Giant Oil and Gas Fields of the Decade 1978–1988

AAPG Memoir 54

Giant Oil and Gas Fields of the Decade 1978–1988

Edited by

Michel T. Halbouty

Proceedings of the
Conference held in Stavanger, Norway
September 9–12, 1990

Michel T. Halbouty, General Chairman
Henrik Ager-Hanssen, Coordinating Chairman

Sponsored by The American Association of Petroleum Geologists and STATOIL

Published by
The American Association of Petroleum Geologists
Tulsa, Oklahoma 74119, U.S.A.

Copyright © 1992
The American Association of Petroleum Geologists
All Rights Reserved
Published 1992

ISBN: 0-89181-333-0

AAPG grants permission for a single photocopy of any article herein for research or noncommercial educational purposes. Other photocopying not covered by the copyright law as Fair Use is prohibited. AAPG participates in the Copyright Clearance Center. For permission to photocopy more than one copy of any article, or parts thereof, contact: Permissions Editor, AAPG, P.O. Box 970, Tulsa, Oklahoma 74101.

Association Editor: Susan A. Longacre
Science Director: Gary D. Howell
Publications Manager: Cathleen P. Williams
Special Projects Editor: Anne H. Thomas
Production: Custom Editorial Productions, Inc., Cincinnati, Ohio

Publisher's Note: AAPG extends its appreciation to the authors, and also to STATOIL for its generous support in underwriting some of the production costs for color illustrations in this volume.

Dust jacket photograph by C.B. Gaynor —

Lobe Falls, south-southeast of Sanaga Sud field, near Kribi, Cameroon. The falls are formed by coast-parallel normal faulting in Precambrian basement. A short distance to the south, near Bwendjo, are outcrops of synrift Aptian-to-Albian arkosic sandstones and conglomerates deposited as debris flows in a lacustrine-to-shallow-marine environment.

About the Editor

Michel T. Halbouty

Michel T. Halbouty has been a member of the petroleum industry since 1930. In a career that has spanned 62 years, he has ardently served the industry and the nation in many capacities. He is in his 62nd year as a member of AAPG, and he served as the Association's 50th President. AAPG has been a material factor in his career, and his devotion to the Association and what it stands for are legendary. Among the many innovations he has initiated for the welfare of AAPG, his prescience led him to establish and implement the AAPG Foundation during his presidency. This volume is a testament to Mr. Halbouty's constant desire to perfect the science of petroleum geology.

Table of Contents

Chapter 1	Introduction *Michel T. Halbouty*	1
Chapter 2	The Point Arguello Field: Giant Reserves in a Fractured Reservoir, California *William E. Mero, Stephen P. Thurston, and Robert E. Kropschot*	3
Chapter 3	West Chalkley, Cameron Parish, Louisiana: A Case for Continued Exploration in Mature Producing Provinces *G. E. Klefstad*	27
Chapter 4	The Reservoir Geology and Geophysics of the Hibernia Field, Offshore Newfoundland *T. J. Hurley, R. D. Kreisa, G. G. Taylor, and W. R. L. Yates*	35
Chapter 5	Geology of Venture, a Geopressured Gas Field, Offshore Nova Scotia *Kenneth J. Drummond*	55
Chapter 6	Mexico's Giant Fields, 1978-1988 Decade *Jose Santiago and Alfonso Baro*	73
Chapter 7	The Tengiz Oil Field in the Pre-Caspian Basin of Kazakhstan (Former USSR)—Supergiant of the 1980s *Nicolai N. Lisovsky, Georgii N. Gogonenkov, and Yuri A. Petzoukha*	101
Chapter 8	The Marlim and Albacora Giant Fields, Campos Basin, Offshore Brazil *Aladino Candido and Carlos A. G. Cora*	123
Chapter 9	The Linguado, Carapeba, Vermelho, and Marimbá Giant Oil Fields, Campos Basin, Offshore Brazil *Paulo Márcio C. Horschutz, Luiz Carlos S. de Freitas, Carlos Varela Stank, Alberto da Silva Barroso, and Wagner Maia Cruz*	137
Chapter 10	El Furrial Oil Field: A New Giant in an Old Basin *Rodulfo Prieto and Gustavo Valdes*	155
Chapter 11	Ceuta-Tomoporo Field, Venezuela *Enrique Ramírez and Fernando Marcano*	163
Chapter 12	The Giant Caño Limon Field, Llanos Basin, Colombia *C. N. McCollough and J. A. Carver*	175

Chapter 13	Takula Oil Field and the Greater Takula Area, Cabinda, Angola *C. T. Dale, J. R. Lopes, and S. Abilio*	197
Chapter 14	Sanaga Sud Field, Offshore Cameroon, West Africa *Robert J. Pauken*	217
Chapter 15	October Field: The Latest Giant under Development in Egypt's Gulf of Suez *Jeffrey J. Lelek, David B. Shepherd, Denise M. Stone, and A. Shawky Abdine*	231
Chapter 16	Villeperdue Field: Exploration of a Subtle Trap in the Paris Basin *Bernard C. Duval*	251
Chapter 17	Barbara Field, Adriatic Sea, Offshore Italy: A Giant Gas Field Masked by Seismic Velocity Anomaly—A Subtle Trap *A. Ianniello, W. Bolelli, and L. Di Scala*	265
Chapter 18	Geochemistry of Oils in the Northern Viking Graben *H. M. Chung, W. S. Wingert, and G. E. Claypool*	277
Chapter 19	The Alba Field: A Middle Eocene Deep Water Channel System in the UK North Sea *G. A. Mattingly and H. H. Bretthauer*	297
Chapter 20	Miller Field: A Subtle Upper Jurassic Submarine Fan Trap in the South Viking Graben, United Kingdom Sector, North Sea *N. M. McClure and A. A. Brown*	307
Chapter 21	Smørbukk Field: A Gas Condensate Fault Trap in the Haltenbanken Province, Offshore Mid-Norway *S. N. Ehrenberg, H. M. Gjerstad, and F. Hadler-Jacobsen*	323
Chapter 22	The Jurassic Snøhvit Gas Field, Hammerfest Basin, Offshore Northern Norway *A. Linjordet and R. Grung Olsen*	349
Chapter 23	Draugen Oil Field, Haltenbanken Province, Offshore Norway *Donald M. J. Provan*	371
Chapter 24	The Geology of Heidrun: A Giant Oil and Gas Field on the Mid-Norwegian Shelf *P. K. Whitley*	383
Chapter 25	The Snorre Field: A Major Field in the Northern North Sea *K. Jorde and G. W. Diesen*	407
Chapter 26	Oseberg Field *Jens Hagen and Benedicte Kvalheim*	417
Chapter 27	The Gullfaks Field *Ole Petterson, Arvid Storli, Eva Ljosland, Ole Nygaard, Ian Massie, and Henrik Carlsen*	429
Chapter 28	Troll Field: Norway's Giant Offshore Gas Field *L. Bolle*	447

Chapter 29	The Suizhong 36-1 Oil Field, Bohai Gulf, Offshore China: Reservoir Delineation by Geophysical Methods *John B. Gustavson and Xin Shi Gang*	**459**
Chapter 30	Iagifu/Hedinia Field: First Oil from the Papuan Fold and Thrust Belt *R. H. Matzke, J. G. Smith, and W. K. Foo*	**471**
Chapter 31	Fortescue Field, Gippsland Basin, Offshore Australia: Flank Potential Realized *J. H. Hendrich, I. D. Palmer, and D. A. Schwebel*	**483**
Chapter 32	Petroleum Geology and Prospects of the Tarim (Talimu) Basin, China *Hu Boliang*	**493**
	The Hydrocarbon Potential of the Norwegian Continental Shelf *Finn Roar Aamodt*	**511**
	The Gorgon Gas Field *L. J. Clegg, M. J. Sayers, and A. M. Tait*	**518**
	Index	**519**

Chapter 1

Introduction

Michel T. Halbouty

*The Michel T. Halbouty Energy Co.
Houston, Texas, U.S.A.*

The success of Memoir 14 and the worldwide interest shown for data on giant fields prompted AAPG to schedule a symposium on giant fields at the end of each subsequent decade. The 1968-78 symposium was held in Houston, Texas, April 1-4, 1979, and the papers were published in AAPG Memoir 30, December 1980.

The Stavanger Conference "Giant Oil and Gas Fields of the Decade: 1978-1988" was held in Stavanger, Norway, September 9-12, 1990, and is a continuation of the Giants of the Decade series.

Scientific studies and projections of future world energy demand indicate that although alternative-energy fuel sources must be actively pursued and developed, there also must be adequate petroleum supplies to bridge the gap. For the international petroleum industry, the years covered by this conference, 1978-88, were complex. They were years of boom and bust. The world's energy consciousness was boosted sharply by the effects of the 1979 Iranian revolution and the resulting embargo, which sent world oil prices to record heights. Global petroleum exploration soon surged, leading to the industry's all-time drilling high in 1981. Then came the oil price collapse in 1985, and the following years were characterized by falling oil prices and drastic budget cuts for exploration and development.

Although exploration dropped sharply during the latter part of the decade, there was a steady flow of giant oil and gas field discoveries. Using the giant field designation criteria of 500 million bbl of oil recoverable for fields in Asiatic Russia, North Africa, and the Middle East, 100 million bbl of oil recoverable for the fields in the remainder of the world, and 3 tcf and 1 tcf of gas reserves recoverable for the same areas respectively, (these definitions exclude heavy oil sands and tar deposits, as well as pervasive ultralight and other nonconventional gas accumulations), it is estimated that at least 182 giant oil and gas fields containing an estimated 140 billion BOE (barrels of oil equivalent) were discovered in 46 countries during the years covered by this conference. Of this number, 33 fields in 17 countries were selected for presentation at the Stavanger Conference.

It is interesting to note that, in the decade 1978-88, 288 giant fields, containing an estimated 330 billion BOE, were found worldwide. Comparison of these figures with those of the preceding decade show that there was an appreciable decline in giant discoveries. It should also be noted that there were more exploratory wells drilled in the latter decade than in the former. The exploratory well totals for the 1978-88 period are misleading, however, because of the great increase in exploratory well drilling in the United States (41,342 more exploratory wells were drilled in the United States than in the 1968-78 decade), where a predominance of small prospects were drilled with relatively small reserves.

The yo-yo fluctuations in oil prices accompanied by a recession in worldwide exploration during the latter part of the last decade added to the drop in giant discoveries.

The majority of the world's prospective petroleum-producing areas currently are nonproductive. Each year a portion of this potential is explored, and it is in these areas that the majority of the future giants of the world will be found.

The papers in this volume discuss the accumulations of vast reserves of oil and gas in, on, and around various types of geological traps. In each case, the type of trap is identified, followed by a discussion of how the trap was formed and found, the age of the reservoir rocks, and, most important, the significance of all these factors to one another and how they were applied to discover the giant field.

During this international symposium, geoscientists from the global petroleum industry described some of the most significant giant field discoveries of the decade and disseminated information of inestimable value, which should be used in searching for the giants yet to be discovered. The presentations covered a selection of giant fields that were found in Venezuela, Brazil, the United States, Canada, Mexico, Colombia, Angola, Cameroon, Egypt, France, Italy, the United Kingdom North Sea, the Norwegian North Sea, Kazakhstan, China, Australia, and Papua New Guinea. The papers focused on how these giant accumulations of petroleum were formed as well as the equally fascinating story of how they were found.

This conference substantiates the importance of giant fields to the global petroleum industry and the world because giant oil and gas accumulations will provide the major portion of our future petroleum reserves. Furthermore, as petroleum explorationists, we know that we must be better prepared than ever before to evaluate and assess the petroleum potential of the world's sedimentary basins in order to search for and find the remaining giant fields waiting to be explored and drilled. We also know that until alternative energy sources can be developed and made readily available, petroleum supplies will remain the mainstay for providing the world's primary fuel demands for decades to come.

I want to acknowledge the help of the many people in the United States and in Norway who assisted in the planning and development of this conference. Without the cooperation and valuable assistance of Henrik Ager-Hanssen, former Senior Vice President of STATOIL, and his staff, it would have been most difficult to present this conference. In addition to thanking all those of my staff who worked on the program, as well as those in AAPG and STATOIL for their efforts in sponsoring this conference, I also want to acknowledge the support of Mobil Exploration Norway, Inc. for underwriting the audiovisual equipment and the Speakers' Center.

I have had the privilege of chairing the preceding two Giant Fields conferences and also serving as Special Editor of the two volumes containing the proceedings of these conferences. I look forward to again serving in both capacities for the conference covering the Giant Oil and Gas Fields of the Decade: 1988–1998, which will be held in the year 1999. At that time I will be 90 years old.

I wish you good reading, good learning, and, above all, good and successful hunting.

MICHEL T. HALBOUTY

Chapter 2

The Point Arguello Field
Giant Reserves in a Fractured Reservoir, California

William E. Mero

*Chevron Overseas Petroleum, Inc.
San Ramon, California, U.S.A.*

Stephen P. Thurston

*Chevron USA
New Orleans, Louisiana, U.S.A.*

Robert E. Kropschot

*Chevron Corp.
San Francisco, California, U.S.A.*

ABSTRACT

Chevron (as operator for its partners, Phillips, Union Pacific Resources, and Impkemix) discovered the Point Arguello oil field in 1981. The discovery well, the Chevron et al. P 0316 #1, was drilled in federal waters 8.5 mi (13.7 km) south of Point Arguello, California. Delineation drilling (by both Chevron and Texaco) has confirmed the discovery of a giant oil field with estimated recoverable reserves in excess of 300 million bbl of oil.

The oil field is located within a small depocenter at the southern edge of the offshore Santa Maria basin. This local depocenter may contain over 15,000 ft (4600 m) of Neogene rocks. The Point Arguello accumulation is trapped in a large north–northwest-trending anticlinal complex and is part of an anticlinal trend of similar Monterey oil discoveries and producing fields within the offshore Santa Maria basin.

The primary reservoir is the middle and upper Miocene Monterey Formation, composed of fractured cherts, porcelanites, siliceous mudstones, and dolostones. Calculated fracture permeabilities range up to 3 darcys. Crestal wells have productive capabilities, after acid, of approximately 6000 BOPD.

Three production platforms have been installed, and connected by oil and gas pipelines to onshore treatment facilities capable of handling 100,000 BOPD and 60 MMCFGD. Thirty-nine development wells have been drilled. Initial production was scheduled for late 1987; however, permitting delays stalled production from this giant oil field until May 1991.

INTRODUCTION

The Point Arguello oil field is located at the southern margin of the Santa Maria basin in federal waters off the California coast (Figure 1). The field is 8 mi (13 km) from the coastline in water depths ranging from 400 to 1200 ft (122 to 335 m). This major offshore oil discovery is a large, complex anticlinal structure whose primary reservoir consists of fractured cherts and porcelanites of the middle and upper Miocene Monterey Formation.

This field was discovered February 27, 1981, on federal Outer Continental Shelf (OCS) Lease P 0316 by a partnership of Chevron, Phillips, Union Pacific Resources, and Impkemix. Following the drilling of the discovery well, seven more exploration and appraisal wells delineated the productive area of the field at over 6000 ac (24.3 km^2) on OCS Leases P 0316, P 0315 (held by Texaco, Pennzoil, Sun, and Koch), and P 0450 (held by Chevron and Phillips). After the initial delineation wells were drilled and three-dimensional seismic surveys conducted, recoverable oil reserves were estimated at some 300 million bbl. Three production platforms have been installed on the three key leases, and 39 development wells have been drilled. A 24-in. oil pipeline and a 20-in. gas pipeline tie platform Hermosa to onshore oil- and gas-processing facilities at Gaviota, a distance of 28 mi (45 km). These facilities, capable of handling production of 100,000 BOPD and 60 MMCFGD, were scheduled to go onstream in 1987 through the Gaviota marine terminal via tanker to refineries in Los Angeles. Permitting delays continue to stall this new significant source of domestic oil production.

EXPLORATION AND DISCOVERY HISTORY

Although California's continental shelf contains 13 offshore Tertiary basins, only two—the Santa Barbara Channel and the offshore Los Angeles basin—had been actively explored and were oil productive prior to the 1981 Point Arguello discovery (Figure 2). California's offshore exploration began in 1896 in the Santa Barbara Channel with the extension of the Summerland field offshore (Yerkes et al., 1969). Over the next 60 years seven more nearshore fields in Los Angeles and Santa Barbara were extended offshore following local lease offerings in coastal waters. The first major California state waters lease sale was held in 1956 and led to 12 new field discoveries and the first offshore drilling and production platform in the Santa Barbara Channel in 100 ft (30 m) of water in 1958. The first significant Federal OCS sale in the Santa Barbara Channel was held in 1968 and resulted in ten new field discoveries (Crain and Thurston, 1987).

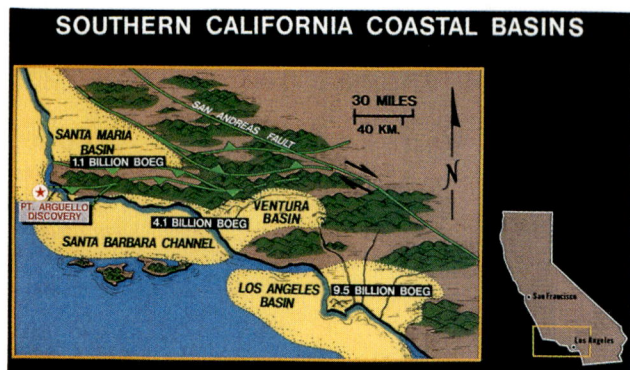

FIGURE 1. Generalized location map of the coastal basins of southern California and the location of the Point Arguello field discovery in the southern Santa Maria basin offshore. The reserves shown represent estimates of the ultimate recovery of oil, and oil equivalent gas (OEG), in barrels, from the Santa Maria onshore, Santa Barbara Channel and Ventura basin combined, and the Los Angeles basin onshore and offshore, respectively. Reserves are tabulated from the 74th Annual Report of the State Oil & Gas Supervisor, 1988, California Division of Oil and Gas, Publication No. PRO6.

FIGURE 2. Generalized location map of California's offshore basins and the location of the Point Arguello field discovery.

Prior to the 1968 sale, the primary offshore exploration target had been sandstone reservoirs in the Eocene, lower Miocene, and Pliocene. The first Federal OCS sale in the Pacific, Sale P-1, was held in May, 1963 and offered leases along the northern and central California coast as far south as the Santa Maria basin. Twenty wells drilled for similar clastic reservoirs were unsuccessful. However, shows of oil were encountered in the Santa Maria, Outer Santa Cruz, Bodega, and Point Arena basins, primarily in the fractured, siliceous shale and chert intervals of the middle and upper Miocene Monterey Formation. The Monterey was not considered a primary reservoir target in the offshore, even though it had been productive since the turn of the century in the onshore Santa Maria basin. The onshore Monterey reservoirs typically produced low-gravity oil at rates too low to be considered commercial in the offshore environment.

As a result of the drilling that followed the 1968 federal OCS sale, P-4, Chevron participated in five new field discoveries in the Santa Barbara Channel (Figure 3). The Monterey fractured shale was tested at rates in excess of 1200 BOPD per test with oil gravities ranging from 12 to 32° API, generally higher than in the onshore Santa Maria basin. These Monterey discoveries (at Hondo, Pescado, Sacate, Santa Clara, and Sockeye) highlighted the Monterey as a significant future exploration target for the next major OCS sale in the Santa Barbara Channel—namely, Sale 48, to be held in 1979 (Figure 3).

Sale 48 (June, 1979) offered 540,000 ac (2185 km²) of the Santa Barbara Channel, including the previously unexplored westernmost portion of the basin (Figure 4). In preparation for Sale 48, the industry drilled the COST (Continental Offshore Stratigraphic Test) 78-164 #1 well in a syncline in the southwestern portion of the Santa Maria basin. This well encountered 9600 ft (2900 m) of Neogene section unconformably overlying Cretaceous rocks. The typical Paleogene section, including the prolific Vaqueros and Gaviota sands of the Santa Barbara Channel, was not present (Figure 5). The Monterey Formation, however, had heavy oil shows and abundant interbedded porcelan-

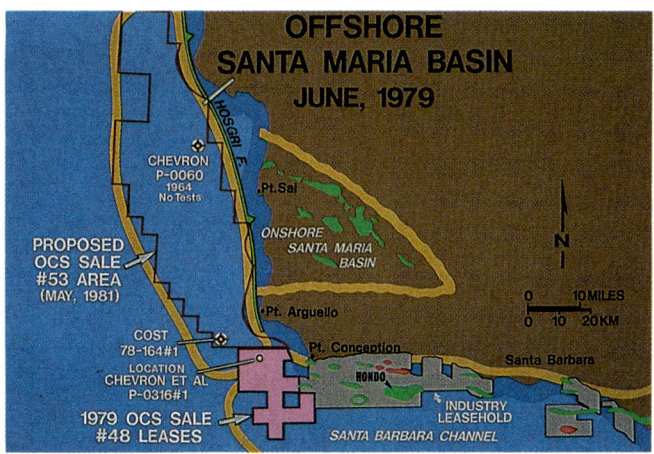

FIGURE 4. Simplified location map of the Santa Maria basin offshore, showing the area of proposed OCS Sale 53 and the location of the newly acquired leases in the June 1979 OCS Sale 48 in the western Santa Barbara Channel.

ites and glassy cherts, which tend to be very fracture-prone. (No tests are conducted in COST wells.)

Chevron's bid maps of the Point Arguello area used in Sale 48 were based on a widely spaced grid of two-dimensional seismic data. These data indicated a series of three large en echelon anticlinal structures at the Monterey level (Figure 6), and a series of smaller structures below the Monterey, perhaps including the Vaqueros sandstone reservoir (Figure 7). The seismic had also suggested that a "paleo high," called the "Amberjack high," perhaps comprised of Upper Cretaceous rocks, was located in block P 0316, and perhaps it was this high that truncated the Eocene section of the Santa Barbara Channel to the east.

In Sale 48, the Chevron group (including Phillips, Union Pacific Resources, and Impkemix) acquired Leases P 0316, P 0317, and P 0318 (P 0316 for $36.5 million) and the Texaco group (including Pennzoil, Sun, and Koch) acquired Lease P 0315 (for $35.3 million). Chevron immediately filed for permits to drill, recognizing that the Point Arguello prospect structure extended into the open acreage planned for OCS Sale 53 to be held in May, 1981. Permits were not obtained until late 1980, and on November 25, 1980, Chevron spudded the P 0316 #1. The well was drilled to 9621 ft (2932 m) and encountered only the Monterey and a thin Point Sal Miocene section above the sands and shales of the Upper Cretaceous. The Vaqueros sands were not present. The Monterey was encountered almost 1000 ft (305 m) low to the 1979 bid mapping. The mud log contained abundant cherts, which became more glassy with depth. Only trace oil shows and moderate-to-poor mud gas shows were logged. However, five successful tests were conducted in the Monterey from the top of the Cretaceous Jalama to the upper Miocene Santa Margarita (Figure 8). Individual DST flow rates ranged from 400 to 2400 bbl of 18 to 22° API oil per day, and the combined total flow for the well was 6580 BOPD and 1680 MCFGD. The testing defined an oil column of at least 1400 ft (425

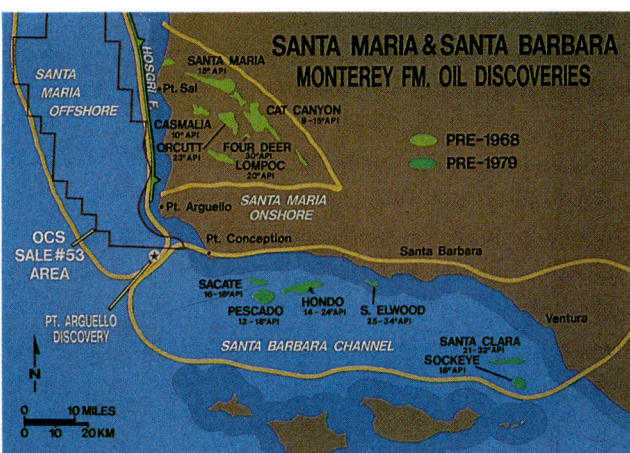

FIGURE 3. Simplified location map of the major oil fields with Monterey Formation reservoirs within the Santa Maria basin and Santa Barbara Channel. Monterey fields discovered prior to the 1968 OCS lease sale are shown onshore, and those discovered prior to the 1979 OCS Sale 48 are shown offshore.

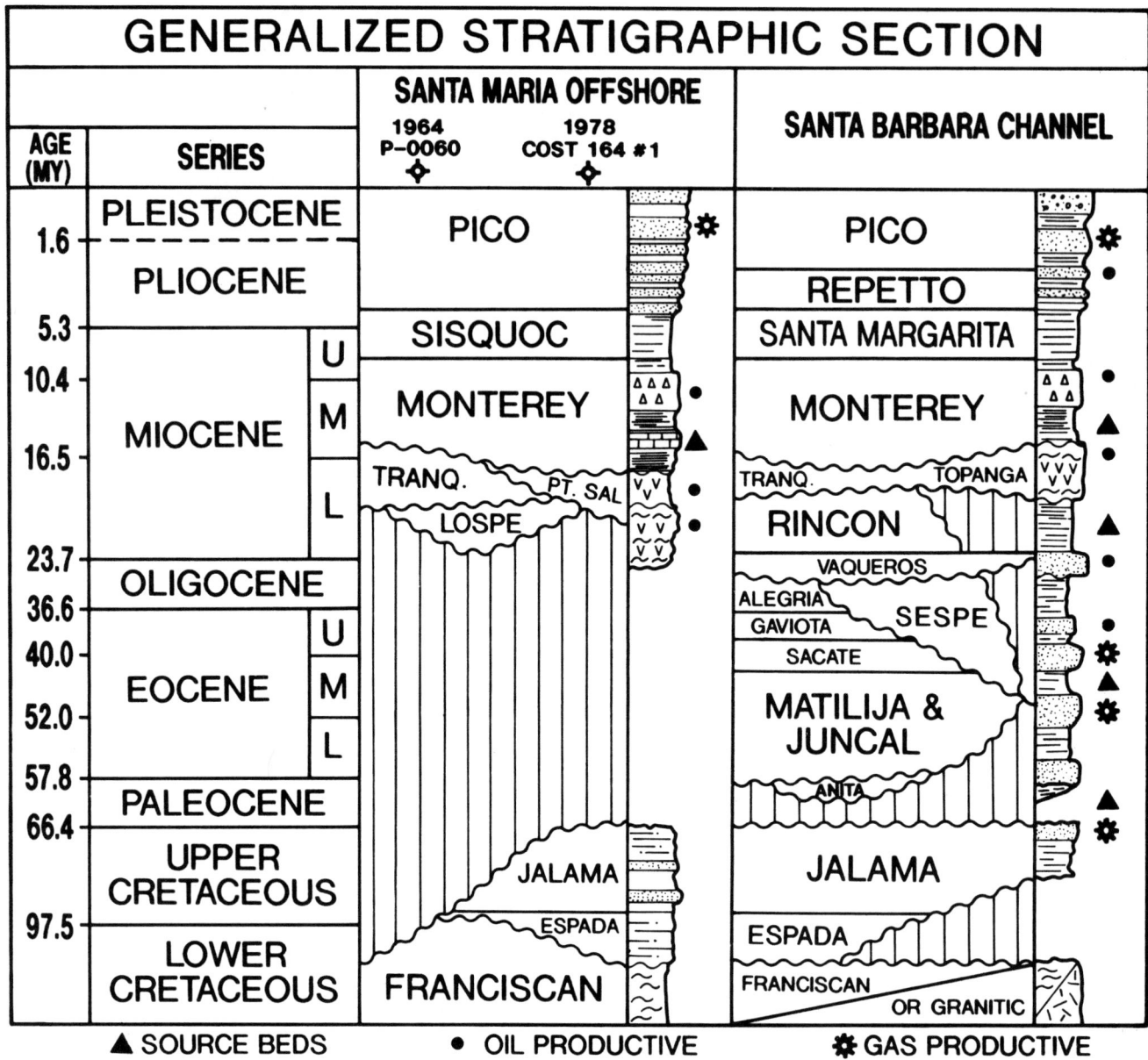

FIGURE 5. Generalized and simplified stratigraphic section for the offshore Santa Maria basin and the Santa Barbara Channel offshore (from Crain and Thurston, 1987).

m) and no oil-water contact. The P 0316 #1 had revealed a major oil field in the fractured Monterey, and with these well data the Chevron group prepared for OCS Sale 53, only three months away.

OCS Sale 53 offered 111 tracts (603,613 ac, or 2443 km^2) for lease at a time when oil prices were over $30/bbl and oil industry interest was high due to the more favored public opinion that searching for domestic oil reserves was important to our nation's security and independence. Sale 53 was the first major lease sale in the entire offshore Santa Maria basin (Figure 9). The only previous sale involving the Santa Maria basin was the 1963 P-1 sale, which leased only a few tracts and caused only one well to be drilled in the northern part of the basin. The Humble P-0060 well was drilled in 1964 and encountered the Monterey Formation with heavy oil shows overlying the lower Miocene Obispo volcanics and reached total depth in the Franciscan (Figure 5). The Monterey was not tested, because it was not considered a viable offshore objective reservoir. Both the COST well and the 1964 P-0060 well had encountered heavy oil shows in the Monterey, but no Miocene, Oligocene, or Eocene sandstone reservoirs. The tests in the P 0316 well were the only real encouragement for the Santa Maria basin offshore sale. Pre-OCS Sale 53 mapping of the Monterey suggested that a major accumulation extended through the Texaco P 0315 lease and into tracts P 0450 and P 0451 (Figure 10). On May 28, 1981, the industry responded to the sale with overwhelming enthusiasm and leased 60 tracts (320,579 ac, or 1297 km^2) for a total bonus

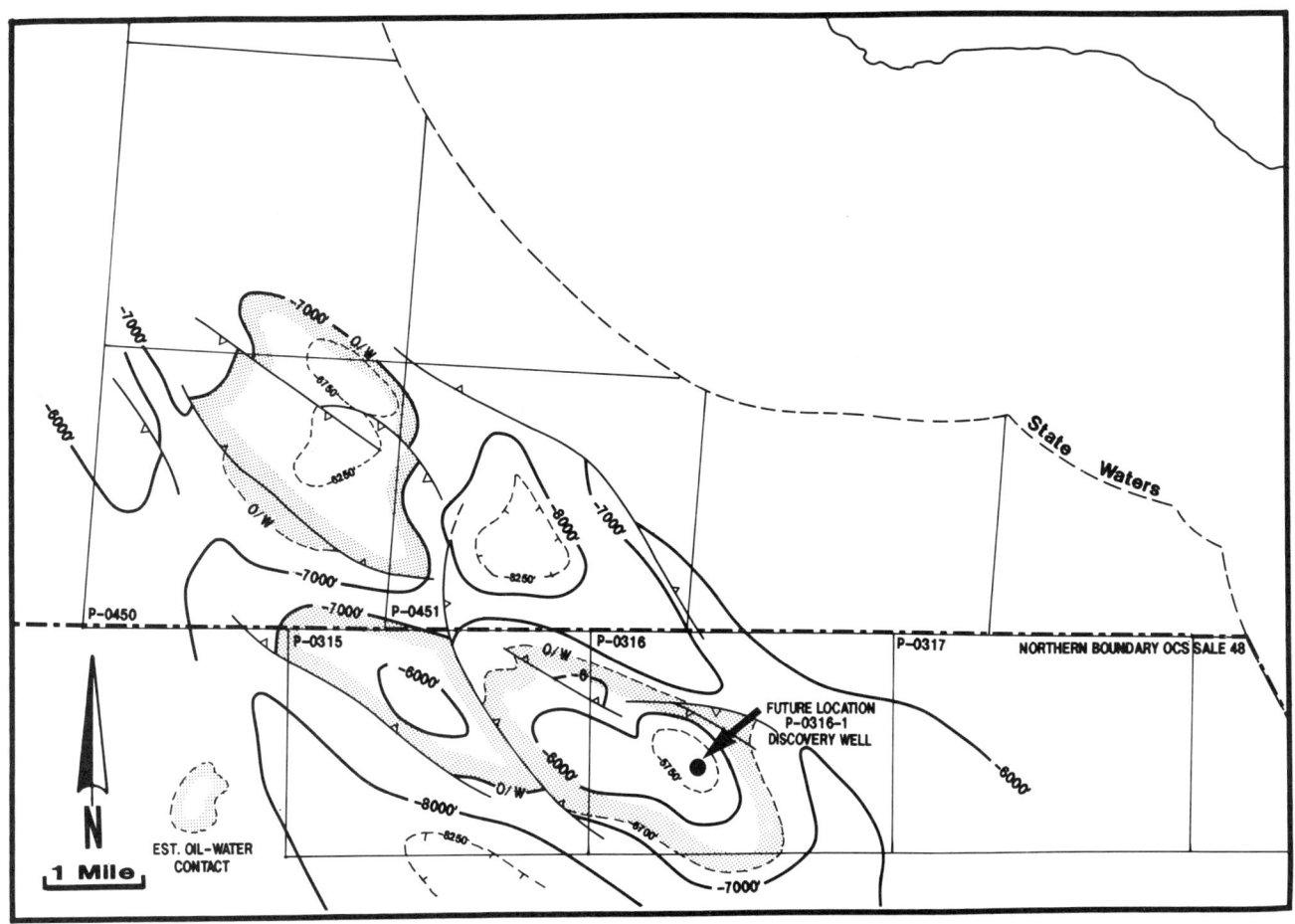

FIGURE 6. Chevron structure map on the top of the Monterey Formation before the June 1979 OCS Sale 48 covering the western Santa Barbara Channel. This interpretation indicated that the culmination of the Point Arguello anticlinal trend was in block P 0316 and that the anticlinal trend extended into the open acreage north of the OCS Sale 48 area. The contour interval is 1000 ft (305 m).

of over $2 billion (Figure 9). The Chevron-Phillips partnership dominated the sale, with high bids on 28 tracts, including $333.596 million for Lease P 0450, the current record for the largest bonus bid for a Federal OCS tract.

Following OCS Sale 53 the Texaco group drilled and tested three wells on the P 0315 lease with combined rates as high as 4200 BOPD (18 to 21° API) from the Monterey. Chevron and Phillips drilled three wells on Lease P 0450 and discovered a new pool of 31 to 33° API light oil which flowed at combined rates as high as 4900 BOPD. By the end of 1982, eight wells had delineated a productive area of nearly 7000 ac (28.3 km²), a maximum oil column of 1800 ft (549 m), and over 300 million bbl of recoverable oil at Point Arguello.

REGIONAL SETTING AND GEOLOGIC HISTORY

The offshore Santa Maria basin is composed of several narrow north–northwest-trending subbasins bounded by basement uplifts (Figure 11). The Santa Lucia high forms the western margin, and the right-lateral oblique-slip Hosgri fault system bounds the eastern margin, of the offshore Santa Maria basin (Hoskins and Griffith, 1971; Crouch et al., 1984). Between the Santa Lucia high and the Hosgri fault is a large basement ridge noted in Figure 11 as the Central Basin high.

The southern boundary of the offshore Santa Maria basin is presumed to be a northeast-trending basement high, informally named the Amberjack high. East of the Amberjack high the stratigraphy is typical of the Santa Barbara Channel, and the Neogene section there overlies thick Paleogene and Cretaceous fore-arc basin deposits (Crain and Thurston, 1987) (Figure 5). West of the Amberjack high, in the Point Arguello field, the stratigraphy resembles the onshore Santa Maria basin where the Neogene section unconformably overlies Jurassic–Cretaceous Franciscan accretionary wedge- and subduction-related deposits. The northwest-trending Point Arguello structure crosses the older northeast-trending Amberjack high and extends over a small depocenter containing up to 15,000 ft (4600 m) of Neogene rocks.

The offshore Santa Maria basin is a Neogene basin that was developed on the accretionary wedge of the convergent margin tectonic system (arc-fore-arc-trench)

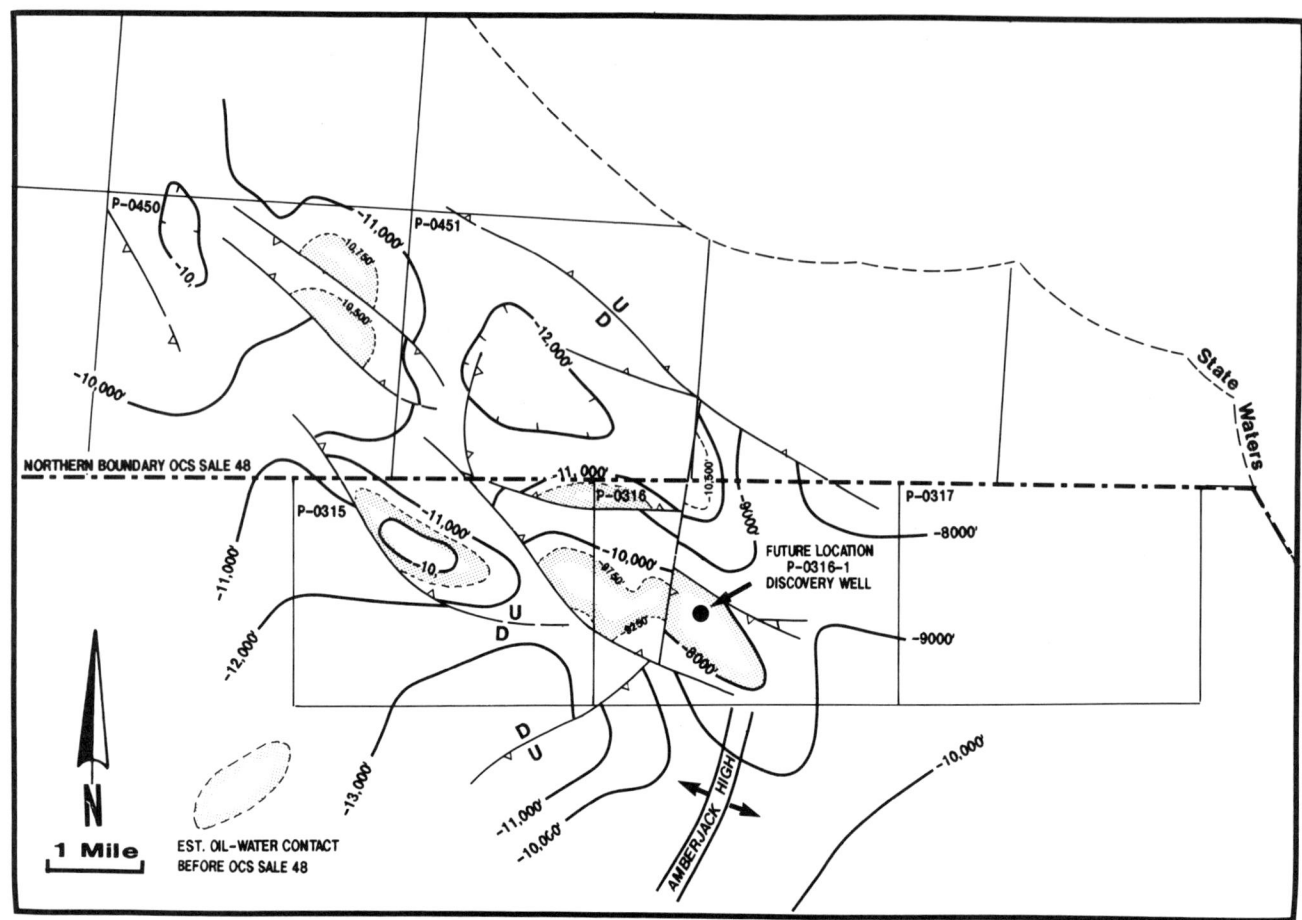

FIGURE 7. Chevron structure map on the "Vaqueros" (pre-Monterey) before OCS Sale 48. This interpretation indicated that the Point Arguello anticlinal trend persisted to pre-Monterey levels that perhaps could contain the Vaqueros sandstone reservoir, a prolific producer in the Santa Barbara Channel to the east. Although no Vaqueros sands were ever found at Point Arguello, some oil was discovered in the lower Monterey (Pt. Sal) on block P 0450.

which had dominated coastal California from the Jurassic to the Oligocene (Dickinson, 1979; Nilsen, 1986; Nilsen, 1987). During the Oligocene, the Pacific-Farallon oceanic spreading center was subducted beneath the North American plate and a transform tectonic margin was established in coastal California (Atwater, 1970; Blake et al., 1978; Dickinson and Snyder, 1979; Page and Engebretson, 1984). This transform margin setting created over 15 separate strike-slip-bounded transtensional (or pull-apart) basins that make up the petroleum basins of coastal California today (Crain and Thurston, 1987). In southern California, early to middle Miocene transform tectonics produced a regional clockwise rotation of the Transverse Ranges crustal block that created the triangular-shaped Santa Maria and Los Angeles basins adjacent to the block (Hornafius, 1985; Luyendyk et al., 1985; Hornafius et al., 1986).

Widespread marine regression and erosion followed the spreading center-trench collision during the late Oligocene, capping the fore-arc margin basins with nonmarine clastics (Nilsen, 1984). The Santa Maria basin began to form in the lower Miocene as evidenced by widespread lower Miocene volcanics and volcaniclastics unconformably overlying Mesozoic fore-arc and accretionary wedge deposits. Rapid basin subsidence continued throughout the middle and upper Miocene, which allowed for thick biogenic deposition in a restricted and anaerobic deep marine environment. The small Neogene depocenter west of the Amberjack high acted as a major hydrocarbon-generating center for the Point Arguello accumulation.

STRATIGRAPHY

Figure 12 is a generalized stratigraphic column of the Cretaceous-Tertiary section in the Point Arguello field as described by Crain et al. (1985). The oldest rocks penetrated are Cretaceous sandstones and shales. Lower Miocene volcanics, volcaniclastics, and fine-grained marine clastics unconformably overlie the Mesozoic section. To the north of Point Arguello, Cretaceous fore-arc rocks have not been penetrated in the offshore Santa Maria, and the lower Miocene volcanics were deposited on the Franciscan. These lower Miocene units grade upward into middle and upper Miocene dolostones, organic mud-

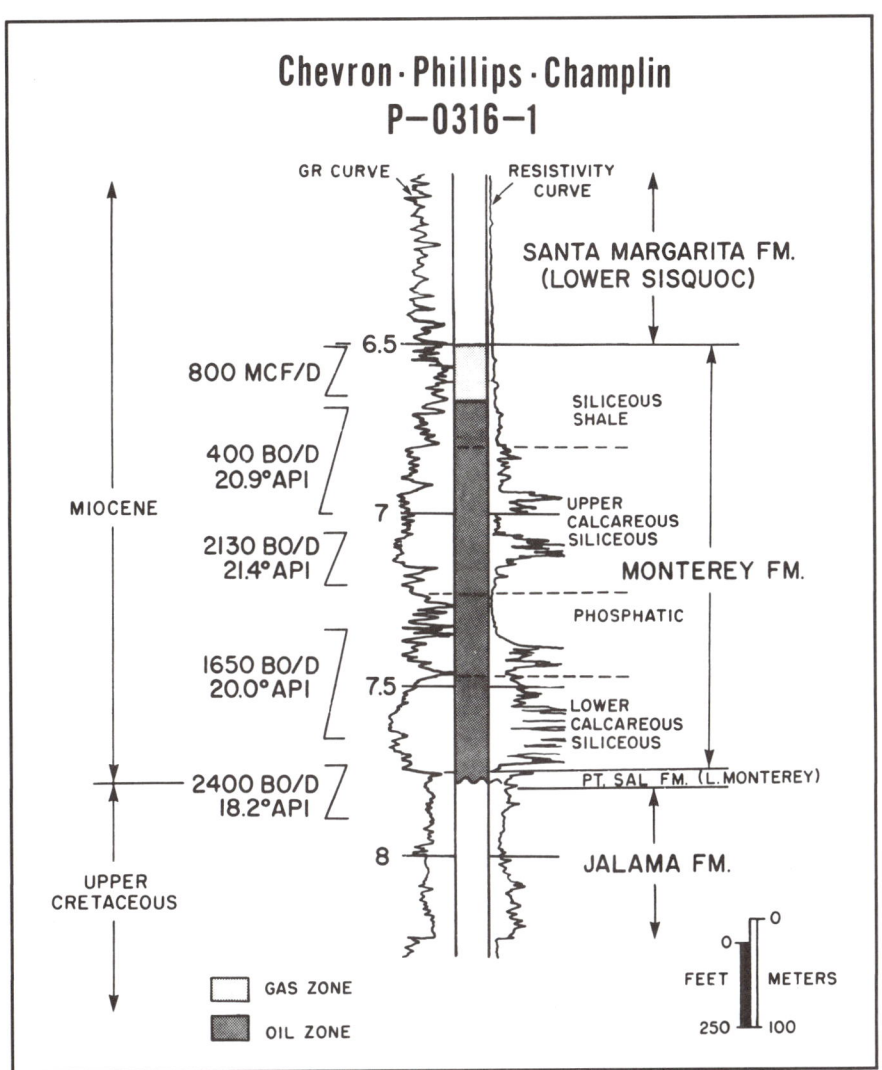

FIGURE 8. Results of electric log and drill-stem tests from the Chevron et al. OCS P 0316 #1, the Point Arguello field discovery well (from Crain et al., 1985). Member subdivisions of the Monterey Formation are from MacKinnon, 1989b. The log is marked in 500-ft (152-m) intervals between 6500 and 8000 ft (1980 and 2440 m).

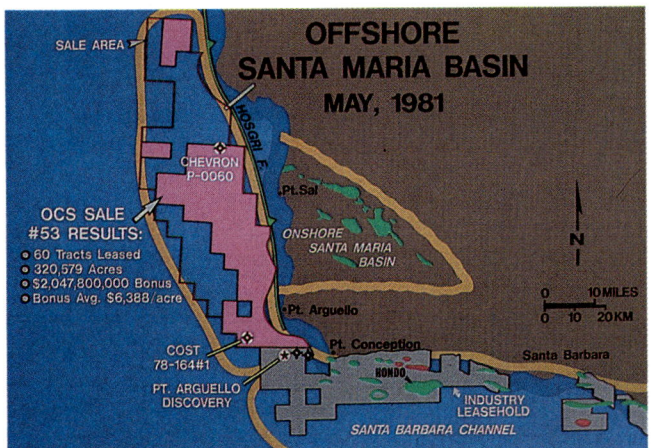

FIGURE 9. Generalized location map of the offshore Santa Maria basin showing the results of OCS Sale 53 held on May 28, 1981.

stones, and siliceous biogenic rocks that were deposited and preserved in an anoxic deep marine basin isolated from significant terrigenous debris until the latest Miocene and Pliocene. By the Pliocene, deep marine sands are seen in the section, followed by progressively shallower marine sands throughout the Pliocene–Pleistocene.

Cretaceous

Espada Formation

Within Lease P 0316, wells have penetrated Lower Cretaceous (Neocomian) dark-brown marine shales and occasional thin, fine-grained arkosic sandstones. These rocks are tentatively assigned to the Espada Formation, after similar rocks mapped onshore by Dibblee (1950). The Espada Formation is assumed to unconformably overlay the metasedimentary subduction melange of the Jurassic–

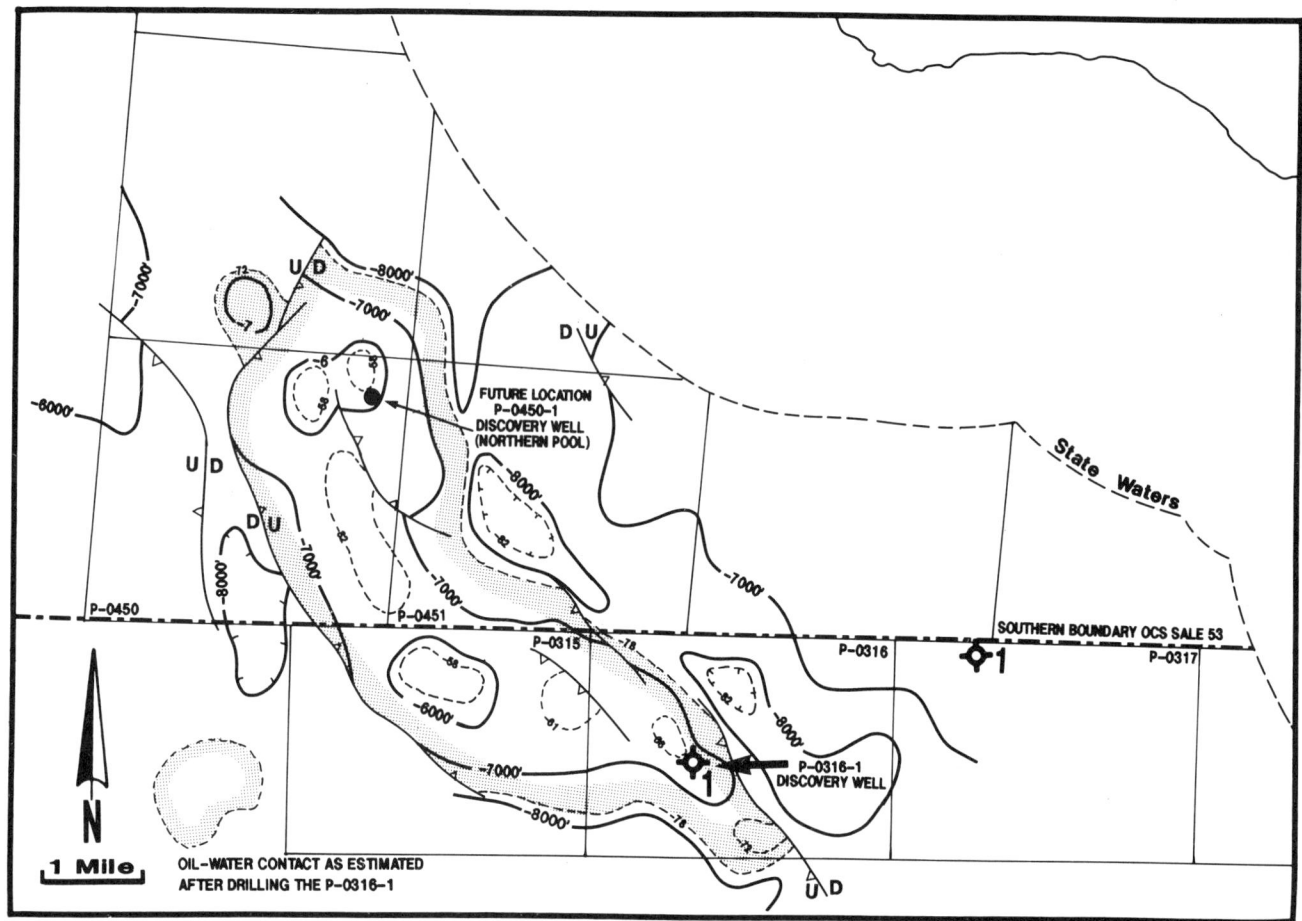

FIGURE 10. Chevron structure map on the top of the Monterey Formation after the drilling of the OCS P 0316 #1 well and before OCS Sale 53 to be held in May 1981. Increased seismic control had aided in optimizing the Monterey structure in the OCS Sale 53 area, and the discovery well provided evidence for the oil-water contact near the structural closure of the field. The contour interval is 1000 ft (305 m).

Cretaceous Franciscan assemblage. The Franciscan is a minor fractured reservoir producing Monterey sourced oil at both offshore Point Pedernales and onshore Santa Maria Valley fields where the Monterey reservoir is in contact.

Jalama Formation

The Chevron et al. P 0316 #1 bottomed in Upper Cretaceous (Turonian to Campanian) dark gray-brown, marine, silty shales interbedded with light buff, fine- to medium-grained, arkosic sandstones unconformably overlain by lower Miocene siltstones. Dibblee (1950) mapped similar Cretaceous rocks onshore as the Jalama Formation.

Tertiary

Lower Miocene

Tranquillion Volcanics. The Tranquillion Volcanics, a marine lower Miocene (Saucesian Stage) series of rhyolitic tuffs, unwelded pumice, tuffaceous volcaniclastics, sandstones, and siltstones, is the oldest Tertiary unit found in the Point Arguello field. The Tranquillion volcanics and their equivalents (Obispo and Lospe) represent the initial extension and opening of the offshore Santa Maria basin. Over 700 ft (213 m) of Tranquillion Volcanics were penetrated in Lease P 0450 and are unconformably overlain by the lower Miocene Point Sal Formation. In the Santa Barbara Channel, the Tranquillion is represented by a thin tuffaceous zone at the top of the Rincon Formation, usually less than 100 ft (30.5 m) thick (Figure 5). The Tranquillion Volcanics are missing by onlap and/or truncation over the Amberjack high (Figure 5).

Point Sal Formation (Lower Monterey). The Point Sal Formation is a lower Miocene (Relizian Stage) limy marine mudstone with interbedded dolostones and minor marine sandstones in places. Onshore, Dibblee (1950) mapped this unit as the lower Monterey. It is roughly coeval to the basal part of the lower calcareous-siliceous member of the Monterey in the Gaviota area as defined by Isaacs (1980). The top of the Point Sal Formation may be as young as middle Miocene (Luisian Stage), and it is transitional to the overlying, more siliceous Monterey

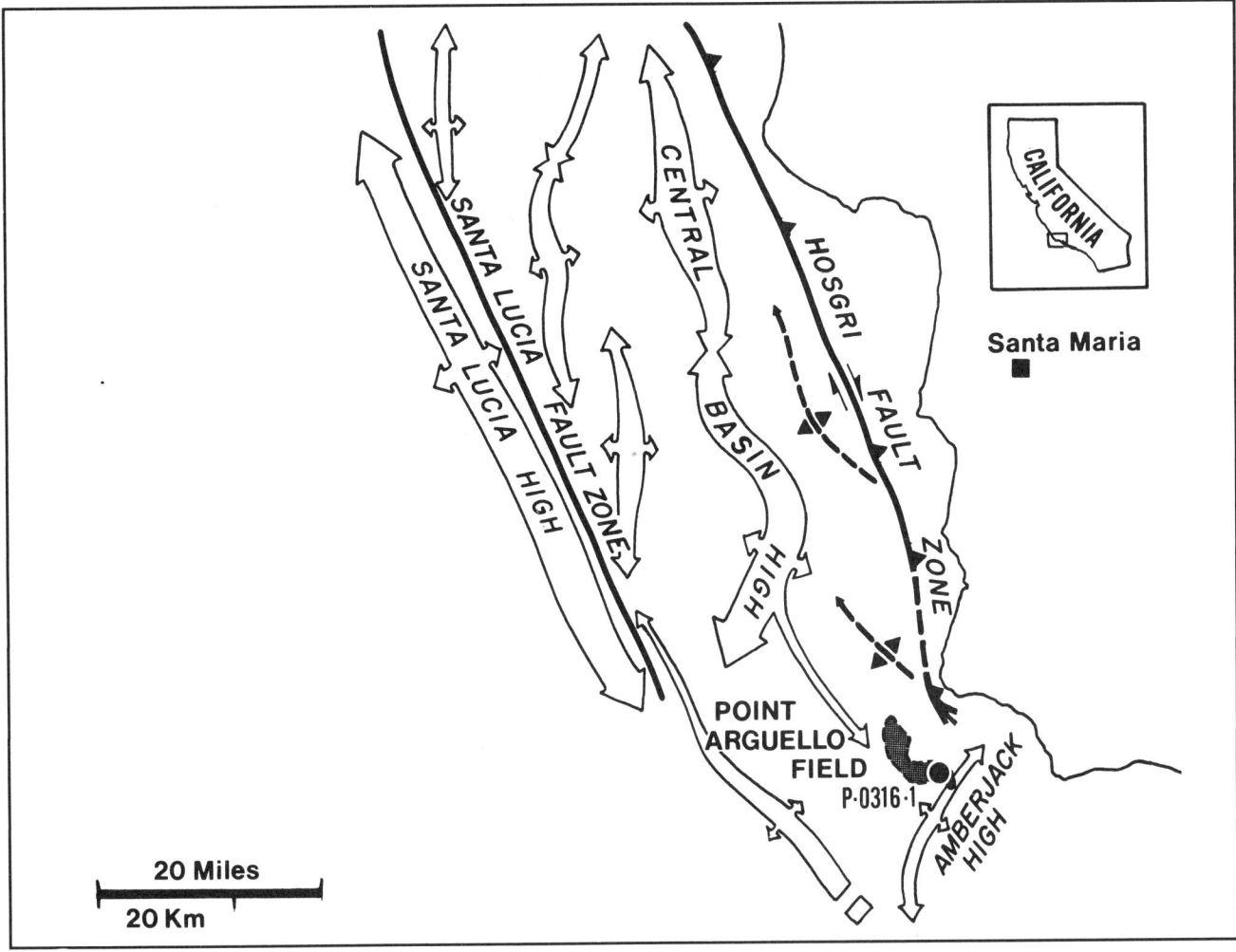

FIGURE 11. Major tectonic elements of the offshore Santa Maria basin with the basement highs and plunge directions shown (modified from Crain et al., 1985).

Formation, diagrammatically shown by the correlation section in Figure 5. The Point Sal Formation can be as much as 2500 ft (762 m) thick in the Point Arguello depocenter, but thins to the southeast onto the Amberjack high (Figure 13).

Fractured dolostones near the base of the Point Sal Formation tested a small amount of 31.7° API oil in the Chevron-Phillips P 0450 #1. The Chevron et al. P 0316 #1 tested a thin Point Sal section in DST #5; however, the thin section of basal Monterey cherts and dolostones included in the test interval may have contributed significantly to the 2400 BOPD in that test (Figure 8). Although not as organic-rich as the Monterey, the Point Sal is also considered an oil-prone source rock. The Point Sal Formation is an important oil producer from sandstone turbidity in several onshore Santa Maria fields (Santa Maria Valley, Casmalia, Orcutt, and Cat Canyon). The Point Sal-lower Monterey Formation within the offshore Santa Ynez unit in the Santa Barbara Channel also contains several oil-prolific deep marine sands (in Isaacs and Garrison, 1983).

Middle and Upper Miocene

The Monterey Formation is both the primary reservoir and the source rock at the Point Arguello field and in the offshore and onshore Santa Maria basins. The Monterey Formation ranges in age from early through late Miocene (Relizian to late Mohnian stages) and in thickness from 2800 ft (853 m) to less than 1000 ft (305 m) at the Point Arguello field. Figure 14 is a Monterey isopach map that shows the Point Arguello depocenter that persisted throughout the middle Miocene; it also demonstrates the thinning onto the Amberjack high.

By 1989, the onshore Santa Maria area had produced 809 million bbl and 792 bcf of Monterey oil and gas. New Monterey discoveries in the offshore Santa Maria basin are estimated to total as much as 2 billion bbl, although much of the oil is low gravity and in deep water, making the fields uneconomic under present conditions. At least 500 million bbl of producible Monterey reserves are estimated to be present in the Santa Barbara Channel (Crain and Thurston, 1987).

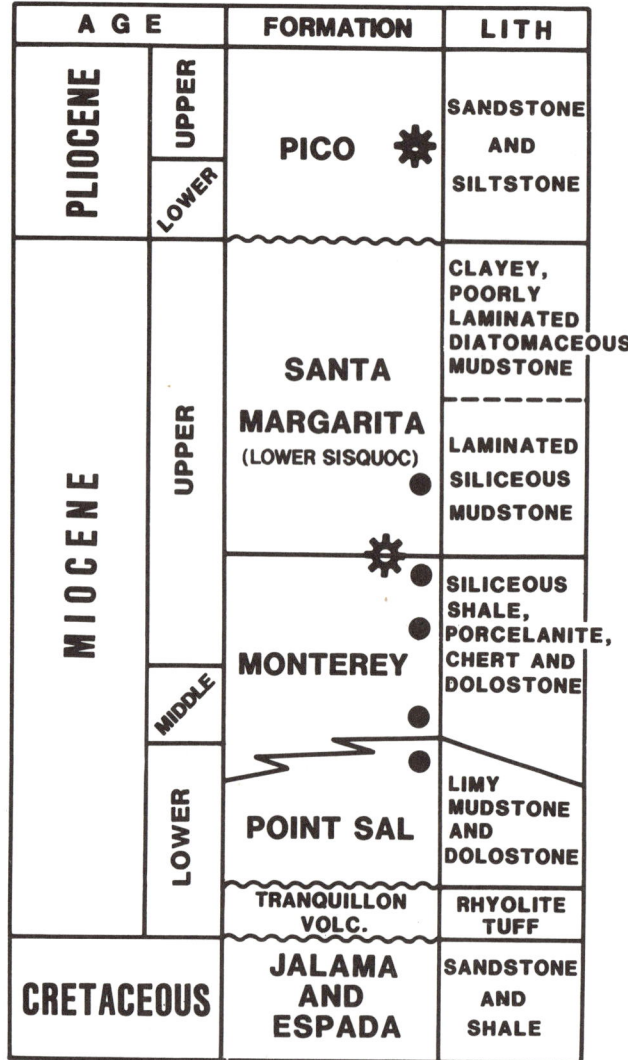

FIGURE 12. Generalized stratigraphic column of the Cretaceous–Tertiary section penetrated by the Chevron et al. exploratory wells at Point Arguello field (from Crain et al., 1985).

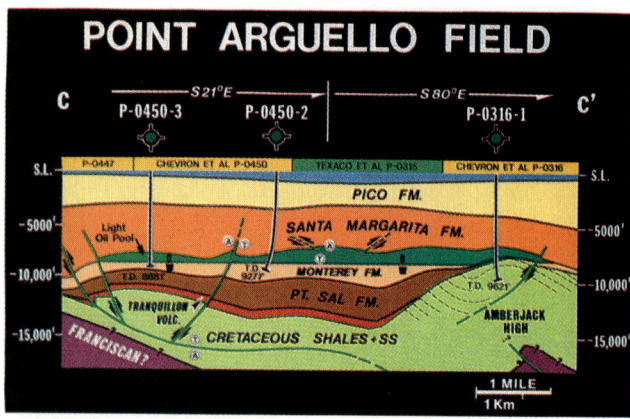

FIGURE 13. Simplified northwest-southeast structural cross section approximately along the axis of the Point Arguello field. Three Chevron et al. exploratory wells are projected into the section. Note the thinning of the Pt. Sal (lower Monterey) and Monterey formations toward the southeast where they unconformably onlap the Cretaceous rocks of the Amberjack high (from Crain et al., 1985).

Middle Miocene transform margin tectonics produced isolated deep marine basins bounded by basement uplifts that created an ocean floor topography very similar to that of the southern California borderland today. The more westerly of the borderland basins were isolated from terrigenous detrital input sourced from the east by the intervening basement uplifts (Figure 15). The Monterey in the offshore Santa Maria basin contains very little terrigenous detritus sourced from the east, whereas in the Cuyama basin to the east, the Monterey is predominantly deep marine turbidite clastics (Lagoe, 1985).

The Monterey Formation in the offshore Santa Maria basin is predominantly a biogenic pelagic deposit comprised of siliceous diatomaceous and foraminiferal oozes thinly laminated and interbedded with organic-rich calcareous and phosphatic mudstones. The middle Miocene climatic and oceanographic conditions favored high plankton productivity, and preservation of the biogenic deposits was aided by the poorly oxygenated conditions in the restricted deep marine environment (Ingle, 1981; Pisciotto and Garrison, 1981; MacKinnon, 1989a). Following burial diagenesis, these biogenic deposits became both an organic-rich source rock (up to 18% TOC) and an excellent fractured reservoir comprised of fracture-prone cherts and porcelanites.

The Monterey Formation in the Santa Maria and western Santa Barbara basins is subdivided into four members: lower calcareous-siliceous, phosphatic, upper calcareous-siliceous, and clayey-siliceous (Isaacs, 1980; Isaacs, 1984; MacKinnon, 1989b). At the Point Arguello field the lower calcareous-siliceous member is primarily composed of thick dolostones, with some cherts, porcelanites, and organic mudstones. The phosphatic member is primarily a highly organic dolomitic phosphatic mudstone, often containing abundant phosphatic nodules and 10 to 18% TOC. This member is characterized by very high gamma ray log response (Figure 8). The upper calcareous-siliceous member is predominantly thin bedded to massive cherts interbedded with thin porcelanites and dolostones. This is the most fracture-prone lithology of the Monterey and the most permeable portion of the fractured reservoir. This member is best identified in E-logs by its lower gamma-ray response, high resistivity, positive SP response, and little or no separation between the neutron and density log curves. The clayey-siliceous member is commonly composed of finely laminated siliceous mudstones interbedded with thin porcelanites and dolostones. The top of this member is the top of the Monterey Formation (M1). In the subsurface, M1 is picked in

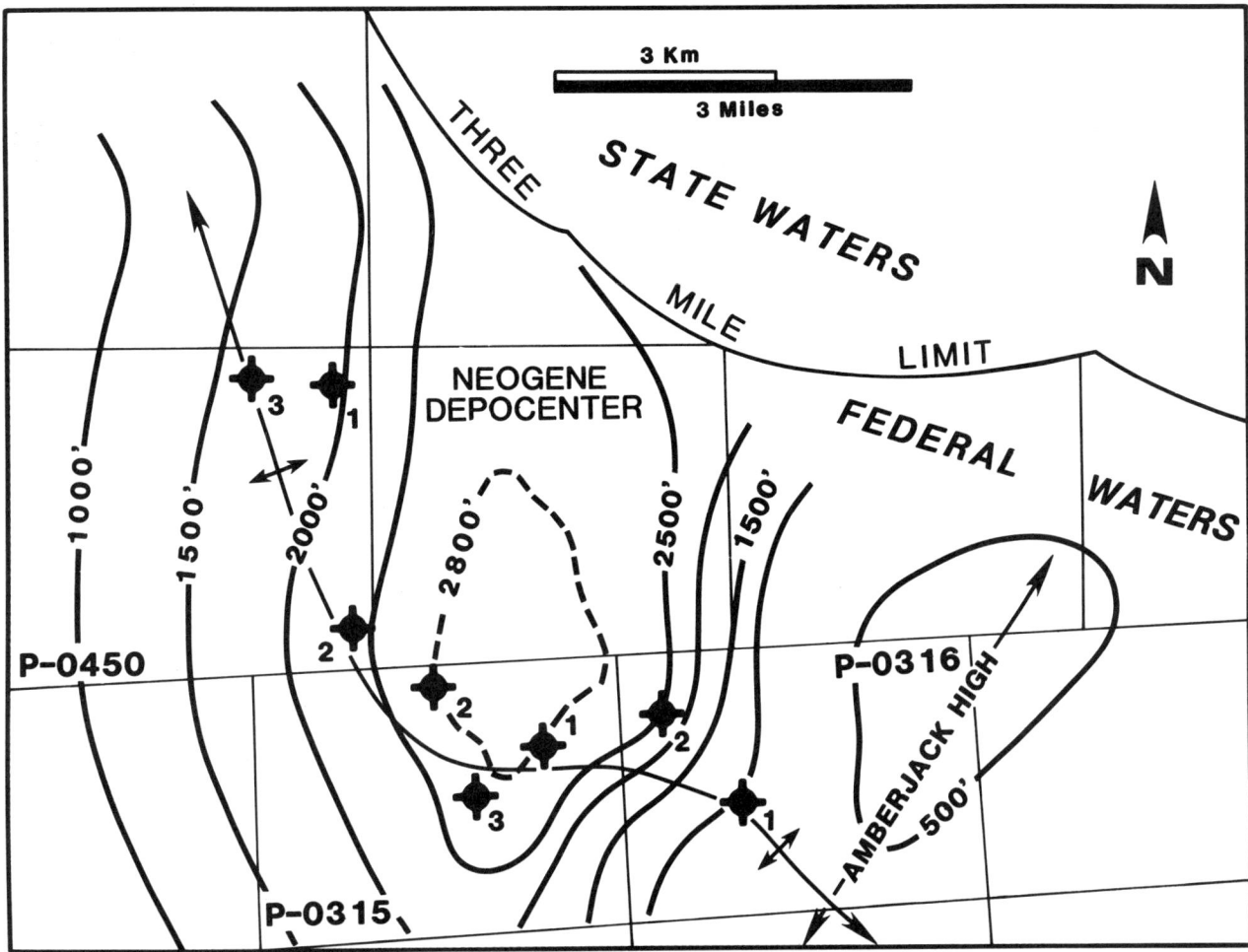

FIGURE 14. Monterey Formation isopach map over the Point Arguello field and the Amberjack high. Note the thinning of the Monterey isopach over the Amberjack high and the location of the Point Arguello depocenter that persisted throughout the Miocene (from Crain et al., 1985).

E-logs at the shift to a higher gamma ray baseline and first thin interbedded siliceous shales and dolostones in contrast to the lower gamma ray response of the more massive diatomaceous mudstones of the overlying Santa Margarita Formation (Figure 8). The M2 marker is within the siliceous shale member and it is generally picked at the first significant increase in the SP and resistivity on the induction log. Ogle et al. (1987) believe this marker may represent a diagenetic change to harder, more fractured lithologies, without regional time significance.

The top of the Monterey is difficult to pick in the coastal outcrops because of the gradational nature of the contact with the Santa Margarita, but onshore it is generally picked between the finely laminated siliceous rocks and the overlying bioturbated massive siliceous mudstones. Chevron dates the Monterey–Santa Margarita (Sisquoc) contact in coastal Santa Maria as upper Miocene near 5.9 m.y.a. with careful diatom biostratigraphy (Dumont, 1989).

Miocene-Pliocene

In the Santa Barbara Channel and offshore Santa Maria basin, the Santa Margarita (lower Sisquoc equivalent) Formation is comprised predominantly of light-gray, diatomaceous, coarsely laminated mudstones and bedded dolostones. At Point Arguello the Santa Margarita is upper Miocene (late Mohnian–early Delmontian) and conformably overlies the Monterey Formation. This section is equivalent to the upper Miocene portion of the Sisquoc Formation as described by Dibblee (1950) and Woodring and Bramlette (1950).

The Santa Margarita (lower Sisquoc) Formation thins over the Point Arguello structure to less than 4500 ft (1372 m) and thickens off the structure to more than 7000 ft (2134 m). A minor unconformity may be present at the top of the Santa Margarita representing a period of nondeposition and mild structural growth in a deep marine environment. A few successful drill-stem tests have been

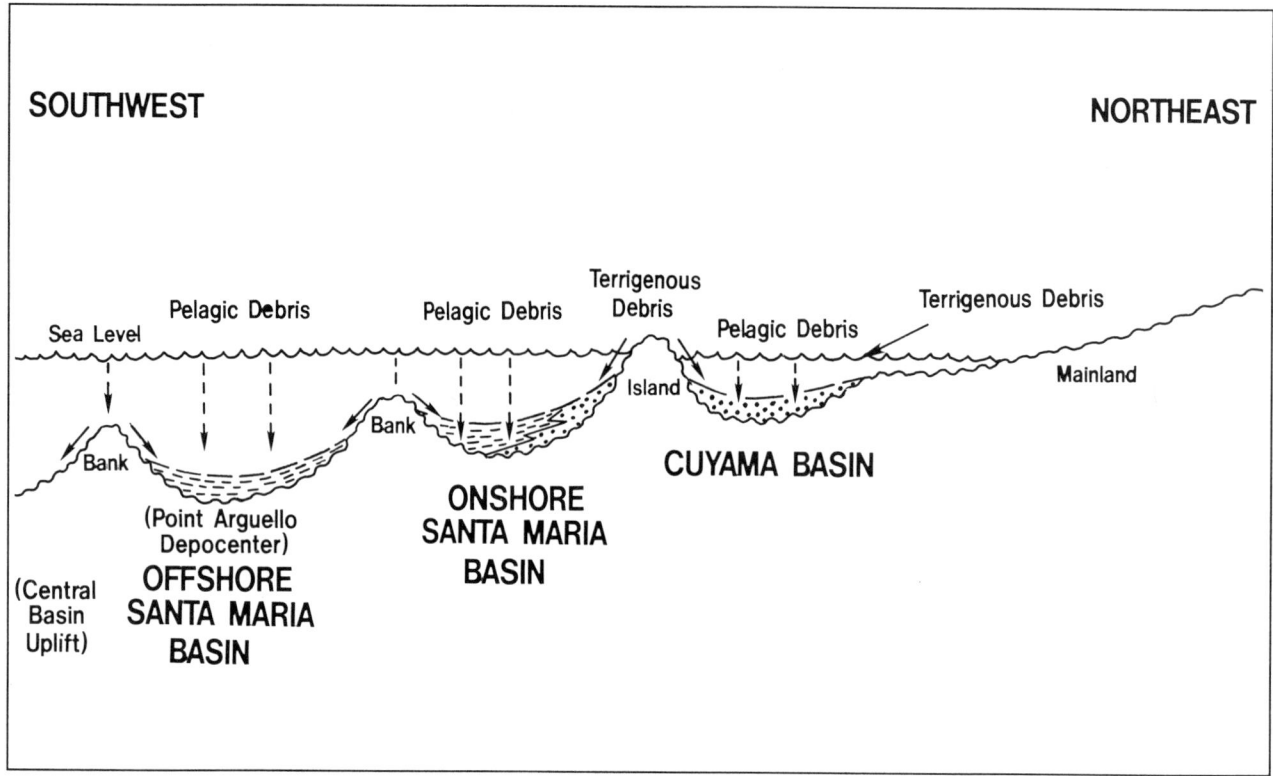

FIGURE 15. Depositional model for middle Miocene Monterey Formation in California's coastal basins. Note that terrigenous detritus (debris) is predominant in the nearshore basins and pelagic detritus in the more outboard basins (modified from Gorsline and Emery, 1959, and Isaacs, 1984).

made in fractured, dolomitic zones near the base of the Santa Margarita, but the long-term producibility appears to be low. Fracturing of the Santa Margarita Formation may be pervasive enough at the undeveloped Rocky Point field to be a commercial reservoir. The lower Sisquoc Formation produces in the onshore Santa Maria Valley and Cat Canyon fields.

Pliocene

Chevron puts all of the Pliocene section within the Pico (upper Sisquoc, Foxen, and Careaga) Formation, a correlation from the Santa Barbara Channel (Figure 12). The Pico Formation in the Point Arguello field area is about 3000 ft (914 m) thick. During the late Pliocene, turbidities deposited channel-like facies of poorly indurated lithic sandstones and pebbly conglomerates across the Point Arguello structure. Most of the Pico Formation is composed of dark gray-brown siltstones and claystones, and is equivalent to the lower Pliocene upper Sisquoc and Repetto and upper Pliocene Foxen and Careaga formations, as commonly mapped in the onshore Santa Maria basin, and the coastal Santa Barbara Channel.

In lease P 0315, Texaco et al. discovered a small gas accumulation from shallow Pliocene sandstones. Seismic amplitude anomalies or "bright spots" can be seen within the Pliocene section. They usually represent noncommercial gassy sands and siltstones.

STRUCTURE

Point Arguello Structure

Figure 16 is a simplified Monterey structure map of the Point Arguello field anticlinal feature. The fold is approximately 9 mi (14.5 km) long and 2 mi (3.2 km) wide. In general, the structure can be divided into two faulted, doubly plunging anticlines separated by a saddle in Lease P 0450.

The locations of seismic lines A-A' and B-B' and structure cross section C-C' are also shown in Figure 16. Line A-A' (Figure 17), through the northern portion of the Point Arguello field, is an east-west-oriented time-migrated seismic line. Complex folding and reverse faulting are shown in Figure 17, as well as the Chevron-Phillips P 0450 #3 and P 0450 #1, the northern pool discovery well. The seismic and well control indicate that thrust faulted structures have formed on the flanks of the Point Arguello structure.

Seismic line B-B' (Figure 18) is a northeast-southwest-oriented three-dimensional user track passing approximately 1800 ft (549 m) northwest of the Chevron et al. P 0316 #1 discovery well for the southern pool and platform Hermosa. Both flanks of the anticline are cut by reverse faults.

Structure cross section C-C' (Figure 13) shows both the structural and stratigraphic relationships between the

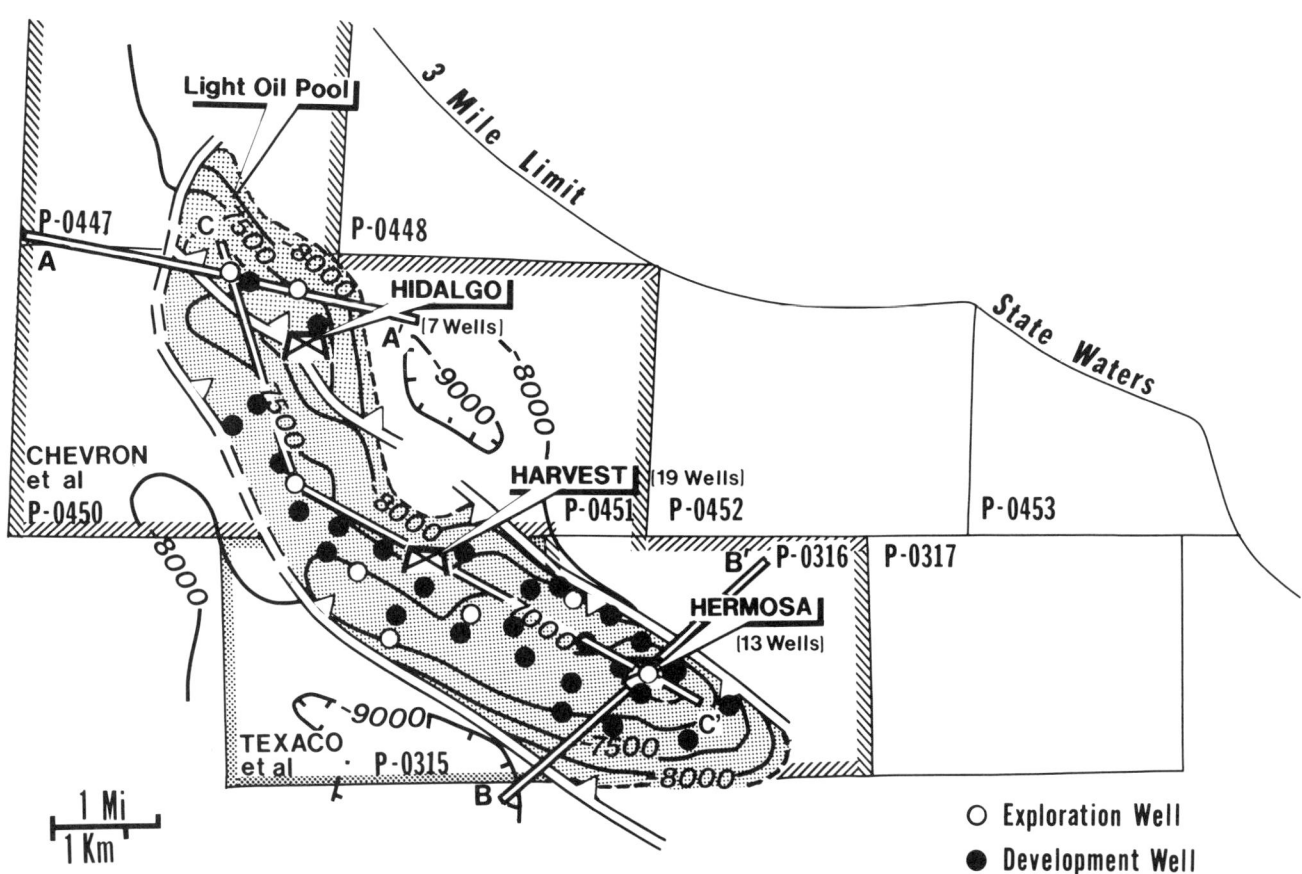

FIGURE 16. Chevron simplified structure map on the top of the Monterey Formation in July 1989 after the interpretation of three-dimensional seismic surveys and the drilling of the 39 development wells from three platforms at the field. Note the location of the light oil (34° API gravity) pool on the northern margin of the field. The contour interval is 1000 ft (305 m).

northeast-trending Amberjack high and the north-northwest-trending Point Arguello field. The stratigraphic onlap and thinning of the Monterey and older Miocene units onto the Amberjack high are illustrated as well as a simplified picture of the cross-faulting between P 0450 #3 and P 0450 #2, which may separate the northern and southern oil pools.

The Amberjack high may have developed as early as the Oligocene and continued to grow through much of the Miocene. As previously discussed, Luyendyk and Hornafius (1987) suggested that the Amberjack high and the associated Neogene depocenter to the north (referred to as the Point Arguello depocenter) evolved during the early Miocene rotation of the western Transverse Ranges. By the latest Miocene and Pliocene this basin extension had ceased and the transform margin underwent compression (Crouch et al., 1984). In the onshore Santa Maria basin, northeast-southwest-directed compression formed the large producing anticlinal structures, and in the offshore this late Miocene compressional event reactivated the Hosgri fault as a southwest vergent low-angle thrust system (Crouch et al., 1984). We believe this initiated the northwest-trending Point Arguello fold trend and superimposed it on the northeast-trending Amberjack high.

Structural growth of the Point Arguello fold continued throughout the Pliocene and in the adjacent Rocky Point field; compressional growth continued into the Pleistocene.

Thinning of the Santa Margarita over the Point Arguello structure indicates that the anticlinal trap began to develop during the late Miocene while middle Pliocene-lower Pleistocene compression gave the Point Arguello structure its present closure.

Trap

The trap is an anticlinal closure enhanced by faulting. Reflection seismic surveys show reverse faults cutting both flanks and the northwest plunge of the Point Arguello structure (Figure 16). Faulting within the structural saddle in block P 0450 (Figure 16) may provide the southern closure for the northern light oil pool.

Seismic line A-A' (Figure 17) demonstrates the northeast-dipping reverse fault that forms the southern boundary of the Point Arguello field. Seismic line B-B' (Figure 18) shows the southwest-dipping reverse fault that breaks the northeast flank and increases the closure

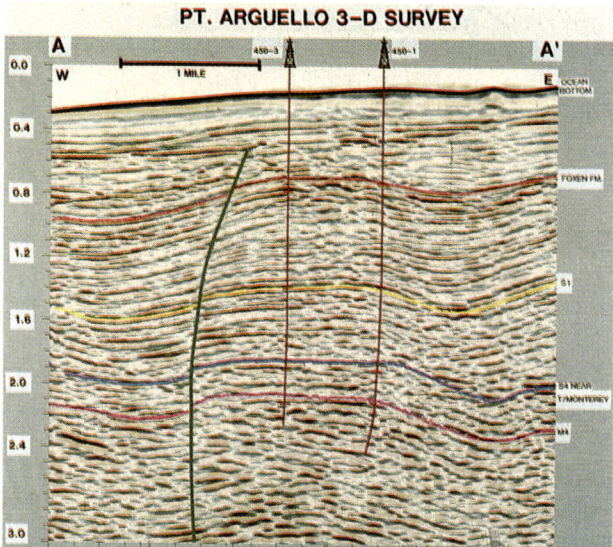

FIGURE 17. East-west time-migrated three-dimensional seismic section passing near the Chevron-Phillips OCS P 0450 #3 and #1 wells. The S4 traverse is within the Sisquoc (Santa Margarita) Formation near the top of the Monterey, and the M4 traverse is within the Monterey Formation at the top of the lower calcareous-siliceous member. See Figure 16 for location.

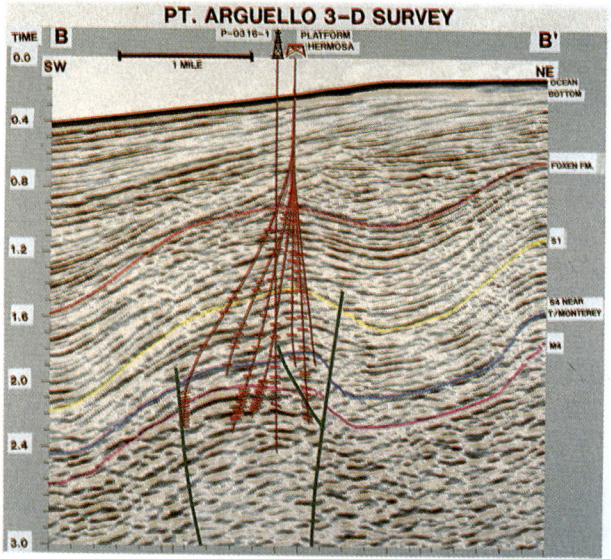

FIGURE 18. Northeast-southwest time-migrated three-dimensional seismic section passing near the Chevron et al. OCS P 0316 #1 well and Platform Hermosa. The S4 traverse is within the Sisquoc (Santa Margarita) Formation near the top of the Monterey, and the M4 traverse is within the Monterey Formation at the top of the lower calcareous-siliceous member. See Figure 16 for location.

of Point Arguello's southern pool. The maximum fault-enhanced closure is about 2400 ft (732 m), and the maximum Monterey hydrocarbon column or "fillup" is approximately 1800 ft (549 m) thick.

Without detailed well control, accurate mapping of the Monterey trap is difficult because of the poor reflection coefficient of the Santa Margarita (lower Sisquoc)-Monterey contact. The early seismic mapping (Figures 6 and 7) was derived from mistakenly traversing the more coherent basal events within the Santa Margarita Formation.

The Santa Margarita (lower Sisquoc) is the overlying seal. Generally, the Santa Margarita siltstones and mudstones are inefficient vertical seals with erratic oil-stained fracture zones often extending several hundred feet above the Monterey reservoir. In the Point Arguello field, strong oil shows in cuttings and core occur below the present oil-water contacts. This suggests that the oil column was originally much deeper or that the oil is still migrating into the reservoir. Fish (1989) suggests that fracturing of the overlying Santa Margarita seal may have caused the oil-water contacts to move upward to their present positions.

Although no seeps are believed to be present over the Point Arguello field, they are commonly associated with known Monterey accumulations (Priestaf, 1979). Geochemical sampling of microseeps in the Point Arguello area has predicted the API gravity of the Monterey in the subsurface with a fair degree of accuracy (Kennicutt and Brooks, 1988). Unless the Santa Margarita–lower Sisquoc seal is thick and unfaulted, a commercial Monterey oil accumulation is a geologically ephemeral phenomenon. Recharging of Monterey reservoirs by present-day oil-generating centers probably helps maintain leaky Monterey accumulations. In the Santa Barbara Channel, over 1 billion bbl of oil have been produced from Pliocene sandstones overlying the Santa Margarita Formation believed to be sourced from the more deeply buried Monterey Formation.

RESERVOIR

General

The southern pool in the Point Arguello field is approximately 6500 ft (1982 m) deep and has an oil column of slightly over 1600 ft (488 m). A small gas cap may exist in the southern pool based on the drill-stem-test in the discovery well, but it has not been confirmed by production. The northern pool, located in a separate faulted anticlinal trap, is approximately 1000 ft (305 m) deeper at 7500 ft (2287 m) and has a maximum oil column of about 650 ft (198 m). No primary gas cap is present at the northern pool.

The principal fractured reservoir facies at the Point Arguello field are the clay-free, glassy cherts and more clayey porcelanites, a microcrystalline siliceous rock with a matte luster like that of unglazed porcelain. The major producing interval is expected to be the upper calcareous siliceous member (Figure 8), composed mainly of cherts and porcelanites.

As the clay content of the rock increases, the reservoir quality decreases. Siliceous mudstones of the clayey-siliceous member and mudstones of the phosphatic member are less profilic reservoirs, but are often oil-prone source rocks. The 1-to-3-ft (0.3-to-1-m) dolomitic beds are usually poorly fractured but in places cause the mudstones and porcelanites to fracture around them. The carbonate-rich, basal calcareous member is not a major reservoir facies at Point Arguello field, simply because it is below the oil-water contact in most places in the field. This member has tested with good oil rates where it is above the oil-water contact on Lease P 0316.

Deposition and Diagenesis

The Monterey Formation at Point Arguello field was deposited in a deep-water anaerobic environment in a subbasin adjacent to the Amberjack high (Figure 15). The finely laminated pelagic deposits of diatomaceous and foraminiferal debris, as well as the organic-rich calcareous and phosphatic mudstones, were well preserved where anoxic bottom environments prevented bioturbation. The Point Arguello subbasin (depocenter) remained isolated from terrigenous detritus until the late Miocene, when the subbasins to the east filled and began to spill fine clastics westward, causing clays to mix with the diatomaceous sediments of the Monterey. This gradual mixing with terrigenous clays occurred at Point Arguello in the upper Mohnian–Delmontian, as evidenced by increased detrital clay content within the clayey-siliceous member (MacKinnon, 1989b). By the late Miocene, oxygen-rich bottom conditions in the Point Arguello depocenter resulted in the bioturbation of the thick-bedded and clayey-diatomaceous Santa Margarita (Sisquoc) Formation. Sand-sized terrigenous detritus did not enter the Point Arguello depocenter until the upper Pliocene.

Diagenesis converted the finely laminated biogenetic-clayey Monterey sediments into cherts, porcelanites, siliceous siltstones, mudstones, and dolostones which comprise the Monterey fractured reservoirs (Isaacs, 1981; Isaacs, 1987). The most significant diagenetic effect is the conversion of the biogenic silica in the diatoms and forams (originally in an opal A phase) to a quartz-phase silica, which is hard and brittle. With increasing time and temperature, the original diatomites were first converted from amorphous silica (opal A) to a more ordered crystalline phase, cristobalite-tridymite (opal CT), as analyzed by x-ray diffraction. Figure 19 is a scanning electron microscope (SEM) photomicrograph of unaltered diatomite illustrating the porous (60 to 70%), low-permeability, low-density prediagenetic deposition. The change from opal A to opal CT occurs within the present temperature range of 105 to 138°F (40.6 to 58.9°C) in the Point Arguello field. Figure 20 is an SEM photomicrograph of porcelanite in the opal CT phase. The microporosity commonly decreases to about 20 to 30% in porcelanites.

The last diagenetic step at Point Arguello was a silica phase change from opal CT to quartz, occurring within the present subsurface temperature range of 150 to 190°F (65.5 to 87.8°C). Figure 21 is an SEM photomicrograph of a glassy chert in the quartz phase. Impermeable but most prone to fracturing, chert zones are the primary fractured reservoirs within the Monterey. The Monterey in the Point Arguello field is entirely within the quartz phase. In other onshore and offshore fields, opal CT–phase porce-

FIGURE 19. Scanning electron microscope (SEM) photomicrograph of unaltered diatomite (silica in opal A phase). This primary biogenic opal A phase material is altered during burial diagenesis to opal CT and eventually to quartz phase silica yielding siliceous shales, porcelanites, and cherts. Bulk density, <1.70–2.04; porosity, 60–70%; velocity, 5800–7800 ft/sec (1800–2400 m/sec).

FIGURE 20. Scanning electron microscope (SEM) photomicrograph of porcelanite. Silica is in the opal CT (cristobalite-tridymite) phase. The depth of the transition from opal A to opal CT phase in the Point Arguello area is shown in Figure 26. Bulk density, 2.24–2.33; porosity, 20–30%; velocity (p), 10,300–13,900 ft/sec (3140–4250 m/sec).

lanites, cherts, and siliceous mudstones also form productive, fractured Monterey reservoirs.

Dolostones occur as beds, lenses, and nodules (Garrison et al., 1984). Thin sections from cores and outcrop samples indicate that the dolostones formed by replacement and pore filling of the siliceous sediments. Close examination of relict laminations within the dolostones and compaction features in the surrounding rocks suggest that the dolostones formed in both the early precompaction and postcompaction stages of diagenesis.

Mudstones containing blebs or fine laminations of phosphatic material occur as erratic thin interbeds

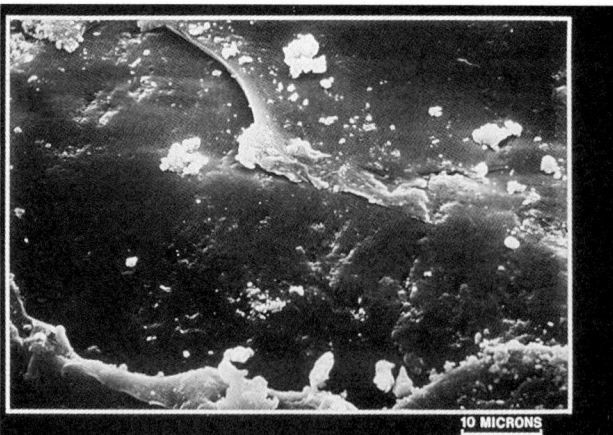

FIGURE 21. Scanning electron microscope (SEM) photomicrograph of glassy chert in the quartz phase. The transition from opal CT to chert phase in the Point Arguello area is shown in Figure 26. Cherts contain little or no terrigenous clay material and can be in either the opal CT or quartz phase. Bulk density, 2.40–2.56; porosity, 0–10%; velocity (p), 14,000–19,000 ft/sec (4270–5800 m/sec).

throughout the Monterey, but are thickest in the phosphatic member. Generally, the mudstones contain the highest organic content (as much as 18% TOC) in the Monterey section (Pytte, 1989). The mudstones are calcareous with nodules and thin interbeds of dolostones, porcelanites, and cherts. Finely laminated siliceous mudstones are more common near the top of the Point Arguello Monterey reservoir within the clayey-siliceous member although thin-bedded cherts, porcelanites, and dolostone nodules are also occasionally present.

Reservoir Character

Crain et al. (1985) reviewed the general porosity and permeability of the Point Arguello field as determined from the initial delineation wells. Fish (1989) reported that limited interference tests suggest good lateral and vertical permeability. Bulk permeabilities have been calculated from drill-stem-test (DST) pressure data. Permeability to oil ranges from less than 1 md/ft (3.3 md/m) to more than 3 darcys/ft (9.8 darcys/m). Permeability measurements on core plugs suggested a matrix permeability of 0.1 md or less for most of the Monterey reservoir facies (Crain et al., 1985).

Variations in fracture and not intercrystalline matrix permeability will control the production rates in the Point Arguello field. It is expected that the decline curves will be similar to the onshore Monterey decline curves: about 45% the first year, 30% the next, and 15% thereafter. The typical Monterey decline rate is assumed by most reservoir engineers to be caused by a dual porosity system. The large and intermediate fracture systems may source the initial flush production. As oils in the major fracture networks are quickly produced, production rapidly declines and then stabilizes as microfractures slowly recharge the large fracture network.

Fracture density varies according to the argillaceous content of the rock. Fracturing is least developed in the mudstones, siliceous shales, and clayey dolostones and is best developed in the glassy cherts, pure dolomites, and porcelanites. Core recoveries from the most productive intervals are poor, forcing us to base our interpretation of the Monterey fracture zones on outcrops, test data, and wireline logs. In the cores, fractures are often parallel and at high angles to the bedding. The natural fractures can be open, partially open, filled with some porosity, or filled with no porosity. Natural tar and heavy oil can be found filling many of the fractures in the coastal outcrops near Point Arguello and Point Conception.

Drilling breaks are most common within the chert- or dolostone-rich intervals, which in some cases are lost circulation zones (Fish, 1989). Dilation breccias (Redwine, 1981; Roehl, 1981) are mapped onshore but have not been cored in the Point Arguello field. However, Fish (1989) reported that wells have occasionally encountered zones of intensely fractured cherts or dolostones up to 50 ft (15 m) thick. These intervals, marked by increased penetration rates and good production rates, have calculated isotropic interclast porosities as high as 25% and may represent fault-associated fracture zones or dilation breccias.

MacKinnon (1989b) summarizes the results of extensive outcrop studies of the fracturing in the Monterey that identified three fracture types and estimated permeability based on fracture spacing and width. The fracture types included small fractures within bedding, brecciated beds in chert-rich intervals, and large fractures that cut several lithologic layers. Estimated permeabilities for fractures within bedding ranged between 56 and 1503 md for brecciated cherts, 2 and 112 md for cherts, and 1 and 56 md for dolostones, mudstones, and porcelanites. When larger multilayer fractures and brecciated zones are mixed together, the net effective fracture permeabilities tend to range between 30 and 300 md (Figure 22). Normal chert-rich zones can have average permeabilities of up to 300 md/ft (984 md/m), but often contain thin intervals of highly brecciated rocks (MacKinnon, 1989b). The unbrecciated layers often have average permeabilities of about 45 md/ft (148 md/m) and are believed to be connected to the erratic fracture zones by occasional high-angle, crosscutting fractures. Porceleneous zones commonly have permeabilities of about 30 md/ft (98 md/m) and average about 2 fractures/ft (6.6 fractures/m). Unfractured porcelanites have permeabilities of less than 1 md/ft (3.3 md/m).

It appears that most fracturing is tectonically induced. Published data from the Santa Barbara Channel's South Elwood field (Belfield et al., 1983) suggest that northeast-southwest tension fractures may be present. Outcrop measurements indicate that fractures occur as conjugate sets of both shear and extension systems, depending on their fold positions.

Total reservoir porosity ranges from 10 to 20%, with fracture porosity averaging between 1 and 2% of the reservoir volume (Crain et al., 1985). Some secondary vugular porosity has been described in both dolostones and cherts. Primary matrix microporosity in the Monterey reservoir ranges from 0 to 15%.

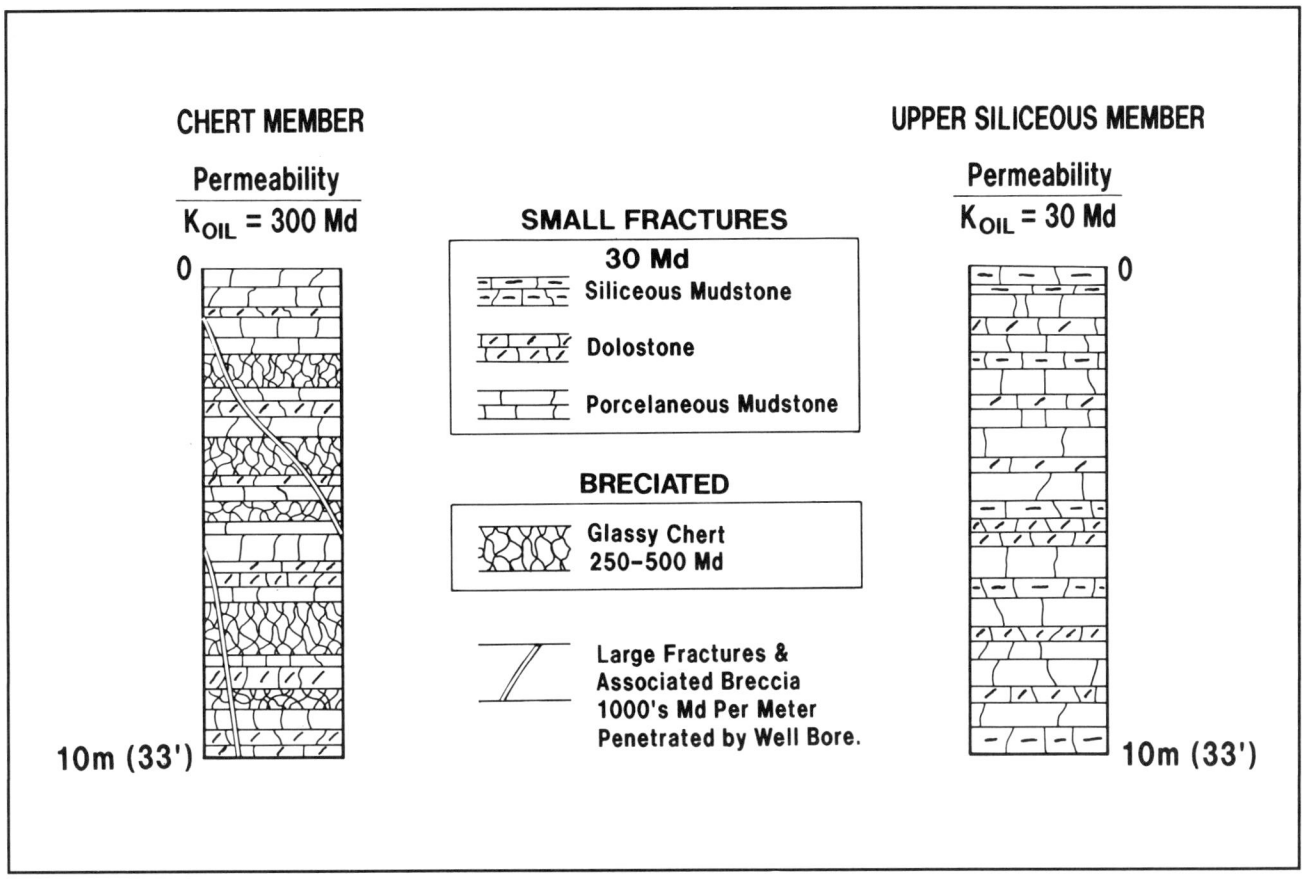

FIGURE 22. Fracture permeability model for the Monterey reservoir at Point Arguello field (from MacKinnon, 1989b). This simplified model suggests that a 33-ft (10-m) section of the clayey-siliceous member of the Monterey, consisting of mudstones, dolostones, and porcelanites, has an average fracture permeability of 30 md. When interbeds of highly fractured glassy cherts and large fractures occur, such as in the chert member (upper calcareous-siliceous), the average permeability of a 33-ft (10-m) interval increases to 300 md.

HYDROCARBONS

Character

The Monterey oil produced from the Point Arguello field is classified as aromatic-intermediate. The southern pool is gravity-segregated, ranging from about 11° API at the base to about 23° API at the top of the Point Arguello field, a segregation common in many Monterey reservoirs.

The API gravity of Monterey oils is roughly inverse to the sulfur content. Sulfur content varies from about 0.8% in the 30° API oils (northern pool) to as much as 5% in the 11 to 23° API oils (southern pool). A plot based on Monterey oil analyses by Orr (1986) and Magoon and Isaacs (1983) illustrates the general relationship between oil gravity and sulfur content (Figure 23).

A representative gas chromatogram is shown in Figure 24. No biodegradation is present. Normal paraffins are well developed with a slight even-carbon preference among the normal paraffins in the C18 to C30 range. The heavier oils in the southern pool contain thermally unstable porphyrins, suggesting an accumulation of thermally immature oil. The carbon isotope ratio (C13/C12) of the whole crude averages approximately −22.7‰ PDB.

The oil is undersaturated with a solution-gas content averaging 400 ft^3/bbl. Gas from the P 0316 lease has a gross heating value of 1183 Btu/ft^3. H_2S is about 0.58 mole %, and methane is around 82 mole %. CO_2 and N_2 average 4% and 0.43%, respectively. Gas from the northern pool has a gross heating value of 1077 Btu/ft^3. H_2S is about 0.05 mole %, and methane averages about 89 mole %. CO_2 and N_2 average 2.6% and 1.24%, respectively (Crain et al., 1985).

Generation

All the hydrocarbons found in the Point Arguello field are believed to have been generated primarily within the Monterey, with a minor contribution from the Point Sal Formation, on the basis of carbon isotopes and biomarker

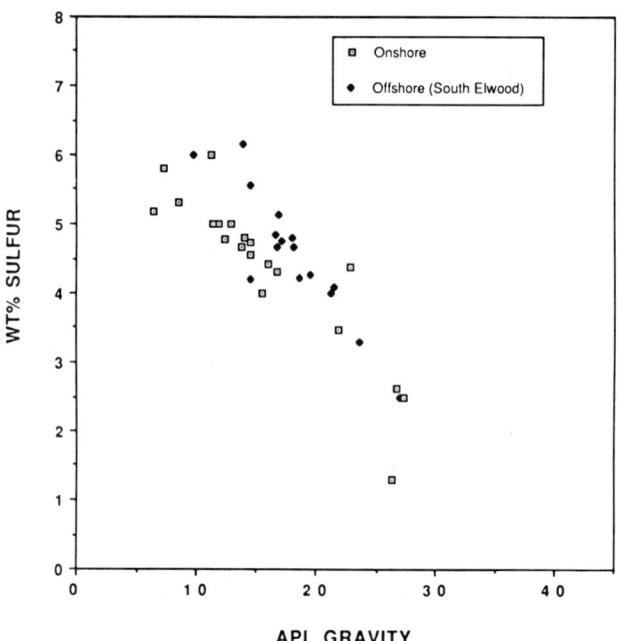

FIGURE 23. Weight percent sulfur versus API gravity for several onshore and offshore oils from Monterey Formation reservoirs.

studies (Crain et al., 1985). Understanding the critical factors that permit the early generation and trapping of large hydrocarbon accumulations of oils with widely diverse API gravities is important to any Monterey exploration program.

The total organic carbon (TOC) content of the Monterey Formation in the Point Arguello field averages about 3%. Phosphatic mudstones vary, but commonly contain between 10 and 18% TOC. The Monterey kerogens are classified as oil-prone high-sulfur type II with a sulfur content of up to 9% in the Point Arguello field (Pytte, 1989) (Figure 25).

Most of the oil-prone Monterey kerogens in the Point Arguello field have an average atomic hydrogen/carbon ratio of about 1.25. Generally, the Monterey section at Point Arguello field has a hydrogen index of 600 to 700 (mg H/g OC) and a hydrocarbon potential of between 160 and 1200 bbl/ac-ft. Of of the kerogens examined, 95 to 98% are sapropelic-amorphous kerogens derived from algal material. Little or no humic organic material is present.

The oil-generating "kitchen" sourcing the Point Arguello field is located in the Monterey depocenter shown in Figure 14. The location of the Point Arguello structure within the generating center is important because it provided a short migration path to the Point Arguello trap, which was necessary for the formation of a large oil accumulation from upper Tertiary source rocks. The higher-than-normal heat flow (Blackwell, 1979) and rapid burial within the Neogene depocenter put early Miocene Monterey source rocks into the oil-generation window by late Miocene to early Pliocene (Figure 26).

In addition, the sulfur-rich Monterey kerogens are cracked into thermally immature, heavy hydrocarbons at a lower maturation level than are normal type II kerogens (Orr, 1984; Orr, 1986). Oil generation from the sulfur-rich Monterey kerogens begins at vitrinite reflectance maturation levels of 0.3 to 0.4% R_o (Petersen and Hickey, 1983) or at a time-temperature index (TTI) value of 3.0 or less (Waples, 1987), a maturation level normally considered thermally immature.

Figure 26 is a generalized geohistory diagram (modified from Waples, 1980) for the Point Arguello depocenter. Typical Monterey kerogen activation energies were used to model the hydrocarbon-generation window after the calculation techniques described by Tissot and Welte (1984). The decrease in heat flow in coastal California since the Oligocene has been modeled by Heasler and Surdam (1985). Using similar thermal-decay assumptions, the present heat flow is calculated at about 88 mW/m², a value compatible with published heat-flow data (Blackwell, 1979). The temperature gradient is about 2.6°F/100 ft (48°C/km), which is similar to the temperature gradient in the nearby COST well (McCulloh and Beyer, 1979).

The geohistory plot shows continuous sedimentation since the deposition of the Point Sal (lower Monterey) Formation. The conversion index (KCl) is a percentage of the kerogen's generative potential that has been converted to hydrocarbons. KCl values of 10 and 90% define the approximate beginning and end of the oil-generating window. Thermal modeling indicates that the Monterey in the Point Arguello depocenter entered the oil window 5 m.y.a. and reached peak oil generation at 4 m.y.a., and portions are within the peak oil-generation window today. From the Pliocene–Pleistocene to the present, high Monterey Formation temperatures could have thermally cracked the lower-temperature, earlier-formed, immature heavy oils. This may be the source of the 30+° API gravity oil found in the Point Arguello and nearby Rocky Point fields. The 30+° API oils could also represent the most thermally mature oils generated from the Monterey at greater depths of burial in the depocenter northeast of the field. Geohistory modeling of the field itself suggests that today the Monterey is within the peak oil window below 9000 ft (2743 m).

FIELD DEVELOPMENT

Production facilities consist of three production platforms, erected on federal leases in OCS waters, and oil- and gas-processing facilities located onshore at Gaviota. These facilities are connected by two industry-built pipelines—one oil, one gas.

Platforms Hermosa and Hidalgo were erected on the Chevron-operated leases, P 0316 and P 0450, whereas Harvest was placed on Texaco's Lease P 0315. Hermosa, measuring 755 ft (230 m) from ocean bottom to the helicopter deck, was set in 600 ft (183 m) of water. The three-deck, 48-slot platform supports oil- and gas-separation facilities and two drilling rigs. Hookup and commission

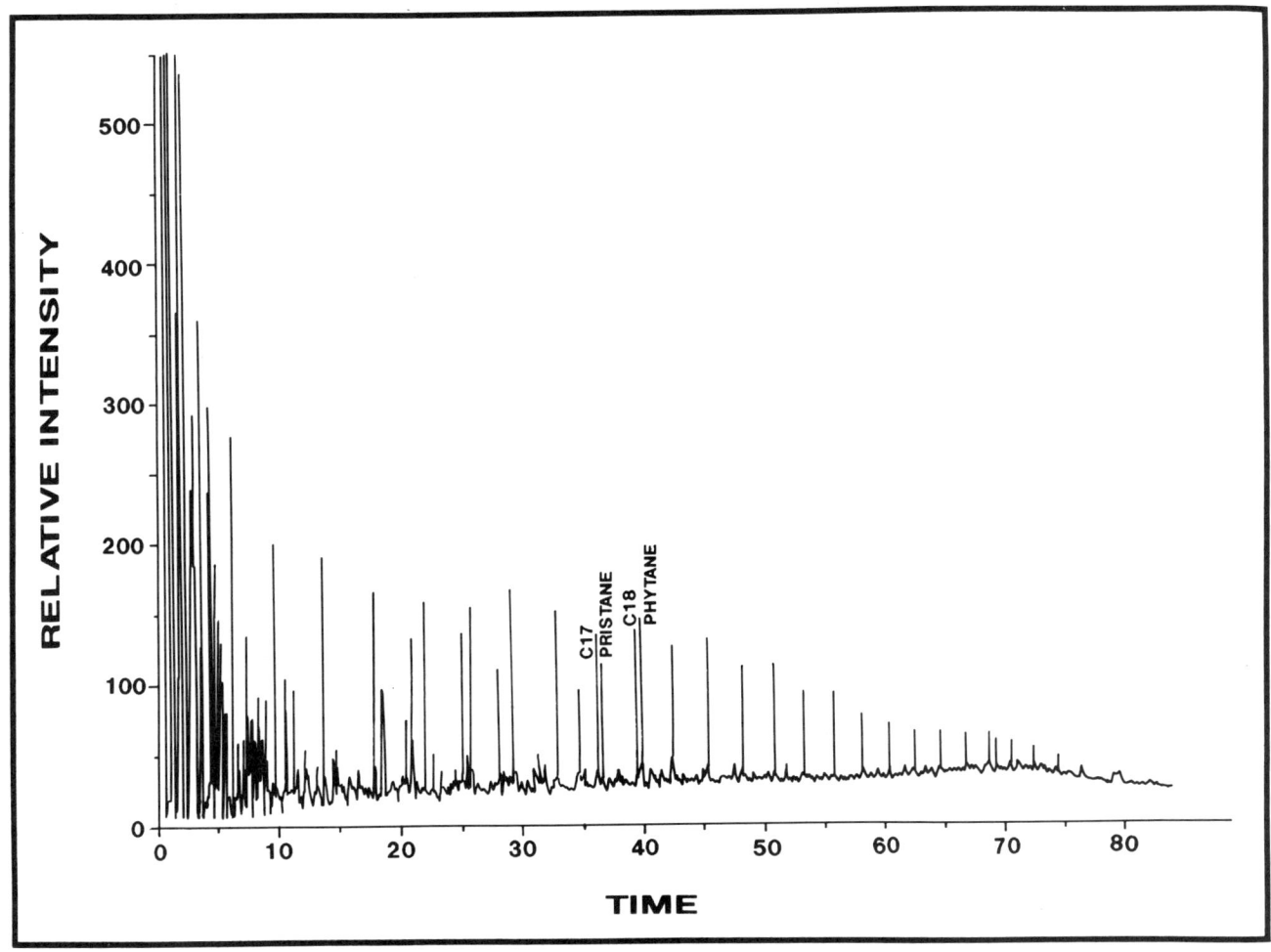

FIGURE 24. Typical gas chromatogram of a Monterey oil sample from the southern pool of the Point Arguello field.

were completed in late 1986, and 13 development wells have been drilled.

Hidalgo was set in 430 ft (181 m) of water in 1987. Seven wells have been completed from this 56-slot platform. The Texaco-operated platform, Harvest, was set in 675 ft (206 m) of water, where drilling of 19 development wells commenced in late 1986.

Further development drilling will depend on well performance and has been suspended on all platforms pending commencement of production. Total production from the two Chevron-operated platforms is expected to peak at about 45,000 BOPD.

Oil and gas from Hidalgo and from Texaco's Harvest will be commingled with production from Hermosa and piped to shore to the Chevron-operated Gaviota Plant for dewatering, desulfurization, and other processing. The Gaviota Oil and Gas Plant has a capacity of 100,000 BOPD and 60 MMCFGD and was completed near the end of 1987.

Total investment by the consortium of 18 oil companies involved in developing the Point Arguello project exceeds $2 billion. The project was scheduled to begin production in late 1987 after dozens of permits were obtained, including one from Santa Barbara County for which more than 160 mitigation measures were satisfied. However, the project is still stalled.[1] Santa Barbara County and the California Coastal Commission must approve the crude oil transportation plan for Point Arguello crude oil, and until a suitable plan for moving the majority of the crude by pipeline is developed, start-up will not be permitted. The interim marine terminal permit for tankering the Point Arguello crude to Los Angeles was allowed to expire without renewal following the oil spill in Prince William Sound in 1989.

---

[1]The Point Arguello field began production on May 27, 1991, under an operating permit that limits the oil flow to the capacity of existing onshore pipelines. The conflict over tankering the oil from the Gaviota Marine Terminal continues. The field produced 10,000 BOPD in June and averaged 35,000 BOPD in October 1991.

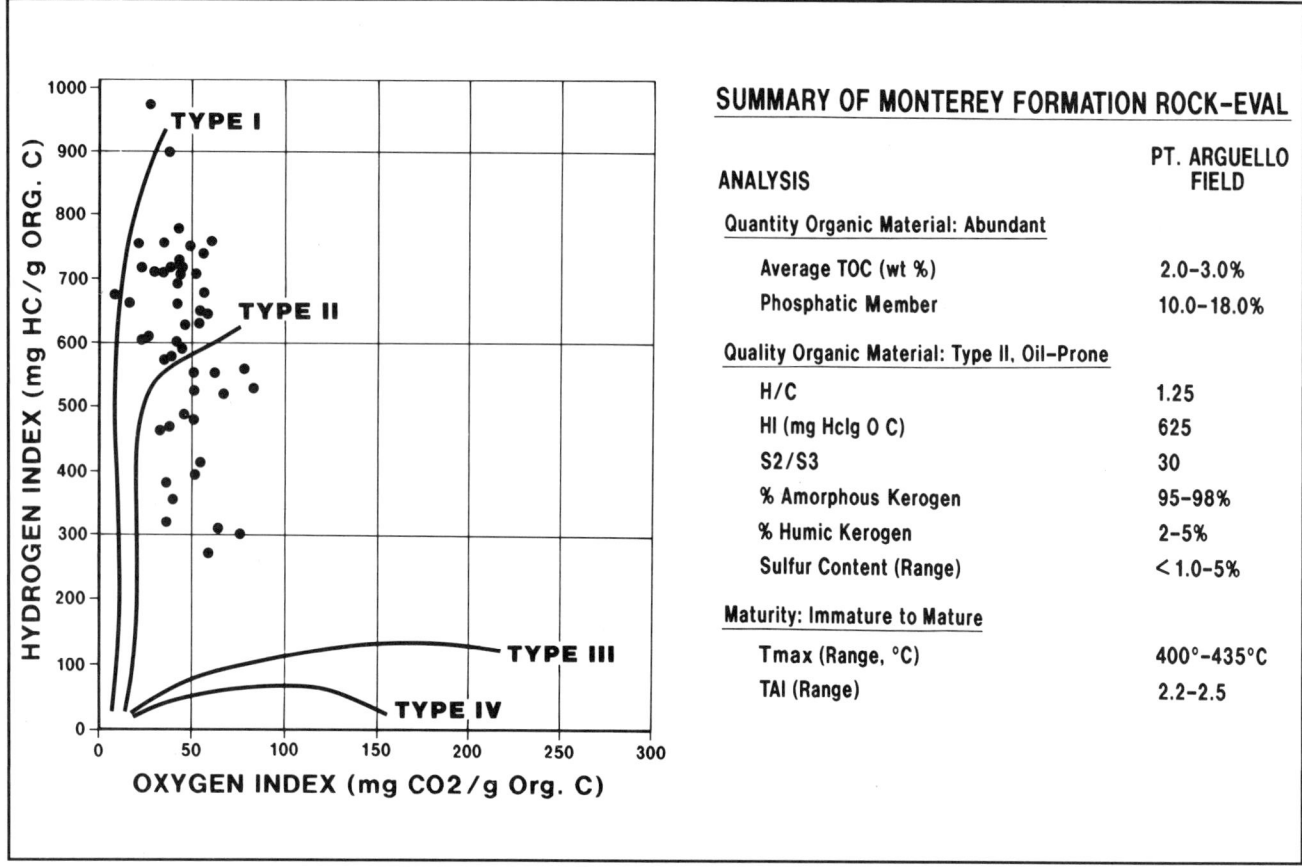

FIGURE 25. Rock-Eval hydrogen and oxygen index plot and a summary of Rock-Eval characteristics of the Monterey Formation in the Point Arguello field and elsewhere in the Santa Maria basin onshore and offshore (from Pytte, 1989).

CONCLUSIONS

The Point Arguello field is part of a regional trend that contains several anticlinal-fault-enhanced traps in close proximity to a major oil-generating center. Monterey sediments are diagenetically altered within this trend due to rapid burial, high heat flow, and high temperature gradients creating fracturable reservoirs of cherts, porcelanites, and siliceous shales. The low clay content of this siliceous Monterey "fairway" produced a high percentage of cherts, the most productive of the fractured reservoirs. The regional Monterey trend is also rich in oil-prone organic matter that converts to hydrocarbons under maturation levels lower than those at which normal type II kerogens convert.

The factors that created the oil accumulation in the Point Arguello field also are common to many fields, throughout Southern and Central California, that produce from the Monterey or Monterey equivalents. These factors are early oil generation, short migration paths, siliceous fractured reservoirs, and structural trap development coincident with oil generation.

Most of the large anticlinal Monterey traps have now been tested except in those California offshore basins closed to exploration. Large remaining Monterey oil accumulations may also be found in stratigraphic-structural traps. The first step in an exploration program will be to identify cherty, low-clay Monterey trends close to or within high-gravity, low-sulfur oil-generating source areas. The seismic response to sharp diagenetic siliceous phase boundaries around the flanks of regional uplifts may identify potential stratigraphic traps formed at the boundaries between opal A and the fracturable CT and quartz phases (Ogle et al., 1987). Because of the leaky nature of the typical Santa Margarita–Sisquoc, a thick overlying section is also required in order to prevent the rapid vertical loss of hydrocarbons from the reservoir. However, large traps may also be formed by near-surface asphaltic seals caused by low temperatures, biodegradation, water washing, and loss of volatiles.

The successful exploration for commercial Monterey accumulations requires the integration of geology, stratigraphy, geochemistry, and geophysical disciplines. Billions of in-place barrels have been discovered in the Monterey during the past decade. Three-dimensional seismic surveys have become a common tool for planning the development of offshore Santa Maria fields. However, velocity determinations for stacking and migration purposes can be extremely difficult with steep dips (over 45°) and rapid lateral velocity changes common over the offshore structures. A layered velocity model from borehole and stacking velocities was constructed, allowing a more successful one-pass three-dimensional migration of seismic events (Fischer, 1987).

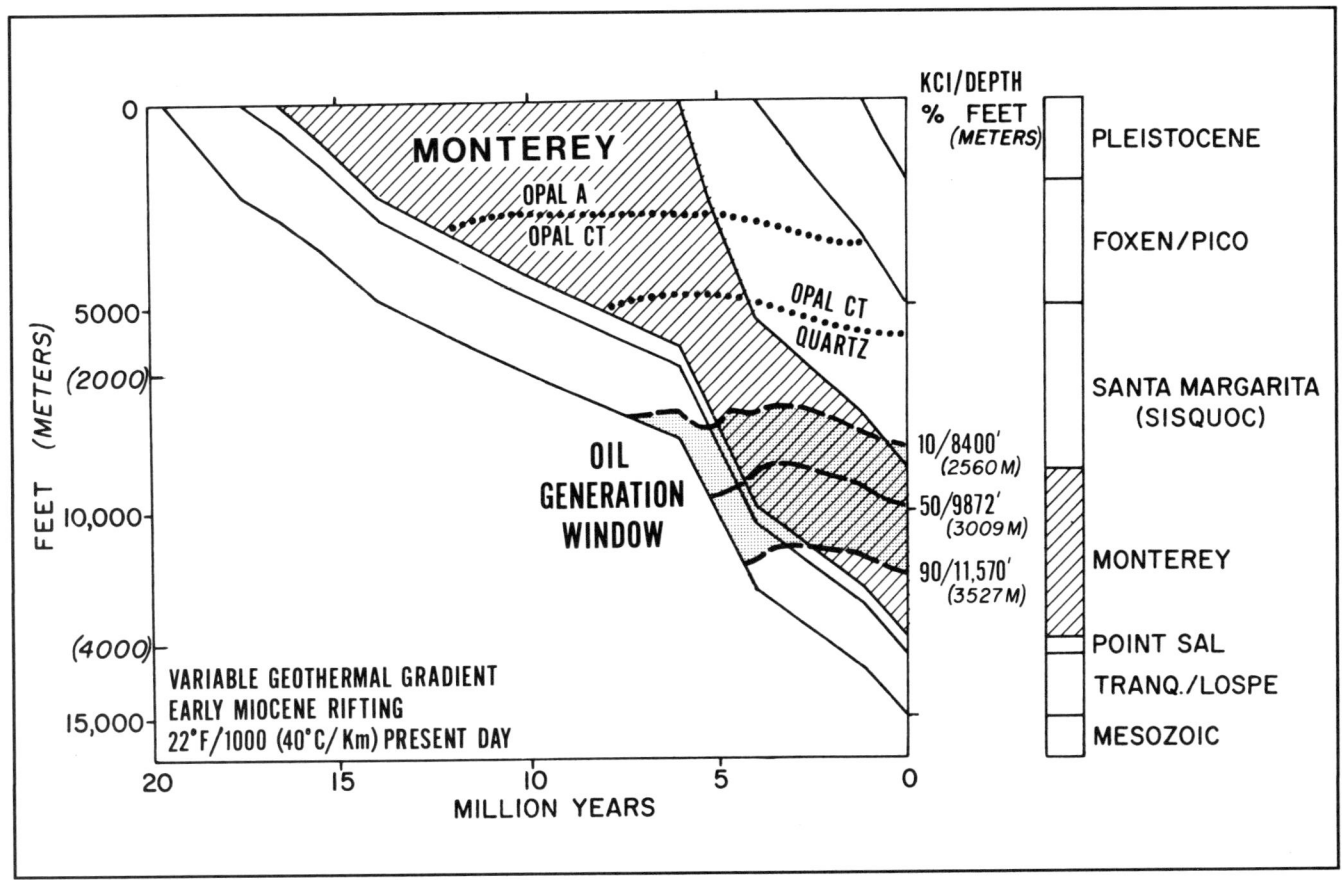

FIGURE 26. Schematic geohistory diagram for the Neogene Point Arguello depocenter located to the northeast of the Point Arguello field (see Figure 14 for location and text for explanation).

The reservoir engineer and the explorationist will play increasingly important roles in expanding our producible reserves from these erratic fractured reservoirs. Horizontal drilling techniques and long-term flowing tests may be necessary to properly evaluate and economically develop many of these Monterey reserves (Skillin, 1989).

SUMMARY

The offshore Santa Maria basin was essentially unexplored until OCS Sale 53 in 1981. Since that sale, 54 exploratory wells have been drilled, resulting in ten new field discoveries. An estimated 1.5 billion bbl of new domestic reserves have been found; however, due to fierce public opposition, only one field is currently producing at some 17,000 BOPD.

The largest discovery in the basin, Point Arguello, contains 300 million bbl of producible oil. Three production platforms and 39 development wells are poised to produce up to 100,000 BOPD.

Farther north along the California coast are six other virtually unexplored basins. It seems inevitable that as U.S. reserves decline and dependence on imports grows, the nation will need to open these basins to exploration.

ACKNOWLEDGMENTS

The authors wish to thank the management of Chevron and Phillips, and our other partners for permission to publish this paper. Mr. Ron Patterson is gratefully acknowledged for assembling the figures at long distance from the authors.

REFERENCES CITED

Atwater, T., 1970, Implications of plate tectonics for the Cenozoic tectonic evolution of western North America: GSA Bulletin, v. 81, p. 3513–3536.

Belfield, W. C., J. Helwig, P. Lapointe, and W. K. Dahleen, 1983, South Ellwood oil field, Santa Barbara Channel, California, a Monterey Formation fractured reservoir, in C. M. Isaacs and R. E. Garrison (eds.), Petroleum generation and occurrence in the Miocene Monterey Formation, California: SEPM, Pacific Section, p. 213–221.

Blackwell, D. D., 1979, Heat flow and energy loss in the western United States: GSA Memoir 152, p. 175–208.

Blake, M. C., Jr., R. H. Campbell, T. W. Dibblee, Jr., D. G. Howell, T. H. Nilsen, W. R. Normark, J. G. Vedder, and

E. A. Silver, 1978, Neogene basin formation in relation to plate-tectonic evolution of San Andreas fault system, California: AAPG Bulletin, v. 62, p. 344–372.

Crain, W. E., W. E. Mero, and D. Patterson, 1985, Geology of the Point Arguello discovery: AAPG Bulletin, v. 69, p. 537–545.

Crain, W. E., and S. P. Thurston, 1987, Geology and oil and gas exploration in California's offshore basins, in M. K. Horn (ed.), Transactions of the Fourth Circum-Pacific Energy and Mineral Resources Conference: AAPG Transactions, v. 4, p. 43–61.

Crouch, J. K., S. B. Bachman, and J. T. Shay, 1984, Post-Miocene compressional tectonics along the central California margin, in J. K. Crouch and S. B. Bachman (eds.), Tectonics and sedimentation along the California margin: SEPM, Pacific Section, v. 38, p. 37–54.

Dibblee, T. W., 1950, Geology of southwestern Santa Barbara County, California—Point Arguello, Lompoc, Point Conception, Los Olivos, and Gaviota quadrangles: California Division of Mines Bulletin 150, 95 p.

Dickinson, W. R., 1979, Cenozoic plate tectonic setting of the Cordilleran region in the United States, in J. M. Armentrout, M. R. Cole, and H. TerBest, Jr. (eds.), Cenozoic paleogeography of the western United States: SEPM, Pacific Section, Los Angeles, p. 1–13.

Dickinson, W. R., and W. S. Snyder, 1979, Geometry of triple junctions related to San Andreas transform: Journal of Geophysical Research, v. 84, p. 561–572.

Dumont, M. P., 1989, The Monterey Formation and biostratigraphy: an overview, in T. C. MacKinnon (ed.), Oil in the California Monterey Formation, Field Trip Guidebook T311, 28th International Geological Congress, American Geophysical Union, p. 28–32.

Fischer, T., 1987, Discovery of the Point Arguello oil field from a geophysical perspective: Geophysics: The Leading Edge of Exploration, v. 6, n. 10, p. 16–21.

Fish, J. L., 1989, Pt. Arguello field formation evaluation of fractured Monterey reservoir, in T. C. MacKinnon (ed.), Oil in the California Monterey Formation, Field Trip Guidebook T311, 28th International Geological Congress, American Geophysical Union, p. 45–49.

Garrison, R. E., M. Kostner, and D. H. Zenger (eds.), 1984, Dolomites of the Monterey Formation and other organic-rich units: SEPM, Pacific Section, Publication n. 41, Los Angeles, 215 p.

Gorsline, D. S., and K. O. Emery, 1959, Turbidity current deposits in San Pedro and Santa Monica basins off southern California: GSA Bulletin, v. 70, p. 279–290.

Heasler, H. P., and R. C. Surdam, 1985, Thermal evolution of coastal California with application to hydrocarbon maturation: AAPG Bulletin, v. 69, n. 9, p. 1386–1400.

Hornafius, J. S., 1985, Neogene tectonic rotation of the Santa Ynez range, western Transverse Ranges, California, suggested by paleomagnetic investigation of the Monterey Formation: Journal of Geophysical Research, v. 90, p. 12,503–12,522.

Hornafius, J. S., B. P. Luyendyk, R. R. Terres, and M. J. Kamerling, 1986, Timing and extent of Neogene tectonic rotation in the western Transverse Ranges, California: GSA Bulletin, v. 97, p. 1476–1487.

Hoskins, E. G., and J. R. Griffith, 1971, Hydrocarbon potential of northern and central California offshore, in Future petroleum provinces of the United States—their geology and potential: AAPG Memoir 15, p. 212–228.

Ingle, J. C., Jr., 1981, Origin of Neogene diatomites around the north Pacific rim, in R. E. Garrison, and R. G. Douglas (eds.), The Monterey Formation and related siliceous rocks of California: SEPM, Pacific Section, Los Angeles, p. 159–179.

Isaacs, C. M., 1980, Diagenesis in the Monterey Formation examined laterally along the coast near Santa Barbara, California: PhD thesis, Stanford University, Stanford, California, 344 p.

Isaacs, C. M., 1981, Porosity reduction during diagenesis of the Monterey Formation, Santa Barbara coastal area, California, in R. E. Garrison and R. G. Douglas (eds.), The Monterey Formation and related siliceous rocks of California: SEPM, Pacific Section, Los Angeles, p. 257–271.

Isaacs, C. M., 1984, The Monterey—Key to offshore California boom: Oil and Gas Journal, v. 82, Jan. 9, p. 75–81.

Isaacs, C. M., 1987, Sources and deposition of organic matter in the Monterey Formation, south central coastal basins of California, in R. F. Meyer (ed.), Exploration for heavy crude oil and natural bitumen, AAPG Studies in Geology #25, p. 193–205.

Isaacs, C. M., and R. E. Garrison (eds.), 1983, Petroleum generation and occurrence in the Miocene Monterey Formation, California: SEPM, Pacific Section, Los Angeles, 228 p.

Kennicutt, M. C. and J. M. Brooks, 1988, Surface geochemical exploration studies predict API gravity off California: Oil and Gas Journal, Sept. 12, p. 101–106.

King, J. D., and G. E. Claypool, 1983, Biological marker compounds and implications for generation and migration of petroleum in rocks of the Point Conception deep-stratigraphic test well, OCS-CAL 78-164 No. 1, offshore California, in C. M. Isaacs and R. E. Garrison (eds.), Petroleum generation and occurrence in the Miocene Monterey Formation, California: SEPM, Pacific Section, Los Angeles, p. 191–200.

Lagoe, M. B., 1985, Depositional environments in the Monterey Formation, Cuyama basin, California: GSA Bulletin, v. 96, p. 1296–1312.

Luyendyk, B. P. and J. S. Hornafius, 1987, Neogene crustal rotations, fault slip, and basin development in southern California, in R. V. Ingersoll and W. G. Ernst (eds.), Cenozoic basin development of coastal California, Rubey Volume VI, p. 259–283.

Luyendyk, B. P., M. J. Kamerling, R. R. Terres, and J. S. Hornafius, 1985, Simple shear of southern California during Neogene time suggested by paleomagnetic declinations: Journal of Geophysical Research, v. 90, p. 12,454–12,466.

MacKinnon, T. C., 1989a, Origin of the Miocene Monterey Formation in California, in T. C. MacKinnon (ed.), Oil in the California Monterey Formation, Field Trip Guidebook T311, 28th International Geological Congress, American Geophysical Union, p. 1–10.

MacKinnon, T. C., 1989b, Petroleum geology of the Monterey Formation in the Santa Maria and Santa Barbara coastal and offshore areas, in T. C. MacKinnon (ed.), Oil in the California Monterey Formation, Field Trip Guidebook T311, 28th International Geological Congress, American Geophysical Union, p. 11–27.

Magoon, L. B., and C. M. Isaacs, 1983, Chemical characteristics of some crude oils from the Santa Maria basin, California, in C. M. Isaacs and R. E. Garrison (eds.), Petroleum generation and occurrence in the Miocene Monterey Formation, California: SEPM, Pacific Section, Los Angeles, p. 201–211.

McCulloh, T. H., 1969, Geologic characteristics of the Dos Cuadras offshore oil field: USGS Professional Paper 679-C, p. 29–46.

McCulloh, T. H., and L. A. Beyer, 1979, Geothermal gradients, in geologic studies of the Point Conception deep stratigraphic test well 1 OCS-Cal 78-164, outer continental shelf, southern California, U. S.: USGS Open-File Report 79-1218, p. 43–48.

Nilsen, T. H., 1984, Oligocene sedimentation and tectonics, California, in T. H. Nilsen (ed.), Fluvial sedimentation and related tectonic framework, western North America: Sedimentary Geology, v. 38, p. 305–336.

Nilsen, T. H., 1986, Cretaceous paleogeography of western North America, in P. L. Abbott (ed.), Cretaceous stratigraphy of western North America: SEPM, Pacific Section, v. 46, Los Angeles, p. 1–39.

Nilsen, T. H., 1987, Paleogene tectonics and sedimentation of coastal California, in R. V. Ingersoll and W. G. Ernst (eds.), Cenozoic Basin Development of Coastal California, Rubey Volume VI, Prentice-Hall, p. 81–123.

Ogle, B. A., W. S. Wallis, R. G. Heck, and E. B. Edwards, 1987, Petroleum geology of the Monterey Formation in the offshore Santa Maria/Santa Barbara areas, in R. V. Ingersoll and W. G. Ernst (eds.), Cenozoic basin development of coastal California, Rubey Volume VI, Prentice-Hall, p. 383–406.

Orr, W. L., 1984, Sulfur and sulfur isotope ratios in Monterey oils of the Santa Maria River basin and Santa Barbara Channel area (abs.): SEPM Annual Midyear Meeting Abstracts, p. 62.

Orr, W. L., 1986, Kerogen/asphaltene/sulfur relationships in sulfur-rich Monterey oils: Organic Geochemistry, Advances in Organic Geochemistry 1985, Part 1, Petroleum Geochemistry, p. 499–516.

Page, B. M., and D. C. Engebretson, 1984, Correlation between the geologic record and computed plate motions for central California: Tectonics, v. 3, p. 133–155.

Petersen, N. F., and P. J. Hickey, 1983, Evidence of early generation of oil from Miocene source rocks, California coastal basins, in C. M. Isaacs and R. E. Garrison (eds.), Petroleum generation and occurrence in the Miocene Monterey Formation, California: SEPM, Pacific Section, p. 226.

Pisciotto, K. A., and R. E. Garrison, 1981, Lithofacies and depositional environments of the Monterey Formation, California, in R. E. Garrison and R. G. Douglas (eds.), The Monterey Formation and related siliceous rocks of California: SEPM, Pacific Section, p. 97–122.

Priestaf, I., 1979, Natural tar seeps and asphalt deposits of Santa Barbara County: California Geology, v. 32, n. 8, p. 163–169.

Pytte, M. H., 1989, Organic geochemistry of the Miocene Monterey and equivalent formations in five California basins, in T. C. MacKinnon (ed.), Oil in the California Monterey Formation, Field Trip Guidebook T311, 28th International Geological Congress, American Geophysical Union, p. 33–36.

Redwine, L. E., 1981, Hypothesis combining dilation, natural hydraulic fracturing, and dolomitization to explain petroleum reservoirs in Monterey shale, Santa Maria area, California, in R. E. Garrison and R. G. Douglas (eds.), The Monterey Formation and related siliceous rocks of California: SEPM, Pacific Section, p. 221–248.

Roehl, P. O., 1981, Dilation brecciation—a proposed mechanism of fracturing, petroleum expulsion and dolomitization in the Monterey Formation, California, in R. E. Garrison and R. G. Douglas (eds.), The Monterey Formation and related siliceous rocks of California: SEPM, Pacific Section, p. 285–315.

Skillin, R. H., 1989, Monterey development—onshore and offshore Santa Maria basins, in T. C. MacKinnon (ed.), Oil in the California Monterey Formation, Field Trip Guidebook T311, 28th International Geological Congress, American Geophysical Union, p. 41–44.

Tissot, B. P., and D. H. Welte, 1984, Petroleum formation and occurrence, Springer-Verlag, New York, 699 p.

Waples, D. W., 1980, Time and temperature in petroleum exploration: application of Lopatin's method to petroleum exploration: AAPG Bulletin, v. 64, p. 916–926.

Waples, D. W., 1987, Predicting thermal maturity, in N. H. Foster and E. A. Beaumont (eds.), Geologic basins I, classification, modeling, and predictive stratigraphy: AAPG Treatise of Petroleum Geology, Reprint Series, n. 1, p.249–282.

Woodring, W. P., and M. N. Bramlette, 1950, Geology and paleontology of the Santa Maria district, California: USGS Professional Paper 222, 185 p.

Yerkes, R. F., H. C. Wagner, and K. A. Yenne, 1969, Petroleum development in the region of the Santa Barbara Channel: USGS Professional Paper 679-B, p. 13–27.

Chapter 3

West Chalkley, Cameron Parish, Louisiana
A Case for Continued Exploration in Mature Producing Provinces

G. E. Klefstad

TXP Operating Company
Houston, Texas, U.S.A.

ABSTRACT

A potential giant gas field has been discovered in the very mature exploration province of south Louisiana, by Transco Exploration Partners (TXP) and Exxon Company U.S.A. The West Chalkley prospect is located in Cameron Parish, Louisiana, and is productive in the upper Oligocene Miogypsinoides (Miogyp) sands. This discovery is in the same producing trend as the prolific South Lake Arthur field, where the Miogyp sands have gas reserves on the order of 1.0 tcf.

The West Chalkley prospect was generated by a combination of trend analysis, subsurface well control, and reflection seismic data. The feature appears to be a faulted anticline separate from the nearest production in the area, Chalkley field, which is located about 1 mi (1.6 km) to the east and was discovered in 1938.

Both TXP and Exxon, working independently, recognized the potential prospect and pursued leasing activities in the area. TXP initiated discussions with the landowner in February 1988 and acquired a 960-ac (3.9-km^2) lease in June. Exxon leased approximately 2100 ac (8.5 km^2) surrounding the TXP lease about one month later. TXP subsequently sold the prospect to Exxon on October 12, 1988. The Exxon #1 Sweet Lake Land & Oil Company was spudded on March 16, 1989, and reached a total depth of 15,600 ft (4755 m) on July 4, 1989. Log analysis indicated a net gas pay zone of nearly 500 ft (152 m) in the 805-ft (245-m) gross productive interval. Testing through perforations near the base of the pay zone yielded flow rates as high as 21.028 MMCFGD and 330 BCPD.

The well was put on production in March 1990 and has produced at rates as high as 45.4 MMCFGD and 740 BCPD. Two successful offsets have been logged, and a third is drilling. At least three more wells are scheduled, with one being a deeper pool wildcat.

INTRODUCTION

A potential giant gas discovery has been announced in south Louisiana. It is probably the largest gas discovery in the onshore United States in the last ten years. The West Chalkley prospect is located in the north-central portion of Cameron Parish only about 18 mi (29 km) from the Gulf of Mexico (Figures 1 and 2). One of the most significant aspects of this discovery is that it is in the very mature south Louisiana exploration province. The first oil discovered in commercial quantities in the state was near the West Chalkley prospect in the fall of 1901. The 1988 production in south Louisiana onshore alone amounted to more than 114 million bbl of crude and condensate and more than 1.1 tcf of gas.

The existing production nearest to the West Chalkley prospect is the Chalkley field production located about 1 mi (1.6 km) to the east. Discovered in 1938, Chalkley field produces from reservoirs as shallow as 8500 ft (2590 m) and as deep as 13,000 ft (3960 m). Cumulative production for Chalkley field through 1988 amounted to 17 million bbl of crude and condensate and 360 bcf of gas. The area has been actively drilled for decades, with 24 wells drilled within a 2-mi (3.2-km) radius, five of which were drilled deeper than 14,600 ft (4450 m).

TREND ANALYSIS

The West Chalkley prospect lies within the Tertiary (lower Miocene–upper Oligocene) producing trend of south Louisiana. It is located downthrown to the regional Oligocene expansion fault system that forms the Camerina embayment. Figure 3 depicts a portion of a biostratigraphic chart pertinent to south Louisiana.

TXP set out to explore for prospects along the expanded Miogyp sand trend in the fall of 1987. The easternmost fields on this trend are S. E. Gueydan and Riceville in Vermilion Parish. Productive from the Miogyp sand and shallower reservoirs, S. E. Gueydan has produced 204 bcf of gas and nearly 7 million bbl of crude and condensate. This field was discovered in 1954, but the Miogyp reservoirs were not found until 1970. A review of S. E. Gueydan determined that the immediate area was fully developed and held by production.

Riceville was discovered in 1958 and has cumulative production of 134 bcf of gas and 4.4 million bbl of crude and condensate, primarily from the Camerina sands. A thick Miogyp sand is present on the west side of the field but has not yet been found productive. After review of the well control and purchase of available seismic, additional exploratory potential was recognized in the Camerina and Miogyp sands. TXP acquired some leasehold and has since sold that prospect to another operator. A Miogyp test is planned for late 1990 or early 1991.

The next significant field and certainly the largest Miogyp producing field is the South Lake Arthur field in Vermilion and Jefferson Davis parishes. South Lake Arthur was discovered in 1955 and has produced 345 bcf

FIGURE 1. Study area map.

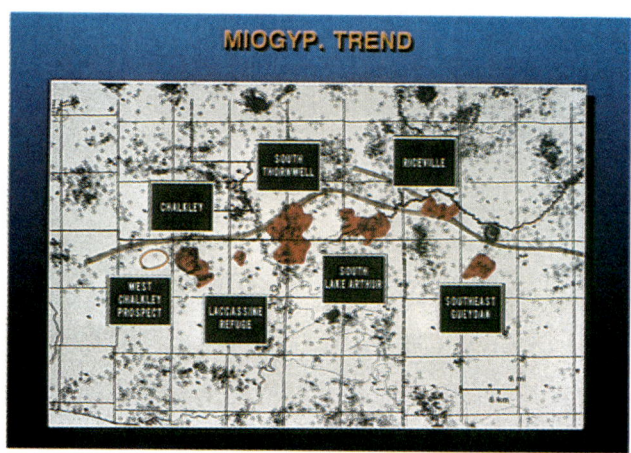

FIGURE 2. Miogyp trend map showing oil and gas fields.

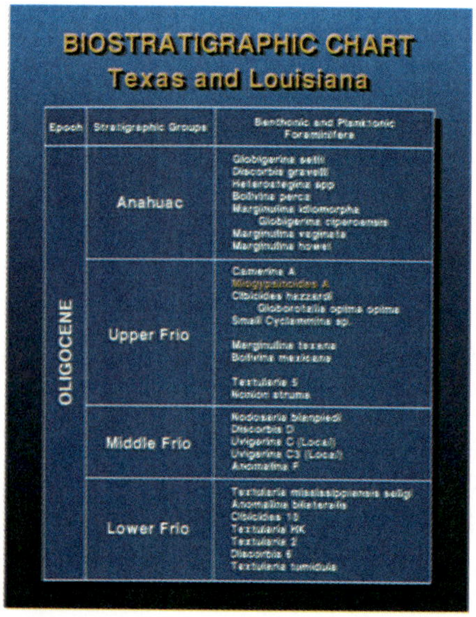

FIGURE 3. Biostratigraphic chart, Texas and Louisiana.

of gas and 2.6 million bbl of condensate, but the Miogyp section was not discovered until the mid-1970s. From that time to the present the Miogyp reservoir has been aggressively developed by several operators. This level of activity has caused lease bonus costs to remain very high. Therefore, this immediate area was eliminated from further exploration by TXP. As of mid-1989, South Lake Arthur had 24 wells completed in Miogyp reservoirs, producing about 240 MMCFGD and nearly 600 BCPD. The Miogyp gas reserves associated with South Lake Arthur are believed to be on the order of 1.0 tcf.

Continuing west, the next major structural complex is the South Thornwell field, discovered in 1942. South Thornwell has produced nearly 700 bcf of gas and 15 million bbl of condensate from Miogyp and younger sands, and a deep test was being drilled. Because so much acreage was held by production and because TXP's seismic database was very poor in the vicinity, the South Thornwell area was bypassed.

The Laccassine Refuge field is the next known structural element heading west along the trend. After some early failures in TXP's attempts to assemble a good seismic grid across the refuge at a reasonable cost, and in the absence of well logs exhibiting the expanded Miogyp sand section, the decision was made not to pursue further activities in the refuge area.

The early analysis of the Chalkley field area of the trend, however, was very encouraging. There were well logs with the Miogyp marker and sands present on the west side of the field, and the acreage in the area appeared to be unleased. In addition, some modern seismic data were available at reasonable prices. The presence of these factors convinced TXP to review the seismic database and integrate existing well control.

STRATIGRAPHIC CONTROL

Figure 4 is a map of the prospect area showing the Miogyp well control. Figure 5 is a stratigraphic cross section through the critical well control in the prospect area. The datum is the Miogyp marker, which is a limey zone present in all wells in the area and is easily correlated. The Miogyp sand is a few hundred feet below the marker.

One of the key wells is the Lynal #2, located on the northwest flank of the prospect. This well was interpreted as seeing the marker but stopping short of the sand section. The other well critical to the prospect was the Mecom #1, which was interpreted as seeing the marker and also seeing some Miogyp sand before faulting into Cibicides hazzardi at total depth. The Mecom well also had a show of gas in the Miogyp sand.

REGIONAL SEISMIC CONTROL

TXP had the 2-by-2-mi (3.2-by-3.2-km) regional seismic grid in the Chalkley area shown in Figure 6. Two regional seismic lines were instrumental in recognizing the potential of the West Chalkley prospect. Both of these lines were recorded 30-fold in 1984 with an asymmetrical split spread and dynamite or a combination of dynamite and air gun as a source. The lines were reprocessed in the fall of 1987, and it was these reprocessed data that were used in the initial interpretation.

The east-west line 9 (Figure 7) tied the Union Oil of California wells to the west and the Mecom #1 well to the east. By projecting the Lynal #2 well onto the north-south line 16 (Figure 8) and mapping a reflection believed to represent the Miogyp marker, it was apparent that one could get high to both the Mecom and Lynal wells.

A data search was initiated, and two brokered lines were purchased. The north-south line 3 had been shot and processed in 1982. TXP reprocessed the data but could not improve on the original. The data were of little value as a result of a great many skips north of the prospect. The second brokered line, line 2, was shot and processed in 1980. This line was also reprocessed by TXP and a significant improvement was obtained. Line 2 confirmed the presence of northeast dip away from the top of the prospect. By use of these data, a structure map of the area (Figure 9) was constructed, mapping a seismic reflection believed to be the Miogyp marker.

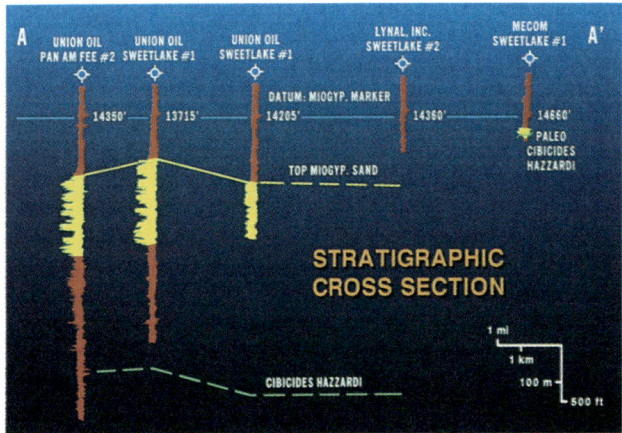

FIGURE 5. Miogyp stratigraphic cross section.

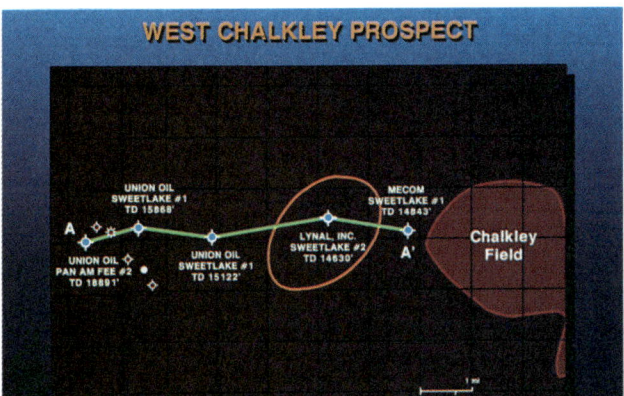

FIGURE 4. West Chalkley prospect area map.

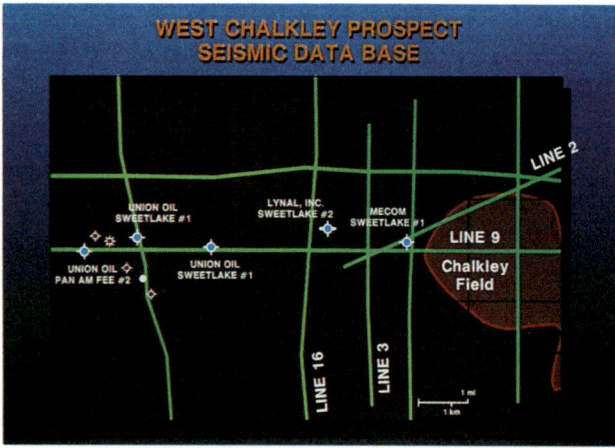

FIGURE 6. West Chalkley prospect, seismic database.

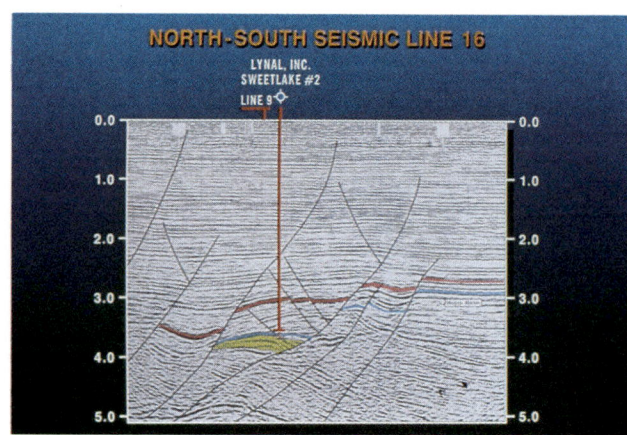

FIGURE 8. Geophysical Pursuit, Inc. north-south seismic line 16.

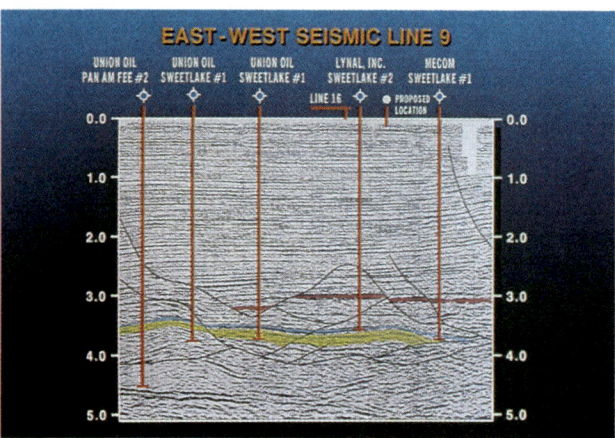

FIGURE 7. Geophysical Pursuit, Inc. east-west seismic line 9.

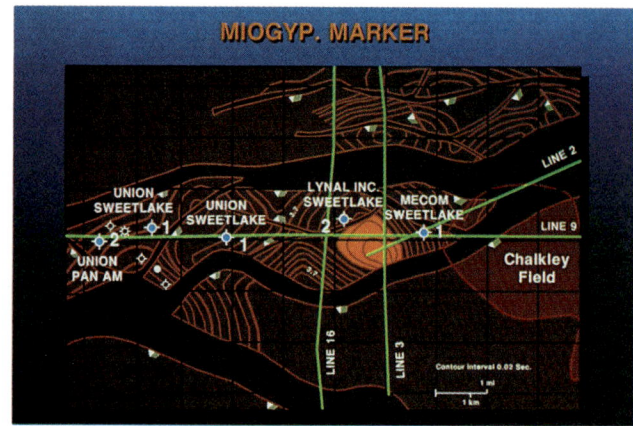

FIGURE 9. Miogyp marker seismic structure map. C.J.M./J.J.S. 6/24/88.

ECONOMICS

The prospect was estimated to have a maximum of 1100 ac (4.45 km²) under closure. The maximum hydrocarbon column was estimated at 335 ft (102 m). The most likely reserves estimates were conservatively placed at 105 bcf gas and 2.84 MMBC. These numbers were based on 700 productive acres (2.8 km²), 100 average net feet (30.5 m) of pay, and a recovery factor of 1500 mcf/ac-ft and 27 BC/mmcf. With an estimated cost of $2.1 million for a 16,500-ft (5030-m) dry hole and a composite probability of success of 20%, the prospect generated very attractive economic parameters.

LEASE ACQUISITION

TXP initiated discussions with the landowner in February 1988 seeking a six-month seismic option covering 1440 ac (5.83 km²) over the prospect. The option would buy enough time, at a nominal cost, to allow participation in a proposed speculative seismic line through the prospect area and shooting and processing of a proprietary seismic line if needed to further determine the optimal drill site.

After several attempts to obtain a seismic option failed, a lease was agreed to in June that covered 960 ac (3.9 km²) situated favorably on the prospective structure, as shown in Figure 10. Exxon, working independently of TXP, also recognized the potential prospect and about one month later acquired leases covering approximately 2100 ac (8.5 km²) surrounding the TXP lease.

PROSPECT SALES

TXP had been deferring discretionary drilling expenditures for some time in response to low product prices. As a result, a significant inventory of prospects had been assembled in all of the areas TXP actively explored, namely the Gulf Coast states of Louisiana, Texas, Mississippi, and Alabama as well as the Rocky Mountain basins. In August 1988, TXP was seeking support from other operators to drill these inventoried prospects. Exxon expressed an interest in the West Chalkley prospect.

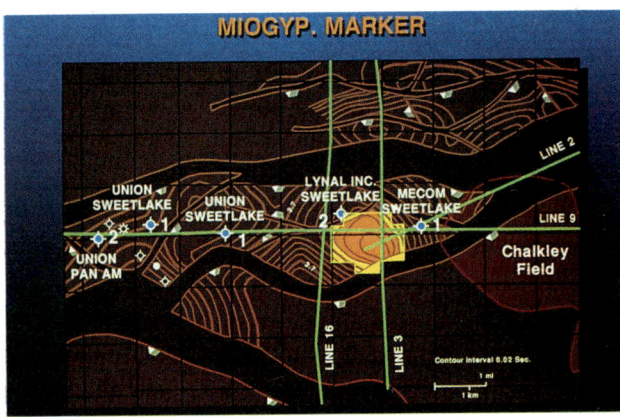

FIGURE 10. Miogyp marker seismic structure map showing TXP lease.

Negotiations commenced and a "farm-out" was signed on October 12, 1988.

DRILLING

The Exxon #1 Sweet Lake Land & Oil Company was spudded on March 16, 1989. Proposed to 15,500 ft (4725 m), drilling operations went smoothly with surface casing set at 2994 ft (913 m) and 9⅝- and 7⅝-in. casings set at 9190 and 13,904 ft (2801 and 4238 m), respectively.

At 13,904 ft (4238 m), six "poor to fair" mud log shows had been encountered, none of which had logged out as pay. However, the log did clearly show the Miogyp marker at 13,700 ft (4175 m), and the Miogyp paleo top came in at 13,470 ft (4105 m). TXP was very excited to be running about 500 ft (152 m) high to their seismic mapping. This led them to suspect a velocity anomaly, which was later confirmed by a velocity survey conducted in the discovery well.

The first log with pay was run on May 23, 1989, to a depth of 14,660 ft (4468 m). The well was then deepened to 14,901 and 15,110 ft (4542 and 4606 m) and was logged each time before the decision to run a 5½-in. casing was made. In running of the 5½-in. liner, it stuck short of total depth of 15,110 ft (4606 m) and was cemented at 14,807 ft (4513 m). A 4.4-in. hole was drilled on to 15,600 ft (4755 m), still in the Miogyp section. The final logs, an induction/short normal and an induction–density–gamma ray, were run on July 3rd and 4th, 1989, and a 2⅞-in. liner was run to 15,595 ft (4753 m).

LOG EVALUATION AND TESTING

Three conventional cores were taken over the course of drilling the wildcat. The amount of conventional core was limited because the core barrel jammed on all three attempts. The first interval cored was from 14,107 to 14,125 ft (4300 to 4305 m), recovering 14.5 ft (4.42 m). Sandstone porosities ranged from 12.2 to 23.4% and permeabilities ranged from 0.35 to 175 md. The second core was taken from 14,152 to 14,160 ft (4314 to 4316 m). Six feet (1.8 m) of silty sandstone were recovered. Porosity and permeability ranged from 10.9 to 24.9% and 0.64 to 53 md, respectively. The third conventional core was taken from 14,901 to 14,932 ft (4542 to 4551 m), recovering 30 of 31 ft (9.1 of 9.4 m). This section exhibited porosities of 12.0 to 14.7%, but permeabilities were very low, ranging from 0.14 to 0.77 md.

In addition to the 2½-in. conventional cores, 154 sidewall cores were recovered from the productive interval and analyzed.

The logs run from 13,904 to 15,110 ft (4238 to 4606 m) included DIL-SFL-sonic, GR-CNL-LDT, EPT, and microlog. A high-resolution dipmeter was also run over this interval, exhibiting dips of about 10° to the northwest in the upper portions of the productive interval. A Repeat Formation Tester was extensively utilized over the productive interval. A total of 52 pressure tests and 11 jug samples were taken. The pressure gradient throughout the interval was measured at 0.15 psi/ft.

The total pay count reached 485 net feet (148 m) of gas pay between 14,063 and 14,868 ft (4286 and 4532 m). The pay sands had an average porosity of 20% and an average water saturation of 31.4%. Figure 11 is the composite 1-in. dual induction log, over the depth interval from 14,000 to 15,000 ft (4270 to 4570 m).

The well was initially perforated from 14,836 to 14,848, 14,854 to 14,860, and 14,863 to 14,866 ft (4522 to 4526, 4527 to 4529, and 4530 to 4531 m). Later perforations were added from 14,670 to 14,674, 14,681 to 14,702, and 14,722 to 14,742 ft (4471 to 4473, 4475 to 4481, and 4487 to 4493 m). The well was then tested at rates of 6.7 MMCFGD and 139 BCPD, with a final tubing pressure of 9522 psi on an 11/64-in. choke. Shut in bottom hole pressure was 12,060 psi. Later tests yielded flow rates of 21.028 MMCFGD and 330 BCPD. Flowing tubing pressure on a 21.5/64-in. choke was 8744 psi with a 1% bottom hole pressure drawdown.

Pressure buildup analysis conducted following the flow tests indicated that the well bore was not damaged. The permeability and static bottom hole pressure were calculated to be 6.81 md and 11,968 psi, respectively. The estimated radius of investigation was 570 ft (174 m) with no reservoir boundaries apparent from the buildup analysis.

The gas and fluid analysis found 0.5 mole % hydrogen sulfide and 2.66 mole % carbon dioxide. Gas-oil ratio was 34,400 standard ft³/bbl. Thermal content was 1075 Btu/ft³. Bottom hole pressure and reservoir temperature were 12,015 psi and 300°F (149°C).

EXPLOITATION

During and subsequent to the drilling of the well, several speculative seismic lines were shot in the area. These lines were purchased following the completion of the well. The suspected velocity anomaly was confirmed with a vertical seismic profile conducted in the discovery. The average velocity to the Miogyp sand at the wildcat was approximately 7800 ft/s (2380 m/s) compared with more than 8100 ft/s (2470 m/s) in the dry hole 3 mi (5 km) to the west.

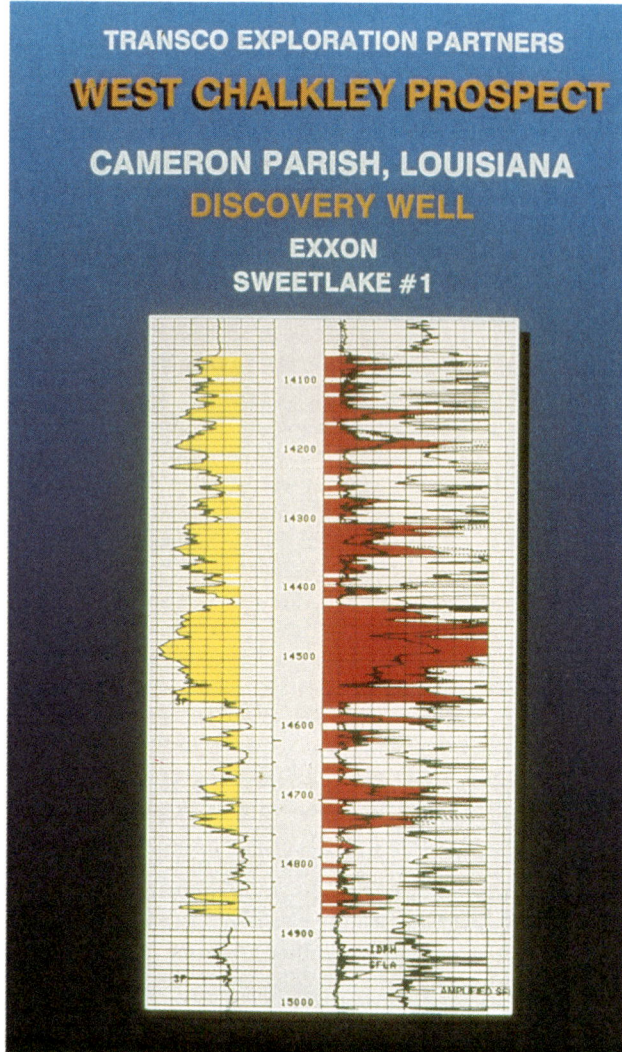

FIGURE 11. Well log. Transco Exploration Partners, West Chalkley prospect, Cameron Parish, Louisiana, discovery well, Exxon, Sweet Lake #1.

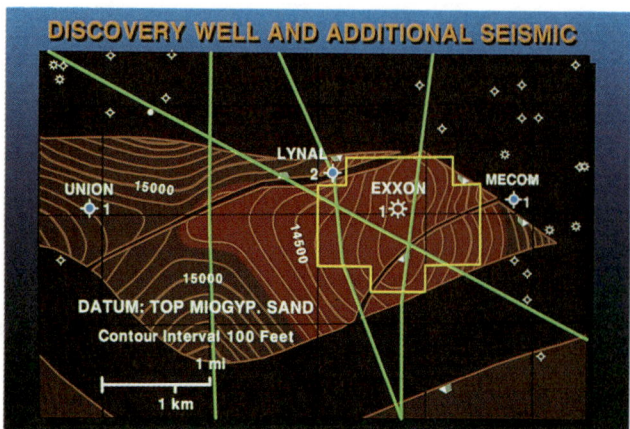

FIGURE 12. Top Miogyp sand structure map. Constructed after discovery well and additional seismic data. G.E.K. 11/3/89.

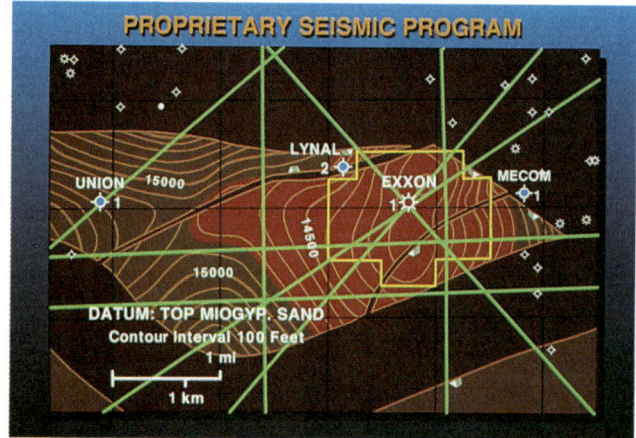

FIGURE 13. Proprietary seismic program map.

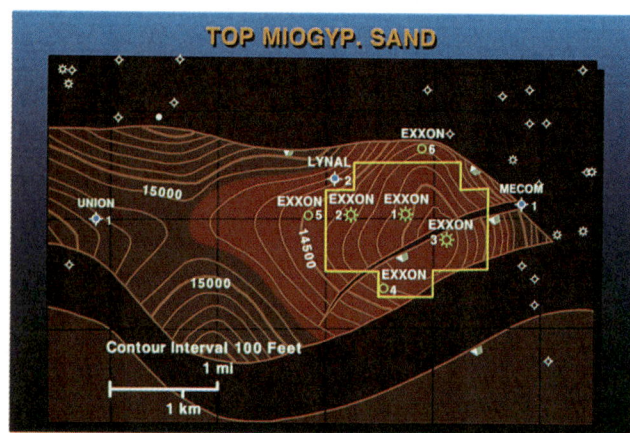

FIGURE 14. Top Miogyp sand structure map. G.E.K. 6/29/90.

Figure 12 is a structure map made on the top of the Miogyp pay sand after integrating the initial well and all of the seismic available at that time. TXP and the operator have jointly acquired approximately 60 mi (97 km) of additional seismic data to further define the structure and assist in determining delineation drilling locations (Figure 13). The seismic acquired was a 240-channel, 60-fold, 165-ft (50-m) group interval.

Because of the hydrogen sulfide and carbon dioxide content of the gas, and to facilitate higher flow rates, the decision was made to replace the initial test string with CRA P-110 chrome alloy tubing. Following that operation, the well has been on production at rates as high as 45.4 MMCFGD and 740 BCPD from perforations in the bottom 115 net feet (35 m) of pay.

Figure 14 is the current structural interpretation incorporating all seismic data and the first three wells drilled. The discovery well has been offset about 2900 ft (884 m) to the west by the #2 Sweet Lake Land & Oil Company. Spudded on January 18, 1990, this well reached total depth of 16,500 ft (5030 m) and encountered 300 net feet (91 m) of gas sand. The #3 Sweet Lake Land & Oil Company is located about 2400 ft (730 m) east-southeast from the discovery. The #3 well was spudded on February 1, 1990, and reached total depth of 16,100 ft (4910 m). This well encountered 246 net feet (75 m) of gas sand. A south offset, the #4 well, is tentatively planned for spudding in September 1990 and is targeted to 16,200 ft (4940 m).

The operator is currently drilling the #5 well as another western step-out. Located approximately 5000 ft (1525 m) west of the discovery well, this well was spudded on June 14, 1990, and is expected to drill to 16,100 ft (4910 m). Spudding of a north offset to the discovery has been planned following completion of the operator's #5 well. The #6 well will be located about 3100 ft (945 m) north-northeast from the #1 well, and its proposed total depth is 16,000 ft (4875 m). Another one or two locations are expected in the near future, including a deeper pool wildcat engineered to be drilled deeper in the productive fault block in an attempt to find additional Miogyp or older sands that may be present. Like the #1 well, each of the wells in the new reservoir is planned to be completed with chrome alloy tubing and as such will be capable of producing at sustained rates of 40 to 60 MMCFGD.

CONCLUSIONS

Accurate reserves estimates are difficult to make at this time given the unknowns of areal extent and net sand distribution, and such estimates will be deferred until more definitive data are collected.

More important than a precise estimate of reserves is the fact that this very significant reservoir lay hidden for 50 years following the discovery of Chalkley field. It is suspected that many explorationists assumed that the Mecom well was on the flank of the Chalkley structure at the Miogyp level. The acquisition of east-west seismic line 9, however, disputed this and suggested the presence of an older buried structure situated west of the old field. This older structure does not manifest itself in datums much younger than the Miogyp.

In the Gulf Coast of the United States, many operators are abandoning exploration efforts in the onshore areas because they are "so mature," preferring instead to focus on the offshore Gulf of Mexico shelf. The offshore shelf certainly remains an attractive exploration area. However, it is the author's contention that, because of the availability of high-quality, low-cost seismic data and the area-wide leasing schedules conducted in the OCS Gulf of Mexico, the offshore patch may not have undergone as much drilling, but it has certainly been more intensely scrutinized and more thoroughly evaluated than many onshore areas.

It is also important to note that the prospect was found without the benefit of a three-dimensional seismic data set, a sophisticated geophysical workstation, or the "alphabet soup technology" of AVO's or HCI's. It was found instead by using basic exploration procedures. By integration of the subsurface well control, a loose regional grid of modern seismic data augmented with reprocessed older data, and fresh geologic thinking, these older, more mature, onshore areas can yield very impressive results.

ACKNOWLEDGMENTS

The author acknowledges the help and support of Transco Exploration Company and many of its past and present employees. Special recognition is due the two outstanding explorationists responsible for generating the prospect, Conrad J. McGarry and J. J. Shelton. The contributions of Keith Vincent and R. L. Brusenhan in lease acquisition and farm-out negotiation are also acknowledged. Thanks go to Susan Downs, Judye Murray, W. B. Miller, and Rayer Chain for their assistance in preparing the manuscript. Thanks also to Geophysical Pursuit Inc. for granting permission to use their seismic data.

Chapter 4

The Reservoir Geology and Geophysics of the Hibernia Field, Offshore Newfoundland

T. J. Hurley
R. D. Kreisa
G. G. Taylor
W. R. L. Yates

*Mobil Oil Canada, Ltd.
Calgary, Alberta, Canada*

ABSTRACT

The Hibernia field is located 315 km (195 mi) offshore and east-southeast of St. John's, Newfoundland. The field was discovered in 1979 and contains an estimated 83 million m^3 (525 million bbl) of recoverable oil. Preproduction investment, which includes the construction of a concrete gravity base structure, is approximately $5 billion (Canadian). Production is scheduled to commence in 1996; the expected plateau production rate is 17,480 m^3 per day (110,000 BOPD). To date, there is no hydrocarbon production from offshore eastern Canada.

Hibernia is located in the northwestern sector of the Jeanne d'Arc rift basin. The trap is a complexly faulted anticline created by rollover into the basin-bounding Murre fault. The integrated geological and geophysical interpretation of the field is based on the results of ten wells and 465 km^2 (180 mi^2) of three-dimensional seismic data.

Berriasian- to Valanginian-aged Hibernia sandstones are the primary reservoirs occurring at an average drill depth of 3720 m (12,200 ft). Sedimentological interpretation of core indicates that the principal reservoirs were deposited as high-bed load distributary channels in a fluvially dominated deltaic complex. Average porosity is 16%; permeability ranges up to 10 darcys. Reservoir sands are interpreted as being elongated and relatively continuous in a southwest-northeast direction. Downdip water injection will be implemented to maximize recovery.

Barremian- to Albian-aged Ben Nevis/Avalon sandstones are the secondary reservoirs and occur at an average drill depth of 2345 m (7700 ft). Core studies indicate deposition within strand plain, transgressive shoreface, and offshore shallow marine environments. The extent of lateral continuity is uncertain due to a combination of thin bed stratigraphy, structural complexity, zones of bioturbation, and calcite cementation.

INTRODUCTION

The Hibernia field is located on the Grand Banks of Newfoundland, 315 km (195 mi) east-southeast of St. John's (Figure 1). Water depth at the site is approximately 80 m (260 ft). Hydrocarbon entrapment is structurally controlled in a broad, complexly faulted, rollover anticline. The field was discovered in 1979 when the Chevron et al. Hibernia P-15 well encountered 58 m of net oil-bearing Upper Jurassic and Lower Cretaceous sandstones. During the 1980–1990 appraisal and evaluation period, Mobil Oil Canada Limited operated the field on behalf of Chevron Canada Limited, Gulf Canada Limited, and Petro-Canada Resources, the working interest owners in the Hibernia project. The development and operation of the field is currently managed by the Hibernia Management and Development Company Limited, a consortium of the four working interest owners.

Approximately 83 million m³ (525 million bbl) of recoverable oil occurs in two Lower Cretaceous, synrift sandstone intervals informally named the Hibernia sandstones and the Ben Nevis/Avalon sandstones. Development will focus on the Hibernia sandstones because 85% of the recoverable oil is contained within this interval. These strata comprise part of a widespread delta/coastal plain regime that occupied the southern and central part of the Jeanne d'Arc Basin during Berriasian–Valanginian times (Figure 2).

Ben Nevis/Avalon production is forecast to commence at the onset of declining production from the Hibernia sandstones. Deposition of the Aptian–Albian portion of this interval is synchronous with structural culmination of the field. A significant degree of uncertainty in reservoir continuity exists due to a combination of thin bed stratigraphy, structural complexity, zones of intense bioturbation, and calcite cementation.

The Hibernia field is located in a harsh physical environment typified by severe winter storms, sea ice, and icebergs. Development will be accomplished by the installation of a fixed production platform, crude oil loading system, and three tankers that will transport the oil to markets. The platform will be a gravity base concrete structure. Production is expected to average 17,480 m³ per day (110,000 BOPD); preproduction investment is approximately $5 billion (Canadian).

REGIONAL GEOLOGY AND STRATIGRAPHY

The Grand Banks is a broad continental shelf partially underlain by a suite of fault-bounded, Mesozoic rift basins (Figure 2). The Hibernia field is located in the northwestern sector of the narrow and elongate Jeanne d'Arc Basin. Prior to the Hibernia discovery, the database for evaluating the Jeanne d'Arc Basin consisted of approximately 25,000 km of conventional seismic reflection data (Sinclair, C-NOPB, personal communication, 1990) and the geologic information from eight well bores (Figure 3A). Within the decade following the discovery of Hibernia, increased exploration activity has resulted in the acquisition of an additional 75,000 km of seismic data (Sinclair, personal communication). Fifty additional wells have been drilled in the basin and eight additional oil fields with significant hydrocarbon accumulations have been discovered (Figure 3B).

The integration of this database with deep seismic reflection data, regional gravity, and regional magnetic surveys has been the basis for numerous publications on the Mesozoic extensional history of the Grand Banks area. Detailed interpretations on the structural and stratigraphic evolution of this rift-basin complex are presented by Hiscott et al. (1990a, 1990b), Tankard and Welsink (1987), Sinclair (1988), Hubbard (1988), Grant et al. (1986a, 1986b, 1988), Jansa and Wade (1975a, 1975b), Amoco and Imperial (1973), Masson and Miles (1984, 1986), and Enachescu (1987, 1988).

The major tectonic elements of the Jeanne d'Arc Basin are the Bonavista platform to the west, the Avalon Uplift to the south, and the Central Ridge complex to the east (Figures 2 and 3). Figure 4 is a lithostratigraphic chart representing the general stratigraphy of the Hibernia field and Jeanne d'Arc Basin. Figure 5 is a generalized geological cross section transverse to the axis of the basin, depicting its structural configuration.

The basin-forming Murre fault, which bounds the Hibernia structure to the west, soles out eastward beneath the basin at approximately 26 km (16 mi) (Tankard and Welsink, 1987). This prominent north–northeast-trending listric fault separates pre-Mesozoic rocks of the tectonically stable Bonavista Platform from basinal Mesozoic strata (Figure 5). Approximately 15 km (9 mi) of Triassic through Lower Cretaceous strata record episodic rifting associated with the opening of the North Atlantic Ocean.

Rifting began during the Late Triassic to Early Jurassic in association with the separation of Africa from North America. The northeast-southwest orientation of the basin complex is inherited from Paleozoic structural fabrics that are exposed onshore in Newfoundland and are apparent in deep seismic reflection data acquired over the Bonavista platform and Jeanne d'Arc Basin (de Voogd et al., 1990). Basement highs, which separate the Mesozoic depocenters, consist primarily of metamorphosed Precambrian and Paleozoic rocks (Jansa and Wade, 1975a, 1975b).

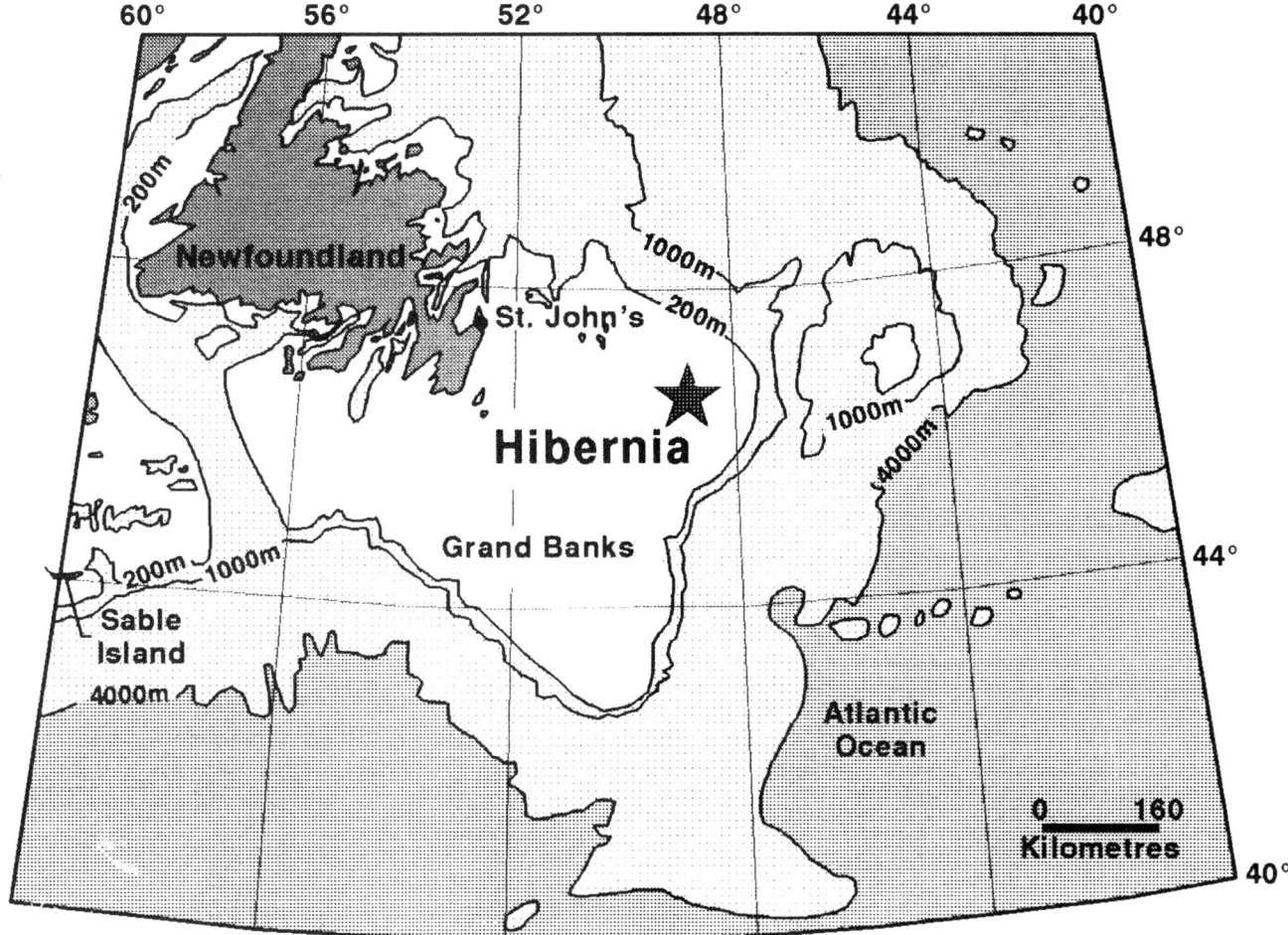

FIGURE 1. Location map of the Hibernia field, offshore Newfoundland.

In the Jeanne d'Arc Basin, incipient Late Triassic through Early Jurassic rift valley sediments are an ascending succession of conglomerates, red beds, and evaporites (Eurydice to Argo formations) which unconformably overlie Paleozoic metasediments. The evaporite complex indicates the first marine invasion into the developing grabens. Subsequent mobilization has created numerous salt structures that were a primary focus in initial basinal exploration.

Tectonic quiescence and normal marine conditions during the Middle Jurassic are represented by a thick (up to 1300 m) succession of shallow marine shales and bioclastic carbonates of the Iroquois Formation and Whale Unit. Oxfordian- to Kimmeridgian-aged organic rich shales are the primary source rock for hydrocarbons in the basin and were deposited in euxinic basin environments prior to a prerift regional doming event. Intense rifting and regional doming of the Avalon Uplift commenced during the latest Jurassic in association with the separation of the Grand Banks from Iberia.

The Avalon Uplift lies to the southwest of the Jeanne d'Arc Basin. Unconformities within the Jeanne d'Arc Basin merge with this peneplain and collectively indicate a 50-60 m.y. (late Kimmeridgian to Albian-Cenomanian) period of episodic uplift and erosion (Grant et al., 1988). Thick sequences of clastic sediments including the Jeanne d'Arc, Hibernia, and Ben Nevis/Avalon sandstones prograded northward into the Jeanne d'Arc Basin during these times. Major hydrocarbon accumulations discovered to date occur within these synrift sandstones in folded and faulted structural traps. Inter-rift periods of basinal subsidence are indicated by shallow marine shales and carbonates that segregate these prominent clastic incursions. Two such carbonates, the B- and A-Marker limestones, generate mappable and relatively continuous seismic reflectors throughout the basin.

Aptian to Albian northeast-southwest extension of the Grand Banks area was associated with the Europe-North America plate break-up (Sinclair, 1988). Culmination of the Hibernia structure was synchronous with this event. Relatively undeformed Albian-Cenomanian shallow marine mudstones bury the faulted and structurally complex synrift sequence. The overlying Turonian Petrel Limestone produces the most consistent seismic reflection in the area.

The stratigraphy of the Hibernia field was previously addressed by Tankard and Welsink (1987) and Arthur et al. (1982). Informal names for the principal reservoirs, namely Hibernia and Avalon, were introduced by Arthur et al. (1982). The stratigraphic nomenclature for the Jeanne d'Arc Basin, however, has not been formalized and since 1982, several different lithostratigraphies have been proposed (McAlpine, 1989; Hiscott et al., 1990a; Sinclair, 1988; Grant et al., 1986a, b). Stratigraphic interpretations

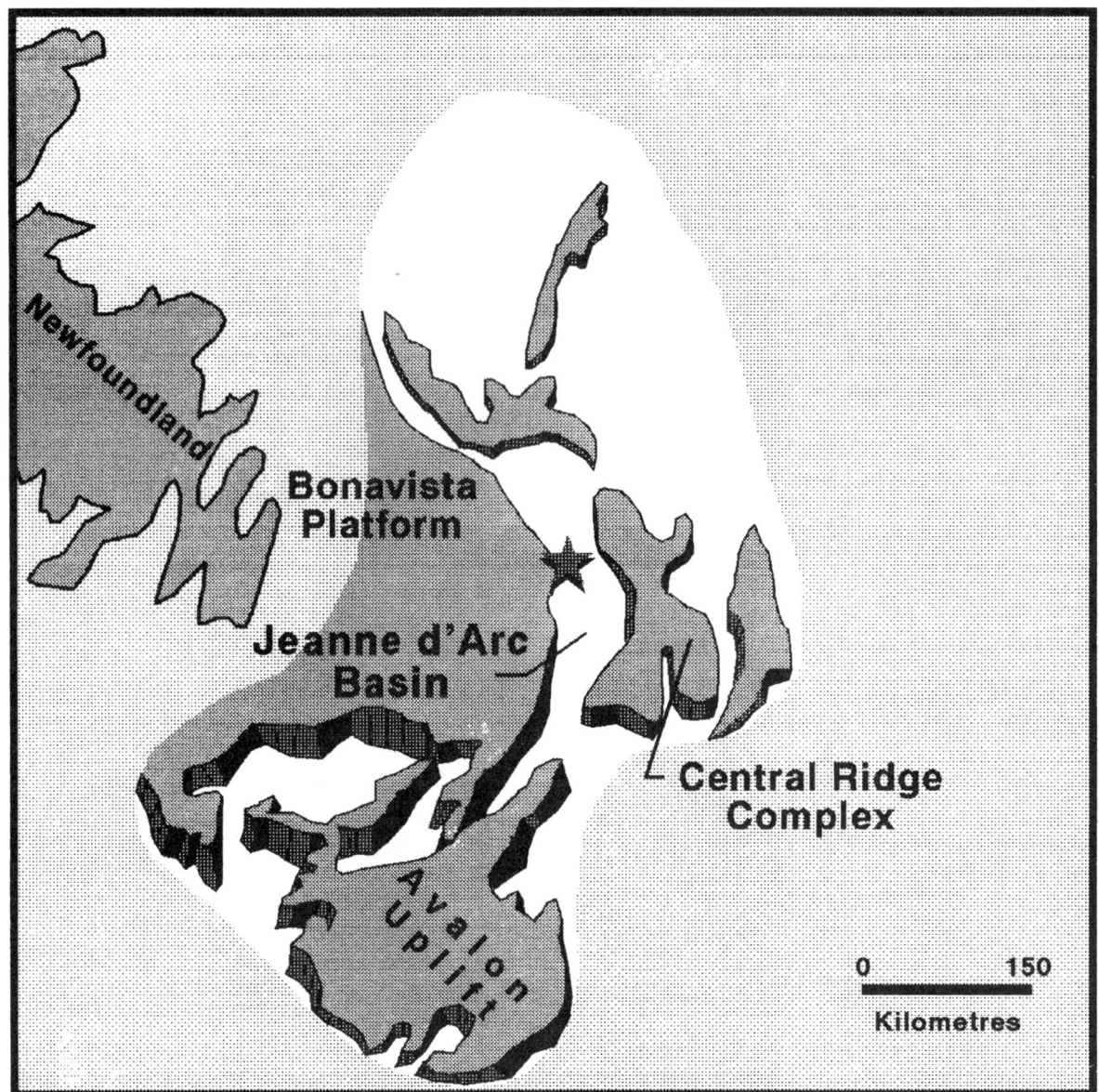

FIGURE 2. Mesozoic rift basins and principal positive areas. The Avalon Uplift was the primary provenance area for siliciclastic sediments deposited in the Hibernia field area and elsewhere in the Jeanne d'Arc Basin.

of the Ben Nevis/Avalon sequence are more varied than the Hibernia interval; Hibernia lithostratigraphic and biostratigraphic assessment has remained essentially unchanged.

Arthur et al. (1982) defined the Avalon reservoirs as Hauterivian-aged sandstones separated by a "pre-Aptian" unconformity from overlying nonreservoir rocks. Sinclair (1988) also interpreted Avalon sandstone truncation by a regional "Aptian" unconformity; sediments that overlie the exposure surface, however, are distinctly different. Sinclair (1988) proposed a marine transgression over the unconformity and deposition of basin-wide, reservoir-quality sandstones of the Aptian–Albian Ben Nevis Formation. The term "Ben Nevis" is now commonly applied to Aptian–Albian-aged sediments in the basin. Sinclair's stratigraphic distinction of this interval, however, hinges upon his recognition and correlation of the "Aptian" unconformity, which we feel is problematic in itself.

In the Hibernia field, reservoir sands occur throughout the Barremian to Albian section. Biostratigraphy indicates a Barremian to early Aptian age for the Avalon sandstones; principal reservoirs are field-wide, "transgressive" sands that overlie a (regional?) mid-Barremian unconformity. Aptian-aged Ben Nevis reservoirs were deposited during structural culmination of the field and are "isolated" sandstones deposited adjacent to an emergent Hibernia structure. These sandstones, where present, do not overlie an unconformity surface but are

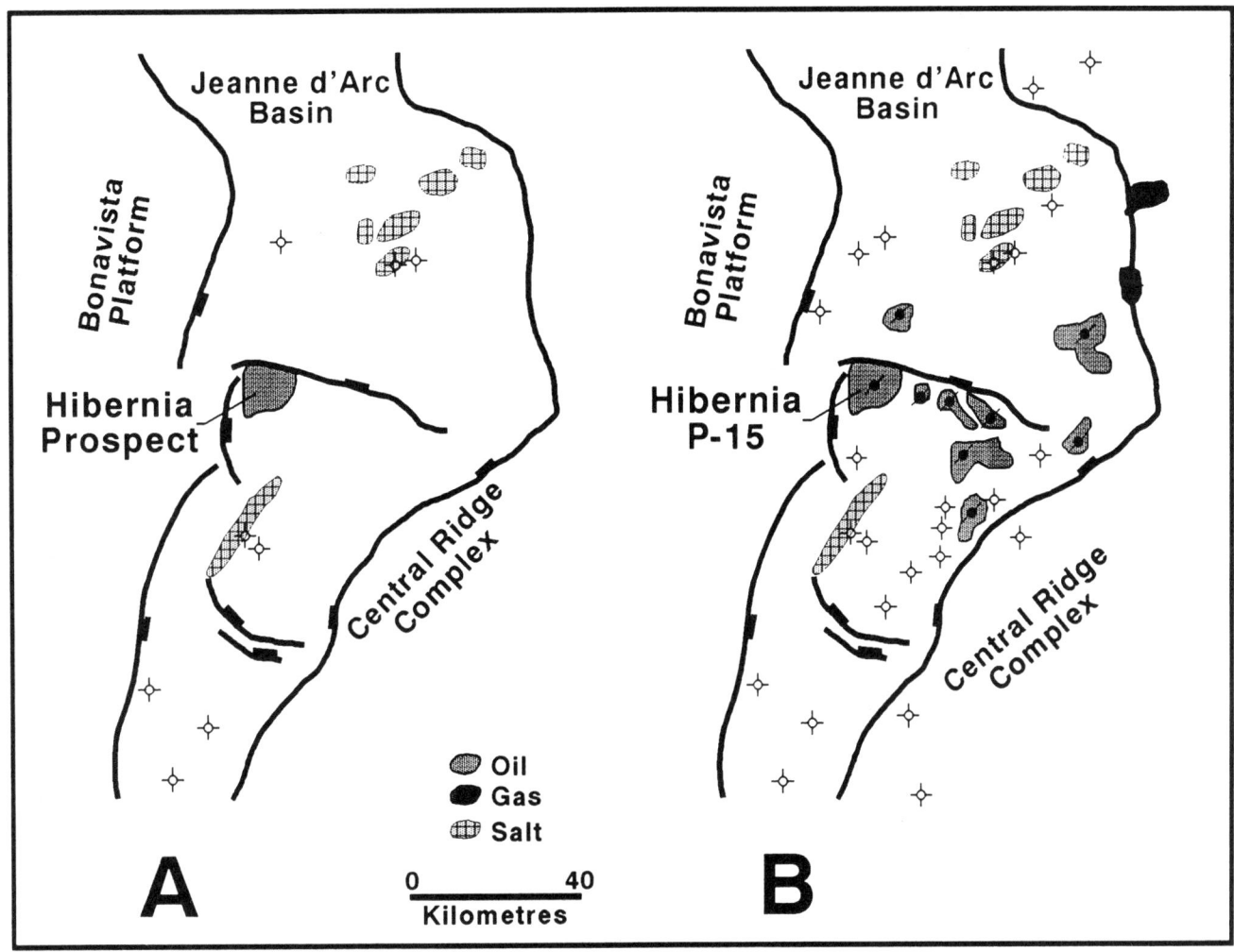

FIGURE 3. (A) Eight wells tested the Jeanne d'Arc Basin prior to the Hibernia discovery in 1979. (B) Since Hibernia's discovery, 50 additional wells have been drilled and eight additional oil fields have been discovered.

coeval with an "Aptian" unconformity localized in the Hibernia field area to the crest of the developing Hibernia anticline.

EXPLORATION HISTORY

In 1959, Mobil Oil Canada initiated acreage acquisition in Canada's eastern offshore by acquiring the mineral rights to 4450 km^2 (1.1 million ac) immediately surrounding and inclusive of Sable Island, offshore Nova Scotia. By 1971, over 809,370 km^2 (200 million ac) had been leased in the Labrador, Grand Banks, and Scotian shelf areas by various oil companies; Drummond (this volume) reviews the exploration history of the Scotian shelf.

Initial drilling activity in the Grand Banks focused on the more readily apparent, seismically definable structures including salt domes, tilted fault blocks, and basement highs. Exploration in this area commenced in 1966 when Amoco and partners drilled the Pan Am a-1 Tors Cove D-52 well in the South Whale Basin. Minor shows of methane in this well were a promising sign of the area's hydrocarbon potential.

The first test in the Jeanne d'Arc Basin was the Amoco Imperial Murre G-57 well, which was drilled in 1971. Reservoir-quality rock was not encountered at this location. In 1972, the Mobil Gulf Adolphus 2K-41 well, the third well drilled in the basin, tested a salt-piercement structure approximately 40 km northeast of the Hibernia prospect. Although not classified as a discovery, a thin Upper Cretaceous sandstone drill-stem tested oil at a rate of 42.6 m^3 per day (268 BOPD). Approximately 40 km south of the Hibernia prospect, the Amoco et al. Egret K-36 well subsequently encountered 236 m (775 ft) of porous, but wet, Lower Cretaceous sandstones as well as potential source rock of late Jurassic age. By 1975, four additional wells were drilled including offsets to the Adolphus 2K-41 well (Adolphus D-50) and the Egret K-36 well (Egret N-46). Thick (wet) Lower Cretaceous sandstones in the southern portion of the basin were reconfirmed by the Egret N-46 well; the porous and productive Upper Cretaceous sandstone in the 2K-41 well, however, was not present in the offset location.

Forty-one wells were drilled within the Grand Banks area prior to the Hibernia discovery; only eight of these were located in the Jeanne d'Arc Basin. None of these

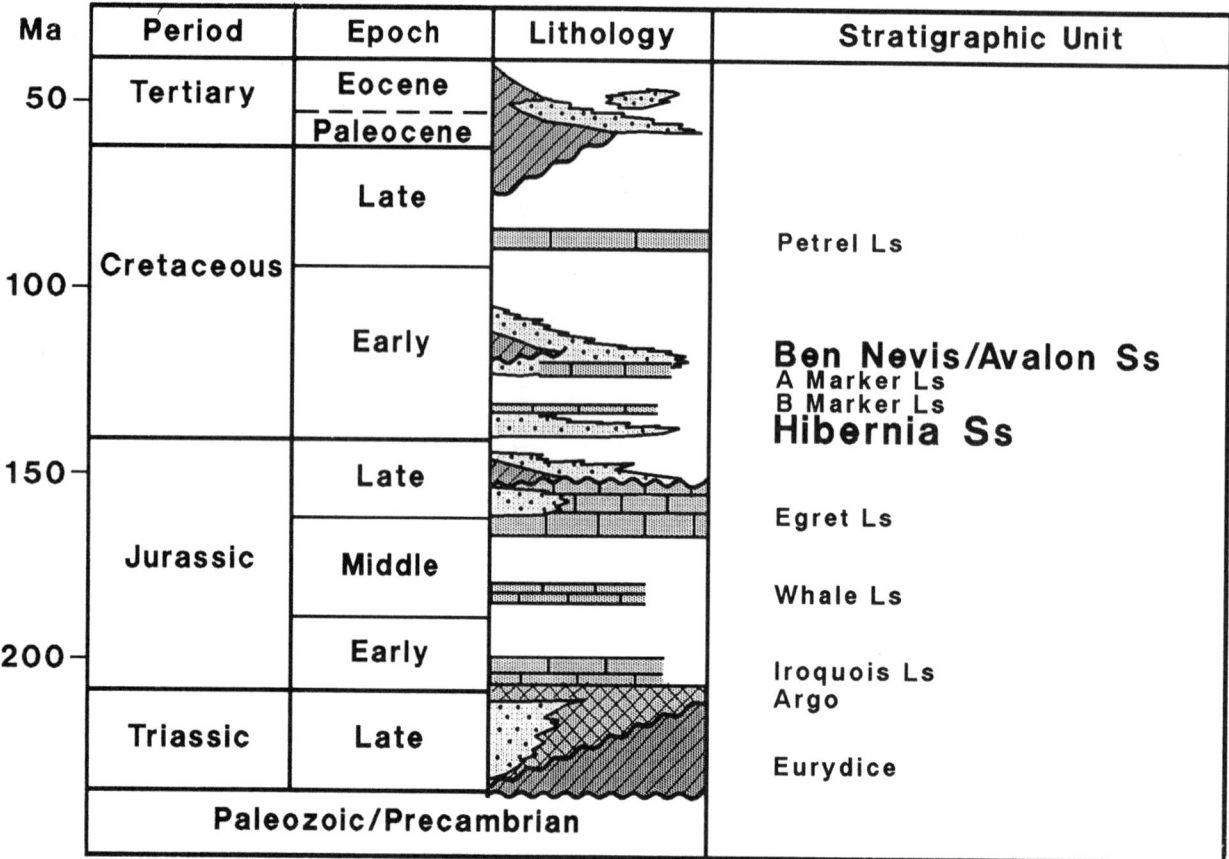

FIGURE 4. Generalized lithostratigraphy of the Hibernia field and Jeanne d'Arc Basin.

wells tested commercially productive hydrocarbons. During this period of exploration, Mobil Oil Canada's Grand Banks land position was reduced, through high grading, from 68,800 to 16,200 km² (17 million to 4 million ac). Included within this 16,200 km² was the untested Hibernia prospect.

In 1979, Mobil Oil Canada and partners negotiated agreements to drill a 4875-m (16,600-ft) Upper Jurassic test on the Hibernia prospect. Drilling of the Chevron et al. Hibernia P-15 well commenced in May 1979. When suspended in January 1980, the potential for commercial oil production from Canada's eastern offshore had been established.

HIBERNIA FIELD APPRAISAL

Nine appraisal wells were drilled from 1980 through 1984 and a three-dimensional seismic survey was acquired over the field in 1980 and 1981. At the time of acquisition, the Hibernia 3D survey was the largest in existence, encompassing 465 km². Approximately 1300 m (4265 ft) of conventional core were recovered from the Hibernia and Ben Nevis/Avalon reservoirs.

Hibernia Sandstone

Seven wells penetrated the Hibernia sandstones. Drill depths to the top of this interval ranged between 3477 and 3913 m (11,407 to 12,838 ft) as defined by B-08 and O-35, the structurally highest and lowest wells, respectively (Figure 6). All wells encountered a thick section of porous and permeable sandstone. Net oil pay ranges from 29 to 68 m (95 to 223 ft). Combined flow rates in the P-15 discovery well exceeded 1670 m³ per day (10,500 BOPD); flow rates from individual drill-stem tests in subsequent wells exceeded 905 m³ per day (5700 BOPD).

An oil-water contact defined in P-15 at approximately 3860 m (12,664 ft) subsea was not disproved by the first appraisal well (O-35), which encountered the top of a porous but wet sandstone sequence at 3888 m (12,755 ft) subsea. A gas-oil interface at 3544 m (11,627 ft) subsea was defined in the third well, B-08. All other wells penetrated the Hibernia sandstone at elevations structurally low to the B-08 gas-oil contact, and this elevation has been assumed to be a field-wide gas-oil interface. Gas is thereby confined to crestal fault blocks located in the north-central portion of the field (Figure 6). Initial descriptions and interpretation of the Hibernia field employed the P-15 and B-08 fluid interfaces as field-wide contacts (e.g., Benteau and Sheppard, 1982; Arthur et al., 1982).

In late 1983, however, the B-27 well encountered water 70 m lower than the P-15 oil-water elevation. This was the eighth well to test the structure, but only the fifth well to penetrate the Hibernia sandstone. An oil-water interface of 3930 m (12,894 ft) subsea was established in this well bore. Subsequently, the K-14 well confirmed this lower oil-water interface. The final appraisal well, C-96,

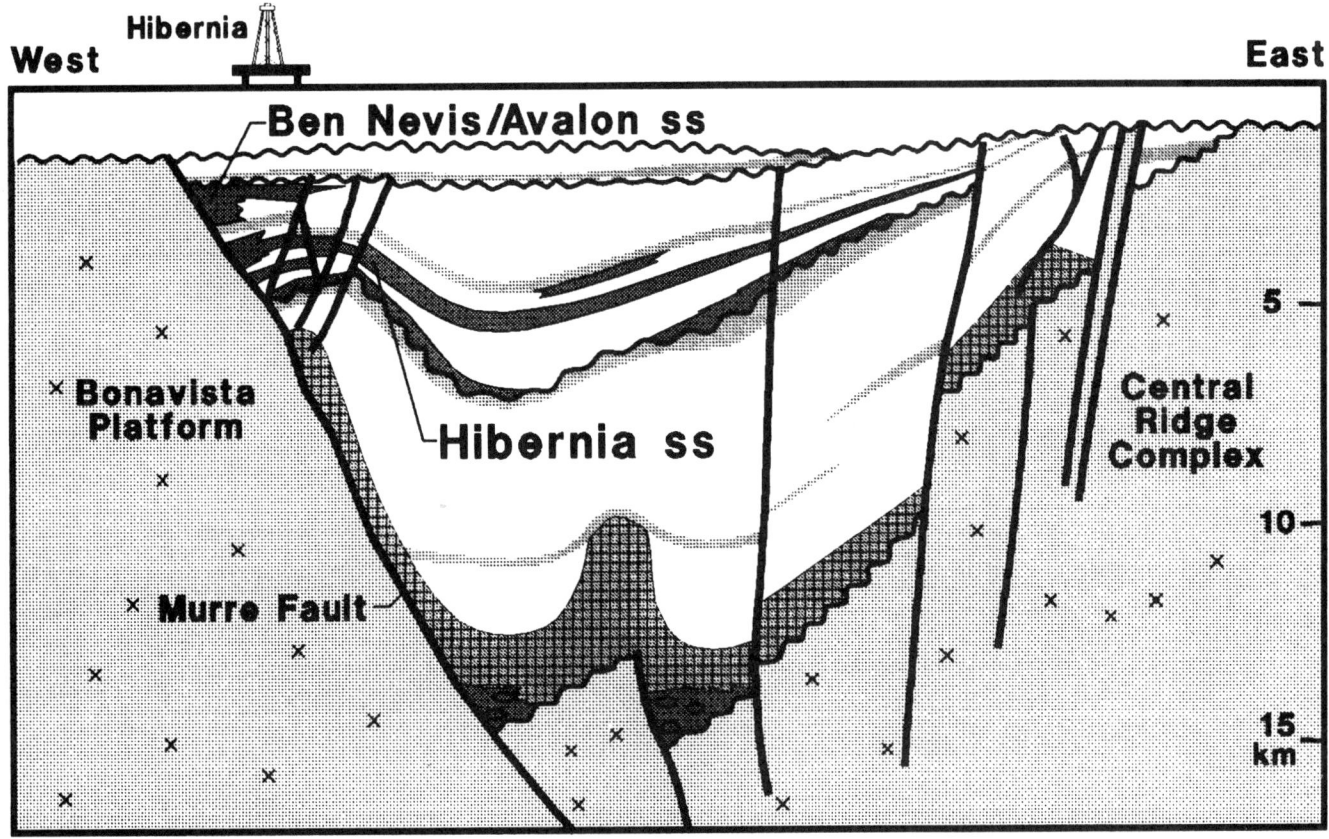

FIGURE 5. Schematic illustration of the rifted Jeanne d'Arc Basin (modified from Tankard and Welsink, 1987). The Hibernia structure is created by rollover into the basin-bounding Murre fault.

was drilled in 1984. This well encountered 58 m (190 ft) of net oil pay and confirmed oil down to 3913 m (12,838 ft) subsea.

Interpretation of the 3D seismic survey during field appraisal identified the P-15 well as being located within a relatively small fault block. The oil-water contact initially defined in P-15 and indirectly verified in O-35 is interpreted as being confined to the individual fault blocks penetrated by these wells. The deeper oil-water contact established by the last three delineation wells is considered to be more representative of the field (Mobil et al., 1985). The presence of multiple hydrostatic pressure systems and the possibility of multiple oil-water contacts, however, have been recognized (C-NOPB, 1986). A vertical oil column of 386 m (1266 ft) exists in the Hibernia reservoir and productive closure encompasses 66.8 km^2 (16,500 ac).

Ben Nevis/Avalon Sandstones

All ten wells drilled within the field penetrated the Ben Nevis/Avalon sandstones. Drill depth to the top of this interval ranged between 2048 and 2432 m (6719 and 7979 ft) in G-55A and B-27, the structurally highest and lowest wells, respectively. Net oil pay ranges from 2 to 101 m (7 to 331 ft). Only the G-55A well failed to encounter hydrocarbons.

A westward-thickening wedge of sediment, seismically defined and anticipated prior to field discovery, was confirmed by initial appraisal drilling. Approximately 600 m (1969 ft) of sand-rich Ben Nevis/Avalon section were encountered in the most westerly G-55A well. In contrast, the B-08 well found 88 m (289 ft) of equivalent sediments. Thickness of this section in the eight remaining wells varies between the G-55A and B-08 extremes.

Excellent productivity was established in the P-15 well. Approximately 183 m^3 per day (1150 BOPD) of 30° API oil was tested from 8 m (26 ft) of net oil pay in one sand-rich interval. The subsequent O-35 well not only confirmed Ben Nevis/Avalon potential but also encountered two prominent, highly productive sandstone intervals. Net oil pay in this well is 101 m (331 ft).

An erosionally thinned B-08 horizon contained 5 m of gross sand and 2 m of net oil pay. The thickest sandstone interval in the field (600 m in G-55A) was found to be porous and structurally high, but wet. This well is apparently fault-isolated from the main structure. Continued appraisal drilling further delineated the field limits by encountering an oil-water contact at 2602 m (8537 ft) subsea in the J-34 well. Log analysis in the subsequent I-46 well supported this contact. No gas cap has been found in the Ben Nevis/Avalon reservoirs, and a vertical oil column of 557 m (1827 ft) has been identified.

STRUCTURE

The Hibernia structure is a rollover anticline located at the intersection of the basin-bounding Murre fault and

42 Hurley et al.

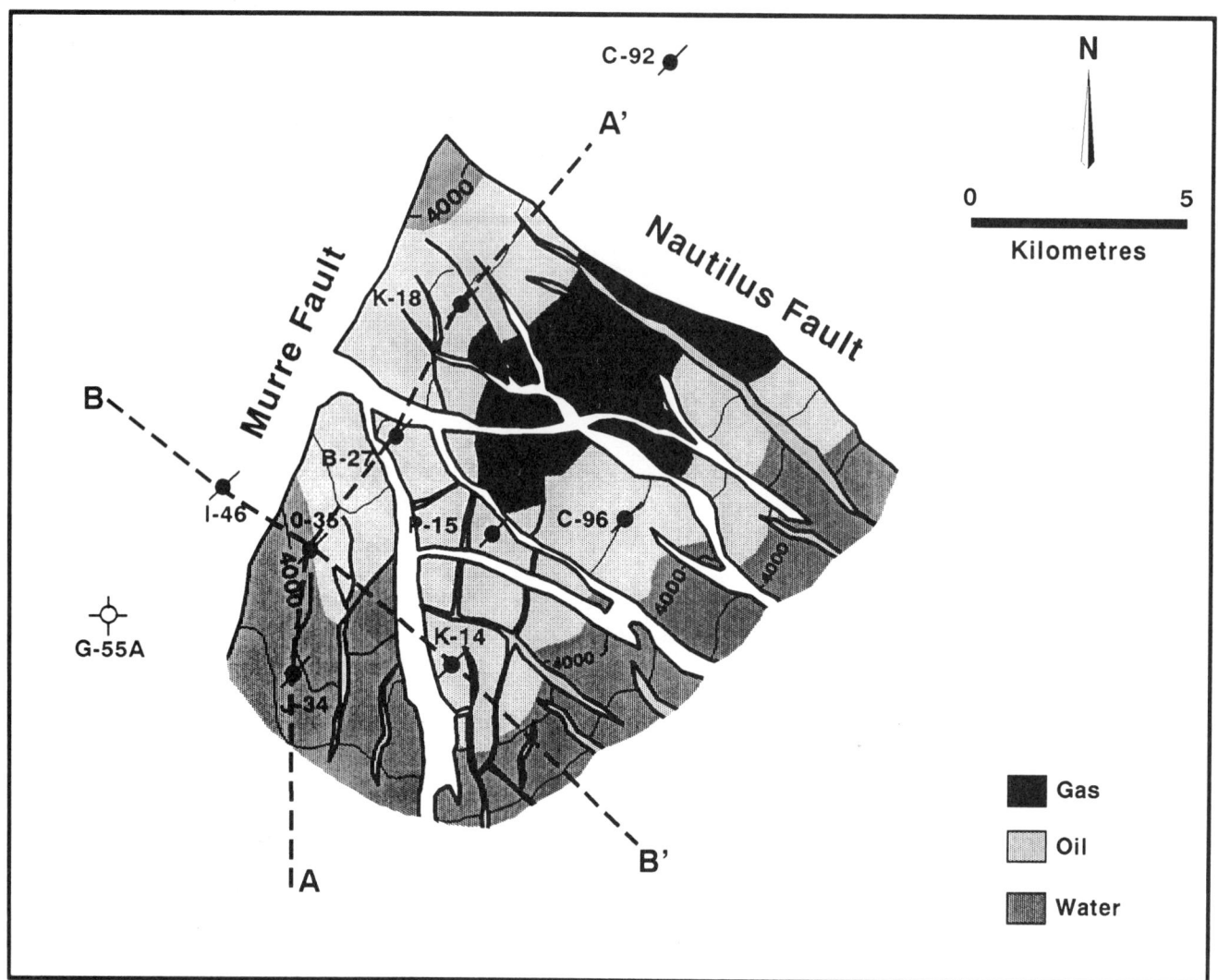

FIGURE 6. Depth structure map on the top Hibernia sandstones (modified from Taylor et al., 1984). Seismic sections A-A' and B-B' are interpreted in Figures 7 and 8.

transbasinal Nautilus fault. Two sets of faults, one subparallel to the Murre and the other parallel to the Nautilus, complicate the structure. Salt diapirism along the Murre fault plane modifies the structure on its western edge (Figure 7). Sixty meters of interbedded salt and dolomite were penetrated in the I-46 well, which stopped drilling approximately 325 m (1066 ft) above the Murre fault plane.

Underlying the anticline, and bounding it to the west, is the listric Murre fault (Figure 7). Subsidence along the Murre fault was likely persistent throughout basinal history; pronounced displacements, however, occurred during Aptian to Albian times. Faults subparallel to the Murre are interpreted as synthetic and antithetic fault sets associated with detachment and rollover.

Bounding the field to the northeast is the Nautilus fault, which offsets beds as young as Barremian by 1500 m (4921 ft) or more (Figure 8). Faults of the Nautilus system are consistent in orientation and spacing and are markedly parallel. Fault plane dips within this system are to both the northeast and southwest. Offsets of the synrift sequence within the Hibernia anticline are measured in hundreds of meters or less, whereas beds of the postrift sequence are offset only infrequently and by much smaller amounts. Most of the displacement is interpreted as having occurred between Barremian and late Albian time.

A distinct difference in the timing of activity along the two fault systems is not clearly evident; offset of one fault set by the other has not been observed. Both master faults extend into Tertiary sediments and both associated fault systems displace late Barremian to early Aptian Avalon reservoirs. Relatively synchronous motion in association with Aptian-Albian structural culmination is indicated.

Several major faults completely offset the principal Hibernia reservoir section. The current depletion strategy proposes field development on a block-by-block basis and assumes that faults (with significant displacement of the reservoir) are sealing. The two different oil-water contacts defined in the Hibernia sandstones support a sealing-fault interpretation. Production and injection wells are planned for each of the principal fault blocks in the field. An illustration of the Hibernia structure schematically

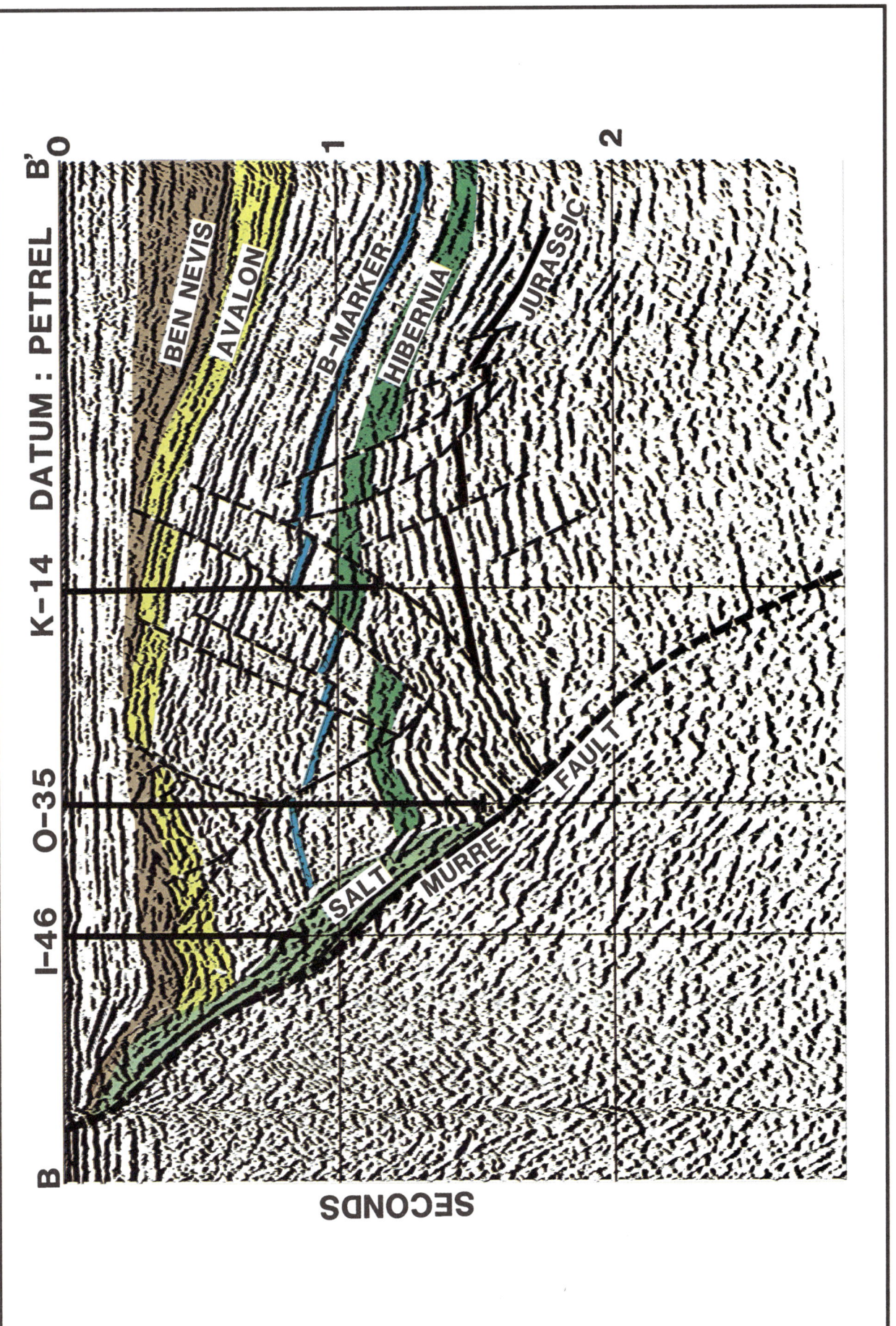

FIGURE 7. East-west transect across the Hibernia Structure illustrating the rollover nature of the anticline, synthetic and antithetic faults, and salt emplacement along the Murre fault plane. Section is flattened on the Petrel Limestone.

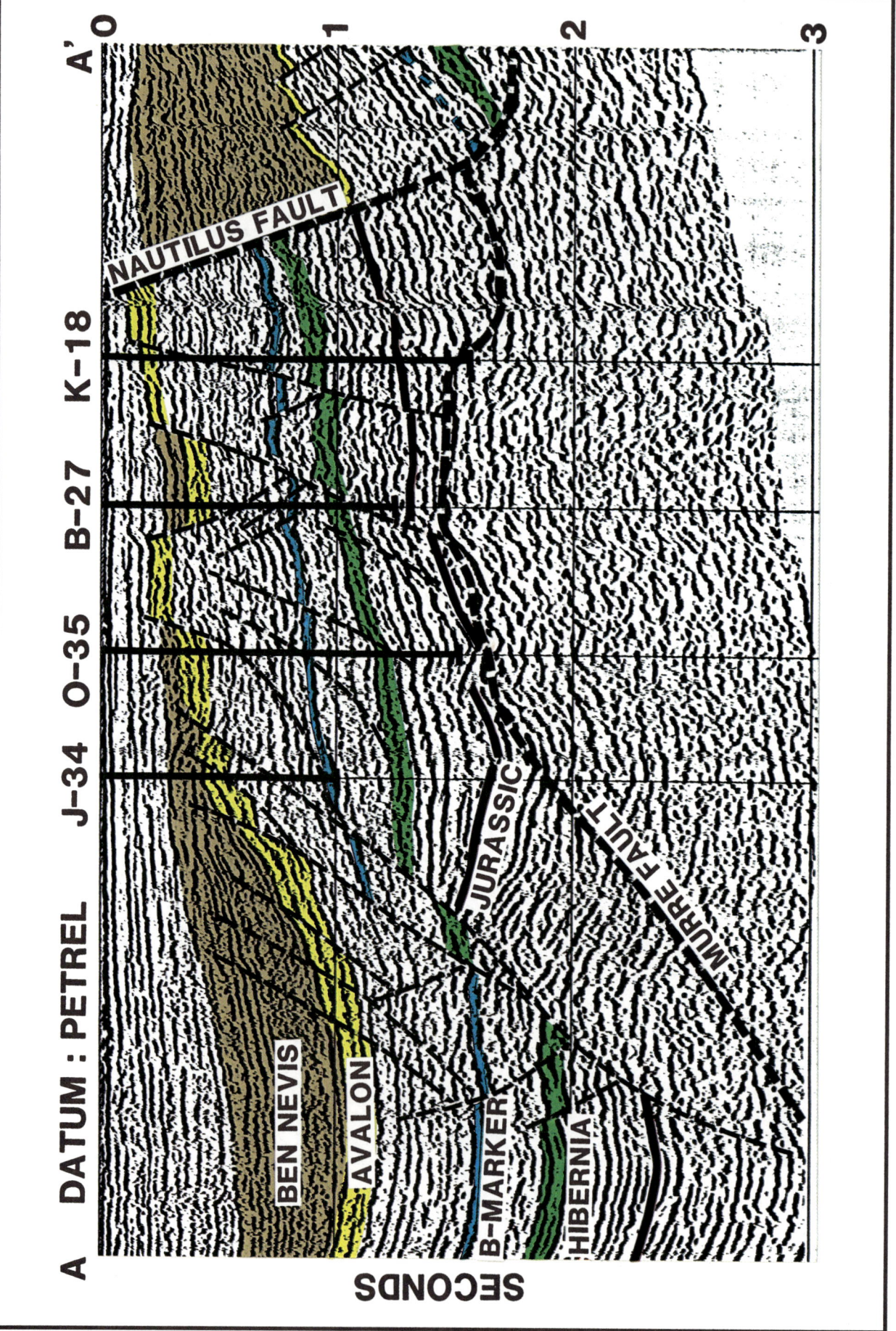

FIGURE 8. North-south transect of the Hibernia field depicting the horst and graben complex created by the system of faults that are strike parallel to the Nautilus fault. Section is flattened on the Petrel Limestone.

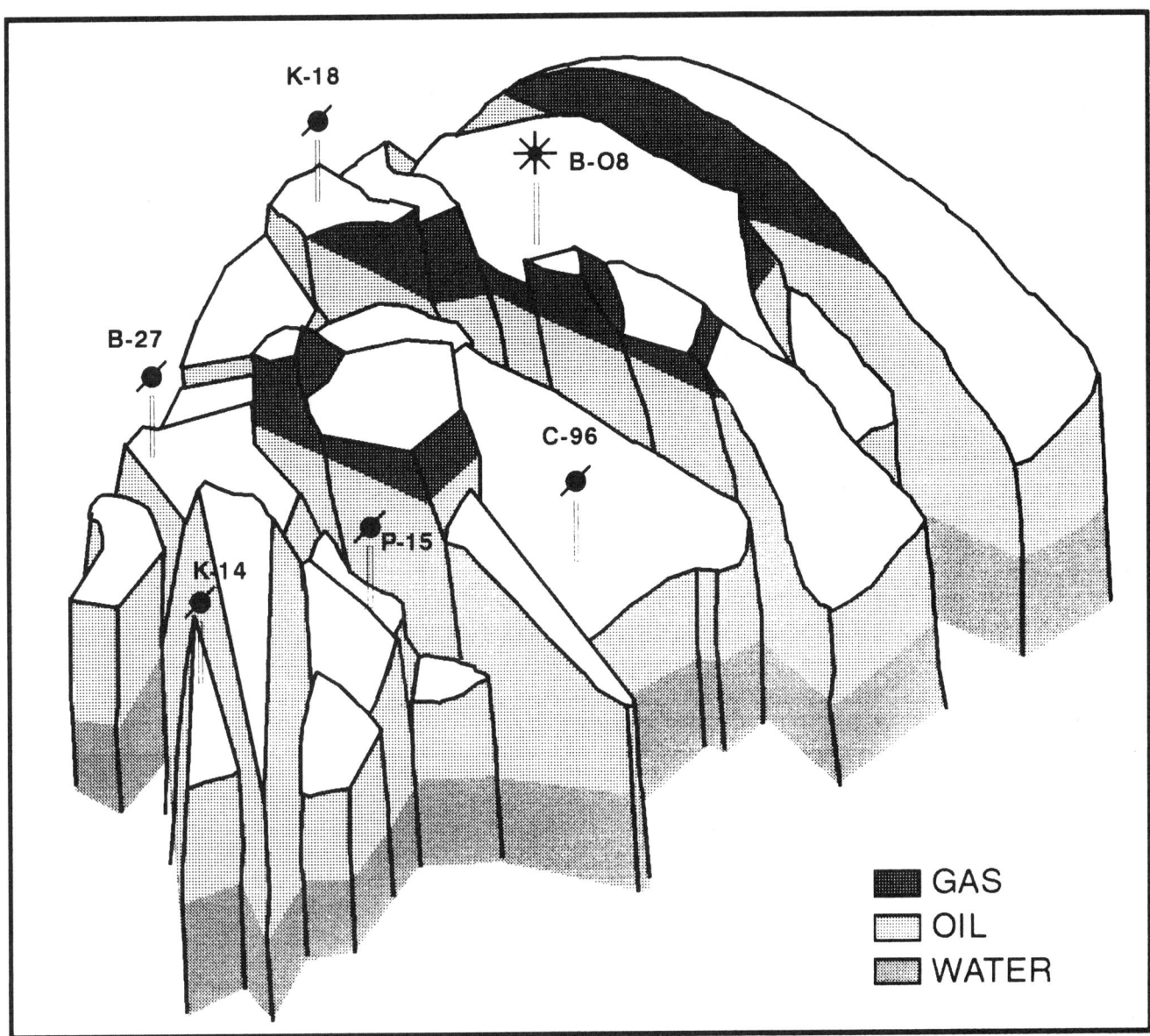

FIGURE 9. Schematic illustration of the fault block configuration and distribution of gas, oil, and water in the Hibernia sandstones. Fifty-eight wells are currently scheduled for full-field Hibernia sandstone development. The production platform will be located above the C-96 fault block.

depicting the fault block configuration and distribution of gas, oil, and water is illustrated in Figure 9.

DEPOSITIONAL MODEL—HIBERNIA SANDSTONES

Biostratigraphic analysis of shale beds above, below, and within the Hibernia sequence bracket the sandstones as Berriasian to Valanginian in age. No biostratigraphically or seismically evident unconformities occur within, or at the base of, this interval. The proposed depositional model and stratigraphic interpretations are based on sedimentological interpretation of 386 m (1266 ft) of conventional core and the integration of this facies analysis with wireline log suites for the seven wells in the Hibernia zone.

The Hibernia interval is characterized by thick, porous sandstones deposited by deltaic distributary channels and interstratified with zones of poorer-reservoir-quality, fine-grained sandstone and mudstone formed in associated overbank, flood basin, bay, marsh, and abandoned channel environments. Rocks deposited in water of normal marine salinity are an important component of these strata, but the delta system was fluvially dominated with marine energy (waves or tides) having very little influence in the distribution of sediments.

Three groups of facies are recognized:

1. Prodelta and delta front
2. Delta plain channel and associated abandoned channel fill, flood plain, bay, and marsh
3. Marine to freshwater bay and crevasse splay.

The over-all vertical sequence of facies reflects initial progradation (prodelta and delta front), delta plain sedimentation (channels and associated facies), and final

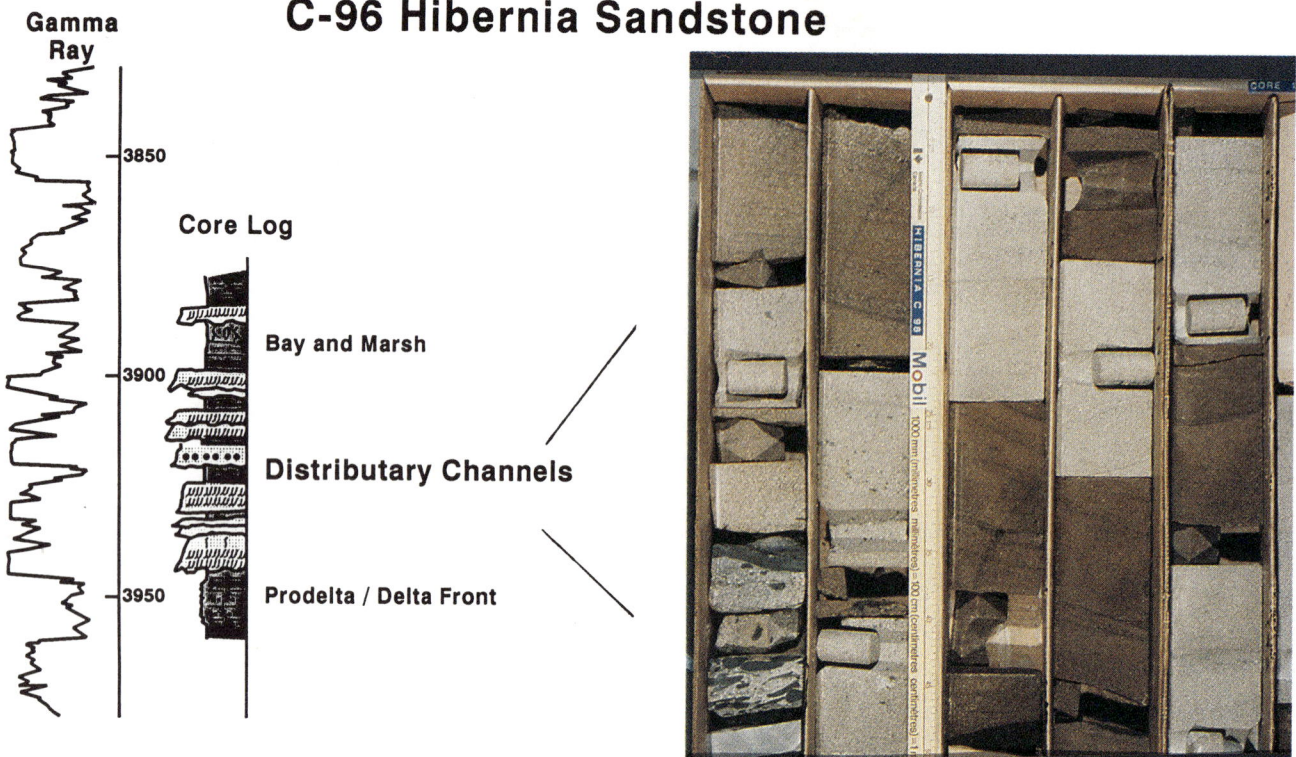

FIGURE 10. Core log of the Hibernia sandstones from the C-96 well. Lag conglomerates and high-angle, large-scale cross-bedding typify the channel belt sandstones.

subsidence and gradual drowning of the delta system (dominantly bay and splay deposits).

Delta Front

In the Hibernia interval, delta front deposits are relatively thin and reservoir quality is typically poor due to fine grain size and carbonate cements. Only a minor amount of core was recovered from the delta front facies.

As seen in core, delta front sands are thin beds of parallel laminated, fine-to-medium-grained sandstone (probably related to land flood events) that are interbedded with shale and overlie prodeltaic black shale. There is little evidence of reworking of these sediments by waves or tides. "Complete" delta front sequences 25 to 35 m thick occur in noncored, upward-coarsening intervals as indicated by gamma ray wireline log response and drill cuttings analysis.

Delta-Plain Sequence

Overlying the delta front is a thick complex of delta plain sediments; the best reservoir quality sandstones are concentrated in this interval. Typically these sandstones are very coarse- to medium-grained, moderately well-sorted quartz arenites characterized by large-scale cross-bedding (10- to 80-cm sets) and subtle vertical grain-size trends. Plant remains and mudstone intraclasts are locally abundant (Figure 10).

Large-scale cross-bedding probably relates to migration of in-channel bars and large bed forms. Locally, superimposed sets of cross-beds diverge about 90° from one another, similar to those formed by modern (Crowley, 1983) and ancient (Cant and Walker, 1976) braided channel-bar deposits. Generally speaking, individual channel sandstones in the upper part of the delta plain show little or no vertical grain-size trends and have corresponding blocky wireline log signatures. These features, plus the dominance of sandstone versus overbank mudstone deposits, indicate deposition in bed load–dominated, low-sinuosity streams, with a subordinate tendency toward somewhat higher-sinuosity mixed-load channels in the lower part of the Hibernia interval.

A relatively minor portion of the delta plain deposits consists of fine-grained sediments. These were formed in a variety of marine-to-brackish and freshwater overbank environments, including levee, flood plain, bay, marsh, and abandoned channel. Horizons interpreted as abandoned channel deposits consist of thinly interbedded mudstones and siltstones that are slightly burrowed and commonly highly contorted and deformed. They abruptly overlie channel sandstones and in turn grade upward into rooted zones and thin coals.

Levee and proximal crevasse splay deposits are represented by ripple and climbing-ripple, cross-laminated, fine-grained sandstones which, in part, overlie channel sandstones. Mudstone and sandy mudstone with rooted horizons, thin coals, and thin-to-medium beds of siltstone to very fine-grained sandstone represent flood plain and freshwater bay and marsh sedimentation, interrupted by crevasse splay deposits.

A medial shale marker and several other thin but important layers of fine-grained sediments formed as marine bay deposits. They consist of silty mudstone with thin beds of very fine-grained sandstone that are highly bioturbated. Because of the bioturbation, most marine bay sediments are massive and have little primary stratification. Marine burrow-types include *Zoophycus*(?), *Chondrites*, and *Planolites*.

Locally, marine bay sediments overlie sharp, irregular, "transgressive" surfaces that include thin pebbly mudstones. Also associated with these transgressive surfaces are thin zones containing storm-wave-induced stratification and wave ripples.

Bay and Crevasse Splay Sequence

The abandonment phase of the Hibernia delta complex is characterized by freshwater to marine bay and marsh facies with numerous thin crevasse splay deposits. These rocks are generally similar to the overbank sediments in the delta plain but lack the associated thick channel sandstones. Instead, sandstones in this interval are thin (about 1 m) and occur at the top of 2- to 5-m thick coarsening-upward sequences. The sandstones are generally massive or weakly laminated. These coarsening-upward sequences are interpreted as bay-filling crevasse splay deposits (essentially small, shallow water deltas) similar to those of the Mississippi delta (Farrell et al., 1984).

Discussion

The Hibernia interval can be viewed as one progradational episode (prodelta through delta plain) that was subsequently transgressed (delta abandonment). The distribution of depositional environments is depicted in Figure 11. Although Figure 11 is somewhat diagrammatic, it is based on cross sections from about the middle of the delta plain sequence (Figure 12).

The channels of the Hibernia system apparently flowed in belts oriented southwest to northeast, roughly parallel to and somewhat to the east of the western boundary fault of the Jeanne d'Arc Basin (Figure 11). This geometry is reflected in both isochore and sand isolith patterns that thin toward the northeast, and by the relative paucity of channel sands in wells near the fault, compared with wells eastward into the basin. Sediment transport direction, as interpreted from wireline dipmeter data, was consistently to the north in channel sandstones of the delta plain sequence (Cox, 1990).

Subsidence along the Murre fault was persistent during deposition of the Hibernia interval, allowing thicker, but not necessarily sandier, accumulations of sediment on the west side of the field. The fine-grained sediments that were deposited within channel belts are dominated by abandoned channel facies, whereas wells near the fault, such as B-27, contain abundant flood basin deposits.

Subsidence along the fault also resulted in the incursion of shallow marine bays into areas along the west side of the field. At the same time, areas on the east side of the field were occupied by channels or freshwater bays. This situation is depicted in Figure 11B, which has been constructed on the basis of cross sections from the upper part of the delta plain. Similar episodes occurred several times during deposition of the Hibernia interval, resulting in several marine horizons that are present only on the west. Rapid subsidence or upstream avulsion of the channel belt at the time of the medial shale marker (stratigraphic datum) produced a widespread but ephemeral marine bay deposit across the entire field.

The final episode of sedimentation in the Hibernia area is represented by increased subsidence, sea level rise, and/or decreased sediment input, which resulted in gradual drowning of the delta system. Crevasse splay and freshwater bay facies seem to be better developed on the eastern side of the field, suggesting that channel belt sedimentation was localized farther to the east. In this case, splays were deposited laterally westward from the channels. Alternatively, they may be frontal splays, feeding directly from the south/southwest.

The ultimate source(s) of the sand in the Hibernia system is not fully understood. Our study supports sediments sourced from highlands to the south (Avalon Uplift); a significant contribution of sand directly from the area of the western boundary fault seems unlikely. Dipmeter studies indicate a northerly transport direction within the channel belt complex; the coarsest grain-size sediments in the system are found in C-96, far removed from a potential western source. In addition, isochore and sand isolith maps thicken to the south while thinning to the northeast (Brown et al., 1989). Thus, it is inferred that the area to the west of the fault was of relatively low relief and provided a "background" source of fine-grained sediment to the Hibernia interval. Source area and resultant channel orientation have important implications for reservoir geometry within the Hibernia field.

RESERVOIR GEOMETRY AND QUALITY

Reservoir sands deposited in linear channel belts of generally low-sinuosity streams are expected to be elongated and relatively continuous in a southwest-northeast direction; reservoir communication parallel to that trend should be excellent (Figure 12). The structural configuration of the field, however, dictates a secondary recovery strategy that, in part, is oriented across this trend. Therefore, the east-west dimensions and interconnectedness between channels and channel belts become important considerations.

The dimensions of the channels (width, depth, meander-belt width) were semiquantitatively determined in order to contribute to an understanding of the lateral communication. Qualitatively, low-sinuosity streams are wide and shallow, whereas high-sinuosity streams are deeper and narrower. Based on the work of Schumm (1968, 1977), sand bodies associated with Hibernia low-sinuosity streams are estimated to range from a few hundred meters to a rare maximum of approximately 1.5 km in width. Typically, they are less than 1 km wide. Higher-sinuosity streams at the base of the Hibernia interval may have been one-fifth to one-tenth the width of the low-sinuosity system. Maximum sand body widths for these higher-sinuosity channels, however, are estimated to be about 1.2 km due to lateral accretion.

The degree of lateral communication that exists due to channel stacking and incisement into pre-existing channel complexes is difficult to evaluate from the limited

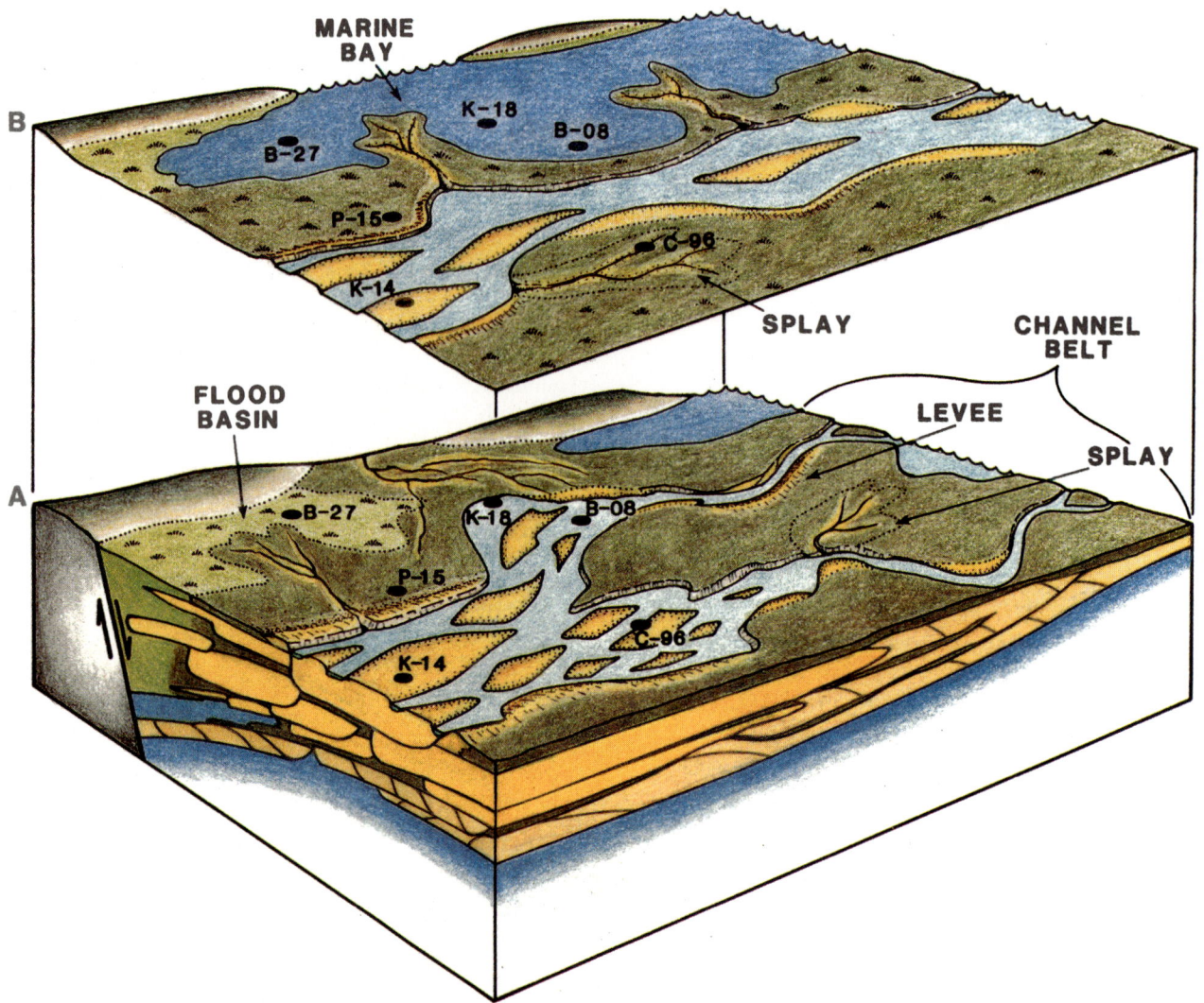

FIGURE 11. Depositional model for the Hibernia sandstones.

data in seven well bores. Preliminary results on extracting stratigraphic information from seismic attribute analysis are encouraging, but more work is needed in this area. Reservoir performance during field depletion will ultimately define the extent of reservoir continuity.

RESERVOIR CHARACTERISTICS

In each of the seven wells that penetrated the Hibernia interval, moderately well-sorted, medium- to coarse-grained, distributary channel sandstones with excellent productivity characteristics were encountered. Measured core porosities range from 15 to 22% and permeabilities range up to 10 darcys in this facies.

Both preserved primary intergranular porosity and secondary dissolution porosity are interpreted within the Hibernia sandstones. To account for the high porosity values at the present depth of burial (3750 m; 12,300 ft), a predominantly secondary origin due to carbonate cement and/or framework grain dissolution has been interpreted by recent petrographic studies (Brown et al., 1989; Abid and Hesse, 1988). Petrographic evidence of cement dissolution includes corroded grain boundaries, oversized pores, inhomogeneity of packing, and remnant cement (Brown et al., 1989). The fraction of the total porosity that is secondary in origin has been estimated to be greater than 80%. Mechanical compaction, syntaxial quartz overgrowths, and minor kaolinite pore-fill are the principal porosity-reducing elements in the Hibernia interval.

Crevasse splay deposits are typically fine-grained sandstones. Although highly productive, this facies does not possess the excellent reservoir quality common to the channel belt sands. Areal extent and fine pore throat size limit the potential floodability of these reservoirs. The core porosity vs. permeability plot (Figure 13A) demonstrates the relationship between high permeability and moderate porosity. A plot of mean grain size vs. permeability (Figure 13B) illustrates an increase in permeability with increasing grain size indicative of the textural control on reservoir quality.

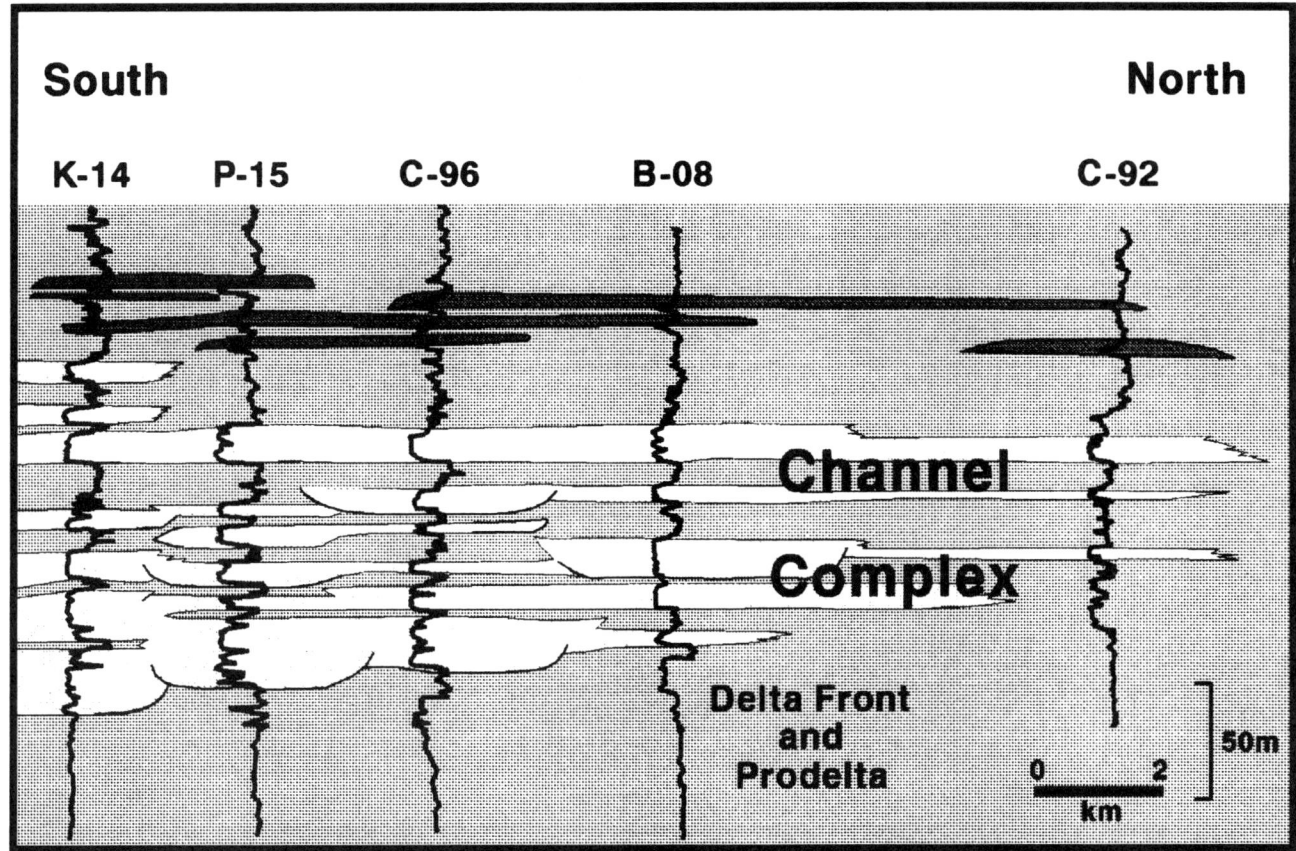

FIGURE 12. North-south stratigraphic cross section of the Hibernia sandstones in the Hibernia field area. Reservoir continuity within the channel belt sandstones is anticipated in the north-south orientation parallel to depositional dip). Laterally discontinuous, fine-grained sandstones deposited in crevasse splays are subordinate Hibernia sandstone reservoirs.

DEPOSITIONAL MODEL—BEN NEVIS/AVALON SANDSTONES

In contrast to the Hibernia sandstones, Ben Nevis/Avalon stratigraphy is complicated by unconformities and structural growth of the field. The distinction between Ben Nevis and Avalon sediments, and the depositional model proposed for these reservoirs, is based on the integration of sedimentology, biostratigraphy, and seismic stratigraphy.

The vertical distribution of depositional facies is recorded and interpreted from 886 m (2906 ft) of conventional core including 450 m and 160 m of continuous core through this interval in the I-46 and J-34 wells, respectively. Core was recovered from Ben Nevis/Avalon sandstones in all wells except P-15. Two horizons have been employed for stratigraphic reconstruction. These are: (1) a Barremian unconformity which is biostratigraphically, lithologically, and/or petrophysically identifiable; and (2) a maximum flooding surface represented by dinoflagellate-rich marine shales. This shale complex caps an over-all transgressive, deepening event on top of the Barremian unconformity.

Within the Ben Nevis/Avalon horizons, two groups of sedimentary structures are predominant: (1) wave- and wave process-associated sedimentary structures; and (2) structures related to rather intense biogenic activity. The dominance of biogenic structures indicates an over-all low-energy depositional setting. In cores, the preserved record of physical depositional processes is largely one of episodic high energy due to marine storms.

Pre-Barremian Unconformity—Avalon Sandstones

Hauterivian-Barremian deposition in the Hibernia field area occurred in low-energy, partially restricted, marine environments. Extensively cored in the I-46 well, this section consists of muddy, burrowed sandstone interstratified with numerous beds (10 to 30 cm thick) of cleaner, less-burrowed sandstones. The sandstone interbeds typically have sharp basal contacts that truncate underlying lithologies and gradually become more burrowed upward. They may contain rip-up clasts and shell debris (especially near the base), horizontal to gently inclined lamination, symmetrical ripples, and rare higher-angle cross-lamination. These beds are interpreted as representing episodic high-energy deposition within an otherwise low-energy environment. They are typical of

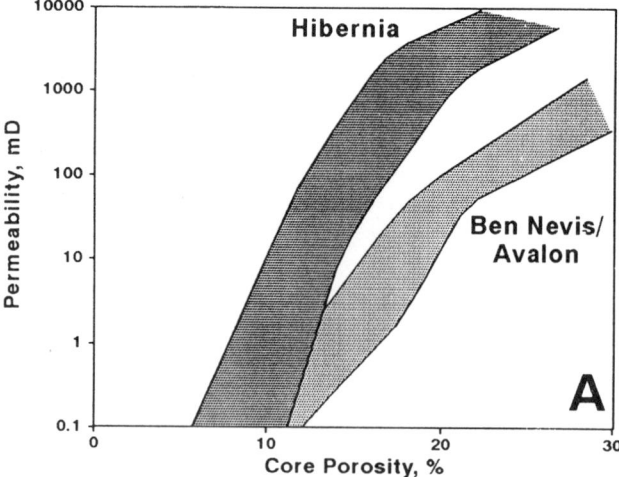
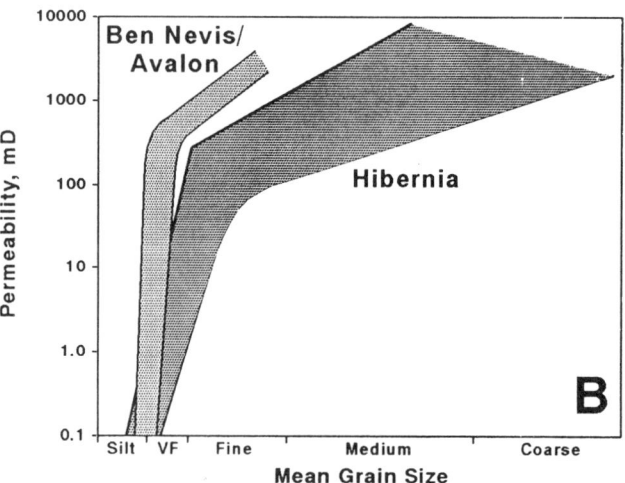
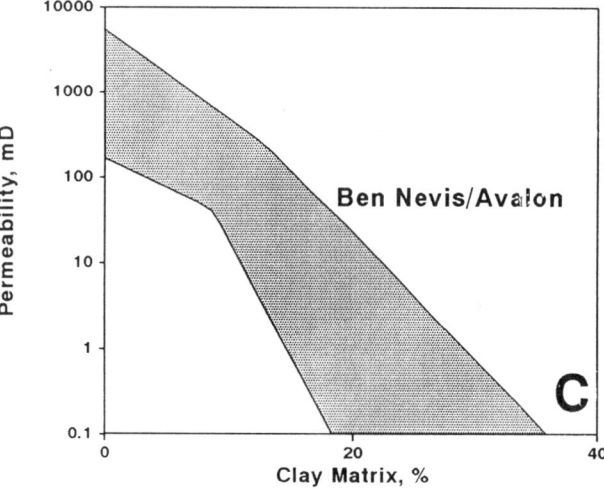

FIGURE 13. (A and B) Comparative plots of porosity vs. permeability and grain size vs. permeability for the Ben Nevis/Avalon and Hibernia sandstones. (C) Clay matrix vs. permeability relationship for the Ben Nevis/Avalon reservoirs.

storm deposits reported in many modern (e.g., Nelson, 1982) and ancient (e.g., Kreisa, 1981; Howard, 1971) sediments. The tops of storm beds are highly burrowed; thin storm beds and interstorm lithologies are also completely bioturbated.

Even though this environment was subjected to periodic marine storms, salinities less than normal marine are inferred. Shells are present but not abundant, and at least some are brackish water forms (S. Root, Mobil-Dallas, personal communication, 1983). Many occur in storm deposits and therefore may have been transported. Burrowing is abundant and many burrow types can be related to marine organisms, but diversity is much reduced from that typical of open-shelf deposits (G. Pemberton, University of Alberta, personal communication, 1983). A partially restricted, shallow bay environment interpretation best fits these data.

Barremian Unconformity

Subaerial exposure of the shallow bay sequence during Barremian time is represented in core (B-27, K-14, C-96) by oxidized and rooted red shales with desiccation cracks and limonite concentrations. Red shales with limonite stain were identified in sample cutting descriptions from the O-35 well. Well bore washouts and sonic log spiking are indicative of this horizon. Barremian palynomorphs *Muderongia simplex* and *M. imparilis* occur above the unconformity. Barremian to Hauterivian palynomorphs *Ctenidodinium elegantulum* and *Phoberocysta* sp., and microfossils *Hutsonia* sp. (3) (Ostracod) and *Choffatella decipiens* (Foraminifera) occur below it (Thompson, 1987). This "red bed" facies occurs within similar aged sediments in wells located approximately 20 km east of the Hibernia field (Mara M-54 and South Mara C-13 wells at drill depths of approximately 2680 and 3115 m, respectively) and is interpreted as representing a regional fall in sea level.

Post-Barremian Unconformity—Avalon Sandstones

Depositional strike of the Avalon sandstone interval was approximately north-south as indicated by the virtually identical vertical successions of lithologies in cored intervals of the B-27 and K-14 wells, which are widely separated (4 km), but nearly aligned north-south.

Reservoir sandstones may include a variety of nearshore facies; the event of major importance in the formation of this complex was a gradual transgression over the mid-Barremian unconformity surface. Each of the four wells from which core was recovered in this interval is characterized by an upward-thinning and upward-fining succession of sandstones overlain by marine shales (Figure 14). Sandstones are comprised of numerous decimeter-scale upward-fining beds that contain the same suite of storm-generated sedimentary structures previously described. The shales contain an abundant and diverse dinoflagellate assemblage. This sequence occurred as the shoreline retreated westward (Figure 15A). The "christmas-tree" pattern of wireline logs through this

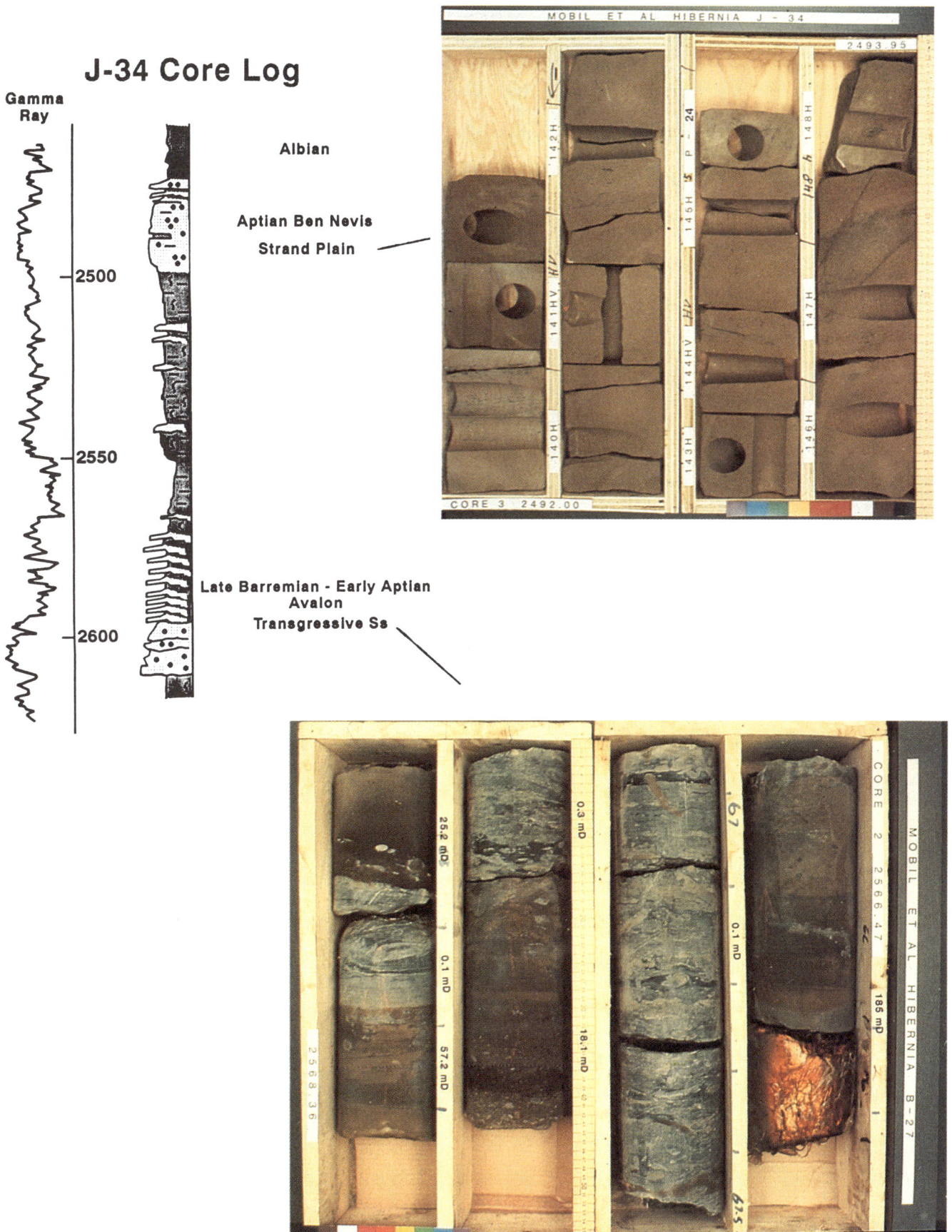

FIGURE 14. Core log for the Ben Nevis and Avalon sandstones in the Hibernia J-34 well. The development of these reservoirs is illustrated in Figure 15.

complex reflects gradually waning energy as marine shelf conditions were established. This over-all transgressive interval is recognized throughout the Hibernia field by correlative wireline log signatures.

Local structural influence on Barremian–early Aptian sedimentation in the Hibernia area was minor compared with such influence during Aptian-Albian times. Conglomerates that overlie coal and red mudstone in the most westerly G-55A well likely represent incipient tectonic activity and alluvial fan sedimentation adjacent to the basin margin. Retreat of the shoreline is indicated by the marine shale that overlies the conglomeratic units in G-55A.

Available sea level curves for this interval suggest that this was a time of global sea level rise (Vail et al., 1977). Subsidence and deepening of the tectonically active Jeanne d'Arc Basin, however, are equally plausible causes for transgressive deposition.

Aptian–Albian Ben Nevis Sandstones

Subsequent to the establishment of open marine conditions across the field area, tectonically influenced sandstone deposition became pronounced. Vast thickness changes of Aptian–Albian sediments occur over the Hibernia structure and stratigraphic relationships of Aptian–Albian Ben Nevis sandstones are complex.

Included within the Ben Nevis sequence is an Aptian-aged, nonburrowed, high-permeability sandstone typified by dominantly horizontal to gently inclined laminae and fine shell material (Figure 14). This sand body was encountered only in the O-35 and J-34 wells (Figure 15B). Age-equivalent sediments westward (purportedly landward) of this sandstone are shallow marine, bioturbated mud and sandy mudstones (I-46 core). Aptian-aged rocks in wells eastward of this O-35/J-34 sand are absent. An unconformable contact between early Aptian and Albian sediments in these eastwardly located wells is biostratigraphically evident, and pronounced angular relationships at this contact have been identified by dipmeter in five of the six locations (Cox, 1988). Early Aptian to Barremian palynomorphs occur below the unconformity, whereas some Aptian as well as Albian and younger forms such as *Xiphophoridinium alatum* and *Rugubivesiculites rugosus* occur above it (Thompson, 1987).

The lateral distribution of Aptian facies and unconformities indicates isolation of the partially exposed Hibernia structure from the mainland. The relatively thick (20 m), high-reservoir-quality O-35/J-34 sandstone is interpreted as a strand plain deposit derived by reworking of the exposed Hibernia structure and deposited adjacent to the western flank of the island in a fault-controlled graben (Figure 15B). Age-equivalent shallow marine, bioturbated mudstone and sandy mudstone recovered in the I-46 core preclude a mainland-attached strand plain interpretation.

Structural development continued into the Albian, producing the horst and graben complex seismically evident today (Figure 15C). Progressive south-to-north burial of the structure is evident by seismic onlap and pronounced thickening of Albian sediments over the southern portion of the structure. The northern edge of the structure, however, remained exposed throughout Albian time (Figure 15C). Erosionally thinned, Barremian to early Aptian Avalon reservoirs in B-08 are overlain by Cenomanian–Turonian sediments; Albian sediments are absent. In contrast, Aptian to Albian sediments are approximately 1500 m thick at the southern edge of the field. Burial and sealing of the structure by shallow marine mudstones were complete by late Cretaceous time.

RESERVOIR GEOMETRY AND QUALITY

The style of sedimentation and depositional processes for the Ben Nevis/Avalon horizons have important implications for both reservoir geometry and reservoir quality. The transgression that apparently moved westward across the Hibernia area imparted a sheetlike geometry to this interval; shoreface sandstones are envisioned as laterally extensive, field-wide deposits.

Modern storm deposits extend over areas of many square kilometers. Thinly bedded, storm-generated sandstones deposited during the transgressive deepening event are expected to be continuous between wells. Vertical communication between those sandstones, however, will be limited by interstratified shales and bioturbated sandy mudstones.

Ben Nevis strand plain sandstones deposited adjacent to the exposed Hibernia structure are anticipated to be elongate and narrow, possibly 5 to 10 km in length and several kilometers wide. The vertical continuity within this reservoir is significantly greater than that of the thinly bedded, heterolithic storm deposits. The areal portion of the field in which both the strand plain sandstones and transgressive sandstones occur is the designated area for future Ben Nevis/Avalon development.

RESERVOIR CHARACTERISTICS

In contrast to the Hibernia interval, the Ben Nevis/Avalon is finer grained and its porosity is dominantly primary in origin. Average porosity in the Ben Nevis/Avalon reservoir sandstones is 21%; permeabilities range up to 3 darcys, but average 100 md.

The most volumetrically significant porosity-reducing agent in the Avalon sandstones is calcite cement. Cementation severely reduces both porosity and permeability in sandstones that initially had good reservoir quality. No definite relationship has been established between depositional facies and calcite cementation; the dominant occurrence, however, is within the "transgressive" Avalon deposits. The effect of porosity reduction by mechanical compaction and quartz overgrowths is comparatively less important than in the Hibernia interval.

In general, biogenic structures dominate the Avalon section. Primary clays introduced by burrowing organisms are a principal cause of low permeability. A plot of percent matrix clay vs. permeability clearly demonstrates permeability reduction with increasing clay content (Figure 13C).

A. **Late Barremian - Early Aptian Avalon Sandstone**

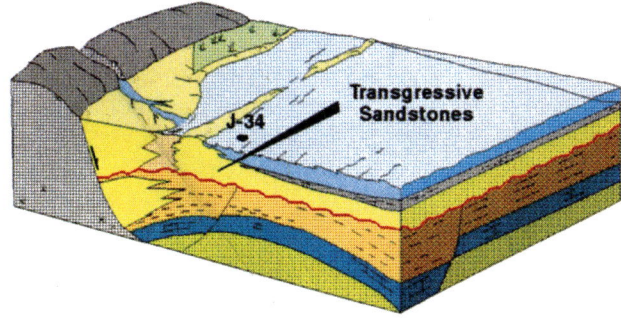

B. **Aptian Ben Nevis Sandstone**

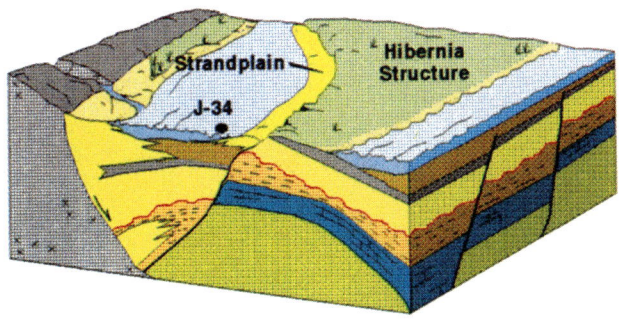

C. **Albian**

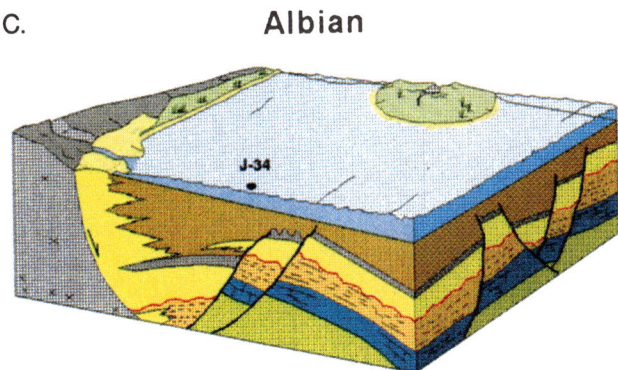

FIGURE 15. Structural and stratigraphic development of the Ben Nevis and Avalon sandstones in the Hibernia field.

Porosity and permeability are markedly enhanced where high-energy events have winnowed fines and/or inhibited burrowing. In many cases, permeability decreases upward in storm beds concomitant with the upward increase in bioturbation due to poststorm repopulation by burrowing organisms. Where storm events have amalgamated, significant lateral and vertical reservoir continuity can be anticipated. However, much of the permeable sandstone within the "transgressive" deepening sequence and underlying shallow bay facies consists of isolated decimeter-scale beds separated vertically by significant permeability barriers. These thin beds can be expected to be continuous between wells, but, because they are so thin, they are extremely susceptible to disruption by minor fault offset or local cementation.

The majority of data in the permeability vs. grain size plot (Figure 13B) lies within the very fine-grained sandstone range where a wide range in permeability is associated with no appreciable change in mean grain size. These sands represent the clay-rich, bioturbated facies common to this shallow marine sequence. High-energy events are represented by very fine to fine-grained sandstones. Permeability increases with grain size in these deposits. A plot of measured core porosity vs. permeability for the Ben Nevis/Avalon sands is compared with an equivalent plot for the Hibernia sandstones in Figure 13A. Table 1 summarizes the characteristics of these reservoirs.

TRAP, MIGRATION, AND SOURCE

Hydrocarbon accumulations in the Hibernia field are structurally controlled. Structural development was essentially complete by late Albian times with only minor displacement on the Murre, Nautilus, and several associated faults continuing into the Tertiary.

Organic-rich source rocks occur in three stratigraphic horizons in the Hibernia area. These are immature Tertiary mudstones, thin coal horizons within the Hibernia sandstone interval, and exceptionally rich (up to 9% organic carbon), locally mature, Late Jurassic shales (Powell, 1985). Geochemical analysis of the Kimmeridgian/Oxfordian shales encountered in the Hibernia K-18 well indicates that these shales are the oil source rocks for the Hibernia structure. The homogeneous composition of all the Hibernia oils, regardless of stratigraphic occurrence, has been interpreted to represent a locally derived source and fault-controlled migration (von der Dick and Meloche, 1986). The onset of hydrocarbon migration occurred in the Late Cretaceous; peak generation occurred in the Oligocene (ibid).

DEVELOPMENT PLAN

The Hibernia field will be developed by the construction and installation of a concrete gravity base structure (GBS) that will support topside facilities including drilling, production and utility systems, and personnel accommodations (Figure 16). The crude oil loading system for the project will include pipelines to transfer crude oil from the production platform to the two offshore loading systems. The Hibernia drilling and production facilities are designed to incorporate two drilling rigs and process facilities for up to 24,000 m^3 (150,000 bbl) of oil, 12,700 m^3 (80,000 bbl) of water, and 7.4 × 10^6 m^3 (263 million ft^3) of gas per day. Water and gas injection facilities are incorporated, and the GBS will contain storage for approximately 200,000 m^3 (1.26 million bbl) of crude oil.

Full-field development of the Hibernia sandstone reservoir is planned. Fifty-eight wells are to be drilled in

TABLE 1. Hibernia field reservoir data.

Reservoir	Hibernia[1]	Ben Nevis/Avalon[2]
Basic Data		
Area	16,500 ac (66.8 km^2)	4,400 ac (17.8 km^2)
Net pay thickness:		
Maximum	68 m (223 ft)	101 m (331 ft)
Average	50 m (164 ft)	46 m (151 ft)
Net pay/gross pay ratio:		
Maximum	0.7	0.4
Average	0.5	0.3
Average porosity	16%	20%
Average permeability	500 md	100 md
Average water saturation	13%	21%
Estimated oil-in-place	1.4 billion bbl (222.6 million m^3)	0.7 billion bbl (111.3 million m^3)
Other		
Gravity of oil (°API)	34	30
Gas cap	Yes	No
Oil–water contact (subsea)	3,930 m (12,894 ft)	2,602 m (8537 ft)
Pressure	5,730 psi (39.51 MPa)	3,870 psi (26.68 MPa)
Temperature	95°C (203°F)	66°C (151°F)
Depletion mechanism	Pressure to be maintained by water injection and gas injection	Pressure to be maintained by water injection

[1]Full field development.
[2]Designated area of development.

the Hibernia sandstones during the period 1996 to 2002. Forty-eight wells are planned to be drilled from the production platform; ten wells will be subsea completions.

Reservoir pressure will be maintained during production by a combination of water injection and gas cap injection into the Hibernia sandstones. Production and injection wells have been located so as to conform to fault block geometries and provide efficient flooding displacement. Exact well numbers and locations will change as the development proceeds and the reservoir characteristics are more clearly defined by production and injection performance.

Average well spacing is approximately 100 ha (250 ac) in the Hibernia sandstones. Production wells are expected to contribute an annual average rate of 1590 m^3 per day (10,000 BOPD). The Hibernia and Ben Nevis/Avalon sandstones will be completed independently and downhole commingling of production from the two formations is not planned. The Hibernia Development Plan proposes an area of Ben Nevis/Avalon development delimited by sufficient reservoir continuity characteristics (Mobil et al., 1985). All 25 of the proposed Ben Nevis/Avalon wells are planned to be subsea completions and production is anticipated to commence at the onset of Hibernia decline.

ACKNOWLEDGMENTS

The authors would like to recognize the geologists and geophysicists of the Hibernia Consortium who were involved in the conception of the Hibernia play and the evaluation of the Hibernia field. Many of the ideas presented in this text hinge upon the initial efforts of Rudi Butot of Mobil Oil Canada. The interpretations presented are those of Mobil Oil Canada and do not necessarily reflect our partners' geological views. We thank Bruce Longmore of Mobil Oil Canada for his review of and

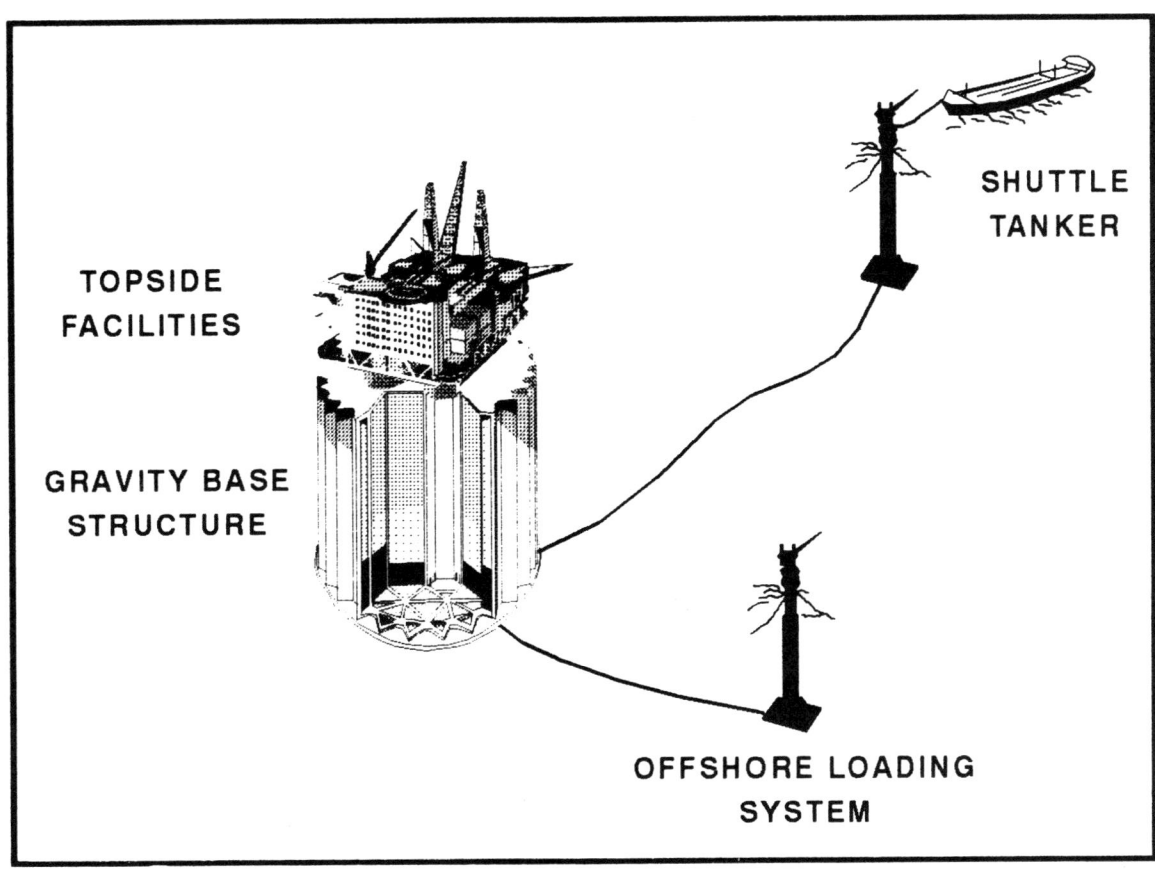

FIGURE 16. Principal components of the Hibernia production system.

improvements to the manuscript, and for his invaluable discussions on the Hibernia project. Regional views and interpretations of Grand Banks geology were greatly enhanced and clarified by numerous discussions with Jock McCracken of Mobil Oil Canada. We also wish to acknowledge the patience and talent of Wayne Jennex and Doug Watt of Mobil Oil Canada for their drafting and Eileen Mayhew for her typing of the manuscript.

REFERENCES CITED

Abid, I., and R. Hesse, 1988, Diagenesis and porosity evolution in Hibernia oilfield, Grand Banks offshore Newfoundland (abs.): Geological Association of Canada Program with Abstracts, v. 13, p. Al.

Amoco Canada Petroleum Company Limited, and Imperial Oil Ltd., 1973, Regional geology of the Grand Banks: Bulletin of Canadian Petroleum Geology, v. 21, p. 479-503.

Arthur, K. R., D. R. Cole, G. G. L. Henderson, and D. W. Kushnir, 1982, Geology of the Hibernia discovery, in M. T. Halbouty (ed.), The deliberate search for the subtle trap: AAPG Memoir 32, p. 181-195.

Benteau, R. I., and M. G. Sheppard, 1982, Hibernia—a petrophysical and geological review: Journal of Canadian Petroleum Technology, v. 21, n. 6, p. 59-72.

Brown, D. M., K. D. McAlpine, and R. W. Yole, 1989, Sedimentology and sandstone diagenesis of Hibernia Formation in Hibernia oil field, Grand Banks of Newfoundland, AAPG Bulletin, v. 73, n. 5 p. 557-575.

Canada-Newfoundland Offshore Petroleum Board, 1986, Decision 86.01: application for approval: Hibernia Canada-Newfoundland Benefits Plan, Hibernia Development Plan, 143 p.

Cant, D. J., and R. G. Walker, 1976, Development of a braided-fluvial facies model for the Devonian Battery Point Sandstone, Quebec: Canadian Journal of Earth Sciences, v. 13, p. 102-119.

Cox, J. W., 1988, unpublished Avalon dipmeter study.

Cox, J. W., 1990, unpublished Hibernia dipmeter study.

Crowley, K. D., 1983, Large-scale bed configurations (macroforms), Platte River Basin, Colorado and Nebraska: primary structures and formative processes: GSA Bulletin, v. 94, p. 117-133.

de Voogd, B., C. E. Keen, and W. A. Kay, 1990, Fault reactivation during Mesozoic extension in eastern offshore Canada; Tectonophysics, v. 173, p. 567-580.

Drummond, K. J., 1992, Geology of Venture, a geopressured gas field, offshore Nova Scotia, this volume.

Enachescu, M. E., 1987, Tectonic and structural framework of the Northeast Newfoundland Margin, in C. Beaumont and A. J. Tankard (eds.), Sedimentary basins and basin forming mechanisms: Canadian Society of Petroleum Geologists, Memoir 12, p. 117-146.

Enachescu, M. E., 1988, Extended basement beneath the intracratonic rifted basins of the Grand Banks of Newfoundland: Canadian Journal of Exploration Geophysics, v. 24, n. 1, p. 48-65.

Farrell, K. M., D. Nummedal, and S. B. Basumallick, 1984, Sedimentation in a transgressed inter-distributary bay on the Mississippi-LaFourche delta lobe: AAPG Bulletin, v. 68, p. 475.

Grant, A. C., K. D. McAlpine, and J. A Wade, 1986a, The continental margin of eastern Canada: geological framework and petroleum potential, in M. T. Halbouty (ed.), Future petroleum provinces of the world: AAPG Memoir 40, p. 177-205.

Grant, A. C., K. D. McAlpine, and J. A. Wade, 1986b, Offshore geology and petroleum potential of eastern Canada: Energy Exploration and Exploitation, v. 4, p. 43-91.

Grant, A. C., L. F. Jansa, K. D. McAlpine, and A. Edwards, 1988, Mesozoic-Cenozoic geology of the eastern margin of the Grand Banks and its relation to Galicia Bank: Proceedings of the Ocean Drilling Program, Scientific Results, v. 103, p. 787-808.

Hiscott, R. N., R. C. L. Wilson, S. C. Harding, V. Pujalte, and D. Kitson, 1990a, Contrasts in Early Cretaceous depositional environments of marine sandbodies, Grand Banks–Iberian Corridor: Bulletin of Canadian Petroleum Geology, v. 38, n. 2, p. 203-214.

Hiscott, R. N., R. C. L. Wilson, F. M. Gradstein, V. Pujalte, J. Garcia-Mondejar, R. R. Boudreau, and H. A. Wishart, 1990b, Comparative stratigraphy and subsidence history of Mesozoic rift basins of North America: AAPG Bulletin, v. 74, n. 1, p. 60-76.

Howard, J. D., 1971, Comparison of the beach-to-offshore sequence in modern and ancient sediments, in J. D. Howard, J. W. Valentine, and J. E. Warme (eds.), Recent advances in paleoecology and ichnology: American Geological Institute Short Course Lecture Notes, p. 148-183.

Hubbard, Richard J., 1988, Age and significance of sequence boundaries on Jurassic and Early Cretaceous rifted continental margins, AAPG Bulletin, v. 72, n. 1, p. 49-72.

Jansa, L. F., and J. A. Wade, 1975a, Paleogeography and sedimentation in the Mesozoic and Cenozoic, southeastern Canada, in C. J. Yorath, E. R. Parker, and D. J. Glass (eds.), Canada's continental margins and offshore petroleum exploration: Canadian Society of Petroleum Geologists Memoir 4, p. 79-102.

Jansa, L. F., and J. A. Wade, 1975b, Geology of the continental margin of Nova Scotia and Newfoundland, in W. J. M. van der Linden and J. A. Wade (eds.), Offshore geology of eastern Canada: Geological Survey of Canada, Paper 74-30, p. 51-105.

Kreisa, R. D., 1981, Storm-generated sedimentary structures in subtidal marine facies with examples from the Middle and Upper Ordovician of southwestern Virginia: Journal of Sedimentary Petrology, v. 51, n. 3, p. 823-848.

Masson, D. G., and P. R. Miles, 1984, Mesozoic sea-floor spreading between Iberia, Europe and North America: Marine Geology, v. 56, p. 279-287.

Masson, D. G., and P. R. Miles, 1986, Development and hydrocarbon potential of Mesozoic sedimentary basins around margins of North Atlantic: AAPG Bulletin, v. 70, p. 721-729.

McAlpine, K. D., 1989, Mesozoic stratigraphy, sedimentary evolution and petroleum potential of the Jeanne d'Arc Basin, Grand Banks of Newfoundland: Geological Survey of Canada Paper, Paper 89-17.

Mobil Oil Canada, Ltd., 1985, Hibernia Development Plan, v. 1.

Nelson, C. H., 1982, Modern shallow-water graded sand layers from storm surges, Bering shelf: a mimic of Bouma sequences and turbidite systems: Journal of Sedimentary Petrology, v. 52, p. 537-545.

Powell, T. G., 1985, Hydrocarbon-source relationships, Jeanne d'Arc and Avalon basins, offshore Newfoundland: Geological Survey of Canada Open File Report 1094.

Shumm, S. A., 1968, River adjustment to altered hydrologic regiment—Murrumbidgee River and paleochannels, Australia: USGS Professional Paper 598, 65 p.

Shumm, S. A., 1977, The fluvial system: Wiley and Sons, 388 p.

Sinclair, I. K., 1988, Evolution of Mesozoic-Cenozoic sedimentary basins in the Grand Banks area of Newfoundland and comparison with Falvey's (1974) rift model: Bulletin of Canadian Petroleum Geology, v. 36, n. 3, p. 255-273.

Tankard, A. J., and H. J. Welsink, 1987, Extensional tectonics and stratigraphy of Hibernia oil field, Grand Banks, Newfoundland: AAPG Bulletin, v. 71, p. 1210-1232.

Taylor, G. G., W. L. Alexander, and D. G. Ward, 1989, Images of the Hibernia Reservoir: A Case Study: Proceedings of the 21st Annual Offshore Technology Conference, Paper No. 5895, v. 1, p. 119-128.

Thompson, L. R., 1987, Cretaceous transgressive and regressive events in the Avalon basin, Grand Banks of Newfoundland: Cushman Foundation for Foraminiferal Research, Special Publication 24.

Vail, P. R., R. M. Mitchum, Jr., and S. Thompson III, 1977, Seismic stratigraphy and global changes of sea level, part 4: global cycles of relative changes of sea level: in Seismic stratigraphy—applications to hydrocarbon exploration: AAPG Memoir 26, p. 83-97.

von der Dick, H., and J. D. Meloche, 1986, Generation, migration and expulsion of hydrocarbons in the Hibernia field: 1986 CSPG Convention Programs and Abstracts, p. 38.

Chapter 5

Geology of Venture, a Geopressured Gas Field, Offshore Nova Scotia

Kenneth J. Drummond

Mobil Oil Canada
Calgary, Alberta, Canada

ABSTRACT

The Venture gas field is a potential giant gas accumulation discovered in 1979 by the Mobil et al. Venture D-23 well drilled by Petro-Canada on a farm-in from Mobil. This well was completed at a total depth of 4945 m (16,224 ft). The field is located on the Scotian Shelf, 5 to 16 km (3 to 10 mi) off the east end of Sable Island, 300 km (186 mi) east of Halifax, Nova Scotia. Water depths across the field vary from 12 to 23 m (39 to 75 ft). Participants in the Venture field include Mobil, Esso (formerly Texaco), Petro-Canada, and Nova Scotia Resources. Venture is the largest discovery on the Scotian Shelf, offshore Nova Scotia, with initial estimates of ultimate gas recovery exceeding 1.5 tcf.

The Venture structure is an elongated east-west anomaly on the downthrown side of a normal east–west-trending fault. Currently five wells define a gas-bearing area of about 3000 ha (7400 ac). In the Venture field, gas occurs in multiple sandstone reservoirs, from Upper Jurassic to Lower Cretaceous age, over a stratigraphic interval of 1600 m (5250 ft). The Venture gas-bearing productive section consists of one to four hydropressured sands and up to 18 geopressured sands, with pressure/depth ratios of up to 0.89 psi/ft. Venture sands exhibit abnormally high porosities for their depths, possibly both preserved primary porosity and secondary solution porosity. The reservoirs are often characterized by high water saturations and very high-salinity formation water.

INTRODUCTION

The Scotian Shelf covers the continental shelf area of Nova Scotia (Figure 1), an area of approximately 155,000 km^2 (60,000 mi^2). To date (September 7, 1990), 122 wells, with a total drilled section of 494,588 m (1,622,664 ft), have been completed at some 87 locations since 1967. This represents one well per 1270 km^2 (490 mi^2) and one structure tested for every 1780 km^2 (690 mi^2). Figure 2 shows the major occurrences of hydrocarbons in the Sable

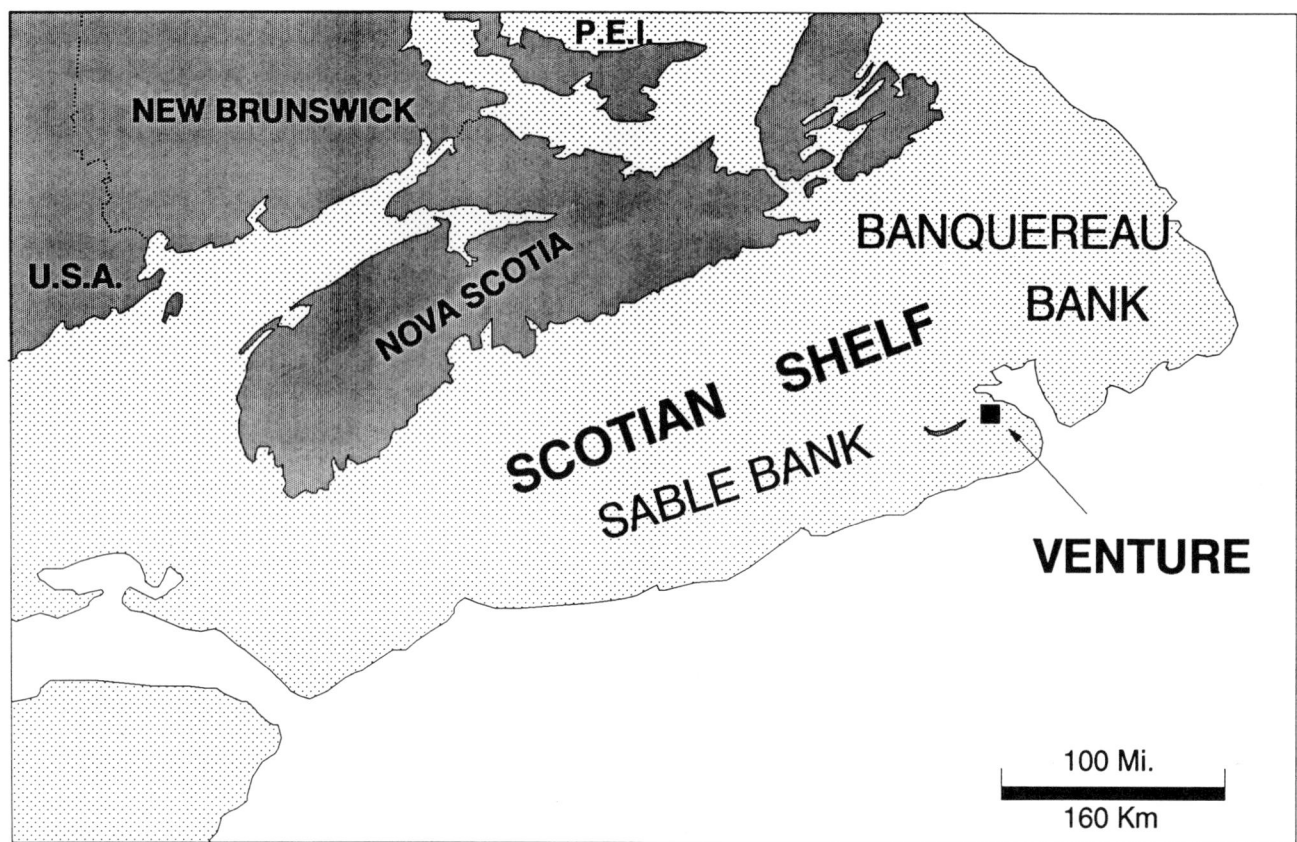

FIGURE 1. Scotian Shelf, offshore Nova Scotia, showing location of Venture gas field.

Island area. Drilling in the Sable Island area has been ongoing since 1967, when Mobil drilled Sable Island No. 1 (Magnusson, 1973) to a total depth of 4604 m (15,106 ft). Only very minor shows of oil and gas were found; however, the test did confirm the presence of a thick sedimentary section capable of trapping hydrocarbons. The first good indication of hydrocarbons occurred in 1969 in Shell Onondaga E-84, the second well drilled on the Scotian Shelf. Although no successful tests were run, log analysis suggested 54 m (177 ft) of gas-bearing Lower Cretaceous sand. The first important discovery with recovery of hydrocarbons was the Mobil-Tetco Sable Island E-48 well, completed October 15, 1971. The well encountered 247 m (810 ft) of net oil, gas, and condensate pay from several Upper Cretaceous sandstones. In 1972, another important discovery well was drilled. Mobil-Tetco Thebaud P-84 encountered a total of 80 m (262 ft) of gas pay in hydropressured sandstones and 15 m (49 ft) of gas pay in Lower Cretaceous geopressured sandstones. The step-out well Thebaud I-94 confirmed the lateral extent of the geopressured gas sand and found all hydropressured sands to be water bearing.

The discovery well for the Venture field was Mobil Texaco Pex Venture D-23, which was suspended as a gas/condensate well on June 16, 1979, at a total depth of 4945 m (16,224 ft). From Sable Island No. 1, abandoned January 2, 1968, to the successful completion of Venture B-43 on April 25, 1982, it has taken the industry 14 years to delineate a potentially viable commercial structure on the Scotian Shelf. The success at Venture and the realization by explorationists of the significance of the geopressured play and the subsequent potential for large reserves led to a resurgence in activity on the Scotian Shelf and several additional discoveries.

REGIONAL SETTING

Figures 3 and 4 illustrate the regional tectonic setting of the Scotian Shelf (Jansa and Wade, 1975; Grant et al., 1986). The LaHave platform, onlapping the Meguma basement complex, was the site of relatively slow shallow water deposition from the Middle Jurassic to the Holocene. The area is bounded along the north by the Orpheus trough, a Jurassic graben, with thick infill of salt and clastics. The shelf sediments shown in the diagrammatic section thicken from a zero line off the coast of Nova Scotia to about 3 km (10,000 ft) at the edge of the basins.

Regional dip across the shelf area is about 30 m/km (158 ft/mi) to the southeast. A tectonic hinge zone occurs in an arc around Sable Island, with regional dip increasing to about 56 m/km (296 ft/mi). The Jurassic hinge line and the associated distribution of carbonate banks controlled the loci of Late Jurassic and Early Cretaceous clastics within the basins. The three main basins on the shelf are the Mohican and Abenaki basins, which are major centers

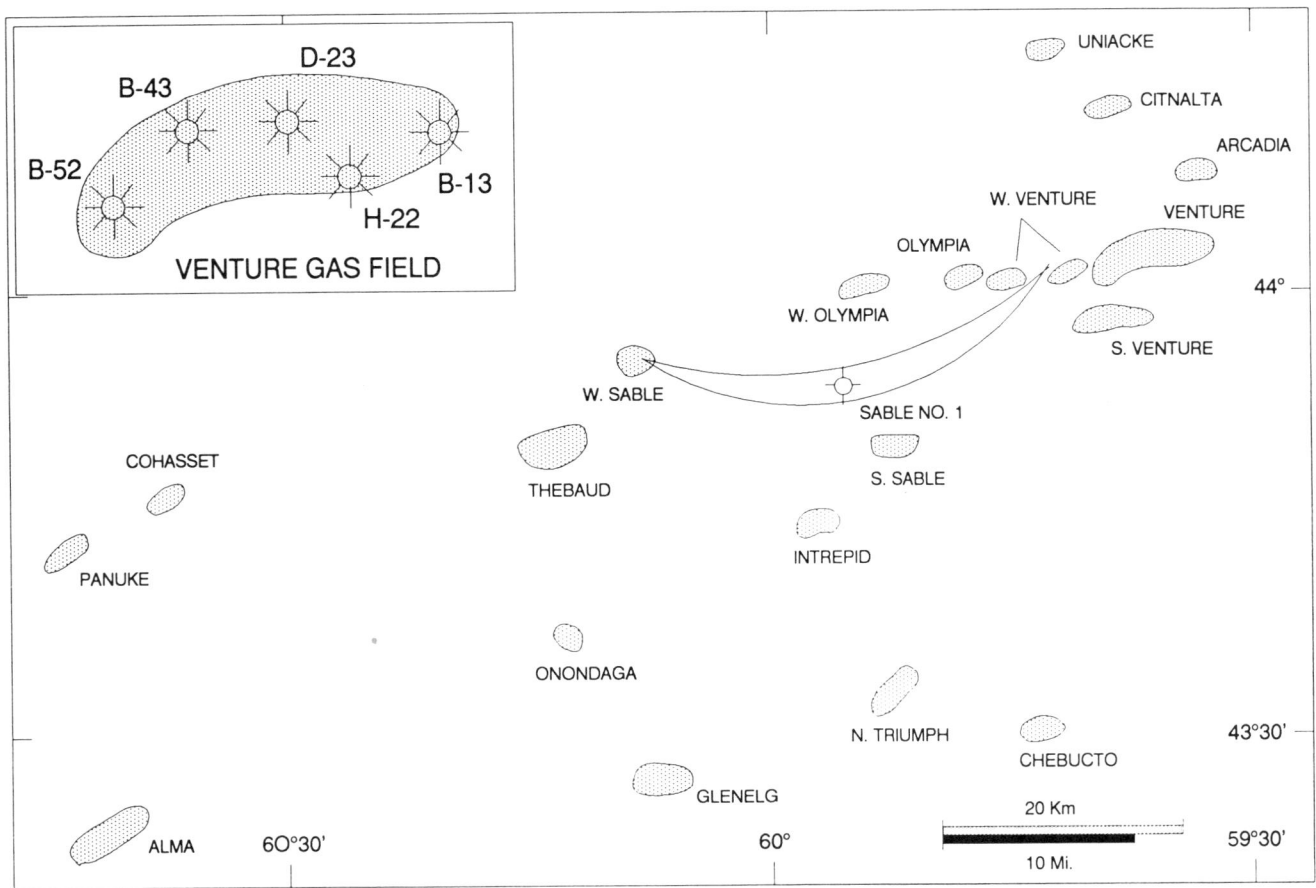

FIGURE 2. Significant discoveries in the Sable Island area, with inset showing the five Venture wells.

of deposition behind the carbonate buildups, and the Sable Basin, which is the major locus of clastic deposition in front of the carbonate trend. A northwest–southeast-trending fault marks the west end of the Abenaki Basin. This general line denotes a major deflection on all regional maps, structure and isopach, through the east Sable area. The Sable Basin is bounded on the southeast by the Sedimentary Ridge Province. Seismic data indicate this to be a distinctive area of elongate diapiric structures that could be salt and/or shale.

The Sable Basin, and the Venture area in particular, are characterized by apparently continuous deposition from the Middle Jurassic to the Tertiary. The postbreakup tectonic history of the Scotian Shelf is in general one of subsidence. The only variation is differential subsidence, with one part subsiding much more rapidly than another. In the Late Jurassic and Early Cretaceous, increased subsidence of the basins occurred, with uplift of the source area to supply the rapid influx of terrigenous clastics. This is the case for the Sable Basin where subsidence and rapid sedimentation have produced the downdropped fault structures along the Late Jurassic–Early Cretaceous hinge line. The Venture gas accumulation occurs in such an environment.

GENERAL STRATIGRAPHIC SECTION

The depositional history of the Scotian Shelf follows the plate tectonic model of Falvey (1974), as indicated on the generalized stratigraphic section (Figure 5) adapted from M. M. Given (1977). On the Scotian Shelf, Triassic and later sediments overlie a pre-Carboniferous basement of Cambrian–Ordovician metasediments and Devonian intrusions. A seaward-thickening wedge of coastal plain sediments over a rifted passive continental margin comprises the stratigraphic section. The Mesozoic-Cenozoic sedimentary thickness in the Sable Island area possibly exceeds 9000 m (30,000 ft).

Four major lithologic sequences comprise the stratigraphic section of the Scotian Shelf:

1. A prerift and rifting sequence of continental clastics, evaporites, and dolomites deposited over a pre-Carboniferous basement of Devonian granites and Cambrian–Ordovician metasediments.

2. Open marine carbonates and clastic equivalents (Abenaki and Mic Mac clastics).

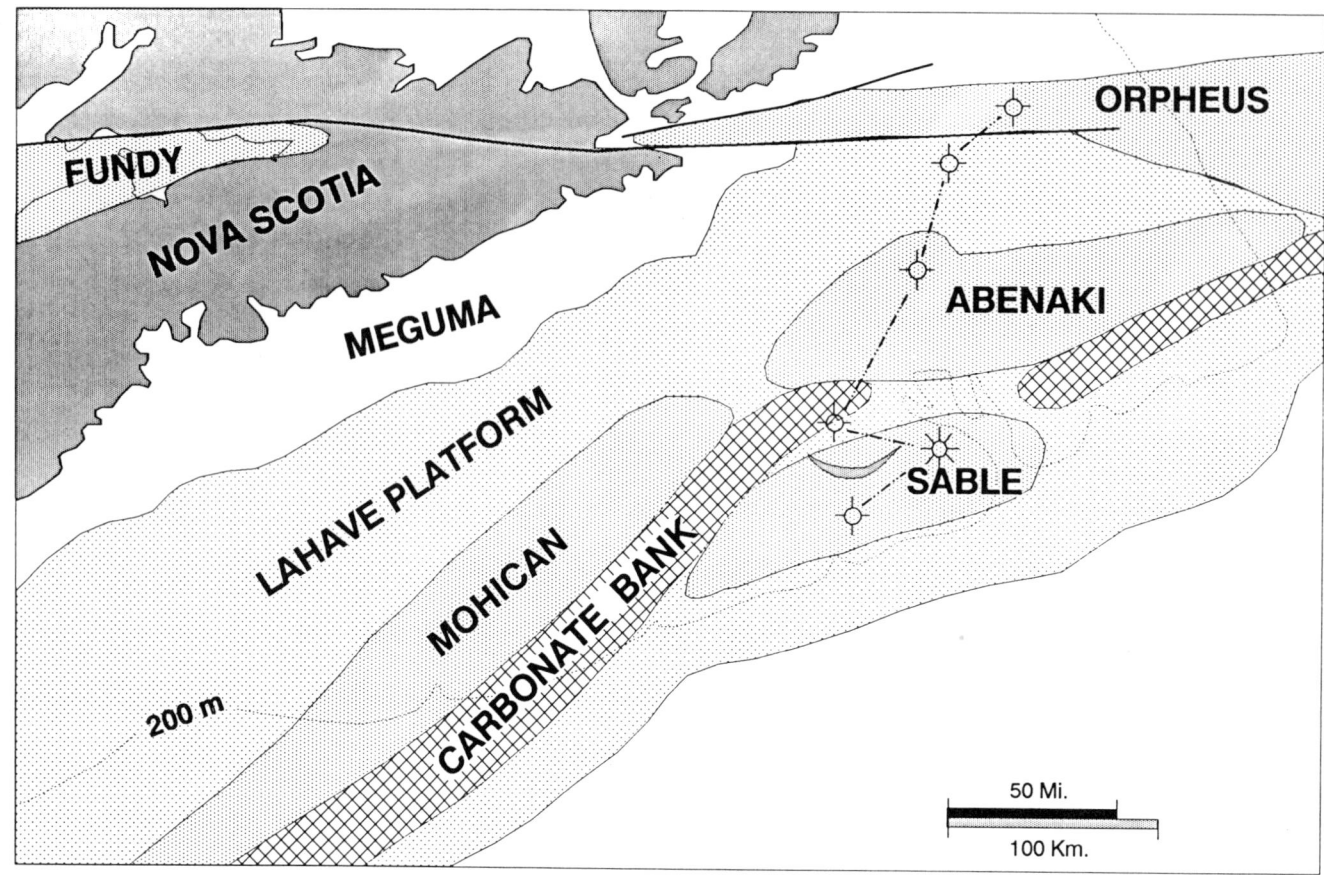

FIGURE 3. Tectonic elements of the Scotian Shelf, showing line of section for Figure 4.

3. Regressive deltaic wedge clastics forming a constructional delta on the continental shelf and slope. Transgressive cycles are represented by the Naskapi and Sable shales.

4. Deep water chalk and shales of the Late Cretaceous marine transgression and early Tertiary regression (outer neritic to bathyal).

Tensional stresses preceding the rifting of the Atlantic margin resulted in the formation of grabens and half-grabens infilled with Permian-Triassic continental red beds (prerift basins). Rifting of the margin was accompanied by Late Triassic to Early Jurassic continental clastic and evaporite deposition. The widespread distribution of evaporites and related clastics was followed by breakup of the continental margin, with a major marine transgression in the Middle Jurassic and the initiation of the drift stage of deposition (postrift). The Middle Jurassic transgression resulted in the deposition of open marine carbonates and associated clastics through Middle to Late Jurassic.

The Middle to Late Jurassic paleogeography shows a carbonate bank, the Abenaki Formation, which developed along the shelf edge, grading basinward to marine shales of the Verrill Canyon Formation, and landward to the deposition of back-bank clastics of the Mic Mac Formation. The shelf edge carbonate formed a tectonic hinge line and is the controlling foundation for subsequent deposition, which is of greatest importance in the Late Jurassic-Early Cretaceous. The Abenaki Formation is composed of three units: the Scatarie, Misaine, and Baccaro members. The initial marine transgression resulted in widespread deposition of the oolitic and bioclastic limestones of the Scatarie member. This was followed by an influx of fine clastics with the deposition of the Misaine shale. The Baccaro member is an areally restricted belt of platform carbonates. This carbonate bank complex developed along the Middle to Late Jurassic shelf edge, with calcareous sands, shales, and carbonate muds of the Mic Mac Formation being deposited in a back-bank shelf area.

The inner shelf area was characterized by a shallow marine deltaic to lagoonal environment. Seaward of the carbonate bank, fine-grained distal or prodelta sediments of the lower Verrill Canyon Formation were being deposited as terrigenous slope deposits of the deeper water outer shelf. There is a break in the shelf edge carbonates in the vicinity of Sable Island. In the Sable Basin area, a downwarping of the shelf and the influx of clastics of the early delta precluded much development of the carbonate bank. There is a greater thickness of Late Jurassic-Early Cretaceous clastics in this region. The provenance of the Missisauga clastics is interpreted to have encompassed a very large area of the Canadian shield, which was drained by an ancestral Saint Lawrence River. This "Proto-Saint Lawrence" would provide detritus from both the Precam-

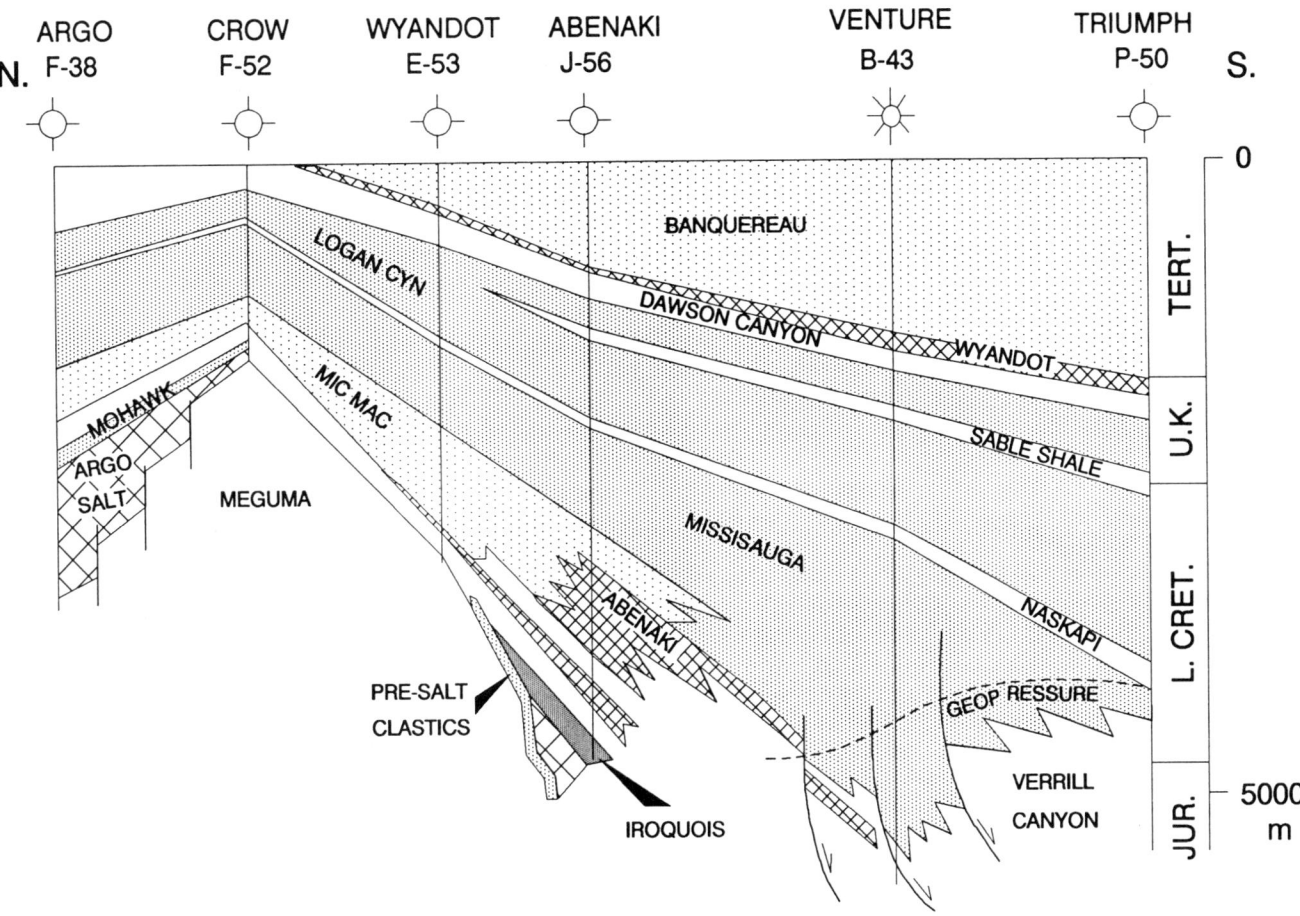

FIGURE 4. Diagrammatic section from Argo F-38 to Triumph P-50.

brian Shield and the Paleozoic metamorphic sequence in Nova Scotia.

GEOPRESSURE

Geopressure is the preservation of a pore-fluid pressure that is greater than that of normal hydrostatic pressure, or hydropressure. A necessary prerequisite for the generation of geopressure is a mechanism for isolating permeable reservoirs from any permeable beds that have continuity to the surface. When the reservoir fluids are cut off from the atmosphere, the fluid pressure in a reservoir then reflects all or part of the superincumbent rock column. Abnormal pore-fluid pressure results where fluids are trapped and cannot escape as increasing pressure is applied. Some of the causes of geopressured pore fluids could be:

1. Compaction of the rock during burial.

2. Released bound water of clays during diagenesis.

3. Thermal expansion of pore water as a result of heating.

4. Generation of hydrocarbons.

To date, 26 wells on the Scotian Shelf (Figure 6) are known to have encountered geopressure. The geopressure trend extends from Cree E-35 on the southwest through Thebaud and Venture to Bluenose on the Sable Bank and Louisbourg and South Griffin on the Banquereau Bank. Geopressure is known in 24 wells on the Sable Bank and in only two on the Banquereau Bank. The top of the geopressure is encountered at depths ranging from 3750 to 4750 m (12,303 to 15,584 ft). In general, the geopressure gets deeper from west to east. The direction of regression is from Bluenose to Thebaud, and geopressure is encountered stratigraphically higher in this direction. Geopressure on the Scotian Shelf is considered to be any pressure/depth ratio (PDR) greater than 0.465 psi/ft. Eighteen of the wells in the Sable Island area have encountered PDRs of 0.6 psi/ft or greater. The depth to the top of this PDR is 3750 to 3800 m (12,303 to 12,467 ft) in Thebaud, increasing to the northeast to 5000 m (16,404 ft) in Bluenose 2G-47. In most wells the PDR goes up in stages, except in Arcadia and Bluenose, where the increase to high geopressure is quite rapid. In Bluenose 2G-47 the PDR increases rapidly from 0.6 to 0.9 psi/ft at about 5000 m (16,404 ft). High PDRs ("hard geopressure") found to date are confined to the Venture area, with depths ranging from 4850 m (15,912 ft) in Arcadia to 5150 m (16,896 ft) in Olympia and Venture. Only three wells have encountered PDRs in excess of 0.9 psi/ft: Bluenose 2G-47, at a

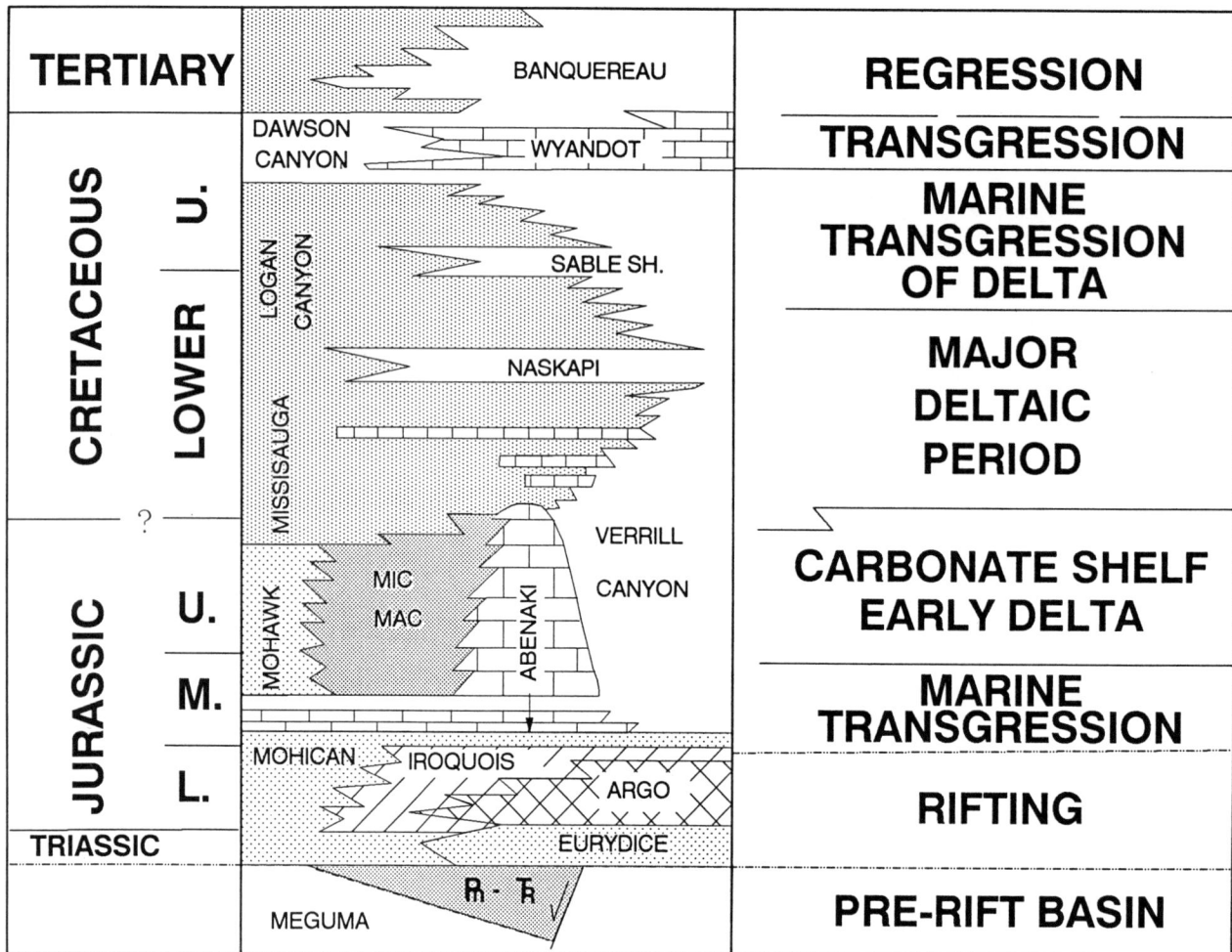

FIGURE 5. Generalized stratigraphic section of the Scotian Shelf (adapted from M. M. Givens, 1977).

depth of 5000 m (16,404 ft); Arcadia J-16, at 5600 m (18,373 ft); and South Venture O-59, at 5840 m (19,160 ft).

The maximum pressure encountered ranges from 0.57 to 0.94 psi/ft. The maximum PDR for many of the wells is probably a function of depth. Many would most likely encounter higher values if drilled deeper. The maximum PDR of the Olympia-Venture trend is 0.89 psi/ft, with higher values to the south, (0.93 psi/ft in South Venture) and to the north (0.94 psi/ft in both Arcadia and Bluenose). Temperature gradients are generally higher over the shelf along the northwest side of the Sable Basin. Local highs occur over the salt domes, such as Primrose, Iroquois, West Sable, and Abenaki. A broad area of lower geothermal gradient occurs to the east and northeast of Sable Island in the general area of most of the fields with geopressured reservoirs. This is the area of thicker sedimentation associated with the Upper Jurassic-Lower Cretaceous Sable delta, and is the main control on the geothermal gradient. It is difficult to make any association between geopressure and temperature in the Sable Island area.

Some of the conclusions concerning geopressure in the Sable Island area are as follows:

- Geopressure in Venture and the Sable Island area occurs within a prograding regressive sequence, associated with growth faulting.

- Venture gas occurs in multiple sand reservoirs with apparent pressure sealing caps.

- Sand grains often have a pervasive coating of chlorite, which is thought to be related in some way to the abnormally high porosities.

- Venture reservoirs are often characterized by rather high water saturations and very high-salinity formation water.

- The shales in the geopressured zone show decreasing shale resistivity and increasing interval transit time with depth, indicative of undercompaction relative to depth of burial.

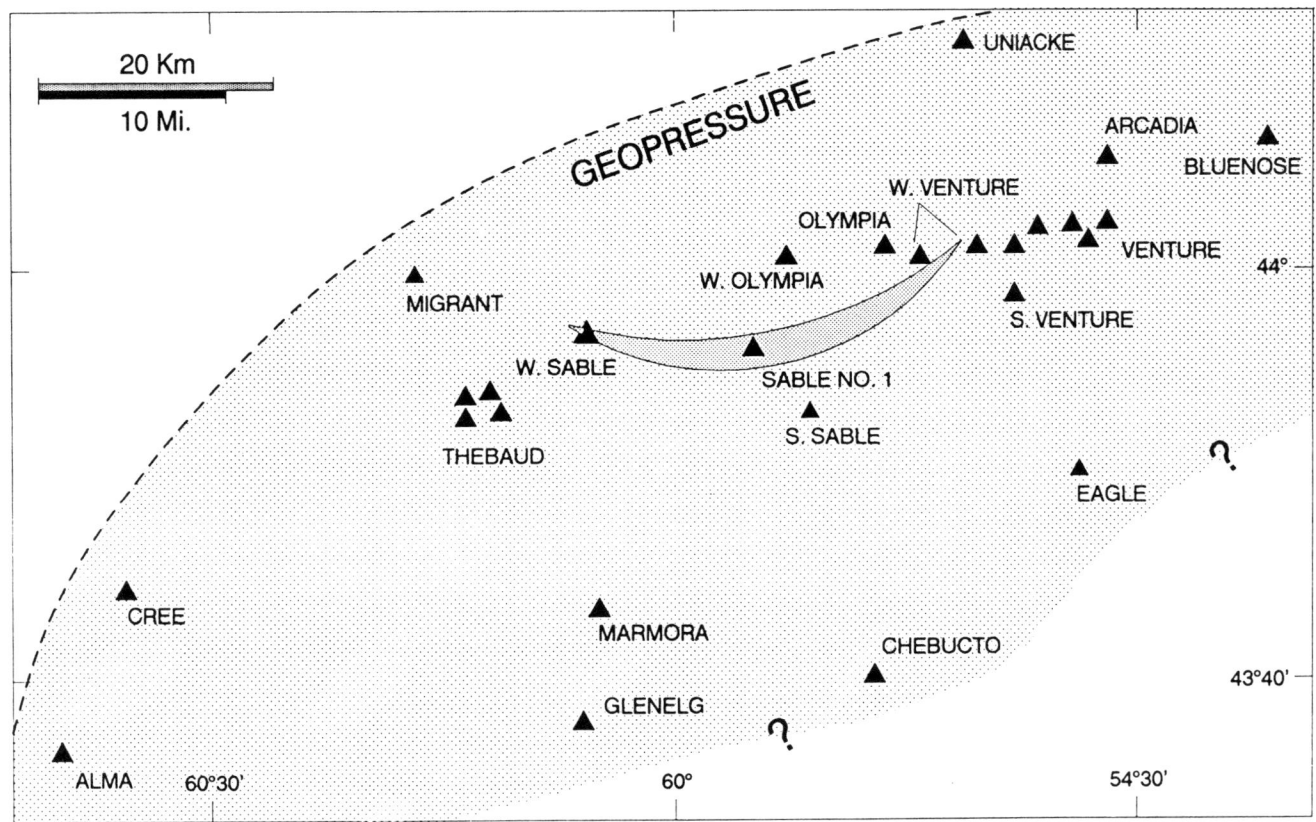

FIGURE 6. Wells with geopressure in the Sable Island area.

♦ Geopressure in the Sable Island area increases in distinct jumps.

Although clay diagenesis and aquathermal pressuring may have played some role, it is concluded that the major cause of geopressure in the Sable Island area is due to compaction. The reservoirs were sealed at a relatively early stage, preserving higher-than-normal porosities; fluids could not escape; and the pore fluids then assumed most of the overburden load, causing the geopressure.

VENTURE FIELD

The Venture field is located on the Scotian Shelf approximately 300 km (185 mi) east of Halifax. It lies a few miles off the eastern tip of Sable Island in relatively shallow water depths of 12 to 21 m (40 to 70 ft). The field currently has five wells drilled and tested—the discovery well (Venture D-23) and four appraisals (Venture B-13, B-43, B-52, and H-22). All wells, although potential gas wells, have been abandoned.

The Venture structure is located in a zone of normal growth faulting along the tectonic hinge zone. The anomaly is a low-relief east-west-trending anticline 10 km long by 2.5 km wide (6.2 by 1.5 mi) and covers an area of about 3000 ha (7400 ac). It is bounded on the north by a down-to-the-basin fault with structural closure resulting from rollover into the fault.

VENTURE COMPOSITE SECTION

Figure 7 is a composite stratigraphic section of the Venture field based on the five wells. The oldest formation penetrated is the prodelta shales of the Jurassic Verrill Canyon Formation. The Verrill Canyon is overlain by the Venture sand sequence, herein considered to be the basal part of the Missisauga Formation. Much of the Venture sand sequence is age-equivalent to the Mic Mac, with the 3A sand possibly being equivalent in time to the top of the Mic Mac. The Missisauga Formation is a regressive sequence of Upper Jurassic-Lower Cretaceous age consisting of interbedded sandstones and shales with minor amounts of carbonate. The sandstones vary in thickness from less than 3 m (10 ft) to more than 60 m (197 ft), and grain size ranges from very coarse to very fine. The section was deposited under shallow water conditions, with limestones representing transgressive periods. The Missisauga Formation is informally divided into the upper Missisauga, the Missisauga shale, and the Venture sand sequence. The thickest section (2830 m; 9285 ft) was penetrated by the B-52 well.

Overlying the Missisauga Formation is the Naskapi Formation of Lower Cretaceous age. The unit is approximately 125 m (410 ft) thick and consists of medium to dark gray silty shale with a minor amount of sandstone. Next in upward succession is the Logan Canyon Formation, which is subdivided into an upper and a lower unit by the Sable shale member. A thickness of approximately 1150 m

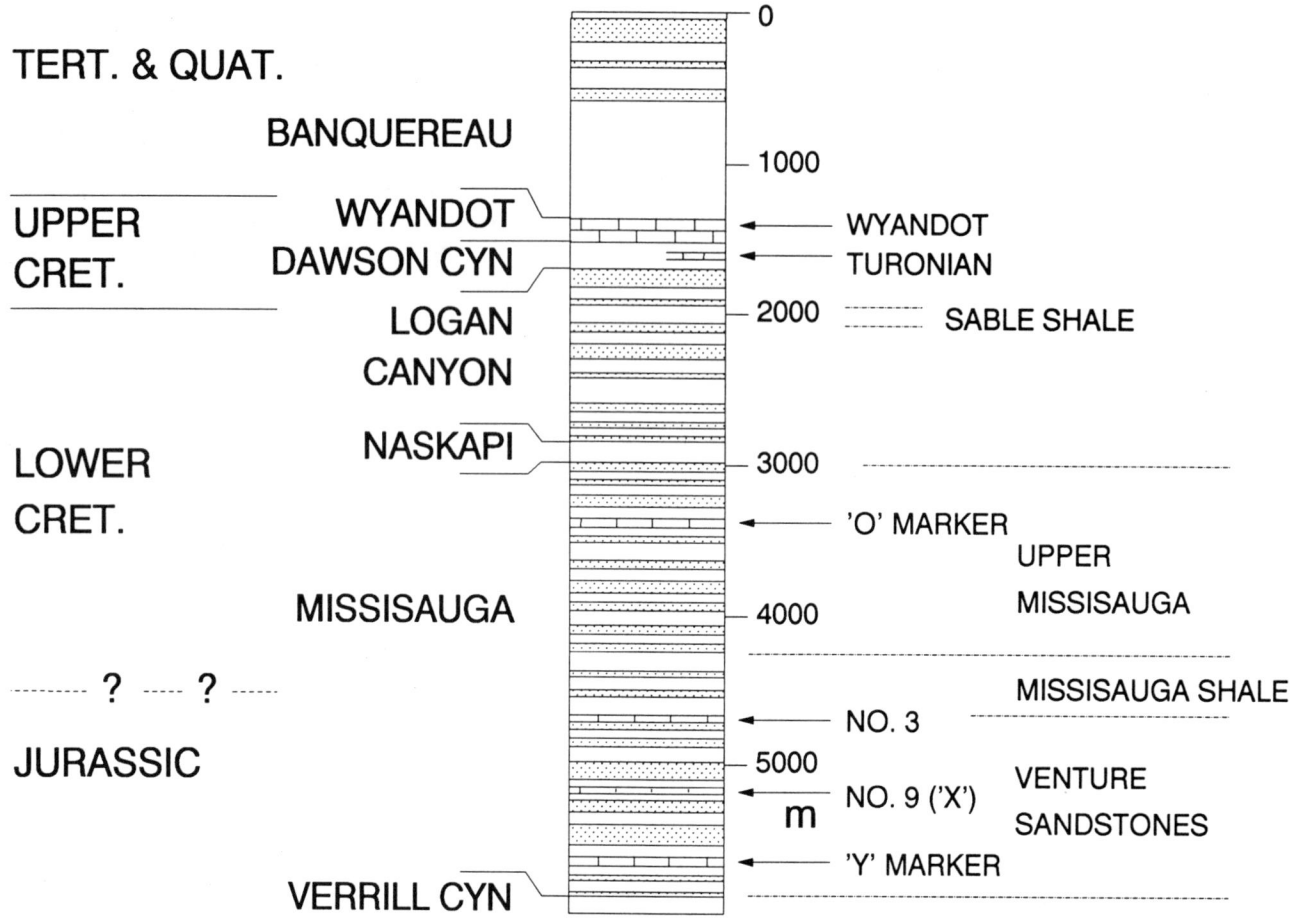

FIGURE 7. Venture composite stratigraphic section, showing seismic markers and subdivisions of the Missisauga formation.

(3773 ft), including 107 m (351 ft) of the Sable member, was penetrated by the Venture wells.

The Dawson Canyon Formation is the final transgressive phase overlying the Logan Canyon Formation. This Late Cretaceous shale section is 150 m (492 ft) thick and forms a correlative marker over a large portion of the Scotian Shelf. It is overlain by some 125 m (410 ft) of chalky limestone of the Wyandot Formation. The uppermost deposits in the Venture field are the sandstones of Late Cretaceous–Tertiary age in the Banquereau Formation.

SEISMIC CONTROL

The Venture prospect was originally mapped from a grid of 12- and 24-fold CDP (common depth point) seismic data shot between 1969 and 1977. Following the 1979 gas discovery at Venture D-23, the prospect was reshot with a series of 60-fold CDP lines totaling 115 km (71 mi). In 1980, an additional 30 km (19 mi) of 98-fold CDP data was recorded across the eastern slope of the Venture prospect. An additional 10 km (6 mi) of 72-fold data was shot across the Venture B-13 well in 1981. In the summer of 1983, a conventional and shallow water program was conducted across the Venture field and the east Sable Island bar. Figure 8 (sand 3A map) shows the seismic sections, which will be discussed. The lines are: line M-311, a regional line from Citnalta to Venture, and Line M-331B, which runs close to the discovery well, Venture D-23.

Line M-311 (Figure 9A) is a regional section that runs north to south from the stable shelf block of Citnalta over the hinge zone and across the Venture structure, about 1 km (0.6 mi) west of B-13. The four wells, Citnalta I-59 and Venture D-23, B-13, and H-22, are projected into the line of section with total depths correlated stratigraphically. The shallow reflections from the Wyandot, Turonian, and "O" markers have been tied to the Venture wells and the Citnalta I-59 gas well located 12.9 km (8 mi) to the north. The deep reflections from zone 3, and the "X" horizon (zone 9 limestone), have been tied to all Venture wells and are mappable only over the Venture fault block. The "Y" horizon, penetrated by B-43, B-52, and H-22, is also mappable only over the Venture fault block. The Jurassic and the Abenaki cannot be identified in the Venture structure. In this interpretation, the lowermost strata are quickly downfaulted, dropping over the hinge line. The Late Jurassic–Early Cretaceous section thickens rapidly, and correlations from one fault block to the next are

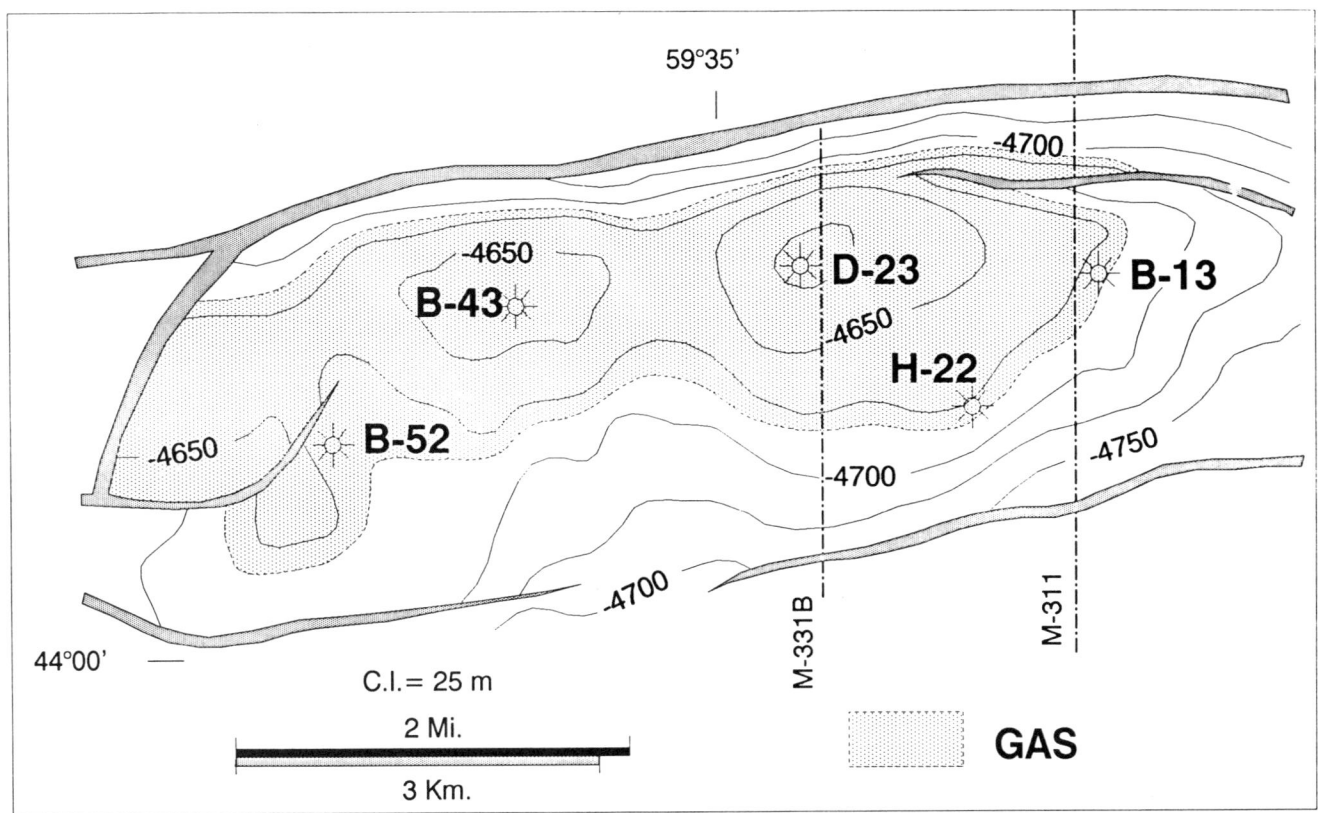

FIGURE 8. Structure on top of the Venture 3A sand, showing locations of seismic lines M-311 and M-331B.

difficult. Shallow gas is indicated to the north of Venture above the main north-bounding fault at a depth of about 0.7 sec.

Line M-331B (Figure 9B) is a section that runs close to the Venture D-23 well site. An enlarged portion of this section clearly shows the distinctive pattern of seismic reflections associated with the Venture structure. The upper reflector is correlated with zone 3. The deeper horizons are limestones, designated "X" (No. 9) and "Y." Shown is a complete loss of character to the south of the Venture structure, probably representing a change of facies into dominantly prodelta shales.

SAND 3A STRUCTURE

The Venture structure, as mapped by the zone 3 seismic reflection, shows an elongated east-west anomaly on the downthrown side of a normal east-west-trending fault. The structure covers approximately 3000 ha (7400 ac). Seismic control and drilling results from the five Venture wells indicate that 2805 ha (6931 ac) could be under structural closure at the No. 3 level. The eastern crest of the prospect shows 80 m (262 ft) of closure between the D-23 and B-13 wells. The western crest of the anomaly shows closure on the order of 40 m (130 ft). The eastern and western crests of the anomaly may have resulted from a deep north–south-trending fault that is indicated on the seismic west of the D-23 well. This deep fault on the seismic appears to be down to the west.

VENTURE SAND SEQUENCE

A composite of the Venture sand sequence as penetrated by the five wells is shown in Figure 10. In Venture, natural gas and condensate are encountered in multiple zones. There are 22 potential gas-bearing reservoirs, as listed in Table 1. Shows of gas were encountered in sands 17 and 18, which, because of depth and low porosity, are not considered to be potential reservoirs. Depths range from 4040 to 5800 m (13,255 to 19,029 ft). The gross thickness of the main geopressured reservoirs is about 1400 m (4593 ft). The Venture sand section is characterized by recurring cycles of marine shale, marine limestone, and shallow marine sandstone. The shales are very dark to black and contain abundant organic (mainly cellulosic) material. Accessory minerals include micas and authigenic pyrite. The Venture sandstones are mainly sublitharenites with minor subarkoses. The sands are generally bimodal: (1) very fine to fine grained, moderately sorted with poor porosity; and (2) more poorly sorted, fine to medium grained, with occasional coarse grains, and with good porosity. In general, the sandstones are clean, with only

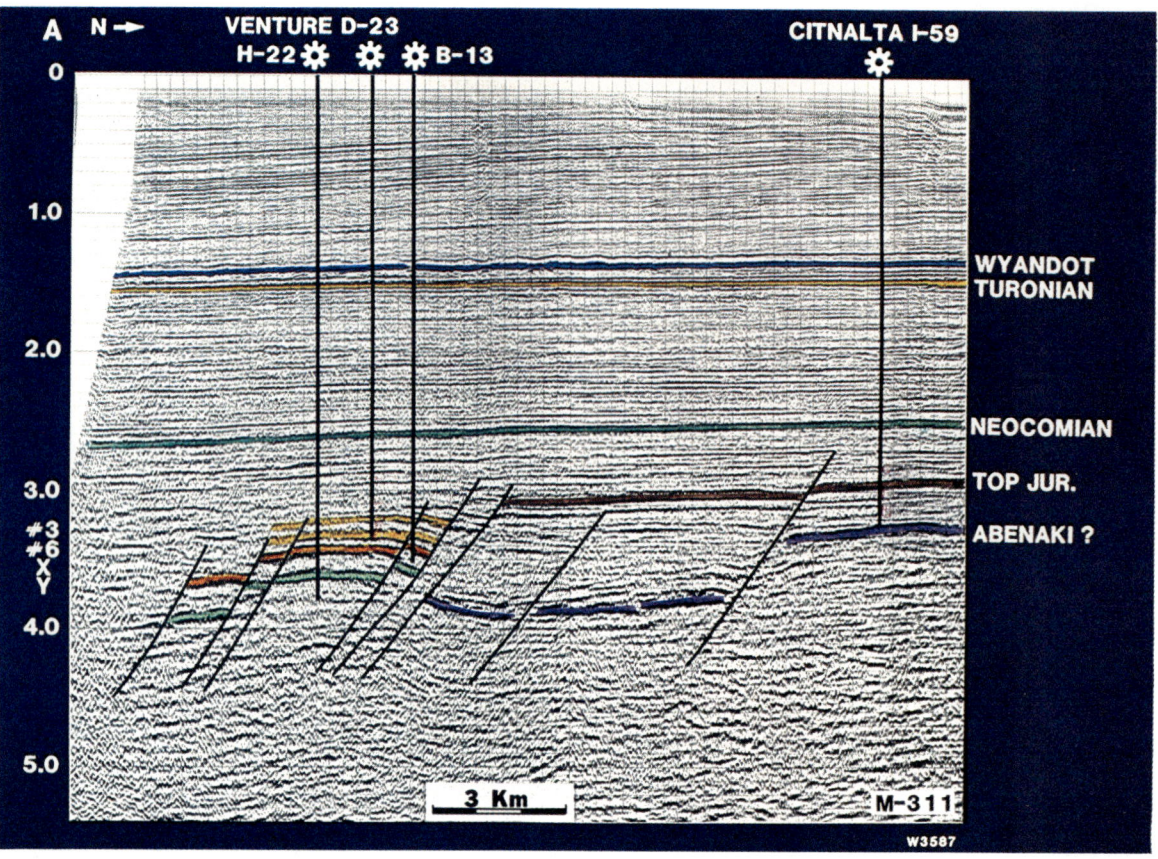

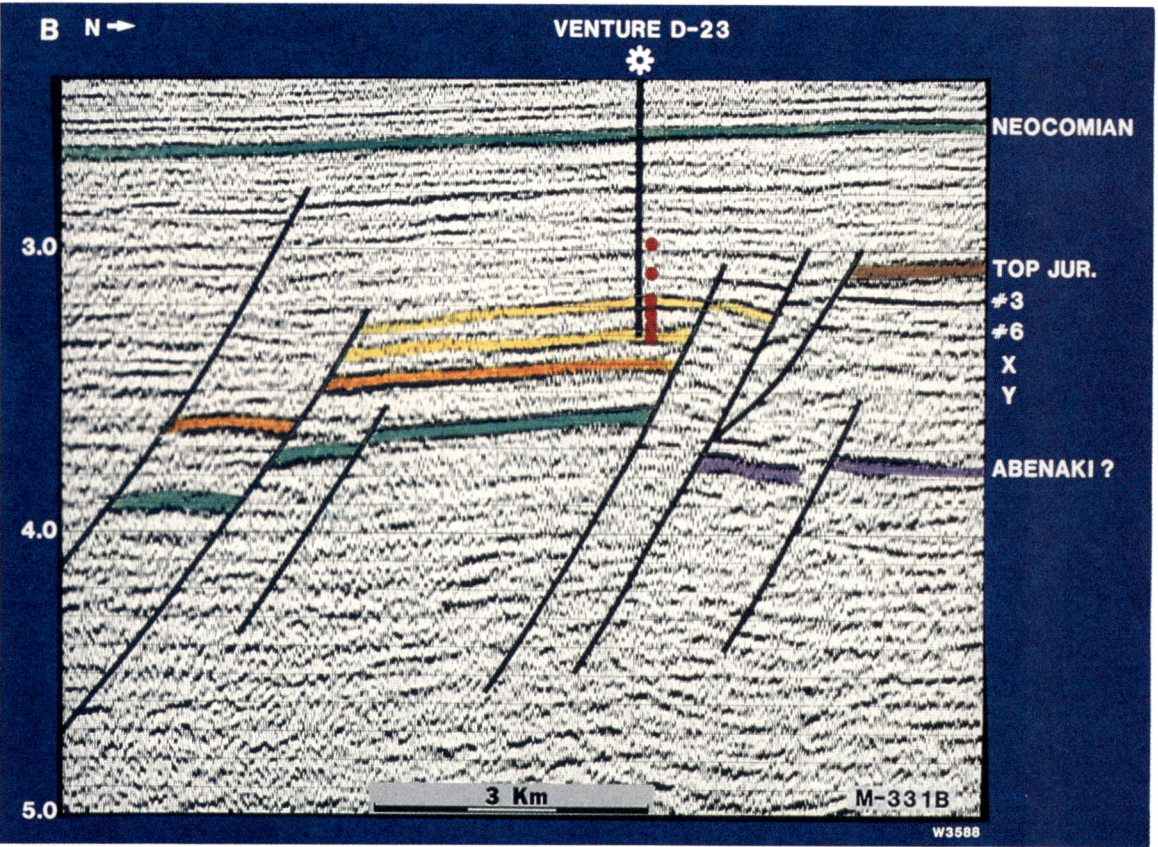

FIGURE 9. North-south seismic sections across the Venture structure: (A) line M-311; (B) line M-331B.

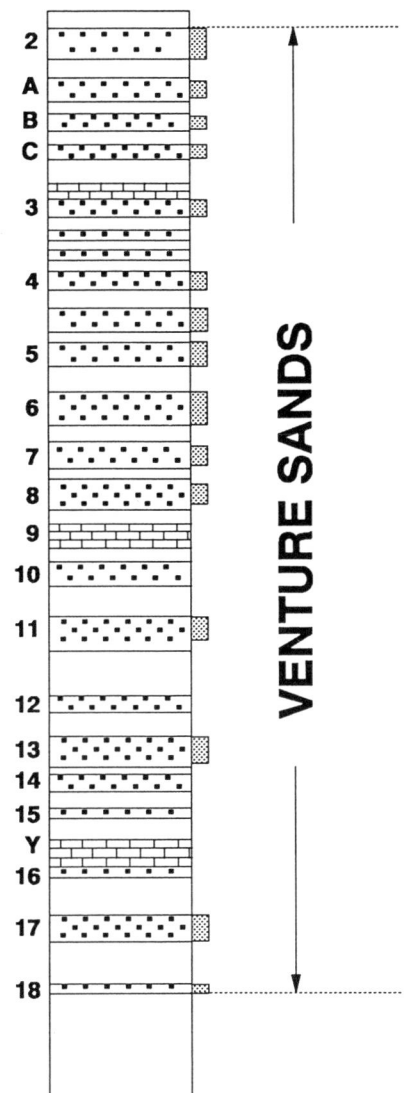

FIGURE 10. Venture sand sequence, showing designation of sands.

Superimposed on the section are the top of geopressure and the occurrence of pressure/depth ratios (PDRs) of 0.6, 0.7, and 0.8 psi/ft. Correlation of pressures among the wells indicates that geopressures are interrelated and controlled stratigraphically. The correlative sands within the geopressured zone of the five wells have similar pressures apparently little related to depth. The first indication of anything above hydropressure occurs below the No. 2 sand. The 0.6-, 0.7-, and 0.8-psi/ft PDRs occur, respectively, at the top of the No. 3 sand, in the No. 4 sand, and below the No. 9 limestone.

PRESSURE AND TEMPERATURE VERSUS DEPTH

The over-all geothermal gradient for Venture B-43 is 1.6°F/100 ft (2.9°C/100 m). Temperatures plotted in Figure 12 are corrected bottom hole temperatures and temperatures from drill-stem tests. The pressure vs. depth plot shows that the geopressured zone is not one continuous zone but is rather a series of geopressured zones each more geopressured than the zone above. It also appears that each jump in PDR is related to an impermeable limestone or calcareous sandstone at the top of each zone. In B-43, the first occurrence of geopressure occurs at the top of sand A at a depth of 4487 m (14,721 ft). In the well there are six different pressure regimes: 0.51 psi/ft over sands A and B, 0.59 in sand C, 0.67 over sand 3, 0.73 in sands 4A through 8, 0.83 in sand 11, and 0.86 in sand 13.

A composite plot of pressure vs. depth for the Venture field (Figure 13) shows pressure measurements in relation to equivalent mud weights. The chart includes all the pressure data from RFT (repeat formation tester) and drill-stem tests to date for the five wells. The data clearly indicate the abrupt pressure changes from one sand package to the next. From 4500 m (14,764 ft) down, the various pressure-depth regimes correspond to: sands A and B, sand C, sand 3, sands 4 through 8, sands 10 and 11, sand 13, and sands 17 and 18.

Plots of shale resistivity (R_{sh}) vs. depth (Figure 14) clearly show a decrease in shale resistivity with increasing pressure. There is also some indication of increased resistivity in the shales immediately above the top of the geopressure. There is a corresponding shift to the right of the delta Δt vs. depth curve, indicating lower velocities in the geopressured shales. The shales appear to be well indurated, as shown in cores from the geopressured zones, but R_{sh} vs. depth and interval transit time vs. depth indicate that there is increased porosity in these shales. In B-43 there is a significant decrease in shale resistivity in the geopressure with a corresponding shift to the right away from the normal trend for delta Δt.

POROSITY VERSUS DEPTH

The composite porosity vs. depth plot for Venture presented in Figure 15 shows a gradual loss of porosity with increasing depth down to about 4300 m (14,108 ft) and then abnormally high porosities in the deeper section. The data are sonic porosity averaged over a 10-m interval

minor argillaceous content. The dominant clay mineral is chlorite. Most important, chlorite often surrounds open pores, either preserving primary porosity or possibly surrounding completely dissolved quartz grains.

VENTURE STRUCTURE

The diagrammatic section presented in Figure 11, which shows the distribution of gas-bearing reservoirs, water sands, and tight sands, indicates good correlation of sands among the five wells. Pressure data also support continuity of reservoirs. Seismic shows continuity of reflectors, characteristic of the gas zones over the entire structure. Results to date indicate that there is continuity of sands and good reservoir quality through all five wells and most likely across the entire Venture structure.

TABLE 1. Reservoir parameters for sands of the Venture field.

Sand	Gross sand m	Gross sand ft	Mean depth m	Mean depth ft	Average porosity (%)	Average SW (%)	Gas gravity	Average pressure (psi)	Average temperature °C	Average temperature °F
1 upper	28	92	4025	13,205	17.1	47.4	0.726	6,273	121	249
2 upper	25	81	4410	14,469	13.8	32.9	0.726	6,881	131	268
2 lower	9	30	4420	14,501	16.4	53.0	0.726	6,881	131	268
A	35	114	4450	14,600	13.7	58.2	0.799	7,368	132	270
B	31	101	4500	14,764	13.6	58.4	0.706	7,716	134	273
C	11	35	4545	14,911	16.6	43.9	0.706	8,934	135	275
3A	23	77	4615	15,141	17.3	53.3	0.712	10,269	137	278
4A	19	62	4720	15,486	14.1	55.3	0.781	11,603	139	283
4B	10	32	4740	15,551	16.3	43.2	0.781	11,632	140	284
4C upper	7	23	4760	15,617	14.9	43.2	0.781	11,654	141	286
4C lower	12	39	4770	15,650	13.7	59.0	0.781	11,654	141	286
4D	8	26	4780	15,682	18.0	41.8	0.781	11,661	141	286
5	11	36	4805	15,764	18.3	44.0	0.853	11,690	142	288
6 upper	15	49	4875	15,994	19.7	36.5	0.749	11,690	144	291
6 middle	19	61	4900	16,076	19.5	35.0	0.749	12,125	144	291
6 lower	10	34	4910	16,109	14.5	56.2	0.749	12,241	144	292
7 upper	2	8	4930	16,175	14.9	53.7	0.711	12,241	146	294
7 lower	21	68	4950	16,240	16.0	45.8	0.711	12,241	146	294
8 middle	11	37	4975	16,322	17.8	53.9	0.820	12,328	147	296
11 upper	3	10	5170	16,962	15.1	58.2	0.716	14,402	152	306
11 lower	14	47	5190	17,028	16.3	52.4	0.716	14,402	152	306
13	29	94	5330	17,487	14.3	50.9	0.692	15,403	157	314
TOTAL/AVG.	353	1156			16.6	47.4	0.749			

for sands with a minimum thickness of 2 m (7 ft). Examination of samples and thin sections suggests that primary porosity is dominant in the upper 2000 m (6562 ft). Between 2000 and 4300 m (6562 and 14,108 ft), porosity development is believed to be both primary intergranular and secondary solution porosity. In Venture B-43, the major reservoir sands have average porosities ranging from 16 to 25%, and are remarkably high, considering their depths (Figure 16). The porosities above the geopressure range from 5 to 8%, and abruptly increase in the geopressure zone to 10 to 25%. In the deeper sands there is a trend toward lower porosities.

DIAGENETIC HISTORY

The Venture sandstones are predominantly sublitharenites, although quartz arenites and lithic arenites are present. The principal detrital components are monocrystalline quartz and polycrystalline quartz, feldspars, micas, chlorite, glauconite, volcanic rock fragments, siliciclastic rock fragments, carbonate rock fragments, and carbonaceous shale fragments. Minor amounts of heavy minerals are also present. Interbedded with the sandstones are dark gray to black organic-rich shales.

A characteristic of the Venture reservoir sandstones having the better porosities is the pervasive occurrence of grain-coating chlorite. Petrographic evidence suggests an early diagenetic origin for most of the chlorite, although late diagenetic chlorite cannot be excluded. Pervasive authigenic chlorite coating of the grain surfaces is associated with the cleaner, more porous reservoirs in the geopressured zone. The early authigenic chlorite has inhibited significant silica cementation. In the shalier sandstones the development of authigenic chlorite is only minor. In these sandstones, quartz diagenesis is often so severe that little intergranular porosity remains. Minor illite occurs, increasing with depth and generally best

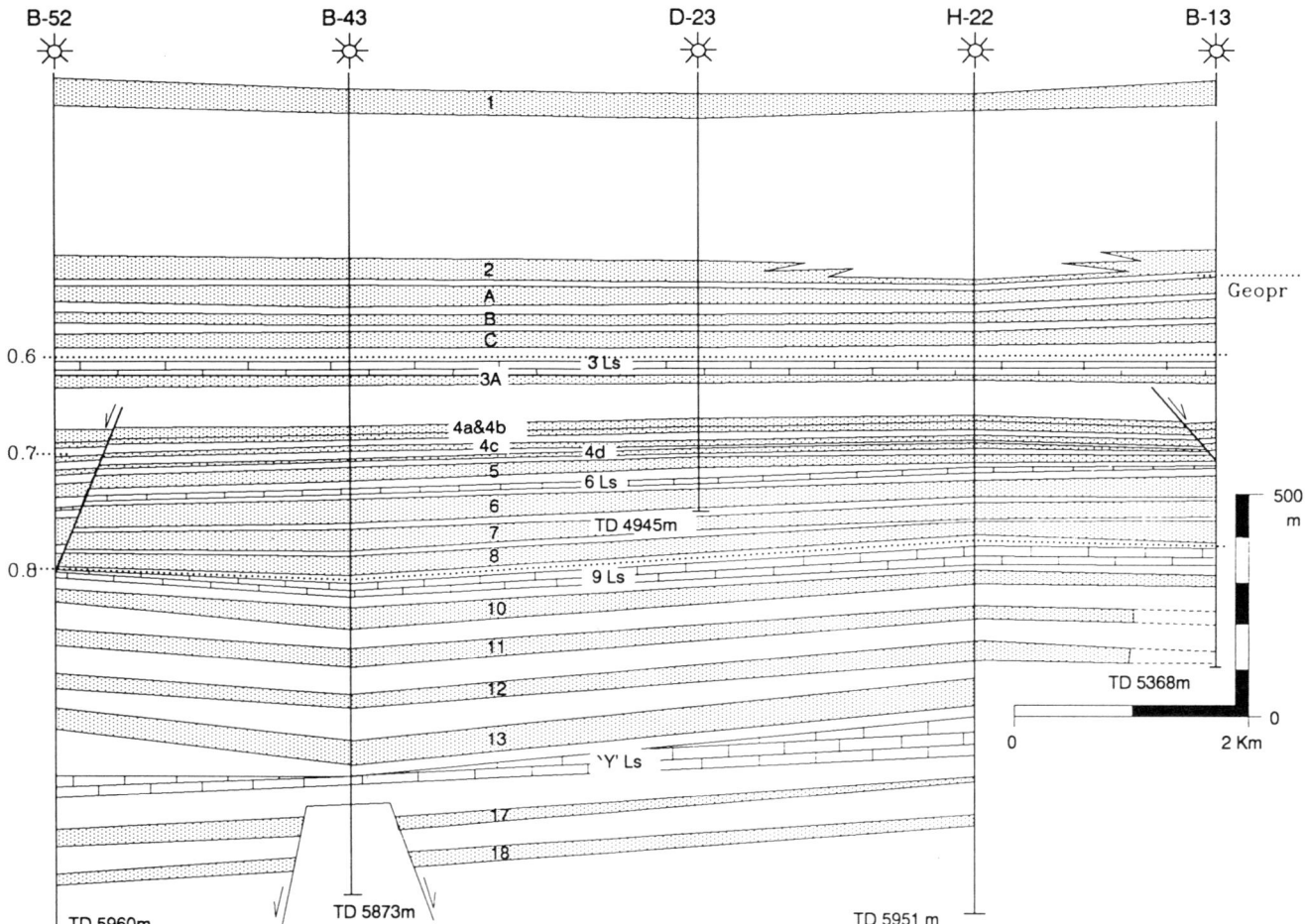

FIGURE 11. Stratigraphic cross section of the Venture field, showing correlation of sands in the five wells, with diagrammatic representation of the faults in B-52, B-43, and B-13. Venture B-13 is not logged over sands 11 and 12.

developed in sandstones with low porosity and permeability. Kaolinite is common in the hydropressured section but generally is not found in the geopressured reservoirs. Pore-filling carbonate cement postdates the authigenic chlorite. The carbonate cement has completely destroyed the intergranular porosity. In other sandstones the carbonate cement fills only a small portion of the porosity, and excellent porosities and permeabilities remain. Grain-coating chlorite characterizes the highly porous sandstone reservoirs. Although the growth of this chlorite has resulted in some reduction of primary porosity, it also appears to have protected these sandstones from significant compaction and quartz cementation. As a result, the presence of chlorite has had a beneficial effect on the porosities of the deeper reservoirs. Leaching of feldspar grains, rock fragments, carbonate grains, and cement has also contributed to the porosities of the better reservoirs.

A photomicrograph of cuttings from 4275 m (14,026 ft) in Venture D-23 shows strong mutual interpenetration between the quartz grains (Figure 17). There is also secondary growth of quartz, with the broken rim of chloritic material defining the edge of the original grain. Under normal compaction, most of the original primary porosity is destroyed by this depth.

Figure 18 shows good development of secondary porosity in the No. 2 sand within the hydropressure zone at a depth of 4436 m (14,554 ft) in B-43, with isolated pockets of porosity from silicate grain dissolution. This is a fine to medium-grained sandstone with 9% porosity and a low permeability of 0.9 md.

A low-magnification view of the No. 6 geopressured sandstone in B-43 from a depth of 4955 m (16,257 ft) shows extensive development of both primary and secondary porosity (Figure 19). This is a fine- to medium-grained sandstone with measured porosity of 28.2% and permeability of 507 md. Framework grains are generally undeformed and rimmed with authigenic chlorite. Freestanding chlorite rims are the remnants from preferential dissolution of unstable host silicate grains.

Secondary porosity is often interpreted as being as abundant in Venture sandstones as preserved primary porosity. It appears that both have played significant roles.

GEOCHEMICAL DATA

The geochemistry of the Scotian Shelf has been discussed in detail elsewhere (Powell and Snowden, 1979; Powell,

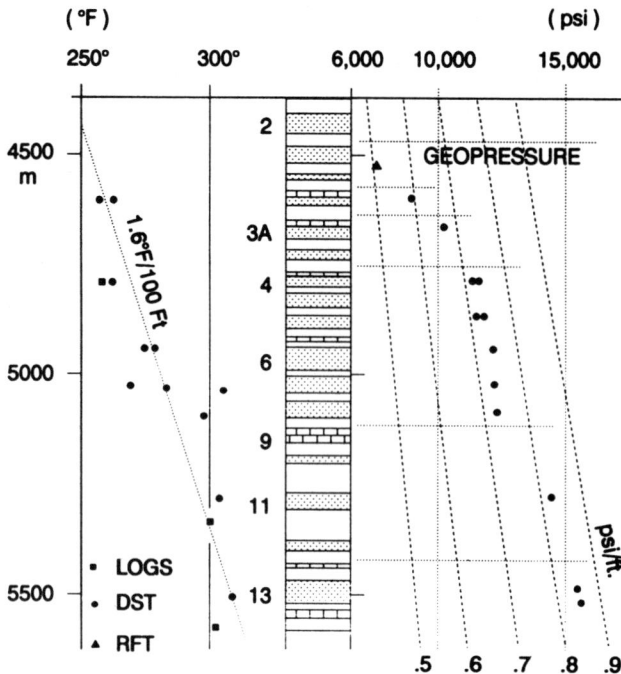

FIGURE 12. Temperature and pressure vs. depth for Venture B-43.

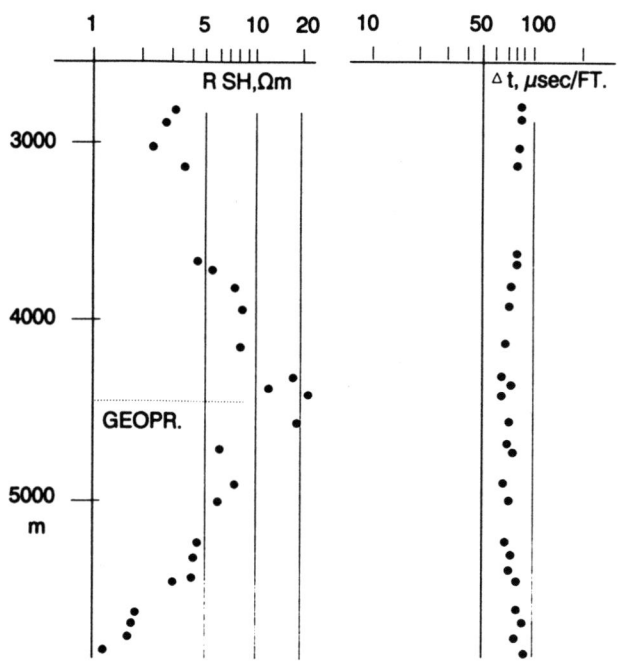

FIGURE 14. Measurements of resistivity and interval transit time for shale in Venture B-43.

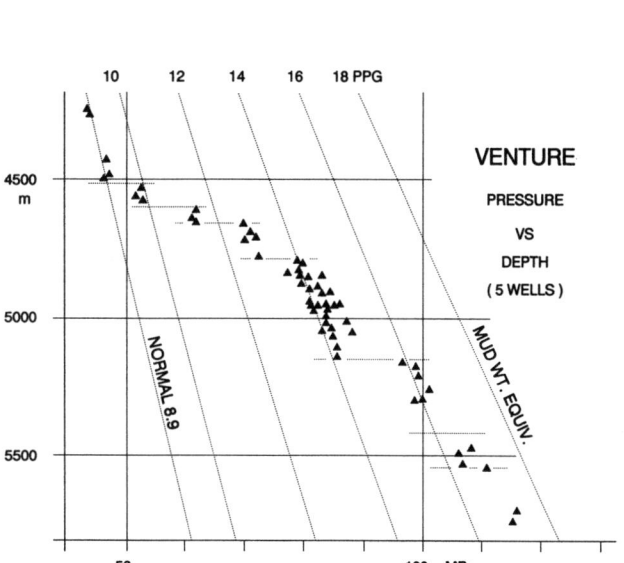

FIGURE 13. Composite pressure vs. depth plot for the Venture field.

1982; and Purcell et al., 1979). Figure 20A shows a plot of vitrinite reflectance values for Venture H-22. In this well the hydrocarbon-generation window is not reached until a depth of about 4600 m (15,092 ft). The maximum R_o value is 1.39 at a depth of 5935 m (19,472 ft), which is the maximum value measured for the Venture field. The Verrill Canyon shales have reached sufficient maturity to generate hydrocarbons. A modified Van Krevelen diagram (Figure 20B) of Rock-Eval pyrolysis data clearly shows type III kerogen to be dominant in the Verrill Canyon

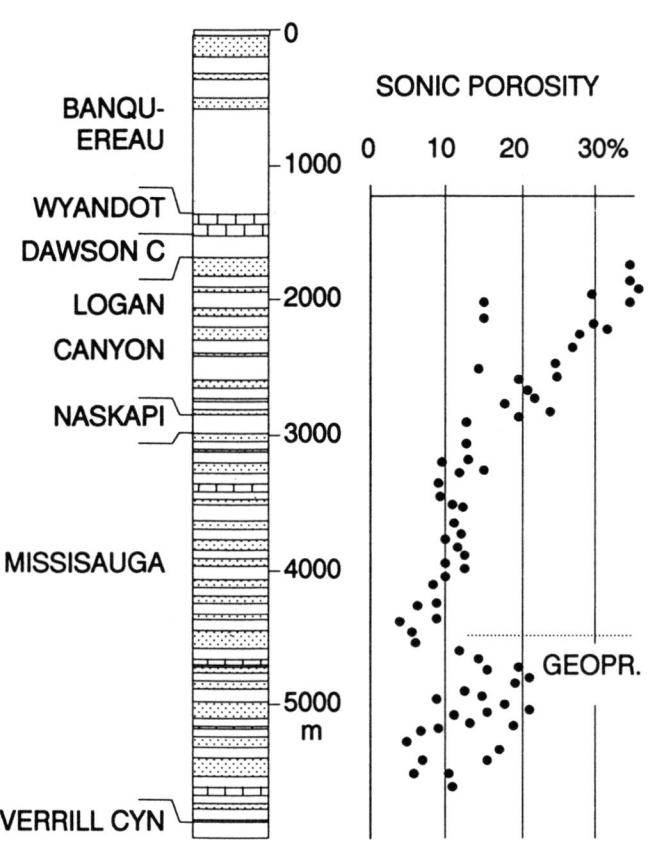

FIGURE 15. Composite plot of porosity vs. depth for the Venture field.

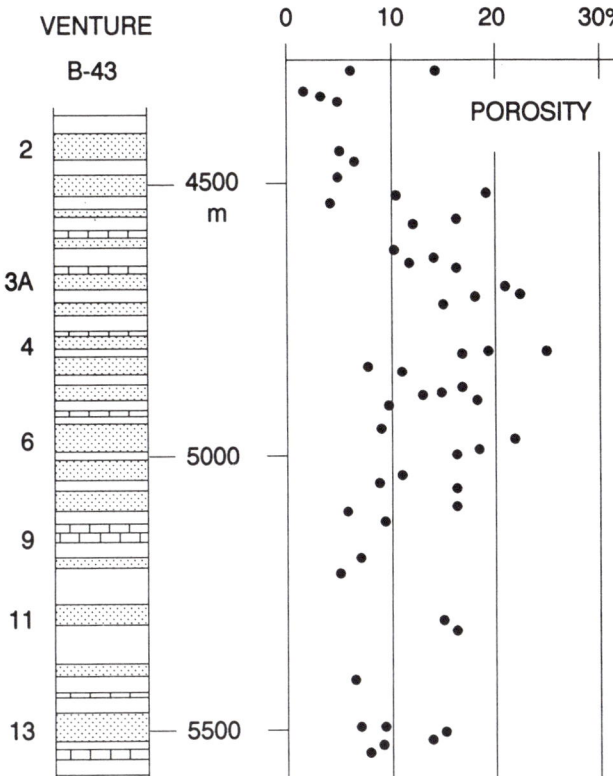

FIGURE 16. Porosity vs. depth plot for Venture B-43.

FIGURE 18. Photomicrograph of No. 2 sand from cuttings at 4436 m (14,554 ft) in Venture B-43, showing good development of secondary porosity.

FIGURE 19. Photomicrograph of No. 6 sand from a depth of 4955 m (16,257 ft) in Venture B-43, showing development of both primary and secondary porosity.

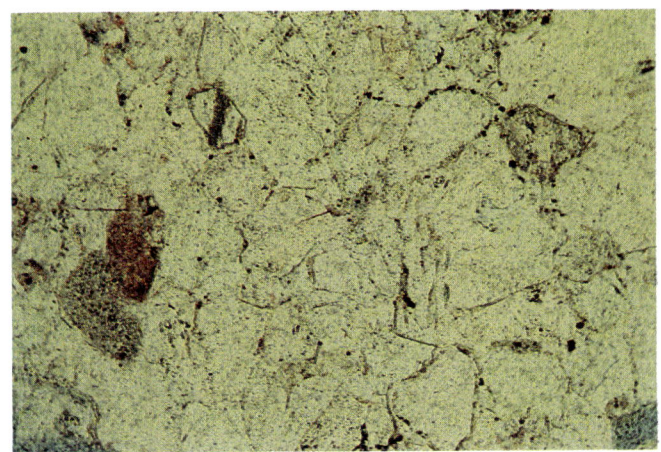

FIGURE 17. Photomicrograph of No. 2 sand from cuttings at 4275 m (14,026 ft) in Venture D-23, showing strong intergrowth of quartz.

shales of Venture H-22 and in the entire Venture area. Type III kerogen is of terrestrial origin and is prone to gas and condensate generation.

RESERVOIR CHARACTERISTICS

All five Venture wells penetrated several significant pay zones. A total of 22 potential gas-bearing reservoirs have been recognized—four in the hydropressured section and 18 in the geopressured zone. Reservoir depths range from 4025 to 5780 m (13,205 to 18,963 ft) subsea. The reservoirs are all sandstones of Late Jurassic to Early Cretaceous age. The sandstones are very fine to coarse grained and poorly to well sorted.

In general, two types of reservoir sandstones are recognized. The first type is more common in the hydropressured interval. These reservoirs are generally characterized by a higher content of shale and dispersed organic matter, significant quartz diagenesis, and the presence of minor amounts of kaolinite. Grain-coating chlorite is poorly developed, and reservoir porosities and permeabilities are generally low.

The second type of reservoir occurs more frequently in the geopressured section. The geopressured reservoirs are generally cleaner and are characterized by the pervasive occurrence of grain-coating chlorite. Authigenic kaolinite and quartz overgrowths are generally absent or

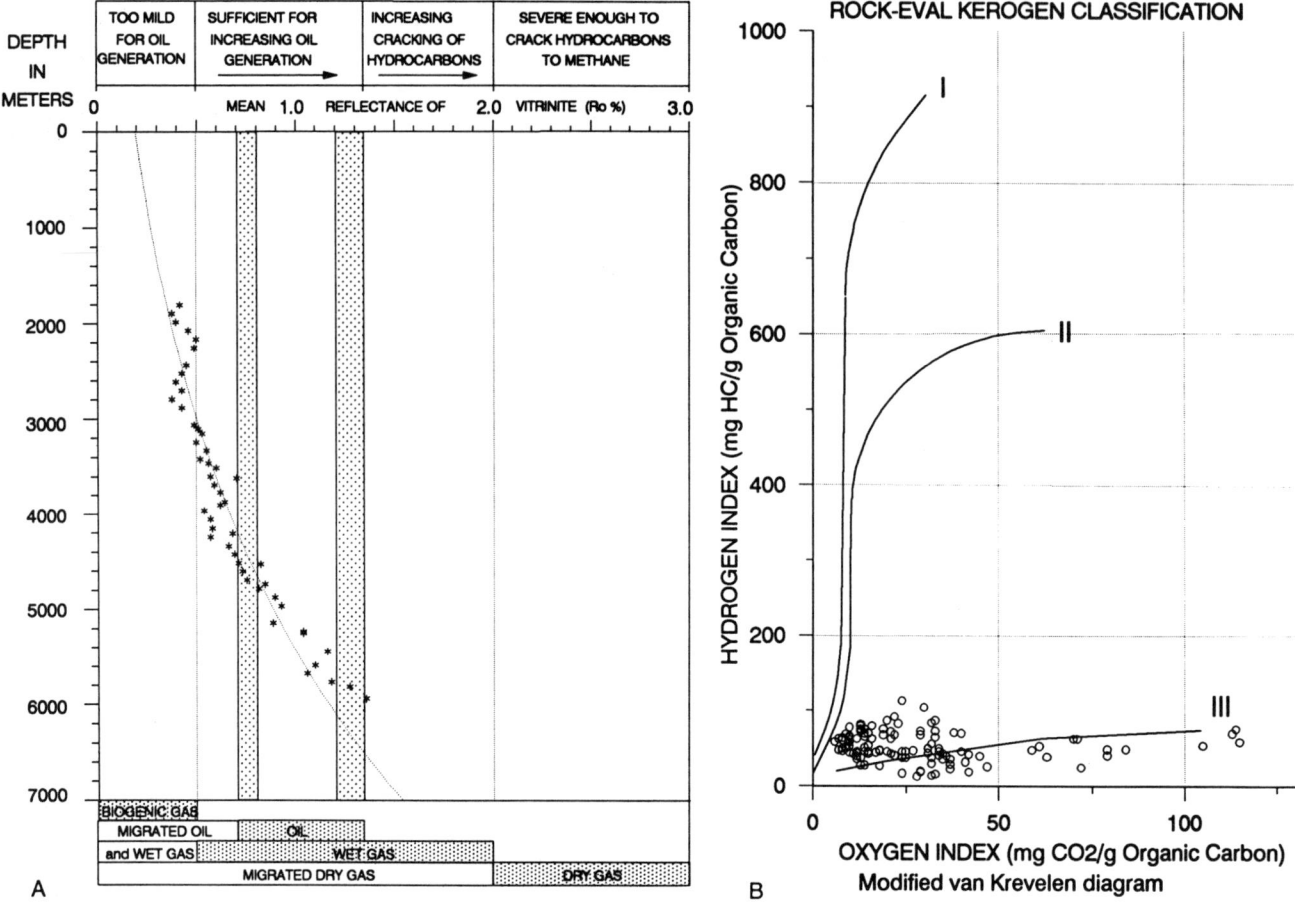

FIGURE 20. Vitrinite reflectance data (A) and modified Van Krevelen diagram (B) for Venture H-22.

poorly developed. The early diagenetic chlorite inhibited quartz diagenesis and resulted in the preservation of high intergranular porosities and permeabilities. There is also a significant contribution of intragranular porosity resulting from the dissolution of feldspars, siliciclastic grains, and carbonate material, as well as microporosity within the chlorite grain rims.

The degree of compaction is highly variable. Some reservoirs have tightly packed fabrics resulting from primarily physical compaction. The best reservoirs show little sign of quartz diagenesis, have open fabrics, and are moderately undercompacted with excellent porosity and permeability.

Table 1 summarizes the parameters of the major reservoir sandstones in the Venture gas field. Average porosities of the sands vary from 14 to 20%, with an overall weighted average of 16.6%. Average water saturations for Venture sands are generally high, ranging from 33 to 59%, averaging about 47%, using a 70% cutoff. Formation water salinities measured on water samples from drill-stem testing range from 95,000 to 202,000 mg/L NaCl.

Drill-stem testing indicates good flow rates for the major sands. Flow rates from sands 2, 3a, 5, 6, 7, and 8 are generally in the range from 5 to 25 MMCFGD. The best flow rate was from the No. 11 sand of well B-52, which flowed at a maximum of 49.2 MMCFGD from the interval 5284 to 5293 m (17,336 to 17,365 ft), with a choke size of 7/8 in., in a multirate extended flow test of 36 hr. The No. 11 sand was cored over the log-corrected interval from 5277.5 to 5290.9 m (17,315 to 17,359 ft). The tested interval includes 4.2 m (14 ft) of sand with average porosity of 20.2% and permeability of 10.8 md. Maximum measured porosity and permeability were 24.2% and 43.2 md, respectively.

Drill-stem testing gave condensate recoveries ranging from 12 to 57 bbl/mmcf, averaging 17 bbl/mmcf. Calculations based on average gas analysis (Table 2) indicated an average liquid content recovery of 52 bbl/mmcf.

SUMMARY

The Venture field is a potential giant gas field on the Scotian Shelf, with gas in multiple sandstone reservoirs. The Venture field is characterized by deep geopressured sandstone reservoirs, with abnormally high porosities for the depth and with both preserved primary and secondary porosity. The geopressure is believed to be due primarily to the effects compaction. The Venture gas field is estimated to contain between 1.5 and 2.0 tcf of recoverable raw natural gas.

TABLE 2. Average gas analysis, Venture field.

Component	Avg. gas composition (mole fraction)
He	0.0001
H_2	0.0000
N_2	0.0100
CO_2	0.0132
C_1	0.8375
C_2	0.0725
C_3	0.0301
iC_4	0.0048
nC_4	0.0079
iC_5	0.0028
nC_5	0.0023
C_6	0.0027
C_7+	0.0161
Total	1.0000

ACKNOWLEDGMENTS

The author would like to acknowledge the work of his colleagues at Mobil Oil Canada, who have made significant contributions to this paper. I also acknowledge the cooperation and assistance provided by our partners Petro-Canada, Esso (formerly Texaco), and Nova Scotia Resources, and would like to thank Mobil Oil Canada and its partners for permission to publish this paper.

REFERENCES CITED

Bradley, J., 1975, Abnormal formation pressure: AAPG Bulletin, v. 59, p. 957-973.

Carstens, H., and H. Dypvik, 1981, Abnormal formation pressure and shale porosity: AAPG Bulletin, v. 65, p. 344-350.

Eliuk, L.S., 1978, The Abenaki Formation, Nova Scotia Shelf, Canada—a depositional and diagenetic carbonate platform: Bulletin of Canadian Petroleum Geology, v. 26, p. 425-514.

Ervine, W., and J. S. Bell, 1987, Subsurface in situ stress magnitudes from oilwell drilling records: an example from the Venture area, offshore eastern Canada: Canadian Journal of Earth Sciences, v. 24, p. 1748-1759.

Falvey, D. A., 1974, The development of continental margins in plate tectonic theory: Australian Journal of Petroleum Exploration, v. 14, p. 95-106.

Fertl, W., 1976, Abnormal formation pressures: Elsevier, New York, 382 p.

Given, M. M., 1977, Mesozoic and early Cenozoic geology of offshore Nova Scotia: Bulletin of Canadian Petroleum Geology, v. 25, p. 63-91.

Grant, A., K. McAlpine, and J. A. Wade, 1986, The continental margin of eastern Canada—geological framework and petroleum potential, in M. T. Halbouty (ed.), Future petroleum provinces of the world: AAPG Memoir 40, p. 177-205.

Hutcheon, I., 1986, The relationship of diagenetic mineral reactions to an overpressured zone on the Scotian Shelf, in Basins of eastern Canada and worldwide analogues (abs.), p. 56.

Hunt, J. M., 1990, Generation and migration of petroleum from abnormally pressured fluid compartments: AAPG Bulletin, v. 74, p. 1-12.

Jansa, J. F., and J. A. Wade, 1975, Geology of the continental margin off Nova Scotia and Newfoundland, in Offshore geology of eastern Canada: Geological Survey of Canada, Paper 77-21, p. 51-105.

Law, B., and W. Dickinson, 1985, Conceptual model for origin of abnormally pressured gas accumulations in low-permeability reservoirs: AAPG Bulletin, v. 69, p. 1295-1304.

Magnusson, D. H., 1973, The Sable Island deep test of the Scotian Shelf, in P. J. Hood (ed.), Earth science symposium on offshore eastern Canada: Geological Survey of Canada, Paper 71-23, p. 253-266.

McIver, N. L., 1972, Cenozoic and Mesozoic stratigraphy of the Nova Scotia Shelf: Canadian Journal of Earth Sciences, v. 9, p. 54-70.

Mudford, B. S., 1988, Modelling the occurrence of overpressures on the Scotian Shelf, offshore eastern Canada: Journal of Geophysical Research, v. 93, p. 7845-7855.

Mudford, B. S., 1990, A one-dimensional two phase model of overpressure generation in the Venture gas field offshore Nova Scotia: Bulletin of Canadian Petroleum Geology, v. 38, p. 246-258.

Mudford, B. S., and M. E. Best, 1989, Venture gas field, offshore Nova Scotia: case study of overpressuring in region of low sedimentation rate: AAPG Bulletin, v. 73, p. 1383-1396.

Plumley, W., 1980, Abnormally high fluid pressure: survey of some basic principles: AAPG Bulletin, v. 64, p. 414-430.

Powell, T., 1982, Petroleum geochemistry of the Verrill Canyon Formation: a source for Scotian Shelf hydrocarbons: Bulletin of Canadian Petroleum Geology, v. 30, p. 167-179.

Powell, T., and L. R. Snowden, 1979, Geochemistry of crude oils and condensates from the Scotian Basin, offshore eastern Canada: Bulletin of Canadian Petroleum Geology, v. 27, p. 453-466.

Purcell, L. P., M. A. Rashid, and I. A. Hardy, 1979, Geochemical characteristics of sedimentary rocks in Scotian Basin: AAPG Bulletin, v. 63, p. 87-105.

Shi, Y., and C.-Y. Wang, 1986, Pore pressure generation in sedimentary basins: overloading versus aquathermal: Journal of Geophysical Research, v. 91, p. 2152-2153.

Chapter 6

Mexico's Giant Fields, 1978–1988 Decade

Jose Santiago
Alfonso Baro

INTRODUCTION

The decade from 1978 to 1988 was prolific in Mexico in the discovery of giant oil fields. Some of these fields can be considered supergiants according to Michel T. Halbouty's classification, including Agave, Giraldas, Paredon, Abkatun, Ek, Ku, Maloob, and Pol.

This paper reports on 21 of these giant oil fields discovered in Mexico during the 1978–88 decade. Eleven of these fields are in the Chiapas-Tabasco area (Agave, Bellota, Cardenas, Giraldas, Iris, Jujo, Luna, Muspac, Paredon, Sen, and Tecominoacan), and ten are in the Sonda de Campeche area (Abkatun, Bacab, Batab, Caan, Chuc, Ek, Ku, Maloob, Pol, and Uech). Original reserves totaled approximately 23 billion bbl of liquid hydrocarbons BOE.

The Chiapas-Tabasco (onshore) and Sonda de Campeche (offshore) areas are adjacent to each other and are similar both geologically and in reservoir behavior (Figure 1). For this reason, the fields from both areas are covered together, with references to regional variations that illustrate the differences among them.

Some of the giant oil fields of southeastern Mexico were described by Santiago (1980).

LOCATION

The division of geologic provinces in southeastern Mexico is shown in Figure 2, where the main producing regions lie. The Sonda de Campeche area is located within the province known as Marina de Coatzacoalcos (Coatzacoalcos Marine), and the Chiapas-Tabasco area is located in the province known as Cuencas Terciarias del Sur Este (Tertiary Basins of the Southeast) and in a small northern region of the province called Sierra de Chiapas.

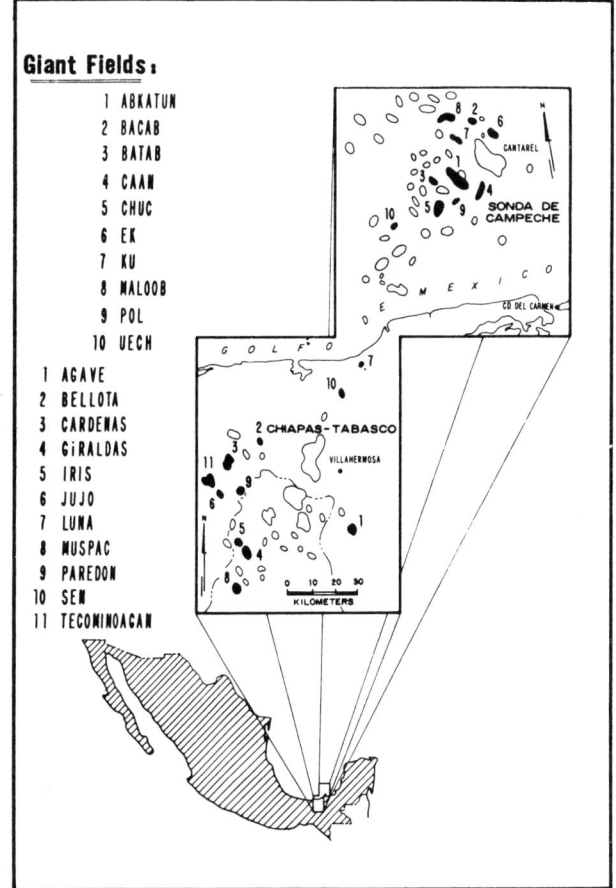

FIGURE 1. Locations of Chiapas-Tabasco and Sonda de Campeche areas, Mexico.

HISTORY OF THE DISCOVERIES

After the 1938 Oil Expropriation, Petroleos Mexicanos (PEMEX), the state-owned corporation in charge of exploration, development, and marketing of Mexico's hydrocarbon resources, successfully continued exploration activities in the southeastern part of the country at the Isthmus Saline basin, Comalcalco basin, and Macuspana basin. Gas and oil were produced from the first two mentioned and dry gas from the third (Figure 3). The production from these basins came from Tertiary sands and amounted to 225,900 BOPD at the beginning of 1972.

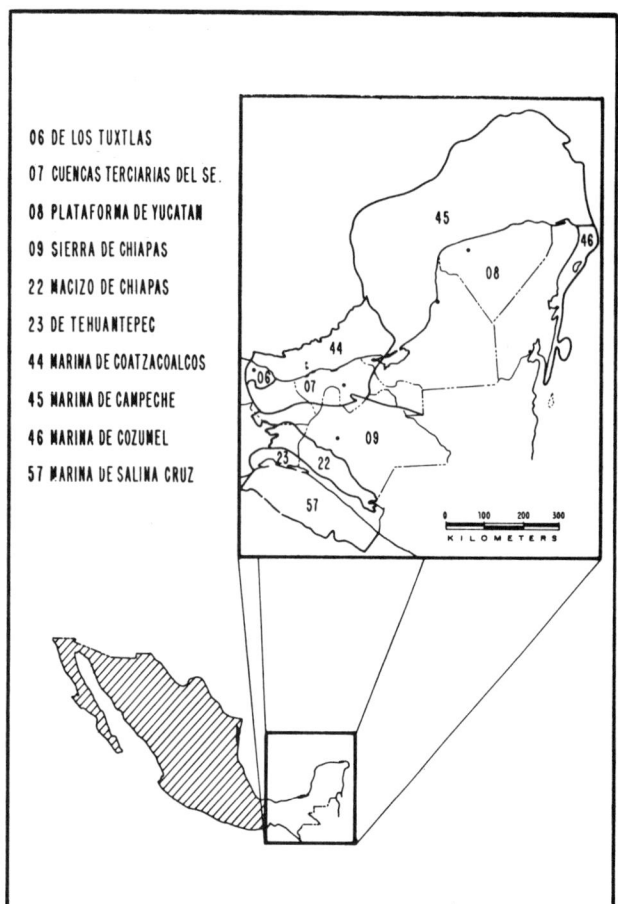

FIGURE 2. Southeast Mexico geologic provinces.

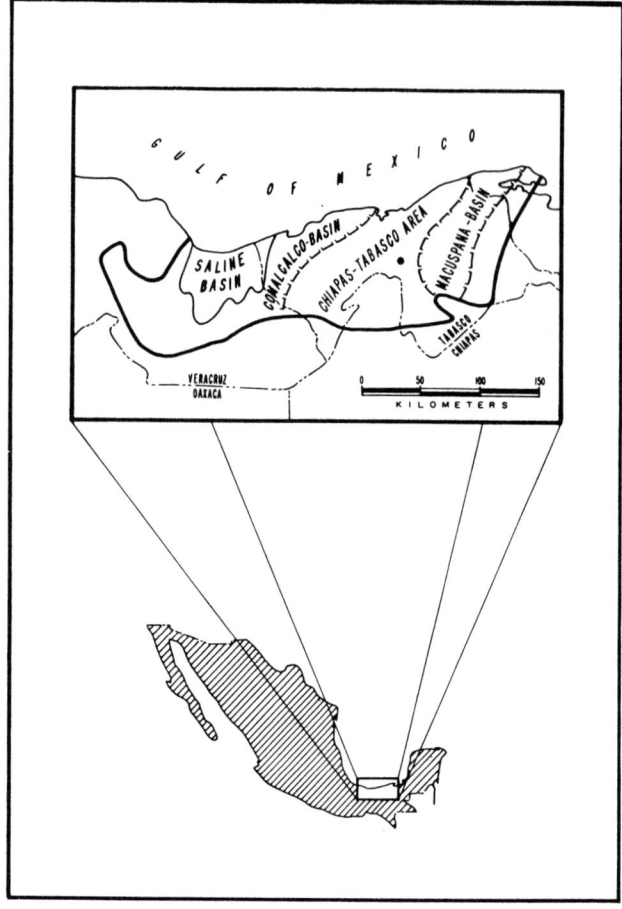

FIGURE 3. Subdivision of Cuencas Terciarias del Sur Este geologic province.

In 1969, the Jalupa No. 3 well was drilled near the present-day A. J. Bermudez field, which at a depth of 2430 m penetrated Upper Cretaceous sediments, represented by marls. The well reached a total depth of 2805 m.

Although by as early as 1960, oil production from Upper Cretaceous formations had been established at the Sierra de Chiapas, the Jalupa No. 3 well was the first to be drilled into Upper Cretaceous sediments within the Cuencas Terciarias del Sur Este Province. Even though the well turned out to be a dry hole, the information obtained was fundamental, because seismic data showed some "deep" reflectors, indicating the possible presence of Cretaceous rocks. From that time on, these areas became prime targets for oil exploration.

Thus, in 1972, the Cactus No. 1 and Sitio Grande No. 1 discovery wells (Figure 4) were drilled close to the township of Reforma.

Cactus No. 1 well, at a depth of 3740 m, produced 2500 BOPD from a calcarenite body of the Upper Cretaceous, and the Sitio Grande No. 1 well, at 4130 m, produced 2800 BOPD from a dolomite formation of the Middle Cretaceous. This marked the beginning of the development of a large area known today as the Chiapas-Tabasco area, which is one of the most important oil-producing areas in Mexico.

Following the Cactus and Sitio Grande discoveries, oil exploration in the area was intensified, using mainly seismic reflection work. A significant number of structures were detected, which, once drilled, resulted in the discovery of several fields with production from various stratigraphic intervals from the Jurassic to the Cretaceous.

Once the strategic characteristics of the Mesozoic from the Chiapas-Tabasco area were known, paleogeographic charts were prepared with the aid of the available geologic information from the Sierra de Chiapas and the Yucatan Peninsula. The paleogeographic charts showed that, toward the Sonda de Campeche, conditions were similar to those found on land. Additionally, as early as the beginning of the century, oil seeps had been found at sea, suggesting the presence of oil in this region.

With this information, a detailed seismic study was carried out offshore of the State of Campeche, covering a surface area of 8000 km^2. More than 30 structures were interpreted. The first of these, known as Chac, was drilled in June 1974. This well penetrated to a total depth of 4934 m and reached Oxfordian-Upper Jurassic sediments. Oil was produced from the 3545–3567 m interval of a limestone breccia of the lower Paleocene. Production amounted to 952 BOPD in March 1976 (Figure 5).

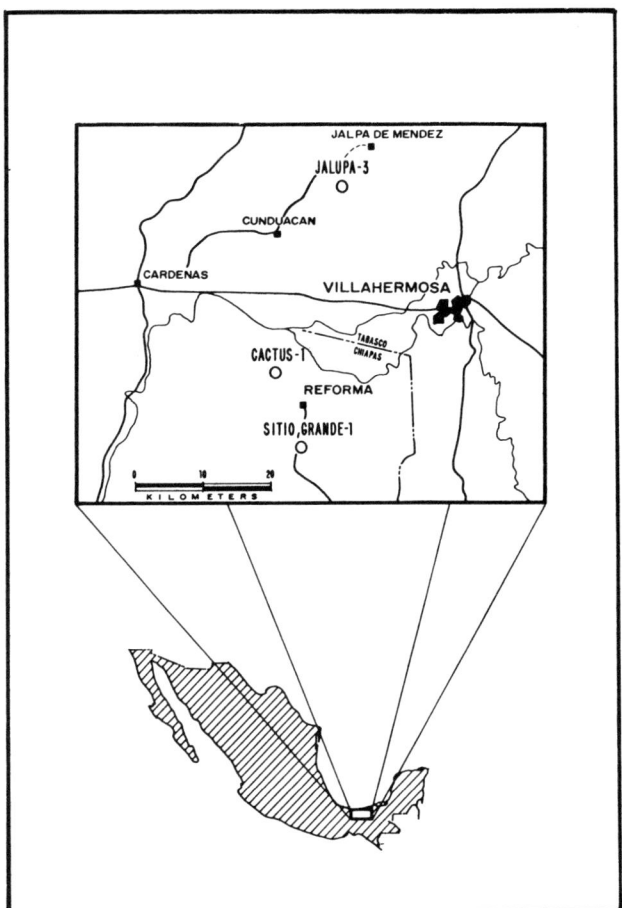

FIGURE 4. Locations of Jalupa No. 3, Cactus No. 1, and Sitio Grande No. 1 wells.

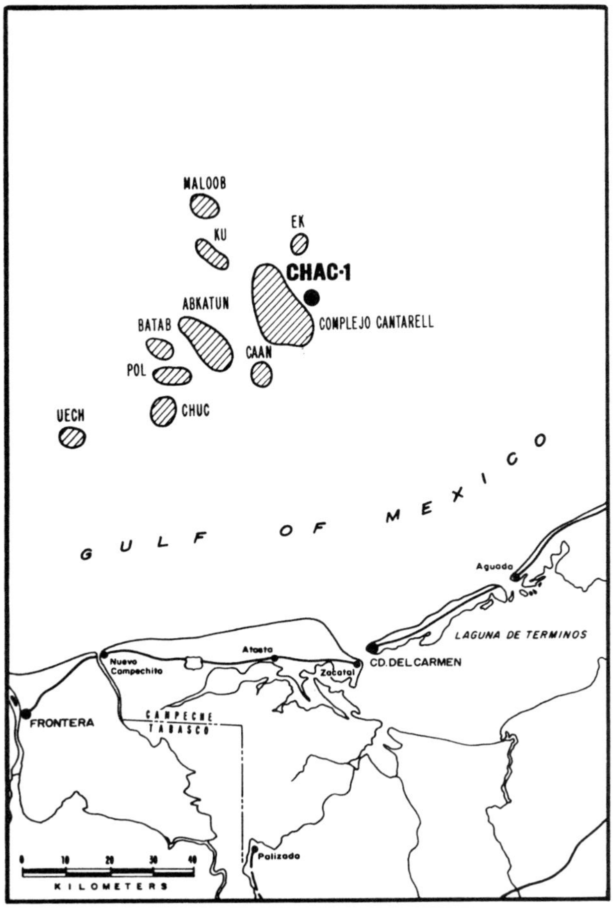

FIGURE 5. Location of Chac #1 well.

REGIONAL STRATIGRAPHY

The sedimentary column found in the Chiapas-Tabasco and Sonda de Campeche areas is supported by a basement made up of metamorphic rocks ranging in age from the Precambrian to the Upper Permian and Lower Triassic, with some activation that reaches the Miocene period (Figure 6). At present, this igneous metamorphic complex is found in the Yucatan No. 1 and 4, Villa Allende No. 1, and Cobo No. 301 wells. (Figures 7 and 8).

Of course, the economic oil potential of this basement is practically nonexistent, but the deposition conditions that it induced as a result of its tectonic evolution allowed the formation of basins that accumulated sediments where hydrocarbons could be generated and stored.

Paleozoic

Up to the present, none of the wells drilled in these areas has reached Paleozoic sediments. Their existence has been shown, however, by the presence of outcrops in the Maya Mountains and in the southeastern part of the State of Chiapas (Figure 9), as well as by some wells drilled in Belize.

Jurassic

During the Mesozoic, the paleogeographic picture was governed by the Yucatan and Chiapas-Oaxaca continents. Initially a continental passive margin basin evolved between these regions, which allowed deposition of salt bodies at a time estimated as Triassic to Jurassic.

The plastic properties of the salt, plus the fact that it was subjected to the influence of at least four tectonic events, affected the folding of the sedimentary column directly through a loading effect. This enhanced the formation of structural and stratigraphic traps, which have appeared at the Saline basin and the Sonda de Campeche area, where hydrocarbons have been produced from the Miocene and the Cretaceous, respectively, in reservoirs associated with salt diapirs.

During the Callovian and the beginning of the Oxfordian, the advance of the seas became more predominant, thus giving rise to the alternate deposition of dolomites and anhydrites in the Chiapas-Tabasco Mesozoic basin. At the Sonda de Campeche, the Oxfordian sediments, which represent the oldest deposits known in this area, are made up mostly of terrigenous deposits.

This terrigenous sequence is made up of sands, silts, and sandstones, alternating with thin strata consisting of

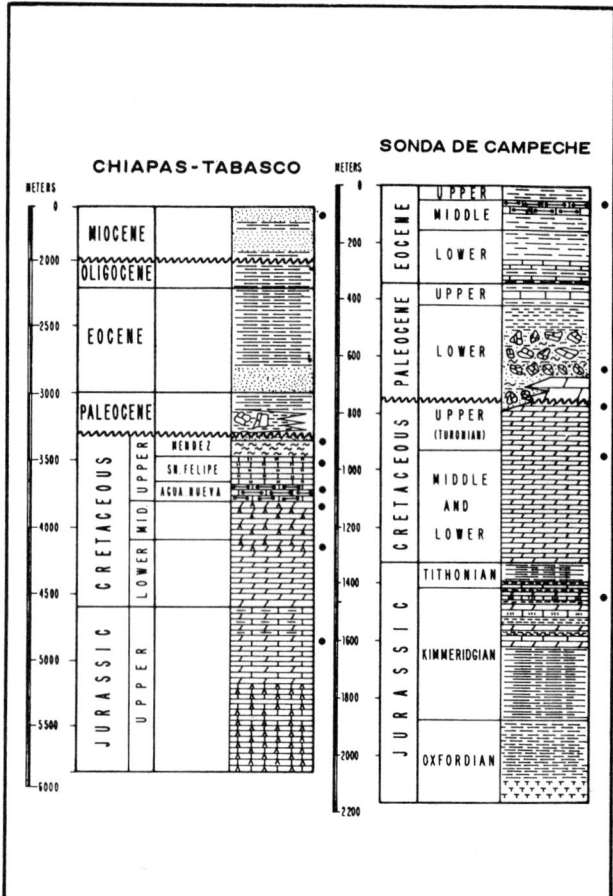

FIGURE 6. Typical geologic column in the Chiapas-Tabasco and Sonda de Campeche areas.

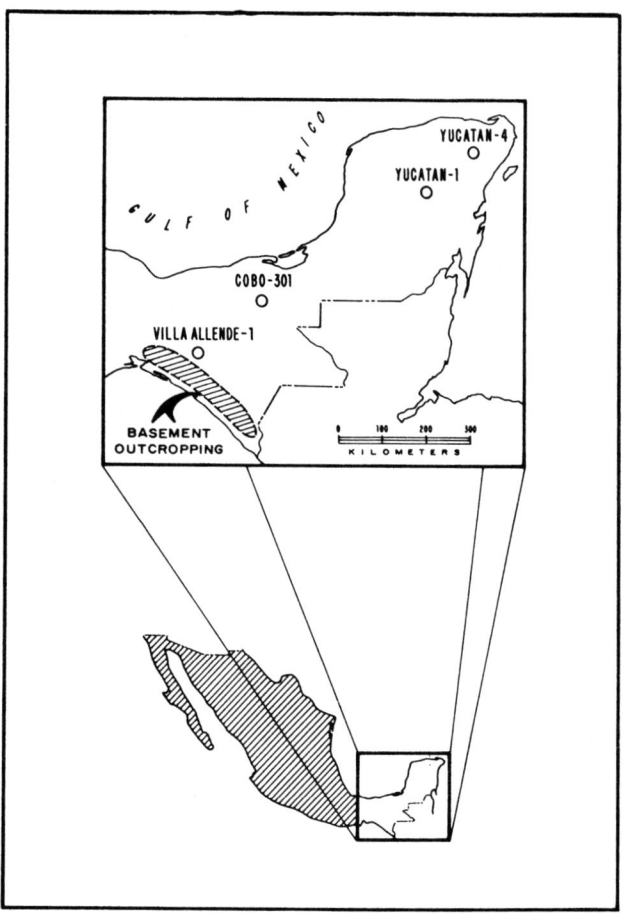

FIGURE 7. Basement outcropping, and locations of Yucatan No. 1, Yucatan No. 4, Cobo No. 301, and Villa Allende No. 1 wells.

carbonated rocks and evaporites, all of which may lead to the assumption that they were deposited in shallow epicontinental seas with an irregular topography.

Recently, production of hydrocarbons was obtained from Oxfordian rocks in the central part of the Chiapas-Tabasco area, where a body of sandstones and shaly limestones with impregnation of light oil was found. For this reason, the discovery of reservoirs in rocks of this age continues to be the primary exploratory objective. The quality of the oil produced was 35° API, and even though in the Chac No. 1 well it was not possible to carry out production tests at this stratigraphic level, at a depth of 4850 m a formation of shaly sandstones and limestones was found with impregnation of light oil.

During the Kimmeridgian, marine transgression continued at the Sonda de Campeche and Chiapas-Tabasco basins (Figure 10).

The lower part of the Kimmeridgian is made up of shaly sediments, which in some parts are bentonitic and sandy with limestone inclusions. In contrast, the upper portion consists of an alternating sequence of oolitic, and partially dolomitized, limestones with layers of dolomites and anhydrites.

Carbonate rocks from the Kimmeridgian constitute one of the main targets for exploration because the oolitic banks, which are characterized by their primary porosity, increased their storage capacity due to dolomitization. This has been amply proven by the prolific fields discovered both on land and offshore.

The best expression of the transgression of the Jurassic is found in the Tithonian. At that time, a large portion of the shelf was covered by deeper seas, with deposits predominantly made up of shaly limestone with inclusions of black shale. Within these sediments a large percentage of sapropelic organic matter can be found as well as lesser amounts of woody and coaly materials. In addition to this, if the degree of thermal maturity is found to be adequate, it must be concluded that the rock formations from the Tithonian period are the most important source rocks for oil and gas at the Sonda de Campeche and Cuenca de Chiapas-Tabasco. This has been verified by means of studies on source rock determination performed by a U.S. company in 1978.

Lower Cretaceous

During the Lower Cretaceous, the sea invaded the continent, covering extensive areas and creating shallow seas. This gave rise to evaporitic deposits, represented by

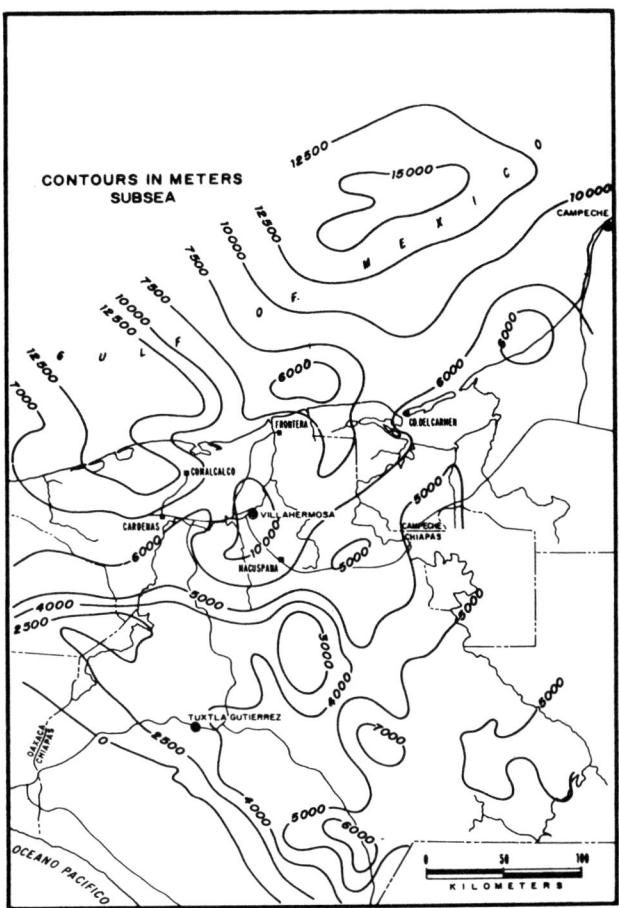

FIGURE 8. Generalized basement contours of the Chiapas-Tabasco and Sonda de Campeche areas.

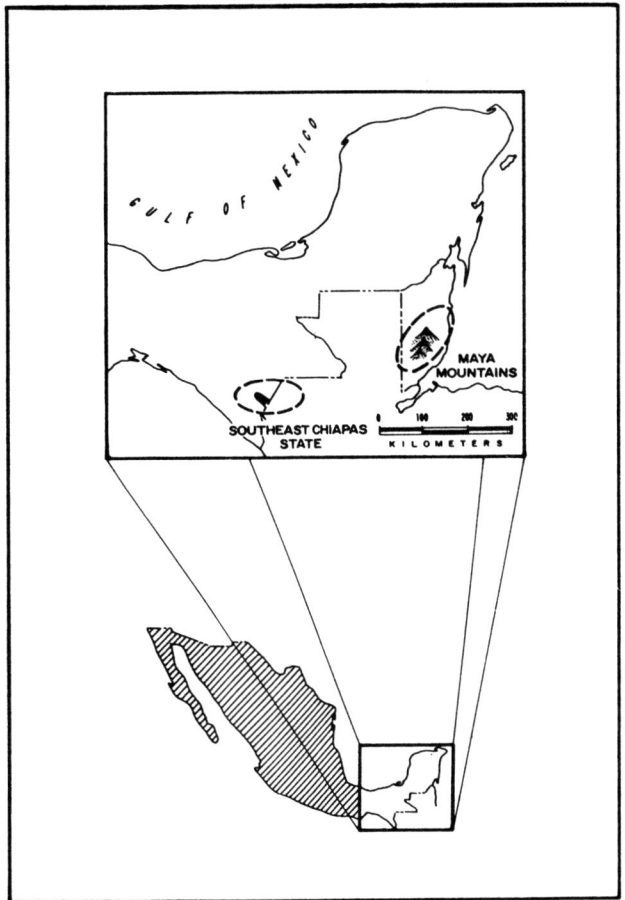

FIGURE 9. Paleozoic outcrop areas.

interbedded dolomites and anhydrites. In contrast, dolomite and shaly limestone appeared on the seacoast. On the offshore portion, shaly limestone together with chert, occasionally interspersed with bentonitic shales, were deposited. Dolomitization of these sediments is predominant in the portions of the basins close to the shelf, where dolomite formations with oil impregnations occurred. Their economic importance is highly significant because a considerable amount of the production at both the Sonda de Campeche and Chiapas-Tabasco areas comes from these rock formations (Figure 11).

Middle Cretaceous

During the Middle Cretaceous, prevailing conditions were the same as those in the Lower Cretaceous. Among the outstanding events was the advance of the sea, which covered all the Yucatan continent as well as most of the Chiapas-Oaxaca continent, thus setting up interoceanic communications. The result of this transgression was the restriction of evaporitic shelves.

Slope and basinal depositions may be found at the Sonda de Campeche, which become dolomitized as they approach the shelf.

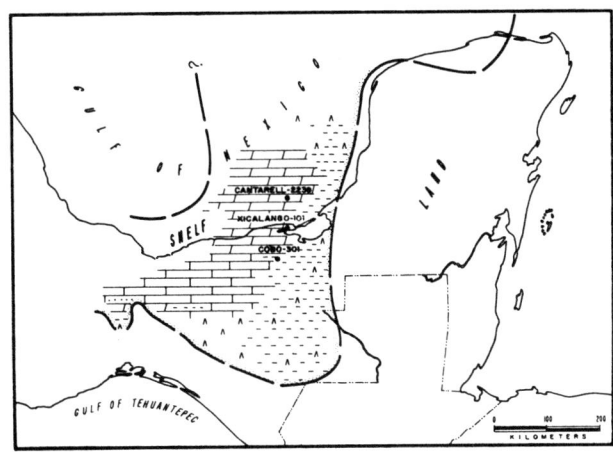

FIGURE 10. Jurassic-Kimmeridgian.

Three well-differentiated facies are found in the Chiapas-Tabasco area: limestone shelves to the south, open sea limestones with dolomitized breccia and dolomites in the central part, and deep water limestones in the north. Their importance as storage rock formations is

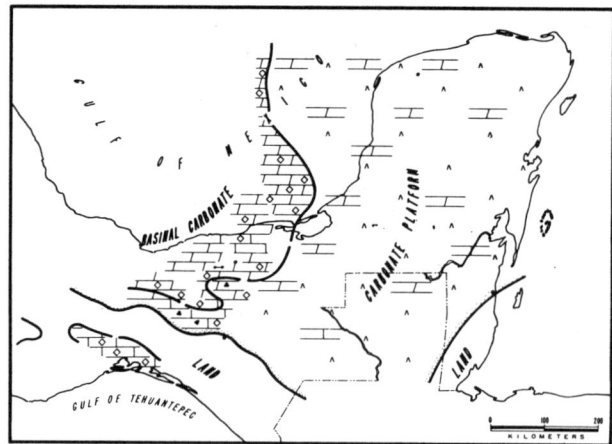

FIGURE 11. Lower Cretaceous.

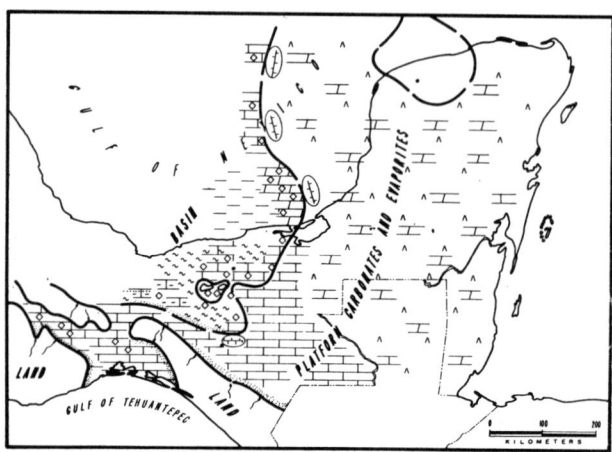

FIGURE 12. Upper Cretaceous.

paramount, because a considerable portion of the oil and gas production obtained offshore, as well as in the Chiapas-Tabasco area, comes from rocks of this age.

Upper Cretaceous

During the Upper Cretaceous, the open sea continued its southward advance over the Middle Cretaceous shelf. At the Sonda de Campeche this interval consists, in general, of dolomites and shaly limestones with chert. Inclusions of dolomitized breccia can also be found, which in the upper part, together with others from the lower Paleocene, make up the main reservoirs in the marine area (Figure 12).

It has been determined that the origin of the breccia is linked to the flow of calcareous detritus coming from the Yucatan shelf and deposited in slopes and basin environments.

In the Chiapas-Tabasco area, the rocks from this period have been well differentiated at Mendez, San Felipe, and Agua Nueva. The Mendez Formation is made up of a continuous sequence of marls, ranging in color from reddish brown to greenish gray. At the upper portion, a body of breccia made up of cream-colored clastic limestone materials can be found. The San Felipe Formation is made up of greenish gray limestone of a bentonitic and shaly nature. The rocks from the Agua Nueva Formation are cream-colored limestones, with brown-amber and white chert and deep brown to black limestone toward the base of the formation.

Cenozoic

The beginning of the Cenozoic, in both the Sonda de Campeche and Chiapas-Tabasco areas, is generally characterized by a change in sedimentation, marked by the contrast between the carbonate rocks from the Cretaceous period and the terrigenous rock formations from the Tertiary. This difference is a result of the evolution of the tectonic phase, which can be correlated with the Laramidic time, whose geologic framework was defined by the Batolito de Chiapas and the Yucatan shelf. The southeastern Tertiary basins were developed between these two regions.

Apparently, at the Sonda de Campeche, paleogeographic conditions at the end of the Mesozoic locally transgressed the beginning of the Paleocene, due to the fact that there was a predominance of slope breccia deposits. These deposits are characterized by the inclusion of bentonitic clays with pelagic microfauna from the lower Paleocene.

Paleocene sediments of the Chiapas-Tabasco area consist of a thick body of bentonitic shales that gradually become reddish brown limonites toward the base, with calcareous breccia inclusions.

At the Sonda de Campeche, Eocene sedimentation is represented mainly by shales with calcarenite and bentonite inclusions. The deposition mechanism for the calcarenites must have been turbidity flows from the neighboring carbonate shelf.

In the Chiapas-Tabasco area, the Eocene column consists of a continuous sequence of shales. The importance of the lower and middle Eocene calcarenites at the Sonda de Campeche may be considered to be secondary from an economic standpoint, because they have proven to be oil and gas producers at the Ku field but remain only potential producers at the Cantarel field.

The Oligocene sequence in both areas is made up of shales, some of which are bentonitic. At the beginning of the Miocene, this sequence, together with the adjacent units, underwent the effects of orogenic phenomena known locally as Chiapaneca, which gave rise to the present-day structural model.

During the Oligocene, there were marked periods of erosion that originated the configuration of a regional unconformity. Therefore, the absence of rocks of this age is common in some parts of the Chiapas-Tabasco and Sonda de Campeche areas.

At the Sonda de Campeche, miocene sedimentation is represented mainly by shales and bentonitic shales, with sand, sandstones, and limestone inclusions, close to the Yucatan shelf. Their potential is evidenced by numerous gas shows in some of the wells drilled at Abkatun, Pol, and Cantarel fields.

In the Chiapas-Tabasco area, sands from the Miocene are important oil and gas producers and traditionally have been producers since before the beginning of production from the Mesozoic.

At the end of the Miocene the sea began to withdraw, which gave rise to increasing sand deposits during the Pliocene and Pleistocene.

REGIONAL STRUCTURAL GEOLOGY

Figure 13 shows the regional tectonic framework in which the various successive events that affected the Chiapas-Tabasco and Sonda de Campeche areas took place. The North American, Caribbean, and Cocos plates are outstanding, and the Cocos is the oceanic plate that goes below the continental crust of the Meso-American trench. The Motagua-Polochic fault system, which is active between the Caribbean and North American plates, crosses the Republic of Guatemala and constitutes the southwestern border of the Chiapas Granitic Massif, within the Mexican territory.

Within the Motagua-Polochic system, in the crushed zone, rocks have been identified whose radiometric ages may lead us to accept the fact that this system has been active at least since the Laramide orogeny, even though its activity may go back to the Permian, with active periods in the Jurassic and Tertiary (Eocene, Oligocene, and Miocene).

Locally, these areas were influenced tectonically by the apparent immobility of the Yucatan shelf, the Motagua-Polochic fault system already described, and the Chiapas Granitic Massif (Figure 14).

In both areas, at least five different deformation events are recorded in the stratigraphic column (Figure 15). Evidence of the first tectonic event, which may be correlated in time with the Nevadian deformation, occurs with the thinning out of the Tithonian carbonates at an approximate thickness of 200 m. As a result of this thinning out due to erosion, the clastic materials, which can be found within the limestones of the Lower Cretaceous, were deposited.

At the end of the Middle Cretaceous, a period of the pre-Laramidic deformations was manifested during which, at the Sierra de Chiapas, a regional horizontal movement thrust fault system became active and an en

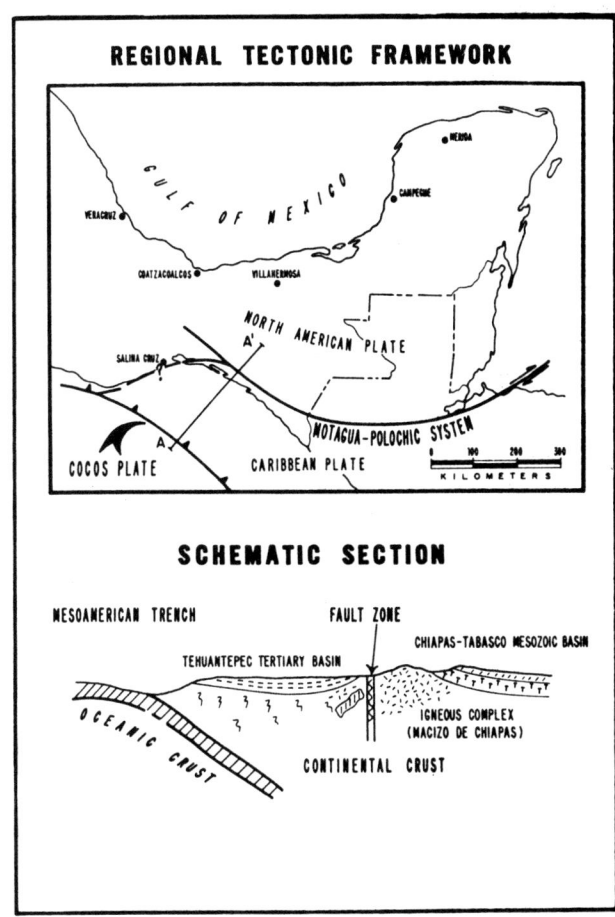

FIGURE 13. Regional tectonic framework and schematic section.

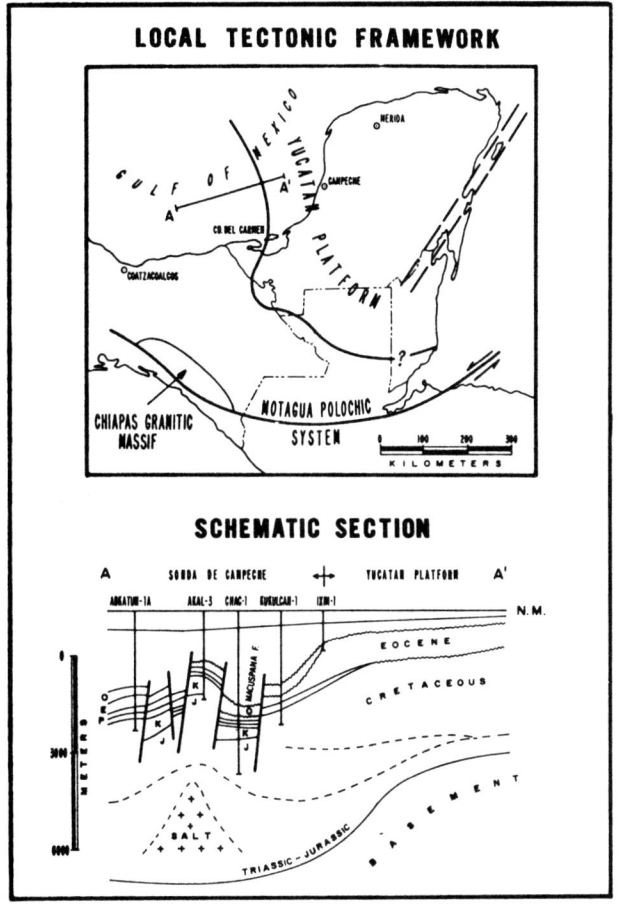

FIGURE 14. Local tectonic framework and schematic section.

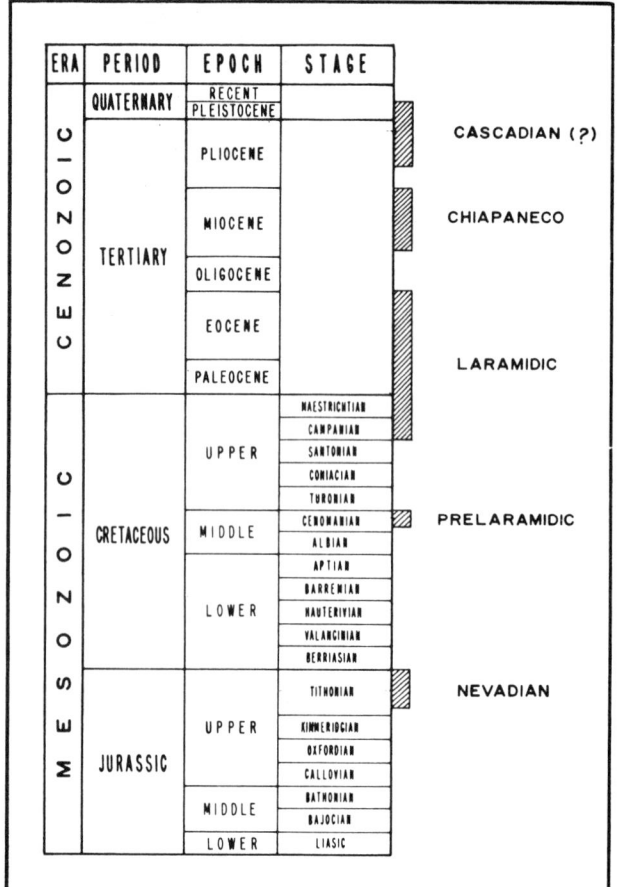

FIGURE 15. Tectonic events in southeast Mexico.

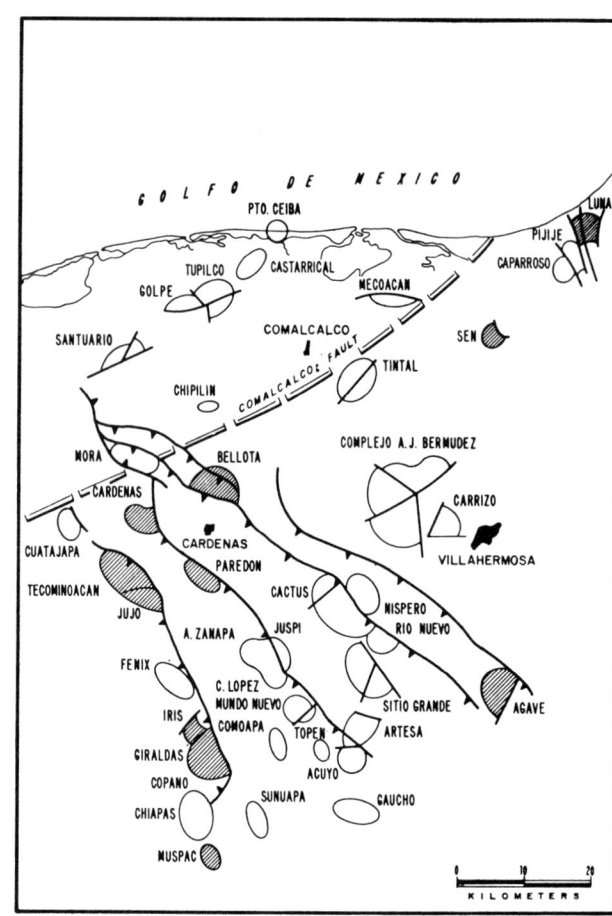

FIGURE 16. Main tectonic aspects of the Tabasco-Chiapas area.

echelon fault system developed. The erosion surface that caused this movement was discordantly covered by the Turonian limestones. This erosion led to the disappearance of all the Middle Cretaceous and a large part of the Lower Cretaceous in some parts of the area.

From the upper Cretaceous to the Eocene, the region underwent a period of transgressional stresses originated by the tectonic system shown in Figures 13 and 14, which may be correlated with the Laramide orogeny and attributed to the movement of the Motagua-Polochic system.

During the Miocene, there was a strong reactivation of deformation of the folded belt, which gave rise to the settling of the en echelon faults as well as to a clear definition of their reverse faults in their outline. Due to the peculiar characteristics of this event, it became known locally as the Chiapaneco Event; it is correlated in time with the Caribbean movement.

During this last-mentioned deformation stage, the structures encompassing the fields of Sonda de Campeche and Chiapas-Tabasco were defined. Regionally, the structures are found at ever greater depths from southeast to northwest in Chiapas-Tabasco and from northeast to southwest in Sonda de Campeche. The most elevated structures can be found toward the Sierra de Chiapas and the Yucatan shelf (Figures 16 and 17).

During the Pliocene and the Pleistocene, the last deformation that affected these areas occurred. It was at this time that tensional stresses were generated within this region, contributing to the formation of numerous normal faults of a listric nature. Some of them constitute a boundary between the Pliocene Comalcalco and Macuspana basins. This deformation could represent the Cascadian Event.

The entire history of the deformation events that evolved in these areas has been affected by a saline tectonism that goes back to Jurassic times, in such a way that the shape of the anticlines is influenced by the saline intrusions. Drilling into the saline mass is a common occurrence, whether it is associated with the surfaces of reverse faults or of the anticlines themselves.

Figures 18 and 19 show seismic portion lines where salt bodies have been found in different shapes within the stratigraphic column in these areas.

Because of the many factors that influenced these areas during their tectonic evolution, they are extremely complex from a structural point of view and therefore quite difficult to interpret.

DESCRIPTIONS OF THE FIELDS

The 21 fields mentioned in the preceding text are described below. These descriptions include basic informa-

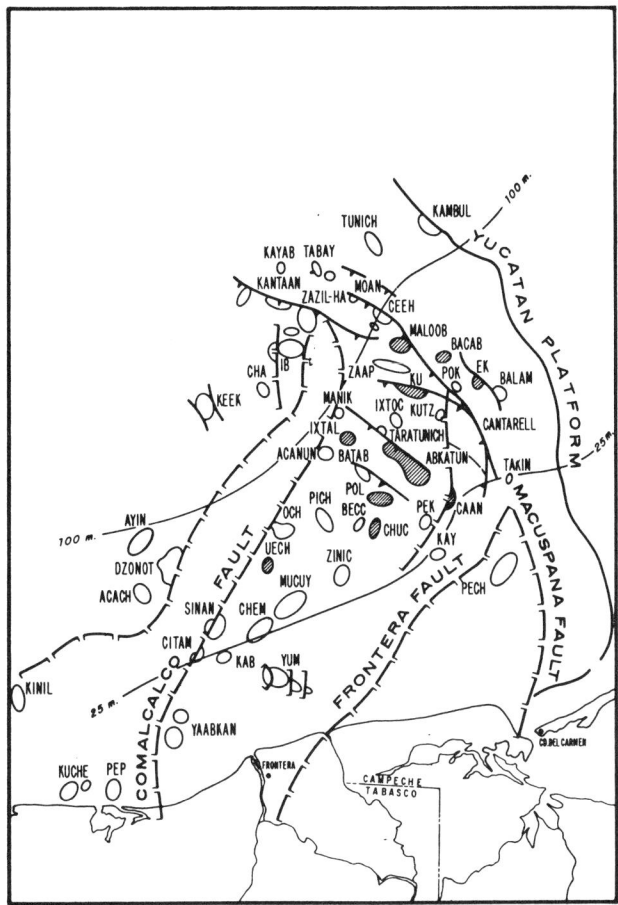

FIGURE 17. Main tectonic aspects of the Sonda de Campeche area.

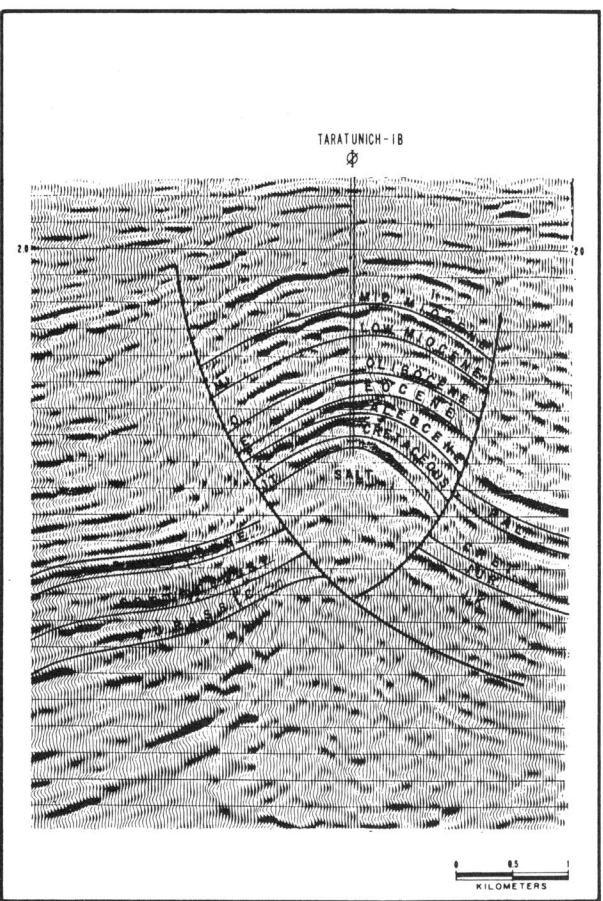

FIGURE 19. Seismic line over the Taratunich structure in the Sonda de Campeche area.

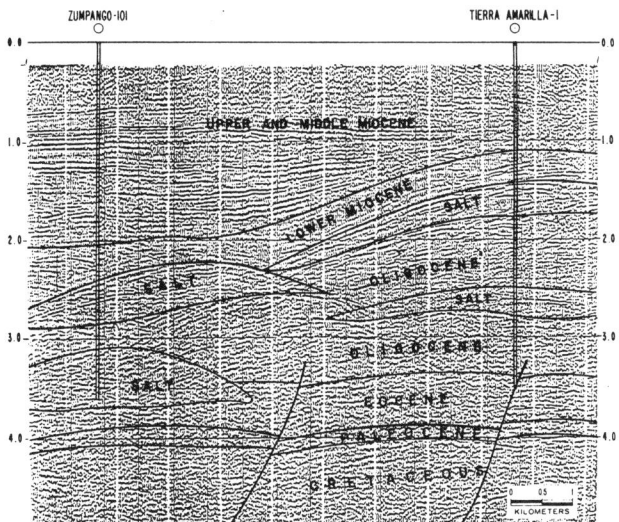

FIGURE 18. Seismic line in the central part of the Chiapas-Tabasco area.

tion on local stratigraphy and structural geology as well as production and characteristic reservoir data, which are complemented with Tables 1, 2, 3, and 4. The locations of these fields are shown in Figures 1, 16, and 17.

Chiapas-Tabasco Area

Agave Field

Location. The Agave field is situated in the southeastern portion of the Chiapas-Tabasco area, 25 km S 12° E of Villahermosa City.

Stratigraphy. At the top of the structure, Pliocene-Pleistocene sediments lie in an unconformity over the Cretaceous rocks, which make up the top of the reservoir. The Upper Cretaceous is represented by mudstone-wackestone, which turns into dolomites to the south. The Middle and Lower Cretaceous are made up of dolomitized limestone.

Structure. The structure is an anticline whose greater axis is oriented from northwest to southeast. It is influenced on its northern flank by a reverse fault and by various small normal faults over the structure that are oriented in different directions (Figures 20 and 21).

Production. Production comes from dolomites and from Upper, Middle, and Lower Cretaceous fractured dolomitized limestones. The discovery well, Agave 1-B, initially produced 1138 BOPD and 2.1 MMCFGD from the 4673-4725 m Lower Cretaceous interval.

TABLE 1. Some characteristics of the Chiapas-Tabasco fields.

Field	Productive area (km²)	Maximum saturated column	Oil-water level (m), subsea	Reservoir rock	Age of the reservoir rock	Average porosity (%)	Water saturation (%)	Initial reservoir pressure (T.P., kg/cm²) × choke size (in.)	Reservoir temp. (°C)	API gravity	Gas/oil ratio (m³/m³)
Agave	48.1	925	4450	Fract. dolomite Fract. dolomite Fract. dol. LS.	Upper Cretaceous Middle Cretaceous Lower Cretaceous	5 5 5	17 17 17	195 × 1/4 87 × 7/8 267 × 1/4	137	41.0	2000 440 325
Bellota	32.1	990	5480	Fract. dolomite Fract. dolomite Fract. dolomite	Middle Cretaceous Lower Cretaceous Jurassic-Kimmer.	3 3 3	13 13 11	163 × 1/2 240 × 1/2 160 × 1/2	148	39.0	400 460 249
Cardenas	41.1	845	5900	Fract. dolomite Fract. dolomite Fract. dolomite	Lower Cretaceous Jurassic-Titho. Jurassic-Kimmer.	3 3 3	17 14 14	132 × 5/8 143 × 5/8 78 × 3/4	159	38.4	264 468 254
Giraldas	36.1	586	5016	Calcarenite and Fract. LS.	Middle Cretaceous	7	13	327 × 5/16	134	39.7	935
Iris	8.4	576	4900 5016	Fract. Limestone	Middle Cretaceous	5	17	300 × 1/4	133	41.4	930
Jujo	35.3	1344	6427	Fract. dolomite Fract. dolomite	Jurassic-Titho. Jurassic-Kimmer.	3 3	12 12	103 × 3/8 196 × 3/8	154	37.6	205 218
Luna	15.0	386	6100	Fract. dolomite	Jurassic-Kimmer.	6	10	360 × 1/2	168	40.8	600
Muspac	17.8	188	2750	Calcarenite	Upper Cretaceous	12	17	180 × 1/2	109	50.6	1646
Paredon	24.0	1270	6090	Fract. dolomite Fract dolomite	Jurassic-Kimmer.	4 4	13 13	158 × 3/8	115	37.5	400 500
Sen	More than 25.0	More than 280	Not determined	Fract. breccia	Upper Cretaceous	7	18	511 × 1/4	155	37.3	460
Tecominoacan	25.0	1300	6175	Fract. dolomite	Jurassic-Kimmer.	3	12	230 × 1/2	147	36.2	159

TABLE 2. Some characteristics of the Sonda de Campeche fields.

Field	Productive area (km²)	Maximum saturated column (m)	Oil-water level (m), subsea	Reservoir rock	Age of the reservoir rock	Average porosity (%)	Water saturation (%)	Initial reservoir pressure (T.P., kg/cm²) × choke size (in.)	Reservoir temp. (°C)	API gravity	Gas/oil ratio (m³/m³)
Abkatun	90	670	3750	Breccia	Pal.-Up. Cretaceous	13	18	116 × 1	110	30	130
				Dolomite	Middle Cretaceous	8	22	110 × 1 1/4	121	29	152
				Dolomite	Lower Cretaceous	9	21	118 × 1 1/4	129	30	120
Bacab	More than 12	More than 185	Not determined	Breccia	Pal.-Up. Cretaceous	12	24	45 × 1/2	120	19	50
Batab	15	300	4720	Breccia	Pal.-Up. Cretaceous	9	27	140 × 2	121	32	138
				Fract. dolomite	Jurassic-Kimmer.	5	26	220 × 1/2	144	32	152
Caan	38	500	3850	Breccia	Pal.-Up. Cretaceous	9	10	95 × 1/4	147	36	42
Chuc	23	420	4020	Breccia	Pal.-Up. Cretaceous	12	20	90 × 2	115	31	120
				Fract. dolomite	Lower Cretaceous	6	24	35 × 2	140	29	140
Ek	More than 11	More than 336	Not determined	Breccia	Pal.-Up. Cretaceous	9	27	Marine test	112	20	
Ku	50	800	3300	Calcarenite	Eocene	19	17	27 × 1/2	80	16	87
				Breccia	Pal.-Up. Cretaceous	8	24	36 × 2	87	20	34
				Dolomite	Middle Cretaceous	5	29	35 × 1	94	22	36
				Fract. dolomite	Lower Cretaceous	5	28	31 × 2	98	22	36
Maloob	30	400	3285	Breccia	Pal.-Up. Cretaceous	9	20	34 × 1	89	20	36
Pol	44	400	4020	Breccia	Pal.-Up. Cretaceous	10	22	69 × 1	122	33	186
				Fract. dolomite	Middle Cretaceous	6	27	72 × 2	127	29	127
				Fract. dolomite	Jurassic-Kimmer.	4	30	89 × 2	138	31	126
Uech	12	400	5100	Fract. dolomite	Jurassic-Kimmer.	15	5	175 × 3 1/2	140	38	300

TABLE 3. Production data for Chiapas-Tabasco fields.

Field	Year of discovery	Depth of the reservoirs (m)	Producer wells drilled to June 1990	Average depth of the wells (m)	Maximum production per day				Cumulative production to Dec. 31, 1988	
					Oil (bbl)	Gas (mmcf)	Month	Year	Oil (10^6 bbl)	Gas (bcf)
Agave	1976	3800–4725	20	4123	56,585	503.2	Jan.	82	94.8	852.5
Bellota	1982	5009–6000	16	5586	43,148	77.1	Aug.	85	49.3	83.7
Cardenas	1980	5115–5960	42	5839	158,689	278.0	Dec.	83	226.7	437.5
Giraldas	1977	4424–4946	23	4812	82,168	353.3	Apr.	82	131.6	905.7
Iris	1979	4454–4990	8	4982	8,435	52.4	Mar.	82	14.5	106.7
Jujo	1980	5081–6435	35	5932	135,996	125.7	Mar.	87	246.3	256.3
Luna	1985	5141–5820	10	5573	29,165	170.0	Sep.	89	2.5	13.5
Muspac	1982	2525–3131	14	3074	20,116	199.1	Nov.	87	12.5	142.1
Paredon	1977	4830–6100	17	5665	51,089	173.1	Mar.	83	129.5	357.8
Sen	1984	4759–5039	2	4920	11,605	35.0	Dec.	89	6.1	18.4
Tecominoacan	1985	5115–6445	23	5924	76,432	113.0	Dec.	85	83.2	87.6

TABLE 4. Production data for Sonda de Campeche fields.

Field	Year of discovery	Depth of the reservoirs (m)	Producer wells drilled to June 1990	Average depth of the wells (m)	Maximum production per day				Cumulative production to Dec. 31, 1988	
					Oil (bbl)	Gas (mmcf)	Month	Year	Oil (mmcf)	Gas (bcf)
Abkatun	1979	3100–3750	55	3471	510,981	369.0	Jan.	83	985.3	772.7
Bacab	1977	3210–	1	3382	Not developed					
Batab	1984	3610–4750	4	4811	11,729	17.1	May	88	5.2	7.6
Caan	1984	3300–3850	4	4295	24,531	25.8	Jan.	88	19.1	19.4
Chuc	1982	3500–4020	16	3987	82,757	85.8	Mar.	89	89.0	75.0
Ek	1980	3075–	1	3411	Not developed					
Ku	1979	2550–3350	25	3024	200,002	150.3	Oct.	89	483.5	240.1
Maloob	1979	2890–3300	9	3480	33,100	148.1	Dec.	89	9.6	12.5
Pol	1980	3730–4150	10	4154	135,571	117.1	Mar.	89	296.6	236.8
Uech	1986	4800–5200	2	5200	5,484	6.1	Aug.	88	0.5	470.2

86   Santiago and Baro

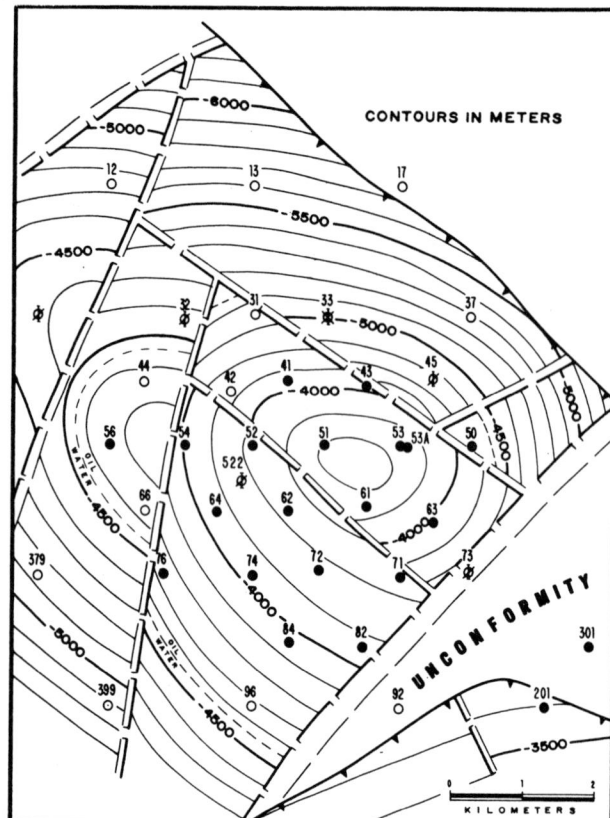

FIGURE 20. Agave field, top of reservoir (Tertiary-Mesozoic unconformity).

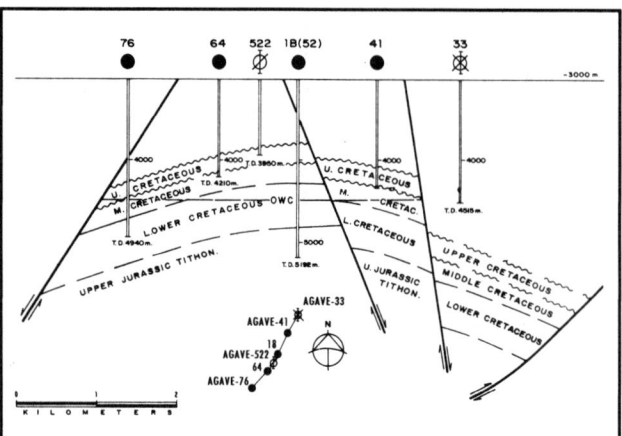

FIGURE 21. Geologic cross section, Agave field.

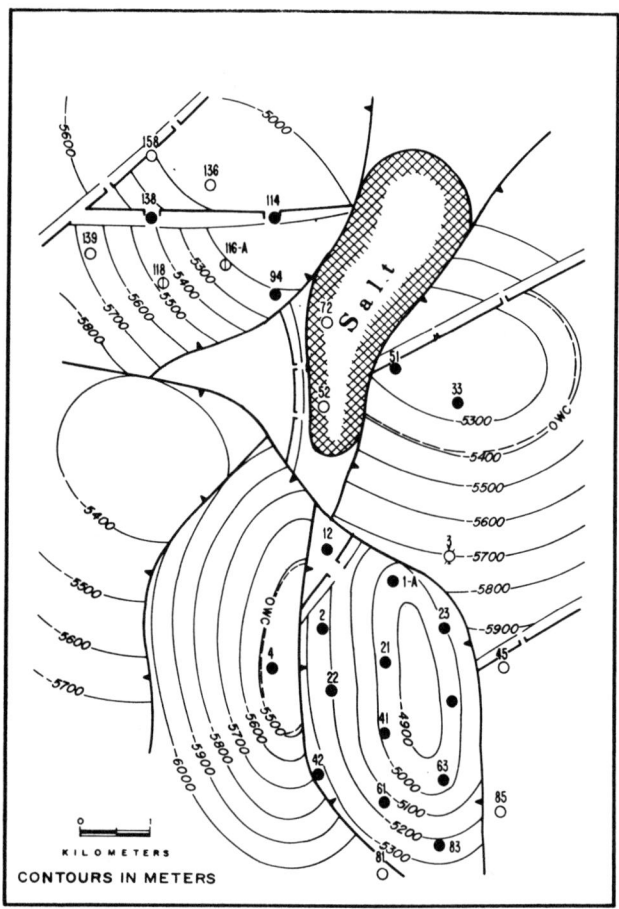

FIGURE 22. Bellota field, top of Middle Cretaceous.

Bellota Field

Location. The Bellota field is 40 km N 71° W of Villahermosa City.

Stratigraphy. The Tertiary sediment top measures approximately 4600 m. The Upper Cretaceous is represented by mudstone-wackestone, and the Middle and Lower Cretaceous by crystalline micro- to mesodolomites. The Jurassic consists of Tithonian shaly mudstone and Kimmeridgian dolomites.

Structure. The Bellota field has a complex structure divided by a salt intrusion at the northern and southern portions. Six reverse faults oriented in different directions are involved in this structure (Figures 22 and 23).

Production. The reservoirs are found in the Middle and Lower Cretaceous and in the Kimmeridgian-Jurassic. Initial production of the Bellota 1-A well was 3730 BOPD and 9.7 MMCFGD from the 5370-5415 m Lower Cretaceous interval.

Cardenas Field

Location. The Cardenas field is 50 km N 84° W of Villahermosa City.

Stratigraphy. The Upper Cretaceous is represented by mudstone and wackestone; the Middle Cretaceous is generally absent because of unconformities. The rest of the column consists of Lower Cretaceous and Upper Jurassic dolomites.

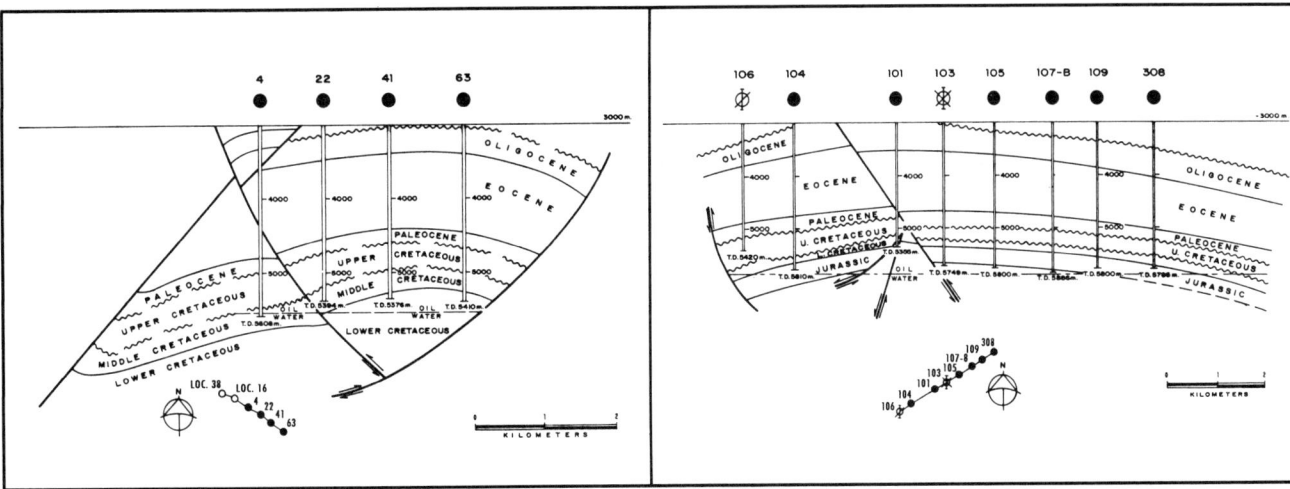

FIGURE 23. Geologic cross section, Bellota field.

FIGURE 25. Geologic cross section, Cardenas field.

Structure. In general terms, the structure of this field is an elongated anticline, oriented from northwest to southeast and influenced by a reverse fault on its southwestern boundary and by various small, normal faults over the structure oriented in different directions (Figures 24 and 25).

Production. The producing rock formations correspond to the Lower Cretaceous and Upper Jurassic. In initial tests, the Cardenas 101 discovery well produced 2943 BOPD, by way of a 1-in. choke, and 4.4 MMCFGD from the 5235–5245 m Lower Cretaceous interval.

Giraldas Field

Location. The Giraldas field is 55 km S 50° W of Villahermosa City, in the southwestern part of the Chiapas-Tabasco area.

Stratigraphy. The Tertiary upper Eocene rocks lie directly on top of Middle Cretaceous rock formations, due to an unconformity. The Middle Cretaceous is represented by mudstone to wackestone and fractured packstone.

Structure. The structure is an elongated anticline with an almost straight north-south orientation, influenced by a reverse fault all along the eastern flank (Figures 26 and 27).

Production. Production comes from Middle Cretaceous rocks. The Giraldas No. 2 well initially produced 1321 BOPD and 7.0 MMCFGD from the 4430–4505 m interval.

Iris Field

Location. The Iris field is 55 km S 56° W of Villahermosa City.

Stratigraphy. The stratigraphic column in this field is quite similar to the one at Giraldas field, taking into account the fact that the regional unconformity causes the Eocene sediments to be placed on top of Upper Cretaceous (Turonian) rock. The Upper Cretaceous rocks are mainly breccia.

Structure. The structure is an elongated anticline with a greater axis going from east to west and influenced by only two normal faults. Owing to the effect of one of them, different impregnation levels of oil-water have been established in two areas of the structure (Figures 28 and 29).

Production. The producing rocks are found in the Middle Cretaceous. The Iris No. 1 well initially produced 724 BOPD and 3.8 MMCFGD from the 4524–4600 m interval.

Jujo Field

Location. The Jujo field is 55 km S 80° W of Villahermosa City.

Stratigraphy. The stratigraphic sequence in the area of this structure is interrupted by an unconformity that puts Upper Cretaceous rocks (mudstone-wackestone) in contact with the underlying Lower Cretaceous dolomites. The Jurassic is dolomitized in both the Kimmeridgian and the Tithonian.

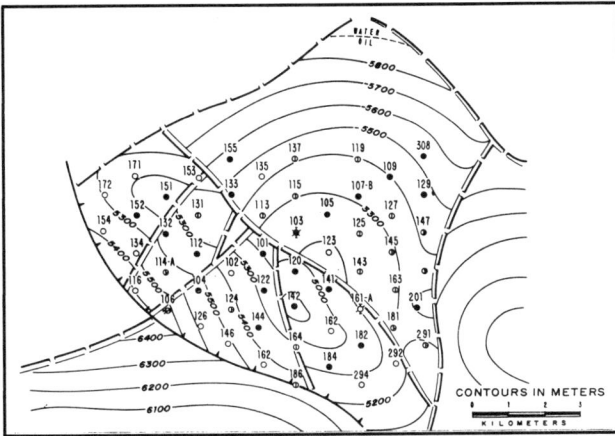

FIGURE 24. Cardenas field, top of Lower Cretaceous.

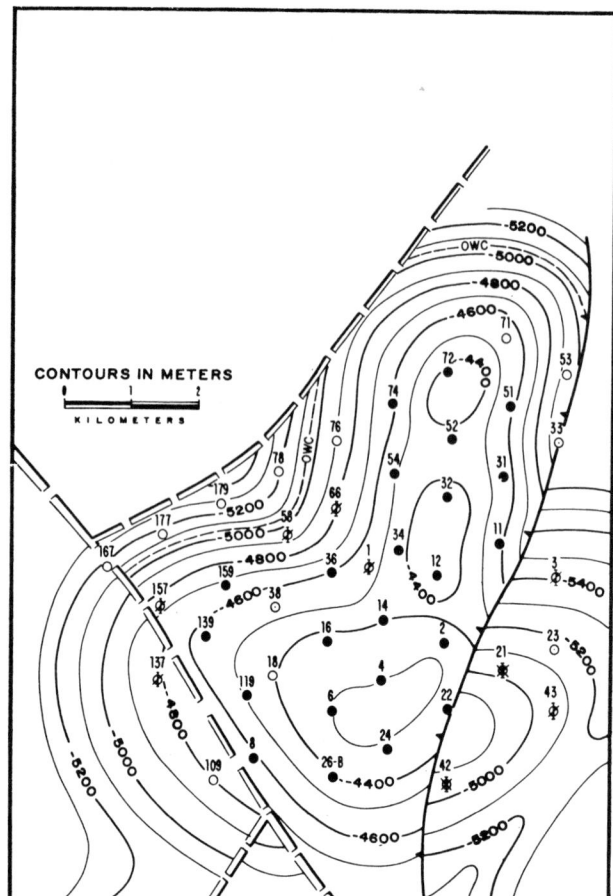

FIGURE 26. Giraldas field, top of Middle Cretaceous.

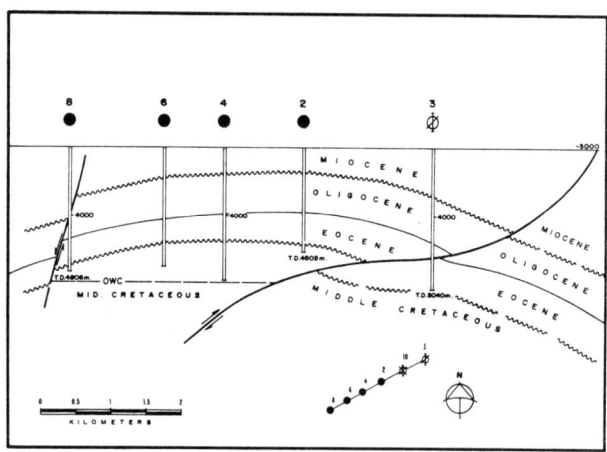

FIGURE 27. Geologic cross section, Giraldas field.

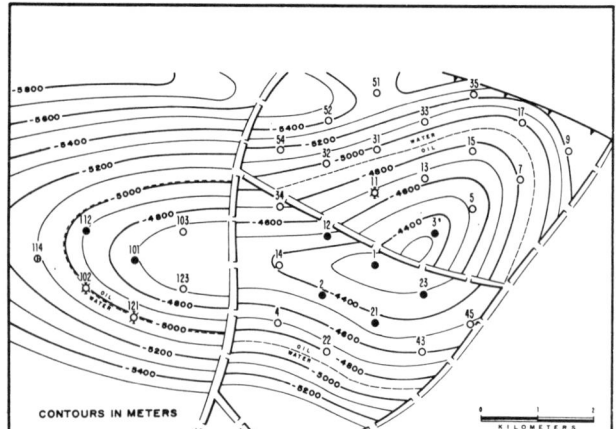

FIGURE 28. Iris field, top of Middle Cretaceous.

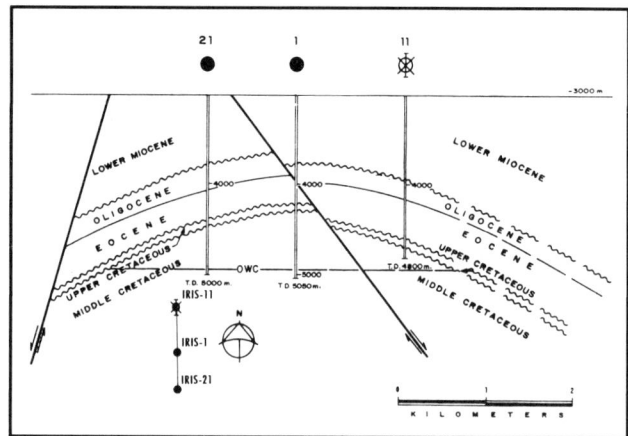

FIGURE 29. Geologic cross section, Iris field.

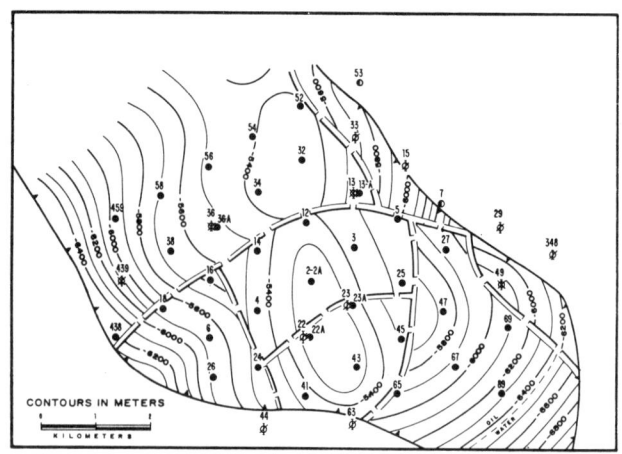

FIGURE 30. Jujo field, top of Jurassic-Kimmeridgian.

Structure. The structure is an elongated anticline oriented from northwest to southeast and is bordered by two reverse faults to the northeast and the southwest. The structure is also influenced by several normal faults that are oriented in different directions (see Figures 30 and 31).

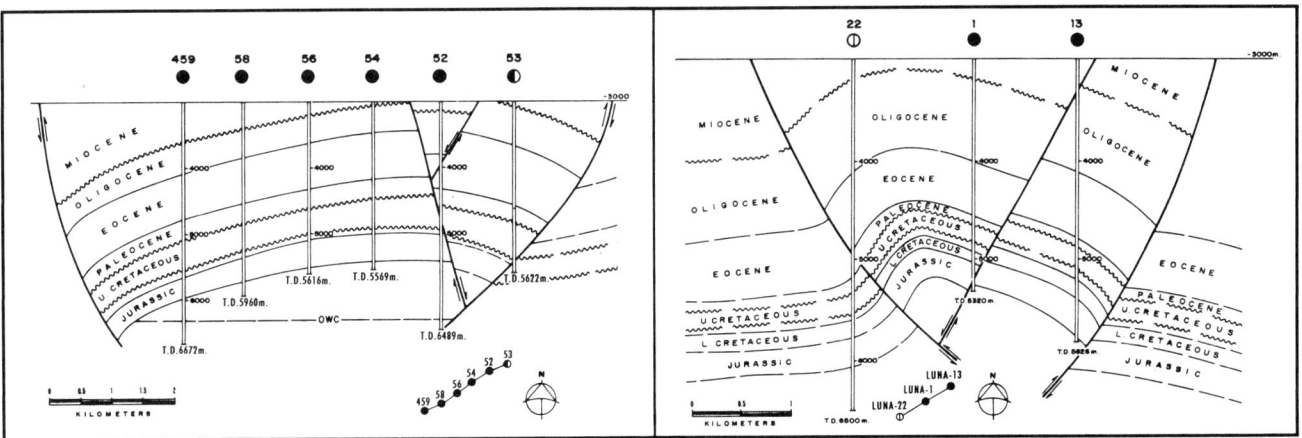

FIGURE 31. Geologic cross section, Jujo field.

FIGURE 33. Geologic cross section, Luna field.

Production. The production rock formations are found in the Upper Jurassic, Tithonian, and Kimmeridgian. The first producer well was Jujo 2-A, with initial production of 13,000 BOPD and 15.3 MMCFGD, by way of a ½-in. choke, from the 5547-5570 m Kimmeridgian interval.

Luna Field

Location. The Luna field is 53 km N 12° E of Villahermosa City, almost on the gulf coastline.

Stratigraphy. The Cretaceous is made up of open sea limestones. The Tithonian-Jurassic formations are slightly dolomitized shaly limestones, and the Kimmeridgian consists of mesocrystalline dolomites.

Structure. The structure is an elongated anticline oriented from northwest to southeast and bordered on its northeast and southwest flanks by reverse faults (Figures 32 and 33).

Production. The producing horizon corresponds to fractured Kimmeridgian dolomites, at a depth range from 5250 to 5680 m. Average production per well is 3638 BOPD and 20.0 MMCFGD. Most wells produce as open holes. The oil-water level has not yet been determined.

Muspac Field

Location. The Muspac field is 70 km S 42° W of Villahermosa City, at the Sierra de Chiapas foothills.

Stratigraphy. The Upper Cretaceous is made up of mudstone and wackestone bioclastic materials.

Structure. The structure is an anticline oriented from northwest to southeast and bordered on the northern flank by a reverse fault and on the south by a normal one (Figures 34 and 35).

Production. Production is obtained from Upper Cretaceous rock formations, with an average of 1140 BOPD and 12.4 MMCFGD per well. The producing rock formations have the highest porosity levels in the area.

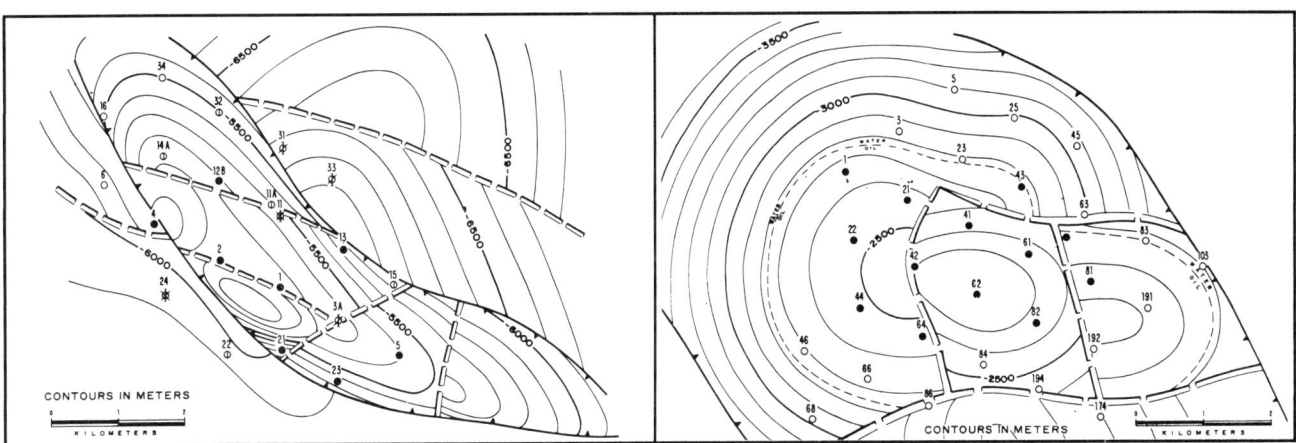

FIGURE 32. Luna field, top of Jurassic-Kimmeridgian.

FIGURE 34. Muspac field, top of Upper Cretaceous.

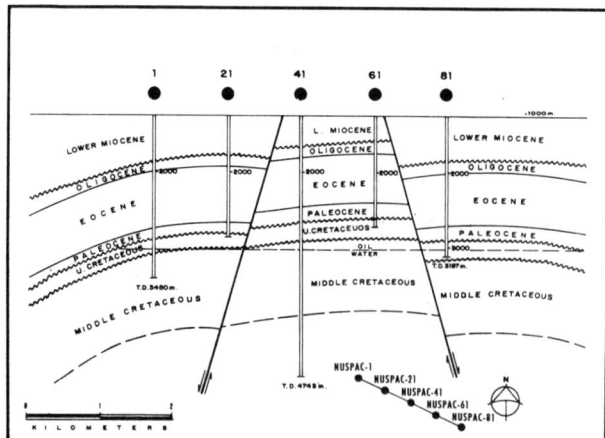

FIGURE 35. Geologic cross section, Muspac field.

Paredon Field

Location. The Paredon field is 45 km S 80° W of Villahermosa City.

Stratigraphy. The Upper Cretaceous is represented by open sea limestones, as opposed to the Middle and Lower Cretaceous, which are represented by dolomites. The Tithonian-Jurassic formations are shaly microcrystalline dolomites, and the Kimmeridgian is made up of mesocrystalline dolomites.

Structure. The structure is an elongated anticline oriented from northwest to southeast and bordered on the northeast by a reverse fault and on the south by a salt intrusion (Figures 36 and 37).

Production. Production comes from Lower Cretaceous and Kimmeridgian-Jurassic fractured dolomites. Production from the discovery well, Paredon 1, was 1554 BOPD and 3.4 MMCFGD from the 5161-5196 m Lower Cretaceous interval.

Sen Field

Location. The Sen field is 34 km due north of Villahermosa City.

Stratigraphy. The Sen field contains Tertiary sediments reaching a thickness of 5000 m. The Upper Cretaceous consists of breccia and limestones, and the Middle Cretaceous is made up of shaly mudstone and wackestone.

Structure. The structure is an anticline oriented from northwest to southeast and divided by three reverse faults with the same orientation (Figures 38 and 39).

Production. The production comes from the Upper Cretaceous. The Sen No. 1 well, flowed by way of a ¼-in. choke and 511 kg/cm² T.P., has average production of 2126 BOPD and 5.5 MMCFGD. The oil-water level has not been defined.

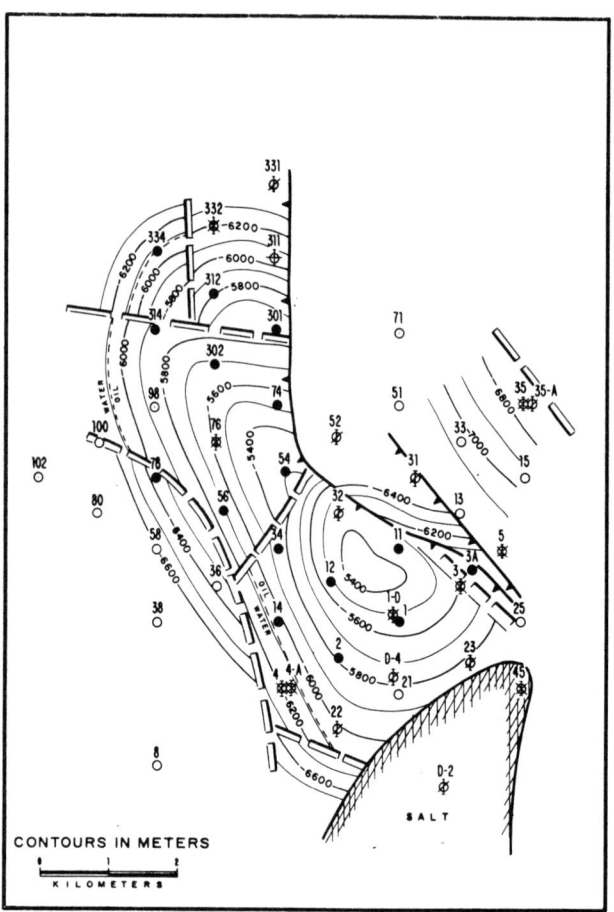

FIGURE 36. Paredon field, top of Upper Jurassic (Kimmeridgian).

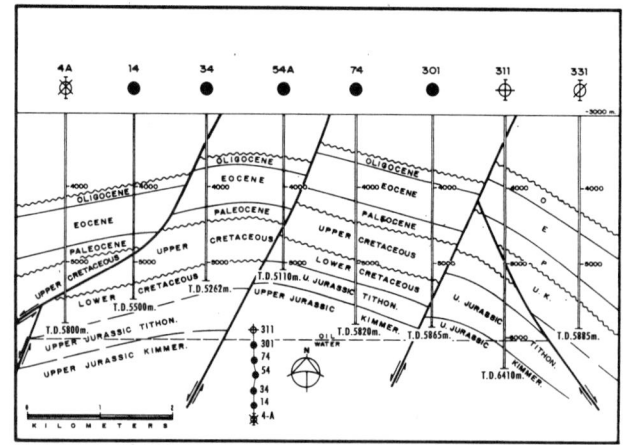

FIGURE 37. Geologic cross section, Paredon field.

Tecominoacan Field

Location. The Tecominoacan field is 60 km S 86° W of Villahermosa City, on the western border of the Chiapas-Tabasco area.

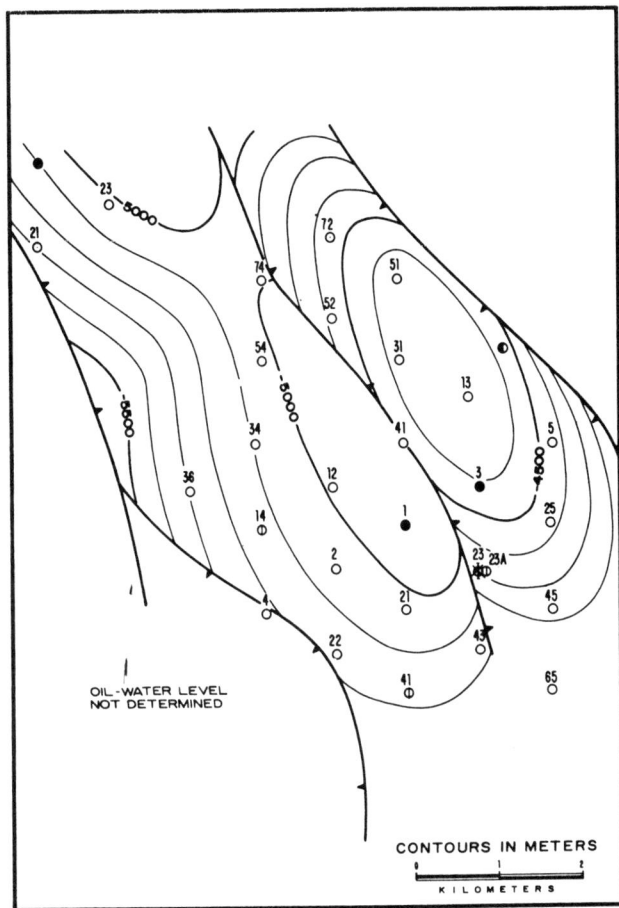

FIGURE 38. Sen field, top of Upper Cretaceous.

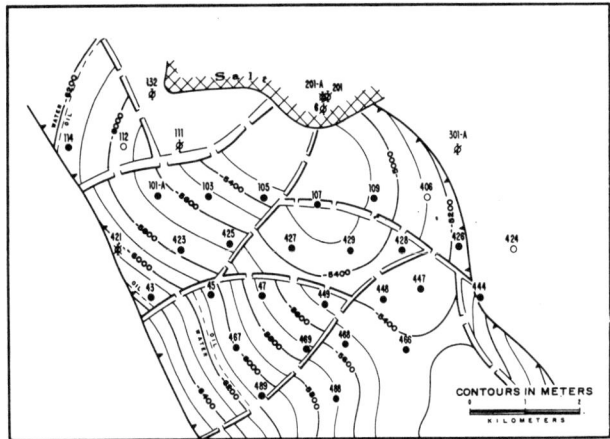

FIGURE 40. Tecominoacan field, top of Jurassic-Kimmeridgian.

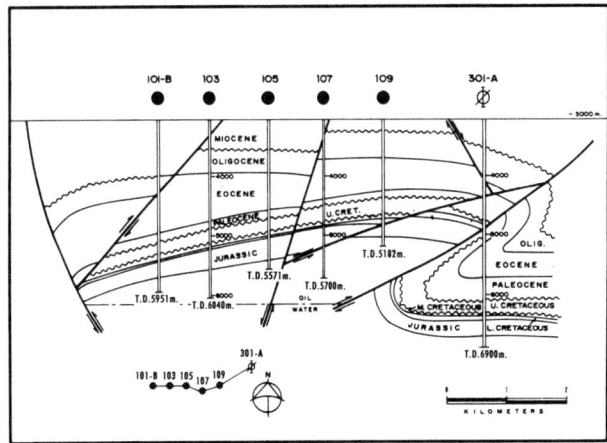

FIGURE 41. Geologic cross section, Tecominoacan field.

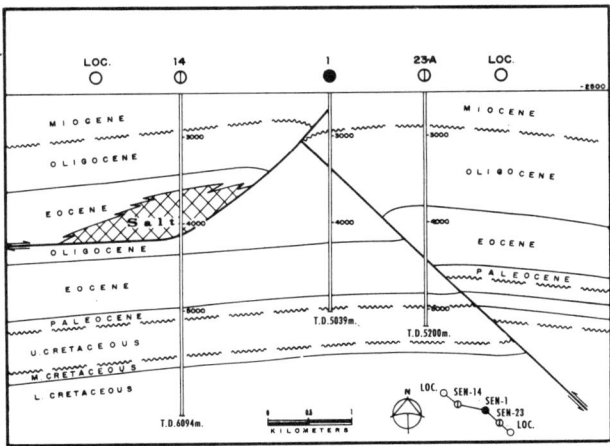

FIGURE 39. Geologic cross section, Sen field.

Stratigraphy. The Upper Cretaceous is represented by open sea limestones. The Middle Cretaceous is absent by unconformity, and the lower Cretaceous and Tithonian are slightly dolomitized shaly limestones. The Kimmeridgian consists of mesocrystalline dolomites.

Structure. The structure is an anticline oriented from northwest to southeast and bordered on its northeastern and southwestern flanks by reverse faults influenced by normal faults at the top. To the north it is truncated by a salt dome, and to the south it borders on the Jujo field as part of the same structure (Figures 40 and 41).

Production. Producing formations are Kimmeridgian fractured dolomites. The discovery well was Tecominoacan 101-B, which initially produced 5887 BOPD and 4.0 MMCFGD.

Sonda De Campeche Area

Abkatun Field

Location. The Abkatun field is 78 km N 28° W of Ciudad del Carmen, Campeche.

Stratigraphy. The Tertiary sequence, which reaches a thickness of 3300 m, consists mainly of shale with thin sandstone inclusions. The Upper Cretaceous-Paleocene breccia are made up of dolomitized calcareous fragments. The Middle and Lower Cretaceous are made up of dolomites with mudstone inclusions.

Structure. The structure is an asymmetric anticline oriented from northwest to southeast and influenced by a reverse fault on its northeastern flank and a normal fault on its southern top, which runs perpendicular to the reverse fault (Figures 42 to 43).

Production. The main reservoir is in the Upper Cretaceous-Paleocene breccia; production is also obtained from the Lower Cretaceous. Initial production from the discovery well was 8500 BOPD and 7.3 MMCFGD. At present there are wells such as the Abkatun 77, which produces 21,160 BOPD and 19.7 MMCFGD.

Bacab Field

Location. The Bacab field is 107 km N 15° W of Ciudad del Carmen, Campeche.

Stratigraphy. The Tertiary top is 3100 m thick. The only well drilled reached 185 m into the Upper Cretaceous-Paleocene breccia without having drilled through it completely.

Structure. According to available seismic information, the structure of the Bacab field is an anticline with an approximate east-west orientation and a reverse fault that influences its northeastern flank (Figures 44 and 45).

Production. At present, the discovery well, Bacab No. 1, is the only well drilled. Its initial test production was 1978 BOPD and 100 MCFGD, from the Paleocene-Cretaceous breccia. Construction of a platform has been scheduled so as to develop this field by 1993.

Batab Field

Location. The Batab field is 85 km N 35° W of Ciudad del Carmen, Campeche.

Stratigraphy. Tertiary sediments are 3600 m thick and consist mainly of shales and thin sandstone inclusions.

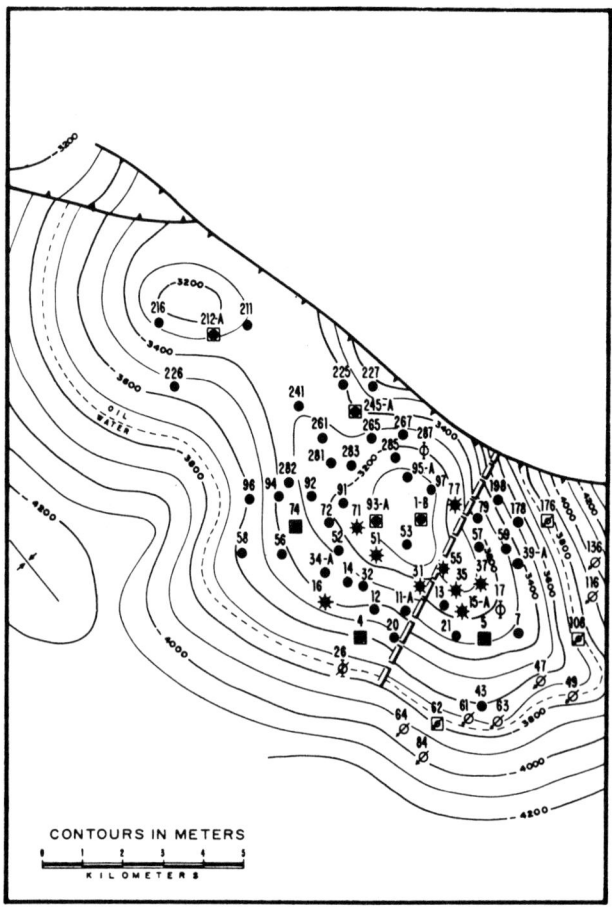

FIGURE 42. Abkatun field, top of breccia, (Lower Paleocene-Upper Cretaceous).

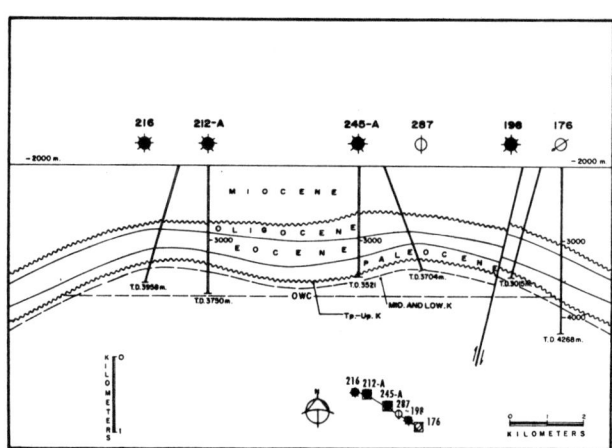

FIGURE 43. Geologic cross section, Abkatun field.

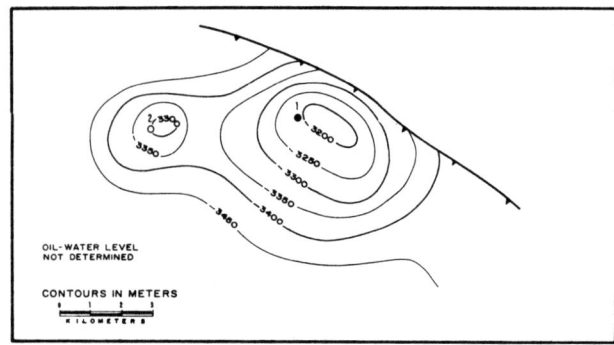

FIGURE 44. Bacab field, top of breccia (Lower Paleocene-Upper Cretaceous).

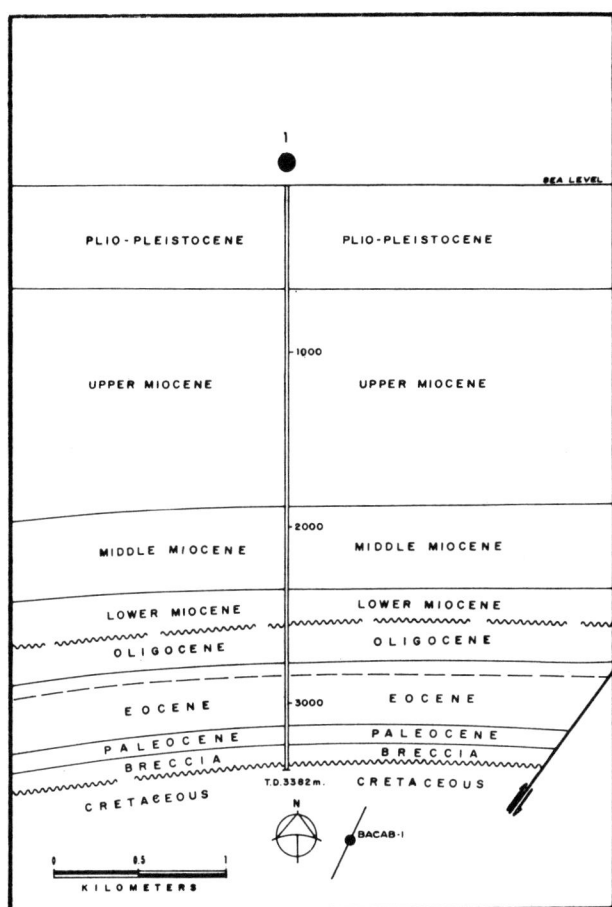

FIGURE 45. Geologic cross section, Bacab field.

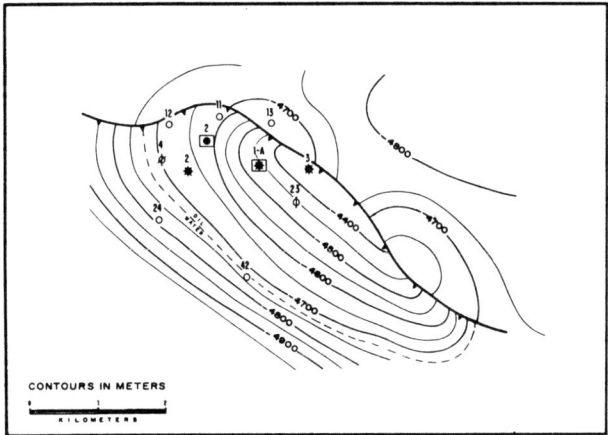

FIGURE 46. Batab field, top of Upper Jurassic (Kimmeridgian).

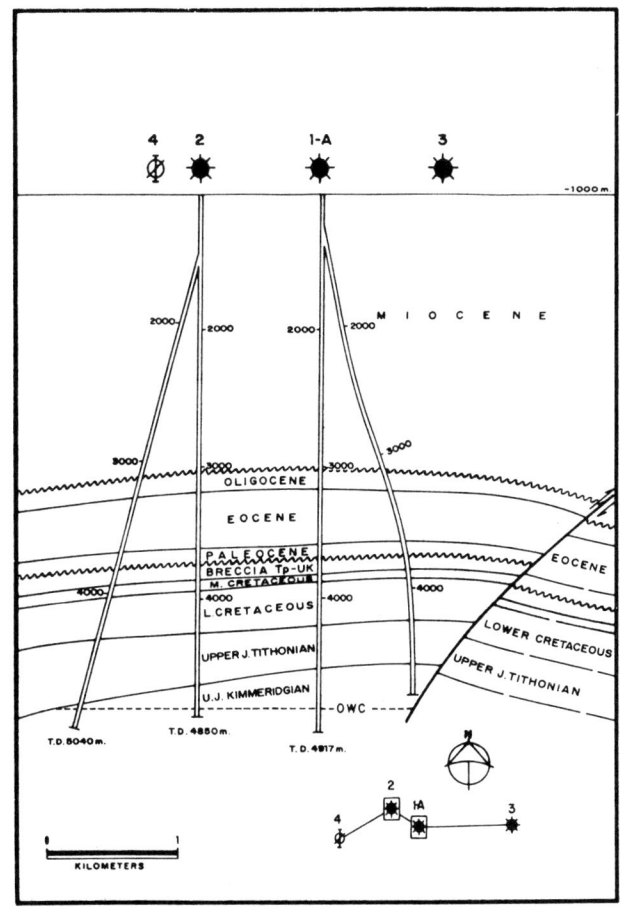

FIGURE 47. Geologic cross section, Batab field.

The Upper Cretaceous-Paleocene breccia reach a thickness of 120 m. Middle and Lower Cretaceous sediments are slightly shaly mudstone and wackestone. The Upper Jurassic-Tithonian is represented by black bituminous shales, and the Kimmeridgian is made up of dolomites with mudstone-wackestone inclusions.

Structure. The structure is an elongated anticline oriented from northwest to southeast and influenced on its northwestern flank by a reverse fault (Figures 46 and 47).

Production. The reservoirs are found within Paleocene-Cretaceous breccia and Kimmeridgian-Jurassic dolomites. Initial production from Batab 1-A was 5718 BOPD and 4.9 MMCFGD from the 4615-4658 m Kimmeridgian-Jurassic interval. Recently Batab No. 23 was tested from the 4110-4235 m interval within the Upper Cretaceous-Paleocene breccia and yielded 12,312 BOPD and 10.4 MMCFGD.

Caan Field

Location. The Caan field is 72 km N 22° W of Ciudad del Carmen, Campeche.

Stratigraphy. The Tertiary column is 3600 m thick. Breccia formations with a thickness of 275 m are similar to the producing ones at the other fields in this area. The remaining part of the Cretaceous is made up of shaly mudstone-wackestone inclusions. Bituminous shales are found in the Jurassic-Tithonian, dolomites in the Kimmeridgian, and limonites, sandstone, and shales in the Oxfordian, with drilling depths of 105, 420, and more than 400 m, respectively.

Structure. The structure is made up of four blocks and four different tops due to the effects of two faults—one reverse fault with a northwest-southeast orientation and one normal fault oriented from north to south (Figures 48 and 49).

Production. Producing rock formations are from the Cretaceous-Paleocene. The Caan No. 1 well initially produced, from the 3645–3675 m interval at a top of 51 kg/cm^2 and by way of a ¼-in. choke, 5439 BOPD and 1.3 MMCFGD.

Chuc Field

Location. The Chuc field is 73 km N 41° W of Ciudad del Carmen, Campeche.

Stratigraphy. A Tertiary sedimentary column 3000 m thick is followed by a Cretaceous-Paleocene breccia 125 m thick and then a sequence of Middle and Lower Cretaceous dolomites. The Upper Jurassic, Tithonian, and Kimmeridgian are quite similar to those described for the Caan field. Under the Kimmeridgian-Jurassic, salt was found in wells 1 and 101. In addition, various thicknesses of salt have been drilled into at wells 3, 4, 11, 14A, 22, 23, 31, 34A, 61, 62, 64, 81, 83, 84, 192, and 193 within Miocene and Oligocene sediments.

Structure. The structure is an anticline oriented from northeast to southwest with two well-defined structural tops. Two normal faults run parallel at the crest, with the same orientation as that of the structure (Figures 50 and 51).

Production. Production comes from Cretaceous-Paleocene breccia and Lower Cretaceous dolomites. The Chuc No. 1 well had the considerable initial production rate of 37,206 BOPD from the 3860–3925 m Lower Cretaceous interval.

Ek Field

Location. The Ek field is 90.8 km N 11° W of Ciudad del Carmen, Campeche.

Stratigraphy. The Tertiary sedimentary column is 3000 m thick and is followed by Cretaceous-Paleocene breccia.

Structure. The structure is an anticline with a northwest-southeast orientation, approximately 6 km long by 3 km wide. It is bordered on the north and east by a reverse fault (Figures 52 and 53).

Production. The discovery well was drilled in 1980. At the beginning of 1990, a fixed platform was installed in order to develop this field. At present, Ek No. 101 well is being drilled. The reservoir is in the Paleocene-Cretaceous breccia.

Ku Field

Location. The Ku field is 110 km N 21°W of Ciudad del Carmen, Campeche.

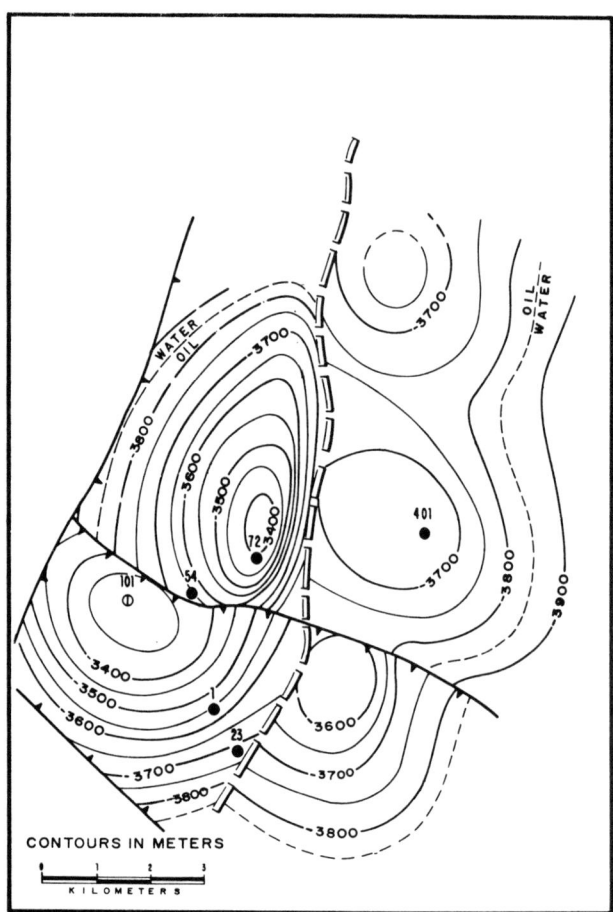

FIGURE 48. Caan field, top of breccia (Lower Paleocene-Upper Cretaceous).

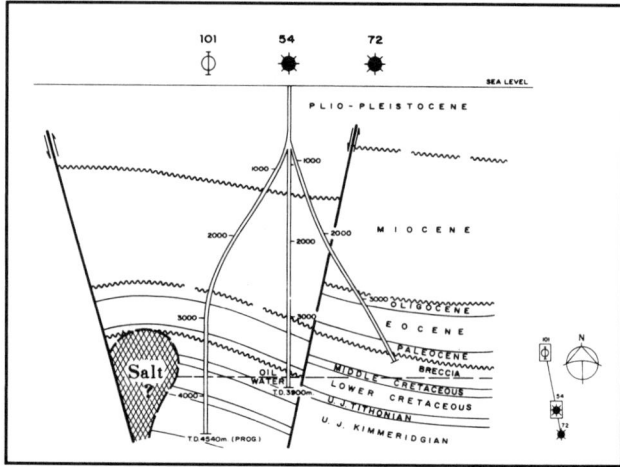

FIGURE 49. Geologic cross section, Caan field.

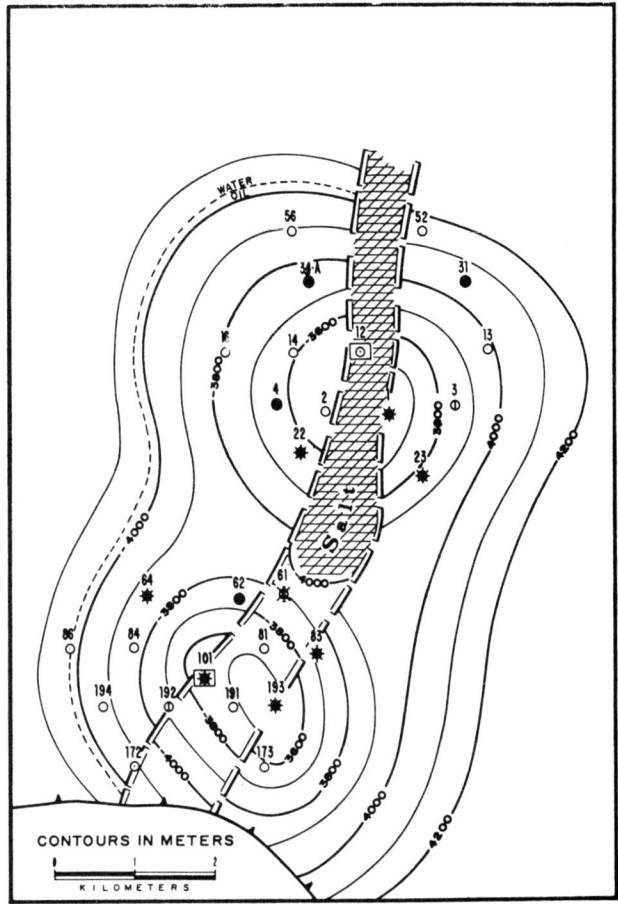

FIGURE 50. Chuc field, top of breccia (Paleocene–Upper Cretaceous).

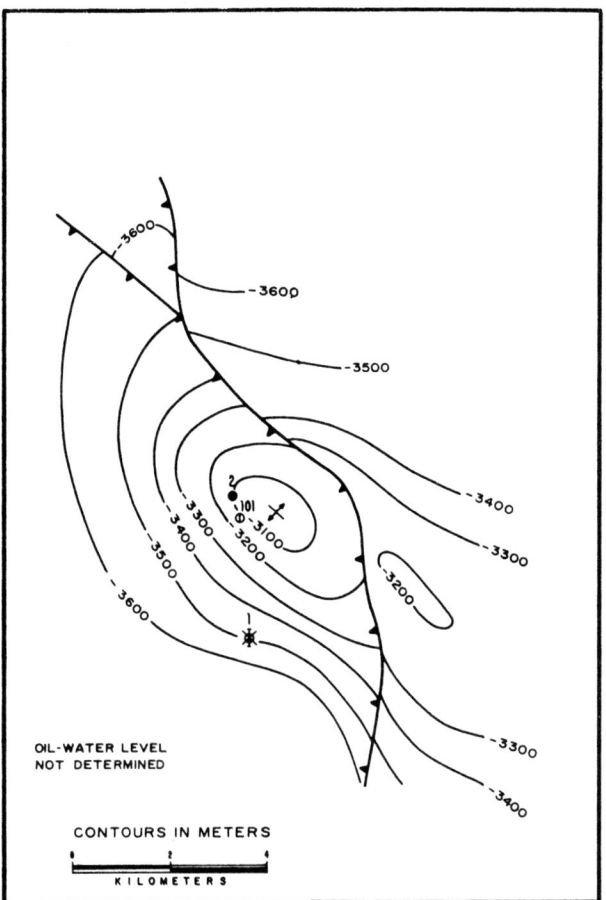

FIGURE 52. Ek field, top of breccia (Lower Paleocene–Upper Cretaceous).

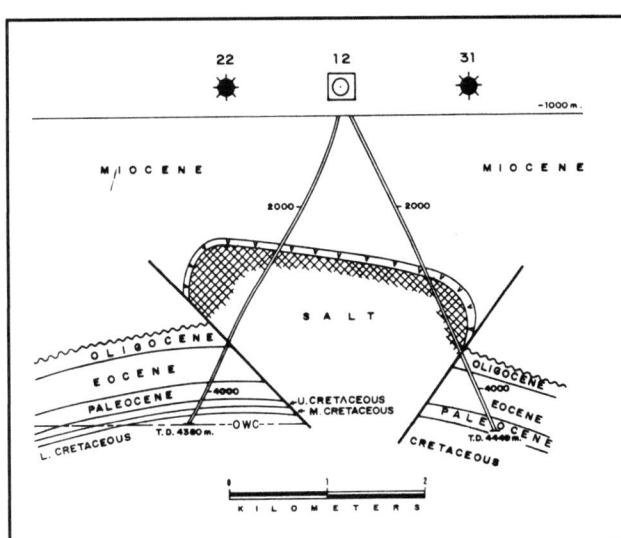

FIGURE 51. Geologic cross section, Chuc field.

Stratigraphy. It is particularly interesting that the Tertiary sequence in this structure contains Eocene calcarenites, with impregnation of hydrocarbons, at an average thickness of 110 m. The Cretaceous is represented by breccia and dolomites.

Structure. The structure is an elongated anticline oriented from northwest to southeast and bordered on the north and east by a reverse fault (Figures 54 and 55).

Production. The main reservoir is at the Upper Cretaceous–Paleocene breccia, with an average thickness of 170 m. Middle and Lower Cretaceous dolomites are also producers, as are Eocene calcarenites in the northwestern part of the structure. (Eocene calcarenites are also present at Cantarel field but have not yet been tested there.) The Ku No. 1 well initially produced 2021 BOPD and 9.1 MMCFGD from breccia. At present, there are wells in this field producing from 14,000 to 15,000 BOPD. The only producing well from Eocene calcarenites is Ku No. 10, which had an initial production of 2866 bbl of heavy oil per day.

Maloob Field

Location. The Maloob field is 125 km N 18° W of Ciudad del Carmen, Campeche.

Stratigraphy. The stratigraphic Cretaceous column of this field is practically the same as that described for Ku field.

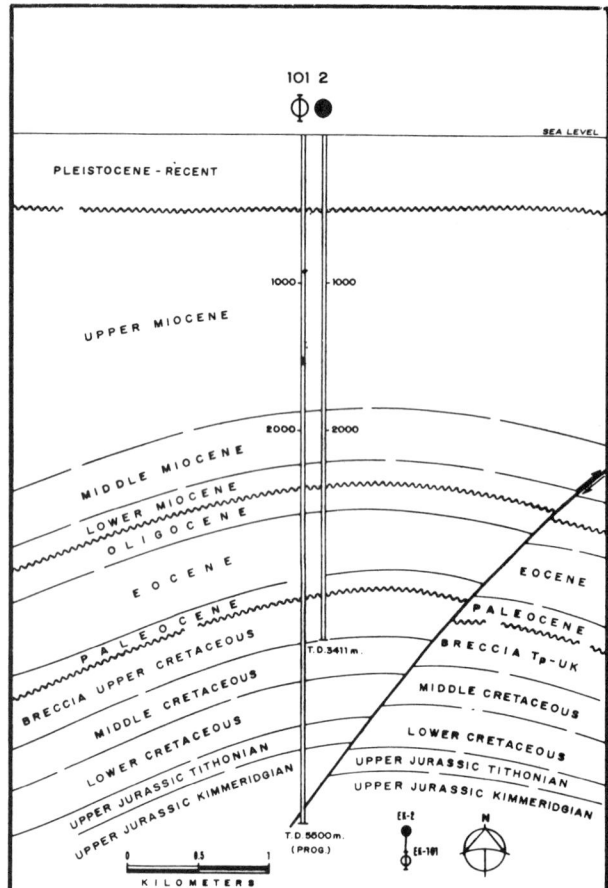

FIGURE 53. Geologic cross section, Ek field.

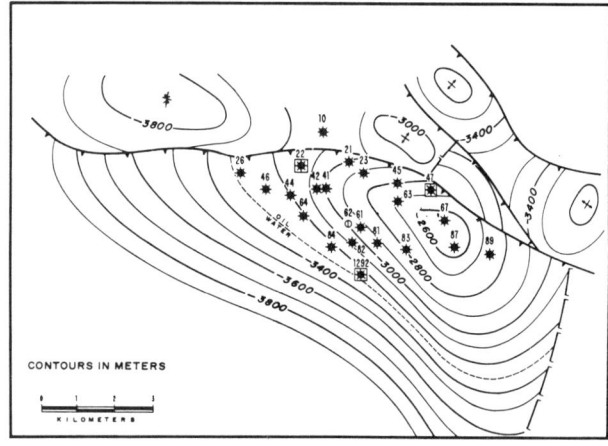

FIGURE 54. Ku field, top of Middle Cretaceous.

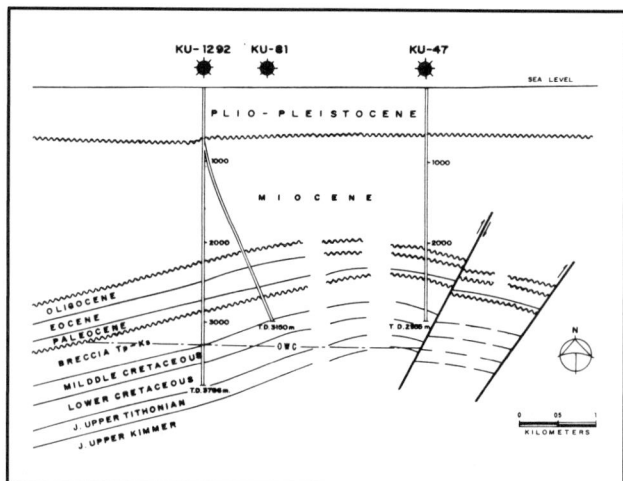

FIGURE 55. Geologic cross section, Ku field.

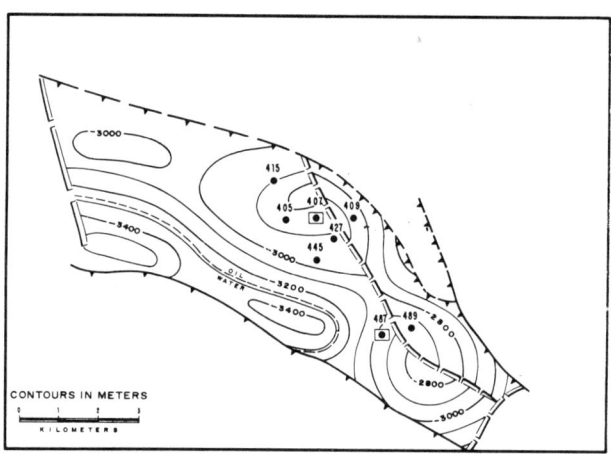

FIGURE 56. Maloob field, top of breccia (Lower Paleocene–Upper Cretaceous).

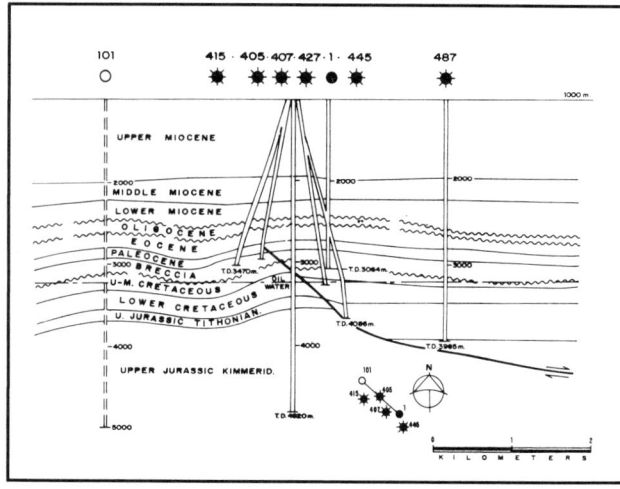

FIGURE 57. Geologic cross section, Maloob field.

Structure. The structure is an asymmetric anticline with its main axis oriented from northwest to southeast, bordered on the north and east by a reverse fault (Figures 56 and 57).

Production. Production comes from Cretaceous-Paleocene breccia. Initially, the Maloob No. 407 well produced 5844 BOPD and 955 MCFGD from a 3040–3107 m interval.

Pol Field

Location. The Pol field is 80 km N 41° W of Ciudad del Carmen, Campeche.

Stratigraphy. The thickness of Tertiary sediments over this structure is 3800 m. The remaining stratigraphy is typically represented by Cretaceous-Paleocene breccia, Middle and Lower Cretaceous dolomites, Tithonian bituminous shales, and Kimmeridgian dolomites.

Structure. The structure is a slightly asymmetric anticline with its main axis oriented from east to west. There are no faulting effects, which is a rare occurrence in this area (Figures 58 and 59).

Production. Production comes from Cretaceous-Paleocene breccia, Middle Cretaceous dolomites, and Kimmeridgian-Jurassic dolomites. Average production per well is about 3900 BOPD and 3.5 MMCFGD for the Kimmeridgian, 9000 BOPD and 19.7 MMCFGD for the Middle Cretaceous, and 6250 BOPD and 6.6 MMCFGD for the breccia, although there are wells that produce up to 19,500 BOPD.

Uech Field

Location. The Uech field is 91 km N 57° W of Ciudad del Carmen, Campeche.

Stratigraphy. Of all the fields in the Sonda de Campeche area, the Uech field exhibits the greatest thickness of Tertiary sediments (up to 4000 m). The Upper Cretaceous is represented by mudstone-wackestone, the Middle and Lower Cretaceous by wackestone-packstone, the Tithonian-Jurassic by bituminous black shale, and the Kimmeridgian-Jurassic by oolitic dolomites, which are quite porous and permeable.

Structure. The Uech field consists of a structural block that has been raised by the effects of two reverse faults oriented from northwest to southeast (Figures 60 and 61).

Production. The producing rock formations are Kimmeridgian oolitic dolomites. Within the Middle Cretaceous, some intervals of dolomites with impregnations have been recorded, although they have not yet been tested. Initial production from the Uech No. 1 well was 9000 BOPD (38° API) and 16 MMCFGD from the 4909–4930 m interval. Figures 62 and 63 show the variations in API gravity for oil found in the Chiapas-Tabasco and Sonda de Campeche areas, respectively.

CONCLUSIONS

In Mexico, in spite of a reduction in the resources allotted to exploration, successes have been achieved in the discovery of giant and supergiant oil fields, both on land and offshore. Attempts are being made to continue the exploration in new areas which offer the possibility of discovering fields that will replace those already under production.

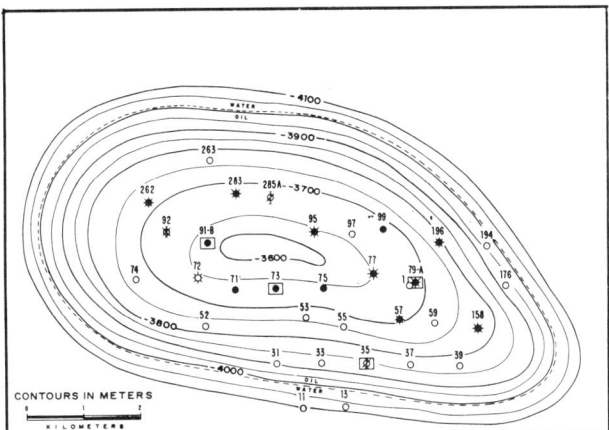

FIGURE 58. Pol field, top of breccia (Paleocene-Upper Cretaceous).

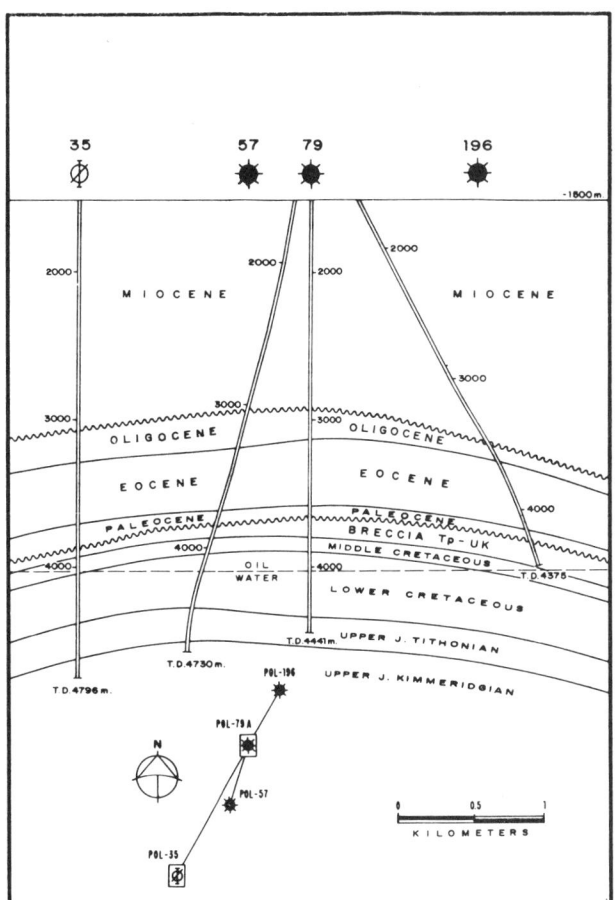

FIGURE 59. Geologic cross section, Pol field.

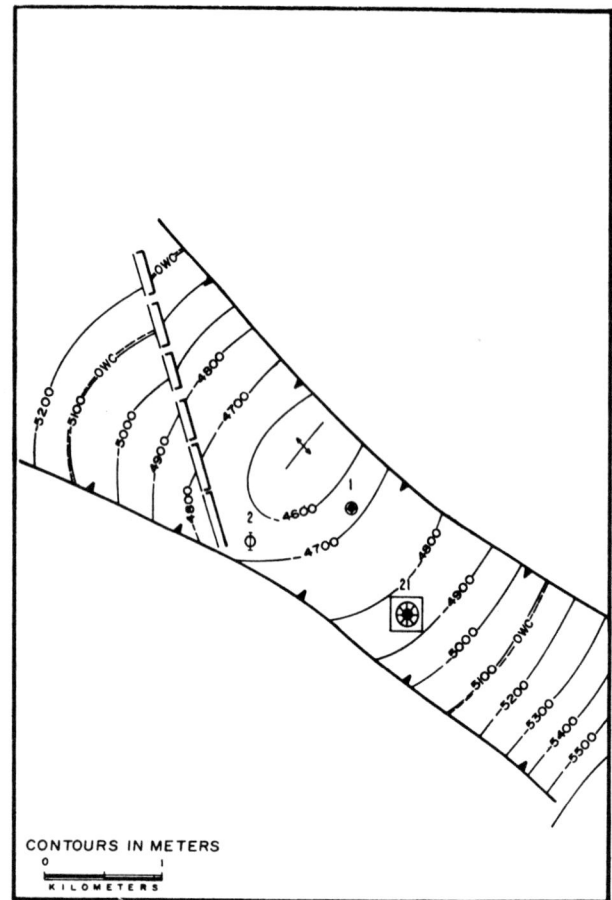

FIGURE 60. Uech field, top of Upper Jurassic (Kimmeridgian).

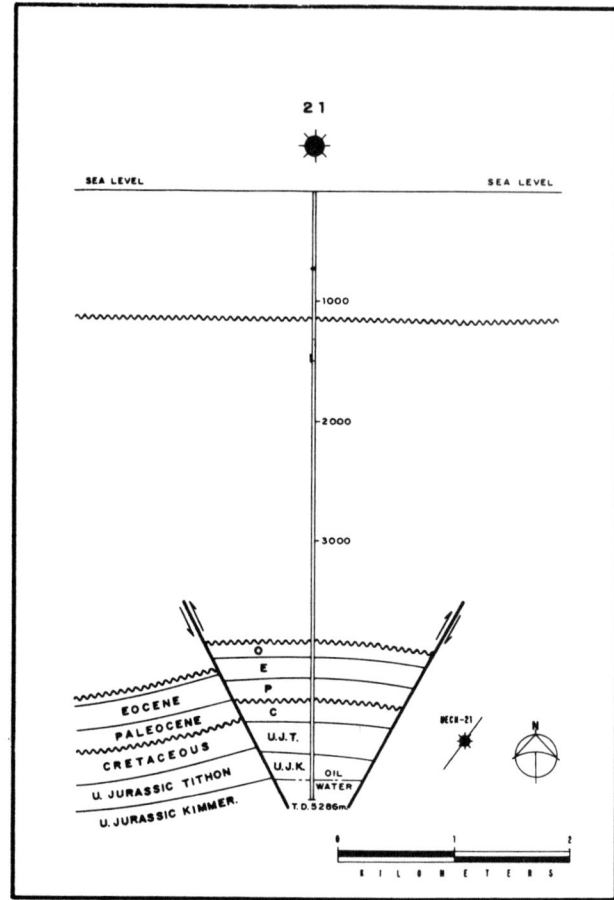

FIGURE 61. Geologic cross section, Uech field.

REFERENCES CITED

Benavides, G. L., 1956, Notas sobre la Geología Petrolera de México, Simposium sobre Yacimientos de Petróleo y Gas, Tomo III, América del Norte, p. 509-510.

Bortolotti, A. C., 1978 (unpublished), IX Excursión Geológica, Petróleos Mexicanos, III Excursión Geológica al Sureste de México, Primera Etapa, Tramo Palenque-Agua Azul, Chiapas, Superintendencia General de Distritos de Exploración, Zona Sur, p. 1-18.

Buffler, T., J. Watkins, F. Shaub, and J. Worzel, 1980, Structure and early geologic history of the deep central Gulf of México, in R. H. Pilger (ed.), The early opening of the central North Atlantic Ocean, Louisiana State University, Baton Rouge, p. 3-16.

Burkart, B., 1978, Offset across the Polochic fault of Guatemala and Chiapas, México: Geology, v. 6. p. 328-332.

Carfantàn, J. C., 1976b, El Prolongamiento del Sistema Polochic-Motagua en el Sureste de México, Una Frontera Entre Dos Provincias Geológicas, Congreso Latinoamericano de Geología 3, Memorias.

Coney. P., 19—, Un modelo tectónico de México y sus relaciones con América del Norte, América del Sur y el Caribe. Revista del Instituto Mexicano del Petróleo, XV, n. 1, p. 6-15.

Dickinson, R., and P. Coney, 1980, Plate tectonics constraints on the origin of the Gulf of México, in R. H. Pilger (ed.), The early opening of the central North Atlantic Ocean, Louisiana State University, Baton Rouge, p. 27-36.

Espinoza, L. L., 1979 (unpublished), IX Excursión Geológica, Petróleos Mexicanos, III Excursión Geológica al Sureste de México, Tercera Etapa, tramo Ocosingo San Cristóbal, Chiapas, Superintendencia General de Distritos de Exploración, Zona Sur, p. 31-37.

Ham-Wong, J. M., 1979 (unpublished), Prospecto Nazareth Chiapas Informe Geológico No. 745, Petróleos Mexicanos, Zona Sur, p. 8-18.

Ham-Wong, J. M., 1986 (unpublished), Geología de la Provincia Geológica de la Sierra de Chiapas, Petroleos Mexicanos, Zona Sureste, p. 185-233.

Jordan, T. H., 1975, The present-day motions of the Caribe Plate: Journal of Geophysical Reserach, v. 80, p. 433-439.

Le Pichon, X., 1978, Sea floor spreading and continental drift: Journal of Geophysical Research, v. 73, p. 3661-3697.

López R. E., 1973, Estudio geológico de la Península de Yucatán: Boletín de la Asociación Mexicana de Geólogos Petroleros, v. XXV, p. 1-22.

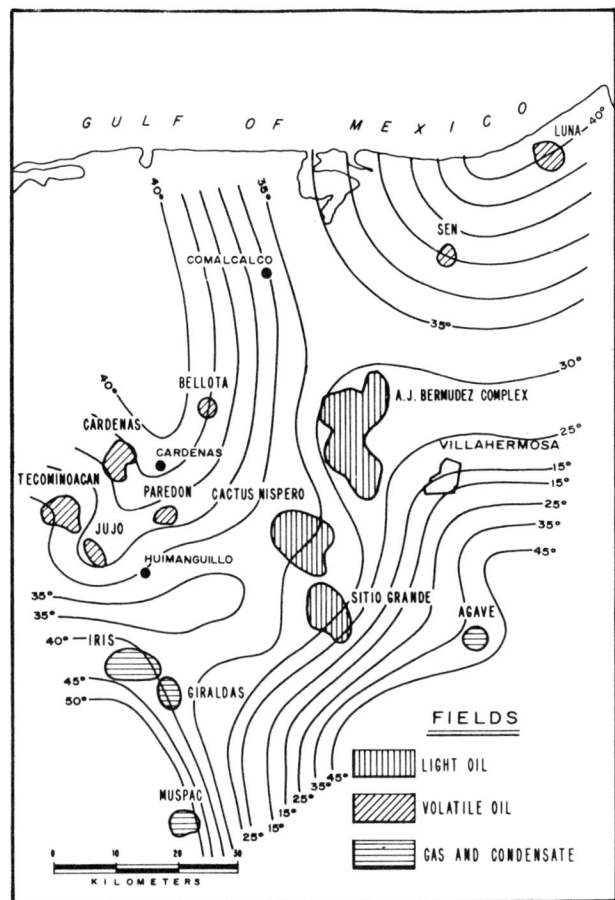

FIGURE 62. Variation in API gravity, Chiapas-Tabasco area.

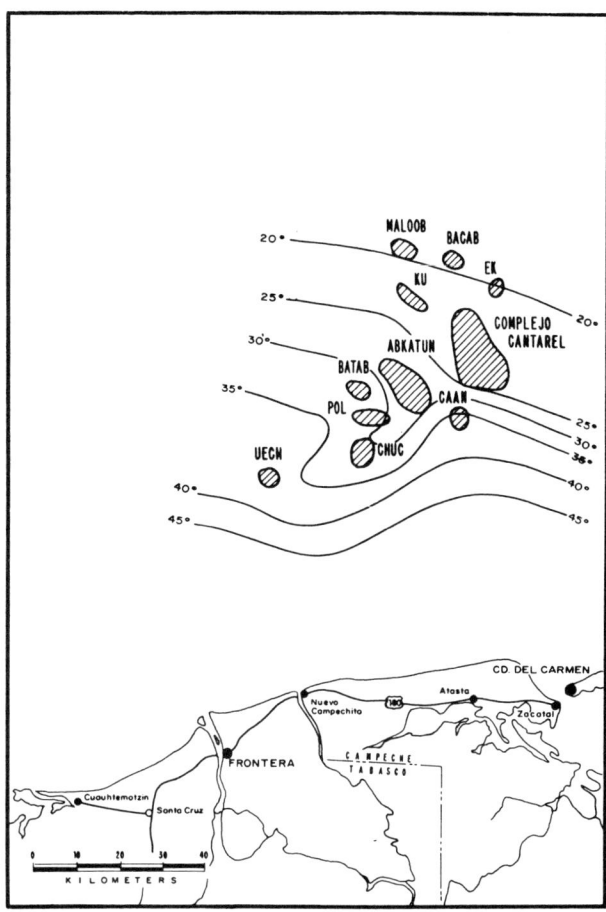

FIGURE 63. Variation in API gravity, Sonda de Campeche area.

Lopez, R. E., 1981, Paleogeografía y tectónica del Mesozoico en Mexico: Instituto de Geología, Universidad Nacional Autónoma de México, v. 5, n. 2, p. 158-174.

Mafait and Dikelman, 1972, Circum Caribbean tectonic and igneous activity and the evolution of the Caribbean plate: GSA Bulletin, v. 83, p. 251-272.

Petróleos Mexicanos (diverse authors), 1964-88 (unpublished), Internal reports and information from Subsurface Geology Dept.

Pindell, J. L., 1985, Alleghenian reconstruction and subsequent evolution of the Gulf of México, Bahamas and Proto-Caribbean: Tectonics, v. 4, n. 1, p. 1-39.

Quezada, M. J. M., 1990, El Cretácico Medio-Superior y el Límite Cretácico Superior-Terciario Inferior en la Sierra de Chiapas: Boletín de la Asociación Mexicana de Geólogos Petroleros, v. XXXIX, n. 1.

Sánchez M. de O., R., 1970 (unpublished), Geología de la Sierra de Chiapas, IX Excursión Geológica, Petróleos Mexicanos, III Excursión Geológica al Sureste de México, Superintendencia General de Distritos de Exploración, Zona Sur, p. 1-19.

Sanchez M. de O., R., 1976b, Geología Petrolera de la Sierra de Chiapas: Boletín de la Asociación Mexicana de Geologos Petroleros, v. XXXI, n. 2 a 6, p. 67-96.

Sanchez M. de O., R. 1986 (unpublished), Marco Geológico Regional de la Zona Sureste de Petróleos Mexicanos, Superintendencia General de Distritos de Exploración, Zona Sureste, p. 7-50.

Santiago A. J., 1980, Giant fields of the Southern Zone México: AAPG Memoir 30.

Vinson L. G., 1962, Upper Cretaceous and Tertiary stratigraphy of Guatemala: AAPG Bulletin, v. 46, n. 4, p. 437-447.

Winker, C. D., and T. Buffler, 1988, Paleogeographic evolution of early deep water Gulf of Mexico and margins Jurassic to Middle Cretaceous (Commanchean): AAPG Bulletin, v. 72, n. 3, p. 318-346.

Chapter 7

The Tengiz Oil Field in the Pre-Caspian Basin of Kazakhstan (Former USSR)—Supergiant of the 1980s

Nicolai N. Lisovsky

USSR Ministry of Oil and Gas Industry
Moscow, Russia

Georgii N. Gogonenkov

Central Geophysical Expedition (CGE),
USSR Ministry of Oil and Gas Industry
Moscow, Russia

Yuri A. Petzoukha

Institute of the Geology and Development of Combustible Fuels (IGIRGI),
USSR Ministry of Oil and Gas Industry, and USSR Academy of Sciences
Moscow, Russia

INTRODUCTION

Oil and gas explorationists have amassed more than a century of experience, and their efforts have led to the discovery of tens of thousands of fields in oil and gas basins throughout the world. However, the number of giants and supergiants among these fields has been extremely small.

The discovery of a supergiant is always a noteworthy event in petroleum geology. It provides a strong impetus not only to the development of the oil-producing district, but also to the economic development of the region and the country.

The Soviet Union is a major producer of hydrocarbons and for many decades has been conducting intensive oil and gas prospecting and exploration. The geographic focus of exploration and the concentration of production have changed significantly over this period (Figure 1). For instance, up until World War II, the Caucasus, chiefly the region around Baku, was the primary producing region. Then several fields, among them the Romashkino supergiant, were discovered in the area between the Volga River and the Urals.

Thanks to advances in exploration, the production level in the Volga-Urals oil and gas province rose to 1.7 billion bbl (240 million MT) per year. The next major advance in the establishment of hydrocarbon reserves occurred after the discovery of the very large oil and gas province in western Siberia, where numerous fields were identified, among them the Samotlor oil supergiant and the Urengoi gas supergiant. With this discovery, annual output in the Soviet Union rose to 4.2 billion bbl (600 million MT) of petroleum and 28 tcf (800 billion m^3) of gas.

The pre-Caspian depression long ago attracted the attention of petroleum geologists. Early exploration was conducted here for relatively shallow reservoirs within reach of what was then state-of-the-art technology. The improvement of the technical level of exploration—above all, the conversion to multiple-fold coverage and digital processing of seismic data—and the drilling of wells to depths of 5 to 6 km made it possible to explore the lower portion of the sedimentary section and to identify several very large gas and condensate fields (Astrakhan, Orenburg, Karachaganak, and others) along the margin of the basin.

This chapter is devoted to the Tengiz oil field in the pre-Caspian basin. By world standards, with in-place reserves estimated at 25 billion bbl (3.5 billion MT) of oil and 46 tcf (1.3 trillion m^3) of gas, this field is the supergiant of the 1980s. Tengiz is located in desert terrain at the

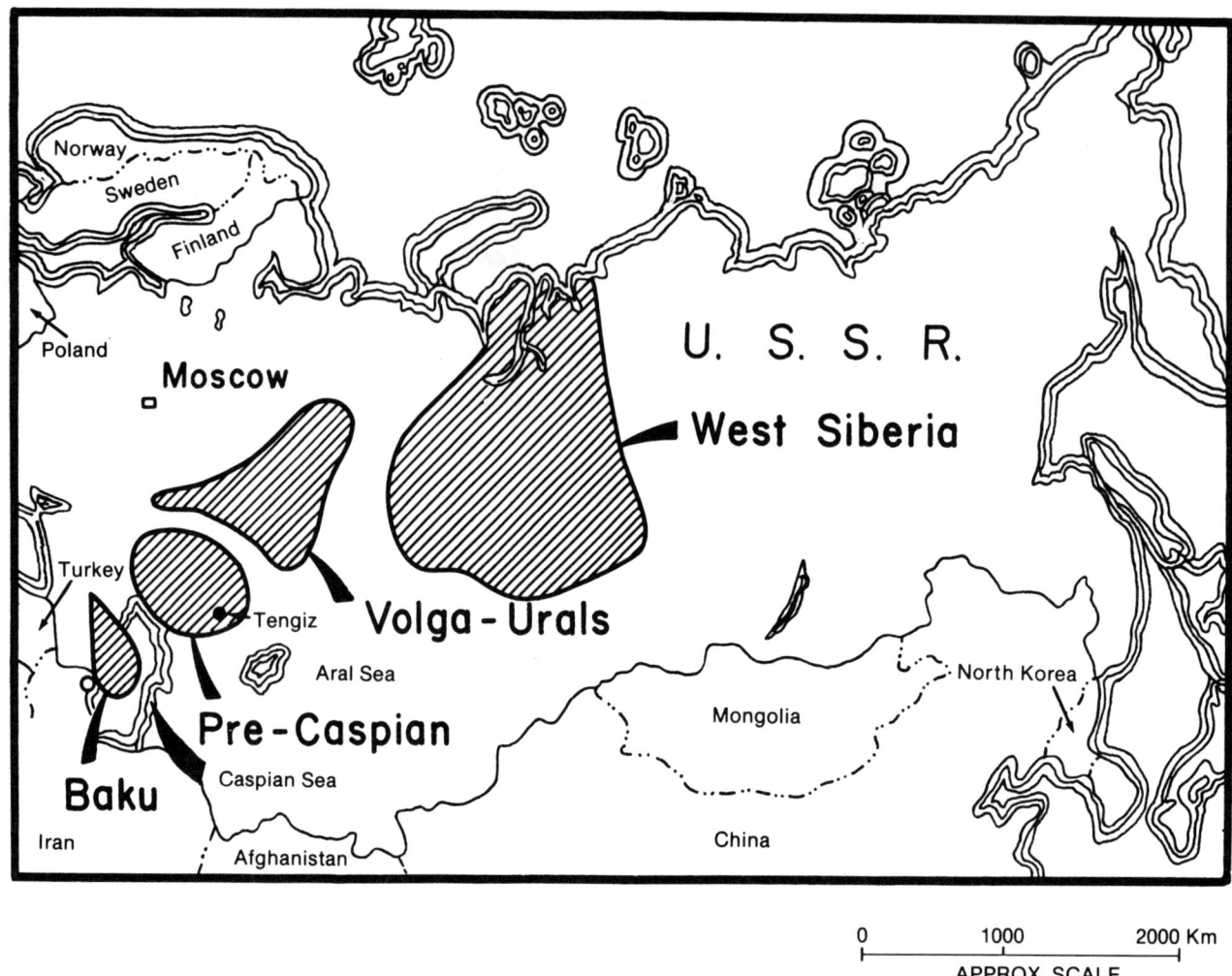

FIGURE 1. Basins (cross-hatched) in which early Soviet oil exploration was conducted, showing the location of the Tengiz field.

edge of the Caspian Sea, where, in addition to extreme seasonal temperature variations, other severe environmental conditions exist.

GENERAL FEATURES OF THE GEOLOGIC STRUCTURE OF THE PRE-CASPIAN BASIN

The pre-Caspian basin is one of the largest depressions in the world, spanning an area of more than 500,000 km² (Figure 2). It is bordered to the north and west by elevated structural elements of the southeastern Russian platform, and to the east, southeast, and south by Hercynides fold belts. The Astrakhan-Aktyubinsk uplift system divides the depression basin into the central pre-Caspian basin and the southeastern basin.

During submergence over the period of hundreds of millions of years from the upper Proterozoic-lower Paleozoic to the present, the pre-Caspian basin has accumulated an enormous sequence of sediments ranging in thickness from 5 km in the marginal zones to 20 to 24 km in the central region.

The sedimentary section of the basin is divided into two megacomplexes by the Permian Kungurian salt formation (Figure 3):

1. The post-salt megacomplex, composed principally of terrigenous deposits of the Upper Permian, Mesozoic, and Cenozoic and considerably deformed by salt tectonics.

2. The pre-salt Paleozoic and upper Proterozoic megacomplex, characterized by a thick carbonate section and, in part, by terrigenous sediments.

According to refraction seismic data, the depth to basement within the pre-Caspian depression varies considerably because of a system of local ridges and depressions up to several kilometers in relief (Figure 4). Prior to the deposition of the Kungurian salt, this relief was largely buried beneath Paleozoic deposits. At that time, the Paleozoic basin was much wider than it is now, and it included extensive areas of the southeastern part of the Russian platform and contiguous Hercynian geosynclinal

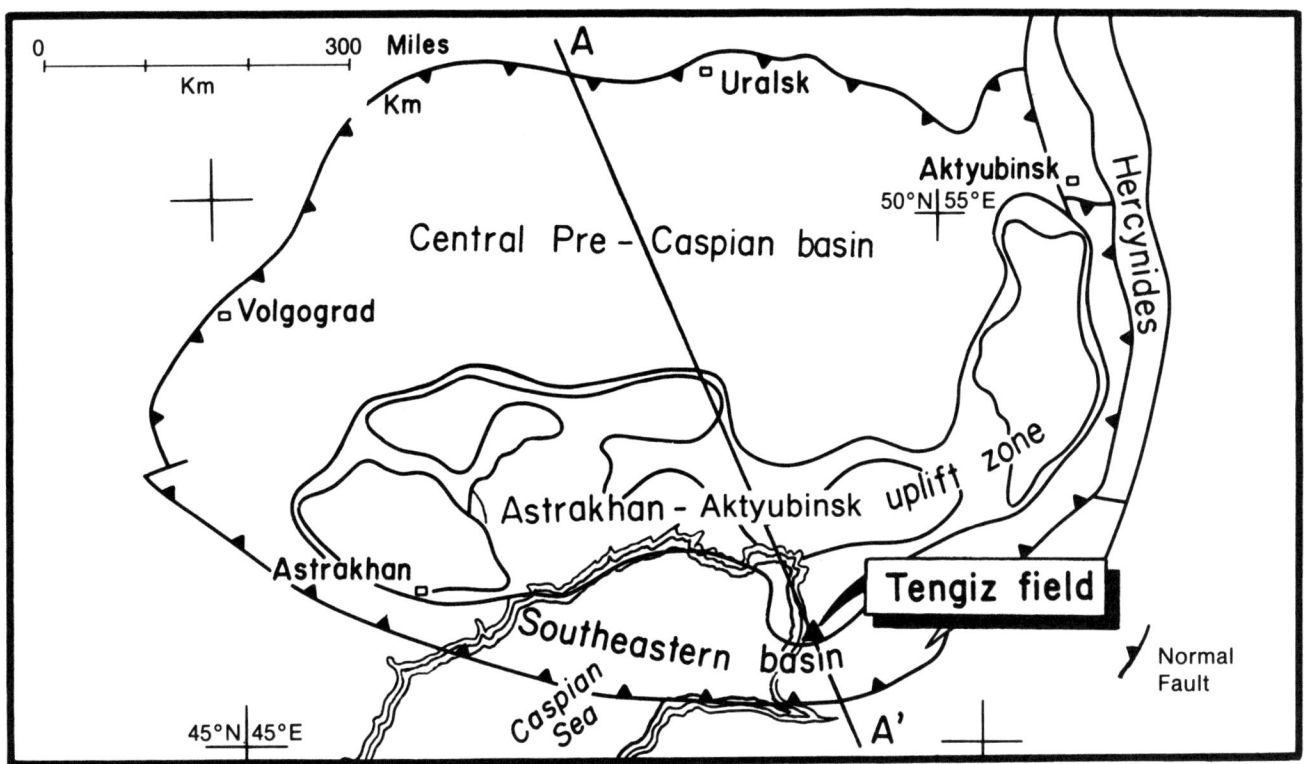

FIGURE 2. Tectonic features of the pre-Caspian basin, mainly just to the north of the present position of the Caspian Sea. Cross section A-A' is shown in Figure 4.

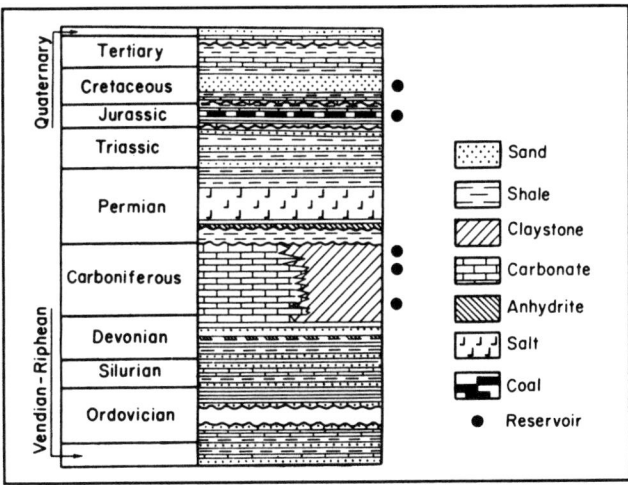

FIGURE 3. Generalized stratigraphic column for the pre-Caspian basin.

zones (Figure 2). As a result, the marginal zones of the basin, which reflect the transition from the shelf portion of the paleobasin to the deep water portion of the basin, have an unusually diverse structure and history of development (Figure 4).

The Paleozoic sediments under the salt are of greatest interest in the pre-Caspian basin from the standpoint of potential oil and gas accumulations. The regional productivity of these sedimentary deposits has been confirmed by the discovery of such very large oil and gas fields as Astrakhan, Karachaganak, and Zhanazhol. The thickness of the Paleozoic deposits in the pre-Caspian basin amounts to nearly half the thickness of the entire sedimentary section, ranging from 6 to 15 km in different areas. Until now, drilling has penetrated only the upper Paleozoic sediments, composed mainly of Carboniferous and Upper Devonian carbonates. Within the Astrakhan-Aktyubinsk uplift zone, the thickness of the Paleozoic deposits decreases sharply to 2 to 3 km (Figure 4). Farther to the south, the thickness of sediments deposited during late Proterozoic-Paleozoic times increased once again to 6 to 8 km (Figure 4).

Characteristically, carbonate zones composed of biogenic shelf limestones almost always occur at the peripheries of large uplifted structural elements. In the southeastern basin of the pre-Caspian depression is the Karaton-Prorva carbonate buildup. It is located on the southern slope of the Guryev arch and runs from north to south along the northeastern shore of the Caspian Sea for more than 150 km. Several high-amplitude (high-relief) structures have been identified within the carbonate buildup, one of which is Tengiz.

HISTORY OF THE FIELD DISCOVERY

The first exploratory wells in the southeastern basin of the pre-Caspian depression were drilled in the late 1960s and

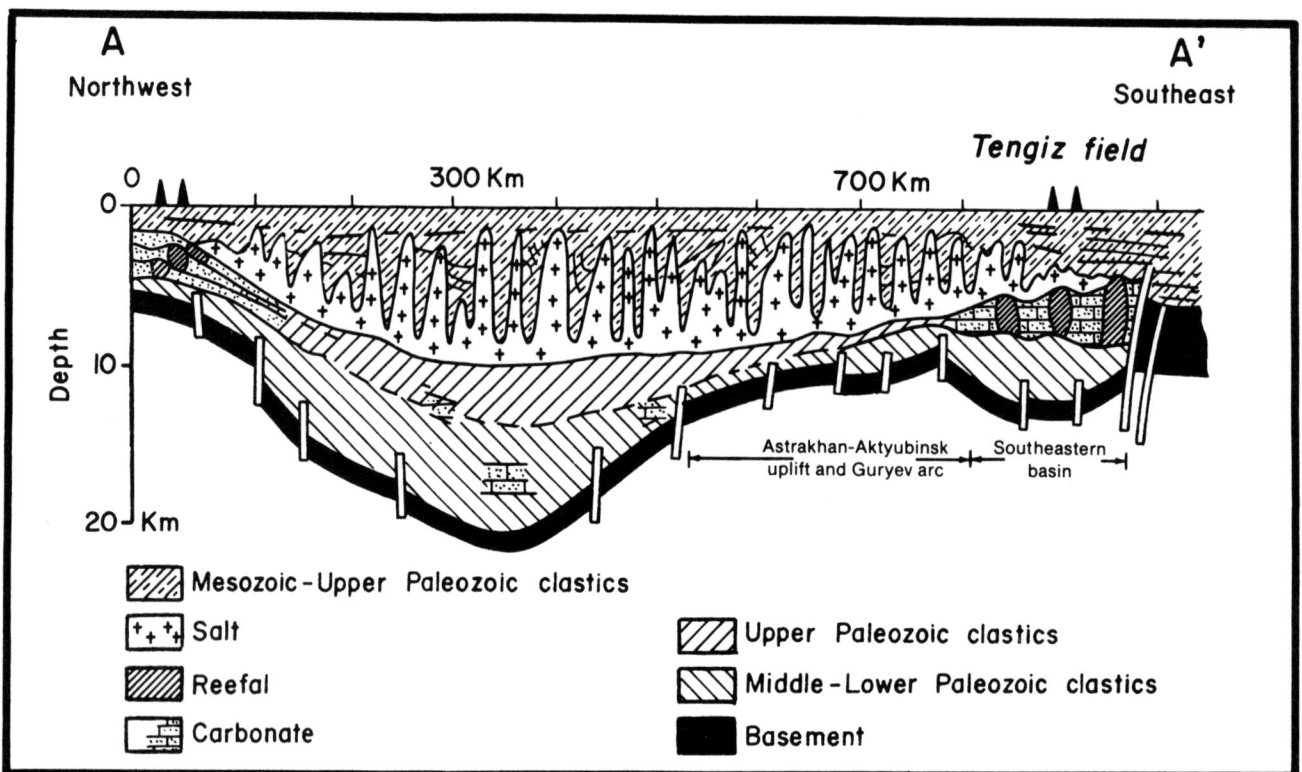

FIGURE 4. Generalized cross section through the pre-Caspian basin, showing the Karaton-Prorva carbonate buildup in the southeastern basin. The Tengiz field is located in this carbonate buildup zone. The location of cross section A-A' is shown in Figure 2.

early 1970s at Karaton. They penetrated Paleozoic deposits under the salt, and, when tested, these deposits produced strong flows of water saturated with hydrocarbon gases. The wells identified the presence of thick biogenic carbonates with good reservoir characteristics and the likely presence of hydrocarbons.

These data inspired optimism, and exploration strategy was directed toward detection of definite trap prospects, including effective seals. Unfortunately, reflection seismic surveys conducted in the 1960s without multiple-fold coverage did not provide accurate structural information on the pre-salt formations.

First Stage

In 1975, following the changeover to multiple-fold seismic surveys utilizing common-depth-point (CDP) stacking, the primary base of salt-reflecting horizon P_1 in the pre-Caspian basin was mapped for the first time. In the process, south of the previously identified Karaton uplift, the Tengiz structure, a major independent uplift with an amplitude of over 1000 m and an area of more than 400 km², was detected at a depth of about 4000 m (Figure 5). Figure 6 shows one of the early seismic sections.

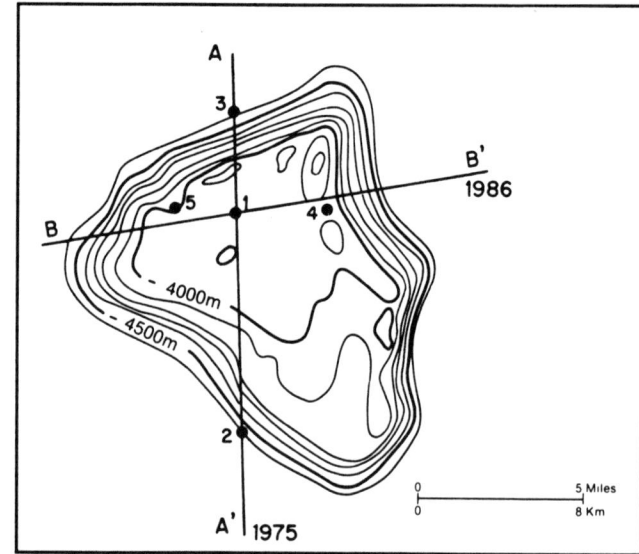

FIGURE 5. Configuration of Lower Permian P_1 base salt in the Tengiz structure. See Figure 8 for stratigraphic position of P_1. Seismic section A-A' (1975 survey) is shown in Figure 6. Section B-B' is the 1986 seismic line shown in Figure 7. Contour interval, 100 m.

The Tengiz Oil Field in the Pre-Caspian Basin of Kazakhstan 105

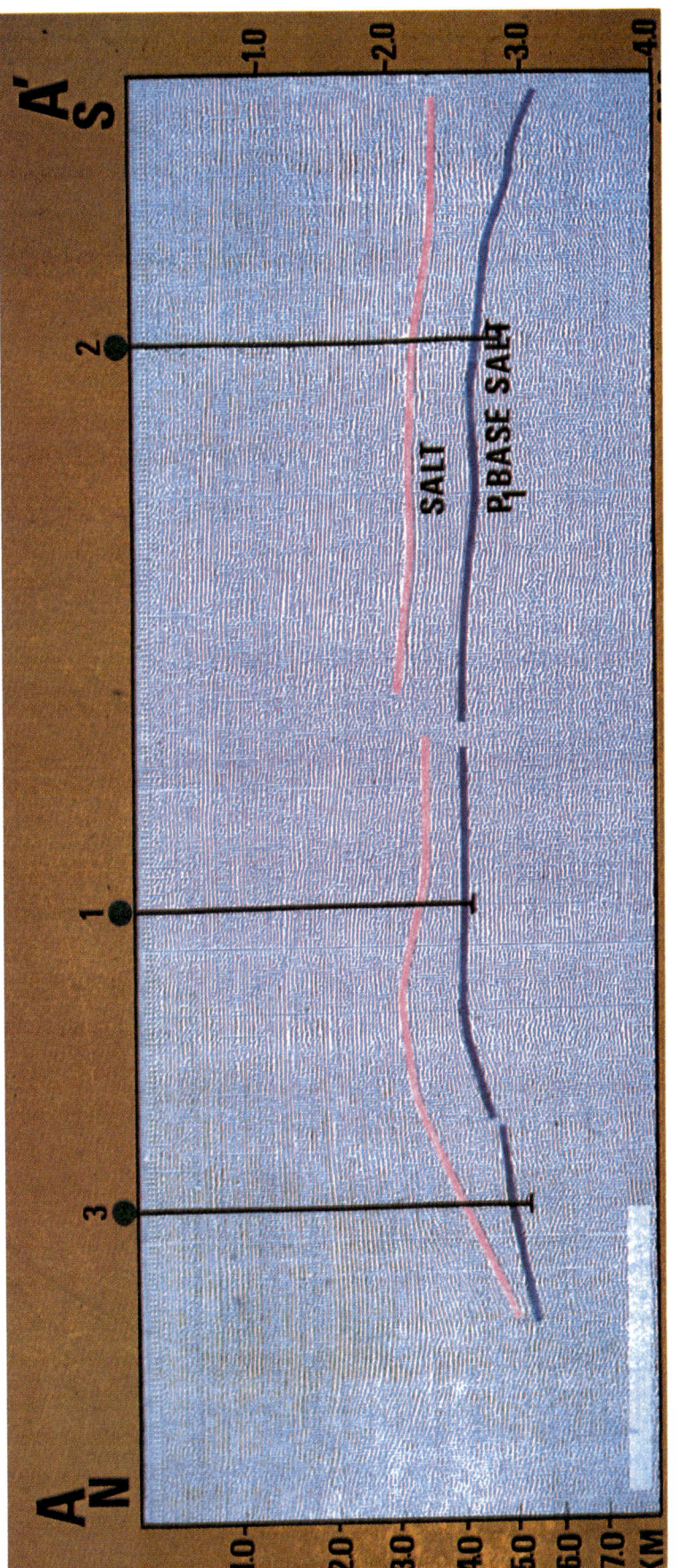

FIGURE 6. Early seismic section of the Tengiz structure. Computer processing of 1975 CDP seismic data identified the Tengiz structure although the quality of the original data was not high. The location of section A-A' is shown in Figure 5. Section length, 25 km.

As shown, the quality of the seismic data—analyzed through only the most primitive computer processing—is not very good. Considerable skill was required on the part of geophysicists to interpret the seismic data and give a three-dimensional picture of the pre-salt Tengiz structure.

Second Stage

The first exploratory well was drilled on the crest of the Tengiz structure in 1979. Testing yielded a significant oil flow and proved the high productivity of middle Carboniferous carbonates capped by Lower Permian clays and the massive Permian salt. Four more wells were drilled immediately in a cross pattern running along the latitudinal and longitudinal directions of the structure (Figure 5).

At the same time, detailing of the field and adjacent territories was started by intensifying seismic field efforts (the coverage factor was increased from 12- to 24-fold, and the maximum offset was raised from 2400 to 3600 m), but the most significant improvement was accomplished by means of more sophisticated computer processing of the seismic data. As a result, there was a sharp increase in the reliability of the seismic data (Figure 7). Deeper reflections from 3.5 to 4.5 sec appeared and revealed a gentler structural configuration in the depth interval from 7.5 to 9 km. Furthermore, the resolution in the top portion of the pre-salt formations became considerably clearer and more differentiated. The nearly planar reflections in the central part of the structure, small ridges 30 to 50 m in size around the edges, and relatively steep rough flanks, judging by the number of diffracted waves, became distinctly visible. Finally, the undulating surface of the salt and the presence of faulting in the post-salt section caused by salt tectonics could be mapped much more accurately.

Analysis of formation velocities based on seismic reflections showed that, in the interval from 4.0 to 7.5 km, the formation velocities varied from 5.0 to 5.7 km/s, corresponding to those of limestones with variable porosity. The analysis also provided an estimate of the velocity in the underlying layer. Its value—5.25 km/s—indicates that this section probably is composed of terrigenous or terrigenous-carbonate rocks.

An integrated interpretation of seismic and gravity data turned out to be extremely informative. Seismic data provided the geometry of the principal density boundaries. The densities of the salt and of the rocks in the post-salt formation were known from drilling. The distribution of rock density within the carbonate body was obtained by using a special program for automatic matching of the observed and computed gravimetric fields in the three-dimensional model of the structure.

The average density of 2.45 g/cm^3, which is 0.17 g/cm^3 less than the average density of limestones, indicates that the carbonate body is characterized by significant porosity. Within the structure, zones of extremely low densities—as low as 2.35 g/cm^3—were identified around the periphery of the carbonate body, and they corresponded with the ridges previously noted on the seismic section.

In the early 1980s, three of the four outpost wells penetrated pre-salt formations and produced significant flows of water-free hydrocarbons (for technical reasons, the fourth outpost well did not produce any flow). Well No. 3 penetrated pre-salt formations and produced the same high-quality oil from a zone 700 m structurally lower than wells 1, 4, and 5 (Figure 5). It became clear that this was indeed a very large field.

Third Stage

The third stage of exploration of the Tengiz field began in 1983. The main feature of this stage was that, for the first time in exploration practice, the decision was made to drill 20 deep wells simultaneously, penetrating the reservoir to depths ranging from 4000 to 5500 m, and to conduct a three-dimensional seismic survey over the entire area of 580 km^2.

In the pay zone of the carbonate section, the presence of numerous zones of lost circulation, high H_2S content, and formation pressure nearly twice the normal hydrostatic pressure created major engineering difficulties and required the development of complex well designs. For these reasons, it still is not possible to penetrate the entire thickness of the carbonate buildup by drilling. The maximum penetrated thickness of the reservoir is 600 m. However, because some of the wells were drilled around the periphery of the structure, it has been possible to study the pre-salt deposits to a depth of 5400 m. Water-free hydrocarbons were obtained in all wells drilled, and, as yet, the oil-water contact has not been reached.

STRATIGRAPHY

The composite stratigraphic section of the Tengiz oil field (Figure 8) provides a generalized summary of the lithology and variability of this carbonate section. In the portion of the section that has been penetrated by drilling, the age determinations and correlations of the rocks are based on paleontological studies of brachiopods, foraminifera, and algae, and on palynological identification of spore types. Age determinations were combined with wireline log suites in order to correlate and compare individual horizons. The stratigraphic column is based on data not only from the Tengiz field, but also from wells in the adjacent area.

Devonian

At present, drilling has penetrated only the Upper Devonian Famennian Stage—37 m at Tengiz and about 400 m at the Karaton location (Figures 9, 10, and 11). These deposits are composed chiefly of a homogeneous biogenic limestone with sparse, thin dolomite interbeds. This section is part of a single shallow water carbonate platform buildup that began in the Upper Devonian and continued into the middle Carboniferous.

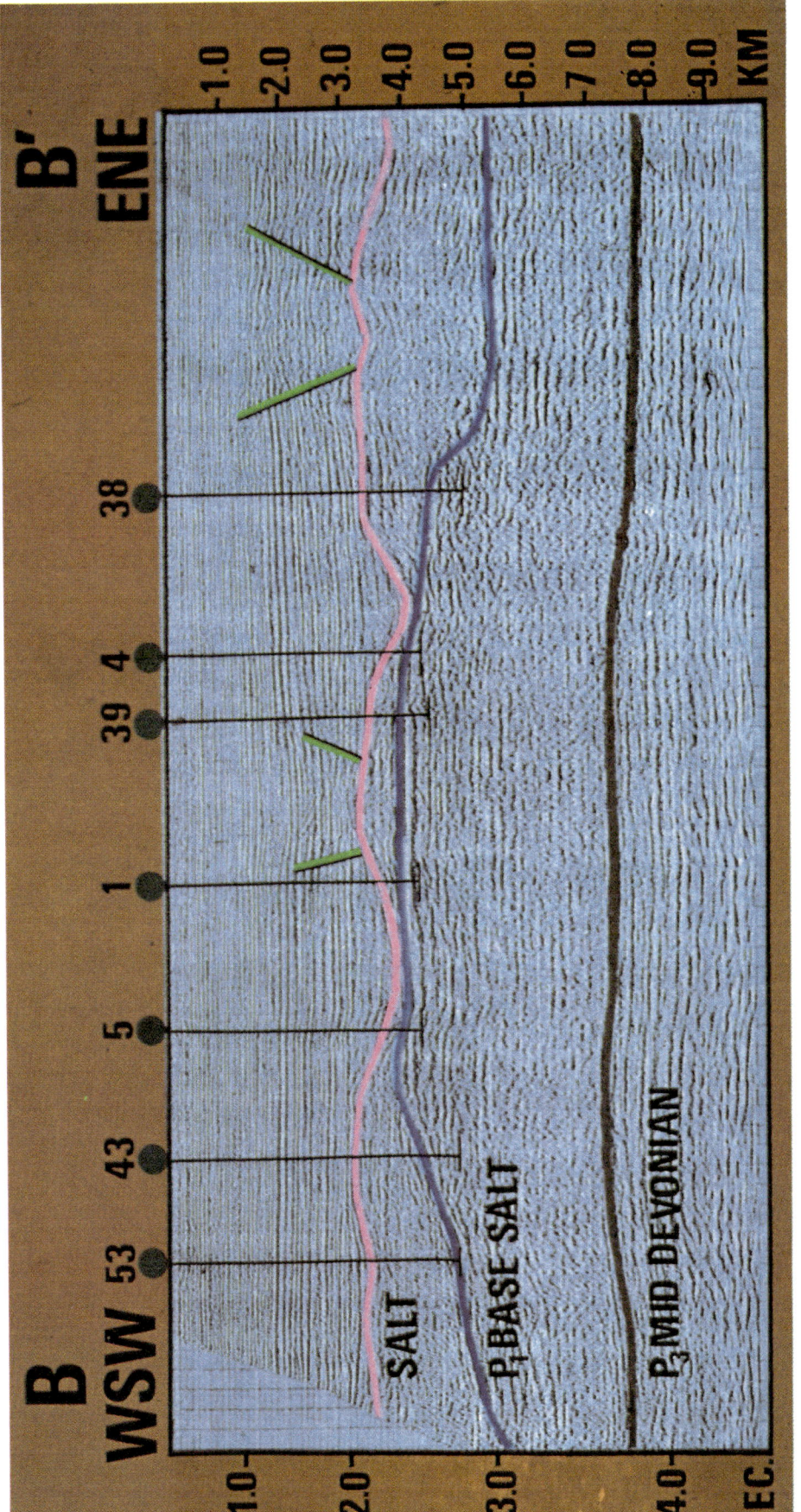

FIGURE 7. Tengiz structure, 1986 seismic section B-B'. The location of section B-B' is shown in Figure 5. Section length, about 25 km.

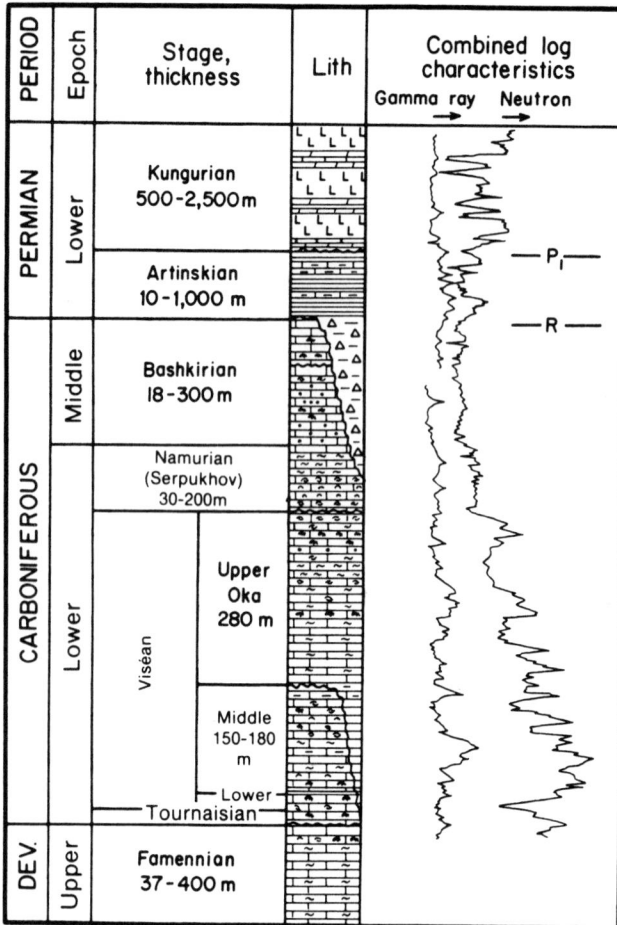

FIGURE 8. Stratigraphic column for the Tengiz structure. P_1 correlates with the base of the Permian salt and with the seismic reflector that was used for the pre-salt mapping that identified the Tengiz structure. R corresponds to a deeper reflector used in subsequent improved seismic surveys. The Namurian Stage is also called the Serpukhov Stage (see Figure 19).

Carboniferous

Lower Carboniferous

Tournaisian deposits rest unconformably on the Upper Devonian and are composed exclusively of gray and brownish gray micritic limestones and interbedded bioclastic limestones containing primary and secondary porosity. The thickness of this stage varies widely, from 30 to 600 m.

The Viséan Stage is subdivided into three substages. The lower Viséan substage is composed of light- and dark-gray limestones. Skeletal and algal-foraminiferal limestones are present in the base of this section, which shoals upward into skeletal packstones and grainstones and is capped by oolitic limestones.

The middle Viséan (Yasnaya Polyana) substage is subdivided into the Bobrikov and Tula formations, which are overlain unconformably by upper Viséan deposits.

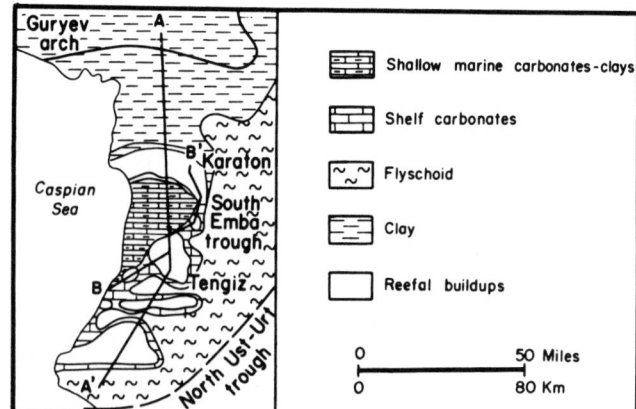

FIGURE 9. Carboniferous lithofacies. Cross sections A-A' and B-B' are shown in Figures 10 and 11.

The Bobrikov horizon is composed of homogeneous algal stromatolite and bryozoan buildups that grade laterally into higher-energy grainstones. In some zones this horizon is overlain by argillaceous mudstones up to 50 m thick and containing calcareous interbeds. The limestones of the Tula Formation are dark-gray, argillaceous, recrystallized, and, in places, silicified, poorly sorted biogenic limestones with an assemblage of foraminifera. The thickness of the middle Viséan substage is 150 to 180 m.

The upper Viséan (Oka) substage is composed of shelf carbonate deposits. It has been penetrated completely at the Karaton, Tengiz, Yuzhnaya, and Korolev fields (Figures 9, 10, and 11). From bottom to top, the Aleksin, Mikhailov, and Venev formations have been identified here. In the lower portion, the Aleksin Formation is composed of limestones that in places are detrital and tend to be coarser-grained in the upper portion. The limestones of the Mikhailov Formation contain richer, more varied fauna. Some wells have penetrated biogenic buildups. At Tengiz, the Venev Formation is composed of biohermal rocks. These are relict-biogenic and, in places, stromatolitic and bryozoan limestones. The formation of these carbonates occurred in the interior portions of the platform system. The thickness of the upper Viséan substage is consistent along the trend at about 280 to 300 m.

The Namurian (Serpukhov) Stage, overlying the Viséan, consists of the Tarussa-Steshnev and Protvin formations. The entire series consists of bioclastic limestones derived from organic buildups and biohermal, algal-bryozoan limestones with an abundance of fauna. The presence of numerous large coral fragments indicates the incipient shoaling of the basin and short-term periods of exposure and erosion. In the Tengiz field, a bed of volcanic ash with fragments of biohermal limestones can be traced along the base of the Protvin Formation. The thickness of the Serpukhov Stage at Tengiz is relatively consistent at 200 m, although in places it is sharply reduced by erosion to 30 m.

Middle Carboniferous

The Bashkirian Stage has been penetrated at the Tengiz, Korolev, and Yuzhnaya fields (Figures 9, 10, and 11). At

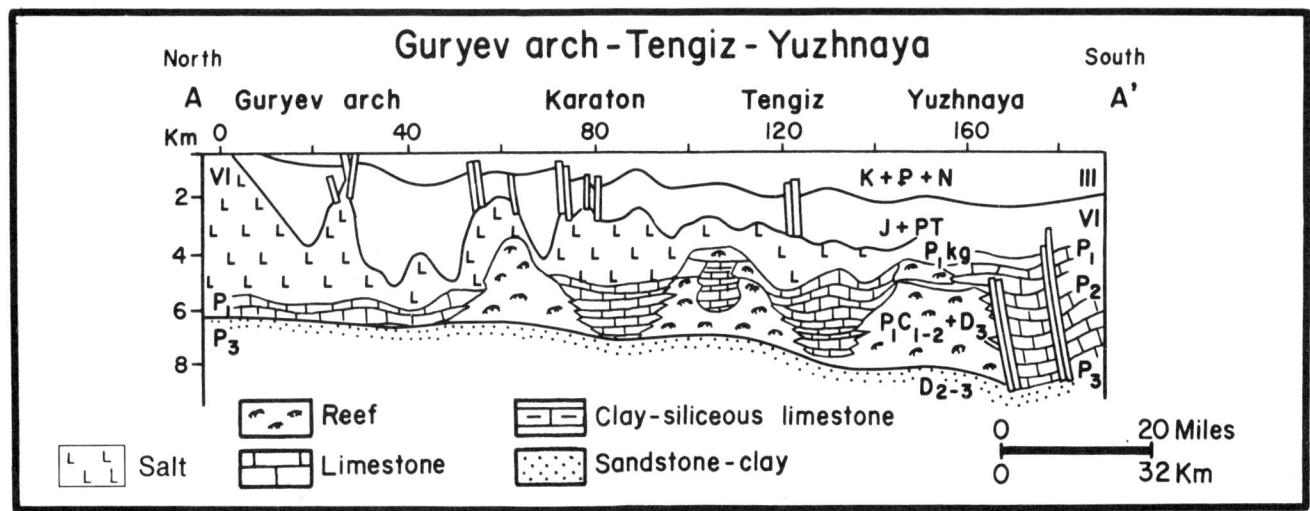

FIGURE 10. North-south regional geologic section A-A'. The location of this section is shown in Figure 9.

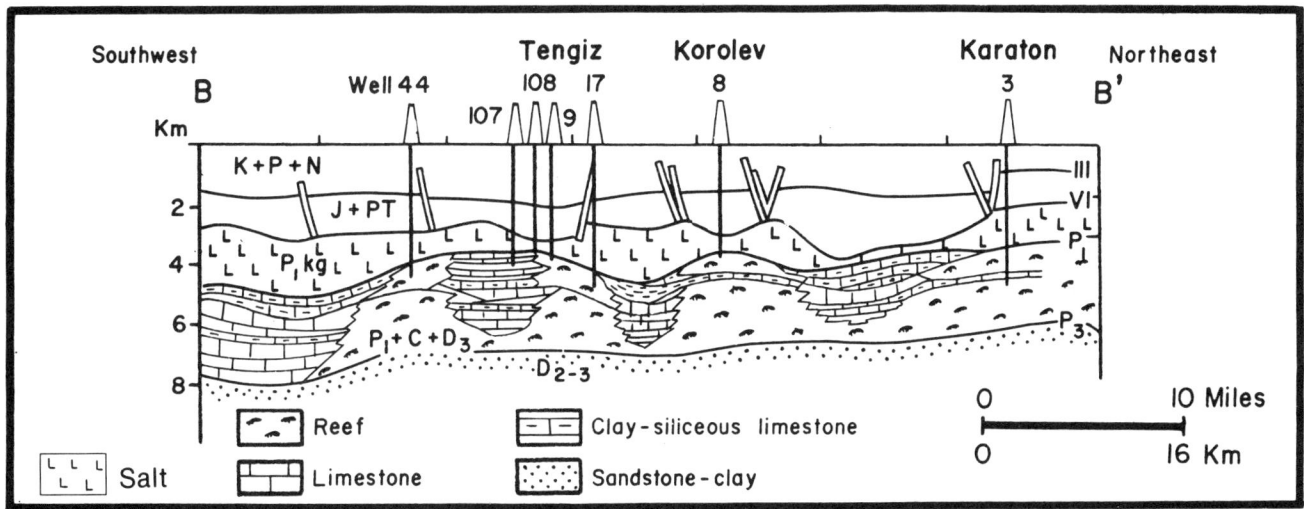

FIGURE 11. Southwest-northeast regional geologic section B-B'. The location of this section is shown in Figure 9.

Karaton, drilling showed that only the lower portion of the Bashkirian Stage was preserved after Early Permian erosion. To the south, the upper Bashkirian substage is preserved in the section at Yuzhnaya.

The deposits of the Bashkirian Stage overlie the Serpukhov Stage, the surface of which is eroded with various degrees of intensity. The lower Bashkirian substage includes the Krasno-Polyana, Severo-Keltmen, and Prikama formations.

In the lower portion, the Krasno-Polyana Formation is composed of bioclastic limestones with an abundance of varied fauna. These deposits are products of the reworking of Oka-Serpukhov biogenic carbonates. From the second half of the Krasno-Polyana Stage, there were favorable conditions for the development of foraminifera and algae. The formation of the carbonate bodies, composed of oolitic and crinoidal limestones, took place under high-energy shoal conditions. The Severo-Keltmen Formation is similar in lithological composition to the Krasno-Polyana Formation. The Prikama Formation is fully represented at Yuzhnaya and is partly eroded at Tengiz. The lower part of this horizon is composed of laminated limestones, and the upper part consists of massive and bioclastic limestones.

Upper Bashkirian deposits were penetrated at Yuzhnaya (Figure 10) and partly penetrated in well 3 of Tengiz (Figure 5). These deposits lie unconformably on the surface of the lower Bashkirian substage and are composed of dark-gray to nearly black, very dense and bituminous micritic limestones. The total thickness of the Bashkirian Stage within the uplift area varies over a wide range, from 18 to 300 m.

Permian

In the area of uplift, Lower Permian deposits lie disconformably on rocks of various ages, ranging from Upper

Devonian to middle Carboniferous. Asselian carbonate deposits of the Lower Permian have been penetrated only at Yuzhnaya. At Tengiz, the Lower Permian is represented by the Artinskian Stage. These deposits are predominantly argillaceous; however, marls and limestones with various clay contents also occur. It should be noted that the top of the upper Artinskian deposits was used to map the surface of the pre-salt Paleozoic complex of the pre-Caspian basin, and is identified as seismic horizon P_1 (Figure 5).

At the Tengiz field, these Permian sediments overlie a thick carbonate buildup 3.5 to 4 km thick in a sharp, angular contact with the steep relief of the unconformity surface. Here the Artinskian Stage is in direct contact with both Lower and middle Carboniferous and Upper Devonian rocks (Figure 8). The thickness of the Artinskian deposits on the crest is minimal, amounting to only 10 to 80 m, whereas the penetrated thickness of the Artinskian deposits exceeds 1000 m on the flanks of the structure and in the synclines and troughs that separate local uplifts.

TECTONICS

Throughout the long history of the formation of the Tengiz carbonate buildup, there were periods in which it was elevated to about sea level and subjected to erosion. This is confirmed by major unconformities found in the pre-salt Paleozoic deposits of the Upper Devonian and Lower and middle Carboniferous, with the most extensive being just above the Carboniferous (Figure 8). These unconformities were due to the Bretonian, Sudetic, and Uralian phases of Hercynian tectonism.

The phases of Alpine tectonism and recent tectonic activity are reflected most significantly in the post-salt complex.

Explicit signs of faulting have not yet been found in the pre-salt carbonate section, except that extensive fracturing has been found in a few wells. It has not been possible to trace reliably any large-displacement faults from presently available seismic data. Previously noted zones of faulting along the northwest, northeast, southeast, and southwest flanks of the structure, which bound a hypothetical large block both at the basement level and in the sedimentary cover, should be viewed as controversial and in need of additional substantiation by further research (Figures 2 and 4).

SEISMOGEOLOGIC MODEL OF THE FIELD

A structural-sedimentary model of the development of atoll-like structures on the platform, which were uplifted relative to the deep water basin, was adopted in the initial stage of the geologic and geophysical study of the Tengiz uplift. Later, a two-tiered tectonic-sedimentary model was proposed. According to this model, shelf-type laminated limestones were deposited in the lower part of the Tengiz carbonate section (Upper Devonian–Tournaisian) and were affected by gentle uplift. Then high-amplitude structures were formed as a result of tectonics and sedimentation (Viséan–Bashkirian).

The seismogeologic model of Tengiz presented below is based on detailed seismic studies. Three time slices from a three-dimensional seismic survey are shown in Figures 12, 13, and 14. The locations of the wells that penetrated these time slices are plotted on these displays. As can be seen, the outer contour of the body is quite tortuous, and its area increases sharply with depth. The configuration of the salt troughs around the periphery of the structure can be recognized. Other details of the structure of the carbonate buildup are shown by vertical seismic sections, the locations of two of which (lines 31 and 520) are shown on the structural map constructed on the top of the pay zone at seismic reflector R (Figure 15).

As can be seen in Figure 16 (line 31), the entire central part of the carbonate body is surprisingly horizontal, whereas the presence of a ridge is distinctly visible on the sides of this structure. This ridge can be traced along three sides of the structure after integration of all the available seismic data. Within the carbonate body, continuous reflectors cannot be mapped. This attests to the absence of internal seals that would impede vertical communication within the carbonate mass. Typically, reflections within the carbonate section almost disappear in zones located under ridges.

A zone of divergent reflections can be seen with varying clarity on the flank of the structure immediately beyond the ridge. Such a zone is distinctly visible in Figure 17 (line 520). Divergent reflections can be traced along the periphery of the carbonate body, and are identified as pre-salt Artinskian deposits. As evidenced by the location of the P_1 and R seismic reflectors, the thickness of these deposits increases down the flank of the carbonate body.

Commercial oil flow tests of the Upper Devonian in well No. 10, which is located on the northeast flank of the structure, support the beliefs that the spillpoint of the structure at the base of the salt, the P_1 reflector, does not limit the accumulation, and that the oil-water contact should be expected to be far deeper than the mapped spillpoint at the P_1 horizon. These data confirm that the zones of divergent reflections identified on seismic sections along the flank of the carbonate body represent reliable sealing facies.

Farther down the flanks of the carbonate body toward its periphery, there is a noticeable increase in the number of reflections on the seismic sections. This suggests a lithologic-facies change in the section and a probable increase in the clay content of the material.

The lower portion of the pre-salt stage, which is identified by reflections at 3.8 to 4.0 sec, shows that this carbonate buildup formed over an extensive uplifted zone substantially greater in area than the Tengiz structure itself. The calculated seismic velocities and a series of parallel reflections indicate the presence of terrigenous deposits that, by analogy with other regions of the eastern Russian platform, are assigned to the Middle Devonian.

Thus, the geologic model of the Tengiz field can be represented as a biohermal structure of the atoll type (Figure 18). It formed on the arch of a gentle Middle

FIGURE 12. A 3D seismic time slice of the Tengiz structure. As shown by the yellow border, the perimeter of the body is tortuous. The locations of wells that reached the depth corresponding to the time slice are shown as green circles.

Devonian uplift during the Late Devonian–Carboniferous. A ridge identified along the edge of the carbonate body is considered to be a reef facies. The flat central part of the body is interpreted as having been a lagoon. In addition to the biogenic carbonates, interbedded argillaceous mudstones, calcareous grainstones, and dark argillites with carbonized plant remains are good indicators of the shoal origin of the deposits encountered in this zone. The

FIGURE 13. A 3D seismic time slice deeper than the one presented in Figure 12, showing that the body is much larger at this depth.

presence of bioclastic limestones both in the lagoonal zone and on the outer flanks beyond the reefs attests to the sedimentation of the erosion products derived from the atoll structure, which at times rose above sea level and was subjected to erosion.

A lithofacies map (Figure 9) constructed on the basis of seismic data and exploration well data gives a general idea of the structure and distribution of lithologies of the pre-salt section of the southeastern pre-Caspian basin. As this map shows, Tengiz is not a solitary structure, but is

FIGURE 14. A still-deeper 3D seismic time slice (compare Figures 12 and 13), showing a still broader structure.

an element of a large carbonate platform within which several massive biohermal bodies formed. The carbonate shelf is replaced by a terrigenous section to the east of the carbonate platform. Within this section there are relatively thin marine carbonate units. The exact locations and potentials of these bodies remain to be evaluated through future studies. The regional A-A' and B-B' structural cross sections (Figures 10 and 11) were derived from regional seismic lines and vividly illustrate the important components of the area.

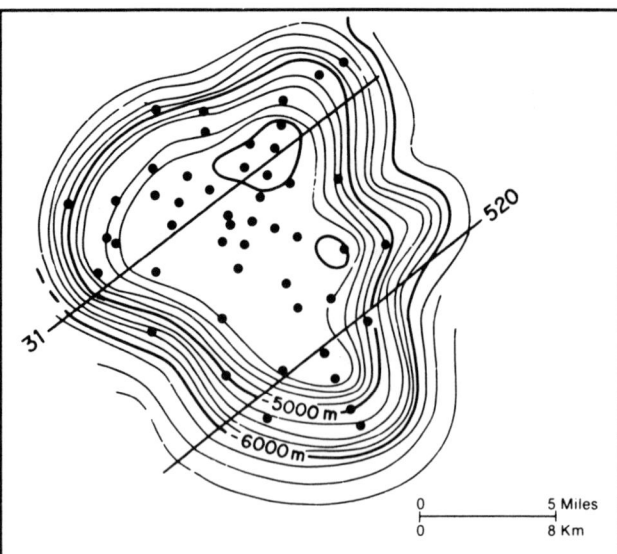

FIGURE 15. Top of Tengiz field reflector R. Seismic lines 31 and 520 are shown in Figures 16 and 17. Contour interval, 200 m.

The A-A' section (Figure 10) intersects, from south to north, the Yuzhnaya, Tengiz, and Karaton biohermal structures, which are distinctly expressed by the relief of the P_1 pre-salt horizon. The base of the carbonate platform, which is identified by the P_3 seismic horizon, rises smoothly from south to north, forming the southern flank of the Guryev arch, which is part of the Astrakhan-Aktyubinsk uplift system.

The B-B' cross section (Figure 11) cuts through part of the extensive ring of structures whose western terminus is covered by the shoal portion of the Caspian Sea. This profile intersects, from east to northwest, a chain of carbonate structures of different amplitudes: the Tengiz, Korolev, and Karaton structures. Commercial oil accumulations have been found in the Tengiz and Korolev structures.

SPECIFIC FEATURES OF THE RESERVOIR ROCK

The pay zone of the Tengiz field, which occurs at depths from 3900 to 5400 m, is confined to a massive carbonate section. Shallow water biogenic limestones comprise the majority of this reservoir. Isolated reefal bioherms with flanking, reworked, and cemented deposits have been identified in some stratigraphic intervals of the section (Figure 19). The textural composition of the limestones of the entire carbonate section is comparatively uniform and has a predominantly bioclastic origin. Nonetheless, significant heterogeneity is observed in the types and properties of the reservoir rocks, both locally and regionally.

The primary texture and porosity of the carbonate reservoir rocks were formed during sedimentation. However, subsequent secondary processes had significant effects on the alteration of the rocks and ultimately led to the present state of the reservoir rocks, with their heterogeneity of pore volume and permeability. Such processes as tectonic fracturing, dissolution and leaching, recrystallization, dolomitization, and silicification are of significance in the reservoir.

Based on the interpretation of core and well log data, one can see that the porosity of the reservoir rocks varies over a wide range, from 1 or 2% to 15 to 25% (Figure 20). Permeability also is nonuniform. On average, the porosity is 6.3%, and the permeability is 10 md. To date, no definite depth dependence has been found for the reservoir properties.

The effects of secondary processes on changes in reservoir properties are zonally manifest. For example, the northern zone of the Tengiz structure is characterized by a predominance of fractured reservoirs throughout the entire section. A zone of sporadic recrystallization also has been identified in this area at depths of 3900 to 5100 m.

A large part of the carbonate section of the Tengiz field, from 3900 to 4600 m, is characterized by intensive leaching and dissolution. It has been determined also that the depths of penetration of secondary diagenetic processes associated with erosion do not exceed 20 to 40 m below the level of regional or local unconformities. In addition to unconformity-related diagenesis, other processes also were involved in the development of secondary characteristics in this carbonate mass.

It is likely that one such factor was the nonuniform stressing of the entire carbonate mass and the resulting rock deformation. When carbonates deform under conditions of nonuniform hydrostatic stress, significant decompaction and an increase in the pore volume of the rocks resulting from microfracturing can occur; then the physical and chemical processes that lead to dissolution, leaching, recrystallization, and other effects can develop more actively. Three main types of reservoir rocks have been identified on the basis of various types of micro- and macro-nonuniformities and the reservoir properties in the matrix of the carbonate formation rocks comprising the Tengiz structure.

The first type of reservoir (type I) includes carbonate rocks in which the entire matrix is permeable and is pervaded by vugs and micro- and macropores. The porosity exceeds 7%, and oil flow occurs via both fractures and the matrix (Figure 21).

The second type of reservoir (type II) includes carbonate rocks in which vugs and micropores are developed in zones along the fractures. These zones are separated from each other by a less-porous matrix. The porosity ranges from 3 to 7%, and permeability is created solely by the fractures (Figure 22).

The third type of reservoir (type III) includes the dense limestones in which the leaching processes have not been effective. In such limestones, the micropores of the matrix are filled with irreducible water, and oil is present only in fractures that cut through the matrix. The porosity is less than 3%, and the matrix is virtually impermeable (Figure 23).

The limited data currently available on the Tengiz carbonate formations allow the identification of only the

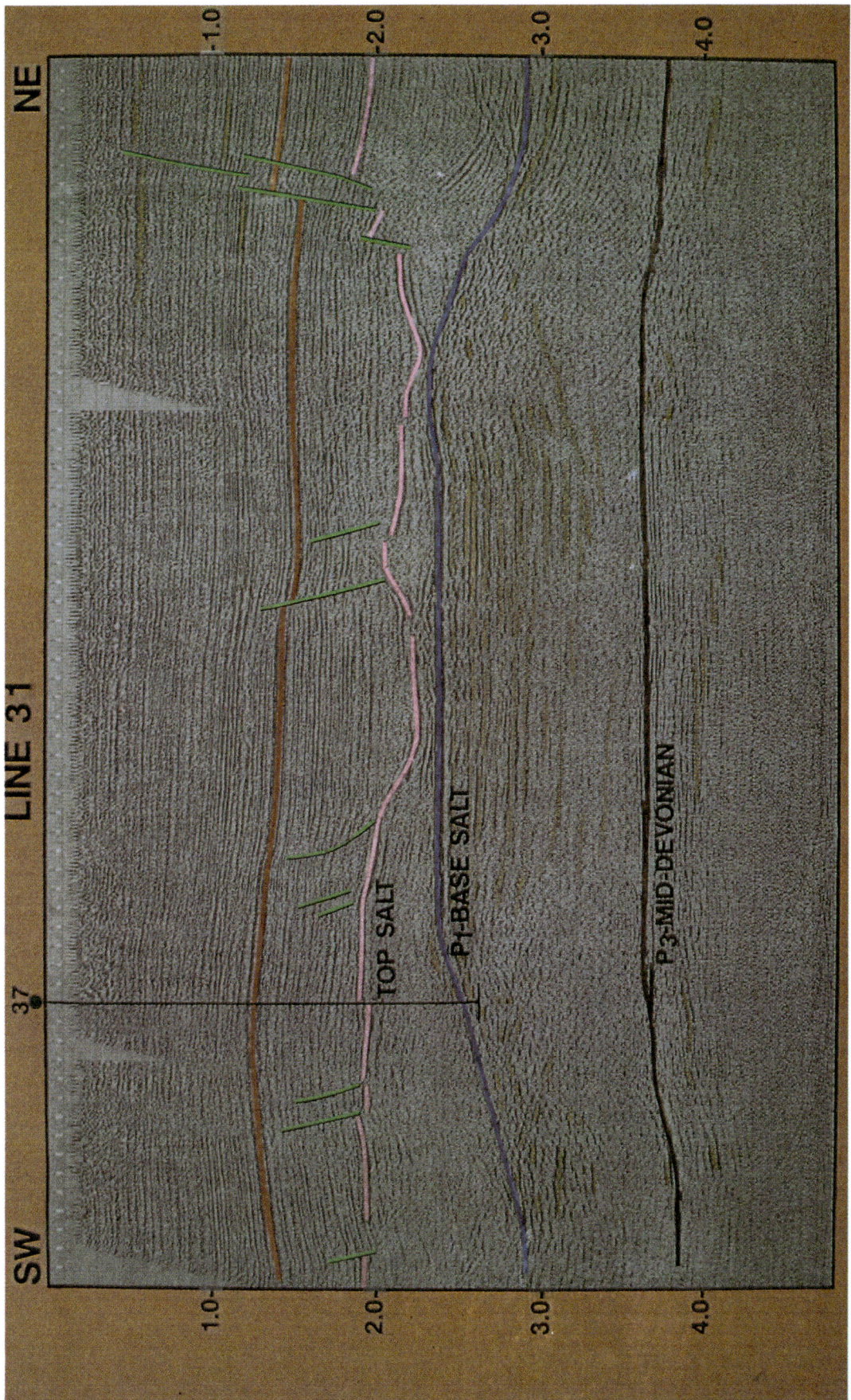

FIGURE 16. West-southwest to east-northeast 2D migrated seismic line 31 over the Tengiz structure. The location of line 31 is shown in Figure 15. This line illustrates the wide, flat top of the structure.

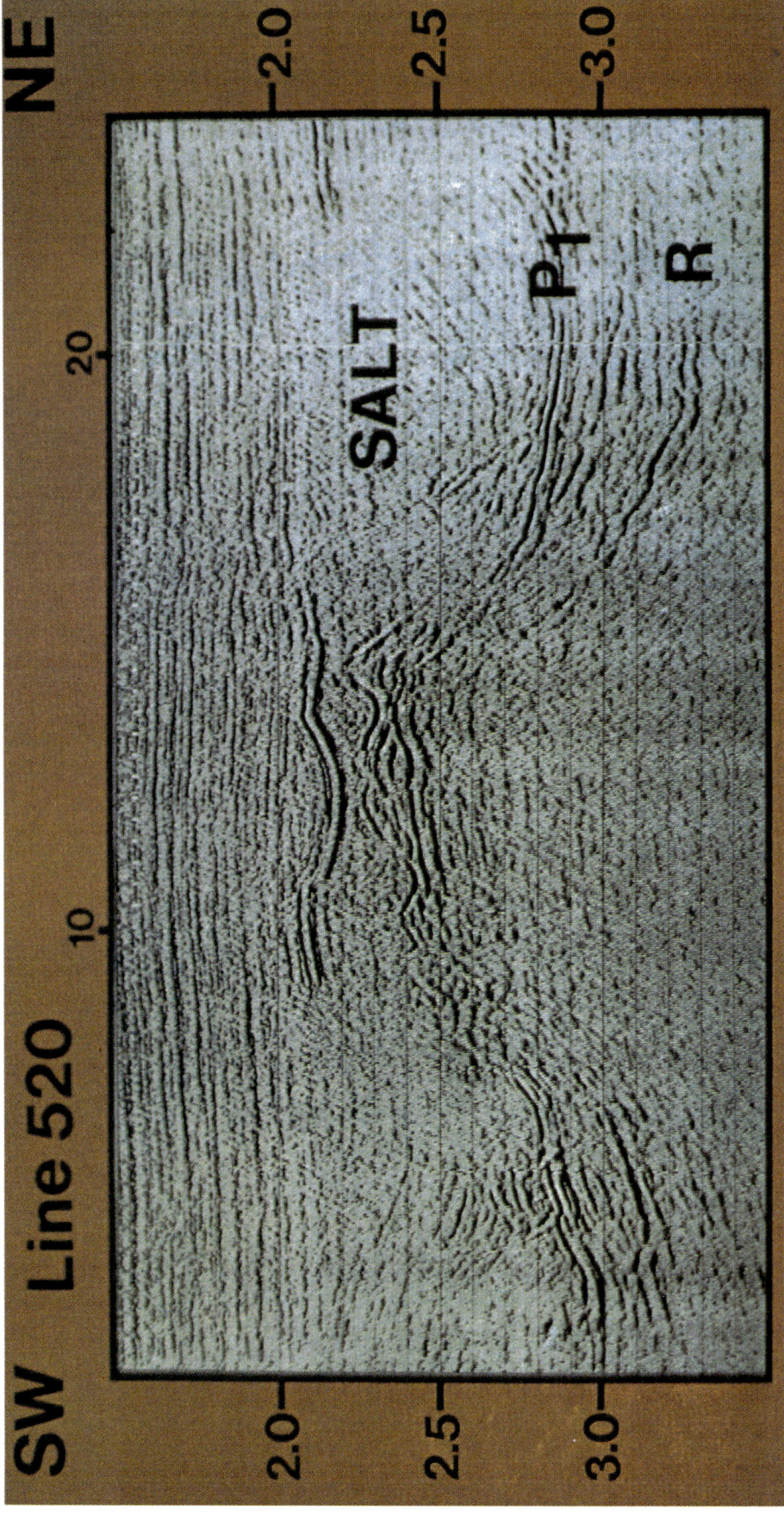

FIGURE 17. Migrated 3D seismic line 520, parallel to line 31 (locations shown in Figure 15). Line 520 shows the zone of divergent reflections on the periphery of the carbonate body.

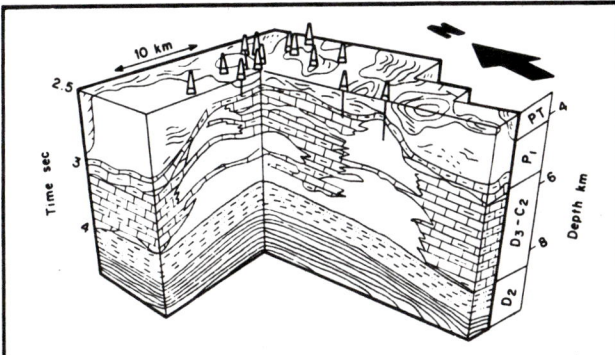

FIGURE 18. Schematic stratigraphic model representing Tengiz field as a biohermal structure of the atoll type. The white areas represent porous, reefal material.

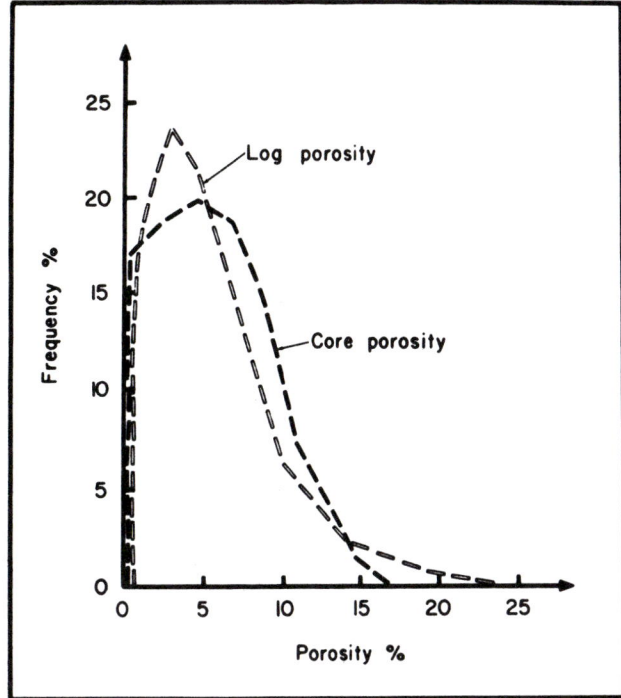

FIGURE 20. Porosity vs. percentage of samples (frequency), based on 8883 log determinations and 892 core sample analyses.

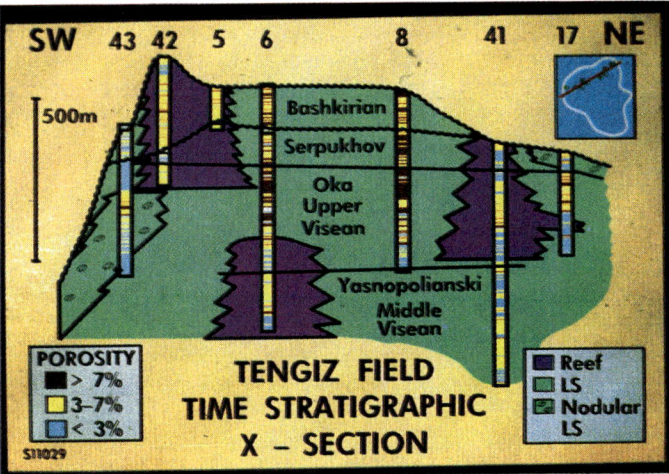

FIGURE 19. Time-stratigraphic cross section of the Tengiz field. The inset at upper right shows the field outline and the location of the section. Width of section, about 18 km.

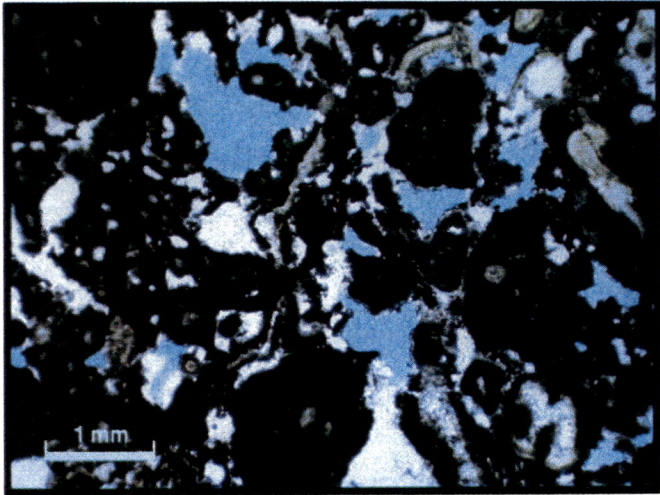

FIGURE 21. Type I reservoir: algal-biolithite carbonate rock with solution-enlarged porosity. The entire matrix is permeable, pervaded by vugs and pores. Porosity, >7%. Fluid flow occurs via both fractures and matrix.

most general features of the reservoirs associated with various stratigraphic complexes.

The Bashkirian reservoir consists primarily of dense biogenic limestones (Figure 8). Secondary processes occurred uniformly throughout the section, leading to dolomitization, silicification, dissolution, and recrystallization. Leaching and fracturing had significant effects in improving the reservoir properties of the rocks. All three reservoir types have been observed in the Bashkirian reservoir, but type II, with a porosity of 3 to 7%, is predominant. In many cases, the primary pore space is filled with solid bitumen. The maximum oil flow rate is 1900 BOPD (270 MT per day).

The Serpukhov reservoir is composed of biohermal limestones. Recrystallization, intense fracturing, and silicification are observed everywhere. The primary pores and fractures often are filled with solid bitumen. High-capacity reservoir rocks with a porosity of more than 7% have been preserved only in elevated sections of the structure. Type II reservoir rocks occur to the north, and type III to the west.

The Oka reservoir has the best reservoir properties. Intense leaching, causing the formation of secondary pores and vugs, is the most characteristic feature of the

FIGURE 22. Type II reservoir: skeletal intraclast grainstone with solution-enlarged porosity. Vugs and micropores developed only along fractures, separated from other porous zones by less-porous matrix. Porosity, 3 to 7%. Fracture permeability only. The scale for Figure 22 is the same as that for Figures 21 and 23.

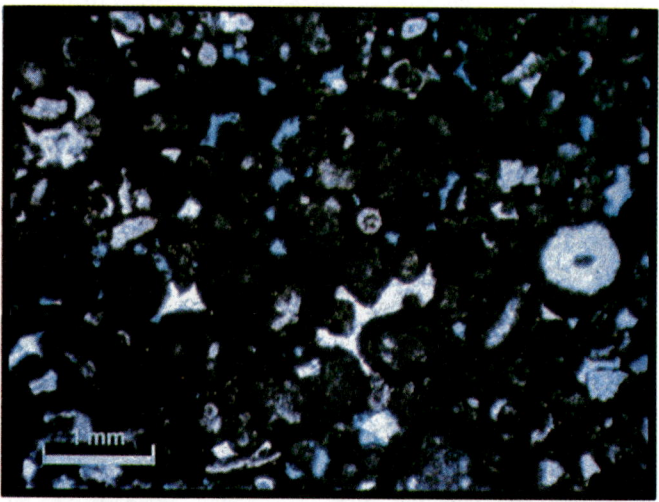

FIGURE 23. Type III reservoir: pisolite grainstone with primary interparticle porosity. This is dense limestone in which leaching has not been effective. Micropores contain irreducible water. Oil is contained only in fractures, which also provide the only permeability. Porosity, <3%.

secondary diagenesis. The vugs are rarely filled with bitumen. Numerous bitumen-filled stylolites are observed in fine-grained recrystallized limestones with a porosity of 3 to 5%. The highest oil flow rates, from 3500 to 8000 BOPD (450 to 1100 MT per day), have been found in this reservoir.

The Yasnaya Polyana reservoir has not been studied very thoroughly to date. Available data indicate low reservoir properties, with reservoir rocks of types II and III being predominant.

The Devonian reservoir has been penetrated by only three wells and is represented by biohermal limestones. The dissolution processes detected in it, with the formation of moldic pores and vugs, provide the basis for assuming that favorable reservoir rocks may be present at depths greater than 5400 m.

RESERVES

The total original volumes of oil and gas in place in the Tengiz field are estimated to be 25 billion bbl (3.5 billion MT) of oil and 46 tcf (1.3 trillion m³) of associated gas. Based on the classification of reserves, the pay zone of the carbonate section is now conditionally subdivided into three zones (Figure 24). Explored (proved) reserves (category C_1) encompass the section from the crest of the structure to a depth of 4700 m. Preliminarily estimated (probable) reserves (category C_2) include the section from 4700 to 5400 m. Hypothetical (possible) reserves (category C_3) are associated with Upper Devonian carbonates and are conjectured to be present from a depth of 5400 m to the still-undetected level of the oil-water contact.

FORMATION FLUIDS

According to the available data, the oil accumulation appears to be a single oil reservoir of the massive type (pressure communication, and similar lithology and structure) that is confined to the Devonian-Carboniferous carbonate stratigraphic section. The initial formation pressure throughout the reservoir is abnormally high, reaching 80 MPa (11,600 psi) at a depth of 4000 m. The temperature ranges from 105 to 116°C (220 to 240°F) at depths of 4000 to 5500 m. There is no gas cap in the pool, and the hydrocarbons are markedly undersaturated. Saturation pressure is about 25 MPa (3625 psi). The main

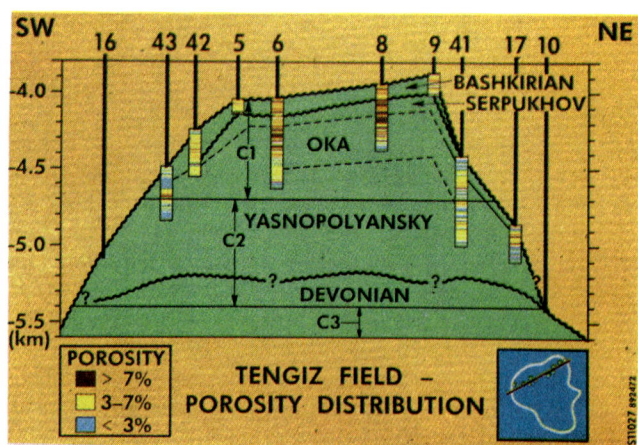

FIGURE 24. Porosity distribution for Tengiz field. The inset shows the field outline and the location of the section. Width of section, about 18 km. The intervals C_1, C_2, and C_3 contain explored (proved) reserves, preliminarily estimated (probable) reserves, and hypothetical (possible) reserves, respectively.

properties of the oil and associated gas are presented in Table 1.

As noted previously, no formation water has been obtained from the pre-salt deposits of the Tengiz field. Formation water with a salinity of 230,000 to 250,000 ppm (270 to 300 g/L) has been produced from Lower Carboniferous limestones at a depth of 4200 to 4300 m at the nearby Karaton well, which is located approximately 30 km northwest of Tengiz. It may be assumed that the salinity of the water in the pre-salt section of Tengiz is approximately the same. The water salinity is 120,000 ppm (130 g/L) in the post-salt Cretaceous deposits at a depth of about 800 m in the Tengiz area.

A sharp difference is observed in the distribution of formation pressures throughout the sedimentary section. The post-salt nonpay zone of Tengiz is characterized by normal hydrostatic pressure. In the pre-salt pay zone, the formation pressures are nearly twice the hydrostatic pressure. It is worth noting that in the nearby Karaton and Yuzhnaya structures, which are nonhydrocarbon-bearing, the formation pressures in pre-salt carbonates are close to hydrostatic pressure. In Tengiz, the development of excess pressures apparently may be related to the same natural process that controlled the maturation of oil-source carbonate rocks, hydrocarbon generation, and the formation of a gigantic oil pool in this enormous isolated body.

GEOCHEMISTRY

The accumulation of organic matter in the carbonate series of pre-salt Paleozoic sediments occurred under littoral-marine conditions. The geochemical environment was reducing or weakly reducing; this contributed to the preservation and further alteration of the organic matter.

Organic matter of marine origin was the precursor of the kerogen for both the Carboniferous limestones and the overlying argillaceous Lower Permian deposits.

The organic carbon content of the Paleozoic carbonates in the Tengiz region is high, generally ranging from 1.2 to 1.4% and occasionally reaching 4%. The oil-generating potential of such rocks is very high. These are type I and type II kerogens with high hydrogen contents, in which the hydrogen-to-carbon atomic ratio ranges from 0.9 top 1.34.

The oils from Tengiz and Korolev are dominantly composed of saturated hydrocarbons, with lesser quantities of aromatic hydrocarbons and small amounts of asphaltenes and polar compounds. Gas chromatograms display certain features that suggest generation from highly mature, carbonate-rich or marly source rocks deposited in an anoxic environment. The high maturity of the oils is indicated by a generally smooth decrease in the abundance of n-alkanes with increasing molecular weight (Figure 25). The oils show a subtle even-over-odd predominance (carbon preference index, 0.97 to 0.99) and a pristane/phytane ratio of 1.02 to 1.22. The content of C_{12} through C_{30} n-alkanes in the oils reaches 80%, and the content of isoprenoids C_{14} through C_{20} ranges up to 20%. All geochemical indices according to gas-liquid chromatography have extremely close values, regardless of the sampling site (Table 2).

Mass-spectrometric analysis of the 200-to-450°C fraction showed that the oil is rich in bi-, mono-, and tricyclic naphthenes. Monocyclic compounds predominate in the aromatic portion, reaching 83%. They probably are derivatives of alkyl benzenes.

The yield of the fraction up to 200°C is about 40%. The paraffin content reaches 70%, of which n-alkanes and iso-alkanes account for 30 and 40%, respectively. In this fraction, naphthenes account for 27%, cyclohexane for 12%, and cyclopentane for 15%. Aromatic hydrocarbons do not exceed 2.5%. Ethyl benzene accounts for 17.4%,

TABLE 1. Properties of oil and gas in the Tengiz field.

Oil	
Specific gravity, °API	46
Viscosity (reservoir condition), cP	0.27
Pour point, °C	−36
Composition, %:	
Paraffins	5.8
Sulfur	0.7
Mercaptan	0.05
Associated gas	
Gas/oil ratio, standard ft³/STB (m³/m³)	2150 (378.6)
Composition, %:	
Methane	41
C_2+	36
H_2S	18
CO_2	5

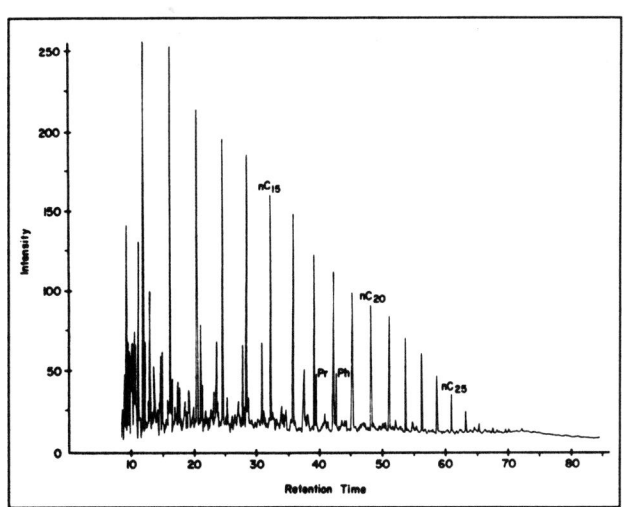

FIGURE 25. Gas chromatogram of oil from Tengiz No. 102 (3996 to 4095 m). The high maturity of the oil is indicated by the generally smooth decrease in n-alkanes with increasing molecular weight.

TABLE 2. Chromatographic ratios of higher alkanes.

n-alkanes/iso-alkanes	4.0–5.0
Pristane/phytane	1.02–1.22
Pristane/nC_{17}	0.29
Phytane/nC_{18}	0.29
(Pristane + phytane)/(nC_{17} + nC_{18})	0.30
(nC_{13} through nC_{15})/(nC_{23} through nC_{25})	3.6

para-xylene for 17.1%, meta-xylene for 45%, and orthoxylene for 22.3%.

Because of their high maturities, the oils show a scarcity of steranes and aromatic steroids. Therefore, in this case, these parameters are unsuitable for deriving a source correlation. In contrast, the oils contain ample concentrations of terpanes. The terpane fingerprints from Tengiz and Korolev oils are very similar. The high maturity of the oils is indicated by the pronounced decrease in the abundance of 17(H)-hopanes from C_{30} to C_{35} and low concentrations of 17(H)-hopanes relative to C_{28} and C_{29} tricyclic terpanes (Figure 26). An "oil-source rock–solid bitumen" correlation reveals that, in most characteristics, the oil composition reflects the genetic type of the argillaceous pre-salt Permian source rocks. However, some features indicate that these oils are derived from a combination of source rocks—i.e., the pre-salt Permian shales and marls and Carboniferous limestones.

The presence of a large quantity of solid bitumens in the pore space of the pay zone is one of the distinctive features of the Tengiz field (Figure 27). Solid bitumens at Tengiz are generally insoluble in common organic solvents. A mixture of orthoxylene, acetone, and ethyl alcohol is commonly used to dissolve them. Solid bitumens contain more than 40% asphaltenes, about 25% resins, and up to 5% sulfur. The hydrogen content ranges from 5.8 to 8%, the hydrogen-to-carbon ratio ranges from 0.85 to 1.2, and the carbon-to-oxygen ratio ranges from 0.5 to 0.12.

According to data from differential thermal analysis and x-ray diffraction analysis, the solid bitumens are classified as transitional between kerites and impsonites. They have high reflectance values of 1.27 to 1.65%, whereas the degree of maturity of the carbonate reservoir rocks is much less according to individual vitrinite reflectance measurements on clay particles, which do not exceed 0.7 to 0.9%.

The solid bitumens encountered in all productive levels of the Tengiz field probably represent residual matter produced during the alteration of the carbonate-rich source rocks in the presence of sulfur, which accelerated the carbonization process.

CONCLUSIONS

1. The Tengiz oil field is the supergiant of the 1980s. It was discovered in the southeastern pre-Caspian basin. It is associated with a large carbonate structure with an amplitude of more than 1000 m and an area exceeding 400 km².

2. The thickness of the pay zone presently exceeds 1500 m. To date, the oil-water contact has not been penetrated. The massive oil accumulation is confined to carbonates from the Upper Devonian to the Lower and middle Carboniferous at depths of 3900 to 5400 m. This eroded carbonate section creates an unconformity surface below which Carboniferous beds, and above which disconformable Artinskian clays and Kungurian salt of the Lower Permian, are in sharply angular contact with the unconformity surface.

3. The limestones throughout the entire carbonate section are of comparatively uniform composition and predominantly bioclastic origin. The primary microstructure of the rocks has been altered significantly by secondary processes. The best reservoir properties are associated with extensive fractures, vugs, and moldic pores.

4. Total original volumes of oil and gas in place are estimated to be 25 billion bbl (3.5 billion MT) of oil and 46 tcf (1.3 trillion m³) of associated gas. The maximum per-well oil flow rates reach 8000 BOPD (1100 MT per day).

5. The initial formation pressure is abnormally high—nearly twice the hydrostatic pressure. The temperature ranges from 104 to 116°C (220 to 240°F) at depths from 4000 to 5500 m.

6. The oil density is about 45° API, and the sulfur content is low. There is no gas cap in the pool, and the oils are markedly undersaturated. The saturation pressure is about 25 MPa (3625 psi). The hydrogen sulfide content in the associated gas is very high, reaching 18 to 22%. The CO_2 content is about 5%.

7. The oil-generating potential of the biogenic Paleozoic limestones of the Tengiz carbonate buildup is very high; the organic carbon content generally ranges from 1.2 to 1.4%, occasionally reaching 4%. Type I and type II kerogens, in which the hydrogen-to-carbon atomic ratio ranges from 0.9 to 1.34, are present.

8. An "oil to source rock" correlation based on chromatographic studies indicates that in most characteristics the oil composition reflects the genetic type of the organic matter of the argillaceous and marly pre-salt Permian source rocks.

9. The presence of a large quantity of solid bitumens in the pore space of the pay zone is one distinctive feature of the Tengiz field. These solid bitumens have high vitrinite reflectance and represent the residual matter from the alteration of the Carboniferous carbonate-rich source rocks in the presence of sulfur.

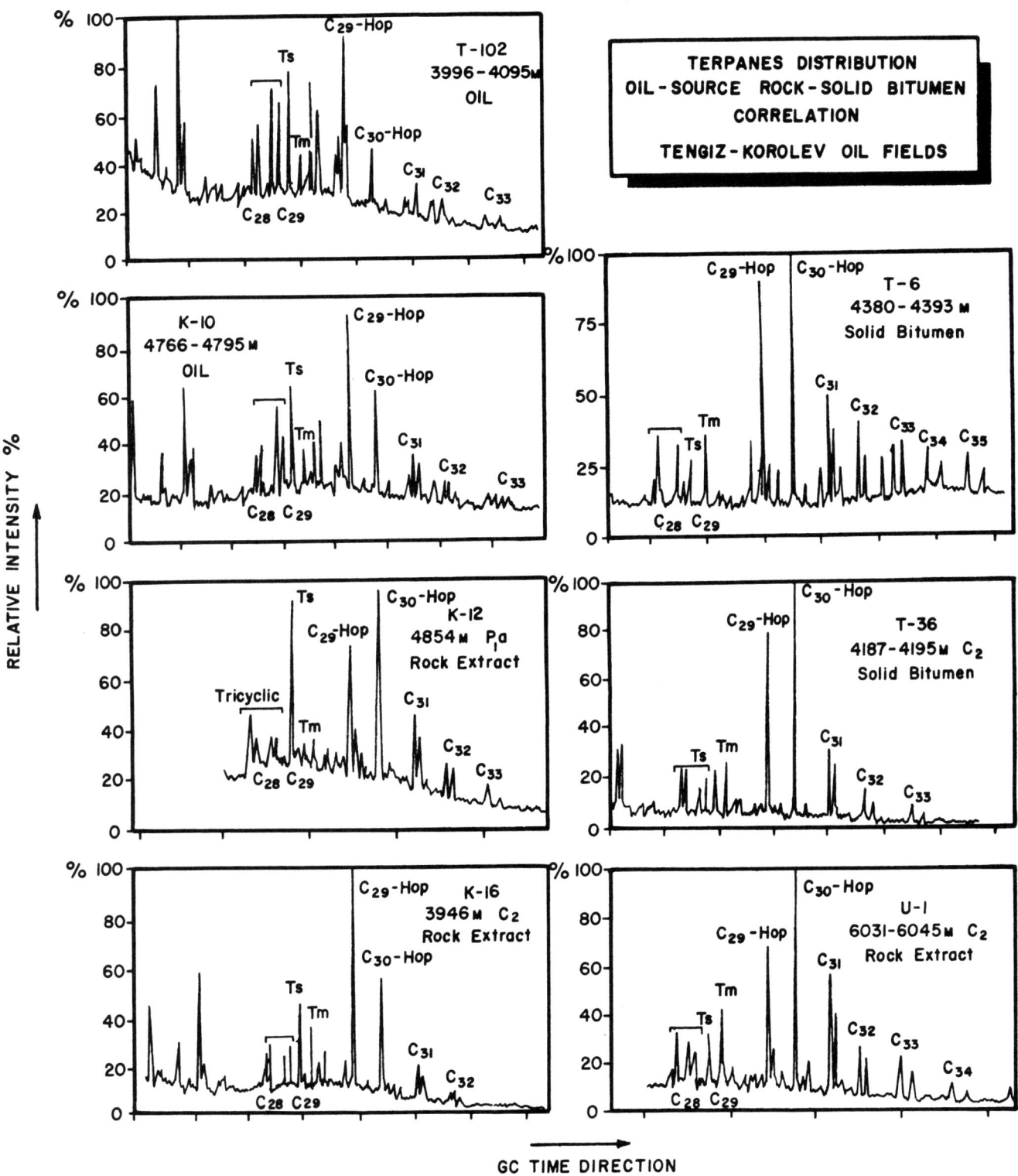

FIGURE 26. Distribution of terpanes, and oil–source rock–solid bitumen correlation, for Tengiz-Korolev oil fields. High oil maturity is indicated (see text).

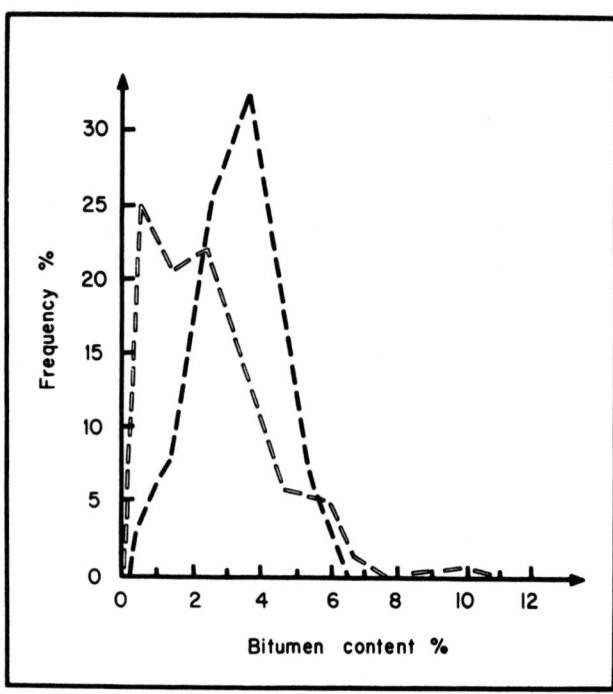

FIGURE 27. Bitumen content vs. frequency. Line of open dashes represents 1157 Lower Carboniferous samples. Line of solid dashes represents 426 "middle" Carboniferous samples.

ACKNOWLEDGMENTS

The authors would like to thank the administration of the USSR Ministry of Oil and Gas Industry for permission to publish this paper, and Chevron Overseas Petroleum, Inc. for providing extensive assistance with the illustrations and in the technical preparation of the text.

Chapter 8

The Marlim and Albacora Giant Fields, Campos Basin, Offshore Brazil

Aladino Candido
Carlos A. G. Cora

Petroleo Brasileiro S/A, Exploration Dept.
Rio de Janeiro, Brazil

ABSTRACT

The Marlim and Albacora fields, which are located offshore Rio de Janeiro State in water depths ranging from 250 to more than 2000 m, cover areas of 350 and 235 km^2 and contain over 14.1 and 4.5 billion barrels of oil in place, respectively. The reservoir rocks of these fields are composed of Albian to Miocene sandstone turbidites.

The Albacora field structure is an anticline feature trending southwest-northeast. The Marlim accumulation combines structural and stratigraphic trapping elements. The reservoir pinches out against marls and shales to the north, west, and south, whereas the eastern, northeastern, and northwestern pool boundaries are defined by normal faults. Albacora is a multiplay oil accumulation whose older reservoirs are located in shallower water depths and whose younger reservoirs are located in deeper water depths. Marlim is a single reservoir oil field but is composed of several pools with different types of oil as a result of the combination of biodegradation and, at least, two phases of oil migration.

The Marlim and Albacora turbidite reservoirs are mostly associated with slope and continental rise deposits. The eastern portions of both fields are dominated by lobe-type sandstones whose successive coalescence resulted in sand bodies with great lateral continuity. To the northwest, these sand bodies become elongated, presenting geometry and facies of typical channel deposits. These channels are thought to be feeders of the lobe system to the southeast.

The development of the Albacora field is being gradually implemented from shallower to deeper waters. It will come on stream in three phases so that each phase will provide information, technology, and cash flow for the subsequent phase. Phase I started in October 1987 and continues presently with a daily production of 6000 m³ of oil and 300,000 m³ of associated gas. In the Marlim field, oil production is projected to start from a pilot system designed to identify the parameters needed to determine the final production systems. This pilot will consist of up to 10 satellite wells located in water depths from 600 to 800 m, with subsea completions, and each well will produce separately to a semisubmersible unit moored at a water depth of 600 m. It is scheduled to start in 1992 and is projected at about 8000 m³ of oil and 600,000 m³ of associated gas per day.

INTRODUCTION

The Marlim and Albacora giant fields are located offshore Rio de Janeiro State, about 110 km from the coast, in the northeastern portion of the Campos Basin—the most prolific petroleum province in Brazil (Figure 1). The water depths in the area of the fields range from 250 to more than 2000 m, and the depths of the pay zones vary from −2400 to about −3200 m. The reservoir rocks are turbidite sandstones deposited during the span of time from Albian to Miocene.

The Albacora field is composed of several vertically stacked accumulations covering an area of about 235 km² and containing a total volume of oil in place of 4.5 billion bbl (Figure 2A). Marlim field is composed of a single Oligocene producing horizon occupying an area of about 152 km². However, when the neighboring proved and potential accumulations are considered, the area increases to 350 km² (Figure 2B). This whole area, known as the Marlim Complex, contains a total estimated volume of oil in place of about 14.1 billion bbl.

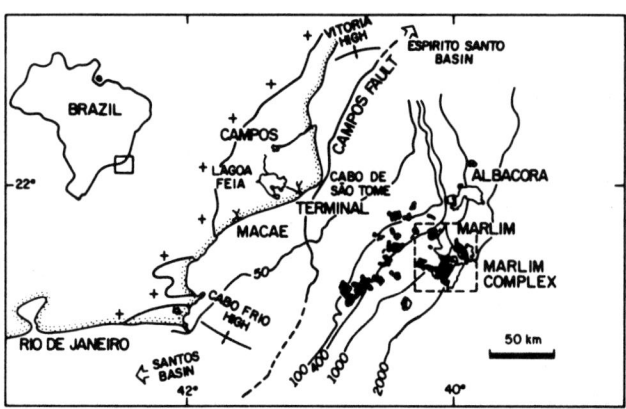

FIGURE 1. Location map of the Marlim and Albacora fields.

GEOLOGIC SETTING—BASIN GEOLOGY

Stratigraphy

The Campos Basin covers approximately 100,000 km² of offshore and onshore adjacent areas, extending roughly from 15 km inland down to places with isobaths greater than 3400 m. It is separated from the southwestern and northeastern adjacent basins by the Cabo Frio and Victória basement highs, respectively (Figure 1). The basin is filled with up to 9000 m of Early Cretaceous to Holocene sediments that are environmentally subdivided into three megasequences, from base to top: (1) a nonmarine megasequence, (2) a transitional megasequence, and (3) a marine megasequence. Figure 3 summarizes the stratigraphic sequences, with accepted stratigraphic nomenclature from Schaller (1973). In the Albacora field, oil accumulations occupy horizons ranging in age from the Albian to the Miocene in the marine megasequence, whereas in the Marlim field there is only one producing interval of Oligocene age. For further details of the stratigraphy, the reader is referred to Schaller (1973), Figueiredo et al. (1985), and Guardado et al. (1989).

Structural Features

The Campos Basin, as are other Brazilian sedimentary basins, is marked by an early Aptian regional unconformity that defines the boundary between two distinct structural styles (Figure 4A). Early Cretaceous block faulting affected mainly the basement volcanics and synrift deposits. Fault blocks are arranged in a series of horsts and grabens bounded by synthetic and antithetic normal faults with throws of up to 2500 m and oriented according to structural lineaments of the adjacent Precambrian shield. Post-unconformity strata are structured by the instability of the late Aptian evaporites, which resulted from the eastward tilting of the basin. This type of tectonism is believed to have controlled both sedimentary

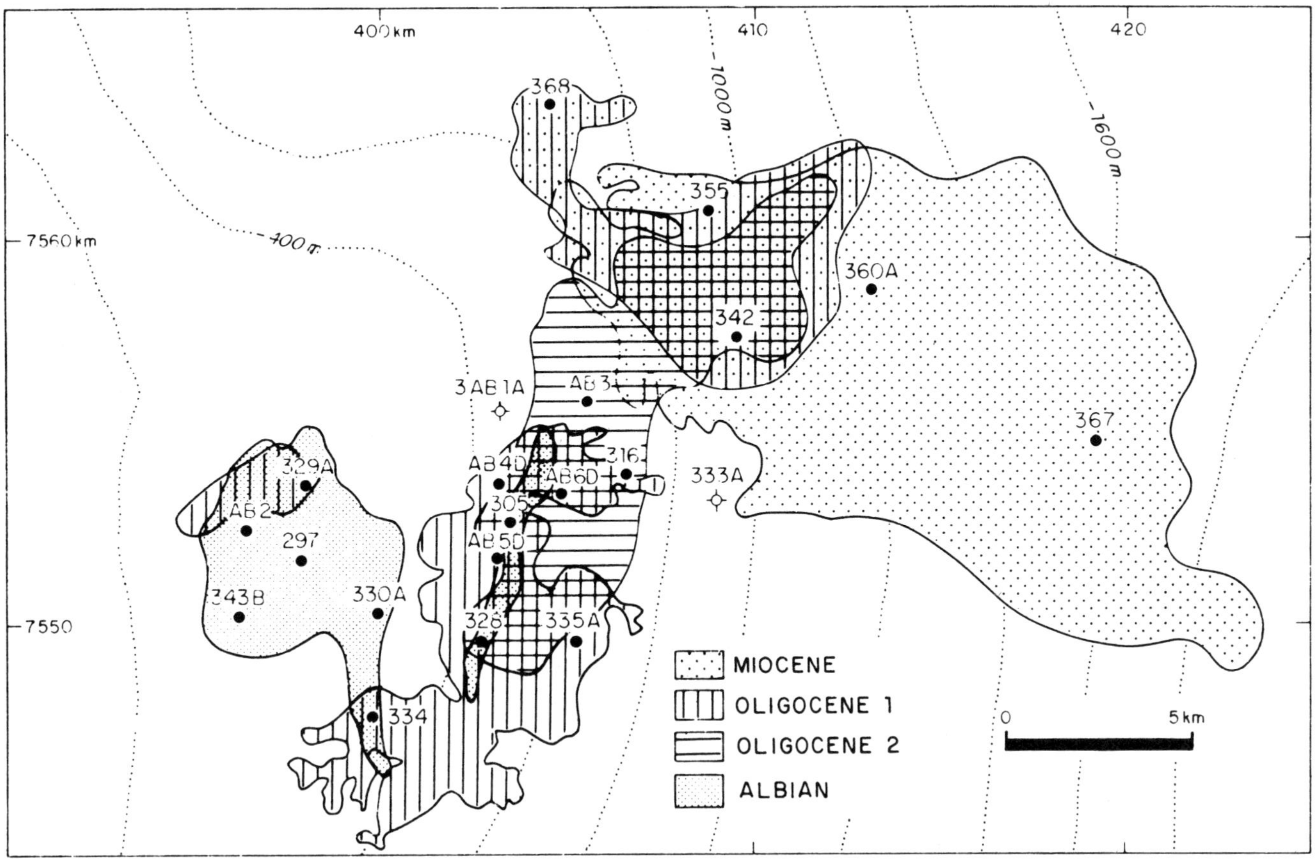

FIGURE 2A. Limits of oil pool in Albacora field.

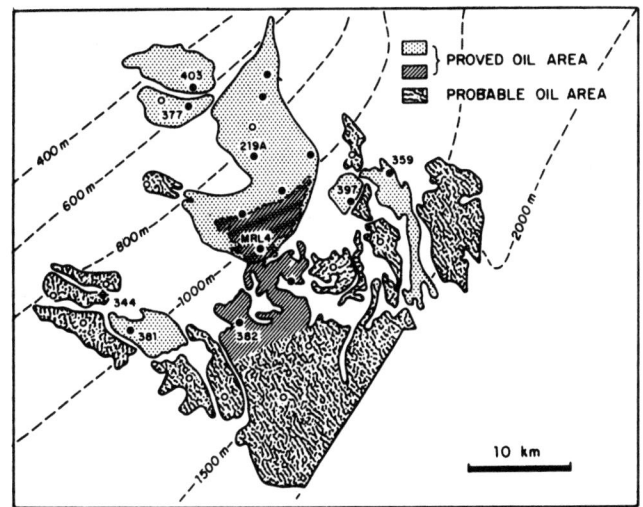

FIGURE 2B. Limits of oil pool in Marlim field.

facies and trap formation for most of the oil accumulations in the basin.

Another significant structural feature is a hinge line delineated by the Campos fault, which is oriented parallel to the basin margin (Figure 1). This hinge line bounds the shallower portion of the basin, where Tertiary sediments rest directly on basement rocks (Figure 4A). Further details can be seen in Ojeda y Ojeda (1982), Figueiredo et al. (1985), and Guardado et al. (1989).

Basin Evolution

The Campos Basin is thought to have originated during the breakup of the Gondwana in the Early Cretaceous. Intense volcanic activity and a tholeiitic basalt spill preceded the inception of rift valleys that were later filled with alluvial and lacustrine sediments. Carbonate deposition occurred, accumulating thick layers of coquinas that constitute oil reservoirs in some areas. With the end of the rift phase, the resulting relief was leveled by a regional unconformity. Postrift deposition started with the formation of the transitional megasequence and ended with the sedimentation of the marine megasequence. During these stages the controlling factors were thermal subsidence, global sea level changes, sediment supply, and paleoclimate. Details on the basin evolution can be found in Figueiredo and Mohriak (1984), Figueiredo et al. (1985), Gamboa (1986), Ojeda y Ojeda (1982), and Guardado et al. (1989).

Petroleum Potential

Most commercial oil production comes from the marine sediments, but geochemical evidence indicates that the

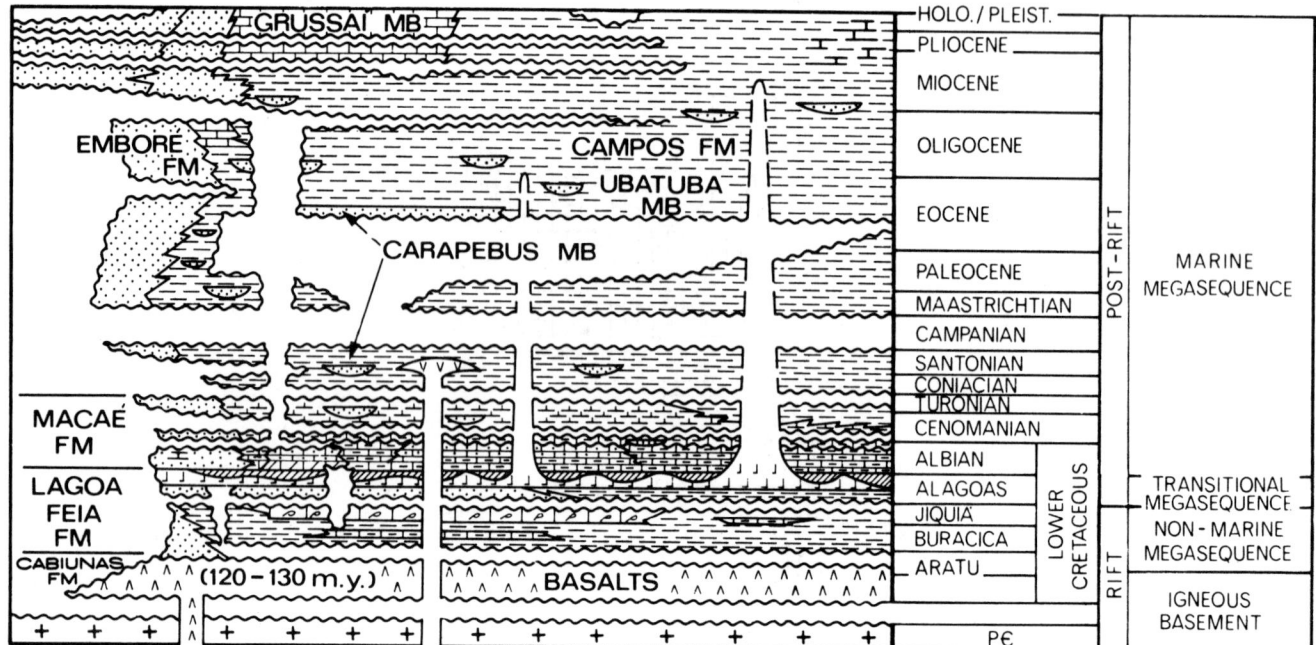

FIGURE 3. Stratigraphic column of Campos Basin (after Guardado et al., 1989).

main source rocks are the late Barremian to early Aptian lacustrine shales of the Lagoa Feia Formation, which are separated from the major reservoirs by a thick evaporitic section (Figure 4A). Therefore, the migration of oil from source to reservoirs required pathways through the salt layers. These pathways are thought to have been developed during the salt diapirism that caused fractures and faulting, further favoring oil movement toward younger reservoirs. Details of oil generation, source rock, and trapping can be found in Figueiredo and Mohriak (1984), Meister (1984), and Guardado et al. (1989).

HISTORY OF DISCOVERY

The discovery of Albacora and Marlim fields opened a new exploration frontier in areas of deep waters, not only for the Campos Basin, but also for other similar Brazilian sedimentary basins. In addition, it constituted a major effort in the research for technological solutions that made it economical to produce oil from deep water areas. The history of these discoveries started as early as 1972, when a seismic survey of the slope regions was executed on the Campos and other Brazilian sedimentary basins. Some of the seismic lines of this 20- to 25-km spacing survey led to the finding of the Marlim field. A regional mapping of the slope and continental rise was carried out in 1981, in water depths up to 400 m. Based on this mapping, the wildcat 1-RJS-219A was proposed to test a faulted anticline with a closure area of 30 km² at the top of the Cretaceous and a huge seismic amplitude anomaly in the Oligocene section that had an area of about 140 km² and was positioned on the flank of the above-mentioned structure. However, the proposed wildcat was not drilled until December 1984. Meanwhile, a detailed grid survey with line spacing of 2 km was shot and interpreted, based on which the well was relocated from a water depth of 660 m to one of 853 m in order to test the seismic amplitude anomaly at a more favorable point. The well reached the final depth of 3607 m in February 1985, within the Albian platform carbonates of the Lower Macaé Formation, but oil was found in the Oligocene target only, at a depth of 2617 m. The pay interval consisted of 75 m of very fine to fine-grained sandstone, with no oil-water contact. The drill-stem test produced 19° API oil at a daily rate of 381 m³, using a ½-in. choke. Following the discovery, four appraisal wells (3-RJS-319A, 3-RJS-325E, 3-RJS-326, and 3-MRL-1) were drilled in the Marlim field during 1985 and 1986, based on the widely spaced two-dimensional seismic survey available. These wells were positioned in water depths varying from 390 to 970 m and, with the exception of the 3-MRL-1 well, were all found to be oil producers.

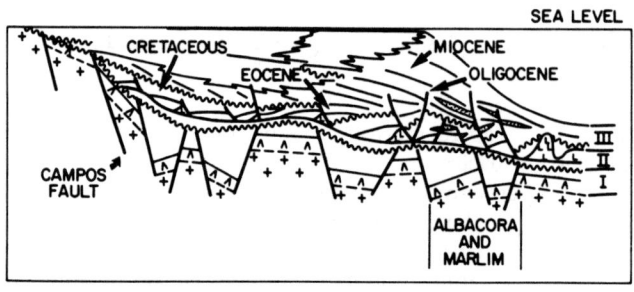

FIGURE 4A. Northwest-southeast diagrammatic cross section of the Campos Basin.

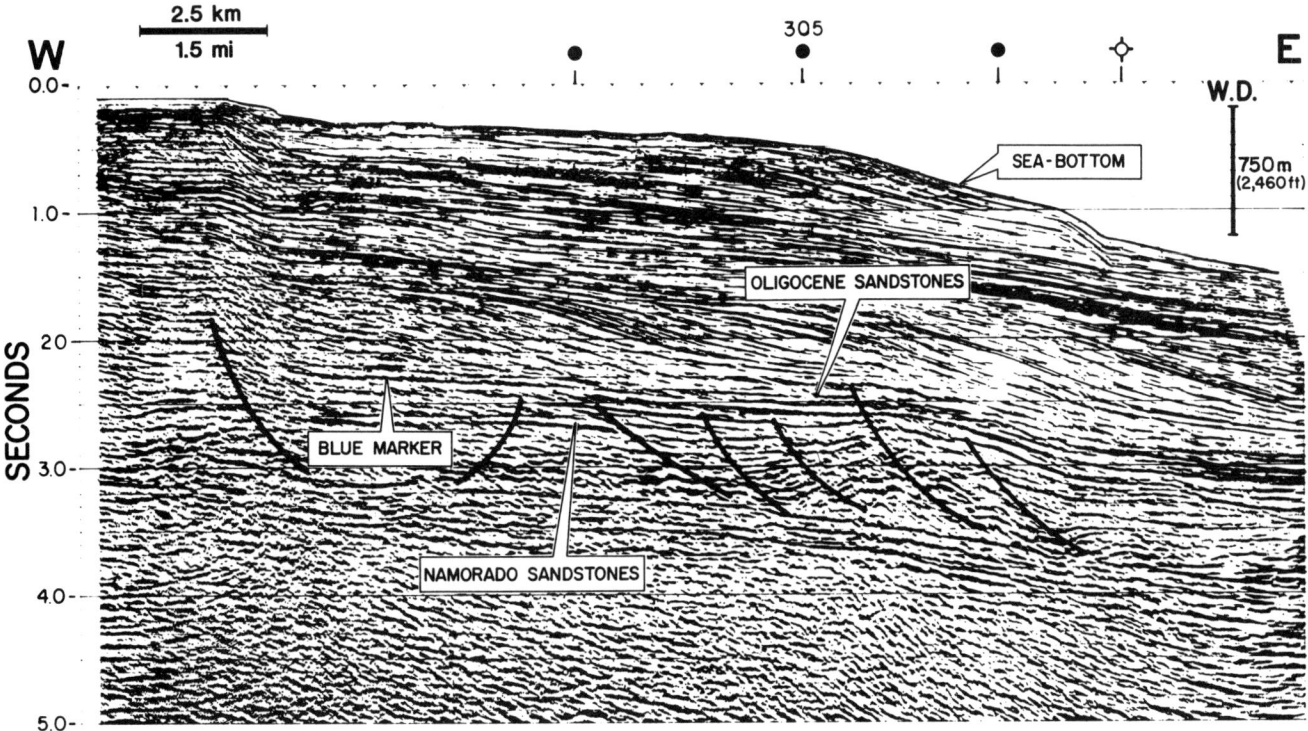

FIGURE 4B. Local signatures of the Albacora area (seismic section from Souza et al., 1989).

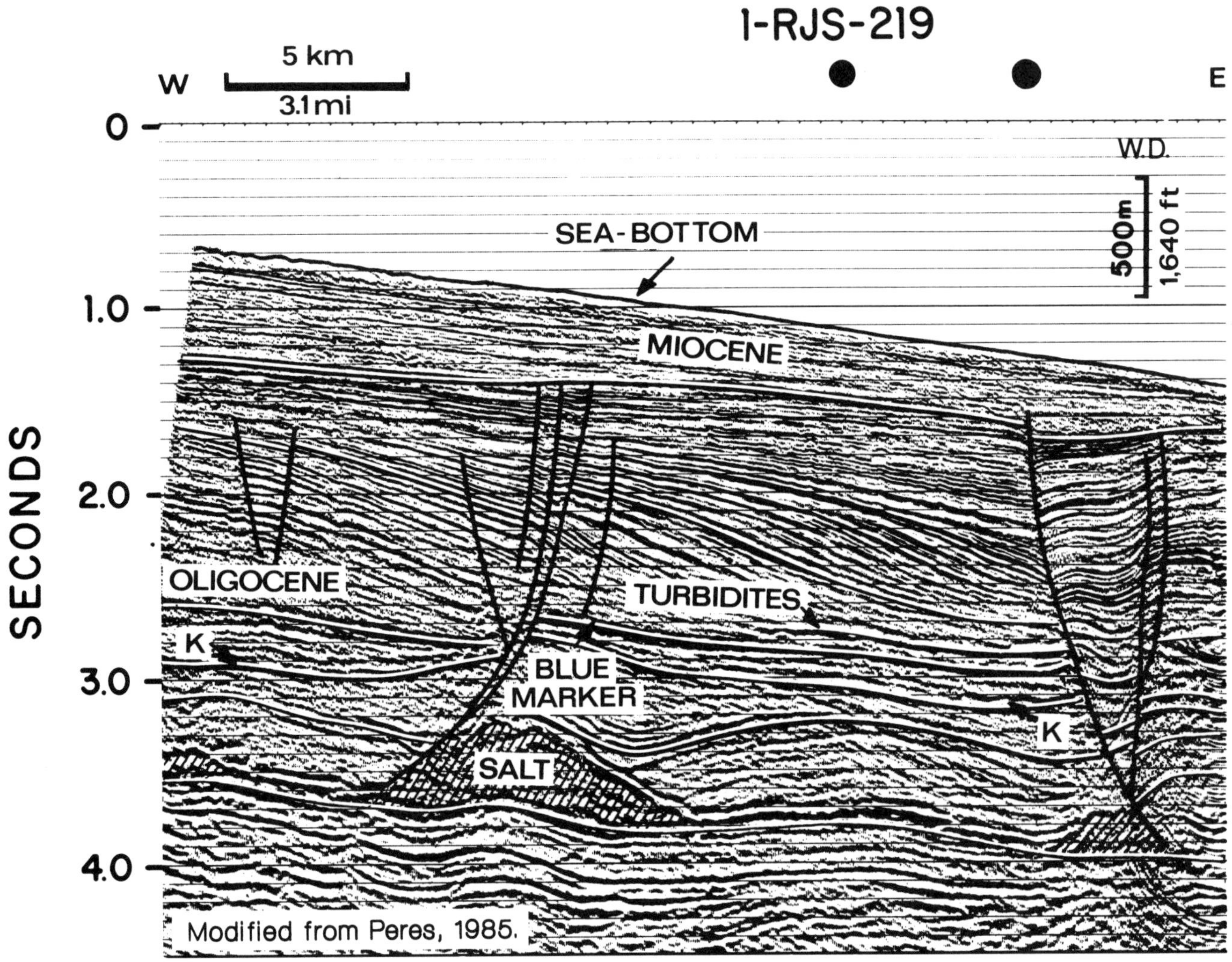

FIGURE 4C. Local signatures of the Marlim area (seismic section from Souza et al., 1989).

The Albacora field was discovered in September 1984 by the wildcat 1-RJS-297, whose approval was based on the above-mentioned seismic survey. This well was located at a water depth of 293 m. Its main targets were the lower Albian high-energy carbonates and a seismic amplitude anomaly interpreted as having been caused by a turbidite sandstone fan in the upper Albian section of the Macaé Formation. The well reached a final depth of 3606 m and cut an oil-impregnated sandstone interval with a thickness of 117 m and a top at −3212 m, which produced 27° API oil at an approximate daily rate of 300 m^3, after acid fracturing. This turbidite sandstone was interpreted as having caused the mapped amplitude anomaly. The well also crossed an interval with sandstone laminations from the Oligocene section whose seismic data indicated thickening toward the east and northeast. This expectation was verified by the wildcat 1-RJS-305 (drilled 5.7 km northeast of well 1-RJS-297), which penetrated an Oligocene sandstone body, thicker than 60 m, that produced 28° API oil at a flow rate of 460 m^3 per day using a ½-in. choke. The well also confirmed oil in the Albian sandstones, but in an accumulation independent of that of the exploratory well 1-RJS-297. Still based on the widely spaced 2D seismic survey, the appraisal well 3-RJS-316 extended the Oligocene sandstones 3.2 km northeast of the exploratory well 1-RJS-305.

In order to improve horizontal resolution and consequently optimize the appraisal and development of the fields, a 3D seismic survey was recorded during 1985, covering about 874 km^2 in the Marlim area and 480 km^2 in the Albacora area. Line spacing of the grid was 75 m, and trace spacing was 12.5 m. Total lengths of the seismic surveys were 19,525 and 6420 km for the Marlim and Albacora areas, respectively. Interpretation of the resulting 3D survey was carried out on an interactive workstation, placing emphasis on amplitude mapping. The highest negative amplitudes are likely to indicate hydrocarbon-bearing reservoirs, whereas lower-amplitude values tend to indicate marls, shales, or even barren sandstones. Stratigraphic elements such as channels and lobe deposits could be detected, and in some cases even the oil-water interface could be depicted (Rosa, 1987).

The interpretation of the Marlim 3D survey revealed clearly the limits of the Oligocene reservoir facies, showing also sedimentary features such as channels and lobes. Two turbidite systems were characterized, one extending beyond the southern limit of the survey area and the Marlim field itself. Based on the these amplitude maps, two appraisal wells have been drilled in the Marlim system (3-MRL-3 and 3-MRL-5), both resulting in oil producers in the Oligocene reservoir as expected. In addition, eight sites were selected, four of which have already been concluded and have resulted in oil producers in the Oligocene.

In the Albacora field, six reservoirs of various ages have been mapped—one Albian, one Eocene, three Oligocene, and one Miocene. Especially in the Oligocene reservoirs, strong reflector patterns disappear laterally over short distances, showing coalescence of the mapped turbidite systems. The Albian and Miocene sandstones present a rather simple stratigraphic pattern revealing continuity throughout the area. Based on these resulting maps, 12 exploratory and four development wells were drilled. Only two of these were dry, one reaching the Oligocene reservoirs below the oil-water contact (3-RJS-333A) and the other hitting a nonreservoir shaly turbidite facies displaying acoustic characteristics similar to those of the reservoir rocks (3-AB-1). Well 1-RJS-342 found gas, and well 3-RJS-355 found oil, in the Miocene reservoir. Based on a 2D seismic survey, wells 3-RJS-360 and 3-RJS-367 extended the Miocene accumulation toward the east, and the wildcat 1-RJS-368 extended the Oligocene pool to the extreme north, of the field. However, this extension well has discovered another independent accumulation.

GEOLOGY OF THE MARLIM AND ALBACORA FIELDS

Structure

The Marlim and Albacora fields, and their relationship with the major structural elements of the Campos Basin, are shown in Figures 4A, 4B, and 4C. Two fault systems are apparent, one affecting the Cretaceous interval only and the other influencing both the Cretaceous and Tertiary layers. Both fault systems resulted from salt collapse, and both are interpreted as having served as pathways for oil migration up to the fields' accumulations.

In the Albacora field, most of the accumulations resulted from the trapping of oil in a general anticline oriented in the southwest-northeast direction and containing a southwestward secondary domic feature (Figure 5). This structure is reflected throughout the Cretaceous and Tertiary strata; however, the Albian reservoirs are also affected by a series of approximately north-south normal faults with throws of up to 300 m. Some of these faults contribute to the definition of the accumulations' boundaries but are not considered restrictive to the flow of fluids (Figure 6). The presence of the secondary domic feature in the area of well 329A resulted in the development of two oil pools independent of each other but having common aquifers for both the Oligocene and Albian reservoirs.

The Marlim accumulation combines structural and stratigraphic trapping elements. The reservoir pinches out against marls and shales to the north, west, and south, whereas the eastern, northeastern, and northwestern pool boundaries are defined by normal faults (Figure 7).

Stratigraphy

The stratigraphic framework of the Marlim and Albacora areas is that defined for the whole basin, as illustrated in Figure 3. However, the Lower Cretaceous Lagoa Feia Formation, which is known for its high potential for hydrocarbon generation, has not been reached by the field wells. The Albian Macaé Formation is composed of a shallow water carbonate section overlain by deep water strata, constituted of shales, marls, calcilutites, and turbidite sandstones. These sandstones are late Albian in age

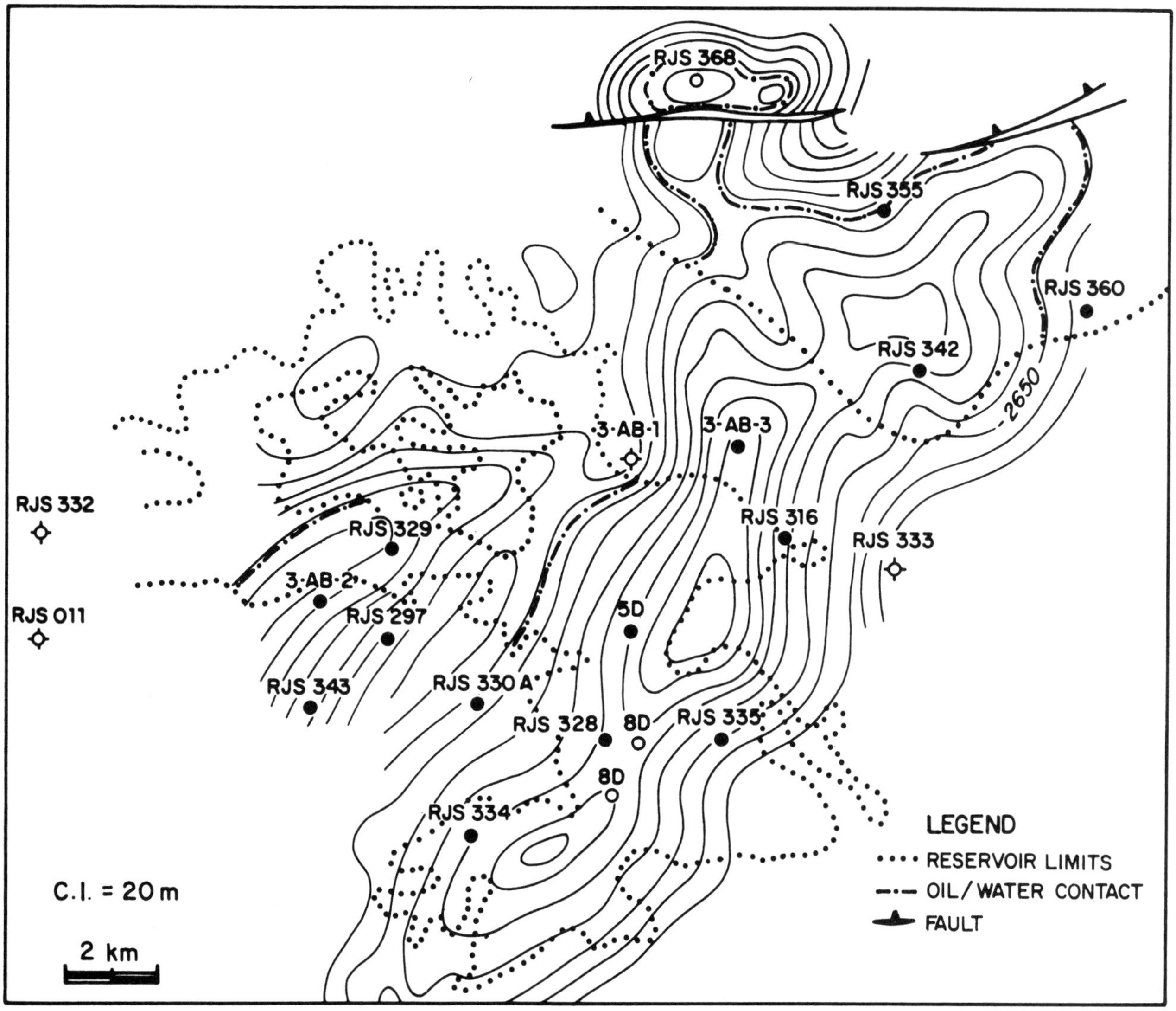

FIGURE 5. Structural contour map on the top of Oligocene 1 reservoir, Albacora field.

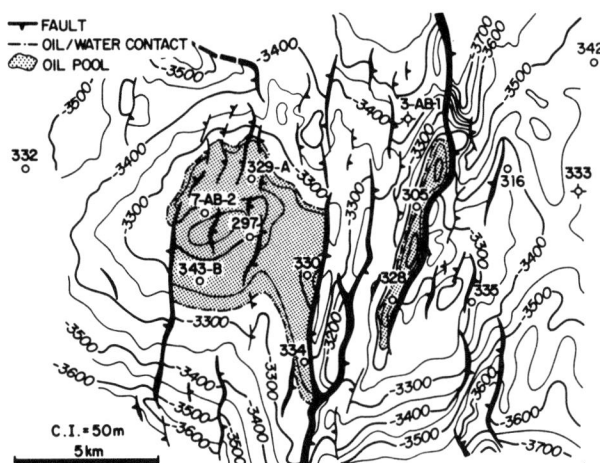

FIGURE 6. Structural contour map on the top of Albian (Namorado) reservoir, Albacora field.

and are the oldest oil reservoirs in the Albacora field. From the end of deposition of the Macaé Formation to the end of the Cretaceous, and during almost all of the Cenozoic, clastic sedimentation persisted in a deep water environment, accumulating a thick sequence of turbiditic sandstones and shales that comprise the Campos Formation. The shaly section is formally designated as the Ubatuba member, whereas the sandy intervals are called the Carapebus member. These sandstones are the main reservoirs in the Albacora field and the only oil producers in the Marlim field. Figures 8 and 9 illustrate the stratigraphic and structural relationships of these sandstones in the fields.

The final stage of basin filling is defined by a prograding sequence of slope and shelf deposits formed by the Ubatuba and Emborê formations, respectively. However, in the area of the fields, the Emborê sediments of Miocene-Pliocene age are not present.

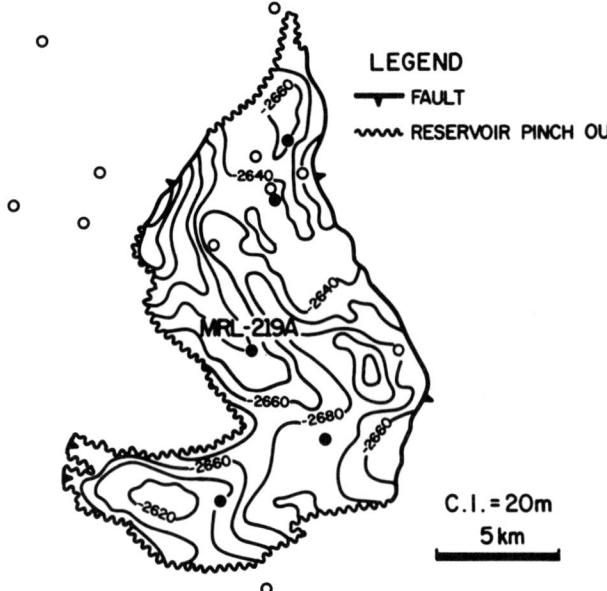

FIGURE 7. Structural contour map on the top of Oligocene reservoir, Marlim field.

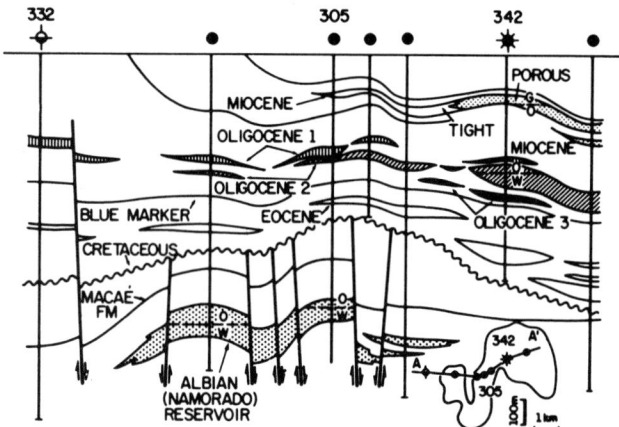

FIGURE 8. East-west geologic cross section of the Albacora field.

RESERVOIR CHARACTERISTICS

The Albacora field is a multiplay oil accumulation with eight different pools. The older reservoirs are located to the southwest, and the younger ones occupy regions in waters that become increasingly deeper until depths of more than 2000 m are reached (Figure 2A). Following this tendency, the oil grades vary from 26 to 19° API, possibly as a result of biodegradation associated with the decrease in reservoir temperatures. Marlim is a single-reservoir oil field, but is composed of several oil pools with different types of oils (26 to 19° API). This difference in oil quality is attributed to the existence of two pulses of hydrocarbon migration. The heaviest oils reached the reservoirs earlier in basin history, when the temperature was low enough

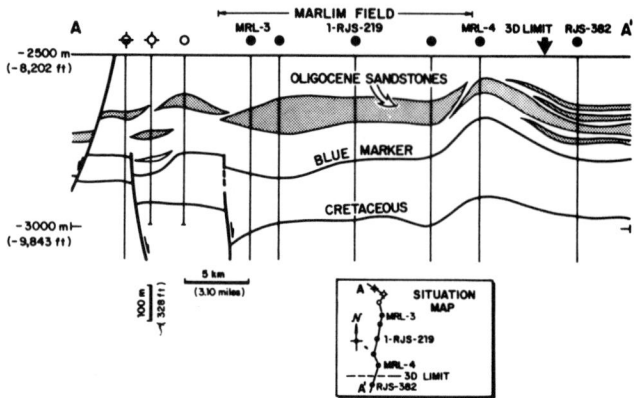

FIGURE 9. North-south geologic cross section of the Marlim field (after Souza et al., 1989).

to cause biodegradation. The lighter oils are thought to be mixtures of these heavier oils with more mature oils of a second migration pulse that reached the reservoirs when they were deeper and when their temperatures were high enough to prevent the action of bacteria.

The other reservoir characteristics of the Marlim and Albacora fields are summarized in Table 1. These characteristics have been derived mainly from seismic, especially amplitude anomaly, well log analysis, and drill-stem tests. Core analyses contributed little to the reservoir description, because recovery was extremely low. In general, the intervals recovered were either shale and marl or reservoir rocks of very poor quality. On the other hand, the best effective permeabilities were obtained through drill-stem tests in sections with no core recovery (Table 1).

In Albacora field as well as in Marlim field, all reservoirs are turbidite deposits mostly associated with the slope and continental rise. The eastern portions of both fields are dominated by lobe-type deposits whose successive coalescence resulted in sand bodies with great lateral continuity, composed mainly of fine to medium-grained massive sandstones (Figure 10). To the northwest, these sand bodies become elongated in the northwest-southeast direction, presenting geometry and facies characteristics typical of channel deposits. These channels are thought to be feeders of the lobe system to the southeast. Typical facies of reworked sandstones by contour currents are detected in the available cores for both lobe and channelized deposits (Figure 11). Special coring techniques using nitrogen permitted rock recovery of 90% in a well of the Marlim field. Routine petrophysical analyses carried out on plugs from this frozen core measured porosities on the order of 30% for both massive and contourite lithofacies, confirming the results obtained by quantitative log analysis. The corresponding permeability value for the massive sandstone was 1200 md; however, for the contourite, this value decreased to only about 630 md. Preliminary modal analyses indicated a proportion of approximately 30% of mainly fresh potassium feldspars. This low fraction of altered fragments, in comparison with other Tertiary reservoirs of the basin, suggests that feldspar-originated diagenetic clays are not expected to pose problems to

TABLE 1. Reservoir characteristics of the Albacora and Marlim fields.

Reservoir Parameter	Albacora						Marlim
	Miocene	Olig. 1	Olig. 2	Olig. 3	Eocene	Albian	Oligocene
Bathymetry (m)	700–2,000	300–650	300–1,100	500–650	900–1,000	200–500	500–1,100
Depth (m)	2,350–2,480	2,500–2,645	2,550–2,645	2,600–2,645	2,830–2,930	3,180–3,260	2,500–2,700
Average net pay (m)	16.8	18.5	16.1	19.0	10.5	29.9	44.5
Porosity (%)	27.8	24.5	23.4	29.0	26.4	17.0	25.0
Water saturation (%)	22.8	20.1	24.9	20.0	18.3	30.7	14.0
Oil gravity (°API)	17.6–25.6	26.0–29.0	24.0–27.0	—	29.7–29.8	25.4–30.6	19.0–20.0
Initial solution GOR (m^3/m^3)	60–70	95.0	97.0	—	118.0	65.0	80.0
Effective permeability (md)	1,460–1,750	280–3,460	433–3,270	—	—	11–71	1,325–5,372
Productivity index (m^3/day per kg/cm^2)	6–37	7–79	9–40	—	28–43	1–6	20–300
Oil in place ($10^6/m^3$)	226.0	121.1	209.6	4.8	14.5	73.4	1,298.5
Oil in place (10^6 bbl)	1,421.54	761.72	1,318.38	30.19	91.20	461.69	8,167.57
Main oil-water contact (water table) (m)	Not known	2,645	2,645	2,645	Not known	3,260	2,740

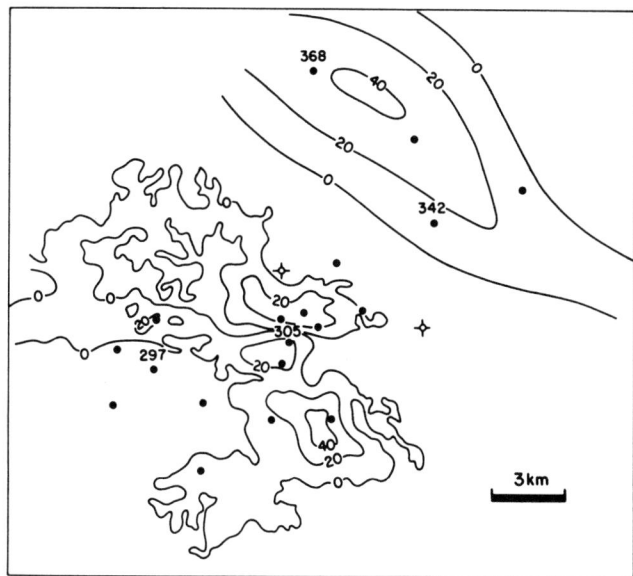

FIGURE 10A. Isopach map of Oligocene 1, Albacora field.

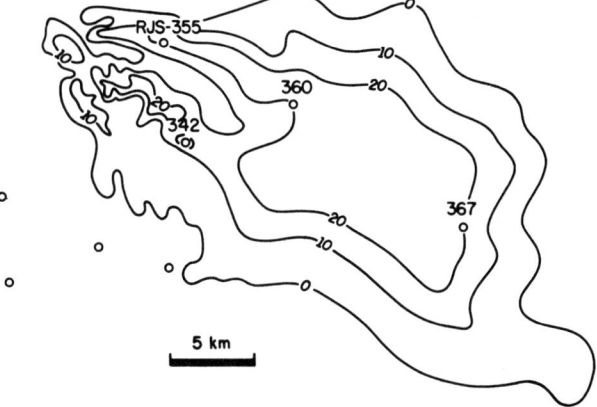

FIGURE 10B. Isopach map of Miocene, Albacora field.

production operations, except the migration of detrital fragments of the sandstone itself.

The Albacora Albian reservoir is similar to those from the Tertiary section in relation to the environment of deposition and sedimentary facies, but is characterized by intense calcite cementation occurring in localized areas of the reservoir as opposed to continuous laminae. Indications of shale are also present in the well logs, and, as the calcite cementation does, these shale layers tend to increase with depth. Because they are both restrictive to the flow of fluids, the influence of the associated aquifer is expected to be reduced to a greater extent than the vertical flow of oil within the reservoir, as evidenced by measured reservoir pressure. Average porosity and water saturation are 17 and 30%, respectively. Permeability varies from zero to 500 md, but generally is less than 100 md. Quantitative log analyses and recovered core samples indicate that both porosity and permeability decrease with depth. This can be seen on plots of core porosity vs.

FIGURE 10C. Isopach map of Oligocene, Marlim field.

depth, indicating that productive capacity is greatest near the top of the reservoir, and that the ability of the underlying associated aquifer to provide driving pressure is reduced (Figure 12).

DEVELOPMENT AND PRODUCTION PROJECTS

The development of the Albacora field is being gradually implemented from shallower to deeper waters. It will come on stream in three phases in such a way that each phase will provide information, technology, and cash flow for the subsequent phase. Figure 13 shows schematically the global development plan and its three stages.

Phase I started in October 1987 and consisted of two parts. Initially, six wells were completed (four in the Albian and two in the Oligocene sandstones), with wet christmas trees in areas with water depths of up to 420 m. These wells were tied to a submarine manifold located at a water depth of 230 m and then to a tanker using a single-point mooring buoy. Presently, eight more wells are being incorporated into the system through a second diver-assisted manifold, six of which came on stream during 1989 (one in the Albian and five in the Oligocene reservoirs), increasing the daily production to about 6000 m^3 of oil and 300,000 m^3 of associated gas.

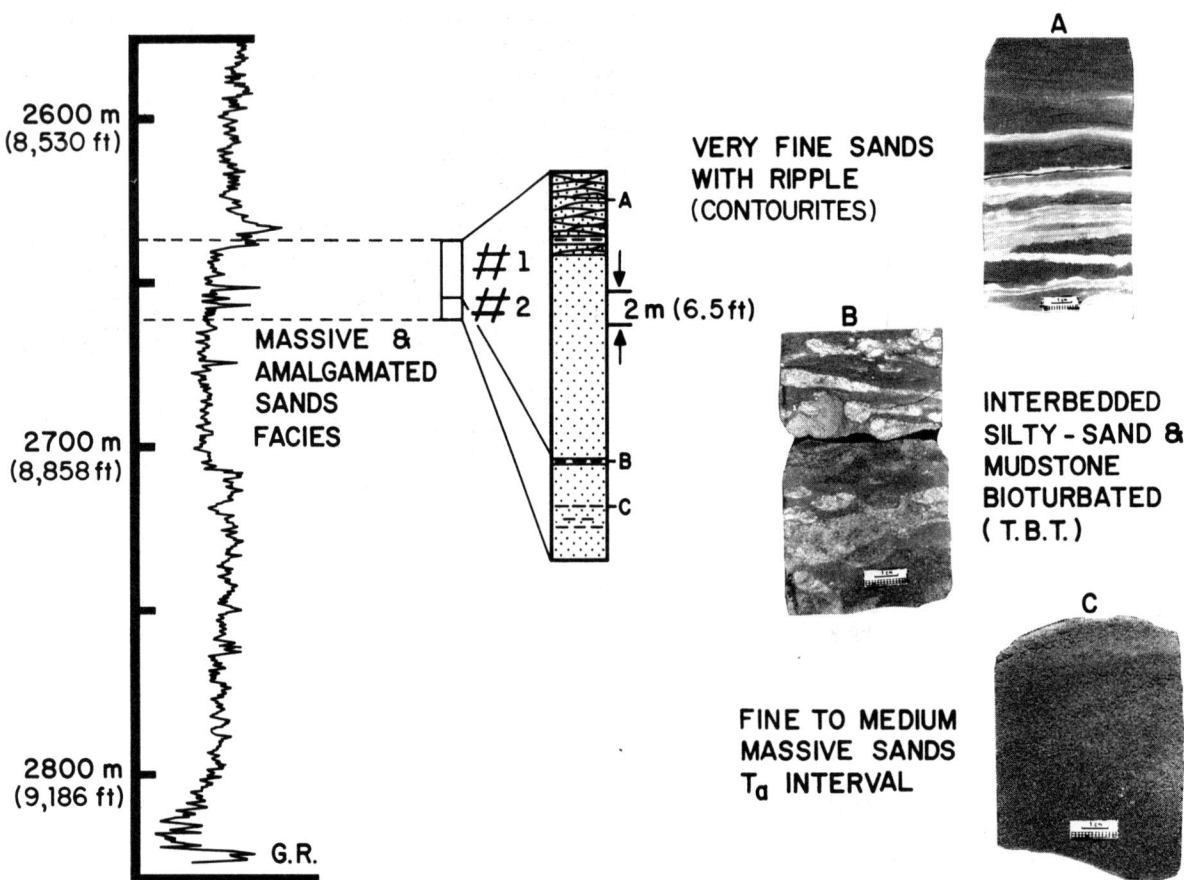

FIGURE 11A. 1-RJS-219A blanket sandstones, lobe system facies, Oligocene, Marlim field (from Guardado et al., 1989).

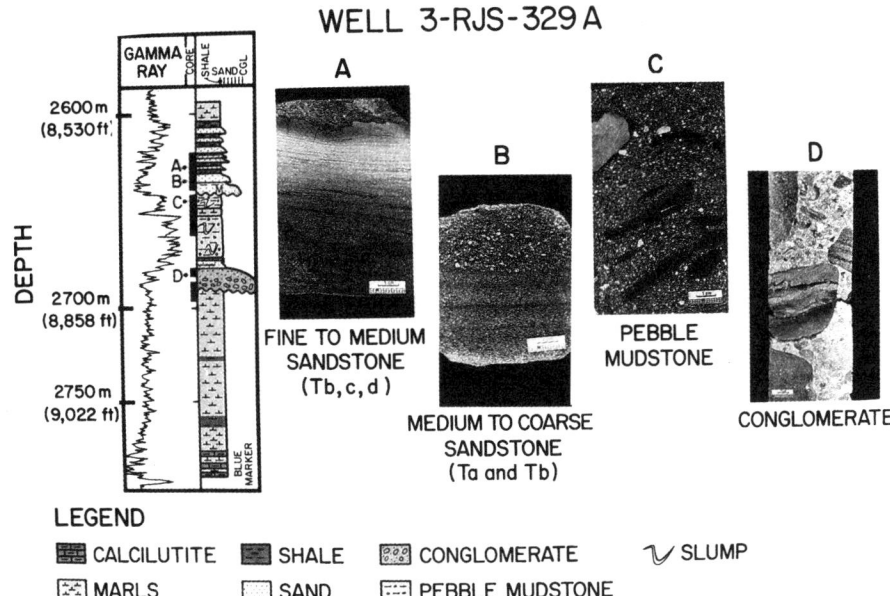

FIGURE 11B. Main reservoir facies, Oligocene, Albacora field.

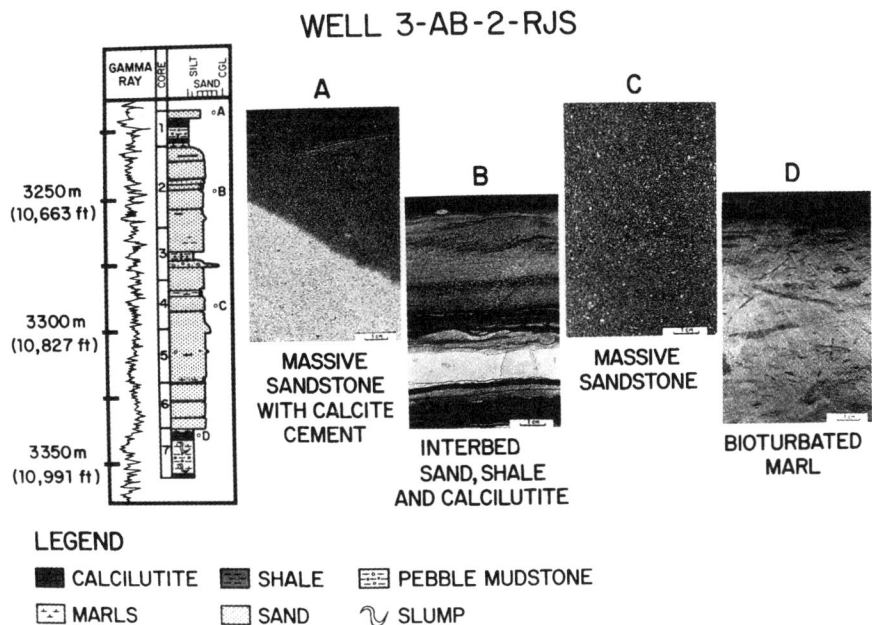

FIGURE 11C. Main facies, Albian, Albacora field (modified from Souza Cruz et al., 1987).

Phase II is scheduled to start in 1992 and is projected to develop the Albian and Oligocene reservoirs entirely. The company plans to complete 116 production wells (46 in the Albian and 70 in the Oligocene reservoirs) connected through submarine manifolds to semisubmersible processing units. Peak production is expected to reach 20,000 m³ of oil and 1.6 million m³ of associated gas per day.

Phase III will conclude the development of the field with the drilling of about 95 production wells in waters deeper than 800 m, with subsea completion and connections to diverless manifolds, and then to semisubmersible

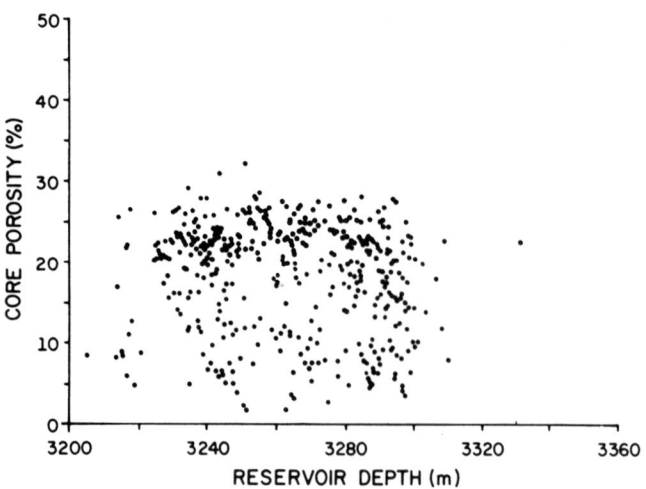

FIGURE 12. Core porosity vs. reservoir depth for the Albian reservoirs.

floating production units. Peak production is projected to reach 50,000 m³ of oil and 5.4 million m³ of associated gas per day.

In the Marlim field, hydrocarbon production is projected to start from a pilot system designed to identify the parameters needed to determine the area's final production systems. This pilot will include the Marlim field itself and three adjacent areas (1-RJS-377, 1-RJS-403, and 4-RJS-413). Starting in 1992, it will consist of up to ten satellite wells located in water depths from 600 to 800 m with subsea completions, where each well will produce separately to a semisubmersible unit moored at a water depth of 600 m. Production is projected at about 8000 m³ of oil and 600,000 m³ of associated gas per day.

The definitive conception of the Marlim field production system foresees the development of the field in two phases. Phase I will consist of two platforms with a total of 32 producing and 20 water-injecting wells with a spacing of 1000 m, positioned in the area of lowest reservoir depths. Commercial production will start in January 1994,

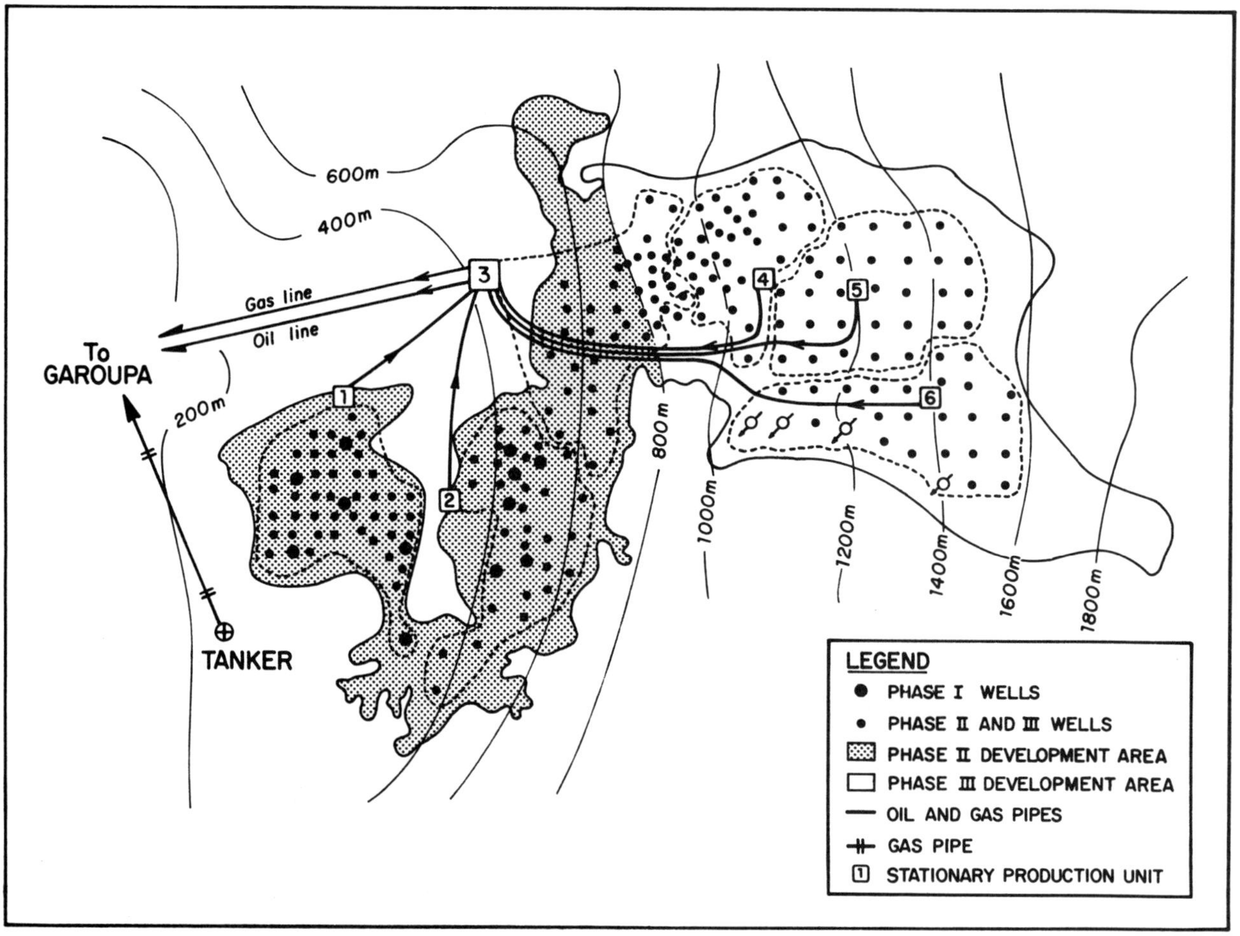

FIGURE 13. Development strategy of the Albacora field.

and the peak production rate is expected to reach 30,600 m³ per day. In Phase II, five additional platforms will be incorporated into the field production scheme in order to complete the global exploitation project. The whole system will consist of 145 wells: 109 producers, 34 water injectors, and two gas injectors. High production rates are projected on the basis of effective permeability and productivity indices obtained during drill-stem testing. Peak production is expected to reach 56,000 m³ of oil per day (Figure 14).

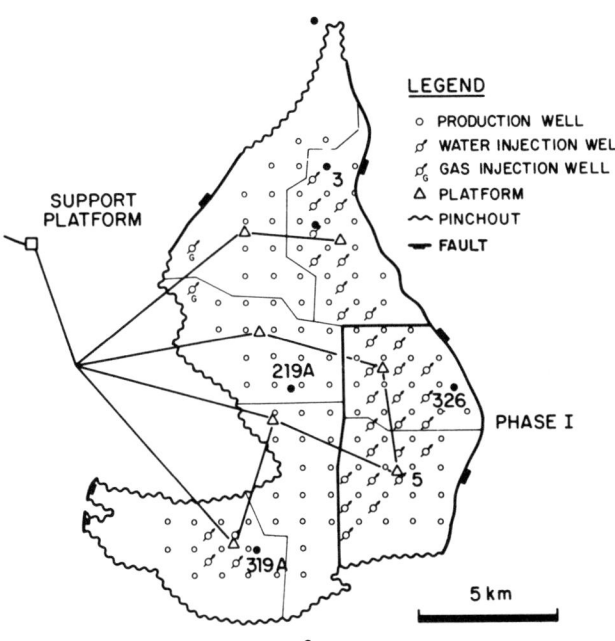

FIGURE 14. Drainage grid, locations of production units, and flowing facilities for the Marlim field.

REFERENCES CITED

Figueiredo, A. M. F., and W. V. Mohriak, 1984, A tectônica salífera e as acumulações de petróleo da Bacia de Campos, in 33° Congresso Brasileiro de Geologia, p. 1380-1394.

Figueiredo, A. M. F., M. Pereira, W. U. Mohriak, and L. A. Trinadade, 1985, Salt tectonics and oil accumulation on Campos Basin, offshore Brazil (abs.): AAPG Bulletin, v. 69, p. 255.

Gamboa et al., 1986, Evidências de variações do nível do mar durante o Oligoceno e suas implicações faciológicas, in 34° Congresso Brasileiro de Geologia, v. 1, p. 8-22.

Guardado, L. R., L. A. P. Gamboa, and C. F. Lucchese, 1989, The petroleum geology of the Campos Basin, Brazil: a model for a producing Atlantic type basin: AAPG Special Basin Series, Divergent passive margin basins, v. 48, p. 3-79.

Meister, E. P., 1984, Historical geology of petroleum in Campos Basin, Brazil: AAPG Annual Meeting, San Francisco.

Peres, W. E., 1985, Interpretaçaõ sismica na area do 1-RJS-219: Rio de Janeiro, PETROBRÁS Internal Report, 5 p.

Ojeda y Ojeda, A. A., 1982, Structural framework, stratigraphy and evolution of Brazilian marginal basins: AAPG Bulletin, v. 66, p. 732-749.

Rosa, A. L. R., 1987, The Albacora field: a case history of seismic amplitude mapping, in Expanded abstracts with biographies, 57th Annual International SEG Meeting, New Orleans, October 11-16, 1987, p. 499-501.

Schaller, H., 1973, Estratigrafia da Bacia de Campos: PETROBRÁS Internal Report, 20 p.

Souza, J. M., J. C. Scarton, and A. Candido, 1989, The Marlim and Albacora fields: geophysical, geological, and reservoir aspects: 21st Annual OTC (Houston, May 1-4) Proc. p. 109-118 (Paper 5894).

Souza Cruz, C. E., S. L. S. Barrocas, and C. J. Appi, 1987, Modelo deposicional dos reservatórios turbidíticos Oligocênicos/Eomiocênicos do Campo de Albacora, Bacia de Campos, Brasil: Boletim Geociências PETROBRÁS, v. 1, n. 2, p. 215-223.

Chapter 9

The Linguado, Carapeba, Vermelho, and Marimbá Giant Oil Fields, Campos Basin, Offshore Brazil

Paulo Márcio C. Horschutz[1]
Luiz Carlos S. de Freitas[1]
Carlos Varela Stank[2,3]
Alberto da Silva Barroso[2,3]
Wagner Maia Cruz[4]

PETROBRÁS
Rio de Janeiro, Brazil

ABSTRACT

About 24 hydrocarbon accumulations have been discovered in the Campos Basin, including four giant fields in shallow to moderate water depths, in the period from 1978 to 1988.

Linguado field is located in the extreme south of the producing area of Campos Basin, in water depths ranging from 95 to 110 m. The pool was discovered by the wildcat 1-RJS-49 in May 1978. The reservoir rocks, which occur between −1700 and −3000 m, are Hauterivian fractured basalts, Barremian pelecypod coquinas, Albian oolitic and oncolitic calcarenites, and, secondarily, Cretaceous turbidite sandstones. The main reservoir is formed by coquinas, which contain 80% of the total original recoverable oil volume, estimated at 130.15 million bbl. The field is located on a regional high, and the accumulation is strongly controlled by stratigraphic and diagenetic factors. High-quality oil (29 to 32° API) is produced through two floating production systems, and the cumulative oil production amounted to 82.25 million bbl as of December 1990.

Carapeba and Vermelho fields are situated at the northern limit of the Campos Basin producing area and, together with the smaller Pargo field, make up the so-called Northeastern Pole of the Campos Basin. Carapeba was discovered in February 1982 by the wildcat 1-RJS-193A and has an estimated recoverable oil volume of 183.81

---

[1]Linguado field
[2]Carapeba field
[3]Vermelho field
[4]Marimbá field

million bbl. Production comes mainly from two Late Cretaceous turbidite sandstone reservoirs. Vermelho field was discovered by the wildcat 1-RJS-241A in December 1982, and its main reservoir is formed by a massive Eocene turbidite sandstone. The estimated recoverable oil volume amounts to 121.55 million bbl. Both Carapeba and Vermelho fields are structural traps associated with the development of subtle anticlines caused by salt tectonics. The fields are gradually being put on stream through five fixed platforms installed in water depths ranging from 70 to 90 m.

Marimbá field, discovered in March 1984 by the wildcat 1-RJS-284 drilled in a water depth of 383 m, is considered the first deep water oil strike in Campos Basin. The field has an estimated recoverable volume of 174.18 million bbl of good-quality (28° API) oil in highly permeable Late Cretaceous turbidite sandstones, trapped under structural-stratigraphic conditions. The Marimbá field has not yet been completely developed, but a floating production system is producing about 22,000 BOPD from four exploratory wells completed in water depths ranging from 383 to 485 m.

INTRODUCTION

Campos Basin is located offshore Rio de Janeiro State, Brazil (Figure 1). Its exploration activities started at the beginning of the 1970s, and the first discovery took place in 1974. To date, about 440 exploratory wells have been drilled and 48 accumulations discovered with a volume of oil in place (VOIP) of approximately 33.73 billion bbl—i.e., 68% of the total discovered in Brazilian territory.

Between 1978 and 1984, about 40 accumulations in shallow to moderate water depths (80 to 600 m) were discovered on the continental shelf of the basin. These include the giant oil fields of Linguado, Carapeba, Vermelho, and Marimbá, with original recoverable oil volumes of 130.15, 183.81, 121.55, and 174.18 million bbl, respectively (Table 1). Together with the giant oil fields of Marlim, Albacora, and Namorado, they are the most important fields in the basin (Figure 2).

The reservoirs are composed of siliciclastics, carbonates, and fractured igneous rocks. They vary in age from the Hauterivian (Early Cretaceous) to the Miocene. The trapping mechanisms involve structural, stratigraphic, diagenetic, and paleogeomorphic elements. The development and production of each field are being undertaken by different methods, which vary primarily with the water depth of the accumulations, and also with the available technology.

Campos Basin is one of the most prolific Atlantic-type passive margin basins in the world. This can be explained by the adequate timing among the generation, migration, and trapping of oil—principally in structures associated with salt tectonics. Great quantities of oil were expulsed from the rift lacustrine shales of the Lagoa Feia Formation during the Tertiary, which migrated to previously formed structures. The salt tectonics provided the routes for and the mechanisms of trapping for the reservoirs structurally above the source rock (Figueiredo and Mohriak, 1984).

REGIONAL GEOLOGY

Campos Basin, located along the coast of southeastern Brazil, has an area of 100,000 km^2, extending to the bathymetric contour of 3400 m. The Victoria high to the north, and the Cabo Frio high to the south, are considered to be the boundaries of the Espirito Santo and Santos basins, respectively.

The origin of the Campos Basin is related to the Early Cretaceous breakup of the Gondwana supercontinent. Its tectonic evolution may be characterized by three distinct phases, from bottom to top, namely: (1) rift, (2) quiescent tectonics, and (3) postrift (drift). In these phases, nonmarine, transitional-evaporitic, and marine megasequences, respectively, were deposited (Asmus and Ponte, 1973; Ponte and Asmus 1978; Chang et al., 1988).

The rift phase started in the late Neocomian with the formation of a rift-valley system between South America and Africa. In this rift, terrigenous and carbonate sediments were deposited from a fluvial-lacustrine environment known as the Lagoa Feia Formation. They were deposited over the volcanic-sedimentary rocks of the Cabiúnas Formation, which are positioned at the section base and considered the economic basement of the basin

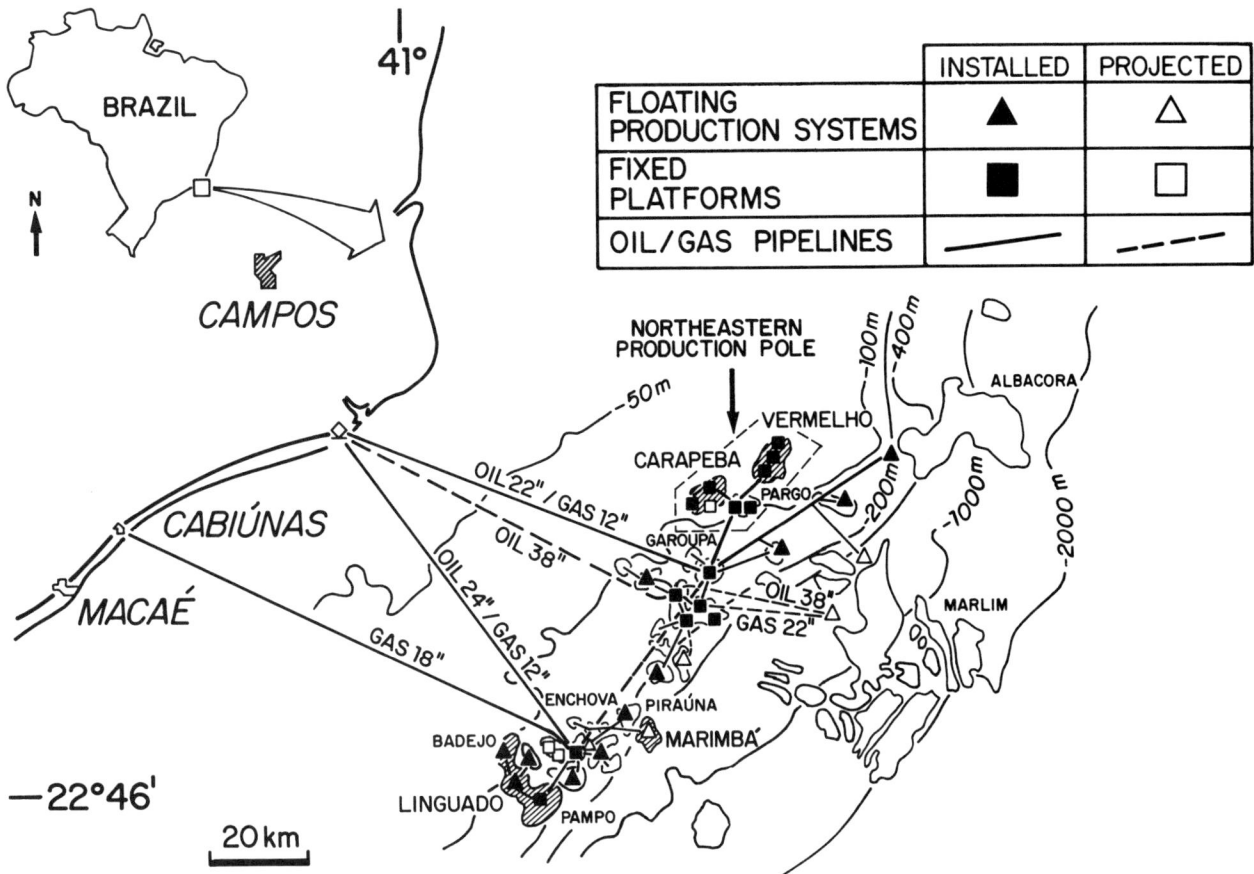

FIGURE 1. Location map of the Campos Basin oil fields.

TABLE 1. Summary of field volumes and production characteristics, December 1990.

Field	Reservoir	VOIP (mm bbl)	Rec. oil (mm bbl)	Reserves (mm bbl)	R.R. (%)	Cumulative production (mm bbl)	Production (BOPD)	Number of productive wells
Linguado	Maastrichtian	18.606	1.861	1.861	10.00	—	—	—
	Albian	167.553	23.896	11.731	15.40	12.165	7,106	3
	Barremian	359.706	104.395	34.759	30.00	69.636	18,610	8
	Neocomian	4.151	—	—	—	0.447	—	—
	Total	550.016	130.152	48.351	—	82.248	25,716	11
Carapeba	Eocene	21.977	1.453	1.453	13.60	—	—	—
	Cretaceous	620.056	182.360	164.786	32.63	17.574	32,984	34
	Total	642.033	183.813	166.239	—	17.574	32,984	34
Vermelho	Oligocene	28.267	1.333	0.767	13.10	0.566	1,044	2
	Eocene	333.312	120.221	100.250	36.07	19.971	43,228	55
	Total	361.580	121.554	101.017	—	20.537	44,272	57
Marimbá	Cretaceous	465.063	174.176	145.733	45.00	28.443	21,946	5

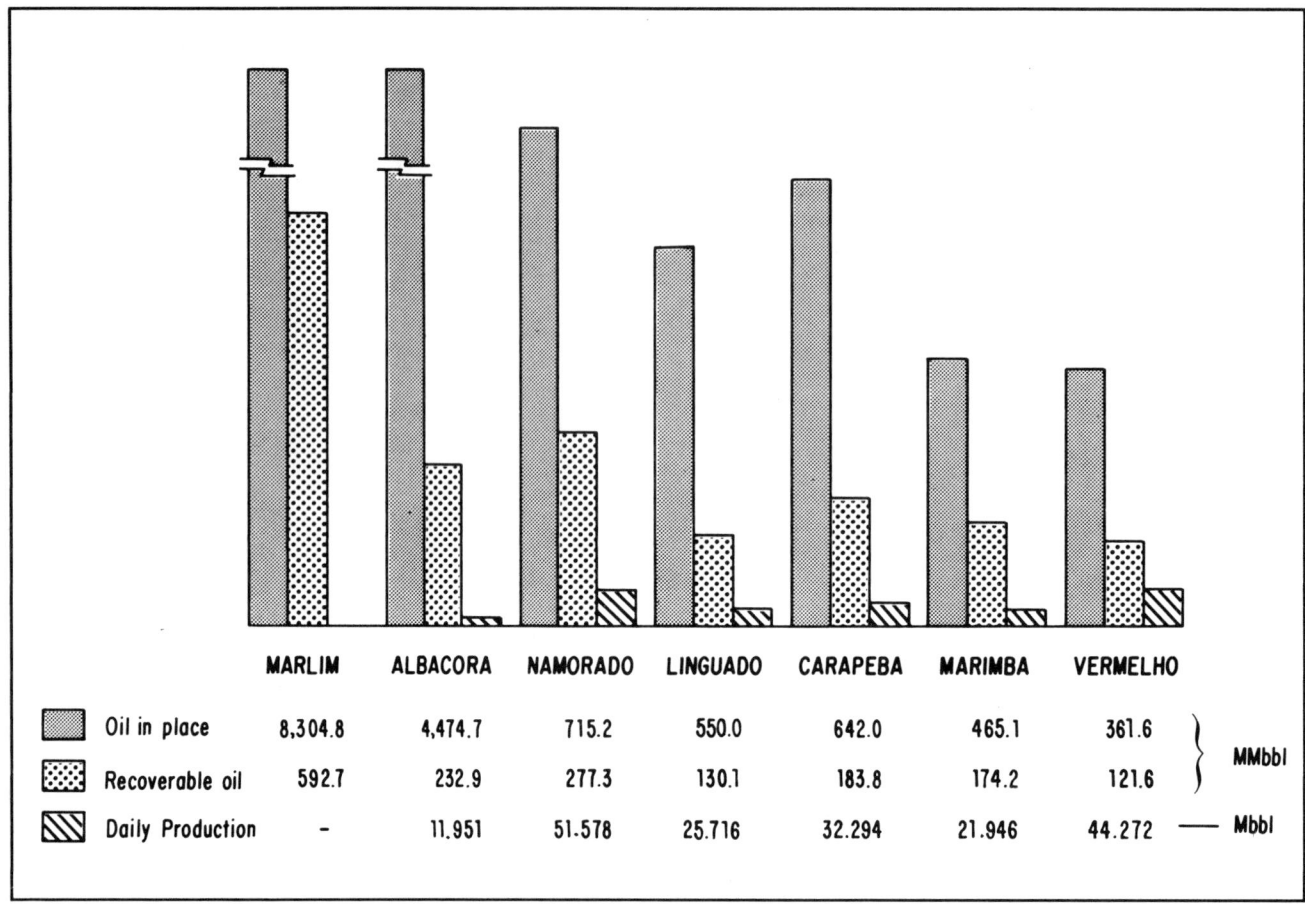

FIGURE 2. Comparative histogram of the seven greatest Campos Basin oil fields.

(Figure 3). The Lagoa Feia Formation is composed of four depositional sequences with a very strong chronostratigraphic relationship: (1) basal clastic sequence, (2) talc-stevensitic sequence, (3) coquina sequence, and (4) clastic-evaporitic sequence (Dias et al., 1988). Among these sequences, the coquina sequence is the most important as a reservoir rock and resulted from an extensive regional deposition of pelecypod valves. The rift phase is characterized by an intense diastrophic tectonism that affected all the pre-Aptian section. Normal faults with throws ranging from tens to hundreds of meters generated a series of horsts and grabens (Figures 4 and 5). Such faults, although very active in this period, were seldom reactivated in the Late Cretaceous and in the Tertiary. This phase finished with a regional unconformity that characterizes the top of the nonmarine megasequence.

The phase of quiescent tectonics (evaporitic-transitional megasequence) is limited to the Aptian, when the South American and African plates started to drift apart and the rift system was invaded by saltwater. The limited circulation of the water body, the restricted environment, and the arid to semiarid climate promoted the formation of extensive siliciclastic dry alluvial fans and evaporitic deposits (anhydrite and halite) widely distributed throughout the basin.

The postrift phase (marine megasequence) is characterized by open sea conditions and is present from the Albian to the Holocene. It is represented by shallow water calcarenites and calcirudites with ramp-type geometry at the lower section (lower Macaé Formation); by calcilutites, marl, and marine-transgressive shales at the intermediate section (upper Macaé Formation); and by marine-regressive turbidite sandstones and shales at the upper part of the section. The adiastrophic tectonism is predominant in this phase due to mobilization of the salt bed resulting from sedimentary overburden, gravitational sliding, and basin tilting toward the eastern direction, which became more intense in the Early Cretaceous. The faults are normal listric, developing associated rollover structures and dying out at the evaporitic layers. Many of these faults reach the Miocene and Oligocene sediments, and many are extended to the actual ocean floor (Figure 5). The Tertiary sedimentation is progradational and was developed concomitantly with the large diastrophic tectonic activity in the adjacent cratonic area, with the uplifting of the Serra do Mar, the formation of the small tafrogenic basins, and the magmatic volcanic episodes in the basin. This activity permitted the influx and storage of great quantities of siliciclastic material on the shallow platform, later transported to the abyssal plain, allowing

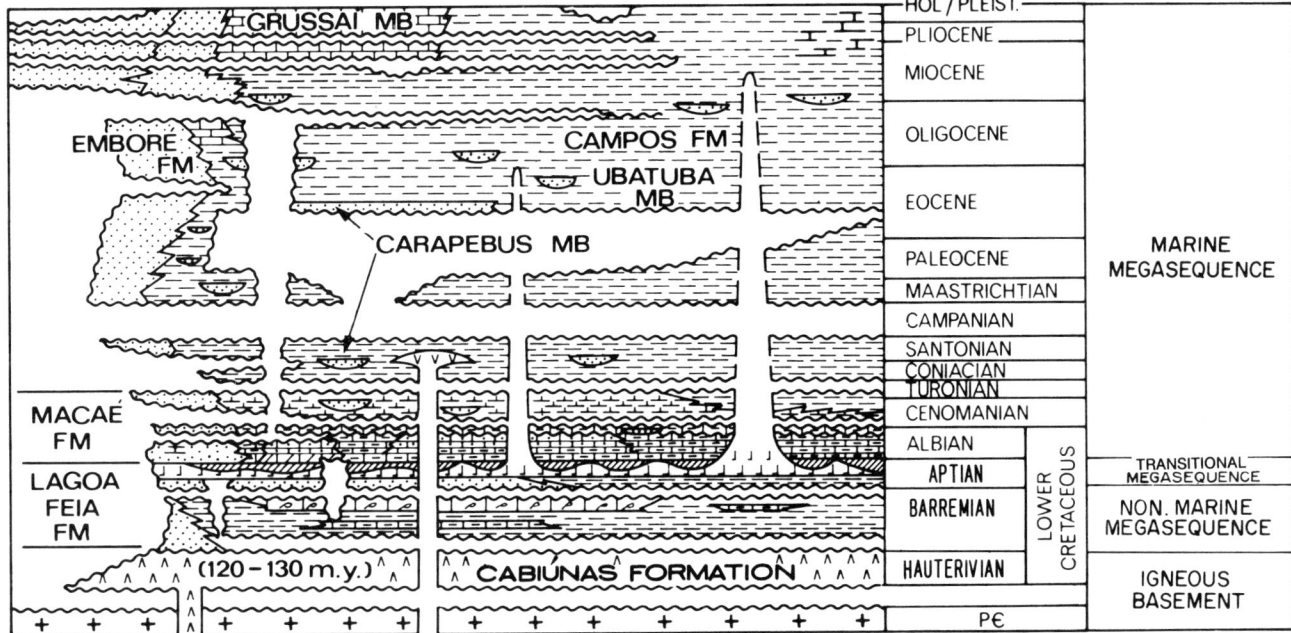

FIGURE 3. Stratigraphic column of the Campos Basin.

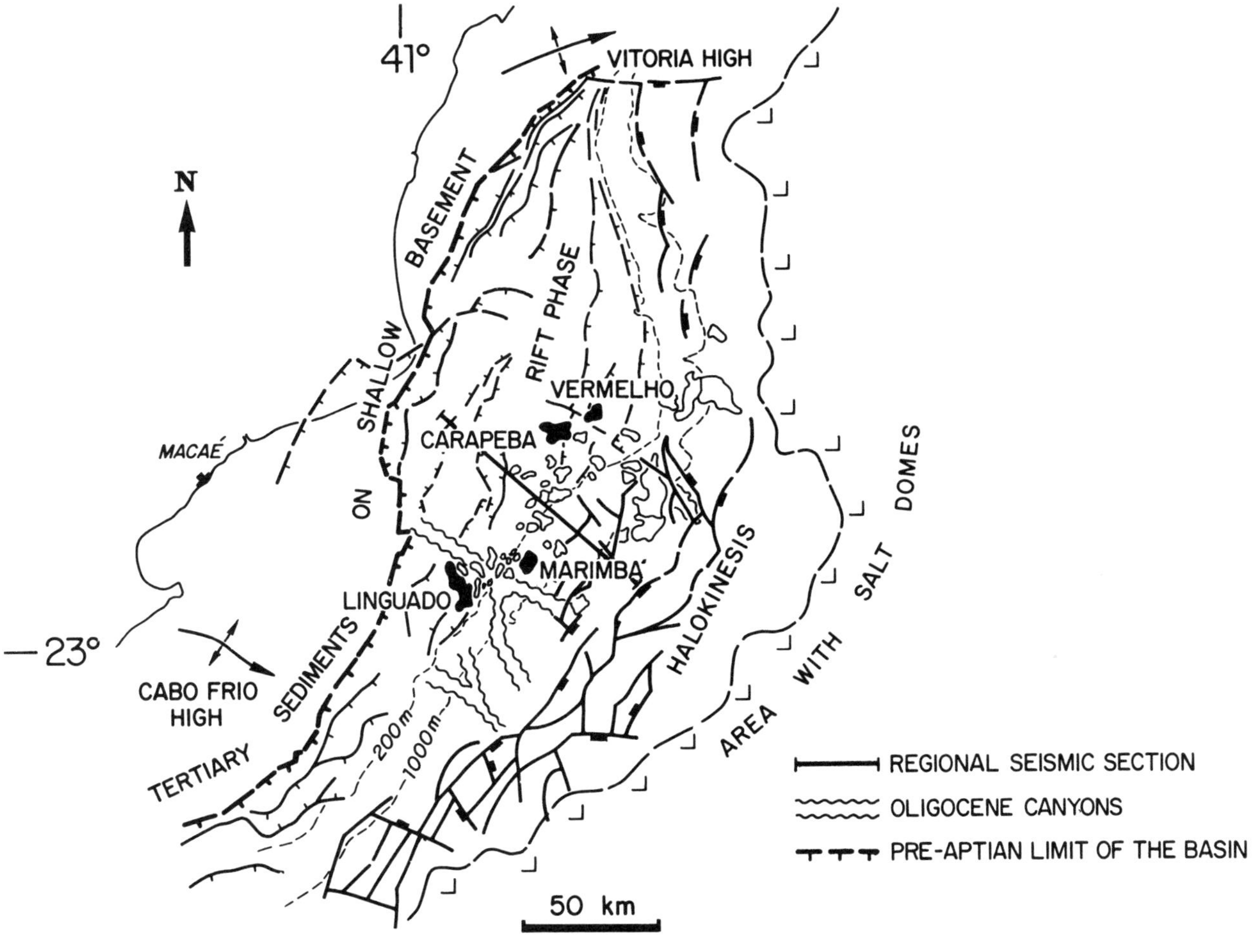

FIGURE 4. Tectonic framework of the Campos Basin.

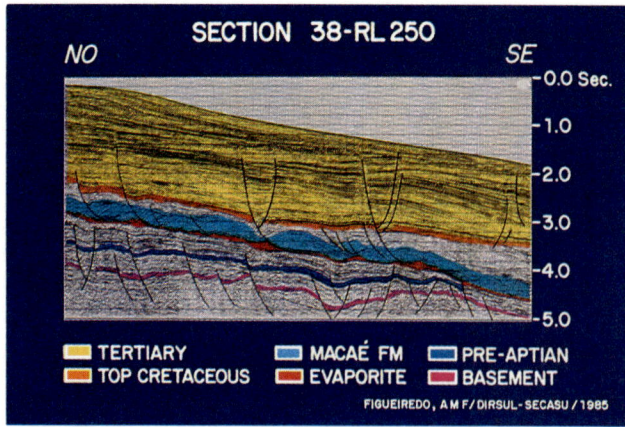

FIGURE 5. Seismic regional section of the Campos Basin.

the formation of great submarine fans that represent the most important reservoirs in the basin.

LINGUADO FIELD

The Linguado field, with an area of 32 km², is located 80 km from São Tomé Cape, in water depths of about 100 m. It was discovered in May 1978 by the wildcat 1-RJS-49, which produces oil from the coquinas of Lagoa Feia Formation, its main reservoir.

The Linguado field is situated at the extreme south of the trend of the Campos Basin accumulations, associated with the Badejo regional high, one of the most prominent structural highs of the basin (Figure 4).

The hydrocarbons are distributed in the Hauterivian (fractured basalts), Barremian (coquinas), Albian (calcarenites), and Maastrichtian (turbidite sandstones) reservoirs. The main reservoir consists of coquinas and is responsible for approximately 65% of the field's VOIP and 80% of its recoverable oil (Table 1).

Barremian Reservoir (Coquinas)

The coquinas were deposited over synchronous highs associated with syndepositional rift normal faults in two depositional phases (Baumgarten, 1985). These phases are separated by siliciclastic pelitic rocks, which represent peripheral river-lake environment facies and are considered electric log markers.

The coquina zonation, based on intercalations of impermeable layers, consists of six zones, including three subzones. Zones I, II, and III contain the upper phase of the coquina deposition. Zones IV, V, and VI contain the lower phase (Baumgarten, 1985). Zone VI, due to its great extension in the area, is the most important and links the Linguado field with the Badejo field to the north and the Pampo field to the south (Figure 6).

Detailed studies of each of the zones have indicated that the coquinas exhibit bank geometry with several shoaling upward cycles. From bottom to top, they are

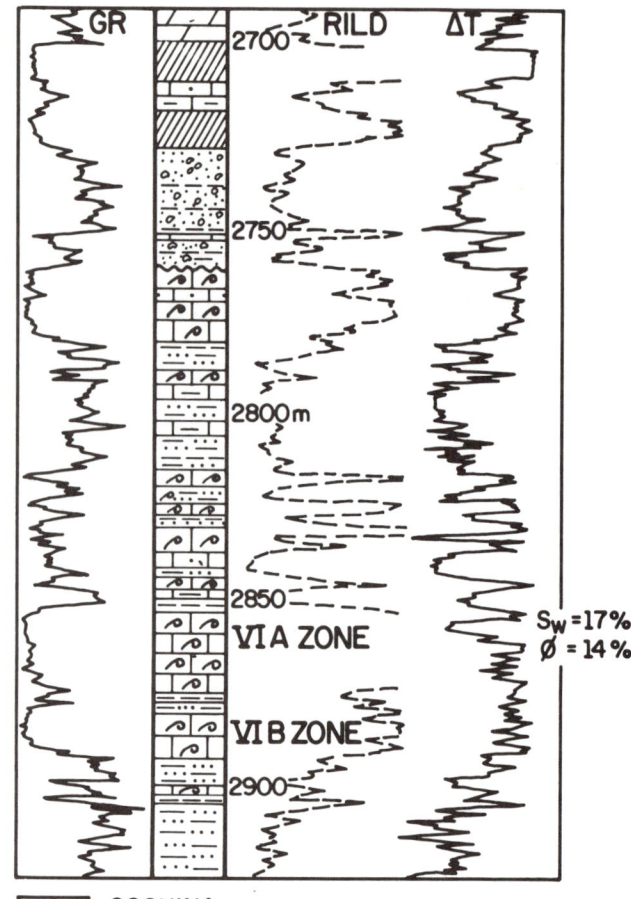

FIGURE 6. Typical section of the Barremian reservoir from the Linguado field.

composed of calcilutites, calcarenites, calcirudites, and calcirudites with calcarenite matrixes. Bioaccumulated carbonates and paleosol horizons occur at random (Carvalho et al., 1986). The calcirudites are the best reservoirs (Figure 7), with average porosity and permeability of 15% and 120 md, at times reaching 20% and more than 1 darcy, respectively (Table 2). Porosity is secondary, predominantly characterized by the interparticle and vugular types (Figure 8), with the moldic type occurring less frequently. The gravity of the oil varies from 29 to 33° API.

The production mechanism for this reservoir consists of a combination of in-solution gas and water-driving.

The structure of the Linguado field at the coquina level is a monoclinal dipping to the east, sectioned by normal faults with throws of up to 50 m (Figure 9).

Besides the faulted structure, stratigraphic and diagenetic factors are effective in trapping oil in the coquinas (Baumgarten et al., 1988). The stratigraphic mechanism is found in the pinch-out of the porous zones in the western direction of the field and in association with the unconformity located at the top of Zone I (Figures 10 and 11). Diagenetic factors have created permeability barriers by cementing with calcite and silica several levels of the

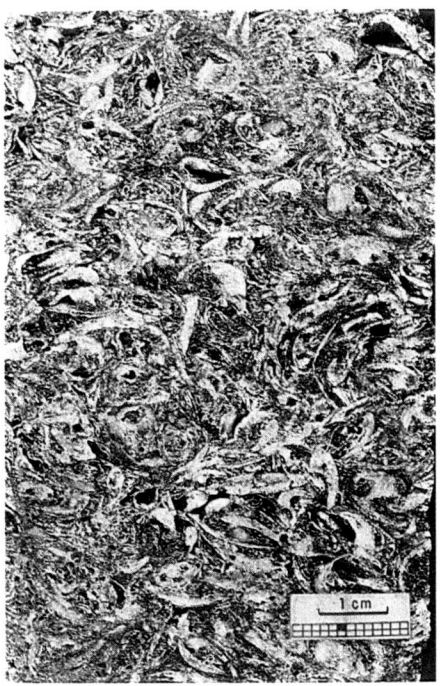

FIGURE 7. Macroscopic view of the coquina, the principal Linguado field productive rock.

coquinas (Figure 12). In Zones I and II the trap is controlled principally by diagenetic factors and, of less importance, structural factors. In Zones V and VI, diagenesis, oriented by structural factors, is mainly responsible for the trapping of petroleum.

Albian Reservoir

The Albian reservoir, containing 31% of the VOIP (Table 1), consists mainly of oncolitic, oolitic, and peloidal calcarenites and calcirudites from the Macaé Formation, deposited in a shallow platform environment. The average porosity and permeability are 20% and 250 md, and the gravity of the oil is 20° API. The structure consists of a smoothly dipping dome, cut by small throw growth faults. The trapping of oil in this reservoir is essentially structural-stratigraphic in nature. The structural component is represented by normal growth faults and the stratigraphic by facies variations in the upper part of the reservoir, where the calcarenites and calcirudites grade into calcilutites to the west and south-southwest of the field. The production mechanism for this field is predominantly water-driving.

TABLE 2. Summary of the characteristics of main reservoirs.

	Field/reservoir			
Item	Linguado/ Barremian	Carapeba/ Cretaceous	Vermelho/ Eocene	Marimbá/ Cretaceous
Datum (m)	−2850	−3000	−2770	−2703
O-W contact (m)	−2943	−3025	−2785	−2754
Area (km^2)	32	35	18	19
Water depth (m)	100	87	80	450
Maximum gross reservoir thickness (m)	361.0	330.0	500.0	114.0
Maximum net pay (m)	105.0	66.0	44.0	43.0
Average porosity (%)	15.0	18.5	24.4	27.0
Average permeability (md)	120	300	700	1700
Water saturation (%)	18.0	39.6	21.4	15.0
Formation pressure (kgf/cm^2)	303.0	313.5	290.7	282.0
API oil gravity	30.0	24.5	23.5	28.0
Gas/oil ratio (m^3/m^3)	164.0	25.0	28.0	85.0
Bubble point (kgf/cm^2)	255.0	75.0	62.0	208.0
B$_o$ (m^3/m^3)	1.49	1.16	1.14	1.29
Reservoir temperature (°C)	85.5	108.0	95.0	112.0
Producing mechanism	Gas in solution and water-driving	Water-driving	Water-driving	Water-driving

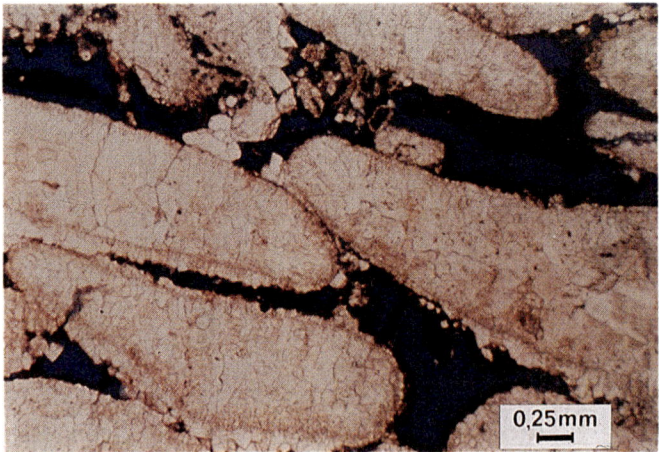

FIGURE 8. Microscopic view of the productive coquina from Linguado field.

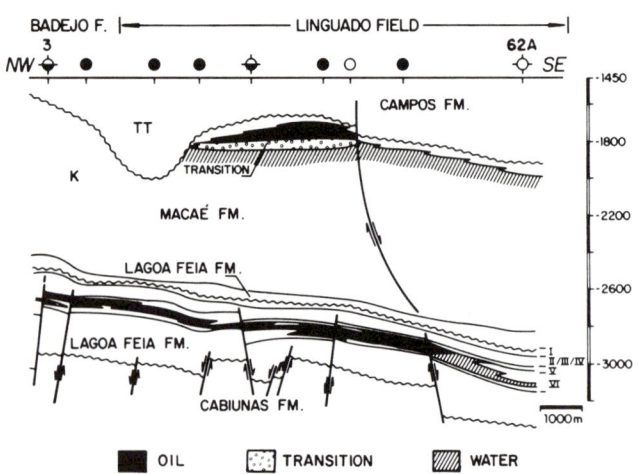

FIGURE 10. Geological section of the Linguado field.

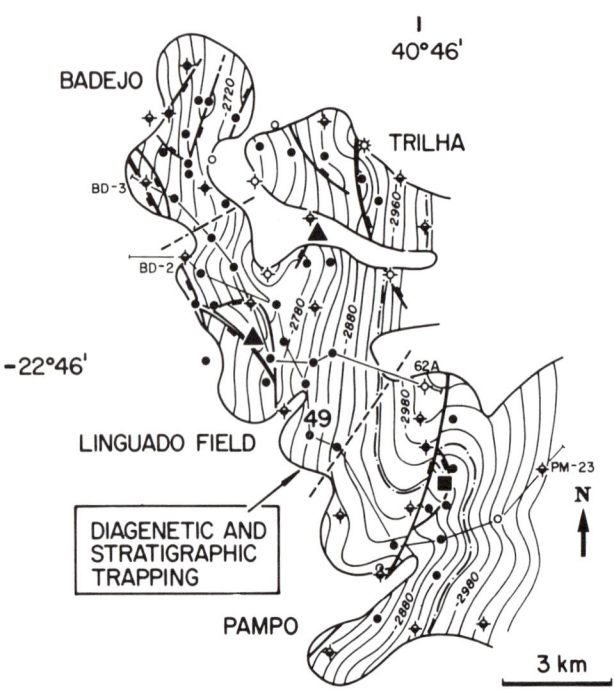

FIGURE 9. Structural map on the top of the Barremian reservoir from Linguado field.

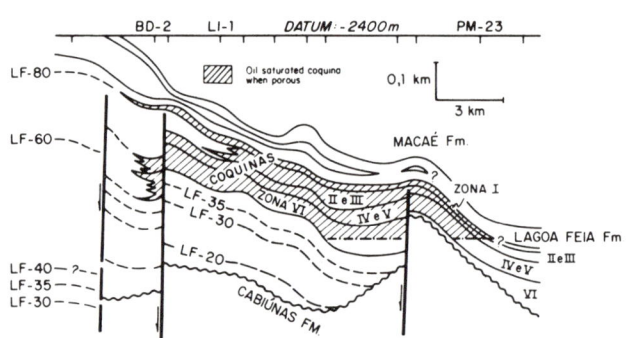

FIGURE 11. Geological section through the coquinas (from Baumgarten, 1988).

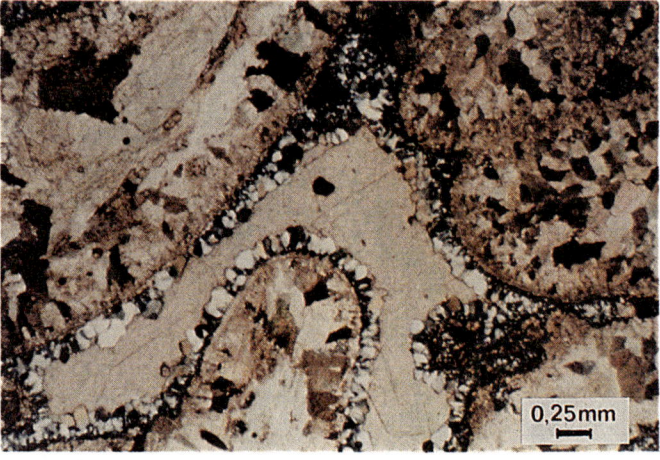

FIGURE 12. Intensive carbonate cementation that causes diagenetic trapping in the coquinas.

Hauterivian Reservoir

The Hauterivian reservoir contains only 1% of the VOIP (Table 1) and consists primarily of fractured vesicular/vugular basalts and secondarily of volcanic breccia from the Cabiúnas Formation. The pore system is represented by fractures, vesicles, and the porosity of the matrix itself, the fractures being the most important. The structure is formed by an elongated semidome in the northeast-southwest direction, cut by normal faults of small throw (up to 40 m). The oil trapping is essentially structural, characterized by faulting and fracturing.

Maastrichtian Reservoir

The Maastrichtian reservoir consists of turbidite sandstones belonging to the Carapebus member of the Campos Formation and contains 3% of the VOIP (Table 1). The average porosity and permeability are 24% and 300 md. The oil is very viscous (18 to 20° API). It is of limited extent (7.62 km^2), with a maximum thickness of 15 m. The trapping of oil is essentially stratigraphic-structural in nature, by shale-out and by dip. To date, this reservoir has not been put into production.

Development and Production

The discovery of Linguado (1978) came after that of Badejo (1975) and Pampo (1977). Recent geological interpretation, supported by pressure data, indicates that these fields are interconnected at the Barremian reservoir level. They form the Badejo-Linguado-Pampo Complex, which has a total area of 97 km^2 and VOIP of 1782.9 million bbl.

The Linguado production began in December 1981 through the wildcat 1-RJS-49. It was put onstream through the early production system of the Badejo-Linguado-Pampo Complex. This wildcat presently has a production rate of 3448 BOPD.

The development studies started using 3D seismic data in 1982, when the field already had ten wells drilled. The good quality of these seismic data has improved the prospects, and still there is the possibility that five more wells may be drilled in the near future.

Presently there are eight wells completed in the coquinas, producing 18,610 BOPD, and three in the calcarenites and calcirudites (Macaé Formation), with a flow of 7106 BOPD. Only one well was completed in the basalts. This well is presently inactive, due to a high water-cut (21.5%). Its cumulative production is 447,000 bbl. The present field production is 25,716 BOPD (Table 1).

The oil is produced through two floating production systems, sent to the fixed Pampo platform, and from there transferred to the central Enchova platform, where, after processing, it is sent to land (Figure 1).

CARAPEBA FIELD

The Carapeba field was discovered in February 1982 by the wildcat 1-RJS-193A. It has an area of 35 km^2, and is located 85 km from São Tomé Cape and 5 km southwest of Vermelho field, in a water depth of 87 m. This field is situated in the northeastern part of the main trend of accumulations of the basin, over a structural high in the central compartment of the basin (Figure 4).

The Late Cretaceous turbidite sandstones of the Carapebus member make up the main hydrocarbon reservoir of this field, containing 97% of the VOIP. Secondarily, oil is accumulated in the Eocene sandstones of the Carapebus member, accounting for 3% of the VOIP (Figure 3 and Table 1).

Cretaceous Reservoirs

The Cretaceous reservoirs correspond to thick, offlapping, coarse-grained to conglomeratic clastic deposits that vary in width from 5 to 10 km and are greater than 300 m thick and more than 20 km long. These deposits resulted from the infilling of a trough, oriented in the northwest-southeast direction, by currents derived from the northwest.

Based on the presence of very persistent shale layers throughout the field, these reservoirs were subdivided into three zones, designated Cretaceous 1 (K1), 2 (K2), and 3 (K3) (Cândido, 1988) (Figure 13).

The reservoirs are basically formed by granular conglomerates and fine to coarse sandstones (Figure 14), which presented average porosities and permeabilities of 17.3% and 414.5 md, and 19.7% and 188.3 md, respectively. The basic composition is subarkosic and its main diagenetic component is calcite, which locally can completely obliterate the rock porosity. Secondarily (never exceeding 5% of the volume of the rock) occur opal, pyrite, and anatase. In the coarse-grained sandstones, the clay fraction (less than 2 μm) varies from 0.3 to 2%, and in the fine-grained sandstones it can reach as much as 10%. The porosity is predominantly of the secondary intergranular type (Figure 15).

A characteristic feature of the Cretaceous reservoirs is the production of water-free oil in intervals with low resistivity. This is caused by the presence of large

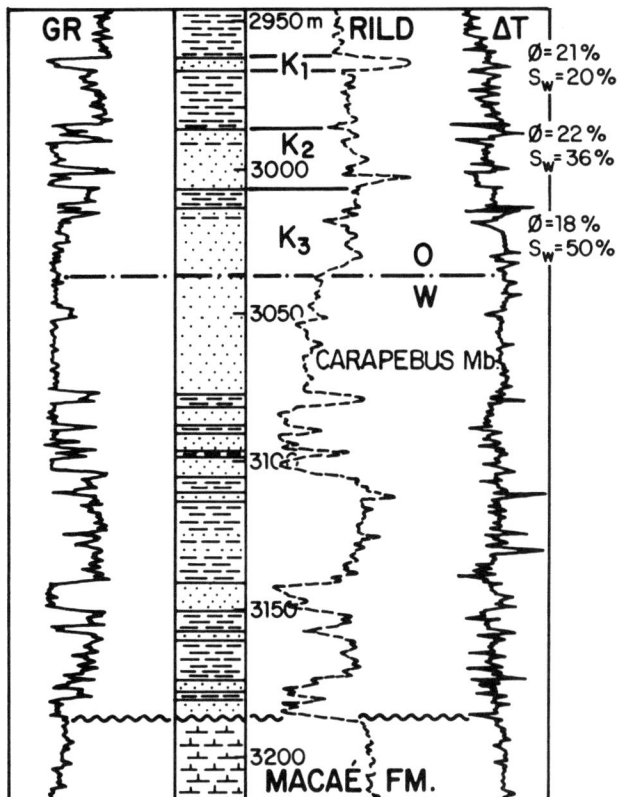

FIGURE 13. Typical section of a Cretaceous reservoir from Carapeba field.

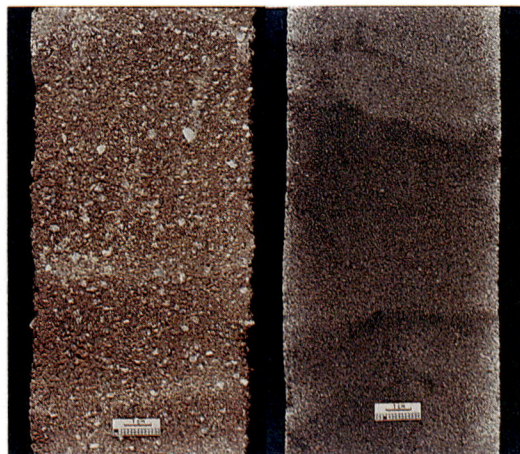

FIGURE 14. Macroscopic views of the productive sandstones from the Carapeba field (from Cândido, 1988).

FIGURE 15. Microscopic view of the productive sandstone from the Carapeba field.

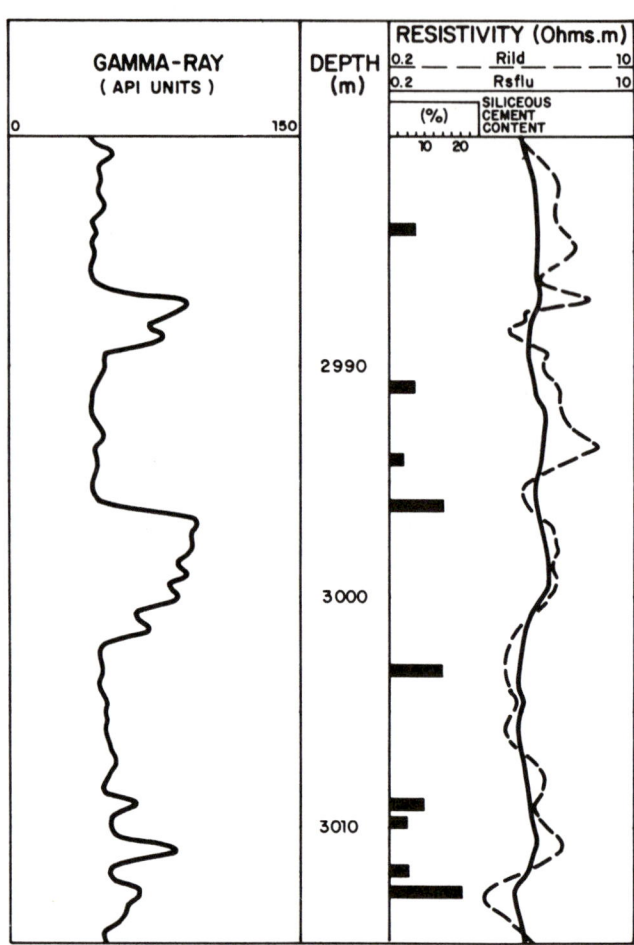

FIGURE 16. Correlation of resistivity and presence of opal cement with microporosity in the Carapeba field reservoirs (from Freitas, 1987).

amounts of connate water held in the microporosity associated with the siliceous material (Freitas, 1987) (Figure 16). The proportion of this microporosity increases with the decrease of grain size, causing deterioration of the quality of the reservoirs from the northwest to the southeast of the field.

The Carapeba field aquifer is of great dimensions. Based on geological data, it is shared with the Pargo field, which will probably provide a very active vertical and lateral water-driving mechanism (Cândido, 1988).

The structure may be defined as being composed of a greater dome elongated in the east-west direction, associated with smaller domes that are cut by normal faults in the north-south direction. These faults dip to the west, and present throws between 20 and 60 m, not restricting the fluid flow (Figure 17). The trap has a predominant structural component, bounded on the southwest by faulting, on the east by pinch-out, on the south by disconformity, and in other directions by structural dip. The oil occurs from −2900 m, with the oil-water contact at −3025 m (Figure 18).

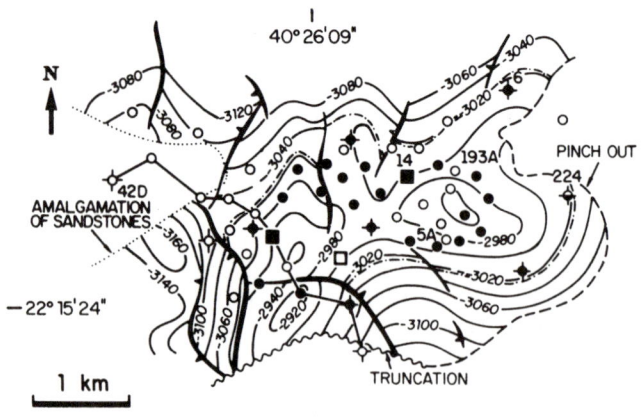

FIGURE 17. Structural map on the top of a Cretaceous reservoir from Carapeba field.

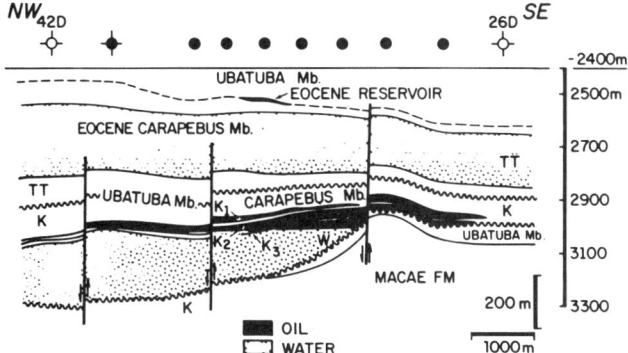

FIGURE 18. Geological section of the Carapeba field.

Eocene Reservoirs

The occurrence of Eocene oil reservoirs is secondary, restricted to sand bodies without oil-water contact, with channel geometry, and less than 10 m thick. They are positioned at the top of a predominantly sandy sequence, characterizing a stratigraphic trapping.

VERMELHO FIELD

The Vermelho field, with an area of 18 km², is located 75 km from São Tomé Cape, close to the Carapeba field, in an average water depth of 80 m. It was discovered in December 1982 by the wildcat 1-RJS-241A and produces oil from the Eocene sandstones of the Carapebus member of the Campos Formation, its main reservoir (Figure 19).

This field has two reservoirs in the Eocene Carapebus member that are designated A, with 2.9% of the total VOIP of the field, and B, the main reservoir, with 89.3% of the total VOIP (Figure 20). The Oligocene Carapebus member has various reservoirs, grouped in two sets, separated by a regional unconformity, holding 7.8% of the VOIP of the field (Table 1). In the Miocene (Embore Formation sandstones), a small oil accumulation was detected, with no commercial interest.

Eocene Reservoirs

The Eocene reservoirs are mainly composed of massive medium-grained sandstones (Figure 21), in beds of 1 to 3 m, in which coarse to fine-grained sandstones are interbedded, resulting in complex bodies of amalgamated

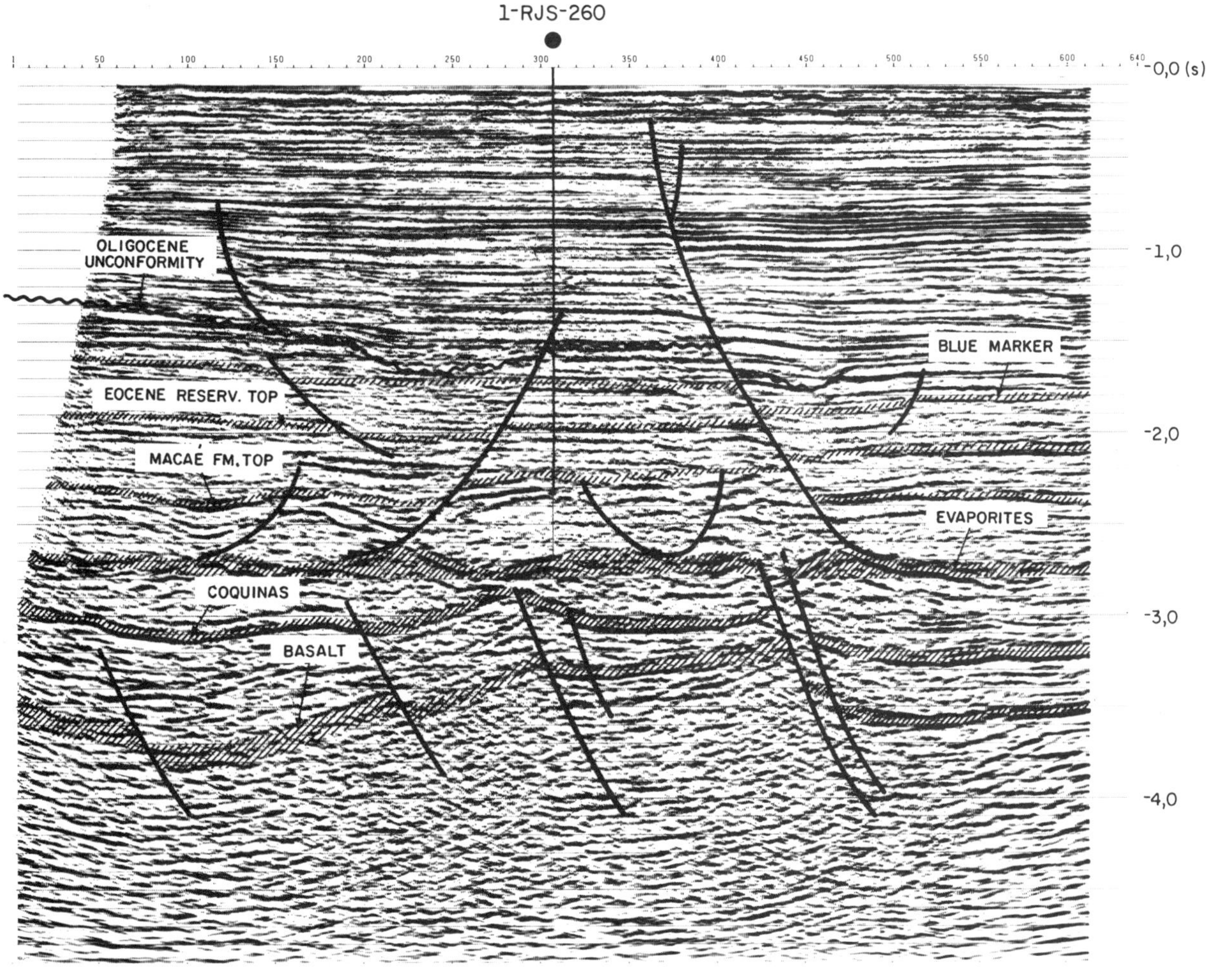

FIGURE 19. Seismic section through Vermelho field.

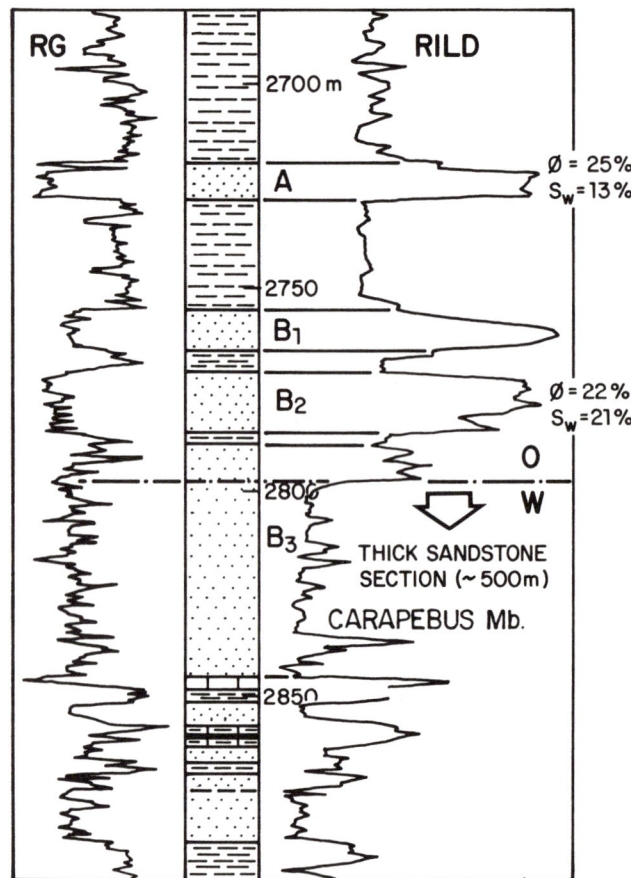

FIGURE 20. Typical section of Eocene reservoirs from Vermelho field.

sandstones with abundant abrupt erosion contacts with multilateral and multistacked geometry. Intercalated with the sandstone bodies are, also very common, bioturbated shales.

The sandstones are classified as submature, subarkosic, and the porosity shows an average value of 24.4% (15.7% intergranular, 3.3% intragranular, and 5.4% microporosity) (Figure 22). The average permeability is 700 md (Table 2). The reservoirs are clean, with 2 to 3% kaolinite, the only significant cement. The microporosity percentages are associated with feldspars, the matrix and clayey fragments, and kaolinite.

Reservoir A is physically separated from reservoir B by a shale bed 40 m thick, and they present different oil-water contacts, at −2752 and −2785 m, respectively.

The oil column in the Eocene B reservoir (Zones B1, B2, and B3) is approximately 60 m and occupies the top of a predominantly sandy section about 500 m thick, whose base marks an Eocene regional unconformity (Figure 20). The aquifer is very large, probably very active, and common for these zones (Barroso and Stank, 1988, 1990).

The shales in Zone B3 are thin and discontinuous, whereas those between Zones B3 and B2, and B2 and B1 (northern part of the field), show good lateral continuity and thicknesses up to 4 m (Barroso and Stank, 1988). The vertical sequence, as seen in cores from Zones B3, B2, and A, is constituted of several normal cycles of up to 5 m, well characterized in the gamma ray logs (Figure 23).

The deposition of this section is related to a middle Eocene lowering of the sea level and to salt movements favoring the displacement of coarse clastic material derived from the shallow platform (Guardado et al., 1986). The deposits represent submarine channeled lobes, formed by braided channels in a prograding sedimentation that results in the formation of a discrete channeled system (Zones B2, B1, and A), and marginal levees (Barroso et al., 1988). Paleontological analyses based on foraminifera confirm a gradual lowering of the sea level during deposition (Barroso and Stank, 1990).

The Vermelho field structure is defined by a horst oriented northeast-southwest and delineated by faults

FIGURE 21. Macroscopic view of the principal productive sandstone from Vermelho field.

FIGURE 22. Microscopic view of the major productive sandstone from Vermelho field.

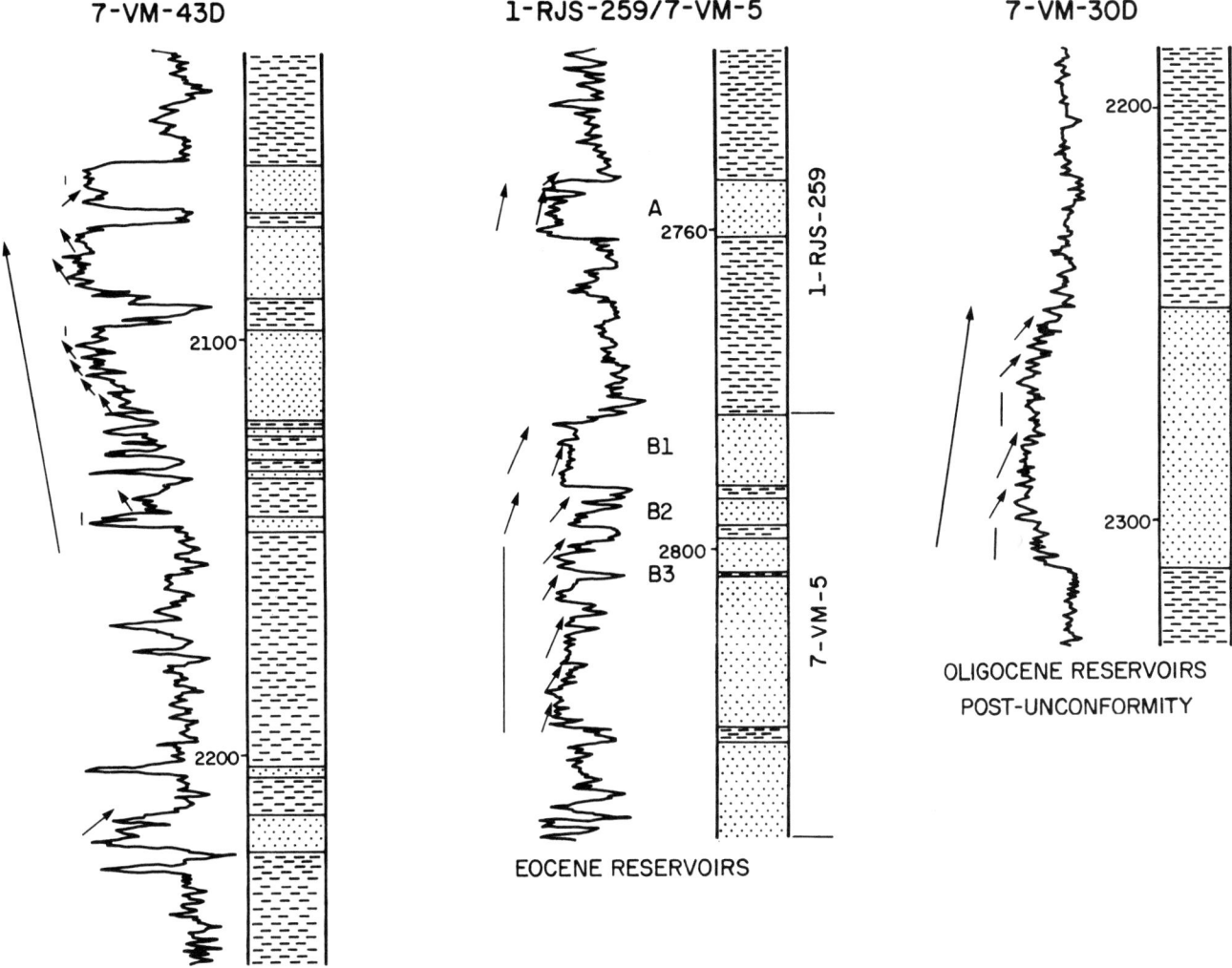

FIGURE 23. Typical gamma ray responses of reservoirs from Vermelho field (from Barroso and Stank, 1990).

with throws between 40 and 250 m. Those of the west and the northwest are antithetic and sealing, and those of the northeast are synthetic and nonsealing (Figures 24 and 25). These faults, originated by the salt movements, are still in motion in the northeast, and represent the main pathway by which hydrocarbons reached the Oligocene and Miocene reservoirs, after the infilling, until the spill-point of the Eocene reservoirs (Barroso and Stank, 1990).

Oligocene Reservoirs

The Oligocene reservoirs are truncated by an upper Oligocene unconformity (Figure 25). The pre-unconformity reservoirs were deposited in a deep neritic (platform) environment and the post-unconformity reservoirs in an upper bathyal (slope) environment (Barroso and Stank, 1990).

The pre-unconformity sandstones exhibit tabular geometry with good lateral correlation, a thickening- and coarsening-upward pattern of sedimentation (Figure 23), a high degree of compartmentation, and multiple contacts between fluids.

The post-unconformity sandstones were deposited on an erosional surface, showing channeled geometry and a finning-upward pattern, representing typical deposits from submarine fan channel-filling (Figure 23). The sandstone bodies in the southern trough are saturated with oil, whereas those in the northern trough, in spite of having good permoporosity characteristics, are saturated with water (Figure 25). The geologic data show that only the southern trough is in contact with the northeast fault, the main route of hydrocarbon migration.

The trapping mechanism is defined principally by the structural component (smooth domes), but stratigraphic and, locally, paleogeomorphic components are also present.

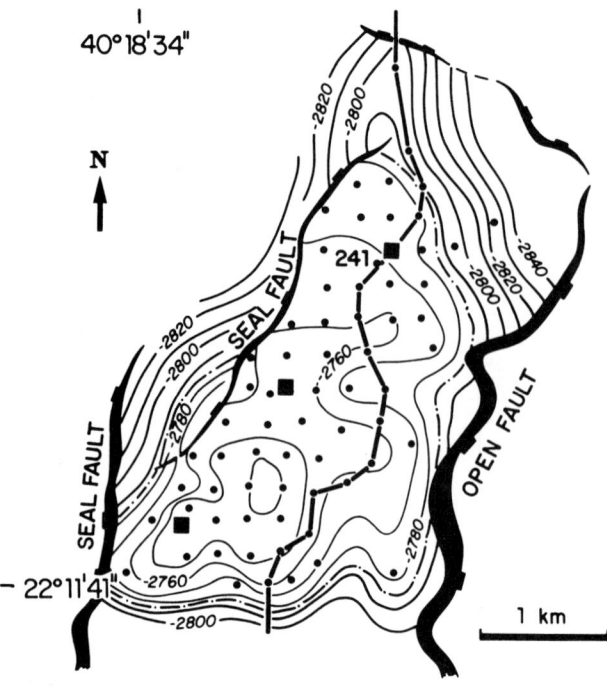

FIGURE 24. Structural map on top of Eocene B reservoir from Vermelho field.

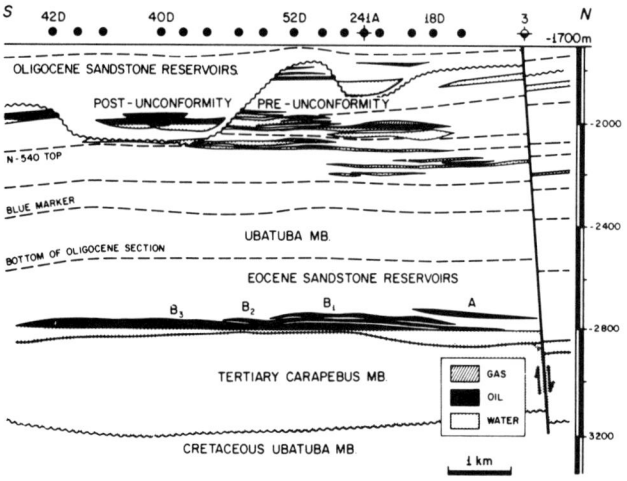

FIGURE 25. Geological section of Vermelho field.

Development and Production of Carapeba and Vermelho Fields

The Carapeba and Vermelho fields comprise, together with Pargo field, the Northeastern Pole of the Campos Basin. It consists of an integrated exploitation project of seven fixed platforms, one for facilities and six for production, with the central one in Pargo. In Vermelho field there are three platforms (21 wells each), and in Carapeba field there are two more (15 and 20 wells), all satellites. From the Pargo platform, the production of the Pole is sent to the Garoupa's central platform, where the petroleum is processed and delivered to land (Figure 1).

This Northeastern Pole was completely delineated before being put onstream and inserted into a unified, previously planned production program, which had an investment of $1.4 billion and an expected return of approximately $6 billion in ten years.

In Carapeba field, between 1982 and 1987, 54 wells were drilled, 19 for exploration and 35 for development. Of these, 36 wells were put into production from two templates. In Vermelho field, between 1982 and 1987, 76 wells were drilled, 10 for exploration and 66 for development. Three existing templates were used to complete 62 wells.

The Carapeba and Vermelho production began in December 1988 and January 1989, respectively, and current production is 32,984 and 44,272 BOPD with watercuts of 0.3 and 3%, respectively (Table 1).

The production wells of these two fields are being equipped with artificial lift semisubmersible pumps. As anticipated by the simulation modeling, the wells are losing flow pressure relatively early, due to the low ratio of solubility. No drilling of any new development wells in the Vermelho field is planned. In the Carapeba field, the implementation of a supplementary production system of approximately 15 wells in the mid-southern area of the field is being studied to achieve better recovery of the reserves.

MARIMBÁ FIELD

The Marimbá field, with an area of 19 km², is located 87 km from São Tomé Cape, in a water depth of 380 to 600 m. The field was discovered in March 1984 by the wildcat 1-RJS-284. It is located in the southern part of the main trend of accumulations of the basin. It is considered the first deep water discovery of the Campos Basin. Among the fields mentioned in this report, Marimbá is the only one that is in the initial phase of development.

The Marimbá field reservoirs are made up of turbidite sandstones, belonging to the Late Cretaceous Carapebus member of the Campos Formation. Based on present available data, the estimated VOIP amounts to 465.06 million bbl (Table 1).

Cretaceous Reservoirs

The Marimbá field reservoirs have a maximum thickness of 250 m, a column of oil of 114 m, and oil-water contact at −2754 m (Figure 26). They consist predominantly of arkosic, medium to coarse-grained sandstones (Figure 27), showing multiple amalgamated bodies with a high sandstone/shale ratio. Secondarily, there are sandstones with indistinct cross- or horizontal-stratification and clayey sandstones. Interbedded bioturbated shales and siltstones are the main dissimilarities of the reservoir.

The sandstones are poorly cemented with calcite, brittle to semibrittle, moderately to poorly sorted, with excellent permoporosity characteristics. Average porosity and permeability values are 27% and 1700 md, respec-

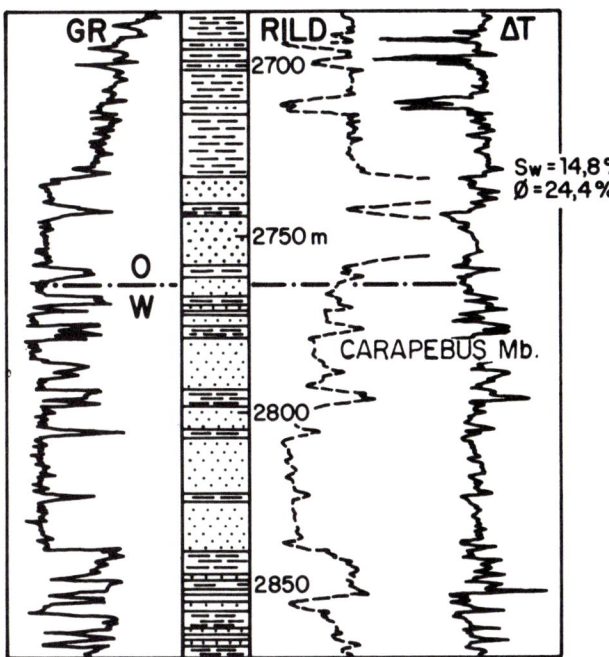

FIGURE 26. Typical section of a Cretaceous reservoir from Marimbá field.

FIGURE 28. Microscopic view of the productive sandstone from Marimbá field.

The abrupt lateral variations of thickness (250 m to zero in 2 km) suggest the coalescence of turbidite lobes, channeled in paleogeomorphic depressions, controlled by faulting and erosion of the underlying Macaé formation, filled with sediments, transported in the northwest-southeast direction.

Based on current knowledge, the main production mechanism of the field is lateral water-driving. The aquifer is limited to the northwestern and southern parts of the field.

The structure is conceived as a tilted fault block in the direction opposite to that of the regional dip, cut by small faults that do not affect the continuity of the reservoir. These faults are parallel to the main fault, which limits the Piraúna field to the northwest and has a throw of approximately 300 m (Figure 28).

The trapping mechanism of the hydrocarbons has a mixed origin, controlled by structural dip and faulting to the northwest and by depositional pinch-out in all the other directions (Figures 29 and 30).

Development and Production

The exploitation of the Marimbá field began in April 1985 when the wildcat 1-RJS-284 was put on stream. This wildcat, in a water depth of 383 m, set a record for subsea completion at that time and was a turning point in the development of deep water oil fields. Next, new records were set with the completion of wells 3-RJS-293D and 3-RJS-294, in water depths of 409 and 413 m in the same field.

Between 1984 and 1990, nine wells were drilled (all oil producers), but presently only five have been completed and produce a total of 21,946 BOPD without water-cutting. The geological data, together with the sharp decline in reservoir pressure, indicate the need to drill new production and water injection wells. Using data from the 3D seismic survey, which proved to be of very good quality, two production and three water injection locations were proposed recently (Figure 31).

FIGURE 27. Macroscopic view of the productive sandstone from Marimbá field.

tively. The porosity is predominantly of the secondary intergranular type (Figure 28). Micaceous minerals can reach 2.5% of the rock volume locally, and the proportion of the clay matrix never exceeds 5%. Lithic fragments, clayey intraclasts, and glauconite are the main accessory constituents.

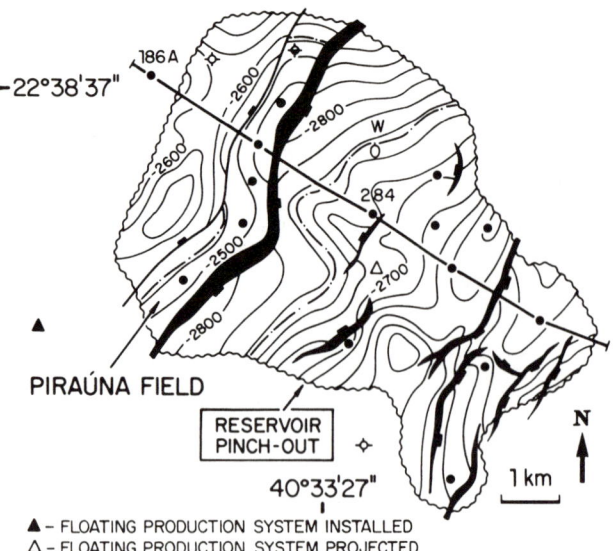

FIGURE 29. Structural map on the top of a Cretaceous reservoir from Marimbá field.

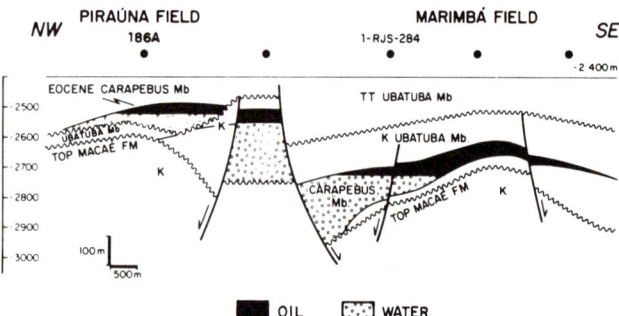

FIGURE 30. Geological section of Marimbá field.

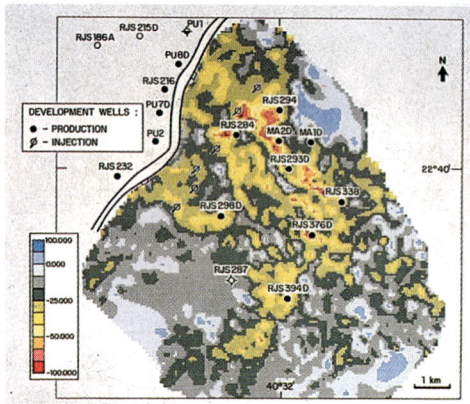

FIGURE 31. Amplitude seismic map on the top of a Cretaceous reservoir from Marimbá field.

The production system already installed in the neighboring Piraúna field was also used for first-stage production from the Marimbá field. Later on, after the expansion of the Marimbá field boundaries, it became necessary to design a system that not only met the water injection project, but also the simultaneous development of both fields, by using an additional rig and increasing the processing capacity from 30,000 to 60,000 BOPD.

The Marimbá field has not yet been totally delimited. It may be expanded with the drilling of new extension wells, which are not foreseen in the near future. The exploration philosophy is based on versatile production systems using semisubmersible drilling rigs adapted for production. This will allow some flexibility in the development, by stages, of costly deep water oil field projects, as additional reservoir knowledge is attained. This field is especially important because it serves as a viable economic laboratory in the development of deep water production technology.

ACKNOWLEDGMENTS

The authors thank PETROBRÁS for the permission to publish this article, and their colleagues for significant discussions and contributions without which this report would not have been possible. Special thanks are due to A. M. F. de Figueiredo, C. S. Baumgarten, and C. A. Tigre for incentive, reading of the originals, and valuable criticisms and suggestions.

REFERENCES

Asmus, H. E., and F. C. Ponte, 1973, The Brasilian marginal basins, *in* A. E. Nairn and F. G. Stehli (eds.), The ocean basins and margins, v. 1, the South Atlantic: New York, Plenum Press, p. 87–133.

Barroso, A. S., and C. V. Stank, 1988, Zoneamento e critérios geológicos para o programa de explotação do reservatório Eoceno (B) do Campo de Vermelho, *in* Annals of the II Seminário de Engenharia de Reservatórios, Teresópolis, v. 3, p. 1490–1519.

Barroso, A. S., and C. V. Stank, 1990, Modelo geológico do Campo de Vermelho, *in* Annals of the IV Seminário de Geologia de Desenvolvimento e Reservatório, Natal, p. 57–72.

Barroso, A. S., C. V. Stank, and F. P. Campozana, 1988, Características dos reservatórios Eocênicos do Campo de Vermelho, *in* Annals of the III Seminário de Geologia de Desenvolvimento e Reservatório, Salvador, p. 226–238.

Baumgarten, C. S., 1985, Evolução estrutural de Pampo, Badejo, e Linguado durante a deposição da Formação Lagoa Feia: Boletim Técnico da PETROBRÁS, v. 28, n. 2, p. 91–101.

Baumgarten, C. S., A. J. C. Dultra, M. S. Scuta, M. V. L. Figueiredo, and M. F. P. B. Siqueira, 1988, Coquinas da Formação Lagoa Feia, Bacia de Campos: Evolução da Geologia de Desenvolvimento, Boletim de Geociências da PETROBRÁS, v. 2, n. 1, p. 27–36.

Cândido, A., 1988, Características dos Reservatórios do Campo de Carapeba, *in* Annals of the III Seminário de Geologia de Desenvolvimento e Reservatório, Salvador, p. 390-401.

Carvalho, M. D., M. Monteiro, and A. M. P. Misuzaki, 1986, Tipo e qualidade da rocha reservatório da Formação Lagoa Feia nos Campos de Badejo, Linguado e Pampo—Bacia de Campos, *in* Annals of the III Congresso Brasileiro de Petróleo, Rio de Janeiro, TT65, p. 1/12, 12/12.

Chang, H. K., R. D. Kowsman, and A. M. F. Figueiredo, 1988, New concepts in the development of east Brazil marginal basins, episodes, v. 11, p. 194-202.

Dias, J. L., J. Q. Oliveira, and J. C. Vieira, 1988, Sedimentological and stratigraphic analysis of the Lagoa Feia Formation, rift phase of Campos Basin, offshore Brazil: Revista Brasileira de Geociências, v. 18, n. 3, p. 252-260.

Figueiredo, A. M. F., 1985, Geologia das bacias Brasileiras, *in* Avaliação de formações no Brasil: Sociedade Brasileira de Pesquisas no Subsolo pelo Método Schlumberger (WEC-Brasil), Chapter 1, p. 1-38.

Figueiredo, A. M. F., and W. U. Mohriak, 1984, Salt tectonics and petroleum accumulations in the Campos Basin, *in* Annals of the XXXIII Congresso Brasileiro de Geologia, Rio de Janeiro, v. 3, p. 1380-1394.

Freitas, L. C. S., 1987, Estudo de reservatório do Membro Carapebus (Cretáceo) no Campo de Carapeba, Bacia de Campos, Estado do Rio de Janeiro, Brasil: Universidade Federal de Ouro Preto, Minas Gerais, Brazil, MSc. dissertation, 230 p.

Guardado, L. R., W. E. Peres, and C. E. S. Cruz, 1986, Depositional model and seismic expression of turbidites in Campos Basin, offshore Brazil: AAPG Bulletin, v. 70, p. 597 (abstract only).

Ponte, F. C., and H. E. Asmus, 1978, Geological framework of the Brazilian continental margin: Geologische Rundschau, v. 67, n. 1, p. 201-235.

Chapter 10

El Furrial Oil Field
A New Giant in an Old Basin

Rodulfo Prieto
Gustavo Valdes

Lagoven, S.A., Petroleos de Venezuela, S.A.
Caracas, Venezuela

ABSTRACT

In 1978, Lagoven, S.A., an affiliate of Petroleos de Venezuela, S.A., started an exploration program aimed at deeper targets on the northern flank of the Eastern Venezuelan basin, where the shallow upper Tertiary section had been explored for more than eight decades. The quality of the seismic data collected was sufficient to give indications of thrust faults with associated structures. From 1978 to 1985, geologic surveying, structural modeling, and interpretation of seismic data were combined to define the first prospect: El Furrial. The discovery well, FUL-1, penetrated 276 m (905 ft) of net oil sand and produced up to 7331 bbl of 26° API oil per day. This is the largest single discovery of medium-gravity oil in the last 25 years in South America. Folding and thrusting of the northern flank of the Eastern Venezuelan basin occurred during the collision of the Caribbean and South American plates. The evolution began at least by the early Paleocene. Cylindrical folds associated with thrusts are aligned in series for a distance of up to 70 km. They constitute a typical foreland overthrusted basin. The reservoir rock is a shallow marine sandstone deposited during the late Oligocene. Gross thickness ranges from 457 to 518 m (1500 to 1700 ft), and porosities range from 11 to 16%. The El Furrial discovery represents an excellent example of the prospectivity of a foreland overthrusted area and also an example of continued successful exploration to pursue deeper objectives in an area already considered mature.

INTRODUCTION

The Eastern Venezuelan basin is the second most important oil province in Venezuela, after the Lake Maracaibo basin. It is located in northeastern Venezuela (Figure 1) and encompasses the Anzoategui-Guarico and Monagas states. The northern limit is the Interior Range, and the southern limit is the Guayana shield. The Eastern Venezuelan basin is divided east-west into two subbasins: the Maturin subbasin to the east, where El Furrial field lies, and the Guarico subbasin to the west (Figure 2). This paper presents the results of an integrated exploratory campaign that led to the discovery of a giant field: El Furrial.

The Eastern Venezuelan basin has been developed over the last 60 years in the traditional fields of Jusepin, Oficina, Orocual, and Quiriquire, where light oil and gas have been produced from the northern flank of the basin and heavy oil from the southern area.

Recent exploratory programs have led to the discovery of deep targets on the northern flank of the Eastern Venezuelan basin, where the shallow upper Tertiary section has produced more than 1.8 billion bbl of oil in the last eight decades.

The discovery of El Furrial gave renewed strength to petroleum exploration for deeper targets on the northern flank of the Eastern Venezuelan basin.

EXPLORATION HISTORY

Oil exploration in the Eastern Venezuelan basin began in the nineteenth century (1890) in the Guanoco area (Figure 3), where commercial production was established in 1913 with the drilling of the Babaui-1 well in the vicinity of Guanoco Lake in the state of Sucre.

Numerous oil seepages along the foothills led to exploratory campaigns aimed at shallow targets in the upper Tertiary section (Pereira and Aymard, 1988). The Quiriquire field, which was discovered in 1928 (Figure 3), has produced over 900 million bbl of oil from Pliocene-Pleistocene alluvial fans. The development of new geophysical reconnaissance surveys (magnetic and refraction) allowed the discovery of the Jusepin field (1936) and the Orocual field (1938). The first seismic reflection survey was carried out during the early 1940s and led to the discovery of the Manresa, Santa Barbara (Figure 3), Mulata, and San Joaquin fields. During the 1960s and 1970s, 30 wells were drilled—mostly to obtain geologic information (Salvador, 1958)—with very little success. Only seven wells penetrated deeper than 3960 m (13,000 ft). At that time, geophysical data did not have enough resolution to identify deeper traps.

In 1978, Petroleos de Venezuela, S.A. started a campaign of exploration for deeper targets. A regional seismic reflection survey of 916 km was conducted using Vibroseis as a source. Although the quality of the data acquired was average to poor, interpretation of these data allowed

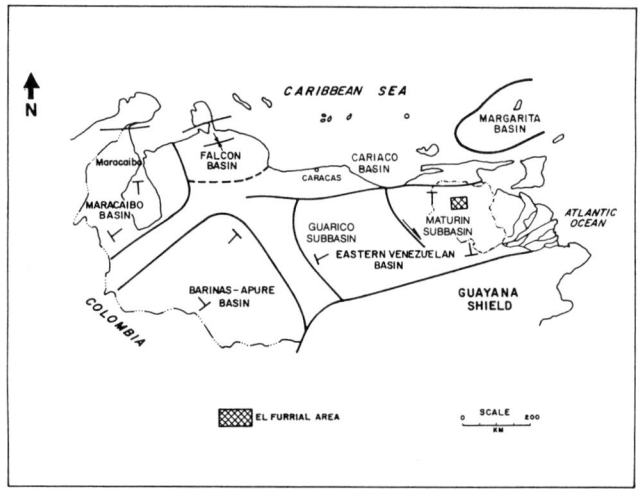

FIGURE 2. Venezuelan oil basins.

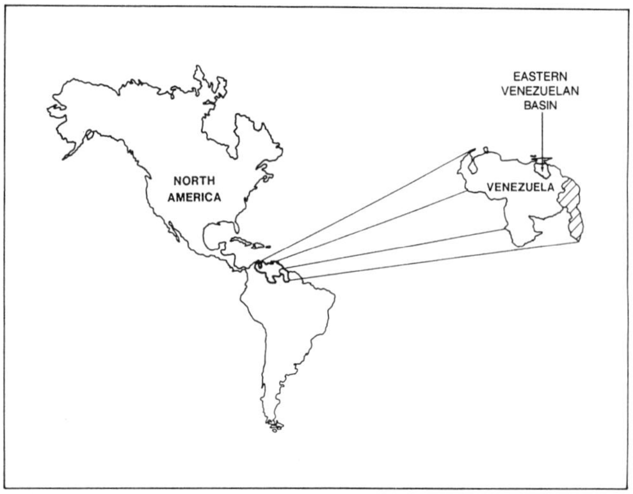

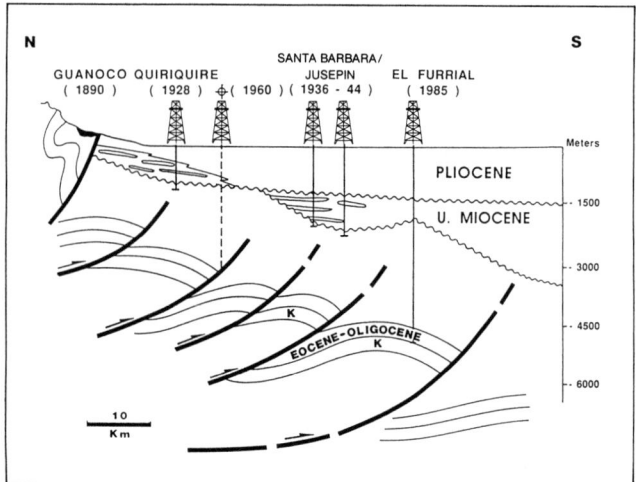

FIGURE 1. Location map showing the Eastern Venezuelan basin.

FIGURE 3. History of exploration in the Eastern Venezuelan basin (after Pereira and Aymard, 1988).

several prospects to be defined in a structurally complex area, one of which was El Furrial field. During 1984 and 1987, 5000 km of new seismic data was acquired using Vibroseis and dynamite in a previously planned testing program for design of the field parameters. Figure 4 shows a seismic line through El Furrial field. Excellent resolution of up to 4.5 sec was obtained.

Parallel with the acquisition of the new seismic data, a drilling campaign was begun with two wells: ORS-52 in the Orocual field and FUL-1 south of the traditional Jusepin field. Excellent results were obtained from the two wells. Condensate was produced from the ORS-52 well, and a giant field with almost 1 billion bbl of oil in place was discovered in a new area called El Furrial.

STRATIGRAPHY

The sequence stratigraphic analysis of the Eastern Venezuelan basin shows the existence of five sedimentary cycles (Prieto et al., 1989; Figure 5; Table 1). These cycles were divided by maximum condensed surfaces (Brown, 1989) as a result of the provinciality of these events. There are several parasequences (Vail et al., 1984) bounded by unconformities or their correlative conformities (not described here).

Cycle I began with a regressive sequence of mostly sandstones and siltstones associated with a fluviodeltaic system on a passive margin (Barranquin Formation), followed by the deposition of mostly deeper water shales of the Garcia Formation (also described as a basal member of the El Cantil Formation; Rosales and Claxton, 1969).

During the Albian, the platform stabilized and biogenic limestones of the El Cantil Formation were formed. A major transgression occurred from the Cenomanian to the Turonian with the deposition of calcareous black shales that constitute the source rocks for most of the oil found in northern Monagas (Querecual Formation).

Cycle II began during the Coniacian with a regressive system. Several submarine fans from this time period have been identified to the north, probably associated with a relative sea level drop and a type I unconformity (Vail et al., 1984). The subsequent rise of sea level formed a retrogradational lowstand wedge of the San Juan Formation capped by the Vidoño Formation without abundant glauconite and high in gamma ray radioactivity. The Vidoño Formation corresponds to a major condensed section (Cabrera and Pizon, 1989).

Cycle III began with a regressive sequence of the Caratas Formation. The tectonic activity probably was initiated during this period. A major (probably eustatic) unconformity resulted from the deposition in the El Furrial area of Oligocene sediments (Naricual Formation) on top of Cretaceous strata. At this time, increased subsidence resulting from the incipient uplifting of the Interior Range created space on the southern flank of the basin. The systems retrograded toward the south, and from the early Oligocene to the middle Miocene an onlap sequence was deposited. The top of this recessional sequence is a condensed section represented by the Areo and lower Carapita formations.

Cycle IV was marked by the sedimentation of the Carapita Formation, which resulted from the arrival in the area of a major river system that deposited a thick shale sequence. Coalescence of several delta systems associated with the Oficina Formation prograded from southwest to northeast.

The upper portion of the Carapita Formation is associated with tectonic pulses of the uplifting of the Interior Range. Molasse-type sand lenses are present in the sequence. The upper Carapita lies unconformably under the La Pica Formation.

Cycle V is bounded by unconformities formed after the uplifting of the Interior Range by the deposition of coarse-grained clastics associated with the tectonic activity in the area. The upper Miocene–Pliocene sediments (La Pica and Las Piedras formations) were deposited in paleolows created by the tectonic activity. The lithology is mostly shale interbedded with sandstones that have been

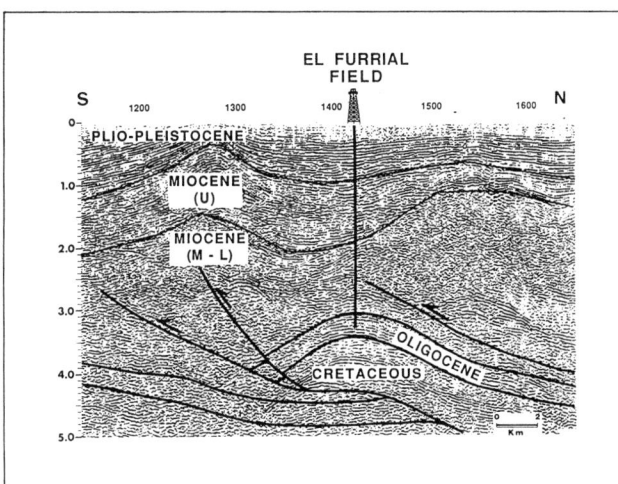

FIGURE 4. Structural interpretation of seismic line NM-84B-05 through El Furrial field.

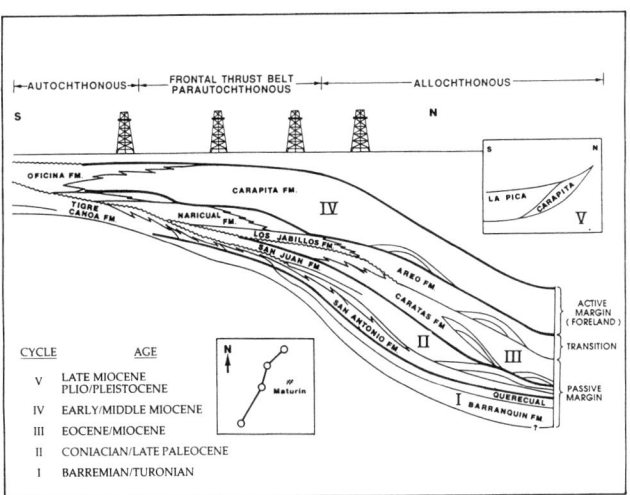

FIGURE 5. Depositional systems in the Eastern Venezuelan basin.

TABLE 1. Sedimentary cycles in the Eastern Venezuelan basin.

Cycle	Age	Formations
I	Barremian-Turonian	Barraquin, El Cantil, Chimana, Querecual (MCS)
II	Coniacian-late Paleocene	San Antonio, San Juan, Vidoño (MCS)
III	Early Eocene-early Miocene	Caratas, Los Jabillos, Naricual, Areo (MCS)
IV	Early Miocene-middle Miocene	Carapita
V	Late Miocene-Pliocene/Pleistocene	La Pica, Las Piedras, Quiriquire, Mesa

productive in the Jusepin, Santa Barbara, and Mulata fields.

Figure 6 shows the stratigraphic chart used in this report.

STRUCTURAL FRAMEWORK

The Eastern Venezuelan basin was formed during the transcollision of the Caribbean plate with the South American plate. It represents a wedge of sediments, deposited during the Cretaceous to the early Tertiary, that was folded and thrusted during the late Tertiary. The transition from a passive margin to a foreland basin started during the Oligocene (Prieto et al., 1989).

Figure 7 presents a schematic profile of the region from the El Pilar fault to the Guayana craton. Three structural provinces can be differentiated:

1. The allochthonous block.

2. The frontal thrust belt or parautochthonous block.

3. The extensive or autochthonous block.

The allochthonous block outcrops in the Interior Range. The main structural features are large-scale concentric folds oriented N70°E. These folds are continuous, measuring up to 70 km long by 5 km wide. Surface work suggests a total shortening for the basin of about 40 km, which represents 28% (Rossi et al., 1987).

The Pirital fault complex represents the southern limit between the allochthonous block and the parautochthonous block. The major structural features found in this area are overthrusted faults identified by seismic data. Interpretation of seismic data shows three main thrust faults characterized by concentric folding that are similar to the thrust faults to the north but lower in amplitude (18 to 10 km in width). Figure 8 indicates the three main thrusts and the location of El Furrial field. The thrusts are mostly blind thrusts that usually end on mud volcanos and that can be followed for many kilometers trending parallel to the main anticlines. The El Furrial trend includes the Carito, Tejero, Boqueron, and El Furrial fields. The chronology of deformation has been interpreted as decreasing in age toward the south and east.

Toward the south, the autochthonous block is characterized by normal faults trending in an east-west direction. Traditional oil fields—the Greater Oficina field, the

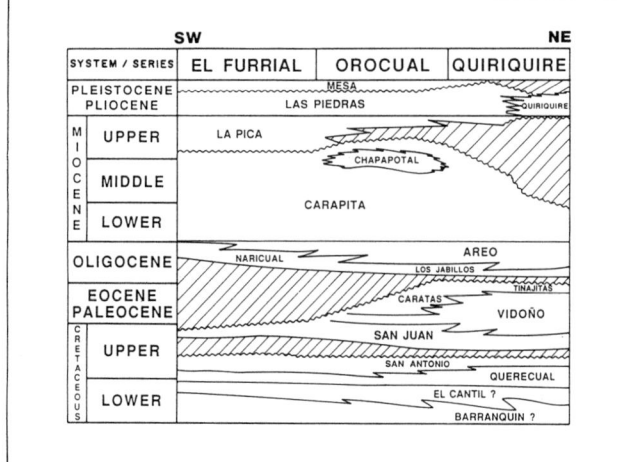

FIGURE 6. Generalized stratigraphic chart for the Eastern Venezuelan basin (see Figure 5).

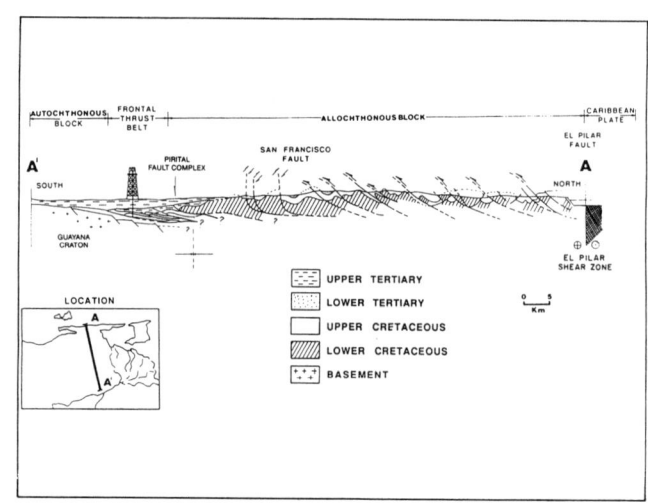

FIGURE 7. Regional cross section through the Eastern Venezuelan basin.

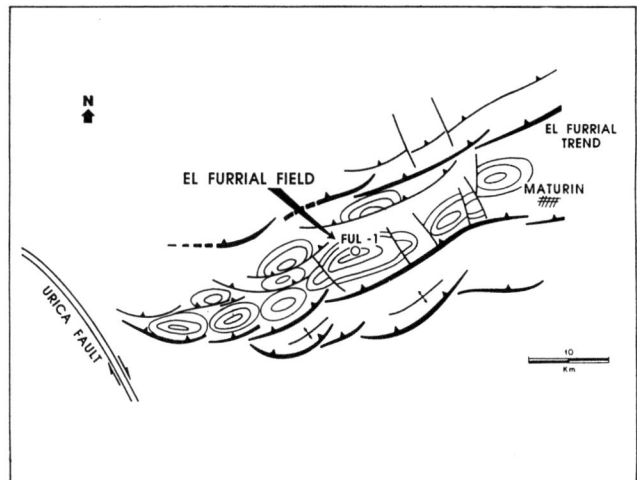

FIGURE 8. Regional tectonic framework of the El Furrial area.

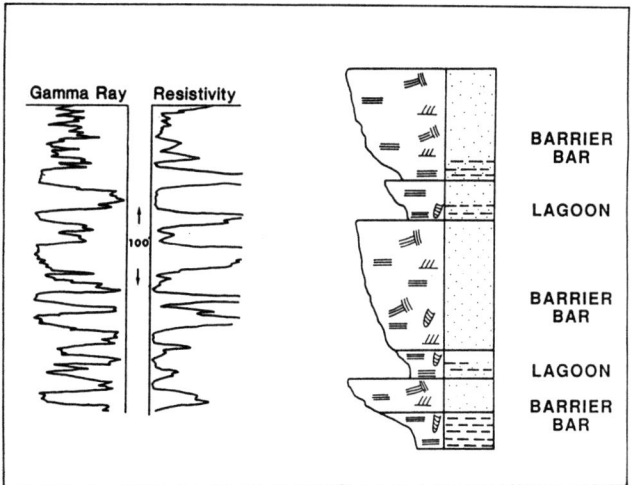

FIGURE 10. Typical lithofacies of the Naricual reservoir sands (after Ghosh et al., 1989).

Anaco field, and the Orinoco tar belt area—are present in this region.

RESERVOIR CHARACTERISTICS

El Furrial field is part of a large anticlinal fold truncated on the south by a major thrust fault and on the north by the Masacua thrust fault. The anticline is 10 km long by 8 km wide and has a vertical closure of 915 m (3000 ft; Figure 9). The most important reservoir rocks found in the field correspond to the Oligocene Naricual Formation.

The discovery well, FUL-1, penetrated the reservoir at a depth of 3960 m (13,000 ft) and found a total pay zone of 276 m (905 ft). The Naricual Formation consists of rocks in a shallow marine environment deposited as a retrogradational sequence that is more marine toward the upper section. The type log and the sedimentological interpretation are shown in Figure 10.

In addition, the lower sequence of the Upper Cretaceous section was drilled by the FUL-4 well, which found 378 m (1240 ft) of net pay sand.

Petrophysical and sedimentological analyses of the reservoir, including the acquisition of 1014 m (3325 ft) of cores in the 17 wells perforated, allowed subdivision of the stratigraphic characteristics of the Naricual Formation into three units: lower, middle, and upper.

The lower unit has the greatest thickness, representing almost half of the formation. Net oil sand thickness is about 58 to 93 m (190 to 305 ft). The middle unit is different from the upper and lower ones owing to a greater number of shale breaks. Net oil sand thickness varies from 29 to 65 m (95 to 213 ft). The upper unit has fewer shale breaks than the other two members. The average porosity in the Naricual Formation varies from 13 to 17%.

The subjacent Cretaceous rocks are composed of sands interbedded with shales and constitute a smaller pay zone varying from 57 to 85 m (187 to 279 ft) thick. These rocks are of lower reservoir quality, exhibiting porosities ranging from 9 to 11% and lower permeabilities (Ghosh et al., 1989).

Table 2 summarizes the reservoir characteristics and petrophysical parameters. Figure 11 is an isoporosity map of the upper unit of the Naricual Formation.

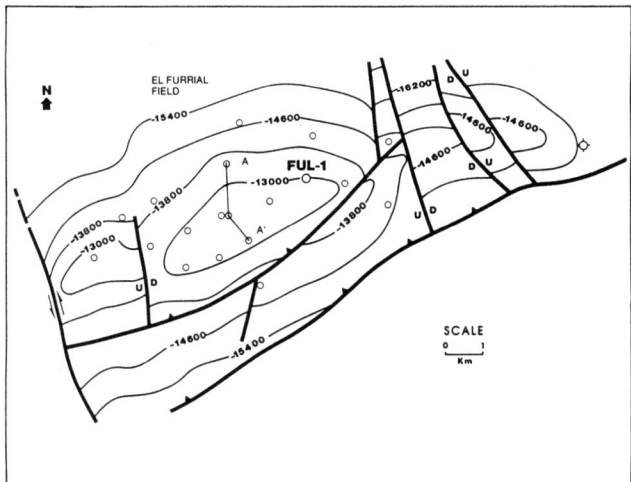

FIGURE 9. Structural map of top of the Naricual Formation. Contour interval = 800 ft.

PRODUCTION DATA

The results of a regional geochemical analysis show that the Upper Cretaceous (Cenomanian–Campanian) Querecual and San Antonio formations are the source rocks responsible for the hydrocarbons in the Eastern Venezuelan basin. Talukdar et al. (1986) have given estimated TOC values of 0.25 to 6.6%, maturities from 0.6 to 1.3% R_o (T_{max} between 440 to 550°C), and a hydrocarbon yield content of 454 mg HC/g TOC. The oil found in El Furrial field is

TABLE 2. Reservoir characteristics and petrophysical parameters for El Furrial field.

Formation	Net oil sand thickness		Porosity (%)	Water saturation (%)	Shale volume (%)
	m	ft			
Upper Naricual	58–93	190–306	14–17	11–17	0.8–4.8
Middle Naricual	29–65	95–214	13–16	11–17	0.5–5.8
Lower Naricual	84–126	275–413	13–17	13–21	0.5–6.1
Cretaceous sequence	57–85	188–279	9–11	17–35	6.8–10.7

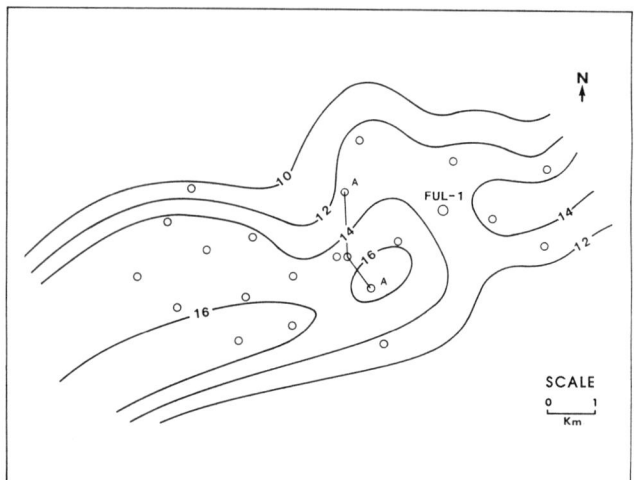

FIGURE 11. Isoporosity map of the upper Naricual Formation. Contour interval = 2%.

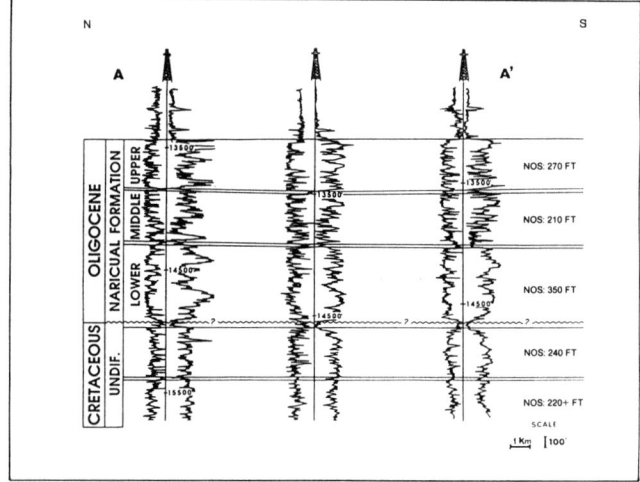

FIGURE 12. East-west stratigraphic cross section correlating three wells in the El Furrial field. See Figures 9 and 11 for location. NOS = net oil sand thickness.

described as mature and was derived mostly from amorphous marine organic matter. Geochemical data indicate that oil was generated from kitchens active from the late Miocene to the Holocene beneath the main thrusts.

Six reservoirs have been defined: four in the Naricual Formation and two in the Cretaceous sequence. A correlation of three wells is depicted in Figure 12.

The discovery well, FUL-1, penetrated the upper Oligocene Naricual Formation at a depth of 3960 m (13,000 ft) and found a total of 430 m (1410 ft) of sandstone and 276 m (905 ft) of net oil pay. Initial production tested 7331 BOPD on a ½-in. choke. The gravity of the oil was 26.5° API.

Seventeen wells had been drilled in the field as of January 31, 1989, with a production rate of 102,000 BOPD and a cumulative production of 49.553 million bbl. Analyses of production data (PVT and pressure) indicated a total OOIP of 4.467 million bbl and primary oil recovery of 893 million bbl.

No oil-water contact has been found, and the drive mechanisms are interpreted as fluid and rock expansion and gas in solution. The average pressure of the productive reservoirs is 11,000 psi, and the bubble point pressure is about 4600 psi. In two wells, FUL-3 and FUL-6, heavy oil was found, indicating the presence of at least two different reservoirs in the area. The tar mat represents the northern and southern limits of the reservoir.

The limits of the field have been established (see Figure 9). To the north and south, the field is bounded by a tar mat at depths of −4756 m (−15,600 ft) and −4512 m (−14,800 ft), respectively, and to the east and west by tear faults.

The information obtained from the exploration program allowed the planning of a development program comprising 43 wells spaced at 1200 m, with an evaluation plan that included logs, cores, fluid analysis, and dynamic testing (RFT, DST, and PVT). The program illustrated in Figure 13 includes 68 completions with a total of 25 dual completions. Each dual completion has a potential production rate of 13,000 BOPD. The total estimated production rate is 180,000 BOPD for 1990, increasing to 220,000 BOPD over a period of four years. The production profile shown in Figure 14 includes a secondary recovery program of pressure buildup starting in 1994.

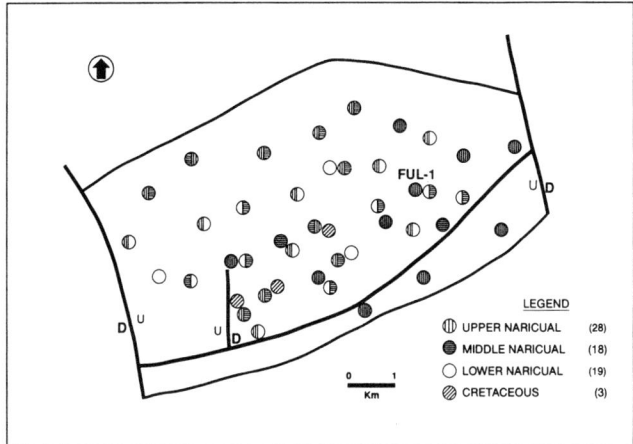

FIGURE 13. Development program for El Furrial field. (Parenthetical numbers in legend denote numbers of completions.)

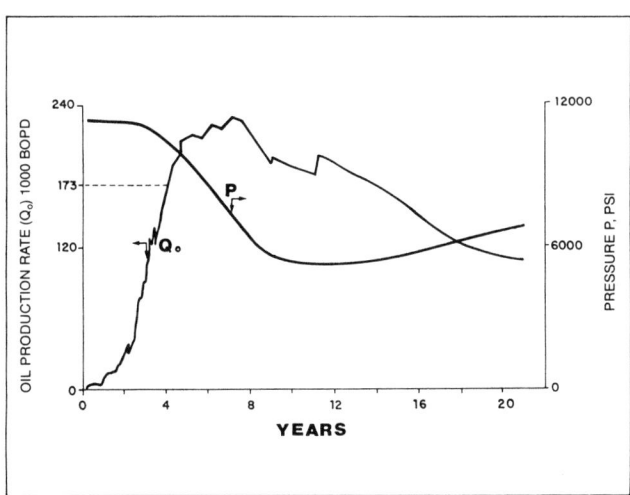

FIGURE 14. Production profile of El Furrial field, including secondary recovery program.

CONCLUSIONS

The discovery of El Furrial field identified a new play in the upper Oligocene section that holds more than 80% of the field reserves.

The Oligocene sandstones were deposited in a nearshore marine environment. Barrier bars are the most important reservoir facies. The Cretaceous sandstones are mostly fluviodeltaic.

An early development plan was designed for El Furrial field to establish a production rate of 220,000 BOPD by 1994.

The El Furrial discovery opened a whole new perspective for the area, with total estimated oil reserves of 5.930 billion bbl for the northern Monagas area and 12.2 billion bbl for the Eastern Venezuelan basin. It reactivated the exploration activity in this area, which had been suspended since 1960.

Further exploration in an area previously thought to be mature led to the discovery of yet another giant field in the Eastern Venezuelan basin. Geophysical data indicate that the structural trend extends toward the northeast. Persistence and new ideas will be required if new giant fields are to be pursued in the adjacent areas.

ACKNOWLEDGMENTS

The authors received support from tl   :tation and Exploration groups from the Geolo   ırtment of Lagoven, S.A. Special thanks go to Ma   , Maria de Salazar, and Daisy Perez de Mejia for tł   al reading of the manuscript.

REFERENCES CITEı

Brown, L. F., Jr., 1989, Seismic/sequence stratigraphy: its role in petroleum exploration: Georgetown, Texas, 61 p.

Cabrera, S., and J. Pizon, 1989, Formacion vidono: ejemplo de una secuencia depositacional: internal report, Corpoven, S.A., unpublished, 30 p.

Ghosh, S., A. Isea, I. Truskowski, and B. Aguado, 1989, Estudio sedimentologico-bioestratiografico de las areas El Furrial-Musipan-Carito, Norte de Monagas: internal report, Intevep, S.A., unpublished, 50 p.

Lagoven, S.A., 1987, Generalized north-south cross-section across Interior Mountain front and Maturin area: unpublished.

Pereira, J. G., and R. Aymard, 1988, Geological integration and evaluation of northern Monagas, Eastern Venezuelan basin: internal report, Lagoven, S.A., unpublished, 35 p.

Prieto, R., G. Hernandez, A. Daal, and A. Gonzalez, 1989, Ephinom: exploracion por hidrocarburos en el norte de Monagas: internal report, Petroleos de Venezuela, 55 p.

Rosales, H., and C. D. Claxton, 1969, The lower Tertiary-Cretaceous of northeastern Venezuela: internal report, Creole Petroleum Corporation, unpublished, 242 p.

Rossi, T., F. Stephan, R. Blanchet, and G. Hernández, 1987, Etude géologique de la Serranía del Interior Oriental (Venezuela): Sur le Transect Cariaco-Maturín: Revue de l' Institut Français du Petrole, n. 1.

Salvador, A., 1958, Northern Monagas oil prospects: internal report, Creole Petroleum Corporation, unpublished, 57 p.

Talukdar, S., O. Gallango, and A. Ruggiero, 1986, Estudio geoquímico regional de la subcuenca de Maturin: Internal Report n. INT 01543,86, Intevep, S.A., Caracas.

Vail, P. R., J. Handerbol, and R. G. Todd, 1984, Jurassic unconformities, chronostratigraphy and sea level changes from seismic stratigraphy and biostratigraphy: AAPG, Memoir 36, p. 129-144.

Chapter 11

Ceuta-Tomoporo Field, Venezuela

Enrique Ramírez
Fernando Marcano

Maraven S.A.
Caracas, Venezuela

ABSTRACT

Ceuta-Tomoporo field is located in the southeastern part of the Maracaibo basin in western Venezuela. It has an area of 320 km² and is producing light- to medium-grade oil. The structural framework is characterized by a major left-slip fault, and oil accumulations are associated with extensive compressional structural features. There are indications that the deposition of Eocene sediments has been controlled by normal faulting in some areas and that a period of shortening altered the character of some faults during the late Eocene, or even later. The main producing intervals are shallow water marine or fluviodeltaic sands of Miocene or Eocene age, with porosities ranging from 12 to 17% and a potential production rate of up to 3500 BOPD per well from depths of about 17,000 ft (5180 m). Most of the oils accumulated in the basin have been derived from the organic-rich, highly oil-prone Cretaceous La Luna source rocks. The reservoirs seem to contain mixtures of hydrocarbons that probably are the results of migration from a common drainage area at different times.

INTRODUCTION

The interactions among the Nazca, Caribbean, and South American plates have controlled the style of the Maracaibo basin and the behavior of its major megastructures. This activity has given rise to the Merida Andes uplift in the east and the Sierra de Perija in the west, where thrusting seems to have been more significant (Figure 1). Since the Paleozoic, the basin has been changed from a back-arc basin to an Atlantic-type passive margin basin to an intracratonic basin.

The Maracaibo basin is bordered by the Oca, Perija, and Bocono major fault systems. The Bocono and Perija fault systems are right-lateral strike-slip faults that have been active since the Paleozoic.

Several locally parallel left strike-slip fault systems oriented roughly north-south are present in the Maracaibo basin, each associated with oil fields. One of these fields is the Ceuta field, which is defined by a major wrench fault structure. The discovery of the field in the 1950s has been followed by important exploration activities, and, through expansion into the neighboring areas, a

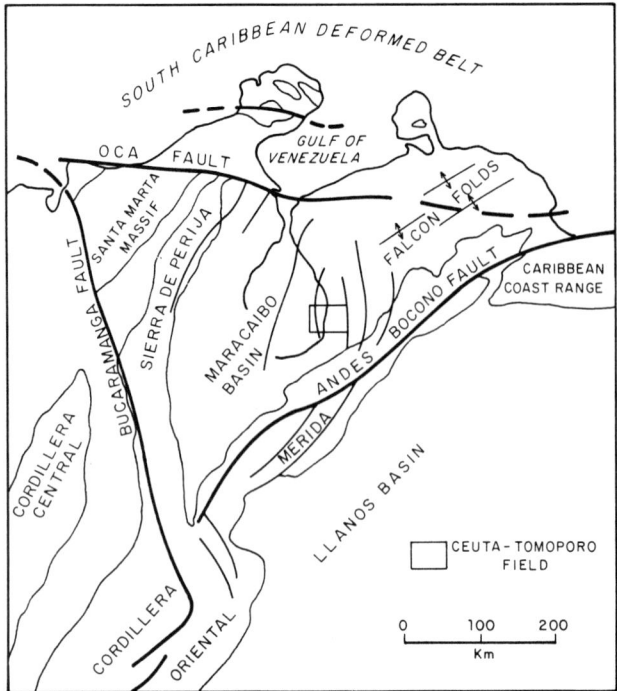

FIGURE 1. Structural map of the Maracaibo basin.

knowledge of the stratigraphy and the characteristics of the various reservoirs has been acquired. The hydrocarbons discovered in either the Tertiary or the Cretaceous reservoirs were generated from the rich marine Upper Cretaceous La Luna source rocks.

The Ceuta field is producing from Miocene and Eocene clastic reservoirs at depths ranging from 15,000 to 17,000 ft. (4570 to 5180 m). In both the Miocene and Eocene layers, multiple reservoirs (in both an areal and a vertical sense) were identified on the basis of fluid distribution. Table 1 gives the production parameters for both the Miocene and the Eocene reservoirs in the Ceuta-Tomoporo field.

STRUCTURAL SETTING

The Ceuta-Tomoporo area is located between two major transcurrent faults in Lake Maracaibo: Barua to the east and Pueblo Viejo to the west. These basement-involving features are subparallel and run roughly north-south (Figure 2). They show left-transpressive displacement related to the clockwise rotational effect produced by the right-lateral Caribbean plate motion, during which a northwest–southeast-oriented compressional effect was responsible for the main deformation in the Ceuta-Tomoporo area (Audemard, 1987).

This event overprinted previous Jurassic rifting associated with the opening of the North Atlantic. One of these normal rift-type faults present in the Ceuta-Tomoporo area was partially reactivated during the Oligocene–Miocene, causing greater deformation in the northern block in contrast to the homocline style in the south (Figure 3). According to the seismic information, the Paleocene and Cretaceous sections show similar deformation, whereas the younger Eocene levels seem to have been affected by later reactivation of these features. The major nonstrike-slip faults that affected the entire Cretaceous–Eocene column are difficult to identify (Figure 4).

On the basis of stratigraphic observations, a hinge zone oriented west-northwest to east-southeast has been postulated in order to explain the rapid thickening of the early and middle Eocene sediments to the east of the Ceuta block (Van Veen, 1972). Along this flexure zone, roughly parallel to the eastern coast of the lake, associated normal faulting has controlled the sedimentation. These faults constitute a down-to-the-basin system oriented west to northwest.

The evolution of the complex structural frame of the Ceuta-Tomoporo area has been pointed out recently by Roberto et al. (1988). According to these authors, at least three major post-Paleocene deformation phases can be recognized:

1. An extensional or divergent wrenching phase related to faults syndepositionally active during the early

TABLE 1. Production parameters for Miocene and Eocene reservoirs in the Ceuta-Tomoporo field.

Cumulative production (MMSTB)	445
Present potential production rate (STB per day)	80,000
Number of wells	309
Oil gravity, average (°API)	30
Oil-gas ratio (std. ft³/STB)	1,230

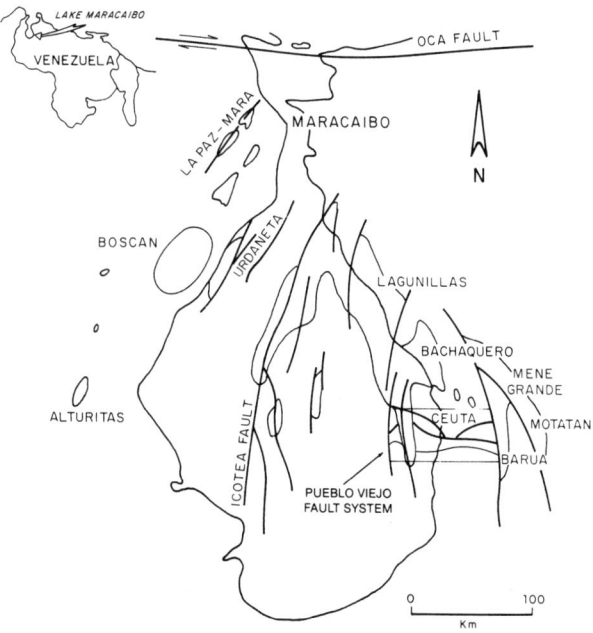

FIGURE 2. Maracaibo basin oil fields.

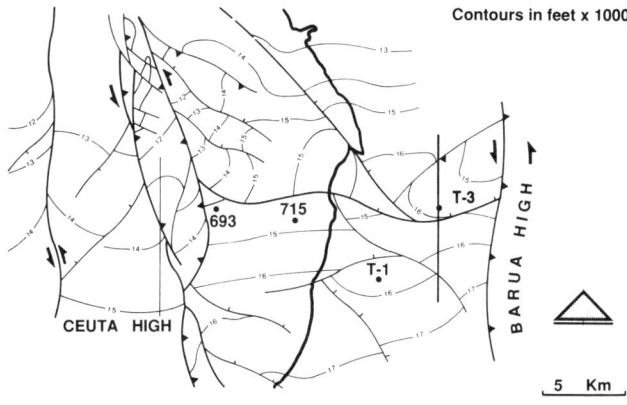

FIGURE 3. Early Eocene structural map.

These authors pointed out that fault geometries and deformational patterns in the Ceuta-Tomoporo area appeared to fit with experimentally derived structural models. The main structural elements that are recognized (Figure 3) are as follows:

1. *Left-lateral strike slip.* Two main north–south-oriented left-lateral strike-slip zones are present: the Ceuta high to the west, and the eastern Barua field. In the Ceuta high, the underlying basement fault behaves as a "scissor" fault with reverse throw in the northern part and normal displacement in the south. The high is bounded by a pair of left-lateral strike-slip zones to the west and by a wrench fault to the east. Two northeast–southwest-oriented en echelon folds bound a pop-up anticline, and they are cut by northwest–southeast-oriented normal faults that control the distribution of the Eocene oil-producing levels.

Eocene, which explains the locally observed differences in the thickness of the Eocene strata.

2. A compressional or convergent phase that accounts for much of the vertical uplifting along the main wrench faults, creating positive flower structures such as the Ceuta and Barua highs. The erosion of thick Eocene intervals represented by a major regional unconformity has been related to this event.

3. Important compressional movements during the Miocene-Pliocene, which were related to the initial growth of the Venezuelan Andes and associated with inversion and locally vertical displacements.

2. *Right-lateral strike slip.* The largest right-lateral antithetic strike-slip fault in the area is a normal basement fault with northward shading that has had a very important dip-slip component. The increased Eocene thickness on the downthrown block suggests important activity during the early Eocene. The fault trace shows two linear segments displaced by a north-south accident, where the right-lateral slip and rotation have created a local pull-apart. The differences in the ratio of strike slip to dip slip seem to have been the cause of the changing attitudes of the associated Riedel-type faults, which behave normally in the west and in a

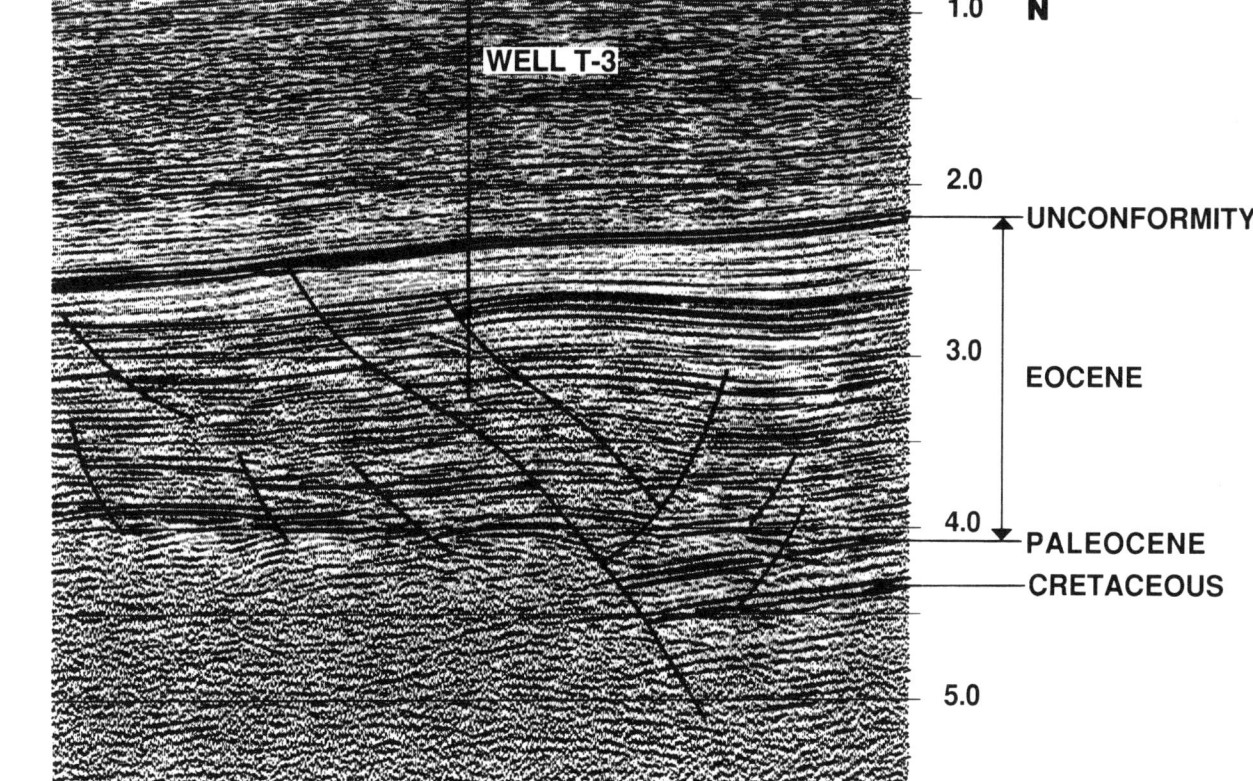

FIGURE 4. Ceuta-Tomoporo seismic section.

reverse manner in the east. This structure provides an ideal trap for the hydrocarbons present in the area.

3. *Folds and reverse faults* are essentially southwest-northeast-oriented folds sometimes bounded by reverse faults. In some areas, these features form important structural trapping elements, which could be confirmed by drilling one of them and finding potential reserves in excess of 100 million bbl. A similar onshore structure in the northern Tomoporo area has proved oil in the upper "B" sandstones (Figure 4).

4. *Normal faults.* The northwest-southeast-oriented normal faults show variations in throw of up to 1000 ft (305 m). Potential hydrocarbon traps exist on both the downthrown and upthrown sides of the fault. The extensional down-to-the-basin faults with west to northwest orientations are very difficult to differentiate from the antithetic wrench faults. Besides, it is possible that these normal faults, as well as the other right-lateral wrench faults, were overtaken by the lateral compressional phases.

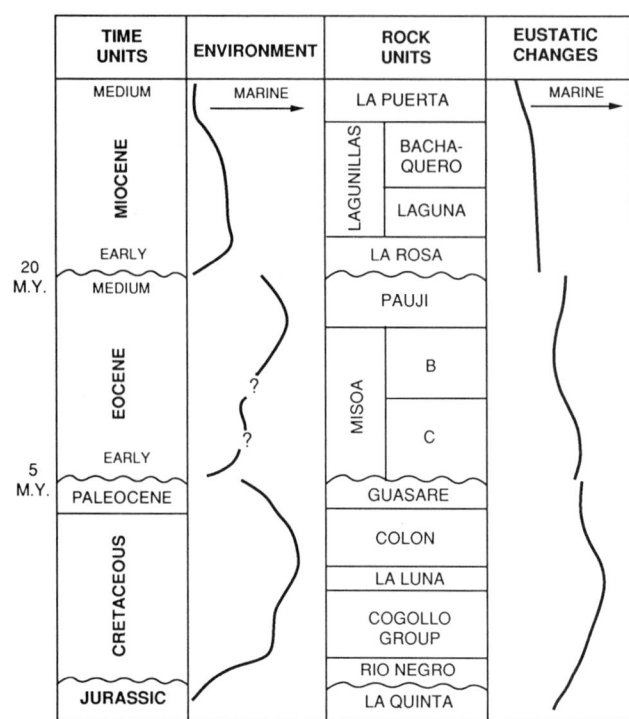

FIGURE 5. Ceuta-Tomoporo stratigraphy.

STRATIGRAPHY

The Maracaibo basin has been developed on a continental crust that is probably upper Precambrian in age. A pile of sediments that had accumulated in the Paleozoic was later deformed and metamorphosed.

The Triassic is represented by a very important regional unconformity, followed by deposition of the Jurassic red beds (La Quinta Formation) in a graben, oriented north-northeast to south-southwest, associated with the Atlantic opening.

Cretaceous sedimentation involved a huge and complete transgressive-regressive cycle. The transgressive event started with a thin, conglomeratic basal sequence, after which the area was covered by thick, shallow platform limestones gathered in the Cogollo Group (Figure 5).

Local movements, probably caused by plate interactions during the Upper Cretaceous, modified the shape of the basin and restricted its connection with the open sea. In this anoxic and very organic-rich marine environment were deposited the limestones of the La Luna Formation, which constitute the main oil source rock of the Maracaibo basin. This unit is overlain by a very thin, very phosphate-rich, glauconitic limestone level called the Socuy Member, which is one of the most conspicuous seismic reflectors in the area and is used as a reference level for Cretaceous mapping. The regressive event is represented by the thick Colon shales.

The Paleocene sedimentary cycle has been developed in a very shallow environment and represents the continuation of the Maastrichtian marine regressive Colon Formation. Deposits consist of a calcareous sequence up to 300 ft (91 m) thick interbedded with local oolitic and shaly layers. This cycle changes gradually to terrestrial environments toward the top, and a short, erosive event of variable intensity, which took place at the end of the period, has been identified.

In the early Eocene, the Maracaibo platform was covered by wide fluvial and deltaic systems characterized by changing flow directions convergent to the basin. These deposits are grouped in the Misoa Formation, and two sedimentary megacycles are recognized: the lower "C" sandstones and the overlying upper "B" sandstones. According to Storey (1980), a stratigraphic hiatus is recognized at the top and base of each sequence. In the basal section, a high-energy sandy level interbedded with thin shales has been identified. Toward the top, the lithology changes to thick sequences of siltstones and shales with occasional thin levels of sandstones, representing low-energy environments. Identification of regional log characteristics allowed this formation to be subdivided into correlatable units within each cycle.

An important tectonic activity took place during the middle part of the lower Eocene. The direction of the Caribbean plate motion changed from northeast to east, and the frontal island arc was passing by the northern part of the Maracaibo basin, controlling both the basin morphology and the direction of the sedimentation of the "B" sandstones. At the base of this cycle, it is possible to recognize seismostratigraphic units downlapping on the "C" sandstones, oriented to the south, and probably linked with the partial erosion of masses located in the north.

Owing to the eastward motion of the Caribbean plate, the Maracaibo basin was tilting gradually in this direction. During the middle Eocene, a marine transgression prograding from the northeast covered most of the eastern zone with thick shale sequences gathered into the Pauji Formation.

In Storey's opinion (1980), this sedimentary unit completes the Eocene sedimentary cycle in the area and includes in its base the uppermost sandstones from the upper "B" megacycle. On the other hand, Schwander (1987) identified four depositional units separated by variable-intensity hiati. The lowermost unit consists of an uncalibrated graben-fill, which is interpreted as containing lower Paleocene marine strata, including turbidite deposits, east of the Ceuta area. The other three are very similar to the Storey units.

Toward the end of the Eocene, very important isostatic movements took place, and most of western Venezuela suffered a gradual and continuous uplift. The exposed areas were strongly eroded, and most of the Pauji Formation was removed. This event continued during the Oligocene, and it constitutes one of the more significant geologic events in the evolution of the Maracaibo basin.

During the early Miocene, a new sedimentary event related to a marine pulse covered the area with a thin sandstone basal sequence known as the Santa Barbara sandstones, which was followed by marine fossiliferous shales. This shale unit, named the La Rosa Formation, also constitutes a producing level in several fields of the Maracaibo basin. From a regional point of view, it onlaps the flanks of the remaining highs. The La Rosa Formation virtually covers the entire Ceuta-Tomoporo area with the exception of the Ceuta high, where it is absent, and the lowermost units of the overlying Lagunillas Formation. The sedimentation of the Lagunillas Formation was restricted to the paleotopographic lows (Figure 6). It represents the transitional changes from marine to continental environments caused by the regression of the sea to the northeast (González de Juana et al., 1980).

The Lagunillas Formation is one of the most important petroliferous units in the Maracaibo basin and the main oil producer in the Ceuta high. According to its lithologic characteristics, it has been subdivided into three minor units. The lowermost unit, the Lagunillas Inferior Member, is a fining-upward sequence that ends with clays and lignites that accumulated in swampy environments. The overlying Laguna Member has been interpreted as a short marine incursion characterized by gray fossiliferous shales and sandy shales. The uppermost unit, the Bachaquero Member, is composed of intercalations of clays, shales, and poorly to very poorly consolidated sandstones the percentage of which increases toward the top, where oil accumulations are common.

These units interfinger eastward with fluvial deposits and alluvial fans, and their evolution has been controlled by tectonic movements related to the Andes uplift, which started during the Miocene-Pliocene.

Since the early Miocene, the Ceuta area has shown a greater stability and has been exposed to frequent shallow overflows. As a consequence, swamps and a variety of platform deposits have been accumulated in the area, congregated in the La Puerta Formation.

GENERATION AND MIGRATION OF HYDROCARBONS

Geochemical studies on the origin of oil in the Maracaibo basin have considered evolution of source rocks, oil typing and oil–source rock correlation, thermal maturity measurements, and reconstruction of the hydrocarbon-generating areas and their evolution through time using thermal modeling methods (Talukdar et al., 1986; Blaser and White, 1984).

It has been established that most of the oils accumulated in the basin have been derived from the organic-rich, highly oil-prone La Luna source rocks. Their thermal maturity at the present time indicates that they have been able to generate the oil and gas in the majority of the basin, remaining immature only in a belt to the west, close to the Perija Range. Petrographic analyses of immature samples have shown amorphous marine organic matter in laminae parallel to stratification regarded as mainly type II (Tissot and Welte, 1984). These rocks have an average organic carbon content greater than 5% and the capacity of generating nearly 30 million bbl of oil per square kilometer.

Geochemical evaluation of Eocene deltaic sediments to the east of the Ceuta area has shown mainly gas-prone type III organic matter with a low potential compared with that of La Luna (Figure 7). The study of oils accumulated in the Eocene reservoirs of the Ceuta area has been based on conventional and special analyses that have shown the origin of the oil to have been marine organic matter.

Comparison of the chromatographic characteristics of the Ceuta and La Luna oils has indicated a clear relationship between them. Moreover, the presence of trace amounts of compounds produced by biodegradation in apparently unaltered oils appears to offer presumptive evidence that earlier-generated oils were later subjected to conditions of biodegradation during a phase of uplifting and erosion at the end of the Eocene and during the Oligocene, as is suggested by geologic modeling (Figure 8).

These unaltered oils in the Ceuta area have been identified as mature and as having resulted from a source rock with a maturity probably corresponding to a vitrinite reflectance of 0.8% or greater. Measurements of vitrinite reflectance and H/C ratios in well samples have been made in order to calibrate the thermal modeling of the area.

EXPLORATION AND DISCOVERY HISTORY

Exploratory drilling in Venezuela started in the southwest corner of the Maracaibo basin, where the Petrolia del Tachira Co. drilled the first producer well in 1883. The operations were carried out by percussion, reaching a depth of 60 m and a production rate of 240 L of oil per day.

The Caribbean Petroleum Co. began drilling the Zumaque-1 well in 1914, following detailed field work near seeps to the east of the Ceuta-Tomoporo area, and discovered the Mene Grande field, which had an initial production rate of 40 m^3 of oil per day from Miocene sediments (Figure 9). A significant exploration activity led to the drilling, in 1922, of the Barrosos-2 well, which was a landmark in the history of the Venezuelan oil industry. This well, out of control, produced 16,000 m^3 of oil per day. Then, in 1924, well M-1 discovered oil in the Eocene layer in the Mene Grande field, the first well in Lake

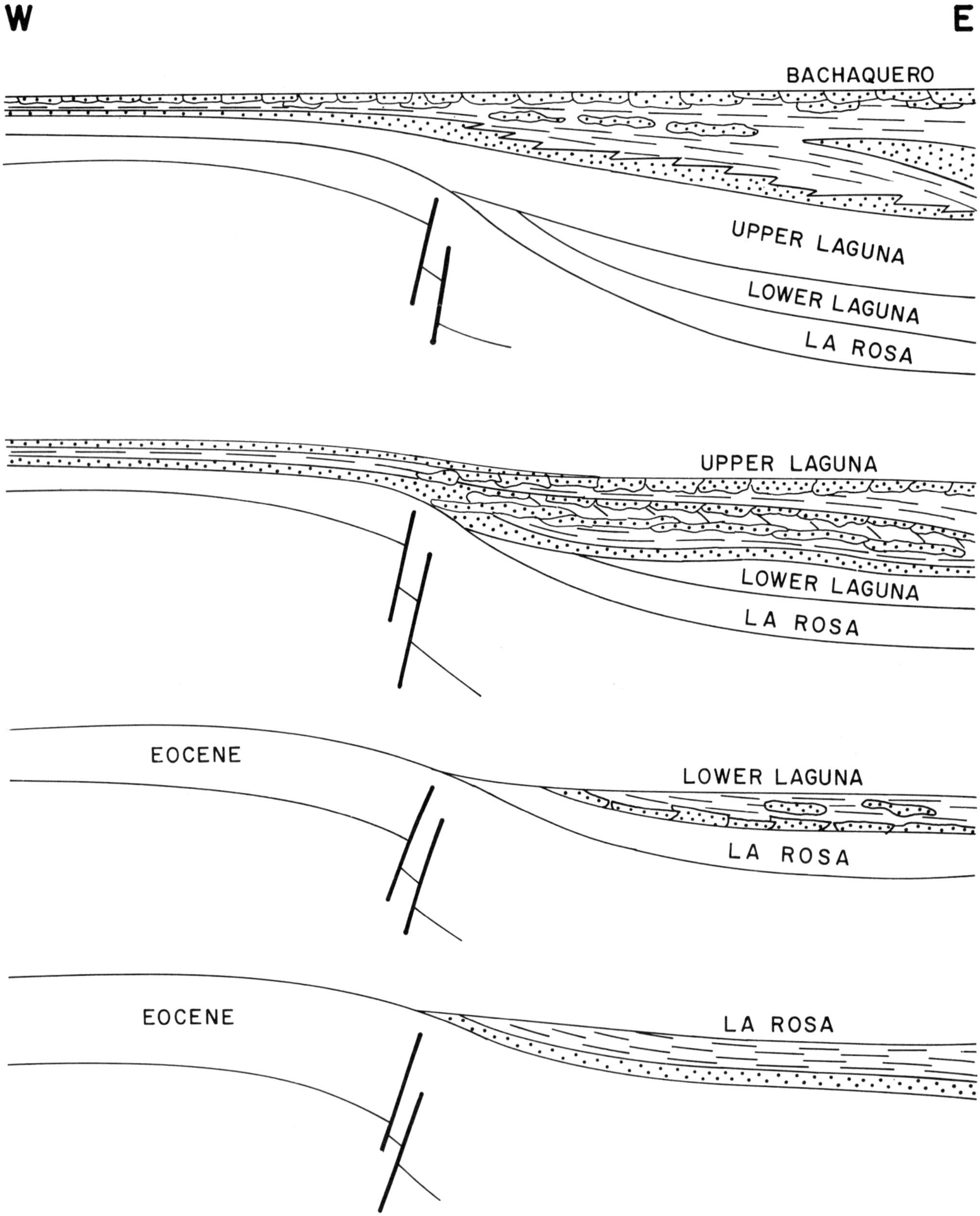

FIGURE 6. Miocene sedimentation on the Ceuta high.

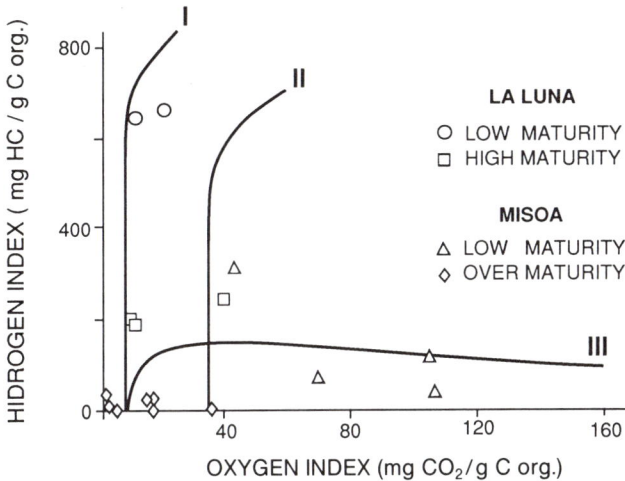

FIGURE 7. Source rock types in the Maracaibo basin (based on Gallango, 1984).

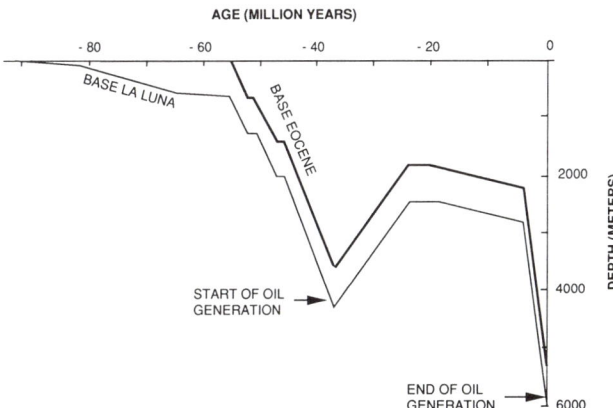

FIGURE 8. Geologic and thermal modeling in the Ceuta area.

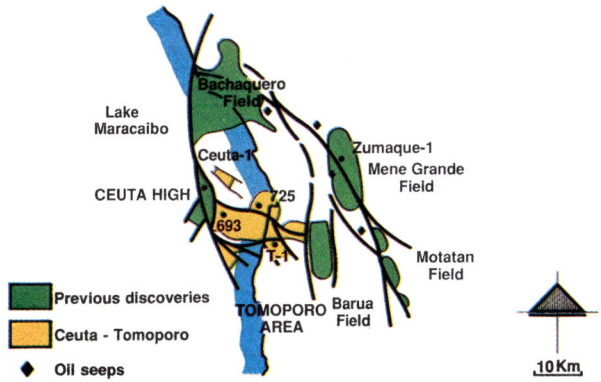

FIGURE 9. Ceuta-Tomoporo discovery.

Maracaibo was drilled, and the first Venezuelan geophysical survey was carried out (using electrical methods).

During 1930, the Venezuela Gulf Co. discovered the Bachaquero field when their Lagunita-1 well blew out. Owing to the successes achieved in the Mene Grande and Bachaquero fields, detailed aeromagnetic and gravimetric studies were performed, and new leads were interpreted in Lake Maracaibo. Then, in April 1956, the Mene Grande Oil Co. drilled the Ceuta-1 well and thereby discovered Ceuta field. Other oil companies continued drilling the area in 1957, extending the field to the north, closer to Bachaquero field.

Ceuta field was operated by Mene Grande Oil Co. until 1976, when the entire oil business was handed over to Venezuelan oil companies under the guide of Petroleos de Venezuela. The Ceuta area was assigned to Meneven S.A., an affiliate of Petroleos de Venezuela, and in 1981 the VLG-693 well discovered new oil accumulations to the east of the traditional Ceuta field, in an area bounded by an important normal fault to the north. The VLG-693 well had an initial production rate of 3100 BOPD from lower "B" Eocene sandstones. Then, in 1982, the VLG-707 well was drilled and produced 4500 BOPD.

In 1983, the Ceuta area was assigned to Maraven S.A., also an affiliate of Petroleos de Venezuela. A new structural model was developed on the basis of the transcurrent tectonic setting, an integration of reservoir engineering data and petrophysical data (V.S.T., 1984) was performed, and the present subdivision of the area was established in order to optimize operational plans.

Meanwhile, exploration activity continued in 1985–86 with the drilling of the VLG-715 well in Lake Maracaibo and the Tomoporo-1X well on land. The Tomoporo-1X well discovered the first commercial accumulation of light oil on the eastern coast of Lake Maracaibo (2000 BOPD). Then, in 1987, the VLG-725 well discovered oil in the hanging wall of the normal east-west fault that formed the northern boundary of the 1981 discovery.

To the west of the Ceuta high, a significant thickness of the "B" Eocene sandstones was removed by erosion during the upper Eocene–Oligocene, and reservoirs were subsequently located in the "C" sandstones, where wells initially produce an average of 3000 BOPD. The first well drilled in this area has produced 5.0 million bbl of oil in five years. Recently, a new discovery was made in the Miocene strata to the south of the traditional Ceuta field, in a separate block.

Even though 309 wells have been drilled in the Ceuta-Tomoporo area (lake and land), exploration has continued and more seismic data have been acquired to the north and south. Also, exploration of deeper objectives, such as untested "C" sands to the east, has been carried out. A special effort is presently being made to focus exploration efforts in the near future on Paleocene and Cretaceous limestones.

MIOCENE RESERVOIRS

The Miocene sequences in the Ceuta field comprise three formations: La Puerta, Lagunillas, and La Rosa. Hydrocarbons are found in the sands of all three of these formations, but in the Ceuta area the main reservoirs are located in the Lagunillas Formation, which has been subdivided into the Bachaquero, Laguna, and Lower Lagunillas members (from top to bottom), and into reservoir units within each member.

Lagunillas Formation

In the Ceuta area, the Lagunillas Formation represents the gradual transgression of the continental environments covering the marine La Rosa Formation.

According to the core studies carried out by Chacartegui (1985), the Lower Lagunillas Member is the result of a complex deltaic activity whereby distributary channels with associated mouth bars, and crevasse splays, became common. The palynologic assemblage suggests the zone of *Echitricolporites maristellae-Psiladiporites minimus*, which represents shoreline/tide-influenced environments from the earliest middle Miocene (Pittelli and Di Giacomo, 1990).

The Laguna Member represents a short, shallow marine pulse. The glauconitic sandstones define complex shoreline bars interbedded with some bay shales. The presence of *Haplophragmoides* sp., *Ammobaculites* sp., *Trochamina* sp., and *Miliammina* sp., and palynologic identification of *Crassoretitriletes vanraadshooveni*, suggest middle Miocene age. The uppermost Bachaquero Member is composed of a very thin basal conglomeratic level followed by fluvial channel sandstones in which point bars and interbedded flood plain deposits are common. According to palynologic identifications of *Echitricolporites spinosus* to *E. mcneillyi*, the age of the Bachaquero Member varies from late Miocene to Pliocene (Pittelli and Di Giacomo, 1990).

Problems in identification and correlation of members and units arise from the rapid changes in thickness and even the disappearance of one or more units. Syndepositional movements caused a complex system of transgression or regression of the Miocene sedimentation on the post-Eocene unconformity, and in places the Lagunillas Formation lies directly on the Eocene Misoa Formation. The possible pinch-out of one or more units, the coalescence and change in log character, or the reduction in thickness of all units makes conventional wireline log correlation impossible. By incorporation of petrophysical and reservoir information, not only major subdivisions of the Lagunillas Formation but also correlations of individual units are possible. The changes in thickness of the Lagunillas Formation have been interpreted as having resulted from a combination of stratigraphic thinning within each unit and the pinch-out of one or more of the units over the rising anticlines. The transgressive La Rosa Formation covers almost the entire area except for the culminations of the two anticlines in the Ceuta high. The La Rosa thinning is the earliest evidence of the presence of the northeast–southwest-trending en echelon anticlines accompanied by normal faulting. The result of these movements is the eventual pinch-out of some units onto the highs or, alternatively, the restriction of the lower parts of the units to the flanks of the highs, which in some cases suggests attractive traps. The key to solving the correlation and recognition of reservoir units has been the use of petrophysical and reservoir engineering data in an interactive working relationship.

Bachaquero Member

The upper part of the Lagunillas Formation comprises the Bachaquero Member, which has been divided by reservoir engineers into upper and lower reservoir units in areas where two contrasting types of performance have been observed. The upper unit is characterized by depletion drive with dry oil production and rapidly declining pressures, whereas the lower unit exhibits high water cuts and high pressures. According to correlation, these two units are separated by a fieldwide correlatable shallow marine shale. The lower unit has been further divided into two units (Figure 10): a sandy upper unit (the middle Bachaquero) and a nonreservoir shaly lower unit (the lower Bachaquero).

Within the individual channel sands of the upper Bachaquero unit, the reservoir quality is generally excellent, with porosities ranging from 20 to 30% in the clean sands. Permeability measurements have shown values up to 1200 md in the cleanest and coarsest parts of these sands. The shales, on the other hand, have acted as barriers to water injection.

The clean, well-sorted sands of the middle Bachaquero unit have porosities ranging from 20 to 30%. The development of stacked channel systems has resulted in good communication, both horizontally and vertically. This is also observed in reservoir performance analyses, which demonstrate that the unit is a single reservoir.

The lower Bachaquero unit is mainly shaly, and only a small part consists of reservoir sands. Most of these sands are water-bearing sands, and only a few have shown signs of oil.

Laguna Member

The Laguna Member is much more variable in both thickness and facies type. In contrast to the Bachaquero, the upper Laguna sands have continually shown high pres-

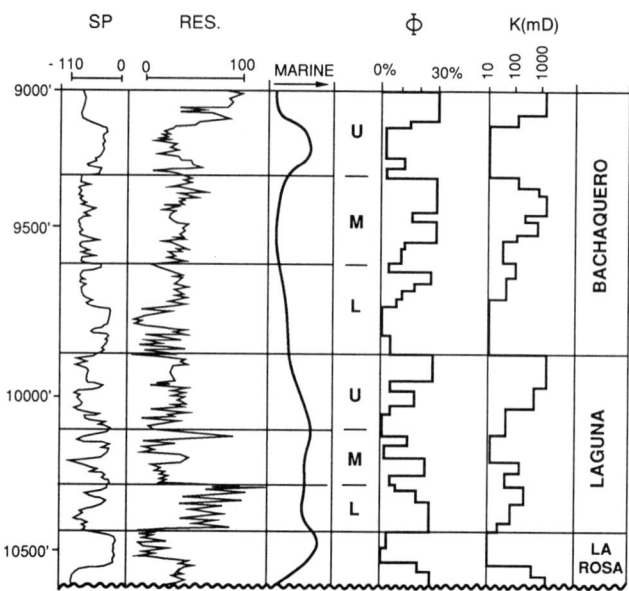

FIGURE 10. Miocene reservoirs.

sures and increasing water cuts, whereas the lower sands have produced dry oil with rapidly declining pressures, indicating drive depletion. Petrophysical evaluation of fluid content in the Laguna Member has greatly aided the recognition of separate units.

The reservoir development in the thick upper Laguna unit is determined by sand quality and continuity in both a vertical and a lateral sense. The variety of lithofacies observed in cores indicates a deltaic to marginal marine depositional environment. The clean channel sands are well sorted and have porosities ranging from 20 to 30% and permeabilities up to 1800 md as measured in cores. The detailed correlation studies have revealed the existence of local shale barriers that within one or more structural blocks may divide the upper Laguna unit into separate reservoirs.

The upper half of the middle Laguna unit consists of a thin, coarsening-upward sequence of interbedded sands and shales that represents a series of coastal barrier bars. The thicknesses of the sand bodies range from 50 ft (15 m) in the main upper channel to 20 to 30 ft (6 to 9 m) in the secondary lower channel. The porosities in the main channel are about 22%, and permeabilities range from 180 to 700 md.

Toward the base of the unit, sudden facies changes occur, resulting in interbedded water-bearing sands and shales. The petrophysical evaluations of the channel sands show porosities of 20 to 30% and an effective permeability of 550 md.

The lower Laguna unit constitutes the basal oil-bearing unit of the Laguna Member and consists of a massive sand sequence. According to stratigraphic correlation sections, it was found to be largely confined to filling-in of paleotopographic lows. Lithologically, it consists of a thick sequence of clean, stacked sands interbedded with thin shale layers. Petrophysical evaluations indicate porosities ranging from 15 to 25%.

La Rosa Formation

The deposits of the La Rosa Formation, which are of earliest Miocene age, represent a regional marine transgression over the Eocene unconformity surface. In the Ceuta field, the La Rosa Formation is divided into the upper La Rosa Shale Member and the lower Santa Barbara Sand Member, which is a clean, porous, oil-bearing sand in part of the Ceuta-Tomoporo field. The La Rosa Shale Member is an excellent marker that is easily recognized on wireline logs and, because of its fossil content, easily recognized during drilling.

EOCENE RESERVOIRS

The B-6 sandstones (lower "B" sands, Misoa Formation) constitute the main producing stratigraphic level in the Ceuta area, except in the Ceuta high and the region to the west, where these sands are absent as a result of erosion.

On the basis of core studies from wells in the southeast quadrant of the area, eight sedimentary units (I to VIII, from bottom to top) have been recognized in these sands. Units I, II, and III (the basal units) are low-energy facies, having been deposited in shallow, brackish water with a high content of organic matter. Biological activity suggests oxygenated water incursions in an environment with restricted flow conditions (Puche et al., 1989). Unit IV shows deposits of low to medium energy, with accumulations of coarse sands and gravels in the distributary channels and clay deposits, characteristic of interdistributary bays. Unit V is an argillaceous section interpreted as deltaic plain deposits with levee splays. Units VI, VII and VIII (the uppermost units) are marine deposits with coastal bars oriented northeast-southwest in which low-energy coastal lagoons are also identified (Chacartegui, 1990).

Because of the highly variable production potentials of nearby wells with identical completion zones, reservoir pressure was closely monitored, as shown in Figure 11. It was noticed that the B-6.0 and B-6.3 subunits showed declines in pressure, whereas the B-6.6 subunit remained at the original pressure. A close check on the log correlation, supported by palynological determinations, showed that thin, shaly layers no greater than 10 ft (3 m) thick separated each of the subunits (Figure 12). A geologic model based on these data, and also supported by reservoir water characterization, was set up for the purpose of development planning.

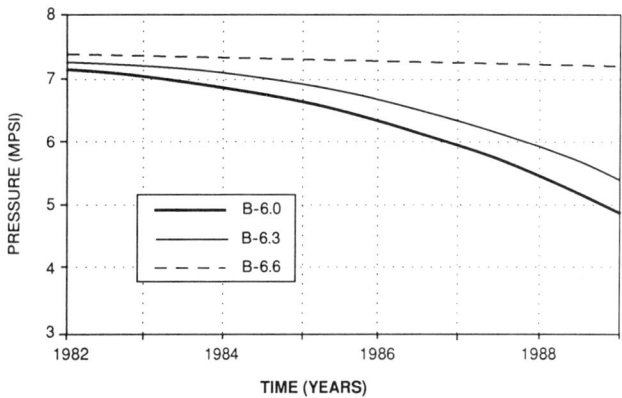

FIGURE 11. Eocene reservoir pressure behavior (after Puche, 1989).

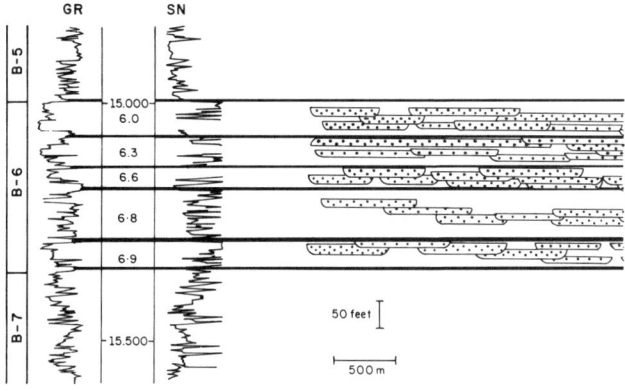

FIGURE 12. Reservoir geologic model and type log for Eocene lower "B" sandstones (after Puche, 1989).

As a result of asphaltene and sand problems, it has been a practice to treat some wells with hydrofluoric or hydrochloric acid, which has resulted in only ephemeral increases in production because of the migration of fine to very fine sediments. It was realized that a detailed study of the sedimentological characteristics was needed.

Core studies indicated a close relationship between productive levels and sedimentological characteristics. Initially, on the basis of grain size as well as mineralogy, texture, and sedimentary structures, eight different rock associations or lithotypes ("facies") were identified. Their characteristics are presented in Table 2.

Porosity measurements in plugs proved that porosities from 10 to 12% in facies S11, S1, and S2 could correspond to very low values of permeability and that porosities from 10 to 17% in facies S and S3 could correspond to very high permeabilities (Figure 13). In addition, petrophysical measurements in plugs, such as grain density and clay correction factor, supported a better interpretation, especially in the porosity values derived from the FDC and CNL logs. Correlation of the cores with the petrophysical logs also indicated a relationship among the facies type, the gamma ray curve, and the irregularities of the hole. In fact, it was discovered that the holes "in front" of facies with good oil saturation, such as S and S3, were circular, with gamma ray measurements lower than or equal to 27 API units, resistivities greater than 40 ohm m, and porosities of about 9 to 16%. On the other hand, in facies with lower or no oil saturation, such as S11, S1, and S2, the holes showed irregular configurations, gamma ray values greater than 27 API units, and porosities from 3 to 15%. It was also shown that ratios greater than 1.6 between shallow and deep DLL values indicated levels of mobile oil.

The coexistence of different types of sedimentary facies within the same sand body indicates that the sedimentary unit does not necessarily correspond to a flow unit and that the petrophysical evaluation of these sand bodies should be based on the type of facies.

These facies have also been recognized in ditch samples, and their correlation with the gamma ray log is a powerful tool for selection of the completion intervals in appraisal and development wells.

RESERVES

The traditional Ceuta field carries ultimate recoverable (proved) oil reserves of 638 MMSTB, including 469 MMSTB in the Miocene reservoirs and 169 MMSTB in the Eocene reservoirs.

As shown in Table 3, new (1980–90) discoveries have added 764 MMSTB (as of December 1990) to the reserves, and most of them (758 MMSTB) are in the Eocene reservoirs. Some of the reserves in the traditional Ceuta field are to be recovered through water and gas injection projects.

Currently defined prospects total 340 MMSTB of light oil, most of which is expected to be found in untested Cretaceous traps and in the Tertiary reservoirs to the north of the Tomoporo area.

REFERENCES CITED

Audemard, F., A. Reymond, J. Maguregui, E. Rodríguez, and J. De Mena, 1987, Distribución y evaluación de las

TABLE 2. Characteristics of lithotypes ("facies") in the Laguna Member of the Lagunillas Formation (from Chacartegui, 1990).

Lithotype code	Description
S	Cross-bedded, coarse-grained, moderately to poorly sorted sandstone, with occasional layers of pebble-size material, composed of quartz (>95%), rock fragments, feldspars, and traces of clay
S3	Cross-bedded, medium-grained, moderately to well-sorted sandstone, composed of quartz (>95%), rock fragments, feldspars, and clay (2%)
S11	Horizontally bedded, fine- to very fine-grained, well-sorted sandstone, with occasional clay laminae, composed of quartz (90%), clays (5%), rock fragments, and feldspars
S1	Laminated, fine- to very fine-grained, well-sorted sandstone, with occasional clay clasts, composed of quartz (90%), clays (8%), rock fragments, and feldspars
S2	Interbedded black shale, gray siltstone, and very fine-grained, occasionally bioturbated sandstone, with current ripples, wavy bedding, and flaser bedding common
SC	Fine-grained, moderately sorted, occasionally bioturbated sandstone, composed of quartz (90%), clays (8%), rock fragments, and feldspars, with flaser bedding, wavy bedding, and current ripples common
H	Coarse-grained, cross-bedded, poorly sorted calcareous sandstone, composed of quartz, rock fragments, feldspars, and marine fossil remains (40%)
L	Black carbonaceous shale, with intercalations infrequent or absent

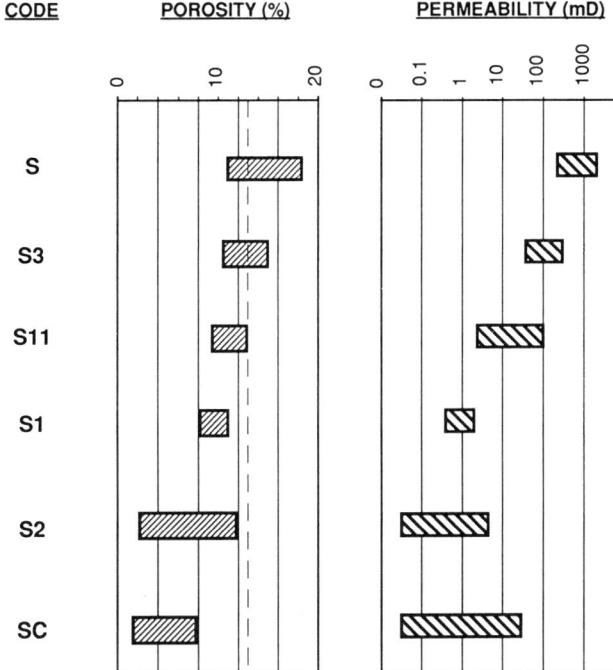

FIGURE 13. Porosity and permeability of Eocene reservoir facies (after Puche, 1989).

TABLE 3. Oil reserves in the Ceuta-Tomoporo field.

Reservoirs	Reserves (MMSTB)		
	Traditional Ceuta field	New discoveries (1980–90)	Total
Miocene	469	6	475
Eocene	169	758	927
Total	638	764	1402

fracturas en las Calizas Cretácicas de Perijá, p. 153, Intevep, Venezuela.

Blaser, R., and C. White, 1984, Source rock and carbonization study, Maracaibo basin, Venezuela: AAPG Memoir 35, p. 229–252

Chacartegui, F., 1985, Sedimentación y diagenésis de la Formación Lagunillas en el área de Ceuta: VI Congreso Geológico Venezolano, v. 1, Venezuela.

Chacartegui, F., S. Herrera, and I. Azpiritxaga, 1990, Estudio sedimentológico y diagenético del Yacimiento B-6, Formación Misoa, Area 8, Campo Ceuta, Lago de Maracaibo: III Simposio Interfilial de Ingeniería de Yacimientos, Maracaibo, Venezuela.

Gallango, O., M. Chin-A-Lien, M. Taheri, and S. Talukdar, 1984, Estudio Geoquimico Regional de la Cuenca de Maracaibo: Intevep Internal Report, Caracas, Venezuela.

Gatrall, M., 1984, Ceuta field, Venezuela: Miocene stratigraphy and use of fluid distribution in correlation: Production Geological/Petrophysical Conference, Venezuela.

Ghosh, S., J. Di Croce, A. Isea, and C. Gonzalez, 1988, Control estratigráfico diagenético-relación entre los ambientes de sedimentación, diagenésis y profundidad de soterramiento en areniscas del Eoceno, cuenca de Maracaibo: Intevep Report 01970,88, v. 1, Venezuela.

González de Juana, C., J. M. Iturralde, and X. Picard, 1980, Geología de Venezuela y de sus cuenczas petrolíferas: Caracas, Ediciones Foninves, p. 1031.

Maraven S.A., 1988, Sometimiento al MEM del modelo geológico y de yacimientos del campo Ceuta, Lago de Maracaibo: Internal Report EPC-12519, Venezuela.

Ordaz, J. de, 1988, Caracterización de hoyo y su efecto sobre producción inicial: Maraven S.A., Internal Report EPC-12562.

PDVSA, 1989, La industria Venezolana de los hidrocarburos: Ed. CEPET, v. 1, Venezuela.

Pittelli, R., and E. Di Giacomo, 1990, Análisis bioestratigráfico-paleoambiental del campo Ceuta (areas 4, 8, Tomoporo, VLG-3725) y Bloque XI: Maraven S.A., Internal Report EPC-12773.

Puche, E., F. Chacartegui, F. Azpiritxaga, I. Brandt, G. Alvarez, F. Ferrer, and O. Suarez, 1989, Estudio multidisciplinario del yacimiento B-6, Formación Misoa del Eoceno, Ceuta SE, Lago de Maracaibo: VII Congreso Geologico Venezolano, v. IV, Venezuela.

Roberto, M., L. Mompart, E. Puche, and F. Scherer, 1988, New oil discoveries in Ceuta area SE Lake Maracaibo, Venezuela: classic petroleum provinces: Memories GSA, v. 17.

Schwander, M., 1987, Tertiary seismostratigraphy in the Eastern Lake Maracaibo: Maraven S.A., Internal Report.

Shell Internationale Petroleum, 1984, Venezuela, Lake Maracaibo–Ceuta field–interim report–field study: Production Development Division, Venezuela Studies Team, SIPM, Report EP-59310, v. 1.

Shell Technical Services, 1985, Venezuela Lake Maracaibo-Ceuta field: reservoir framework report of the Miocene and Eocene formations: The Hague, The Netherlands, v. 1.

Smith, F. D. (Coordinator), 1963, Resumen geológico exploratorio—campo Ceuta, in Aspectos de la industria petrolera en Venezuela: Primer Congreso Venezolano de Petroleo, v. 1, p. 71–76.

Storey, T. P., 1980, Eocene and Oligocene transgressive and paleogeomorphological phases: Maraven S.A., Internal Report EPC- 6363.

Tissot, B. P., and P. H. Welte, 1984, Petroleum formation and occurrence (2d ed.): New York, Springer-Verlag, p. 699.

Talukdar, S., O. Gallango, and M. Chin-A-Lieu, 1986, Generation and migration of hydrocarbons in the Maracaibo basin, Venezuela: an integrated basin study: Org. Geochemical., v. 10, p. 261–279.

Van Veen, F. R., 1972, Ambientes sedimentarios de las Formaciones Mirador y Misoa del Eoceno Inferior en la cuenca del Lago de Maracaibo: IV Congreso Geológico Venezolano, v. 2, p. 1077–1104, Venezuela.

V.S.T. (S.I.P.M.) Maraven (EyP), 1984: Technical visit of G. Alvarez, Ceuta field, Caracas, Venezuela.

Chapter 12

The Giant Caño Limon Field, Llanos Basin, Colombia

C. N. McCollough

Bakersfield, California, U.S.A.

J. A. Carver

*Occidental International Exploration & Production Co.
Bakersfield, California, U.S.A.*

ABSTRACT

After 40 years of sporadic exploration that yielded negative or marginal results, the Llanos basin of eastern Colombia was thrust to the forefront of world attention by the discovery of the giant Caño Limon field in July 1983.

This discovery was the culmination of an intensive three-year exploration effort by Occidental International Exploration and Production Co. involving 4000 km of dynamite seismic, 20 stratigraphic tests from 1300 to 3500 ft (395 to 1065 m) deep, and 12 exploratory wells.

Before Occidental entered the area, 61 exploratory wells were drilled with meager results—namely, two fields with total reserves of about 20 million bbl of light oil and one field with reserves of 90 million bbl of 13.6° API oil, none of which was commercial.

The Llanos basin was known for its abundant excellent reservoir sands, and opinions varied as to whether there was adequate source rock. The major problem had been definition of traps. Except for the very young folding along the Andean front, the known structural traps were sparse and subtle.

Most of the exploration had been done in the western part of the basin near the basin deep or in the Andean foothills. Occidental took a very large acreage position east of the area of past exploration efforts and found an exception to the small fault closures known elsewhere in the basin.

This exception, the Caño Limon area, is dominated by major early Tertiary northeast-southwest strike slip faulting. Concurrent folding in combination with fault sealing formed the Caño Limon field and other much smaller fields in the area.

The Caño Limon field, encompassing 8821 ac, contains an estimated 1.800 billion BOIP, which 1.066 billion bbl are expected to be recovered with the very strong natural water drive.

The bulk of the oil is in deltaic sands of Eocene Mirador, with additional reservoirs in the Upper Cretaceous. Average porosity of the Mirador is about 25%, permeability about 5 darcys, and water saturation 23%. Individual well flow rates have exceeded 20,000 BOPD. The average oil gravity is 29.5° API, with a gas-oil ratio of 8 ft^3/bbl (0.2 m^3/bbl) and a sulfur content of 0.41%. Current production is about 230,000 BOPD.

INTRODUCTION

The month of June 1983, was eventful for oil exploration in Colombia. An elite panel of eight Colombian geologists was called back from jobs overseas to assess the oil situation in the country. The focus was the Llanos basin, where the reserves of Intercol's (Exxon's) deep Arauca field discovery were suspected of being closer to 3 million bbl than to the 200 million bbl initially estimated, and the commerciality of Ecopetrol's Apiay discovery was being questioned.

On June 21, 1983, the panel presented its findings to the President of Colombia. The next day *El Tiempo*, the leading newspaper, carried the story under the by-line "Petroleum Expectations in the Llanos Collapse."

Ironically, unbeknown to the panel, Occidental's Caño Limon-1 was at the time drilling ahead in the Cretaceous, after having penetrated 174 ft (53 m) of net oil pay in the Eocene Mirador sands in what was to become a giant field. Occidental's current reserve estimates are 1.05 billion bbl of 29.5° API oil. Suddenly, the unpopular Llanos basin became the hottest exploration play in the western hemisphere, if not the world.

Although the geologic panel reached the logical conclusion based on the information available to it, this incident illustrates that new frontiers are new frontiers precisely because their potential has not been generally recognized, and that for those with imagination and a willingness to take risks, new frontiers and opportunities still exist.

GEOGRAPHIC SETTING

The Llanos basin covers an area of about 200,000 km^2 in the low-lying grassy plains east of the Eastern Cordillera (Andes) in Colombia (Figures 1 and 2). The basin is cut by numerous easterly flowing jungle-lined rivers feeding the major Meta River, which flows northeast to join the Orinoco River. To the east, the grasslands give way to Amazonian jungle and in the north, near Caño Limon, large lily pad swamps remain wet even through the dry season.

Heavy rainfall virtually prohibits land activity for eight months of the year. In the four-month dry season, roads become deep in dust. The region is sparsely settled, with the two main towns being Villavicencio in the southwest and Arauca in the north. Cattle ranching is the only significant legal agricultural activity.

EXPLORATION HISTORY

Llanos exploration has been very cyclical (Figure 3). The first well in the basin was the San Martin-1, drilled by Shell Oil along the southern basin edge in 1944.

The well tested heavy oil and was followed by the San Martin-2 through -6 wells, but the area proved subcommercial and was abandoned.

In the south central portion of the basin, Shell also drilled the wildcats Chaviva-1 in 1946 and Voragine-1 in 1947. Neither was successful, and exploratory drilling in the basin ceased for more than a decade.

In 1958, Intercol drilled five stratigraphic tests in the east central portion of the basin. During 1959–62, six wildcats were drilled in the Llanos, three of them by Texaco on the Guavio anticline in the foothills belt. The Guavio-1, drilled in 1960, reportedly yielded a total of 554 bbl of 39° API oil from the Tertiary on two drill-stem tests, but two delineation wells, Guavio-2 and Guavio-3, failed to establish commerciality. The other tests during the period were dry holes, ending the second phase of Llanos exploration in 1962.

Chevron launched the third phase of Llanos exploration in 1969 with its Castilla-1 wildcat, which tested heavy oil (about 14° API) from the Cretaceous. Chevron subsequently has developed the field and is trucking production to Bogota. Ironically, the Castilla-1 is only 1.5 mi (2.4 km) from San Martin-1.

The Castilla strike sparked a surge of exploration that lasted until 1978. During this period, operators drilled 25 exploratory wells, mostly in the foothills fold-thrust province. Exploration in this area was not commercially successful, although some wells—most notably, Phillips' 1127-1X, Chevron's Yali-1, and Ecopetrol's Tauramena-

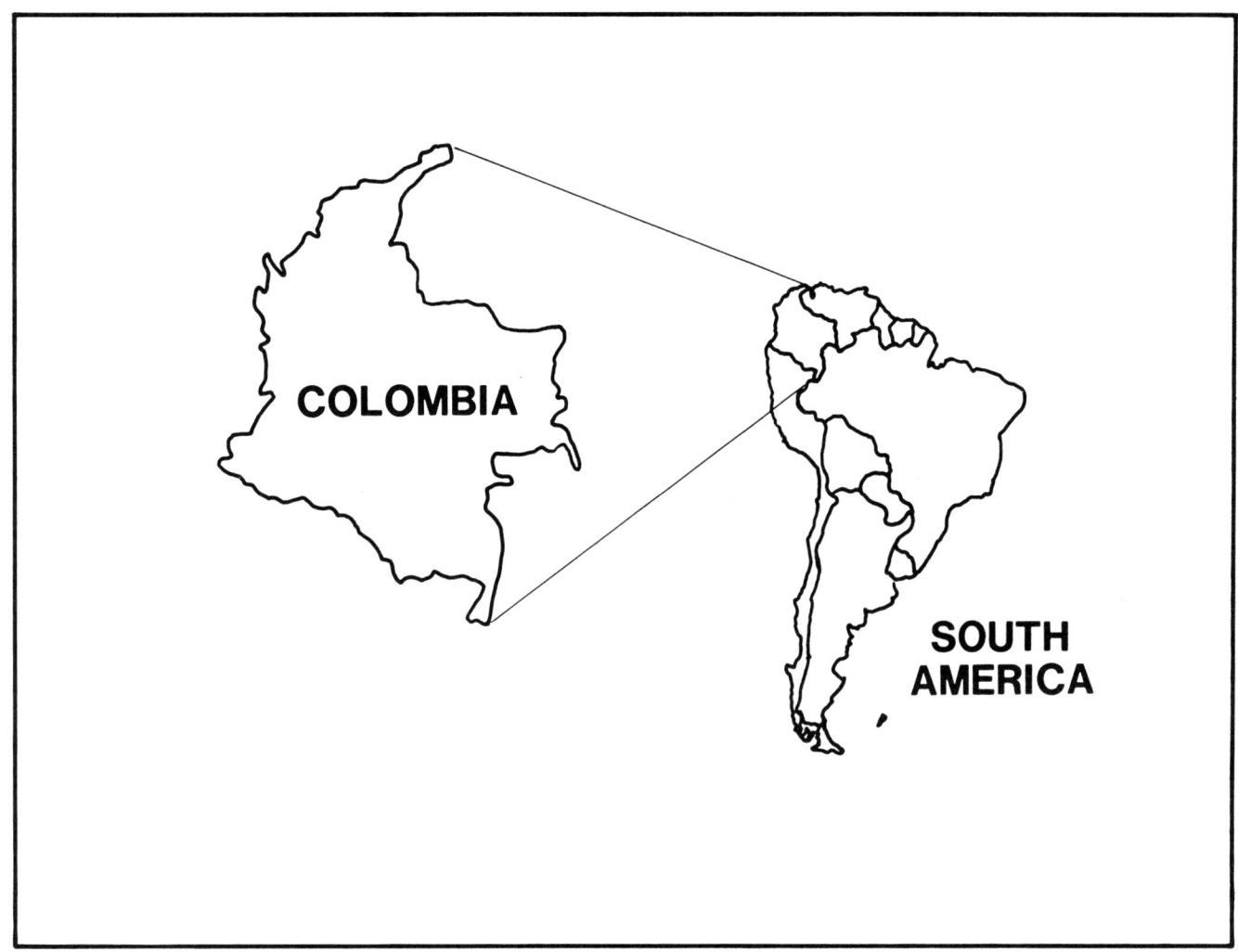

FIGURE 1. Colombia, South America.

2X—encountered significant shows. Exploratory drilling in this area was hindered by the depths of the objectives and by problems created by sealing of the Leon shale.

It was during this period that Elf Aquitaine began its exploration program in the central Llanos, drilling in 1974 the Trinidad-1 discovery well that tested about 1600 bbl of 32° API oil per day from the Mirador.

Exploration declined during 1976–79, totaling only two wells in 1977 and one in 1978. None of these wells was successful, although Elf Aquitaine's Fortaleza-1A encountered subcommercial shows.

Exploration activity increased somewhat in 1979, with Elf accounting for three out of the four wells spudded that year. Elf drilled and completed the Trinidad-2 confirmation well and discovered the Cano Garza field.

Intercol spudded the Arauca-1 well in 1979, completing it as a discovery well at depths of 18,090 to 18,140 ft (5514 to 5529 m) in 1980. This well tested 1941 BOPD (41° API), and the Arauca-2 well tested a total of 4970 BOPD (34 and 41° API) from Mirador and Cretaceous intervals in 1981. At the time of discovery and confirmation, the Arauca discovery was hailed as a major find, with recoverable reserve estimates as high as 200 million bbl. Subsequent delineation drilling was disappointing, and reserve estimates have been downgraded to about 7.5 million bbl.

Llanos exploration accelerated in 1980, primarily due to increased activity by Ecopetrol, Elf Aquitaine et al., and Intercol, along with the start of Occidental's exploration program. A total of 35 exploratory and confirmation wells and 37 stratigraphic tests (20 by Occidental; 17 by Intercol) were drilled during 1980–82.

Much of Intercol's activity focused on a heavy oil play in the southern portion of the basin that yielded four noncommercial discoveries during 1981–82. Ecopetrol discovered the Apiay field on the margins of the foothills belt in 1981, testing 4509 bbl of 24.6° API oil per day from the Cretaceous.

Occidental's Caño Limon-1 discovery, which tested oil from two Mirador zones at a rate of 10,690 BOPD (about 31° average gravity), was announced in July of 1983. Subsequent drilling has established the discovery as a giant oil field, which Occidental estimates will ultimately produce more than 1 billion bbl of oil.

The intense exploration activity in the Llanos basin following the Caño Limon discovery failed to find additional large fields. In 1989, six years after the Caño Limon-

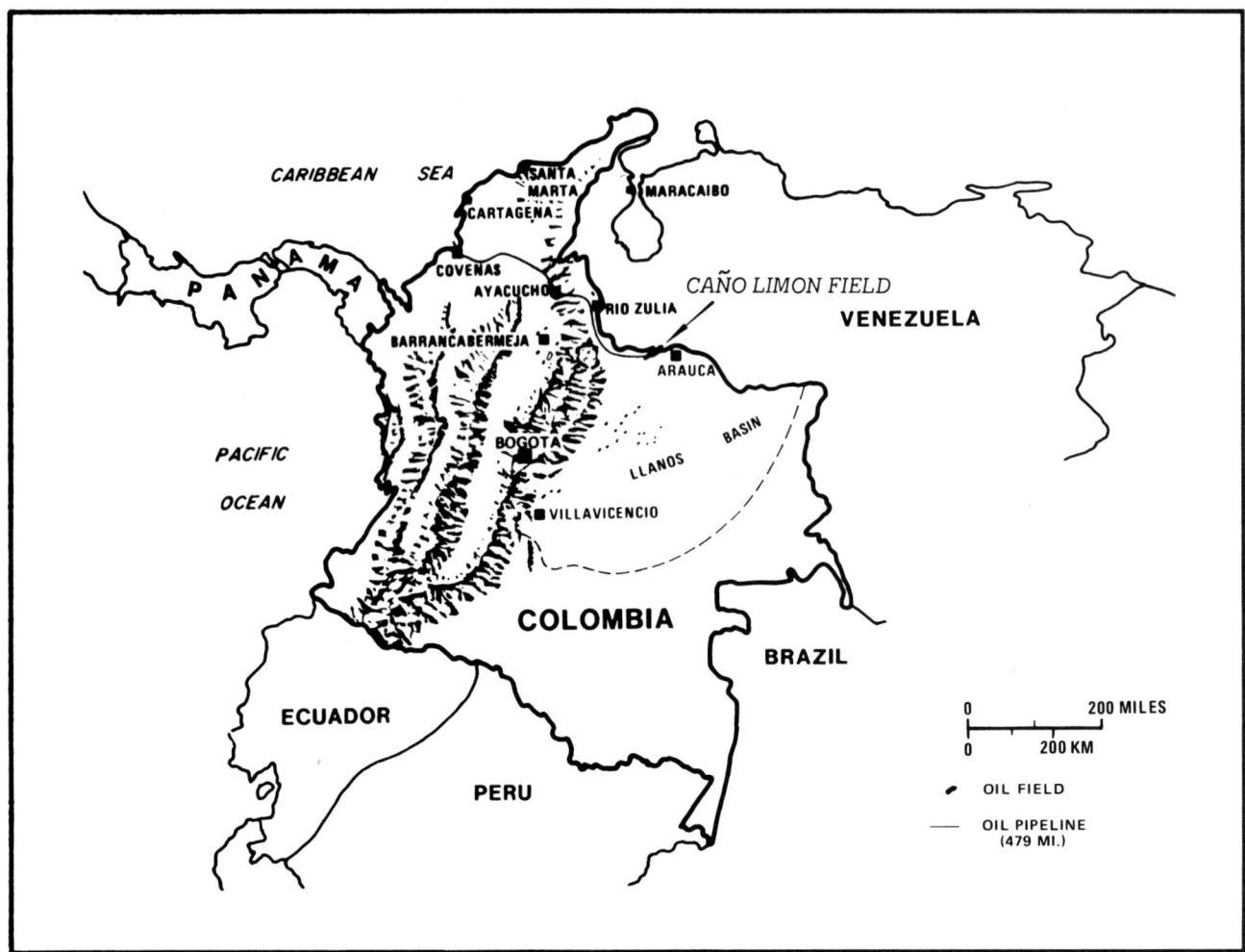

FIGURE 2. Index map of the Llanos basin, Colombia.

1 success, exploration activity has moderated, although about 50 small fields have been found in antithetic fault traps. Counting the Trinidad and Caño Garza fields, which were found earlier, there are now 45 of these fields with reserves of less than 10 million bbl and seven fields with reserves between 10 and 20 million bbl. However, at the time of this writing there is a major new play developing in the fold-thrust belt led by the British Petroleum Cusiana discovery.

BASIN SETTING

The Llanos basin is one of a long chain of sub-Andean foreland basins extending from Venezuela to Argentina (Figure 4). These basins, which lie east of the Cordillera (Andes), are characterized by a fold-thrust belt at the mountain front, a basin deep immediately to the east, and a shallowing onto the granitic shield at the eastern margin (Figures 5 and 6).

The maximum depth of the Llanos basin is estimated to be at least 30,000 ft (9150 m), with the sediment package thinning to zero at 500 km to the east, where the granitic Guyana shield is exposed.

BASIN EVOLUTION

Until the Andean orogeny in the late Miocene, the Llanos basin and the Magdalena basin to the west were one. The basin deep was in the area now occupied by the Eastern Cordillera. With the uplift of the Eastern Cordillera during the Andean orogeny of late Miocene to Holocene, the basin was bisected, with the larger eastern portion becoming the Llanos basin.

An earlier component of the basin evolution occurred during the Paleozoic and early Mesozoic. A large failed rift is now recognized trending northeast-southwest through the northern portion of the basin. The depth of this graben is not known, but seismic reflectors are seen to at least 5 sec, representing an estimated 28,000 ft (8537 m). The sediments in the graben are mainly Cambrian-Ordovician with possible Devonian in some portions. In limited areas the Paleozoic is overlain by Jurassic-Triassic red beds. As yet no information that explains the Permian-Carboniferous hiatus is available.

A major unconformity at the base of the Cretaceous puts Early Cretaceous on Paleozoic in the deeper western part of the basin and progressively younger Cretaceous on older Paleozoic and ultimately on the granite basement

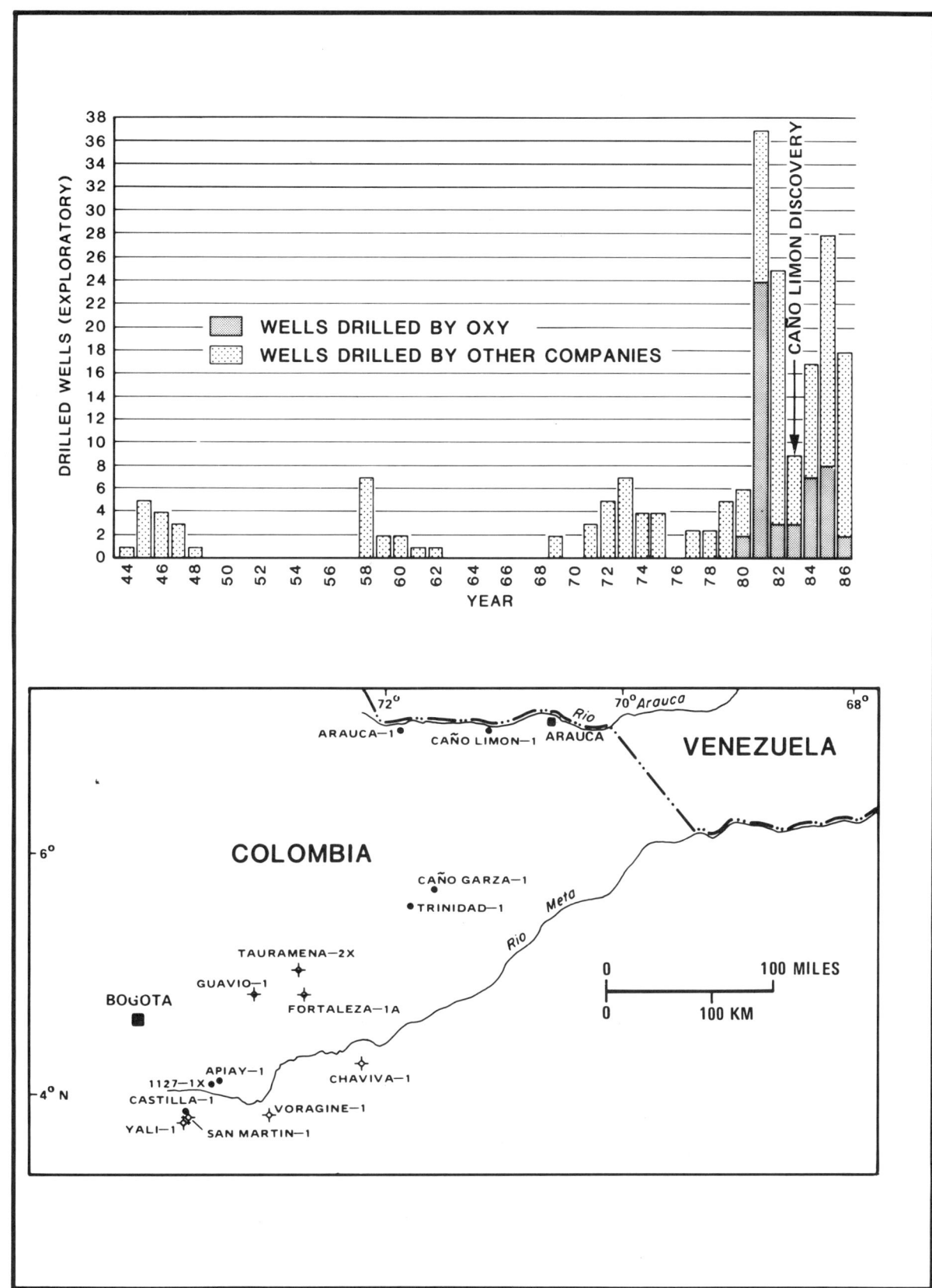

FIGURE 3. Drilling history of the Llanos basin, and location map (Oxy = Occidental International Exploration & Production Co.).

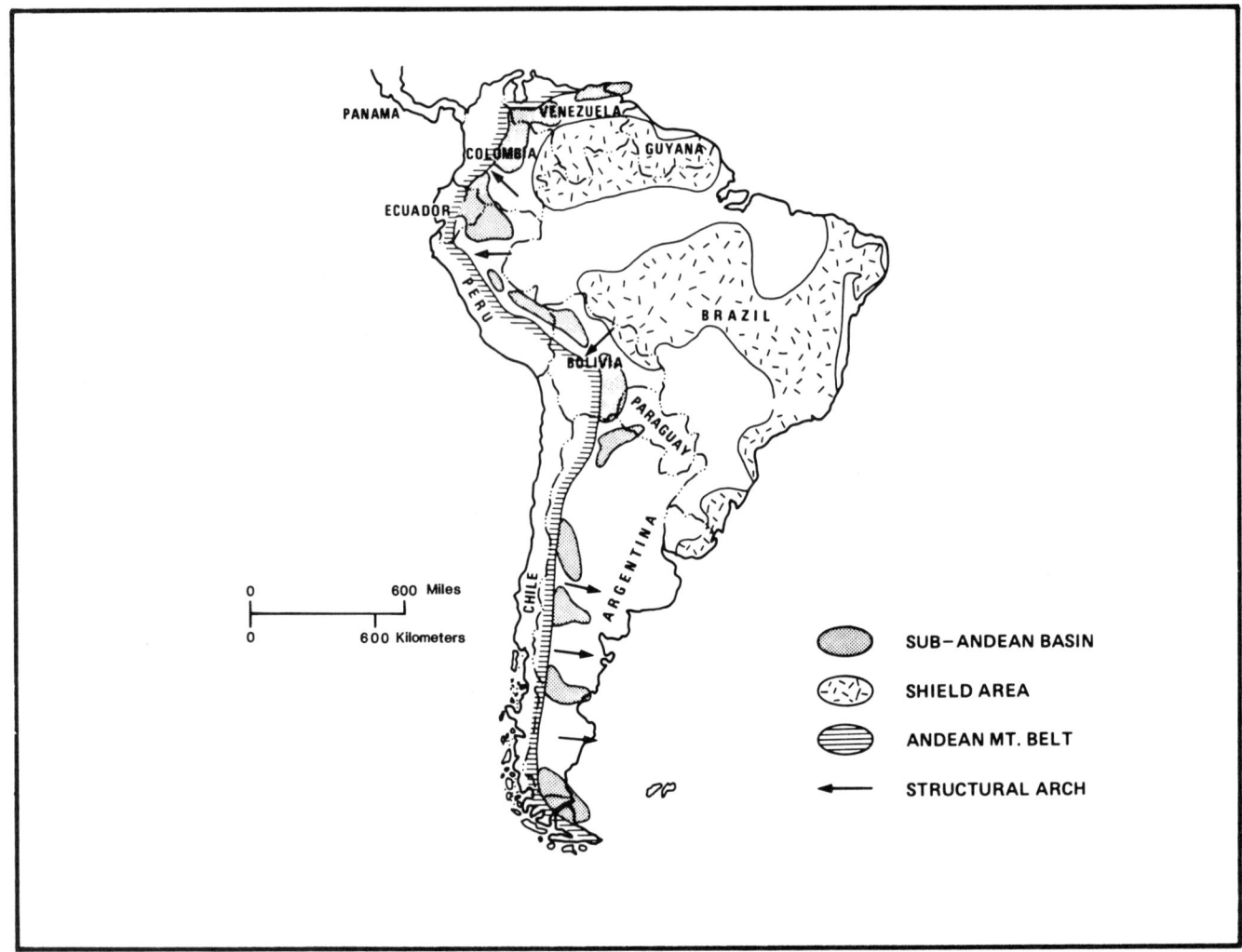

FIGURE 4. South American sub-Andean basins.

to the east. In the southeastern Llanos, a Paleozoic subbasin preserved a considerable thickness of Cambrian-Ordovician sandstones and shales overlain by Tertiary rather than Cretaceous sediments.

STRATIGRAPHY

The stratigraphy of the Paleozoic is not well known (Figure 7). The few wells drilled into Paleozoic sediments have found mainly siltstones and shales with a few sandstones. Paleozoic sediments over the major part of the basin are well indurated, but cannot be called metamorphic. Sandstones with porosities up to 22% have been seen, but the more usual values are in the 10 to 12% range. Whether adequate oil source rocks exist is a question.

With the exception of minor carbonates in the Paleozoic and Cretaceous, the Llanos basin is filled with sands, shales, and siltstones. Prior to the Andean orogeny, probably all sediments entering the basin were derived from the granitic Guyana shield to the east and southeast. A western source, the Central Cordillera, contributed sediments to the Magdalena portion of the basin. After the onset of the Andean orogeny, a probable dual eastern and western source was present that later became a totally western source as basinal sediments, now exposed in the Eastern Cordillera, were eroded and carried eastward.

Cretaceous units ranging in age from Albian or possibly older to Campanian are predominantly alluvial, deltaic, turbiditic, and basal transgressive sands. Shales are a minor constituent to the east, but the basinal shaley equivalents to the continental/deltaic sands thicken into the basin and contain turbidites, as might be expected. Cretaceous sands are generally of very good reservoir quality.

The Paleocene Cuervos-Barco shale-sand units are present in the western portion of the basin but terminate along an eastern pinch-out, the position of which is controversial. The Barco sands are productive in Intercol's deep Arauca field.

Another controversial unit is the Eocene Mirador sand. It is widely believed that the Mirador sand in outcrops in the Eastern Cordillera and Merida Andes of Venezuela is not the same unit that is the main producing sand in Caño Limon and other Llanos fields. There is growing

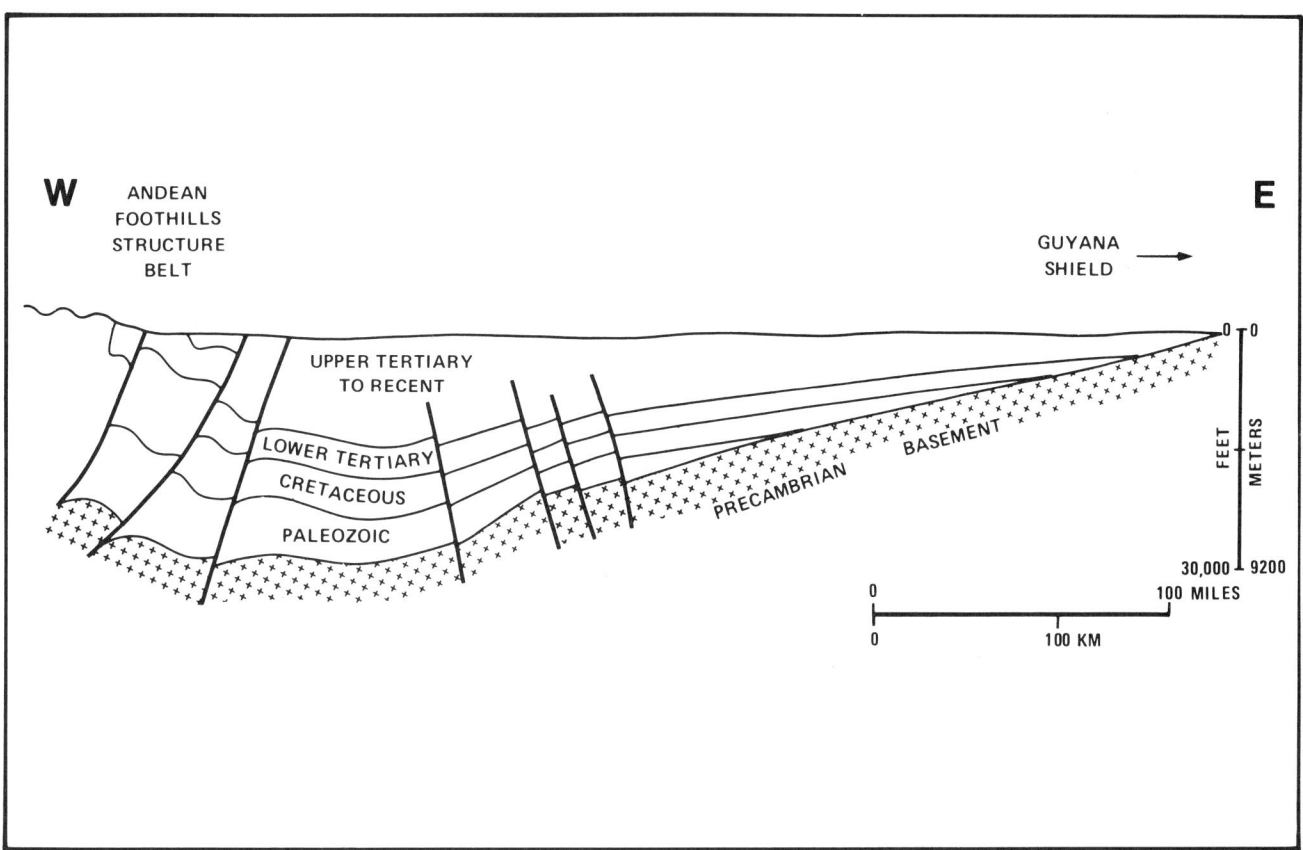

FIGURE 5. Geological section of the Llanos basin.

support for the idea that the "true Mirador" of the outcrops is stratigraphically deeper and pinches out to the east somewhere in the western part of the basin. The other Mirador, which is present over most of the Llanos, may be stratigraphically shallower and in some areas may overlap the "true Mirador." At least a portion of the higher "Mirador" may more correctly be a basal sand member of the Carbonera Formation. The Oligocene-Eocene boundary occurs within the upper part of the "Mirador" Formation at Caño Limon on the basis of palynology. This zone is now being called Lower Carbonera/"Mirador" in the Caño Limon area.

In any event, the "Mirador" is present over essentially all the Llanos basin as a major regressive unit. It appears to be a series of stacked deltaic and alluvial sands with few shales. It is an excellent reservoir ranging in thickness from about 200 to over 500 ft (61 to 152 m). In the remainder of this paper, the terms Mirador and Lower Carbonera/"Mirador" both refer to this unit.

The Carbonera Formation is Oligocene in the lower part and appears to be Miocene in the upper part. It incorporates three to four shaley transgressive units and the same number of regressive sandy units. The regressive sandy units are predominantly deltaic with coal swamp deposits and thin coal beds that give the formation its name. Within the prodelta and basinal shales are found some thin sands that are probably turbidites.

Oil has been found at several levels of the Carbonera, but mainly in the lower sands just above the Mirador.

Overlying the Carbonera is the Miocene Leon shale, a varicolored claystone with thicknesses from 560 to 785 ft (170 to 239 m). In the northern Llanos near Caño Limon, it contains numerous sand interbeds that make its upper boundary difficult to pick. This is the most ubiquitous shale of the basin but to date seems to have no economic significance because no oil is known to be trapped in the sands immediately beneath it.

Above the Leon shale is a thick easterly thinning wedge of sands and shales of the Miocene to Holocene Guayabo Formation. No hydrocarbon potential is known to be present in this formation.

STRUCTURE

The primary deterrent to exploration in the Llanos basin has been that, except for the mountain front, structural activity has seemed to be virtually nonexistent. Early reconnaissance seismic, and in some areas reasonably closely spaced seismic grids, failed to define any interesting structure. Several wells were drilled along the fold-thrust mountain front belt, but without commercial success. Wells were generally deep, and drilling was difficult and expensive. Several never reached their objectives.

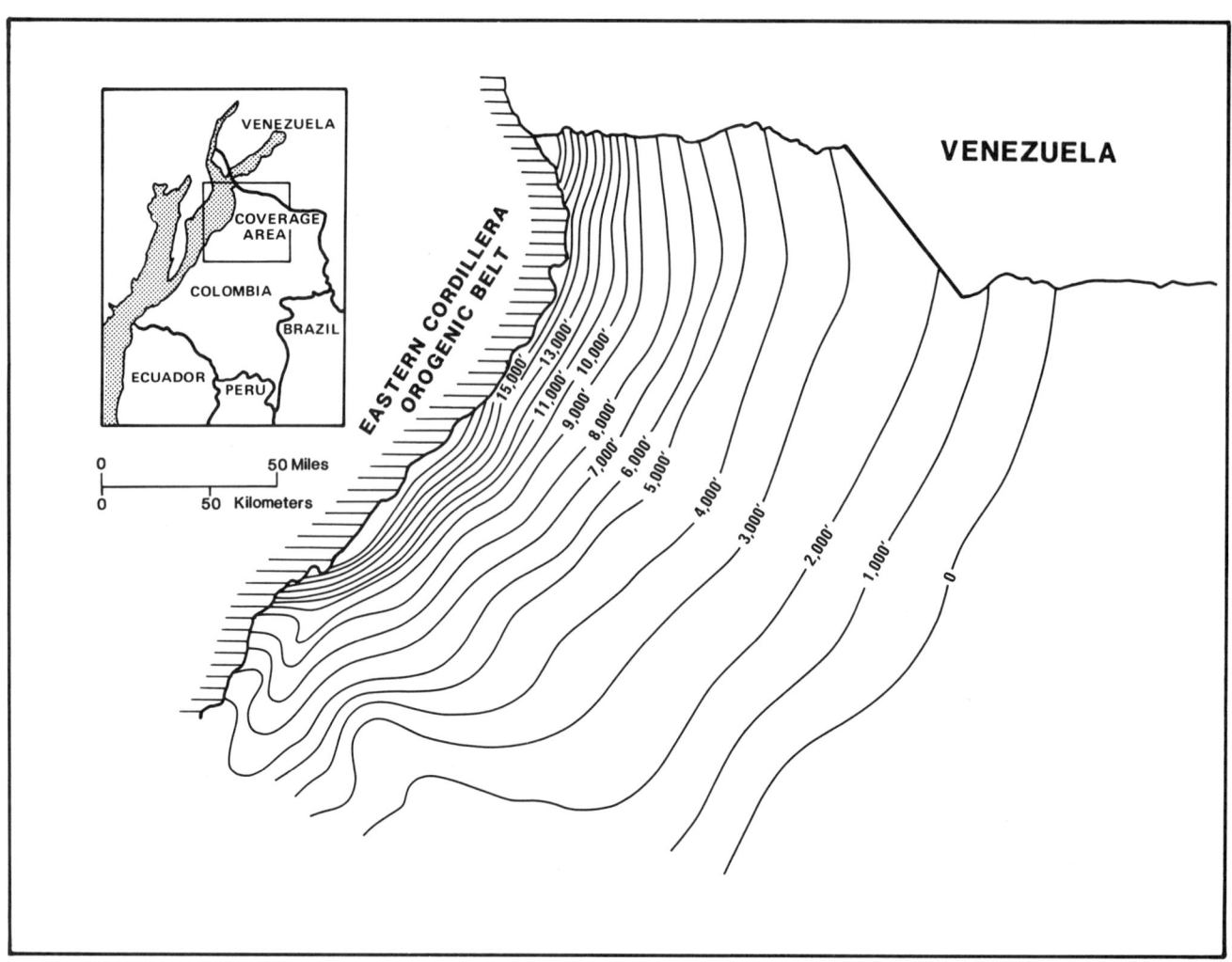

FIGURE 6. Tertiary isopach map of the Llanos basin (contour interval = 1000 ft).

Small amounts of oil were found, but often it was degraded. Fresh waters and porosity destruction by diagenesis were also common.

Subtle up-to-the-basin (antithetic) faults were recognized on seismic in the west central part of the basin. The first potentially commercial light oil discovery, Elf Aquitaine et al. Trinidad-1, was made in 1974 on a structural closure against one of these faults. Since then more than 52 small oil fields have been found trapped by antithetic faults in this basin.

These antithetic faults are very young and appear on many seismic lines to come essentially to the surface (Figure 8). The authors believe they were formed too late to trap oil from the main migration from Cretaceous sources and are being sourced by younger or more shelfward shales, possibly still generating. Most of these traps do not appear to be filled to capacity, although it is difficult to determine what the capacities of these structures really are.

The major strike-slip fault system responsible for the trapping structures in the Caño Limon area is still being studied with regard to the nature and timing of the tectonic forces involved.

OIL SOURCE

Until the discovery of Caño Limon there was a belief that the Llanos basin had a limited oil source. As suggested above, the problem may have been timing rather than quantity of source. It seems likely that the main source of oil was marine shales of the Cretaceous, most of which are now uplifted in the Eastern Cordillera. Based on estimated sediment thickness and geothermal gradients, it seems likely that oil was generated and expelled from the Cretaceous source during the Oligocene. Any traps in the basin that were present at that time should be filled. Because there were apparently insufficient traps, much of this oil migrated to the margins of the basin and was degraded or lost. Heavy oil and oxidized oil residues in a large number of wells along the southern and southeastern margins of the basin testify to this thesis.

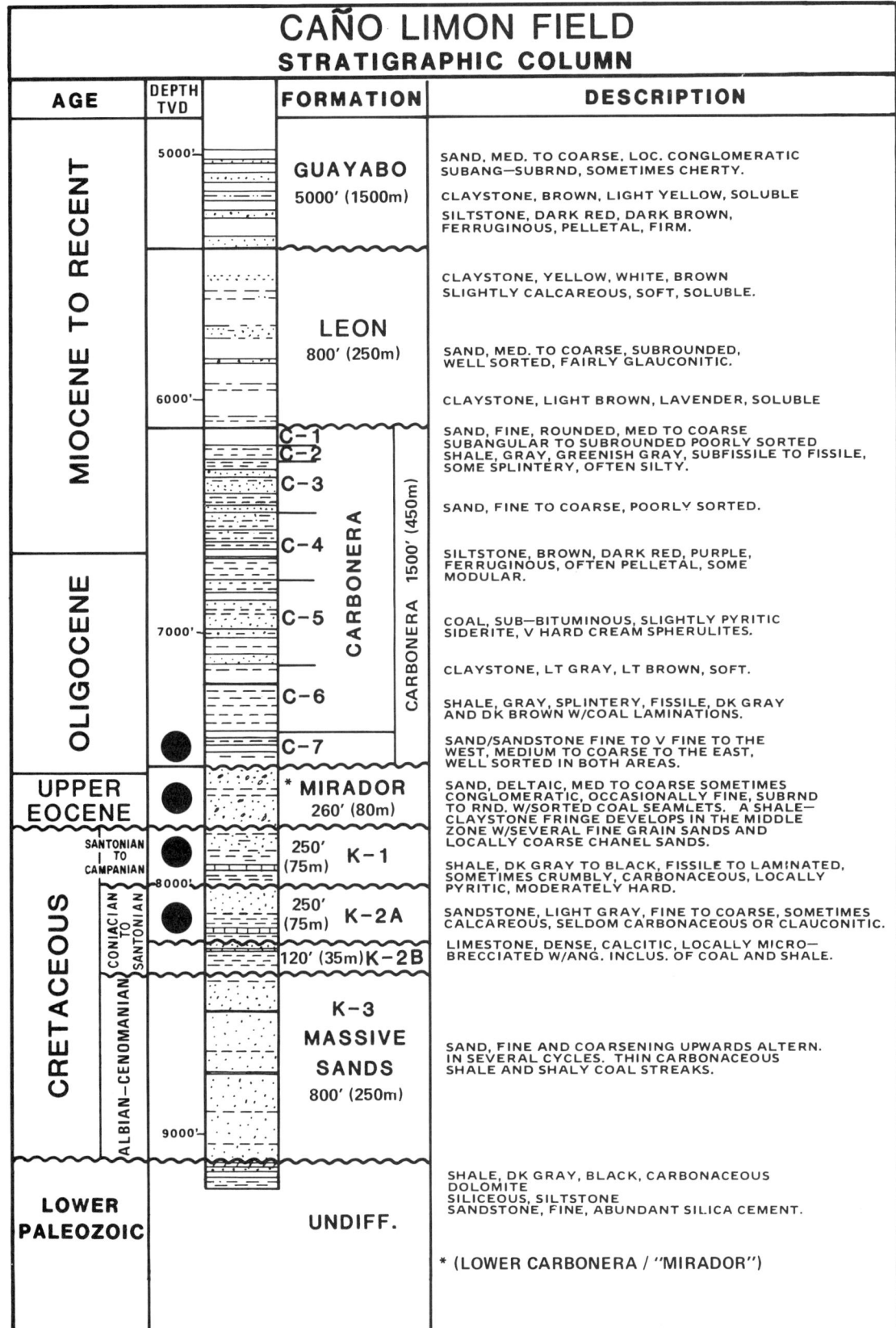

FIGURE 7. Stratigraphic column for the Caño Limon field.

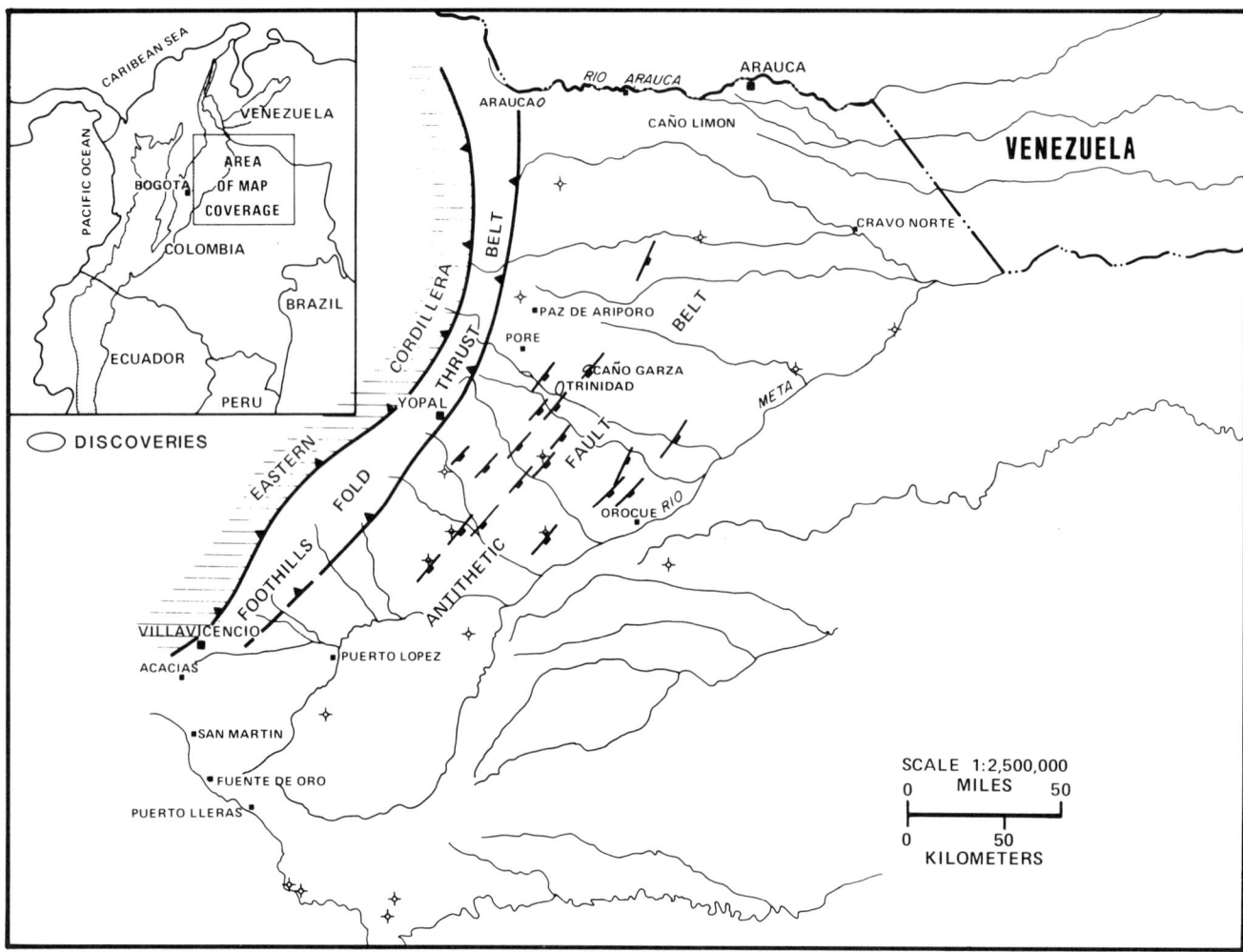

FIGURE 8. Structural elements of the Llanos basin.

The antithetic fault traps were not present at this time. The oil in these accumulations is most likely sourced from younger shales—possibly the basinal equivalents of the Mirador, the Cuervos, or even the Carbonera. An alternative possibility is that this late-generated oil is from Cretaceous beds in more shelfal positions.

This difference in sources is supported by geochemical data showing that oils from early traps such as Caño Limon and Ecopetrol's Apiay field in the southern Llanos are from a strongly marine source, whereas oils from the antithetic fault traps have a much more terrestrial origin. This later source may indeed be limited, or perhaps it is only that the young structures found so far have limited capacity.

OCCIDENTAL'S EXPLORATION PROGRAM

When Occidental entered the Llanos basin in 1980, much of what we know today was unknown. The factors in the decision to explore an unpopular basin were as follows:

1. It was apparent that excellent reservoirs were everywhere.

2. Sealing shales appeared adequate in number and competency, although in the eastern parts of the basin this was a concern.

3. Contrary to some published reports, it was deduced that there probably had been plenty of mature source in the pre-Andean basin deep.

4. Despite the indications at that time, it was reasoned that no basin of this size could completely escape structural deformation throughout its history. Also, Occidental had had the experience in Peru of having found some 500 million bbl of oil in a very subtle drape-type structure along the hinge line 100 km east of the basin deep.

5. Although not proved, the Arauca arch was believed to exist at the Colombia-Venezuela border. It was postulated that this could provide the necessary structural and stratigraphic complications and migration focus for oil accumulations.

6. There was some 90,000 km² that was totally unexplored.

7. Ecopetrol, the national oil company, had an enlightened approach to attracting foreign exploration while protecting the national interests.

On these bases, Occidental signed association contracts on five blocks of 2.5 million ac (1 million ha) each—a total of 12.4 million ac (5.0 million ha) (Figure 9). Within the selected areas there existed a small amount of seismic, three stratigraphic tests, and one exploratory well (La Heliera-1), whose location was determined at least partly by how far up the Rio Casanare the barges transporting the rig were able to navigate.

The commitment for the first contract year was to drill four stratigraphic tests on each block. At the end of the first year one-half of the acreage had to be relinquished.

In order to locate the stratigraphic tests most effectively and to prepare for the 50% relinquishment, Occidental rushed to start recording a large seismic program. The effective date of the contracts was July 1, 1980, and by August a program of 1100 km of seismic on the major rivers was begun. Portable dynamite crews started the land program shortly thereafter.

Because one of the potential accumulation types was a heavy oil belt like the Orinoco belt in Venezuela, many of the stratigraphic tests were drilled along the shallow eastern portions of the blocks.

The results of the first year seismic and stratigraphic drilling program were as follows:

1. Seven of the stratigraphic tests found shows of oil including dead residuals along the shallow eastern margin of the basin.

2. Several antithetic faults for potential traps were found in the east central part of the basin.

3. A new type of structure—drape folds over granite knobs—was found in the northeastern areas (Figure 10).

4. A new type of large early faulting was seen on seismic lines in the area southwest of Caño Limon (Figure 11).

In the second year additional seismic was recorded and a commitment of ten exploratory wells was com-

FIGURE 9. Original Occidental blocks in the Llanos basin.

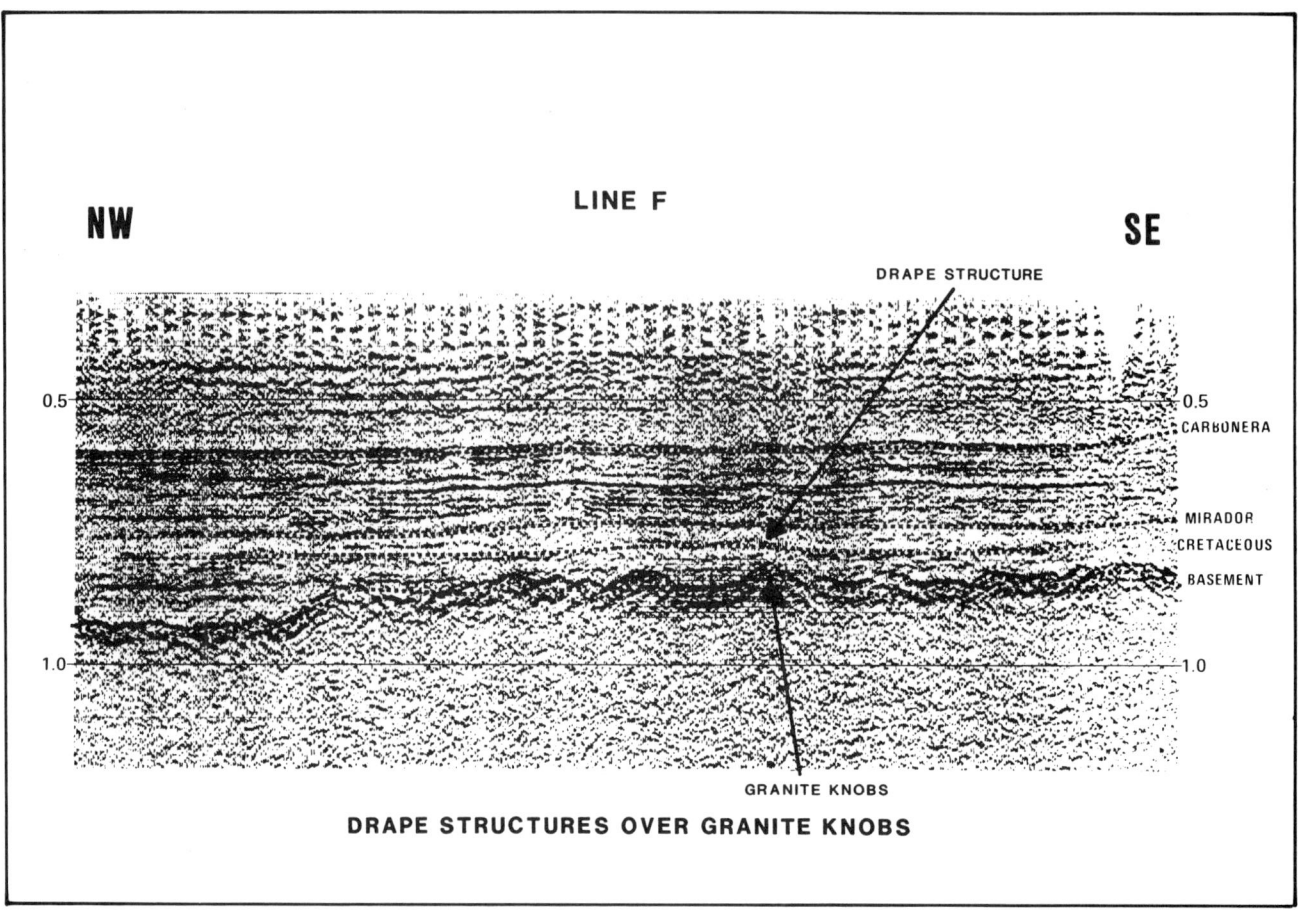

FIGURE 10. Seismic section through the eastern Llanos basin, showing drape structures over granite knobs.

pleted. Additional encouragement resulted—in particular, recognition that the faults in the Caño Limon area formed a major system and were probably strike-slip movements.

The possibility of folding associated with the major fault system began to spark enthusiasm, and a third small seismic program was recorded in the Caño Limon area to check several indications of folding. In particular, line 61.7 was shot along the south (downthrown) side of the main northeast-southwest fault now called the Caño Limon fault. This was done to check a very small rollover that later was drilled as the Caño Verde discovery, but farther to the northeast a larger rollover was found that was to be the Caño Limon discovery (Figure 12).

It was now late 1982, and the oil industry had begun to slump as a result of falling oil prices. Occidental, like most other companies, was reducing exploration expenditures worldwide, and compounding its problems were the strains of the merger with Cities Service. The decision was made to farm out 50% of the whole area for the drilling of the Caño Limon well. Exploration enthusiasm for the prospect was high, but from a business standpoint it appeared the options might be to farm-out or drop.

Thus began Occidental's largest farm-out effort. In total, 73 contacts were made with 59 companies, with presentations to many of them.

In the meantime, it was found that a well location could not be constructed in the swampy area overlying the prospect. A site was found on the natural levee of the Caño Agua de Limon, and a 43-km road from Troncal was started while equipment made its way to the drill site to raise the site elevation 2 m.

Because a deviated well would have to be drilled, it was realized that a seismic line from the drill site to the apparent crest of the structure was necessary for proper well planning. The interpretation at that point, based on a one-line rollover, was a small four-way dip closure; but it was apparent that the structural configuration was not really known (Figure 13). So, in addition to a northwest-southeast line from the drill site to the target, a second line was shot NNE-SSW across the apparent crest of the structure. These two lines showed that it was not a four-way closure, but was closed against the major Caño Limon fault (Figure 14). Thus it became necessary to believe that the fault would form a seal.

As the end of the dry season approached, it was clearly necessary to get a rig in before heavy rains closed the road. Management agreed to go ahead provided that farm-out attempts were vigorously continued. The rig was trucked in as the rains began. The trucks carrying the last load, the main mud pit, became stuck short of the

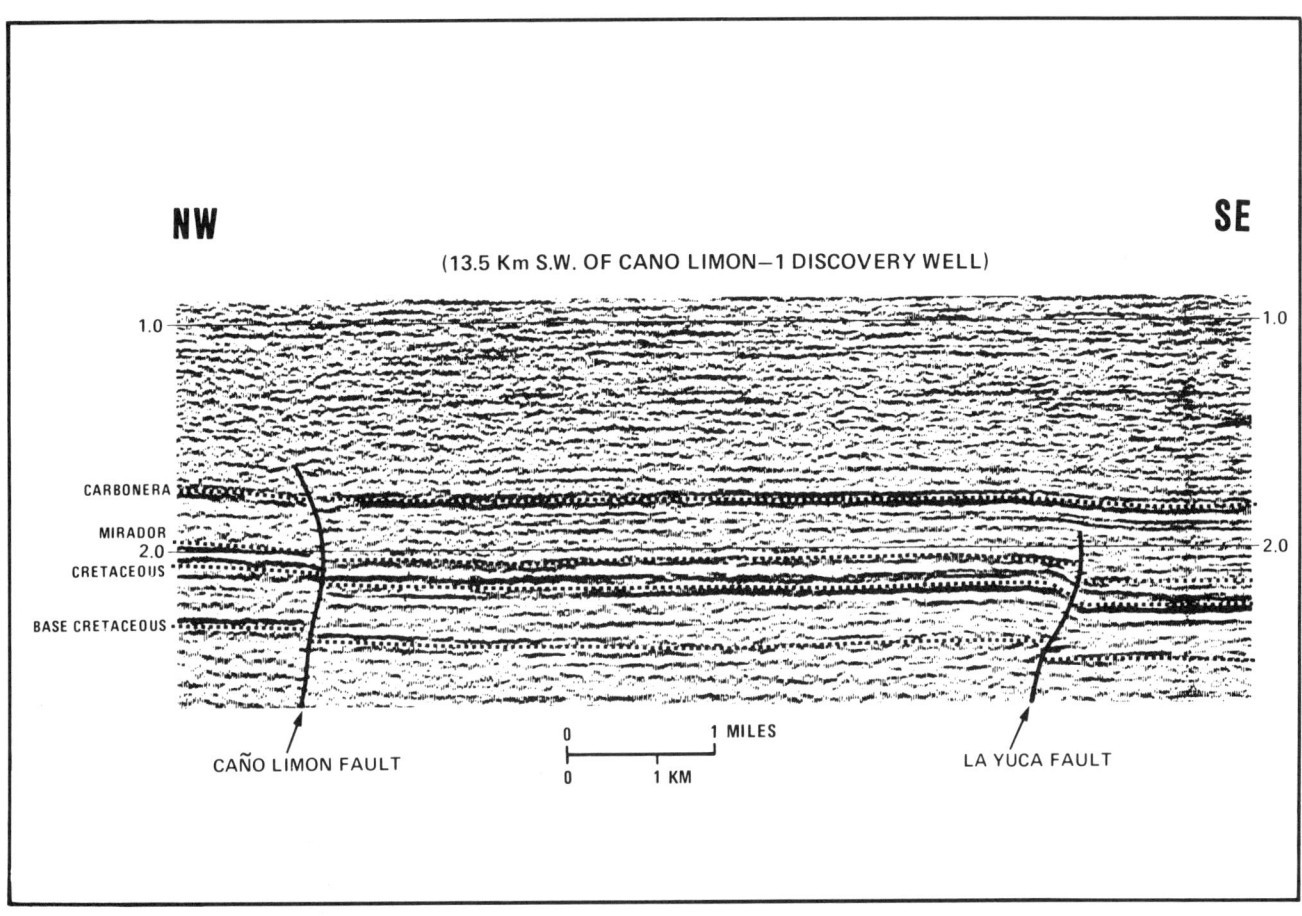

FIGURE 11. Seismic section through area 13.5 km southwest of Caño Limon field, showing Caño Limon and La Yuca faults.

location, and the mud pit was dragged by bulldozers to the Caño Limon and onto a barge for transportation to the drill site. The well was spudded on April 20, 1983, while farm-out attempts continued.

The Mirador sands were penetrated with no gas shows and only traces of fluorescence on a few sand grains. It did not look good, but the generally low-GOR sands in Peru had had similar shows in production wells.

Occidental was surprised by the first log, which showed 160 ft (49 m) in two massive sands with resistivities up to 2500 ohm m (Figure 15). There was no doubt that a significant discovery had been made. The well tested at a combined rate of 10,690 BOPD from the two zones—one zone on a 1⅞-in. choke and the other on a 9/16-in. choke. The API gravity varied from 33.4° at the top of the Mirador to 29.8° at the base. No oil-water contact was present in the Mirador in this well.

The rig was skidded on the same location and Caño Limon-2 was directionally drilled to a target location on the west flank of the structure. The oil-water contact was penetrated and commerciality was granted by Ecopetrol on an Occidental reserve estimate of 64 million bbl of oil.

Drilling was suspended until the following dry season when equipment could again enter the area to build locations. Three locations were built: one for Caño Limon-3 on the southeast flank of the structure, and two for wildcats on nearby structures updip to the northeast, La Yuca and Matanegra (Figure 16).

The first indication that a major oil-field could be present came in Caño Limon-3, where the oil-water contact was found to be lower than in Caño Limon-2 and also lower than the syncline between the Caño Limon and La Yuca structures. The La Yuca-1 well soon added support to the idea that the two structures were in a common oil accumulation. Oil was found down to a major sealing shale in the Cretaceous K-1 zone, and pressure measurements indicated and oil-water contact at the same depth as in Caño Limon-3. The interconnection was subsequently proved by La Yuca-3, which found oil in the syncline.

Matanegra-1 found oil in the Mirador, K-1, and K-2 sands, adding a major fault-bounded segment to the Caño Limon field. This fault wedge is not in communication with the Caño Limon—La Yuca portion of the field, as shown by a lack of pressure response and different oil-water contacts across the Caño Limon fault.

By the end of 1989, 50 wells had been drilled in the field, of which 47 were producers, two were downdip pressure observation wells (Matanegra-3 and La Yuca-22), and one was abandoned and replaced because of mechan-

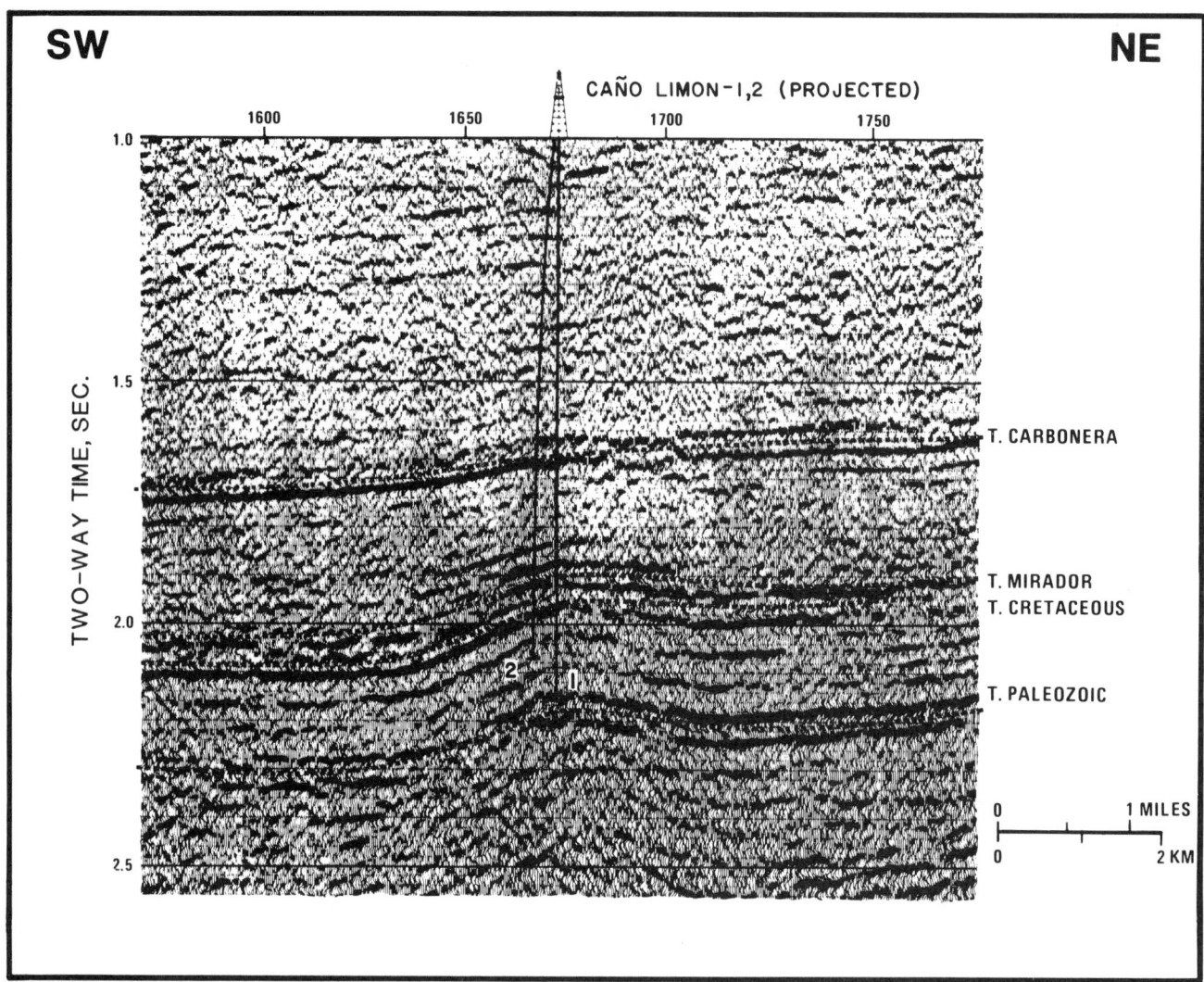

FIGURE 12. Seismic section through the region of the Caño Limon-1 and -2 wells.

ical problems (Matanegra-2) (Figure 17). A total of 75 producers may be needed.

CAÑO LIMON FIELD

Structure

The Caño Limon structure resulted from strike-slip movements of the Caño Limon and La Yuca faults during the Eocene and early Oligocene. Minor folding may have occurred just prior to Mirador deposition or during early Mirador time, because there is some thinning of the Mirador over the crestal areas. The main folding clearly took place after Mirador and before Carbonera, because there is marked thinning of basal Carbonera units over structural highs (Figures 12 and 18). Although folding of the Caño Limon structural complex terminated early in Carbonera time, movement on the major faults continued longer, as shown by thickening of higher Carbonera zones in down-thrown blocks and fault cuts in some of these higher zones.

Lateral offset on the order of 2 to 4 km on the major faults is estimated from offsets in the thickness trends within specific Mirador and Carbonera units and offsets of certain areally limited Cretaceous marker beds.

Stratigraphy

Cretaceous sands of Albian to Campanian age rest on Cambrian-Ordovician sandstones, siltstones, and shales over most of the Caño Limon area (Figures 7 and 19). In a few places (the Arauquita-1 well and the Guafita area in Venezuela), Triassic sand-shale red beds are preserved beneath the base Cretaceous unconformity.

The Cretaceous has been divided into four members, K-1, K-2A, K-2B, and K-3, in the Caño Limon area, but it now appears that this K-1 is not the K-1 as defined by Ecopetrol on a regional basis but is the regional K-2. The regional K-1 is apparently not present. Whether it is miss-

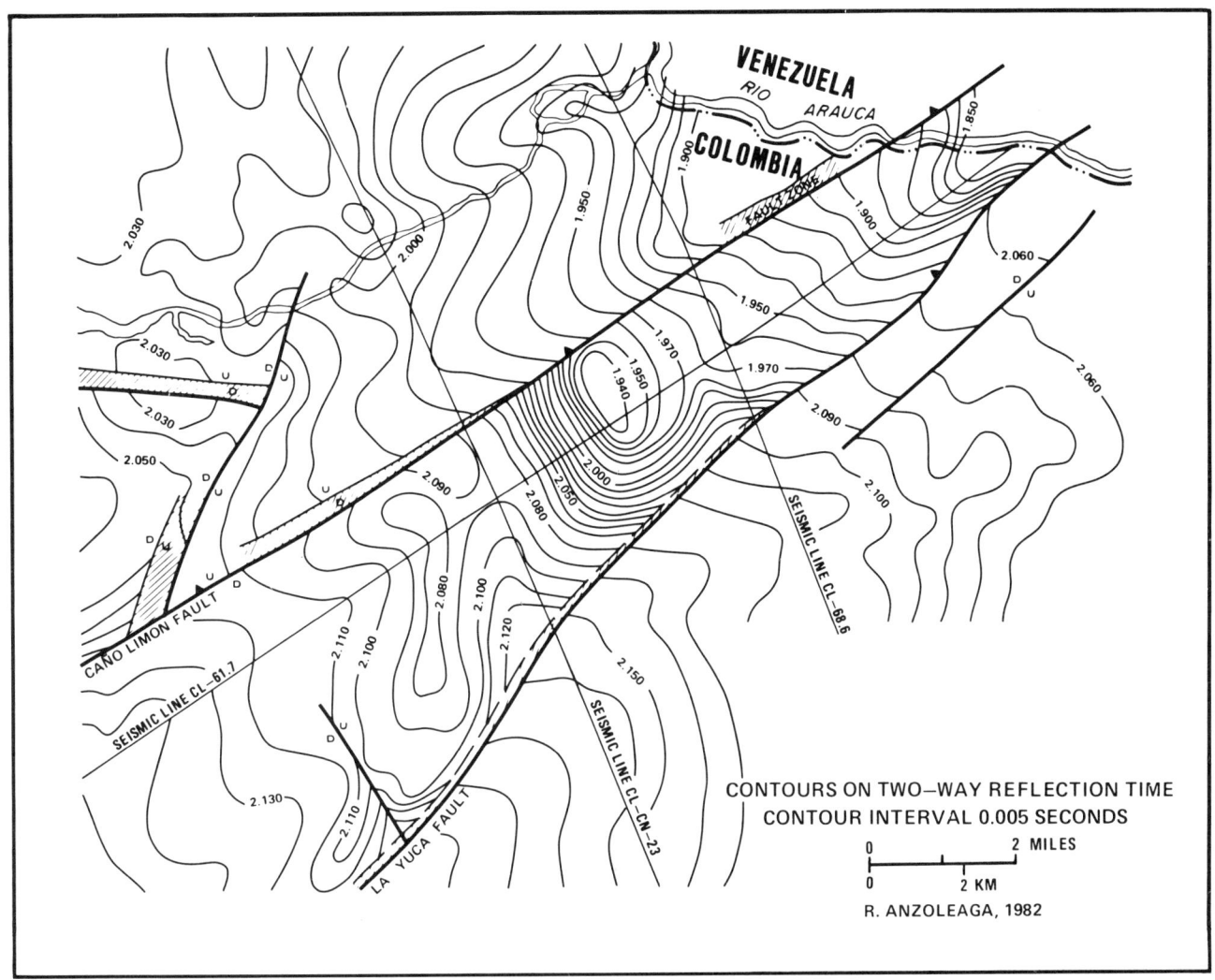

FIGURE 13. Structural map on the top of the Upper Cretaceous in the Caño Limon field when the decision was made to drill.

ing by nondeposition or truncation is not yet known. Its absence confirms that the Arauca arch or high was present in Late Cretaceous or early Tertiary time.

The K-3 basal member is a massive sand unit of stacked continental braided stream deposits with an average thickness of 800 ft. Oil has been found in the K-3 only in the Caño Rondon-1 well, a small discovery 33 km southwest of Caño Limon.

The K-2 members of Santonian to Coniacian age are marine-dominated regressive sequences (deltaic) containing several thin calcareous sandstones and shelly limestones. The K-2A member averages 250 ft (76 m) in thickness. It is a major reservoir sand in the Matanegra fault block where it has produced at a rate of 10,784 BOPD. It has also produced oil in the Caño Rondon-1 well. The K-2B averages 120 ft (37 m) in thickness but has not yet been found productive.

The K-1 member falls within the Santonian-Campanian age interval. It is a more marine sequence of prodelta shales, thin limestones, and shelf sands with deltaic sands near the top. This interval is productive in the La Yuca and Matanegra areas and has produced at rates of up to 8000 BOPD in La Yuca-4. The average thickness of the K-1 is 250 ft (76 m).

The Mirador zone of Eocene age is mainly a river-dominated delta system but has several marine shales, delta fringe sands, and probably wave-dominated delta sands. The river-dominated delta channel sequences are often massive sand bodies with excellent permeability and good continuity in the channel direction. Continuity normal to the channel direction is expected to vary. Pressure interference tests between wells have all shown rapid response when equivalent Mirador zones were open, and no response when the zones were not the same. An injectivity test into the downdip La Yuca-22 well produced pressure responses in Caño Limon-6 after 7 hr of injection (a distance of 2.4 km) and in La Yuca-11 after 11 hr of injection (a distance of 1.6 km).

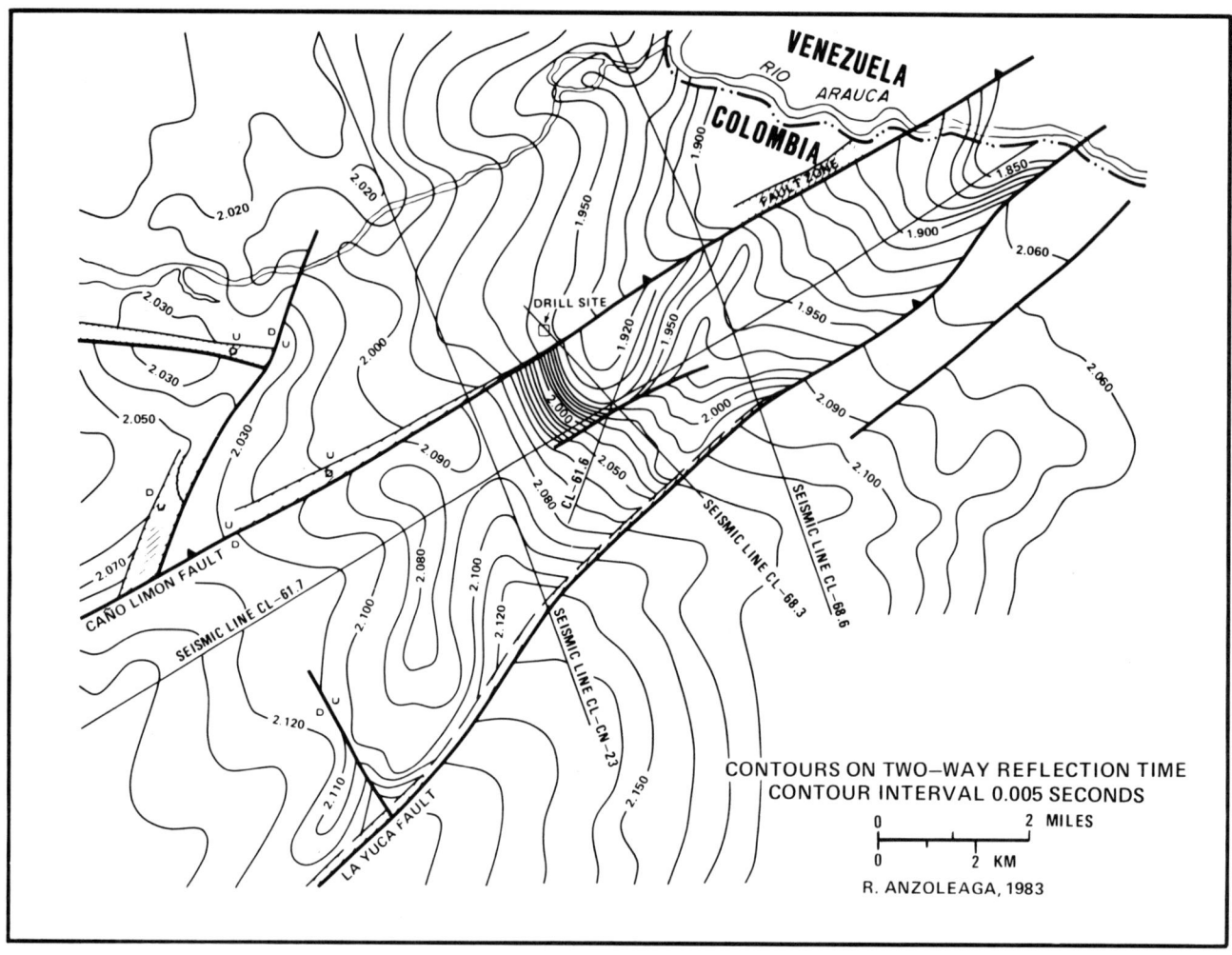

FIGURE 14. Structural map on the top of the Upper Cretaceous in the Caño Limon field when the discovery well was drilled.

The average thickness of the Mirador zone is 260 ft (79 m). Approximately 80% of the Caño Limon reserves are in this reservoir.

The basal sands of Carbonera C-5 zone are productive on the flanks of the Caño Limon and La Yuca structures. These sands are quite discontinuous.

Field Characteristics

Caño Limon has been described as "the kind of field geologists and engineers dream of." Some of the attributes that give it this reputation are:

1. In the Mirador, which holds 80% of the reserves, the sequences of deltaic channel and fringe sands results in excellent to good reservoirs with several continuous shales to reduce major fluid crossflow and assist in reservoir management capability (Figure 18). Cretaceous reservoirs are also good to excellent.

2. Permeabilities very widely depending on facies but average 1145 md from core analysis (under overburden pressure) and range from 2 to 15 darcys on pressure buildup analyses.

3. Flowing well rates can exceed 20,000 BOPD.

4. The oil gravity averages 29.5° API, and the oil is low in sulfur (0.4%).

5. The gas-oil ratio is about 8 ft^3/bbl (0.2 m^3/bbl), resulting in a formation volume factor of 1.05, or virtually no shrinkage.

6. Formation waters are fresh (100 to 200 ppm NaCl) and noncorrosive in the water leg, and are only slightly more saline within the oil accumulation.

7. A strong natural water drive obviates the need for pressure support by water injection. Occidental reser-

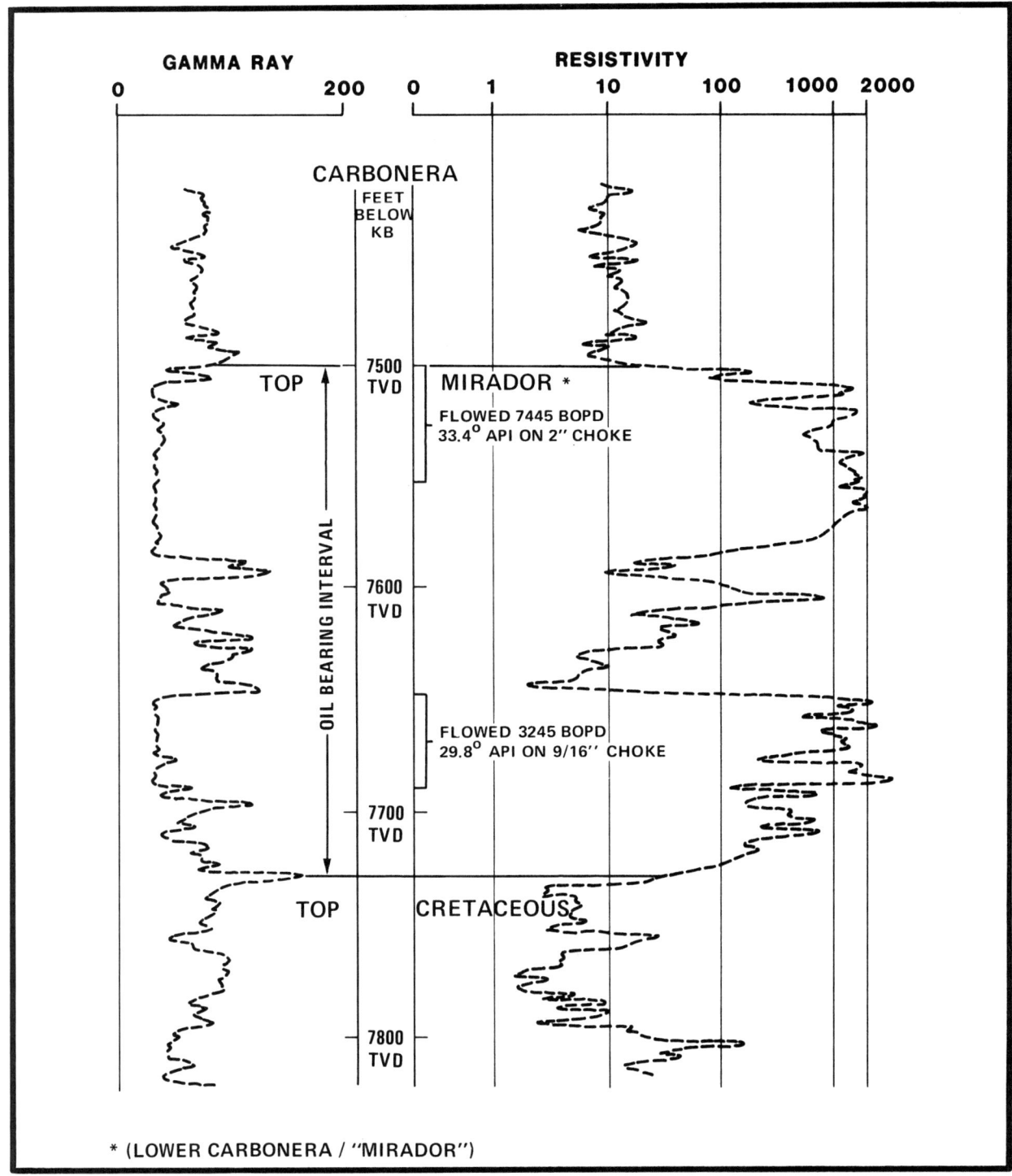

FIGURE 15. Gamma ray and resistivity profiles of the Lower Carbonera/"Mirador."

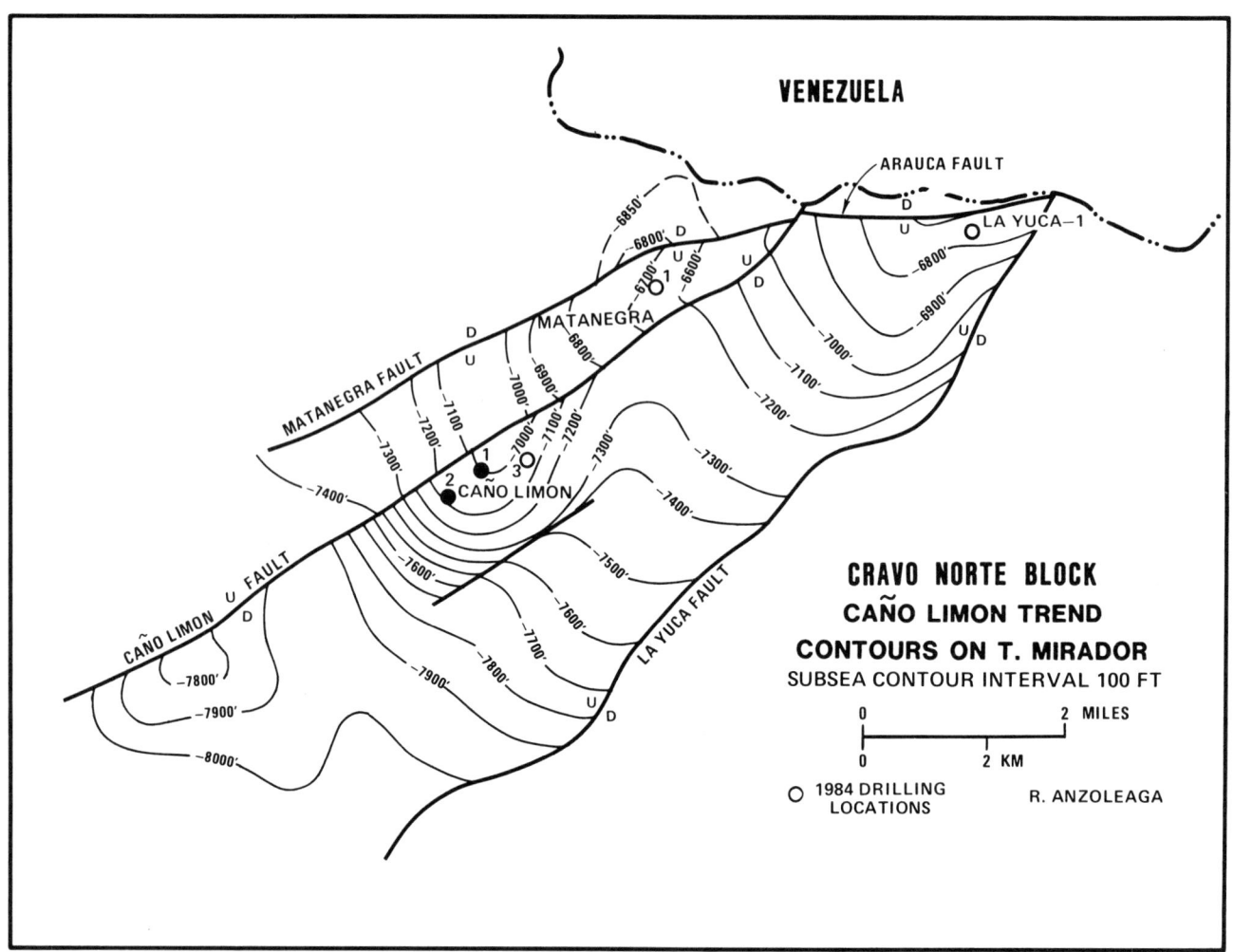

FIGURE 16. Structural map of the Caño Limon trend in the Llanos basin, showing 1984 Caño Limon, La Yuca, and Matanegra drilling locations.

voir engineers believe that over-all recovery will exceed 50%.

8. Wells are very easy to drill, and depths are moderate, with the top of the reservoir averaging 7500 ft (2290 m) and total depths being near 8000 ft (2440 m).

9. Seismic resolution is good. A 3D survey has been recorded to detail structural and stratigraphic relationships.

IN RETROSPECT

Looking back at the Occidental Llanos basin exploration play, the factors that appear to have been important in the discovery of the Caño Limon field are:

1. Recognition that the basin had all the prerequisites for big oil except known early traps, and that such traps could be associated with the Arauca arch or could exist in the vast areas to the east that lay totally unexplored.

2. Willingness to accept a very high level of risk in a basin that conventional wisdom said was lacking in potential.

3. Perseverance in the face of adversity as long as the essential requirements for oil were in sight.

ACKNOWLEDGMENTS

The authors thank the managements of Occidental International Exploration & Production Company and of its partner companies in Colombia—Ecopetrol, Shell, and Repsol—for permission to publish this paper.

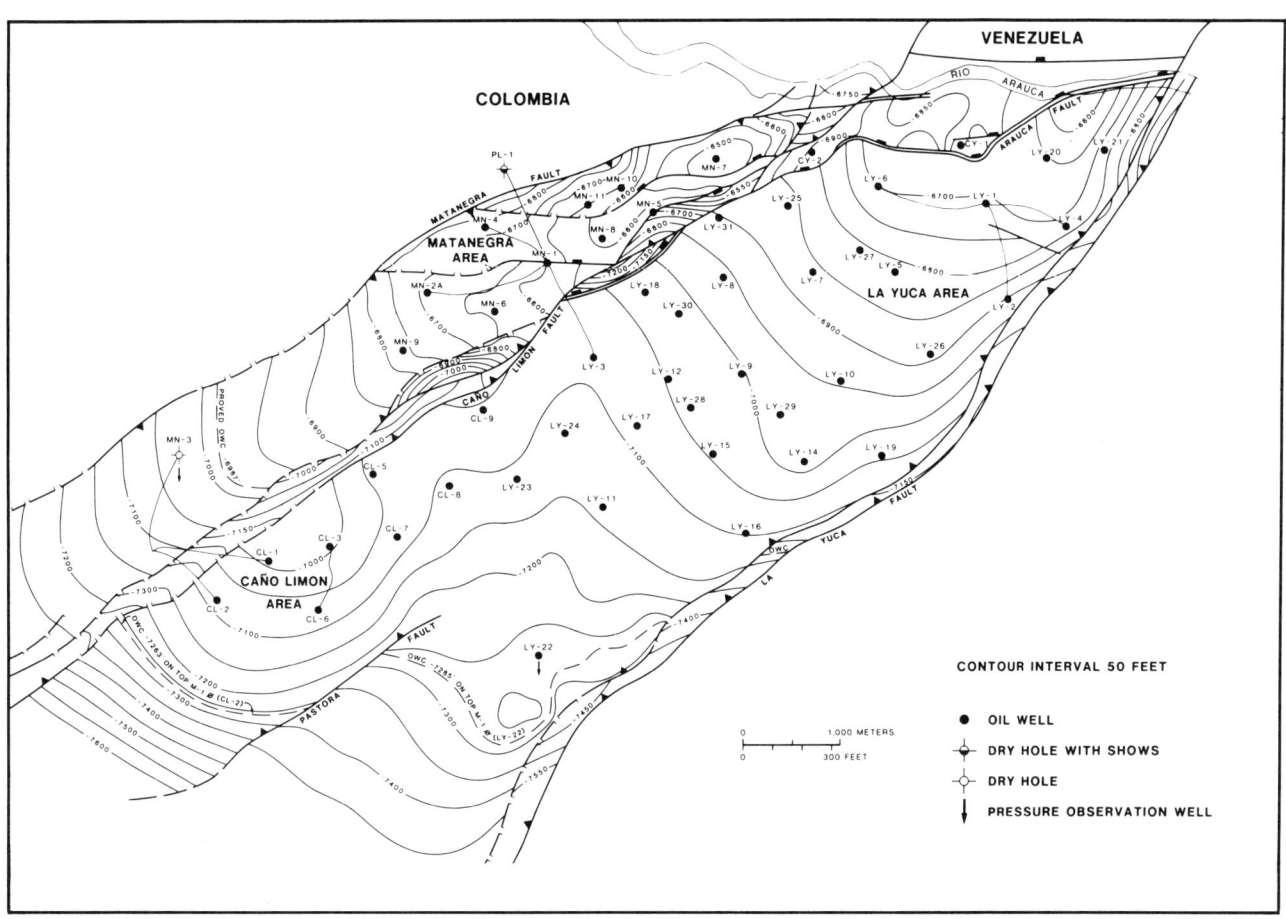

FIGURE 17. Structural map on the top of the Lower Carbonera/"Mirador," showing wells drilled by the end of 1989.

REFERENCES CITED

Cleveland, M. N., 1988, The role of geological core interpretation in reservoir management of Caño Limon field, Colombia: Tercer Congreso Colombiano de Petroleo Memoirs, v. 1, p. 165–182.

Cleveland, M. N., and J. Molina, 1989, Deltaic Caño Limon field, Colombia (in press), to be published in Springer-Verlag Sandstone Petroleum Reservoirs.

Gabela, V. H., 1985, Campo Caño Limon, Llanos orientales de Colombia: II Simposio Bolivariano Memoirs, v. 1, p. 1–29.

McCollough, C. N., 1988, Geology of the super giant Caño Limon field and the Llanos Basin, Colombia, Transactions: Fourth Circum-Pacific Energy and Mineral Resources Conference.

McCollough, C. N., in progress, Caño Limon Field, Colombia: AAPG Atlas of Oil and Gas Fields.

Molina, J., 1988, Condiciones hidrodinamicas de los Llanos septentrionales: Tercer Congreso Colombiano de Petroleo Memoirs, v. 1, p. 183–198.

Parker, C. A., 1986, Caño Limon reservoir properties suggest high recovery factor: Oil and Gas Journal, May 12, p. 55–58.

Rivero, Ramon T., J. G. Dominguez, J. A. Slater, and C. L. Hearn, 1988, Caño Limon field, Colombia: the latest giant oil reservoir in South America: Oil and Gas Journal, July, p. 874–880.

Robertson Research, 1983, The northern Llanos of Colombia, hydrocarbon potential and stratigraphic control, v. 1, 2, and 3.

Taylor, J. B., 1984, Occidental Caño Limon discovery in the Llanos of Colombia: new ideas, new methods, new developments: Proceedings of the Southwestern Legal Foundation Exploration and Economics of the Petroleum Industry, v. 22, New York, Matthew Bender & Company.

Tillman, R. W., 1985, Sedimentology and reservoir properties of Occidental de Colombia, Caño Limon No. 3, Llanos basin, Colombia: Technical Report No. 131, p. 65.

Valderrama, R., 1982, Desarrollo de facies de la Cuenca de los Llanos orientales de Colombia: Simposio "Explora-

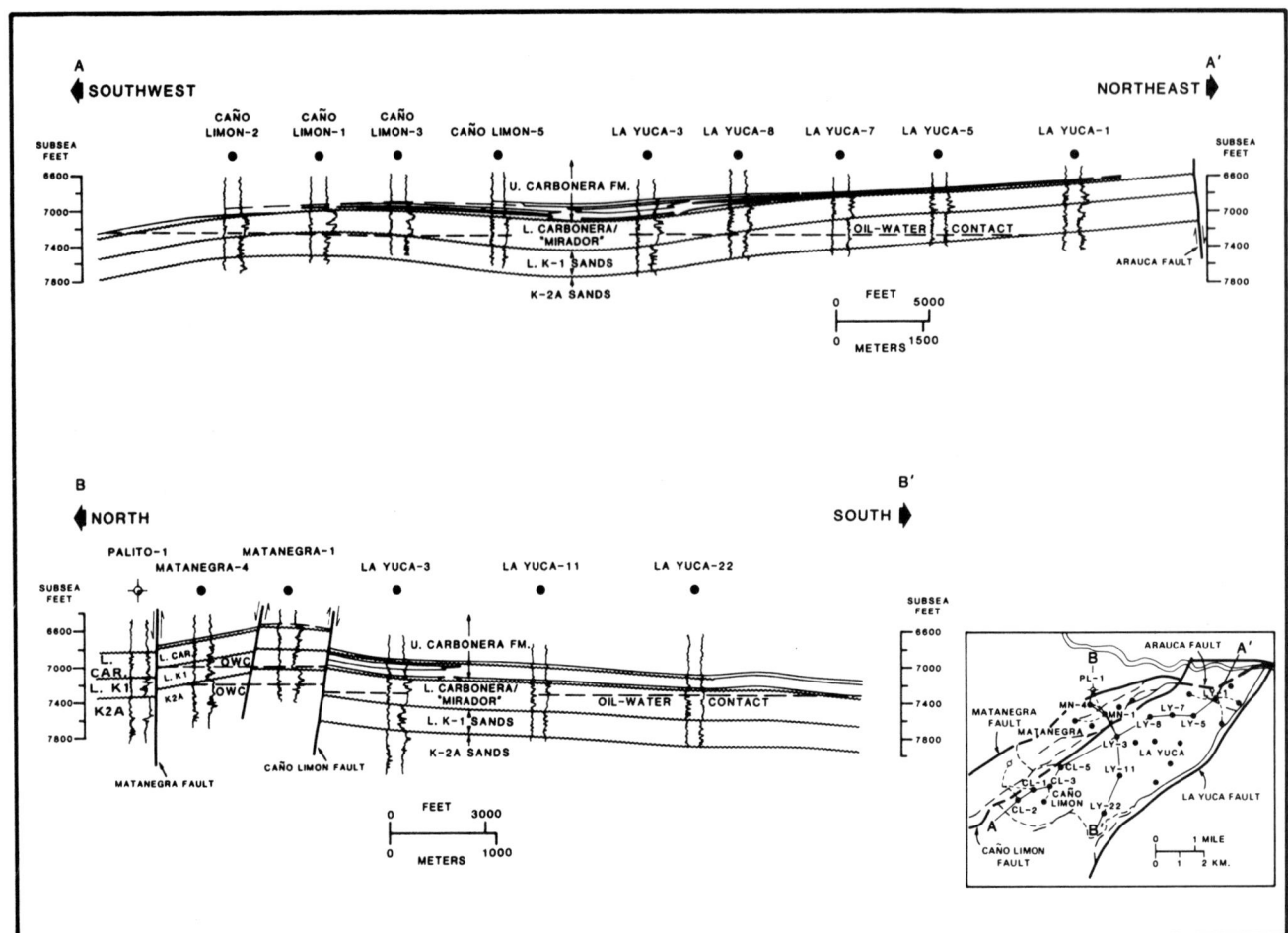

FIGURE 18. Cross sections through Caño Limon field.

cion petrolera en las cuencas subandinas de Venezuela, Colombia, Ecuador y Peru", Asociacion Colombiana de Geologos y Geofisicos de Petroleos, Bogota.

Zambrano, E., E. Vasquez, B. Duval, M. Latreille, and B. Coffinieres, Paleographic and petroleum accumulation history of western Venezuela: Congreso Geologico Venezolano, v. 1, Memoir 4, Dir. Geol., Bol. Geol., Special Publication No. 5, p. 483–552.

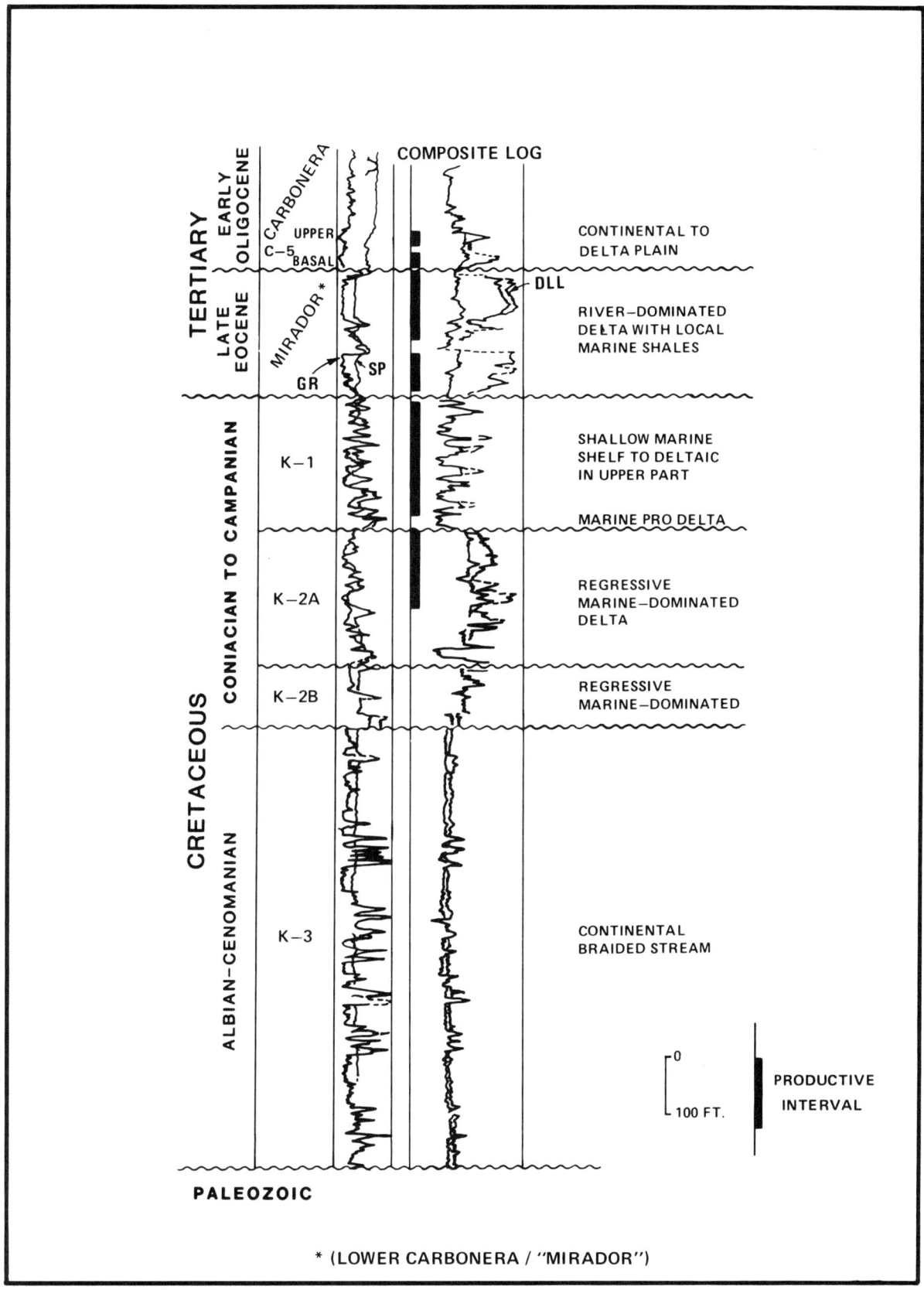

FIGURE 19. Sedimentary sequence for the Caño Limon field.

Chapter 13

Takula Oil Field and the Greater Takula Area, Cabinda, Angola

C. T. Dale
J. R. Lopes

Chevron Overseas Petroleum, Inc.
San Ramon, California, U.S.A.

S. Abilio

Sociedade Nacional de Combustiveis de Angola (SONANGOL)
Luanda, Angola

ABSTRACT

The Greater Takula area comprises four offshore fields that produce oil predominantly from Upper Cretaceous reservoirs. They are located approximately 40 km (25 mi) west-northwest of the Malongo terminal, Cabinda, Angola, in water depths between 50 and 75 m (165 and 230 ft). Current total production level is approximately 200,000 bbl of 32° API oil per day. The largest field is Takula, where oil was first discovered in rocks of Barremian age in 1971. However, it was not until 1979 that the giant oil accumulation in the Cenomanian section was tested. An estimated 2060 MMSTB of OOIP (original oil in place) was present in Takula field. By 1983, appraisal and exploratory drilling had found the prolific Greater Takula trend of rollover anticlinal structures striking northwest to southeast through Takula. These structures host Wamba, Takula, Numbi, and Vuko fields. Over 80% of the original oil in place is held in the Cenomanian Vermelha sandstone reservoirs. These reservoirs are a stacked sequence of nearshore sandstones deposited in a coastal environment. The Greater Takula area is now in a mature stage of development, having produced over 250 MMSTB from an OOIP of 3450 MMSTB.

INTRODUCTION

The Greater Takula area is located 40 km (25 mi) west-northwest of the Malongo Terminal, Cabinda, Angola, in water depths of 50 to 75 m (165 to 230 ft). Approximate geographic coordinates are 5°15'S and 11°50'E. This complex of four fields is situated in the Cabinda concession of Angola in the Lower Congo Basin. The concession is operated by Cabinda Gulf Oil Company (CABGOC), a Chevron subsidiary, on behalf of the partnership, which includes SONANGOL (Sociedade Nacional de Combustiveis de Angola) (51%), CABGOC (39.2%), and AGIP (9.8%). Prior to the Takula discovery in 1971, the bulk of the oil produced in Cabinda came from the giant pre-salt (pre-Aptian) pools of Malongo North and Malongo West (Figure 1). These fields produce from both Toca carbonates and Lucula clastics of the early continental rift sequence (Figure 2). The first well in the Takula area, drilled in 1971, was also drilled to test the pre-salt section. The 44-1X exploratory well drilled an untested horst structure. It was successful and tested 5600 bbl of 32° API oil per day from the Barremian Toca carbonates. This Toca accumulation was areally limited and it was not until 1979 that the 57-2X well showed the potential of the untested Cenomanian Vermelha sandstones. By 1983, exploratory and appraisal drilling had confirmed the prolific trend of salt-cored rollover anticlines. These structures host Wamba, Takula, Numbi, and Vuko fields, which produce primarily from the Cenomanian Vermelha sandstones. The other producing reservoirs are the Aptian Pinda sandstones and the Barremian Toca carbonates. These fields are in a mature stage of development. At the present time 104 wells produce approximately 200,000 BOPD in the Greater Takula area. A water flood project, scheduled for start-up in late 1990, will increase oil recovery at Takula and Numbi.

EXPLORATION HISTORY

The earliest indications that Cabinda was a potentially rich oil province were oil seeps and asphaltic outcrops in the onshore areas. These encouraging signs led, in 1954, to the formation of Gulf Oil Corporation's first geologic field parties to the then Portuguese colony of Cabinda. On November 22, 1957, the Cabinda Gulf Oil Company was awarded its original concession covering 7267 km² of onshore and offshore acreage out to the 30-m isobath.

On the concession's anniversary in 1966, additional offshore acreage extending to the 200-m isobath was added. This addition and a small onshore relinquishment in 1960 (655 km²) resulted in a concession area of 10,116 km². The concession is presently divided into three areas:

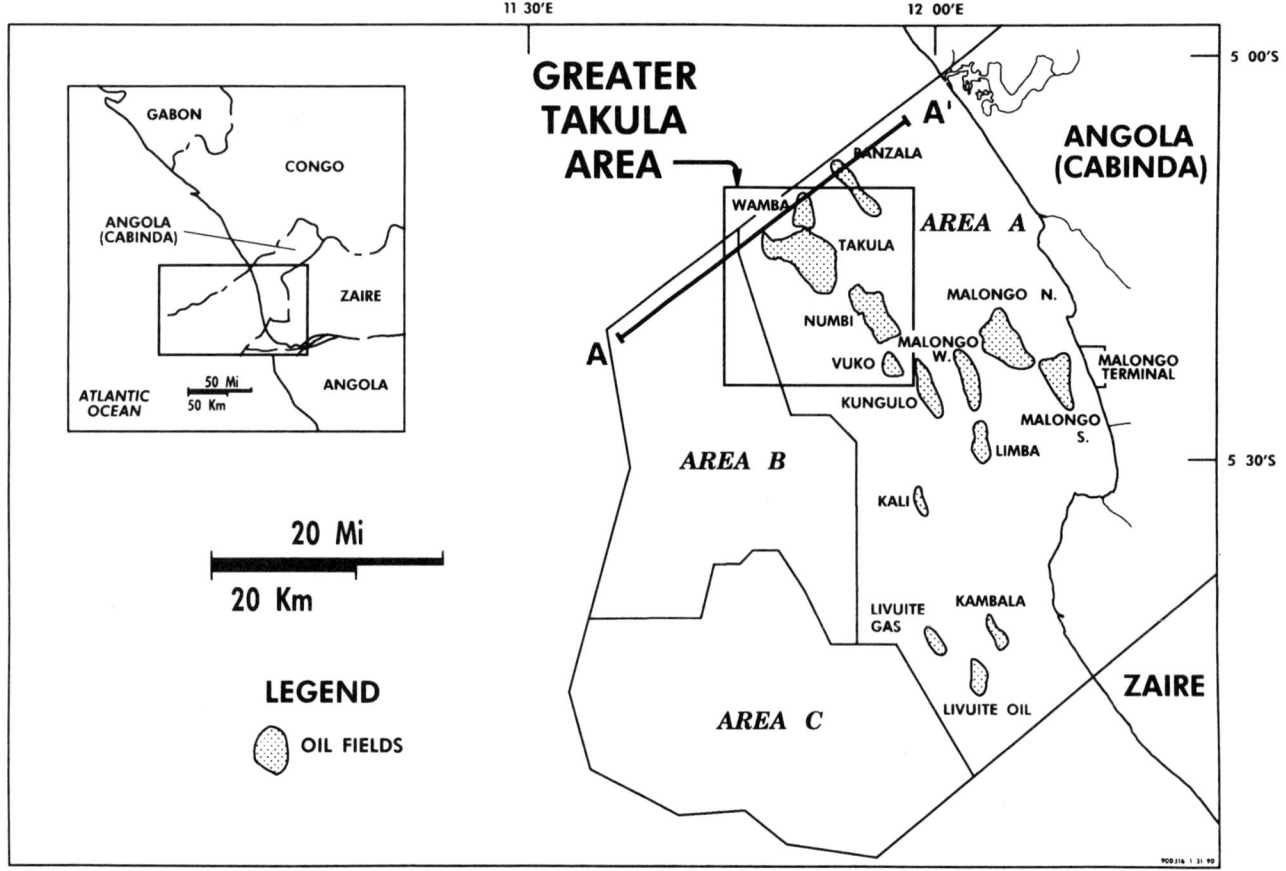

FIGURE 1. Location map showing the regional position of Cabinda, the Greater Takula area, other producing fields, and the location of the regional seismic line A-A'.

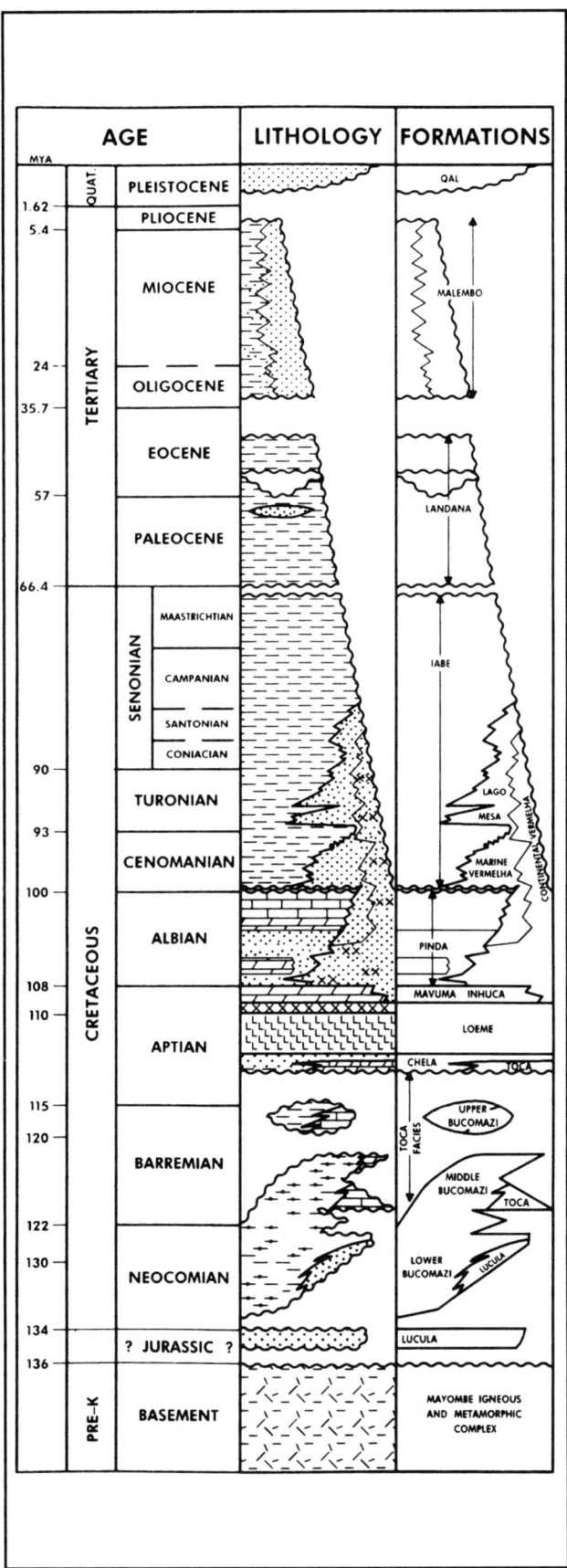

FIGURE 2. Generalized stratigraphic column for Cabinda.

A, B, and C. All current production comes from area A fields (Figure 1).

Early exploratory work concentrated onshore. Gravity, magnetics, seismic, and outcrop surveys led to minor discoveries. These were encouraging enough that in 1966 the first offshore well was drilled on the basis of marine geophysical work. The well was completed in the Cenomanian Vermelha sands of the Limba field and produced 350 BOPD using a ¼-in. choke (see stratigraphic section, Figure 2). This success led to an intensification of the offshore oil search. By late 1966, two additional Cenomanian fields, Malongo North and South, had been discovered. These fields formed the basis of the concession's first export production in late 1968. Exploration work continued, and in 1967 the first oil test from the deeper pre-salt (pre-Aptian) synrift section was made at Malongo North. By 1970 the giant pre-salt pool of Malongo West had been discovered (950 MMSTB original oil in place). The Malongo West and Malongo North fields produce from both carbonate and clastic rocks of early Cretaceous age.

The search for similar pre-salt fields turned to the northern part of the concession. In late 1971, the first Takula field well, 44-1X, was drilled on a postulated basement-cored horst structure. Basement was not penetrated, but pre-salt carbonates tested a total of 5600 bbl of 32° API oil per day from two separate zones. Oil shows were noted in the post-salt Cenomanian Vermelha section, only hinting at the large post-salt reserves that were later to be found.

The formation of the People's Republic of Angola, of which Cabinda is a province, resulted in a hiatus in drilling operations until the concession was renegotiated in 1977. The Cabinda Gulf Oil Company (CABGOC) became the operator and a 49% partner with the Angolan state-owned oil company SONANGOL.

On July 11, 1979, drilling in Takula field recommenced with the 44-4X well designed to test both pre- and post-salt reservoirs. Although the pre-salt rocks were essentially unproductive, sands of the Albian Pinda Formation flowed 1675 BOPD on a ¾-in. choke from two zones between 1495 and 1529 m (4905 and 5018 ft). The Vermelha was encountered between 1037 and 1201 m (3402 and 3940 ft), and good shows were noted. However, they were not tested because of the silty nature of the interval.

In early 1980, drilling was undertaken to further evaluate the Pinda Formation and also the untested Vermelha in a structurally favorable position (57-2X). Although the Pinda did not produce, Vermelha sands between 971 and 1038 m (3186 and 3404 ft) flowed at a cumulative rate of 4544 BOPD. This well proved the commerciality of Takula field and initiated field development. The Vermelha reservoirs, which contained over 65% of the field's OOIP, cover an area of approximately 53 km² (13,000 ac) and have a vertical closure of about 135 m (450 ft) (Figure 3). Follow-up exploratory drilling in the vicinity of Takula field resulted in the discovery of Wamba field in mid-1982, Numbi field in late 1982, and Vuko field in late 1983.

The complex of structures known as the Greater Takula area contained an estimated OOIP of 3450 MMSTB. To date, slightly more than 250 MMSTB have been produced. Current production rate is about 200,000 BOPD. To map

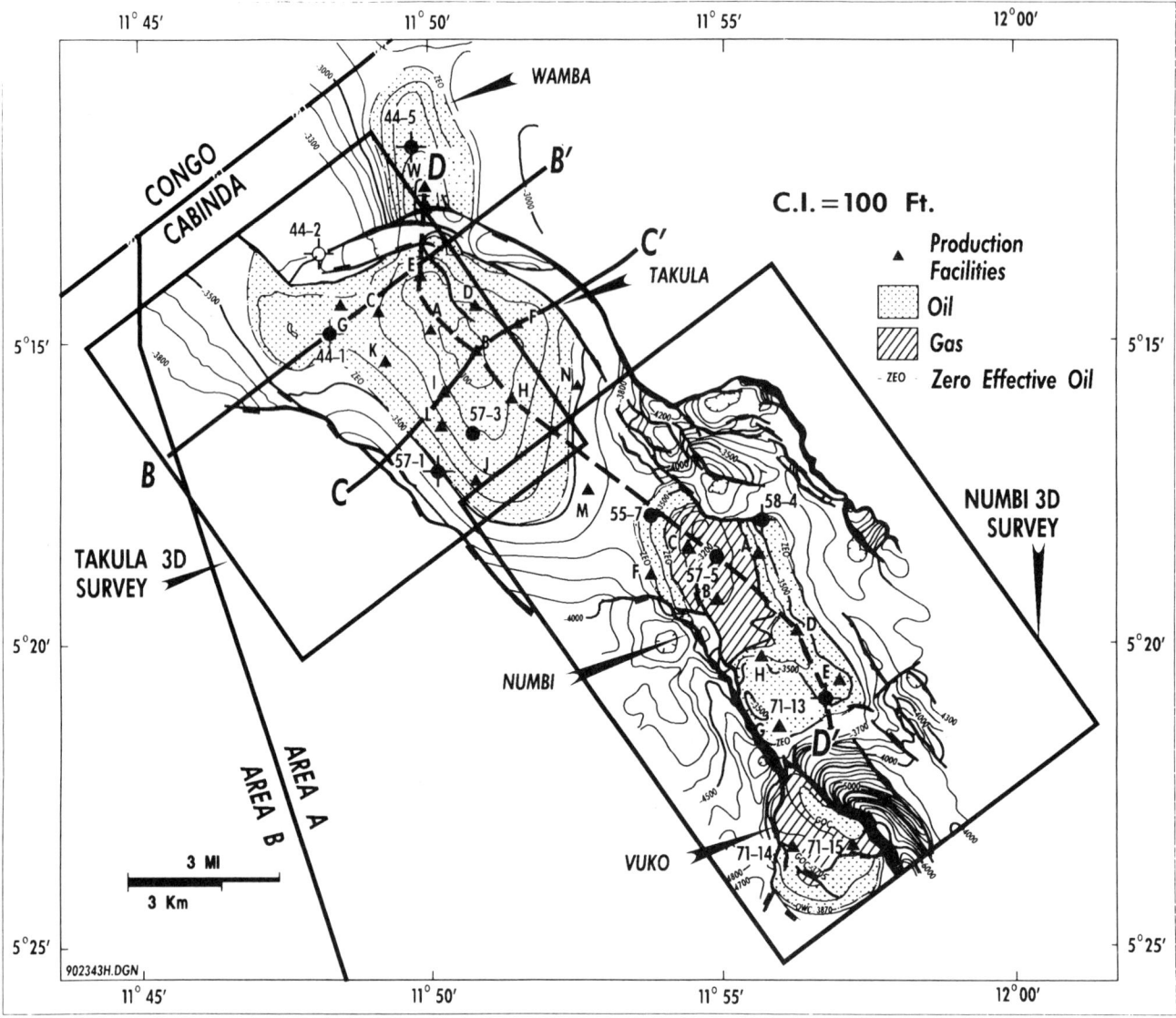

FIGURE 3. Depth structure map on the top of the Vermelha Formation, Greater Takula area, showing production jacket locations, seismic line B-B′, Vermelha cross-section locations C-C′ and D-D′, and the areas of 3D seismic coverage.

effectively the numerous reservoirs in the area, two three-dimensional seismic surveys have been acquired. In 1984 the Takula 3D survey was shot, followed in 1987–88 by the Numbi 3D survey (Figure 3). Both are used primarily to support development drilling and waterflood pressure maintenance programs.

REGIONAL GEOLOGY

Angola lies midway along the length of the West African margin of the South Atlantic rift. The rift developed in the Late Mesozoic as plate tectonic activity caused the separation of South America from Africa. The tectonic elements and sedimentary sequences observed along the continental margins of both Angola and Brazil developed in response to this event (Belmonte et al., 1965; Asmus and Ponte, 1973; Campos et al., 1974; and Lehner and DeRuiter, 1977). In general, the protocontinental margin responded to this rifting by the development of a series of terrestrial-lacustrine and later marine basins that first formed in latest Jurassic time. The Cabinda concession is located in the Congo Basin, which is flanked to the north and south by the Gabon and Cuanza basins, respectively.

The structure and stratigraphy of Cabinda can be divided into two major geologic units separated by the Loeme salt sequence of Aptian age (Caflisch, 1979; Brice et al., 1982; and McHargue, 1991). These units have distinct tectonic styles and contrasting terrigenous and marine clastic sedimentation. A generalized stratigraphic column (Figure 2) and regional seismic line A-A′ (Figure 4) illustrate the important structural characteristics and seismic expression of the geologic units.

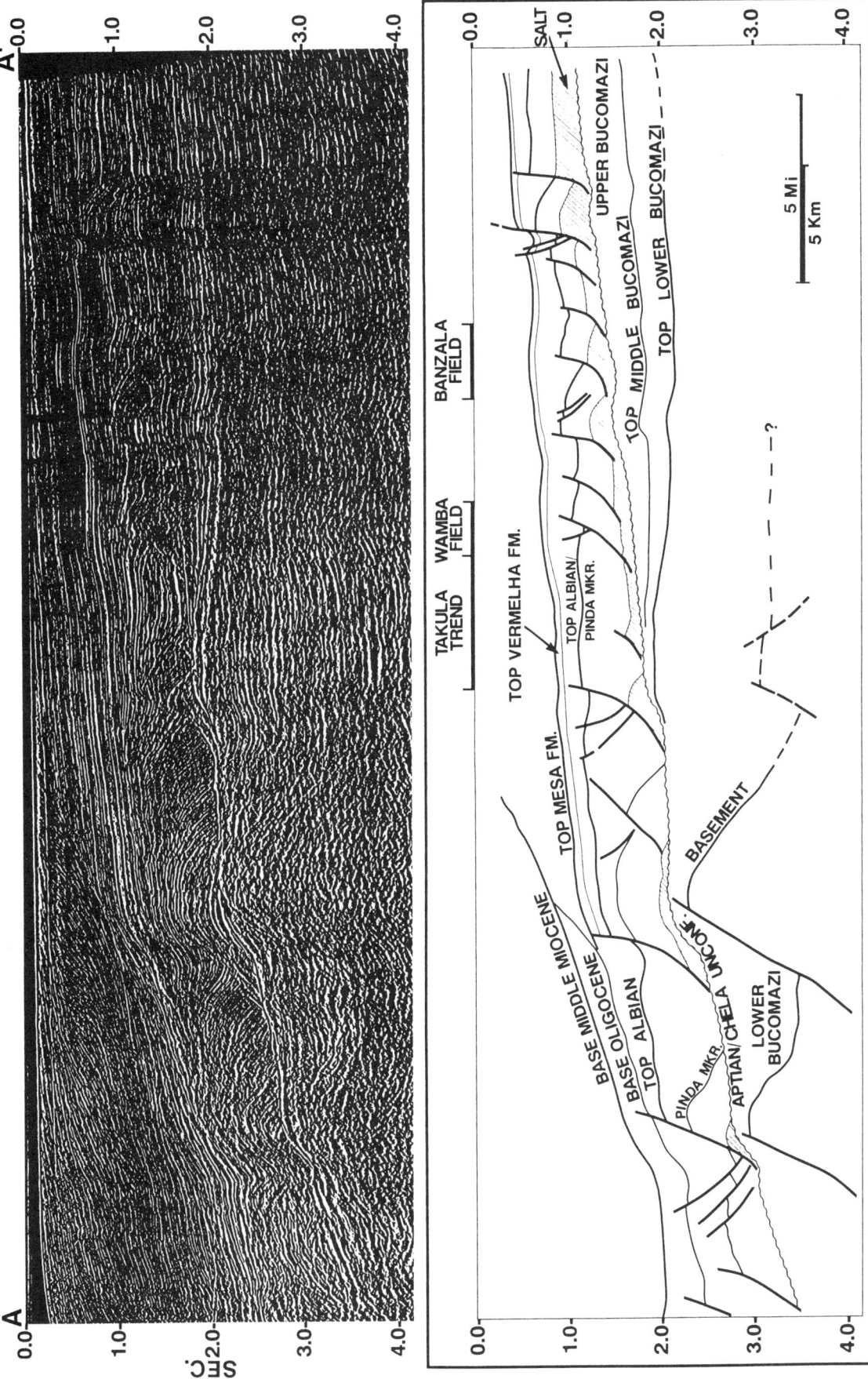

FIGURE 4. Regional seismic line A-A' and interpretation. Note the distinct unconformities separating the pre-salt (pre-Aptian) from the post-salt section and the erosional base of Oligocene.

Pre-Salt Period

The pre-salt period spanned earliest Cretaceous (possibly latest Jurassic) to mid-Aptian. During prerift time, gentle tectonism influenced the deposition of thin continental clastics. The active early-rift structures were dominantly southwest–northeast-dipping basement (Precambrian to Paleozoic) involving half-grabens segmented along the strike into compartments or subbasins by southwest-northeast transfer zones (Gibbs, 1987). This segmentation influenced both detailed structural development and sediment distribution. Most basement-involved tectonism occurred in the early rifting phase with reactivation and partial inversion of some half-grabens continuing throughout pre-salt and into post-salt time.

The dominant pre-salt basin-fill is the lacustrine Bucomazi Formation. It is subdivided into three informal members: upper, middle, and lower. The lower Bucomazi is the thickest unit within the Bucomazi Formation (up to 1000 m). It is made up of clastics deposited during the early active rift period. They are dominantly lacustrine silts and shales of Neocomian age, some of which are rich in organic material (up to 5% TOC). Sand-rich facies within this interval were sourced from exposed igneous and metamorphic basement. Locally called the Lucula sands, they were deposited around basement highs as well as within the basins. Lucula sands are important reservoir rocks in the pre-salt fields outside of the Greater Takula area.

An end-Neocomian unconformity, resulting from renewed rifting, usually separates the lower from the middle Bucomazi. Immediately subsequent to this tectonism, regional subsidence led to uniform anoxic lacustrine deposition. These conditions promoted the deposition of an extensive, laterally continuous, organic-rich sequence (up to 20% TOC) that provides the primary source of hydrocarbons in the Takula subbasin. In the shallower areas of the lake system, these organic- and carbonate-rich shales grade laterally into shallow water shales and algal carbonates of the Toca Formation.

Toward the end of the Barremian, the lake system became oxygenated and deposition of organic-rich shales ceased. The top of the organic shales marks the base of the upper Bucomazi member. At the basin margins an unconformable relationship is seen between the upper and middle Bucomazi members.

In shoal areas, algal carbonates of the Toca Formation continued to be deposited. The carbonates are found as time equivalents to both the upper and middle Bucomazi. They are important reservoir rocks in the Takula field and were the first Takula reservoirs to be discovered and tested. Porosity and permeability are secondary and probably resulted from subareal or phreatic-zone exposure.

The Barremian ended with isostatic uplift and minor structural readjustment followed by extensive erosion of the rift sediments. The resultant pervasive unconformity is readily recognized on both seismic and well data. Renewed subsidence and erosion during the Aptian allowed the accumulation of the sandstones and shales of the Chela Formation on this unconformity.

Aptian Salt Deposition and Structural Role

Evaporite deposition marked the onset of marine incursion into the basin. The Aptian Loeme Salt Formation is locally over 1000 m (3300 ft) thick. It is composed of halite interbedded with numerous, sometimes thick, bands of potash salts. At least four major cycles of potash deposition have been recognized within the formation. Toward the top of the halite/potash section, thin clastics become common and the salt grades upward into a regionally extensive anhydrite bed. This anhydrite represents basinwide lower salinity and marks the top of the Loeme Formation. The regional cross section shows that the salt section acts as a décollement zone for the faulted post-salt structures (Figure 4). Salt thickness is highly variable, but no salt section in excess of 1000 m has been drilled. The present thickness of the salt is directly related to both halokinesis and the structural evolution of the basin margin since the salt was deposited.

Areas where no salt is present are often called "salt windows." These windows provide significant conduits for the migration of hydrocarbons from pre-salt source rocks into much younger post-salt reservoirs.

Post-Salt Period

In Cabinda, open marine conditions prevailed in early Albian time and produced the sequence of continental shelf clastics and carbonates of the Pinda Formation. The lowest unit of the Pinda section is mostly limestone and dolomite resulting from the early marine transgression. Pinda sediment thickness, passive margin subsidence, and the movement of underlying salt caused the Albian shelf to break into a series of listric normal fault–bounded structures generally oriented northwest-southeast (parallel to the ocean margin). The Pinda structures evolved as rollover growth features with bounding listric faults that sole out into the salt sequence. Pinda stratigraphy developed in response to these structural features. In general terms, the carbonate shelf consisted of a relatively narrow shelf margin, a broad shelf area, and a shoreward coastal complex. The complex interaction of associated facies tracts with the continuously evolving growth structures of the rotating and sliding Pinda blocks produced a stratigraphy of intercalated calcareous shelf sandstones, dolomites, shales, and sandstones. Horizontal continuity and facies can vary dramatically on a local scale because of these complexities. The Pinda is a secondary reservoir in the Greater Takula area.

The predominantly carbonate-clastic sedimentation of the Albian gave way to dominantly siliciclastic deposition in Cenomanian times. The Vermelha reservoir sandstones were deposited in a nearshore/shoreface environment along a trend through the producing fields of Wamba, Takula, Numbi, Kungulo, and Limba. Relative sea level remained more or less constant with respect to these structures for much of the Cenomanian, yielding a thick (approximately 168 m, or 550 ft) stacked sequence of nearshore deposits. To the west of this trend, deeper water sedimentation of silts and shales prevailed, whereas to the east increasingly nonmarine red-bed sedimentation occurred. The degree of motion on the structure forming listric faults attenuated with time. The structures involving the Vermelha section are broader and less faulted than those involving the Pinda section.

The remainder of the Upper Cretaceous sedimentation was dominated by a marine transgression. This section is composed primarily of the shales of the Iabe Formation with regionally extensive beds of dolomite, sandy dolomite, limestones, and sands of the Mesa, Lago, and Azul formations. These were deposited as the coastline migrated eastward across the area.

During the Late Cretaceous and Tertiary, passive continental margin subsidence continued, resulting in a regional tilt toward the west. This, combined with a significant lowering of eustatic sea level, produced a major unconformity at the base of the Oligocene section. Marine transgression with minor regressions continued in conjunction with regional basin tilting to the west throughout the late Tertiary. Massive amounts of sediment were supplied into the basin during the Tertiary. The Miocene had the highest sedimentation rate, resulting in thicknesses greater than 3050 m (10,000 ft). These rocks are primarily shales with associated turbidites and large cut and fill channel systems.

The structural style of this period generally involves reactivation and accommodation features whose grain sizes are controlled primarily by the underlying Cretaceous structures overprinted with the effects of gravity sliding and compaction.

GREATER TAKULA AREA FIELDS

Structure

The structural and general large-scale stratigraphic features are best shown by seismic data. Two 3D seismic surveys have been acquired over the structures and cover 360 km^2 (Figure 3). A dip-oriented seismic line through Takula field and a time-slice mosaic at the Vermelha reservoir level illustrate the major elements (Figures 5 and 6).

The post-salt section consists of fault-controlled extensional rollover anticlines. They are generally salt cored and are geometrically constrained by faults soling into the Aptian salt section, which acts as a décollement.

The pre-salt section is difficult to map in detail because of poor seismic reflectivities, peg-leg multiples, and complex structure. Underlying Takula field is a local fault-controlled horst block complex. This high area creates a hinge line in the overlying post-salt section. The structures of Wamba, Takula, Numbi, and Vuko fields all lie along this hinge zone fairway on the west side of the Takula subbasin within the Lower Congo Basin.

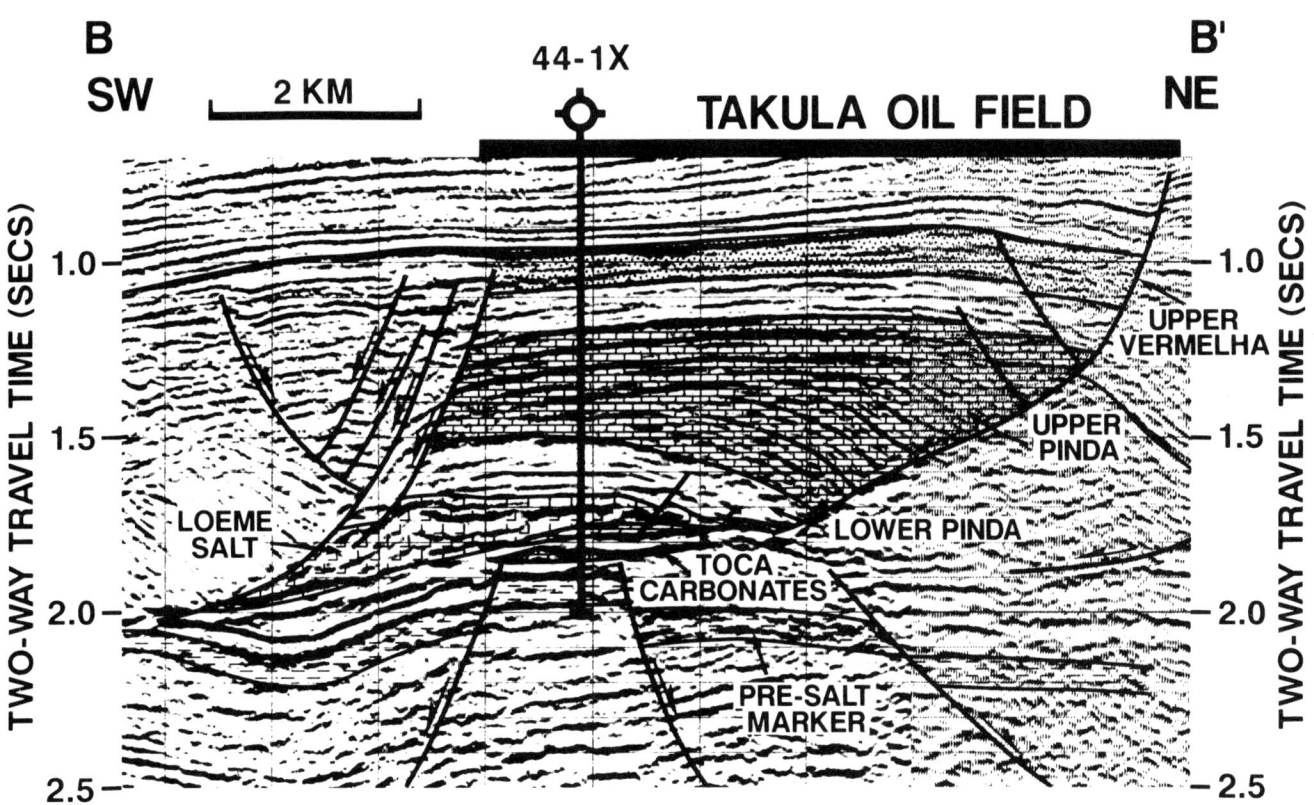

FIGURE 5. Dip-oriented seismic line B-B' through Takula field (see Figure 3 for location).

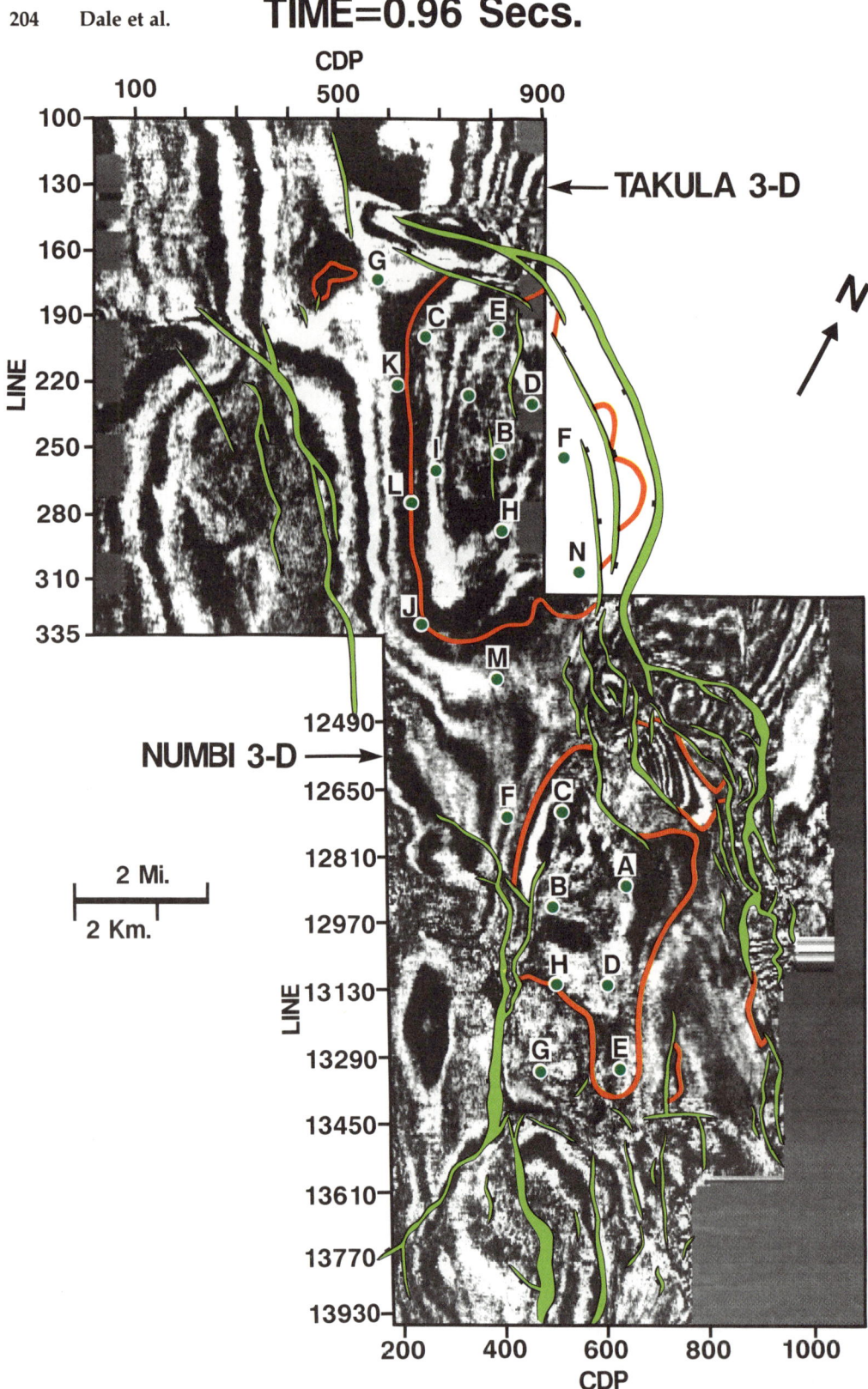

FIGURE 6. Time-slice mosaic at 0.96 sec two-way travel time from the Takula and Numbi 3D seismic surveys. The structure contour on the top of Vermelha, and significant faults, are shown (see Figure 3).

The position of the western-bounding post-salt fault system was controlled in part by the presence of the hinge area. The steepened west flank of the pre-salt high encouraged extensional downslope movement in the post-salt section as it was deposited (see also Figure 4). This resulted in more growth in the geologic section to the west of the hinge line. The amount of salt underlying the Pinda and Vermelha sections is also markedly different on either side of the hinge zone. In the Takula area the salt remains as elongate ridges. In between these ridges there are areas with little or no salt. Farther east the salt body is more intact, with almost complete preservation of a maximum of 1000 m of salt close to the coastline. To the west, less and less salt underlies the structures. The large amount of structural movement has led to both salt dissolution and smearing along the décollement surface.

The productive pre-salt section at Takula is from a high horst block that exposed Barremian carbonates during development of the Aptian unconformity. The result was enhanced porosity and permeability. This reservoir is relatively limited areally and covers only 485 ha (1200 ac). Only a minor structural closure is present, which cannot account for the hydrocarbon column present. As discussed below, updip stratigraphic closure is evident from well data.

The post-salt section accounts for 95% of the Greater Takula area reserves. The rollover anticlines involve the Pinda and Vermelha formations. The seismic dip line illustrates structural tightening and steepening with depth as well as stratigraphic thickening into the structure-bounding listric normal fault. The Vermelha Formation is the least faulted and has the greatest areal closure. The oil-bearing structural closures cover 107 km² (26,500 ac) and held an estimated 2.8 billion bbl of OOIP (Figure 3). The map and 3D seismic time-slice mosaic (Figure 6) combined with the dip-oriented seismic section (Figure 5) show the separations between the fields to be controlled by structural saddles or faults. Individual pools have independent and different gas-oil and oil-water contacts. Vuko and Numbi both have gas caps whereas the Takula and Wamba fields do not. This is related to hydrocarbon-migration pathways and the reservoir spillpoints.

Stratigraphy and Reservoir Geology

Reservoir characteristics for the Toca, Pinda, and Vermelha formations have been determined from cores and well logs. Core-derived petrophysical parameters have been tied to the formation-specific log responses. Thus realistic extrapolation of log data to a field-wide formation evaluation is possible for each stratigraphic interval.

Core descriptions have defined sedimentary lithofacies. The textural, mineralogical, and petrophysical properties of each facies have been determined and related to each rock type and its abundance. These data aid our understanding of the distribution of the reservoir rocks and their properties.

The following reservoir descriptions constitute a summary of the core and log work that has been done.

Toca Reservoir

In the Greater Takula area the pre-salt Toca carbonates have been developed only at Takula field. The Toca reservoir consists of two thick (up to 130 m, or 425 ft) carbonate units informally known as the upper and lower members of the Toca Formation.

The two carbonate units are independent reservoirs separated by an argillaceous section that acts as a permeability barrier (Figure 7). These reservoirs support oil columns in excess of 107 and 122 m (350 and 400 ft), respectively. Structural mapping shows a vertical closure of approximately 91 m (300 ft) at the top of the reservoir (Figure 8). This fact and well data support a stratigraphic seal to the east as the reservoir-quality carbonates grade laterally into their basinal shale equivalents. Production history indicates an effective porosity of 16 to 20% and local permeabilities up to 600 md. These reservoirs currently produce about 12,500 BOPD, and the seven producing wells cumulatively had produced 28 million bbl by 1990.

The individual reservoir units are difficult to resolve using seismic data (Figure 5). However, the reservoir section as a whole is recognizable, as is the erosive pre-Chela unconformity.

The Toca carbonates were deposited in a shallow lacustrine environment either on top of or around locally exposed or shoal areas. These predominantly algal carbonates have been diagenetically altered since deposition. The correlation section (Figure 7) illustrates the degree of local variability within the reservoir and the stratigraphic trapping mechanism provided by the basinal shale equivalents of the carbonates to the east. The rapid lateral changes in lithofacies control the heterogeneous nature of the reservoir properties. The carbonates consist primarily of recrystallized limestones, locally argillaceous, dolomitic, or partially silicified. Textures vary from wackestones to grainstones with fragments of algae, algal stromatolites, ostracods, and pelecypods. Minor amounts of very fine to medium crystalline interlocking dolomites are present.

Seismic and well data confirm that the pre-Chela unconformity played an important role in the development of diagenetic Toca porosity and permeability. The Barremian carbonates were exposed to either subareal or phreatic-zone erosion and dissolution, which resulted in the development of vugular and moldic porosity. In addition, both fenestral and fracture porosity have been recognized in the reservoir facies. The fractured, vugular rock volumes make a significant contribution to the overall productivity of the reservoir.

Pinda Reservoir

The Albian Pinda Formation consists of a transgressive sequence of continental shelf clastics and carbonate rocks. The environment of deposition spanned nearshore and tidal flat settings through the outer shelf. Although generally deposited in shallow water, the stratigraphy of the Pinda Formation is closely related to its structural development. Both carbonates and clastic lithofacies offer reservoir-quality rocks that often are limited in extent.

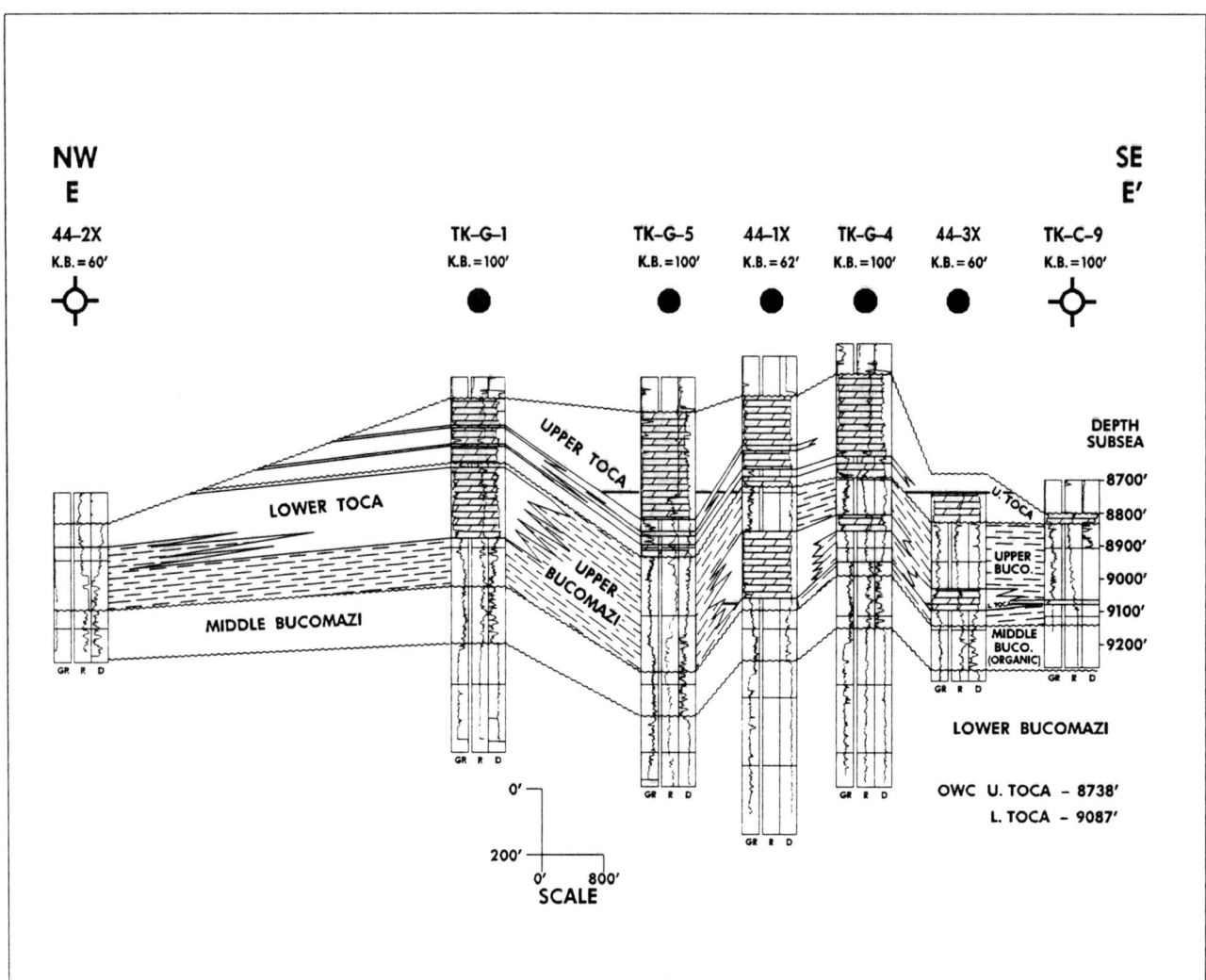

FIGURE 7. Structural-stratigraphic correlation section E-E' through the Takula field Toca reservoir (see Figure 8 for location).

The sandstones and dolosandstones provide the best reservoirs and were generally deposited in nearshore bars and shoals as well as shoreface and channel facies. Porosities and permeabilities in the sandstones average 22% and 150 md, respectively. The limestone facies generally contribute to the reservoir when the porosity and permeability have been enhanced by dissolution and/or dolomitization. Typical porosity values average between 10 and 20%, although permeabilities are relatively low (on the order of 10 to 20 md).

The Pinda Formation at Takula field has been informally divided into upper and lower members. A correlation cross section (Figure 9) and structure map on the top of Pinda reservoir (Figure 10) illustrate the general characteristics of the upper Pinda reservoirs.

The productive intervals consist primarily of dolomitic sandstones and microcrystalline dolomite. There are four separate upper Pinda reservoir units, which have independent oil-water contacts. They have an average porosity of 20% and an average permeability of 20 md.

The lower Pinda is currently drained by only two wells. Lithologically the reservoir is similar to the upper Pinda, although the sandstone facies tend to be more homogeneous. Three separate reservoirs have been identified, with a gross oil column of 235 m (780 ft).

To date the Pinda reservoirs have produced 21 million bbl of oil from an estimated OOIP of 400 MMSTB. Additional development drilling is planned in the near future.

Vermelha Reservoir

The Cenomanian Vermelha sandstone reservoir accounts for over 80% of the OOIP in the Greater Takula area. Not surprisingly, our most abundant database is for this reservoir; over 1200 m (4000 ft) of core have been collected.

The effective reservoir rock consists primarily of very fine to coarse-grained, clean, consolidated to poorly consolidated arkosic quartzose sands. Good intergranular

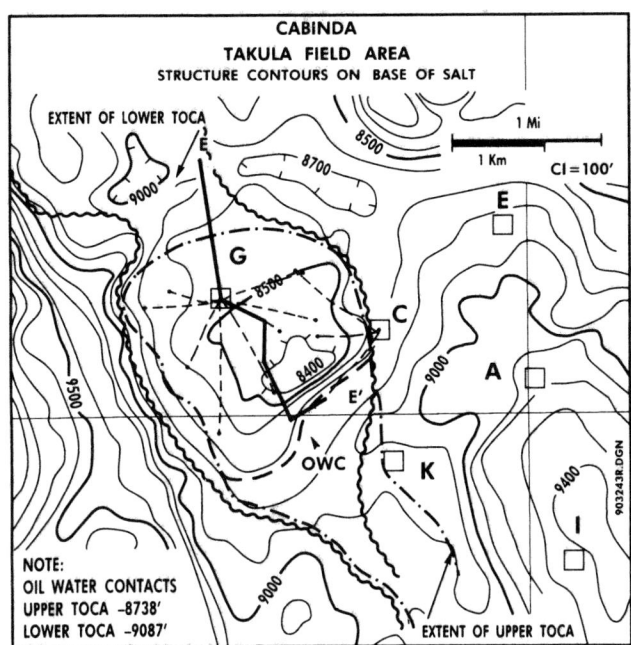

FIGURE 8. Map showing the depth structure on the Toca carbonate and the stratigraphic extent of the reservoirs. Locations of cross section E-E', production jackets, and well penetrations are indicated.

porosity averages about 25% and is as high as 35%. Permeability and water saturation average 1 darcy and 21%, respectively. These primary reservoir rocks grade laterally into well-cemented dolomite sandstones with porosities of approximately 10%. Nonreservoir facies within the Vermelha are mostly dolomitic siltstones and argillaceous dolostones with minor amounts of shale. The Vermelha reservoir rocks formed in a coastal marine environment in Cenomanian time. Relative sea level remained more or less constant with respect to the developing rollover structures, yielding a thick stacked sequence of nearshore deposits.

Dip- and strike-oriented correlation cross sections C-C' and D-D' (Figures 11 and 12) show the relationships within the formation. Based on log characteristics and reservoir production data, the reservoir has been divided into 11 informal units (S1 to S11). The sedimentary packages within this framework are relatively continuous laterally although local heterogeneities are present.

The Vermelha reservoir is divided into upper and lower reservoir units at the top of the S8 sandstone (Figures 11 and 12). They are separated by a thick section of carbonates and shales throughout the Takula field. This permeability barrier supports the two separate oil-water contacts observed but must pinch out off structure because pressure communication is evident between the two reservoirs. The lower Vermelha is oil bearing only in Takula field and consequently holds only about 5% of the Vermelha reserves.

The best reservoir rocks were deposited in foreshore (beach), tidal channel, and offshore marine bar environments. Lower-porosity, lower-permeability rocks that can act as seals were deposited in either shallow lagoonal/tidal flat or deeper water offshore environments.

Figure 13 shows core photographs of the tidal channel and foreshore/beach facies. These facies have average porosities over 25% and horizontal permeabilities in excess of 1 darcy. Vertical permeabilities tend to be lower than horizontal permeabilities. The foreshore/beach facies comprises very well sorted, fine to very fine sandstones that are horizontally bedded. Thin, coarsening-upward sequences that repeat many times over through some intervals are observed. The porosities and permeabilities of the good reservoir sandstones generally depend on the abundance of dolomite cement (Figure 14). Petrographic examination reveals a very fine, well-sorted sandstone composed primarily of quartz and feldspar framework grains and 4% heavy minerals. The dolomite cement composes 14% of the sample. The porosity and permeability are approximately 28% and 1.5 darcys, respectively, and are controlled primarily by the dolomite cement.

The tidal channel facies consists of medium sorted coarse to very fine sandstones. Grain-size variations were determined by the energy of the depositional environment. The primary sedimentary structures are crossbedding and horizontal bedding, which are generally visible in the finer sandstones. Occasionally the sandstones are massive with some bioturbation. The framework grain lithologies are very similar to the foreshore facies. Again, dolomite is the major cement and pore-occluding material (Figure 15). The more poorly sorted intervals generally have lower porosities and permeabilities. The porosity and permeability average 20 to 25% and 0.5 darcy, respectively, for this facies. A section of core description from the 57-3 well illustrates a typical vertical section through the reservoir (Figure 16). Observed sedimentary structures, lithologies, and interpreted environments of deposition are shown. The major distinction between reservoir and nonreservoir rock is in the average grain size and the degree of reworking through bioturbation. A thick, stacked lagoonal sequence is evident between 3351 and 3376 ft (1021 and 1029 m). Such sequences can act as vertical permeability barriers between reservoir sandstone units.

Lower-porosity, lower-permeability, nonreservoir-quality rocks were mostly deposited in the lower-energy environment of lagoonal or tidal flat settings. These rocks are generally siltstones, sandy dolomudstones to dolo-wackestones, and shales. The units can be laminated or bioturbated and often contain plant fragments and shell remnants. Porosity varies between 5 and 15%, and permeability ranges from 0.1 to several md.

Reservoir Fluid Properties

The crude oils produced in the Greater Takula area have low-viscosity and high-temperature pour points. Table 1 summarizes the characteristics of the crudes by reservoir in Takula field.

The formation water in both the pre- and post-salt sections is saline, with an equivalent NaCl concentration of 20.8%. The reservoir temperatures are closely modeled

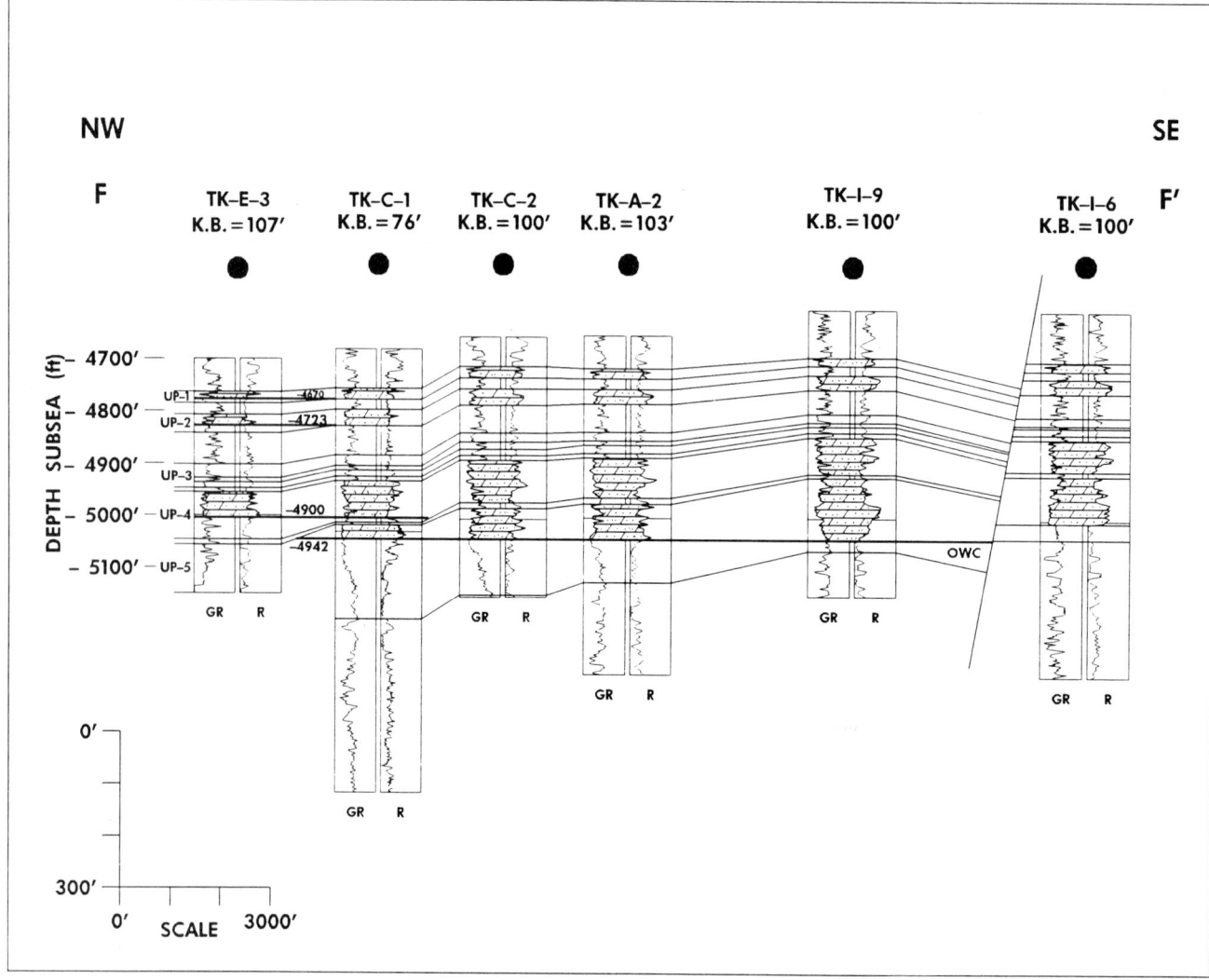

FIGURE 9. Structural-stratigraphic correlation section F-F' through the Takula field Pinda reservoir (see Figure 10 for location). Effective reservoir-quality rock is indicated within the correlation framework (patterned area between logs).

by a temperature gradient of 2.7°C/100 m (1.5°F/100 ft) and an annual mean surface temperature of 27°C (80°F).

Source Rocks and Hydrocarbon Migration

All the crudes produced in the Greater Takula area were sourced from the prolific source rocks of the pre-salt middle or "organic" Bucomazi Formation. These source rocks contain type 1 lacustrine kerogens. They are extremely rich; some samples contain in excess of 20% TOC and have an average source potential index of 46 tons HC/m^2 (Demaison and Huizinga, in press). This is one of the richest source rock intervals found worldwide. Directly underlying Takula field, geochemical analysis indicates that the source rock is marginally mature but passes quickly into the generative oil window. Therefore, it is adequate that oil migration only has to take place over short distances to charge overlying structures. The Chela unconformity sandstones make an excellent conduit and gathering system for oils migrating vertically from the pre-salt section. The sands are laterally extensive, and structural dip is more or less monoclinal to the west. When salt is not present, post-salt faults (which sole into a zone about the unconformity) can act through time as conduits into the post-salt structures. If a continuous salt layer is present between the source rocks and post-salt reservoirs, oil-migration pathways may not exist for charging the structures. The Greater Takula area structures are ideally situated to receive hydrocarbons generated in the adjacent basins. The structures all lie above "windows" in the salt and are bounded by faults that sole into the Chela unconformity.

Development Program

The discovery of the Vermelha oil pool at Takula field in 1980 triggered the large development project that is still ongoing. By December 1982 the first production from

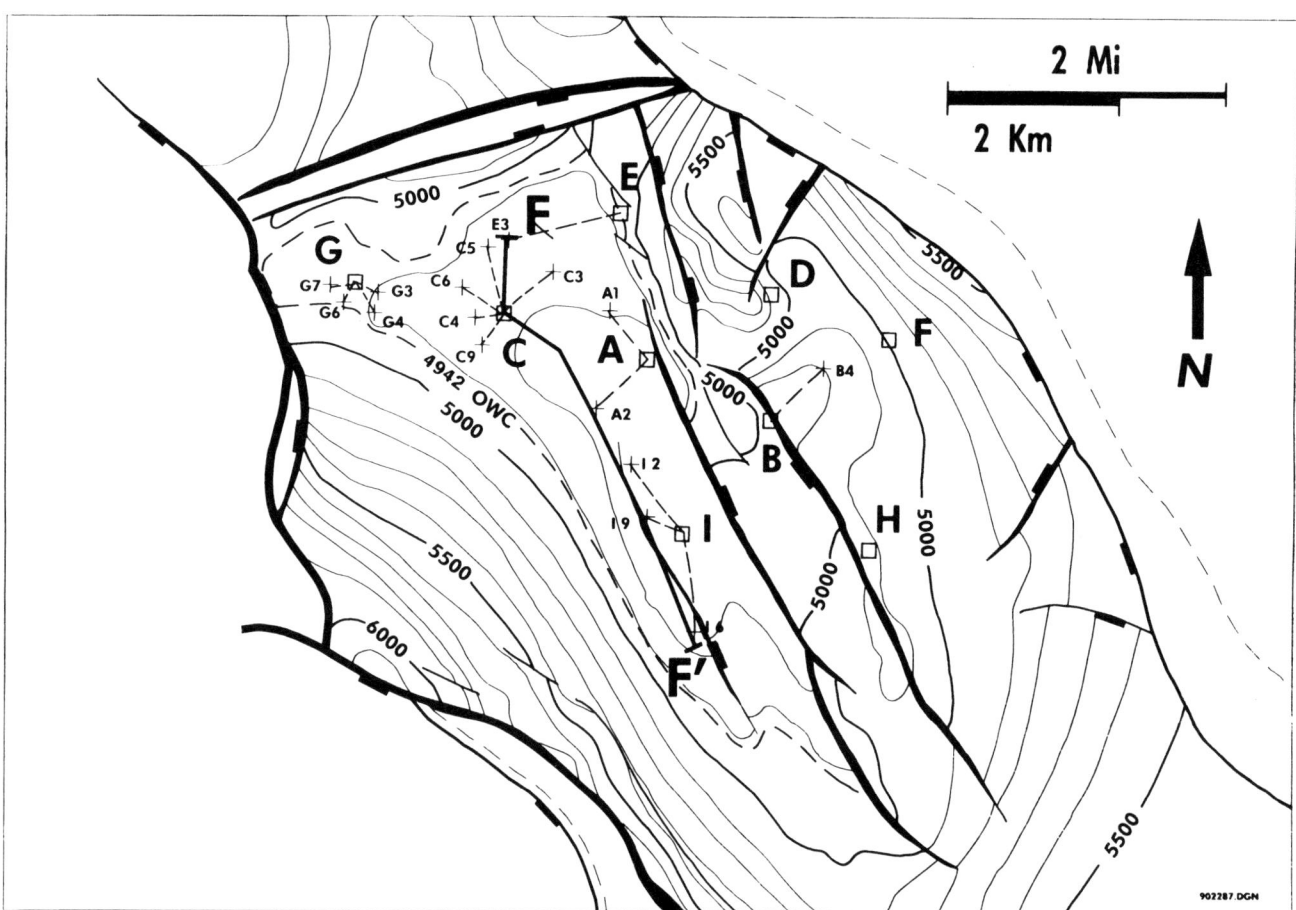

FIGURE 10. Map showing the depth structure on the top of the upper Pinda reservoir. Locations of cross section F-F', production jackets, and well penetrations are indicated (contour interval = 100 ft).

Takula was onstream. Exploration efforts in the intervening time led to three important discovery wells along the structural trend southeast and north of Takula. The discoveries of Numbi and Wamba occurred in 1982 and Vuko in 1983. With these discoveries, the major producing trend of the Greater Takula area had been found. Throughout the 1980s, development drilling work continued. Two 3D surveys were shot to cover the area. In 1984 the Takula survey was acquired and in 1987 the Numbi/Vuko portion was added to provide detailed structural mapping of the producing horizons (see Figure 3). Stratigraphic variation and associated petrophysical parameters were incorporated from log and core analysis.

At the end of 1990, 75 wells were producing from Vermelha reservoirs, ten wells from the Pinda, and seven from the Toca. Initial production from all these wells was from primary solution and/or expansion gas cap drive. As rates decline, gas lift is used to stimulate the wells. Primary recovery estimates for the Vermelha are about 25% of the OOIP. With primary recovery alone, the peak production output would already be passed. To boost the ultimate recovery, secondary recovery through waterflooding is planned. Reservoir simulation has been used to help predict reservoir performance limits. This analysis indicates that in excess of 35% of the OOIP may be recoverable from the Vermelha using secondary recovery methods (Figure 17). Thirty water injection wells were completed by mid-1990. The water injection platform (WIP) was commissioned in late 1990. This central injection platform will also service the Numbi field through flow lines (Figure 18).

Production and Export Facilities

The large areal extent of the fields, the depth of the Vermelha reservoir, and environmental considerations dictated the production scheme to be used. Figure 18 shows a schematic of the facilities planned for completion by mid-1991. By year end 1990, all the facilities except the Takula "O" production jacket were installed.

The production jackets are four-pile structures that can support up to nine wells used for either production or water injection. The jackets are connected back to the central gathering stations and the artificial lift or water injection platforms. A total of 25 production jackets are planned: 14 for production alone, five solely for water injection, and five in the Numbi field for joint water

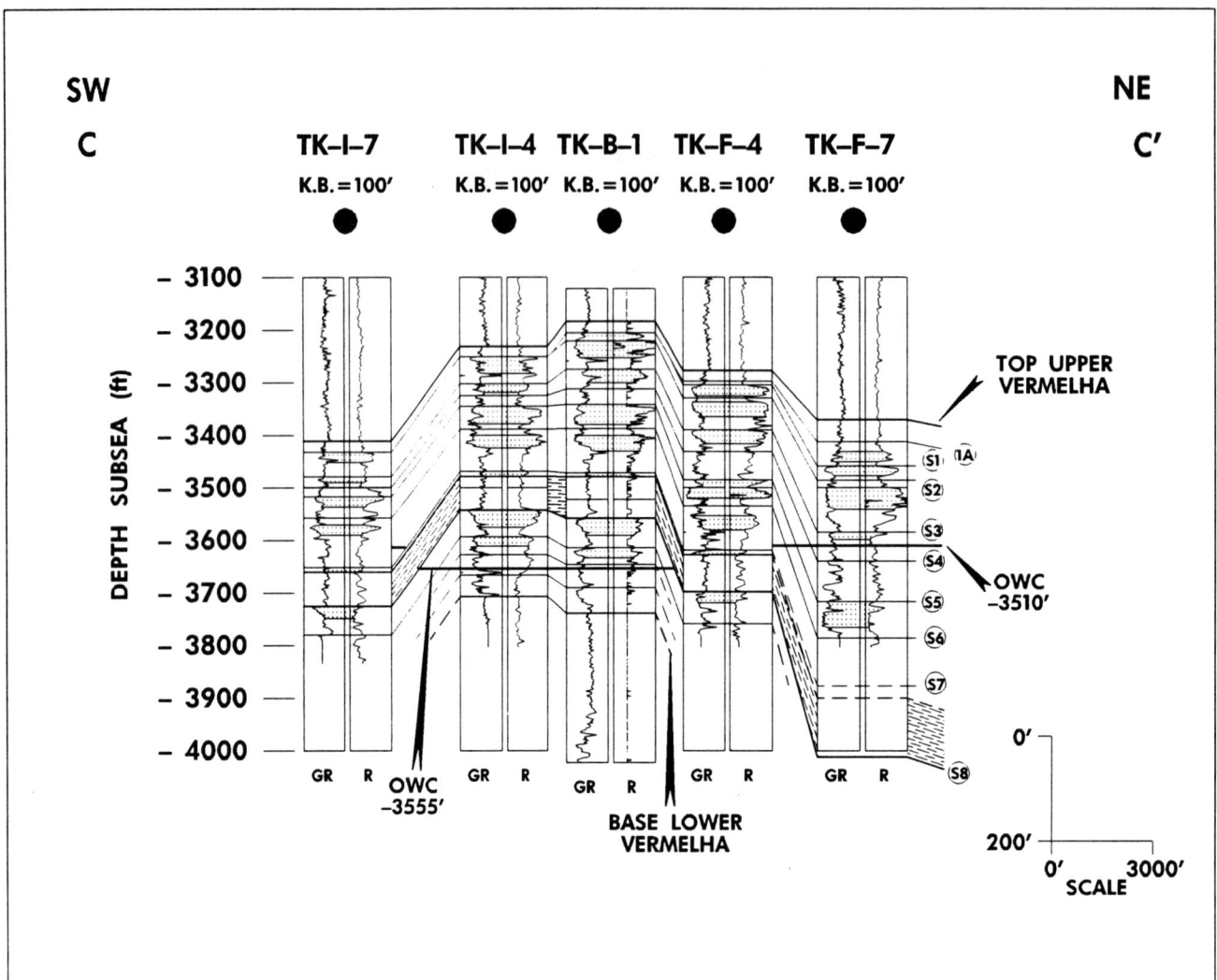

FIGURE 11. Structural-stratigraphic correlation section C-C' of the Vermelha reservoir in a dip orientation through Takula field (see Figure 3 for location). Effective reservoir-quality rock is indicated within the correlation framework (patterned area between logs).

injection/production. By the end of 1990, these facilities were supporting 92 producing and 30 water injection wells. The water injection facilities have the capacity to handle in excess of the 350,000 BWPD planned for Vermelha injection. With the gathering systems completed, the facilities will have the capacity to handle production rates of approximately 400,000 BOPD. To improve operations efficiency, offshore crew accommodations are planned for 1991. This will alleviate transportation logistics to and from Malongo Base 30 miles west. Crude from Vuko field is moved to the Malongo fields via flowlines and then into the export line through Malongo terminal. Undoubtedly, additional reserves both in and around the fields will be added through step-out drilling and smaller satellite fields. The Greater Takula area infrastructure will enable expedient development of such reserves into the export stream.

The majority of the crude produced in the Greater Takula area is exported via offshore loading facilities. Two SPMs (single point moorings) are utilized. One berth is continuously occupied by a storage tanker, the Chevron Ocean, with a capacity of 2.2 million bbl. The second berth is used for export loading of the stored crude when weather conditions prevent side-by-side export from the storage tanker.

ACKNOWLEDGMENTS

The authors would like to thank SONANGOL (Sociedade Nacional de Combustiveis de Angola), Chevron, and AGIP for permission to publish this paper. Interpretations presented herein are those of the authors and do not necessarily correspond with the views of all parties within

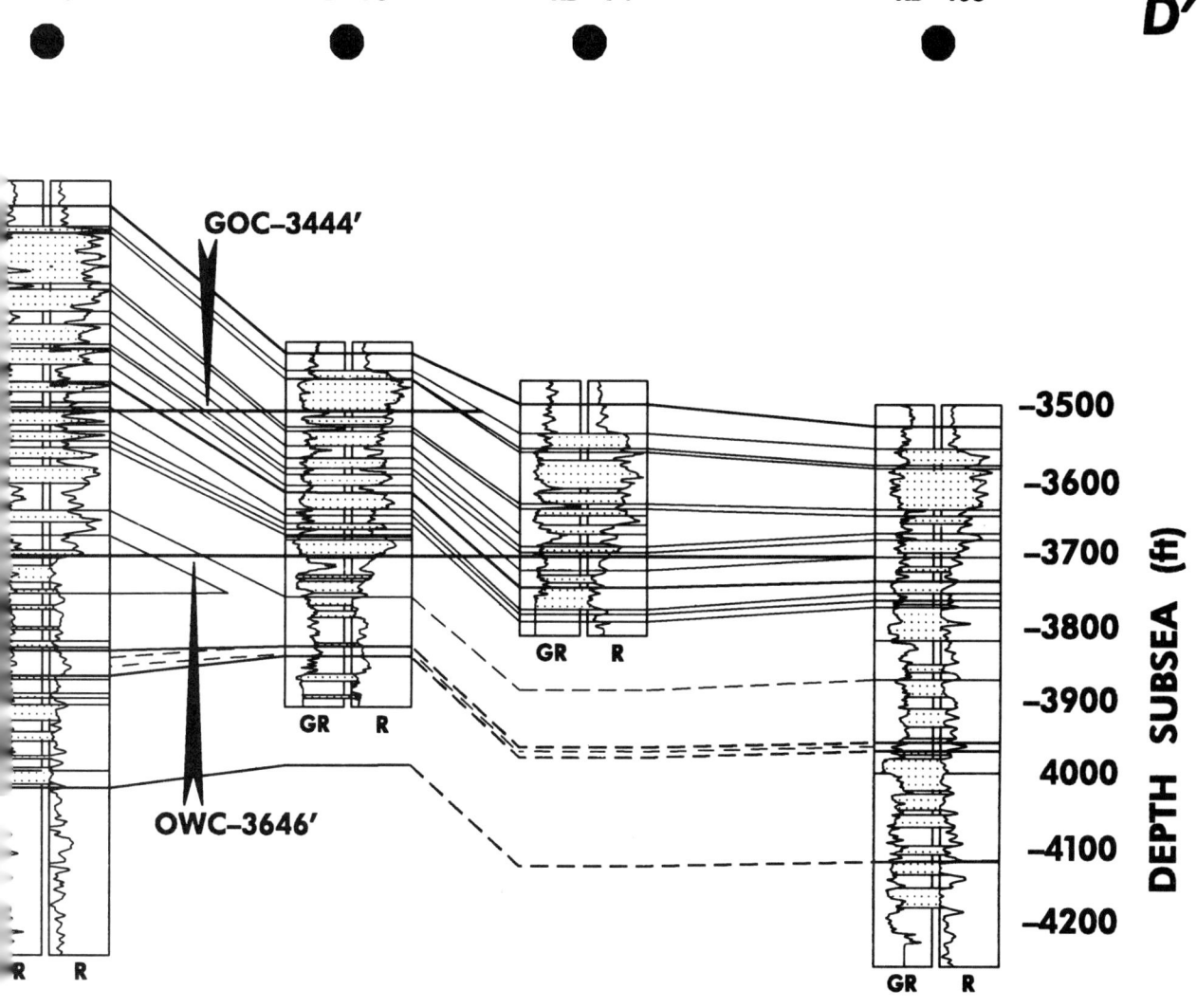

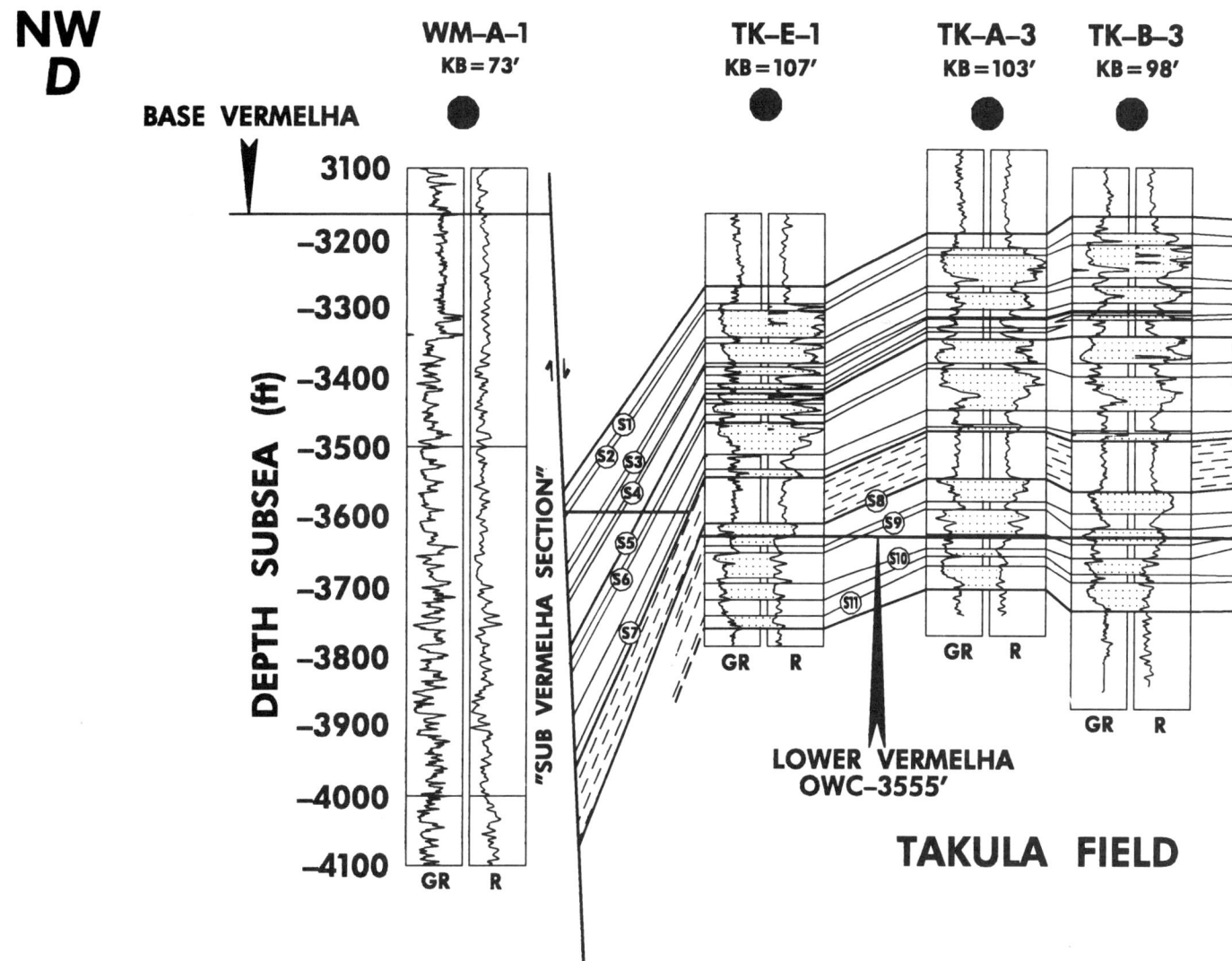

FIGURE 12. Structural-stratigraphic correlation section D-D' of the Vermelha reservoir in a strike orientation through Wamb Figure 3 for location). Effective reservoir-quality rock is indicated within the correlation framework (patterned area between logs

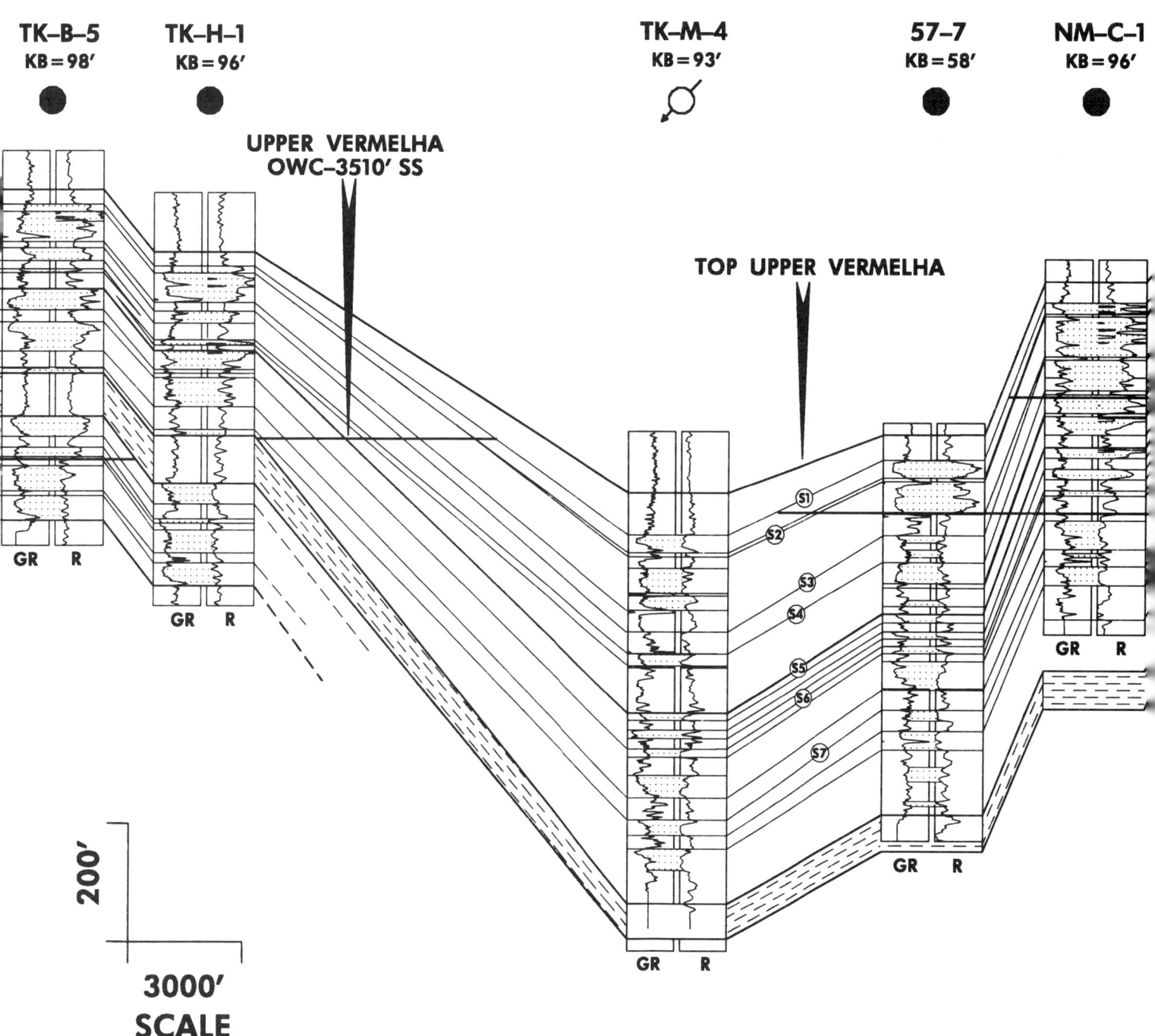

a, Takula, and Numbi fields (see

FIGURE 13. Core photographs of the tidal channel and foreshore/beach facies from the 57-3X well location: (A) 3385.6–3386.2 ft, S-5 interval, tidal channel. Grain size generally decreases upward. Large-scale cross-bedding is prominent in the coarser sandstone, and small-scale cross-bedding or ripples are visible in the finer sandstone. (B) 3390–3390.5 ft, S-5 interval, tidal channel. Grain-size varieties, cross-bedding, and horizontal bedding indicate varying depositional energies common to tidal channel settings. (C) 3632–3633 ft, S-8 interval, upper shoreface and overlying foreshore/beach. A sharp contact is evident between the two units. The foreshore/beach is very well sorted, horizontally bedded, very fine sandstone underlain by bioturbated argillaceous siltstone of the upper shoreface.

212 Dale et al.

FIGURE 14. Photomicrographs from 3380.5 ft in well 57-3X. This example of the foreshore or beach facies clearly shows the very well sorted, very fine grained sandstone. Framework grains are quartz (Q), feldspar (F), heavy minerals (black), and mica (M). Small crystals of dolomite (arrows) are the dominant cement. Quartz and feldspar overgrowths are detectable. The evenly distributed pore space (blue) averages 28%, and measured permeability is 1.5 darcys. Note that the thin sections have been impregnated with blue-dyed epoxy and stained with alizarin red-S in order to distinguish calcite (which takes on a pink tint) from dolomite (which is not reactive). The high-magnification area is from the center of the low-magnification plate.

FIGURE 15. An example of the tidal channel facies. The moderate sorting is evidenced by framework grain size diversity. Once again, dolomite (arrows) is the major pore-occluding material. Quartz and feldspar overgrowths are less abundant. Some of the large rounded pores (blue) are attributed to framework grain dissolution. Porosity, 16%; measured permeability, 21 md. The high-magnification area is from the center of the low-magnification plate.

the joint venture. We would like to thank the Chevron and Gulf employees past and present whose contributions and criticism made this paper possible. In particular, T. A. Anderson, L. P. Caflisch, E. L. Couch, J. E. Ellithorpe, C. J. Harrison, K. W. Knutson, E. P. Maxwell, and P. J. Vita offered contributions, encouragement, and review. Finally, we acknowledge the tremendous efforts of the Malongo facility personnel who make the project work.

REFERENCES CITED

Asmus, H. E., and F. C. Ponte, 1973, The Brazilian marginal basins, in A. E. Nairn and F. G. Stehli (eds.), The ocean basins and margins, v. 1, The South Atlantic: New York, Plenum, p. 87–133.

Brice, S. E., M. D. Cochran, G. Pardo and A. D. Edwards, 1982, Tectonics and sedimentation of the South Atlantic

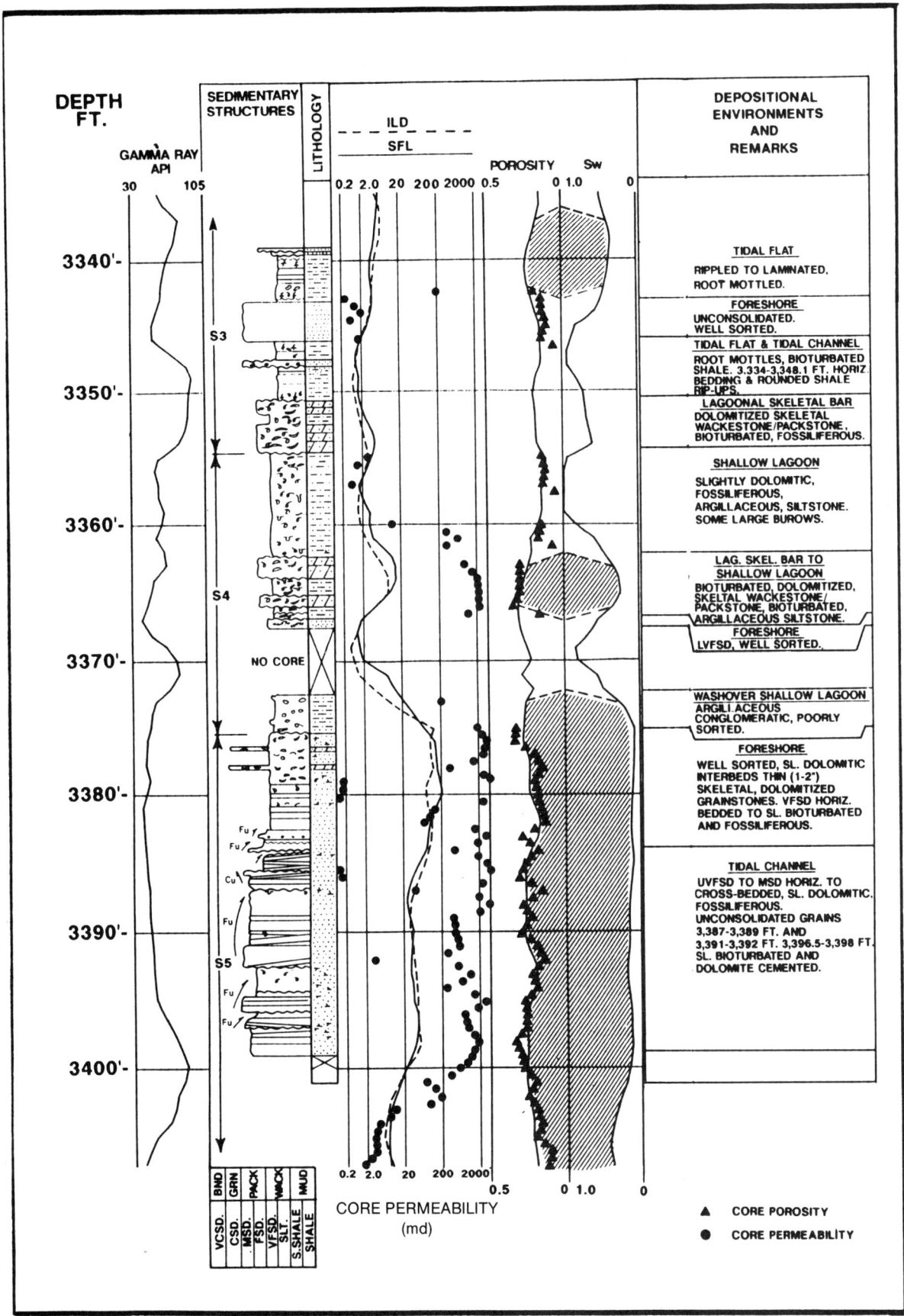

FIGURE 16. Section of core description from 3339 to 3399 ft in well 57-3X for parts of Vermelha sandstone units S3, S4, and S5. The hydrocarbon log showing net effective reservoir is included for comparison.

214 Dale et al.

TABLE 1. Reservoir fluid properties at Takula field.

Reservoir	API	GOR @ bubble point (scf/bbl)	Bubble point pressure (psia)	Reservoir datum depth (ft subsea)	Viscosity (cp)	Original reservoir pressure (psi)	Formation volume factor (rbbl/stb)	Reservoir temperature (F)
Toca	33.8	761	3715	8726	0.41	4303	1.44	210
L. Pinda 2	33.3	420	2077	6373	1.16	3051	1.27	186
L. Pinda 1	33.3	420	2077	5934	1.16	2850	1.27	179
U. Pinda	34.2	407	1140	4842	1.15	2277	1.27	165
L. Vermelha	29.8	131	595	3356	3.87	1553	1.14	143
U. Vermelha	33	307	1065	3211	1.75	1525	1.19	133

Note: All data are at original reservoir conditions.

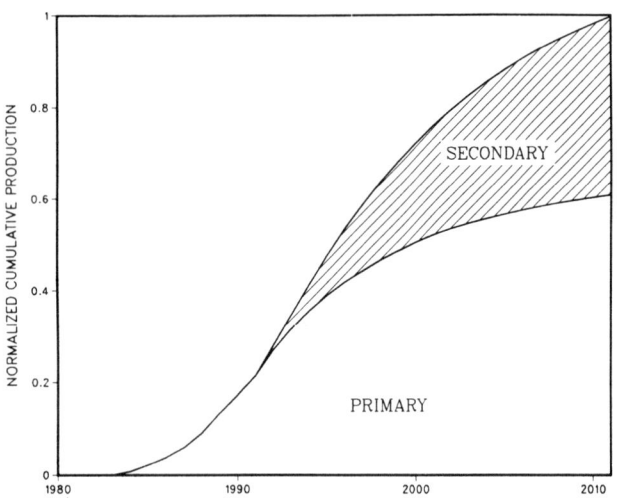

FIGURE 17. Anticipated production profile for the Vermelha reservoirs at Takula and Numbi, assuming primary solution gas drive and secondary waterflood.

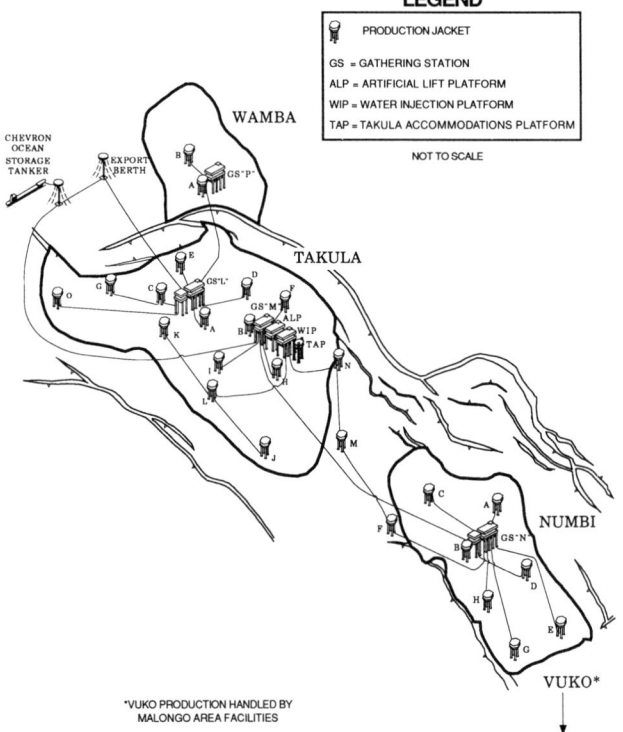

FIGURE 18. Schematic showing the production facilities used in the Greater Takula area.

rift sequence: Cabinda, Angola, *in* J. S. Watkins and C. L. Drake (eds.), Studies in continental marine geology: AAPG Memoir 34, p. 5-18.

Belmonte, Y., P. Hirtz, and R. Wenger, 1965, The salt basins of the Gabon and the Congo (Brazzaville)—tentative paleogeographic interpretation, in salt basins around Africa: The Institute of Petroleum, London, p. 55-74.

Caflisch, L., 1979, Oil exploration in pre-salt and post-salt sequence of West Africa—results in Cabinda: AAPG Annual Convention, Houston.

Campos, C. W. M., F. C. Ponte, and K. Miura, 1974, Geology of the Brazilian continental margin, *in* C. A. Burk and C. L. Drake (eds.), The geology of the continental margins: New York, Springer Verlag, p. 447-461.

Demaison, G., and B. J. Huizinga, in press, Genetic classification of petroleum basins: AAPG.

Gibbs, A., 1987, Development of extension and mixed-mode sedimentary basins, *in* M. P. Coward, J. F. Dewey, and P. L. Hancock, (eds.), Continental extensional tectonics: Geological Society Special Publication No. 28, p. 19-33.

Lehner, P., and P. A. C. DeRuiter, 1977, Structural history of Atlantic margin of Africa: AAPG Bulletin, v. 61, p. 961-981.

McHargue, T. R., 1991, Tectonostratigraphic development of proto-South Atlantic rifting in Cabinda, Angola: a petroliferous lake basin, *in* B. J. Katz and B. R. Rosendahl (eds.), Lacustrine basin exploration—case studies and modern analogues, AAPG Memoir.

Tissot, B. P., G. Demaison, P. Masson, J. R. Deltril, and A. Combaz, 1980, Paleoenvironment and petroleum potential of Middle Cretaceous black shales in Atlantic basins: AAPG Bulletin, v. 64, p. 2051-2063.

Chapter 14

Sanaga Sud Field, Offshore Cameroon, West Africa

Robert J. Pauken

*Mobil New Exploration Ventures Co.
Dallas, Texas, U.S.A.*

ABSTRACT

The Sanaga Sud field is located 10 km northwest of the coastal town of Kribi in the central portion of the Douala basin. The Sanaga Sud A-1 discovery well was drilled in 1979 to test an eroded, paleotopographic high block composed of Lower Cretaceous sands and shales and overlain by Upper Cretaceous shales. The structure contained a prominent near-horizontal seismic amplitude event.

Drilling results showed that the amplitude event was a gas-water contact. Two appraisal wells, Sanaga Sud A-2 and A-3, were drilled in 1981. All three wells tested gas and condensate. Gross pay thickness averages 250 m, with an average porosity of 23% and an average permeability of 142 md. Total recoverable hydrocarbons for the field are estimated to be 900 bcf of gas and 4.5 million of bbl of condensate.

INTRODUCTION

The Sanaga Sud gas field is located in the central Douala basin, offshore Cameroon, northwest of the coastal town of Kribi and southwest of Mt. Cameroon (Figure 1). The northern Douala basin lies north of the Sanaga River and extends to the Cameroon volcanic line. The southern Douala basin lies between the Sanaga River and the Corisco arch, near the Gabon/Equatorial Guinea border. The basin is bounded on the east by basement outcrops that occur close to the shoreline.

Most of the Sanaga Sud field lies in the H-38 permit area, which is under concession to the Mobil/Total/Ocelot/Damson group of partners. The H-38 concession, following a recent partial relinquishment, presently covers 583 km². A small portion of the field lies in the Londji permit, the northern subdivision of a larger concession area referred to unofficially as the Kribi block. The Kribi block is the remaining portion of the H-17 concession originally granted to Total and Mobil. The three subdivisions of the Kribi block are La Lobe' in the west and Londji and Batanga in the east. The Kribi block has an areal extent of 1094 km² and has had 15 wells drilled within its boundaries (Figure 2).

Ownership of the H-38 concession is as follows: Mobil, 54.1%; Total, 18.0%; Ocelot, 15.6%; and Damson

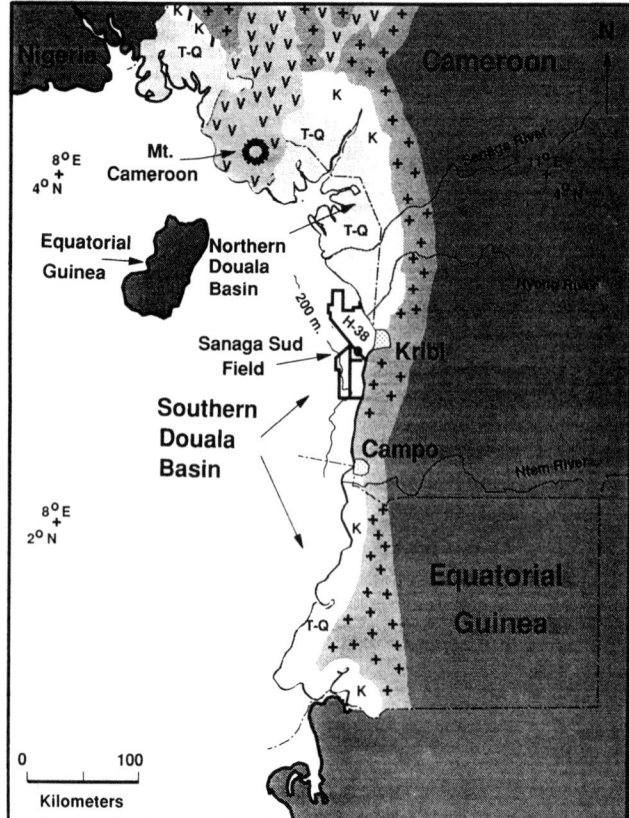

FIGURE 1. Location map for the Sanaga field in the Douala basin.

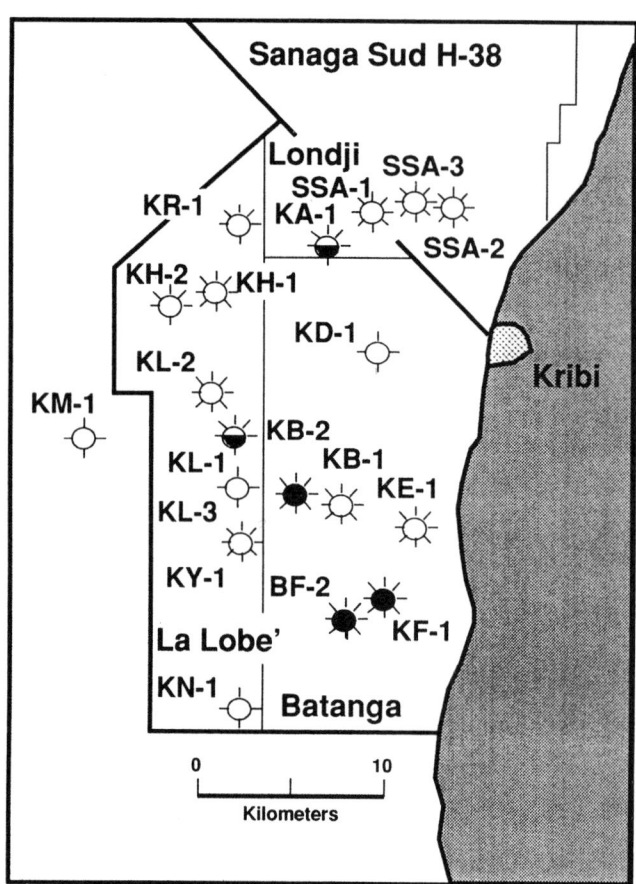

FIGURE 2. Map showing concession outlines and exploration wells in the Kribi block.

12.3%. Ownership of the Londji portion of the field is: Mobil, 25%; Total, 25%; and Société Nationale des Hydrocarbures, 50%.

The entire field encompasses an area of 1210 ha (2990 ac) under closure and at present is defined by a 1-km grid of seismic, one discovery well, and two appraisal wells. The field is presently shut-in.

REGIONAL GEOLOGY AND BASIN DEVELOPMENT

The Douala basin is one of several divergent margin basins along the West African coast. Many of these basins, along with their counterparts on the conjugate margin of South America, contain significant accumulations of oil and gas. The basins along the southwest African margin were formed as the result of the rifting of Africa and South America during the breakup of Gondwanaland.

Continental rifting along the southwestern African margin began during the Late Jurassic and lasted until the onset of seafloor spreading in Albian-Cenomanian time. During this time interval, thick sequences of Lower Cretaceous, fluvial-lacustrine and restricted marine clastic sediments and evaporites were deposited in the developing basins. These sediments ultimately became the source and reservoir facies for a large number of significant hydrocarbon accumulations in the various basins of the South Atlantic.

In the Douala basin, the onset of the drift phase in late Albian time was accompanied by tectonic uplift and faulting. Structural movement, combined with a relative lowstand of sea level, resulted in severe erosion of the uplifted fault blocks along the margin of the rift. Subsequent postrift subsidence along with a major rise in eustatic sea level resulted in the late Albian erosional surface (unconformity) being covered with thick onlapping pelagic shales of the transgressive systems tract. These shales became the seal facies for the gas accumulation in the Sanaga Sud field. During the deposition of the uppermost portion of this shale sequence, the deepest portion of the Lower Cretaceous section began to reach sufficient thermal maturity for oil generation to begin. Maturation continued to migrate upward into the stratigraphically younger uppermost Lower Cretaceous and Upper Cretaceous sections during the deposition of the overlying Tertiary shale and carbonate sequence.

The regional cross section shown in Figure 3, oriented perpendicular to the coastline, depicts the regional structural and stratigraphic setting of the area in which the Sanaga Sud field is located. The Sanaga Sud structure is shown buried under a relatively thin Upper Cretaceous and Tertiary sedimentary veneer, but contains

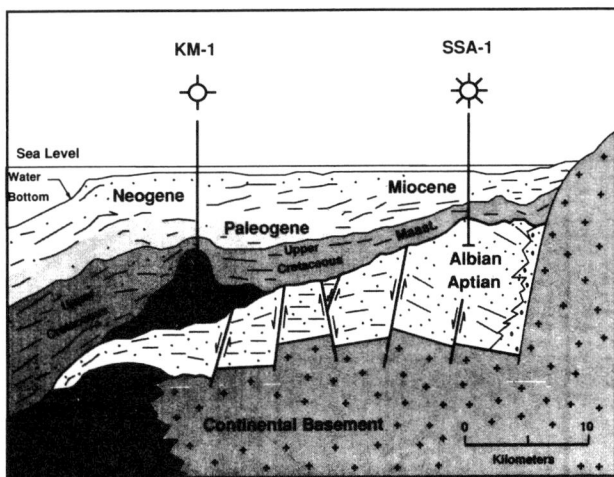

FIGURE 3. Regional east-west cross section through the Douala basin at about 3°N latitude (see Figure 2 for well locations).

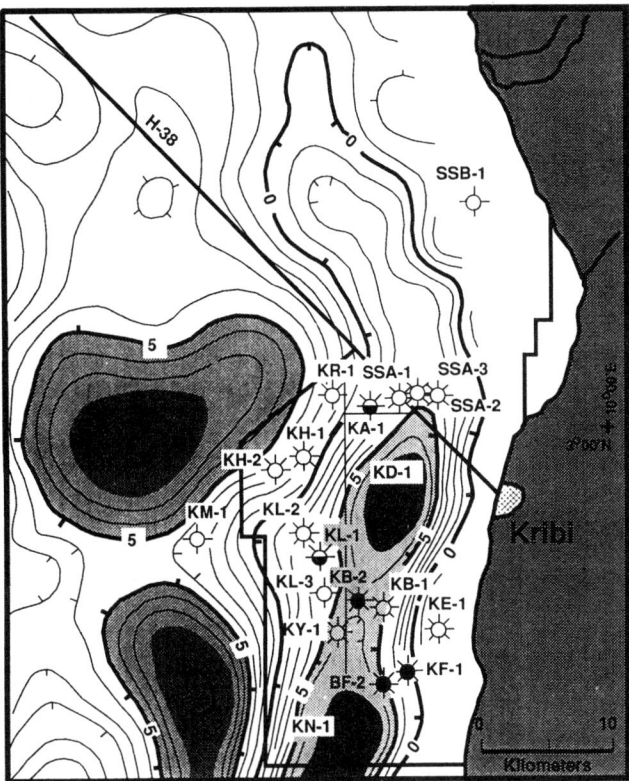

FIGURE 4. Filtered residual Bouguer gravity map in the Kribi–Sanaga Sud area.

a thick Lower Cretaceous section. Salt, known to be present seaward of the field, may extend farther shoreward than shown, but direct evidence for this has not been observed.

Reservoir and source facies consist of Lower Cretaceous continental, marginal marine, and open marine clastic sediments. These sediments are sealed by Upper Cretaceous marine shales. This sequence of lithologies, combined with Lower Cretaceous structural development, has resulted in a geologic setting conducive to generation, migration, and entrapment of hydrocarbons. Only the Lower Cretaceous synrift and Upper Cretaceous postrift sequences are considered to be thermally mature and therefore prospective sources of hydrocarbons in this part of the basin.

The Douala basin is the northernmost of the West African salt basins. The known occurrence of salt extends northward from the Walvis ridge to just south of 3°N latitude, where salt was encountered in the Kribi Marine-1 well (Figure 2). The age of the salt is generally assumed to be Aptian, because this is the age of the salt in offshore Gabon, Congo, and Angola to the south. The salt may have contributed to structural development seaward of the Sanaga Sud field. An example might be the Kribi-H feature.

GRAVITY

Because of the lack of deep resolution in the seismic data, basement topography and the location of the thickest sediment accumulations were estimated by mapping and modeling Bouguer gravity. The residual gravity map presented in Figure 4 depicts a coast-parallel trend of gravity minimums 6 km west of the coastline. These gravity lows are interpreted as containing thick accumulations of synrift sediment. Gravity modeling in the area indicates the presence of 8 to 10 km of sedimentary fill.

The strong gravity gradient between these lows and the prominent gravity highs to the west, which are thought to be buried volcanic seamounts, is interpreted as being the location of transitional crust associated with the continent/ocean boundary. This gradient follows the coastline at a distance of about 10 to 15 km in the Kribi–Sanaga Sud area.

EXPLORATION HISTORY

Exploration in the Kribi area was initiated in 1969, in the H-17 permit, with the drilling of Kribi Marine-1 as an anticlinal salt feature. Following this, there was a 10-yr drilling hiatus until 1979, when the Mobil/Total consortium drilled the Sanaga Sud A-1 discovery well in the H-38 concession (Figure 2). This stimulated a flurry of drilling activity, which peaked in 1981, when ten wells were drilled, including the two Sanaga Sud appraisal wells. Table 1 lists the wells drilled in the Kribi–Sanaga Sud area from 1969 to the present.

No exploration drilling has taken place since 1982. Since that time, Ph-40, the Campo block, has been relinquished and H-38 has been reduced in area.

STRATIGRAPHY

Exploration drilling results, geophysical interpretations, and application of rift theory in the Kribi–Sanaga Sud area

TABLE 1. Exploration well results in the Kribi–Sanaga Sud area.

Well name	Operator	Date	TD (m)	Results
Kribi Marine-1 (KM-1)	Mobil	4-69	4215	Dry
Sanaga Sud A-1 (SSA-1)	Mobil	9-79	2479	Gas and condensate
Kribi B-1 (KB-1)	Total	6-80	1420	Gas and condensate
Kribi E-1 (KE-1)	Mobil	8-80	1373	Gas
Kribi A-1 (KA-1)	Mobil	10-80	3475	Gas and oil show
Kribi H-1 (KH-1)	Total	10-80	2624	Gas and condensate
Kribi D-1 (KD-1)	Mobil	12-80	2463	Dry
Kribi L-1 (KL-1)	Total	1-81	3294	Gas and oil show
Sanaga Sud A-2 (SSA-2)	Mobil	2-81	2464	Gas and condensate
Kribi N-1 (KN-1)	Mobil	3-81	1295	Gas shows
Kribi F-1 (KF-1)	Mobil	5-81	1691	Oil and gas
Kribi R-1 (KR-1)	Mobil	7-81	3126	Gas show
Kribi L-2 (KL-2)	Total	7-81	3530	Gas
Sanaga Sud B-1 (SSB-1)	Mobil	8-81	1598	Dry
Kribi L-3 (KL-2)	Total	9-81	2838	Dry
Sanaga Sud A-3 (SSA-3)	Mobil	10-81	2081	Gas and condensate
Kribi Y-1 (KY-1)	Total	11-81	2316	Gas
Kribi B-2 (KB-2)	Mobil	1-82	1922	Oil and gas
Kribi H-2 (KH-2)	Mobil	4-82	3105	Gas shows
Batanga F-2 (BF-2)	Mobil	6-82	1479	Oil and gas

allow some important interpretations to be made about the tectonodepositional history of this portion of the Douala basin. Observation of seismic data, combined with gravity and magnetics modeling, indicates the presence of a significant untested sedimentary thickness below that penetrated by the deepest wells.

It is hypothesized that the older, undrilled portion of the stratigraphic section of the basin consists of Late Jurassic to Early Cretaceous continental deposits, possibly containing lacustrine source rock facies. Overlying the Jurassic is an undrilled thickness of continental to marginal marine clastics and evaporites of Aptian–Albian age. The diapiric salt drilled in the Kribi Marine-1 well, considered to be Aptian in age, is probably the stratigraphic equivalent of the Ezanga salt of the Gabon basin. The salt appears to have been deposited seaward of the main Lower Cretaceous depocenter.

The drilled portion of the Lower Cretaceous section becomes more marine upward. The lithofacies consist of coarse to fine clastics that become generally finer-grained westward, away from the paleoshoreline.

The Upper Cretaceous is primarily Maastrichtian pelagic shales, which prograded seaward over the Lower Cretaceous escarpment during a major transgressive event. These shales onlap a major unconformity developed between the Lower and Upper Cretaceous sections. The amount of time represented by this unconformity is estimated to be approximately 26 million years (Figure 5). During this time period, two relative lowstands of sea level appear to have contributed to the development of the top-Albian unconformity.

The Tertiary section also consists of dominantly Miocene marine shales, with minor amounts of carbonates. Another significant unconformity exists between the Paleocene and the Miocene in this area. The Paleocene is absent in some wells.

DEPOSITIONAL ENVIRONMENTS

The drilled portion of the Lower Cretaceous section records an upward transition of depositional environments from a continental to restricted marine setting to open marine shelf and deep water marine environments (Figure 6). This interpretation is based on studies of assemblages of fossils, including foraminifera, calcareous nannofossils, and palynomorphs. A second deepening-upward cycle is shown above the top Albian unconformity.

The Aptian–Albian syntectonic sequence present in the coast-parallel rift system in the Kribi-Sanaga Sud area was deposited by a variety of gravity flows. These gravity flows originated updip, not far from the paleoshoreline, and prograded seaward as submarine fans and fan-deltas. Syntectonic motion along the boundary faults of the rift reversed the primary depositional dip and resulted in dipping and thickening of the beds toward the continent by late Albian time (Figure 3).

A depositional model was constructed for this part of the basin by using various types of core analyses, seismic stratigraphic techniques correlated with gamma ray logs from the various wells, and the fossil-based paleoenvironmental interpretations discussed above. Sedimentation in the Sanaga Sud area was dominated by syntectonic, coalescing, and prograding submarine fans and fan-deltas deposited concurrently with the development of the continental rift system (Figure 7). Cores from all three Sanaga Sud wells and adjacent wells reflect this mode of deposition. The gamma ray response in all three wells is domi-

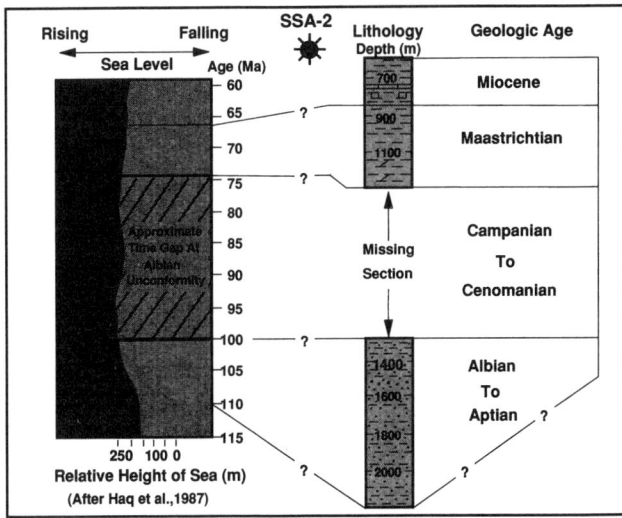

FIGURE 5. Age range of missing section relative to sea level at the Albian unconformity.

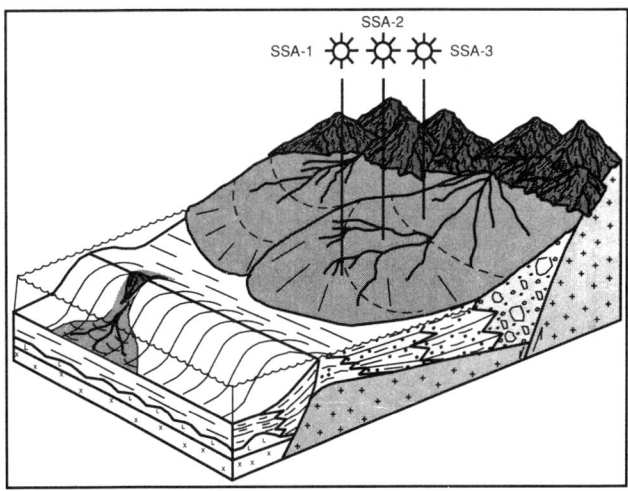

FIGURE 7. Block diagram depicting coalescing submarine fan deposition and facies relationships in the area of the Sanaga Sud field.

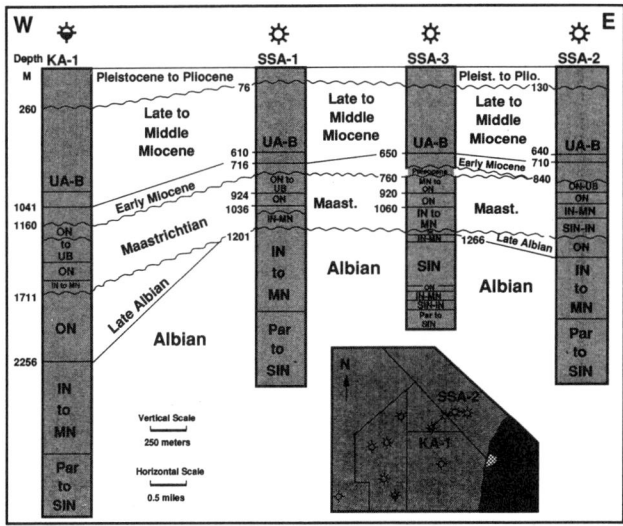

FIGURE 6. Paleoenvironmental interpretation of the major stratigraphic units in the Sanaga Sud wells. Environments include paralic (Par), shallow inner neritic (SIN), inner neritic (IN), middle neritic (MN), outer neritic (ON), upper bathyal (UB), and upper abyssal (UA).

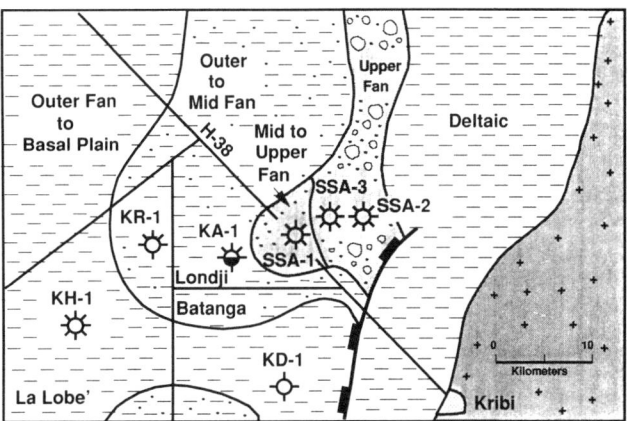

FIGURE 8. Lower Cretaceous lithofacies distribution in the Kribi–Sanaga Sud area.

nated by cylindrical to bell-shaped character, which is typical of channel-fill deposits that tend to become progressively finer grained upward. These observations, combined with the presence of planktonic foraminifera, argue for a submarine fan origin for the stratigraphic section in this region.

LITHOFACIES DESCRIPTIONS

Petrologic and petrographic studies of the Sanaga Sud cores were conducted by Fontaine (1982), Larson (1983), and R. J. Moiola of the Mobil Dallas Research Laboratory. The paragraphs below summarize the findings of the first two studies. The results from the Mobil Research Laboratory were utilized in Larson's and Fontaine's studies.

The Lower Cretaceous sandstone reservoir facies are classified as subarkoses and sublitharenites following the classification of Folk (1974). Accessory minerals include muscovite, garnet, kaolinite, and chlorite. Calcite and dolomite cements are significant locally. Individual sandstone units are up to 100 m in thickness and have porosities ranging between 20 and 25%.

Generally, the various lithofacies, as defined by core and log analyses, interfinger in a seaward direction from proximal conglomerates to sandstone and finally to shale, and represent upper fan, midfan, and outer fan settings, respectively (Figure 8). Minor disruptions of this pattern are due to different sources and rates of sediment supply, local slumping, and differential erosion of the rotated fault blocks. Four generalized lithofacies are recognized and are displayed graphically in Figures 7 and 8. Although the lithofacies interpretations rely heavily on gamma ray log patterns, these interpretations were also calibrated with core data where available. Figure 9 shows the depths and

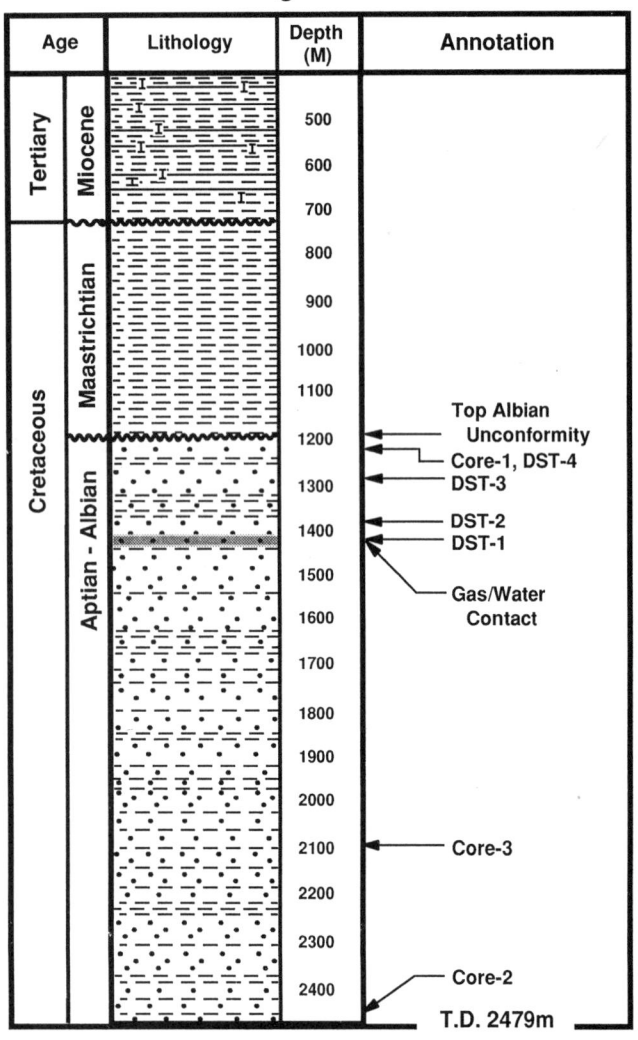

FIGURE 9. Stratigraphy, lithology, coring intervals, and drill-stem tests in the three Sanaga Sud wells: SSA-1, SSA-2, and SSA-3.

stratigraphic levels of coring and drill-stem testing. All cores reflect a progradational submarine fan mode of deposition.

Lithofacies 1 consists of conglomeratic sandstone containing large clasts composed of shale and metamorphic rock fragments. The clasts range up to 10 mm in diameter (Figure 10A). This rock type is present in cores 1 and 2 from the SSA-2 well (Figure 9) and represents deposition in the upper midfan or upper fan environment (Figure 8). Some intervals consist of pebbly sandstone with clasts up to 10 mm in diameter (Figure 10A).

Lithofacies 2 is represented by sandstones without the large clasts of lithofacies 1. These sandstones are typically subarkoses and litharenites. Slabs of these cores are photographically featureless and are not illustrated.

Lithofacies 3 consists of dominantly turbiditic shale units with cross-bedded silt and sand interbeds of varying thickness. In some cases the sand, silt, and shale cycles display the classic Bouma sequence with the "A" division missing (Figure 10B). This is interpreted as reflecting turbidite deposits in a distal fan setting.

Lithofacies 4 typically is a sandstone containing thin laminae of opaque organic material. The sandstone fraction has the composition of lithofacies 2.

PETROGRAPHIC ANALYSIS OF RESERVOIR QUALITY

Based on geologic and petrophysical data from the various wells within and adjacent to the Sanaga Sud field, the best reservoir character is found in the midfan facies (Figure 8). In this facies the sands tend to be cleaner and better sorted. The over-all sand/shale ratio is extremely high, locally exceeding 70% sand. Reservoir quality decreases away from the midfan facies. In the upper fan facies, toward the east, the sands contain a wider variety of clastic debris and are more poorly sorted (Figure 10A).

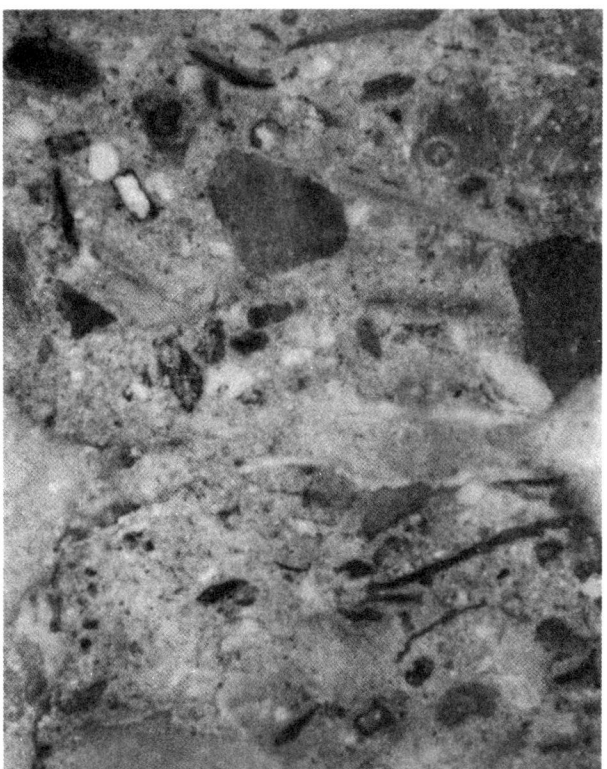

FIGURE 9. (continued)

Seaward of the midfan facies, the sandstones are finer-grained and contain more clay, mica, and coal fragments.

A thin-section petrographic study of cores from the Sanaga Sud wells indicated that primary porosity was the main contributor to the total observed porosity. Because of the relatively shallow depth of the reservoir units in the field area, post-depositional reduction of the primary porosity does not appear to have been significant. Diagenetic effects observed in the cores included microporosity, with kaolinite books partially filling pore spaces (Figure 11A), fracturing (Figure 11B), general dissolution (Figure 11C), and feldspar dissolution (Figure 11D).

PALEONTOLOGY

Paleontological data from the Lower Cretaceous section are generally sparse, poorly preserved, and often lacking in index fossils; therefore, biostratigraphic correlation is difficult. The best fossil assemblages are found in the upper portions of the Lower Cretaceous section. This may be a reflection of improved marine circulation at that time. Paleoenvironmental interpretations, based on fora-

FIGURE 10. Slabbed cores from the Sanaga Sud A-2 well: (A) core 3; (B) core 2.

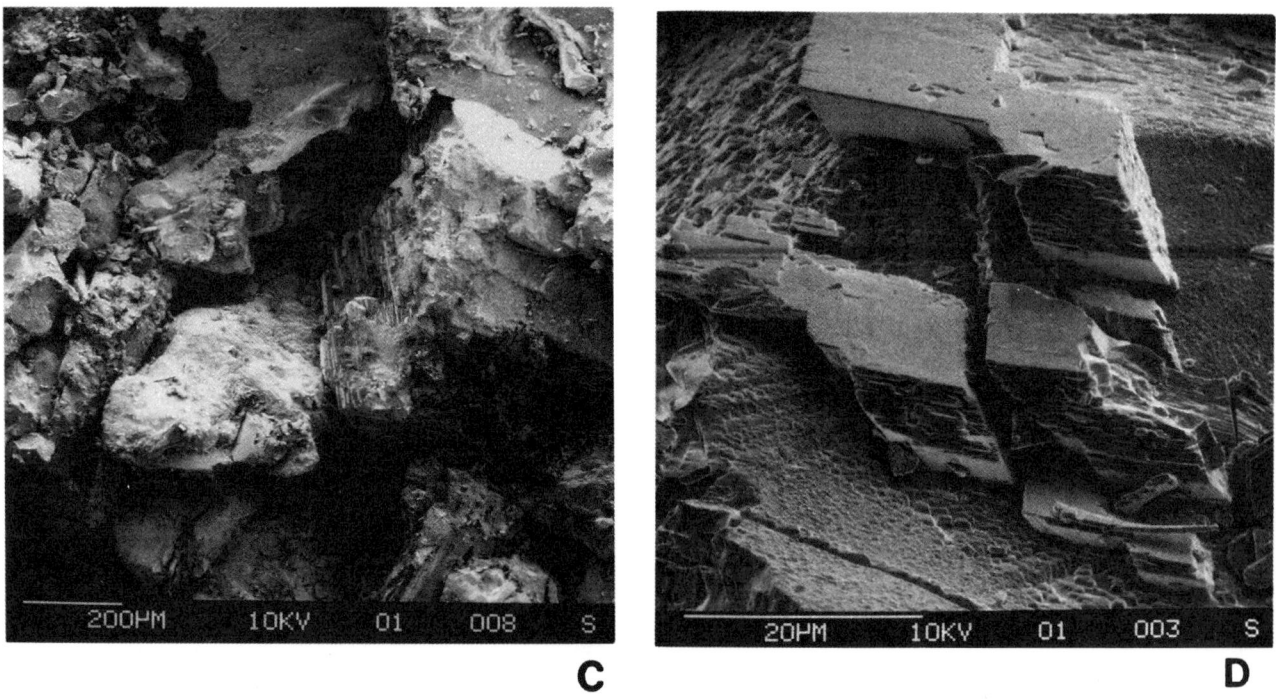

FIGURE 11. SEM photomicrographs depicting types of porosity found in the Sanaga Sud field: (A) microporosity; (B) fracturing; (C) general dissolution; (D) feldspar dissolution.

minifera and calcareous nannofossils, are considered to be reasonably accurate.

STRUCTURE AND TRAP CONFIGURATION

Following deposition of Aptian–Albian synrift sediments, there was a major uplift of the rifted margin, probably associated with the rift-to-drift transition. Major motion on the boundary fault(s), located close to the present shoreline, broke up the continental shelf into a series of fault blocks that dip into the boundary fault. Tectonic uplift and erosion of this broken-up shelf beveled the tilted fault blocks and formed several deeply incised canyons that probably followed faults oriented northeast-southwest at a high angle to the coastline. This produced a terrain of canyons and spurs that were mapped on seismic (Figures 12 and 13). The shaded portion of the map presented in Figure 12 corresponds to the Lower Cretaceous shelf during late Albian time. The numerous structural closures shown on this map were the exploration targets drilled during the early 1980s.

Figure 14 shows a well-calibrated seismic-based east-west cross section that depicts strata inclined counter to the primary depositional dip. Dipmeter logs from the SSA-1 well show a 45° southeast structural dip in the strata immediately underlying the Albian unconformity. At the SSA-2 well, the structural dip under the Albian unconformity ranges between 20 and 25° southeast. Not all of the rotated fault blocks in the area have the same magnitude and azimuth of dip. This is at least partially due to structural accommodation effects resulting from differences in orientation and throw on the various faults.

Paleotopographic features on the cross section were subsequently buried by transgressive marine shales, which in the Sanaga Sud area were dominantly Maastrichtian in age. These deep water shales are the seal facies for the oil and gas accumulations trapped in porous and permeable Lower Cretaceous sandstones beneath the Albian unconformity in the Kribi–Sanaga Sud area.

The hydrocarbon-bearing portion of the structure has 390 m of vertical closure above the gas-water contact. The area prospective for gas and condensate production is 2.5 km wide and 7.2 km long. Critical east closure is provided by late Albian shales beneath the unconformity (Figures 14 and 15). This is why a small portion of the accumulation does not follow the 1.5-sec contour along the southeastern margin of the field.

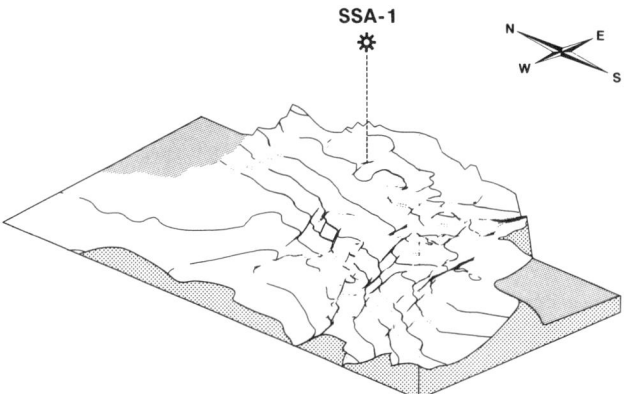

FIGURE 13. Isometric plot of the time map shown in Figure 12.

FIGURE 12. Seismic time map on the top-Albian unconformity surface.

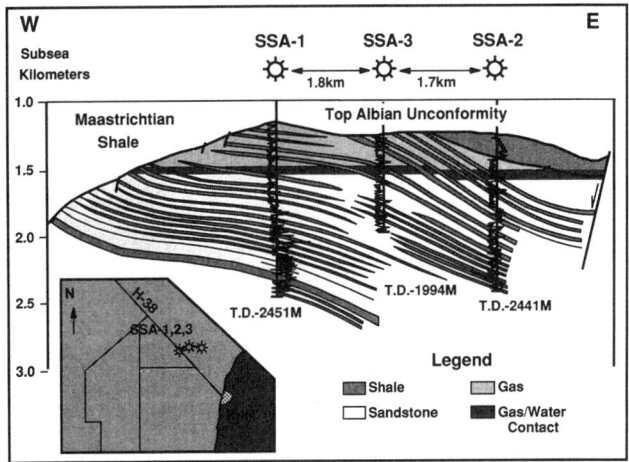

FIGURE 14. Structural cross section through the Sanaga Sud wells, showing relative thicknesses of sandstone and shale correlated with gamma ray log response, the gas-water contact, and the gas accumulation.

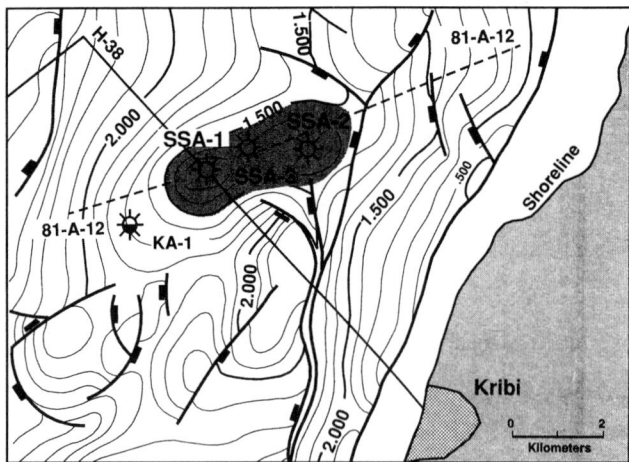

FIGURE 15. Detailed structure map of the Sanaga Sud gas field, showing the locations of the three wells and the gas-water contact at 1.5 sec.

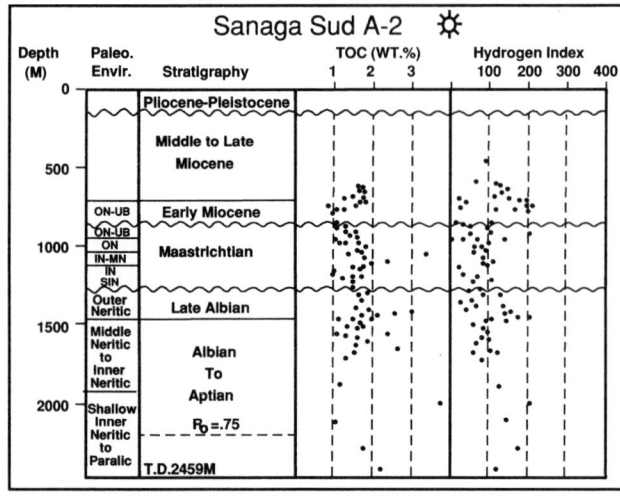

FIGURE 16. Hydrogen index and paleobathymetry plotted against depth in the SSA-2 well.

GEOCHEMISTRY

Various geochemical analyses have been conducted on rock and hydrocarbon samples from the Kribi–Sanaga Sud area. These analyses were designed to establish local and regional organic richness and maturation trends and to determine which stratigraphic intervals were potential sources for discovered oil and gas accumulations.

Oil characterization data, geochemical analyses of Lower Cretaceous sediment samples, and basin modeling results suggest two potential source areas for the discovered hydrocarbons: (1) the thick Lower Cretaceous section underlying or near the various rotated tilt-blocks and (2) the equivalent section seaward of the escarpment at the Albian unconformity. The structural attitude of the Lower Cretaceous in both of these areas would allow migration of generated hydrocarbons into the observed traps, if appropriate conduits existed.

Mature Lower Cretaceous source rocks underlying, downdip, or near the structures could have generated hydrocarbons that simply migrated updip into the traps. More mature, downdip Lower Cretaceous-sourced hydrocarbons would have migrated along reservoirs, faults, or the Albian unconformity. Mature Upper Cretaceous shales basinward of the faulted field and onlapping the Albian unconformity could also have sourced hydrocarbons in addition to the more mature Lower Cretaceous sections.

In general, gas-prone source rocks are common in the Lower Cretaceous, and oil-prone rocks are rare. Geochemical data from the SSA-2 well are used to represent the geochemical character of the sediments underlying and near the gas accumulation in this field. Plots of TOC and hydrogen index show that the relatively immature kerogen samples had hydrogen indices less than 300, which is typical of type III kerogen (Figure 16). Kerogen type is also indicated by pristane-phytane ratios determined from sample averages for the three wells: SSA-1 (3.9), SSA-2 (2.9), and SSA-3 (4.0).

In order to determine the stratigraphic depth from which the gas and condensate were likely to have been generated, temperature gradients were compared with vitrinite reflectance (R_o) profiles constructed for the SSA-2 and other wells. The SSA-2 R_o plot shows a fairly uniform trend of increasing reflectance with depth (Figure 17). By use of age, depth, and appropriate erosional estimates from regional interpretations, a burial profile with a constant geothermal gradient of 1.77°F/100 ft (32.26°C/km) was used to construct a theoretical R_o curve that closely matched the observed reflectance curve (Figure 17). Following sensitivity studies, the above geothermal gradient was selected to model the thermal maturity of the section underlying the SSA-2 well through time.

Recent modeling of stress-induced vertical tectonic motion by Cloetingh and Kool (1990) suggests that, in passive margin settings, the effects of vertical tectonics have a tendency to overpower the contributions of eustatic sea level oscillations. Certainly, local degrees of tectonism also have to be considered in this context. The older portion of the time gap represented by the top-Albian unconformity (100 to 90 Ma) was a period of relative sea level rise (Figure 5). This implies that local and/or regional tectonic uplift was a significant process in the development of this portion of the unconformity. This also provides justification for assuming some degree of erosional removal of previously deposited sediment.

According to the Tissot et al. (1987) type III kerogen kinetics model, maximum changes in fractional conversion occurred near the total depth of the SSA-2 well (Figure 18) at about 50 Ma. The amounts of potential erosional removal at the Albian and the other two unconformities were determined by estimating the sedimentation rate for the stratigraphic section immediately underlying each unconformity and multiplying it by the time gap represented by the unconformity. It was not possible to determine objectively the relative contributions of subsidence, nondeposition, and uplift for each unconformity. Consequently, the degree of thermal maturation depicted in

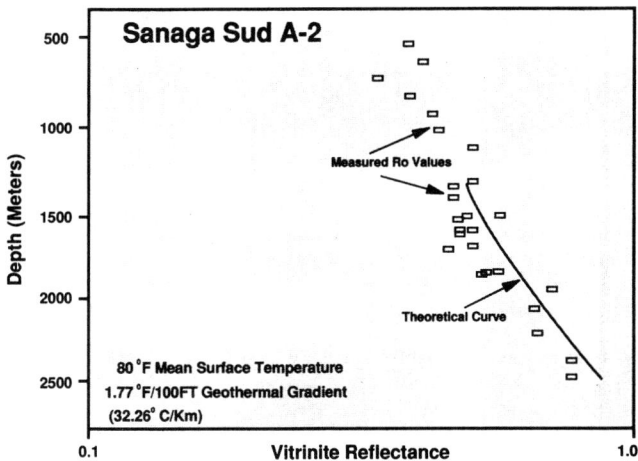

FIGURE 17. Theoretical and measured vitrinite reflectance for the SSA-2 well.

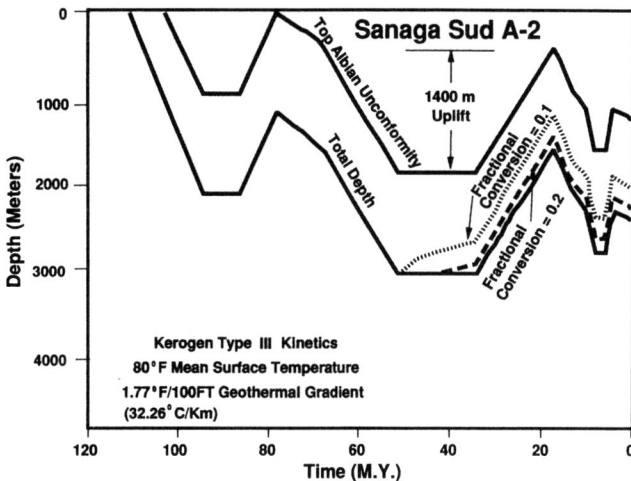

FIGURE 18. Isomaturity profile for the SSA-2 well, based on type III kerogen kinetics.

Figure 18 assumed that subsidence, nondeposition, and erosion were equally divided.

Figure 19 shows that, according to type III kerogen kinetics, little fractional conversion has taken place from sediments located stratigraphically close to the Albian unconformity at the SSA-2 well. A maximum of about 30% of the type III kerogen in rocks located close to the total depth of the well would have been converted to hydrocarbons. The sections underlying the well, or the stratigraphically equivalent section downdip to the drilled part of the structural block, would have had a higher percentage of conversion, given the same assumptions cited above.

GEOPHYSICS

Numerous seismic surveys were shot in the Kribi-Sanaga Sud concession areas during the late 1970s and early 1980s. The line spacing was 1 km in most areas. Most of the data are rated fair to good but generally lack resolution below about 2.5 sec. Reprocessing of these data would probably improve resolution and could lead to additional play opportunities in the existing concession areas.

One technique that has shown promise for delineating hydrocarbons trapped in Lower Cretaceous structures in this area is seismic inversion. Experimental seismic inversion processing was conducted by the Mobil geophysical staff for a line crossing one of the Albian fault-block structures in the southern part of the Batanga concession in 1985. The color display (not included here) shows a dramatic decrease in interval velocity between the hydrocarbon-water contact and the unconformity at the top of the Albian. This result emphasizes the utility of inversion technology for depicting porosity preservation and the presence of trapped hydrocarbons in this area.

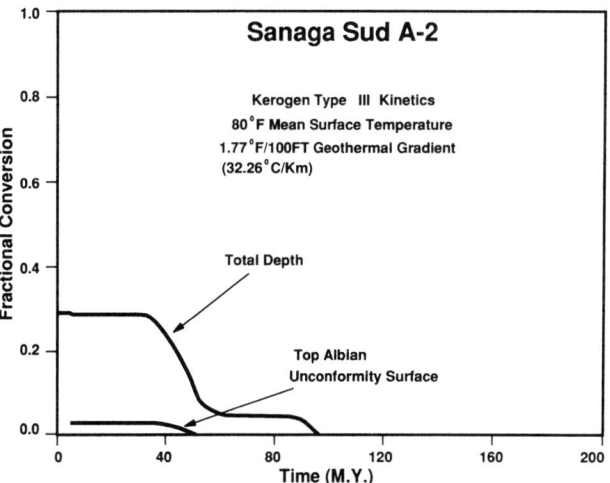

FIGURE 19. Fractional conversion of type III kerogen in the Sanaga Sud field area, based on type III kerogen kinetics and data from the SSA-2 well.

COMPLETION RESULTS FOR THE SANAGA SUD WELLS

Table 2 lists the important petrophysical reservoir properties encountered in the three wells drilled in the Sanaga Sud field, along with a variety of drilling information, including the coring program.

The Sanaga Sud prospect (Figure 20) was drilled to test a structural closure at the top-Albian unconformity. A prominent seismic "flat event" was recognized below the unconformity at about 1.5-sec, two-way travel time, and was interpreted as a hydrocarbon-water interface. Areally mapped closure on the Sanaga Sud structure is 1615 ha (3990 ac). Vertically mapped closure is 329 m (Figure 15).

TABLE 2. Petrophysical reservoir properties and drilling information for the three wells drilled in the Sanaga Sud field.

Item	Well SSA-1	Well SSA-2	Well SSA-3
Water depth (m)	17.1	16.0	17.0
Total depth (m)	2479	2459	2018
Latitude	3°00′N	3°01′N	3°01′N
Longitude	9°51′E	9°52′E	9°52′E
Spud date	8/27/79	12/26/80	8/31/81
Core-1 depth range (m)	1220–1239	1507–1516	1263–1272
Core-2 depth range (m)	2460–2479	1581–1590	1272–1280
Core-3 depth range (m)	2118–2119	1829–1838	1344–1363
Core-4 depth range (m)	—	—	1473–1491
Log top pay (m subsea)	1163	1436	1221
Gas-water contact (m)	1491	1553	1503
Log gross pay (m)	331	118	282
Net gas pay (m)	261	90	183
Net/gross ratio	0.79	0.76	0.65
Average porosity (%)	25.8	17.3	20.4
Average SW (%)	17.0	18.3	17.4
Average permeability (md)	181	112	133

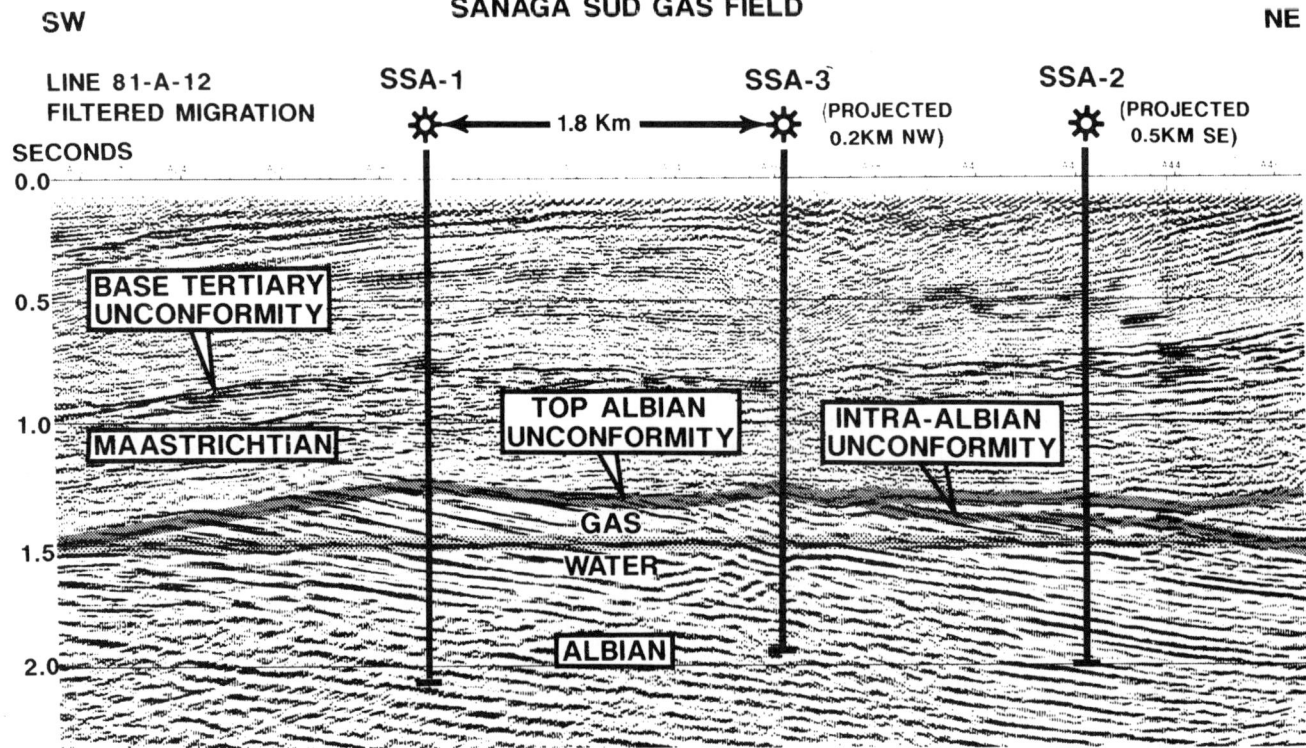

FIGURE 20. Seismic line 81-A-12, oriented northeast-southwest through the Sanaga Sud field, showing the positions and total depths of the three wells and the gas-water contact (line position is shown in Figure 15).

The Sanaga Sud A-1 well encountered 261 m of net pay with 83% gas saturation. Four drill-stem tests averaged 11 MMCFGD and 85 BCPD.

Sanaga Sud A-2, the first appraisal well, had 90 m of net pay with 82% gas saturation. The four drill-stem tests averaged 10 MMCFGD and 51 BCPD.

The Sanaga Sud A-3 well, the second appraisal well, had 183 m of net pay with 83% gas saturation. Four drill-stem tests averaged slightly over 8 MMCFGD and 32 BCPD.

RESERVOIR POTENTIAL AND RESERVES

The Albian sandstone reservoir section of the Sanaga Sud accumulation ranges in depth from 1163 to 1553 m. The gas-water contact is tilted toward the east. In the SSA-1 well it is at 1491 m, whereas in the SSA-2 well it is 62 m lower, at 1553 m (Figure 12).

The gas is relatively free of nonhydrocarbon gases. The initial condensate content, as indicated by stabilized flow tests, ranges up to 8.2 STB/mmcf of gas (Table 3). Condensate recovery over the productive life of the field is estimated to be 5 STB/mmcf of gas.

The reserve estimates currently accepted by Mobil for this field are 900 bcf of gas and 4.5 million bbl of condensate, within a reservoir volume of slightly over 1 million ac-ft.

Subdividing this by area, the Londji concession has 210 bcf of gas and 1.675 MMBNGL, and the H-38 concession has 690 bcf of gas and 5.500 MMBNGL. The proved and probable distributions of these numbers are shown in Table 4.

FIELD DEVELOPMENT AND PRODUCTION

The partners in the blocks have discussed plans for an LNG plant, but no action has been taken. As a result, no plans have been made for further appraisal drilling or development of this field. The flow capabilities cited above should sustain deliverability of the gas reserves over a production period of 20 yr.

ACKNOWLEDGMENTS

In listing the major contributors to this study, my apologies are extended to any that have been omitted. The

TABLE 3. Drill-stem test results for wells SSA-1, SSA-2, and SSA-3.

Well	Test No.	Depth (m)	Gas (MMCFGD)	Condensate (BCPD)	Choke size (in.)
SSA-1	1	1482–1485 1462–1472	5.3	38	11/32
SSA-1	2	1418–1420 1422–1429	12.6	100	19/32
SSA-1	3	1330–1344	10.8	75	19/32
SSA-1	4	1222–1241 1330–1344	15.2	125	19/32
SSA-2	1	1557–1563	2.8	—	3/4
SSA-2	2	1510–1518	15.6	57	3/4
SSA-2	3	1457–1466	7.1	32	3/4
SSA-2	4	1474–1483	13.7	113	3/4
SSA-3	1	1498–1505	4.0	—	32/64
SSA-3	2	1467–1476	8.9	48	48/64
SSA-3	3	1340–1349	10.3	46	32/64
SSA-3	4	1244–1253	9.9	34	32/64

TABLE 4. Proved and probable reserves by concession area.

Concession	Proved gas (bcf)	Probable gas (bcf)	Proved condensate (MBNGL)	Probable condensate (MBNGL)
Londji	125	85	625	1050
H-38	550	140	2750	2750

intent is to recognize those who contributed to the technical aspects of this paper and not necessarily all who may have contributed to this successful exploration play. The Mobil report covering the Kribi–Sanaga Sud area evaluation, which led to Mobil's acreage acquisition, was authored by H. Aves, R. F. Dyar, P. D. Jackson, S. M. Kiger, G. M. Patterson, A. Telles, B. Wildgoose, J. G. Winston, and C. L. (Barton) Gaynor. The preparation of this paper was significantly aided by J. C. Cooke, R. E. Devlin, G. J. Gail, W. E. Loomis, and the Mobil drafting department, particularly K. C. Bourland. I express my sincere gratitude to the management of Mobil Oil Corp., Mr. P. Orieux of TEP-CAM, Mr. I. J. Bond of Ocelot Industries, Mr. C. W. Dye of Damson Oil Corp., and the Société Nationale des Hydrocarbures for their permission to publish this paper.

REFERENCES CITED

Aves, H., R. F. Dyar, C. L. (Barton) Gaynor, P. D. Jackson, S. M. Kiger, G. M. Patterson, A. Telles, B. Wildgoose, and J. G. Winston, 1978, Cameroon Kribi block evaluation: Mobil Report SEP TM-658, 16 p.

Cloetingh, S., and H. Kool, 1990, Intraplate stresses and continental margin stratigraphy: new constraints on the relative contributions of tectonics and eustacy to the record of sea level changes: AAPG Bulletin, v. 74, n. 5, p. 630.

Folk, R. L., 1974, Petrology of sedimentary rocks: Austin, Texas, Hemphill Publishing Company, 170 p.

Fontaine, R. X., 1982, Lower Cretaceous reservoir trend study, Sanaga-Kribi-Campo offshore Cameroon: Mobil Producing Cameroon Report CGR-7, 23 p.

Larson, P. A., 1983, Sanaga Sud, a Lower Cretaceous (Cameroon offshore) core description and interpretation: Mobil Applied Stratigraphy Report, 14 p.

Moiola, R. J., 1982–83, Miscellaneous unpublished Mobil Research Laboratory technical service reports and letters.

Tissot, B. P., R. Pelet, and P. Ungerer, 1987, Thermal history of sedimentary basins, maturation indices, and kinetics of oil and gas generation: AAPG Bulletin, v. 71, p. 1445–1466.

Chapter 15

October Field
The Latest Giant under Development in Egypt's Gulf of Suez

Jeffrey J. Lelek

Gulf of Suez Petroleum Company
Cairo, Egypt

David B. Shepherd

Amoco Production Company
Houston, Texas, U.S.A.

Denise M. Stone

Amoco Production Company
Tulsa, Oklahoma, U.S.A.

A. Shawky Abdine

Gulf of Suez Petroleum Company
Cairo, Egypt

ABSTRACT

October field, the third largest oil field in Egypt, produced over 420 million bbl of oil from its discovery in 1977 until January 1991. It is the northernmost giant oil field in the Gulf of Suez rift basin. Forty-five wells from eight platforms in approximately 58 m (190 ft) of water drain over 3703 ha (9150 ac) of the October producing trend. Recent reservoir discoveries demonstrate the viability of continuing exploration along this oil-rich trend.

This structurally trapped field is a complex of rotated fault blocks typical of rift basins worldwide. A northwest-trending normal fault with a throw of approximately 1220 m (4000 ft) has trapped the largest of several oil accumulations, which has a 333-m (1092-ft) oil column on the upthrown eastern side. Severe seismic multiple problems resulting from thick Miocene evaporites restrict definition of the highly productive prerift section. These same evaporites serve as the ultimate seal throughout the Gulf of Suez. Approximately 95% of field reserves are within Carboniferous to Lower Cretaceous massive Nubia Sandstones, with remaining reserves in more lenticular Upper Cretaceous sands, basal Miocene rift-fill Nukhul Formation clastics, and the Asl Member of the Miocene Upper Rudeis Formation.

Several distinct reservoir accumulations exist, with the deepest and most significant original oil-water contact at -3558 m (-11,670 ft) subsea. October field oil gravities range from 14 to 39° API. The main source rock unit for the field is postulated to be the Campanian Brown Limestone Member of the Sudr Formation. Typical reservoir parameters for the Nubia Formation are: porosity, 17%; permeability, 236 md; and net pay thickness, 137 m (450 ft). The expected recovery factor is 45%. A field-wide reservoir study is currently in progress to address reservoir management of the October area. GUPCO (the joint operating company for Amoco Production Company and the Egyptian General Petroleum Company) operates October field, which from all reservoirs produced approximately 170,000 BOPD at the beginning of 1991.

INTRODUCTION

Like most things in Egypt, oil exploration has a long and impressive history. Most of Egypt's petroleum production is clustered in the small Gulf of Suez rift zone, within which the October field is the northernmost giant oil field (Figure 1). It is also the third largest field in the country, following only El Morgan and Belayim fields.

Throughout the world, cratonic rift basins harbor significantly larger reserves than those of convergent and divergent margin rift basins (Figure 2). Among cratonic rift basins, the North Sea is the most prolific, with over 50 billion bbl of oil reserves. The Sirte (Libya) and North German-Polish basins are next in importance. The Gulf of Suez contains approximately 11 billion bbl of known reserves, more than 90% of Egypt's total. There are approximately 40 oil fields in Egypt, and seven of them have reserves greater than 250 million bbl.

The 350-km- (220-mi-) long Gulf of Suez represents the northern terminus of the East African Tertiary rift system, a 7000-km- (3800-mi-) long megafeature that splits much of the African continent. Figure 3 shows this Tertiary rift system along with Africa's older Triassic to Cretaceous rifts and other cratonic basins. Although hydrocarbons have been found in the Cretaceous rift system—notably in Sudan, Chad, and Niger—the Gulf of Suez Tertiary rift contains essentially all of the production within East Africa.

Egypt produced approximately 5 billion bbl of oil from 1911 to January 1991. Figure 4 shows historic trends of production, which was 870,000 BOPD in early 1991.

October field is a series of rotated fault blocks typical of the Gulf of Suez rift. Like many fault-block fields, October field is not as simple as a first glance would indicate (Zahran, 1986; GUPCO, 1990). It consists of at least six separate reservoirs in four stratigraphic horizons, with different hydrocarbon types and various reservoir drive systems. The main field area, producing primarily from the Nubia Formation, accounts for the bulk of reserves and is the focus of this paper.

REGIONAL TECTONIC AND STRUCTURAL SETTING

The Gulf of Suez rift is a northwest elongated structural depression 350 km (220 mi) long, ranging in width from 52 km (32 mi) in the north to 90 km (56 mi) in the south. Gulf waters with an average water depth of 50 m (164 ft) occupy only the central portion of the rift; the remaining flank portion is exposed onshore. Numerous authors have published excellent articles on regional tectonics (El Shazly, 1977; Patton et al., in press) and the structural development of the Gulf of Suez (Sellwood and Netherwood, 1984; Khalil and Meshref, 1988; Soliman et al., 1988). Figure 5 (Hempton, 1987) depicts the major tectonic elements in the Middle East, two of which are critical to an evaluation of the October field area.

The Tethyan system forms the boundary between the Arabian-African plates and the Eurasian-Turkish plates (Argyriadis et al., 1980; Dercourt et al., 1986). This plate margin diverged and opened to form the Neotethys Sea in a time-progressive manner from east to west. Opening began in the Oman area during the Permian, and migrated westward through Iran, Iraq, and the present-day Mediterranean Sea. In the area of Egypt, the opening of the Neotethys commenced in the Jurassic. Subsequently, the Neotethys closed, imparting an east-west structural grain to the northern Gulf of Suez during the Upper Cretaceous to Eocene.

The second and most important tectonic system is the Tertiary rift system of East Africa (Figure 3). Time-progressive from south to north in this portion of the system, rifting began in the southern Red Sea during the lower Eocene to middle Oligocene. Igneous intrusives appeared in the Gulf of Suez during the Oligocene, and

October Field: The Latest Giant under Development in Egypt's Gulf of Suez   233

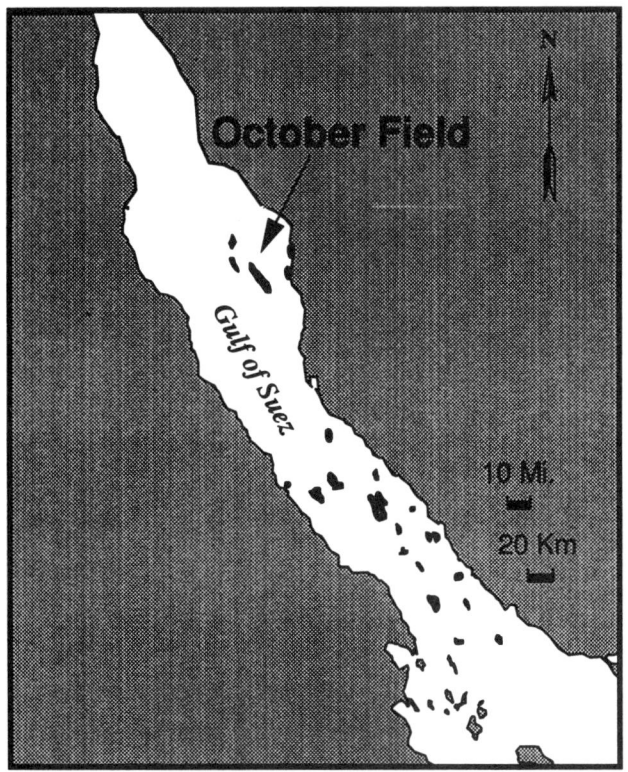

FIGURE 1. Location map for the October field complex (three black offshore areas indicated by arrow), the northernmost significant offshore production in the Gulf of Suez.

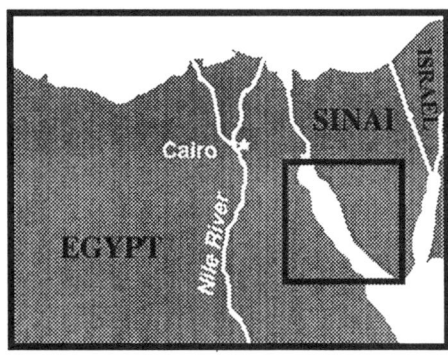

FIGURE 2. Worldwide rift basin reserves. Estimated hydrocarbon reserves discovered to date are plotted for the world's producing rift basins. The Gulf of Suez ranks seventh.

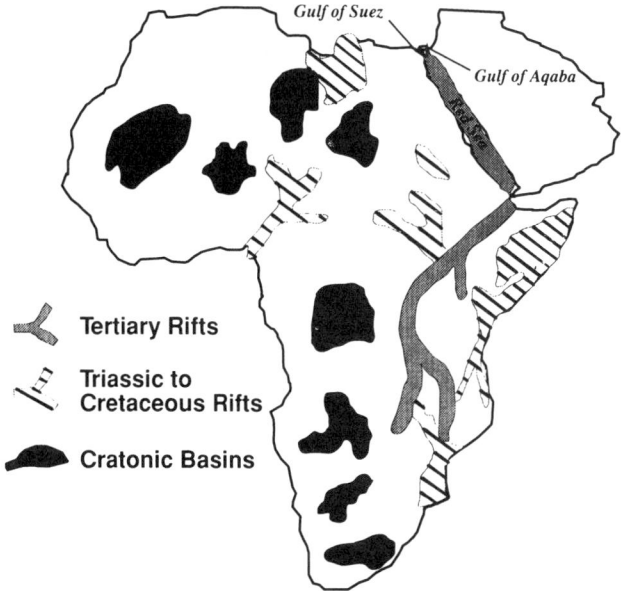

FIGURE 3. African rift basins. The Gulf of Suez represents the northern terminus of the East African Tertiary rift system, shown extending from Egypt south to Tanzania. Triassic to Cretaceous rifts, unrelated to the Tertiary system, are also shown.

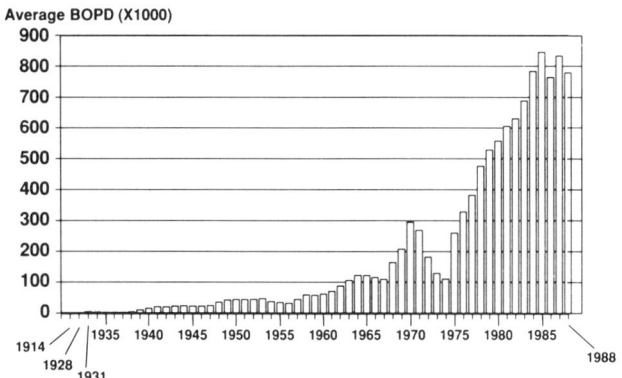

FIGURE 4. Egyptian oil production history. Average daily production increased dramatically beginning in the late 1960s. The current (early 1991) production of approximately 870,000 BOPD is expected to remain constant for at least several years.

major rifting began there about 22 m.y.a. during the lower Miocene. Fifteen m.y.a., the direction of opening changed slightly from an east-northeast direction of spreading to a more northeasterly direction, coincident with the start-up of the Gulf of Aqaba strike-slip system (LePichon and Gaulier, 1988). Nine million years ago, this largely halted horizontal extension in the Gulf of Suez, which henceforth underwent thermal subsidence and evaporite deposition.

Figure 6 summarizes regional tectonic events, placing them in a time frame relative to the deposition of reser-

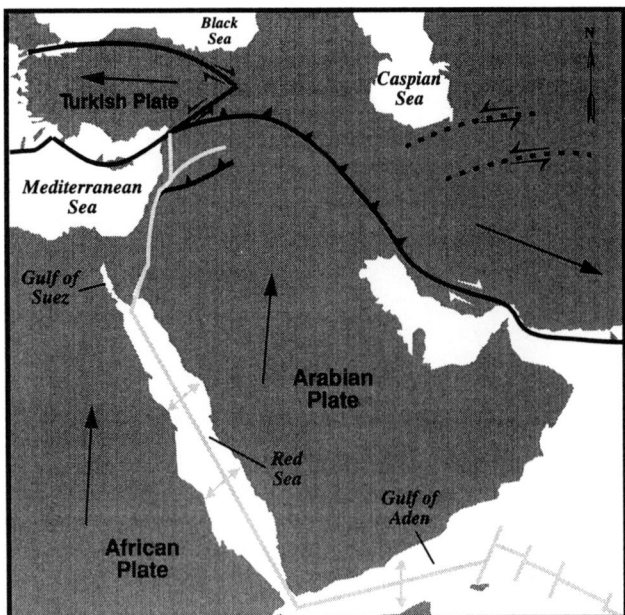

FIGURE 5. Middle East tectonics. The Tethyan compressional system is shown as dark-colored thrusts and wrench faults. The East African Tertiary extensional system is shown as light-colored spreading axes and strike-slip faults. Relative major plate motions are indicated by the large arrows. (After Hempton, 1987.)

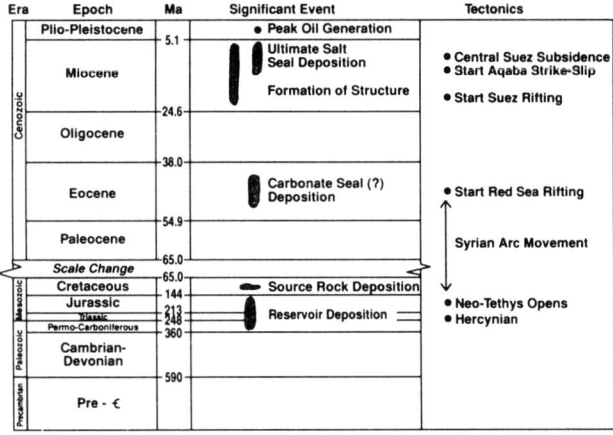

FIGURE 6. Tectonic timetable. Regional deposition of October field's major reservoir and source rock in the Mesozoic predated rifting. Semiregional deposition of the evaporite seal, as well as structural development and hydrocarbon migration, occurred in the Cenozoic as a direct result of rift extension and subsidence.

voirs, source rocks, and seals, and relative to the time of peak oil generation, all of which are intimately related to, or caused by, the tectonics. With this relational concept in mind, the specifics of October field geology can best be appreciated.

Studies of structural styles in both outcrop and the subsurface have enhanced our understanding of the region (Robson, 1971; El-Tarabili and Adawy, 1972; Chenet and Letouzey, 1983; Patton, 1984; Angelier, 1985; Colletta et al., 1986 and 1988; Jackson et al., 1988; Moretti and Colletta, 1988; Vigano and Patton, 1988; Coffield and Schamel, 1989; Morley et al., 1990). The dominant northwest-southeast fault trend (310 to 340°) was labeled "Clysmic" after the ancient Roman city of Clysma, where this trend was first recognized in outcrop. The Clysmic trend lies parallel to the geographic strike of the gulf and perpendicular to the inferred regional extension that formed the basin. This trend is often interrupted by three other common fault trends termed "north oblique" (350 to 030°), "northwest oblique" (280 to 310°), and "cross" (050 to 075°). The relative occurrence of these faults is shown graphically by a rose diagram compiled by analysis of aerial photographs in conjunction with data from the literature (Figure 7). Interactions among the clysmic and oblique faults allow transfer of displacement between major fault systems and produce the characteristic zig-zag shapes of faults in the region. These fault patterns are easily identified on Landsat images of the basin margins (Figure 8). The tilted updip corners of these fault blocks are often favorable traps for oil or gas in the subsurface.

Landsat photographs (Figure 8) also show large wadis, or intermittent desert stream systems, which transport large volumes of sediment to the marine environment. The wadis that today feed the Gulf of Suez existed throughout the Miocene; in fact, some have not changed geographic position since the lower Miocene, when they were responsible for supplying sediment that ultimately formed several important synrift productive reservoirs. Miocene Nukhul production as well as the newly discovered Miocene Asl production at October field are related to such synrift depositional systems.

Another structural feature in the Gulf of Suez, also common in other cratonic rifts, is the presence of regional

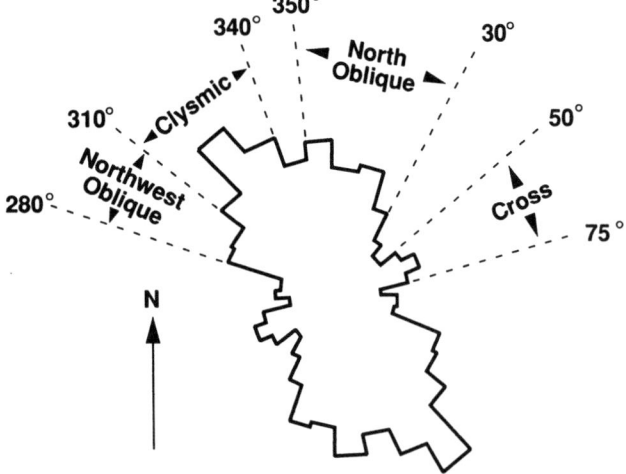

FIGURE 7. Frequency distribution rose diagram of Suez rift fault trends. The clysmic trend is dominant, followed by the frequently occurring north oblique, northwest oblique, and cross trends (Patton, personal communication, 1991).

FIGURE 8. Landsat image of Suez rift. Shoreline and outcrops show clysmic trending structures that zig-zag due to offset by oblique changes in fault strike. Large wadis, intermittently active since the Miocene, build outward into the Gulf and were responsible for sediment transport of many synrift reservoirs.

dip domains. Three major domains exist in the Gulf of Suez (Figure 9), with October field lying within the middle Gulf domain where dips are to the northeast. The linear zones where regional dips change are called "accommodation zones." In many cases, accommodation zones appear to have localized synrift sand deposition and provided structural complexity, both of which enhance reservoir trapping potential, as in the case of El Morgan field.

The generally east-west-trending accommodation zone just north of El Morgan field has been studied by Colletta et al. (1988), who constructed a block diagram showing the geometry of this area (Figure 10). Several major oil fields, including the supergiant El Morgan, are located immediately north and south of this zone. Here, as throughout the East African rift zone, as major rift faults approach the accommodation zone they bifurcate several times, bend in predictable directions, and decrease their throw through interaction or displacement transfer between faults. This results in numerous trapping geometries involving multiple reservoirs.

Morley et al. (1990) describe the interactions and resulting geometries of fault displacement transfer zones in extensional regions, which are critical to an understanding of structural style in the Gulf of Suez. Transfer zones are subdivided, according to the attitudes and directions of throw of their major faults, into synthetic and conjugate zones. Synthetic transfer zones consist of faults that dip in the same direction and link to exchange displacement laterally along strike. Conjugate transfer zones consist of faults that dip opposite to one another; they are further categorized as convergent and divergent

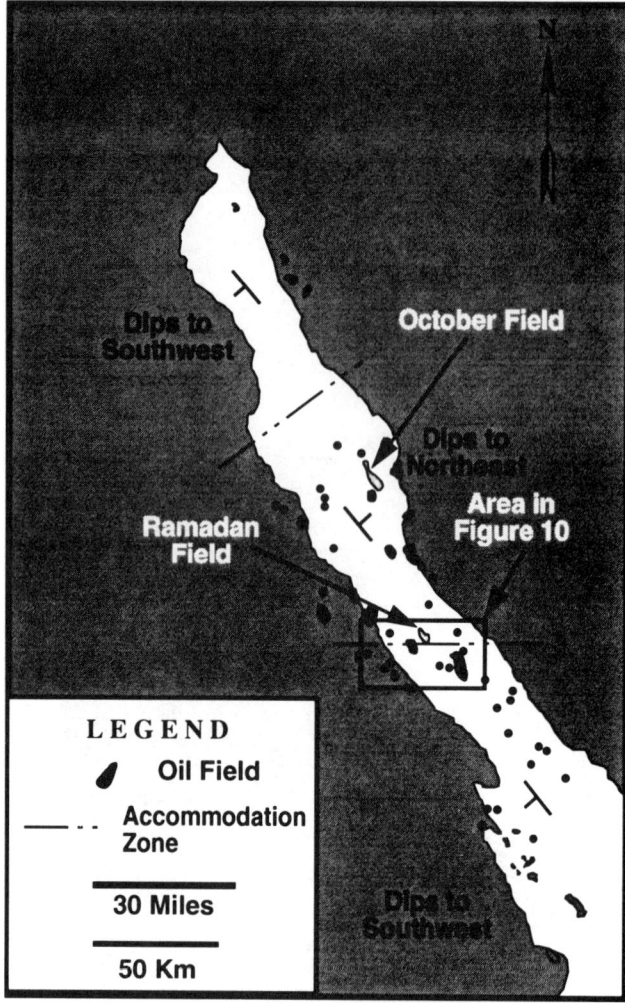

FIGURE 9. Gulf of Suez dip domains. Three mega-scale dip domains exist in the Gulf; October field dips to the east within the central dip domain. (After Moustafa, 1976.)

systems. In convergent conjugate transfer zones, faults dip toward each other (full graben geometry), whereas in divergent conjugate transfer zones, faults dip away from each other (horst geometry). These conjugate transfer zones may approach each other and lose displacement, overlap one another, completely overlap (collateral), or lie in line (collinear). With respect to regional dip domains, the three areas of dominant dip reflect synthetic transfer zones of varied geometry, separated by complex conjugate transfer zones.

The northern Gulf of Suez accommodation zone is poorly delineated by sparse well control and over-all poor-quality seismic. The structure within October field might be subtly related to this cross-gulf phenomenon. As exploration continues north from October field, the definition and importance of this large-scale accommodation zone will receive much study.

Poor seismic data quality, owing to multiple interference from thick, continuous evaporite sequences deposited late in the Miocene and shallow in the basin, is a challenging problem for explorationists. Identification of both prospective pre-Miocene and synrift structures from seismic has been based primarily on drape and folding of the younger Miocene evaporite sequences. Pre-Miocene objectives are nonetheless difficult to predict because of thickening and thinning of synrift sequences, which reduce the apparent relief associated with underlying structure. The difficulty of defining these structures is evidenced by the number of dry holes drilled for both exploration and development purposes. Such dry holes are located throughout the Gulf of Suez, including the area immediately surrounding the October field.

STRATIGRAPHY AND PRODUCING HORIZONS

The Gulf of Suez basin went through three major depositional periods: prerift, synrift, and postrift sag (Figure 11). The prerift section consists of early Paleozoic to Oligocene (pre-Miocene) clastics and carbonates, which directly overlie crystalline basement. The synrift and postrift sag sections consist of Miocene to Holocene sediments, dominated by clastics and evaporites that unconformably overlie the pre-Miocene. From 1600 to 2000 m (5250 to 6560 ft) of pre-Miocene sediments were deposited before the Gulf of Suez rift developed. The synrift and postrift graben fill section of Miocene age is generally 1000 to 2500 m (3300 to 8200 ft) thick. Oil from October field is produced from four stratigraphic horizons: the Cretaceous Nubia Formation, the Upper Cretaceous Nezzazat Group, the Miocene Nukhul Formation, and the Miocene Asl Member of the Upper Rudeis Formation (Table 1; Figures 16 and 13). The focus of this paper is the prerift Nubia Formation, which contains most of the reserves in October field.

The early Paleozoic to Cretaceous Nubia Formation overlies basement and is the oldest sediment in the field area. The 184-1 well (Figure 16), which penetrated basement, shows a gross Nubia thickness of 850 m (2800 ft). Other wells in the field generally penetrated only the upper third of the Nubia section, typically reaching total depth just below the oil-water contact. The Nubia is overlain by the Upper Cretaceous Nezzazat Group.

The Nezzazat Group consists of the Matulla, Wata, and Raha formations, which are predominantly interbeds of limestone and sandstone. Oil is produced from the Matulla and Wata, with reservoirs commonly in local dolomites and channel sandstones. The Matulla Formation exhibits the best lateral reservoir continuity. The Wata offers some upside potential, but the sands are interpreted as being more discontinuous than those in the Matulla. The average initial flow rate of the seven wells that have produced from the Nezzazat in the field area is 5100 BOPD, with minimum and maximum rates of 3400 and 10,000 BOPD, respectively. Production decline is often rapid as a result of limited reservoir continuity.

Above the Nezzazat Group is the Sudr Formation, a widespread chalk that is significant to exploration. The Brown Limestone Member at its base is a regional radioactive marker bed and the probable major source rock for oil in the Gulf of Suez. The top of the Sudr Formation

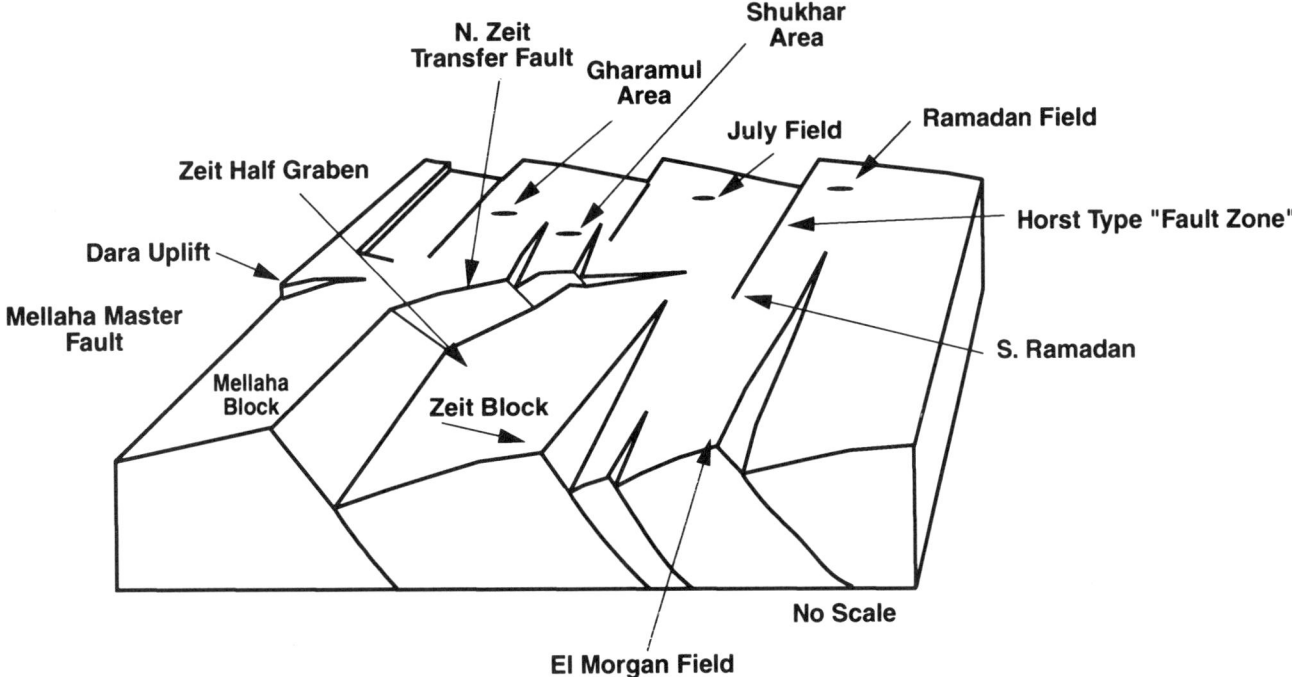

FIGURE 10. Block diagram of central Gulf of Suez accommodation zone area. Prevalent regional dips are to the southwest south of the accommodation zone and to the northwest north of the accommodation zone. Major faults bifurcate and curve as they approach the accommodation zone. The July, Ramadan, and El Morgan giant fields are shown. (From Colletta et al., 1988.)

marks the erosional unconformity between the Cretaceous and the Tertiary.

Locally, the oldest Tertiary sediments thin westward toward the crest of the main field. These sediments include carbonates of the Eocene Thebes Formation and red beds and shales of the Oligocene Abu Zenima Formation (Figure 11). Unconformably overlying these sediments is the 1000-to-2500-m (3300-to-8200-ft) Miocene synrift and postrift sag section. From the base upward, this consists of the Nukhul/Lower Rudeis synrift section, the Upper Rudeis/Kareem/Belayim/South Gharib/Zeit postrift sag section, and post-Miocene sediments. Up through the Kareem is a combination of continental to deep marine deposits of sandstone, shale, and limestone, with minor amounts of anhydrite.

Alluvial fan or fan-delta sandstones of the Nukhul Formation are productive in the F and D platform areas of the October trend (Figure 16), where structurally influenced depositional and erosional patterns controlled the distribution of reservoir-quality sediments. The Nukhul Formation represents the first period of deposition following the initiation of rifting in the Gulf. Because the Nukhul Formation was deposited downthrown to major faults as a result of upthrown fault-block erosion, Nukhul distribution and reservoir quality are functions of proximity to local horst blocks. Generally, Nukhul sediments are texturally and compositionally immature, consisting of conglomerates, sandstones, limestones, and shales.

Locally, Lower and Upper Rudeis marine shales and marls overlie the Nukhul Formation. They generally serve as the sealing lithology juxtaposed against the Nubia reservoir at the main trapping clysmic fault. Within the Upper Rudeis is the Asl Sandstone, the most recent addition to the list of producing horizons in the October trend, productive on the downthrown side of the bounding clysmic fault. The Asl is a fluvial-alluvial mixed sandstone-limestone and shale unit. Lateral facies changes over short distances observed in both outcrop and the subsurface make interpretation within this unit a challenge. Some investigators interpret these sands as being related to talus cones and fan-deltas deposited during and following fault rejuvenation and rotation of fault blocks associated with the mid-Clysmic event (Gawthorpe et al., 1990). Deposition was controlled by rift paleotopography.

Overlying the middle to late Miocene Rudeis units are the Belayim, South Gharib, and Zeit formations, which represent a marine evaporite sequence of salt, anhydrite, and shale that serves as the ultimate regional hydrocarbon seal in the Gulf of Suez (Fawzy and Abdel Aal, 1986). Thickness of this section ranges from 600 to 1200 m (2000 to 4000 ft). Significant isopach expansion is observed in this section downthrown to major rift normal faults.

Post-Miocene sediments are primarily thick sandstones with interbedded limestones and shales. Thick post-Miocene sediments downthrown to major faults

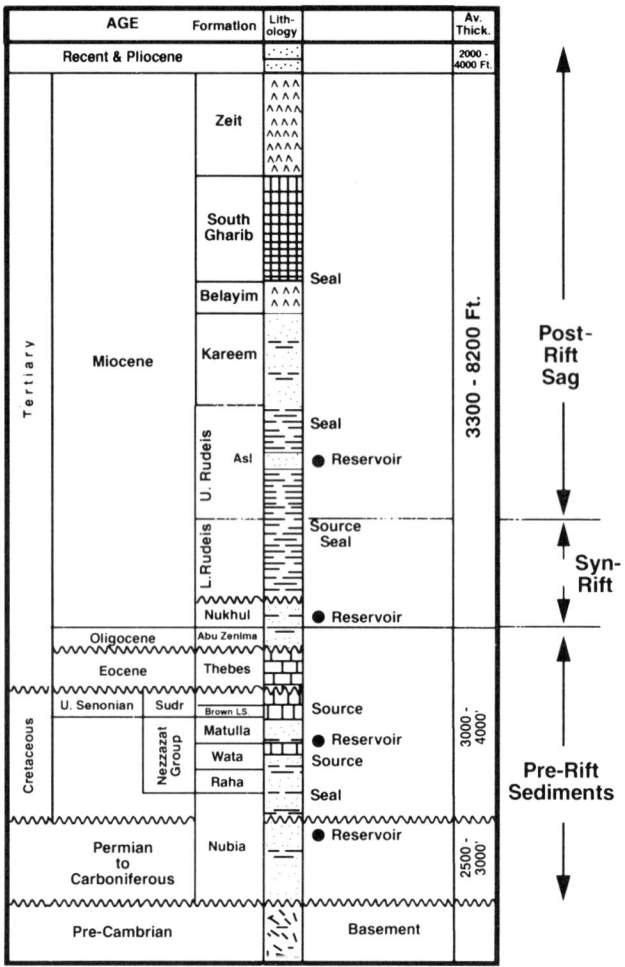

FIGURE 11. Stratigraphic column. Regionally continuous clastics dominate the pre-Miocene, with the exception of the Eocene limestone section. Miocene rifting and subsidence resulted in laterally variable clastics, carbonates, and evaporites. Extensive evaporites provide a very effective reservoir seal.

reflect reactivation of basement blocks. This section ranges in thickness from 900 to 1500 m (3000 to 5000 ft).

HISTORY OF EXPLORATION AND DISCOVERY

The history of oil discovery within the Gulf of Suez is traced in Figure 14. Although oil from seeps was used throughout history, Gemsa field was discovered in 1886 and was put into commercial production in 1911. The first surge of exploration in the Gulf of Suez basin began in the 1920s and led to discovery of the Ras Ghareb field on the central Gulf's west coast. Discovery of the giant Belayim land field in 1955 initiated production on the eastern margin of the gulf basin, starting the race for exploration in offshore areas. Cope (now Petrobel) conducted the first marine seismic survey in 1955, which marked the emergence of the oil industry in the Gulf of Suez. Despite a limited ability to define deep subsurface structure with conventional seismic techniques, success came quickly. The first Amoco well in the Gulf discovered the giant El Morgan field in 1964 (Figure 14). The Gulf of Suez Petroleum Company (GUPCO), a joint Amoco and Egyptian General Petroleum Company operation, was established in 1965 to operate and develop Amoco-held concessions. The El Morgan field's peak production rate of 300,000 BOPD in 1970 set the stage for aggressive exploration programs in the region and the emergence of the Gulf of Suez as a major oil-producing province. Discovery of the giant July and Ramadan fields in 1974 confirmed the prolific nature of the trend and added the massive Nubia sandstone to the list of proved reservoirs.

Amoco obtained the October concession in 1963, but exploration was delayed from 1967 to the end of 1975 because of the Arab-Israeli conflict. In 1976, Amoco acquired a large seismic survey in the area, confirming the presence of a large northwest–southeast-trending anticline bounded by a down-to-the-west normal fault. The primary target for the October structure was the Nukhul Formation, which produced at the onshore Abu Rudeis field 10 to 15 km (6 to 9 mi) to the east (Figure 15). The massive Nubia sandstone was a major secondary target. The first well in the October structure, GS 195-1 (now labeled the October A-1 well, drilled vertically beneath the A platform), began drilling in 1977 and penetrated the Nubia at 3308 m (10,850 ft) subsea, nearly 610 m (2000 ft) high to prediction (Figures 16 and 17). The Nubia was productive, with 165 m (541 ft) of net oil pay that tested at a rate of 4500 BOPD. The well penetrated a nearly normal Miocene and pre-Miocene section, but the Nukhul reservoir was absent. The first delineation well, GS 195-2 (now labeled the vertical October B-1 well), was drilled about 1.5 km (1 mi) to the northwest and penetrated the Nubia slightly lower at 3395 m (11,140 ft) subsea, with 90 m (297 ft) of net pay (Figures 12 and 17).

In 1978, the third delineation well, GS 185-1 (now labeled the vertical October C-1 well), was drilled about 5 km (3 mi) northwest of GS 195-1 (Figures 12 and 17). The well penetrated the Nubia at 3453 m (11,328 ft) subsea, with 59 m (193 ft) of net pay. The GS 185-1 also penetrated 21 m (70 ft) of net pay in lower Senonian sands, a new reservoir for the area. A common oil-water contact in the Nubia was recognized in the three wells at 3557 m (11,670 ft) subsea, suggesting a large and continuous accumulation. Later in the year, the GS 173-1 delineation well (now labeled the vertical October D-1 well) was drilled 6 km (4 mi) northwest of the GS 185-1, encountering a wet Nubia section at 3890 m (12,764 ft) subsea, downthrown relative to the Nubia discovery wells (Figure 18). However, the Nukhul section penetrated by the well had 32 m (106 ft) of net oil pay that flowed at a rate of 4900 BOPD.

Initial Nubia reservoir development in the main field consisted of 22 successful wells from the A, B, and C platforms. In 1988 and 1989, the G and H development platforms were installed to improve recovery of crestal reserves. To date, ten additional development wells have been successfully drilled along the trapping fault west of the A and B platforms (Figure 17). Currently, in the main October field area, 27 wells produce from the Nubia at an average per-well rate of 4500 to 5000 BOPD.

TABLE 1. Geologic and petrophysical parameters for October field reservoirs.

| | Reservoir | | | |
Parameter	Nubia	Nezzazat	Nukhul	Asl
Reserves (%)	93	2	4	1
Number of producing wells	30	7	2	2
Formation age	Cretaceous/Permian–Carboniferous	Cenomanian-Senonian	L. Miocene	L. Miocene
Lithology	Quartz SS	Carbonate/SS	Conglomerate	Calcareous SS
Depositional environment	Fluvial channel	Marginal marine	Alluvial	Alluvial/fluvial
Average gross thickness [m (ft)]	701 (2300)	366 (1200)	0–76 (0–250)	76 (250)
Average net pay thickness [m (ft)]	137 (450)	24 (80)	30 (100)	40 (130)
Average porosity (%)	16	25	30	11
Average permeability (md)	236	N.A.	N.A.	120
Average water saturation (%)	17	33	27	19
Original OWC (ft TVDSS)	−11,670 (main field)	Unknown	Unknown	Unknown
Original productive area [ha (ac)]	2792 (6900)	2590+ (6400+)	809+ (2000+)	485 (1200)

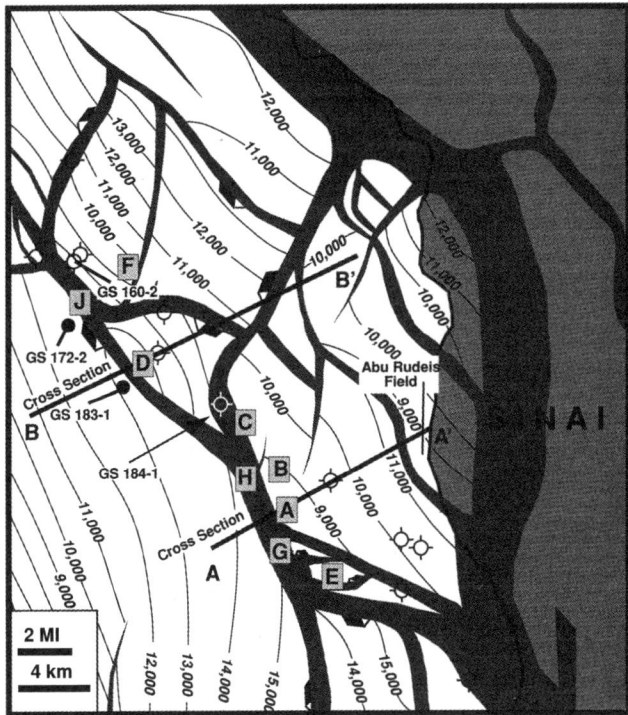

FIGURE 12. October area pre-Miocene structure map. This map shows major pre-Miocene structural elements (faults with displacement greater than 300 m, or 1000 ft), including northeast-dipping fault blocks bounded by major extensional faults that converge to form the October Nubia trap. Significant wells are also shown. The first four exploratory wells, 195-1 (A-1), 195-2 (B-1), 185-1 (C-1), and 173-1 (D-1), were platform locators for the A, B, C, and D platforms, respectively. Contours show depth to pre-Miocene, measured in feet below sea level.

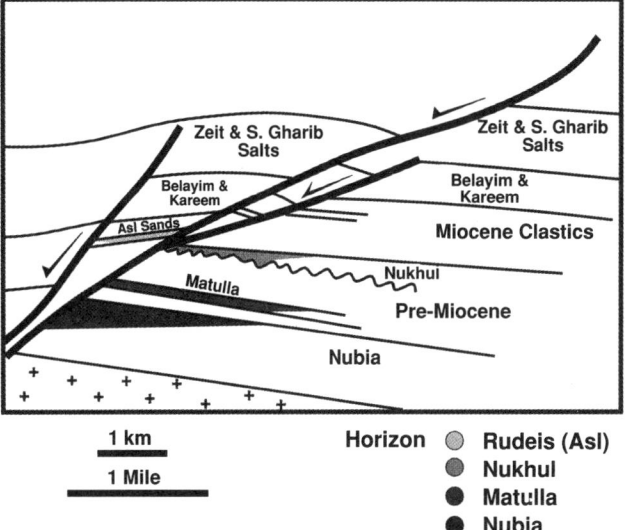

FIGURE 13. Conceptual dip cross section showing the distribution of reservoirs in October producing trend. The main reservoir area currently produces from the Nubia and Matulla. The D platform area currently produces from a Nukhul and a Rudeis reservoir. The J platform area currently produces from a Nubia and a recently discovered Rudeis reservoir.

Numerous other exploratory tests were drilled on the October trend between 1979 and 1988, but with limited commercial success. Although accumulations in the Nubia were increasingly hard to find, lower Senonian sands were often found to be oil-bearing. In the past few years, Nubia wells with declining production rates have been

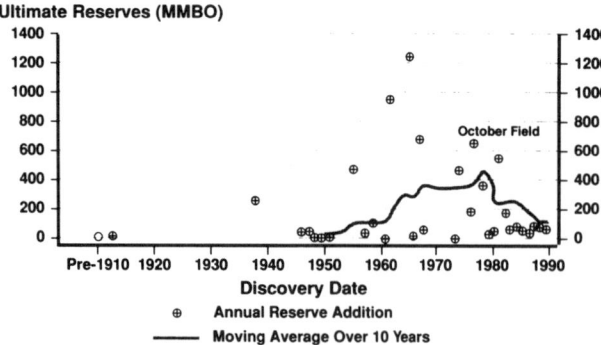

FIGURE 14. History of discoveries within the Gulf of Suez. Individual field reserve estimates are shown as points; the line traces a 10-yr moving average, showing a typical basin progression. (Amoco Basin Analysis Group, with data from Petroconsultants.)

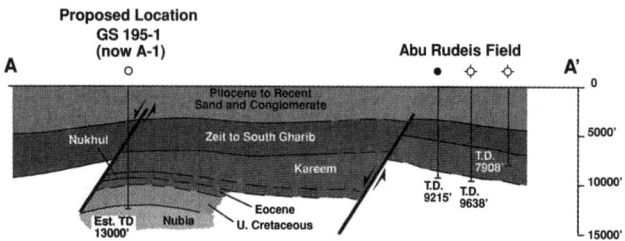

FIGURE 15. Structural cross section from October structure to Abu Rudeis field. This pre-October field discovery cross section (circa 1976) shows October to be similar to the Abu Rudeis structure with potential Nukhul and Nubia reservoirs.

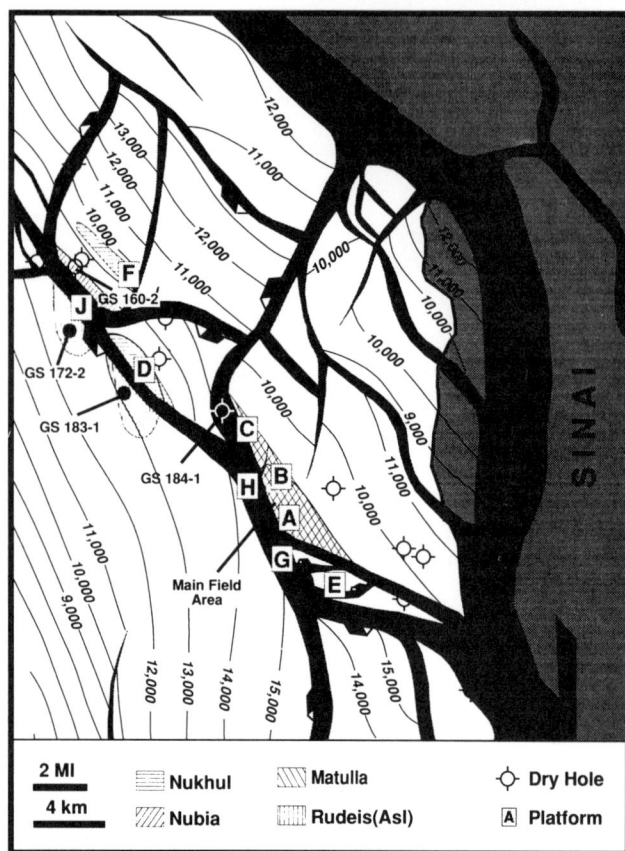

FIGURE 16. Distribution of reservoirs in October trend. The main field area is located in the south, whereas five smaller accumulations (two Nukhul reservoirs, two Asl reservoirs, and one Nubia reservoir) have been discovered in the North October area. Contours show depth to pre-Miocene, measured in feet below sea level.

recompleted in lower Senonian sands with encouraging results. Currently, seven wells produce from the lower Senonian at an average per-well rate of 1600 to 2000 BOPD. Additional development of the lower Senonian reservoir in the main Nubia reservoir area is currently under evaluation.

In 1989, another productive Nubia fault trap was discovered in the North October area. The GS 172-1 well (now labeled the vertical October J-1 well) was drilled 5 km (3 mi) northwest of the October D-1 well immediately updip from the GS 160-2 well (Figure 16). It penetrated the Nubia at 3268 m (10,723 ft) subsea with 77 m (254 ft) of net oil pay that tested at a rate of 7880 BOPD. This discovery led to the drilling of the GS 183-1 well, a Nubia test located updip and west from the October D-1 well (Figure 18). GS 183-1, a directional well, penetrated a wet Nubia section, tight Senonian and truncated/eroded Nukhul, but found productive Miocene Asl sands (Lower Rudeis Formation) downthrown to the major trapping fault. The Asl discovery consisted of 37 m (122 ft) of net oil pay that tested at a rate of 20,170 BOPD. In 1990, the J platform was placed at the GS 172-1 discovery location for subsequent Nubia reservoir development. The first well drilled from the J platform, GS 172-2, drilled to the south-southwest, encountered another Asl reservoir with 49 m (160 ft) of net oil pay that tested at 10,370 BOPD (Figure 16). Evaluation of these exciting new Asl reservoirs was in progress in early 1991.

OCTOBER FIELD STRUCTURE

The October field trend is situated along an overlapping synthetic transfer zone that extends north-south through the October area. Pre-Miocene fault blocks dip to the northeast and are bounded by down-to-the-west normal faults. The October field trend is believed to be an offshore down-faulted extension of the Nezzazat Mountain Range that outcrops to the south along the Sinai Coast for 20 km (12 mi). The main Nubia reservoir is situated at the juncture of two major fault systems (Figure 16). The trap consists of a two-way fault-controlled closure on a northeast-dipping pre-Miocene block. A splay of the clysmic fault system to the northwest creates a trap for a reservoir in the Miocene Nukhul clastics situated on the hanging wall of the main October fault. The northernmost production along the trend occurs in a smaller but significant Nubia fault trap currently under development. The

October Field: The Latest Giant under Development in Egypt's Gulf of Suez   241

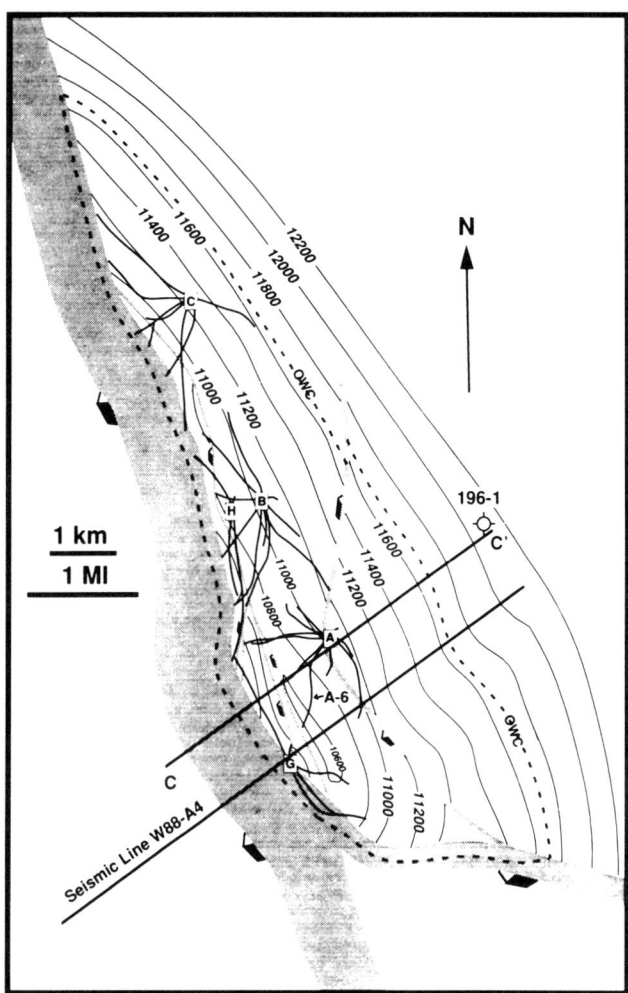

FIGURE 17. Nubia structure map of main reservoir area. Five production platforms drain the northeast-dipping fault block. Several small adjustment faults splay from the trapping fault to complicate and flatten crestal relief due to loading and drag from the overlying Miocene section. Contours represent feet below sea level.

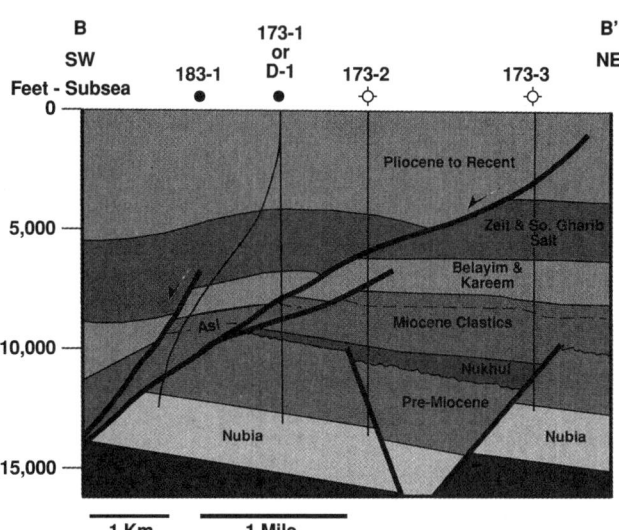

FIGURE 18. Structural dip cross section B-B' through the D platform. The Nukhul reservoir develops on the hanging wall to the main October fault block. The GS 173-1 (now D-1) well found the Nubia to be low and wet, but penetrated oil-bearing Nukhul clastics. The GS 183-1 well, an updip test of the Nubia, was successful, penetrating oil-bearing Lower Rudeis Formation Asl sands, within the hanging wall. (After A. Hamzy, modified by M. Hsu, Amoco Internal Report.)

style of this trap is remarkably similar to that of the main reservoir.

A structure map of the main reservoir (Figure 17) shows that the Nubia dips gently to the northeast (N57°E) at an average of 9°. The main reservoir has maximum dimensions of 13 km (8 mi) in length and 3.4 km (2 mi) in width, with a total surface area of 25.9 km² (6400 ac). The October A-6 development well encountered the Nubia in the structurally highest position to date at 3225 m (10,578 ft) subsea. With the original oil-water contact at 3558 m (11,670 ft) subsea, the total height of the oil column penetrated to date is 333 m (1092 ft). The structure is complicated by several small faults of 25- to 90-m (80- 300-ft) displacement that control relief but have limited impact on reservoir performance. These faults represent adjustments to loading from the overlying Miocene clastics and evaporite sequences. Along the crest of the structure, every new well shows previously unrecognized faulting.

Figure 19 shows a dip structural cross section (C-C') of the Nubia through the initial discovery well and several development wells. It shows the rotated pre-Miocene fault blocks overlain by thick Miocene clastics and evaporites. Current interpretation shows that faults generally trend parallel to the trapping fault.

The Nubia reservoir is sealed by the fault-related juxtaposition of Lower Rudeis shales with the Nubia. The size of the accumulation could be limited by either source potential or fault seal. Clearly, Lower Rudeis shales effectively seal the 333-m (1092-ft) oil column. The Nukhul and Thebes formations, underlying the Lower Rudeis shales on the hanging wall, would not be effective seals and may represent the spillpoint of the structure. Migration of additional oil would be spilled to the next favorable trapping position, such as the Nukhul reservoir at the D platform, which is also sealed by juxtaposition of Lower Rudeis shales.

The magnitude of displacement along these predominantly dip-slip clysmic faults at the pre-Miocene horizon is remarkable (Figure 16). As much as 1250 m (4100 ft) of displacement is transferred along the northwest oblique fault system from the Feiran field, and as much as 1000 m (3280 ft) of displacement is transferred along the clysmic trending fault system from the Belayim marine field. Because a complete pre-Miocene section has not been penetrated downthrown to October field and is not adequately defined on seismic, the exact amount of dip-slip displacement along the bounding fault is unknown. The dip angle of these faults ranges from 45 to 55°.

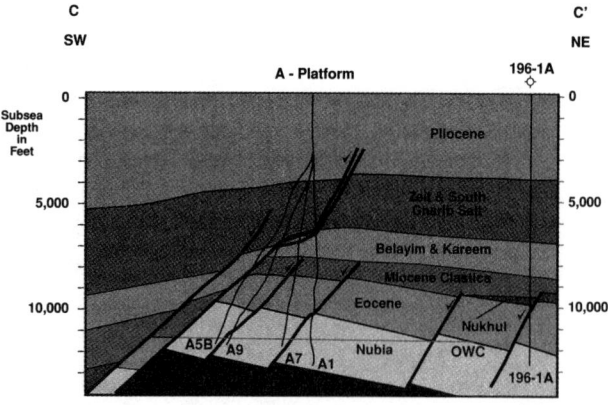

FIGURE 19. Structural cross section C-C' through main Nubia reservoir. The tilted pre-Miocene fault block is bounded on the west by a major fault system. Juxtaposition of Lower Rudeis shales across the trapping fault from the Nubia effectively seals a 333 m (1092 ft) column of oil. The Miocene section drapes the escarpment and shows growth downthrown due to syndepositional movement on the west-bounding faults.

Figure 20, a depth-migrated seismic profile (after multiple suppression) oriented in the dip direction through the crestal G platform, shows two large basin expansion faults. The westernmost fault cuts the Miocene evaporite sequences and produces major expansion into the basin. The fault immediately to the east is the trapping fault for the Nubia reservoir, with significant displacement at the pre-Miocene level, but no displacement above the Belayim horizon. Seismic suggests displacement on the trapping fault of 1250 m (4100 ft) near the G platform, whereas correlation of the Lower Rudeis near the C platform indicates displacement of 1037 m (3400 ft). Displacement within the pre-Miocene, calculated from the Lower Rudeis, would be underestimated by the thickness of the Nukhul section, which overlies the pre-Miocene on the hanging wall but is absent on the footwall.

Recent development of crestal reserves revealed complex propagation of the trapping fault in the B platform area. Figure 16 shows that the two major faults diverge near the H platform, where the westernmost fault assumes a northwest oblique strike and the bounding fault for the reservoir turns to north oblique and back to clysmic orientation. The western fault system is the dominant feature to the north and provides much of the structural complexity associated with the recent discoveries in the North October area.

Also remarkable is how little displacement is shown on these faults at the Belayim horizon, indicating little late stage rejuvenation in this area. The top of the Belayim Formation is generally the deepest seismic horizon that can be reliably mapped in the gulf basin; locally, it represents the base of salt and the top of Miocene synrift clastics. Much of the displacement on these fault systems is taken up by dramatic drape of the evaporites over the fault escarpment, as well as significant thickening or growth of the Miocene synrift clastics on the hanging wall. In the main reservoir area, Miocene evaporites drape the fault escarpment, whereas in North October the evaporites roll into the fault, producing multiples that have approximately the same dip as the underlying pre-Miocene section, greatly inhibiting seismic recognition. Extensive drape of the evaporites produces seismic multiples with opposing dips to the primary reflections, making structural interpretation easier. Multiple suppression techniques developed by Amoco, as well as seismic contractors, are often helpful, but the number of dry holes drilled for exploration and even development indicates the importance of close interaction between geologists and geophysicists.

NUBIA CHARACTERISTICS

The Nubia Formation is subdivided into six major stratigraphic units based on lateral correlation of electric logs and petrophysical properties. These units, from top to bottom, are the Transitional Nubia, MN, MI, MII shale, MIIA, and MIIB zones (Figure 21). The section is approximately 850 m (2800 ft) thick and gently increases in thickness across the October area toward the northeast. It consists of a series of vertically stacked massive sandstones distinctly interbedded with massive to thin lenticular shales. The Nubia records the prerift transition from a continental to a marginal marine depositional environment.

Generally, the six stratigraphic subdivisions are comprised of 80% sandstone and 20% shale. They show good lateral continuity in both the strike and dip directions (Figure 22) over a 24-km (15-mi) distance from the J platform area in the north to the G platform area in the south. Below the MII shale, correlation quality tends to decline. There, the MIIA and MIIB zones are more randomly interbedded, and correlation is more interpretive.

The Transitional Nubia is stratigraphically the highest oil-bearing reservoir unit in the Nubia section. It is composed of Cenomanian thin calcareous sandstone and shale interbeds, the tops of which are delineated fieldwide by a distinct thin porous zone distinguished on logs (Figure 21). Because of the large aerial extent of oil-bearing Transitional Nubia in the October complex (3400 ha, or 8400 ac, above the oil-water contact), the Transitional Nubia contains approximately 25% of the main field oil in place. This is impressive given its moderate thickness, ranging from 35 to 60 m (115 to 200 ft).

Below the Transitional Nubia are the massive MN and MI sandstones. Together they range in thickness from 100 to 135 m (330 to 440 ft) and are interpreted as fining-upward stacked fluvial channels. They have excellent reservoir quality and contain over 50% of the total Nubia reserves in the field. In the section above the MII shale, the net-to-gross pay ratio is an impressive 0.9. Sandstones are dominantly medium-grained with some coarse-grained and fine-grained intervals. Good porosity and good vertical and horizontal permeability exist. Petrologic studies reveal that the sandstones are moderately to well-sorted quartz arenites with accessory minerals of basement-derived tourmaline and zircon. Authigenic kaolinite and quartz overgrowth cement act as primary

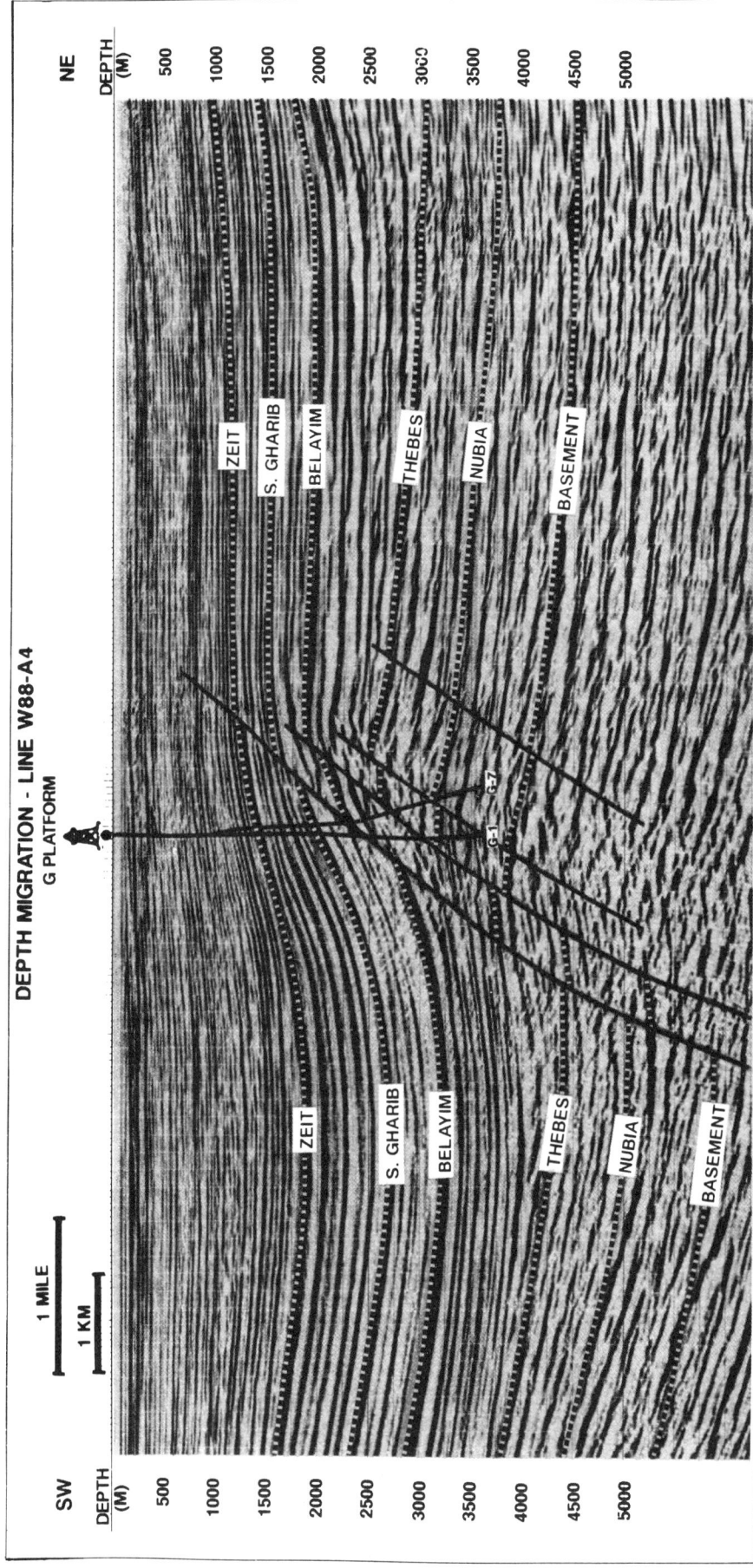

FIGURE 20. Depth migrated seismic line W88–A4. An acquisition test line through the crestal G platform area of the main field area, with multiple suppression processing, offers some of the best seismic recognition of pre-Miocene anywhere in the Gulf. The structural description is similar to cross section C–C' located 1 km to the north. Onlap of the Nukhul onto the rift block to the east of the bounding faults indicates early rotation on the October structure. The position of the pre-Miocene on the hanging wall is implied.

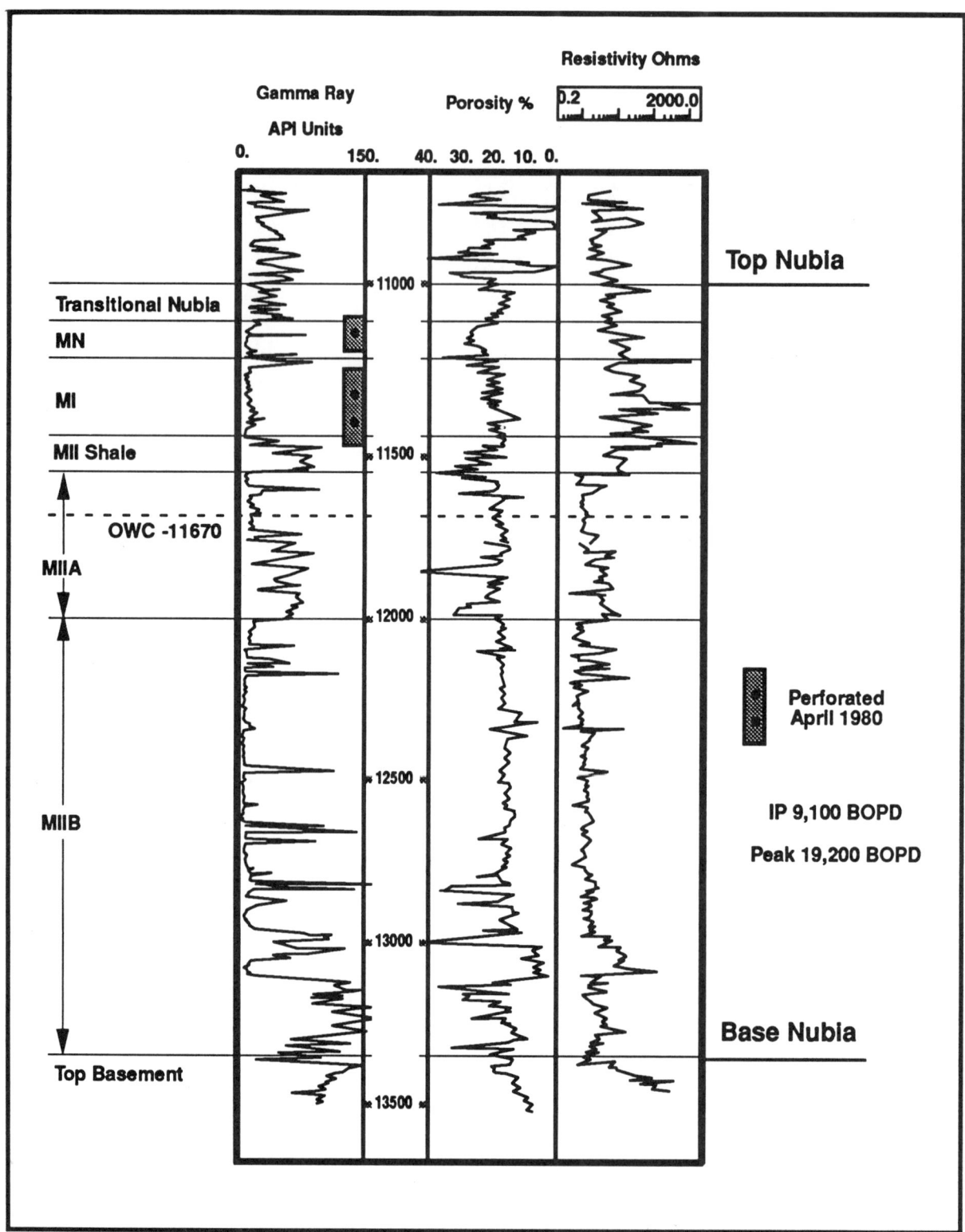

FIGURE 21. Nubia reservoir type log, composited from two wells, representing a typical well in the structure. The depth track indicates measured depth in feet.

porosity-reducing agents. The MN and MI sandstones are separated vertically by the thin MI shale marker.

Beneath the MN and MI zones is the MII shale. It is a reddish brown, nonproductive unit ranging in thickness from 18 to 27 m (60 to 90 ft). Nubia production history shows it to act as a fieldwide vertical permeability and pressure barrier, isolating the light oil-bearing Nubia above and below it. The position of the MII shale with

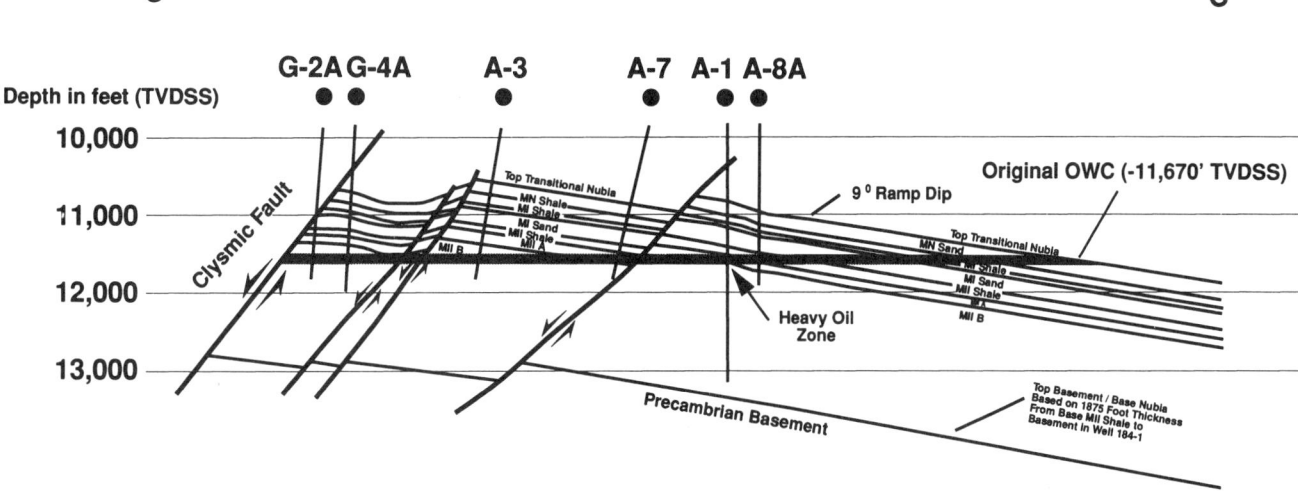

FIGURE 22. Dip cross section of Nubia reservoir subdivisions. The section shows stratigraphic units of the reservoir dipping gently to the east, their juxtaposition across faults, and the heavy oil zone angularly crossing the units.

respect to the oil-water contact must be considered when targeting either section with development wells.

The MIIA and MIIB zones are the deepest productive units in the field and serve as reservoirs only in the crestal area. Limited well penetrations through these two zones, owing to their positions below the field oil-water contact, limit their interpretation. They were probably deposited by alluvial and fluvial systems draining the basement terrain from a westward direction. Of the two, the MIIB zone has the best reservoir quality. This is demonstrated by the MIIA and MIIB net-to-gross pay ratios of 0.3 and 0.8, respectively.

Continuous sedimentation from Carboniferous through late Paleozoic time caused reworking and maturation of the sediments in these continental depositional systems. Increased bedding organization upsection, and increased textural and compositional maturity, reflect the transition from a continental to a marginal marine depositional environment.

SOURCE ROCKS, GENERATION, AND MIGRATION

The Campanian Brown Limestone Member of the Sudr Formation (Figure 11) is interpreted as the dominant regional source rock for oil in Gulf of Suez fields, including October field. Other source rocks of lesser significance are thin-bedded calcareous and siliceous shales of the Eocene Thebes and Miocene Rudeis formations, and the Cretaceous Nezzazat Group.

The Campanian Brown Limestone Member is a marine unit deposited regionally during southward transgression of the Tethys Sea. It ranges in thickness from 25 to 70 m (80 to 230 ft) (Chowdhary and Taha, 1987). Geochemical analyses performed on both well cuttings and outcrop samples from the Gulf of Suez rift show that total organic carbon (TOC) averages 2.6 wt % for the Campanian limestone, with a maximum of 3.2%. It typically contains amorphous type II kerogen. TOC values as high as 21% have been documented in Campanian Brown Limestone samples collected from underground phosphate mines in southeastern Egypt (Robison, 1986). Rock-Eval pyrolysis shows excellent hydrocarbon generation potential with S2 ranging from 1400 to 7000 ppm. Average TOC values for the Thebes and Nezzazat formations are 2.3 and 1.5 wt %, respectively.

Independent analyses of Gulf of Suez oils conclude that they are genetically related to one source (Zein El-Din and Shaltout, 1982; Rohrback, 1983). Even-carbon paraffin distribution and high sulfur content suggest that the source is a limestone rather than a shale (Lewan, personal communication). Whole oil chromatograms show October field Nubia and Nukhul light oils to be compositionally similar (Figure 23).

Well data show that the geothermal gradient varies in the October field area from 2.2 to 2.7°C/100 m (1.5 to 1.8°F/100 ft), depending on structural position—i.e., structurally high areas tend to be cooler and basinal areas warmer. The average depth to the top of the oil window is 3050 m (10,000 ft) based on basin subsidence studies (Chowdhary and Taha, 1987). Structural interpretations show that the Campanian Brown Limestone is in the oil window in basinal areas surrounding October field.

Although the exact generation area for October field oil is unknown, three adjacent subbasins (Figure 24) may have played roles in sourcing the oil for October field. These are (1) the subbasin to the southwest, downthrown to the main clysmic or bounding fault; (2) the area to the south, down-structure from the E platform; and (3) the downdip area to the east-northeast of the October trend.

In the case of the subbasin to the southwest, the clysmic fault seals the Nubia reservoir from the oil-water contact (OWC) to the crest of the structure. At the crest,

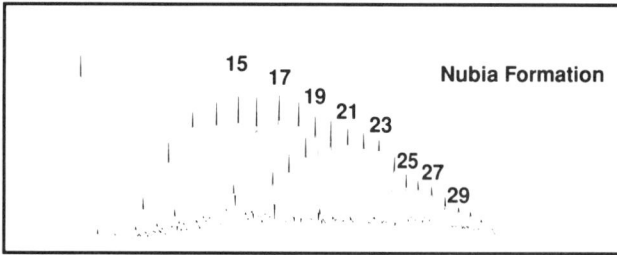
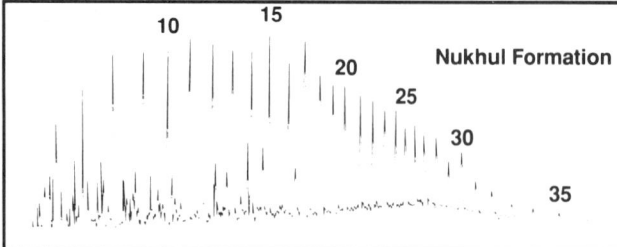

FIGURE 23. Whole oil chromatogram of Nubia and Nukhul oils. These oils are interpreted as originating from the same source rock. Present-day compositional differences reflect timing of expulsion, migration path, and possible late stage alteration.

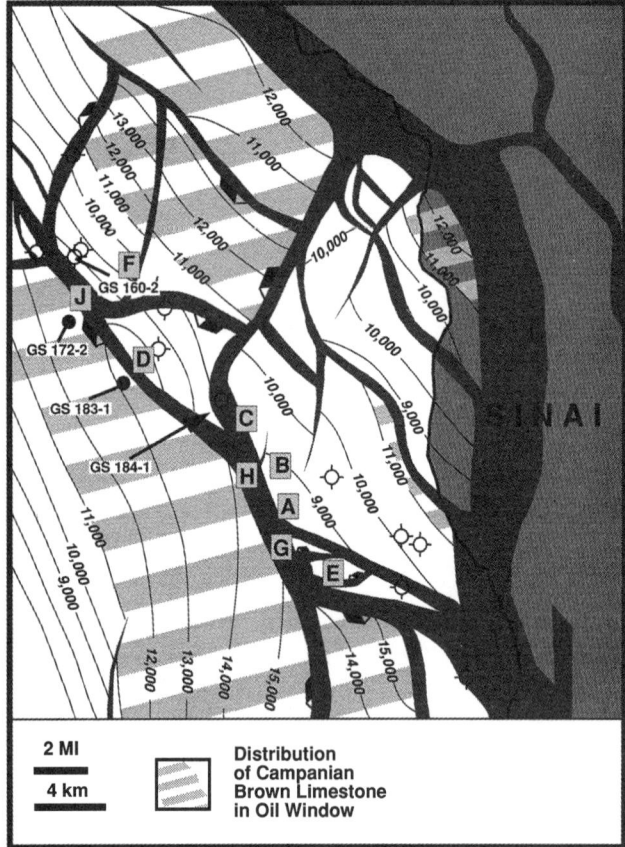

FIGURE 24. Hydrocarbon source basin map. The Campanian Brown Limestone lies in the oil window in several basins surrounding October field. Migration pathways are complicated and uncertain. The Lower Rudeis could also be in the oil window to the west of the main reservoir, offering a more direct migration pathway into the Nubia. Contours indicate depth to pre-Miocene in feet below sea level.

Miocene Rudeis shales and marls are directly juxtaposed against oil-bearing Nubia sandstones. Cross-clysmic fault, sub-OWC oil migration may have occurred with oil entering a fault block and then migrating vertically within the Nubia. This assumes that the position of the Nubia OWC or the volume of oil contained in the Nubia is limited primarily by a change in seal capacity somewhere around the perimeter of the field.

The area downdip to the east-northeast of October probably plays a major role in sourcing the three significant oil fields that flank this basin—October, Ras Budran, and Abu Rudeis fields. Similar to the southern subbasin down-structure from the E platform, complex faulting of the source interval suggests tortuous updip migration into surrounding fault blocks. Oil must migrate across nonsealing faults, and even within fault blocks migration is controlled by stratigraphically influenced permeability pathways.

OIL CHARACTERISTICS

Although the Nubia reservoir in the main field area contains both light and heavy oils (Table 2), only the light oil is currently being produced. The properties of light and heavy oils are important in understanding reservoir behavior. Heavy oil is largely immobile, acting as a permeability barrier and restricting the natural water drive. The light oil reservoir forms a wedge ranging in thickness from 270 m (900 ft) at the crest to zero at the eastern dip closure. The average gravity is 27° API and generally decreases with depth (Figure 25).

Below the light oil is a heavy oil zone 45 m (150 ft) thick. It is laterally continuous throughout the field, although it does thin downdip to the east. The zone crosscuts the stratigraphic units of the east-dipping Nubia reservoir (Figure 22). More than ten years of RFT (repeat formation tester) pressure data show that the heavy oil zone acts as a permeability barrier to upward aquifer encroachment into the overlying light oil. Compositionally, it contains intergranular bitumen and pyrobitumen interpreted as having been formed by de-asphalting.

PRODUCTION AND PRESSURE HISTORY

Approximately 411 million bbl of oil was produced from the Nubia Formation at October field from October 1977, when the field came on line, through December 1990. Figure 26 traces the production history of the field. The total field production rate reached 100,000 BOPD in 1982 and has exceeded that level ever since, with an early 1991 rate of about 170,000 BOPD (115,000 BOPD from Nubia). Water breakthrough began in early 1982 and slowly increased to an early 1991 rate of approximately 20,000 BWPD, which equates to a Nubia water-oil ratio of 17%. The producing gas-oil ratio is currently flat at 338 standard ft³/STB.

Nubia reservoir pressure has steadily declined from an original value of 5506 psi at the 3429-m (11,250-ft) standard subsea datum to an early 1991 value of 2621 psi at datum (Figure 27). A hydrocarbon column of 335 m (1100 ft) exists with significant vertical hydrocarbon com-

TABLE 2. Characteristics of light and heavy oils from Nubia reservoir, October field.

Characteristic	Light oil	Heavy oil
Gravity (° API)	27	13–18
Bubble-point pressure (psi)	1683	540–1740
Solution GOR (standard ft^3/STB)	279	134
Density at 4000 psi and 265°F (g/cm^3)	0.7915	0.89
Viscosity at 4000 psi and 265°F (cp)	1.007	36.8
Primary drive mechanism	Edge water	—

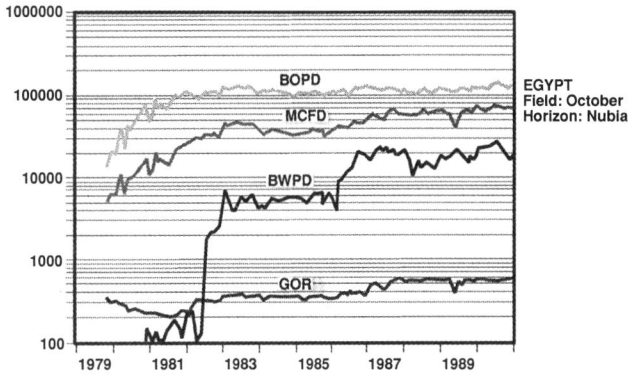

FIGURE 26. Nubia reservoir production history. The top two curves on the graph show average daily oil (BOPD) and gas (MCFD) production since production began in 1979. The third curve shows water (BWPD) production, which began in late 1980, with significant water breakthrough beginning in early 1982. Water breakthrough gas-oil ratio (GOR), the fourth curve, has ranged between 300 and 600 SCF/STB since 1979.

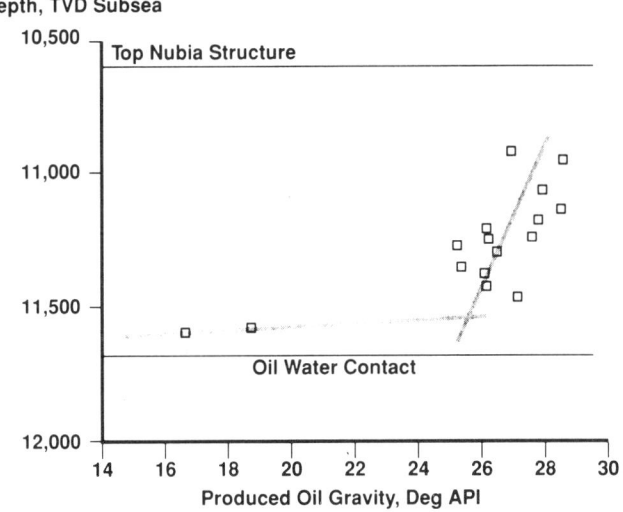

FIGURE 25. Depth vs. API gravity. Vertical compositional differences (of oil) within the field are shown by the decrease in API gravity with increasing depth. The two deep data points reflect the heavy oil leg near the oil-water contact. Depths are in feet below sea level.

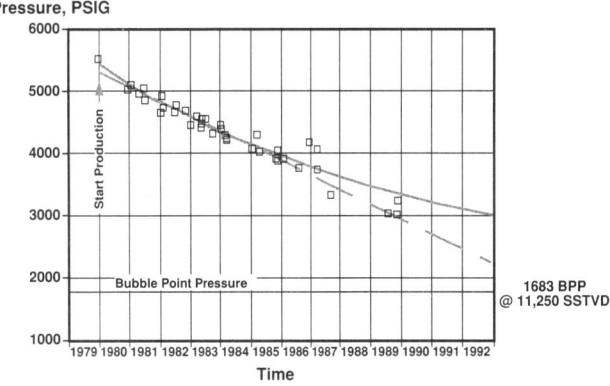

FIGURE 27. October field pressure history, showing a typical decline.

positional differences. Pressure measurements received from RFT tests throughout field development are plotted against subsea depth in Figure 28. These data show the vertical pressure gradient of 0.25 to 0.30 psi/ft within the reservoir column above the MII shale, as well as the decline in reservoir pressure through time. The RFT vs. depth plot also highlights the pressure differential across the oil-water contact and associated heavy oil zone, supporting the hypothesis that vertical water drive through the heavy oil zone is ineffective. Water drive is believed to be from the east, where the eastward-thinning heavy oil zone approaches the top of the formation near the oil-water contact. Because of this edge water drive, recent development drilling has favored the west-bounding fault face to maximize ultimate recovery.

Bubble-point pressure exhibits a large vertical variance in the field, decreasing dramatically with depth (Figure 29). At the top of the Nubia structure it is approximately 2400 psi; at the oil-water contact it is close to 1400 psi; and at the traditional datum of 3429 m (11,250 ft) it is 1683 psi. In 1990, the approach of the reservoir pressure toward the bubble-point pressure (Figure 30) became a great concern, initiating an intensive joint field study at Amoco's Houston office and the Cairo office of GUPCO. As the study progressed, it was determined that falling of the reservoir pressure below the bubble point would not be devastating to reservoir management. A final reservoir management plan for the October field Nubia had not yet been formulated at the time this paper was submitted.

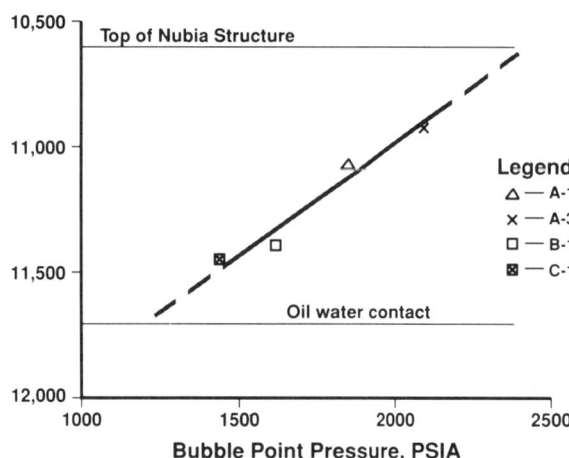

FIGURE 28. October field pressure trends. Pressures from various RFT logging runs are plotted against depth, with dots indicating individual point tests, and lines indicating individual logging runs. Logging dates are shown above the RFT lines. Reservoir pressure is shown approaching the bubble-point pressure through time. Depths are in feet below sea level.

FIGURE 29. Depth vs. bubble-point pressure. Bubble-point pressure decreases dramatically as depth increases, suggesting the reservoir will first encounter its bubble point at the structural crest. Depths are in feet below sea level.

CONCLUSIONS

The giant October field is a complex of gently rotated fault blocks typical of rifts throughout the world. It contains both light and heavy oils in six separate reservoirs in four stratigraphic horizons. The Nubia Formation is the main reservoir, containing over 90% of the reserves. Major transverse and clysmic faults complicate development drilling and production plans.

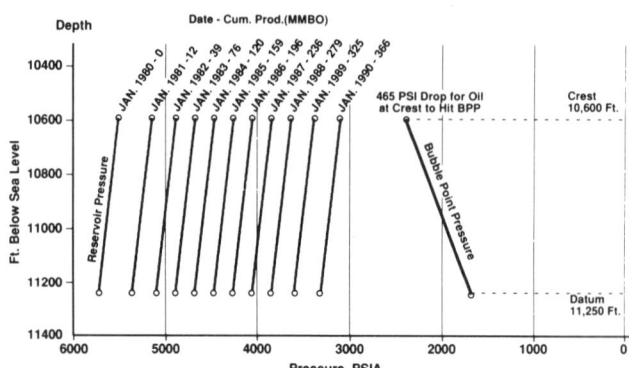

FIGURE 30. Reservoir pressure history. Internal reservoir pressures at yearly intervals are plotted on this depth vs. pressure plot; pressure increases to the left. The number following each date indicates cumulative production as of that date, in millions of barrels. The bubble-point pressure is also shown, decreasing with depth.

Exploration efforts as well as subsequent development efforts at October field were hampered by the shallow salt section, which provides an excellent hydrocarbon seal but creates severe seismic multiple problems. During drilling, the position of the structural crest was extremely difficult to discern. This crestal resolution remains a major objective, along with continuing evaluation of the non-Nubia reservoirs.

Continuing in-depth reservoir studies are normally necessary to ensure an optimum reservoir development and management program, especially in big fields where remedial/incremental work can add significant economic value through production rate increases and reserve growth. In the case of October field, a complete reservoir study, completed eight years after field discovery, identified certain problems, recommended a specific plan of action, and made a host of performance predictions. Only four years later, with additional well and production data, and new discoveries adjacent to the original "field," different problems arose, which required another in-depth study. The optimum way to conduct such reservoir studies is to dedicate an interdisciplinary team to the identification and solution of the highest-impact problems. For October field, this team included geologists, geophysicists, petrophysicists, and reservoir engineers in both Houston and Cairo, as well as researchers in Tulsa and field personnel in the Ras Shukheir Egypt office. This type of well-planned, intensive effort pays out handsomely in terms of optimum reservoir management, maximum oil recovery, and maximum present financial worth.

ACKNOWLEDGMENTS

The authors would like to thank all the creative and dedicated people who have influenced exploration and development of the October trend. In particular, we thank Dave Smith, Peter Manoogian, and Joe Ader for their contributions, Ron Nelson for his constructive review, and Amoco Production Company and the Gulf of Suez

Petroleum Company for their permission to publish this paper.

REFERENCES CITED

Angelier, J., 1985, Extension and rifting: The Zeit region, Gulf of Suez: Journal of Structural Geology, v. 7, p. 605-612.

Argyriadis, I., P. C. De Garciansky, J. Marcoux, and L. E. Ricou, 1980, The opening of the Mesozoic Tethys between Eurasia and Arabia-Africa: Mem. du Bureau de Recherches Geologiques et Minieres, v. 115, p. 199-214.

Chenet, P. Y., and J. Letouzey, 1983, Techtonique de la zone comprise entre Abu Durba et Gebel Nezzazat (Sinai Egypte) dans le contexte d' evolution du reft de Suez: Bulletin Centres Recherches Explor.-Prod., Elf-Aquitaine, v. 7, p. 201-215.

Chowdhary, L.R., and S. Taha, 1987, Geology and habitat of oil in Ras Budran field, Gulf of Suez, Egypt: AAPG Bulletin, v. 71, n. 10, p. 1274-1293.

Coffield, D. Q., and S. Schamel, 1989, Surface expression of an accommodation zone within the Gulf of Suez rift, Egypt: Geology, v. 71, p. 76-79.

Colletta, B., I. Moretti, P. Y. Chenet, C. Muller, and P. Gerard, 1986, The structure of the Gebel Zeit area, a field example of tilted block crest in the Suez rift: Institut Francais du Petrole, Ref. I.F.P. 34, p. 547-560.

Colletta, B., P. Le Quellec, J. Letouzey, and I. Moretti, 1988, Longitudinal variation of the Suez rift structure (Egypt): Tectonophysics, v. 153, p. 221-233.

Dercourt, J., et al., 1986, Geologic evolution of the Tethys belt from the Atlantic to the Pamirs since the Lias: Tectonophysics, v. 123, p. 241-315.

El Shazly, E. M., 1977, Geology of the Egyptian region, in A. E. M. Narin, F. G. Stehli, and W. H. Kanes (eds.), The ocean basins and margins, the eastern Mediterranean, 4A, p. 379-344.

El-Tarabili, E., and N. Adawy, 1972, Geologic history of the Nukhul-Baba area, Gulf of Suez, Egypt: AAPG Bulletin, v. 56, p. 882-902.

Fawzy, H., and A. Abdel Aal, 1986, Regional study of Miocene evaporites and Pliocene-Recent sediments in the Gulf of Suez: Egyptian General Petroleum Corp., Seventh Exploration Conference, 1984, p. 49-74.

Gawthorpe, R. L., J. M. Hurst, and C. P. Sladen, 1990, Evolution of Miocene footwall-derived coarse-grained deltas, Gulf of Suez, Egypt: implications for exploration: AAPG Bulletin, v. 74, p. 1077-1086.

GUPCO, 1990, October oil field: AAPG Treatise of Oil and Gas Fields, p. 27.

Hempton, M. R., 1987, Constraints on Arabian plate motion and extensional history of the Red Sea: Tectonics, v. 6, p. 687-705.

Jackson, J. A., N. J. White, Z. Garfunkel, and H. Anderson, 1988, Relations between normal-fault geometry, tilting, and vertical motions in extensional terrains: an example from the southern Gulf of Suez: Journal of Structural Geology, v. 10, p. 155-170.

Khalil, B., and W. Meshref, 1988, Hydrocarbon occurrences and structural style of the southern Suez rift basin: Egyptian General Petroleum Corp., Ninth Exploration and Production Conference.

LePichon, X., and J. M. Gaulier, 1988, Rotation of the Arabian and the Levant fault system: Tectonophysics, v. 153, p. 271-294.

Moretti, I., and B. Colletta, 1988, Fault-block tilting: the Gebel Zeit example, Gulf of Suez: Journal of Structural Geology, v. 10, p. 9-19.

Morley, C. K., R. A. Nelson, T. L. Patton, and S. G. Munn, 1990, Transfer zones in East African rift systems: AAPG Bulletin, v. 74, p. 1234-1253.

Moustafa, A. M., 1976, Block faulting in the Gulf of Suez: Egyptian General Petroleum Co., Fifth Exploration Seminar, 1976, p. 36.

Patton, T. L., 1984, Surface studies of normal-faulting geometries in the pre-Miocene stratigraphy, west central Sinai Peninsula: Egyptian General Petroleum Corp., Sixth Exploration Seminar, 1982, p. 437-452.

Patton, T. L., A. R. Moustafa, R. A. Nelson, and A. S. Abdine, in press, Tectonic evolution and structural setting of the Suez rift: AAPG Special Publication, Interior Basins.

Robison, V. D., 1986, Organic geochemical characteristics of the Late Cretaceous-Early Tertiary transgressive sequence found in the Duwi and Dakhla formations, Egypt: Ph.D. dissertation, University of Oklahoma.

Robson, D. A., 1971, The structure of the Gulf of Suez (clysmic) rift, with special reference to the eastern side: Journal of the Geological Society, v. 127, p. 445-503.

Rohrback, B. G., 1983, Crude oil geochemistry of the Gulf of Suez: advances in organic geochemistry, in M. Bjorøy et al. (eds.), New York, John Wiley, p. 39-48.

Sellwood, B. W., and R. E. Netherwood, 1984, Facies evolution in the Gulf of Suez area: sedimentation history as an indicator of rift initiation and development: Modern Geology, v. 9, p. 43-69.

Soliman, S. M., F. Berthelot, N. Lyberis, and X. LePichon, 1988, Subsidence in the Gulf of Suez: implications for rifting and plate kinematics: Tectonophysics, v. 153, p. 249-270.

Vigano, P. L., and T. L. Patton, 1988, Mid-Clysmic Event, Gulf of Suez rotational deformation associated with a deep crustal detachment fault (abs.): AAPG Bulletin, v. 72, p. 1029.

Zahran, M. E., 1986, Geology of October oil field: EGPC Eight Exploration Seminar, p. 13.

Zein El-Din, M. Y., and E. M. Shaltout, 1982, Geochemical correlation of crude oils, Gulf of Suez region, Egypt: AAPG Bulletin, v. 66, p. 646.

Chapter 16

Villeperdue Field
Exploration of a Subtle Trap in the Paris Basin

Bernard C. Duval

TOTAL Compagnie Francaise des Pétroles
Paris, France

ABSTRACT

The Villeperdue oil field is located in the Paris basin less than 100 km east of Paris. The first well was drilled in 1959 and tested some oil, but it was not until 1982 that exploration was resumed and the field proved commercial.

The trap is not obvious from seismic data because of velocity variations in the Tertiary and in the Cretaceous chalk. It is a combination of stratigraphic, structural, and diagenetic features. The structure is a western-plunging nose, the eastward updip closure being controlled mainly by variation of permeability and possibly influenced by gentle faults and pressure barriers. The Villeperdue field has a producing surface area of about 65 km^2 and a 60-m oil column.

The reservoir is an oolitic limestone of Middle Jurassic (lower Callovian) age. Its average thickness is about 30 m and its approximate depth is 1850 m (1650 m subsea). Two main units, linked to specific oolitic microfacies, can be identified. Detection of porosity over the field, and consequently delineation, are therefore dependent on detailed facies and sedimentological studies.

The generation of oil from Liassic source rocks started during the Cretaceous, and the migration upward into the lower Callovian carbonate was made possible by Tertiary faulting.

The current production of 12,000 BOPD comes from 100 wells and is enhanced by water injection. Horizontal drilling has given positive results and will be used more in future programs.

The play concept has been successfully applied on a regional basis and has resulted in several other discoveries, although minor in size.

INTRODUCTION

Although the reserves of the Villeperdue oil field fall short of 100 million bbl, which serves as a limit for qualification as a giant field, Villeperdue is one of the two largest fields discovered in the Paris basin (the other one is Chaunoy). It has an interesting history, which began with modest shows of hydrocarbon. These shows were not fully understood for many years, but nevertheless led to the discovery of a subtle but large stratigraphic trap that is partly controlled by structure.

Elaborate seismic time–depth-conversion techniques, detailed lithological analyses of the carbonate reservoir formation, and a reconstruction of the over-all reservoir sedimentological model have facilitated the understanding of the trapping mechanism and the determination of the extension of the field area that can be produced economically.

BACKGROUND HISTORY OF OIL EXPLORATION: REGIONAL AND PETROLEUM GEOLOGY

The Paris basin is a large cratonic sag that is bordered by the Massif Armoricain of Brittany to the west, the Ardennes-Brabant axis to the north and northeast, the Massif des Vosges to the east, and the Massif Central to the south. It contains a Mesozoic and a Tertiary section (Figure 1). The central part of the basin is almost perfectly circular, as can be seen in an isobath map of the Middle Jurassic on which a few important faults are represented (Figure 2). The Villeperdue field is located about 100 km east of Paris.

The first petroleum licenses in the Paris basin were allocated in 1951, and, although several early wells found good indications of oil, it was not until 1958 that the first sizeable oil field was discovered by Petrorep in Middle Jurassic carbonates at Coulommes-Vaucourtois. By scanning the years of discovery of the main fields producing from Middle Jurassic carbonates, one can see that Villeperdue was discovered more than 20 years after a first cycle of Middle Jurassic discoveries took place on anticlinal traps, of which Coulommes (Tilloy and Dardenne, 1960), Chailly (CEP and RAP, 1960), and Saint-Martin de Bossenay (Bouché et al., 1960) were the most prominent (Figure 2).

It must be pointed out that discoveries also were made in reservoirs other than Middle Jurassic—for instance, Lower Cretaceous sandstones such as Châteaurenard, which was found in the early days of exploration (Manguy, 1960), and Triassic Keuper sandstones such as Chaunoy, which is a major discovery that was made in 1983 by Essorep (Morelot and Pages, 1989).

The old fields were found mostly on fairly simple low-relief anticlinal structures with analog seismic, single-fold coverage. The subsequent exploration period during the 1960s, 1970s, and 1980s was not successful for two reasons:

1. The price of crude oil was low, and exploration consequently slowed down, during part of that period.

2. Improved seismic acquisition and processing technologies were not adequate to overcome the main difficulties of poor velocity control and frequent, unpredictable velocity variations in the Tertiary and Cretaceous carbonate-rich overburden. Cretaceous sediments in the Paris basin are largely made up of chalk, and the Tertiary also contains high-velocity limestones.

Figure 3 shows the regional lithostratigraphic column of the Paris basin, and more particularly the various res-

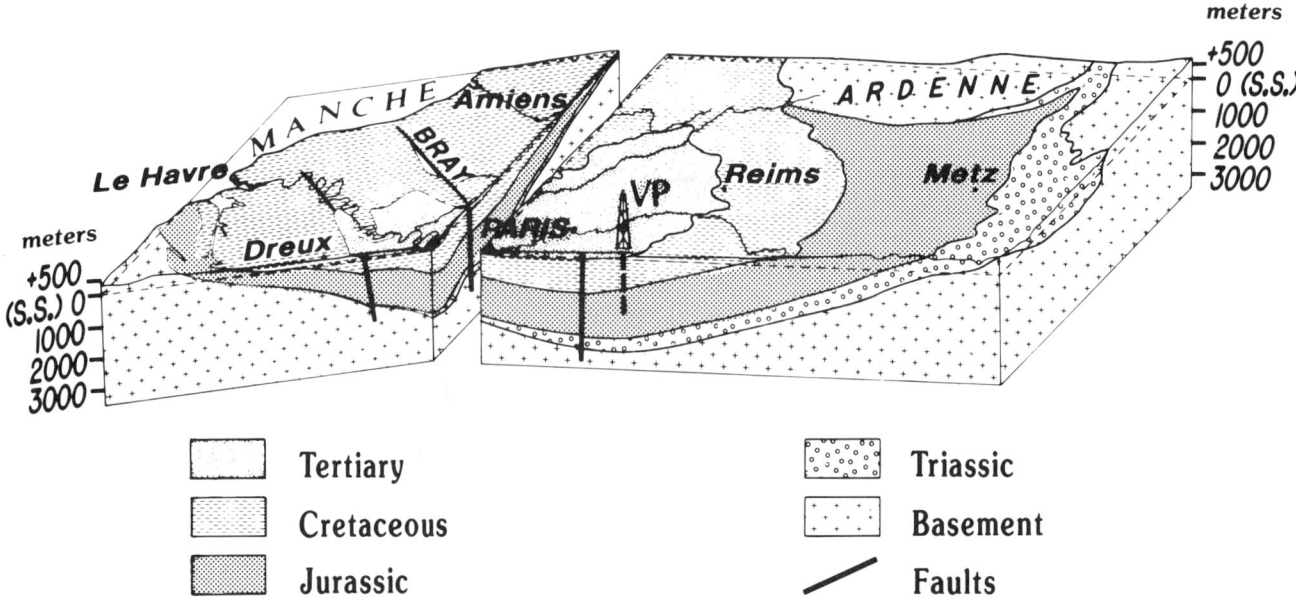

FIGURE 1. Regional block diagram of Paris basin.

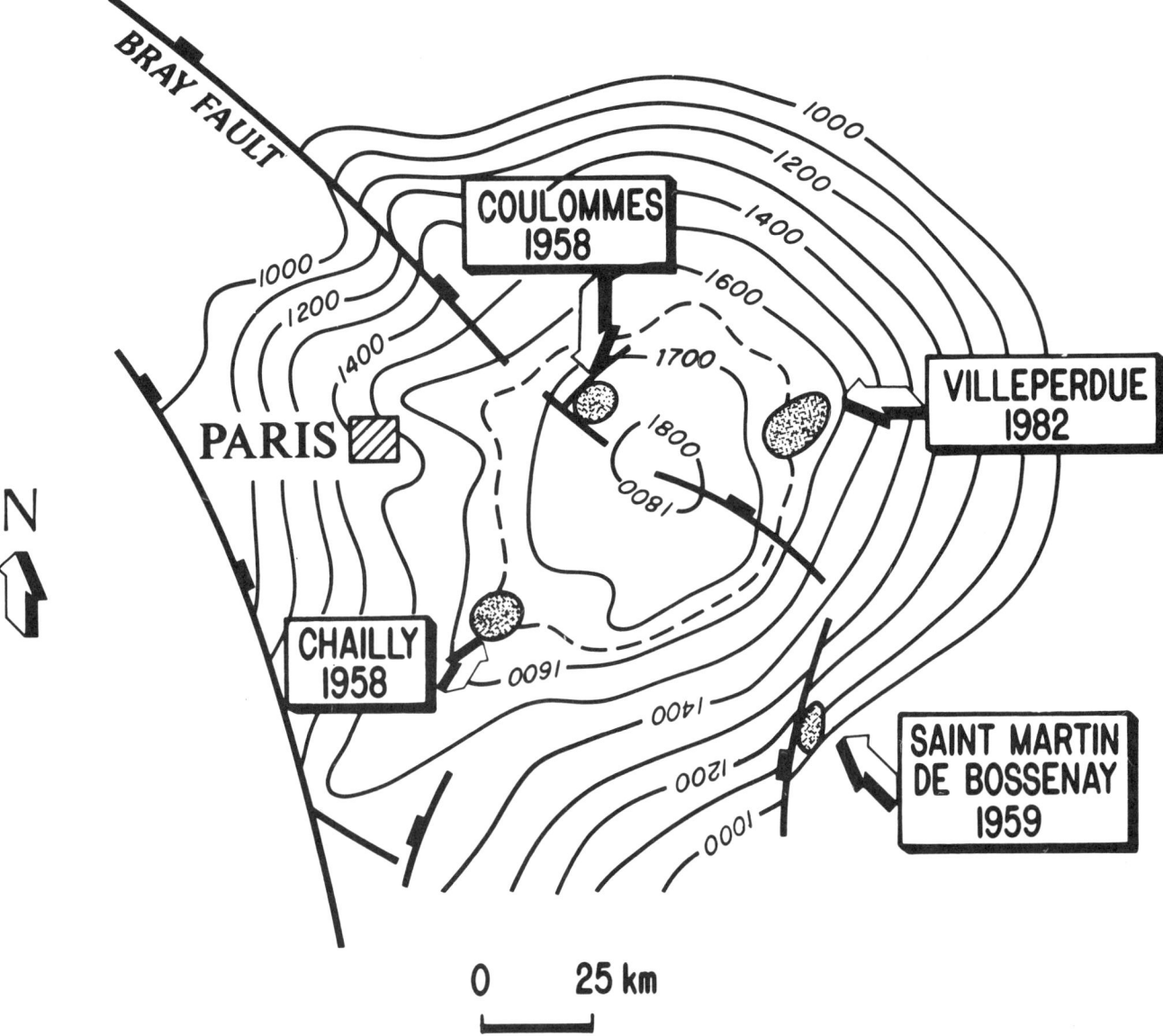

FIGURE 2. Regional contour map of top of Middle Jurassic limestone, showing locations of main Dogger oil fields with years of discovery. Contour interval, 100 m.

ervoir units of Triassic, Lower and Middle Jurassic, and Lower Cretaceous ages and the source rocks of Lower Jurassic age. The latter contain several shaly marine episodes that were deposited in a confined, anoxic silled basin. Four intervals of good to excellent type II source rocks (Tissot and Welte, 1984) have been identified within a Liassic interval with a thickness of 350 m or more. The hydrocarbon yields measured by Rock-Eval pyrolysis are generally in excess of 4 kg per ton of rock (Espitalié et al., 1984). The four intervals are:

1. Hettangian and Sinemurian source rock at the bottom (70 m).

2. Pliensbachian source rock (15 m).

3. Lower Toarcian source rock, locally known as "schistes carton" or "paper shale" facies (65 m), with pyrolitic yields in excess of 10 kg of oil per ton of rock.

4. Upper Toarcian source rock (30 m).

The maximum burial depth of the lower Liassic at the center of the basin is approximately 2800 m, but the maturity needed for these formations to be in the oil window has been reached at a depth of about 2000 m (Tissot, 1971). Because these Liassic source rocks have never been buried deep enough to reach the gas window, it is not surprising that the oil is undersaturated. Another point of interest is that all the oil fields of economic significance are located inside the limits of the Liassic "hydrocarbon kitchen," which lies at the center of the Paris basin, just east of Paris (Espitalié et al., 1986).

Among the several reservoir rocks present throughout the basin, only a lower Callovian carbonate formation, called the "Dalle Nacrée," or, literally, "nacreous flagstone," will be discussed here because it is the producing horizon in the Villeperdue oil field. The basal Oxfordian shales form the seal of the Villeperdue oil accumulation. Particular attention should also be paid to the Cretaceous chalk because it plays an important role with respect to in-depth structural interpretation.

DISCOVERY OF VILLEPERDUE AND SUBSEQUENT EVOLUTION OF ENTRAPMENT CONFIGURATION

Montmirail-2 (MT-2), the first well to discover oil in this part of the Paris basin, was drilled by Régie Autonome des Pétroles, a French state company, in 1959. After producing minor amounts of oil in 1959–60, the MT-2 well was shut down as noncommercial by the standards of the day. Following a detailed study of all available geophysical and well data, with a strong feeling that some exploration problems in this sector of the basin were still unsolved, TOTAL Compagnie Francaise des Pétroles asked for and was awarded a license in 1980, which included the location of the MT-2 shut-in well. TRITON then entered into a farm-in agreement with TOTAL and, to fulfill it, undertook a two-phase program:

1. MT-2 was re-entered and put back into production at the very modest rate of 35 BOPD with a 35% watercut.

2. Following a new Vibroseis seismic survey in 1981, further interpretation led to a structural map very similar to the one that was used until September 1983 (Figure 4). Two wells were drilled, and the second one, Villeperdue-1 (VPU-1), 3 km south of MT-2, was completed as an oil producer in December 1982.

After this encouraging result, TOTAL (to which operatorship had returned) and TRITON implemented a field delineation and development program, starting with a regular grid 1.2 km wide, followed by deviated wells at a later stage. It is notable that one of the first delineation wells, VPU-4, produced at an initial rate of 430 BOPD, which was an encouraging result when compared with the average well productivity of the Paris basin.

The execution of such a program became a difficult exercise. First, this sector of the Paris basin is only very

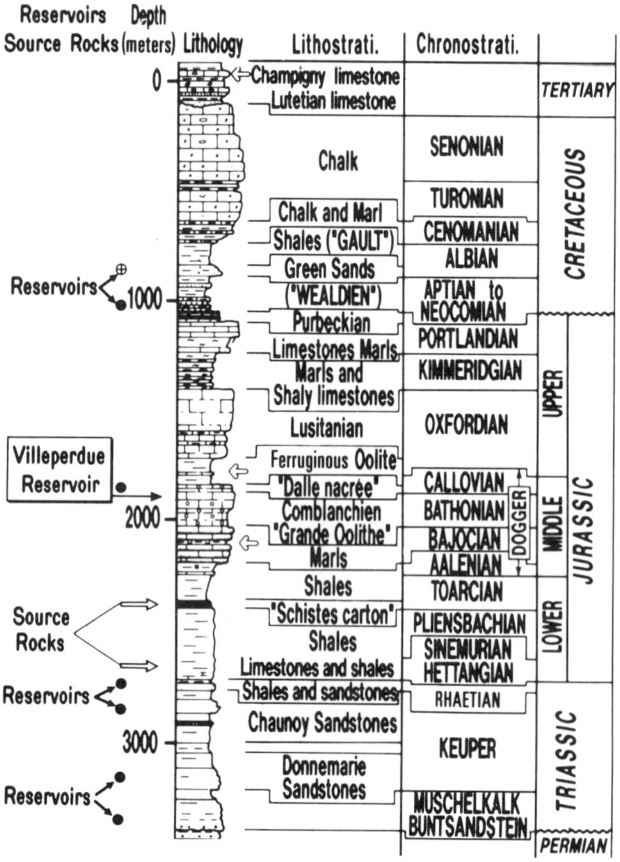

FIGURE 3. Generalized lithostratigraphic column of Paris basin, showing source rocks and reservoir units.

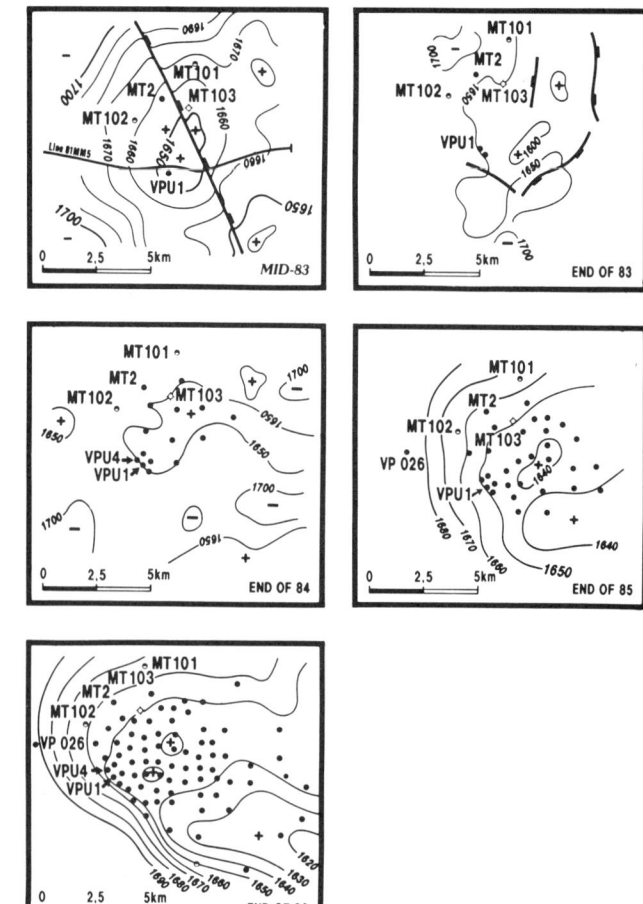

FIGURE 4. Contour map of top of Middle Jurassic, showing how the structural image of the Villeperdue field evolved across time. Contour depths in meters.

weakly structured. The structures are mostly low-amplitude noses or anticlines with some faulting. They result from the interaction of three tectonic phases (BRGM, 1980; Pomerol, 1989; Trémolières, 1981):

1. An extensional Liassic phase that resulted in north-south synsedimentary faulting.

2. An extensional Neocomian phase followed by a slight Albian compressional period.

3. A compressional phase, still going on, driven by repercussions of the Pyrenean and Alpine orogenies. This phase has produced a series of strike-slip faults, the most remarkable one being the Bray fault, which crosses transversely the whole basin (Figures 1 and 2) and the main structural features observed in the area of Villeperdue.

Second, lateral changes of seismic velocities in the overburden, particularly in the Cretaceous chalk, hamper the reliability of structural interpretations. These problems were not suspected in the early stages of exploration of the Villeperdue accumulation.

Figure 5 shows a west-east seismic section across the field, which has been compressed to illustrate the point. There is an obvious link between the time structure at the Cenomanian level (just below the chalk) and the Dogger (Middle Jurassic objective) below, and, as a general rule, the structural attitude in time at the top of the Cenomanian has been found to be a fair guide to the structure in time of the Dogger below. Even if the highest

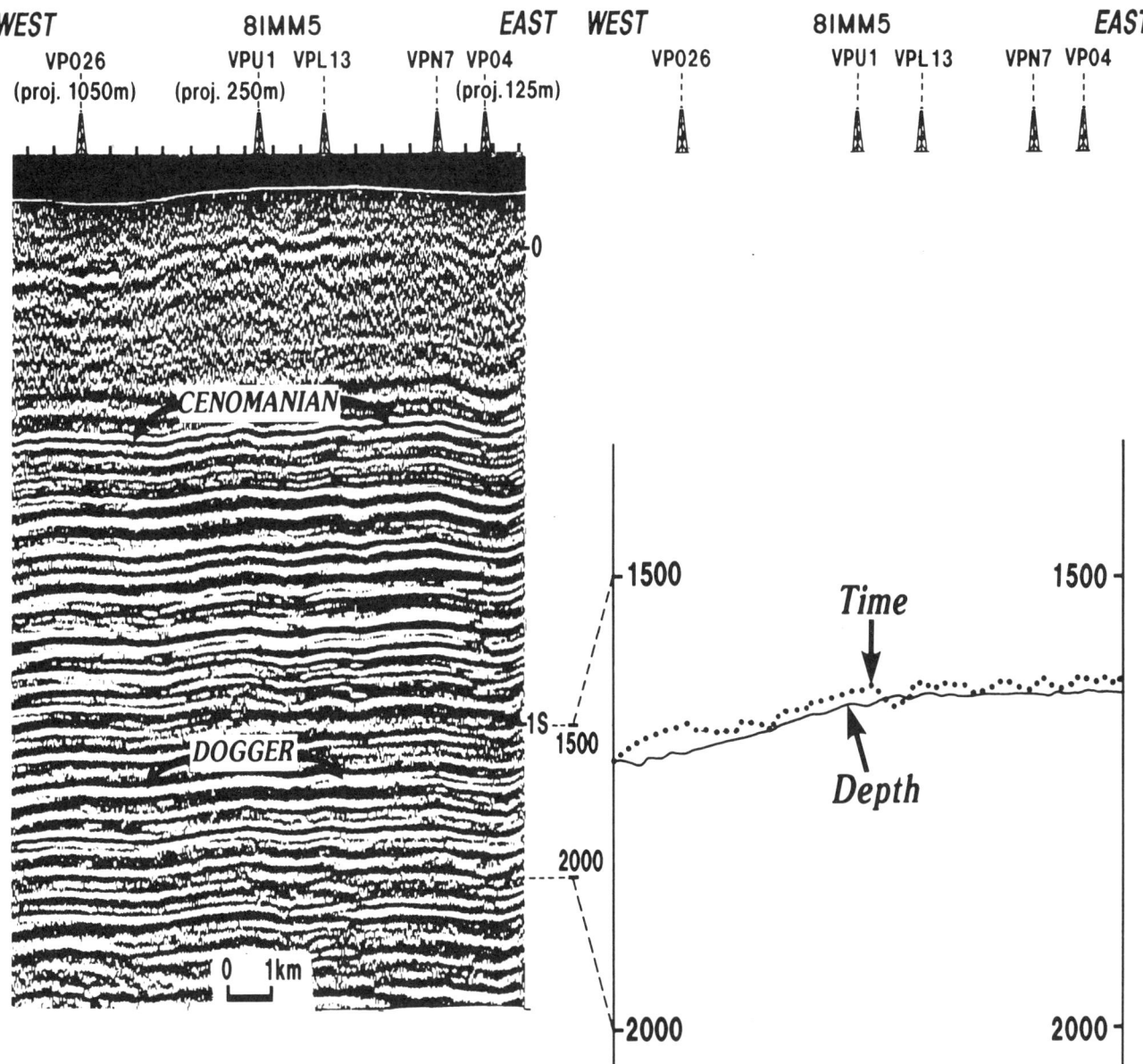

FIGURE 5. West-east seismic line, left, and close-up of depth-converted Dogger horizon, right (depths in meters). Velocity pseudostructure (VP026) is shown vs. neighboring true structure (VPU1).

points of the Cenomanian are not strictly above the highest points of the Dogger, experience has shown that the shift tends to be very slight. The main problem with structural mapping in the Paris basin has always been the acquisition of accurate depth-conversion maps. To overcome this difficulty, the structure at the top of the chalk and the thickness variations between the top of the chalk and the top of the Middle Jurassic were examined and modeled (it is important to note that velocities are usually lower below surface topographic highs of the chalk). In the case of Figure 5, one can see bumps on the line, but, because of strong velocity variations in the upper chalk, as illustrated by the sonic logs of two wells located on the section (Figure 6), once time-depth conversion has been made, some bumps simply disappear, as shown in Figure 5 and the included close-up. This explains why the structural images drawn through the years are found to be so different (Figure 4).

In interpretations made prior to September 1983, the structure appears as a small anticlinal closure that is transversally faulted. In the interpretation made at the end of 1983, the main fault has disappeared, new small faults are drawn, and there remains only a vague gentle structural closure. In the interpretation made in 1984, there is no longer an anticline, and an east-west axis begins to take shape. In April 1985, a large nose appears in the interpretations, but its limits and orientation are still uncertain. Finally, in 1986, after many more wells have been drilled, the structure is interpreted as a broad, gentle nose dipping toward the northwest. Updip, toward the southeast, there appears to be no structural closure, a fact that leads to the conclusion that the entrapment must be partly stratigraphic. This geometry was in no way suspected at the start of the exploration program.

LITHOLOGY, SEDIMENTOLOGY, AND POROSITY DISTRIBUTION OF LOWER CALLOVIAN CARBONATE RESERVOIR

The oil in the Villeperdue oil field is hosted in a carbonate formation locally called the "Dalle Nacrée." It is of Upper Jurassic, lower Callovian age. It is a bioclastic, oolitic limestone up to 30 m thick. Generally, it is not a porous or permeable unit in the Paris basin, but in some areas it can develop favorable petrophysical characteristics that are considered more as local anomalies than as regionally predictable conditions.

The principal calcareous elements are various types of oolites, the remainder being oncoids, pellets, and fossil fragments, mainly from echinoderms, molluscs, bryozoans, and foraminifera. Quantitative petrographic analysis allowed the vertical percentage variations to be determined, which helped to define the three main units, whose vertical distribution is shown in Figure 7. Three types of oolites are dominant (Figure 8). The radial type is

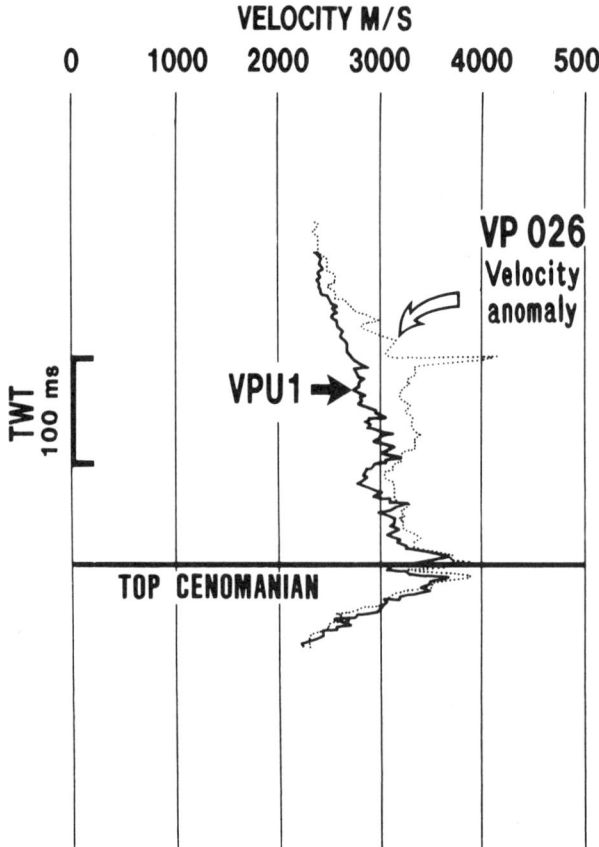

FIGURE 6. Velocity anomaly in the Senonian chalk, shown by portions of sonic logs of VPU1 and VP026.

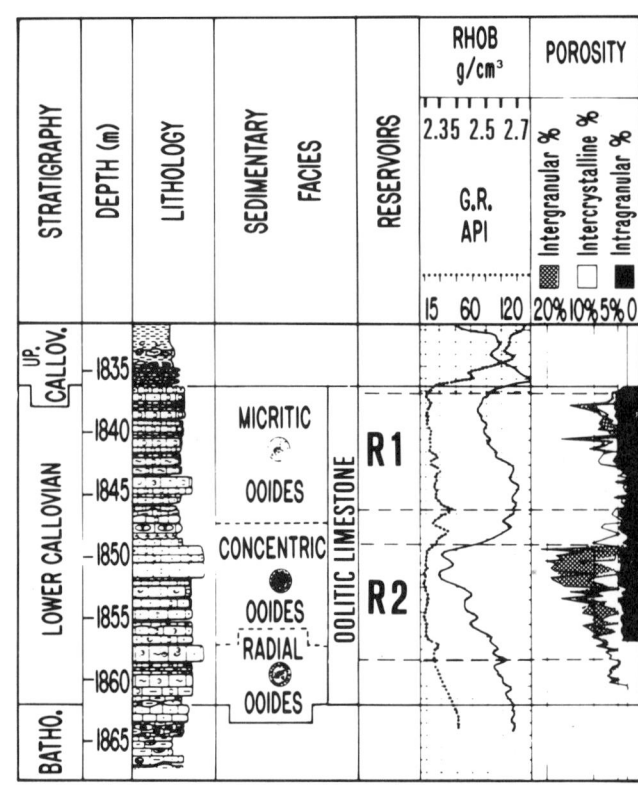

FIGURE 7. Vertical distribution of the three main types of oolitic facies in the lower Callovian.

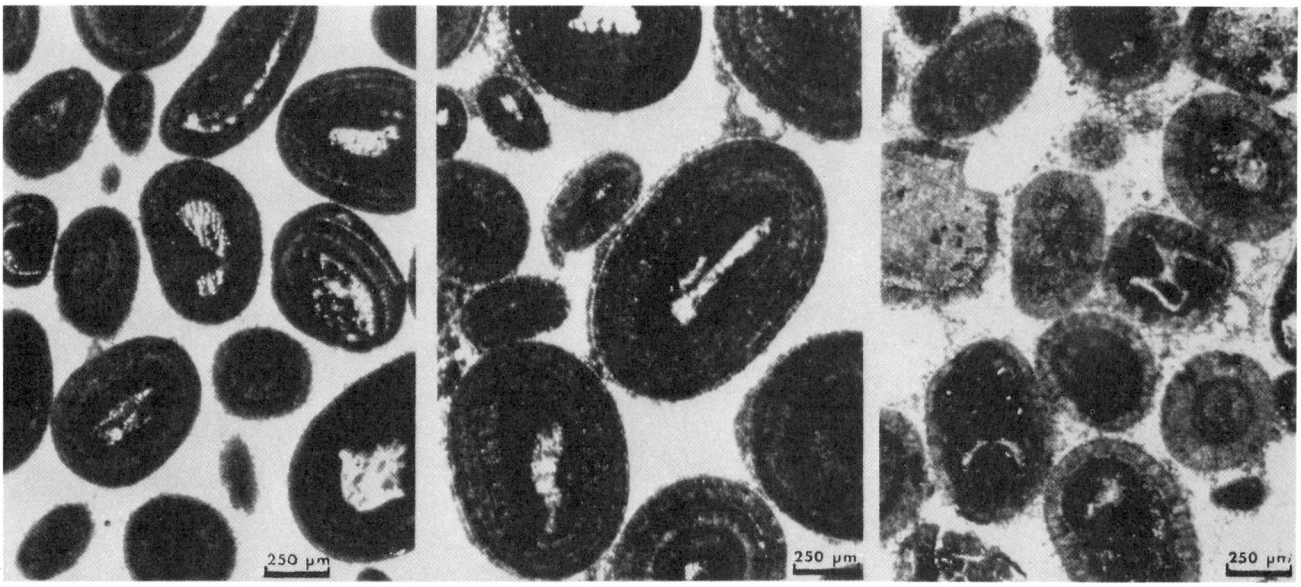

"Micritic" "Concentric" "Radial"

FIGURE 8. The three different types of oolitic microfacies.

predominant in the lower unit, or "sole," which is tight, the grains being embedded in a micritic matrix, whereas intergranular porosity is better developed between micritic and concentric oolites in what are called, respectively, the "R1" and "R2" units. The R1 unit has an average net thickness of 8 m and a fairly uniform porosity of about 12%, but its permeability is poor, ranging from 2 to 10 md. The R2 unit has an average net thickness of about 12 m. It is much less homogeneous than the upper (R1) unit, with porosity ranging from 8 to 20% and permeability varying from 5 to 100 md.

The crossing over from one unit to the other is often very abrupt. The variation in oolite types is thought to be indicative of changing environments, which is supported by evidence for an apparent upward reduction in water depth throughout the unit, an associated reduction in the rate of oolitic sedimentation, and an upward increase of the oolite growth stages. The cement is mostly calcitic, and is rarely dolomitic.

Particular emphasis has been placed on the description and understanding of the sedimentary structures that are visible in the "Dalle Nacrée" lower Callovian reservoir unit. Because these structures are intricate, they will not be discussed in detail here, and we will limit ourselves to presenting the interesting features that have a demonstrated bearing on the distribution of the entrapped oil. Two important observations have been made:

1. Whereas the sole of the over-all reservoir unit displays several indications of wave influence together with possible biological mounding features, the R1 and R2 reservoir units display sequences of dip values arranged into bell-shaped statistical distributions, with dips increasing up to 35° (Figure 9), and a general progradation trend toward the east-southeast (Figure 10). This depositional setting was documented from cores and from dipmeter data. One can even draw the elongated envelopes of each sequence with a given dip distribution and confirm the east-southeast directions indicated earlier (Figure 11). They are interpreted as stackings of oolitic sand waves or bars, and some local progradations in an opposite direction may be interpreted as tidal effects. Apart from these observations, additional evidence to support the depositional model of sand waves and bars can be drawn from the isopachs of R1 and R2, which display elongated forms.

2. The important consequence affecting the shape and size of the field is that the best reservoirs, which correspond to the micritic and concentric oolites, decrease in thickness updip, toward the southeast, being partly replaced by radial oolite types. The eastward thinning of porous units is visible in the section shown in Figure 10.

Active research is still being conducted on these reservoirs, and our conclusions are still tentative. However, the economic consequences of reservoir distribution are quite obvious on the net pore volume map presented in Figure 12, which shows clearly the decrease in reservoir porosity updip to the southeast, along the axis of the structural nose. This combination of a structural nose together with an updip permeability barrier in the Callovian carbonates is what is responsible for the trap geometry of the Villeperdue oil field.

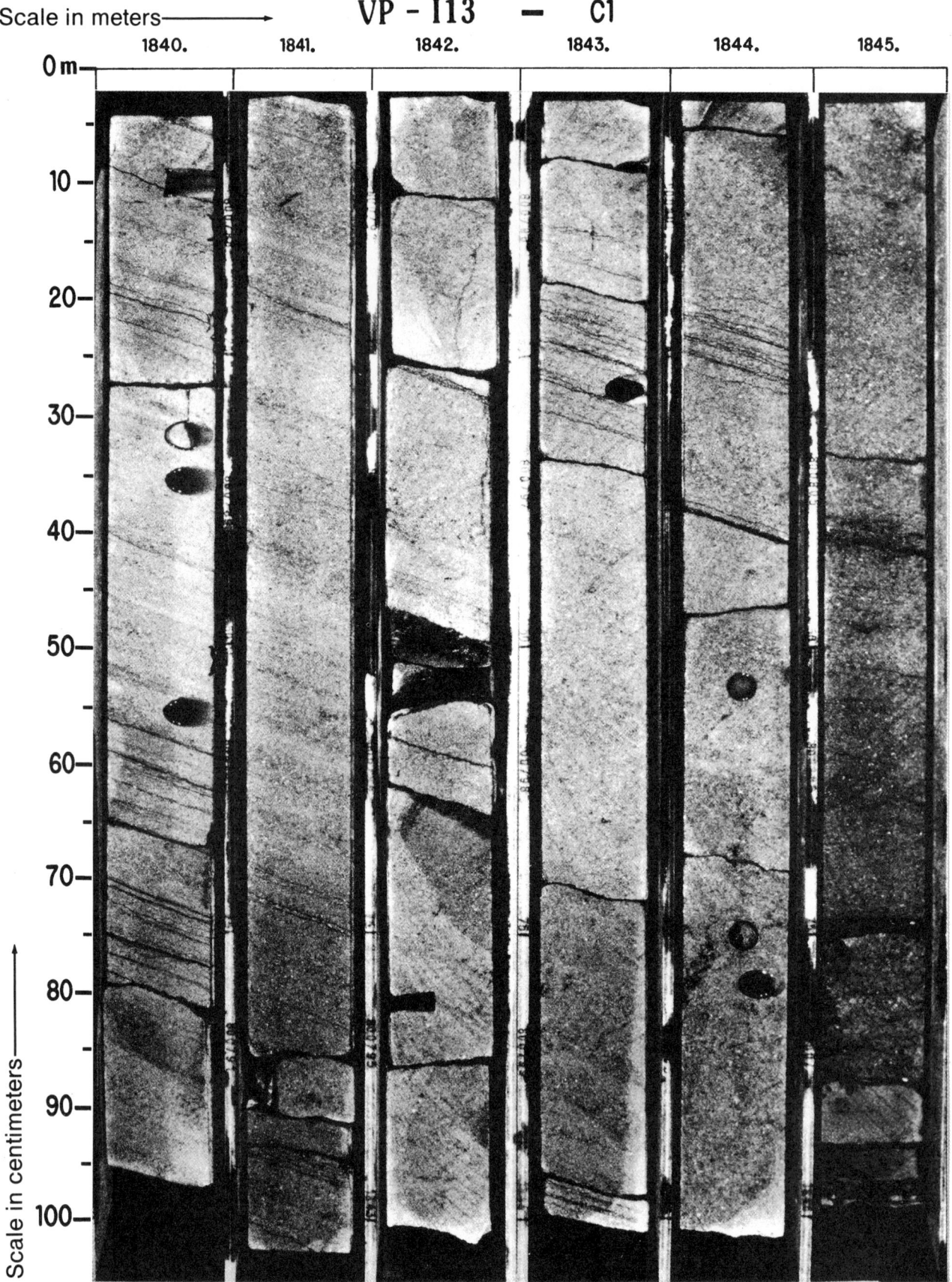

FIGURE 9. Cores of R1 and R2 units, showing evidence of sedimentary dips.

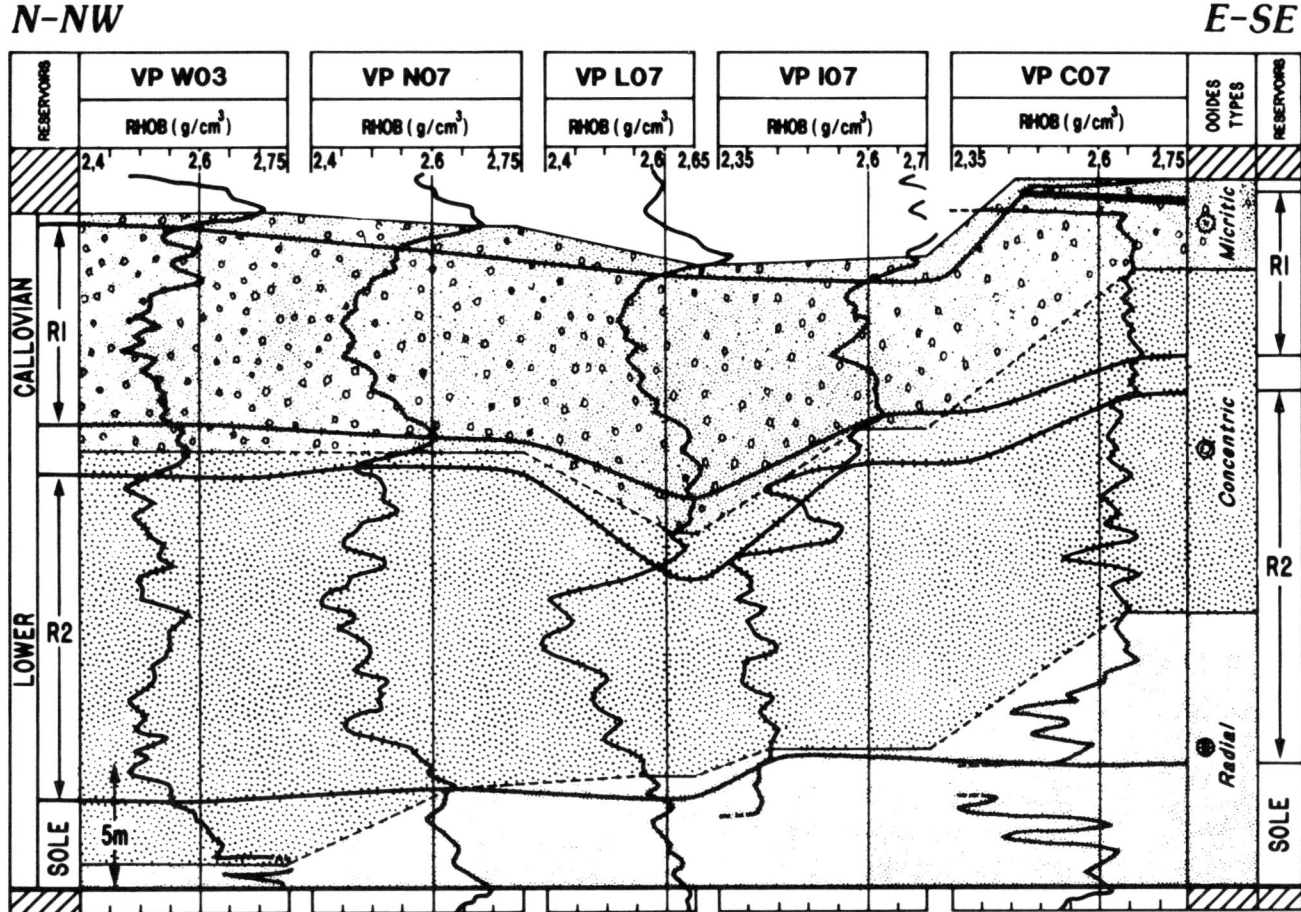

FIGURE 10. North-northwest–south-southeast profile across the field, showing the geometry of the reservoirs in relation to sedimentary units.

MIGRATION HISTORY AND TRAPPING MECHANISM

Villeperdue oil field is not far from the deepest part of the Paris basin, where the depth of effective hydrocarbon expulsion from Liassic source rocks is about 2400 m subsea (Espitalié et al., 1986). Oil generation in the "hydrocarbon kitchen" started during the Lower Cretaceous and is still active today. The structural grain itself, as discussed above, results from Cretaceous and Tertiary tectonic phases, and it is proposed that vertical migration of oil up to Middle Jurassic carbonate reservoirs in most fields (Coulommes, Chailly, Saint-Martin) occurred along subvertical faults activated during the Tertiary (Figure 13).

Faults are obvious at the top of the Dogger reservoir in the Coulommes and Hautefeuille fields (Hautefeuille is a smaller field immediately north of Villeperdue) but have not been proved at Villeperdue, where only some deep-seated faults can be observed on the seismic. They can eventually reach up to the Dogger, but by this time they are without seismically noticeable throw.

The sealing mechanism in the reservoir is complex, probably resulting from the subtle interaction of sedimentology, faulting, hydrodynamics, and porosity change. The reservoir could well be in hydrodynamic communication, through faults, with the underlying "Oolithe Blanche." This lower reservoir formation of Middle Jurassic age is charged by fresh water from outcrops to the east and southeast, and so it has been speculated that the Villeperdue reservoir could be affected by a slow invasion of fresh water through faults. In this case, water circulation would, perhaps, be responsible for the porosity loss observed toward the eastern part of the field. The most likely hypothesis, however, is that lateral variations of reservoir quality permitting entrapment are linked to facies variations, with some likely diagenetic effect related to sedimentologic types. Thus, it is suggested that it is partly because of its restricted areal development in porosity and permeability that the Callovian "Dalle Nacrée" reservoir at Villeperdue has entrapped the existing oil field (Figure 14). Geochemical analyses of the oils demonstrate that they are not biodegraded and thus are unlikely to have been affected by meteoric water invasion.

PRODUCTION DATA

By the end of 1989, the proved recoverable reserves amounted to 66 million bbl. Field development started in 1984, and enhancement of production by water injection

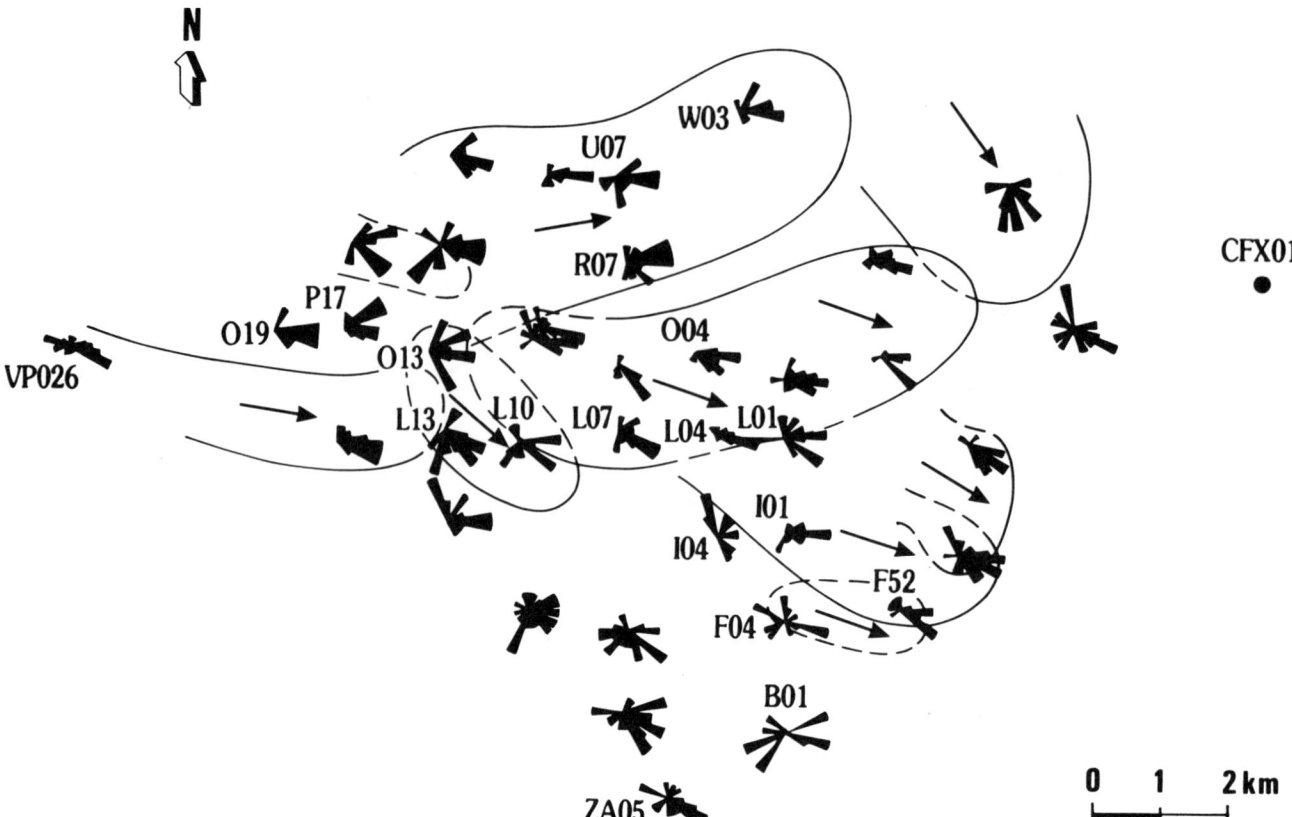

FIGURE 11. Azimuth diagram of sedimentary dips in R2 reservoir unit (data from dipmeter and cores).

closely followed the beginning of development. A total of 150 wells had been drilled by the end of 1989, of which 100 are now producers and 20 are water injectors. The envelope of the existing field is fairly large and would approximately coincide with the shape and size of the city of Paris. Production capacity varies from one well to another, the best producing well having an output of over 600 BOPD. At the beginning of 1990, the daily production rate of the entire field was a little less than 12,000 BOPD, and the total cumulative production had reached 22 million bbl by the end of 1989.

As can be expected in such a field, water-cut is increasing regularly. The oil column is 60 m high. The oil has a gravity of 36° API, and the GOR is about 3.5 volume per volume. The reservoir pressure is 178 bars at 1650 m subsea, and the reservoir temperature is 72°C. The bubble point is about 6 bars. After a first success with this new technique, further development programs are planned, including several horizontal wells, in order to increase the productivity of the field.

PERSPECTIVES AND CONCLUSION

Since the discovery of Villeperdue, additional smaller discoveries have been made in the same general area. In all cases, the trap style requiring complex seismic velocity and permeability barrier considerations in exploration methods has been similar to that found at Villeperdue. The Fontaine-au-Bron and Hautefeuille pools were put into production in 1986 and 1987, respectively. Sancy-lès-Provins, the most recent discovery, about 10 km to the south of Villeperdue, followed in 1990, with a high initial well productivity, and delineation is under way. This pool was found on a structural nose parallel to the nose of Villeperdue, and could be a geologic analog. So far, however, the reserves appear to be much smaller that those at Villeperdue.

The conclusions are as follows:

1. The exploration and production of Middle Jurassic reservoir plays in the Paris basin are largely controlled by the complex geology of carbonate reservoirs in combination with the low-amplitude geometry of structural elements. Petrophysical and sedimentological studies can facilitate the understanding and prediction of porosity trends in the reservoir unit.

2. Nearly all obvious anticlinal structures in the Paris basin have been tested. Those that are inside the Liassic "hydrocarbon kitchen" are oil-productive or have given major shows. Additional discoveries will be made in more subtle traps, of which Villeperdue is a good example, with its combination of a structural nose dipping into the "kitchen" and an updip permeability barrier in the reservoir.

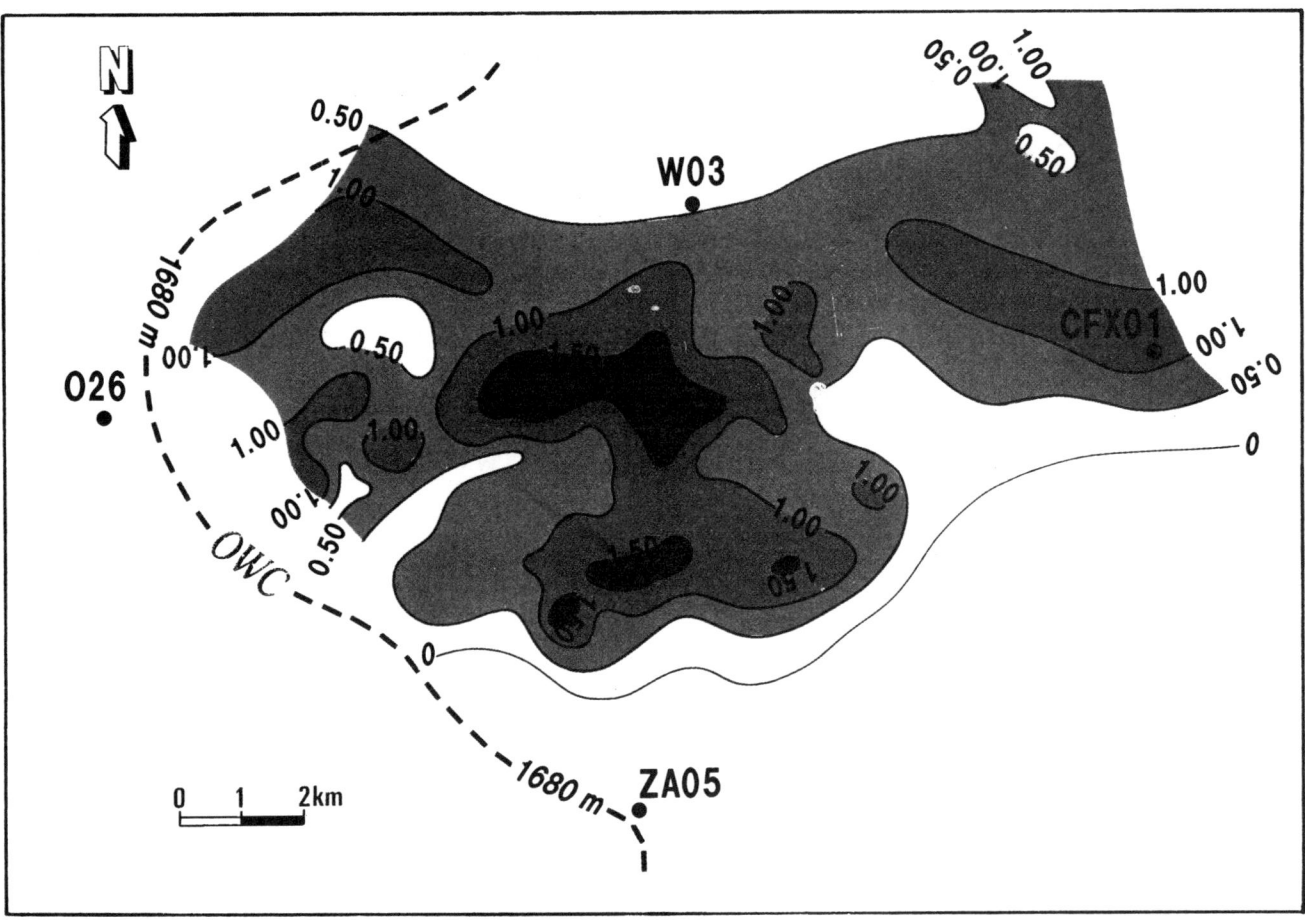

FIGURE 12. Net pore volume map of R2 reservoir unit (porosity vs. thickness in meters; cutoff porosity, 12%).

3. Because of very low-amplitude folding and the nature of the overburden, future discoveries must rely on the continued evolution of improved seismic techniques and, more specifically, on the careful monitoring of seismic velocities in the Cretaceous chalk.

4. The approach must incorporate a systematic and thorough re-evaluation of all existing seismic and well data as well as a continued reappraisal of the geologic model, in the light of new seismic and drilling.

ACKNOWLEDGMENTS

The author wishes to thank TOTAL Compagnie Francaise des Pétroles and TRITON France for permission to publish this paper, and gratefully acknowledges the help received from P. Bousquet, C. Goy, P. Lossel, and P. Pouclée of TOTAL in discussing the various aspects of Villeperdue field and gathering data and illustrations. Comments and advice were kindly provided by C. Cramez of TOTAL and G. Demaison, Consultant.

REFERENCES CITED

Allen, J. R. L., 1980, Sand waves: a model of origin and internal structure: Sedimentary Geology, v. 26, p. 281–328.

Arbin, P., and M. Cassoudebat, 1991, Synthèse géologique et pétrolière du bassin Parisien: Rapport interne TOTAL-CFP, 2 vol. en cours d'édition.

Arbin, P., and F. Euriat, 1989, Des réservoirs convoités: les calcaires du Dogger du bassin Parisien in Les roches au service de l'homme. Géologie et Préhistoire du bassin Parisien: Bull. Inf. des Géologues du Bassin de Paris, Mémoire hors série n. 7.

Bouché, P., J. Follot, J. M. Pagezy, and J. C. Henberger, 1960, Gisement de Saint-Martin-de-Bossenay: Revue de l'IFP, v. 15, n. 6, Editions Technip: les gisements du bassin de Paris.

BRGM, 1980, Synthèse géologique du bassin Parisien: Mémoires 101, 102, and 103. BRGM ed.

CEP and RAP, 1960, Champ de Chailly-Chartrettes: Revue de l'IFP, v. 15, n. 6.

Espitalié, J., F. Marquis, and I. Barsony, 1984, Geochemical logging, in K. J. Voorhees (ed.), Analytical Pyrolysis,

262 Duval

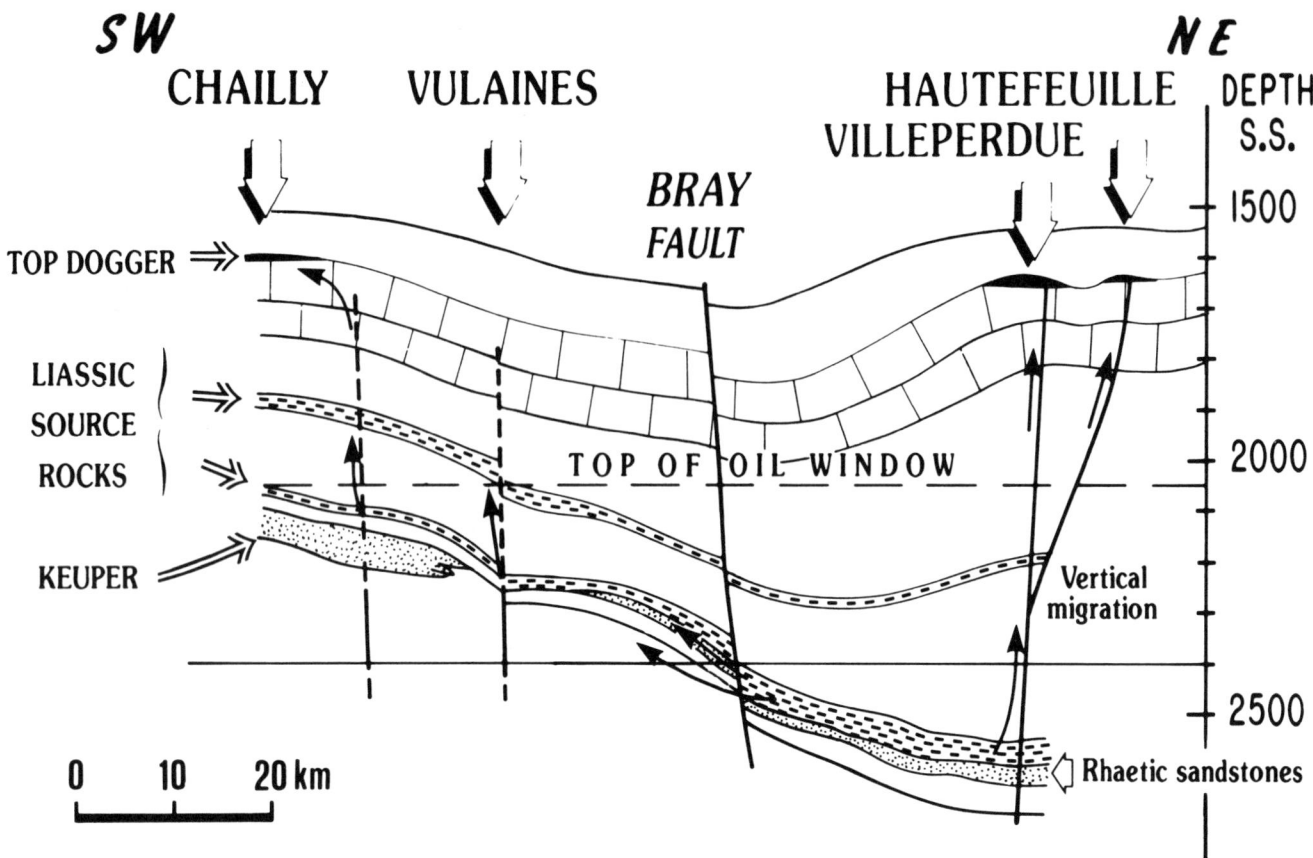

FIGURE 13. Migration pathways from Liassic source rocks to Dogger reservoirs.

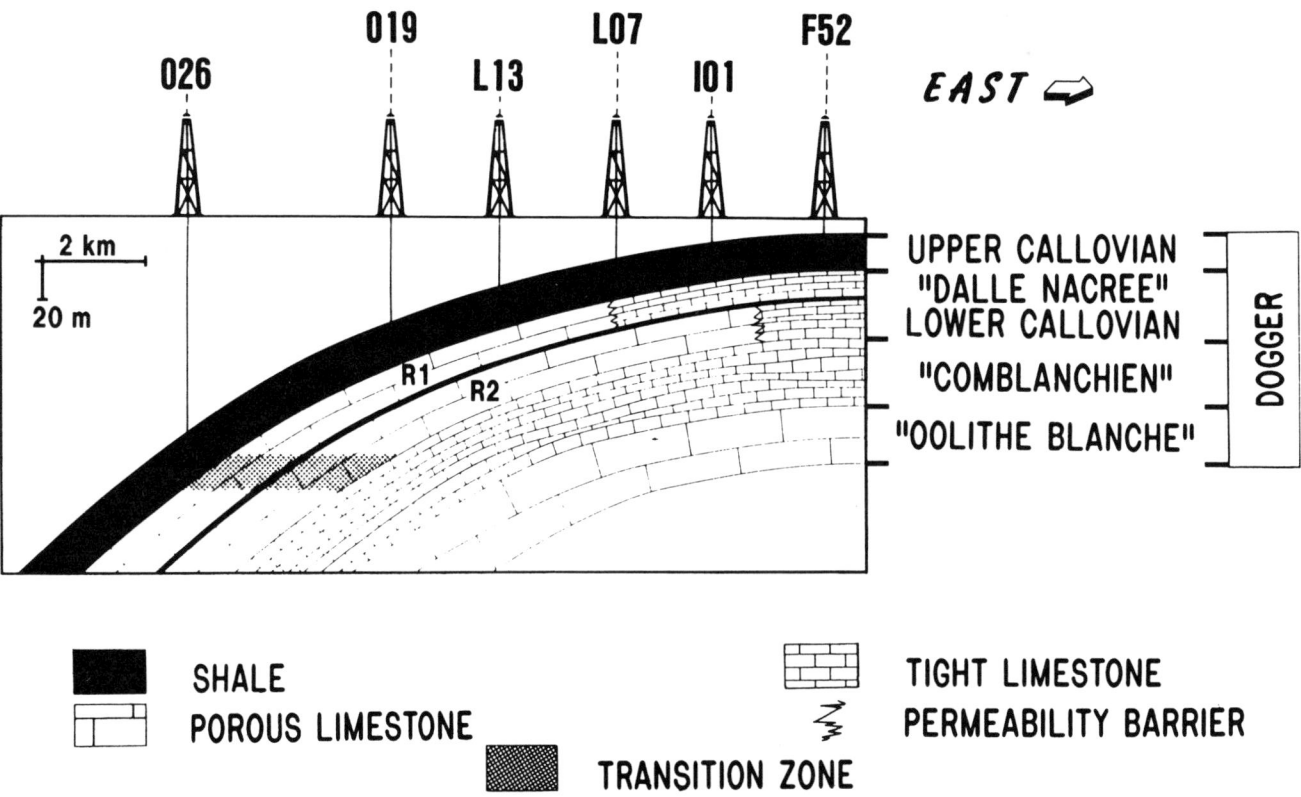

FIGURE 14. West-east schematic cross section illustrating the updip permeability barrier in the Villeperdue field. Note extreme vertical exaggeration.

Techniques and Applications: Butterworth, p. 276-304.

Espitalié, J., F. Marquis, and L. Sage, 1986, Organic geochemistry of the Paris basin, in J. Brooks and K. Glennie (eds.), Geology of NW Europe: London, Graham and Trotman, p. 71-86.

Fitzgerald, M., and E. Mousset, 1987, Triton France details its successes in the Paris basin: Oil and Gas Journal.

Floquet, M., P. Laurin, P. Laville, D. Marchand, J. C. Menot, A. Pascal, and J. Thierry, 1989, Les cycles sédimentaires bourguignons d'âge Bathonien terminal-Callovien: Bull. Centres Rech. Explor.-Prod. Elf-Aquitaine, 13/1, p. 133-165.

Handford, R. C., 1988, Review of carbonate sand-belt, deposition of ooid grainstones and application to the Mississippian reservoir, Damme field, southwestern Kansas: AAPG Bulletin, v. 72, n. 10, p. 1184-1199.

Hoecker, C., K. M. Eastwood, J. C. Herweijer, and J. T. Adams, 1990, Use of dipmeter data in clastic sedimentological studies: AAPG Bulletin, v. 74, n. 2, p. 105-118.

Laville, P., R. Cussey, J. Durand and M. Floquet, 1989, Faciès, structure et dynamique de mise en place de dunes oobioclastiques au Callovien inférieur en Bourgogne: Bull. Centres Rech. Explor.-Prod. Elf-Aquitaine, 13/2, p. 379-393.

Manguy, V., 1960, Gisement de Châteaurenard: Revue de l'IFP, v. 15, n. 6.

Monchaux, A., and P. Tremolieres, 1985, Le gisement de Coulommes-Vaucourtois. Historique et données nouvelles: Pétrole et Techniques, n. 313, p. 19-27.

Morelot, G., and L. Pages, 1989, Objectif Trias: à la recherche des gisements pétrolifères du bassin Parisien in les roches au service de l'homme. Géologie et préhistoire du bassin Parisien: Bull. Inf. des Géologues du Bassin de Paris, Mémoire hors série n. 7.

Pomerol, C., 1989, L'évolution du bassin Parisien in dynamique et méthodes d'étude des bassins sédimentaires: Editions Technip, p. 165-178.

Pommier, G., 1985, Le développement d'un gisement complexe: Montmirail-Villeperdue: Extrait des Annales des Mines.

Purser, B. H., 1975, Sédimentation et diagénèse précoce des séries carbonatées du Jurassique moyen de Bourgogne: Thèse d'Etat, Orsay, 450 p.

Tilloy, R., and M. Dardenne, 1960, Gisement de Coulommes Vaucourtois et le périmètre de Dammartin en Goële: Revue de l'IFP, v. 15, n. 6.

Tissot, B. P., 1971, Origin and evolution of hydrocarbons in early Toarcian shales, Paris basin, France: AAPG Bulletin, v. 55, p. 2177-2193.

Tissot, B. P., and D. H. Welte, 1984, Petroleum formation and occurrence (2nd Ed.): Berlin, Heidelberg, New York, Tokyo, Springer-Verlag, 699 p.

Trémolières, P., 1981, Mécanismes de la déformation en zone de plateforme. Méthode d'application au bassin de Paris (2e partie): Revue IFP, v. 36, n. 5, p. 579-593.

Chapter 17

Barbara Field, Adriatic Sea, Offshore Italy
A Giant Gas Field Masked by Seismic Velocity Anomaly—A Subtle Trap

A. Ianniello
W. Bolelli
L. Di Scala

AGIP S.p.A.-S. Donato Milanese
Milan, Italy

ABSTRACT

Barbara gas field, discovered in 1971, is located in the northern sector of the Adriatic offshore. It is a gentle anticline, involving Quaternary clastic sediments, shaped by carbonate Mesozoic morphology.

The presence of shallow gas pockets at the crest of the structure distorts the seismic signal to such an extent that structural reconstruction is not immediate.

Seismic attribute analysis provides, however, a key to the understanding of the seismic anomalies and is a valuable tool for the reconstruction of the real structural configuration of the field.

The appraisal history of the field illustrates how the progressive understanding of the above-mentioned complications helped upgrade the reserves from an initial value of 10 billion m^3 of gas to 40 billion m^3, making Barbara the most important Italian gas field of the decade.

INTRODUCTION

The Barbara gas field is located in the northern part of the Italian Adriatic continental shelf in 70 m of water, very near the median line with Yugoslavia and at a distance of 50 km from the coast (Figure 1).

The field produces from the Quaternary sands, which are shaped as a gentle anticline and capped by shales. The main gas-bearing sequence extends from 1000 to 1400 m in depth, and the field covers an area of 70 km^2.

Recoverable reserves are estimated to be 40 billion m^3 of gas. Development started in 1980 and is still going on with the completion of the development wells from the sixth platform. A seventh platform is foreseen in the near future. By the end of 1990, the total number of development wells will be 72.

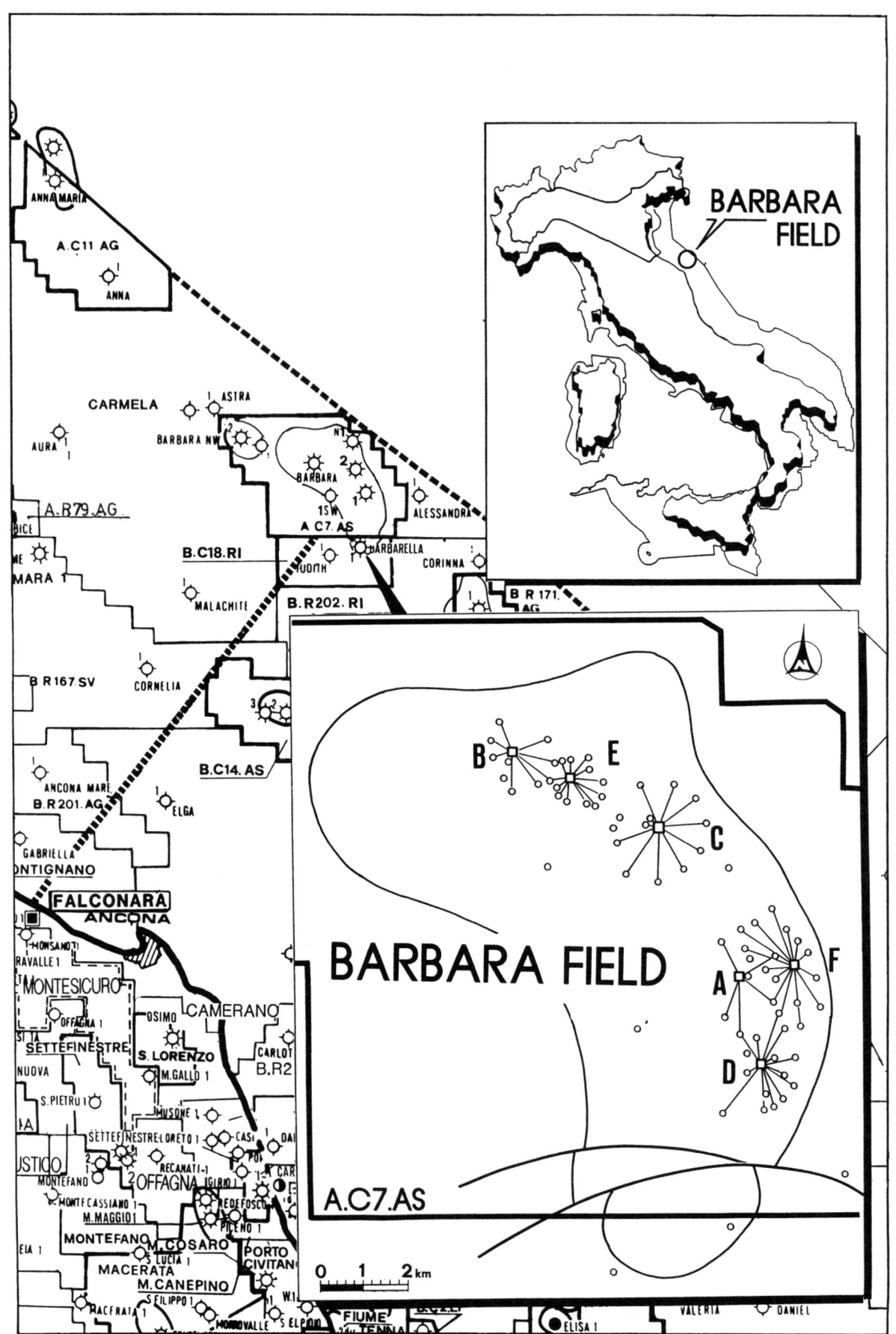

FIGURE 1. Location map for Barbara field.

Production started in 1981. Average present production is 8.3 million m³ per day. Cumulative production will reach 8.4 billion m³ by the end of 1990.

HISTORY

The Barbara field was discovered in 1971 under exploration permit A.Rll.AS. The permit was allocated in October 1967 to AGIP (51%) and Shell (49%) as a result of the opening of the exploration activity in Adriatic Zone A.

The discovery well, Barbara 1, was located on what was considered a not very large structural high having a NNW-SSE trend on the time maps. The well found a total of 320 m of gross pay subdivided into 13 separate levels (Figure 2).

Subsequent to the discovery, several appraisal wells were drilled to define the structural configuration of the field (Figure 3), which was considered a minor one (10 billion m³ of gas reserves) because its depth reconstruction was reflecting the shapes of the time maps.

These maps presented one main anticlinal feature in the southern part of the concession and a minor one to the north, separated by a syncline. Additional minor undulations complicated the general picture.

The first appraisal wells were located in the early 1970s: Barbara NW 1 in 1971, Barbara 2 and Barbara SW 1 in 1972, and Barbara 3 in 1973. Subsequently, concession A.C7.AS was awarded in November 1974, and during 1976 AGIP acquired 100% of the participating interest.

In 1979 a new detailed seismic survey was shot, followed by a new structural interpretation that mainly confirmed the previous one. In order to define additional reserves, two more appraisal wells were drilled on the two minor undulations in the northern part of the concession (Barbara N 1 and Barbara NW 2). Both wells found gas, but Barbara N 1 started to indicate the possibility of a unique gas-water contact with the wells located to the south.

Doubts were then raised on the structural configuration of the field and on the presence of the syncline separating the Barbara N 1 culmination from the Barbara 1 positive trend. Barbara N 2 was therefore spudded into the syncline present on the time maps. The well encountered the reservoirs higher than expected and demonstrated that the syncline was in fact related to a pull-down effect caused by shallow gas layers that distorted the seismic signal and the amplitude of the reflections, limiting the bright spot effect. The boundary of the bright spot area was taken initially as the actual boundary of the field, the extension of which was therefore underestimated.

Additional appraisal wells (Barbara N 3 and N 4) subsequently confirmed the final structural configuration of the field (Figure 4).

In the meantime, improvement in logging technology and data acquisition through coring and testing enhanced the potential of the silty sequence, which proved to have a good production capacity.

All these factors have contributed to an increase in the estimated reserves of the field to the present level of 40 billion m³.

GEOLOGY

The Barbara field lies in the Adriatic foreland area, about 30 km northeast of the Apennine "thrust belt front," at the southwestern boundary of the Istrian platform (Figure 5).

During the Late Triassic–early Liassic, a "tidal flat complex" environment prevailed in the area with deposition of a dolomitic sequence. From the early Liassic onward, the occurrence of oolitic and reefoid facies indicates the evolution toward an "open shallow platform" environment.

Deposition of shallow marine carbonates continued until Cenomanian time (Cellina Limestone) and, apart from a hiatus episode between the Late Cretaceous and early Paleocene, until mid-Eocene, with deposition of the Peschici Limestone (Figure 6).

Emersion of the platform occurred during the Oligocene and Miocene, while in the surrounding Adriatic area the deposition of basinal flysch, and subsequently of the Messinian evaporitic sequence, was taking place.

During the lower Pliocene a new sedimentation cycle started in the whole Adriatic area with subsidence and deposition of mainly argillaceous sediments. Subsequently, from mid-Pliocene to Pleistocene, a turbidity phase took place with deposition of sands, prograding from the northwest toward the Barbara area.

QUATERNARY DEPOSITIONAL ENVIRONMENT

Well log and core analyses integrated with seismic interpretation suggest that the Quaternary sedimentary clastic sequences encountered in the Barbara field are the result of the aggradation of a deep water turbidite system that reflects the action of the relative sea level changes on the progradation of coastal deposits. The observed vertical trend shows a coarsening- and thickening-upward tendency resulting from the repetition of:

♦ Relative lowstand phases with large-scale erosion of underconsolidated shelf deposits and basinal deposition of thick-bedded turbiditic sand lobes having a north-south depositional trend. The main source for the sands is furnished by the Tertiary outcrops surrounding the eastern part of the Po valley. Only minor and mainly muddy sediments are generated from progradational wedges originated on the eastern Apennine margin.

♦ Relative highstand phases characterized by an increase of progradation sequence in the shelf area with generation into the basin of only thin, shaly deposits.

THE TRAP

The Barbara gas field is a result of the entrapment of biogenic gas, generated by the Pliocene–Quaternary shaly sequence, within sand beds deposited in a deep water en-

Text continues on p. 272

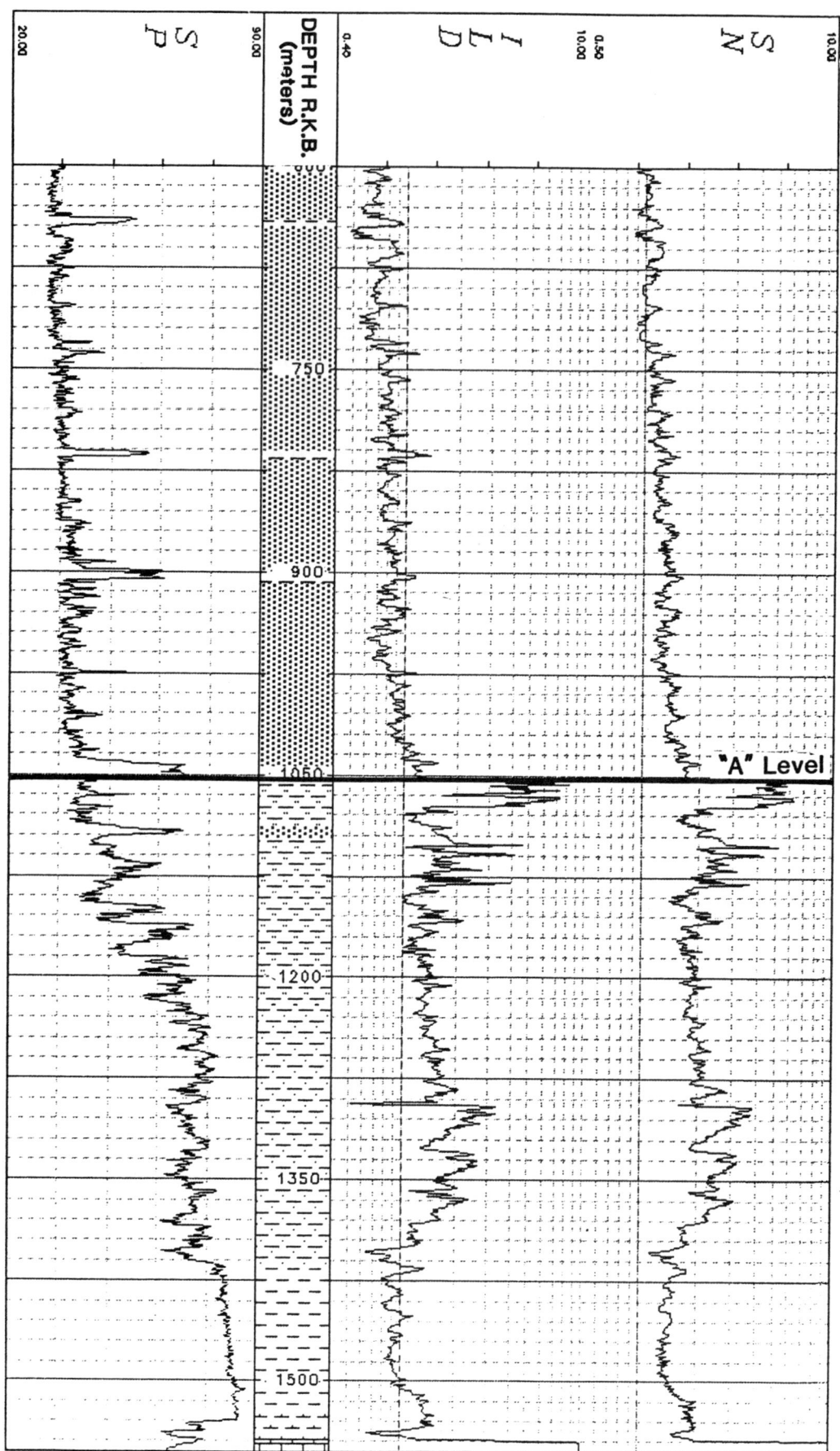

FIGURE 2. Type log for Barbara 1 well.

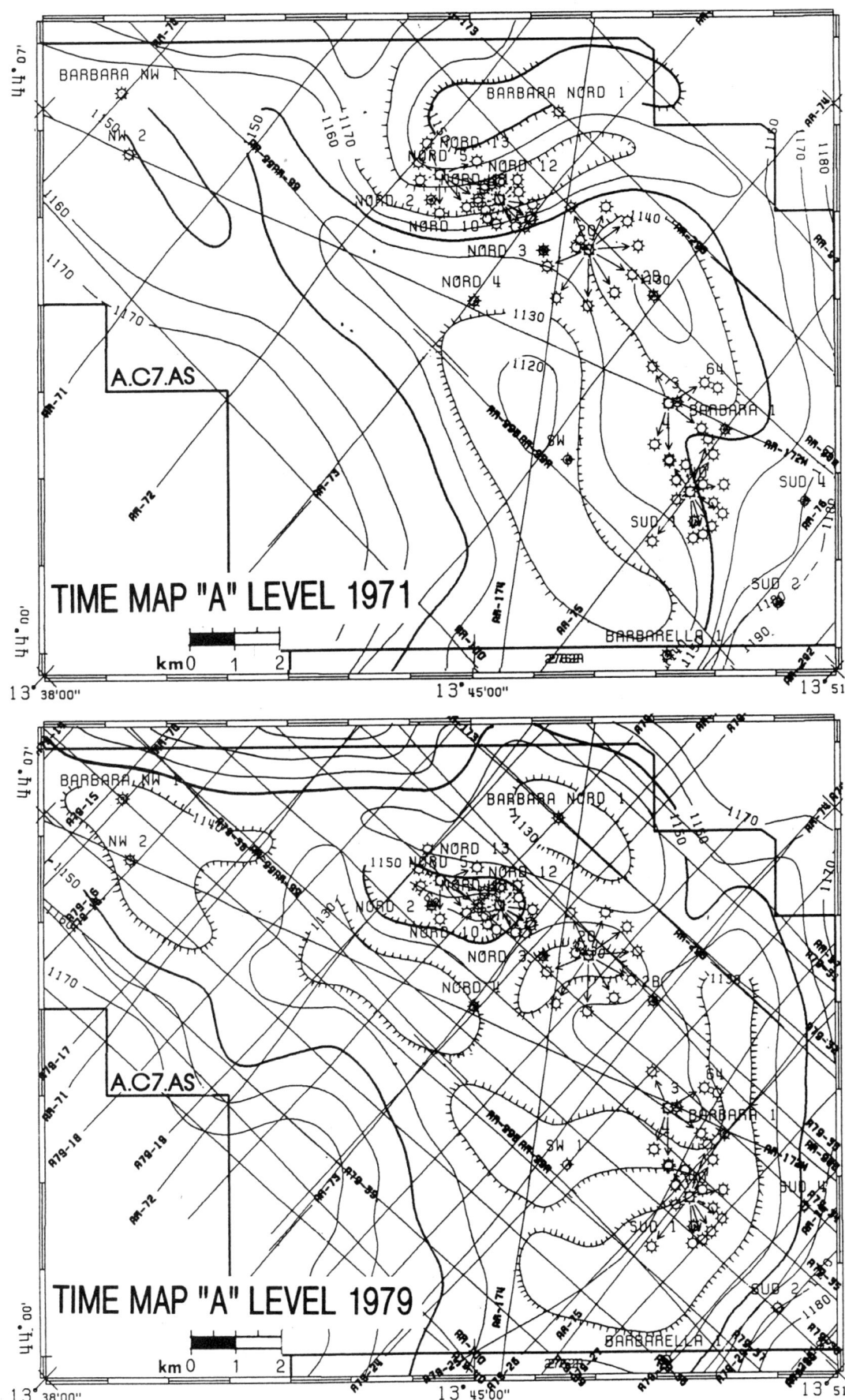

FIGURE 3. "A" level time maps for 1971 and 1979.

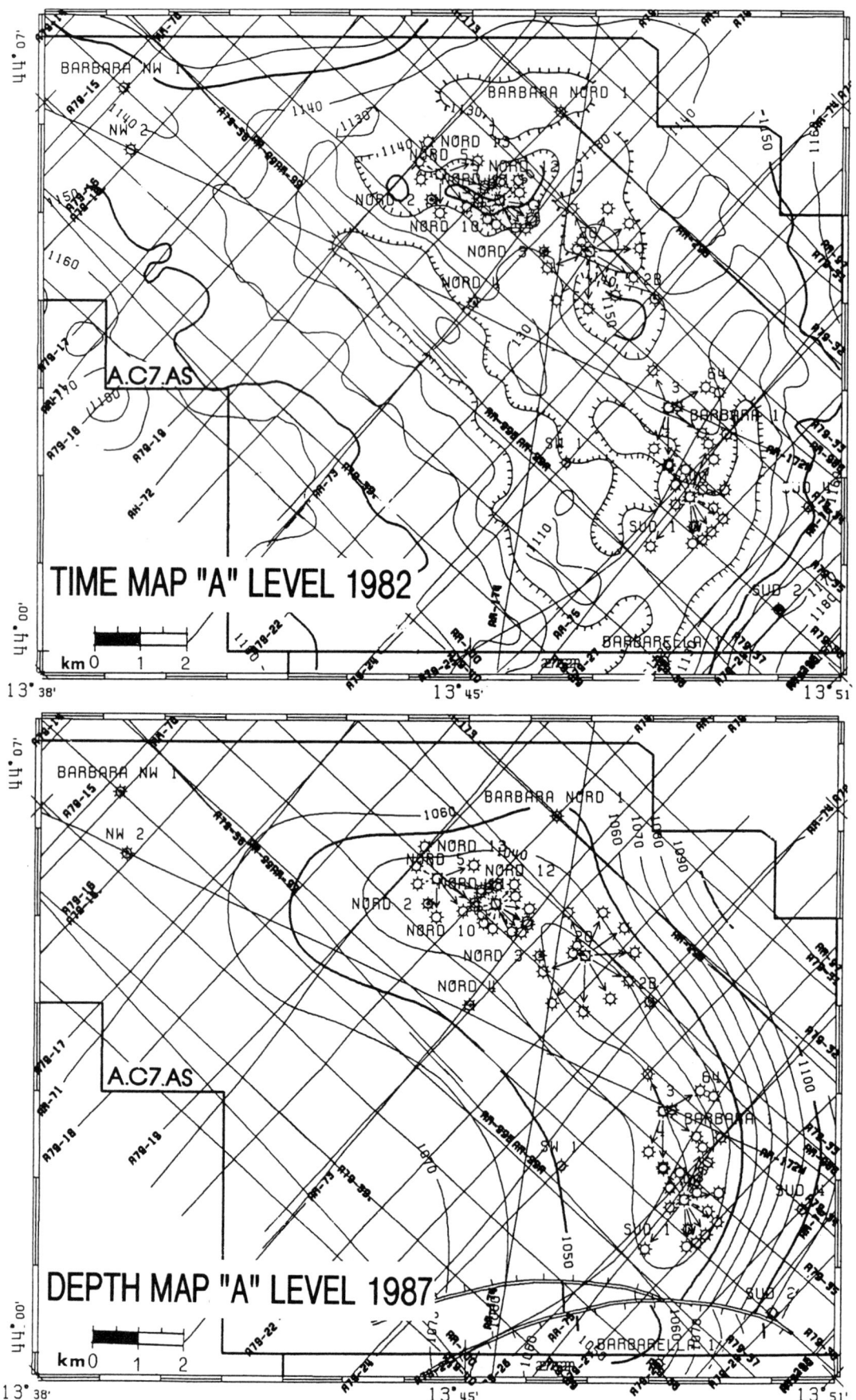

FIGURE 4. "A" level time maps for 1982 and 1987.

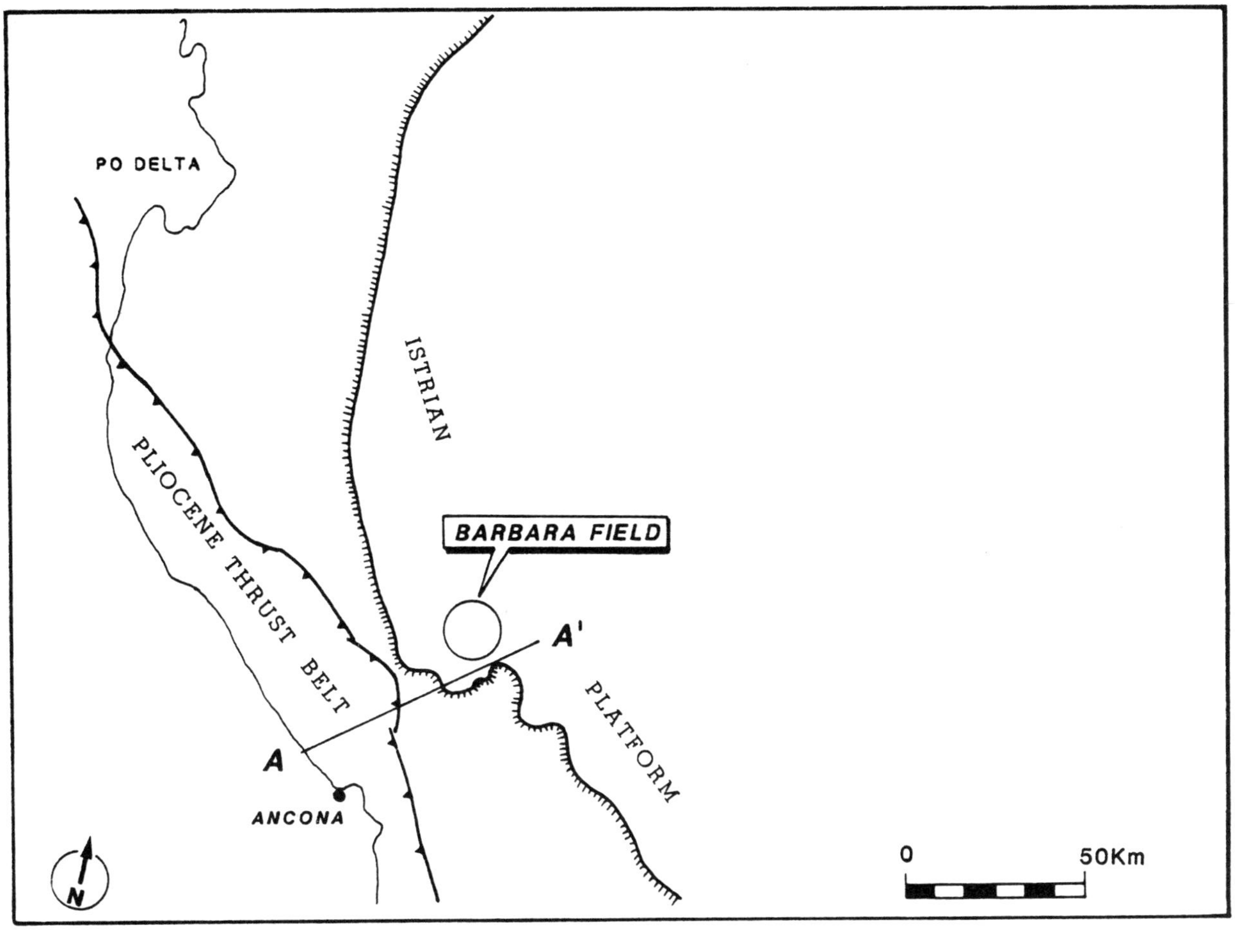

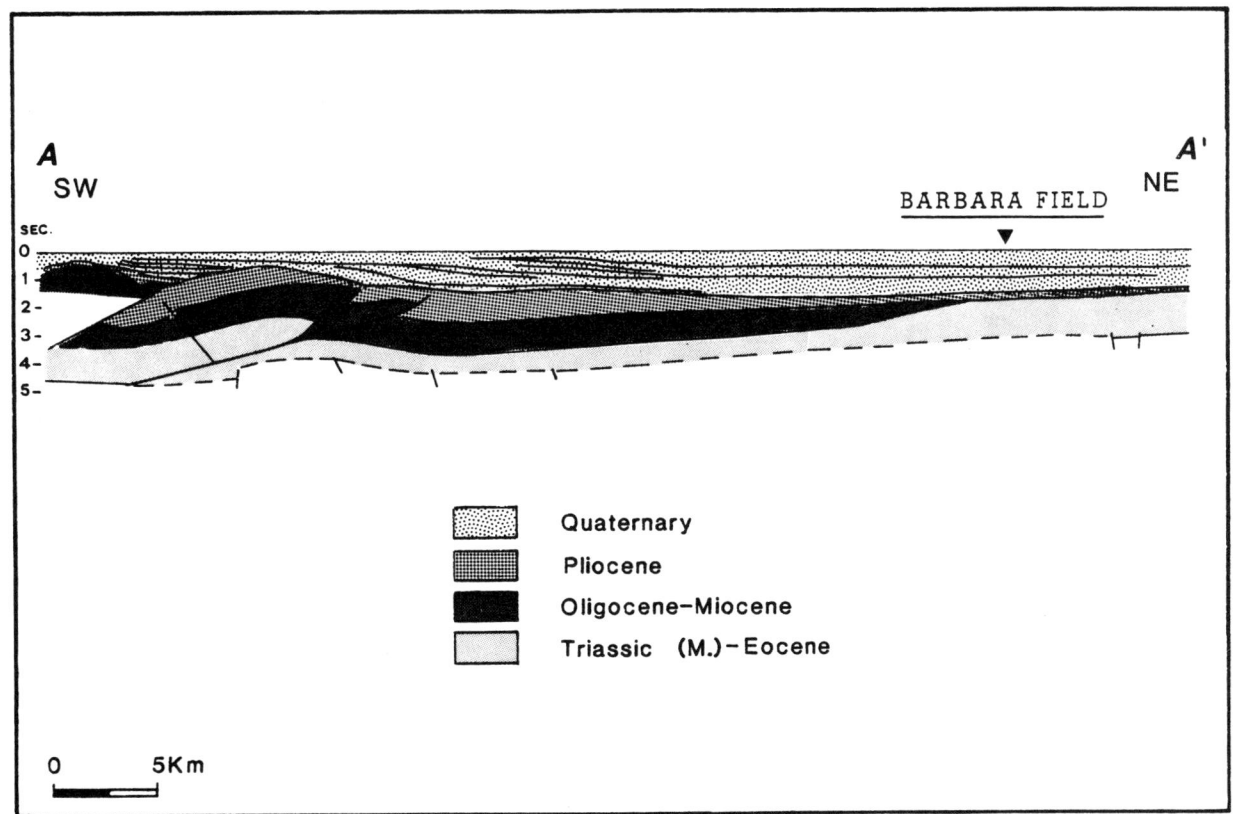

FIGURE 5. North Adriatic area tectonic sketch.

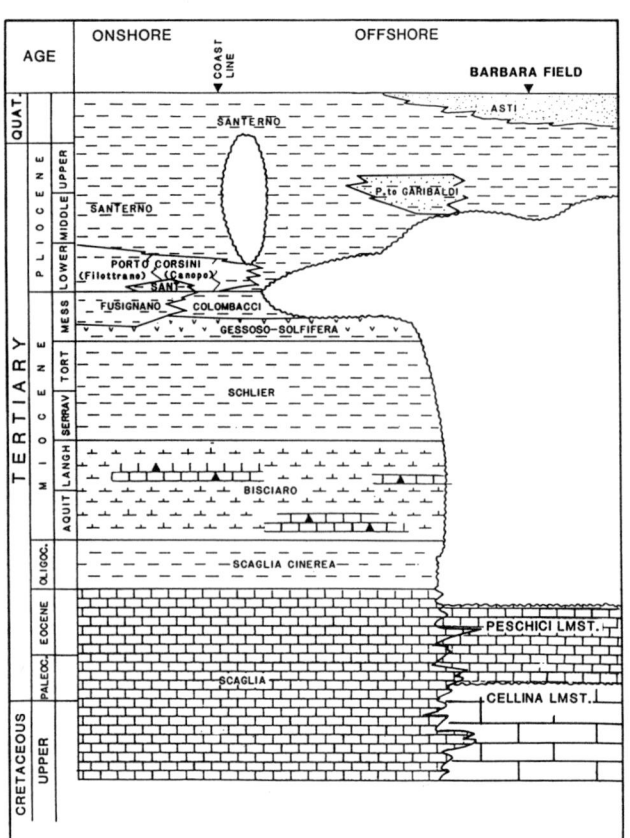

FIGURE 6. Generalized stratigraphy of the Barbara field area.

vironment (Figure 7). The trapping mechanism is related to the draping of the Quaternary and Pliocene sequences overlying thick Mesozoic carbonates cut by subvertical normal faults downthrown to the east. Such a mechanism has allowed the creation of a broad and gentle anticline covering an area of 70 km² and having a vertical closure of 60 m. The trap is not filled to the spillpoint. The gas has migrated into the reservoirs from the basin on the west.

The main gas pools are concentrated in the lower Quaternary and are capped by well-developed shales. The shales, however, are almost absent in the shallower sequence, which is composed mainly of sands and where most of them have an average thickness on the order of 1 m.

Nevertheless, some gas, trapped by these shales, distorts the ray paths and the reflection time of seismic data and creates a low-velocity anomaly on the crest of the structure. The areal extension of such low velocity in the shallow sequence, reconstructed using well data, has the same shape as the syncline related to the pull-down effect of the time maps and corresponds to the highest part of the depth-contoured map (Figure 8).

GEOPHYSICS

The Barbara appraisal case history serves as a valuable aid to definition of other similar situations, especially by utilizing seismic attributes, interactive workstations, and modeling. These techniques, in fact, now allow us to recognize easily those cases where pull-down effects, signal attenuation, and bright spots are related to gas accumulations in both shallow and deep reservoirs.

When applied to the Barbara field, the amplitude, phase, frequency, and relative acoustic impedance seismic data respond in the following ways (Figure 9).

The instantaneous amplitude or reflection strength display shows the magnitude of the change in acoustic impedance at the reflecting interface. Higher values are associated with small amounts of gas in shallow sediments, whereas the productive levels and the underlying carbonates present an abrupt lowering of amplitude owing to a loss of energy through the shallow accumulations.

The instantaneous phase display emphasizes the configuration and reflection continuity. The high continuous signals of the Pleistocene sequence appear disrupted in the gas-affected area.

The instantaneous frequency is sensitive to bed spacing and absorption effects. The display shows a typical "low-frequency shadow" that corresponds to the gas-bearing levels and to the sequence immediately beneath them.

The relative acoustic impedance generated through the inversion of the seismic trace discriminates quite well the low velocity of the gas-bearing sands, because it is not affected by the distortion created by the shallower gas accumulations.

PETROPHYSICAL AND THERMODYNAMIC PARAMETERS

The Barbara gas-bearing sequence can be subdivided, from a reservoir point of view, into three different lithotypes: one, in the upper part, composed of metric to decimetric sand beds; a second one composed of decimetric to centimetric silty layers; and a third one made of centimetric interbedded sands and shales.

Petrophysical parameters have been obtained through laboratory analyses of core data (Table 1). For average porosity values, which range from 29 to 31% for the thickest sands, there is good correlation between values from log data and those from core data.

Water saturation averages 35% in the sands up to 1.5 m thick, but log information for the thin sandy layers gives a value of 75%. Such log indications are unacceptable, because tests run to verify the presence of movable water have resulted in production of dry gas.

Net/gross ratio is a critical parameter, mainly because intervals that logs indicate to be shaly have, during testing, demonstrated good, and sometimes very good, production rates. Net/gross ratio varies from 27% in the silty layers to 95% in the upper sands.

Permeability is directly dependent on sand quality. This parameter, which varies generally from 5 to 100 md, can be as high as 1 darcy in the highest-quality sands.

Text continues on p. 276

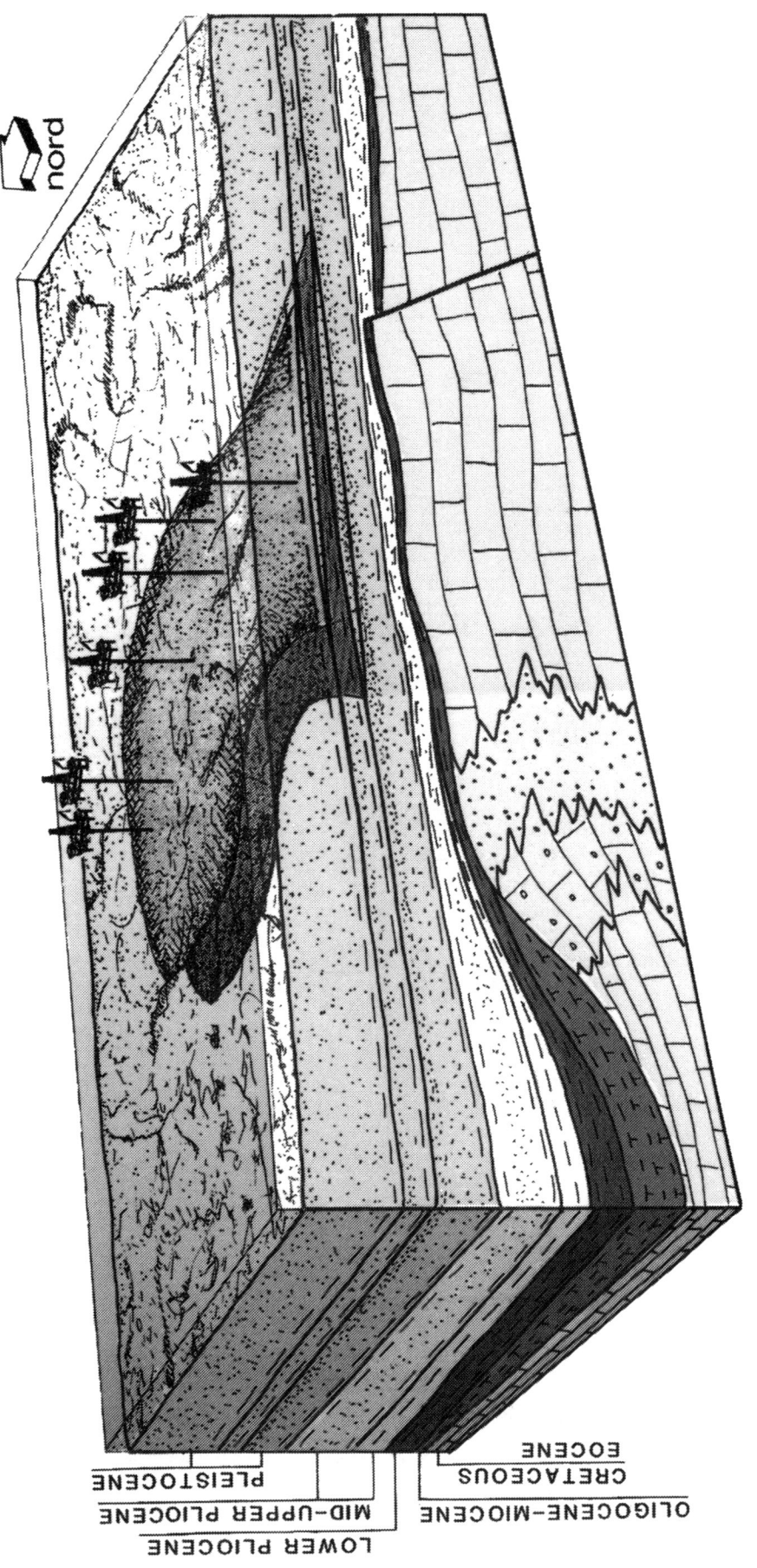

FIGURE 7. Geologic model of Barbara field.

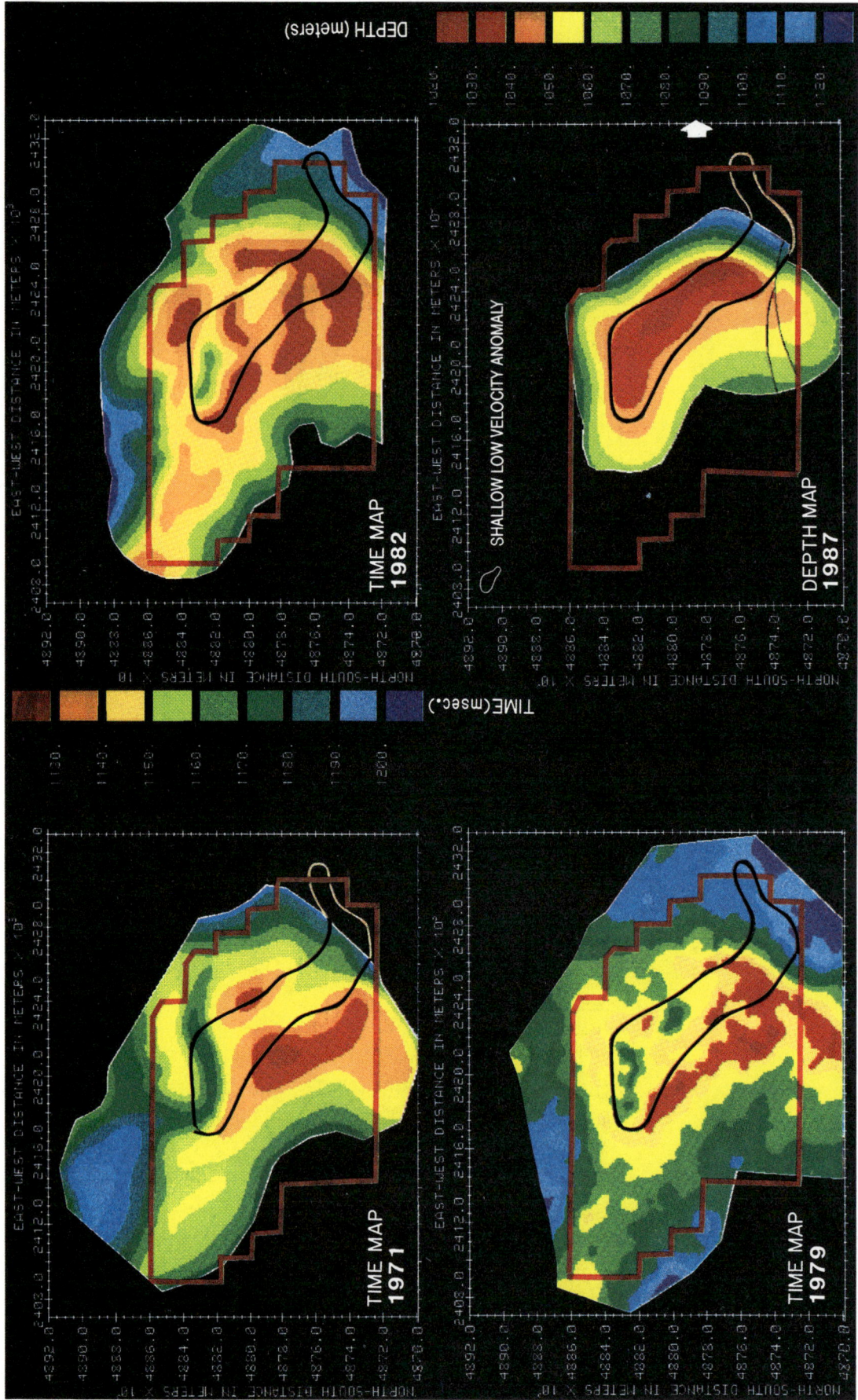

FIGURE 8. Evolution of "A" level maps, showing shallow low-velocity anomaly.

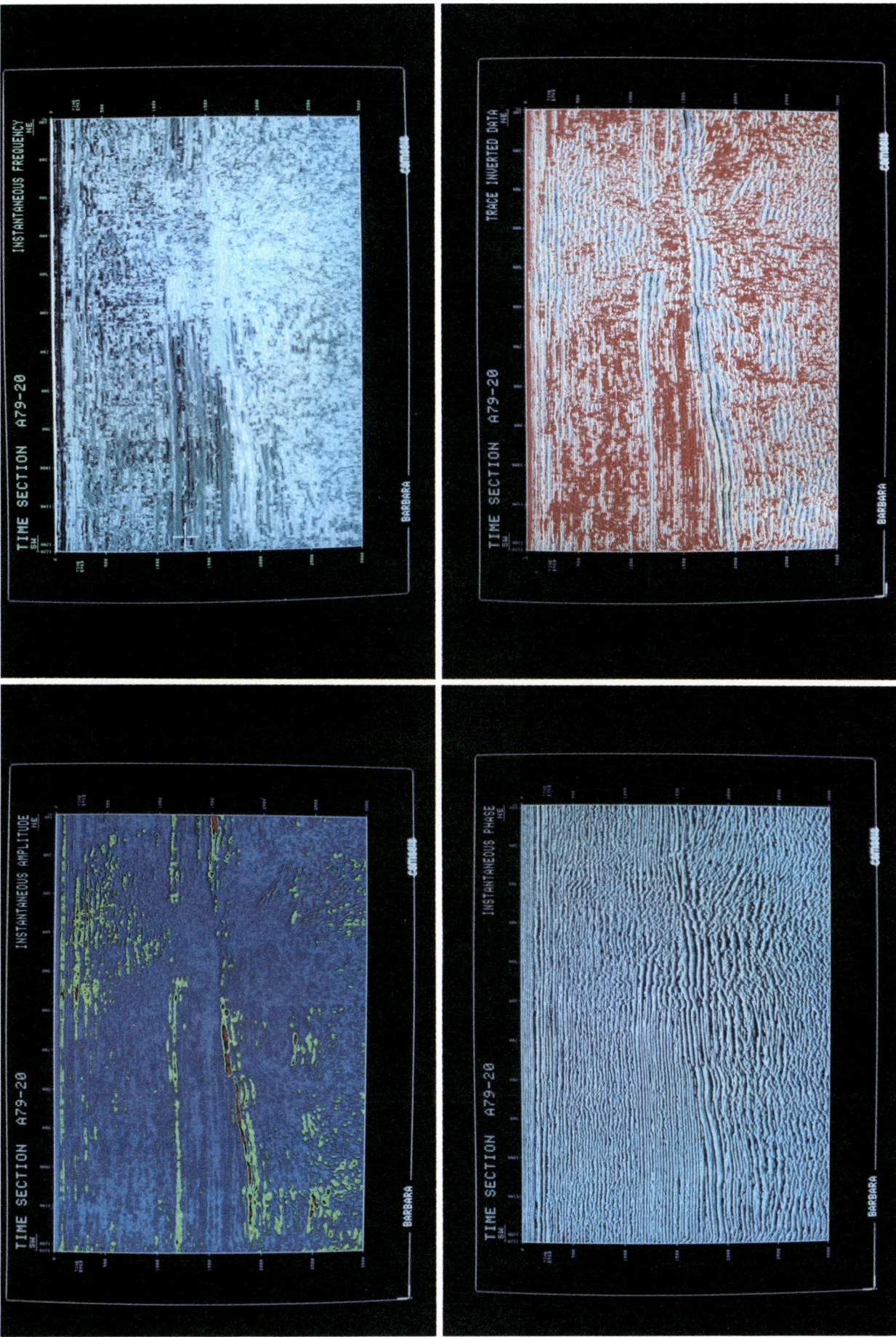

FIGURE 9. Seismic attributes of line A79-20, crossing Barbara gas field.

TABLE 1. Reservoir and fluid parameters for Barbara field.

Reservoir parameters

Average porosity	29 to 31%
Average water saturation	35%
Net/gross ratio	27 to 95%
Permeability	5 md to 1 darcy[1]
Initial static pressure	110.7 kg/cm^2 at −1045 m
	157.6 kg/cm^2 at −1350 m
Temperature	44.6°C at −1045 m
	56.1°C at −1350 m

Fluid parameters

Gas composition	99.5% methane
Gas density[2]	0.558
Water salinity	31.5 g/L
Water density	1.053 g/cm^3 at 15°C

[1]Permeability varies generally from 5 to 100 md, but can be as high as 1 darcy in the highest-quality sands.
[2]Density of air = 1.

BARBARA DEVELOPMENT SCHEME

Barbara field is presently producing from five platforms called Barbara A, B, C, D, and E with a total of 57 development wells. The sixth platform (Barbara F) is already set, and 15 additional wells are being completed; production from these wells is scheduled to begin during 1991.

Almost all wells have dual completion with sand control. The long string generally has an open hole gravel pack, whereas the short one has an inside casing gravel pack. This technique was adopted because of the sand problems encountered in the first production wells. The sand, in fact, was limiting both production capacity and well life. This approach allows:

- Avoidance of frequent workovers.

- High production rates and high drawdown during the depletion phase of the field.

- Avoidance of the risk of sand in the strings caused by shut-in and subsequent reopening of the wells.

The average production rate per well is 145,000 m^3 per day, with peaks of 250,000 m^3 per day. Gas is transported through a 14-in. sea line from the six platforms to the T terminal, which is linked through a bridge to the C platform. The gas reaches the Falconara treatment plant onshore through a 24-in. sea line. In addition, two more 3-in. sea lines link the T terminal to the shore.

CONCLUSIONS

The Barbara field, which started as a low-interest discovery in the early 1970s, has been transformed into the second largest Italian gas field as far as reserves are concerned. Its contribution to 1990 AGIP domestic gas production amounted to 18% of the total. Its role as an important producer will be maintained well into the 1990s. Moreover, the technical difficulties encountered during the appraisal of the field have provided several keys that have enhanced the prospectivity of similar traps over a wide area of the northern Adriatic.

ACKNOWLEDGMENTS

The authors wish to express their thanks to the AGIP management, who authorized the present publication, and to their colleagues at the AGIP laboratories and technical departments, who have consistently contributed to the understanding of this gas field.

Chapter 18

Geochemistry of Oils in the Northern Viking Graben

H. M. Chung
W. S. Wingert
G. E. Claypool

Mobil Research and Development Corp.
Dallas, Texas, U.S.A.

ABSTRACT

The compositions of crude oils in the northern Viking graben area reflect the influence of both the types of organic matter and the thermal maturity of Jurassic source rocks in the major basin depocenters, referred to as the Oseberg and Troll kitchens. Forty-four oil samples were analyzed and classified primarily on the basis of chemical composition, but with recognition of constraints on origins imposed by geographic and stratigraphic locations. Five groups are recognized: (1) Snorre oils, (2) Statfjord-Gullfaks oils, (3) Brent-Oseberg Beta-Veslefrikk oils, (4) Oseberg Alpha-Troll oils, and (5) Agat and Blocks 35/8 and 35/11 oils. The compositions of oils in the first three groups reflect the large relative proportions of marine organic matter and the lower temperatures of expulsion characteristic of oils derived from the Draupne Formation. The compositions of oils in the last two groups reflect source rocks having more terrestrial organic matter and higher temperatures of expulsion.

The initial phase of hydrocarbon expulsion and migration from the Draupne Formation in the northern Viking graben began about 56 m.y.a. in the Troll kitchen and about 45 m.y.a. in the Oseberg kitchen. With continued subsidence and heating of Jurassic source rocks, gases and condensates were expelled, beginning about 40 m.y.a. in the Troll kitchen and about 30 m.y.a. in the Oseberg kitchen. This secondary phase of hydrocarbon generation has mixed with and displaced some of the earlier accumulations.

INTRODUCTION

The northern Viking graben in the North Sea is a prolific petroleum province with several giant accumulations such as Brent, Gullfaks, Oseberg, Snorre, Statfjord, and Troll fields. Total recoverable reserves in these giant and nearby satellite fields are reported to be about 9.7 billion bbl of oil and about 54 tcf of gas (Carmalt and St. John, 1986).

The most important source rock in the north Viking graben is the Upper Jurassic (Oxfordian/Kimmeridgian/Tithonian) Draupne or Kimmeridge Clay Formation. In addition, the Upper Jurassic (Callovian/Oxfordian) Heather Formation and the Middle Jurassic Brent Formation are considered as source rocks for gas and condensate (Kirk, 1980; Goff, 1983; Cornford, 1984; Dahl and Speers, 1985; Thomas et al., 1985).

Most oil fields in the northern Viking graben are located on the highs of rotated fault blocks near the edge of the graben system. Structural and stratigraphic traps are generally filled to spillpoint and are sealed by Lower Cretaceous or Upper Jurassic claystones or marls. Possible source rocks in the immediate vicinity of many oil fields are thermally immature or marginally mature, implying that oil and gas were generated in the deeper parts of the graben system and migrated to shallower depths. Middle Jurassic sandstones are the most important reservoirs and the most important regional migration conduits. In the northern Viking graben, two major depocenters are viewed as the main sites of hydrocarbon generation (Thomas et al., 1985): one is west of Oseberg field (the Oseberg kitchen), and the other is east of the Gullfaks field (the Troll kitchen).

The bulk of the petroleum accumulations in the northern Viking graben has been derived from marine shales of the Humber/Viking Group that are now deeply buried in the axial parts of the graben system (Cornford, 1984; Thomas et al., 1985). Jurassic rock samples from the graben deeps are generally unavailable. Where limited samples are available, the organic matter is thermally overmature. Therefore, in order to determine the source regions, the pathways, and the timing of migration for different petroleum accumulations, it is necessary to rely more heavily on regional patterns of variation in hydrocarbon composition than on documented specific oil-source rock correlations.

Although North Sea crude oils are generally similar, compositional differences occur in regionally significant patterns. In addition, some fields contain oil undersaturated with gas, whereas other fields contain dominantly gas. A variety of factors can explain these patterns of compositional variation in the petroleum accumulations of the northern Viking graben. Such factors include: (1) contribution of hydrocarbons from (stratigraphically) different source rocks, (2) contribution of hydrocarbons from source rocks at different stages of thermal maturity, and (3) contribution of hydrocarbons from source rocks that exhibit regional variations in organic facies.

The purpose of this paper is to document the patterns of variation in the geochemistry of the oils and to interpret the possible geologic and geochemical factors that control this variation. This interpretation is developed within a framework that requires knowledge or assumptions regarding (1) the influence of source material and thermal maturity on certain diagnostic geochemical properties of crude oils and natural gases, (2) regional and stratigraphic variation in the composition of organic matter in the possible source rocks, (3) the general burial history and timing of oil generation and expulsion from the main source rock depocenters, and (4) the main pathways of migration from the source areas to the accumulations.

Improved understanding of the geologic and geochemical factors controlling regional variations in crude oil composition will enable a more detailed and valid reconstruction of the timing and pathways of petroleum migration in the northern Viking graben. Such reconstruction should enhance the ability to evaluate undrilled prospects.

GEOLOGIC FRAMEWORK

The Viking graben is located at the northern margin of the northern North Sea Permian-Triassic rift system. The graben is bordered by the Tampen spur to the northwest and the Horda platform to the east (Figure 1). Development of the North Sea graben system began during Permian time and reached its maximum intensity during intermittent tectonic events (locally referred to as Cimmerian events) that occurred between Late Triassic and Early Cretaceous (Ziegler, 1980, 1981). These tectonic movements resulted in deep axial graben systems bounded by rotated fault-block complexes. Rifting died out progressively during the Late Cretaceous and was accompanied by rapid deposition of Mesozoic sediments. The Mesozoic sediments are overlain by gently downwarped Tertiary sediments, comprising a total sedimentary section of up to 9 km.

The Jurassic period was characterized by an over-all rise in sea level, with periodic emergence and erosion associated with Cimmerian events (Karlsson, 1986). At the beginning of the Middle Jurassic, the main Cimmerian updoming of a large area south of the Viking graben took place and the concomitant erosion of this area supplied clastics to the basin toward the north leading to deposition of the most important reservoir rocks, the Brent Group. With continued rise of the sea level, the Brent delta was progressively drowned, bringing the return of marine conditions and deposition of the Heather Formation. Intermittent but progressively more stable and widespread anoxic conditions are recorded in deposition of the Upper Jurassic Draupne Formation (equivalent to the Kimmeridge Clay). According to Thomas et al. (1985), this organic-rich source rock is a time-transgressive unit within a single marine system, and the depositional and organic facies vary regionally and temporally according to extent of anoxic bottom-water conditions. In the southern Viking graben (e.g., at the Gudrun field) the basin was anoxic as early as the Oxfordian, whereas in the northern Viking graben (e.g., at the Troll field) the basin was largely aerobic until latest Jurassic (Volgian/Ryazanian) time. In addition, anoxic conditions prevailed more frequently in

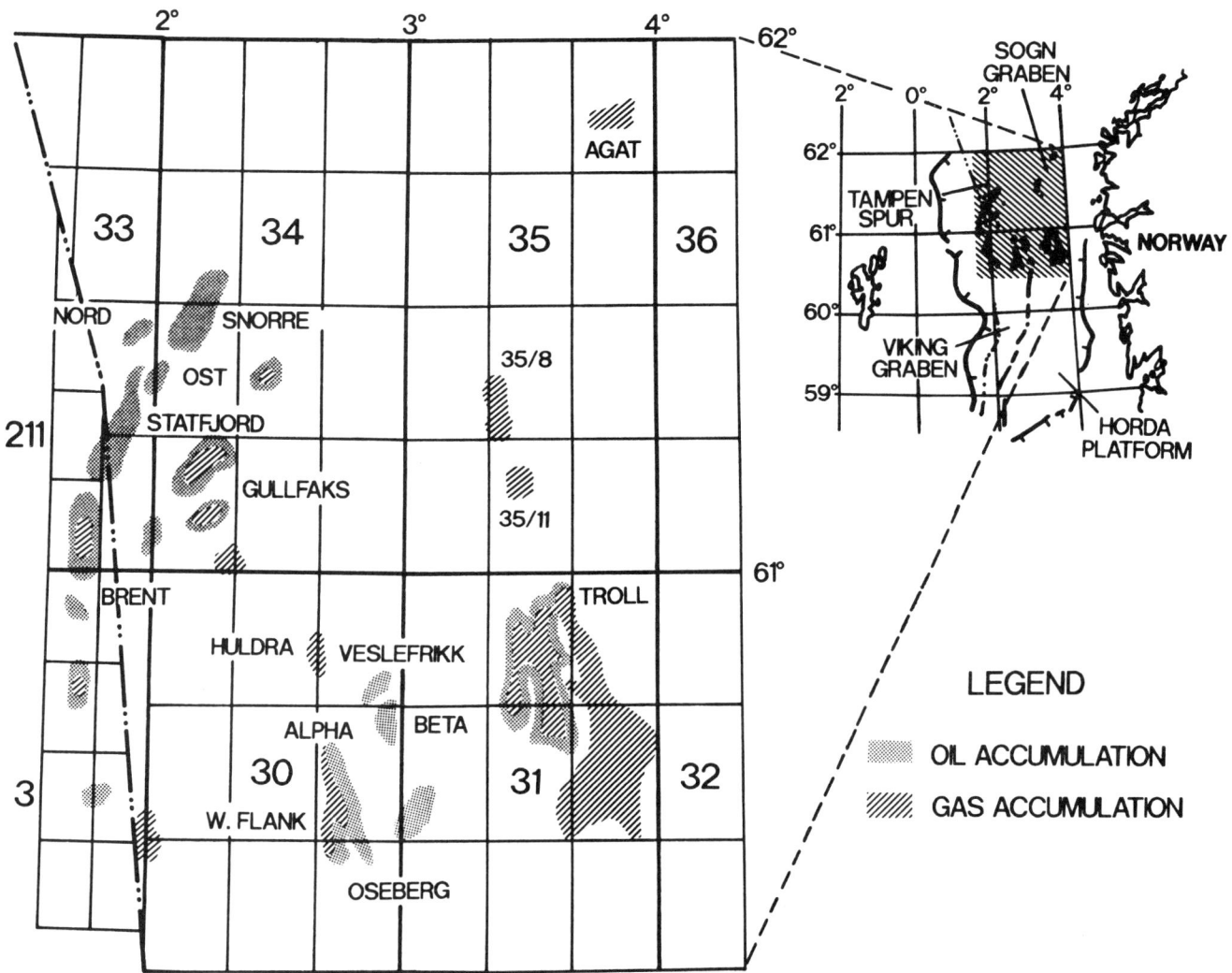

FIGURE 1. Structural features and oil fields in the northern Viking graben.

the deepest graben centers, compared with the transitional areas or on the graben flanks (Cornford et al., 1986).

In the Early Cretaceous, rapid deepening of the basin and opening of broader seaways to the Norwegian-Greenland sea and the Tethys resulted in circulation of freer water and in reestablishment of oxic conditions (Doligez et al., 1987). The Cromer Knoll Group was deposited during this time. This followed a period of erosion due to the latest Cimmerian movement that removed a variable thickness of the Upper Jurassic rocks and resulted in the Base Cretaceous unconformity. The claystones and marls of the Late Cretaceous Shetland Group, the Early Cretaceous Cromer Knoll Group, and the Upper Jurassic Viking Group form the seals of reservoirs in the region.

RESERVOIR ROCKS AND SAMPLE LOCATIONS

The prospective strata of the northern Viking graben are subdivided into the Lunde Formation, the Statfjord Formation, the Dunlin Group, the Brent Group, the Viking/Humber Group, and the Cromer Knoll Group (Figure 2). Major oil fields of the northern North Sea are located on both sides of the Viking graben (Figure 1). Our study concentrates on the fields on the Tampen spur (Brent, Statfjord, Gullfaks, Snorre) on the west side, and on the fields on the Oseberg-Troll complex (Oseberg Alpha, Oseberg Beta, Oseberg Gamma, Oseberg Kappa, Oseberg Omega, Huldra, Veslefrikk, and Troll) on the east side, of the Viking graben. In addition, we have included oil samples from the Agat field and from Blocks 35/8 and 35/11.

The basic geologic information (field and rock unit names) for the 44 crude oil or condensate samples used in this study are given in Table 1. Sample locations are shown in Figure 1. The major producing intervals of the various fields sampled are indicated in Figure 2.

GEOCHEMISTRY OF JURASSIC SOURCE ROCKS

To provide a background for the interpretation of crude oil compositional differences, and to establish possible

PERIOD	SERIES	STAGE	ROCK UNIT NAME		FIELD OR BLOCK
CRETACEOUS	LOWER	BARREMIAN	CROMER KNOLL		Agat
		HAUTERIVIAN			
		VALANGINIAN			
		RYAZANIAN			
JURASSIC	UPPER	VOLGIAN	VIKING/ HUMBER	DRAUPNE/ KIMMERIDGE CLAY	Statfjord Nord Troll
		KIMMERIDGIAN		HEATHER	
		OXFORDIAN			
		CALLOVIAN	BRENT	TARBERT	35/8, 35/11, Statfjort Ost, Statfjord
	MIDDLE			NESS	35/11, Oseberg Beta, Oseberg Omega, Gullfaks, Statfjord
		BATHONIAN		ETIVE	35/11, Veslefrikk, Huldra, Oseberg Beta, Brent, Oseberg Zeta, Oseberg Alpha, Statfjord Nord
				RANNOCH	
		BAJOCIAN		BROOM/OSEBERG	35/11, Statfjord
	LOWER	TOARCIAN	DUNLIN		Statfjord
		PLEINSBACHIAN			
		SINEMURIAN			
		HETTANGIAN	STATFJORD		Oseberg Kappa, Statfjord, Brent
TRIASSIC	UPPER	RHAETIAN	LUNDE		Snorre, Gullfaks

FIGURE 2. Stratigraphic nomenclature chart showing producing intervals of oil fields.

general oil-source rock relationships, we have compiled analyses of samples of the Draupne and Heather formations. These samples are mainly from locations on the flanks of the grabens where the organic matter is thermally immature to marginally mature. For over 170 rock samples that have more than 2 and 1% TOC for the Draupne and Heather formations, respectively, the Pr/Ph (pristane/phytane) ratios are plotted against the HI (hydrogen indices) values in Figure 3, showing an inverse relationship between the two geochemical properties. The Heather Formation samples have Pr/Ph ratios larger than those of the Draupne Formation samples (avg. 2.51 vs. 1.33). The Heather Formation samples also tend to have HI values smaller than those of the Draupne Formation samples (avg. 242 ± 109 mg/g C vs. 376 ± 145 mg/g C).

Carbon isotopic ratios of kerogens are plotted against HI values of source rocks in Figure 4. Both carbon isotopic ratios become isotopically lighter with increasing HI values. The samples with larger HI values (>350 mg/g C), most of which are the Draupne samples, have lighter carbon isotopic ratios of kerogens. In contrast, the samples with smaller HI values (<350 mg/g C), most of which are the Heather samples, have the opposite.

The following generalizations are based on the interrelationships among HI values, Pr/Ph ratios, and carbon isotopic ratios:

1. Thermally immature samples of both the Draupne and Heather formations contain organic matter that shows a wide range of apparent hydrogen content and implies oil-generating character (HI of 500 to 600 mg/g C).

2. A good correlation occurs between the carbon isotopic ratio and the HI for both the Draupne and Heather formations, such that $\delta^{13}C$ is about $-25‰$ when HI is <100 mg/g C, and about -27 to $-29‰$ when HI is >400 mg/g C.

3. There appears to be an inverse relationship between Pr/Ph and HI for the Draupne and Heather formations. The Draupne Formation samples have low Pr/Ph ratios (<2) at all levels of HI, whereas the Heather Formation samples have a wide range of Pr/Ph ratios at HI levels below 300 mg/g C, and lower Pr/Ph values at higher levels of HI.

Based on the observed differences in the organic matter, we can conclude that the Draupne Formation should generate oil with a Pr/Ph ratio of about 1.5 and a carbon isotopic ratio of about $-29‰$. In contrast, the Heather Formation should generate oil with a Pr/Ph ratio in the

TABLE 1. Sample locations and geochemical data for oils from the northern Viking graben.

No.	Field	Block	Formation	S (%)	API (°)	Pr/Ph	SAT (%)	ARO (%)	NSO (%)	%20S (%)	DS/DR	Oil (‰)	SAT (‰)	ARO (‰)	NSO (‰)	Gasoline (‰)
1	Agat	35/3	Cromer Kn.	0.06	50.3	2.83	78.2	18.6	3.2	48.7	2.36	-25.63	-26.39	-25.14	-24.26	-25.01
2	35/8	35/8	Tarbert	0.02	43.2	2.68	71.6	32.6	4.8	53.7	5.58	-26.74	-27.41	-26.12	-25.94	-25.96
3	35/11	35/11	Oseberg	0.03	40.2	2.92	71.2	25.4	3.4	54.0	4.60	-26.74	-27.42	-26.49	-26.35	-25.88
4	35/11	35/11	Etive	0.02	40.4	2.84	68.8	26.4	4.8	52.3	5.01	-27.03	-27.61	-26.66	-26.55	-26.02
5	35/11	35/11	Ness	0.03	42.7	2.88	69.2	26.5	4.3	52.8	4.93	-26.95	-27.63	-26.63	-26.58	-26.07
6	35/11	35/11	Tarbert	0.02	40.8	2.47	63.1	29.6	7.3	52.0	3.96	-27.25	-28.13	-26.89	-26.72	-26.66
7	Troll	31/2	Sognefjord	0.00	49.5	2.71	NA	NA	NA	48.2	2.97	-26.30	NA	NA	NA	-25.47
8	Troll	31/2	NA	0.25	27.6	2.10	46.0	34.4	19.5	NA	NA	-27.67	-28.61	-27.82	-27.76	-25.73
9	Troll	31/2	Sognefjord	0.02	28.0	NA	45.9	37.7	16.5	47.7	2.66	-27.77	-28.68	-27.93	-27.72	-25.75
10	Veslefrikk	30/3	Etive	0.08	39.2	1.38	59.4	32.1	8.5	49.3	2.88	-29.00	-29.25	-28.29	-28.15	-28.81
11	Huldra	30/2	Etive	0.01	42.8	1.72	80.1	17.7	2.3	54.7	6.30	-27.81	-28.71	-27.33	-27.36	-26.48
12	Oseberg Beta	30/6	Etive	0.14	36.6	1.53	56.6	33.1	10.2	49.0	2.27	-29.06	-29.37	-28.60	-27.99	-28.95
13	Oseberg Beta	30/6	Etive	0.13	37.2	1.52	56.1	33.1	10.8	48.1	2.35	-29.08	-29.26	-28.52	-27.81	-28.78
14	Oseberg Beta	30/6	Etive	0.07	38.2	1.44	62.9	30.2	6.9	47.3	2.60	-29.15	-29.41	-28.36	-27.91	-28.92
15	Oseberg Beta	30/6	Brent	0.06	38.3	1.44	63.1	28.8	8.1	48.1	2.66	-29.37	-29.27	-28.58	-28.00	-29.11
16	Oseberg Beta	30/6	Etive	0.12	38.2	1.43	63.4	28.8	7.8	48.4	2.80	-29.27	-29.52	-28.52	-27.99	-28.98
17	Oseberg Beta	30/6	Etive	0.04	37.3	1.36	62.9	28.8	8.3	47.0	2.63	-29.04	-29.22	-28.39	-27.84	-28.82
18	Oseberg Beta	30/6	Ness	0.04	37.9	1.42	63.3	29.6	7.0	48.4	2.57	-29.09	-29.27	-28.48	-27.92	-28.81
19	Oseberg Alpha	30/6	Etive	0.29	34.8	1.52	53.1	34.6	11.9	49.5	2.90	-28.22	-29.03	-28.08	-27.86	-27.07
20	Oseberg Alpha	30/6	Brent	0.25	33.6	1.53	55.0	33.0	11.5	48.5	3.07	-28.04	-28.85	-27.86	-27.66	-27.13
21	Oseberg Gamma	30/6	Brent	0.31	34.3	1.52	49.4	36.4	13.8	48.0	2.61	-28.28	-29.06	-28.04	-27.89	-27.19
22	Oseberg Omega	30/9	Ness	0.13	34.2	1.77	57.5	34.4	8.1	48.6	2.69	-28.00	-28.82	-27.94	-27.61	-26.43
23	Oseberg Omega	30/9	Ness	0.07	40.5	1.79	66.2	27.4	6.4	40.5	1.51	-27.71	-28.30	-27.21	-27.54	-26.98
24	Oseberg Omega	30/9	U. Ness	0.02	55.5	2.25	79.1	15.2	5.7	49.8	3.14	-27.28	-28.76	-27.01	-26.39	-26.51
25	Oseberg Kappa	30/6	Statfjord	0.11	40.8	1.40	65.5	28.4	6.1	48.9	1.86	-29.04	-29.71	-28.50	-27.75	-28.21
26	Oseberg Kappa	30/6	Statfjord	0.08	35.4	1.63	71.6	24.7	3.7	51.4	2.88	-28.24	-29.00	-27.77	-27.33	-26.80
27	Gullfaks	34/10	Brent	0.48	28.8	1.52	49.5	36.0	14.0	46.6	1.49	-28.77	-29.05	-28.86	-28.69	-27.18
28	Gullfaks	34/10	Brent	0.41	28.4	1.47	48.9	37.6	13.1	44.0	1.65	-28.81	-29.63	-28.86	-29.18	-27.28
29	Snorre	34/7	Lunde	0.00	38.7	1.63	61.1	31.4	7.5	46.2	1.90	-30.24	-30.40	-29.65	-29.27	-29.97

TABLE 1. (Continued)

No.	Field	Block	Formation	S (%)	API (°)	Pr/Ph	SAT (%)	ARO (%)	NSO (%)	%20S (%)	DS/DR	Oil (‰)	SAT (‰)	ARO (‰)	NSO (‰)	Gasoline (‰)
30	Statfjord	33/9-9	Statfjord	0.22	39.6	1.61	62.0	25.1	12.7	43.5	1.19	-30.11	-30.30	-29.81	-29.41	-29.68
31	Statfjord	33/9-9	Statfjord	0.20	40.2	1.38	60.1	29.2	9.9	44.5	1.21	-30.06	-30.13	-29.56	-28.93	-29.54
32	Statfjord Nord	33/9	Etive	0.28	36.9	1.44	55.6	30.3	13.3	44.7	1.59	-29.09	-29.72	-28.92	-28.65	-28.71
33	Statfjord Nord	33/9	Draupne	0.31	37.0	1.42	54.2	30.1	15.1	45.0	2.39	-28.30	-28.96	-27.71	-27.56	-28.27
34	Statfjord Ost	33/9	Tarbert	0.25	36.0	1.49	55.1	30.4	14.3	45.7	1.48	-29.18	-29.46	-28.82	-28.62	NA
35	Statfjord Ost	34/7	Brent	0.21	35.7	1.44	57.5	35.2	7.3	44.8	1.77	-29.25	-29.54	-28.84	-28.34	-28.58
36	Statfjord	33/9	Dunlin	0.41	34.6	1.56	58.3	26.6	14.4	45.0	1.36	-29.10	-29.65	-28.94	-28.78	-28.40
37	Statfjord	33/9	Brent	0.38	35.3	1.51	57.3	29.9	12.5	43.7	1.65	-29.11	-29.64	-28.78	-28.66	-28.54
38	Statfjord	33/9	Statfjord	0.32	34.8	1.34	57.1	31.0	11.8	43.7	1.43	-29.20	-29.67	-28.92	NA	-28.46
39	Statfjord	33/9	Ness	0.22	38.0	1.53	54.4	31.3	14.0	44.9	1.50	-29.35	-29.73	-28.95	-28.70	-28.72
40	Statfjord	33/9	Tarbert	0.28	36.4	1.57	56.8	31.4	11.7	45.4	1.55	-29.20	-29.68	-28.88	-28.68	-28.76
41	Statfjord	33/12	Statfjord	0.45	33.4	1.39	54.1	30.3	15.5	43.6	1.66	-29.02	-29.61	-28.74	-28.64	-28.46
42	Statfjord	33/12	Broom	0.33	37.2	1.39	57.4	31.4	11.1	46.8	1.67	-29.12	-29.55	-28.76	-28.58	-28.57
43	Brent	211/29	Brent	1.03	36.8	1.32	58.3	28.8	12.8	46.0	2.31	-28.75	-29.16	-28.26	-28.26	-28.38
44	Brent	3/4	Statfjord	0.20	35.7	1.56	54.2	32.0	13.7	47.9	3.48	-28.97	-29.30	-28.18	-28.00	-29.14

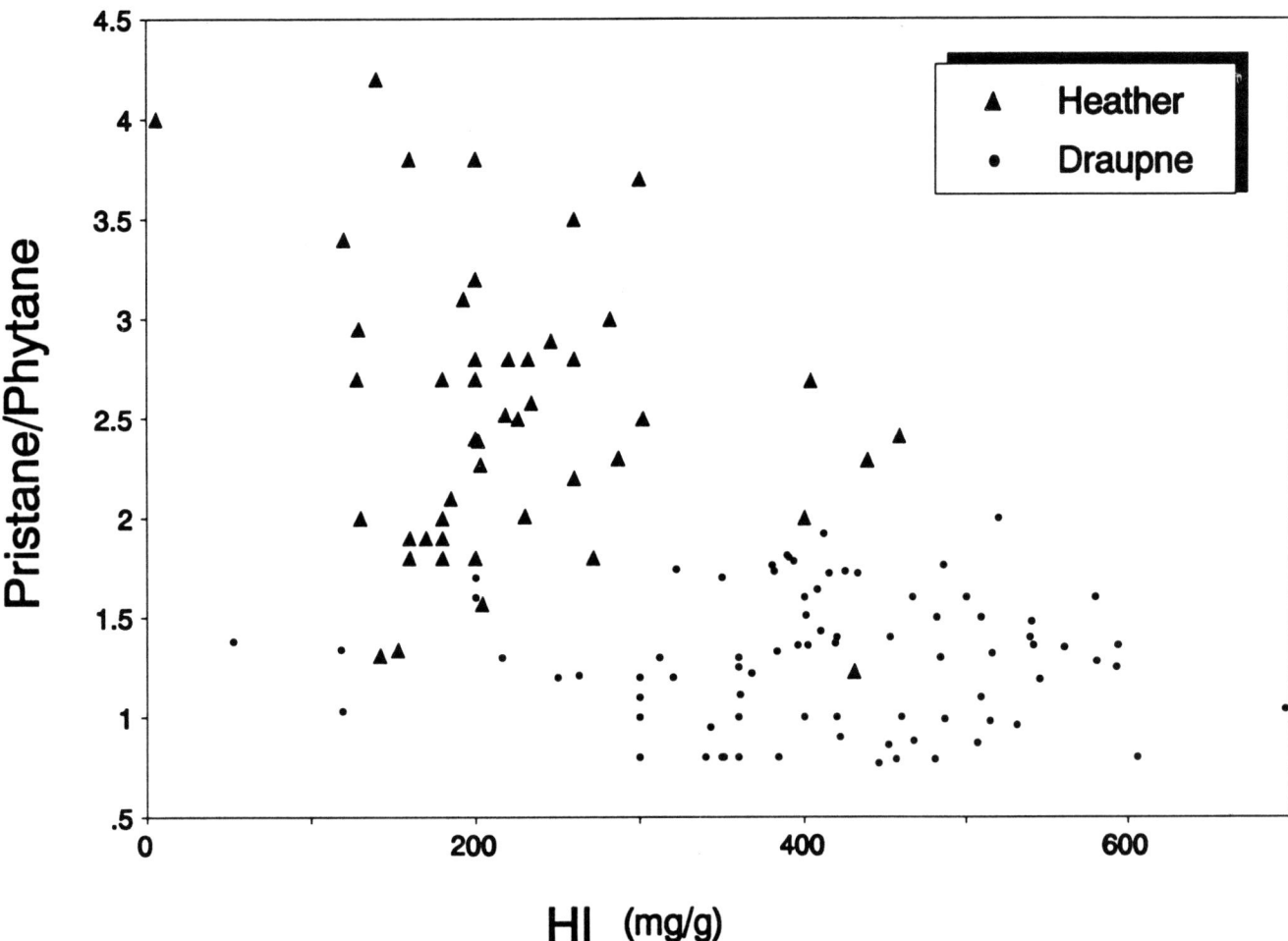

FIGURE 3. Pristane/phytane (Pr/Ph) ratio vs. hydrogen indices (HI) for Upper Jurassic source rock samples.

range of 2 to 3, and a carbon isotopic ratio of about −28 to −26‰.

Compositional Effects of Depositional Environment

The contrasting organic geochemical characteristics of the Draupne Formation and the Heather Formation are well established (Cooper and Bernard, 1984; Field, 1985; Thomas et al., 1985). On average, organic carbon contents and HI values for the immature to marginally mature Draupne Formation samples are about two times larger than comparable values for the Heather Formation samples. These differences in organic geochemical characteristics are believed to reflect mainly differences in the environment of deposition. The samples of the Draupne Formation probably represent rocks deposited under anoxic marine conditions, whereas the samples of the Heather Formation represent rocks deposited under dysaerobic or intermittently anoxic conditions (Thomas et al., 1985; Knudsen et al., 1987). The main effect of anoxic depositional conditions on these Upper Jurassic rocks was enhanced preservation of marine organic matter, which subsequently was altered by intense, early (microbial) diagenetic processes. Under dysaerobic or intermittently anoxic depositional conditions, marine organic matter was less efficiently preserved. As a result, terrestrially derived components (which are more resistant to oxidation during deposition and early diagenesis) make up a larger proportion of the preserved organic matter in the rocks deposited under dysaerobic conditions.

In addition to organic carbon contents and HI values, the effects of differences in depositional environment are also reflected in other properties of the organic matter, such as Pr/Ph ratio, carbon isotopic composition, and organic sulfur content. The enhanced preservation of marine organic matter results in lower Pr/Ph ratios (e.g., 0.9 to 1.5) as a result of (1) chemically reductive degradation of isoalkane and alkane precursors (also leading to increased relative abundance of even-numbered C_{24} to C_{32} n-alkanes) and (2) increased input of acyclic isoprenoid microbial lipids as a consequence of more intense diagenetic processes (ten Haven et al., 1987). Higher Pr/Ph ratios (2 to 4) are associated with poorer preservation of marine organic matter and result in an increased proportion of terrestrially derived organic matter. Marine organic matter of Jurassic age is relatively more ^{13}C-

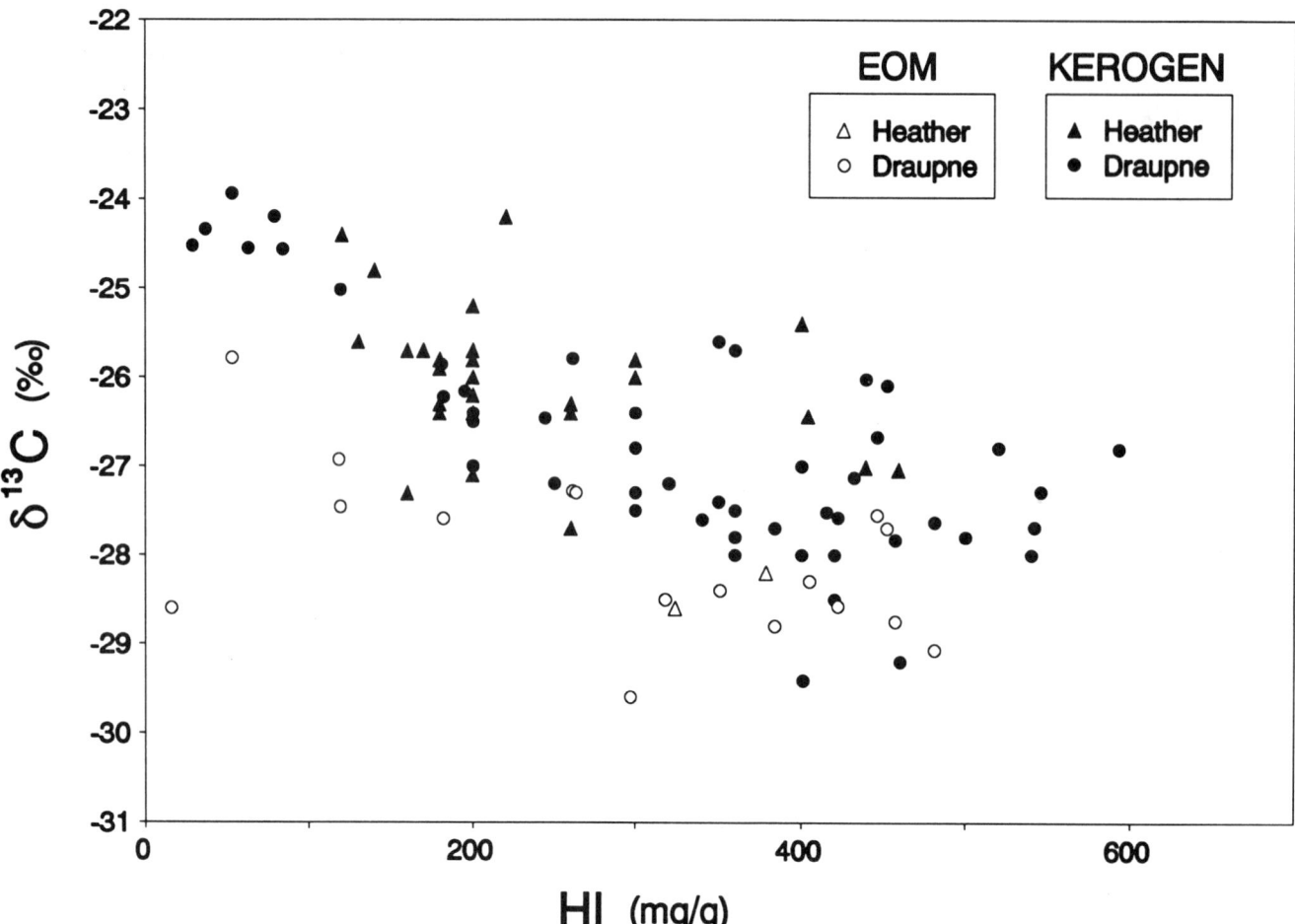

FIGURE 4. Carbon isotopic ratios of kerogens and extracts (EOM) vs. hydrogen indices (HI) for Upper Jurassic source rocks.

depleted, or isotopically lighter ($\delta^{13}C$ of kerogen = $-30‰$), than terrestrial organic matter ($\delta^{13}C$ of kerogen = $-25‰$). In addition, the more intense diagenesis associated with enhanced preservation of marine organic matter results in higher levels of H_2S generation, some of which may be incorporated into the organic matter (Orr, 1986).

Petroleum Generation and Expulsion

The composition of the organic matter in the Upper Jurassic source rocks also changes with increasing depth of burial and temperature. The HI value decreases with increasing degree of conversion of the solid organic matter to petroleum. The composition of the extractable bitumen becomes more like that of petroleum as the extractable components reflecting original biological input to the sediment are diluted by the generated petroleum hydrocarbons. In addition, the hydrocarbon retention capacity, which is directly related to porosity, also decreases with increasing depth of burial and compaction. Expulsion takes place when the quantity of petroleum generated exceeds the retention capacity of the source rock (Ungerer et al., 1985; Durand, 1988). At the depths where hydrocarbons are generated, shale porosities are in the range of about 2 to 4%, and hydrocarbon retention capacities are about 5 to 10 kg/ton. For rich source rocks such as those of the Draupne Formation, with original oil-generating potentials of about 40 kg/ton, migration thresholds are reached and expulsion can occur when 12 to 25% of the kerogen has been converted to hydrocarbons. For leaner source rocks such as those of the Heather Formation, with original oil-generating capacities of 10 kg/ton, expulsion may require 50 to 100% conversion of kerogen. Thus source rocks with high original contents of oil-generating organic matter will expel oil earlier, at lower temperatures, and at lower degrees of conversion of kerogen to petroleum. Conversely, rocks with low contents of oil-generating organic matter may not expel hydrocarbons until higher temperatures and more advanced degrees of kerogen conversion are achieved. At higher temperatures, a greater proportion of the hydrocarbons retained in the source rocks will be converted to light hydrocarbons and gas. Because the Draupne Formation contains greater amounts of marine organic matter, it should expel lower-maturity oils that have geochemical characteristics of less terrestrial organic matter influence. The Heather Formation, by contrast, should expel only higher-maturity oils that have geochemical characteristics of more terrestrial organic matter influence.

The source potential of Middle Jurassic coals of the Brent Group has not been directly investigated as a part of our study. We assume that these rocks are mainly a source for gas and that high temperatures are required for expulsion, either because the kerogen does not react and generate hydrocarbons until high temperatures are reached or because of a high expulsion threshold requiring cracking of liquid hydrocarbons to gas before expulsion can take place.

GEOCHEMISTRY OF CRUDE OILS

As a part of this study, 44 crude oil and condensate samples from the northern Viking graben area have been analyzed. The major oil occurrences (Brent, Statfjord, Snorre, Oseberg) are typical North Sea oils with low to moderate sulfur contents, medium API gravities, and naphtheno-paraffinic hydrocarbon compositions (Cornford, 1984). Oils with higher API gravities and lower sulfur contents occur in the Agat field and Blocks 35/8 and 35/11. In the shallow reservoirs of the Gullfaks and Troll fields, the oils have higher sulfur contents and lower API gravities. The geochemical data, detailed molecular (biological marker) data, and carbon isotopic data are given in Table 1.

The oil and condensate samples from the northern Viking graben can be grouped on the basis of compositional similarity. These groupings generally correlate with geographic location, and to a lesser extent with stratigraphic or structural setting of the reservoir within the field. The migration pathways and drainage areas of the fields are determined by the source rock–carrier bed–reservoir rock geometries at the time of petroleum generation and migration. By evaluating differences among the oils in terms of the possible influences of source and thermal maturity factors, we can test various migration hypotheses and determine whether they are consistent with observed patterns of crude oil composition. The ultimate goal is to reconstruct the petroleum generation, expulsion, migration, and accumulation history in the northern Viking graben.

Thermal Maturity of Oils

In this study, biomarkers are used primarily to assess levels of thermal maturity of crude oils. In particular, we emphasize the degree of C_{29}-sterane isomerization, expressed as % 20S. This property has previously been used as a maturity indicator for North Sea crude oils and source rocks (Mackenzie and McKenzie, 1983; Mackenzie et al., 1984). It is less sensitive to differences in the nature of the source material than other maturity-sensitive biomarker properties (Rullkötter and Marzi, 1988). In addition, the relative amounts of the diasterane biomarkers are also used (Mackenzie et al., 1982).

Sterane isomerization (% 20S) is plotted against diasterane/regular sterane ratio (DS/RS) in Figure 5. With some exceptions, oils in the Tampen spur area (Brent, Statfjord, Snorre, Gullfaks fields) are less mature than oils from the east side of the Viking graben. As a group, oils in the Tampen spur area have the least mature sterane distributions (% 20S = 45.1). Oils from the Kappa and Omega fields at Oseberg, which we refer to as Oseberg West Flank, are generally more mature than the oils from the Tampen spur area (% 20S = 47.8). Oils from the Agat, Troll, Veslefrikk, and Oseberg fields also have more mature sterane distributions (% 20S = 48.3) than the oils from the Tampen spur area. As a group, oils from the Huldra field and from Blocks 35/8 and 35/11 have the most mature sterane distributions (% 20S = 53.2).

Source Characteristics of Oils

The Pr/Ph ratios and carbon isotopic compositions of the source rocks reveal differences in source characteristics. Similarly, these properties are used as cursory measures of an oil's source character. Additional information is derived from the carbon isotopic compositions of the volatile hydrocarbons.

Pr/Ph and Carbon Isotopic Ratios of Whole Oils

We have used Pr/Ph and carbon isotopic ratios of whole oils as the initial bases for grouping oils in terms of common or similar source rocks. The effects of thermal maturity on these properties are minor, and interpretation can be readily adjusted for such effects, if necessary.

The Pr/Ph ratios are plotted against the carbon isotopic ratios of whole oils in Figure 6. Most oils from the Tampen spur, Veslefrikk field, and Oseberg Beta field have Pr/Ph ratios in the range of 1.3 to 1.6 and carbon isotopic ratios of whole oils in the range of −30.2 to −28.3‰. They are the typical North Sea oils (Cornford, 1984). The oils from the Oseberg Alpha, Oseberg West Flank (Kappa, Omega), and Troll fields have Pr/Ph ratios that are generally in the range of 1.5 to 2.3, and carbon isotopic ratios of oils from −29.0 to −26.3‰. Finally, the oils in the Agat field and Blocks 35/8 and 35/11 have extreme Pr/Ph ratios and extreme carbon isotopic ratios (>2.5 and −27.3‰, respectively). This wide variation in basic geochemical source characteristics is inconsistent with the oils from the northern Viking graben being derived from a single source rock.

The Tampen spur area oils have geochemical source characteristics that suggest derivation from the Draupne Formation. By contrast, most of the oils from the east side of the Viking graben have source characteristics that indicate varying degrees of influence of terrestrial organic matter in their source rocks.

Carbon Isotope Profiles

In order to characterize further the differences among these oils, we have determined the carbon isotopic distribution within the oils as a function of molecular size or boiling point. This plot of carbon isotopic ratio vs. distillation fraction number is referred to as the isotope profile. Isotope profiles of oils characterize volatile hydrocarbons, which are the most abundant portion of reservoired hydrocarbons. In addition, isotope profiles of oils are consistent properties that reflect source characteristics as

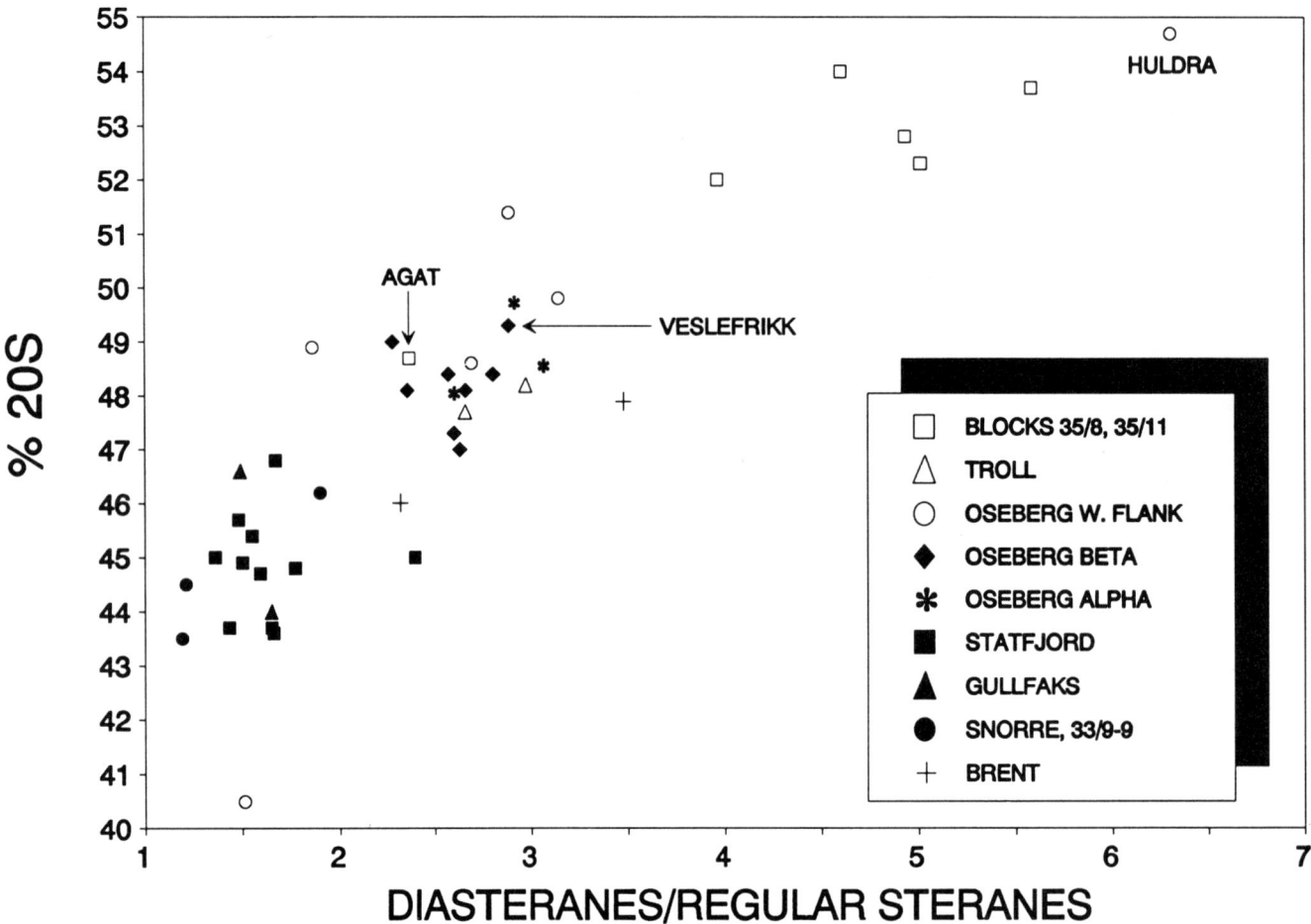

FIGURE 5. Degree of isomerization of C_{29}-sterane (% 20S) vs. ratio of diasterane to regular sterane (DS/RS).

well as thermal maturity of the oils from the northern Viking graben (Northam, 1985). Isotope profile data are given in Table 2.

The basic principle of the isotope profile technique is shown in Figure 7, using three representative isotope profiles of oils A, B, and C from the northern Viking graben. The isotope profile of oil A exhibits the most negative carbon isotopic ratios among the three oils, and a relatively flat or gentle slope. Oil A has the isotope profile of a typical marine shale–derived oil. The isotope profile of oil B is isotopically heavier than the isotope profile of oil A, but both isotope profiles are similar in shape. Oil B could be a more mature variety of oil A, or oil B could be derived from a different facies of the source rock that is responsible for oil A. Oil C has a steeply sloped isotope profile and is isotopically the heaviest among the three oils. The isotope profile of oil C is characteristic of deltaic oils such as the Tertiary oils from the Niger basin. The characteristic isotope profiles of deltaic oils in the northern Viking graben probably represent the influence of terrestrial organic matter and expulsion at advanced degrees of thermal maturity.

Based on these general guidelines for interpreting isotope profiles, we will now evaluate the evidence for possible genetic relationships among the northern Viking graben oils.

Oils in the Northern Viking Graben

The Brent and Statfjord Fields

Two samples from the Brent field and seven samples from the Brent and Statfjord formations in the Statfjord field have similar isotope profiles, which are typical of most North Sea oils (Figure 8). The Brent and Statfjord accumulations in the Statfjord field, with the exception of the Statfjord accumulations from well 33/9-9 (samples 30 and 31), are referred to here as the typical Statfjord oils. The oils in these fields may have originated partly from local depocenters in the East Shetland basin to the west of the fields (Kirk, 1980; Goff, 1983).

The oils from the Statfjord Ost and Nord fields have geochemical properties that are similar to those of the typical Statfjord oils (Table 1). However, the oils from the Statfjord Nord field (sample 33) have slightly heavier isotope profiles and carbon isotopic ratios of nonvolatile hydrocarbon fractions (Tables 1 and 2).

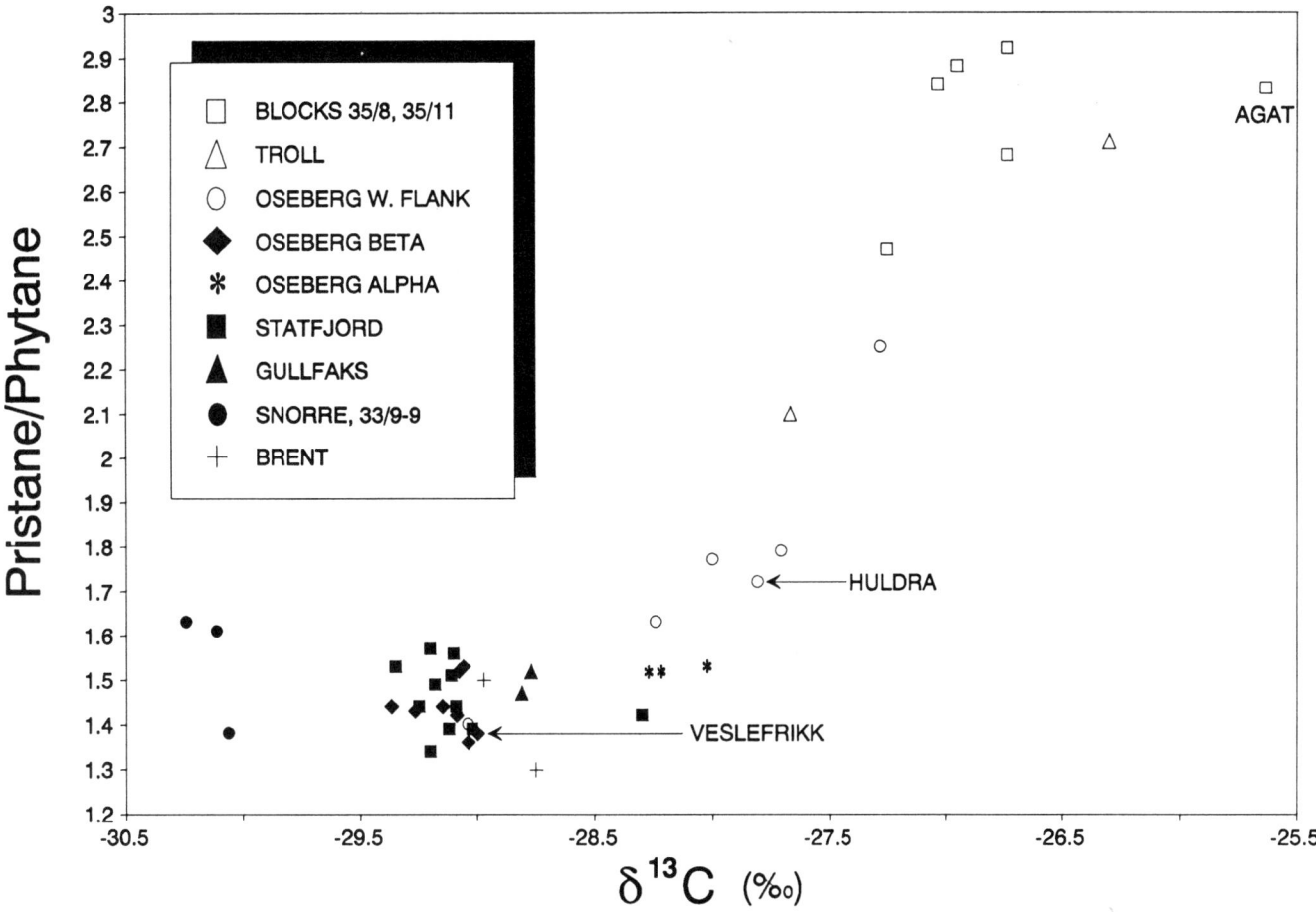

FIGURE 6. Pristane/phytane (Pr/Ph) ratios vs. carbon isotopic ratios of whole oils.

Statfjord Reservoir in Statfjord Field and Lunde Reservoir in Snorre Field

The oils from these reservoirs have isotope profiles that are isotopically lighter than those of the typical Statfjord oils (Figure 8). The Statfjord accumulation in the Statfjord field has been reported as completely separate from the overlying Dunlin and Brent accumulations (Roberts et al., 1987). This is consistent with our analyses of two samples from well 33/9-9, which show one oil type in the Statfjord Formation and a second oil type in the Dunlin/Brent Formation.

Gullfaks Field

Gas chromatograms of oils from the Gullfaks field show a lack of n-alkanes, indicating that the oils are biodegraded. Because n-alkanes are isotopically lighter than other hydrocarbons (Zaback, 1987), the isotope profiles of the Gullfaks oils are isotopically heavier than those of the typical Statfjord oils (Figure 8). Biodegradation also explains why the oils have higher sulfur contents (0.45% vs. 0.34%) and lower API gravities (28.6° vs. 35.7°) than the typical Statfjord oils (Table 1).

Oseberg Beta, Oseberg Zeta, and Veslefrikk Fields

The Oseberg Beta, Oseberg Zeta, and Veslefrikk oils have virtually identical geochemical compositions (Table 1). The geochemical data also show these Oseberg Beta oils to be similar to the Brent oils but slightly more mature than the typical Statfjord oils. The isotope profiles of Oseberg Beta and typical Statfjord oils are similar (Figures 8 and 9), suggesting that both were derived primarily from the Draupne Formation.

Oseberg Alpha, Alpha North, and Gamma Fields

The oils from the Oseberg Alpha, Alpha North, and Gamma fields are the same, and they are described as Oseberg Alpha oils. These oils are different from the Oseberg Beta oils. Differences in the carbon isotopes of the two groups are revealed by their isotope profiles (Figure 9). Oseberg Alpha oils are about 1 to 2‰ heavier than Oseberg Beta oils. The biomarkers (% 20S), however, show that both groups have the same maturity (Table 1).

Differences between the Oseberg Alpha and Beta oils have also been observed by others (Dahl et al., 1987). The sedimentary cover was minimal (20 to 30 m) in the Ose-

TABLE 2. Isotope profile data for oils from the northern Viking graben.

No.	Field	1 (%)	2 (%)	3 (%)	4 (%)	5 (%)	6 (%)	7 (%)	8 (%)	9 (%)	10 (%)	11 (%)	12 (%)
1	Agat	2.20	3.16	11.71	16.24	11.76	8.62	6.72	6.51	6.78	6.93	5.70	13.68
2	35/8	0.69	2.38	7.11	9.17	6.70	4.91	5.05	5.05	5.20	7.14	6.52	40.07
3	35/11	0.29	1.20	5.25	7.41	6.23	5.92	5.45	5.26	5.20	6.91	7.99	42.88
4	35/11	0.55	1.43	5.43	7.67	6.39	5.84	5.17	5.03	5.74	6.70	6.47	43.57
5	35/11	1.29	2.42	7.66	8.39	6.28	5.60	4.90	4.83	5.47	6.19	5.99	40.97
6	35/11	1.20	2.17	6.87	7.54	6.19	5.42	5.47	4.19	5.02	6.45	5.70	43.80
7	Troll	0.84	1.80	20.04	24.33	16.54	12.44	7.92	5.98	4.47	0.00	0.00	5.63
8	Troll	0.00	0.49	2.30	3.58	3.48	3.04	3.77	4.23	4.47	6.68	5.73	62.23
9	Troll	0.00	0.32	2.67	3.40	3.90	3.48	3.19	3.73	5.17	6.56	6.23	61.34
10	Veslefrikk	1.77	2.36	4.37	4.75	4.58	4.59	4.14	4.07	4.90	4.14	6.20	54.13
11	Huldra	0.48	1.50	6.59	9.23	7.77	6.65	5.68	5.48	6.62	6.25	5.94	37.80
12	Oseberg Beta	0.21	4.43	1.44	4.91	3.98	4.00	4.60	4.75	4.80	5.28	6.04	55.57
13	Oseberg Beta	0.28	1.20	4.14	4.16	4.66	5.76	4.01	4.12	4.78	5.99	5.68	55.23
14	Oseberg Beta	1.24	2.03	3.95	4.66	4.66	4.35	4.08	4.16	4.73	5.33	5.88	54.94
15	Oseberg Beta	1.39	1.83	3.64	4.59	5.23	3.81	4.23	4.16	5.06	7.68	6.04	52.34
16	Oseberg Zeta	1.23	1.68	3.55	4.59	4.31	4.38	3.86	4.15	4.88	5.55	5.95	55.87
17	Oseberg Beta	1.09	1.94	4.34	4.68	4.47	4.39	4.17	4.50	5.64	4.52	5.34	54.90
18	Oseberg Beta	1.43	2.12	4.04	4.80	4.60	4.55	4.20	4.11	4.61	5.43	5.60	54.51
19	Oseberg Alpha	1.43	1.51	3.64	4.00	4.32	4.16	4.38	4.19	5.27	5.93	6.67	54.50
20	Oseberg Alpha	0.59	1.25	3.74	4.05	3.14	5.93	3.99	4.34	5.47	6.12	6.48	54.90
21	Oseberg Gamma	0.76	1.94	3.17	4.91	3.51	4.40	3.93	4.82	4.76	6.16	7.28	54.39
22	Oseberg Omega	0.39	1.41	4.03	4.68	4.19	4.82	4.53	4.06	0.00	0.00	0.00	71.79
23	Oseberg Omega	0.82	1.71	4.07	4.85	4.82	4.65	4.72	4.98	6.22	7.86	8.97	46.34
24	Oseberg Omega	4.67	5.55	13.63	14.79	11.20	9.40	7.42	6.45	6.62	5.92	7.35	7.00
25	Oseberg Kappa	0.99	3.22	5.14	5.78	5.30	5.33	4.89	4.77	5.26	5.83	6.07	47.42
26	Oseberg Kappa	0.47	1.17	2.62	4.61	3.98	3.23	3.15	3.79	3.17	3.64	5.94	64.23
27	Gullfaks	0.28	0.46	1.64	3.03	4.11	2.79	3.72	4.02	4.62	6.01	6.39	62.94
28	Gullfaks	0.23	0.49	1.38	3.07	3.88	3.50	3.25	4.13	6.16	6.57	5.40	61.95
29	Snorre	1.46	2.14	4.73	5.27	5.15	5.05	4.61	4.45	5.17	5.74	5.75	50.47
30	Statfjord	2.90	2.62	4.52	5.34	6.13	5.15	5.33	3.64	5.62	6.22	6.12	46.38
31	Statfjord	3.23	3.21	3.97	5.89	5.42	5.50	4.92	6.31	3.19	5.36	6.43	46.54
32	Statfjord Nord	1.93	1.65	2.48	5.27	6.38	4.03	5.08	4.42	4.13	6.00	5.78	52.84
33	Statfjord Nord	1.91	2.24	3.91	4.42	4.93	4.90	4.34	4.15	4.51	5.18	5.86	53.62
35	Statfjord Ost	0.48	1.39	3.54	4.47	4.67	4.71	4.38	4.25	4.87	5.88	5.91	55.46
36	Statfjord	1.98	1.79	4.04	4.46	4.94	4.20	4.86	2.92	5.87	5.48	6.23	53.22
37	Statfjord	1.82	2.26	3.59	5.61	5.96	4.86	4.61	4.25	5.31	5.25	6.61	49.86
38	Statfjord	0.91	1.70	3.62	3.96	5.26	4.68	4.93	4.72	6.30	3.43	7.28	53.20
39	Statfjord	2.53	2.16	3.66	5.18	5.31	3.99	4.09	5.84	2.93	8.12	5.48	50.70
40	Statfjord	2.56	4.41	6.14	2.93	4.98	5.39	3.74	5.08	6.66	6.52	3.68	48.21
41	Statfjord	2.35	2.12	3.92	4.97	4.80	3.04	4.58	6.80	5.98	5.74	8.01	47.65
42	Statfjord	0.87	2.20	4.31	4.63	4.99	4.66	4.04	4.55	4.59	5.46	7.53	52.19
43	Brent	2.21	2.03	4.35	5.27	5.85	6.94	3.70	4.04	5.11	5.39	5.44	49.67
44	Brent	0.00	2.13	3.46	4.06	4.67	3.70	2.16	6.40	2.78	4.72	5.94	59.98

TABLE 2. (Continued)

No.	Field	1 (‰)	2 (‰)	3 (‰)	4 (‰)	5 (‰)	6 (‰)	7 (‰)	8 (‰)	9 (‰)	10 (‰)	11 (‰)	12 (‰)
1	Agat	-25.56	-25.06	-24.91	-25.08	-25.35	-25.46	-25.66	-25.70	-25.75	-25.81	-25.92	-26.11
2	35/8	-26.85	-26.11	-25.95	-25.96	-26.15	-26.48	-26.68	-26.78	-26.85	-26.92	-27.00	-26.82
3	35/11	-27.03	-26.14	-25.68	-26.02	-26.18	-26.39	-26.52	-26.67	-26.72	-26.73	-26.94	-27.05
4	35/11	-27.00	-26.10	-26.00	-26.03	-26.36	-26.61	-26.77	-26.90	-26.95	-26.99	-27.11	-27.29
5	35/11	-27.14	-26.29	-26.04	-26.10	-26.39	-26.63	-26.81	-26.90	-27.11	-27.02	-27.15	-27.28
6	35/11	-27.98	-27.00	-26.67	-26.65	-27.04	-27.30	-27.44	-27.51	-27.55	-27.51	-27.67	-27.63
7	Troll	-24.00	-24.30	-25.16	-25.72	-26.31	-26.75	-27.05	-27.29	-27.51	NA	NA	-27.98
8	Troll	NA	-25.08	-25.37	-25.96	-26.54	-26.85	-27.13	-27.27	-27.47	-27.57	-27.71	-28.10
9	Troll	NA	-25.07	-25.48	-25.96	-26.50	-26.84	-27.09	-27.29	-27.46	-27.59	-27.78	-28.13
10	Veslefrikk	-29.74	-29.00	-28.85	-28.77	-28.96	-29.08	-29.09	-29.16	-29.20	-29.17	-29.25	-28.89
11	Huldra	-26.21	-25.90	-26.36	-26.56	-27.06	-27.46	-27.71	-27.88	-27.97	-28.03	-28.14	-28.36
12	Oseberg Beta	-29.55	-29.02	-28.86	-28.98	-29.23	-29.33	-29.36	-29.33	-29.43	-29.45	-29.48	-29.09
13	Oseberg Beta	-29.37	-28.79	-28.74	-28.81	-29.04	-29.13	-29.19	-29.26	-29.27	-29.26	-29.34	-28.95
14	Oseberg Beta	-29.82	-28.92	-28.95	-28.90	-29.10	-29.18	-29.27	-29.24	-29.27	-29.30	-29.33	-28.98
15	Oseberg Beta	-30.45	-29.28	-29.14	-29.09	-29.34	-29.42	-29.48	-29.49	-29.47	-29.52	-29.64	-29.12
16	Oseberg Zeta	-29.81	-29.12	-29.00	-28.96	-29.20	-29.36	-29.39	-29.42	-29.45	-29.44	-29.50	-29.15
17	Oseberg Beta	-29.48	-28.55	-28.86	-28.78	-29.00	-29.12	-29.15	-29.16	-29.26	-29.30	-29.32	-29.00
18	Oseberg Beta	-29.15	-28.66	-28.79	-28.82	-29.01	-29.14	-29.21	-29.23	-29.27	-29.29	-29.33	-29.01
19	Oseberg Alpha	-27.12	-27.01	-26.89	-27.22	-27.76	-28.17	-28.30	-28.27	-28.39	-28.38	-28.44	-28.48
20	Oseberg Alpha	-27.17	-26.90	-26.96	-27.29	-27.43	-27.63	-27.80	-28.00	-28.10	-28.16	-28.21	-28.30
21	Oseberg Gamma	-27.18	-27.03	-27.06	-27.28	-27.59	-27.82	-28.05	-28.14	-28.23	-28.26	-28.46	-28.44
22	Oseberg Omega	-26.55	-26.05	-26.28	-26.56	-26.97	-27.29	-27.60	-27.83	NA	NA	NA	-28.31
23	Oseberg Omega	-27.81	-27.18	-26.97	-26.99	-27.33	-27.58	-27.68	-27.76	-27.69	-27.58	-27.51	-27.88
24	Oseberg Omega	-26.08	-26.33	-26.45	-26.56	-27.00	-27.37	-27.68	-27.91	-28.04	-28.06	-28.21	-28.49
25	Oseberg Kappa	-28.40	-27.83	-28.09	-28.31	-28.60	-28.82	-28.95	-29.04	-29.09	-29.10	-29.20	-29.26
26	Oseberg Kappa	-26.33	-26.03	-26.49	-26.98	-27.44	-27.67	-27.87	-27.89	-27.98	-28.11	-28.21	-28.68
27	Gullfaks	-26.86	-26.71	-26.78	-27.40	-27.92	-28.26	-28.63	-28.63	-28.76	-28.76	-28.78	-29.07
28	Gullfaks	-26.58	-26.80	-27.03	-27.39	-27.95	-28.24	-28.50	-28.63	-28.64	-28.70	-28.87	-29.09
29	Snorre	-31.04	-30.19	-29.99	-29.95	-30.14	-30.26	-30.30	-30.35	-30.34	-30.34	-30.35	-30.06
30	Statfjord	-30.92	-30.22	-29.66	-29.70	-30.10	-30.26	-30.33	-30.33	-30.32	-30.32	-30.30	-29.95
31	Statfjord	-31.01	-30.10	-29.56	-29.53	-29.70	-29.86	-30.15	-30.15	-30.23	-30.23	-30.30	-29.85
32	Statfjord Nord	-29.37	-28.89	-28.73	-28.70	-28.79	-29.06	-29.25	-29.25	-29.28	-29.28	-29.38	-29.16
33	Statfjord Nord	-29.52	-28.80	-28.22	-28.31	-28.34	-28.50	-28.65	-28.65	-28.47	-28.47	-28.50	-28.30
35	Statfjord Ost	-29.20	-28.15	-28.50	-28.64	-28.88	-29.09	-29.16	-29.25	-29.17	-29.26	-29.26	-29.27
36	Statfjord	-29.25	-28.67	-28.30	-28.49	-28.58	-28.72	-29.16	-29.16	-29.33	-29.33	-29.30	-29.13
37	Statfjord	-29.36	-29.08	-28.30	-28.70	-28.90	-29.07	-29.20	-29.20	-29.10	-29.10	-29.40	-29.13
38	Statfjord	-29.27	-28.65	-28.07	-28.82	-29.14	-29.24	-29.17	-29.17	-29.21	-29.21	-29.45	-29.29
39	Statfjord	-29.61	-28.86	-28.49	-28.89	-29.08	-29.30	-29.34	-29.34	-29.30	-29.30	-29.50	-29.32
40	Statfjord	-29.76	-28.70	-28.58	-29.14	-29.20	-29.30	-29.38	-29.38	-29.45	-29.45	-29.60	-29.38
41	Statfjord	-29.27	-28.71	-28.20	-28.67	-28.95	-29.08	-29.05	-29.05	-29.04	-29.04	-29.36	-29.24
42	Statfjord	-29.36	-28.95	-28.45	-28.69	-28.93	-29.09	-29.04	-29.04	-29.10	-29.10	-29.23	-29.26
43	Brent	-29.42	-28.59	-28.29	-28.46	-28.84	-29.00	-28.97	-28.97	-29.02	-29.02	-29.10	-28.85
44	Brent	NA	-30.81	-29.20	-29.09	-29.22	-29.21	-29.23	-29.23	-29.37	-29.37	-29.57	-28.74

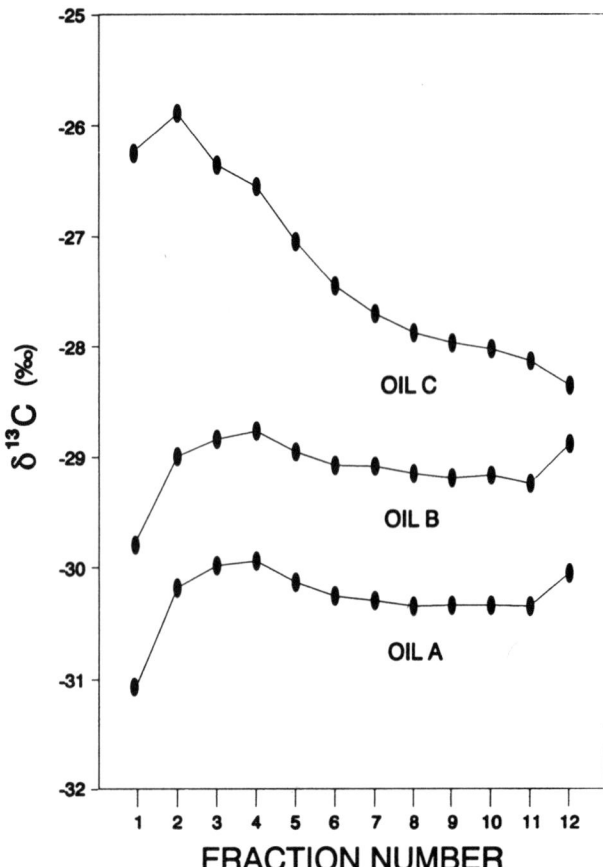

FIGURE 7. Isotope profiles of oils A, B, and C. Source characteristics and effects of thermal maturity for these isotope profiles are discussed in the text.

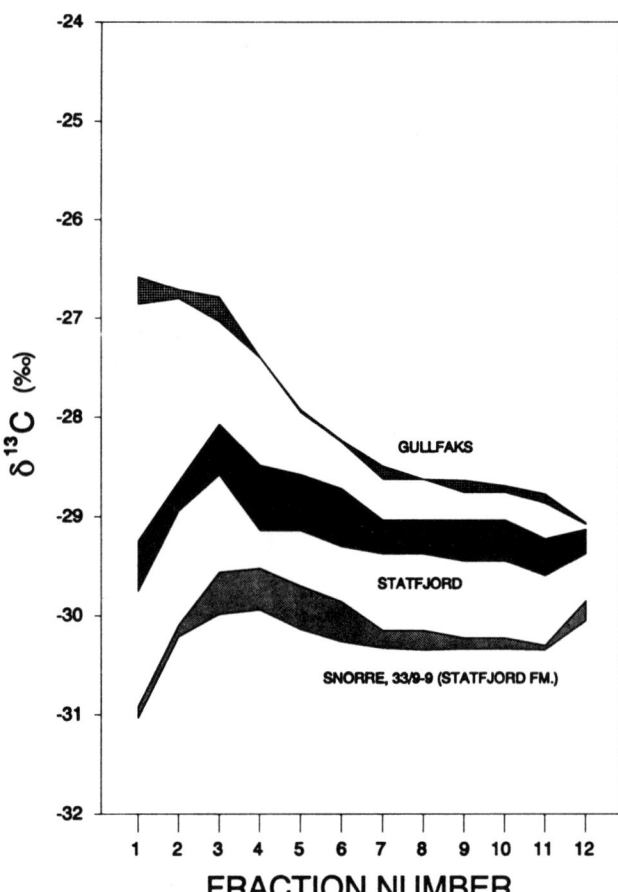

FIGURE 8. Ranges of isotope profiles of oils from the Snorre field, the Statfjord Formation in well 33/9-9, the Statfjord field, and the Gullfaks field.

berg Alpha structures during the time of significant oil migration, whereas a thickness of up to 1000 m of sediments was present in the Oseberg Beta. Structural traps in the Beta field, being the first formed, were the first available to retain early migrated hydrocarbons. Therefore, the Oseberg Beta oils must be less mature than the Oseberg Alpha oils. This argument is supported by carbon isotopic data; the Oseberg Beta oils are isotopically lighter than the Oseberg Alpha oils.

Oseberg West Flank

The structurally complex area to the west of the Oseberg Alpha field, comprising Oseberg Omega, Kappa, Theta, and Delta, is collectively called the Oseberg West Flank in this report. We have analyzed five oils from the Oseberg Omega and Oseberg Kappa fields.

The isotope profiles of three of the Oseberg West Flank oils are similar (Figure 9). Based on the shapes of the isotope profiles, hydrocarbon accumulations in the Oseberg West Flank were generally derived from source rocks similar to those of the Oseberg Alpha oils.

Two anomalous oils occur in the Oseberg West Flank. The oil from the Omega field (sample 23) is from a deeper reservoir than other oils from the Oseberg West Flank. It has, however, less mature biomarker properties (the least mature oil from the northern Viking graben). A second anomalous oil is from the deepest reservoir of the Oseberg Kappa field (sample 25). Unlike the other Oseberg West Flank oils, this oil has an isotope profile similar to those of the Statfjord oils. These two anomalous oils may represent accumulations derived from local source rocks.

The Oseberg West Flank oils probably represent accumulations with variable timing and hydrocarbon sources in this structurally complex area. Most of the Oseberg West Flank oils, however, have isotopically heavier gasoline-range hydrocarbons than those of the Oseberg Alpha oils.

Huldra Field

The biomarker and other geochemical properties show that the Huldra field contains the most thermally mature oil in the northern Viking graben (Figure 5), suggesting that the Huldra oil was expelled most recently from the graben center. The isotope profile of the Huldra oil (Figure 10) is identical to those of the Oseberg West Flank oils (Figure 9). The Huldra oil contains a larger abundance of

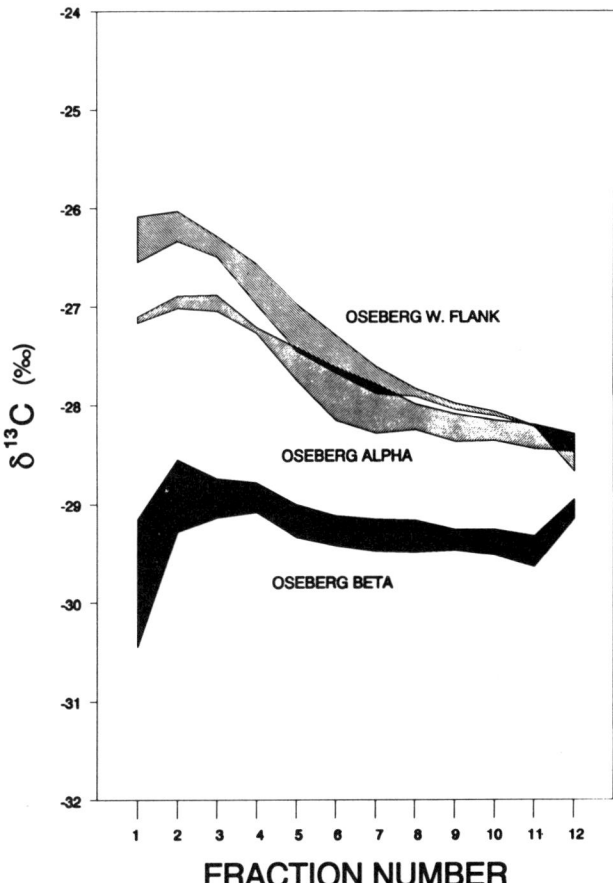

FIGURE 9. Ranges of isotope profiles of oils from the Oseberg Beta (Zeta) field, the Oseberg Alpha (Gamma) field, and the Oseberg West Flank.

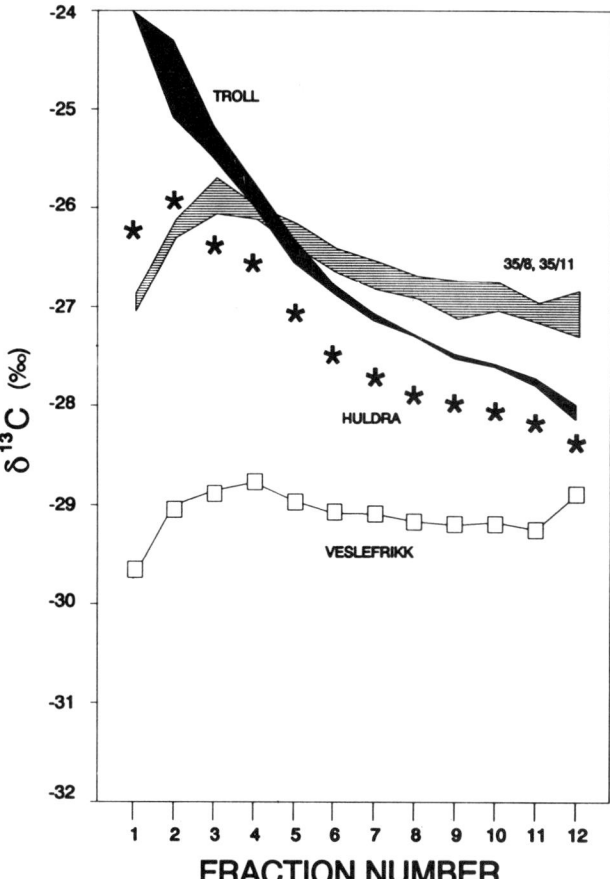

FIGURE 10. Ranges of isotope profiles of oils from the Veslefrikk field, the Huldra field, the Troll field, and Blocks 35/8 and 35/11.

isotopically heavy gasoline-range hydrocarbons (combined distillation fractions 3 and 4) than the nearby Oseberg Alpha and Oseberg Beta oils.

Troll Field

The Troll field is a major gas accumulation underlain by a thin (only 4 to 28 m) oil rim (Gray, 1987). Oils from the Troll field are biodegraded.

The isotope profiles of three Troll oils are genetically related to the Oseberg West Flank or Huldra oils, with a shift in carbon isotopic ratios (generally less than 1‰) due to biodegradation (Figure 10).

Blocks 35/8 and 35/11

The isotope profiles of the four oils from Blocks 35/8 and 35/11 are shown in Figure 10. One exception (sample 6, not shown in Figure 10), is isotopically lighter than other oils in the same area (Table 2). Extremely heavy carbon isotopic ratios of isotope profiles suggested that the oils in Blocks 35/8 and 35/11 were not derived from the same source of typical North Sea oils.

Agat Field

The light oil sample from the Agat field is less mature than the oils from the Huldra field and Blocks 35/8 and 35/11 (Table 1). It also has the heaviest isotope profile among oils in the northern Viking graben (Table 2). Compared with the typical Statfjord oils, however, this oil is isotopically heavier by more than 4‰ in all oil fractions. Such a large shift in the carbon isotopic ratios of oils cannot be attributed to thermal maturity alone. Therefore, the Heather Formation or a terrestrial facies of the Draupne Formation must be invoked to explain the distinct chemical and carbon isotopic composition of this oil. The Agat oil is genetically similar to those in Blocks 35/8 and 35/11.

Summary of Oil Classification

The oils from the northern Viking graben can be grouped in terms of the carbon isotopic ratio of the gasoline-range hydrocarbons against that of the aromatic (ARO) fractions (Figure 11). The carbon isotopic ratio of the gasoline-range hydrocarbons represents the shape of an isotope profile as a single number. The carbon isotopic ratio of the aromatic fraction reflects mainly the source characteristics

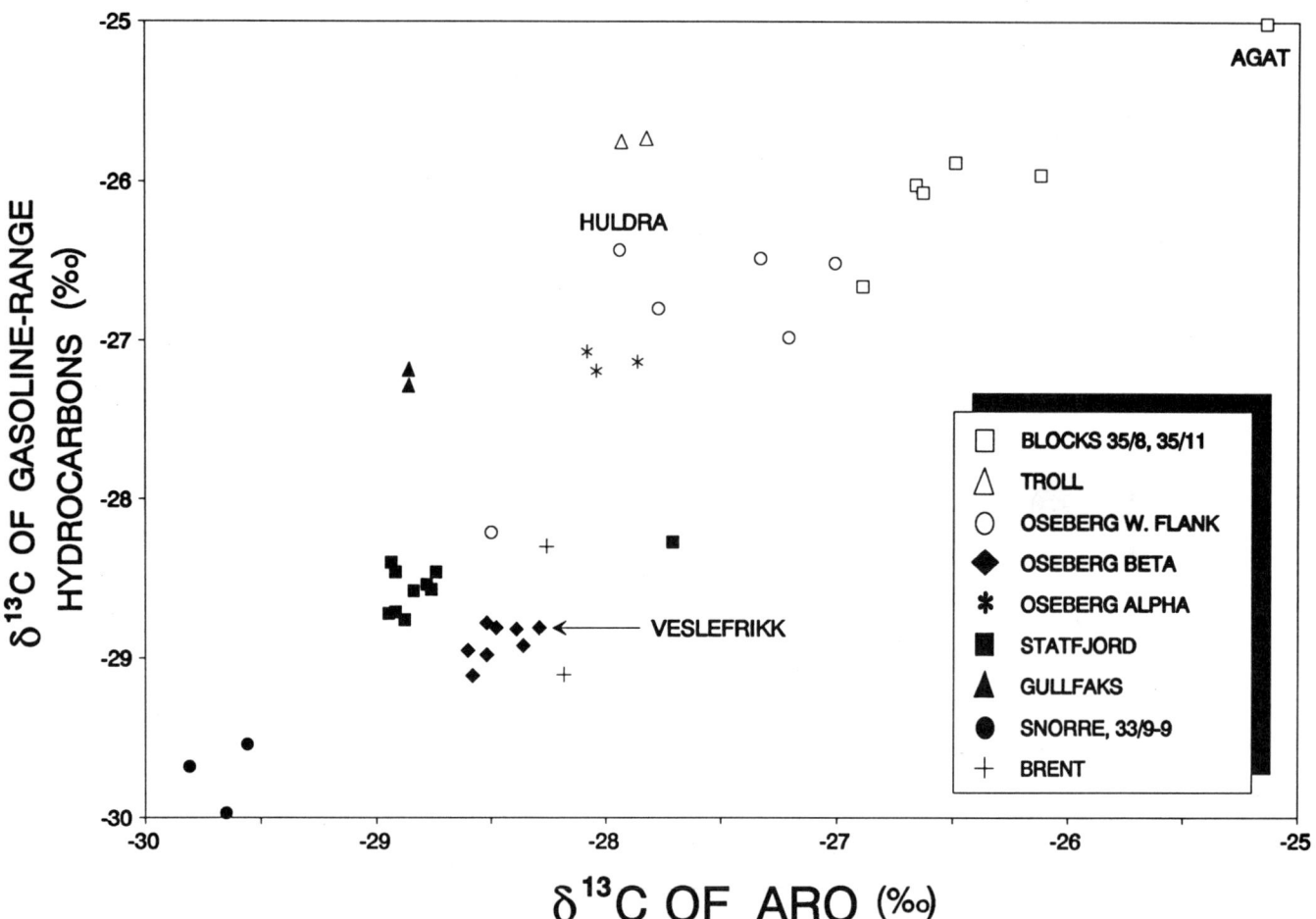

FIGURE 11. Carbon isotopic ratio of gasoline-range hydrocarbons (distillation fractions 3 and 4) vs. carbon isotopic ratio of aromatic fraction (ARO) of oils.

and is less influenced by biodegradation. Figure 11 suggests five major oil types in the northern Viking graben, as follows.

Snorre Oils. The oil from the Lunde Formation in the Snorre field (sample 29) and two oils from the Statfjord Formation in the Statfjord field (samples 30 and 31) belong to this oil type. As a group, these oils have the most negative carbon isotopic ratios and slightly less-mature biomarker distributions. They were probably derived from the Draupne Formation in the drainage area located northeast of the Snorre field (Hollander, 1987).

Statfjord-Gullfaks Oils. These Middle Jurassic oils on the west side of the Viking graben are typical North Sea oils, presumably derived from the Draupne Formation. The isotope profiles of the Gullfaks oils differ from those of the Statfjord oils as a result of biodegradation.

Brent-Oseberg Beta-Veslefrikk Oils. The oils from the Brent, Oseberg Beta, and Veslefrikk fields are similar to the Statfjord-Gullfaks oils, but have biomarker distributions indicating more advanced thermal maturity. The Brent field is located on the west side of the Viking graben and drains the Oseberg kitchen, whereas the Oseberg Beta and Veslefrikk fields are on the east side of the Viking graben and drain the Troll kitchen.

Oseberg Alpha-Troll Oils. The oils from the Huldra, Oseberg Alpha, Oseberg West Flank, and Troll fields are grouped together as a common oil type and have isotopically heavy gasoline-range hydrocarbons. The Troll oils are biodegraded and have an isotopic shift that is similar to that observed for the Gullfaks oils. The Oseberg West Flank oils have variable chemical and isotopic compositions but are included together in this group because their properties on average resemble those of the Oseberg Alpha–Troll oils.

Agat and Blocks 35/8 and 35/11 Oils. The Agat and Blocks 35/8 and 35/11 oils are grouped together on the basis of both geochemical similarity and geographic location. As a group, these oils have the heaviest carbon isotopic ratios and the highest Pr/Ph ratios. The Blocks 35/8 and 35/11 oils have the most mature biomarker distributions (next to the Huldra oil).

Timing and Direction of Expulsion and Migration of Hydrocarbons

As discussed previously, source rocks with low contents of oil-generating organic matter such as the Heather Formation expel hydrocarbons at more advanced stages of thermal maturity than source rocks with high contents of oil-generating organic matter such as the Draupne Formation. The Middle Jurassic Brent coals are also believed to have a high expulsion threshold, due to less reactive kerogen and hydrocarbon adsorptive properties of the rock. Source rocks that expel hydrocarbons at advanced stages of thermal maturity generate gaseous hydrocarbons and condensates. Thus hydrocarbons derived from the Brent and Heather formations should have characteristics associated with both advanced thermal maturity and terrestrial influence.

Consideration of kinetic models, generative capacities, and expulsion thresholds for Jurassic North Sea source rocks (Mackenzie et al., 1987; Mackenzie and Quigley, 1988) suggests that the Draupne Formation should have expelled oil at temperatures of about 120°C, the Heather Formation should have expelled light oils, condensates, and gas at temperatures of about 130 to 150°C, and the Brent Formation should have expelled gas at temperatures greater than 150°C. This implies that at a given locality where the rocks are undergoing continuous burial, the Draupne Formation should have expelled hydrocarbons about 15 million years earlier than the Brent Formation, based on a maximum thickness of Upper Jurassic rocks in the Viking Graben of about 700 m (Thomas et al., 1985), a thermal gradient of about 32°C/km, and a heating rate of about 1.2°C/m.y. In this report, hydrocarbons derived from the Brent and Heather formations at more advanced stages of thermal maturity are referred to as "secondary hydrocarbons," as opposed to the "primary hydrocarbons" derived from the Draupne Formation at earlier stages of thermal maturity.

Possible source rock units are buried about 700 m deeper in the Troll kitchen than in the Oseberg kitchen. This means that hydrocarbon generation is more advanced, and should have started about 12 million years earlier, in the Troll kitchen than in the Oseberg kitchen. The Viking graben system is not symmetric but is a series of asymmetric half-grabens (Harding, 1984; Rosendahl, 1987). As a result of this asymmetry, the main direction of hydrocarbon migration out of the graben deeps is focused away from the steep side of the graben. The major directions of migration are northwestward out of the Oseberg kitchen and southeastward out of the Troll kitchen. Migration pathways of the primary and secondary hydrocarbons from the Troll and Oseberg kitchens are shown in Figure 12.

The oil accumulations in the Tampen spur area are derived mainly from the Draupne Formation in the Oseberg kitchen. The Snorre, Statfjord, and Gullfaks oils are primary hydrocarbons expelled from the Draupne Formation in the Oseberg kitchen at about 120°C, probably 40 to 45 m.y.a. Secondary hydrocarbons, which were expelled from Brent coals in the Oseberg kitchen at about 160°C (30 m.y.a.), have displaced oil from the North Alwyn fields and partially from the Brent field (Karlsson, 1986).

On the east side of the Viking graben, the Draupne Formation expelled primary oils earlier (55 to 60 m.y.a.) from the Troll kitchen than from the Oseberg kitchen. Secondary hydrocarbons began to be generated from the Troll kitchen about 35 to 45 m.y.a., and have since displaced most of the primary oil from the Troll field. The Huldra, Veslefrikk, and Oseberg complex fields receive hydrocarbons mainly from the Troll kitchen, but with some possible contributions from the Oseberg kitchen. The Huldra field has the largest proportion of secondary hydrocarbons. The primary hydrocarbons that were originally in the Huldra field have probably been displaced into the Veslefrikk, Oseberg Beta, and Oseberg Alpha fields. The Oseberg Alpha field has secondary hydrocarbons originated from the Troll kitchen through the Huldra field and also from the Oseberg kitchen through the Oseberg West Flank. By contrast, the Oseberg West Flank structures have received hydrocarbons from the Oseberg kitchen.

The displacement mechanism of primary hydrocarbons is supported by studies of diagenetic minerals in the Brent sandstones (Glassman et al., 1989). The results of potassium-argon dating of illites in the Brent sandstones at the Huldra field suggest that oils entered the reservoir about 58 m.y.a., whereas gas entered the reservoir about 38 m.y.a. In contrast, at the Veslefrikk field, oil entered the reservoir about 31 m.y.a.

The Quadrant 35 oils (Agat and Blocks 35/8 and 35/11 oils) differ from the other oils in having source characteristics (Pr/Ph and carbon isotopic ratios) that suggest more affinities with terrestrial organic matter. The shapes of the isotope profiles (Figure 10) are not consistent with these oils containing large proportions of secondary hydrocarbons. Based on these geochemical data, the oils in this area were probably derived from the Heather Formation or from a more oxidized facies of the Draupne Formation.

Migration pathways for hydrocarbons in Blocks 35/8 and 35/11 are difficult to determine. If the Troll kitchen were responsible for the bulk of the hydrocarbons, we would expect geochemical characteristics of oils indicative of their derivation from the Draupne Formation. Because of the different source characteristics of the Agat and Blocks 35/8 and 35/11 oils, indicating greater influence of terrestrial organic matter on oil composition, we propose that lateral organic facies changes in both the Draupne and Heather formations may explain the origin of the oils in this area. The drainage area of these hydrocarbons is probably the Sogn graben to the north. In this northern portion of the Viking graben system, both the Draupne and Heather formations may have been deposited under intermittently anoxic conditions, resulting in less content of oil-generating organic matter.

CONCLUSIONS

We have characterized 44 oils from the northern Viking graben in terms of the probable type of source rock and thermal maturity at the time of expulsion. Five general groups of geochemically similar hydrocarbon occurrences

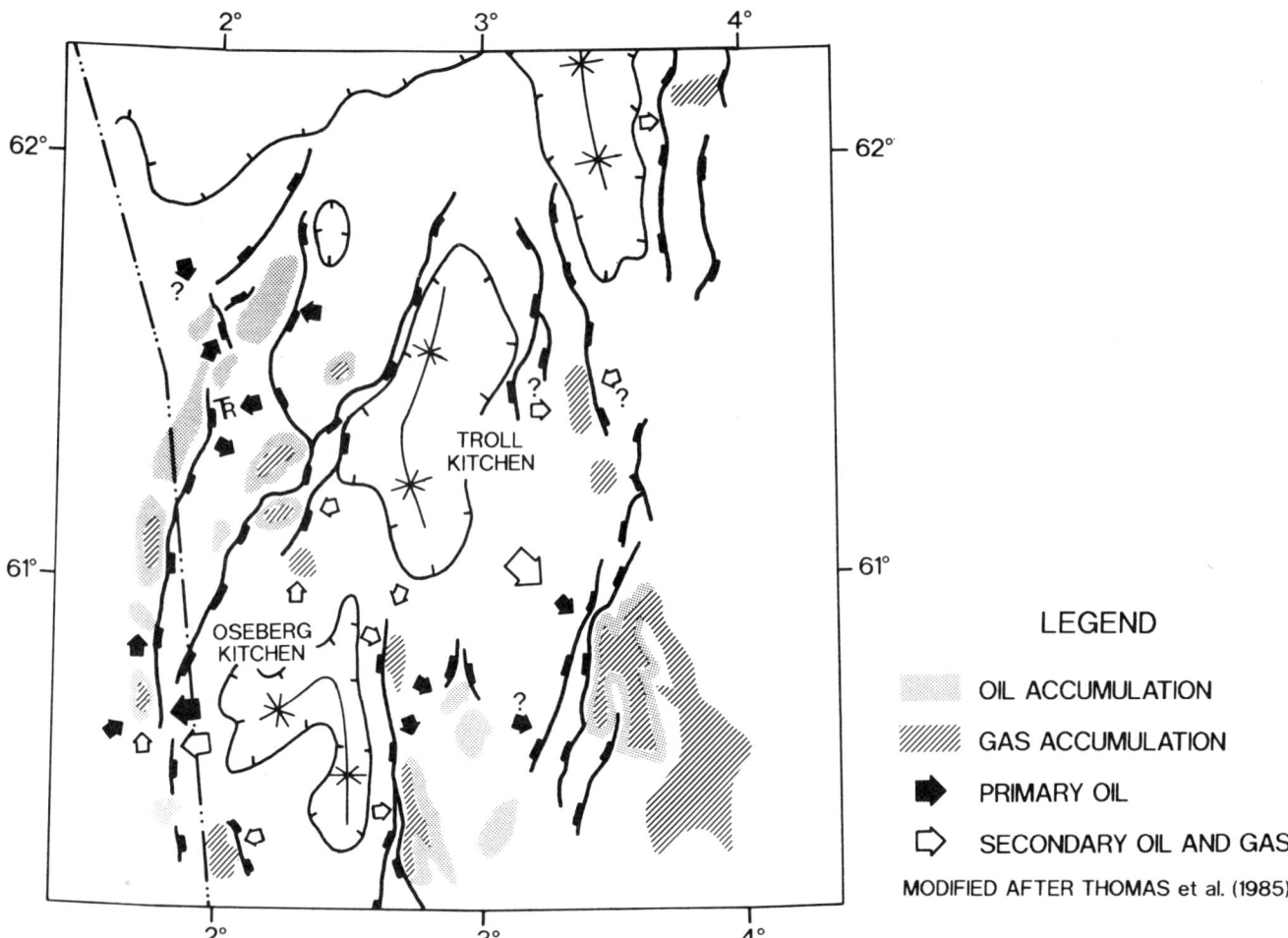

FIGURE 12. Migration pathways of primary and secondary hydrocarbons generated from the Troll and Oseberg kitchens.

are recognized: (1) Snorre oils, (2) Statfjord-Gullfaks oils, (3) Brent-Oseberg Beta-Veslefrikk oils, (4) Oseberg Alpha-Troll oils, and (5) Agat and Block 35/8 and 35/11 oils. The differences among these groups of oils indicate a complex history of generation, expulsion, migration, and accumulation. A possible reconstruction of some aspects of this history has been developed that is consistent with the observed geochemical differences among the oils and with the regional and stratigraphic distribution of types of source rocks and with their inferred burial histories.

The Snorre oils are found in the Snorre field and in the Statfjord reservoirs of well 33/9-9 of the Statfjord field. These oils were derived from the Draupne Formation, probably buried in local depocenters to the north of the Snorre field.

The Statfjord-Gullfaks oils were derived from the Draupne Formation in the Oseberg kitchen, and possibly also in local depocenters in the East Shetland basin.

The Brent, Oseberg Beta, and Veslefrikk fields constitute a third group derived primarily from the Draupne Formation. The Brent field is on the west side of the Viking graben and is currently receiving hydrocarbons being generated in the Oseberg kitchen. The Oseberg Beta and Veslefrikk fields contain oils that were originally generated from the Draupne Formation in the Troll kitchen and that were displaced out of the Huldra field by late-generated hydrocarbons.

The Oseberg Alpha-Troll oils contain isotopically heavy, late-generated hydrocarbons that originated in the Troll kitchen. Most of the earlier-generated oil has been displaced from the Troll structure.

Finally, the Agat field and Blocks 35/8 and 35/11 oils are derived from Upper Jurassic source rocks that are probably buried in a part of the graben system north of the Troll kitchen. The source rocks in this part of the Norwegian offshore apparently contain more terrestrial organic matter.

Primary hydrocarbons in the northern Viking graben were generated from the Draupne Formation during two major episodes beginning about 56 m.y.a. in the Troll kitchen and 45 m.y.a. in the Oseberg kitchen. With continued subsidence and heating of the Jurassic source rocks, gases and condensates (secondary hydrocarbons) were generated, mainly from the cracking of nonexpelled liquid hydrocarbons in the Heather Formation, and possibly from less reactive kerogen in the Brent coals. Secondary hydrocarbons began to be expelled from the Troll kitchen about 40 m.y.a. and from the Oseberg kitchen about 30 m.y.a., and have mixed with and displaced the thermally less mature primary hydrocarbons from some of the accumulations.

ACKNOWLEDGMENTS

We thank J. D. Cline, J. W. Snedden, and C. C. Walters for their critical reviews. We also thank Mobil management in Dallas, London, and Stavanger for their permission to publish this paper.

REFERENCES CITED

Carmalt, S. W., and B. St. John, 1986, Giant oil and gas fields, *in* M. T. Halbouty (ed.), Future petroleum provinces of the world: AAPG Memoir 40, p. 11–53.

Cooper, B. S., and P. C. Bernard, 1984, Source rocks and oils of the central and northern North Sea, *in* G. Demaison and R. J. Murris (eds.), Petroleum geochemistry and basin evaluation: AAPG Memoir 35, p. 303–314.

Cornford, C., 1984, Source rocks and hydrocarbons of the North Sea, *in* K. W. Glennie (ed.), Introduction to the petroleum geology of the North Sea, Chapter 9: Blackwell, p. 171–204.

Cornford, C., C. E. Needham, and L. de Walque, 1986, Geochemical habitat of North Sea oils and gases, *in* A. M. Spencer (ed.), Habitat of hydrocarbons on the Norwegian continental shelf: Graham & Trotman, p. 39–54.

Dahl, B., and G. C. Speers, 1985, Organic geochemistry of the Oseberg field (I), *in* B. M. Thomas et al. (eds.), Petroleum geochemistry in exploration of the Norwegian shelf: London, Graham & Trotman, p. 185–196.

Dahl, B., E. Nysaether, G. C. Speers, and A. Yukler, 1987, Oseberg area—integrated basin modelling, *in* J. Brooks and K. W. Glennie (eds.), Petroleum geology of north west Europe, v. 2: London, Graham & Trotman, p. 1029–1038.

Doligez, B., P. Ungerer, P. Y. Chenet, J. Burrus, F. Bessis, and G. Bessereau, 1987, Numerical modelling of sedimentation, heat transfer, hydrocarbon formation and fluid migration in the Viking Graben, North Sea, *in* J. Brooks and K. W. Glennie (eds.), Petroleum geology of north west Europe, v. 2: Graham & Trotman, p. 1039–1048.

Durand, B., 1988, Understanding of hydrocarbon migration in sedimentary basins (present state of knowledge): Organic Geochemistry, v. 13, p. 445–459.

Field, J. D., 1985, Organic geochemistry in exploration of the northern North Sea, *in* B. M. Thomas et al. (eds.), Petroleum geochemistry in exploration of the Norwegian shelf: Graham & Trotman, p. 39–57.

Glassman, J. R., R. A. Clark, S. Larter, N. A. Briedis, and P. D. Lundegard, 1989, Diagenesis in the Bergen High area, North Sea: relationships to hydrocarbon maturation and fluid flow, Brent sandstone: AAPG Bulletin, v. 73, p. 1341–1360.

Goff, J. C., 1983, Hydrocarbon generation and migration from Jurassic source rocks in the East Shetland basin and Viking Graben of the northern North Sea: Journal of Geological Society of London, v. 140, p. 445–474.

Gray, D. I., 1987, Troll, *in* A. M. Spencer et al. (eds.), Geology of the Norwegian oil and gas fields: Graham & Trotman, p. 389–401.

Hollander, N. B., 1987, Snorre, *in* A. M. Spencer et al. (eds.), Geology of the Norwegian oil and gas fields: Graham & Trotman, p. 307–318.

Harding, T. P., 1984, Graben hydrocarbon occurrences and structural style: AAPG Bulletin, v. 68, p. 333–362.

Karlsson, W., 1986, The Snorre, Statfjord and Gullfaks oil fields and the habitat of hydrocarbons on the Tampen spur, offshore Norway, *in* A. M. Spencer (ed.), Habitat of hydrocarbons on the Norwegian continental shelf: Graham & Trotman, p. 181–197.

Kirk, R. H., 1980, Statfjord field, a North Sea giant, *in* M. T. Halbouty (ed.), Oil and gas fields of the decade 1969–78, AAPG Memoir 30, p. 95–116.

Knudsen, K., D. Leythaeuser, B. Dale, and S. R. Later, 1987, Variation in organic matter quality and maturity in Draupne Formation source rocks from the Oseberg field region: Organic Geochemistry, v. 13, p. 1051–1060.

Mackenzie, A. S., and D. McKenzie, 1983, Isomerization and aromatization of hydrocarbons in sedimentary basins formed by extension: Geological Magazine, v. 120, p. 417–528.

Mackenzie, A. S., and T. M. Quigley, 1988, Principles of geochemical prospect appraisal: AAPG Bulletin, v. 72, p. 399–415.

Mackenzie, A. S., S. C. Brasell, G. Eglinton, and J. R. Maxwell, 1982, Chemical fossils: the geological fate of steroids: Science, v. 217, p. 491–504.

Mackenzie, A. S., C. Beaumont, and D. P. McKenzie, 1984, Estimates of the kinetics of geochemical reactions with geophysical models of sedimentary basins and applications: Organic Geochemistry, v. 6, p. 875–884.

Mackenzie, A. S., I. Price, D. Leythaeuser, P. Muller, M. Radke, and R. Schaefer, 1987, The expulsion of petroleum from Kimmeridge Clay source-rocks in the area of the Brae oilfield, UK continental shelf, *in* J. Brooks and K. Glennie (eds.), Petroleum geology of north west Europe, v. 2: Graham & Trotman, p. 865–878.

Northam, M. A., 1985, Correlation of northern North Sea oils: the different facies of their Jurassic source, *in* B. M. Thomas et al. (eds.), Petroleum geochemistry in exploration of the Norwegian shelf: Graham & Trotman, p. 93–99.

Orr, W. L., 1986, Kerogen/asphaltene/sulfur relationships in sulfur-rich Monterey oils: Advances in Organic Geochemistry, v. 10, p. 499–516.

Roberts, J. D., A. S. Mathieson, and J. M. Hampson, 1987, Statfjord, *in* A. M. Spencer et al. (eds.), Geology of the Norwegian oil and gas fields: Graham & Trotman, p. 319–340.

Rosendahl, B. R., 1987, Architecture of continental rifts with special reference to East Africa: Annual Review of Earth and Planetary Sciences, v. 15, p. 445–503.

Rullkötter, J., and R. Marzi, 1988, Natural and artificial maturation of biological markers in a Toarcian shale from northern Germany: Organic Geochemistry, v. 13, p. 639–645.

ten Haven, H. L., J. W. de Leeuw, and J. Rullkötter, 1987,

Restricted utility of the pristane/phytane ratio as a paleoenvironmental indicator: Nature, v. 330, p. 641-643.

Thomas, B. M., P. Moeller-Pedersen, M. F. Whittaker, and N. D. Shaw, 1985, Organic facies and hydrocarbon distribution in the Norwegian North Sea, in B. M. Thomas et al. (eds.), Petroleum geochemistry in exploration of the Norwegian shelf: Graham & Trotman, p. 3-26.

Ungerer, P., A. Chiarelli, and J. L. Oudin, 1985, Modelling of petroleum genesis and migration with a bidimensional computer model in the Frigg sector, Viking Graben, in B. M. Thomas et al. (eds.), Petroleum geochemistry in exploration of the Norwegian shelf: Graham & Trotman, p. 121-129.

Zaback, D. A., 1987, Classification of crude oil based on stable carbon isotopes: M. S. thesis, the University of Texas at Dallas.

Ziegler, P. A., 1980, Northwest European basin: geology and hydrocarbon provinces, in A. D. Miall (ed.), Facts and principles of world petroleum occurrences: Canadian Society of Petroleum Geologists, Memoir 6, p. 653-703.

Ziegler, P. A. 1981, Evolution of sedimentary basin in northwest Europe, in L. V. Illing and G. D. Hobson (eds.), Petroleum geology of the continental shelf of northwest Europe: Institute of Petroleum, London, p. 521.

Chapter 19

The Alba Field
A Middle Eocene Deep Water Channel System in the UK North Sea

G. A. Mattingly
H. H. Bretthauer

Chevron U.K. Ltd.
London, England

ABSTRACT

The end of the appraisal stage of the Alba field has opened the way for the first development of a middle Eocene reservoir in the North Sea.

The Alba field is located in the Witch Ground graben between the Fladen Ground spur to the north and the Renee ridge to the south, entirely in UKCS (U.K. continental shelf) Block 16/26. In 1984, oil was discovered in a sand within the middle Eocene Alba Formation at a depth of 6100 ft (1860 m) subsea. Seventeen subsequent wells, including sidetracks, have been drilled to appraise the discovery. This drilling indicates that the Alba field is a stratigraphic trap covering an area of approximately 3400 ac (1375 ha).

The sands represent a brief interruption in the hemipelagic sedimentation that dominated this part of the Witch Ground graben during the middle Eocene. Sediment was supplied intermittently from a shelf area to the northwest, into a deep-water environment. Well correlations, seismic mapping, and core analyses indicate that these sands were deposited from turbidity flows as part of a constructional channel/levee complex, within a mud-rich, shelf-sourced, submarine channel system. The cap, updip, and lateral seals of the reservoir are shale.

The reservoir is predominantly a homogeneous, very fine to fine-grained, unconsolidated sand. The average reservoir porosity is 35% and the average permeability is 2.8 darcys. Oil in place is estimated to be 1.1 billion bbl of 20° API biodegraded crude oil.

INTRODUCTION

The Alba field lies approximately 140 mi (225 km) northeast of Aberdeen, Scotland, in Block 16/26 in the UK sector of the North Sea (Figure 1). It was discovered in 1984 when well 16/26-5 encountered middle Eocene oil-bearing sands in a section that was entirely shale in previous wells in the block (Figure 2). Data from 17 subsequent appraisal wells, combined with core data and seismic data, indicate that these sands were deposited as part of a deep water channel system.

Biostratigraphic analyses indicate that the sands were deposited in a restricted, poorly oxygenated, deep water basin. These analyses have also identified "events" that aid in subdividing the middle and late Eocene section and assessing the relative ages of the sands in each well throughout the field. The stratigraphic variation of the sands indicates that there were multiple episodes of sand input into the area from possibly more than one direction.

The model as currently perceived will evolve as more data become available through drilling and the interpretation of a new 3D seismic data set.

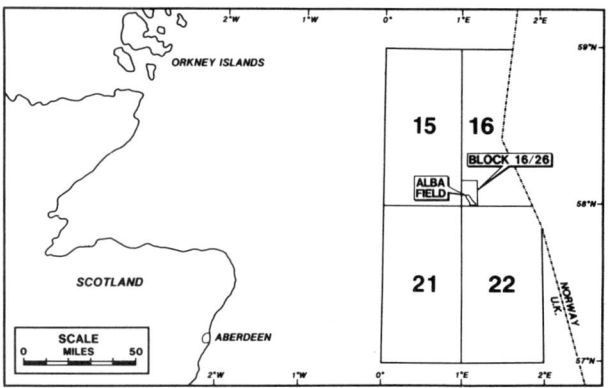

FIGURE 1. Location map for the Alba field within UKCS Block 16/26.

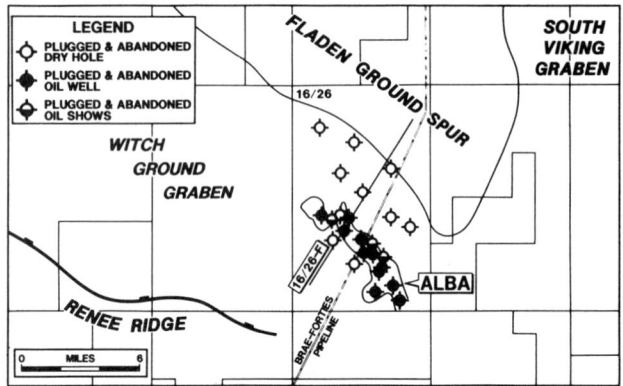

FIGURE 2. Map showing the major structural features in the area surrounding Alba field. Wells in Block 16/26 are shown with their status in the Alba Formation.

EXPLORATION HISTORY

Block 16/26 was awarded, under Production License P.213, to the Sea Search Group as part of the fourth round of UK licensing in 1972. Chevron operates the block, with Amerada Hess, Aran, Baytrust/Unilon, Clyde, Conoco, Fina, Oryx, Phillips, and Santa Fe/Transocean as the other licensees.

Initial exploration efforts in Block 16/26 were directed at Jurassic and Lower Cretaceous sands, which appeared to be trapped by pinch-out and/or truncation against the Fladen Ground spur in the northern part of the block (Figure 3).

The fifth well drilled in the block, 16/26-5, shifted exploration emphasis to include shallow Tertiary objectives. The well encountered 112 ft (34 m) of an oil-bearing sand within the middle Eocene Alba Formation, referred to as the Nauchlan Member (Latham and Mattingly, in press).

A total of eight 2D seismic surveys consisting of 2324 km of data was shot over the block from 1978 to 1987. A 3D survey covering the Alba field was acquired in 1989. The 3D survey contains data covering 5000 km subsurface over an area of 105 km^2.

GEOLOGIC SETTING

A relative lowering of sea level during part of the middle Eocene in the Witch Ground graben enabled terrigenous clastic sediments to be supplied to a deep water basinal environment (Harding et al., in press). This produced a rapid basinward progradation of a clastic shelf complex from the north and northwest. The amount of deltaic and shallow marine sediments being supplied to the shelf break and subsequently into the deep basin was increased and produced the turbiditic sediments that form the reservoir for the Alba field.

STRATIGRAPHY

The Tertiary system in the North Sea has been classified by Deegan and Scull (1977) and by the Robertson Group (1987). These early classification schemes do not provide the detail necessary to differentiate among the numerous Eocene sands that are being encountered as companies pursue Tertiary oil prospects in the North Sea. Drilling results in the late 1980s, particularly in Block 16/26, necessitated a further subdivision of the Eocene section.

A lithostratigraphic classification of the Eocene, proposed by Latham and Mattingly (in press), divides the Eocene into three formations—from oldest to youngest, the Orkney, Lothian, and Alba formations. The Alba Formation contains two sandstone members—the Brioc and Nauchlan members—with the Nauchlan being the main reservoir for the Alba field (Figure 4).

The Alba Formation is middle to late Eocene in age and occurs within a claystone-dominated stratigraphic section that is bounded at the top by the Base Oligocene Unconformity and at the base by an intra-middle Eocene seismic sequence boundary referred to as the Blue Marker (Figures 5 and 6). The Base Oligocene Unconformity

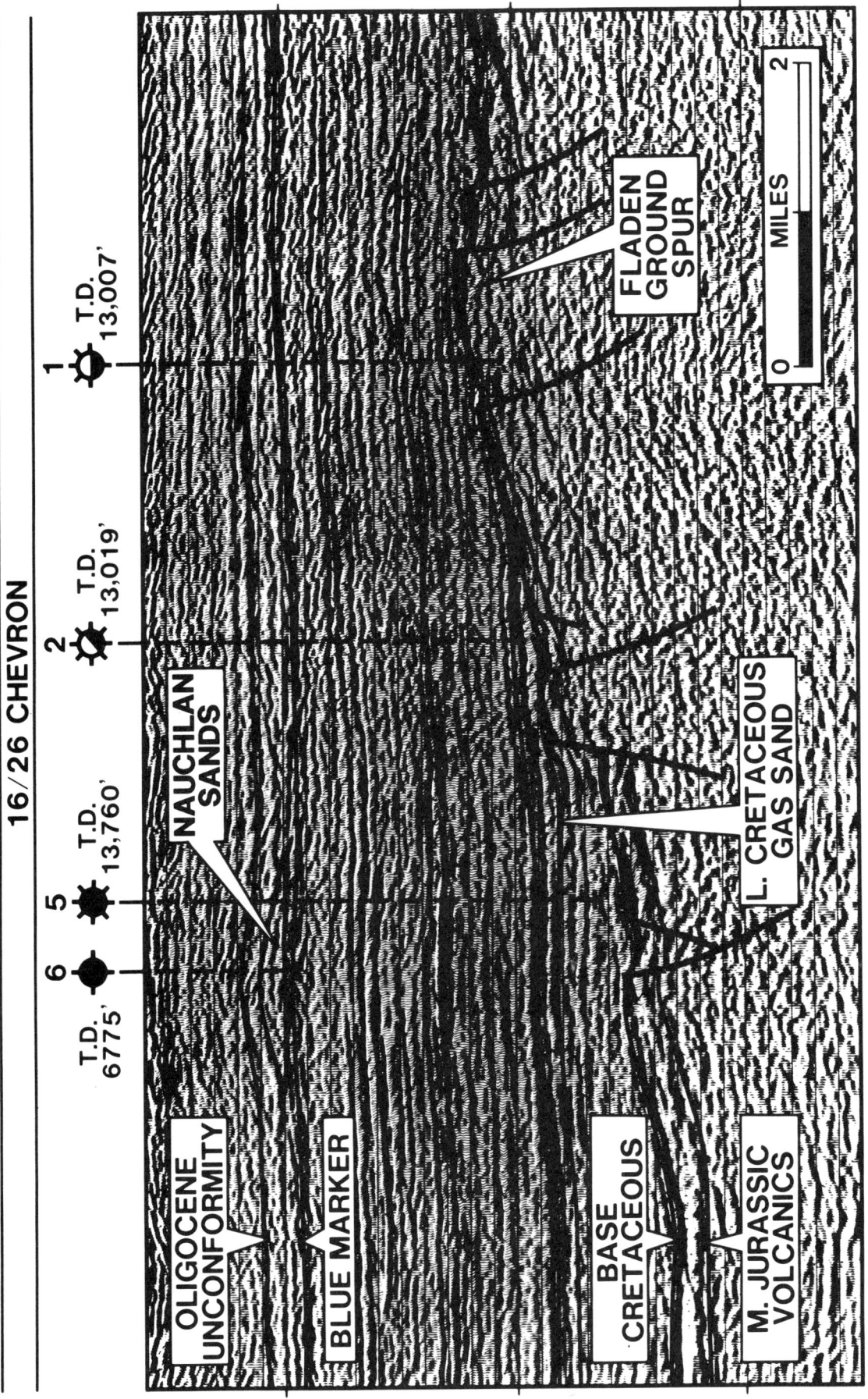

FIGURE 3. Seismic line showing the Nauchlan sands in the Alba field and their position relative to the Fladen Ground spur. The location of this seismic line is shown in Figure 2.

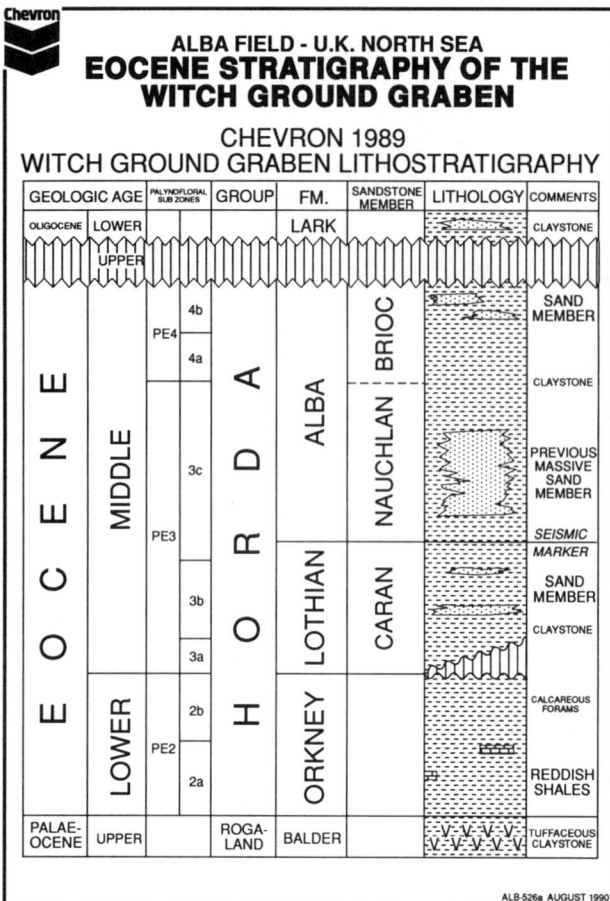

FIGURE 4. Proposed Eocene stratigraphy of the Witch Ground graben, integrating the lithostratigraphy and the biostratigraphic subzones and events of the Robertson Group (Latham and Mattingly, in press).

represents a major hiatus with part of the late Eocene section absent. The Blue Marker is a lithostratigraphic and chronostratigraphic event that occurs just below a downhole increase in the foraminiferal species, *Spiroplectammina spectabilis*, and just above a downhole increase in the palynological species, *Diphyes ficusoides*. Both of these biostratigraphic events are recognized on a regional scale (Robertson Group, 1987).

Between the Blue Marker and the Base Oligocene Unconformity there is a lithostratigraphic and chronostratigraphic event that is represented away from the field as a seismic sequence boundary. This event is defined by a downhole increase in the palynological species, *Systematophora placacantha*, which marks a regionally recognized subzone (Robertson Group, 1987), and by the occurrence of the foraminiferal species, *Cibicides* c.f. *ungerianus*. The occurrence of *C.* c.f. *ungerianus* is a short-lived event that is recognizable over the area. This lithostratigraphic event occurs between the top of the Nauchlan Member and the Base Oligocene Unconformity and is used as the stratigraphic datum for cross sections through the field.

The microfauna within the claystones of the Alba Formation in and around Block 16/26 is dominated by agglutinated foraminifera and a distinct lack of calcareous benthonic foraminifera. This is interpreted as indicating deep marine sedimentation in a basin with restricted circulation.

Biostratigraphic zonation of the Eocene is complicated by the lack of diversity and extinction/evolution events. Within the Alba Formation, zonation relies on increases and decreases in abundance of particular species that in some cases are used to define subzonal tops. The original zonation scheme (Robertson Group, 1987) defined approximately 20 biostratigraphic events within what is now the Alba Formation. Of these events, only a few appear to be of correlative value.

Correlations (Sandstone Relationships)

Within the envelope defined by the Base Oligocene Unconformity and the Blue Marker, the stratigraphic position of the Nauchlan Member varies along the axis of the field from northwest to southeast. The main sands of the Nauchlan Member are those that are greater than 100 ft (30 m) thick and are mapped on seismic data (shown as the stippled sands on the cross sections, Figures 5 and 6). These are the sands that have been mapped to produce the reservoir maps of this report. The stratigraphic variations, combined with differing sand thicknesses, are used to define four different areas of the field.

The stratigraphic variations of the Nauchlan Member in the northwest end of the Alba field are shown by the correlation between wells 16/26-5 and 16/26-6. The stratigraphic positions of the top and the base of the thin Nauchlan Member in well 16/26-5 are high compared with those in well 16/26-6 (Figure 5). The Nauchlan sand in well 16/26-5 is interpreted as being younger than the Nauchlan Member in well 16/26-6. This interpretation is based on the high stratigraphic position of the base of the sand and the occurrence of biostratigraphic events beneath the sand in 16/26-5 that occur above the Nauchlan Member in other wells in the north end of the Alba field. In addition, a shale section characterized by low resistivity is present beneath the sand in 16/26-5. This shale, which is time-equivalent to part of the Nauchlan Member in other wells, is present in all wells that have no Nauchlan sand. Although the sand in 16/26-5 is defined as Nauchlan, it must be younger than the main Nauchlan sands in 16/26-6 to allow for this shale to be developed.

In contrast to well 16/26-6 in the northwest end of the Alba field, the Nauchlan Member in the middle of the field exceeds 300 ft (90 m) in thickness, and the base and top of the sand are stratigraphically high in the section. This change is shown by the correlation of wells 16/26-6 and 16/26-7 (Figure 5).

At the south end of the Alba field in well 16/26-8, the top and base of the Nauchlan Member are stratigraphically low in relation to well 16/26-7. The interval between the base of the Nauchlan Member and the Blue Marker is thin, and there is a thicker sequence from the stratigraphic datum to the top of the Nauchlan sand. This is com-

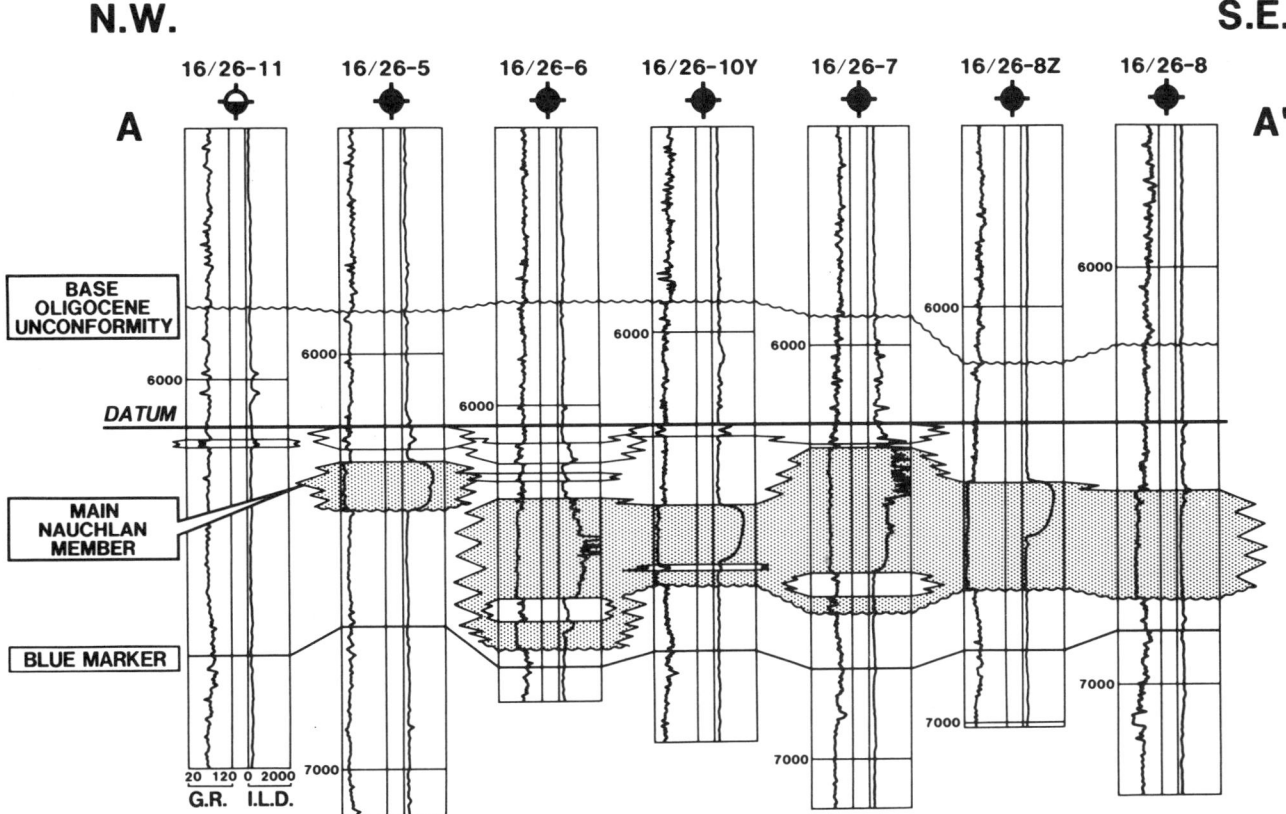

FIGURE 5. Axial stratigraphic cross section with the datum being a lithostratigraphic log correlation supported by two biostratigraphic events. This cross section has no horizontal scale. The stratigraphic location of the main Nauchlan Member sands varies throughout the field.

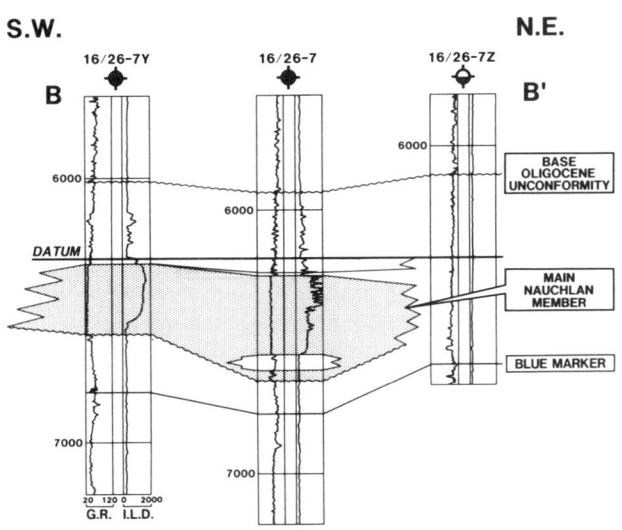

FIGURE 6. Stratigraphic cross section with no horizontal scale, normal to the axis of the field, with the same stratigraphic datum as in Figure 5. The section shows the abrupt boundary of the sand between wells 16/26-7 and 16/26-7Z.

mon in wells with a stratigraphically low Nauchlan Member.

Erosion into the shale underlying the Nauchlan Member is not evident from logs over some of the Alba field. The stratigraphic section from the base of the Nauchlan Member to the Blue Marker is of similar thickness in wells 16/26-10Y, 16/26-7, and 16/26-8Z, denoting equal or no erosion at the base of the reservoir over this part of the field.

Evidence for erosion at the base of the Nauchlan Member does exist in other parts of the field, but erosion does not appear to have gone deep into the section. The sands in wells 16/26-6 and 16/26-8 sit stratigraphically low at the base, which could be due to the erosion of the underlying claystone. The low stratigraphic positions of these sands are interpreted as having resulted, at least in part, because these sands are older than sands stratigraphically higher in the section. Normal to the axis of the field, the correlation of the base of the Nauchlan Member in 16/26-7 is shown to be lower than in 16/26-7Y (Figure 6). Because 16/26-7 is in the center of the channel system, this difference could be due to either erosion or the sand being older. The relationship of the base of the Nauchlan

Member to the underlying shales, particularly on the flanks of the field, is uncertain because of the paucity of well data. The amount of erosion at the base of the Nauchlan Member is still under review.

The sand and shale sequence above the Nauchlan Member in wells 16/26-10Y and 16/26-6 is correlated as being the lateral equivalent of the upper part of the Nauchlan Member in well 16/26-7 (Figure 5). These correlations in the central part of the field indicate that mud was being carried downslope and deposited as levees flanking the distributary channels. This mud formed the levees that represent at least some of the boundaries and lateral seals of the reservoir.

The Alba Formation thins from 16/26-7 to 16/26-7Z on the flank of the field (Figure 6). A structural low exists beneath 16/26-7 at both the Blue Marker and the base of the Nauchlan horizons. The concave-upward shape at the base of the sand and the abrupt boundaries imply a channel-like geometry of the sand distribution. The Nauchlan Member appears to end abruptly on the northeast and the southwest flanks of the field.

Facies

The Nauchlan Member is composed of three sandstone facies: structureless sands, thinly bedded parallel to subparallel laminated sands, and rippled sands. These facies combine to form the sequences observed in Alba cores.

The structureless sand facies occurs vertically throughout the Nauchlan Member in every well cored to date in Alba. The structureless sections can be several feet thick and can either be abruptly terminated at the top of the unit or grade up into a rippled or water escape sequence from 3 to 6 in. (75 to 150 mm) thick.

The laminated sands that have been observed in cores represent very thin depositional units of light gray sandstone interbedded with medium gray sandstone. The medium gray sandstones appear to be slightly finer-grained and are interpreted as being the top of a single depositional unit. The light gray sandstones commonly exhibit erosional bases into the underlying medium gray units. The depositional units represented by these two sands vary in thickness from one-half to several inches. The genesis of these structures is currently being investigated. These units show the amalgamated nature of the Nauchlan sand as well as the scouring effects of a turbidite flow into the sediment over which the flow passed.

The sands that appear to be rippled occur at the top of a structureless sequence of sands and may represent sediment deposited from the traction part of a waning turbidity flow.

Some of the primary sedimentary structures in the reservoir have been distorted by water escape processes. These processes probably occurred in at least two phases. In sandstone sets that are rippled, the primary phase of water escape is evident because it distorts these structures. Just after deposition, water escape caused minor distortions of sedimentary structures that appear in cores as faint dish structures. The second phase occurred after burial of the sands and appears to have been a very violent event that severely distorted or obliterated primary sedimentary structures. This episode is represented in cores by large "pipes" that are several feet in length and 6 in. (150 mm) wide, passing vertically through the Nauchlan sand. This dewatering event could be associated with the injection of the Nauchlan sand into the shales that surround the reservoir.

The reservoir sand facies is underlain and overlain by claystone-dominated sequences in which three facies are recognized. The first consists of medium dark gray, thick, poorly bedded, fissile to subfissile, laminated, and extensively bioturbated claystones.

The second claystone facies consists of claystones identical to the first claystone facies but with interlaminated thin horizons of buff-colored dolomitic/sideritic silty shales.

The third claystone facies consists of interbedded sands and claystones. The claystones are similar to those in the first facies. The interbedded sands provide thin reservoir units above the Nauchlan sand and also make up part of the Brioc Member. These sands are similar in character to those of the main part of the Nauchlan Member, but occur as thinner beds. Examination of cores indicates that at least some of the thin sands within the Brioc interval have been injected into the claystones above the Nauchlan Member as dikes and sills. The lateral extent and frequency of these injected sands, as well as the communication within them, are unknown. Therefore, the producibility of this interval is uncertain.

Depositional Processes

A possible interpretation of the different stratigraphic positions of the Nauchlan Member sands in Alba wells is amalgamation of sinuous channels as suggested by wells 16/26-6, 16/26-10Y, and 16/26-7. These wells, drilled in the thickest part of the Alba Formation, have Nauchlan sands with vastly differing thicknesses and stratigraphic positions (Figure 7).

The channel system that deposited the Nauchlan Member appears to have been active in the area through the middle Eocene. The variability in the stratigraphic position of the Nauchlan Member suggests that at any one time sand was being deposited in only a small part of the channel. The locus of sand deposition seems to have moved laterally through time.

If the younger sands in well 16/26-7 are not present to the northwest, the sands in the 16/26-6 and 16/26-7 areas of the field could be parts of different depositional systems with different sediment input pathways (Figure 7).

Onlap of strata onto the Base Oligocene Unconformity is clearly shown by seismic data, indicating that by the end of the Eocene a large bathymetrically positive channel system had developed on the sea floor. This is largely responsible for the mounding features observed in seismic data at the top of the Nauchlan Member and the Base Oligocene Unconformity over the Alba field.

STRUCTURE

The Alba field consists of three separate hydrocarbon accumulations, each with a different oil-water contact.

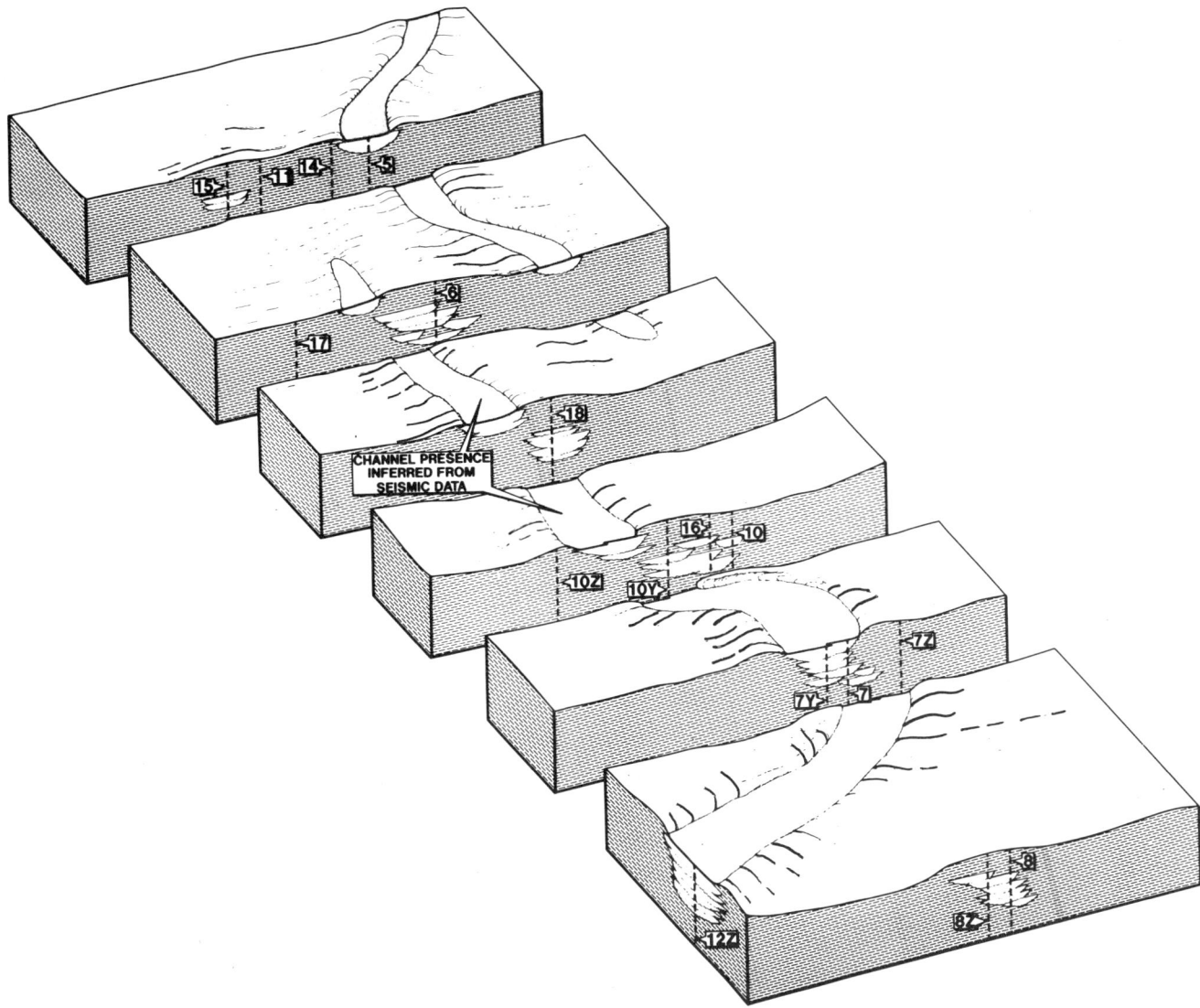

FIGURE 7. Diagrammatic split block diagram designed to have a top surface that approximates the stratigraphic datum of the stratigraphic cross sections shown in Figures 5 and 6. The thickness and lateral extent of the Nauchlan sands are shown to change within the field, particularly in the northwest-to-southeast direction.

The two smallest accumulations lie in the north and northwest of the main field. The accumulations tested by wells 16/26-15 and 16/26-5 have mapped structural separation between them and the main field. Both wells are interpreted as having encountered thin water legs at the base of the Nauchlan Member.

The top and base of the Nauchlan Member are mapped using 2D seismic data. The limits of resolution of the data result in sands of different ages being mapped as one unit within the main field. Therefore the top and base structure maps do not adequately express the stratigraphic variability of the sands.

Regional work on Tertiary faults in the Witch Ground graben indicates that faulting is present in the Alba field. Faulting appears to be most prolific in the Miocene–Oligocene section with a subordinate number of faults cutting through the Alba Formation. The period of most active fault movement is dated by the truncation of the faults by the mid-Miocene Unconformity.

Reservoir Structure

The top and base of the main sands of the Nauchlan Member are mapped within the zero sand line of the field, as determined from seismic data. The mapping of the Nauchlan Member was confined to this area because no time equivalents of the sand are discernible away from the field.

Several faults with displacements of generally less than 100 ft (30 m) have been interpreted on the 2D seismic data. These faults do not impact fluid levels within the reservoir and have been omitted from the maps for simplicity.

The structure at the top of the Nauchlan Member is a plunging anticline down the axis of the field from the northwest to the southeast (Figure 8). This geometry is at least in part due to the differential compaction of the sand-prone channel fill relative to the mud-prone flanks of the channel system. It results in the top of the reservoir

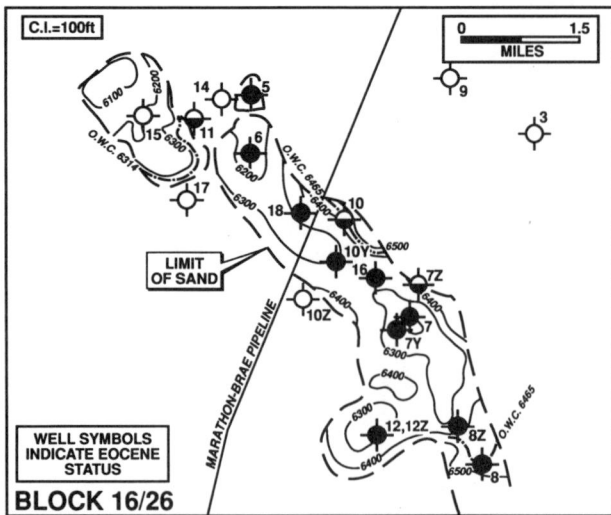

FIGURE 8. Top Nauchlan sand depth structure map, simplified by removing small faults.

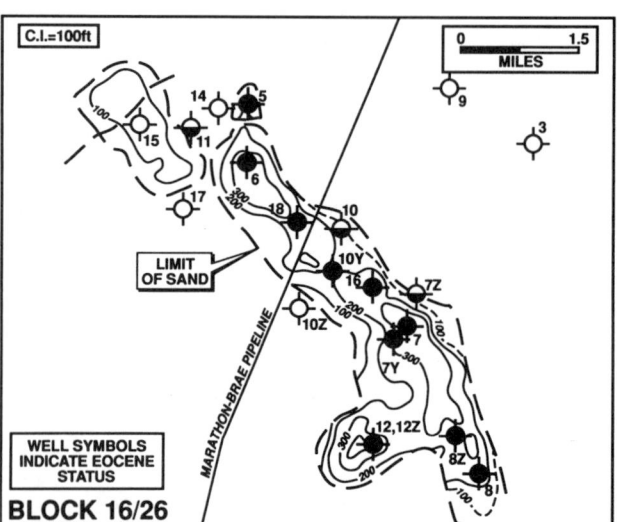

FIGURE 10. Main Nauchlan sand gross interval isochore, made up almost entirely of sand. The intersection of some of the contours with the limit of sand demonstrates that parts of the field have been mapped with vertical lateral boundaries.

intersecting the oil-water contact in only two areas of the main field—a small area on the northeast flank around well 16/26-10, and an area at the far southeast end of the field.

The structure at the base of the Nauchlan Member is a narrow syncline plunging and bifurcating to the southeast (Figure 9). The concave geometry normal to the axis of the field and the gentle plunge result in a water leg across almost the entire field. The geometry at the base of the Nauchlan Member is the result of a preexisting bathymetric low and localized erosion and/or loading of the sand into the poorly compacted shale beneath, or a combination of both.

Subtracting the top from the base structure results in a narrow, elongated feature in which the gross interval thickness increases into the middle of the field (Figure 10).

This thick area trends down the axis of the field, bifurcating in the southeast.

RESERVOIR CHARACTERISTICS

The Nauchlan Member rock properties and the nature of the fluids within the reservoir give contrasting expectations regarding development. The rock properties (unconsolidated sands in excess of 300 ft, or 90 m, thick, vertical and horizontal permeabilities of approximately 2.8 darcys, and porosities of 35%) define an ideal reservoir. The 20° API gravity of the Alba crude oil and the areal extent and thickness of the water leg, however, could reduce the productive efficiency of the reservoir.

Petrographic Description

The Nauchlan sands are classified as subarkoses within the classification scheme of Pettijohn et al. (1972). They are very fine to fine-grained, micaceous, and variably argillaceous. The grains are moderately well-sorted, subangular to subrounded in shape, and partly corroded.

The detrital mineralogy is dominated by monocrystalline quartz with subordinate polycrystalline quartz, lithic fragments, feldspars, micas, glauconite, heavy minerals, and detrital clay. Quartz makes up 80 to 90% of the rock. The sands within the Brioc Member have a detrital mineralogy similar to that found in the Nauchlan Member.

Authigenic clay minerals include kaolinite, chlorite, and indeterminable "illitic" types. Clays occur in the reservoir in amounts from a trace to 5%. They are not present in sufficient amounts to form effective cements or significantly reduce permeability.

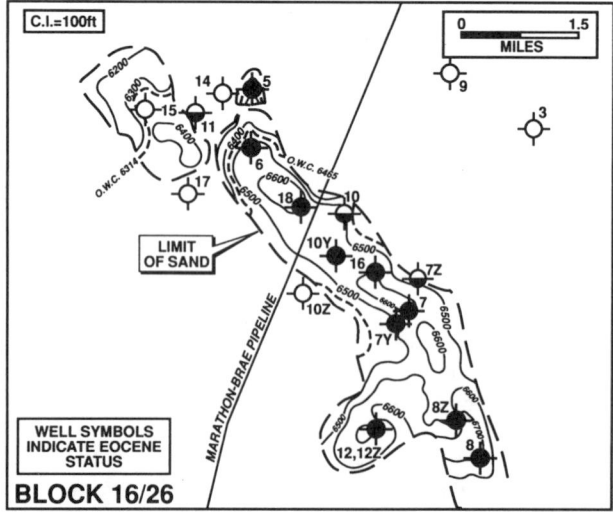

FIGURE 9. Base Nauchlan depth structure map, simplified by removing small faults.

FIELD DEVELOPMENT

The Alba field will be developed by means of two platforms designed as standard North Sea jackets. The first platform will be set in the north end of the field in the vicinity of the 16/26-6 well. Deviated well bores and horizontal wells, both of which will be gravel packed, will be used in production. Water injection is planned to maintain reservoir pressure from the point of initial production. First oil is expected in 1993.

CONCLUSIONS

The Alba field is a stratigraphic trap in which the reservoir is composed of several sands that vary in stratigraphic position throughout the field and collectively comprise the Nauchlan Member. This is interpreted to indicate variations in sediment source directions and timing of sand deposition. These variations in sediment sourcing and timing of deposition appear to be related to distinct areas within the field.

Initial geologic analysis involved subdivision of the Eocene sequence into three formations. Detailed work then focused on the Alba Formation. Changes in frequency distribution of particular fossil species have aided in subdividing the Alba Formation into two members and identifying chronostratigraphic markers within it. This basic stratigraphic framework has increased the understanding of the depositional processes that transported and deposited the Nauchlan Member and should provide a tool for better reservoir management.

ACKNOWLEDGMENTS

We would like to thank Chevron management and Block 16/26 participants for permission to publish this paper. Block 16/26 participants have added constructive comments over the years and have agreed to place this information in the public domain. However, the thoughts put forward in this paper are those of the authors and do not necessarily reflect the participants' interpretations. The list of people who have worked on the field is long, but two people—Karl Thompson and Mark Lunsford—deserve specific mention. Kevin Flanagan has contributed constructive ideas and criticisms. Special thanks go to Bill Terrell for his unending support of this project and for allowing the various ideas to surface.

REFERENCES CITED

Deegan, C. E., and B. J. Scull, 1977, A proposed lithostratigraphical nomenclature for the central and northern North Sea: Rep Institute Geologic Science No. 77/25 (Bulletin Norwegian Petroleum Directorate No. 1), 36 p.

Harding, A. W., T. J. Humphrey, A. Latham, M. K. Lunsford, and M. H. Strider, in press, Controls on Eocene submarine fan deposition in the Witch Ground Graben.

Latham, A. L., and G. A. Mattingly (in press), Lithostratigraphy of the Eocene in the area around UKCS Block 16/26.

Pettijohn, F. J., P. E. Potter, and R. Siever, 1972, Sand and sandstone: Berlin, Springer-Verlag, 618 p.

Robertson Group, 1987, Central Graben North Sea, stratigraphy, structure and petroleum geology of the Paleocene and Eocene: The Robertson Group Press, 5 volumes (available by purchase only)

Chapter 20

Miller Field
A Subtle Upper Jurassic Submarine Fan Trap in the South Viking Graben, United Kingdom Sector, North Sea

N. M. McClure
A. A. Brown

BP Exploration[1]
Glasgow, U.K.

ABSTRACT

The Miller field is situated at a depth of 4 km in the South Viking graben, some 270 km northeast of Aberdeen. The field was discovered in 1983 in previously relinquished U.K. license blocks 16/7b and 16/8b. The discovery can be attributed to a detailed understanding of the regional sedimentological and seismic velocity models, which predicted the presence of submarine fan sediments within a structural nose some 10 km from the graben margin sediment source.

The Miller field reservoir comprises up to 100 m of oil-bearing Upper Jurassic, Brae Formation turbidites. Core, electric log, and well test analyses show that the turbidites have excellent reservoir qualities. Porosity ranges from 12 to 23%, and net/gross ratios are typically greater than 0.75. The reservoir is highly productive, with permeability typically ranging from 50 to 1200 md.

The reservoir fluid is an undersaturated, CO_2-rich, sour, 37.5° API oil sourced from the Kimmeridge Clay Formation, which overlies and interfingers with the reservoir. The oil is trapped by a subtle combination of structural and stratigraphic mechanisms.

Thirty development wells will be drilled from a single platform. A total of ten wells (five producers and five injectors) will be predrilled through a template prior to platform installation. The first of these wells was spudded in March 1989 in preparation for first oil production in early 1992. Ultimate recovery is estimated to be approximately 300 MMSTB of oil and 0.57 tcf of associated gas.

---

[1]Now BP Exploration Europe

INTRODUCTION

Miller field is an oil and gas field located in the South Viking graben in the northern North Sea (Figure 1). It is 270 km northeast of Aberdeen, in U.K. license blocks 16/7b and 16/8b. Discovered in March 1983 by British Petroleum (BP), the field lies at a depth of about 4 km and covers an area of 45 km^2 (11,100 ac, or 17.4 mi^2). It is a broad structural drape over an accumulation of submarine fan sand lobes of the Brae Formation, which form the reservoir. These sand lobes interdigitate with basinal organic-rich mudstones of the Kimmeridge Clay Formation, which sourced the hydrocarbons and form the cap rock.

The Miller field extends considerably beyond BP's block 16/7b into the Conoco-operated block 16/8b, and was the largest hydrocarbon discovery made by BP in the North Sea during the years 1978–88. It has also proved to be one of the largest UKCS discoveries made by any company during that decade.

Seven appraisal wells drilled by BP and Conoco before the end of 1986 confirmed the existence of a giant oil field with a maximum oil column of 110 m above an oil-water contact at 4090 m subsea. This field has an estimated oil in place of about 640 million bbl and an initial GOR of about 1900 standard ft^3/bbl.

Ultimate recovery is estimated to be 300 million bbl of oil and 0.57 trillion standard ft^3 of associated solution gas, which contains 14 to 20 mol% CO_2 and 80 to 800 ppm H_2S. The crude oil, which has API gravity of 35.7 to 38.5°, is undersaturated by about 2350 psi at a virgin reservoir pressure of 7250 psia.

The development plan was devised and completed during 1984–87, and, at the same time, studies were undertaken to assess the likely performance of the reservoir. The development plan called for predrilling of ten wells from a template so that production operations could commence as soon as possible after the arrival of the seabed-supported production platform. The first template well was spudded in March 1989, and first oil production is expected in early 1992.

The results from appraisal wells indicate that there has been a progressive decline in pressure of about 850 psi from the virgin reservoir pressure to the present pressure. This is believed to have resulted from post-1983 production in the South Brae field, which is located in reservoirs of similar age just 3 km to the southwest. Pressure communication between Miller and the upper South Brae reservoir units occurs via the underlying aquifer. To maintain reservoir pressure above the bubble point and to replace voidage, it is intended that half of the Miller template wells and several of the platform wells be used for the injection of seawater.

In the exploration phase, Miller field was regarded as an eastern extension of the larger "Brae" oil field. However, following the drilling of 16/8b-3, Conoco's view that the pool was separate from the Brae complex led them to name the accumulation "Miller," in recognition of pioneering contributions to science by the Scottish geologist Hugh Miller (1802–1856).

The larger "Brae" oil accumulation is now known to comprise at least three separate fields with almost identical oil-water contacts (at about 4090 m subsea): South Brae, Central Brae, and Miller. However, whereas the Brae field reservoirs are dominated by gravels and conglomerates derived from the adjacent fault scarp, Miller reservoirs are predominantly sandy. A structural saddle forms the separation between the Miller and South Brae fields, whereas reservoir pinch-out provides stratigraphic separation between Central Brae and Miller.

Adjacent to the north, and also containing reservoirs in the Brae Formation, are the North Brae and East Miller fields (Figure 2). Thus, all the Upper Jurassic fields in the Brae-Miller area are related stratigraphically, and their distinctive reservoirs are aprons and lobes of a submarine fan complex that evolved through time, changing in character and sediment entry points along the actively faulted margin of the South Viking graben.

The purpose of this paper is to document BP's current understanding of the Miller field and to bring up to date the evolving sedimentological model of the Brae Forma-

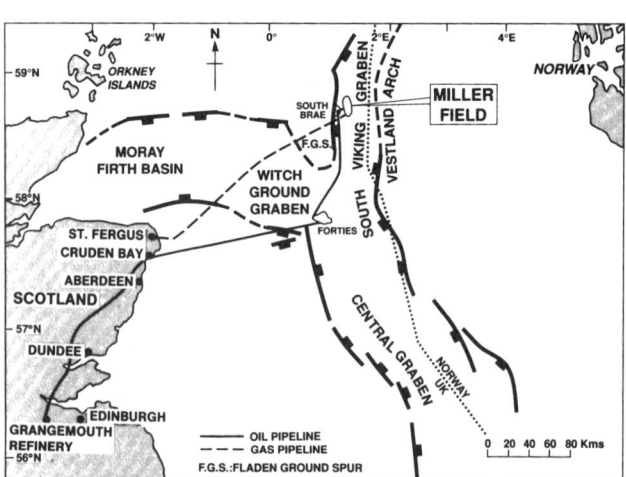

FIGURE 1. Location of Miller field and oil and gas pipeline routes.

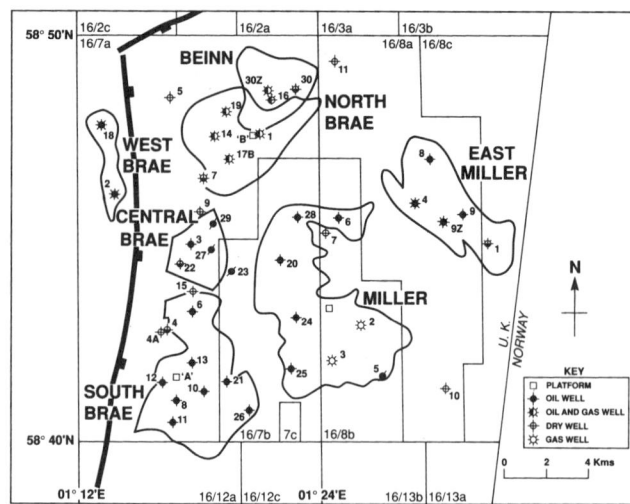

FIGURE 2. Fields in the Brae-Miller area.

tion reported in earlier studies (Turner et al., 1987; Stow et al., 1982; Harms et al., 1981). The information presented here is based on extensive 3D seismic coverage of the Brae-Miller area and on access to high-quality data from more than 30 exploration and delineation wells in the vicinity of Miller field.

EXPLORATION HISTORY

Block 16/7 was originally awarded to Pan Ocean (later merged with Marathon) in the third U.K. licensing round in 1970. Significant hydrocarbon discoveries were made at North Brae, West Brae, Central Brae, and South Brae between 1975 and 1977. In 1976, the eastern part of block 16/7 and the western part of the adjacent block 16/8 were relinquished by Marathon and Shell, respectively. Neither of the relinquished areas had been drilled.

The relinquished blocks were reawarded in the seventh licensing round in 1980 to BP as operator for 16/7b and to Conoco as operator for 16/8b. Current equity holders in Miller field are BP (operator), 40%; Conoco, 30%; Enterprise, 18%; and Santa Fe, 12%.

BP initially became interested in the 16/7b block in 1979, after their regional geologic studies of the Upper Jurassic in the South Viking graben area suggested that:

♦ The Brae Formation (as later defined) in block 16/7a was likely to be the proximal expression of a submarine fan system.

♦ Middle Jurassic sands could have been deposited across the Fladen Ground spur and much of the East Shetland platform.

♦ With increasing submergence concomitant with the opening South Viking graben, these sands were visualized as having sloughed across the western boundary fault during the Upper Jurassic, plunging down the paleoslope and spreading across the basin plain as a fan, thus providing a reservoir target in a deep water setting.

At the time of the application, BP had a small grid of about 200 km of 2D seismic data over the Brae area, together with the results of early South Brae and Central Brae wells. The original exploration play envisaged a low-relief nose extending southeastward from Central Brae with separate small four-way dip closures at the base of the Cretaceous. Although the top of the Brae Formation could not be mapped directly, the high areas were interpreted as sand accumulations in the Upper Jurassic, whereas the bounding lows, interpreted as isopach thins, were thought to reflect the greater degree of compaction associated with mudstones, which might provide a lateral seal. The trap was therefore assumed to be at least partly stratigraphic. The original license application map is shown in Figure 3, with the current field outline superimposed.

BP acquired block 16/7b in the UKCS seventh round in 1980 and initiated a sedimentological study to review

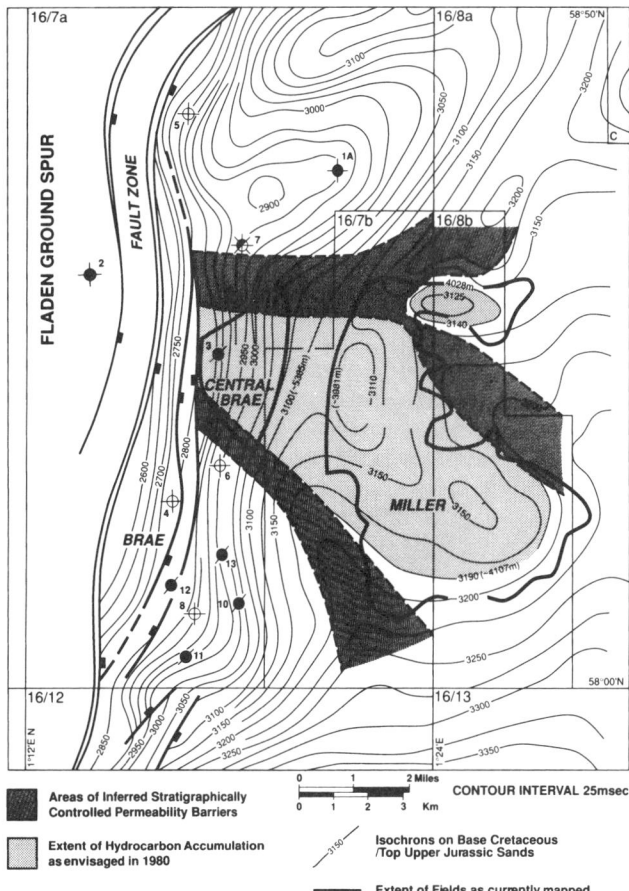

FIGURE 3. Map attached to BP's application for block 16/7b in the UKCS seventh licensing round. The current Central Brae and Miller field outlines have been added for comparison. This application derived substantially from in-house studies by R. G. Widdowson and S. Edrich.

possible depositional models for the Upper Jurassic in the Brae field and surrounding areas. These models included:

1. Subaerial and marginal marine coalescing fan deltas (Harms et al., 1981).

2. Fault-controlled slope-apron small fans (Stow et al., 1982).

3. Fault-controlled proximal area of a more extensive submarine fan.

It was concluded on the basis of BP's wider well control that neither model 1 nor model 2 was tenable, but that model 3, which proposed a system of small, tectonically controlled submarine fans, was likely. Data from well 16/7-3 indicated that thick sandstone turbidite sequences were being deposited during the Late Jurassic and/or Early Cretaceous.

The early exploration of Miller field was based on three 2D seismic surveys, totaling about 600 km, acquired by both operators between 1981 and 1983. The top of the

Brae Formation could not be resolved on these data, and top reservoir mapping was based on picking the base Cretaceous seismic event and adding a uniform Kimmeridge Clay isopach. Morphological highs and lows on the base Cretaceous surface were used to predict reservoir and nonreservoir facies, respectively.

During the depth conversion phase, the important observation was made that the interval velocity of the Cretaceous section showed significant lateral variation, with a decreasing trend from the Brae fields eastward. This amplified the subtle low-relief two-way time closures seen on the relatively flat southeast-trending nose and extended the Central and South Brae field closures eastward into block 16/7b.

BP's first well on the 16/7b block, well 16/7b-20, was spudded in November 1982 and completed in March 1983. It was followed by Conoco's well 16/8b-2, which was drilled between February and May of 1983. Both wells were targeted at low-relief closures mapped at the base of the Cretaceous. They both confirmed the presence of excellent-quality Brae Formation reservoir sands, which substantiated the submarine fan hypothesis. Similar gross pay zones of about 60 m were encountered in both wells, which flowed significant quantities of oil and gas on test. Data indicated similar types of fluids, with a common oil-water contact and a common pressure regime.

The discovery of thick, deep water fan lobe sands in Miller field has wider significance for the exploration for giant petroleum fields, because it has established the viability of a play type now known to be present locally along the South Viking graben in the Toni, Tiffany, and Thelma discoveries (Harris and Fowler, 1987). By implication, there seems to be no reason why similar sands should not be present intermittently in front of Mesozoic synrift margins throughout the world, although conjunction with rich source rocks is, of course, critical for success.

DELINEATION HISTORY

Appraisal of the main part of the accumulation by BP and Conoco took place rapidly between February 1983 and July 1984. During this period, wells 16/7b-21, 16/7b-23, 16/7b-24, 16/7b-25, 16/7b-26, 16/8b-3, and 16/8b-5 were drilled to confirm the initial discovery and to delineate the field from South Brae, Central Brae, and North Brae.

Well 16/8b-3 confirmed an oil-water contact between 4090 and 4095 m subsea, which was similar to those at South Brae and Central Brae. At first, the accumulation was thought to be part of a giant Brae field, but Conoco had named their 16/8b discovery "Miller" on the assumption that it was separate from the Brae field discoveries to the west.

The apparent relationship between Miller field and the Brae fields became more complex when BP wells 16/7b-21 and 16/7b-26 confirmed the extension of the South Brae field into block 16/7b, but it has been established that they were drilled west of a saddle that provides structural separation between the oil legs in South Brae and the southwestern part of Miller.

The 16/7b-23 well was drilled to test the eastward extent of Central Brae and to examine the relationship between Miller and Central Brae. There is no mapped structural separation, but palynological data show that the main Central Brae reservoir is older than the oil-bearing interval at Miller. The upper part of the Miller reservoir shales out westward between wells 16/7b-20 and 16/7b-23, demonstrating that the northwestern boundary of Miller is stratigraphically constrained.

The final phase of field appraisal took place between October 1984 and 1986, when three additional wells—16/7b-28, 16/8b-6, and 16/8b-7—were drilled to delineate the northern part of the field. These wells, which were drilled after South Brae came into production in July 1983, showed indications of pressure depletion from original virgin reservoir conditions, suggesting pressure communication via the aquifer between Miller and South Brae. The continuing effect of South Brae production was later confirmed when well 16/8b-5 in the southeast portion of the field was reentered and further pressure depletion was observed.

Appraisal wells drilled in Central Brae after South Brae production started did not show the same depletion as the Miller wells, again suggesting that Central Brae was distinct from Miller and South Brae.

The field appraisal results were incorporated into a development plan for the Miller field, and the Annex B application for development was approved by the Department of Energy in 1988.

STRATIGRAPHY

The regional stratigraphy of the South Viking graben is summarized by the stratigraphic column in Figure 4. The lithostratigraphic nomenclature used in Figure 4 follows the recommendations of the joint Norway/U.K. Lithostratigraphic Nomenclature Committee (Deegan and Scull, 1977).

The Miller reservoirs are Upper Jurassic in age and form part of the Brae Formation as defined by Turner et al. (1987). Over 80 exploration, appraisal, and production wells have now been drilled in the Brae-Miller area. Many of these wells have been extensively cored within the Upper Jurassic reservoir interval. A selection of these data, to all of which BP has access through various partnerships, has aided the construction of a depositional model for the Brae and Kimmeridge Clay formations.

BP's regional understanding of reservoir development and correlation relies largely on in-house studies of Jurassic sequence stratigraphy and tectonics (Hayward and Rattey, 1990). This approach has permitted consistent interpretation of the stratigraphic development of the area with reference to all available seismic, biostratigraphic, and sedimentological data.

The reservoir geology of individual Brae fields is discussed further in Turner et al. (1987), Roberts (1991), Stephenson (1991), and Turner and Allen (1991).

Regional Geologic Setting

Subregional studies of the Brae-Miller area (Wood and Hall, in press) clearly demonstrate the strong influence of structural setting and tectonic activity on sediment distribution patterns and stratigraphic development.

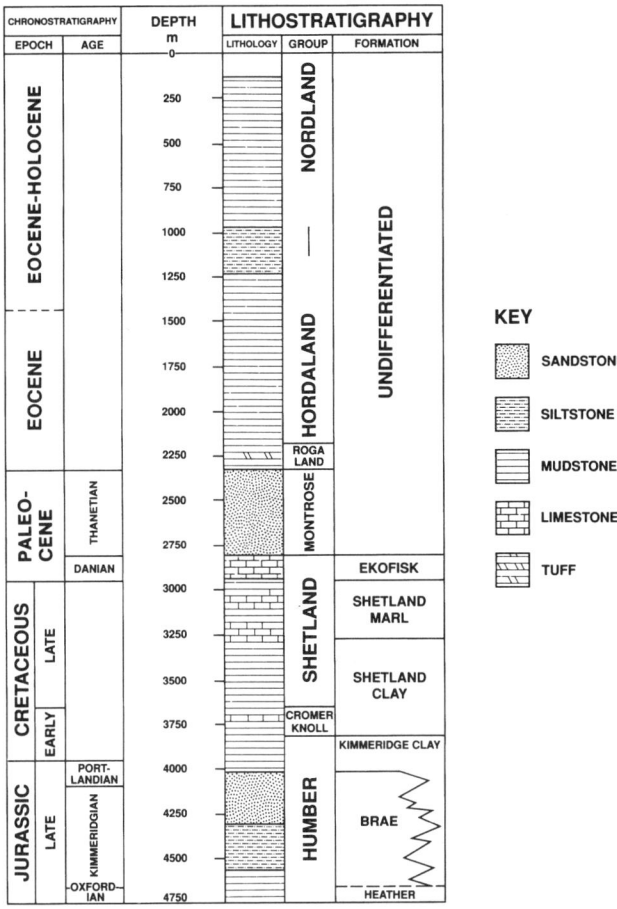

FIGURE 4. Stratigraphic column for the Brae-Miller area.

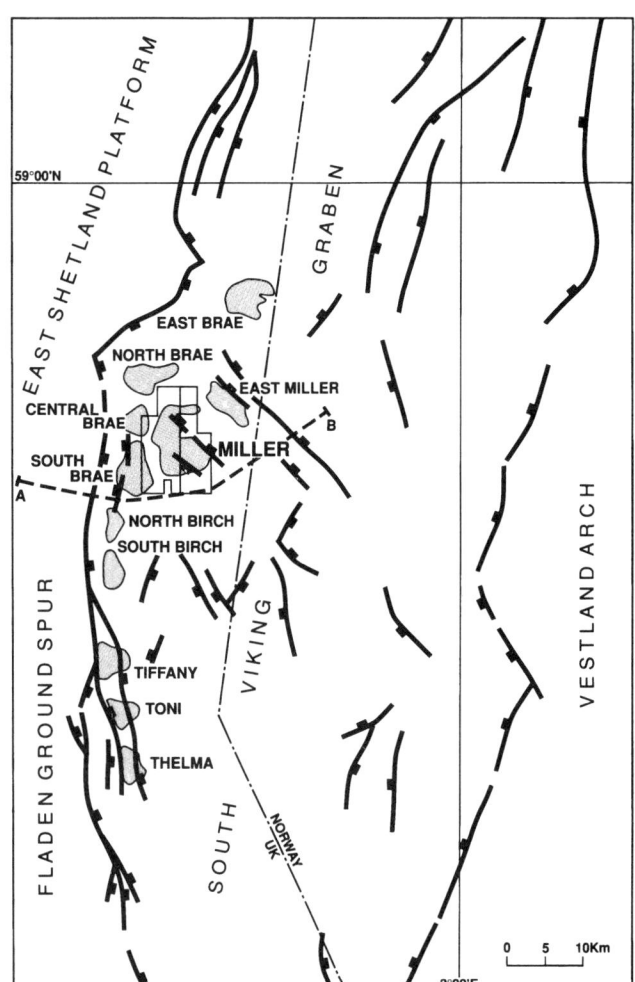

FIGURE 5. Structural outline of the South Viking graben, showing Upper Jurassic fields. Section A-B is shown in Figure 6.

The Miller field lies in the southern part of the South Viking graben, which in this area is a north–south-trending half-graben that is fault-bounded to the west against the Devonian "basement" of the Fladen Ground spur (Figure 5). The graben rises gently eastward to the Vestland arch, but is cut by several deep-seated northwest–southeast-trending faults oblique to the main boundary fault. This northwest-to-southeast lineation is the dominant trend of faults within the Miller field, and is coincident with the structural nose over which Miller has developed.

The east-west regional seismic line presented in Figure 6 shows the position of Miller field within the graben, and Figure 7 is a geologic interpretation of part of this section that indicates the present distribution of Upper Jurassic.

Figure 8 is a chronostratigraphic summary that indicates the fluctuating distribution of reservoir facies with time, relating this to BP's sequence stratigraphy (Hayward and Rattey, 1990) and to the geologic time scale (Haq et al., 1987).

Rifting in the South Viking graben was active from at least the Early Jurassic, and extensional movement continued intermittently on the western boundary fault until the Tertiary, when there was a brief phase of inversion before activity ceased.

Following deposition of epicontinental sediments in the South Viking graben during the Early to Middle Jurassic (Harris and Fowler, 1987), the northward-prograding deltas of the Bathonian were inundated during a major transgression in the early Callovian; this may have coincided with a phase of extension on the fault that forms the western boundary of the Viking graben (Figure 7). A second large-scale phase of extensional faulting, which began in the late Oxfordian, controlled deposition of the Brae Formation, building out as a clastic wedge that thins eastward from the boundary fault. The crustal extension that began in the late Oxfordian reached a climax during the Kimmeridgian and waned through the Portlandian to Ryazanian.

The western boundary fault, comprising a series of en echelon faults with associated basement terraces, has itself been cut back by submarine erosion, in places forming an escarpment. The close proximity of sediment source to receiving basin, together with active extensional tectonics, provided the ideal setting for the development of numerous, often overlapping, small-scale (approximately 5 to 10 km wide), gravel-dominated submarine fans along this fault-defined basin margin. Well data show the suc-

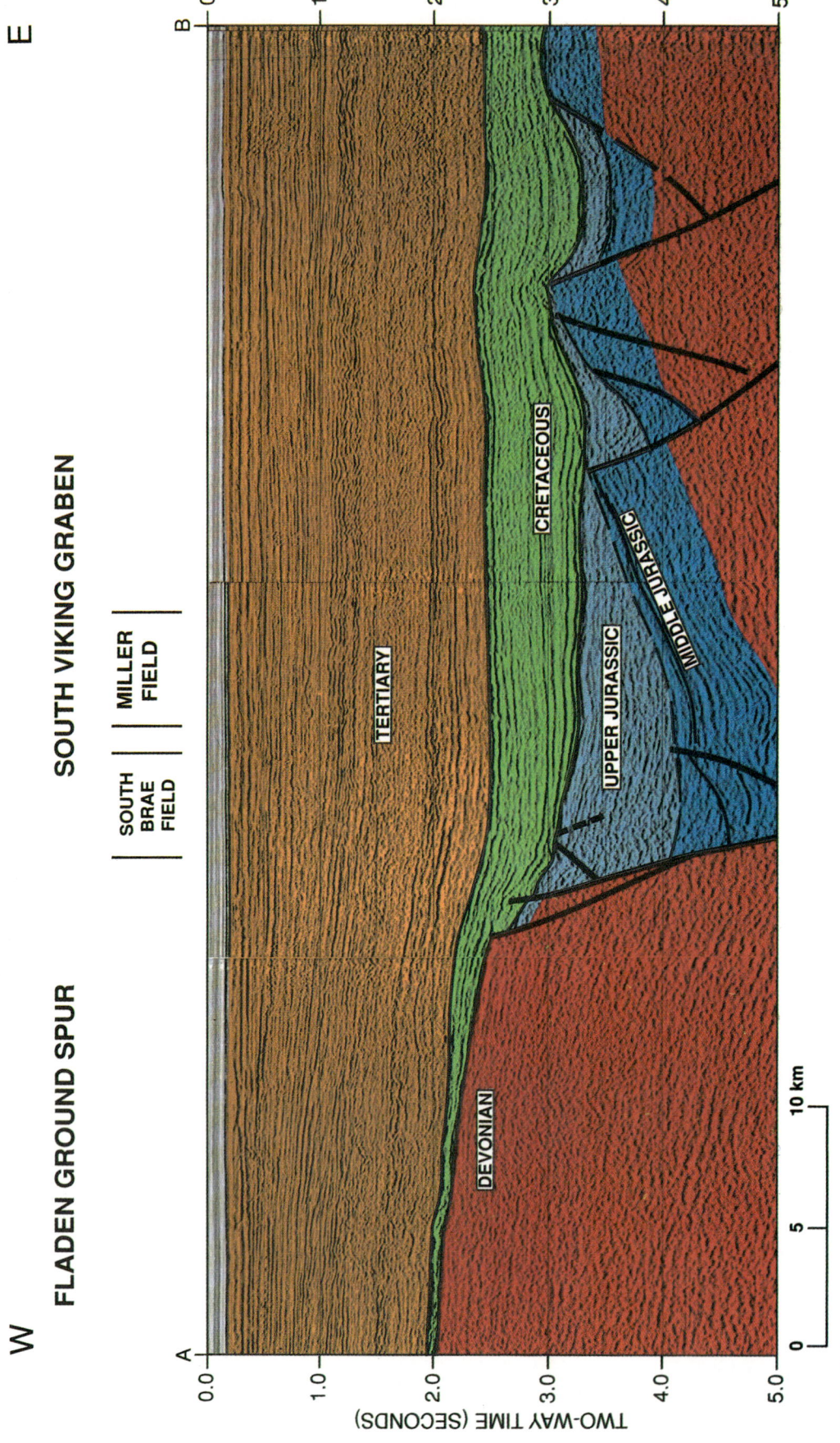

FIGURE 6. Regional seismic section close to the South Brae and Miller fields within the South Viking graben (line of section shown in Figure 5).

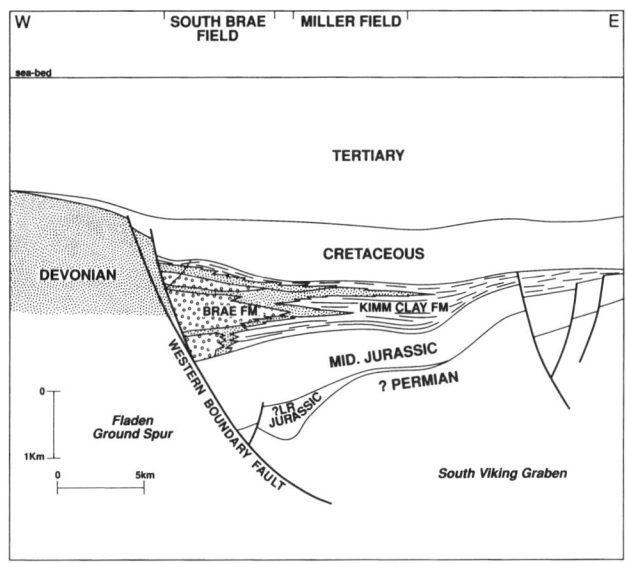

FIGURE 7. Geologic cross section across the western part of the South Viking graben (based on part of seismic section shown in Figure 6).

cession to be dominated by medium- to large-scale (100 to 150 m thick) fining-upward sequences. Interbedded mudstones represent the hemipelagic background sedimentation preserved during periods of quiescence and low clastic input. Occasional laterally continuous shales may indicate that basin subsidence was sometimes interrupted. In general, the gravel-rich fans are confined to a narrow belt (about 10 km wide) extending along the basin's actively faulted margin.

Rates of hinterland uplift and basin subsidence decreased with time. A relative rise in sea level, probably associated with footwall collapse at the basin edge, led to the development of a wider shelf area. This had a profound effect on the nature and geometry of the submarine fan system. The apron-fringing, gravel-dominated fan complex gave way to a major sand-rich lobate fan system. Fan development occurred basinward of the apron system, bypassing early topography near the faulted margin that had been infilled by the earlier gravel-rich system. These younger sand-rich systems form the main reservoir succession within the Miller field.

A well-exposed Tertiary analog of this type of sand-rich fan depositional system occurs in the Tabernas basin

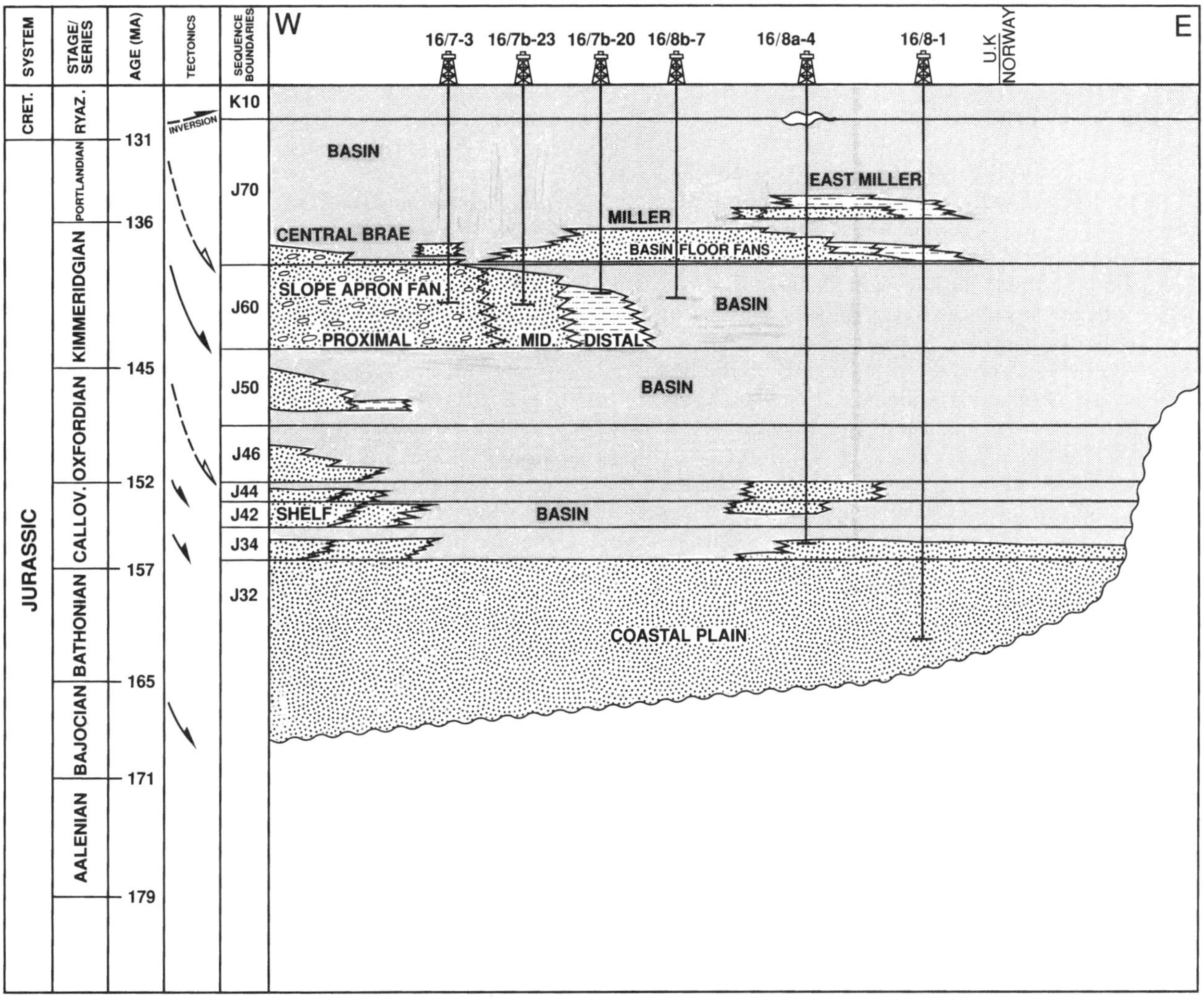

FIGURE 8. Chronostratigraphic summary of the Jurassic in the South Viking graben.

in southern Spain (Kleverlaan, 1989). Similar depositional systems in Baja California have been described by Busby-Spera (1988). Less-accessible outcrops in a complete analog of the entire Brae-Miller system, also of Upper Jurassic age, occur in eastern Greenland (Surlyk, 1987).

Submarine degradation of the Fladen Ground spur eventually starved the South Brae–Miller area of its supply of coarse clastic material and marked the end of major submarine fan deposition in the late Portlandian. The final stages of submarine fan deposition within the area are recorded by the gradual retreat of the sand-rich fan with back-filling of its principal conduits within South Brae. After this, coarse clastic input into the graben continued, but the dominant sediment entry point switched to the north, where thick Brae Formation reservoirs were deposited at North Brae field (Figure 2). Distal equivalents of these sediments occur over the northern part of Miller field as thin sands within the Kimmeridge Clay Formation.

In the Brae area, the main rifting phase had been completed by the Ryazanian. Sea level continued to rise, placing the Brae-Miller area even more distal from potential sources of clastic material. Decreased rates of crustal extension in the Cretaceous, together with high global sea level, allowed thick mudstones to onlap across the western boundary fault onto the Old Red Sandstone, and coarse clastic deposition in the graben ceased.

Biostratigraphic Framework

A biozonation of the Miller reservoir has been established using data from cored wells in this field. Figure 9 indicates how this scheme permits a correlation between Miller and South Brae fields that is in general accord with that proposed for the Brae fields by Riley et al. (1989). It also is consistent with the sequence stratigraphy established by Hayward and Rattey (1990).

The Miller field biozones are based mainly on the first downhole occurrence of particular dinoflagellate cysts, and a color-coded series of biozone tops has been set up (Figure 9).

Two important markers recognized within the oil-bearing interval are the *Oligosphaeridium patulum* "black marker" (at the top of the Kimmeridgian) and the *Cribroperidinium longicornis* "pink marker," which mark the tops of units 2B and 2C, respectively. A third marker, the "light brown marker" (*Gonyaulacysta jurassica*), correlates with the top of the unit 3 reservoir interval. The pink and light brown markers are both within the Kimmeridgian.

Facies Descriptions

Within the Miller field, four main lithofacies have been recognized in the submarine fan sediments. These lithofacies form a subset of those in the Brae Formation as a whole (Turner et al., 1987). Conglomeratic facies are not present in Miller field, although small, rounded pebbles are seen, more commonly at deeper levels in the reservoir. The four lithofacies are:

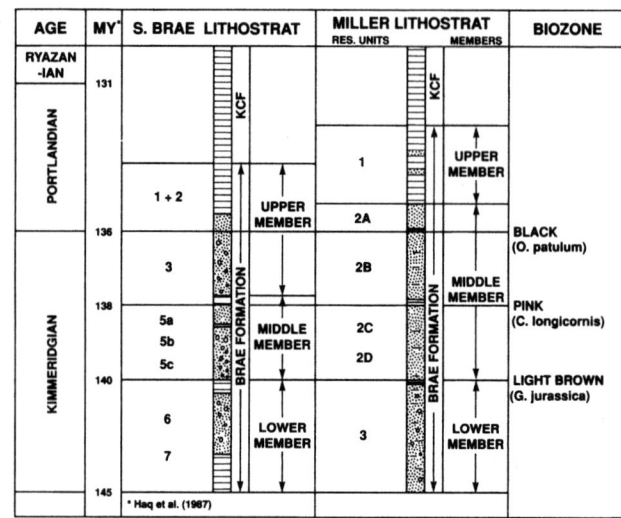

FIGURE 9. Correlation between Miller and South Brae fields, as indicated by biostratigraphic markers.

1. *Granular sandstones.* These are the dominant facies of the Miller reservoir, comprising medium-grained, structureless sandstones forming thick beds. These sandstones were deposited by high-density turbidity currents in what were probably unconfined or partly confined lobes and channels, somewhat resembling a sandy braided river system.

2. *Interlaminated mudstones and fine-grained sandstones.* Also referred to as "tiger stripes" (Stow et al., 1982), these consist of rapid alternations of mudstones, siltstones, and very fine- to fine-grained sandstones (Bouma sequences), and are thought to have been deposited by low-density classical turbidity currents in relatively quiescent areas of the mid- to outer fan.

3. *Argillaceous sandstones.* Medium- to fine-grained sandstones with abundant argillaceous laminae and some carbonaceous plant debris. Bed thickness ranges from 5 to 100 cm. These sands were deposited by high-density turbidity currents. The increased argillaceous content suggests a more marginal setting within the midfan and inner fan, away from the main axis of sand input.

4. *Mudstones.* Massive mudstones with occasional siltstones and very fine-grained sandstone laminae (<5 mm thick). These represent dilute turbidite deposition on the outer fan and basin plain.

Depositional Model

The Brae Formation has been divided into three reservoir units (Figure 10), each recording a phase in the sedimentary evolution of the Miller area (Figure 9). Unit 3, the lowest unit (see Figure 10), forms the main reservoir in

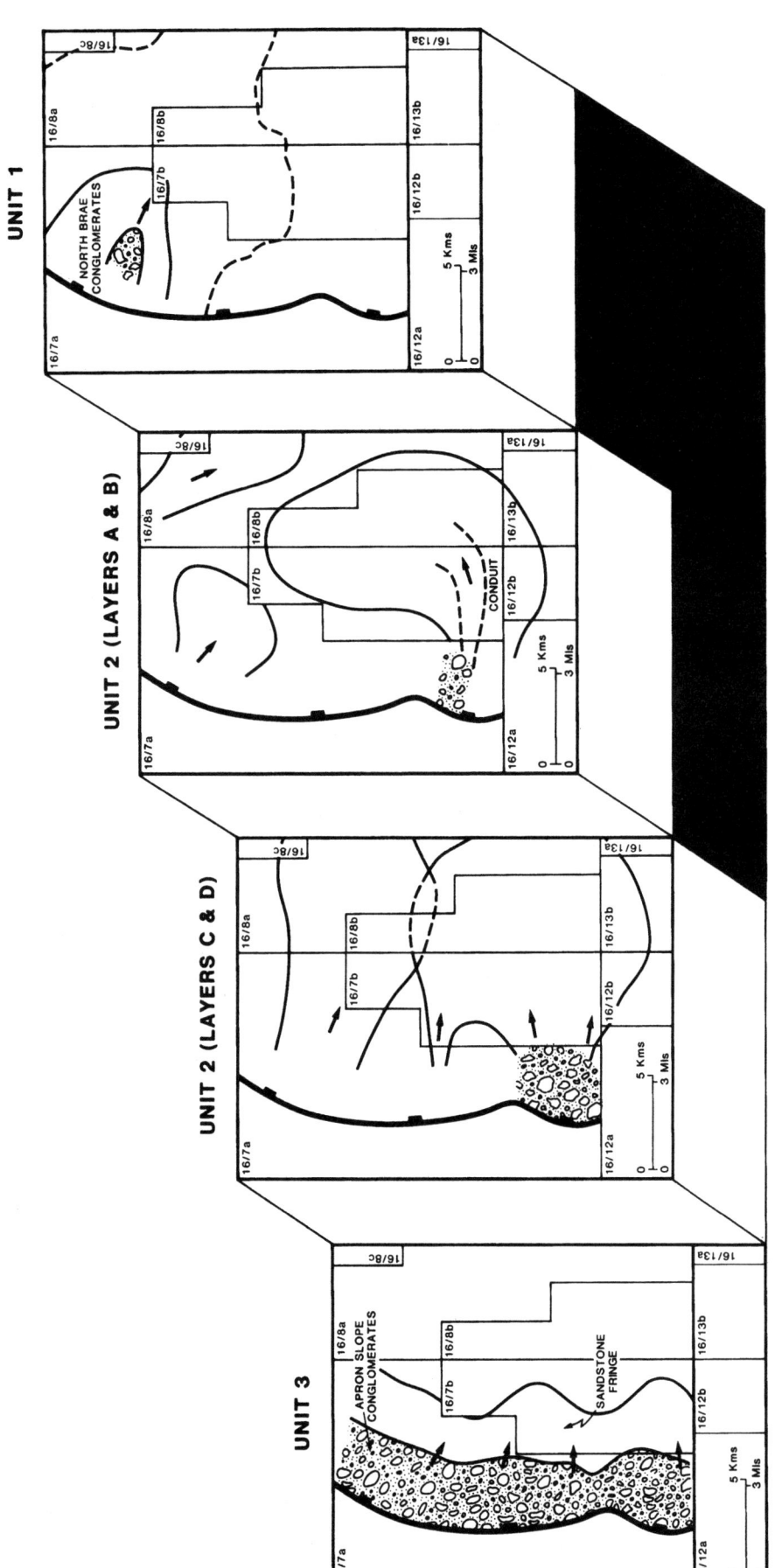

FIGURE 10. Evolution of submarine fan reservoir facies in the Brae-Miller area.

South Brae. It comprises a series of coalescing, low-efficiency, apron-fringing fans up to 3000 m thick and laid down close to the basin margin fault. It was deposited in the late Oxfordian during a period in which the western margin of the basin was subsiding more rapidly than it could be filled with sediment. Proximal gravels and conglomerates in the west pass eastward into massive sandstones with pebbles and then into a succession of distal thin-bedded turbiditic sandstones and shales that underlie the Miller reservoir.

The sands that comprise the lowermost part of unit 2 (units 2D and 2C) are the distal expressions of the narrow wedge of fining-upward gravels (unit 3) that thin and fine upward within a few kilometers as they pass eastward. They are thought to reflect a general decrease in the rate of sediment supply and fault activity along the graben margin, probably accompanied by hinterland retreat and the development of a wider shelf along the basin edge.

This phase gave way to the development of a sand-dominated lobate fan system in the Miller area, although fining-upward apron-fringing fan sequences continued to be deposited in South Brae. This resulted in the deposition of units 2B and 2A, which form the bulk of the oil-bearing reservoir in Miller field.

Isopachs suggest that the Miller sand was sourced through a major southeastward-trending conduit in the South Brae area (some 1.0 to 2.5 km wide). At the eastern end of the South Brae conduit, the submarine fan sediments were deflected northward and transported axially into the basin across blocks 16/7b and 16/8b.

Unit 2A is present only intermittently over the field, and its distribution may have been controlled by the topography of the main unit 2B fan surface. Penecontemporaneous submarine fans that were active in North Brae and East Miller can be distinguished petrographically. The northwestern fan contains characteristic calcareous bioclasts, whereas the East Miller fan also contains abundant sponge spicules. The Miller fan, by contrast, contains very little bioclastic material.

The final stage of coarse clastic deposition in the Brae Formation is represented by the unit 1 sandstones. By this time, the southern feeder channel had ceased to be active, and deposition had switched to a new entry point to the northwest of Miller field in the North Brae field area. Unit 1 in Miller is represented by thin sandstones and shales believed to be the distal equivalent of the North Brae system.

Unraveling the complex submarine fan depositional history of the Brae-Miller area was based largely on the study of well data. Until recently, lobes could not be resolved on seismic data. The large volume of 3D data now available allows detailed seismostratigraphic mapping, and is revealing depositional features that broadly support the initial model.

STRUCTURE

The regional geologic setting of the Brae-Miller area within the South Viking graben has been described above. The Miller field forms a southeast-plunging nose that rises northwestward toward Central Brae. Dip closure is present to the north, south, east, and southwest, whereas closure to the northwest is provided by stratigraphic pinch-out (Figure 11).

In the southwest, Miller is structurally separated from South Brae by a gentle north–south-trending syncline. Another gentle syncline separates Miller from Shell's East Miller accumulation to the northeast.

Miller field is a fairly low-relief, four-way dipping feature with an oil-water contact at 4090 m and a maximum hydrocarbon column of about 110 m. Within the field, several faults are present at top reservoir level. The major faults generally trend from northwest to southeast, parallel to the plunge of the structure, and they cut both the top of the reservoir and the base of the Cretaceous. This trend is similar to the trend in the South Viking graben east of Miller. In the northern part of the field, a northwest-trending graben reduces the area above the oil-water contact.

The major intrafield faults have throws of up to about 140 m, less than the gross reservoir thickness of 250 to 300 m. In addition, some synsedimentary faulting has been observed. These faults produce only a slight inflection at the base of the Cretaceous, with thickening of the Kimmeridge Clay on the downthrown side. Interpretation of the pressure decline in response to South Brae production suggests that faults within the reservoir are not sealing.

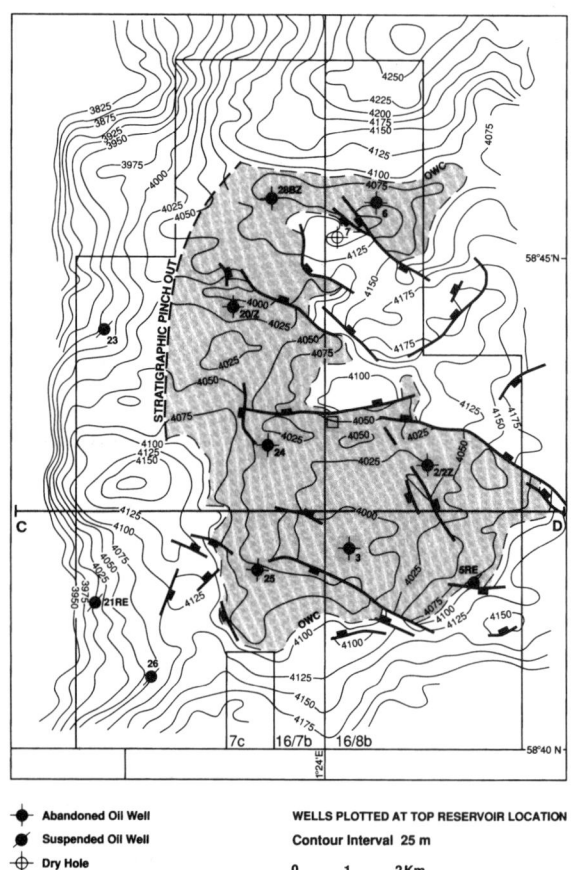

FIGURE 11. Depth map for the top of the Brae reservoir. Section C-D is shown in Figure 12.

Seismic Interpretation

The structure has been mapped at top reservoir level using part of the Brae-Miller 3D survey recorded and processed between 1985 and 1987. The total survey comprises 12,400 km of subsurface data, of which about 4000 km was used for the interpretation of Miller. These data were tied to synthetic seismograms from 18 exploration and appraisal wells within and adjacent to the field, and the interpretation was carried out using a seismic workstation. A seismic section through the center of the field is shown in Figure 12.

The top of the reservoir (top of the Brae Formation) can be mapped directly by picking a trough on zero phase data that corresponds to an increase in acoustic impedance between the Kimmeridge Clay Formation and the higher-velocity Brae Formation. The amplitude of the top reservoir event varies, however, and is more difficult to pick over the northern part of the field.

A two-way time map (Figure 13) over the field shows the southeast-plunging nose, with independent dip closure of only 30 msec over the crest of the structure near well 16/8b-3. Structural closure is amplified in depth (Figure 11) by the average velocity to the top of the reservoir, which decreases over the crest of the field. The general west-to-east decrease in average velocity (Figure 14) from the Brae area to Miller field, which was recognized early in the exploration phase, is attributed to an increase in detrital clay content of formations within the Upper Cretaceous Chalk Group.

Intrareservoir and seismostratigraphic mapping have been attempted, but in view of the inherent complexity of the depositional system they are critically dependent on data quality for success. In general, the south of the field is easier to map than the north. Stratigraphic features that have been recognized include downlaps indicative of the western pinch-out edge of the unit 2B/unit 2A fan (Figure 12).

RESERVOIR GEOLOGY

The Upper Jurassic sequence in Miller field has been subdivided into three main units, designated units 1, 2, and 3, which are shown in Figure 9. These units can be broadly related to the upper, middle, and lower units of the Brae Formation identified in the Brae fields.

Unit 3 is generally below the oil-water contact and comprises thick, often pebbly sands interbedded with shales, and the base is not penetrated over most of the Miller area. Unit 3 is equivalent to much of the reservoir interval in Central Brae.

Unit 2 contains the main Miller accumulation. It has been further subdivided into four reservoir zones: units 2A, 2B, 2C, and 2D (Figure 15). The bulk of the reserves are within units 2A and 2B, with only a small volume in unit 2C. Shales are developed between units, and they appear to act as barriers to vertical fluid flow and pressure communication over significant areas of the field.

Units 1 and 2A are equivalent to the upper member of the Brae Formation and form the reservoir in North Brae field.

To date, reservoir zonation has been based on biostratigraphic and log correlation, with additional input from static pressure measurements in the appraisal wells. Subsequent wells have been particularly significant in that they show differential depletion between reservoir layers caused by production from the South Brae field, which commenced in 1983. As Miller field moves through the development phase, further refinement of the reservoir zonation, based on production data, may be anticipated.

The sandstones are quartzose in composition, with very minor amounts of detrital feldspar and mica. Quartz cement forms the major diagenetic phase, occurring as syntaxial overgrowths on the detrital grains.

Locally, calcite cements formed carbonate doggers during later stages of diagenesis. The best-quality reservoirs occur in the granular sandstones, where average horizontal permeability is 120 md, with a typical range of 50 to 1200 md. Over-all average porosity is about 14%, with a range of 12 to 23%, predominantly primary porosity. Net/gross ratios are high, typically greater than 0.75.

Reservoir quality is controlled primarily by grain size and sorting, with the coarser-grained, moderately sorted to well-sorted sandstones providing better reservoirs than the shalier, more poorly sorted sands. In the former, the principal mode of quality reduction is the development of quartz overgrowths, whereas in the latter, compaction is more significant.

No discernible change has been observed in the cement phases above and below the oil-water contact, implying that cementation preceded oil migration across the area.

Descriptions of the individual reservoir units are given below.

Unit 3

The unit 3 reservoirs are the oldest reservoirs present, being early Kimmeridgian in age. They comprise interbedded sands and shales, typically with a net/gross ratio of 0.47 and an average porosity of 12%. They are believed to be distal turbidites of the Central and South Brae apron fans. Sand content decreases consistently eastward.

Unit 3 sands are oil-bearing within structural closure under the crest of the field in wells 16/8b-2, -3, and -5 down to an oil-water contact at 4303 m subsea. Water saturation is high because of a high clay content. Reservoir pressure in unit 3 may be as much as 1000 psi higher than in the main accumulation, from which it is separated by a sealing shale. Thus, to avoid mechanical complications, this unit will not be developed at present.

Unit 2D

Unit 2D is the deepest reservoir zone within the main sand-rich submarine fan in Miller, and, lying below 4090 m subsea, is in the water leg of the main accumulation. Thickness ranges from 50 to 150 m. In the north of the field, a small, probably isolated oil leg with a contact at 4172 m subsea was found in unit 2D below the main pool in well 16/7b-28.

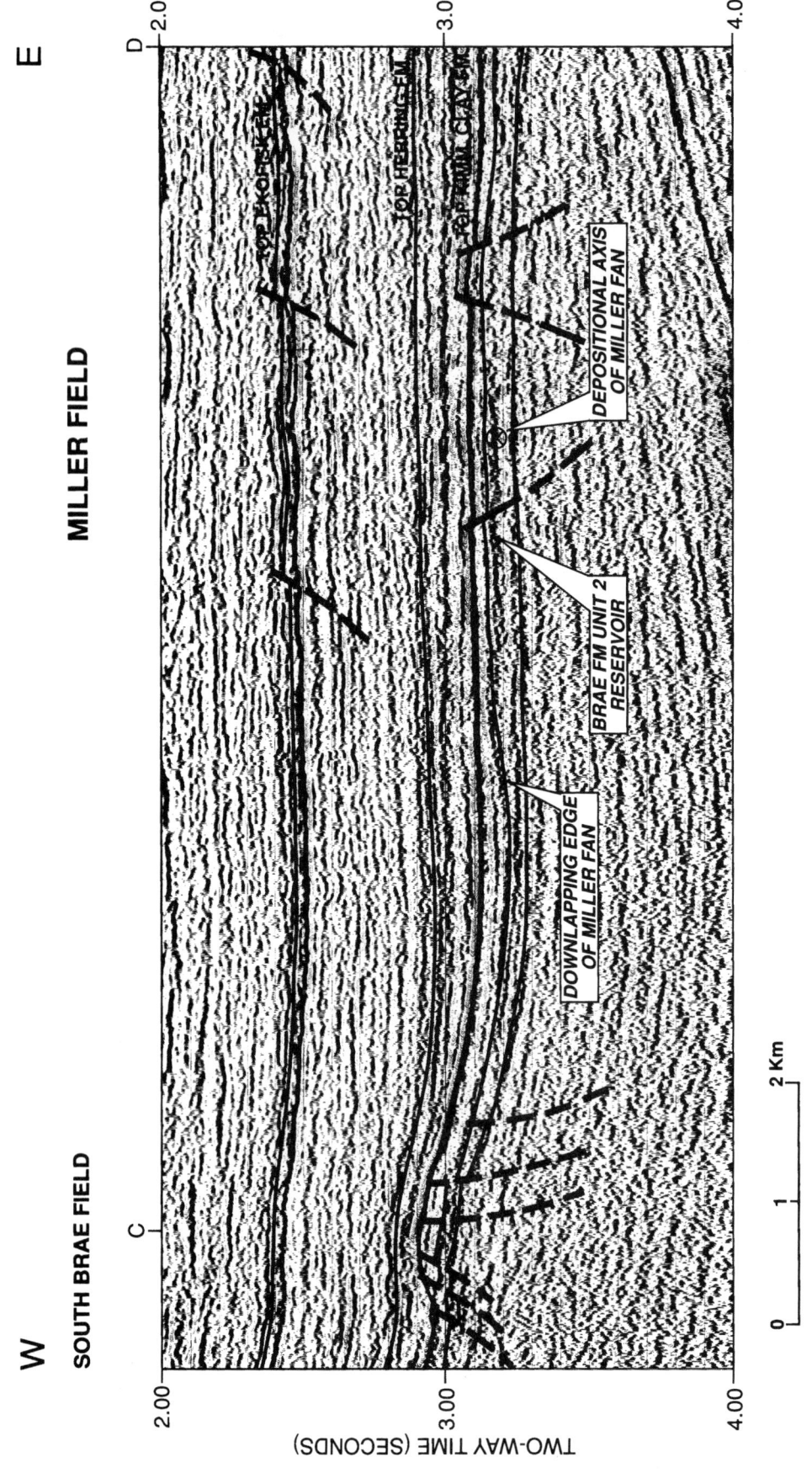

FIGURE 12. Seismic section across the South Brae and Miller fields (line of section shown in Figure 11).

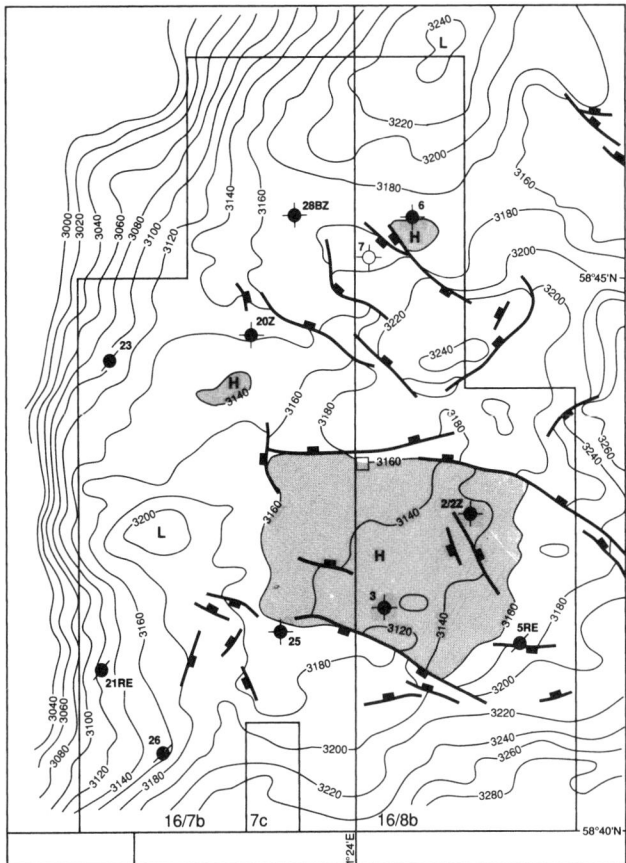

FIGURE 13. Two-way time map for the top of the Brae reservoir (contour interval = 20 m/sec).

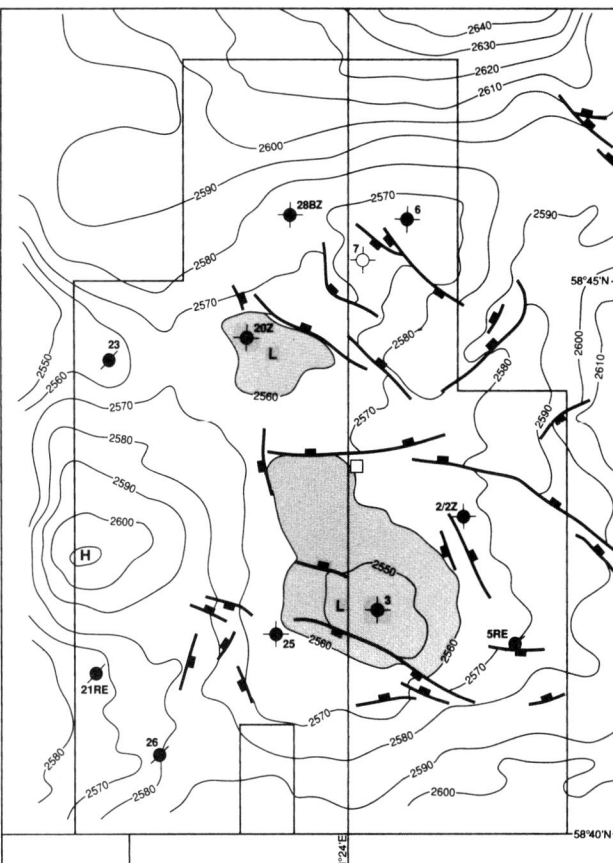

FIGURE 14. Average velocity map for the top of the Brae reservoir (contour interval = 10 m/sec).

Unit 2D comprises predominantly medium-grained sandstones with occasional thin beds of shale. Its net/gross ratio is the highest in the south central part of the field (0.68 to 0.78), decreasing northward toward well 16/7b-20 (0.36). Farther north, sand content increases again (0.57 to 0.70). Average porosities range from 12 to 15%, and permeabilities range from 20 to 40 md.

This unit is believed to represent two major, partially overlapping lobes. The main one is sourced from the southwest via South Brae, and the other is sourced from the north via North Brae (Figure 10). Well 16/7b-20, with poor sand development, would lie in an interlobe or a peripheral setting.

Unit 2C

As does unit 2D, unit 2C lies below the oil-water contact over much of the field area, although it does contain oil in the crestal area. It comprises mainly medium-grained sands and is interpreted as being bilobate with a depositional origin similar to that of unit 2D (Figure 10). It is 40 to 80 m thick. Net/gross ratios range from 0.24, in the interlobe area between wells 16/7b-20 and -28, to 0.93, in well 16/8b-3 in the center of the field. Porosities range from 12 to 17%. Average test permeabilities range from 2 to 160 md.

Unit 2B

Unit 2B, which is between 50 and 100 m thick, contains the bulk of the Miller reserves and is separated from unit 2C by a shale break that acts as a pressure barrier, as indicated by differential pressure depletion. Unit 2B consists mainly of medium-grained sandstone and forms the thickest part of the sand-rich fan in Miller field. Net/gross ratios exceed 0.85 over most of the field, but well 16/7b-28 in the north has a substantially lower net/gross ratio (0.61), which is interpreted as reflecting a marginal to outer-fan setting. Average porosities range from 13 to 19%, and average test permeabilities range from 30 to 600 md.

The depositional model envisaged for unit 2B is shown in Figure 10. Unit 2B is believed to have formed as a single major lobe fed from the southwest via a conduit through the upper part of the South Brae conglomerates. Well 16/7b-26 in South Brae is interpreted as having encountered unit 2B within the feeder channel, and the entry point into the Miller fan is believed to be just south of well 16/7b-25.

Most of Miller field lies within the center of the unit 2B fan, and the northern and eastern limits of the fan appear to be beyond the structural spillpoint at 4090 m subsea. In the northwest, however, the pinched-out edge of the fan provides stratigraphic separation from the older Central Brae reservoirs.

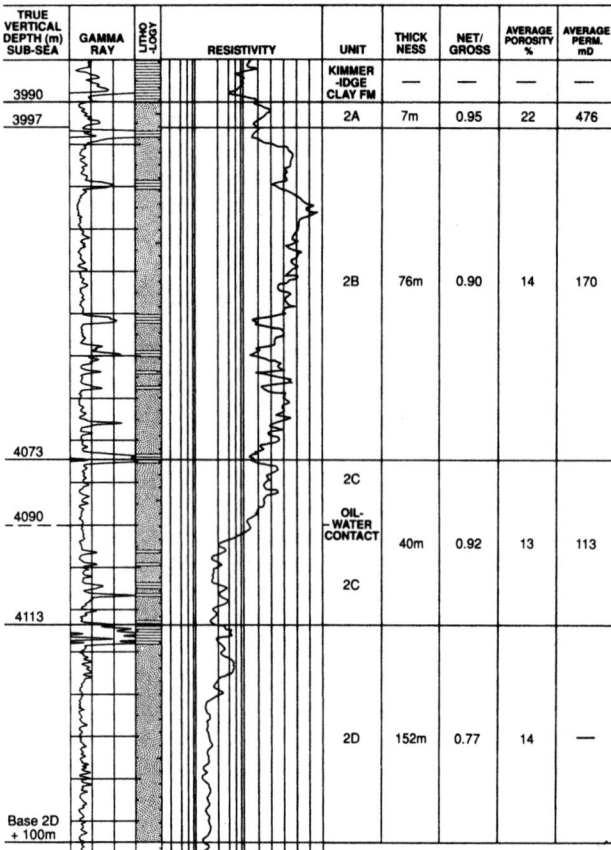

FIGURE 15. Petrophysical summary of Miller reservoir units in well 16/8b-3.

Unit 2A

Unit 2A is thin (up to 10 m) and is not present over all parts of the field. Its distribution may be related to the topography of the underlying main unit 2B fan lobe. Reservoir quality of unit 2A is excellent, with average porosities of 14 to 22% and net/gross ratios of 0.67 to 0.98. Test permeabilities of 250 md have been measured.

Unit 1

Over most of Miller field, unit 1 is predominantly shaly, although interbedded turbidite sands were deposited over the northern part of the field after clastic deposition through the South Brae channel system had ceased. The major entry point shifted north, and unit 1 appears to have been sourced via North Brae (Figure 10).

Thin oil legs are present in wells 16/7b-28 and 16/8b-6 down to an oil-water contact at 4035 m subsea. Net/gross ratios for unit 1 reservoirs are typically 7 to 19%; average porosities range from 12 to 20%, and test permeabilities of 30 to 60 md have been measured.

SOURCE ROCK, HYDROCARBONS IN PLACE, AND RESERVES

The source of oil in Miller field is the Kimmeridge Clay Formation, which interfingers with and overlies the Upper Jurassic reservoirs in the South Viking graben. Oil generation began in the early Tertiary to the northeast of Miller field, and continues to this day.

The current estimate of the mean volume of oil in place in Miller field is 640 MMSTB. Of this, the volume of recoverable oil is estimated to be around 300 MMSTB. In addition, the associated gas that will be produced is 0.57 trillion standard ft^3.

DEVELOPMENT

Thirty development wells are currently planned for Miller field (Figure 16), with the first having been spudded in March 1989. Twenty of these wells will be producers, supported by ten peripheral water injectors. The development plan calls for ten of these wells (five producers and five injectors) to be predrilled through a seabed template over which a single platform will be positioned during 1991. The bulk of the reserves lies in the southern part of the field, and this is the area currently under development in the template drilling phase. The northern area, which will require closer well spacing, will be developed from the platform.

It is intended that the field be put into production in early 1992, and that an average plateau rate of 113,000 STB of 37.5° API oil per day be reached within one to three months. The plateau period is expected to last just under four years, with a field life of about 13 years. The platform will have a single process train designed to handle 120,000 STB per day, and a water injection facility capable of handling 300,000 bbl of water per day.

As illustrated in Figure 1, the Miller platform will be linked by an 18-in. pipeline to South Brae, and oil will then be exported via Marathon's Brae/Forties pipeline through the Forties field. From there it will join BP's Forties pipeline system for distribution to Kinneil and the Grangemouth refinery.

Gas from Miller field has high H_2S and CO_2 contents (H_2S up to 800 ppm; CO_2 up to 20 mol%) and will be exported separately via a separate sour gas trunk pipeline to a landfall at St. Fergus. The total length is 240 km, and the line has been designed to allow spare capacity for additional tie-ins. The entire Miller gas export stream has been sold to the North of Scotland Hydro Electricity Board to fuel a power station at Peterhead.

CONCLUSIONS

Miller field is a giant oil and gas field containing 640 million bbl of oil in place, giving recoverable reserves of about 300 MMSTB of oil and 0.57 trillion standard ft^3 of gas. Its reservoir is the Brae Formation, a Kimmeridgian-Portlandian sand-rich submarine fan sequence deposited in the South Viking graben. This sequence has been subdivided on the basis of core evidence and gamma ray logs

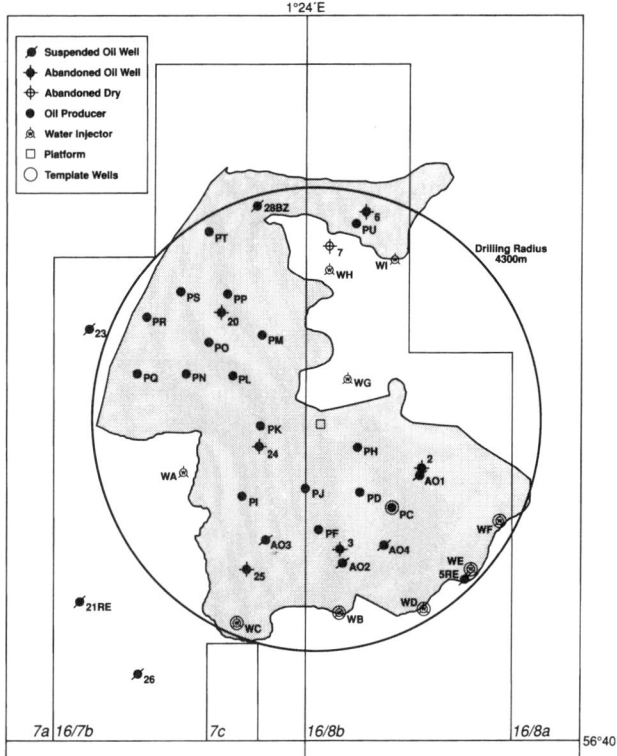

FIGURE 16. Locations of development wells in Miller field.

into four units that provide a basis for correlation within the field. The lithostratigraphic correlation is supported by biostratigraphic and pressure evidence that indicates widespread lateral continuity within the reservoir.

The depositional process is interpreted as having involved submarine transport of sands sourced on the Fladen Ground spur, through conduits in the Brae fields area, to deposit interdigitating lobes of sand-rich submarine fans in the basin.

The structure of the field is one of topographic relief, reflecting the compaction and minor faulting that accompanied and immediately followed deposition. The top of the reservoir can be mapped directly from the change in acoustic impedance between the Kimmeridge Clay Formation and the Brae Formation. The field will be developed by installing a single fixed platform that will have 20 production wells and ten water injectors. The oil and gas will be delivered to Scotland in separate streams by pipeline.

ACKNOWLEDGMENTS

This paper was derived from studies undertaken by staff members at BP Exploration as part of the ongoing development of Miller field. The authors are indebted to all of their colleagues who provided assistance, comments, and suggestions that contributed to the preparation of the manuscript. The Chairpersons and Boards of Directors of BP and their partners, Conoco, Enterprise, and Santa Fe, are thanked for their permission to publish this paper.

The views expressed herein are based on data available to BP and are not necessarily coincident with those of BP's partners in the field.

REFERENCES CITED

Busby-Spera, C., 1988, Development of fan-deltoid slope aprons in a convergent-margin tectonic setting: Mesozoic, Baja California, in W. Nemec and R. J. Steel (eds.), Sedimentology and tectonic settings: Blackie and Son, p. 419–429.

Cazzola, C., F. Fonnesu, E. Mutti, G. Rampone, M. Sonnino, and V. Vigna, 1981, Geometry and facies of small, fault-controlled deep-sea fan systems in a transgressive depositional setting (Tertiary Piedmont basin, NW Italy), in F. Ricci Lucchi (ed.), IAS Excursion Guidebook, 2nd European Regional Meeting, Bologna, p. 7–56.

Deegan, C. E., and B. J. Scull, 1977, A standard lithostratigraphic nomenclature for the central and northern North Sea: Bulletin of the Norwegian Petroleum Directorate, v. 1, p. 1–36.

Haq, B. U., J. Hardenbol, and P. R. Vail, 1987, Chronology of fluctuating sea levels since the Triassic: Science, v. 235, p. 1156–1166.

Harms, J. C., P. Tackenberg, E. Pickles, and R. E. Pollock, 1981, The Brae oilfield area, in L. V. Illing et al. (eds.), Petroleum geology of the continental shelf of northwest Europe: London, Institute of Petroleum, p. 352–357.

Harris, J. P., and R. M. Fowler, 1987, Enhanced prospectivity of the Mid-Late Jurassic sediments of the South Viking Graben, northern North Sea, in J. Brooks and K. W. Glennie (eds.), Petroleum geology of north west Europe: London, Graham and Trotman, p. 879–898.

Hayward, A. H., and R. P. Rattey, 1990, Depositional sequence stratigraphy of a failed rift system. The Middle Jurassic to Early Cretaceous basin evolution of the North Sea, in Proceedings of EAPG Conference, Copenhagen, 1990.

Kleverlaan, K., 1989, Three distinctive feeder-lobe systems within one time slice of the Tortonian Tabernas fan, SE Spain: Sedimentology, v. 36, p. 25–45.

Normark, W. R., 1970, Growth patterns of deep-sea fans: AAPG Bulletin 54, p. 2170–2195.

Normark, W. R., 1978, Fan valleys, channels and depositional lobes on modern submarine fans: character for recognition of sandy turbidite environments: AAPG Bulletin 62, p. 912–931.

Riley, L. A., M. J. Roberts, and E. R. Connell, 1989, The application of palynology in the interpretation of Brae Formation stratigraphy and reservoir geology in the South Brae field area, British North Sea, in J. D. Collinson (ed.), Correlation in hydrocarbon exploration: London, Graham and Trotman, for Norwegian Petroleum Society, p. 339–356.

Roberts, M. J., 1991, The South Brae field, block 16/7a, U.K. North Sea, in I. L. Abbotts (ed.), United Kingdom oil and gas fields commemorative volume: Geological Society Memoir 14, p. 55–62.

Rooksby, S. K., 1991, Miller field, blocks 16/7b, 16/18b, U.K. North Sea, in I. L. Abbotts (ed.), United Kingdom oil and gas fields commemorative volume: Geological Society Memoir 14, p. 159-164.

Stephenson, M. A., 1991, The North Brae field, block 16/7a, U.K. North Sea, *in* I. L. Abbotts (ed.), United Kingdom oil and gas fields commemorative volume: Geological Society Memoir 14, p. 43-48.

Stow, D. A. V., C. D. Bishop, and S. J. Mills, 1982, Sedimentology of the Brae oilfield, North Sea: fan models and controls: Journal of Petroleum Geology, v. 5, p. 129-148.

Surlyk, F., 1987, Slope and deep shelf gully sandstones, Upper Jurassic, east Greenland: AAPG Bulletin 71, p. 464-475.

Turner, C. C., and P. J. Allen, 1991, The Central Brae field, block 16/7a, U.K. North Sea, *in* I. L. Abbotts (ed.), United Kingdom oil and gas fields commemorative volume: Geological Society Memoir 14, p. 49-54.

Turner, C. C., J. M. Cohen, J. R. Connell, and D. M. Cooper, 1987, A depositional model for the South Brae oilfield, *in* J. Brooks and K. W. Glennie (eds.), Petroleum geology of north west Europe: London, Graham and Trotman, p. 853-864.

Vail, P. R., J. Hardenbol, and R. G. Todd, 1984, Jurassic unconformities, chronostratigraphy and sea-level changes from seismic stratigraphy and biostratigraphy, *in* J. S. Schlee (ed.), Interregional unconformities and hydrocarbon accumulation: AAPG Memoir 36, p. 129-144.

Wood, J. L., and S. Hall, in press, The tectonostratigraphic model for the deposition of submarine slope apron fans and basin floor fans during the Late Jurassic-Early Cretaceous of the South Viking graben.

Chapter 21

Smørbukk Field
A Gas Condensate Fault Trap in the Haltenbanken Province, Offshore Mid-Norway

S. N. Ehrenberg
H. M. Gjerstad
F. Hadler-Jacobsen

Statoil
Stavanger, Norway

ABSTRACT

Smørbukk field, discovered in 1984, is situated approximately 230 km west of the Norwegian mainland where the water depth is 250 to 300 m. The field lies at the crest of a southeast-dipping fault block, bounded on the west by a major normal fault and on the north by an east–west-trending graben that transects the crest of the fault block. Hydrocarbons are contained in four Lower to Middle Jurassic sandstone formations deposited in tidally influenced nearshore and braid delta front environments. Multiple reservoir zones with various hydrocarbon contacts and gas-oil ratios are separated by transgressive shale units that form vertical pressure barriers. The fluids are mainly rich gas condensates (GOR, 1500 to 1800 standard m^3/standard m^3), but volatile oil (GOR, 470) is present in one zone.

Permeability varies as a function of facies-dependent primary sand quality and depth-dependent diagenetic alteration. The sandstones are heavily quartz cemented and extensively illitized, reflecting their present maximum burial depth of 3800 to 4400 m MSL. In the better zones, however, sufficient primary intergranular macroporosity survives to give core permeability measurements ranging from 10 to 1000 md.

Total in-place reserves are estimated to be 106 billion standard m^3 (3.7 tcf) of gas and 90 million standard m^3 (566 million bbl) of condensate/oil. However, only three wells have been drilled in the field, and further appraisal drilling is needed to confirm these estimates. Simulation studies have not yet been carried far enough to provide realistic estimates of total recoverable reserves.

INTRODUCTION

Smørbukk field (Figure 1) is located mainly in block 6506/12, approximately 230 km west of the mainland where the water depth is 250 to 300 m (Figure 2). Block 6506/12 was awarded in 1984 under PL (production license) 094, presently consisting of the following companies: Statoil (operator), 50%; Mobil Development Norway A/S, 15%; Norsk Agip A/S, 10%; Neste Petroleum A/S, 10%; Conoco Petroleum Norge A/S, 10%; and Norsk Hydro Produksjon A/S, 5%. The undrilled southern portion of the field, with an estimated 24% of the total in-place reserves, extends into block 6506/11, under PL 134, consisting of Statoil (operator), 50%; Norsk Agip A/S, 30%; Conoco Petroleum Norge A/S, 10%; and Enterprise Oil Norge Ltd., 10%.

Preliminary details regarding Smørbukk field can be found in Aasheim et al. (1986). The sources and compositions of hydrocarbons in Smørbukk and other Haltenbanken fields are discussed in Heum et al. (1986).

Smørbukk's in-place reserves are estimated to be 90 million standard m^3 of condensate/oil and 106 billion standard m^3 of gas. Estimates of the proportions of these volumes that are recoverable are not currently available.

Although this paper concerns only Smørbukk field, it is relevant also to provide a few facts concerning Smørbukk Sør field, a complexly faulted domal structure just southeast of Smørbukk (Figure 2). Here the hydrocarbons are mainly light oil, rather than gas condensate, and 88% of the reserves is reservoired in the Garn Formation, which contains only a minor proportion of the total reserves at Smørbukk field. Because the oil from Smørbukk Sør field will probably be produced much sooner than the gas condensate from Smørbukk field, production simulation has been carried farther for Smørbukk Sør, and estimates of recoverable reserves are therefore available. The reserves of Smørbukk Sør field are estimated to be 77 million standard m^3 of condensate/oil in place, of which 26 million m^3 (34%) is recoverable, and 38 billion standard m^3 of gas in place, of which 21 billion m^3 (55%) is recoverable. The oil would be produced by reinjection of associated gas, with production tentatively planned to begin in 1996 and continue through 2010.

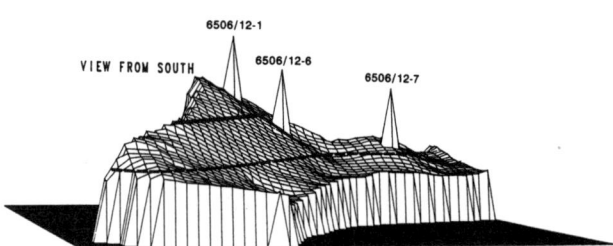

FIGURE 1. Perspective view of Smørbukk field at the top of the Garn horizon, showing well locations, gas-water contact in the Garn Formation (solid line), and projected position of oil-water contact in the underlying Tilje Formation (dashed line).

EXPLORATION AND DISCOVERY HISTORY

The first Haltenbanken exploration licenses were awarded in 1980 for blocks 6507/12 and 6507/11 following the Fifth Concession Round. In 1982, blocks 6407/1, 6407/2, and 6507/10 were also awarded. Prior to licensing of the Smørbukk block in 1984, eleven wells had been drilled in Haltenbanken, resulting in three discoveries: gas condensate in 6407/1-2 at South Tyrihans field, gas and oil in 6407/1-3 at North Tyrihans field (Aasheim and Larsen, 1984; Larsen et al., 1987), and gas in 6407/2-2 and 6507/11-1 at Midgard field (Ekern, 1987).

A seismic grid shot in 1979 and covering block 6506/12 with 2-km spacing was evaluated by Statoil in 1980 and reevaluated in 1983. In March 1984, following the Eighth Concession Round, block 6506/12 was awarded under PL 094.

The discovery well 6506/12-1 was spudded in August 1984 and was plugged and abandoned in February 1985, having proved gas condensate in five zones. Subsequently, three more wells have been drilled on the structure (Figure 3): 6506/12-4, drilled in August 1985, which showed hydrocarbons to be absent in the northern part of the Smørbukk structure; 6506/12-6, drilled in August 1986, immediately downdip from well 6506/12-1; and 6506/12-7, drilled in August 1987, which tested the eastern extent of the field. During the same period, wells 6506/12-3, 6505/12-5, 6406/3-3, and 6506/12-8 were drilled to discover and evaluate Smørbukk Sør field.

In 1984, a second seismic survey was shot, infilling the lines of the 1979 survey, and the previous data were reprocessed to match the new data, giving a combined grid with 1-km spacing. In 1988, a 3D seismic survey was shot over Smørbukk field, including 5696 km of 3D coverage with 500-m spacing plus 568 km of additional 2D coverage.

STRATIGRAPHY

The stratigraphy of Haltenbanken is summarized in Figure 4. Pre-Triassic strata have not been penetrated by any Haltenbanken wells, but thick sequences observed on seismic data in local fault-bounded basins probably consist mainly of Devonian to Carboniferous strata. The oldest strata that have been penetrated so far at Haltenbanken are a thick sequence of Middle to Upper Triassic age from block 6507/12, east of Smørbukk field (Figure 2). These strata can be traced seismically with little thickness variation throughout most of Haltenbanken. The section contains 220 m of continental shale and sandstone, overlain by two marine evaporite formations, each 400 m thick and separated by 500 m of dolomite- and anhydrite-rich shale. The upper evaporite is overlain by a mud-dominated lacustrine to fluvial red bed sequence 500 m thick that grades upward into 150 m of sandier "gray beds." This color change and the presence of overlying coal-rich strata indicate transition from arid to humid climatic conditions.

The oldest unit penetrated at Smørbukk field is the Upper Triassic to Lower Jurassic Åre Formation (Figure 5). None of the four Smørbukk wells reached the base of the

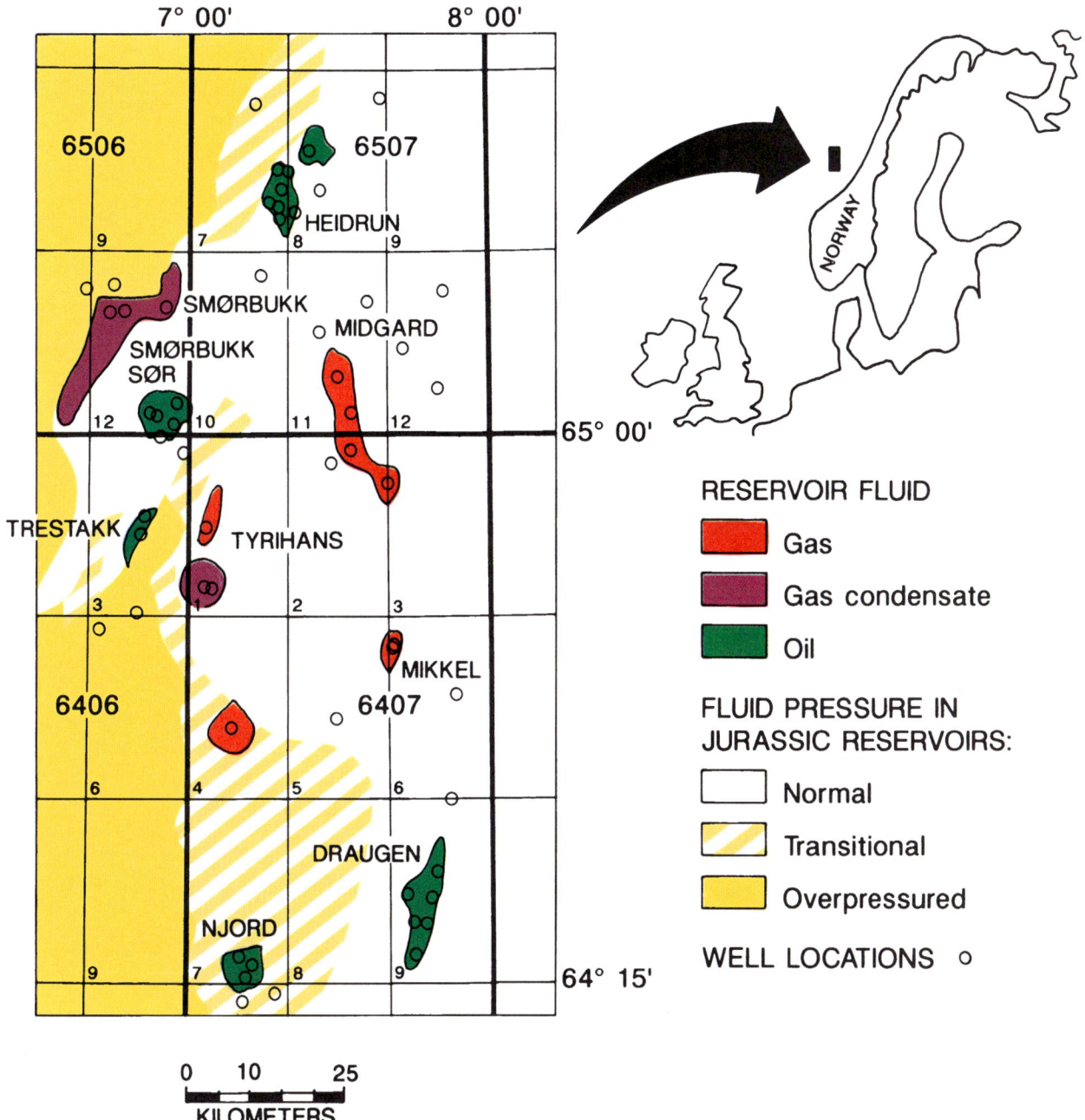

FIGURE 2. Map of Haltenbanken, showing license blocks, well locations, outlines of hydrocarbon discoveries, and areas where the Jurassic reservoir section is overpressured.

formation, but seismic data indicate a thickness of about 500 m. The Åre Formation consists of sandstone, shale, and coal, representing a delta plain environment. The coals are important source rocks (Heum et al., 1986). The sandstones are interpreted as point bar and crevasse splay deposits, commonly show fining-upward trends, and do not appear to be correlative between wells. In all four Smørbukk wells, a marine shale bed 12 m thick occurs at the top of the formation.

The overlying Tilje Formation varies from 119 to 151 m in thickness in the Smørbukk wells. It consists mainly of tidally influenced nearshore marine sandstone interlayered with thinner zones of offshore shale and bioturbated siltstone that tend to be readily correlative among all wells. In Smørbukk field, the Tilje Formation is divided into three units, labeled 1, 2, and 3 from bottom to top (Figure 6):

Tilje 1 is mainly burrowed, fine-grained sandstone (lower shoreface to offshore environment) with occasional medium-grained sandstone layers 20 to 60 cm thick (storm deposits).

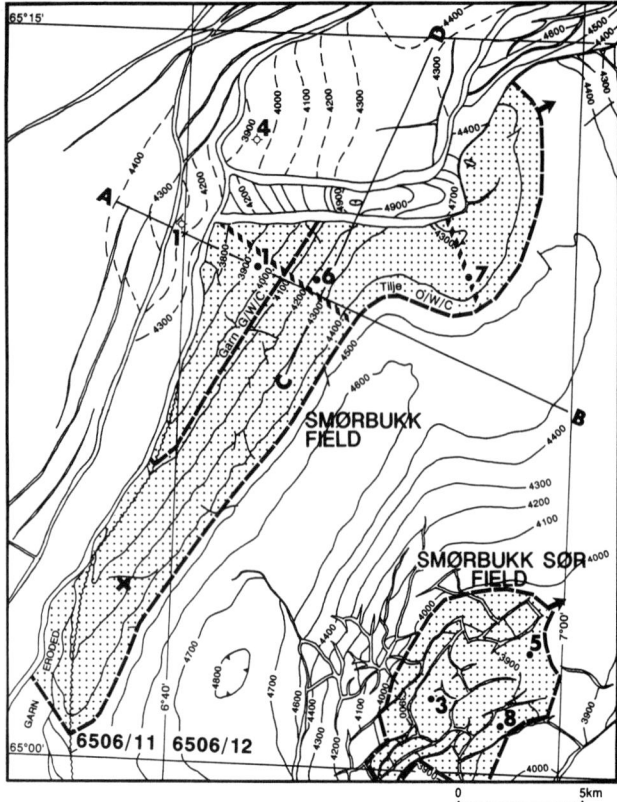

FIGURE 3. Structural contour map of the top of the Garn Formation in block 6506/12 (contours in meters MSL), showing well locations, hydrocarbon-water contacts, spillpoints (arrows), locations of seismic cross sections in Figure 9 (lines A-B and C-D), and locations of cross sections used for simulation of production from the Tilje Formation (diagonally dashed lines).

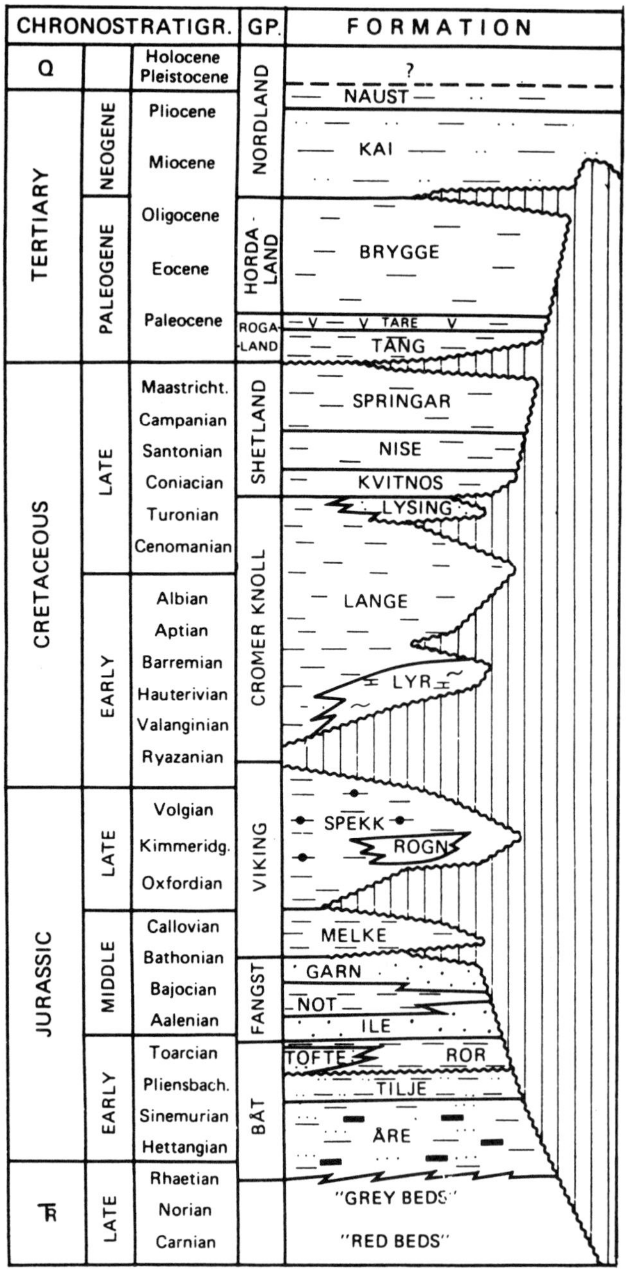

FIGURE 4. Generalized time and lithostratigraphic section extending from the Halten terrace (left) northward to the Nordland ridge area (right).

Tilje 2 is a classic regressive sequence beginning with a silty shale at the base (minor transgression), overlain by a coarsening-upward fine- to medium-grained sandstone (a tidal bar complex with minor tidal channels), and capped by a fine-grained sandstone and an organic-rich silty shale (overbank deposits in an interdistributary bay).

Tilje 3 is quite heterogeneous. The lower portion (subunits 3.1 and 3.2) consists of fine- to medium-grained sandstones (tidal bars and channels with subordinate storm layers) with intervening silty shales. Subunit 3.3 is a very fine-grained, micaceous sandstone with hummocky cross-stratification (interpreted as a storm bed). At the top of the Tilje Formation, subunit 3.4 is an upward-coarsening sequence of medium-grained, cross-bedded sandstones (delta front).

Major trangression of the entire Haltenbanken area resulted in deposition of the Ror Formation, an over-all upward-coarsening sequence of marine shales with storm-deposited sandstone beds in its upper portion. Around Smørbukk field and in areas farther north, however, Ror deposition was interrupted by the localized outpouring of a fan delta of coarse-grained, very poorly sorted, cross-bedded sands, forming an areally restricted unit named the Tofte Formation (0 to 84 m thick in the Smørbukk wells). This fan delta appears to have been deposited very rapidly into deep water in response to tectonic uplift of the Sklinna high immediately to the west. Ror shales occur both under and over the Tofte Formation. The thickness of the upper Ror unit varies from 53 to 73 m in the Smørbukk wells and from 0 to 21 m in the lower Ror unit, decreasing toward the north and west, nearer the source of the Tofte sandstones (Figure 6).

Following the Ror transgression, the culminating Middle Jurassic regression of the Haltenbanken area produced the Fangst Group, including the Ile, Not, and Garn formations. These units were deposited during the same major sea level lowstand as the Brent Group of the northern North Sea.

The Ile Formation is a series, 60 to 82 m thick, of tidally influenced nearshore marine sands and thinner bioturbated shale/siltstone intervals similar to the underlying Tilje Formation. In Smørbukk field, the Ile Formation is divided into three units, which are described as follows, from bottom to top (Figure 6):

Ile 1 is a coarsening-upward sequence of fine- to medium-grained sandstones with thinner shaly interlayers (interpreted as a tidal bar complex).

Ile 2 consists of generally finer-grained, more bioturbated sandstones with more interlayered shale/siltstone (interpreted as the more distal, lateral equivalent of Ile 1).

Ile 3 consists of medium-grained sandstones having an erosive contact on Ile 2 (interpreted as a subtidal channel complex), overlain by a fine-grained, extensively burrowed sandstone/siltstone unit (intertidal flat environment).

Marine transgression of the top of the Ile Formation is recorded in a conglomerate lag bed 10 cm thick.

The Not Formation is a marine shelf deposit 24 to 34 m thick that is similar to the Ror Formation. It consists mainly of shale at the base, grading upward into highly bioturbated siltstone and eventually into fine-grained sandstone at the top. The transition to the overlying Garn Formation is a sharp erosional contact.

The Garn Formation has a uniform thickness of 41 to 44 m in the Smørbukk wells. Over-all, this unit may be characterized as a fan delta or braid delta front deposit.

Garn 1 consists entirely of massive to cross-bedded, mainly medium-grained sandstone. The presence of minor mud laminations and herringbone cross-laminations suggests tidal reworking in a delta front setting, but much of the sand may also have been deposited by braided streams.

Garn 2 consists of bioturbated, clearly marine sandstone (recording minor transgression).

Garn 3 is an upward-coarsening, cross-bedded sandstone, also showing indications of tidal influence. A transgressive lag conglomerate is present at the top of the formation.

The section overlying the Garn Formation consists dominantly of shale (Figures 4 and 5). The Melke Formation, consisting of 117 to 282 m of shale and siltstone, is the equivalent of the Heather Formation of the northern North Sea. It was deposited in open marine conditions, and is moderately organic-rich (generally 1 to 4% TOC). The overlying Spekk Formation, which consists of 8 to 73 m of deep marine shale, is the equivalent of the Draupne

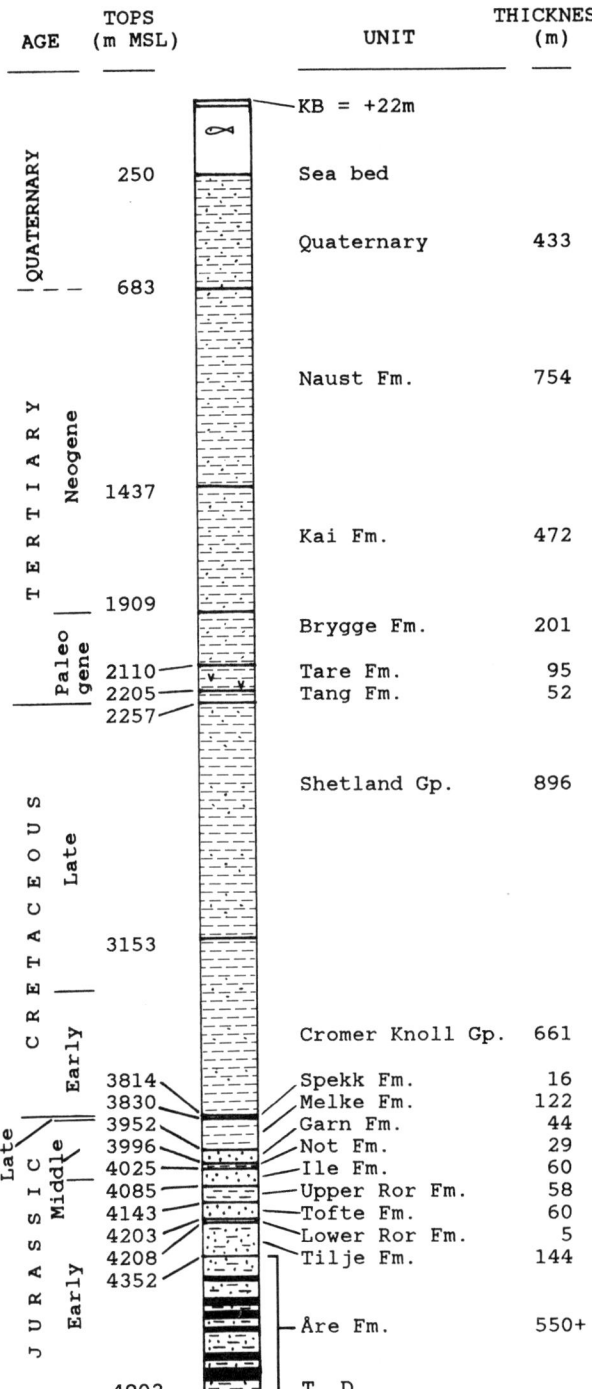

FIGURE 5. Stratigraphic column for the Smørbukk field discovery well, 6506/12-1.

In well 6506/12-4, the lower Ror unit is missing entirely, and the Tofte shows its most proximal development, as interpreted from the gamma ray log. To the south in well 6506/12-1, coarse, proximal sand still makes up most of the formation, but the upper and lower parts of the section consist of fine-grained, highly bioturbated sandstones. The proportion of this distal facies increases in well 6506/12-6. Farther east in well 6506/12-7, and farther south in Smørbukk Sør field, the Tofte Formation has shaled out.

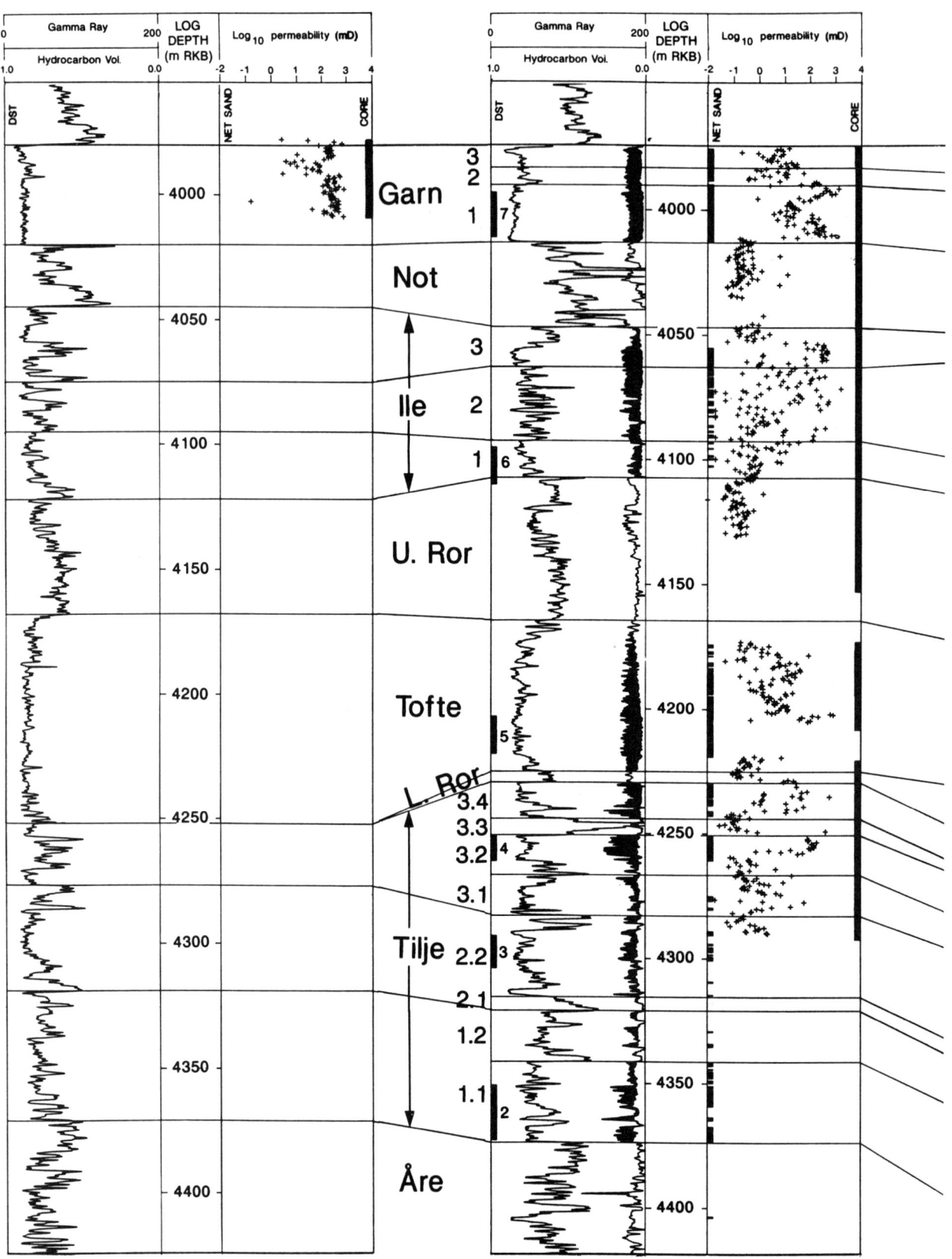

FIGURE 6. Correlation of reservoir zones between Smørbukk wells, showing core permeability measurements, net pay, hydrocarbon and water saturation volumes, and locations of drill-stem tests (figure plotted by A. Fylling).

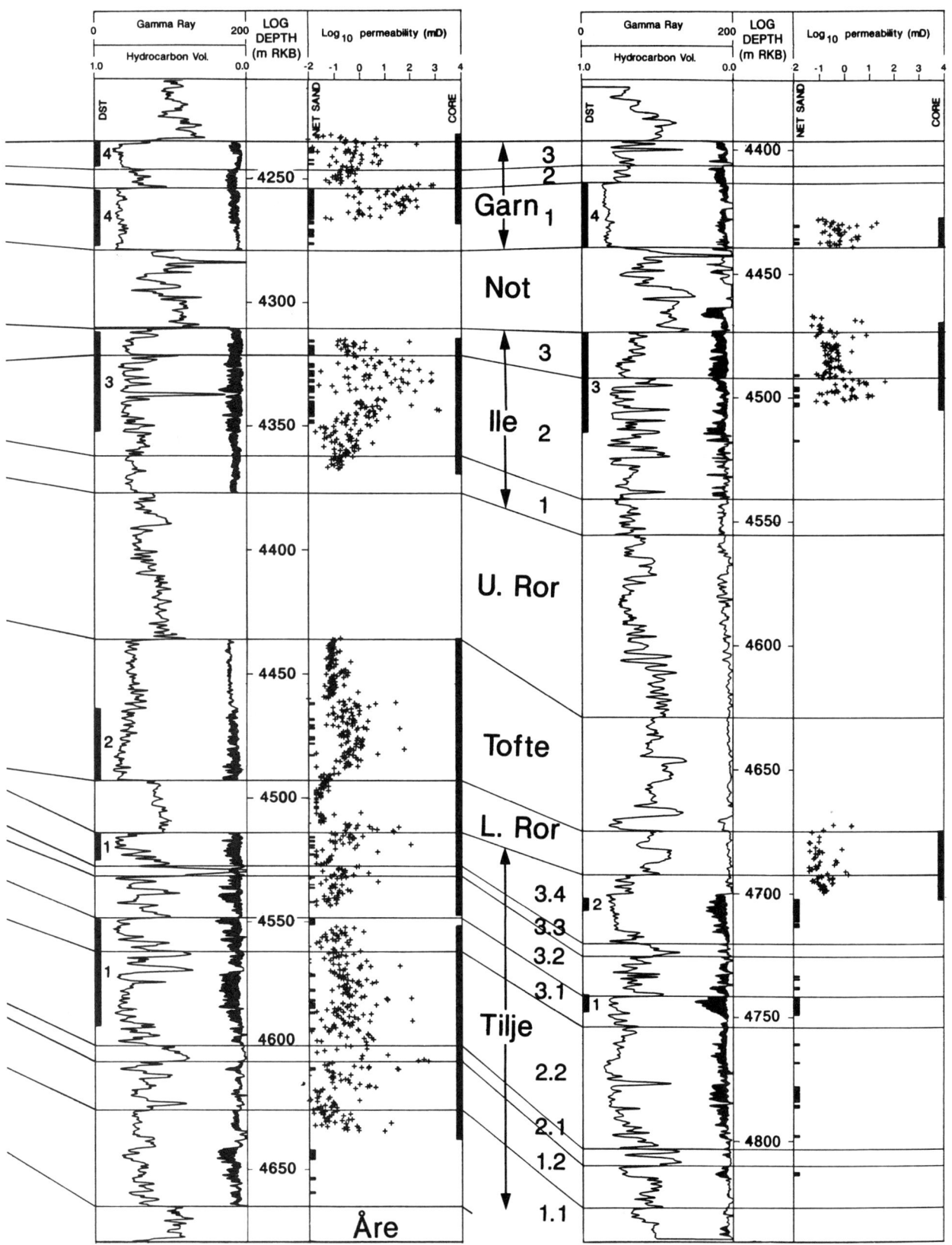

Shale of the northern North Sea and is correspondingly highly radioactive and an important source rock (generally 5 to 8% TOC). The Melke-Garn contact is an unconformity over many of the higher structures, and the Spekk Formation is bounded by regional unconformities. In the Smørbukk wells, the Cretaceous Cromer Knoll and Shetland groups, which are 661 to 777 and 869 to 922 m thick, respectively, consist mainly of shale with occasional turbiditic sandstones.

Tertiary deposition followed a regional Upper Cretaceous unconformity. The Tang Formation, consisting of 40 to 100 m of marine shale, is overlain by the tuffaceous Tare Formation, 70 to 106 m thick, which is equivalent to the Balder Formation of the northern North Sea. The overlying Brygge Formation consists of 168 to 372 m of marine shale with minor amounts of sandstone and limestone. Following a regional mid-Oligocene unconformity, the Kai Formation was deposited, consisting of 317 to 472 m of westward-prograding marine shale and subordinate sandstone and limestone. Finally, beginning in the late Pliocene with the Naust Formation and continuing through the Quaternary, all of Haltenbanken was covered by a thick sequence of alternating gray shale and poorly sorted sand, which is 1162 to 1243 m thick in the Smørbukk wells.

TECTONIC SETTING

The main tectonic elements of the mid-Norwegian shelf are illustrated with reference to the thickness of post-Jurassic sediments in Figure 7. The cross section in Figure 8 shows the general structural configuration in the Haltenbanken area. The Trøndelag platform is separated from the deep Vøring and Møre basins to the west by a wide zone of predominantly down-to-the-west normal faulting. In the Haltenbanken area, this zone broadens to its widest extent of about 80 km and is referred to as the Halten terrace. To the north the Halten terrace passes into the narrower Dønna terrace, and to the south the displacement is concentrated within the narrow Klakk fault complex. The Halten terrace occupies an intermediate structural level between major down-to-the-west fault zones forming its western and eastern limits—the northern extension of the Klakk fault complex on the west, and the Vingleia and Bremstein fault complexes on the east. The fault zones tend to be associated with major footwall uplifts on their eastern sides—the Nordland ridge along the Revfallet fault complex, and the Frøya and Sklinna highs along the Klakk fault complex.

The tectonic history of Haltenbanken can be summarized by the following chronological sequence of events:

1. Probable Early Permian rifting is recorded by basement blocks trending north-northwest and onlapped by Triassic seismic reflectors. Regional thermal subsidence followed in the Late Permian to Early Triassic, with deposition of locally thick strata (up to 2 sec TWT).

2. From the Middle Triassic to the earliest Jurassic, mainly continental strata were deposited over both the Halten terrace and the Trøndelag platform, including two evaporite formations, red beds, and overlying coaliferous delta plain clastics (Åre Formation). This period was tectonically quiescent, except for minor uplift and faulting along the Nordland ridge and the Føya high. Positive relief is indicated by the observation that the evaporites apparently consist of anhydrite over these features and halite over the rest of Haltenbanken.

3. In the Early Jurassic, a relatively minor tectonic phase resulted in down-to-the-west growth faulting, oriented in a north-northeast direction, throughout the Halten terrace. At this time, progradation of a delta front sand complex (the Tilje Formation) was followed by regional transgression and deposition of marine shale (the Ror Formation). During deposition of the Ror Formation, uplift in the vicinity of the Sklinna high shed a local apron of coarse sand (the Tofte Formation) southwestward into the Smørbukk area and also farther to the north over part of the Dønna terrace.

4. Middle Jurassic regression produced a major coastal sand complex (the Fangst Group) broadly equivalent to the Brent Group of the northern North Sea. This was an interval of nearly total tectonic quiescence.

5. A second major transgression, beginning toward the end of the Middle Jurassic and continuing into the Early Cretaceous, produced marine shale with moderate organic content (the Melke Formation), followed unconformably by a rich source rock (the Spekk Formation, equivalent to the Draupne Formation of the northern North Sea). Intense northwest–southeast- to east–west-oriented regional extension began with the earliest Melke deposition and continued through the Early Cretaceous. This episode created the tilted fault blocks and horsts that contain most of the hydrocarbon accumulations of Haltenbanken, including Smørbukk field. On the Halten terrace, most of these structures appear to have formed by detachment into the two Triassic evaporite layers. However, a few of the larger faults, possibly including the western Smørbukk boundary fault, are high-angle basement faults.

6. During the Cretaceous, thick marine shales with thin carbonate and sand beds were deposited over the Halten terrace. These units thin to drastically condensed sequences over the Trøndelag platform. Through the Tertiary, thick marine shale units accumulated over both the Halten terrace and the Trøndelag platform.

7. From the late Pliocene (approximately 3.0 Ma) through the Quaternary, the Halten terrace and Trøndelag platform underwent rapid subsidence and progradation from the southeast of about 1 km of poorly sorted, glaciomarine clastics (Naust Formation and overlying Pleistocene/Quaternary). This depositional episode coincided with the creation of western Norway's spectacular scenery by isostatic uplift of 1 to

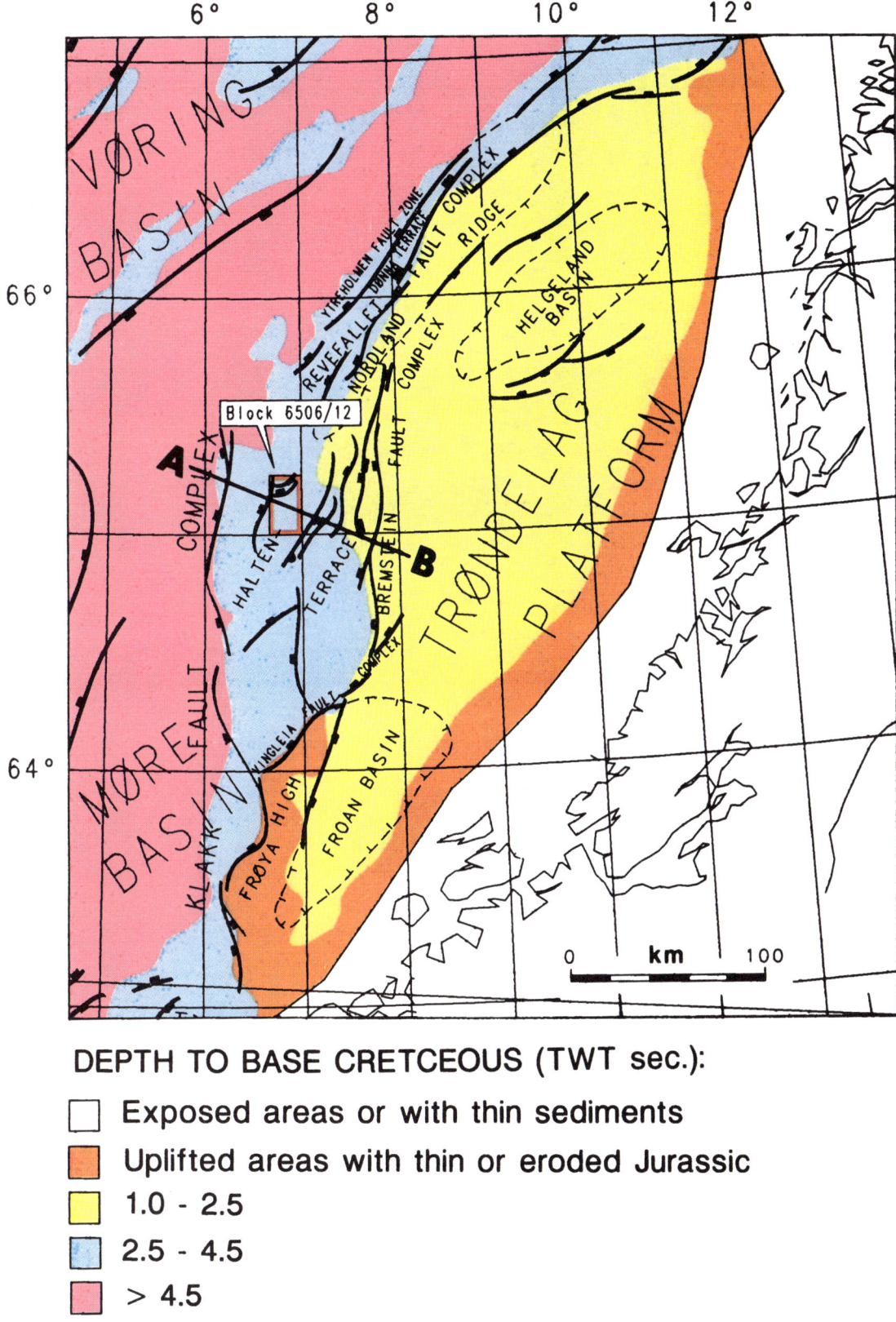

DEPTH TO BASE CRETCEOUS (TWT sec.):

- ☐ Exposed areas or with thin sediments
- ■ Uplifted areas with thin or eroded Jurassic
- ■ 1.0 - 2.5
- ■ 2.5 - 4.5
- ■ > 4.5

FIGURE 7. Map showing principal tectonic elements of mid-Norwegian shelf and thickness of post-Jurassic strata. Line A-B shows the location of cross section in Figure 8.

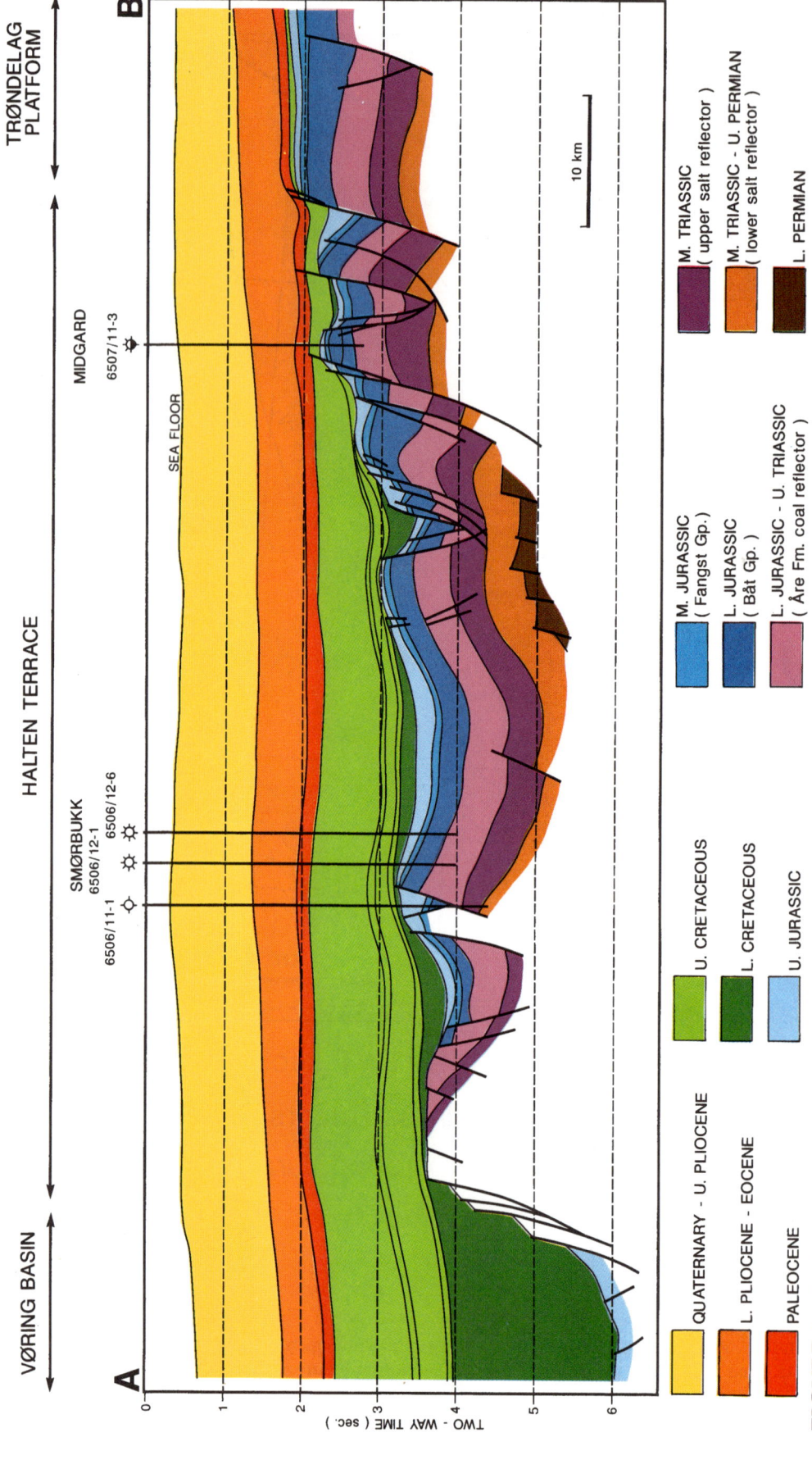

FIGURE 8. Regional east-west structural cross section (A-B) of Haltenbanken area. Location of section is shown in Figure 7.

2 km, a phenomenon that is at present only poorly understood. Uplift of the mainland also resulted in westward tilting of the Trøndelag platform and truncation of the entire post-Paleozoic section along its eastern margin.

This late depositional episode resulted in drastic deepening and heating of the underlying strata throughout Haltenbanken, with consequent acceleration of hydrocarbon generation and migration. As a result, most of the hydrocarbons filling Haltenbanken fields were generated after 3.0 Ma, although modeling indicates that modest amounts began accumulating in several structures, including Smørbukk, as early as the Late Cretaceous.

Another important effect of the late Pliocene–Quaternary deposition was the generation of overpressure in the Jurassic reservoir section throughout the western part of the Halten terrace. In the eastern part of Haltenbanken, overpressures were generated in the thick Cretaceous shale sequence, whereas pressures within the underlying Jurassic sand sequence were relieved by communication through shallower sandy units farther east. Here the overpressured shale sequence provides an effective seal to hydrocarbons in the underlying normally pressured sandstones. In the western part of Haltenbanken, however, faulting has sealed the Jurassic sand sequence from pressure communication to the west, resulting in overpressuring and consequent loss of hydrocarbons from many otherwise prospective structures due to fracturing of the overlying shale. The line separating the eastern province of normally pressured Jurassic reservoirs from the western overpressured province coincides with the western and northern fault boundaries of Smørbukk field (Figure 2).

STRUCTURE

Smørbukk field lies at the crest of a southeast-dipping fault block, bounded on the west by a major normal fault and on the north by an east–west-trending graben (Figure 3). The western boundary fault is tentatively interpreted as a high-angle basement fault (Figure 9, line A-B). Throw on this fault diminishes northward, being transferred to the Revfallet fault zone to the northeast. The displacement partitioning between the Revfallet fault zone and the Smørbukk boundary fault causes a warp or ramp between them. The graben forming the northern boundary of Smørbukk field appears to be a result of extension in the postbasement cover caused by warping of this ramp (Figure 9, line C-D).

Hydrocarbons are trapped in multiple zones of mainly regressive sandstone bodies separated by transgressive shale units that form vertical pressure barriers. Gas condensate is trapped in the uppermost reservoir unit, the Garn Formation, by the overlying marine shale of the Melke Formation. To the northwest, the trap is formed by the western boundary fault, which brings the reservoir zones in contact with Upper Jurassic to Lower Cretaceous shales. To the north, the reservoirs are faulted against Lower Cretaceous shales filling the east-west crestal graben.

Closure on the top of the Garn horizon measures 700 m vertically and covers an area of 120 km^2 (6 km wide by 30 km long). For all reservoir intervals, the structural spillpoints are located at the northeasternmost limit of the field (Figure 3).

The Smørbukk fault block continues north of the crestal graben, but well 6506/12-4 revealed that the reservoirs are highly overpressured here and do not contain hydrocarbons. The reservoirs within Smørbukk field are of normal pressure, and so it is presumed that the overpressure in the northern block is to blame for the absence of hydrocarbons. A consistent correlation between the occurrence of hard overpressure and a lack of hydrocarbon accumulations has been noted in structures throughout the western part of the Halten terrace. Modeling studies (Ungerer et al., 1987) indicate that the lack of hydrocarbons in the northern structure results from rupturing of the overlying shale section caused by overpressure in the reservoir section.

The 1988 3D seismic survey of Smørbukk field showed that the reservoir intervals within the hydrocarbon accumulation are not cut by any faults that can be expected to form barriers to production. Several minor faults were mapped, but these faults generally have about 10 msec of throw and are of limited extent in map views.

RESERVOIRS

The reservoir section has been divided into 18 zones, ten of which contain potentially economical hydrocarbons. Figure 5 shows the correlation of these zones among the four wells. Figure 10 shows the reservoir zones in an east-west structural cross section, with the interpreted hydrocarbon distributions that have been used for calculating proven, probable, and maximum (possible) reserves, respectively.

Results of drill-stem tests in wells 1, 6, and 7 are summarized in Table 1. In several cases (particularly the tests in the Garn and Ile formations in wells 1 and 6), calculated test permeabilities are dramatically lower than arithmetic averages of core measurements over the same intervals. In the Garn test in well 1, this discrepancy appears to have resulted at least partly from retrograde condensation caused by pressure drawdown well below the dew point in a zone extending about 30 m out from the well bore. In general, however, it is suspected that core permeabilities may have been artificially enhanced by the effect of air drying on fibrous, pore-bridging illite.

Figure 11 summarizes interpretations of pressure communication vertically between zones and laterally between wells, based on RFT and DST pressure measurements. For all zones, a lateral barrier is indicated between the western part of the field (including wells 6506/12-1 and 6506/12-6) and well 6506/12-7 to the east. No faults capable of accounting for such a barrier could be identified on the 1988 3D seismic survey, and so it has been tentatively suggested that the barrier could have resulted from fractures or minor faults along which permeability has been reduced by carbonate or quartz cementation. Because of this barrier, reserve estimates have been made separately for the western and eastern parts of the field, assuming that the barrier is a vertical zone of zero thick-

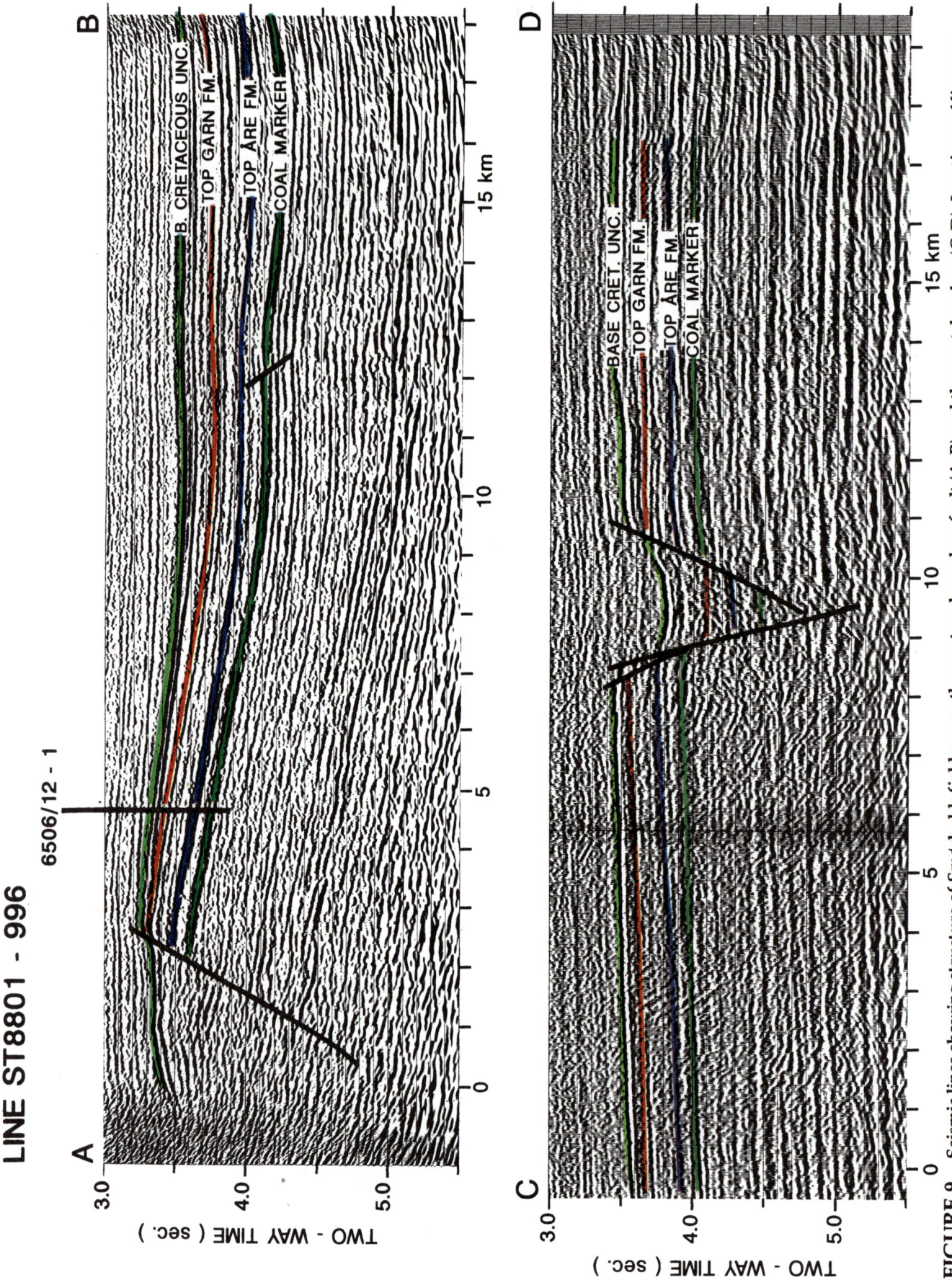

FIGURE 9. Seismic lines showing structure of Smørbukk field across the western boundary fault (A-B) and the crestal graben (C-D). Locations of lines are shown in Figure 3.

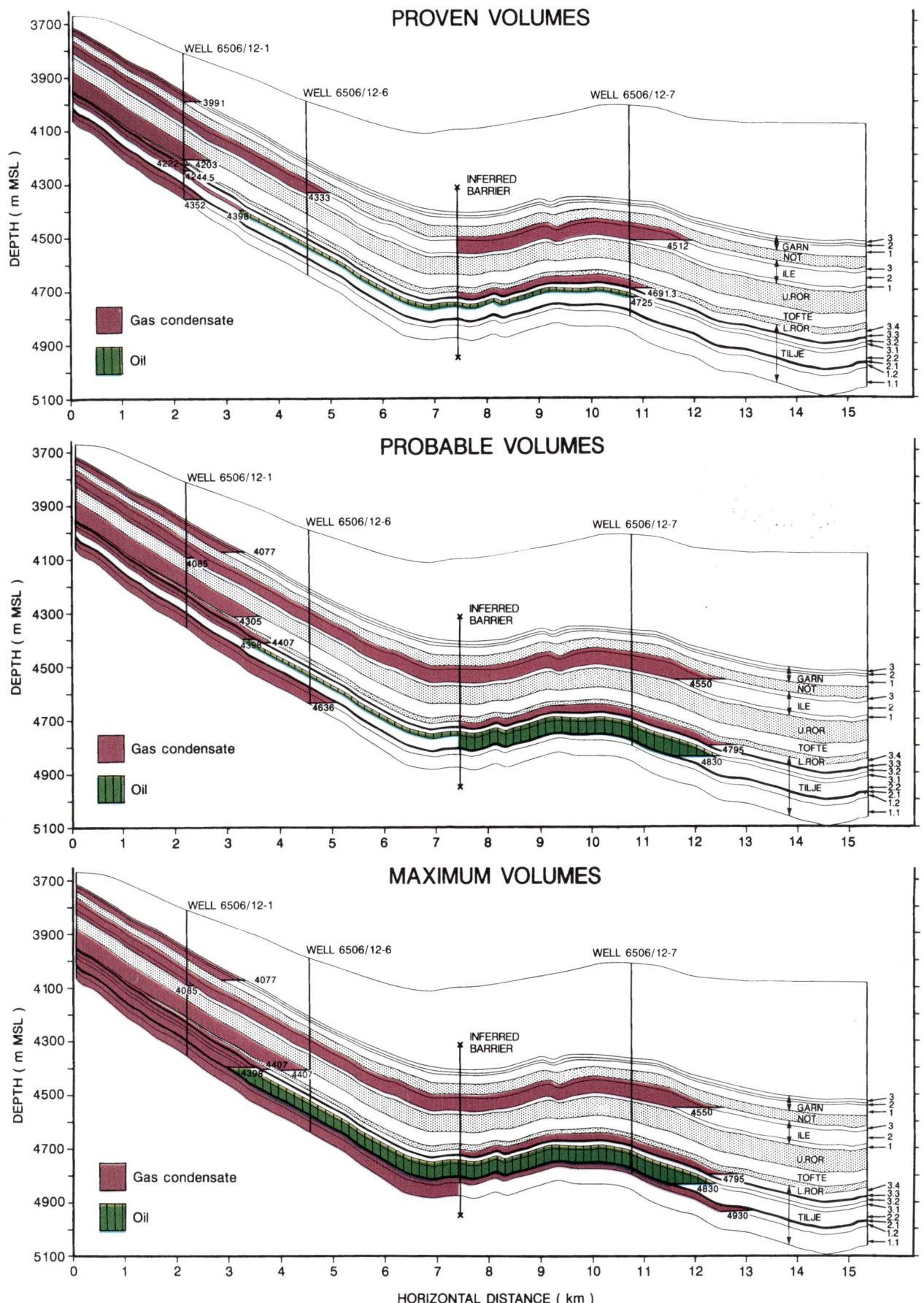

FIGURE 10. East-west structural cross section of Smørbukk field, showing extent of proven, probable, and maximum hydrocarbon volumes.

TABLE 1. Summary of drill-stem test results for Smørbukk field.

Zone	Well	DST	Fluid type	Choke Diameter (mm)	Rate (std. m³/d) Gas	Rate (std. m³/d) Condensate	Test permeability (md)	Core permeability (md) Arithmetic	Core permeability (md) Geometric	Comment number
Garn	12-1	7	Gas condensate	32	727,000	511	14	112	26	1
	12-6	4	Water	—	—	—	7	57	8	
	12-7	4	No flow	—	—	—	0.0	2	1	
Ile	12-1	6	No flow	—	—	—	0.2	145	22	2
	12-6	3	Gas condensate	21	596,000	322	5	88	4	
	12-7	3	Gas condensate	159	110,000	70	2	6	3	
Tofte	12-1	5	Gas condensate	25	584,000	377	9	—	—	3
	12-6	2	No flow	—	—	—	—	3	1	
Tilje 3.4	12-6	1	Water	—	—	—	—	—	—	
	12-7	2	Gas condensate	79	145,000	200	7	—	—	
Tilje 3.2	12-1	4	Gas condensate	24	603,000	620	298	70	19	
Tilje 3.1	12-6	1	Oil	10	58,000	142	6	10	1	4
	12-7	1	Oil	127	230,000	520	385	—	—	
Tilje 2.2	12-1	3	No flow	—	—	—	0.2	—	—	5
Tilje 1.1	12-1	2	Gas condensate	25	692,000	517	40	—	—	

Comments:
1. Low test permeability compared with core permeability at least partly reflects retrograde condensation near the well as pressure was drawn down below the dew point. Core permeabilities include only plugs within the interval thought to have contributed to test flow.
2. Test interval did not include the much better-quality upper Ile zones 2 and 3.
3. Core permeability measurements are available for nearly all of the Tofte Formation except the tested interval (arithmetic average, 21 md; geometric average, 1.6 md).
4. Production is interpreted as having come only from the upper 3 m in the lower of the two perforated intervals (top of Tilje 3.1). The upper perforated interval (Tilje 3.4) produced only water.
5. Reservoir quality appears to be much better in the untested lower part of the Tilje 2.2 zone in well 1.

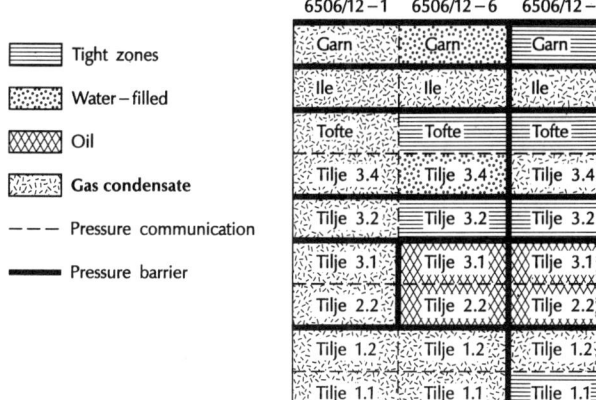

FIGURE 11. Schematic representation of interpretations regarding pressure communication both vertically between different reservoir zones and laterally between different well locations in Smørbukk field.

ness oriented as shown in Figure 10. This is an obvious oversimplification, but was chosen as the simplest configuration consistent with the available information.

DIAGENESIS AND RESERVOIR QUALITY

The relationships described here are based partly on a regional study of diagenesis and reservoir quality in the Garn Formation reported by Ehrenberg (1990) and Ehrenberg and Nadeau (1989). This regional study involved cores from 16 wells, including the Smørbukk wells 6506/12-1 and 6506/12-4. The conclusions for the Garn Formation also provide a basis for understanding the best-quality sandstone intervals within the Ile, Tofte, and Tilje formations, but need to be modified to take into account the generally poorer primary sand quality in these units, including over-all finer grain size and higher detrital clay content in the Ile and Tilje formations, and very poor sorting and abundant detrital clay in the Tofte Formation. Another difference is that preservation of anomalously high porosity by chlorite is observed within certain zones in the Tilje Formation, but not in the Garn Formation.

The discussion in this section is illustrated by photomicrographs of representative thin sections from the various Smørbukk reservoirs (Figures 12 and 13). Figure 12A and B show the contrast between typical samples from the Garn Formation in well 6506/12-1, where the unit has rather good permeability, and in well 6506/12-7, where the unit is tight. Figure 12C and D illustrate the range of variation in reservoir quality within the Garn Formation in well 6506/12-6. Figure 12E and F show the analogous range in reservoir quality within the Ile Formation in well 6506/12-6. Figure 13A and B show typical samples from the Tofte Formation in well 6506/12-1, where moderate permeability exists, and in well 6506/12-6, where the unit is tight. Figure 13C and D show the contrast between a porous, chlorite-coated sandstone (C) and a tight, chlorite-poor sandstone (D) in the Tilje Formation in well 6506/12-1. Figure 13E and F show a similar pair of samples from the Tilje Formation in well 6506/12-6.

Permeability in the Smørbukk reservoirs varies mainly as a function of facies-dependent primary sand quality (sand texture and composition at the time of deposition), but has also been strongly affected by diagenesis. Thus, the wide fluctuations in permeability over short vertical distances, which are especially apparent in the Tilje and Ile formations, correlate with varying grain size and detrital clay matrix content, reflecting alternating depositional facies. The more uniform permeability profiles of the lower part of the Garn Formation reflect more constant depositional conditions. The significance of variations in primary sand quality can be appreciated visually by comparing Figure 12C and D, which show good- and poor-quality sandstones, respectively, from the Garn Formation in well 6506/12-6. Figure 12E and F show the same type of comparison for two samples from the Ile Formation in well 6506/12-6.

In the Garn Formation in wells 6506/12-1 and 6506/12-4, porosity and permeability have been reduced by quartz cementation (filling about 16 vol. % of the rock). Quartz cement is believed to have been supplied by development of stylolites along primary clay laminations throughout the formation. Permeability has been further reduced by luxuriant growth of authigenic illite, a phenomenon that displays regional correlation with a burial depth of 3.7 km below the sea floor—the approximate depth to the top of the Smørbukk structure at the top of the Garn horizon. Illite growth consumed both earlier authigenic kaolinite and detrital K-feldspar, resulting in conspicuous secondary porosity within the molds of dissolved feldspar grains (Figure 12A, B, and D). Molds of dissolved feldspars make up about 5% of the total rock volume in the Garn Formation in wells 6506/12-1 and 6506/12-4.

Permeability variations in the Garn Formation can be understood by using modal analyses of thin sections to classify the pore system into effective and ineffective types of porosity. The larger pores between grains (the intergranular macroporosity) are thought to represent the potentially well-interconnected portion of the pore system accessible to the type of fluid flow measured in a permeability test. When measured by point counting of thin sections, "macroporosity" is defined as pores with cross sections larger than approximately 20 μm (areas of blue epoxy entirely free of any mineral matter). Macropores within molds of dissolved feldspars (the intragranular macroporosity) are characteristically much larger than the surrounding intergranular pores and commonly contain, or are surrounded by, abundant microporous materials (authigenic clay and skeletal remnants of dissolved feldspar) that tend to cut off the secondary macropores from the intergranular macropore system. It thus appears questionable whether these secondary pores are available to fluid flow and whether their hydrocarbon content is displaceable under normal production operations. Microporosity can be estimated by subtracting the point-counted macroporosity from the helium porosity measured on the plug from which the thin section was cut. The microporosity is assumed to be occupied mainly by residual water saturation. "Effective porosity" is therefore thought to be approximately equivalent to the point-counted intergranular macroporosity, whereas the sum of the intragranular macroporosity and the microporosity gives a measure of the "ineffective porosity."

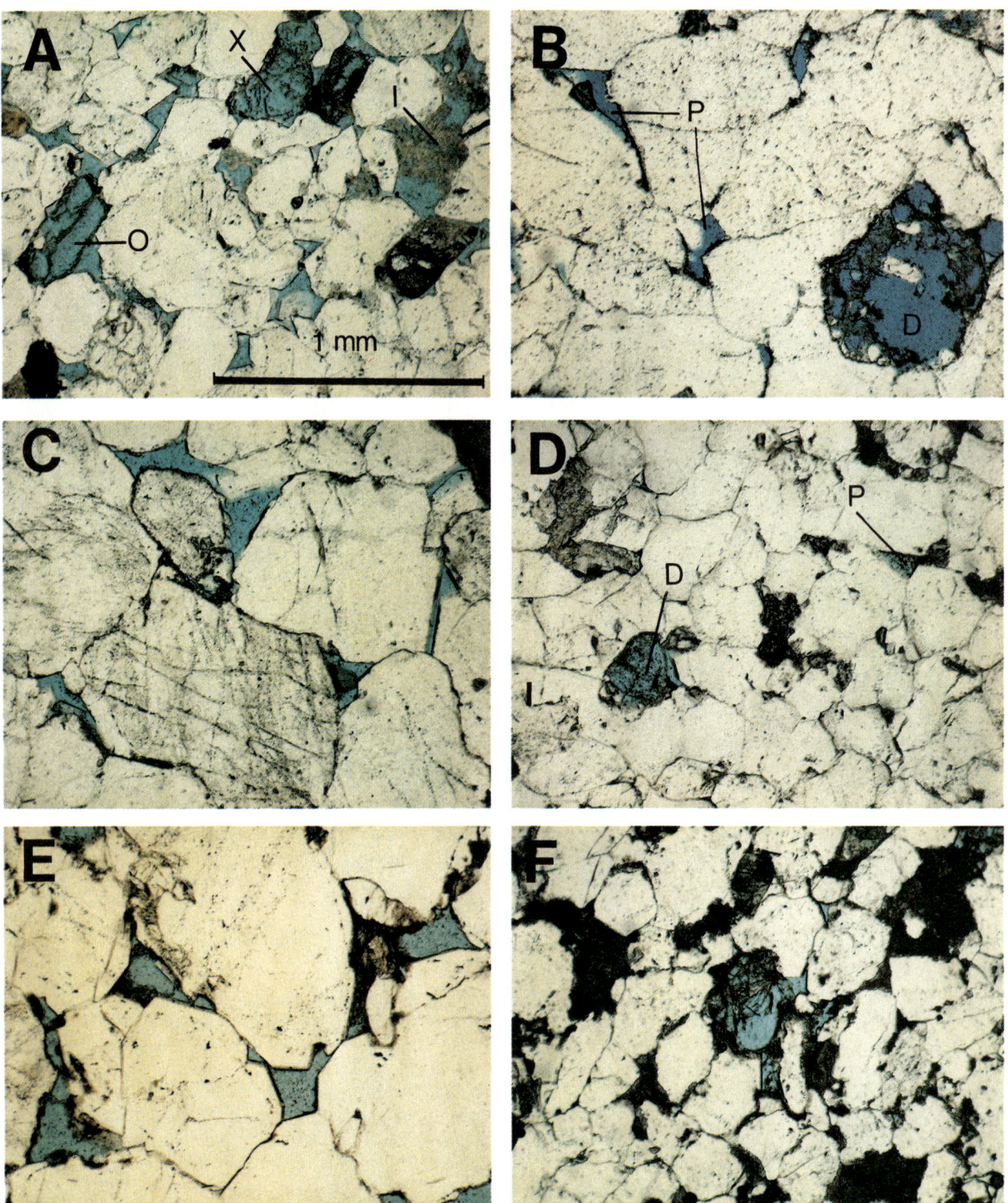

FIGURE 12. Photomicrographs of thin sections from Garn and Ile formations. All views have the same magnification (see micrograph A). Log depths are given to facilitate comparison with permeability-depth plots in Figure 6.)

(A) Garn Formation, well 6506/12-1: core depth, 3982 m RKB (log depth, 3984); porosity, 13.7%; permeability, 22 md. Abundant intergranular macroporosity (blue) appears to be preserved primary porosity that has been much reduced by compaction and quartz cementation. Clusters of authigenic illite ("I") clog many intergranular pores. Conspicuous secondary porosity is developed in molds of dissolved potassium feldspars. In some cases, the secondary porosity consists mainly of macroporosity ("O"), but elsewhere it is mainly microporosity ("X").

(B) Garn Formation, well 6506/12-7: core depth, 4436 m RKB (log depth, 4436); porosity, 10.1%; permeability, 2.2 md. Compared with the rock shown in A, this typical sample from well 7 has much less primary intergranular macroporosity ("P"),

Figure 14 shows how the relative abundances of porosity types in the Garn Formation vary between well 6506/12-1 and other locations having higher and lower average core permeabilities. Regionally, the Garn Formation shows relatively limited variation in primary sand quality, and so variations in average permeability from well to well are thought mostly to reflect differences in diagenetic grade resulting from different degrees of thermal exposure. Figure 14 illustrates how the regional trend of decreasing permeability with increasing thermal maturity is related to decreasing total porosity and effective porosity.

The Garn Formation in well 6506/12-4 is not represented in Figure 14, but has a composition nearly the same as that in well 6506/12-1, reflecting nearly identical depositional, burial, and thermal histories in the two locations. In wells 6506/12-1 and 6506/12-4, quartz cementation and illitization have not completely occluded the pore system, and the rock still contains about 5% intergranular macroporosity. Reasonably good connection of this macroporosity is attested to by the average core permeabilities (based on all routine core analyses) of 108 md in well 6506/12-1 and 216 md in well 6506/12-4. Nevertheless, about 60% of the pore system is comprised of ineffective porosity (microporosity and macroporosity within molds of dissolved feldspars).

The Garn Formation in well 6506/12-3 in Smørbukk Sør field provides an example with higher permeability. The main difference between wells 6506/12-1 and 6506/12-3 is that 6506/12-3 has higher porosity resulting from less quartz cement, although effective porosity accounts for a higher proportion of the total porosity. The Smørbukk and Smørbukk Sør examples may be further compared with a core from a much shallower depth (well 6407/6-3 from Mikkel field) where the pore system, except for mechanical compaction, has been little altered by diagenesis.

An example of very low permeability is found in well 6407/4-1 (average 0.5 md). Here the average total porosity is only a few percent lower than in well 6505/12-1, but intergranular macroporosity has been nearly eliminated. Although the burial depths in wells 6506/12-3, 6506/12-1, and 6407/4-1 are nearly the same, decreasing reservoir quality correlates with a progressive increase in thermal maturity, as measured by the equivalent vitrinite reflectance at the top of the Garn Formation. Another factor could be inhibition of diagenesis by oil in well 6506/12-3, but wells 6506/12-1 and 6407/4-1 both contain gas condensate, which modeling studies indicate to have been emplaced during the early Tertiary—long before illitization and quartz cementation are thought to have occurred. Thermal maturity is expected to increase with depth in the Smørbukk structure downflank from well 6506/12-1, and so the situation in well 6407/4-1 gives an idea of the general tendency to be expected toward wells 6506/12-6 and 6506/12-7 and in the deeper, underlying sandstone reservoirs. Average core analyses in the Garn Formation give permeability values of 30 md at 4249 m relative to kelly bushing (RKB) in well 6506/12-6 and 1 md at 4433 m RKB in well 6506/12-7, which are consistent with this predicted tendency.

Figure 15 shows plots of depth vs. all routine core analyses from wells 6506/12-1, -6, and -7. Although there is considerable scatter, reflecting variation in primary sand quality within each unit, an over-all decrease in reservoir quality with increasing depth is apparent. The main cause of this decrease is thought to be increasing infilling of intergranular pores by quartz cement with higher temperature and thermal maturity, as supported by the regional results reported by Ehrenberg (1990). Two additional factors that could have contributed to the decrease are: (1) slightly more distal facies development and consequently poorer over-all primary sand quality in the direction of depth increase, from well 1 to well 6 to well 7; and (2)

---

but retains abundant secondary porosity within dissolved grains ("D"). Comparison between A and B gives an idea of the reasons for the over-all poorer reservoir quality in well 6506/12-7 than in well 6506/12-1, both in the Garn Formation and generally in other units as well.

(C) Garn Formation, well 6506/12-6: core depth, 4260 m RKB (log depth, 4262); porosity, 10.0%; permeability, 40 md. This sample is from Garn zone 1. Comparison with photomicrograph D shows how the better reservoir quality of zone 1 is related to better primary sand quality (coarser grain size and less detrital clay) and consequently a better degree of preservation of primary intergranular macroporosity.

(D) Garn Formation, well 6506/12-6: core depth, 4241 m RKB (log depth, 4243); porosity, 6.6%; permeability, 0.1 md. This tightly quartz-cemented, relatively fine-grained sandstone is typical of the more distal upper portion of the Garn Formation (Garn zones 2 and, to a lesser extent, 3). Content of detrital clay and prevalence of bioturbation also tend to be greater than in zone 1, represented by the examples in A, B, and C. There is nearly no primary intergranular macroporosity ("P"); most macroporosity is contained within molds of dissolved grains ("D").

(E) Ile Formation, well 6506/12-6: core depth, 4332 m RKB (log depth, 4335); porosity, 16.9%; permeability, 646 md. Many sandstones from the Ile Formation have relatively high permeabilities due to excellent primary sand quality. However, thin interlayering with finer-grained, shaly sandstones is characteristic and may be a problem for lateral continuity.

(F) Ile Formation, well 6506/12-6: core depth, 4360 m RKB (log depth, 4363); porosity, 11.3%; permeability, 0.6 md. This sample is typical of the relatively impermeable lower part of Ile zone 2. Poor quality reflects tight quartz cementation and abundant detrital clay matrix (dark areas). Similar sandstones are also thinly interlayered with better rock, such as in E, throughout the more permeable upper part of Ile zone 2.

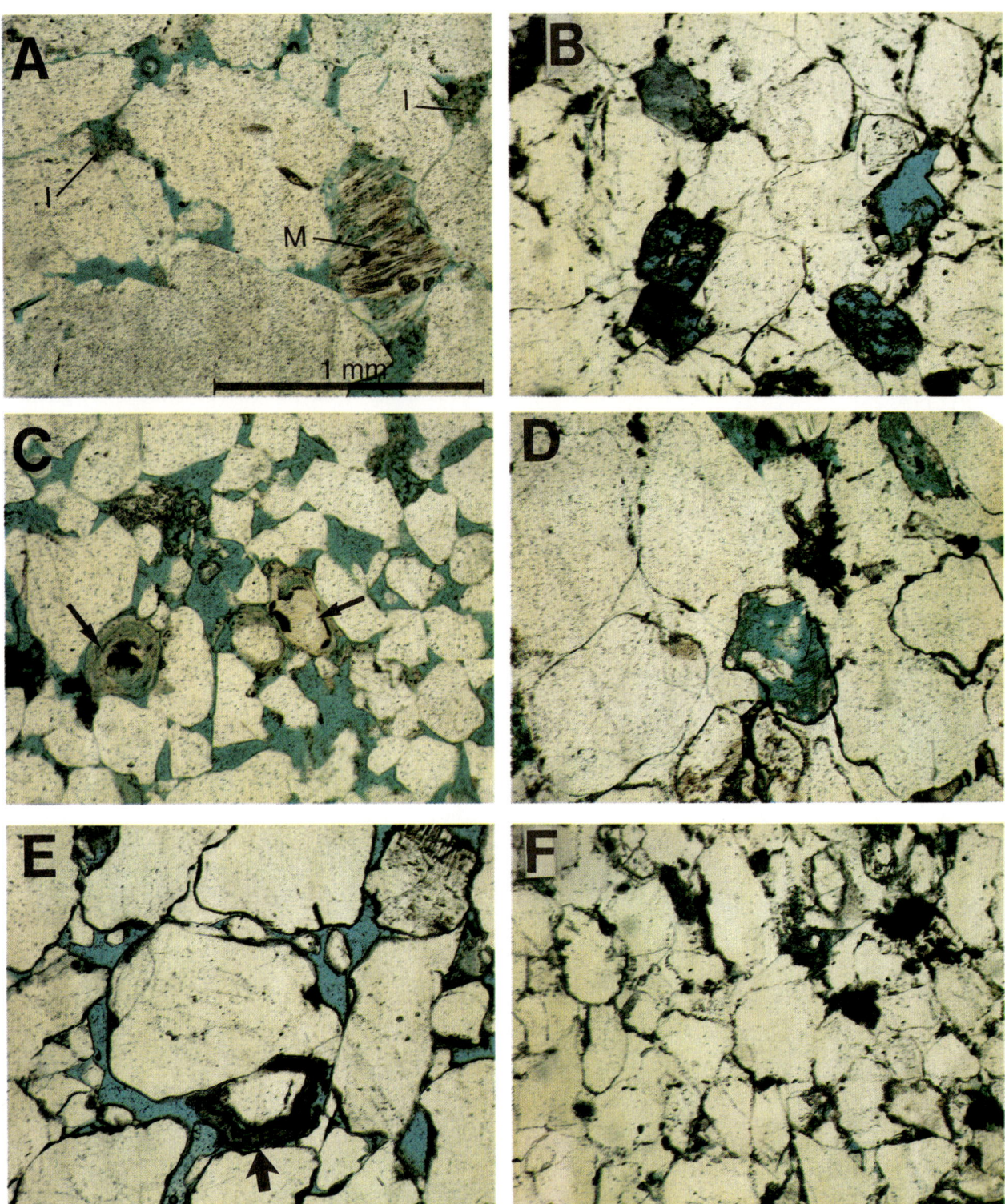

FIGURE 13. Photomicrographs of thin sections from Tofte and Tilje formations. All views have the same magnification (see micrograph A). Log depths are given to facilitate comparison with permeability-depth plots in Figure 6.

(A) **Tofte Formation,** well 6506/12-1: core depth, 4184 m RKB (log depth, 4184); porosity, 13.2%; permeability, 40 md. Despite abundant quartz cement, permeabilities in the Tofte Formation in well 1 mainly range from 1 to 40 md. Tofte sand composition is relatively immature compared with the Garn Formation, exhibiting poor sorting, abundant expanded mica grains ("M"), and clusters of detrital clay ("I") that have been largely recrystallized to diagenetic illite.

(B) **Tofte Formation,** well 6506/12-6: core depth, 4486 m RKB (log depth, 4486); porosity, 8.9%; permeability, 0.3 md. In contrast to the situation in well 6506/12-1, quartz cementation and illitization have eliminated virtually all intergranular macroporosity in the Tofte Formation in well 6506/12-6, resulting in permeabilities mostly <1 md.

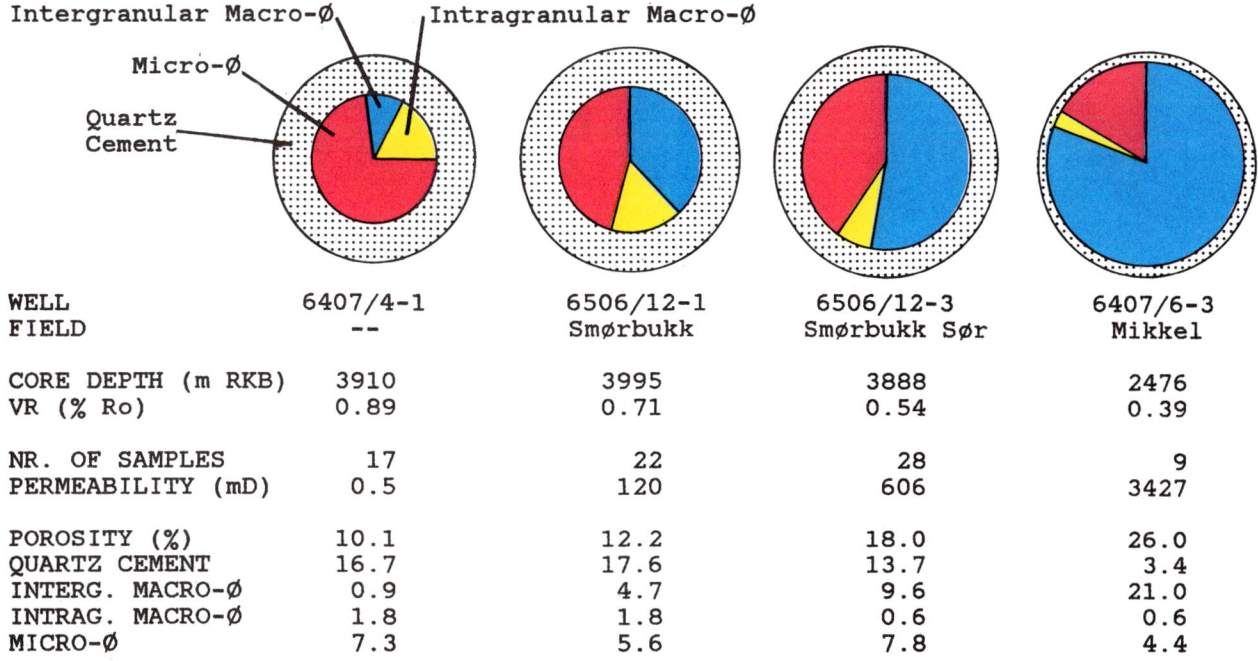

FIGURE 14. Relative abundances of different porosity types in the Garn Formation in Smørbukk field compared with those in other Haltenbanken wells. Each diagram represents the average composition of samples from a particular well analyzed by point counting of thin sections (300 counts). Further data are given in Ehrenberg (1990). Porosity and permeability values are averages of routine core analyses of the same samples. Areas of different fields are proportional to percentages of bulk rock volume. The sum of porosity and quartz cement (area of outer circle) remains relatively constant, indicating that variation in total porosity is mainly a function of quartz cement abundance. With decreasing total porosity, there is a dramatic decrease in the ratio of "effective porosity" (intergranular macroporosity) to "ineffective porosity" (macroporosity within dissolved grains plus microporosity). As shown in Figure 15, the situation shown here for well 6506/12-1 represents the best-case scenario for Smørbukk field.

inhibition of diagenesis in the crestal part of the field (well 1) resulting from earlier filling with hydrocarbons (probably lower Tertiary, as opposed to late Pliocene, downflank).

An exception to the over-all tendency of decreasing reservoir quality with depth in Figure 15 can be seen in a relatively small percentage of the samples from the Tilje Formation, which have anomalous porosities of 20 to 28% at a depth of about 4250 m (well 12-1) and 14 to 22% at depths of 4500 to 4700 m (wells 12-6 and 12-7). In these samples, porosity appears to have been preserved by the presence of early authigenic, grain-coating chlorite, which has inhibited quartz cement growth. Petrographic study of these porous samples reveals that for the most part

---

(C) Tilje Formation, well 6506/12-1: core depth, 4256 m RKB (log depth, 4258); porosity, 23.9%; permeability, 110 md. This sample is from a high-porosity, high-permeability zone about 6 m thick, surrounded by sandstones of generally much poorer quality (Figure 5). Abundant intergranular macroporosity and the near absence of quartz cement in such sandstones appear to have resulted from the presence of grain-coating authigenic chlorite, which inhibits quartz cement growth. Chlorite ooids (arrows) are conspicuous minor features of the chlorite-coated sandstones, indicating that Fe-rich clay was available for accretion around sand grains at the time of deposition. The presence of the chlorite is therefore believed to have been controlled by geochemical conditions at and near the sediment-water interface.
(D) Tilje Formation, well 6506/12-1: core depth, 4238 m RKB (log depth, 4240); porosity, 6.2%; permeability, 1.2 md. Clean, low-porosity sandstones, such as the one in this example, are interlayered with high-porosity rocks, such as that in C. The difference here is heavy quartz cementation, reflecting thinner or less-continuous chlorite coatings.
(E) Tilje Formation, well 6506/12-6: core depth, 4606 m RKB (log depth, 4609); porosity, 14.7%; permeability, 193 md. In well 6506/12-6, Tilje porosities tend to be lower than in well 6505/12-1, apparently reflecting generally heavier quartz cementation. Even the most chlorite-rich sandstones, such as the one in this example, contain considerable amounts of quartz cement. Chlorite ooids (arrow) are conspicuous minor features.
(F) Tilje Formation, well 6506/12-6: core depth, 4596 m RKB (log depth, 4599); porosity, 4.6%; permeability, 0.2 md. Chlorite-poor, clean sandstones in the Tilje Formation in well 6606/12-6, such as that shown here, are tightly cemented.

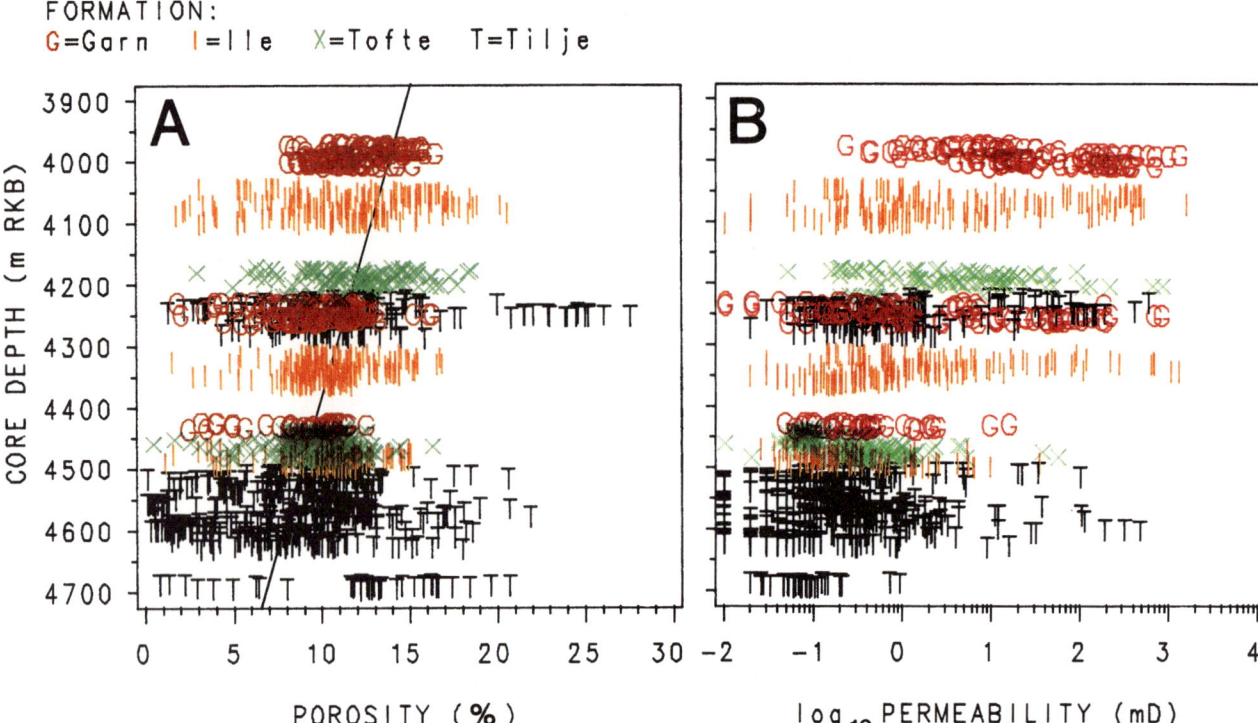

FIGURE 15. Plots of porosity and permeability vs. core depth for Smørbukk wells 6506/12-1, -6, and -7. All routine core analyses are plotted with symbols identifying the various formations. Different clusters of points for the same formation correspond to the three wells (shallowest, 6506/12-1; intermediate, 6506/12-6; deepest, 6506/12-7). The diagonal line in the depth-porosity plot is the regional trend of decreasing average core porosity with depth reported by Ehrenberg (1990). Both porosity and permeability show overall deterioration with depth in Smørbukk field, whereas the considerable scatter at each depth is believed to reflect variation in primary sand quality. The highest-porosity samples from the Tilje Formation at all depths are samples in which quartz cement growth has been inhibited by early authigenic, grain-coating chlorite.

they are unbioturbated sandstones containing minor amounts of quartz cement and having abundant coatings of iron-rich chlorite (Figure 13C and E). Adjacent lower-porosity zones in the same cores are shale layers, bioturbated matrix-rich sandstones, or clean sandstones with abundant quartz cement overgrowths and minor or discontinuous chlorite coatings (Figure 13D and F).

Measurements of residual gas and water saturations and capillary pressure data for Smørbukk sandstones reflect the strong influence that diagenesis has had on petrophysical properties. Residual gas saturations (Figure 16) are high (30 to 65%) in comparison with the range typical of North Sea reservoir rocks (30 to 40%), possibly reflecting the abundance of "vuggy" secondary porosity in the Smørbukk sandstones. Mercury injection data (Figure 17) show a wide range of results, but for most samples the "average" pore throat diameter (vertical axis) is less than about 10 μm, and 20 to 60% of the total porosity is trapped behind pore throats with diameters smaller than 1 μm.

HYDROCARBONS

The general characteristics of Haltenbanken source rocks and hydrocarbons as described by Heum et al. (1986) also apply to block 6506/12. The compositions of hydrocarbons in the various Smørbukk reservoirs, and their phase behavior, are shown in Table 2 and in Figures 18 and 19, respectively. The gas condensates are relatively rich, especially in the Tilje Formation (maximum liquid drop-out during constant volume depletion, 10 to 32%). The reservoir oils are relatively volatile (GOR, 440 to 494 standard m^3/standard m^3). The gas condensate in the Garn Formation is approximately at its dew point, but conditions in the other reservoirs are 65 to 120 bars below saturation pressure. The CO_2 contents of the reservoir fluids (3 to 5 mol %) conform to the regional trend of increasing CO_2 partial pressure with increasing temperature reported by Smith and Ehrenberg (1989) for clastic reservoirs of the Norwegian continental shelf and the U.S. Gulf Coast.

The migration and accumulation of hydrocarbons in block 6506/12 have been investigated with basin modeling computer programs by the Institut Français du Pétrole. This work includes several 2D sections through Smørbukk and Smørbukk Sør fields and through the northern overpressured structure (Ungerer et al., 1987), and a 3D model of Smørbukk Sør field (Forbes et al., 1991). The results show that hydrocarbons began to accumulate as early as the Eocene, but that the vast majority of the total hydrocarbon expulsion (85%) took place within the

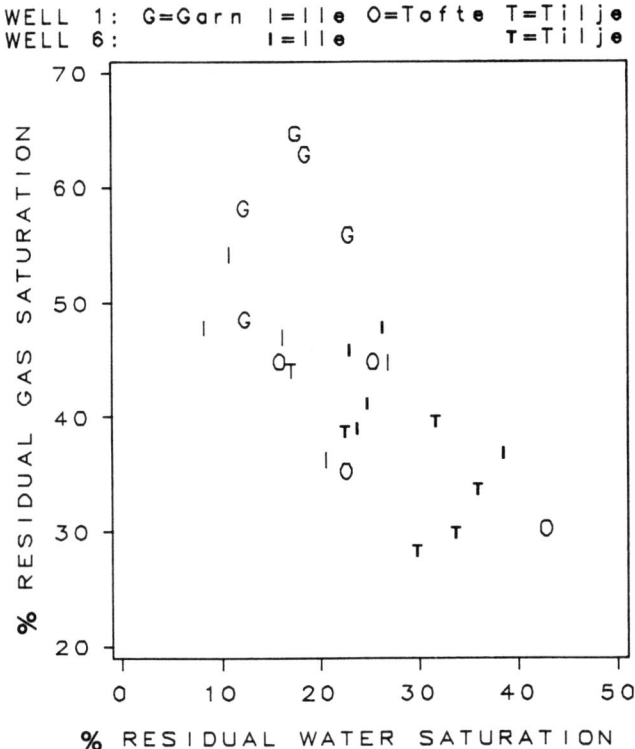

FIGURE 16. Data for residual gas and water saturations for Smørbukk reservoir rocks. Plotted symbols identify well and formation of each sample.

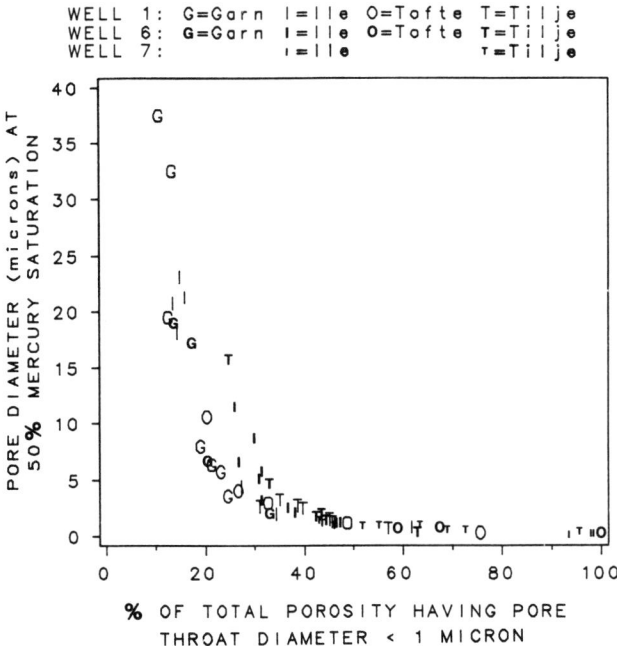

FIGURE 17. Mercury injection capillary pressure data for Smørbukk reservoir rocks. Plotted symbols identify well and formation of each sample.

last 5 million years. The total amount expelled was several times the volume of the present accumulations, the excess mainly spilling northeastward into the drainage basin of Heidrun field.

The basin modeling results also suggested that most of the hydrocarbons were generated from the coals of the Åre Formation, with relatively minor contributions from the more oil-prone source rocks of the Spekk Formation. However, organic geochemical studies performed by Statoil, comparing source rock extracts and Smørbukk reservoir fluids (including measurements of pristane, phytane, hopane, sterane, and carbon isotope compositions), indicate that the fluids had a mixed marine/terrigenous source and strongly support their derivation from both Åre Formation and Spekk Formation source rocks.

RESERVES

In-Place Reserves

In-place hydrocarbon volumes are listed for each reservoir zone in Table 3. Major uncertainties concern the geometry of the barrier between the western and eastern parts of the field and the extension of reservoir properties to the south in the western part of the field, as far as 19 km away from well control. Drilling of another well has been proposed to address the latter problem.

Reservoir Simulation

During 1986, studies of several production simulation models were performed, including radial single-well and full-field studies. However, the results of these studies are now regarded as invalid because: (1) they were based on earlier in-place reserve estimates that were about 2.6 times the current estimates; (2) only data from well 6506/12-1 were used, whereas wells 6506/12-6 and -7 have given a much less optimistic picture of reservoir quality and continuity; and (3) production from the Tilje Formation was not included in the models. However, one conclusion from the modeling that appears to remain valid is that gas expansion should be used as the main driving mechanism in production of the Garn, Ile, and Tofte reservoirs, because gas reinjection would require too many wells and excessive pressures.

In 1990, 2D simulations were performed for individual zones of the Tilje Formation along the two cross sections shown in Figure 3. Each cross section was divided into 30 cells (each 250 m wide for the western cross section and 150 m wide for the eastern cross section). Each cell consisted of five reservoir zones (Tilje 1.1, 1.2, 3.1, 3.2, and 3.4), with no vertical communication among zones. Production from each reservoir zone was simulated independently, and so it is somewhat misleading simply to sum up the recovery from all zones to estimate a total production profile for the Tilje Formation. A more realistic simulation would need to model the simultaneous production of several if not all zones from a single completion. Nevertheless, a summation of the zones gives an

TABLE 2. Recombined reservoir fluid compositions (mol %) used for calculation of reserves in different reservoir zones.

Sample Reservoir	A Garn	B Ile*	C Tofte**	D Tilje 3.4	E Tilje 3.2/3.1/2.2	F Tilje 3.2/3.1/2.2	G Tilje 3.2/3.1/2.2	H Tilje 1
Fluid	Gas/c.†	Gas/c.	Gas/c.	Gas/c.	Gas cap	Oil leg	Oil leg	Gas/c.
Field part	West	West	West	East	West	West	East	West
Well	12-1	12-6	12-1	12-7	12-1	12-6	12-7	12-1
DST	7	3	5	2	4	1	1	2
Nitrogen	0.53	0.60	0.62	0.50	0.57	0.42	0.48	0.56
Carbon dioxide	3.30	3.34	3.47	5.38	4.13	4.56	4.04	5.13
Methane	72.98	74.16	69.95	62.84	66.88	55.77	57.41	67.92
Ethane	7.68	7.90	8.85	9.32	8.96	8.79	9.28	9.93
Propane	4.10	4.15	4.89	5.29	4.98	5.45	5.62	5.34
Iso-butane	0.70	0.71	0.87	0.87	0.89	1.02	1.00	0.88
n-butane	1.42	1.44	1.72	1.92	2.00	2.34	2.22	1.80
Iso-pentane	0.54	0.53	0.63	0.67	0.77	0.90	0.83	0.62
n-pentane	0.67	0.66	0.73	0.87	0.94	1.10	1.05	0.64
C_6 fraction	0.85	0.81	0.82	1.07	0.89	1.37	1.35	0.86
C_7 fraction	1.33	1.20	1.47	1.70	1.65	2.27	2.21	1.16
C_8 fraction	1.33	1.15	1.29	1.78	1.29	2.48	2.59	1.16
C_9 fraction	0.78	0.63	0.99	1.09	1.03	1.42	1.49	0.78
$C_{10}+$ fraction	3.79	2.71	3.70	6.70	5.02	12.11	10.43	3.22
GOR (standard m³/standard m³)	1458	1807	1508	920	920	440	494	1508
g/cm³ at 15°C	0.798	0.784	0.801	0.818	0.809	0.821	0.820	0.795
API gravity	46	49	45	42	43	41	41	47
Maximum CVD % cond.	15	10	12	33	23	—	—	13
Dew/bubble point (bar)	447	388	382	420	410	358	359	386
Reservoir pressure (bar)	467	476	473	488	474	480	494	478
Reservoir temperature (°C)	143	155	150	164	152	156	165	154

*This composition is also used for Ile east.
**This composition is also used for Tilje 3.4 west.
†Gas/c. = Gas/condensate.

approximate, although probably optimistic, idea of the total possible recovery from the Tilje Formation (Table 4). The "probable" in-place volumes were used for these models (Table 3).

Two different cases were studied for each cross section:

1. Production by pressure depletion, using the black oil simulator ECLIPSE (also cross-checked using the MORE simulator).

2. Reinjection of separator gas in the highest-permeability zones (Tilje 3.2 in the west, which contains gas condensate, and Tilje 3.1 in the east, which contains oil), using the compositional simulator MORE.

Results are presented in Table 4. Recovery efficiency with pressure depletion is low because of high residual gas saturation (40% used for all zones) and high residual oil saturation (25% used for zone 3.1). Reinjection in the gas condensate zone 3.2 in the west gives higher oil recovery, because the recycling process substitutes lean gas for wet gas. Reservoir pressure is maintained above saturation pressure, and so no liquid dropout occurs in the reservoir. Reinjection in the oil zone 3.1 in the east gives higher oil recovery, because pressure is maintained above the minimum multicontact miscibility pressure for most of the plateau period, in effect resulting in replacement of oil by gas in the residual saturation.

Uncertainties Regarding Effects of Phase Changes on Recovery

In planning of field development, it will be important to consider that pressures near production wells will fall

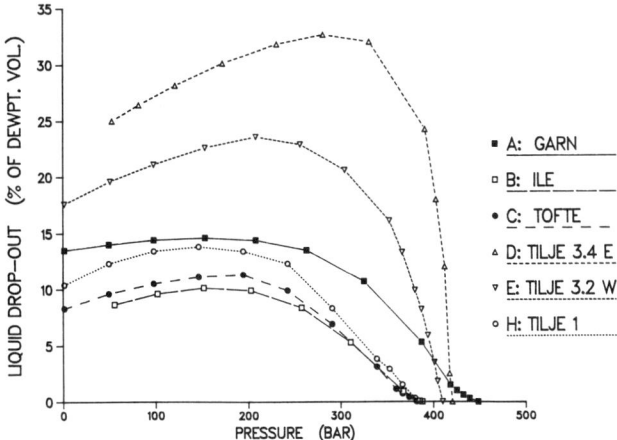

FIGURE 18. Retrograde liquid dropout from Smørbukk field gas condensates during constant-volume depletion at reservoir temperature. These curves show the amounts of liquid forming (as a percentage of the gas volume at dew point) as pressure is decreased at constant volume and temperature.

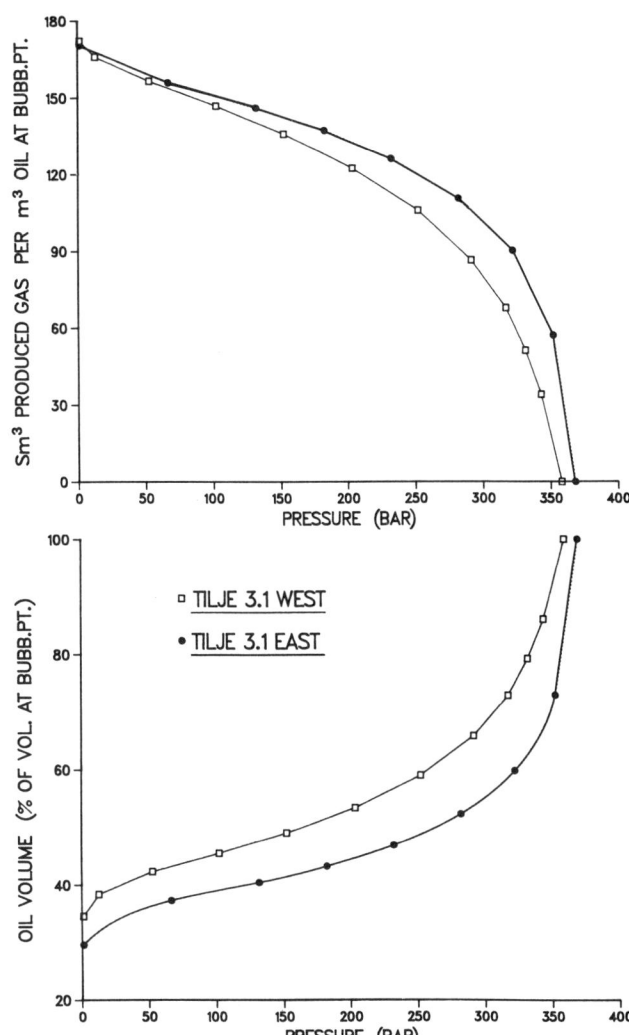

FIGURE 19. Gas and oil volumes during differential vaporization of Smørbukk field oils. These curves show the volumes (relative to the oil volume at bubble point) of gas produced (upper plot) and oil remaining (lower plot) as pressure is decreased at constant volume and temperature.

below the saturation pressure of the reservoir fluid, with resultant precipitation of condensate and reduction of gas permeability. This would occur almost immediately in the Garn Formation, where the difference between the ambient and saturation pressures is only 20 bars, and would eventually occur in the other zones as well, where the differences are from 73 to 135 bars (Table 2). Analysis of DST results shows that in five of the tests the well pressure fell below the saturation pressure of the reservoir fluid, but that in only one test (DST 7 in the Garn Formation in well 6506/12-1) was a two-phase gas condensate region likely to have been developed to any appreciable distance into the formation.

It has been recommended that in future wells, production testing should be carried out over long time intervals and at various production rates in order to study the effect of within-formation condensation on well productivity. Methods of compensating for the "skin effect" produced by near-well condensation—hydraulic fracturing and acid treatment, for example—might be explored.

The effect of condensate precipitation on well productivity could also be studied experimentally by performing imbibition gas-oil relative permeability analyses. The analogous problem of within-formation gas evolution from the volatile oil in the oil zones of the Tilje Formation could be modeled by drainage gas-oil relative permeability experiments. In such experiments, it would be important to consider the variation of phase composition with progressive gas condensation or oil evaporation, because this affects the gas-oil surface tension and therefore the relative permeability. Matching the phase compositional range realistically would be a challenging and nonroutine experimental problem.

Recoverable Reserves

At present, no realistic estimates of total recoverable reserves can be given for Smørbukk field. For the past several years, total recoverable reserves have been quoted as being approximately 75 billion standard m^3 of gas and 27 million standard m^3 of condensate/oil, but in light of recent reevaluation these values now appear to have been based on insufficient data and to be overly optimistic. Estimation of realistic recovery factors is difficult, because modeling studies done to date have not adequately considered the marginal to poor reservoir quality of the Smørbukk reservoirs and the effects on well productivity of within-formation condensation and evaporation resulting from pressure drawdown.

TABLE 3. Estimates of in-place hydrocarbon volumes for Smørbukk field.

Zone	Smørbukk west							Smørbukk east							Total Smørbukk						
	Gas (10^9 std. m³)			Oil (10^6 std. m³)			Gas (10^9 std. m³)			Oil (10^6 std. m³)			Gas (10^9 std. m³)			Oil (10^6 std. m³)		
	Min.	Prob.	Max.	Min.	Prob.	Max.	Min.	Prob.	Max.	Min.	Prob.	Max.	Min.	Prob.	Max.	Min.	Prob.	Max.
Garn 3	1.5	2.3	2.3	1.0	1.6	1.6	0	0	0	0	0	0	1.5	2.3	2.3	1.0	1.6	1.6
Garn 2	0.7	1.3	1.3	0.5	0.9	0.9	0	0	0	0	0	0	0.7	1.3	1.3	0.5	0.9	0.9
Garn 1	3.0	5.4	5.4	2.1	3.7	3.7	0	0	0	0	0	0	3.0	5.4	5.4	2.1	3.7	3.7
Total Garn	5.2	9.0	9.0	3.6	6.2	6.2	0	0	0	0	0	0	5.2	9.0	9.0	3.6	6.2	6.2
Ile 3	8.6	14.1	14.1	4.7	7.8	7.8	0.4	0.6	0.6	0.2	0.3	0.3	9.0	14.7	14.7	4.9	8.1	8.1
Ile 2	11.5	22.2	22.2	6.4	12.3	12.3	1.1	1.6	1.6	0.6	0.9	0.9	12.6	23.8	23.8	7.0	13.2	13.2
Ile 1	0	0.5	0.5	0	0.3	0.3	0	0	0	0	0	0	0	0.5	0.5	0	0.3	0.3
Total Ile	20.1	36.8	36.8	11.1	20.4	20.4	1.5	2.2	2.2	0.8	1.2	1.2	21.6	39.0	39.0	11.9	21.6	21.6
Tofte	11.3	15.5	19.8	7.5	10.3	13.1	0	0	0	0	0	0	11.3	15.5	19.8	7.5	10.3	13.1
Tilje 3.4	1.5	4.2	4.2	1.0	2.8	2.8	1.9	4.9	4.9	2.1	5.3	5.3	3.4	9.1	9.1	3.1	8.1	8.1
Tilje 3.2	3.7	7.4	7.4	4.0	8.0	8.0	0	0	0	0	0	0	3.7	7.4	7.4	4.0	8.0	8.0
Tilje 3.1 Gas	0.8	0.8	0.8	0.8	0.8	0.8	0	0	0	0	0	0	0.8	0.8	0.8	0.8	0.8	0.8
Oil	3.0	4.2	4.2	6.9	9.5	9.5	1.9	5.0	5.0	3.9	10.1	10.1	4.9	9.2	9.2	10.8	19.6	19.6
Tilje 2.2 Gas	0	0	1.3	0	0	1.5	0	0	0	0	0	0	0	0	1.3	0	0	1.5
Oil	0	0	3.9	0	0	8.9	0	3.3	3.3	0	6.7	6.7	0	3.3	7.2	0	6.7	15.6
Tilje 1.2	0.3	1.1	2.3	0.2	0.7	1.6	0	0	0.7	0	0	0.5	0.3	1.1	3.0	0.2	0.7	2.1
Tilje 1.1	4.2	11.3	16.9	2.8	7.5	11.2	0	0	0	0	0	0	4.2	11.3	16.9	2.8	7.5	11.2
Total Tilje	13.5	29.0	41.0	15.7	29.3	44.3	3.8	13.2	13.9	6.0	22.1	22.6	17.3	42.2	54.9	21.7	51.4	66.9
Total	50.1	90.3	106.6	37.9	66.2	84.0	5.3	15.4	16.1	6.8	23.2	23.8	55.4	105.7	122.7	44.7	89.5	107.8

TABLE 4. Results of 2D Simulation of Tilje Formation.

		Smørbukk west				Smørbukk east				Total Smørbukk*			
		Gas produced		Oil produced		Gas produced		Oil produced		Gas produced		Oil produced	
Zone	Case**	Volume (10^9 std. m³)	Recovery (%)	Volume (10^6 std. m³)	Recovery (%)	Volume (10^9 std. m³)	Recovery (%)	Volume (10^6 std. m³)	Recovery (%)	Volume (10^9 std. m³)	Recovery (%)	Volume (10^6 std. m³)	Recovery (%)
3.4	P	0.08	2	0.06	2	0.37	8	0.52	10	0.45	5	0.58	7
3.2	P	2.15	29	1.56	20	—	—	—	—	—	—	—	—
	R	—	—	4.15	52	—	—	—	—	—	—	4.15	52
3.1 + 2.2	P	0.29	6	0.88	9	1.24	15	4.33	27	0.29	3	0.88	26
	R	—	—	—	—	—	—	7.39	44	—	—	7.39	
1.2	P	0.25	23	0.16	23	—	—	—	—	0.25	23	0.16	23
1.1	P	1.18	10	0.70	9	—	—	—	—	1.18	10	0.70	9
TOTAL		—	—	—	—	—	—	—	—	2.17	5	13.86	27

*Totals are based on values for gas reinjection in Tilje 3.2 west and Tilje 3.1 east.
**P = pressure depletion; R = reinjection of separator gas.

The simulation results in Table 4 provide reasonable maximum recovery estimates for the Tilje Formation, but similar studies have not been carried out for the other reservoirs. Absolute maximum values for these reservoirs can be calculated very simplistically by taking average values of the available residual gas measurements for each formation (Figure 16) and assuming 100% recovery efficiency for all nonresidual gas and no loss of liquid resulting from condensation within the formation. This method yields recovery factors of 40 to 60% for the Garn, Ile, and Tofte formations. It is, of course, unrealistic to assume 100% sweep efficiency in these very heterogeneous reservoirs, but the basis for a meaningful evaluation is not presently available.

FIELD DEVELOPMENT PLAN

Further appraisal drilling is necessary before any decisions can be made regarding commerciality and field development. The work referred to in the preceding section indicates that the bulk of Smørbukk field must be produced as a gas field by pressure depletion, but that oil in the Tilje 3.1 and 2.2 reservoirs could be produced most efficiently by gas reinjection.

Exploitation of Smørbukk's gas reserves must await the establishment of a regional gas transport system for Haltenbanken. Smørbukk's production will be phased in at such time that Smørbukk becomes the most attractive alternative for maintaining the regional gas production profile. Oil from the Tilje Formation could be produced sooner, probably as a satellite contribution to the Smørbukk Sør field, where oil production is tentatively planned to begin in 1996 and continue through 2010.

CONCLUSIONS

Smørbukk field is a structurally simple fault trap in which rich gas condensate and subordinate volatile oil are trapped in a complex sequence of Lower to Middle Jurassic sandstone reservoirs. Reservoir complexity results from depositional heterogeneity and diagenetic reduction of effective porosity, with consequent occlusion of pressure communication both vertically between zones and laterally within zones.

The optimal exploitation of Smørbukk field's considerable in-place reserves will be a major technological challenge. Success in meeting this challenge will depend in large part on how skillfully problems associated with well productivity and reservoir continuity can be anticipated and compensated for, and on the extent to which future advances in enhanced recovery technology can be applied to Smørbukk's difficult reservoirs. The danger is that high residual hydrocarbon saturations, extensive within-formation condensation caused by pressure drawdown, and widely variable reservoir continuity could result in low over-all recovery efficiency. The desired result will be an optimum balance between maximization of profits and maximization of recovery efficiency. Because production of oil probably must await development of Smørbukk Sør field and production of gas condensate must await establishment of a regional gas transport system, there is time for further study.

The name "Smørbukk" means literally "butter ram" (a fat male goat) and is taken from an old Norwegian folk tale in the classic collection by Asbjørnsen and Moe (1982). In this tale, a plump lad called Smørbukk (Figure 20) outwits a group of trolls who are planning to eat him for dinner, kills them, and carries off their gold and silver. Analogies between this folk tale and reality can be drawn in several ways. When it finally comes time for the dramatic story of Smørbukk field to be acted out, the companies of PL 094 must use their technological wits to ensure that they are cast in the role of the clever fat boy, who defeats the trolls (potential production problems), rather than in the role of the trolls, who fail to feast on the succulent fat boy.

NOTE ADDED IN PRESS

As discussed above, a major uncertainty concerning estimation of reserves in Smørbukk field has been the lack of well control at the southern end of the field. This problem was addressed during the summer of 1991 by the drilling of well 6506/11-2 at the location marked "X" in Figure 3. The results substantiate the reserve estimates given in Table 3. The Tilje 1, Tilje 3, and Ile reservoirs were found to be filled with gas condensate in apparent pressure communication with the same zones in well 6506/12-1.

ACKNOWLEDGMENTS

This report has benefited from contributions by the following Statoil personnel: A. Fylling on petrophysical evaluation; B. T. Larsen on tectonic setting and structure; A. Dalland on stratigraphy; E. Vik on exploration history; K. K. Meisingset on reservoir fluid compositions; E. Siring on plotting of mercury injection data; R. Sørbel on seismic interpretation; O. Dyrnes on DST results; and O. Harstad on reservoir simulation.

REFERENCES CITED

Aasheim, S. M., A. Dalland, A. Netland, and A. Thon, 1986, The Smørbukk gas/condensate discovery, Haltenbanken, in A. M. Spencer (ed.), Petroleum geology of the northern European margin: Norwegian Petroleum Society, Graham and Trotman, London, p. 299–305.

Aasheim, S. M., and V. Larsen, 1984, The Tyrihans discovery—preliminary results from well 6407/1-2, in A. M. Spencer (ed.), Petroleum geology of the northern European margin: Norwegian Petroleum Society, Graham and Trotman, London, p. 285–291.

Asbjørnsen, P. C., and J. Moe, 1982, Samlede Eventyr, første bind, Den Norske Bokklubben, p. 430–435.

Ehrenberg, S. N., 1990, Relationship between diagenesis and reservoir quality in sandstones of the Garn Formation, Haltenbanken, mid-Norwegian continental shelf: AAPG Bulletin, v. 74, p. 1538–1558.

Ehrenberg, S. N., and P. H. Nadeau, 1989, Formation of diagenetic illite in sandstones of the Garn Formation,

FIGURE 20. Illustration of "Smørbukk" by T. Kittelsen (from Asbjørnsen and Moe, 1982).

Haltenbanken, mid-Norwegian continental shelf: Clay Minerals, v. 24, p. 233–253.

Ekern, O. F., 1987, Midgard, in A. M. Spencer et al. (eds.), Geology of the Norwegian oil and gas fields: Norwegian Petroleum Society, Graham and Trotman, London, p. 403–410.

Forbes, P. L., P. M. Ungerer, A. B. Kuhfuss, F. Riis, and S. Eggen, 1991, Compositional modeling of petroleum generation and expulsion: Trial application to a local mass balance in the Smørbukk Sør field, Haltenbanken area, Norway: AAPG Bulletin, v. 75, p. 873–893.

Heum, O. R., A. Dalland, and K. K. Meisingset, 1986, Habit of hydrocarbons at Haltenbanken (PVT-modelling as a predictive tool in hydrocarbon exploration), in A. M. Spencer (ed.), Petroleum geology of the northern European margin: Norwegian Petroleum Society, Graham and Trotman, London, p. 259–274.

Larsen, V., P. O. Mørkeseth, and S. M. Aasheim, 1987, Tyrihans, in A. M. Spencer et al. (eds.), Geology of the Norwegian oil and gas fields: Norwegian Petroleum Society, Graham and Trotman, London, p. 403–410.

Smith, J. T., and S. N. Ehrenberg, 1989, Correlation of carbon dioxide abundance with temperature in clastic hydrocarbon reservoirs: relationship to inorganic chemical equilibrium: Marine and Petroleum Geology, v. 6, p. 129–135.

Ungerer, P., B. Doligez, P. Y. Chenet, J. Burrus, F. Bessis, E. Lafargue, G. Giroir, O. Heum, and S. Eggen, 1987, A 2D model of basin scale petroleum migration by two-phase fluid flow application to some case studies, in B. Doligez (ed.), Migration of hydrocarbons in sedimentary basins: Institut Français du Pétrole, p. 415–455.

Chapter 22

The Jurassic Snøhvit Gas Field, Hammerfest Basin, Offshore Northern Norway

A. Linjordet

Statoil
Harstad, Norway

R. Grung Olsen

Statoil
Stavanger, Norway

ABSTRACT

The first well offshore northern Norway was drilled in 1980, and the Snøhvit field, which is the largest gas find and the first oil discovery in the area, was discovered in 1984. The Snøhvit field is in the southwest Barents Sea in the center of the Hammerfest Basin and straddles blocks 7120/6, 7121/4, and 7121/5. Water depth is approximately 300 m. The field covers 90 km² (22,200 ac) and has a gas column of 124 m (407 ft) overlying a 14 m (46 ft) thick oil leg. This oil leg makes Snøhvit the only significant oil find on the Barents Shelf to date.

The reservoir consists of Lower to Middle Jurassic sandstones deposited in a transgressive coastal to inner shelf sequence. Three wells, each situated in a separate fault block, define the Snøhvit field. The wells have common fluid contacts. Reservoir properties are fair, with a porosity of 15% and permeability from 200 to 500 md in the main reservoir. The water saturation in the gas zone averages 10%, and varies from 3 to 26%.

A burial history involving uplift, erosion, and renewed burial during the Tertiary has influenced the distributions of oil and gas in the reservoir and the positions of the fluid contacts.

The most likely estimate of gas in place is 160 billion standard m³ (5.6 tcf), and the gas has a recovery factor of 70%. The oil in place is estimated at 73 million standard m³ (450 million bbl). Because of the thin oil leg and the areal distribution of the oil in the reservoir, it is not thought economically feasible to develop the oil with present-day technology. Various development scenarios have been studied for the Snøhvit gas, but currently there are no firm development plans.

INTRODUCTION

The Snøhvit field, discovered in 1984, is located in the southwestern Barents Sea, about 130 km (80 mi) off the Norwegian coast, northeast of Tromsø in northern Norway (Figure 1). Water depth in the area is 290 to 350 m (951 to 1148 ft). The field covers about 90 km² (22,200 ac) and straddles three licenses. The field name comes from the fairy tale "Snow White and the Seven Dwarves," because in the early exploration phase one large structure and seven smaller ones surrounding it were identified.

The three production licenses, PL 097, PL 099, and PL 110, were awarded in the eighth and ninth license rounds in 1984 and 1985. Statoil is operator for PL 099 and PL 110, and Norsk Hydro is operator for PL 097. Nine companies have interests in the Snøhvit field. The companies in addition to the operators are Total Marine Norsk a.s, Conoco Norway Inc., Elf Aquitaine Norge A/S, Esso Norge a.s, Amerada Hess Norwegian Exploration A/S, Fina Exploration Norway, and Deminex (Norge) A/S.

From Late Permian to Middle Jurassic, the Hammerfest Basin was part of a regional intracratonic basin, with deposition of marginal carbonates and thick clastic sequences (Sund et al., 1984). The structural elements in the Hammerfest Basin are the results of the late Cimmerian tectonic phase (Olsen and Hansen, 1987). The main subsidence in the area took place during the Cretaceous.

EXPLORATION HISTORY

Seismic data in the Barents Sea were first acquired in 1970 and made available to oil companies in 1977. In total, more than 400,000 km of seismic has been acquired offshore northern Norway. Exploration drilling started in 1980, when limited areas in the southwestern part of the Barents Sea were awarded to groups with Norwegian oil companies as operators. One of the first wells had good oil shows in the background gas and in the core samples. The discoveries that followed—Askeladd (gas) in 1981 (Olsen and Hansen, 1987); Albatross (gas) in 1982; and Snøhvit, which tested both oil and gas in 1984—encouraged the oil industry.

Since 1980, 44 wells have been drilled in the northernmost Norwegian exploration area, including 37 wildcats. The deepest water depth of any well has been 475 m (1558 ft), and the deepest well was drilled to 5200 m (17,060 ft). Nineteen of the 37 wildcats made discoveries, but only nine of these wells discovered significant volumes of hydrocarbons, mainly gas (Foyn, 1989).

No significant hydrocarbon discoveries have been made in the Hammerfest Basin since 1984. Tertiary uplift, erosion, hydrocarbon expulsion, and leakage may explain some of the dry prospects drilled since 1984.

DISCOVERY HISTORY

A total of six wells have been drilled in the three Snøhvit licenses. In 1984, the objective of the discovery well 7121/4-1 was to test the Middle/Lower Jurassic Stø Formation for hydrocarbon accumulations (Figure 2). This well was located on the crest of the largest of three fault blocks and found the largest offshore northern Norway gas/oil discovery to date, with a gas column of 108 m (354 ft). The well was tested at four levels, the two deepest of which were in the Triassic. The first and lowest test was an aquifer test. The second was a 3-m (11-ft) gas test with an initial gas rate of 391,000 standard m³ per day (13.8 MMCFGD); this test ended, after a few hours, with a water-cut of 92%. The third test was perforated across the gas-oil contact in the Lower Jurassic Nordmela Formation. The shaly Nordmela has poor reservoir characteristics, and the interpretation is that the gas in the gas cap assisted the flow of oil into the well at a rate of about 88 standard m³ per day (554 BOPD) and a high GOR of 1000 standard m³/standard m³ (5614 ft³/bbl). The fourth test had a perforation interval of 32 m (105 ft) in the lowest and best part of the Stø Formation. The well produced gas at a rate of 839,500 standard m³ per day (29.6 MMCFGD) with a GOR of 7830 standard m³/standard m³ (43,958 ft³/bbl).

The gas-oil and oil-water contacts were not determined in well 7121/4-1, because the oil was located in a shaly lithology and because of the increasing shale content toward the base of the Nordmela Formation. In the second Snøhvit well, 7120/6-1, the oil rim is situated in the best and lowest part of the Stø Formation. From testing and log analysis, the oil-water contact was interpreted as being at 2418 m MSL (7933 ft subsea). A water test below the oil-water contact produced water at a rate of 409 standard m³ per day (2570 BWPD). In the oil rim test, a peak rate of 1526 standard m³ per day (9600 BOPD) was obtained before the water coned in and gave a 60% water-cut after four hours. A test in the middle, shaly part of the Stø Formation did not flow to surface, whereas a gas test in the upper part of the Stø Formation produced 1.183 million standard m³ per day (41.8 MMCFGD).

The main objective for the easternmost well in the field, 7121/5-1, was to test for additional oil below the oil-water contact found previously, but there was no evidence of an oil-water contact below 2418 m MSL (7933 ft subsea). The oil-water contact was found close to the top of the Nordmela Formation, and the gas-oil contact was determined to be at 2404 m MSL (7887 ft subsea) from static pressure measurements and log analysis. The oil leg was consequently 14 m (46 ft) thick. The well was tested extensively in a 3-m (10-ft) perforation interval. The initial oil rate was 232 standard m³ per day (1460 BOPD). After reperforation, the oil rate went up to 430 standard m³ per day (2705 BOPD) and then slowly decreased to 360 standard m³ per day (2264 BOPD). The GOR increased from 120 standard m³/standard m³ (674 ft³/bbl) to 460 standard m³/standard m³ (2582 ft³/bbl) during ten days of testing after the reperforation.

The two Snøhvit appraisal wells were successful in testing the huge gas cap, whereas the oil leg tests gave disappointingly low oil rates. The eastern and northwestern parts of the field were delineated by these two wells.

In the western part of the Snøhvit field, there is an area (a gas chimney) where the seismic data are clearly affected by gas (Figure 3). The seismic reflectors sag and lose continuity, and so the western extent of the field has

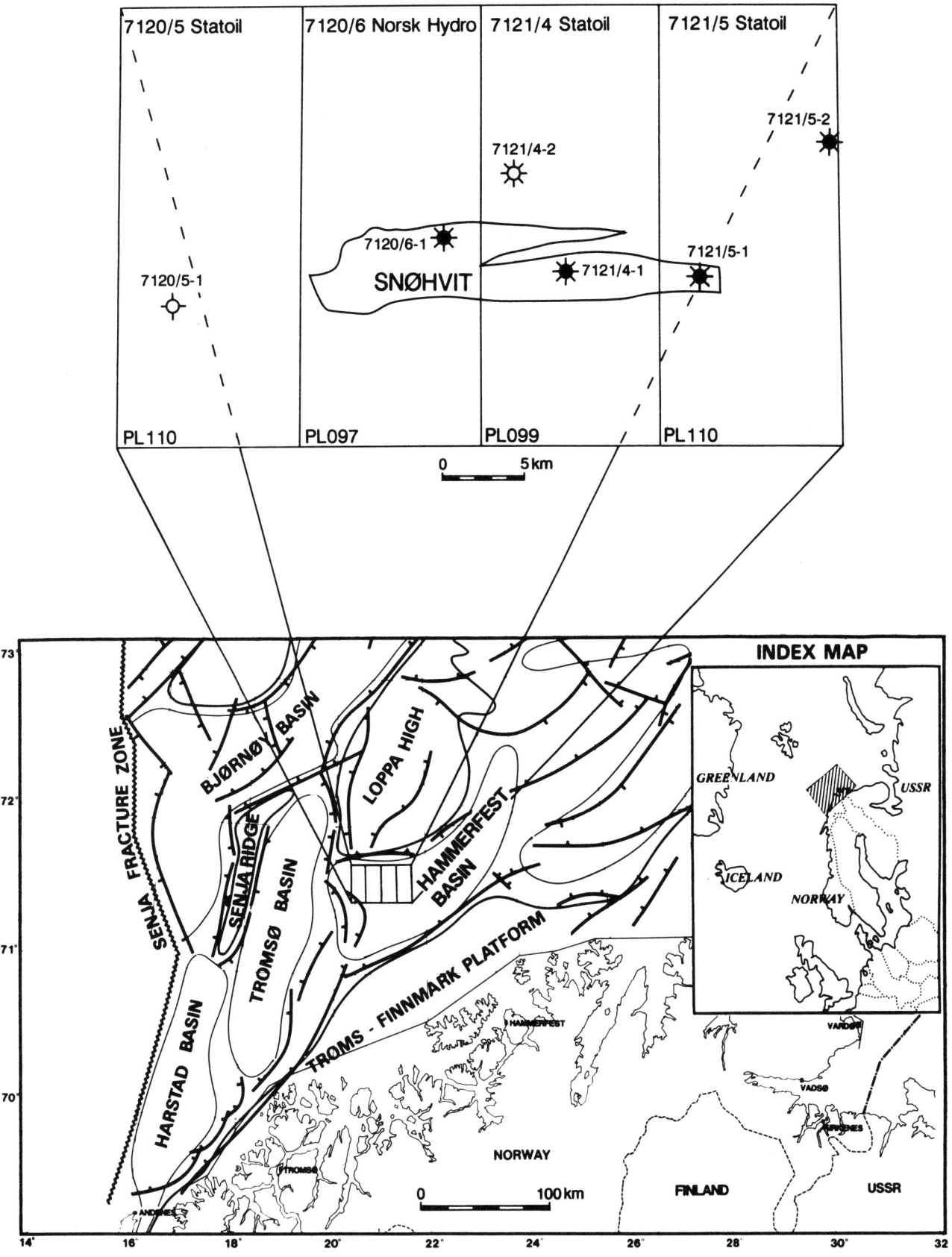

FIGURE 1. Location map for Snøhvit field.

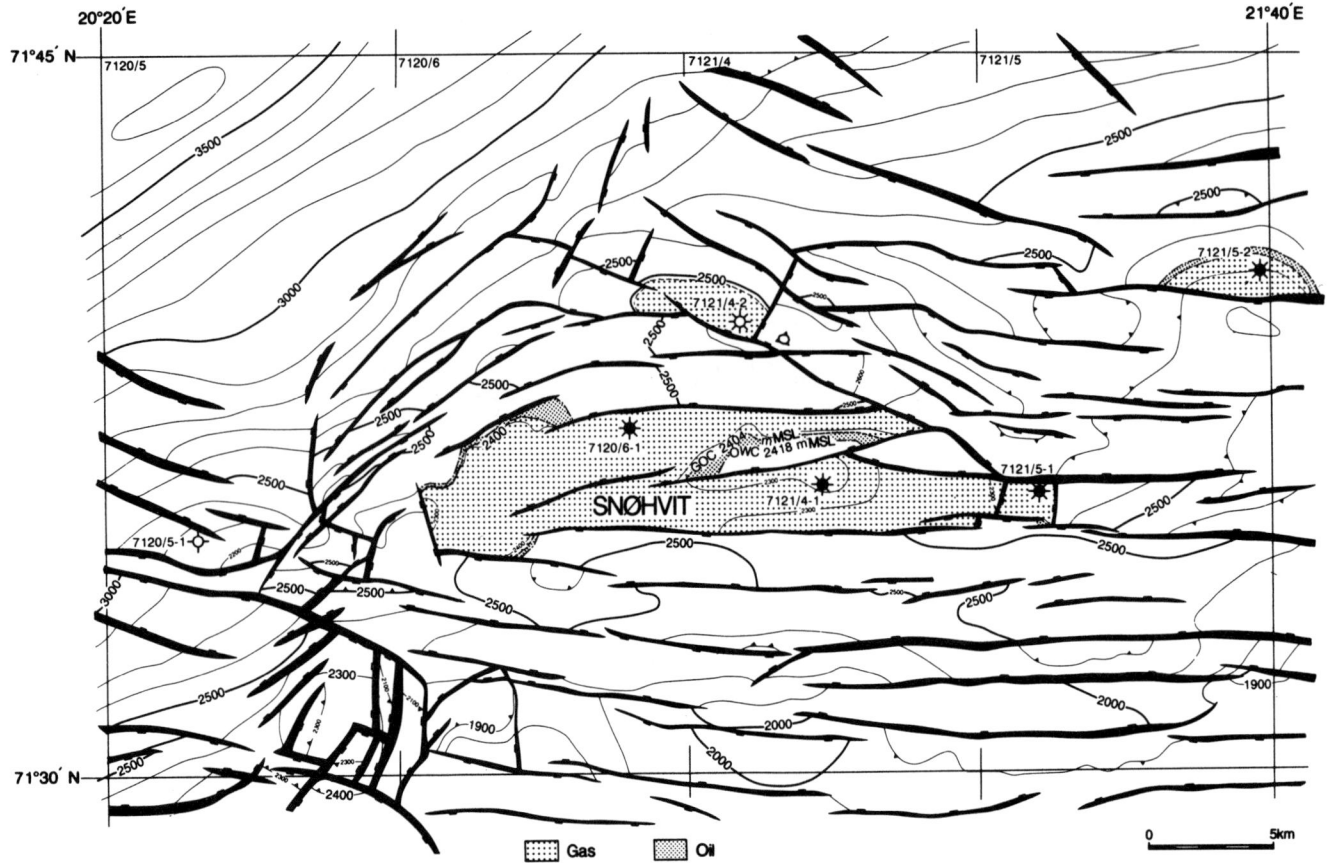

FIGURE 2. Structural depth map on the top of the Stø Formation.

not been well defined. A structure west of the gas chimney was tested by well 7120/5-1 in 1985, but it was dry. The current interpretation is that about 15% of the proved Snøhvit hydrocarbons in the west are located in an area where the seismic is affected by gas.

The two other wells that have been drilled in the Snøhvit licenses were located on separate, smaller structures. They both made discoveries in the Stø Formation. Well 7121/4-2 tested a fault block with about 7 billion standard m³ (247 mmcf) of gas initially in place. Well 7121/5-2 also found a gas cap overlying an oil rim only 6 m (20 ft) thick. The well discovered 4 billion standard m³ (141 mmcf) of gas and 11 million standard m³ (69 million bbl) of oil. Because of technical problems, well 7121/5-2 was not tested.

STRATIGRAPHY AND DEPOSITION

The Hammerfest Basin contains 5000 m (16,400 ft) of strata above the basement in the area of the Snøhvit field (Figures 4 and 5). The deepest Snøhvit well, 7121/5-1, reached into the Upper/Middle Triassic sandy shale.

Triassic

The Middle to Upper Triassic strata are characterized by a lower sequence of interbedded shales and sandstones that are occasionally carbonaceous and contain coal fragments, overlain by a shaly and silty unit that has increasingly more interbedded sandstones upward. The fossil assemblages are impoverished in most of the Triassic sequence, and fossils with ages older than Rhaetian have not been identified. These sediments are interpreted as having been deposited in a deltaic environment.

Jurassic

The Lower to Middle Jurassic strata consist mainly of sandstones interbedded with thin shale layers, deposited in a shallow marine to coastal plain environment with fluctuating coastlines.

The lower part, the Tubåen Formation, has thick sandstone bodies with thin shale beds, which are in part carbonaceous, and cores from Snøhvit wells contain thin coal layers. These sediments are interpreted as representing an estuary. A small gas accumulation is found in the upper part, at the crest of the Snøhvit structure.

The conformably overlying Nordmela Formation has silty shales and very fine grained sandstones in the lower part, overlain by fine-grained sandstones. The conformably overlying sand-dominated unit, the Stø Formation, consists of fine- to coarse-grained sandstones. The Stø and Nordmela formations are described in detail in the section on reservoir characteristics.

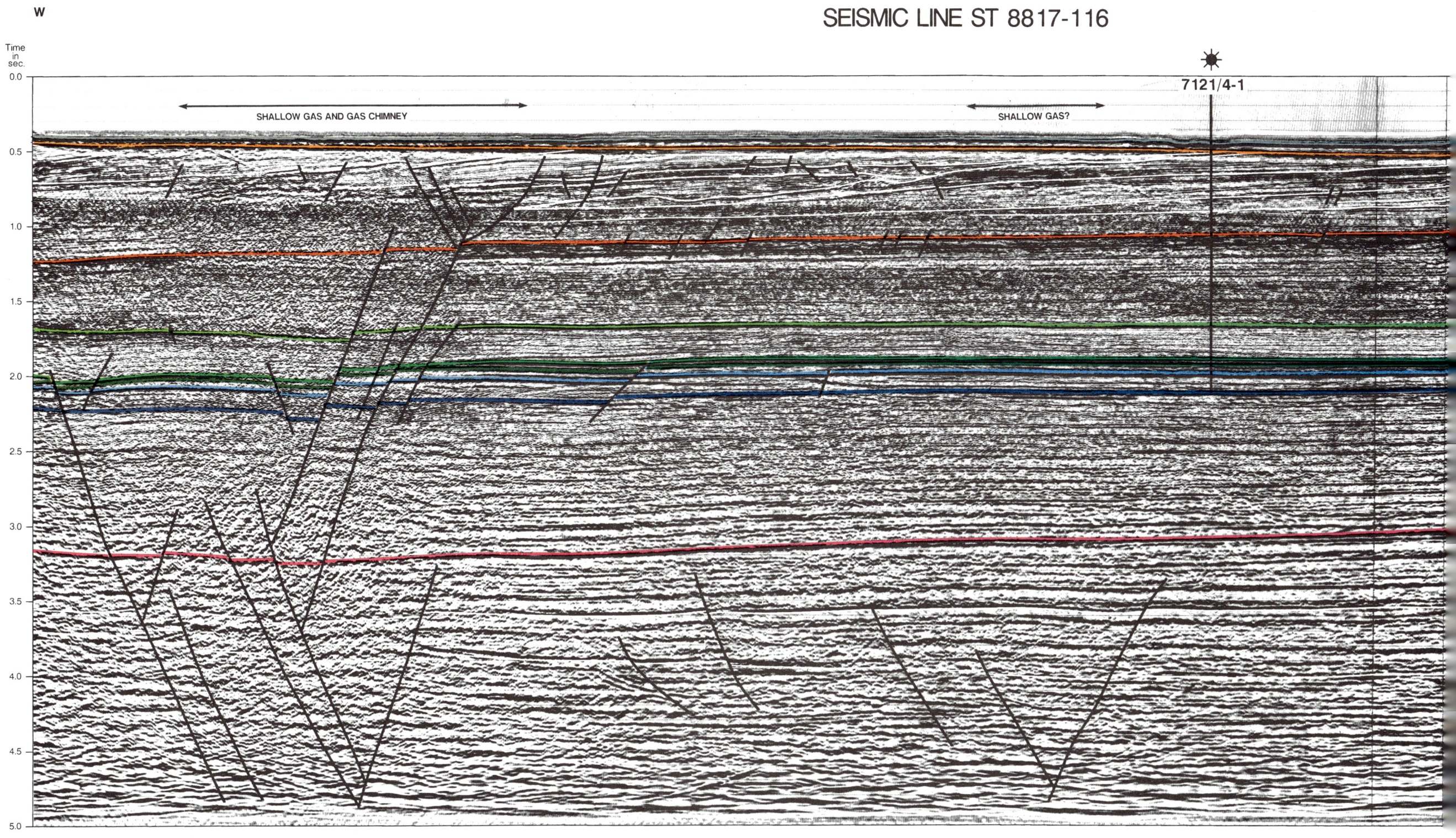

FIGURE 3. East–west-oriented seismic line ST8817, showing gas-affected seismic.

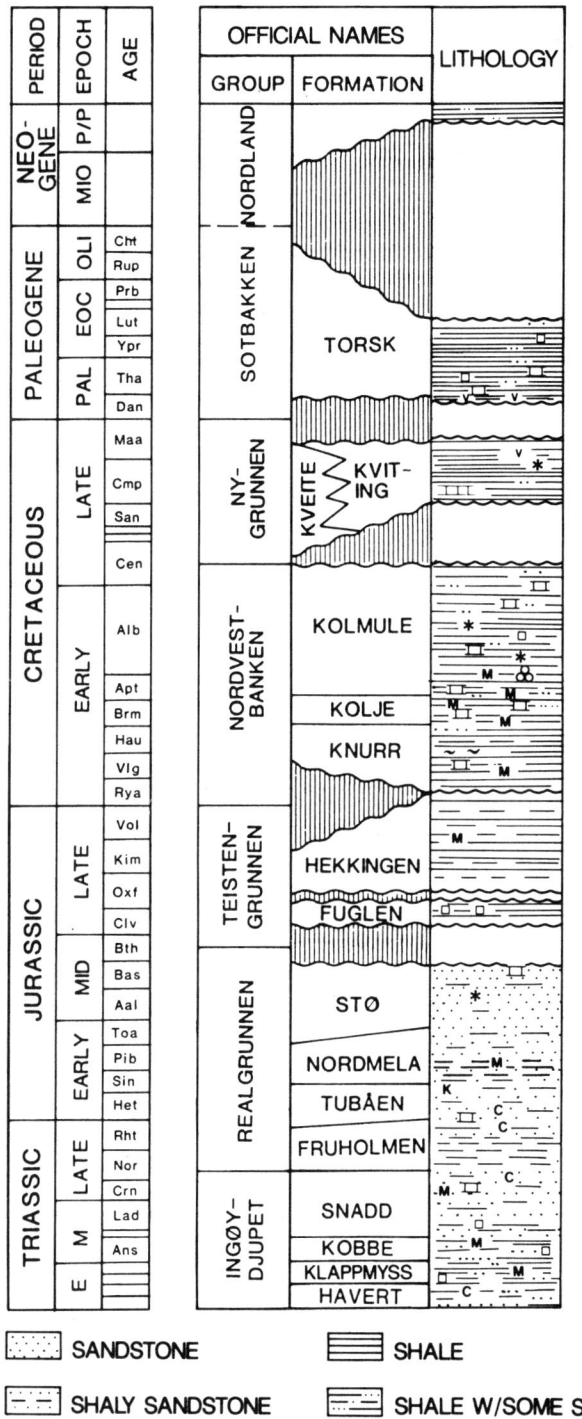

FIGURE 4. Stratigraphy of the Hammerfest Basin (adapted from Worsley et al., 1988).

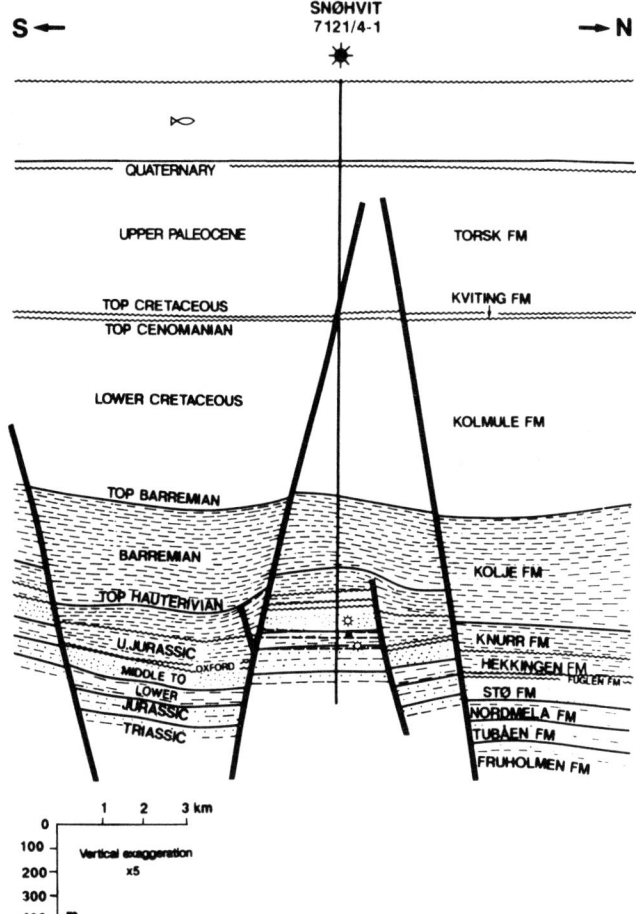

FIGURE 5. North-south geologic section through the discovery well 7121/4-1.

The seismic and sedimentological data indicate that these sediments were deposited under stable tectonic conditions. The upper part, however, was probably influenced by the central doming in the Hammerfest Basin. There is a major hiatus between the Middle and Upper Jurassic, and Bathonian strata are missing in five of the six wells drilled in the Snøhvit licenses. Above this hiatus, the Upper Jurassic consists of shales that are very organic-rich in the upper part. The depositional environment was a marine shelf with anaerobic bottom water conditions. The organic-rich shales are good source rocks. These shales also form the cap rocks for the Snøhvit reservoir. The total thickness of the Upper Jurassic shows a westward increase from 67 to 100 m (220 to 328 ft) across the field.

Cretaceous

There is a hiatus between the Upper Jurassic and Lower Cretaceous strata and another between the Lower and Upper Cretaceous rock units. The Cretaceous strata consist mainly of claystones with thin sandstone and siltstone stringers in the middle part. Limestone and dolomite stringers occur mainly in the lower and middle parts, whereas limestones interbedded with claystones are found in the upper part. The claystones are generally gray, but red claystones occur in the middle and lower parts. Traces of tuff and tuffaceous claystones occur in the middle and uppermost parts of the sequence. The Cretaceous sediments were deposited in a marine shelf environment. The total thickness is approximately 1200 m (4000 ft). There is another hiatus between the Cretaceous and Tertiary strata.

Tertiary

Paleocene to Eocene claystones containing stringers of sand, siltstone, limestone, and dolomite, and traces of tuff, occur at the base of the Tertiary sequence. The depositional environment was a marine shelf with restricted bottom water conditions. The thickness is about 600 m (2000 ft).

There is a major hiatus in the upper part of the Tertiary sequence, and the Pliocene to Pleistocene sequence is represented by only 100 m (328 ft) of soft claystones.

STRUCTURAL GEOLOGY

The post-Caledonian development of the southwestern part of the Barents Sea probably started with deposition of terrestrial sediments in fault-bounded extensional basins (Sund et al., 1984).

The Caledonian structural trend observed in regional fault systems on the Norwegian mainland is from northeast to southwest, but only a few offshore faults related to this system have been mapped.

After post-Caledonian rifting in a north-south system, a thick salt sequence of Permian age was deposited in the paleo-Tromsø Basin, and the paleo-Loppa high was partly eroded along the eastern margin of the basin (Figure 1). Marginal carbonates and clastics were deposited in the Hammerfest Basin.

Fault activity ceased during the Middle Permian, and the entire area became part of a regional intracratonic basin until the Middle Jurassic. The only indication of structural movement is an Early Triassic unconformity on the Loppa high.

The Snøhvit field is situated in the center of the Hammerfest Basin, which is a late Kimmerian trough trending east-northeast and lying among the Loppa high to the north, the Tromsø Basin to the west, and the Troms-Finnmark platform to the south. The basin is symmetrical, but widens and deepens westward. It was affected by a doming parallel to the basin axis that was active from the Middle Jurassic to the early Barremian. The main subsidence of the basin occurred along the north and south margins (Figure 6). The dominant east-west-trending fault system in the central part of the basin was formed by flexural extension related to the doming. A majority of these faults dip toward the basin axis, where horsts and grabens formed along the crest of the dome. Because of this geometry, the Hammerfest Basin can be divided into northern and southern provinces with respect to hydrocarbon generation and migration.

In the late Paleocene–early Eocene, subsidence and sedimentation occurred in the western part of the Barents Sea, whereas uplift and erosion affected the area after the early Eocene opening of the Norwegian-Greenland Sea.

The updoming in the central part of the Hammerfest Basin created a series of east-west-oriented normal faults. The Snøhvit accumulation occurs in three of these fault blocks, with the major part in a large, east-west-oriented horst that dips gently to the west. The maximum throw on the southern east-west fault is 200 m (656 ft) (Figure 2). This horst was tested by well 7121/4-1. The rest of the

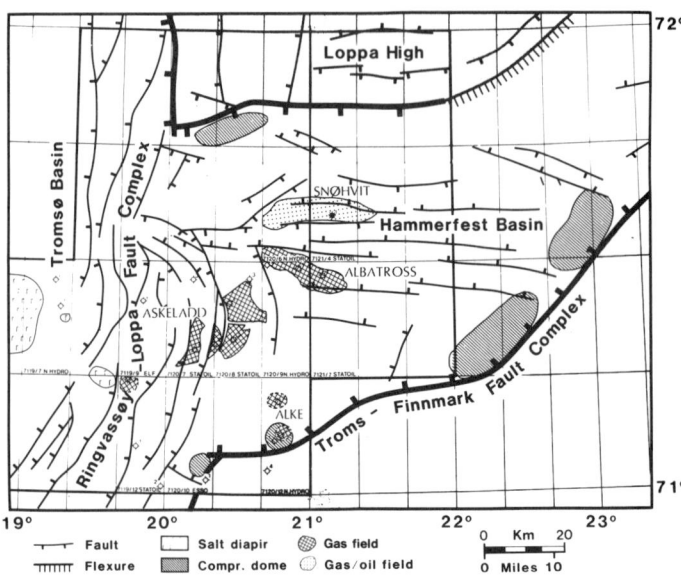

FIGURE 6. Structural elements of the southwestern Barents Sea (from Sund et al., 1984).

Snøhvit reserves are situated in two smaller "semihorsts": one to the east, penetrated by 7121/5-1; and one to the northwest, tested by 7120/6-1.

RESERVOIR CHARACTERISTICS

Mineralogy

The sandstones of the Lower to Middle Jurassic Nordmela and Stø formations are generally well sorted and medium to fine-grained, and exhibit various degrees of rounding. Some sandstone units are poorly sorted and have grain sizes ranging from fine to coarse. The sandstones of the Nordmela Formation are generally finer in grain size than those of the Stø Formation. The detrital mineralogy of the Stø and Nordmela formations consists primarily (80 to 90%) of monocrystalline quartz. The other minerals observed are polycrystalline quartz, chert, mica, potassium feldspar, plagioclase, metamorphic and sedimentary rock fragments, and heavy minerals. Quartz is the dominant cement. Other cementing minerals are ankerite, dolomite, calcite, kaolinite, illite, and pyrite. In the Stø Formation, the intergranular porosity has been reduced by pressure solution overgrowth (Figure 7). In the lower part of the Stø Formation, there are abundant stylolites that developed as a result of compaction. The stylolites probably originated from thin mud drapes (Figures 8 and 9).

Sedimentology

The major portion of the hydrocarbons in Snøhvit field is encountered in the Stø Formation, with about 10% in the Nordmela Formation (Figures 10 and 11).

The Nordmela Formation consists of alternating very fine to fine-grained sandstones, siltstones, and mudstones,

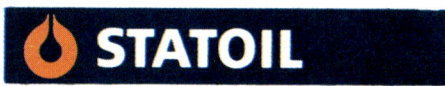

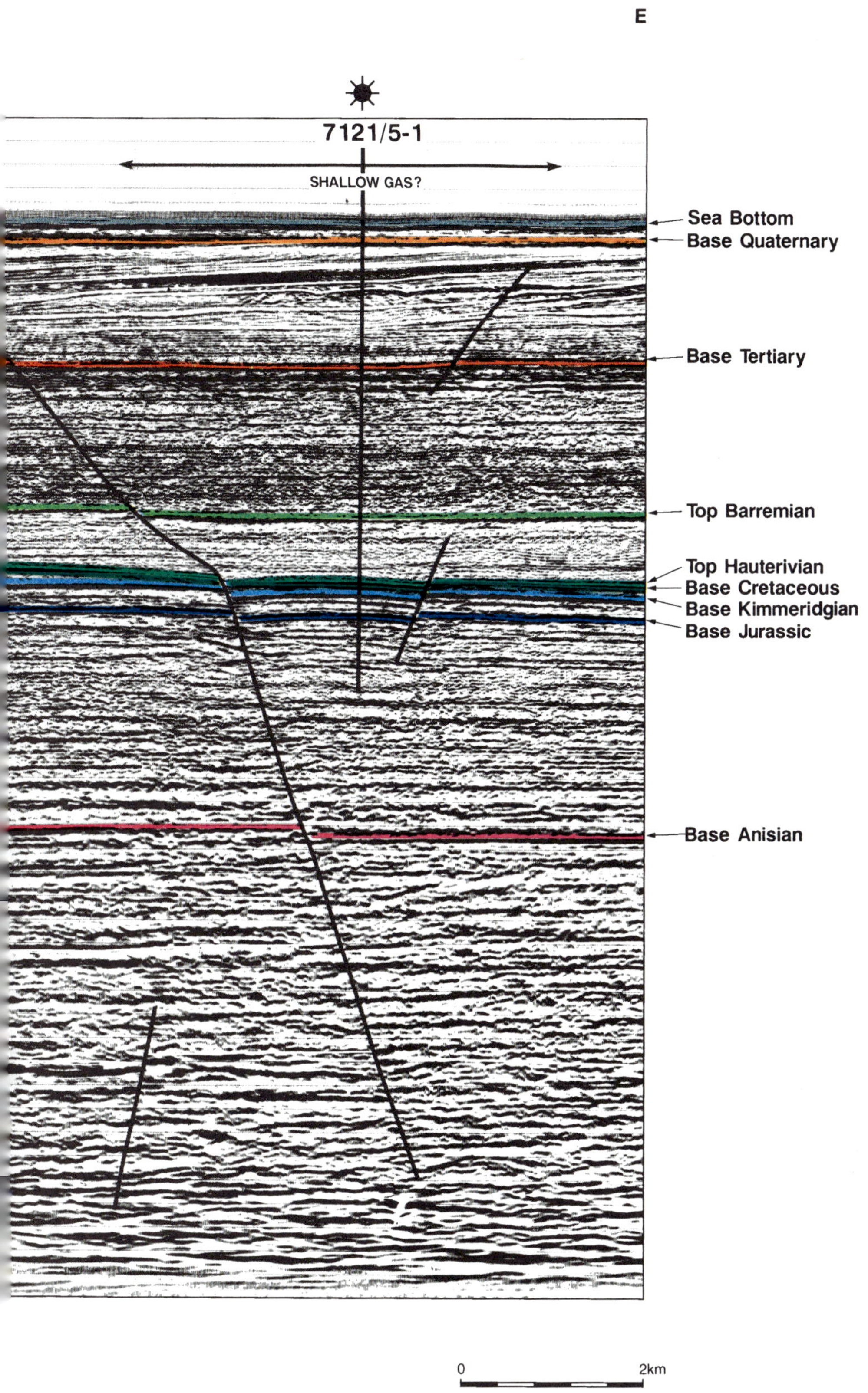

E
7121/5-1
SHALLOW GAS?
Sea Bottom
Base Quaternary
Base Tertiary
Top Barremian
Top Hauterivian
Base Cretaceous
Base Kimmeridgian
Base Jurassic
Base Anisian
0
2km

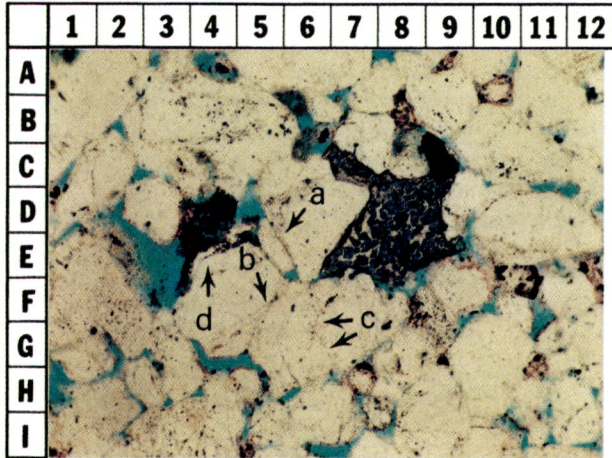

FIGURE 7. Thin section of sample 3.1, from a depth of 2366 m RKB, layer S1, well 7121/5-1. This sample is a well-compacted, fine- to medium-grained, well-sorted sandstone. It consists mainly of monoquartz and displays (a) long, (b) concavo-convex, and (c) sutured grain contacts. The diagenetic effects on the Stø Formation also include an early syntaxial quartz overgrowth phase (d), which has been partly lost by compaction. The open primary intergranular porosity has been reduced by compactional features and various cement types. Scale bar = 100 μm. Plane polarized light.

FIGURE 8. Stylolites in the Stø Formation, well 7121/4-1, layer S3. Stylolites occur every 3 to 4 cm. The stylolites are interpreted as having started from thin mud drapes, and show various degrees of continuity.

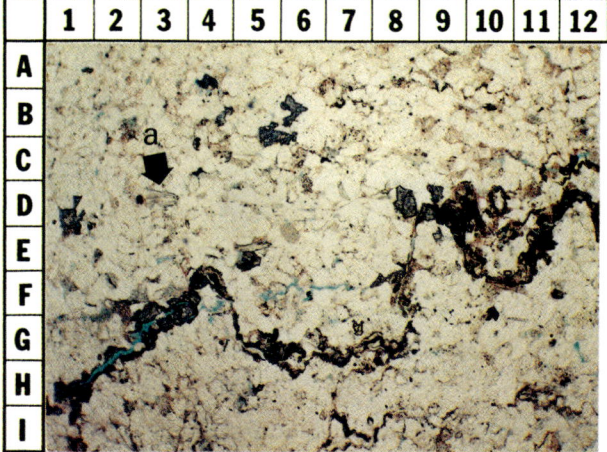

FIGURE 9. Thin section of sample 17.1, from a depth of 2371 m RKB, layer S1, well 7121/5-1. This sample, comprising well-compacted silt to fine-grained, moderately sorted sandstone, consists of monoquartz and some rare muscovite mica flakes (a). A late stage stylolitization phase has caused distinct pressure solution of framework grains, in a sinus-shaped pattern across the picture. Insoluble detritus has been concentrated along the stylolite junction. Scale bar = 500 μm. Plane polarized light.

accompanied by thin coal and coaly shales, in both fining-upward and coarsening-upward sequences. The formation is divided into two layers, N1 and N2. Layer N2 has very poor reservoir characteristics and forms the cap rock of the gas discovery in the Tubåen Formation, which was tested by well 7121/4-1. Layer N1 has poor to moderate reservoir characteristics. The Nordmela Formation was deposited in a coastal or delta plain environment with active distributary feeder channels and possibly tidal channels. There was more marine influence in N2 than in N1, and both layers become more marine westward toward well 7120/5-1. The formation thickness increases from 60 to 105 m (197 to 344 ft) westward across Snøhvit.

The Stø Formation consists of thick sandstones alternating with thin shales and mudstones. The sandstones were deposited in an over-all shallow marine setting, whereas the shales mark transgressive events in an offshore environment. The formation thickens from 70 to 95 m (230 to 312 ft) from west to northwest (Figure 12). For reservoir evaluation purposes, the Stø Formation has been divided into five reservoir layers that are labeled S5 to S1 from bottom to top (Figure 10).

Layer S5, which is the thickest layer of the formation, has the best reservoir properties and correlates easily across the field. It increases in thickness from east to west.

This layer is dominated by medium- to fine-grained, cross-bedded, and intensively bioturbated sandstones. Most beds show fining-upward trends. In intervals where bioturbation has not completely disturbed the primary sedimentary structures, the sandstones show a domi-

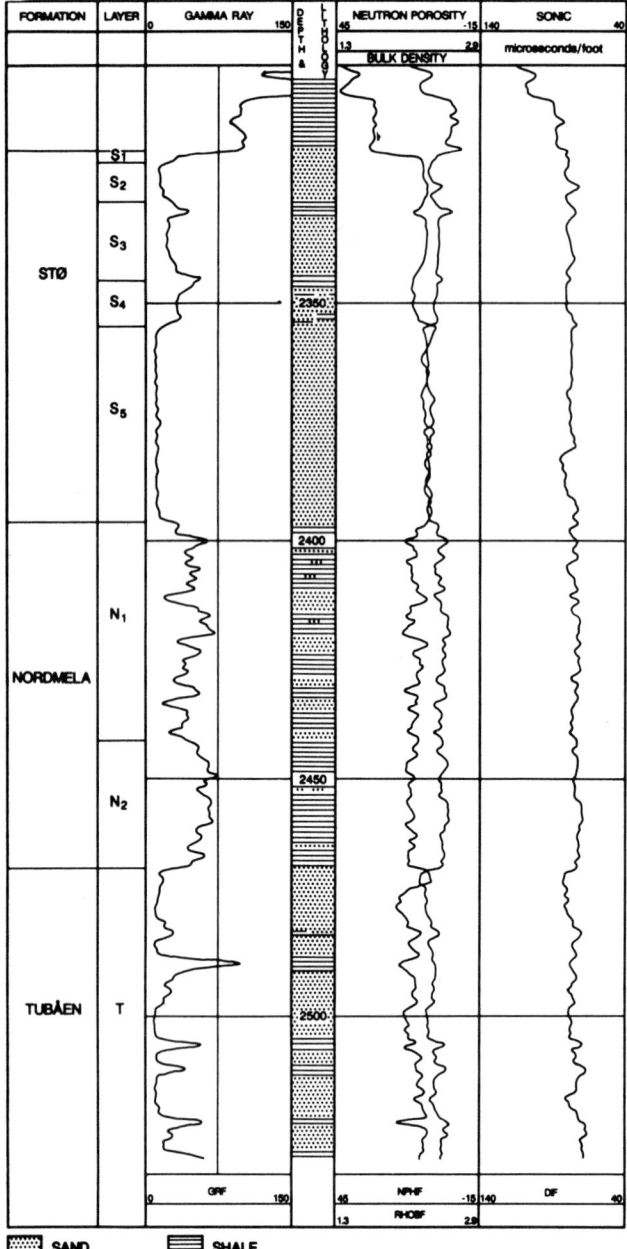

FIGURE 10. Reservoir layering in well 7121/4-1, Snøhvit field.

nance of planar cross-stratification, but associated current ripple lamination and cross-stratification are also present (Figure 11). In wells 7121/5-1, 7121/5-2, and 7121/4-2, flaser bedding and scattered plant material are relatively common. Shell fragments also are common. The sandstones become more bioturbated south of Snøhvit, in the Albatross field. Common trace fossils are *Diplocraterion, Ophiomorpha, Planolites, Skolithos, Chondrites,* and *Arenincolites.*

Layer S5 was deposited in a large, open marine embayment with barriers, lagoons, tidal deltas, and estuaries in the area of the Snøhvit field (Figure 13). Fluvial (delta plain) input probably occurred from both the north and the southeast, but has not been penetrated by any well. Tidal delta and shoreface deposition dominated in the area where the wells were drilled.

The tidal delta deposits are distinguished from normal shoreface deposits by the presence of planar cross-stratification (indicating bar migration), a low degree of bioturbation, an alternation of current ripple lamination and cross-stratification (indicating fluctuating currents), flaser bedding (episodic sand deposition), and associated plant material (closer to the shoreline). Owing to both wave agitation and tidal action, this marine environment was dominated by sand deposition that resulted in relatively uniform primary reservoir properties.

Layer S4 consists of fine- to medium-grained, intensively bioturbated sandstones with reservoir qualities poorer than those of the underlying layer S5. Layer S3 comprises an alternation of bioturbated, very fine grained sandstones and siltstones together with thinner mudstones and shales. Two of the three shale units are present throughout the area. The reservoir quality of the interval is generally poor, and the thin shales represent barriers to vertical flow of fluids.

Layers S3 and S4 show that, during deposition, the delta plain feeder channels retreated, probably as a result of a relative rise in sea level. A lower to middle shoreface, occasionally transgressive, offshore environment existed, with dominance of bioturbated, fine-grained sandstones and siltstones and with vertical and areal variations in reservoir properties (Figure 14).

Layer S2 consists of fine- to coarse-grained, cross-bedded and current ripple laminated sandstone intervals often arranged in fining-upward sequences in the Snøhvit area. However, in the vicinity of well 7120/5-1 to the west, and in wells to the south, layer S2 consists of fine-grained, bioturbated sandstones. The reservoir quality is good. In layer S2, the current ripple and cross-bedded sedimentary structures and the fining-upward character in the Snøhvit area indicate fluvial and/or tidal channel deposits. Farther south and in the western part of Snøhvit, the sandstones were probably deposited in a lower to middle shoreface environment.

Layer S1, the uppermost layer of the Stø Formation, is present in wells 7121/4-1, 7121/4-2, 7120/6-1, and 7120/5-1. It consists of fine-grained, cross-bedded sandstones with a moderate to high degree of bioturbation, often with a basal conglomerate (in wells 7121/4-2 and 7120/6-1). The reservoir quality is poor. The marine bioturbation and the types of sedimentary structures indicate deposition in a middle to upper shoreface environment. The basal conglomerate represents a transgressive, reworked deposit.

During deposition of layers S1 and S2 (Figure 15), the delta plain prograded from both the north and the south toward the open marine embayment, as a result of a drop in sea level. As a consequence, sandstones of fluvial origin rest directly on marine sandstones, and a complete marine to marginal marine to continental transition is missing. This drop in sea level is thus marked by a sequence boundary. The upper part of layer S2 is dominated by current ripple laminated and cross-bedded sandstones of fluvial origin. A probable tidal influence on the channels is observed in wells 7121/4-1 and 7121/5-1. In the central part of the area, outside well control, marine conditions

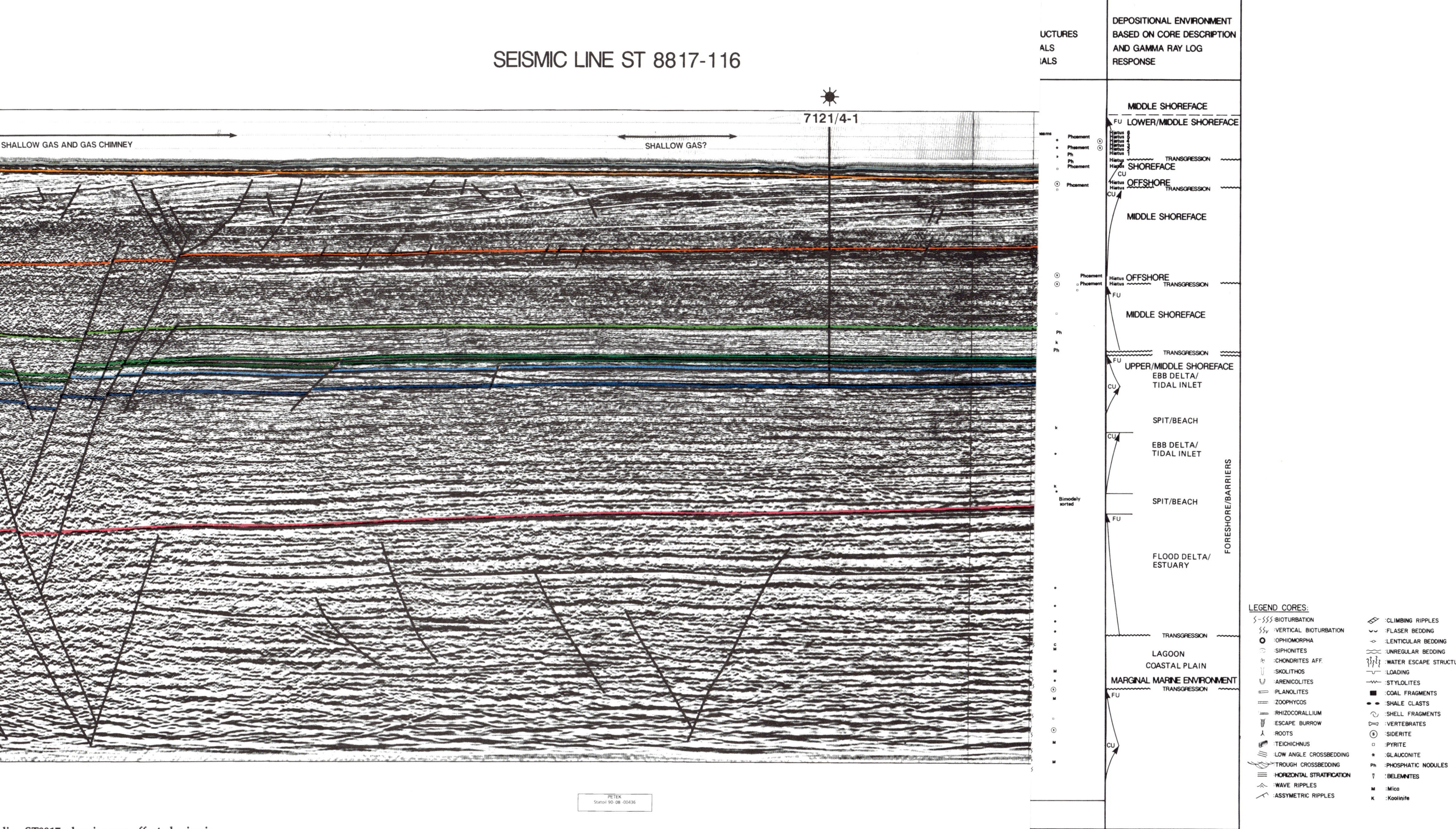

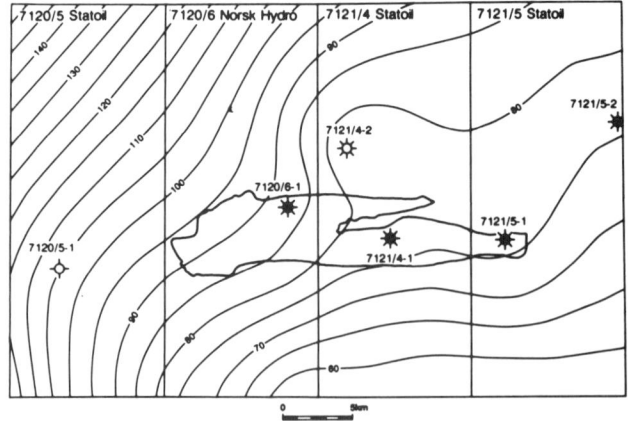

FIGURE 12. Stø Formation isochore (in meters).

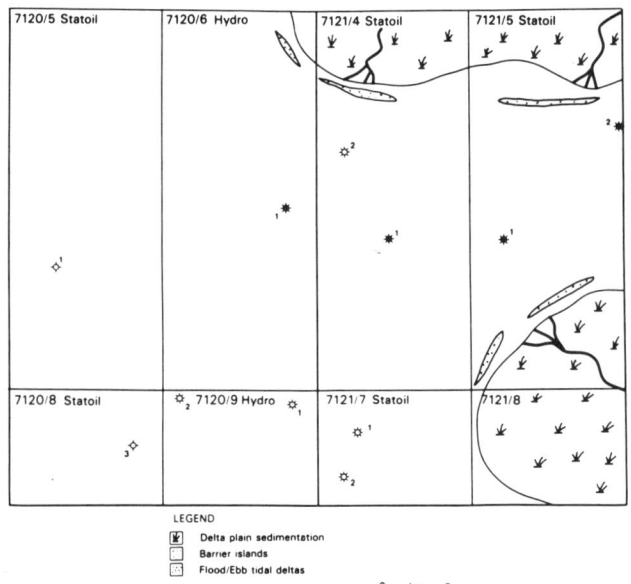

FIGURE 14. Depositional environment of layers S3/S4.

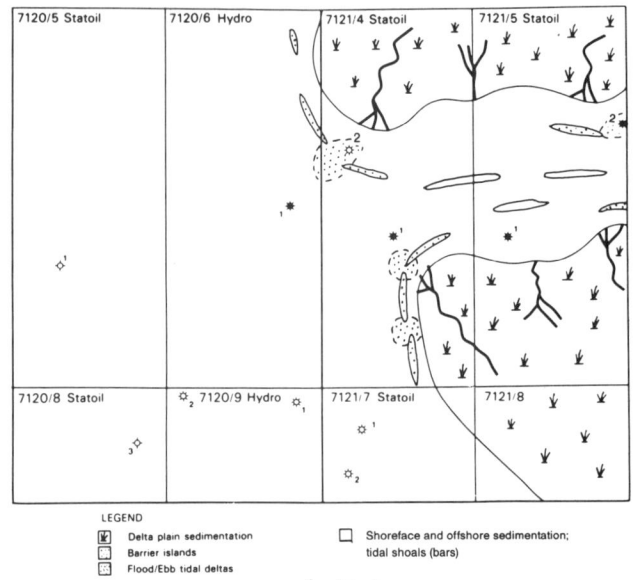

FIGURE 13. Depositional environment of layer S5.

may have existed (Figure 15). Well 7120/5-1 and wells to the south are clearly different from the other wells because of their distal, marine positions, where shoreface deposition continued.

Aquifer

The minor gas accumulation in the Tubåen Formation at the structural crest of the Snøhvit field proves that the lower part of the Nordmela Formation is a barrier to vertical flow. This means that gas production wells in that area will not undergo any aquifer influx from the underlying Tubåen Formation.

Water from the regional aquifer may therefore enter the Snøhvit reservoir only by way of the downflank reservoir rocks to the west or to the east. The lateral extension of the reservoir rocks must be very large, because the same rock is observed in all other wells in the Hammerfest Basin. The available area/volume of the aquifer thus appears to be very large. However, the fault pattern allows only continuous aquifer communication across the faults in an irregular pattern. The fault pattern defining the main Snøhvit horst leaves only a narrow entrance of about 1.5 km (4900 ft) for the western aquifer. The eastern aquifer has to cross two faults that both appear to be sand/shale juxtaposed. This will also restrict the possible aquifer response during a production periods.

Reservoir Properties

The Snøhvit reservoir exhibits two classes of reservoir rock quality. Layers S2 and S5 of the Stø Formation are the best reservoir rocks. They have porosities of about 17% and core permeabilities ranging from 150 to 500 md (Table 1). Layers S1, S3, S4, and N1 have average porosities of about 12% and permeabilities less than 20 md (less than 0.01 md in the shale layers). The three shaly layers included in layer S3 are considered to be barriers to vertical flow during production periods. The average net/gross ratio is 0.89, varying between 0.23 and 1.00. The irreducible water saturation determined from log interpretation and core plug capillary measurements is often less than 10%, and the lowest average value is 2.4% in layer S5. The low water saturation is probably a result of the erosion and uplift, and later reburial, described in the section on erosion and uplift, which resulted in at least two stages of expulsion (drainage processes) of the original water in the reservoir.

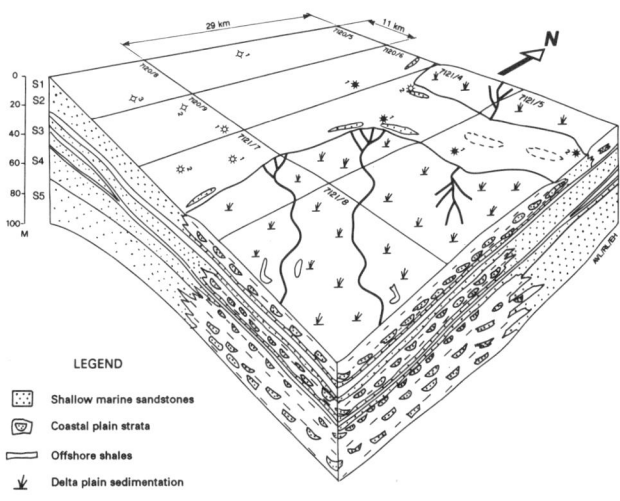

FIGURE 15. Depositional environment of layers S1/S2 with block diagram of the underlying units.

GEOCHEMISTRY

Hydrocarbons

The Snøhvit field contains both gas and oil. The gas cap contains more than 80 mol % methane, 5% CO_2, and 2% N_2. PVT analysis shows that there will be a small liquid dropout (<1%) in the reservoir during production. The liquid will not flow in the reservoir and will not be produced. The initial GLR is 9000 standard m³/standard m³ (50,000 ft³/bbl). The gas deviation factor (Z) is 0.936, and the gas expansion factor (E) is 229 vol./vol.

The Snøhvit oil is medium light, with a stock tank density of 856 kg/standard m³ (33.8° API), a high pour point (20°C), a low wax content (5%), and an above-average aromatic content of C_{10+} (>30%). Figure 16 shows mass fragmentograms of the oil from DST 3 in well 7121/4-1. The initial flash GLR is 138 standard m³/standard m³ (776 ft³/bbl).

TABLE 1. Snøhvit field reservoir summary.

Trap (tilted fault blocks)	
Depth to crest, top of Stø Formation	2280 m MSL (7480 ft subsea)
Lowest closing contour	2470 m MSL (8104 ft subsea)
Area	90 km² (22,200 ac)
Gas-water contact	2404 m MSL (7887 ft subsea)
Oil-water contact	2418 m MSL (7933 ft subsea)
Gas column	124 m (407 ft)
Oil column	14 m (46 ft)
Pay zone	
Formations	Stø and Nordmela
Age	Early/Middle Jurassic
Gross thickness	130–200 m (427–656 ft)
Net/gross ratio, average (range)	0.89 (0.23–1.0)
Porosity, average (range)	15% (10–18%)
Water saturation, average (range)	10% (3–26%)
Permeability, average (range)	200 md (20–500 md)
Temperature at 2404 m MSL	92.8°C (200°F)
Oil properties	
Density, ρ_o	856.3 kg/std. m³ (33.8° API)
Formation volume factor, B_o	1.329 Rm³/std. m³ (1.329 rb/STB)
Gas-oil ratio, GOR	138.0 std. m³/std. m³ (775 ft³/bbl)
Viscosity, μ_o	0.585 mPa · s (0.585 cp)
Bubble point, P_b	267 bar (3872 psi)
Pressure gradient (reservoir), dP/dD	0.071409 bar/m (0.315762 psi/ft)
Compressibility (reservoir), c_o	1.733×10^{-4} bar^{-1} (1.195×10^{-5} psi^{-1})

TABLE 1. (Continued)

Gas properties

Gas gravity, γ (1 = air)	0.7485
Gas deviation factor, Z	0.936
Formation volume factor, B_g	0.00436 Rm³/std. m³ (0.00078 rb/std. ft³)
Expansion factor, E	229 vol./vol.
Gas-liquid ratio, GLR	9000 std. m³/std. m³ (50,000 ft³/bbl)
Viscosity, μ_g	0.02423 mPa · s (0.02423 cp)
Dew point, P_{dew}	267 bar (3872 psi)
Pressure gradient (reservoir), dP/dD	0.019843 bar/m (0.087744 psi/ft)
Compressibility (reservoir), c_g	2.84×10^{-3} bar^{-1} (1.958×10^{-4} psi^{-1})
Condensate density, ρ_c	749.6 kg/std. m³ (57° API)

Water properties

Density ρ_w	1120 kg/std. m³ (69.9 lb/ft³)
Formation volume factor, B_w	1.025 Rm³/std. m³ (1.025 rb/STB)
Viscosity, μ_w	0.40 mPa · s (0.40 cp)
Pressure gradient (reservoir), dP/dD	0.1082 bar/m (0.478445 psi/ft)
Compressibility (reservoir), c_w	4.0×10^{-5} bar^{-1} (2.73×10^{-6} psi^{-1})

Volumes

Gas initially in place, GIIP	160 billion std. m³ (5.6 tcf)
Recoverable gas (70% recovery)	110 billion std. m³ (3.9 tcf)
Oil initially in place, STOIIP	73 million std. m³ (450 million bbl)

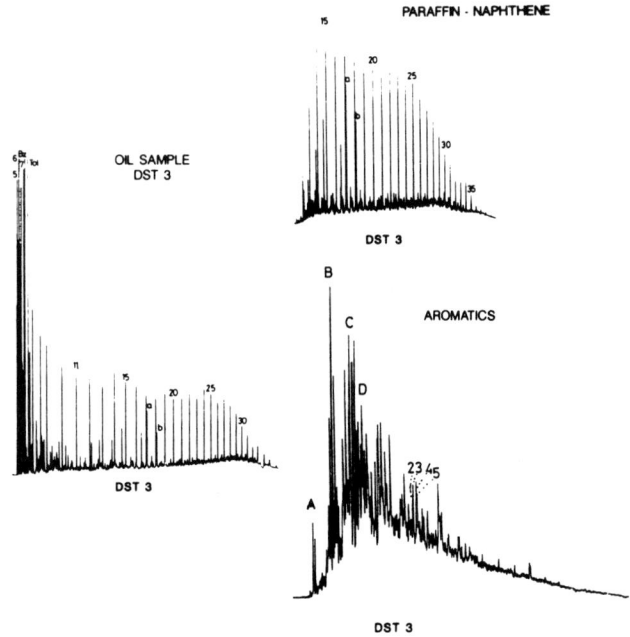

FIGURE 16. Mass fragmentograms of oil from DST 3, well 7121/4-1.

The initial flash oil formation volume factor is 1.329 vol./vol., and the oil viscosity is 0.585 mPa · s (cp) at a pressure of 264 bar (3828 psia) and a reservoir temperature of 93°C (200°F). It is noteworthy that the oil viscosity increases to 2.19 mPa · s when the pressure is reduced to atmospheric pressure.

Source Rocks

There are three possible source rocks: the Late Jurassic anaerobic shales of the Hekkingen Formation, the Early Jurassic Nordmela Formation, and the Triassic shales (Table 2). The Hekkingen Formation is the best source rock penetrated by the Snøhvit wells, and it has a very good potential for light oil, condensate, and gas. The Nordmela Formation is a clay-rich, terrestrial-deposited source rock. It contains negligible coal and has potential for generating waxy hydrocarbons and gas. The wax in the Snøhvit oil was probably generated from the Nordmela Formation. The Triassic shales have mainly a gas potential. In the Hammerfest Basin, these shales are bur-

TABLE 2. Characteristics of Snøhvit source rocks.

Source rock	Age	Thickness m	ft	TOC (%)	Bound hydrocarbon (mg HC/g rock)	Hydrocarbon index (mg HC/g TOC)	Kerogen type	Potential hydrocarbons
Hekkingen Formation	Late Jurassic	10-50	33-164	8-20	10-62	100-475	II/III	Oil/gas
Nordmela Formation	Early Jurassic	10-15	33-49	1-4	2-23	130-250	III/II	Waxy oil/gas
Triassic shales	Triassic	0-60	0-197	2-8	55-40	200-590	III/II	Gas/minor oil

ied deeply and have generated considerable volumes of gas.

Maturation

The Hekkingen and Nordmela formations are immature/early mature in most of the Hammerfest Basin, but to the northwest they are in the oil window. In the Tromsø Basin, farther west, both formations are in the gas window. The Triassic shales have been buried relatively deeper in the Hammerfest Basin, with most of the shales in the gas window.

Migration/Filling of the Reservoir

The three source rocks in the Hammerfest Basin have generated both gas and oil. PVT simulations have shown that the Snøhvit oil could not have released gas of the same volume as the Snøhvit gas cap of today. Our interpretation is that either the Snøhvit reservoir was filled with oil and gas simultaneously, or that the reservoir was partly filled with gas before the oil entered the trap.

The Snøhvit hydrocarbons are probably a mixture of the three source rocks. The migration took place in the Late Cretaceous to early Tertiary. Our interpretation is that it occurred partly vertically through the overlying strata and partly "horizontally" through reservoir sands and fault planes, as shown in Table 3.

LEAKAGE FROM THE RESERVOIR

The Snøhvit reservoir is not completely filled. The oil-water contact is almost 50 m (164 ft) above the structural spillpoint. The estimates of the volumes of generated and migrated hydrocarbons suggest that the reservoir should be filled with hydrocarbons, implying that leakage from the reservoir has occurred. Another observation supporting leakage from the reservoir is that the seismic has been affected by a gas chimney in the western part of the Snøhvit field (Figure 3). In addition, some less-severe leakage has been mapped north of well 7121/4-1.

The leakage theory is also supported by geochemical and petrophysical analyses, which show high hydrocarbon saturation below the oil-water contact. In well 7121/5-1, extraction of organic matter with separation between nonhydrocarbons and extracted aromatics (that have not vaporized since the core was cut) shows a gradually decreasing concentration to 19 m (62 ft) above the gas-oil contact (Figure 17). These high concentrations of aromatics in the gas zone are interpreted as traces of the oil that filled these pores earlier. This suggests that oil leakage has taken place from the Snøhvit reservoir. For the other two Snøhvit wells, the geochemical data are not as clear, but

TABLE 3. Interpretations of migrations from the Snøhvit source rocks.

Source rock	Location	Time of primary migration	Dominating type of migration
Hekkingen Formation	Mainly the area northwest of Snøhvit	Middle Tertiary	Oil, horizontally, via reservoir sands and fault planes
Nordmela Formation	Mainly the area northwest of Snøhvit	Middle Tertiary	Oil, horizontally, via reservoir sands and fault planes
Triassic shales	A position relatively deeper than the other source rocks in the Snøhvit area	Late Cretaceous-early Tertiary	Gas, vertically, through sands and shales

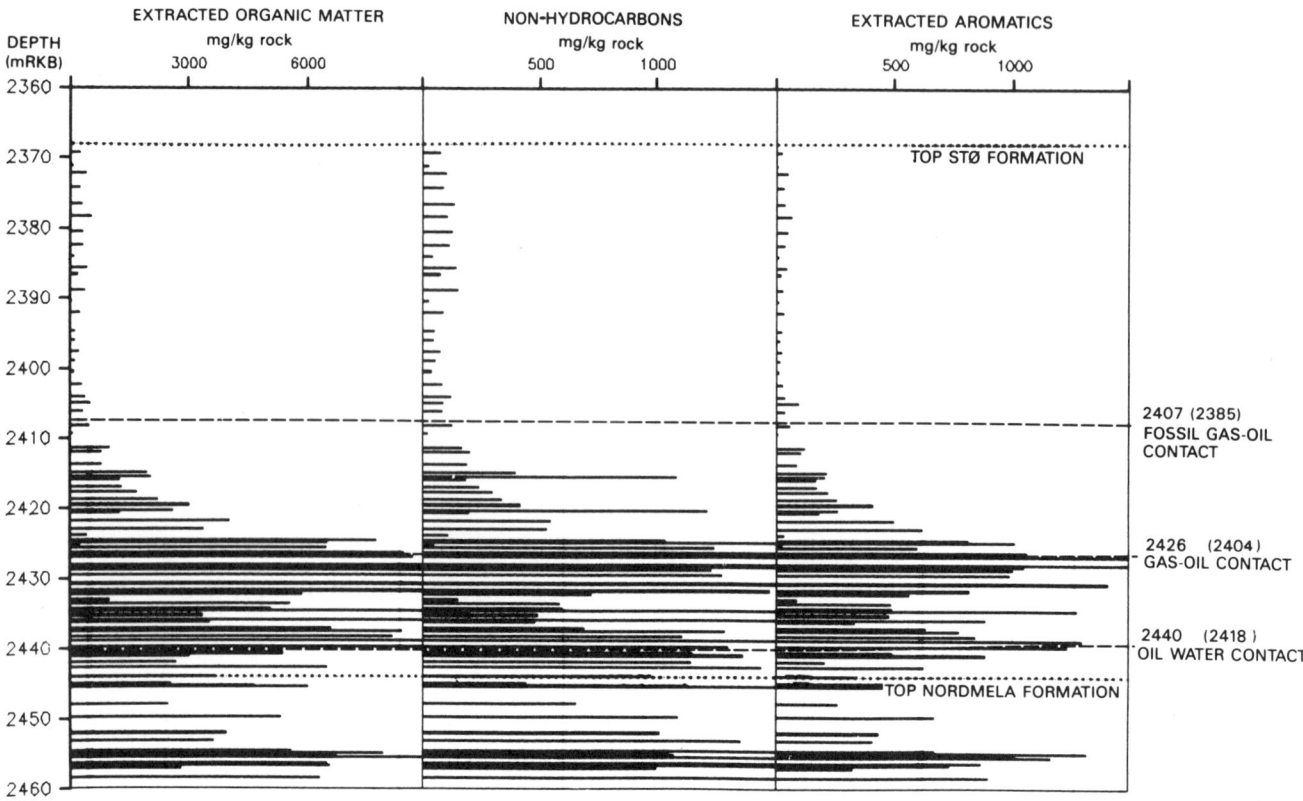

FIGURE 17. Geochemical analysis of well 7121/5-1.

both show similar gradually decreasing concentrations of extracted aromatics into the gas column. The fossil gas-oil contact is interpreted as being at 2385 m MSL (7825 ft subsea).

SEAL

The Snøhvit field has two shale formations sealing the hydrocarbons: the Fuglen Formation and the Hekkingen Formation.

Lithology

The lower part of the Upper Jurassic shales, the Fuglen Formation, which is 7 to 30 m (23 to 98 ft) thick, is a mudstone interbedded with thin limestones giving characteristic log responses (gamma ray, sonic, and density). The shales are dark to light brown, are occasionally micromicaceous, are rich in pyrite nodules, and contain pyritized wood fragments from 1 to 100 mm in length. In well 7120/6-1, 13.5 m of core was recovered from the Fuglen Formation. About 1% of the shales are silty, based on a visual description under the microscope.

The Hekkingen Formation, with a thickness of 60 to 80 m (97 to 262 ft), consists of brownish gray to very dark gray shale and claystone with thin interbeds of limestone/dolomite and siltstone. The color becomes darker, and the organic content increases, toward the base. Microscopic pyrite and mica are present. This formation, which is firm to hard, increases in thickness northward to more than 100 m (328 ft).

Brittleness

X-ray diffraction analyses of core samples (wells 7121/4-2 and 7120/6-1) from the Fuglen Formation, and of sidewall cores from the Fuglen and Hekkingen formations, show a mineralogical composition of about 35% quartz, 15% pyrite and siderite, 15% kaolinite, 20% carbonate, 5% plagioclase, and 10% mica. The ratio of the combined cement and quartz content to the content of the softer kaolinite and mica is used as an estimate of brittleness. The higher the cement content (pyrite, siderite, dolomite, and calcite) relative to the kaolinite and mica content, the higher the brittleness of the shales. Of the two cap rocks, the Fuglen Formation is the softer rock.

Rock Strength

The core samples had not been correctly preserved between the time they were collected and the mechanical tests performed five years later. The samples had dried out, and thus the measurements probably did not give an accurate estimate of the rock strength. Nevertheless, most of the deformational tests of the Fuglen shale samples required triaxial pressures of 50 to 100 bar (725 to 1450 psi) before the samples fractured. This is surprising, even considering the poor preservation of the cores, and suggests a relatively high mechanical strength.

Pore Pressure

An increased pore pressure in the cap rocks is indicated by an increased rate of penetration and increased gas content in the mud. In addition, decreases in the drilling exponent and in the shale density are observed in the area.

There is a relationship between the gamma ray log on the one hand and the gas content in the mud and the pore pressure on the other—i.e., increases in gamma ray readings are accompanied by increases in gas content and pore pressure. The highest pressure is interpreted as occurring near the highest gamma ray readings. For the Snøhvit reservoir, this means that the best seal, in terms of pore pressure, is in the lowest part of the Hekkingen Formation. This is one of the reasons for interpreting the Snøhvit field as having two cap rocks acting as seals for the hydrocarbons. The typical Snøhvit cap rock pore pressure is displayed in Figure 18, together with the typical gamma ray, resistivity, and sonic logs. When the top of the Hekkingen Formation is entered, the gamma ray and resistivity readings increase whereas the sonic reading decreases. The Hekkingen Formation exhibits the highest gamma ray readings of the two Upper Jurassic shale formations.

Permeability

Two measurements of permeability were made on unpreserved samples from the Fuglen Formation. Attempts were made to restore the samples after the five years of unfavorable storage. The corrected permeability values were 0.000037 and .00015 md.

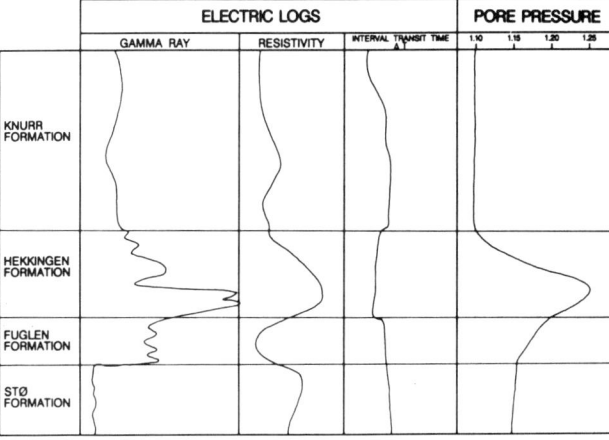

FIGURE 18. Typical pore pressure and log character vs. depth in the Snøhvit cap rocks.

Geochemical Analysis

The Hekkingen Formation is the best source rock in the area. This has complicated the geochemical analysis of the possible leakage of light hydrocarbons through the seal. For the heavier hydrocarbons, several analyses have shown that there has been no vertical migration of heavier hydrocarbons into the strata above the reservoir.

Borehole Stress and Fault Sealing Capacity

The major stress directions and their relative importance affect the sealing capacity of major, bounding faults. In general, faults that are oriented 90° to the horizontal ("extensional") stress direction are most likely to fail in sealing. The greatest horizontal stress in a borehole occurs where the largest horizontal stress direction is tangent to the borehole. Consequently, this is also the location where the borehole wall will be the most unstable and where the deepest breakouts will occur. If a well is drilled through strata with anisotropic stress, the borehole will become oval as a result of the increased incidence of breakouts. The ovalization of the boreholes in geologic strata deposited after the time of hydrocarbon migration gives information on the paleoregional stress. The ovalization, or the breakouts in the boreholes, has been recorded by a four-arm caliper tool. The recorded variations in hole shape vs. depth for wells 7120/5-1, 7121/4-1, and 7121/5-1 are organized and interpreted below.

The Snøhvit reservoir is bounded by three east–west-oriented normal faults and several smaller faults oriented in various directions (Figure 2). Several of the faults cut the section from Triassic (or older) to Tertiary. The gas- and oil-filled Stø Formation is fault-juxtaposed against the Lower Cretaceous Knurr and Kolje formations (Figure 5).

The trapping of the Snøhvit hydrocarbons is interpreted as having occurred in the early Tertiary. As a result of regional uplift and erosion, only 500 m of Tertiary strata and about 100 m of Quaternary strata overlie the Cretaceous.

Ideally, this interpretation should have been performed on strata deposited after hydrocarbon migration and after the regional uplift and erosion. The exact times of these events are not known. Because some faults cut into the lower Tertiary, it is acceptable to use the borehole breakout information from the lower Tertiary to evaluate the sealing capacity of the Snøhvit faults.

For reference, the upper Albian interval is shown in Figure 19. For the upper Albian interval, the S_h (horizontal stress) azimuths are north-south. Regarding the sealing capacity of the fault planes, it can be concluded that the Snøhvit reservoir at that time was in a favorable anisotropic stress situation that closed off the east–west-oriented, major faults. However, this was before the time of hydrocarbon migration and is of minor importance in explaining the leakage from the Snøhvit reservoir. In the lower Tertiary Torsk Formation, however, it is noticeable how different—almost opposing—the greatest horizontal stress directions are over a distance of 20 km (Figure 20). Only four borehole breakouts occur in well 7121/4-1, whereas there are 60 in well 7120/5-1.

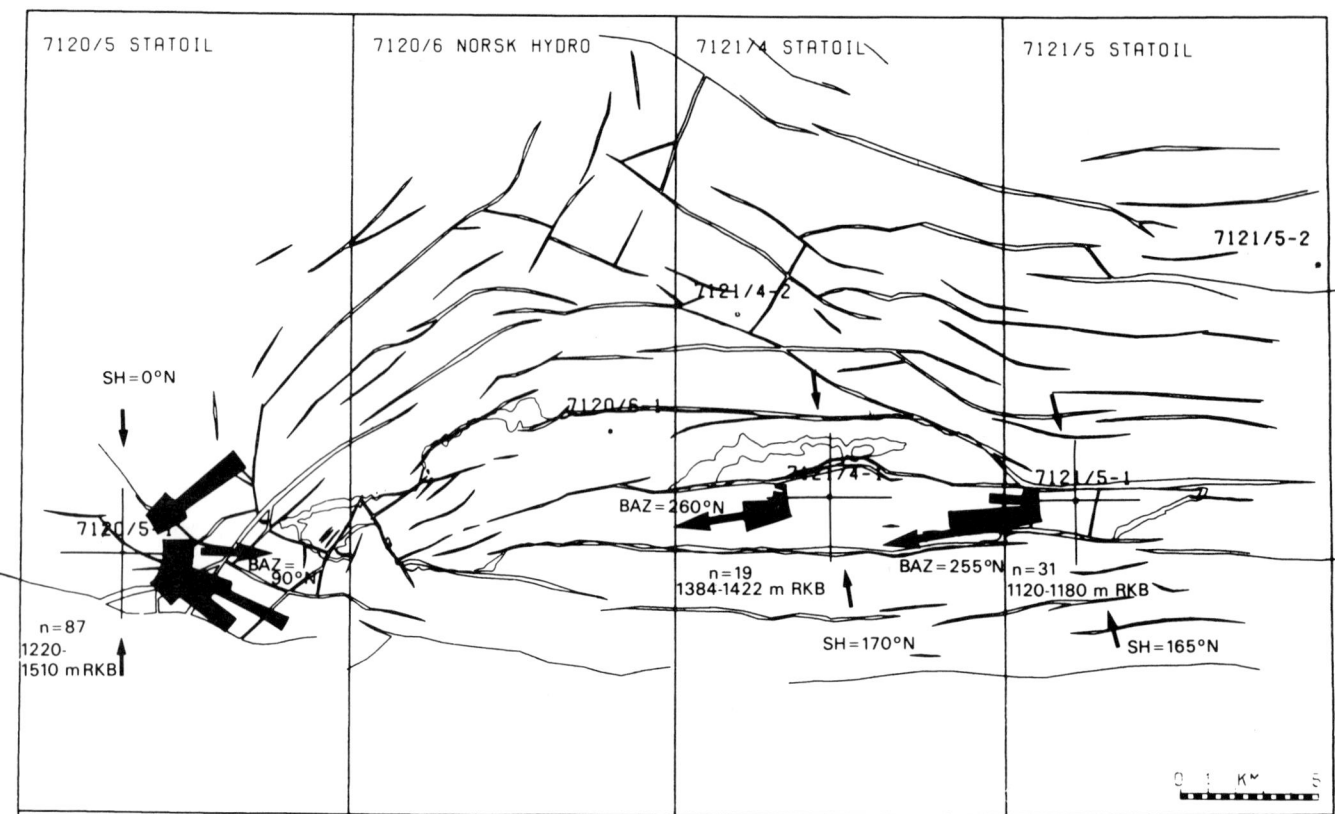

FIGURE 19. Borehole breakouts in upper Albian Kolmule Formation.

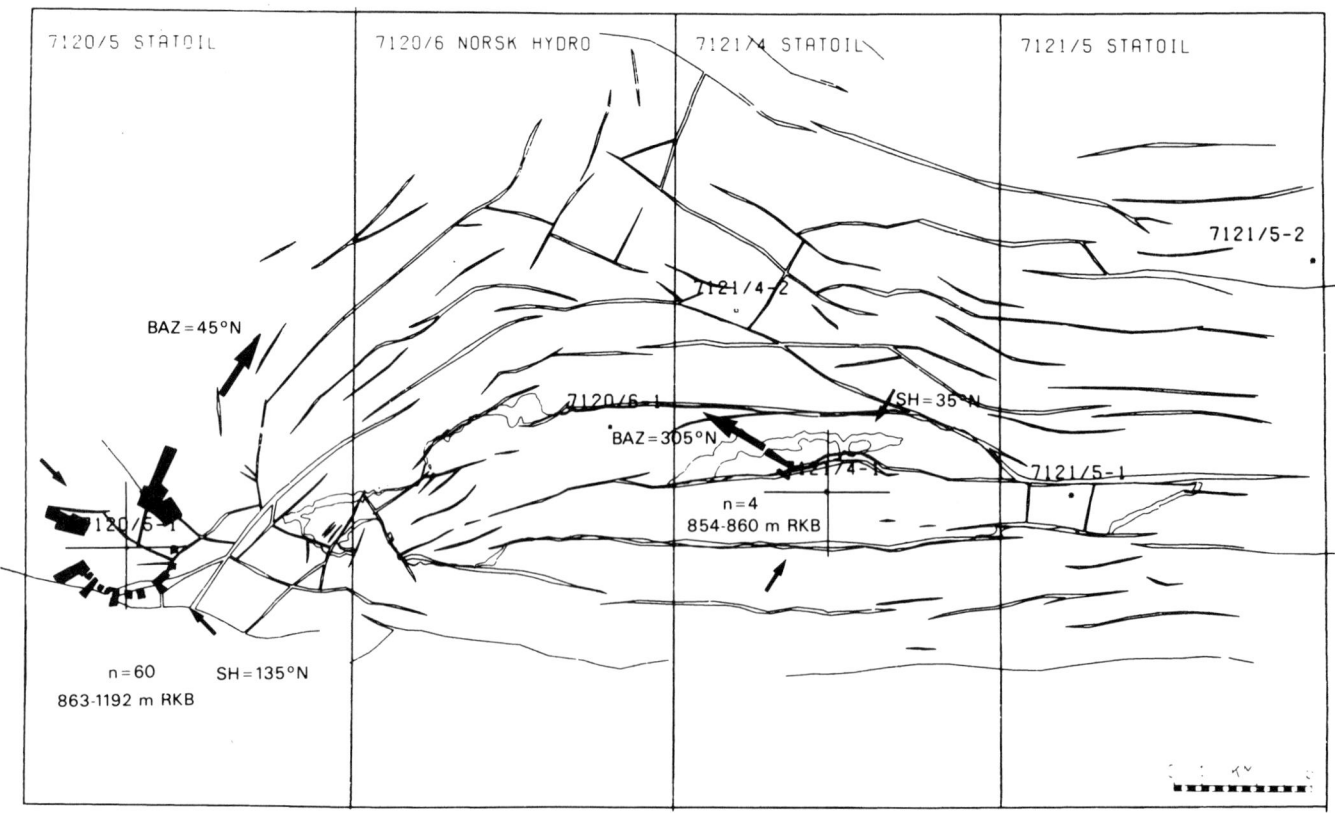

FIGURE 20. Borehole breakouts in Paleocene-Oligocene Torsk Formation.

Figure 20 shows borehole breakouts in only the two wells to the west. The population of breakouts is lower in well 7121/4-1 than in well 7120/5-1. This means that the central and eastern areas of the Snøhvit structure have been subjected to a more isotropic stress field than the western area. In other words, the anisotropic horizontal stress observed in well 7120/5-1 has been released between wells 7120/5-1 and 7121/4-1, perhaps along the weakest fault planes. For 7120/5-1, the borehole breakout azimuth is 45°N and the maximum compressional horizontal stress azimuth is 135°N.

The information from well 7120/5-1 in Figure 20 has been used to analyze the faults. They can be divided into three groups on the basis of the three dominating fault plane orientations (Figure 21). The azimuth of each fault plane normal (the fault pole) is plotted on a frequency diagram in Figure 22 and shows the direction of downfaulting. The plotting of the fault poles is combined with the interpreted, postmigration, anisotropic stress from well 7120/5-1. The corresponding compressional and extensional stress regimes are shown in Figure 22. It can be seen that it is the faults in group 3 that are oriented in the extensional stress regime. Consequently, according to this analysis, the faults in group 3 are the faults that will have the poorest sealing capacities.

A dry prospect was tested by well 7120/5-1. The normal to the fault northeast of the well location (group 3, labeled fault p in Figure 22) is oriented in the extensional stress regime, very close to the azimuth for maximum extensional, horizontal stress.

One of the westernmost faults bounding the Snøhvit reservoir (group 3, labeled fault q in Figure 22) is also situated close to the maximum extensional stress. It is therefore natural to propose that (some of) the leakage has occurred vertically through the p and q fault planes. These faults are in the areas where the seismic is affected by leaked gas (Figure 21).

EROSION AND UPLIFT

Several observations point to the conclusion that much of the sequence in the Hammerfest Basin has been buried at depths greater than its present depth. Shale densities are higher than expected for today's depth and overburden, suggesting that the compaction of the shale has resulted from a previously deeper burial. Log measurement of bulk density in well 7121/4-1 at the top of the reservoir (2300 m Rotary Kelly Bushing, or RKB) yields a value of 2.51 g/cm^3, whereas just below 1000 m RKB, the shale density is 2.31 g/cm^3 (Figure 23). These are the example values used in Figure 24. The Sclater and Christie (S & C) curve shows a normal development of density vs. depth from a study of the central North Sea Basin. The two example densities equate to burial depths of 1900 m (2.31 g/cm^3) and 3200 m (2.51 g/cm^3), instead of 1000 and 2300 m, respectively, suggesting that the top of the Snøhvit reservoir has been buried 900 m deeper than it is today.

The sonic log transit time is generally lower than in the North Sea (Figure 23). At 2300 m, it is 80 μsec/ft,

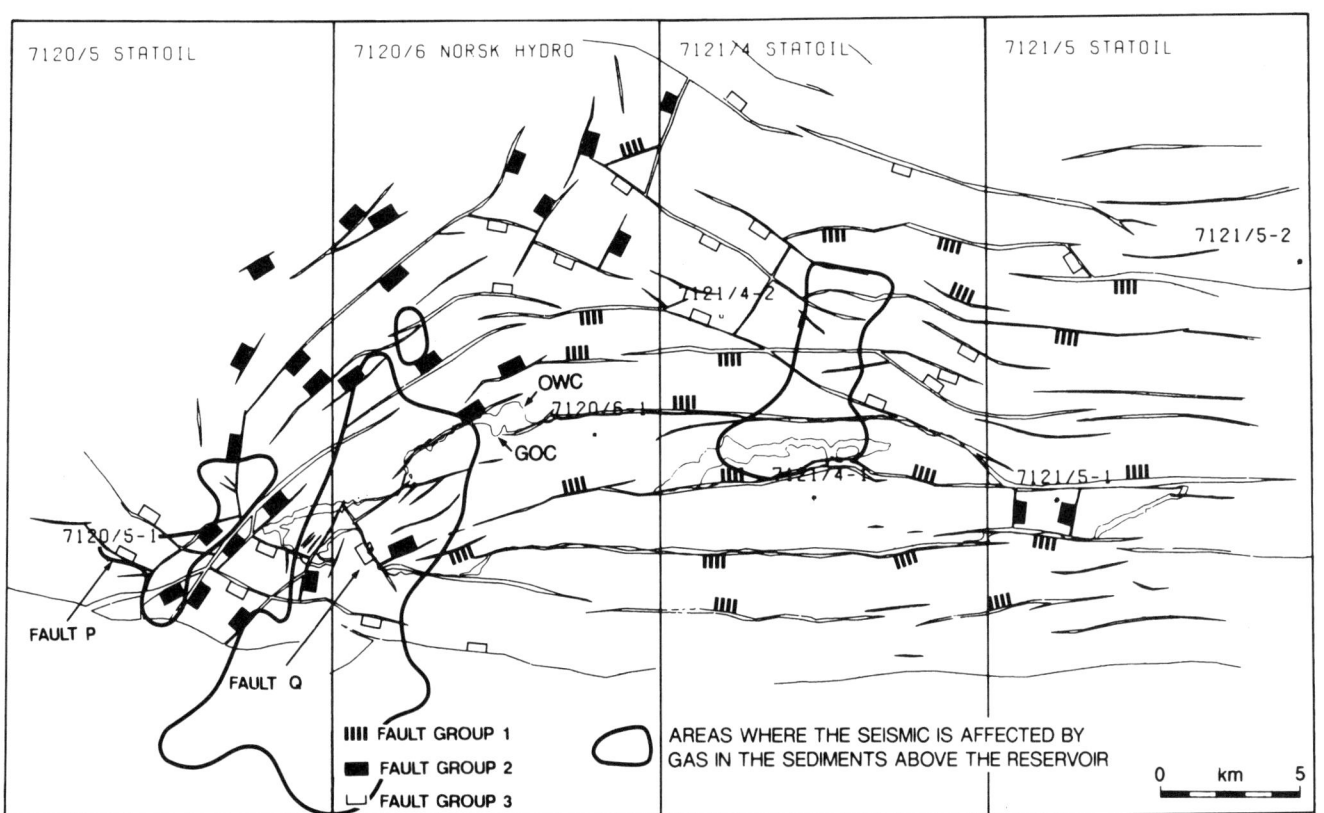

FIGURE 21. Fault groups 1, 2, and 3.

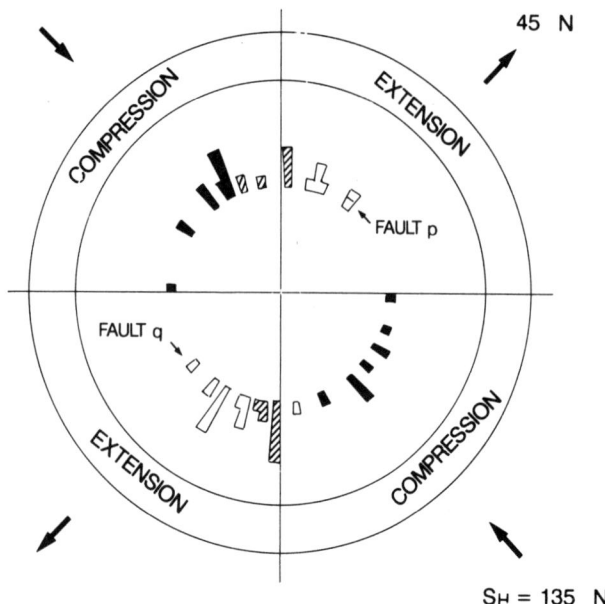

FIGURE 22. Fault pole plot for the Snøhvit area combined with horizontal, postmigration stress directions interpreted from the Torsk Formation in well 7120/5-1.

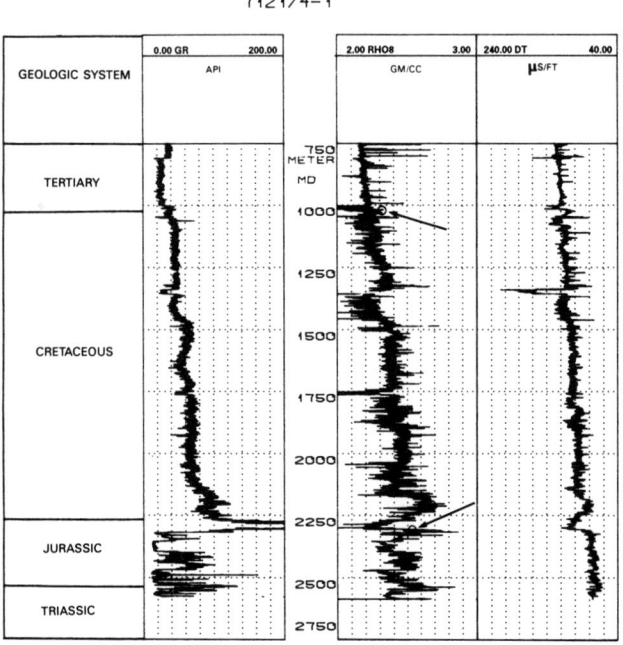

FIGURE 23. Sonic and bulk density logs for well 7121/4-1.

whereas in the northern North Sea, in the Jurassic at 2300 m, the typical value is 90 to 110 μsec/ft. This also indicates that the reservoir has been buried deeper than it is today.

Stø Formation stylolites, with an average spacing that varies between 2 and 40 cm, increase in abundance with depth. This indicates a previously greater depth within the Hammerfest Basin.

There has been a reduction in porosity due to compaction, through the processes of quartz grain dissolution and overgrowth cementation (Figure 7). Studies have shown that these processes started at a depth of about 2600 m, with a maximum depth of burial of about 3000 m, indicating a former burial at least 700 m deeper. This has been inferred from temperature information, fluid inclusion studies, isotopic studies, and clay mineral transformation studies done by Norsk Hydro.

Vitrinite reflectance vs. depth for well 7121/5-1 is shown in Figure 25. By applying $R_o = 0.2$ as the surface value, 1100 m of erosion can be inferred.

The observations presented above strongly indicate that the Snøhvit field has been buried 1000 m (3280 ft) deeper than it is today. The figure of 1000 m is taken as an average of the indicators listed above and gives only the correct order of magnitude of the previously deeper burial.

Burial History

Within the stratigraphic section of the Snøhvit field, there is a major unconformity between the Paleocene and Pliocene–Pleistocene strata. This means that the information given above has to be used to interpret the burial history for the 50 million years of missing section. Figure 26 presents a burial history of the Stø Formation, based on well 7121/4-1, offering the interpretation that the initial depth of burial was 1000 m greater than today's depth, followed by uplift, erosion, and subsequent reburial.

Effect on the Reservoir Fluids

Hydrocarbon migration took place in the early Tertiary, before uplift and erosion. During uplift, the reservoir pressure would have been reduced, allowing the gas cap to expand and to force the gas-oil contact downward, thus leaving a fossil gas-oil contact in the gas zone. Considerable volumes of oil must then have been spilled out through one or more spillpoints, probably at the major faults.

Subsequent renewed burial affected the positions of the fluid contacts. Increased pressure transformed parts of the gas cap to solution gas in the oil, and the aquifer pressure forced the position of the oil-water contact upward. In Figure 27, an east-west cross section shows the three main steps in the interpretation described above. The interpretation concludes that the Snøhvit reservoir must have contained more oil than it does today.

Figure 27 also shows an interpretation of the changes in water saturation vs. depth that occurred during the

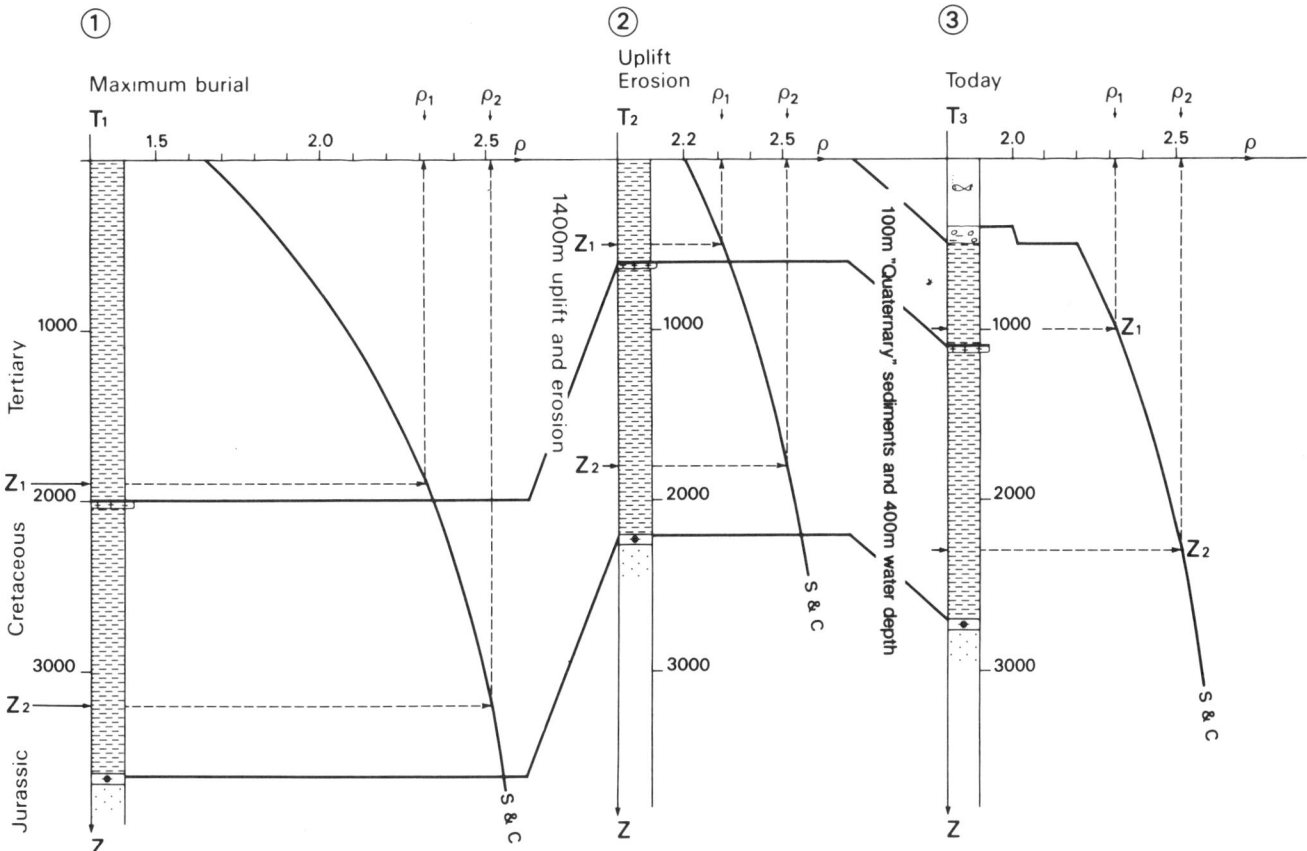

FIGURE 24. Principle for calculating uplift and erosion in the Hammerfest Basin. The Tertiary is estimated to have had a maximum thickness of 2000 m. By comparison of the bulk densities on the S & C curve, one can see that the Z_1 sediments, previously buried to 1900 m, occur at 1000 m today. For the Z_2 sediments, the depths are 3200 m and 2300 m, indicating 900 m of uplift. Today, there are 500 m of Tertiary sediments above Z_1; this suggests 1400 m of eroded section (from L. N. Jensen, personal communication).

three main events described above. The high residual hydrocarbon saturation below the oil-water contact observed in the Snøhvit wells supports this interpretation.

FIELD DEVELOPMENT

The hydrocarbon volumes in place in the Snøhvit field are 160 billion standard m³ (5.6 tcf) of gas and 73 million standard m³ (450 million bbl) of oil. Various sensitivity estimates show that the in-place volumes may vary by ± 20%. The sensitivities relate to uncertainties in depth conversion, migrated positions of bounding faults, two-way time correction in the gas-affected area, the petrophysical interpretation, and possible residual oil within the gas cap. Figures 28 and 29 show the areal distributions of proven gas and oil, respectively.

The 14 m thick oil zone is presently not considered profitable to develop. The recorded low flow rates caused by gas and/or water coning have so far left no hope for development of oil production. Recently, however, horizontal wells have been successful in producing thin oil zones in the North Sea.

Development of the Snøhvit gas has far greater potential, and reservoir simulations indicate that a 70% recovery factor for gas is realistic. As yet, there are no firm gas development plans for Snøhvit, but various development scenarios are being studied. These scenarios include fixed/floating platforms or subsea completion, all with the gas being piped to shore to a dedicated LNG plant. The studies continue with the development of new technology. The trigger for development will be an agreement on sale of the Snøhvit gas. There are expectations that the sleeping Snøhvit will be awakened in the late 1990s.

CONCLUSIONS

The Snøhvit field, discovered in 1984, is the largest gas find and the first oil discovery in the Norwegian Barents Sea. The field straddles blocks 7120/6, 7121/4, and 7121/5 in a water depth of 300 m. It covers an area of 90 km² (22,200

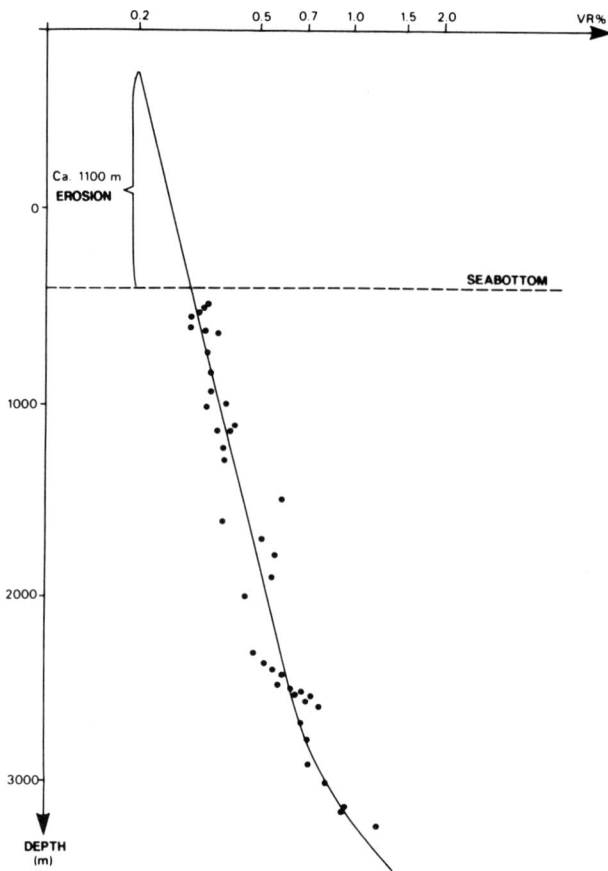

FIGURE 25. Vitrinite reflectance profile for well 7121/5-1.

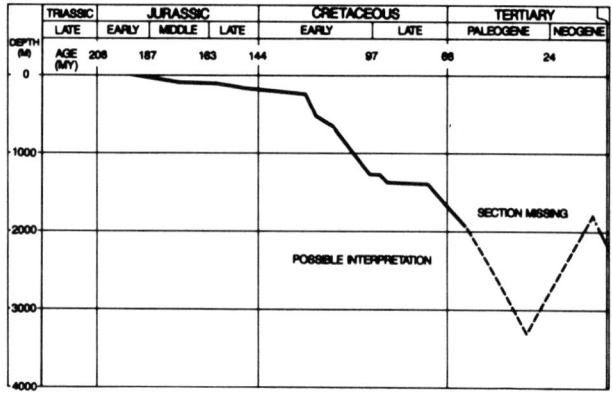

FIGURE 26. Interpreted burial history for the Middle/Lower Jurassic Stø Formation.

ac) and has a gas column of 124 m (407 ft) overlying a 14 m (46 ft) thick oil leg.

The reservoir consists of Lower to Middle Jurassic sandstones deposited in a transgressive coastal to inner shelf sequence. Three wells, each situated in a separate fault block, define the Snøhvit field. The wells have common fluid contacts. Reservoir properties are fair, with a porosity of 15% and permeability ranging from 200 to 500 md in the main part of the reservoir. The water saturation in the gas zone is low, varying from 3 to 26% and averaging 10%.

A burial history involving uplift, erosion, and reburial in the Tertiary has influenced the positions of the fluid contacts and the distributions of oil and gas in the reservoir. The cap rocks show good vertical sealing capacity. The observed leakage is interpreted as having occurred along the fault planes.

The most likely estimate of gas in place is 160 billion standard m^3 (5.6 tcf), and the gas has a recovery factor of 70%. The oil in place is estimated at 73 million standard m^3 (450 million bbl). Because of the thin oil leg and the areal distribution of the oil in the reservoir, development of the oil with present-day technology is not thought to be economically feasible. Various development scenarios have been studied for the Snøhvit gas, but currently there are no firm development plans.

ACKNOWLEDGMENTS

The authors would like to thank the Snøhvit owners for permission to publish this paper. The following Statoil personnel contributed work for the licensees of which parts have been used in this paper, and also contributed to numerous discussions about the Snøhvit field: O. Skarpnes, L. N. Jensen, R. Johansen, R. Lundstrøm, J. Skagen, L. O. Olsen, B. Moltu, and P. E. Eliassen. A. D. Hughes of Redwood Corex did thin section analysis. P. Rowe of Total contributed suggestions for improving the English. Thanks go also to T. Oliversen and E. Heitman of Statoil, who drew the figures, and to A. M. Spencer for reviewing the manuscript.

REFERENCES CITED

Foyn, R., 1989, Geologiske forutsetninger og forventninger til resurspotensialet, in Naturdatakonferanse, Harstad, September 26-28, 1989, p. 17-21.

Olsen, R. G., and O. K. Hansen, 1987, Askeladd in A. M. Spencer et al. (eds.), Geology of the Norwegian oil and gas fields: Graham and Trotman, p. 419-428.

Sclater, J. G., and P. A. F. Christie, 1980, Continental stretching. An explanation of the the post-mid-Cretaceous subsidence of the central North Sea Basin: Journal of Geophysical Research, v. 85, p. 3711-3739.

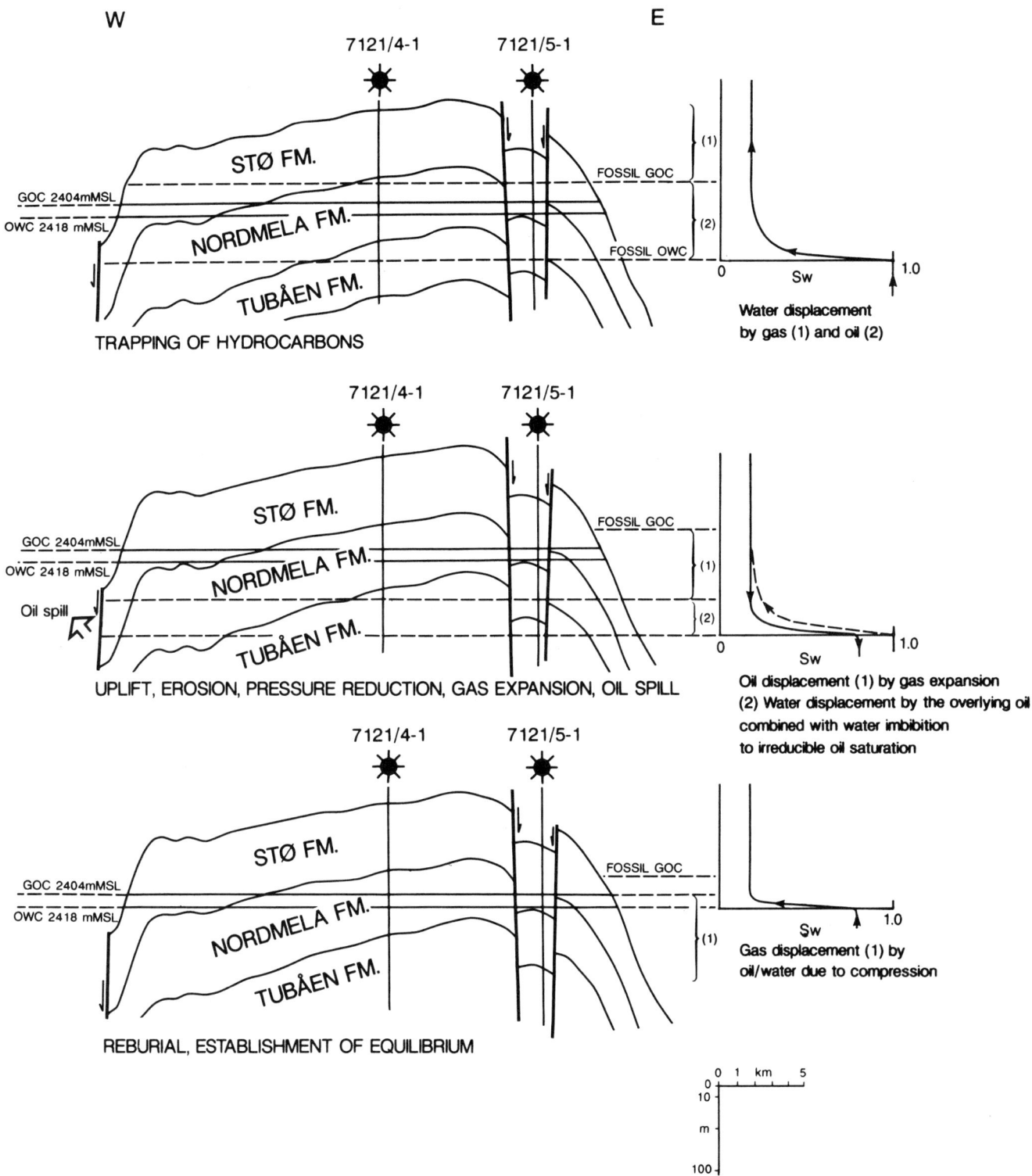

FIGURE 27. East-west cross section through the Snøhvit field, showing the effects of burial, uplift, erosion, and reburial.

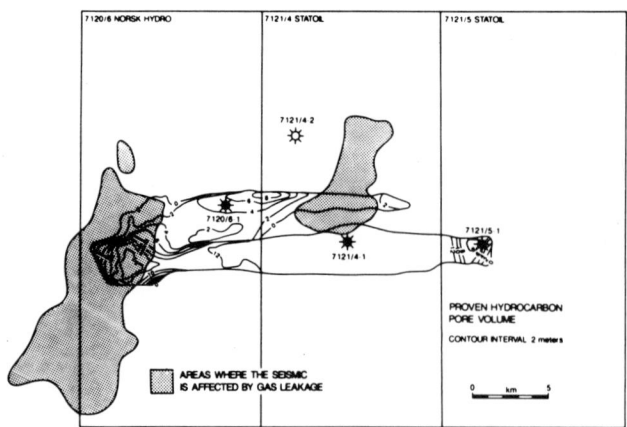

FIGURE 28. Iso-gas map of Stø and Nordmela formations.

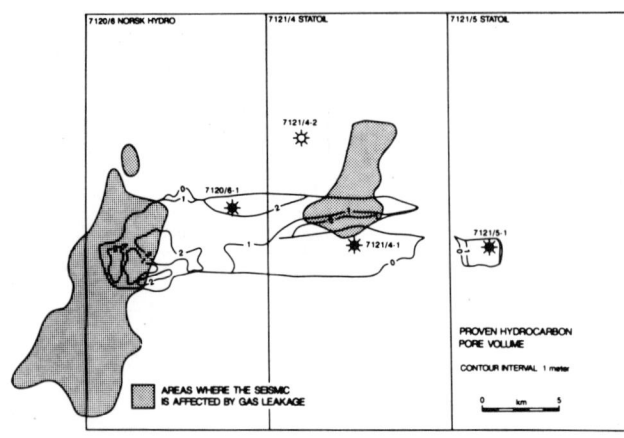

FIGURE 29. Iso-oil map of Stø and Nordmela formations.

Sund, T., O. Skarpnes, L. Nørgård Jensen, and R. M. Larsen, 1984, Tectonic development and hydrocarbon potential offshore Troms, northern Norway: AAPG Memoir 40, p. 616–626.

Worsley, D., S. E. Kristensen, and R. Johansen, 1988, A lithostratigraphic scheme for the Mesozoic and Cenozoic succession offshore Norway north of 62° N: Norwegian Petroleum Directorate.

Chapter 23

Draugen Oil Field, Haltenbanken Province, Offshore Norway

Donald M. J. Provan

A/S Norske Shell
Tananger, Norway

ABSTRACT

The Draugen oil field lies in block 6407/9 in the Haltenbanken oil and gas province. The field is located 150 km off the coast of Norway and 200 km south of the Arctic Circle, in water depths of 240 to 280 m.

The Draugen field was discovered in 1984 by well 6407/9-1. Five additional exploration/appraisal wells and 2D seismic assisted in delineating the reservoir

The field is hosted by a low-relief north–south-trending anticline measuring some 20 by 6 km. The reservoir lies at a depth of 1600 m subsea and has an oil column of 40 m. The main reservoir is the Late Jurassic Rogn Formation sands, interpreted as a shallow marine sand bar. A separate, smaller accumulation has also been proved in Middle Jurassic Garn Formation sands in the western part of the field.

The estimated STOOIP is 180 million standard m^3. Recoverable reserves are estimated at 67 million standard m^3 of oil and some 3 billion standard m^3 of associated gas.

The field will be developed with a concrete gravity base structure and offshore loading. The initial development plan calls for six oil producers and six subsea water injectors. The platform will be installed in the summer of 1993, with first oil shortly thereafter. The planned plateau production rate is 14,300 standard m^3 of dry oil per day.

INTRODUCTION

The Draugen is a figure of Norwegian folklore—a sea creature that appears as a warning of ill fate to come.

The Draugen oil field is located in block 6407/9 in the Haltenbanken area some 160 km northwest of Trondheim in water depths of 240 to 280 m (Figure 1). The field lies 200 km south of the Arctic Circle and 400 km from existing facilities in the North Sea.

The production license (PL 093) was awarded in March 1984 to a group consisting of A/S Norske Shell (operator), Statoil, and BP Petroleum Development of

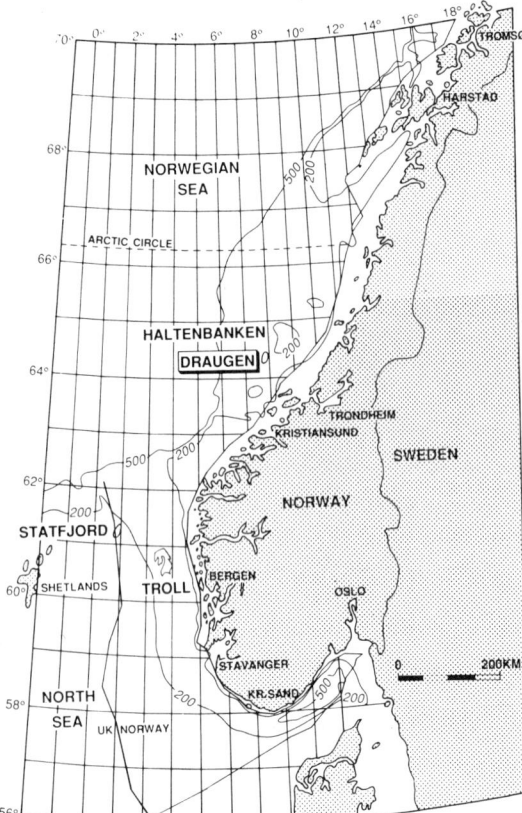

FIGURE 1. Location map. The Draugen field lies in the Haltenbanken area 150 km off the coast of Norway in water depths of 240 to 280 m. Note the distance to existing fields in the North Sea. The operations base will be sited in the town of Kristiansund. Bathymetry in meters.

Norway A/S. Current equity shares are 21, 65, and 14%, respectively.

The field was discovered by well 6407/9-1, which was spudded in June 1984. Five more exploration and appraisal wells drilled through 1986 allowed further definition of the structure and reservoir quality.

The plan for development and operation of Draugen field was approved by the Norwegian authorities in December 1988. The plan calls for crestal oil production from a fixed concrete platform, with the crude being transported via an offshore loading buoy and shuttle tankers. Pressure support will be provided by water injectors drilled from two subsea templates. First oil is scheduled for August 1993.

EXPLORATION AND APPRAISAL

The discovery well 6407/9-1 was spudded in June 1984 to test a structure recognized on the basis of regional seismic data acquired in the period 1973–77. The well encountered the Late Jurassic Rogn Formation sandstone at a depth of 1596 m subsea. Logging established a net oil column of 39 m with an oil down to (ODT) at 1636 m subsea. During production testing, the well flowed 1350 standard m^3 of 40° API oil per day.

Well 6407/9-2 was spudded in November 1984 as an outstep well 4 km to the north, whereas well 6407/9-3 appraised the crestal part of the field in May 1985. The wells confirmed the excellent reservoir quality of the Rogn sandstone and established a common oil-water contact (OWC) at 1638 m subsea. Well 6407/9-4 tested the western flank of the field; the Rogn formation was absent through pinch-out, but a small oil accumulation was proved in the deeper Middle Jurassic Garn Formation sandstone. This unit flowed oil with properties similar to those of the Rogn accumulation at a rate of 1000 standard m^3 per day; the OWC was at the same depth. The well was deepened to evaluate a closure at Early Jurassic level, but the objective Ile Formation sandstone was found to be water-bearing.

In September 1985, well 6407/9-5 tested the southern extension of the field. The final well, 6407/9-6, was drilled in 1986, west of the crestal part of the field. All six wells were extensively cored, providing detailed sedimentological and petrophysical input to field development planning.

GEOLOGIC SETTING

Structure

The regional structure of the Haltenbanken area has been described by Bukovics et al. (1984) and Bukovics and Ziegler (1985). The Draugen field is located near the western margin of the Trøndelag platform, an eastward-dipping monocline abutting the Norwegian mainland (Figures 2 and 3). The north–south-trending Bremstein fault complex separates the Trøndelag platform from the deeper Halten terrace.

In contrast to the relatively undeformed Trøndelag platform area, the Halten terrace is characterized by a series of rotated fault blocks, some of which host sizeable hydrocarbon accumulations (e.g., the Njord, Midgard, Heidrun, and Smørbukk fields). According to Caselli (1987), the Halten terrace was formed as a pull-apart basin in a regime characterized by dextral, strike-slip faulting. Faulting was initiated in the Early Jurassic, culminating in widespread tectonic activity in the Late Jurassic and Early Cretaceous. Rifting and the formation of oceanic crust during the early Eocene completed the separation of Greenland and Fennoscandia.

Regional Stratigraphy

The stratigraphy of the Mesozoic and Cenozoic interval offshore mid- and north Norway has been summarized by Dalland et al. (1988).

Only one well has penetrated Paleozoic rocks on the mid-Norwegian shelf. The overlying Triassic and Jurassic sediments comprise a thick package of coarse clastics deposited during periods of rapid erosion of the neighboring highlands, alternating with shales and silts deposited during periods of quiescence.

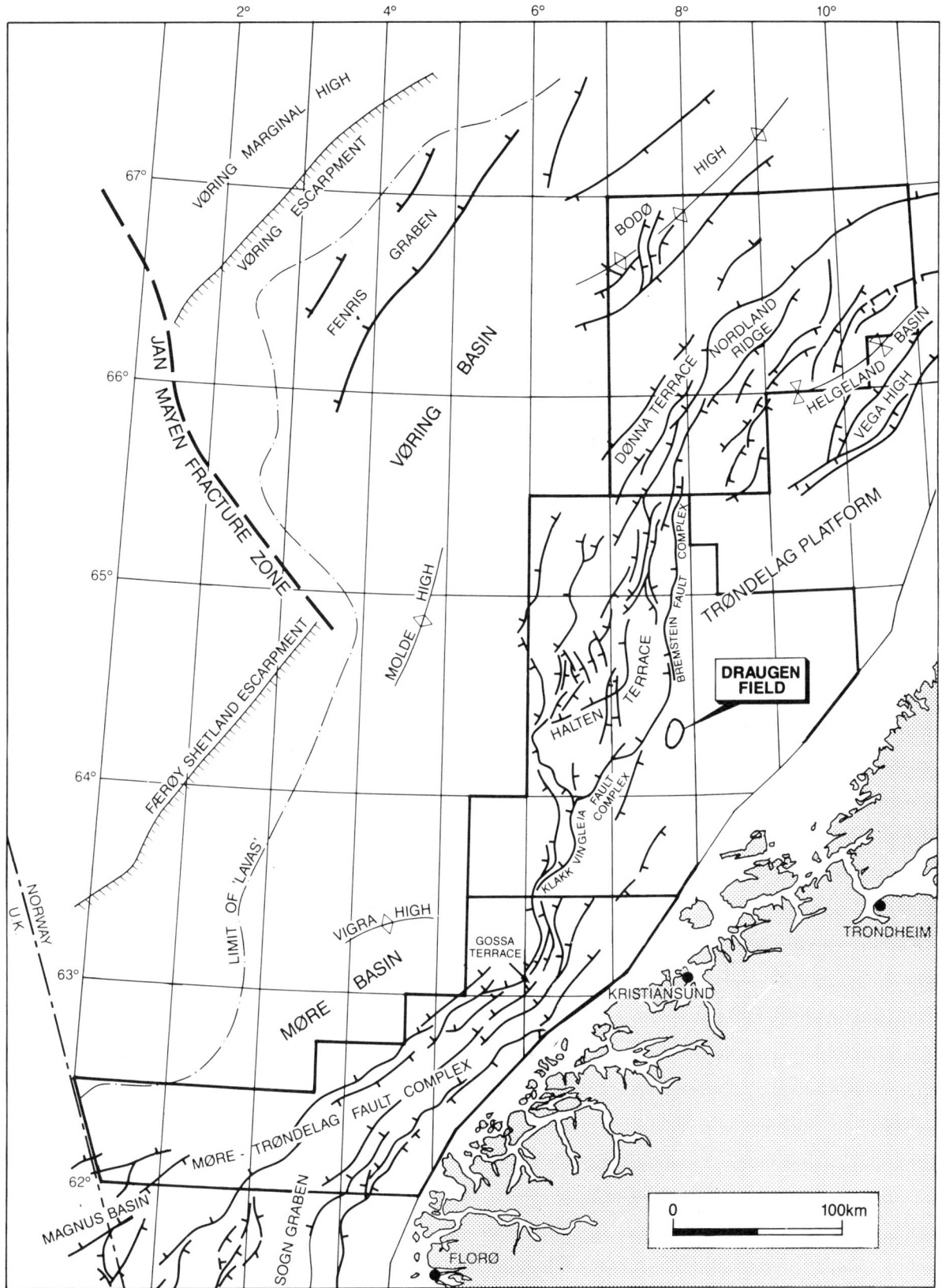

FIGURE 2. Structural elements. The Draugen field is located near the western edge of the relatively undeformed Trøndelag platform. The majority of the oil and gas fields in the area are hosted by rotated fault blocks within the deeper Halten terrace.

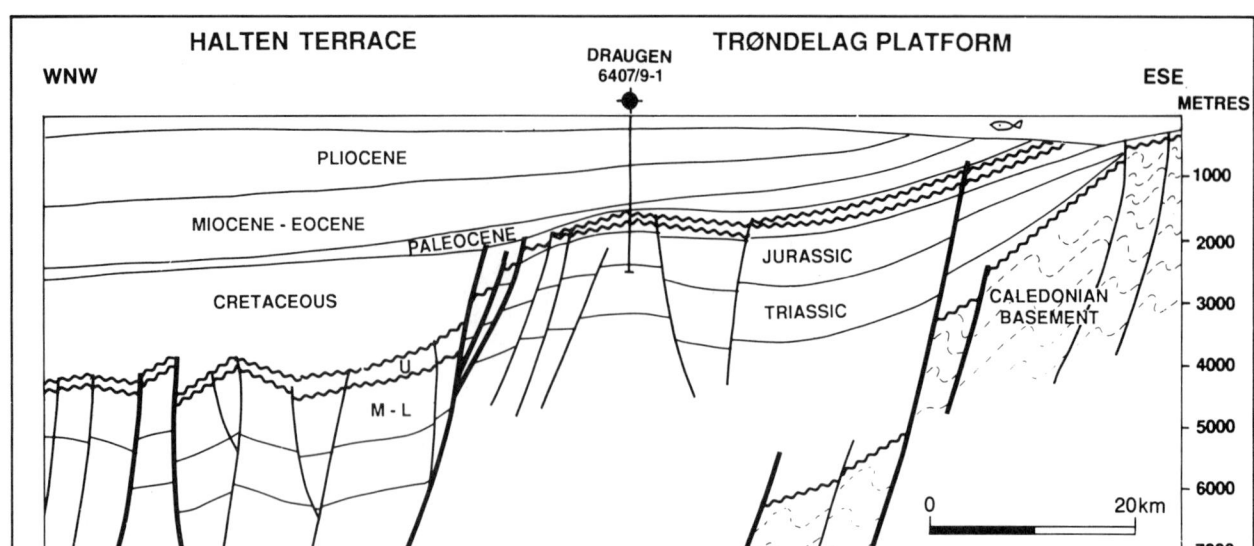

FIGURE 3. Regional cross section. The north-south-trending Bremstein fault complex separates the Jurassic and older sediments of the Trøndelag platform from the downthrown Halten terrace with its rotated fault blocks. Objective reservoirs occur in the Jurassic and Triassic sequence below the regional base Cretaceous unconformity.

Late Triassic sedimentation is characterized by a thick sequence of continental sandstones, siltstones, and shales, locally interbedded with evaporites (Figure 4). Fluviatile to deltaic conditions in the Early Jurassic allowed the deposition of the widespread Åre Formation, consisting of sandstones, shales, and coals. The succeeding Tilje Formation (Sinemurian-Pliensbachian), comprising coastal to deltaic sandstones with minor shales, forms an important reservoir in the Haltenbanken area.

Open marine conditions were initiated toward the end of the Early Jurassic with the deposition of the Ror Formation shales. The late Toarcian-early Bajocian Ile Formation comprises shallow marine sands deposited under the influence of wave and tidal processes. A renewed transgression at the end of the Bajocian resulted in the deposition of the Not Formation shales.

The succeeding Garn Formation (Bajocian-Bathonian) is dominated by upper shoreface and fluviatile sandstones. This unit forms an important reservoir at Draugen and elsewhere in the Haltenbanken area. Open marine conditions returned during the Callovian-Oxfordian with the deposition of the Melke Formation claystones. During this period, widespread tectonic activity and block faulting resulted in a regional unconformity and the emergence of the Trøndelag platform.

In the Draugen area, the Melke Formation is apparently missing, and the Garn Formation is unconformably overlain by organic shales of the Spekk Formation (Oxfordian-Ryazanian). These shales were deposited under stagnant bottom conditions and form the main source rock in the region.

The Rogn Formation sands (Oxfordian-Kimmeridgian) occur as a lens within the Spekk Formation. This unit is interpreted as a shallow marine sand bar, derived by erosion of Middle and Late Jurassic clastics exposed elsewhere on the Trøndelag platform.

The topography resulting from Jurassic block faulting had a major influence on sediment thickness during the Cretaceous and Cenozoic. Only a relatively thin wedge of marine shales and silts, interrupted by periods of subaerial erosion, accumulated on the Trøndelag platform. In contrast, continued subsidence, associated with the reactivation of regional faults, allowed the deposition of considerable thicknesses of sediments on the neighboring Halten terrace.

RESERVOIR GEOLOGY

Structure and Accumulation Conditions

The current structural interpretation is based on a 1 by 1 km grid of good-quality seismic data (1192 km) acquired in 1984, supplemented by infill crestal lines shot in 1986 (Figure 5). During 1990, a 3D survey was acquired over the entire field to assist in field development and the location of development wells.

The 2D data were processed to zero phase reflectivity and acoustic impedance (Figure 6). Depth conversion of the tops of the Rogn and Garn horizons was performed using average-velocity maps between sea bottom and the horizon in question.

The Draugen structure is a low-relief north-south-trending anticline measuring some 20 by 6 km (Figure 7), with the crest at 1600 m subsea and maximum vertical closure of some 50 m. The main accumulation is separated from a southern extension by a low structural saddle.

Several relatively minor northeast-southwest-trending normal faults with throws generally less than 10 to 15 m traverse the structure. Most of the faults die out at the base of the Rogn Formation.

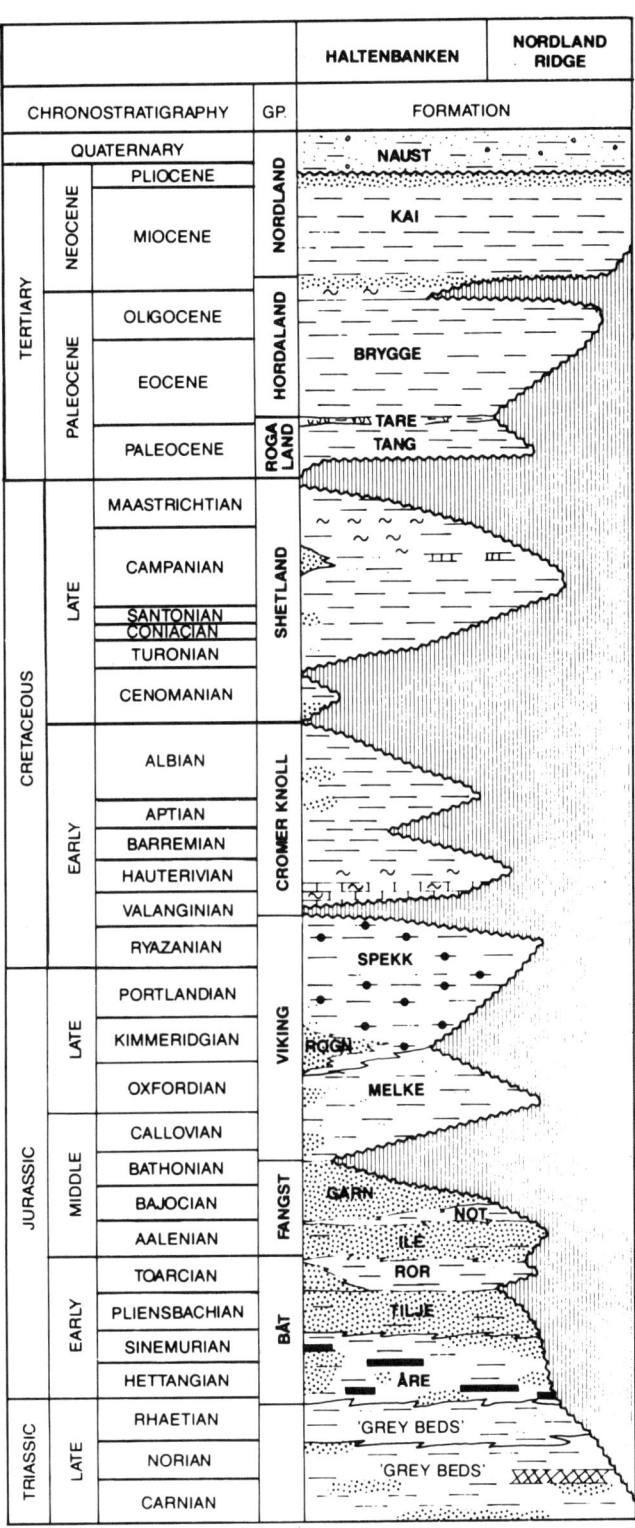

FIGURE 4. Regional stratigraphy. Objective reservoirs are found in the Jurassic sequence, with alternating sands and shales/siltstones forming excellent reservoir/seal pairs. Source rocks comprise the Early Jurassic Åre Formation coals and the Late Jurassic–Early Cretaceous Spekk Formation organic shales.

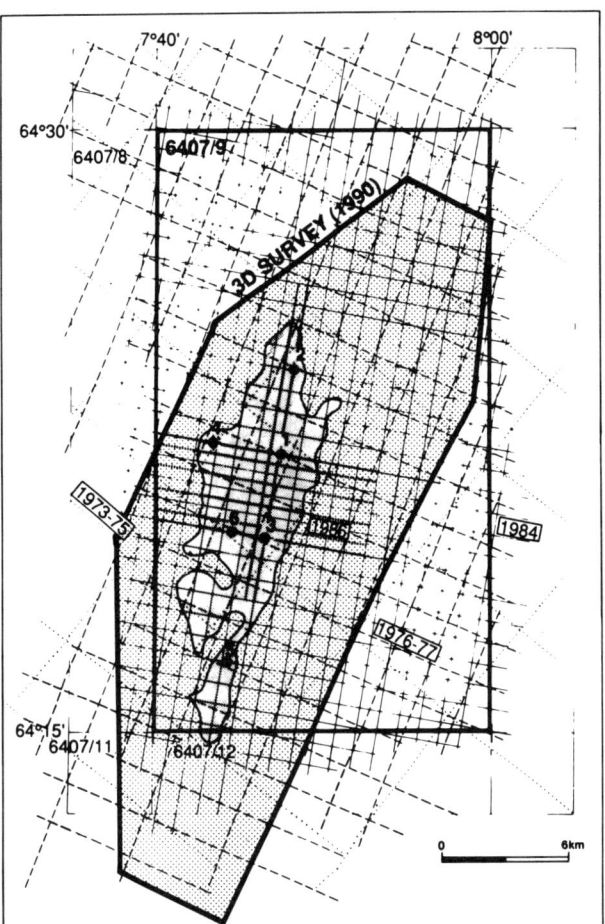

FIGURE 5. Seismic cover. 2D seismic data from a series of surveys acquired in the period 1973–86. A 3D survey was shot in 1990 to assist in development planning and the siting of development wells.

The main reservoir, the Rogn Formation sands, contains an estimated 92% of the oil in place. The Spekk Formation forms the cap rock as well as the seat seal. On the flanks, the accumulation is sealed by dip closure to the north and south, and by a combination of pinch-out and dip closure to the east and west (Figures 8 and 9).

The underlying Middle Jurassic Garn Formation sandstones are water-bearing over the greater part of the field, but in the west the unit rises above the field OWC of 1638 m subsea to form a separate accumulation.

RESERVOIR CHARACTERISTICS

Reservoir-quality sands in the Draugen field are found in the Rogn and Garn formations. The underlying Lower Jurassic Ile Formation (argillaceous, micaceous sandstones interbedded with claystones) was found to be fully water-

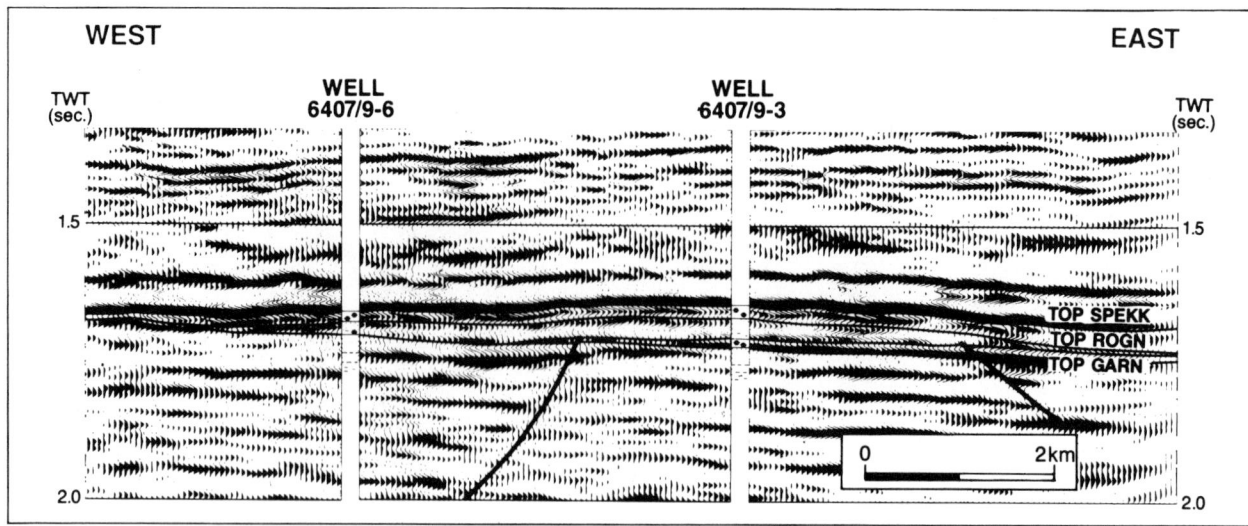

FIGURE 6. 2D seismic line acquired in 1984, processed to band-limited acoustic impedance. The Rogn Formation reservoir forms a subtle moundlike feature, pinching out to the west and east, below the regional base Cretaceous unconformity (top Spekk reflector).

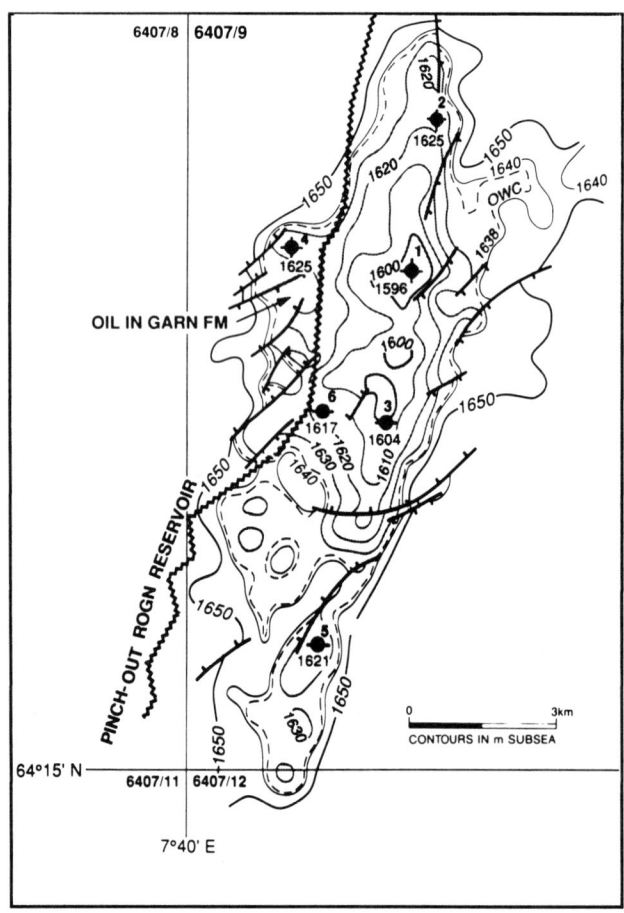

FIGURE 7. Depth map, top of reservoir. The Draugen field is hosted by a low-relief anticline with the crest at 1600 m subsea and maximum closure of some 50 m. To the west, the main Rogn Formation reservoir pinches out, and the underlying Garn Formation reservoir rises above the field OWC at 1638 m subsea.

bearing, but may be used for temporary storage of produced gas.

Rogn Formation

The mineralogy and reservoir geology of the Rogn Formation has been reviewed by Goesten and Nelson (in press), and Van der Zwan (1989, 1990) has described the palynostratigraphy and palynofacies of the unit.

The Rogn Formation forms a coarsening-upward sequence subdivided into two main units, units I and II, each with field-wide extent (Figure 10).

Unit II, forming the transition from the underlying Spekk Formation shales, ranges in thickness from 10 to 20 m along the crestal part of the field, thinning to zero in the west and east (Figure 8). The lower part consists of silty claystones and sands, grading upward into strongly bioturbated fine-grained sandstones. Porosities range from 26 to 30%, and permeabilities from 700 to 2000 md, in the upper part of the unit.

Unit I consists of unconsolidated fine- to medium-grained sandstones. Sorting is moderate, with a relatively high content of dispersed coarse grains. The sands are generally massive and bioturbated, although low-angle cross-bedding is developed locally (Figure 10). Over the greater part of the field, the unit ranges from 20 to 40 m in thickness. The unit pinches out to the west, where it is absent in well 6407/9-4, and is interpreted from seismic as thinning to the north and east. Reservoir quality is excellent, with porosities in the range 27 to 32% and permeabilities up to 10 darcys.

The Rogn Formation sands, like those of the underlying Garn Formation, are predominantly quartz arenites, although arkosic and lithic arenites are developed occasionally. Only limited compaction and diagenesis have taken place. Diagenetic processes are restricted to early burial diagenesis (with the development of local calcite,

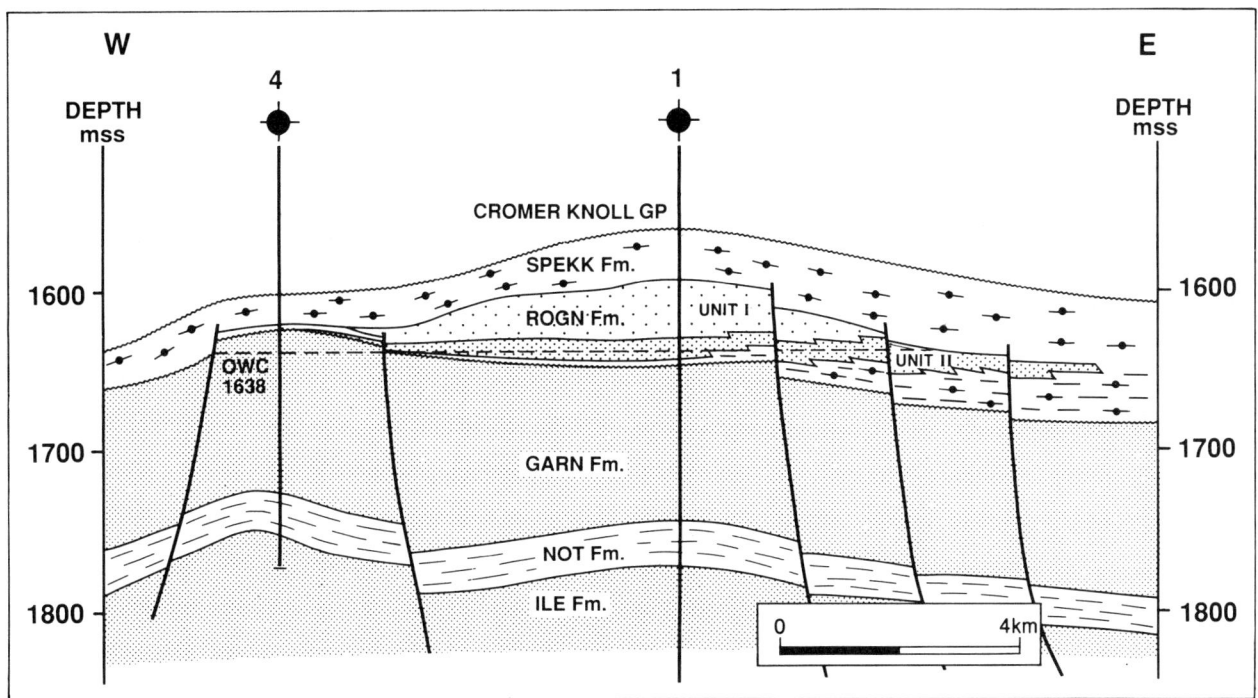

FIGURE 8. East-west cross section. Note the pinch-out of the Rogn Formation reservoir to the west and east, and the separate accumulation in the Garn Formation in the western part of the field.

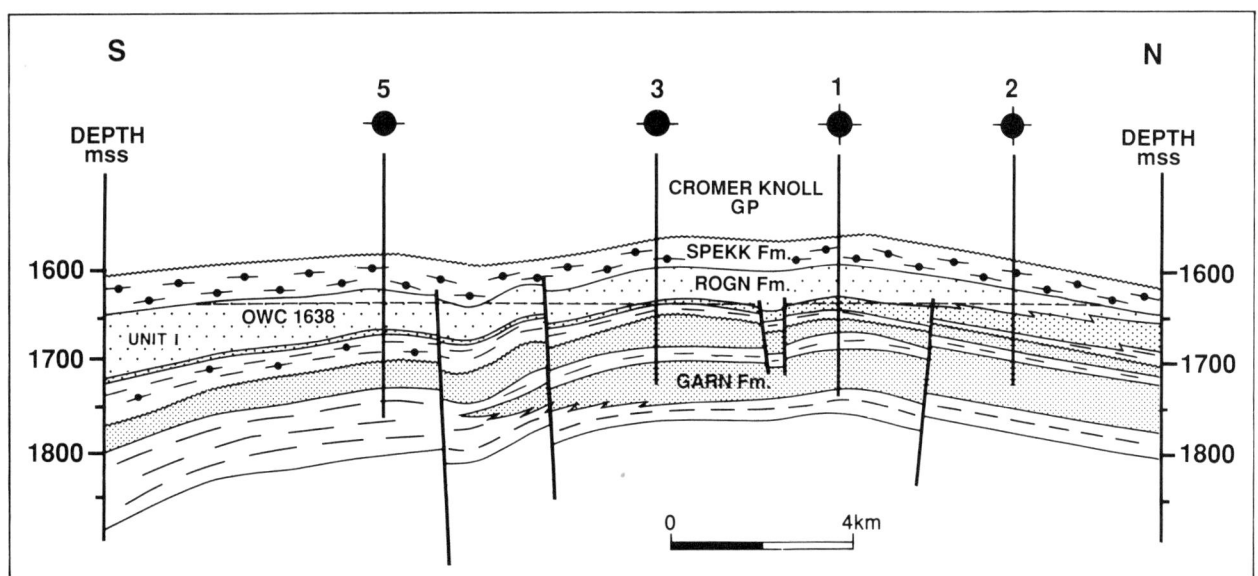

FIGURE 9. North-south cross section. The Draugen accumulation is sealed by dip closure to the north and south. Unit I of the Rogn Formation is mapped as thickening to the south, providing communication with the aquifer. Apart from this area, aquifer support during production is expected to be limited.

siderite, and pyrite cements) and extensive feldspar leaching as a result of influx of fresh water rich in carbon dioxide.

Palynofacies assemblages in unit II of the Rogn Formation point to deposition in a medium-energy open marine environment (Van der Zwan, 1990). The fine micaceous sands and silts were probably deposited in the lower part of an offshore marine sand bar. The palynofacies evidence suggests that unit I was deposited in a medium- to high-energy environment (Van der Zwan, 1990). Along with evidence from the bimodal sand grains, bioturbation, and occasional cross-bedding, this suggests that the unit represents the upper part of a shallow marine sand bar (Figure 11). The bar is thought to have been sourced from Middle Jurassic sediments, locally exposed to erosion on the Trøndelag platform.

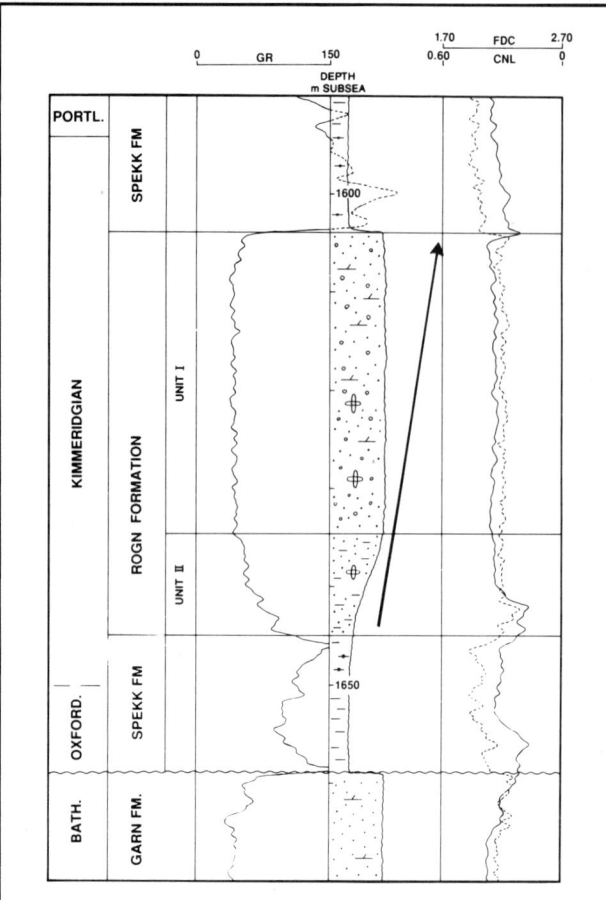

FIGURE 10. Rogn Formation, log of well 6407/9-3. The Rogn forms the main reservoir at Draugen, containing some 92% of the oil in place. The reservoir was deposited as a shallow marine sand bar within the Spekk Formation shales, which form both seat and cap seal. Fine-grained micaceous sands (unit II) grade upward into the fine- to medium-grained unconsolidated and frequently bioturbated sands of unit I.

The closest ancient analogs to the Rogn formation as developed at Draugen are found among the Upper Cretaceous shelf sandstones of the U.S. Western Interior seaway and are summarized by Boyles and Scott (1982). The sandstones were deposited as elongate bodies, tens of kilometers offshore from the contemporaneous coastline. Processes controlling their deposition include tides, storms, and longshore currents. Coarsening-upward sequences, tabular cross-bedding, and local bioturbation are characteristic features of these sand bodies. In terms of facies distribution and over-all dimensions, the Duffy Mountain sandstones (up to 18 m thick, 8 to 16 km wide, and 50 km long) come closest to the Rogn Formation as encountered at Draugen.

Garn Formation

The Middle Jurassic Garn Formation comprises several coarsening-upward sequences and is interpreted as a progradational sequence of shallow marine to coastal deposits. This unit maintains a relatively constant thickness of some 110 m across Draugen.

The Garn Formation has been subdivided into two major units, HI and HII (Figure 12). The uppermost unit, HI, consists of silty shales with thin, bioturbated sand layers at the base, passing upward into fine-grained cross-bedded sandstones. Porosities in the sandier section average about 36%. Permeabilities range from 4000 to 7000 md in the central part of the field but decrease toward the north.

Only the upper part of unit HII has been cored. From logs, the unit is interpreted as consisting of two coarsening-upward sequences, with basal silty shales passing upward into coarse-grained, moderately to poorly sorted massive sands at the top. The sands are best developed in the north central part of the field, where porosity averages some 29% and permeabilities range between 5000 and 10,000 md. This unit is absent in well 6407/9-5 in the south.

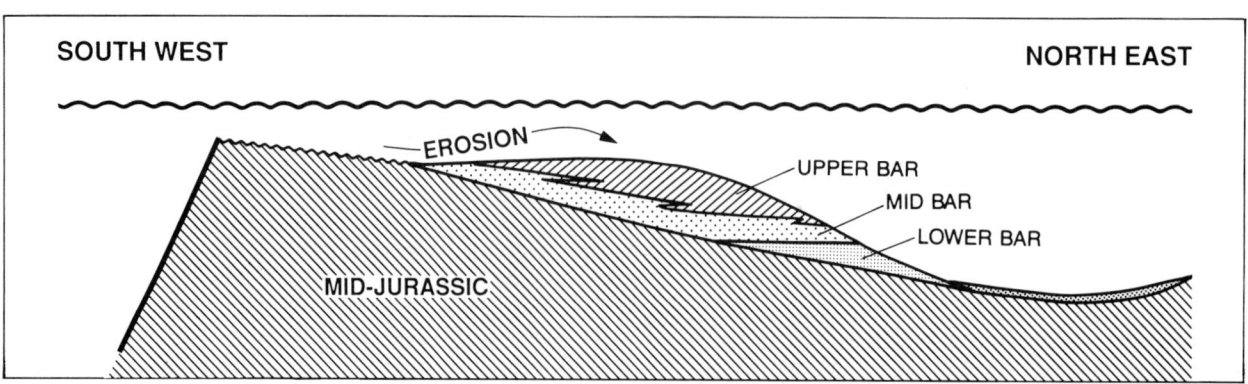

FIGURE 11. Rogn Formation, depositional model. Sand was derived by erosion of Middle Jurassic sediments locally exposed on the Trøndelag platform and redeposited as a shallow marine sand bar. (Modified from Van der Zwan, 1990.)

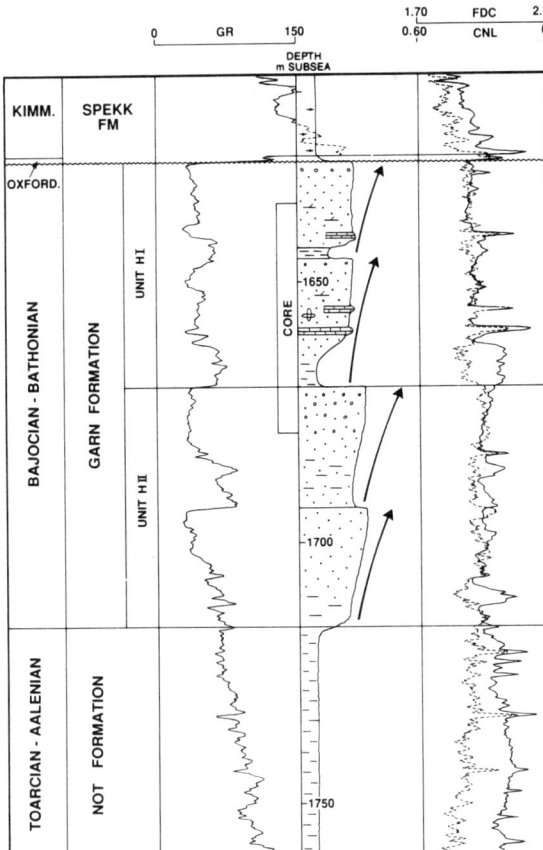

FIGURE 12. Garn Formation, log of well 6407/9-4. The Garn hosts a smaller, separate accumulation in the western part of the field. The unit comprises several coarsening-upward sequences, deposited in a shallow marine to coastal environment.

TABLE 1. Reservoir fluid parameters for Draugen field.

Hydrocarbon type	Undersaturated oil
Formation pressure, bara	165
Formation temperature, °C	71
Datum depth, m subsea	1630
Oil gravity, °API	40
Solution GOR, std. m³/std. m³	61
Bubble point, bara	59
Oil viscosity, mPa · s	0.67
Water salinity (NaCl equivalent), ppm	37,500

FLUID PROPERTIES AND VOLUMETRICS

The Draugen field contains undersaturated 40° API oil with a low dissolved gas content (Table 1). The field is hydrostatically pressured, with a formation pressure of 165 bara (2392 psia) at datum 1630 m subsea.

The estimated STOOIP in Draugen is 179.4 million standard m³ (1128 million STB), with 165.7 million standard m³ (1042 million STB) in the Rogn Formation reservoir and 13.7 million standard m³ (86 million STB) in the Garn (Table 2). The expectation of reserves is 67.5 million standard m³ (425 million STB).

DEVELOPMENT PLANNING

General Considerations

Extensive reservoir simulation studies indicate that optimum development of the Rogn Formation reservoir at Draugen should be based on pressure support through water injection. This is dictated by the low intrinsic reservoir energy, which is again a function of the limited connected aquifer, the highly undersaturated crude, and the hydrostatic pressure. Additional factors are the permeability profile (with the highest permeabilities at the top of the reservoir) and the favorable water-oil mobility ratio (0.6).

The flat structure and thin oil column, together with the presence of bottom water, imply that early water production will be difficult to avoid and will have to be accounted for in the development plan. Production wells will be located crestally to minimize the effect of bottom water coning.

The Rogn Formation pinches out rapidly to the east and west, at or near the OWC of the accumulation (Figure 8). To avoid injection into the oil column, the water injection wells will be located in the north and south of the field, where injection into the water leg is feasible.

Development Plan

Salient features of the Draugen development plan are given in Table 3 and Figures 13 and 14. The basic plan consists of a centrally located concrete monocolumn with integrated deck. The platform has ten conductors and several spare J-tubes. Initially, five platform oil producers will be drilled to drain the central portion of the field. Additionally, appraisal wells 6407/9-3 and -5 (currently suspended) will be recompleted as subsea oil producers in the central and southern parts of the field, respectively (Figure 13).

Water injection will be done by means of two subsea template/manifold systems located axially in the north and south of the field. Each template will serve three injectors.

Crude export will employ an offshore loading buoy and shuttle tankers. Pending construction of a permanent gas offtake system in the Haltenbanken area, the produced associated gas will be reinjected. Initially, gas will be injected by means of a subsea gas injector into the deeper Ile Formation sands in a water-bearing structure

TABLE 2. Volumetrics for Draugen field.

	Volume (million std. m³)					
	STOOIP			Recoverable reserves		
Reservoir	Low	Expectation	High	Low	Expectation	High
Rogn Formation	155.8	165.7	175.5	58.5	67.2	75.9
Garn Formation	12.6	13.7	14.9	0.0	0.4	0.8
TOTAL	168.4	179.4	190.4	58.5	67.6	76.7

TABLE 3. Development plan for Draugen field.

Facilities	Concrete gravity base structure with integrated deck
	Two water-injection template/manifold systems
	Offshore loading buoy
Wells	5 platform oil producers
	2 subsea oil producers
	6 subsea water injectors
	1 subsea gas injector
Plateau rate	14,300 std. m³ per day (90,000 BOPD)
First oil	1993
Investment cost	10.8 billion Nkr (1989)

proved by well 6407/9-4 in the western part of the field (Figure 8). Injection is planned for some three years of production in this structure. In the event that permanent gas export facilities are not available at the end of this period, further injection could be performed in the Rogn Formation in the southern part of the field (area to be drained by subsea producer 6407/9-5).

CONCLUSIONS

Interpretation of 2D seismic data and the results from six exploration and appraisal wells have allowed delineation of the reservoir geometry and establishment of the geologic model for the Draugen oil field, offshore mid-Norway.

The main reservoir is the Rogn Formation sandstone, deposited as a shallow marine sand bar on the Trøndelag platform in the Late Jurassic. A smaller accumulation occurs in the Middle Jurassic Garn Formation sandstone. Extensive petroleum engineering studies point to pressure support by water injection as the optimum production scheme. Production wells will be located in a crestal position in the central part of the field, whereas the water injection wells will be sited near the northern and southern extremities.

Water depth at Draugen is 240 to 280 m, and the field lies 150 km offshore and some 400 km from existing infrastructure. These factors were instrumental in the decision to develop the field with a concrete monotower with integrated deck, combined with offshore loading of the produced crude. Water injection will be performed by means of two subsea template/manifold systems. Pending construction of a permanent gas offtake scheme for the region, the produced associated gas will initially be reinjected into a deeper water-bearing reservoir in the western part of the field.

The Draugen field is due to go onstream in 1993 as the first oil or gas field on the mid-Norwegian shelf.

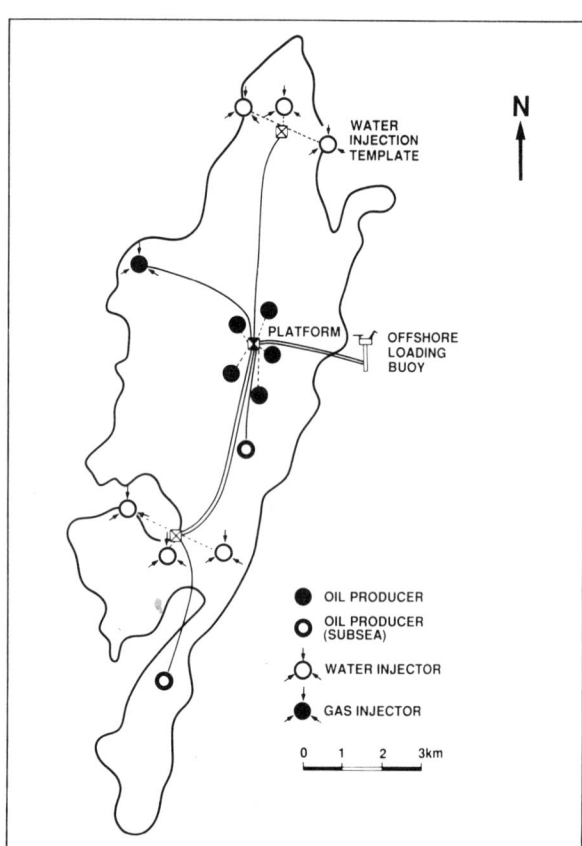

FIGURE 13. Development plan for the Draugen field, consisting of a central drilling/production and accommodation platform, two subsea water-injection template/manifold systems, and an offshore loading buoy. Produced gas will be initially reinjected into a deeper, water-bearing reservoir in the western part of the field.

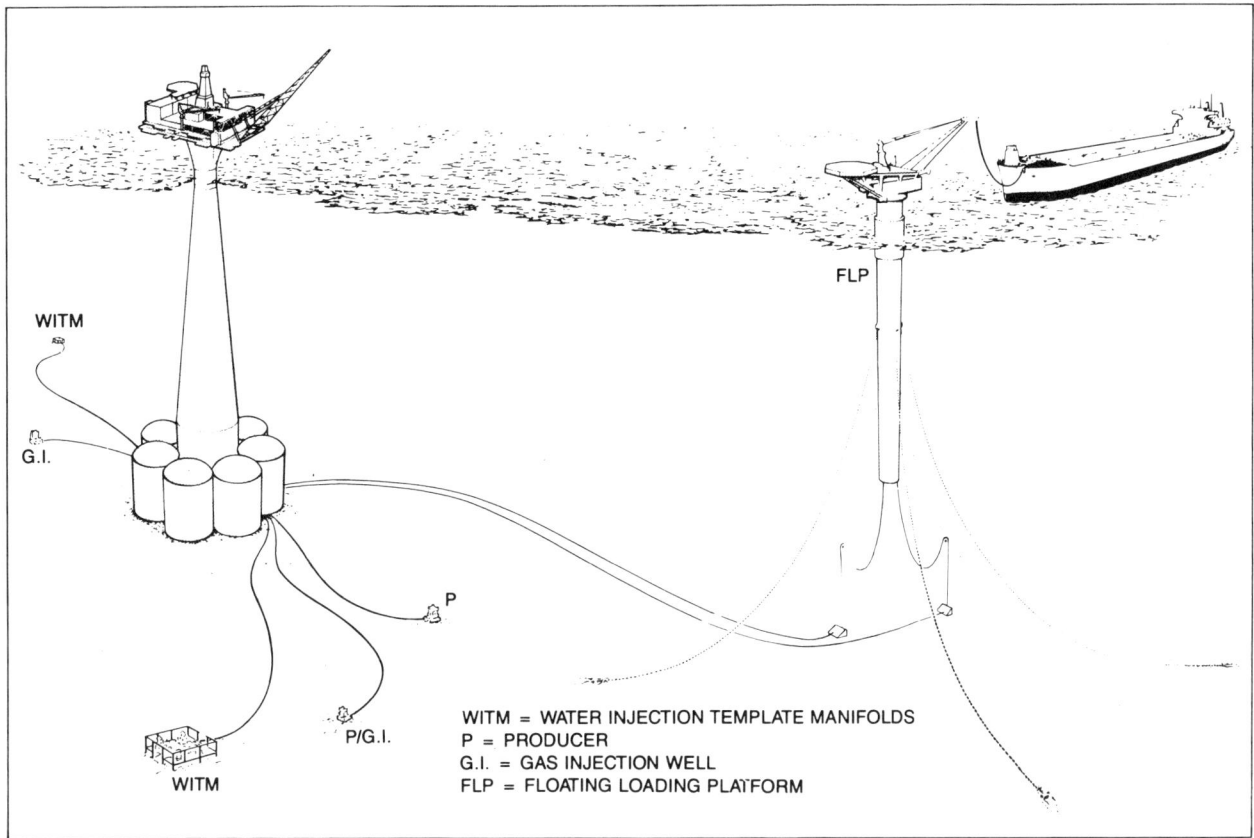

FIGURE 14. Schematic overview of the Draugen field, showing the concrete monocolumn platform in water depths of 260 m, offshore loading buoy and shuttle tankers for crude export, water-injection template/manifold systems, and subsea oil production and gas injection wells.

ACKNOWLEDGMENTS

The author thanks the partners in the Draugen license, Statoil and BP Petroleum Development of Norway, and the management of Norske Shell and Shell Internationale Petroleum Maatschappij, for permission to publish this paper. Grateful acknowledgment is also extended to the author's colleagues at Norske Shell and to the Draugen project team for their assistance in the preparation of the manuscript and the illustrations.

REFERENCES CITED

Boyles, J. M., and A. J. Scott, 1982, A model for migrating shelf-bar sandstones in Upper Mancos Shale (Campanian), Northwestern Colorado: AAPG Bulletin, v. 66, n. 5, p. 491–508.

Brenner, R. L., 1978, Sussex Sandstone of Wyoming—example of Cretaceous offshore sedimentation: AAPG Bulletin, v. 62, n. 2, p. 181–200.

Bukovics, A., and P. A. Ziegler, 1985, Tectonic development of the mid-Norway continental margin: Marine and Petroleum Geology, v. 2, n. 1, p. 2–22.

Bukovics, C., E. G. Cartier, N. D. Shaw, and P. A. Ziegler, 1984, Structure and development of the mid-Norway continental margin, in A. M. Spencer et al. (eds.), Petroleum geology of the north European margin: London, Graham and Trotman, p. 407–423.

Caselli, F., 1987, Oblique-slip tectonics mid-Norway Shelf, in J. Brooks and K. Glennie, (eds.), Petroleum geology of north west Europe: London, Graham and Trotman, p. 1049–1063.

Dalland, A., D. Worsley, and K. Ofstad, 1988, A lithostratigraphic scheme for the Mesozoic and Cenozoic succession offshore mid- and northern Norway: Norwegian Petroleum Directorate, Bulletin n. 4.

Ellenor, D. W., and A. Mozetic, 1986, The Draugen oil discovery, in A. M. Spencer et al. (eds.), Habitat of hydrocarbons on the Norwegian continental shelf; proceedings of an international conference: London, Graham and Trotman, p. 313–316.

Fruit, D. J., and R. Douglas-Elmore, 1988, Tide and storm-dominated sand ridges on a muddy shelf: Cottage Grove Sandstone (upper Pennsylvanian), northwestern Oklahoma: AAPG Bulletin, v. 72, n. 10, p. 1200–1211.

Goesten, M. J. B. G., and P. H. Nelson, in press, Draugen field, Haltenbanken area, offshore Norway: Contribution to the AAPG Field Atlas.

Harris, N. B., 1989, Reservoir geology of Fangst Group (Middle Jurassic), Heidrun field, offshore mid-Norway: AAPG Bulletin, v. 73, n. 11, p. 1415-1435.

La Fon, N. A., 1981, Offshore bar deposits of Semilla Sandstone Member of Mancos Shale (Upper Cretaceous), San Juan basin, New Mexico: AAPG Bulletin, v. 65, n. 4, p. 706-721.

Van der Zwan, C. J., 1989, Palynostratigraphical principles as applied in the Jurassic of the Troll and Draugen fields, offshore Norway, *in* Correlations in hydrocarbon exploration: Norwegian Petroleum Society, Graham and Trotman, London, p. 357-365.

Van der Zwan, C. J., 1990, Palynostratigraphy and palynofacies reconstruction of the Upper Jurassic to lowermost Cretaceous of the Draugen field, offshore mid Norway: Review of Palaeobotany and Palynology, v. 62, p. 157-186.

Chapter 24

The Geology of Heidrun
A Giant Oil and Gas Field on the Mid-Norwegian Shelf

P. K. Whitley

Conoco Norway Inc.
Stavanger, Norway

ABSTRACT

The Heidrun field is located in the Haltenbanken region on the mid-Norwegian continental shelf in water depths of approximately 350 m. It lies 190 km from shore. The field was discovered in 1985, and appraisal drilling has subsequently confirmed a giant Cimmerian structure with hydrocarbons trapped in Jurassic sandstone reservoirs. The field has 750 million bbl of recoverable oil reserves, with associated gas reserves of 0.45 tcf and free gas reserves of 1.32 tcf. The accumulation extends over 37 km^2 and was unitized in 1989 with 75% of the field in block 6507/7 and 25% in block 6507/8.

Block 6507/7 was acquired by Conoco Norway and partners in 1984 in the eighth round of licensing offered by the Norwegian government. The acquisition was based on extensive in-house exploration regional studies, which identified a Jurassic rift element in Haltenbanken and the potential for the generation of liquid hydrocarbons.

The Heidrun structure is a large southwest-plunging horst block on the southwest flank of the Nordland ridge and was formed during the Cimmerian extensional tectonic phase in the Late Jurassic–Early Cretaceous. The Heidrun reservoirs are severely truncated at the northern edge of the structure and are sealed by Cretaceous shales.

The Heidrun Jurassic reservoir rocks, the Fangst Group, and the Tilje and Åre formations were deposited on the southeastern flank of the developing northeast Atlantic rift domain. Despite an over-all transgressive regime, the interval was characterized by a high, coarse, clastic influx from the elevated rift shoulders. The shallow depth of burial (less than 2500 m) has limited compaction effects. Reservoir quality in the clean Fangst sands is somewhat enhanced by dissolution, and the sands exhibit maximum permeabilities higher than 10 darcys and porosities in excess of 30%.

The primary source of the petroleum is the Upper Jurassic Spekk Formation, which is mature in the downdip areas 5 to 15 km southwest and west of Heidrun. The Åre Formation coaly beds are also potential source rocks in the same downdip areas. Therefore, fairly long petroleum migration paths are inferred.

A 5000 km high resolution 3D seismic survey was acquired in 1986. State-of-the-art processing has produced a data set far superior to previous 2D data. The discovery well and appraisal wells were extensively cored through the Middle and Lower Jurassic. The 3D data set forms the basis for detailed geologic and reservoir models. These models were used in the application for development consent, submitted to the Norwegian government in late 1989 by the Heidrun unit owners.

The current development plan proposes that exploitation of the Heidrun field will be carried out using an innovative concrete tension leg platform, offshore oil loading, and transportation of the gas to shore by pipeline for onshore use.

INTRODUCTION

The Heidrun field, located in the Haltenbanken region (Figure 1), was discovered in 1985. It has subsequently proved to be a giant field, with 750 million bbl of recoverable oil reserves, 0.45 tcf of associated gas, and 1.32 tcf of primary (free) gas. It lies approximately 190 km off the coast of mid-Norway, in 350 m of water, and is to date the most northerly oil field in Europe.

The field was unitized in 1989, with 75% of the field in block 6507/7 and 25% in block 6507/8. The current unit owners and tract participants are: Conoco Norway Inc. (CNI), 25.8%; Statoil, 50.75%; Neste OY, 9.85%; Conoco Petroleum Norge A/S, 9.85%; Norsk Hydro Produksjon a.s, 2.5%; and Det Norske Oljeselskap a.s (DNO), 1.25%.

CNI is the designated operator for the development stage. Statoil will assume operatorship within nine months after installation of the main Heidrun platform.

This paper describes Heidrun in two phases: the exploration phase, and the appraisal and development phase.

PHASE I—EXPLORATION

Overview

The Haltenbanken area was opened for initial (fifth round) license applications by the Norwegian government in 1980; early drilling results encouraged further exploration (Campbell and Ormassen, 1987). In 1983, nine blocks were offered under the eighth round, and were fully evaluated by CNI. In the awards, CNI was appointed operator for block 6507/7.

The drilling of the discovery well 6507/7-2 in 1985 marked the culmination of an intensive exploration analysis of the Haltenbanken region by CNI, which commenced in the late 1970s and came to fruition in the early 1980s.

CNI Exploration Team

A team of experienced geologists and geophysicists, both Norwegian and international, was assembled in the early 1980s to evaluate the Norwegian continental shelf north of 62°N, and to identify areas of petroleum potential. This team was augmented by organic geochemistry and reservoir diagenesis experts from the Conoco Research Division in Ponca City, Oklahoma.

The exploration strategy was multidisciplinary and drew heavily on seismic stratigraphy, as developed in Payton (1977). This technique caused a revolution in thinking within Conoco in the late 1970s, with the realization that single-block evaluations would be optimized if a regional approach to the geologic and geophysical interpretation were undertaken.

A working environment was created within CNI that encouraged the application of modern exploration techniques, including seismic stratigraphic analysis, organic maturation studies, and state-of-the-art computer mapping. The combination of an experienced team, the free exchange of geologic ideas, and the application of diverse modern technology led to a successful exploration approach that is continuing within Conoco.

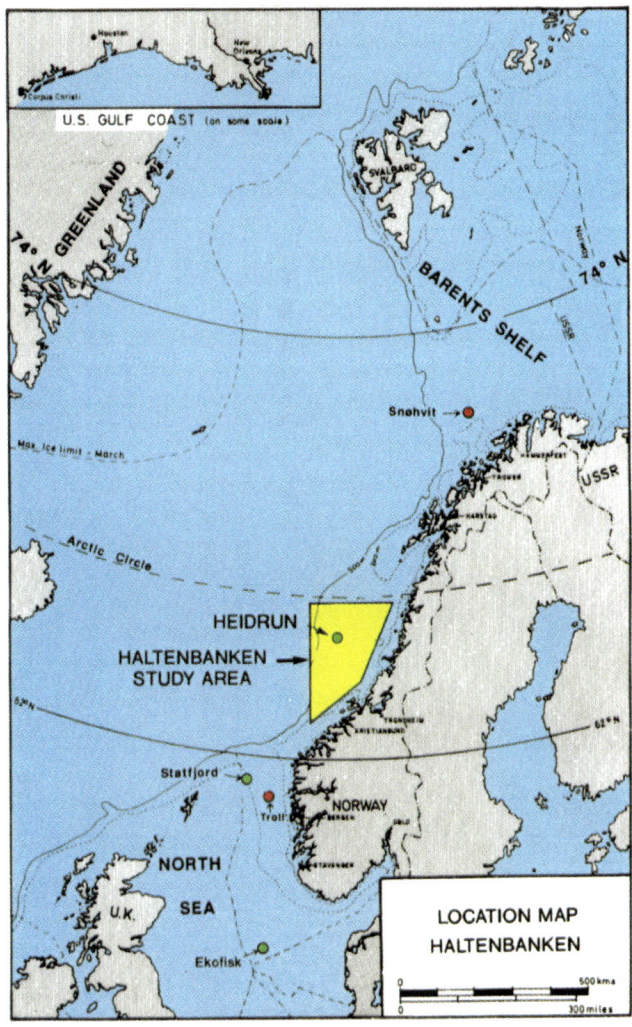

FIGURE 1. Location map for Heidrun field.

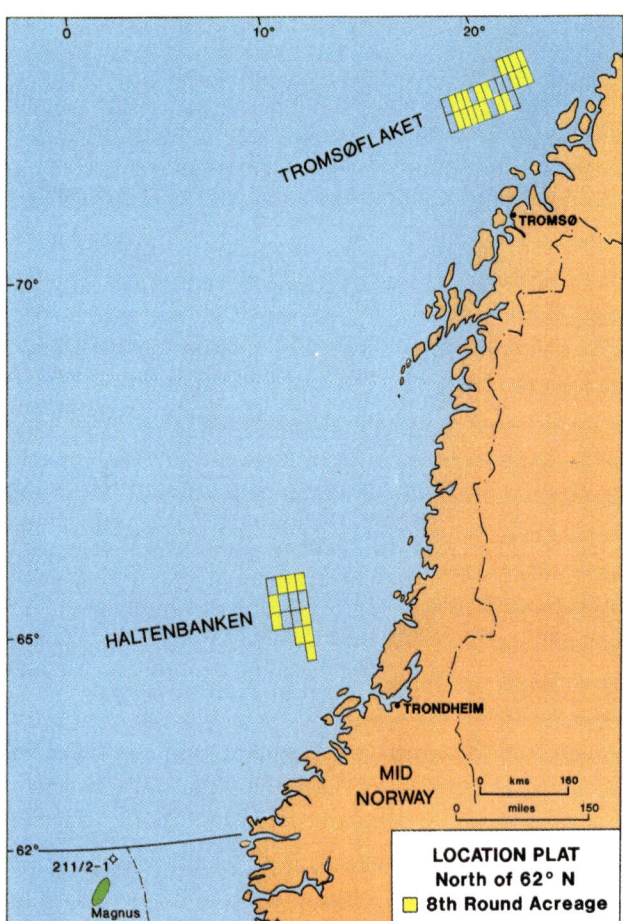

FIGURE 2. Exploration status north of 62°N in 1980.

Regional Studies in the 1980s

During this early 1980s study, the most northerly well drilled in the North Sea was the Møre Basin well, 211/2-1, drilled in 1978 by BP in the U.K. sector (Figure 2). This well encountered a thick Cretaceous section lying unconformably on Triassic sediments with the prospective Jurassic section absent. The seismic data from the mid-1970s indicated that the Møre Basin subsided rapidly during the Cretaceous and Tertiary (Jørgensen and Navrestad, 1981). The later Cimmerian tectonic phase was not as pronounced as it was in the Viking graben, and the Jurassic section was interpreted as being absent as a result of erosion or nondeposition. The interpretation of the seismic data at that time therefore indicated that the Viking graben, with its prospective Jurassic rift section, lay south of 62°N, and this cast some doubt on the more northerly exploration prospectivity.

Regional seismic data acquired by the Norwegian Petroleum Directorate (NPD) in Haltenbanken, and interpreted by Jørgensen and Navrestad (1981), provided renewed encouragement. This work interpreted a preserved Jurassic rift section within the Haltenbanken passive continental margin. CNI's previous worldwide experience with passive continental margins had not yielded high-potential exploration plays (e.g., Baltimore Canyon). However, CNI recognized the significance of a preserved rift section within a passive continental margin, such as those found offshore Ivory Coast and Newfoundland. Therefore, Haltenbanken was upgraded and identified as prospective, and regional studies of Haltenbanken offshore were intensified.

As more regional Haltenbanken seismic was interpreted, a picture of the paleogeographic evolution of the northwest Norwegian continental shelf was assembled by CNI.

These studies were based on 15,520 km of seismic data acquired from the NPD and various operators between 1978 and 1983. The geologic studies focused on the plate tectonic evolution of the mid-Norwegian continental shelf during the Mesozoic and Cenozoic periods. The early drilling in Haltenbanken between 1981 and 1983 yielded invaluable stratigraphic data that served as input for the regional studies. The structural features of Haltenbanken were identified, and the paleogeographic/tectonic evolution was described.

Tectonic Development

Basic to the early regional evaluation was construction of a geologic plate tectonic model for Haltenbanken (Figure 3). The proto-Atlantic rift evolution began in the late Paleozoic, after the intense compressional phases associated with the Caledonian suture and orogeny and strike-slip displacement of Scandinavia and Greenland.

Lithospheric extension and mantle upwelling along the ancient Caledonian suture zone between Laurentia and Baltica occurred in the late Paleozoic. Extensional tectonics were initiated in the Permian-Triassic with downwarping, rifting, erosion, and clastic infill. The rifting culminated in the Late Jurassic with the Cimmerian tectonic phase, when there was widespread footwall uplift and erosion and tilting of the Jurassic fault blocks, initiating the classic trap geometry.

During the Cretaceous, basin infill and subsidence ensued, followed by the initiation of seafloor spreading and continental drift of Greenland from Norway in the Paleocene-Eocene.

Structural Features Map of Haltenbanken

A structural features map (Figure 4) was assembled on the basis of seismic stratigraphic analyses of the NPD seismic data and plate tectonic reconstruction. This area is composed of seven subbasins. Heidrun is located in the northern part of the Halten terrace, marginal to the Nordland ridge. A reactivated northeasterly Caledonian trend is prominent, with a pronounced Cimmerian northerly trend also present in the Halten terrace.

The predominant structural features in Haltenbanken are the Trøndelag platform, the Nordland ridge, the Halten terrace, and the Vøring Basin.

The Heidrun field is located on the southernmost extension of the predominantly southwest-northeast-trending Nordland ridge, which is an intensely faulted high that separates the Helgeland Basin from the more westerly Vøring Basin. South of the Nordland ridge, the area plunges into the less intensely faulted Halten terrace, where numerous oil and gas fields have been discovered.

Paleogeographic Evolution of Haltenbanken and the Northwest Norwegian Shelf

Overview

A series of paleogeographic maps have been extracted from Doré (1990) and Doré and Gage (1986). Heidrun field is highlighted as a red dot on all these maps. The Triassic/Jurassic maps (Figure 5) show a narrow proto-Atlantic rift 100 to 250 km wide and trending north-northeast. This rift underwent frequent marine transgressions that, coupled with rapid influx of coarse, clastic sediments from the margins and the intrabasinal highs, resulted in the deposition of excellent reservoir rocks.

A major rise in sea level and a marine transgression in the Late Jurassic caused the rift to be inundated and a

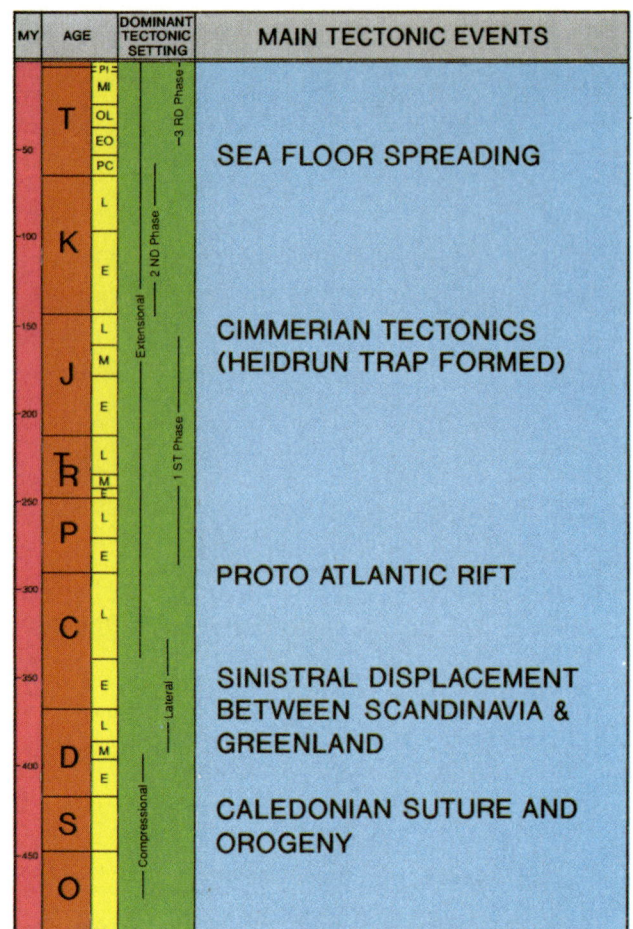

FIGURE 3. Tectonic development of Haltenbanken (after Bukovics and Ziegler, 1983).

"hot shale" to be deposited in a narrow restricted rift, in a setting that was ideal for the deposition of rich organic source rocks.

Continued subsidence and basin fill in the Cretaceous-Tertiary provided an effective seal for the basin and sufficient overburden for thermal maturation (Figure 6).

Triassic (Excepting Rhaetian)

A eustatic lowstand in the Triassic brought the return of continental conditions to the Haltenbanken area, and red siltstones, shales, and thin sandstones accumulated in a fluvial setting. The earliest Triassic section penetrated to date in the area (well 6507/12-2) is of probable Middle Triassic age. Two thick halite sequences of Middle and Late Triassic age are present in the same well. They probably represent incursions from the predominantly marine area of the Barents shelf, along arms of the Atlantic rift system. The precise distribution of these units over the Haltenbanken regions is unknown, although there is good evidence of salt-related tectonics in blocks 6407/5 and 6407/6.

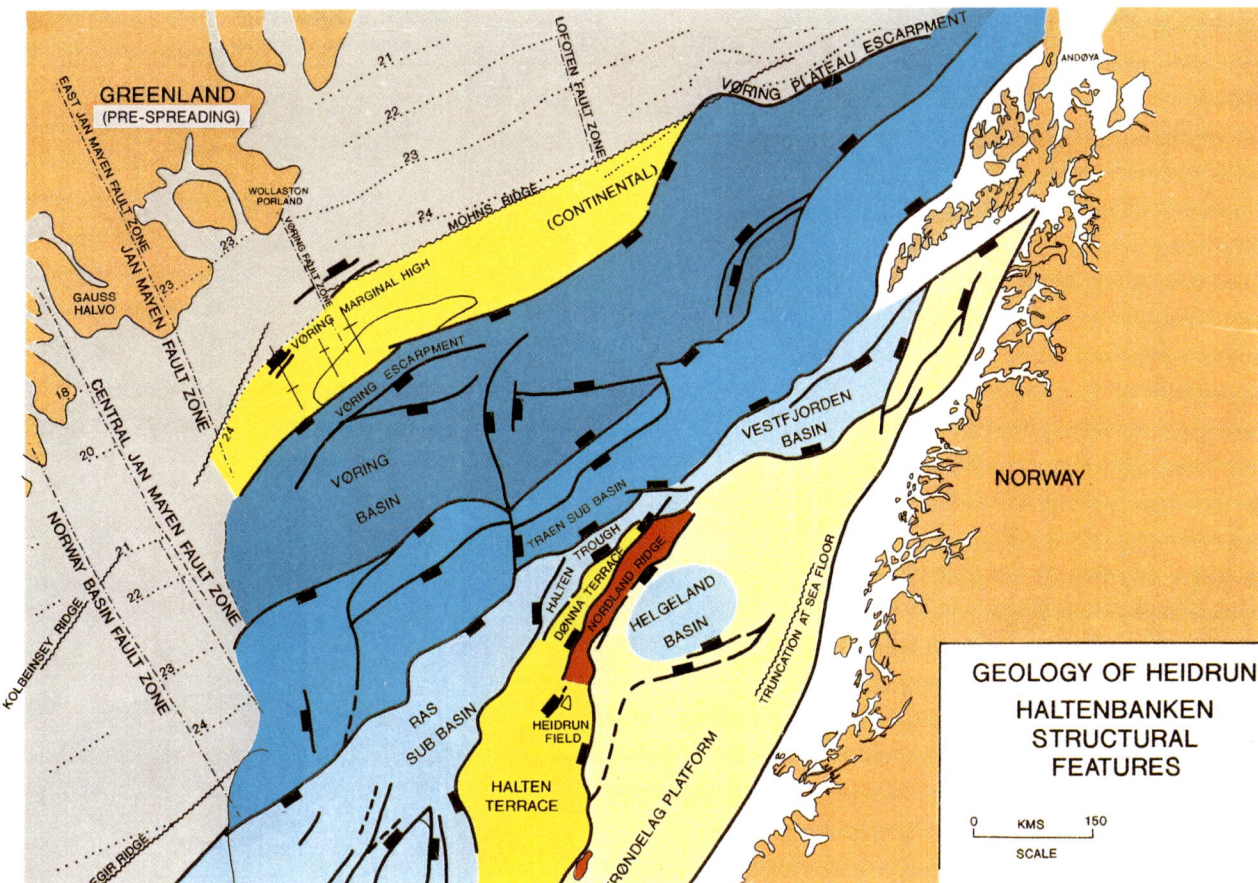

FIGURE 4. Structural features of Haltenbanken.

Rhaetian and Lower Jurassic

A change from arid to more humid climatic conditions and a gradual rise in sea level brought paralic, swampy conditions to the area, resulting in the deposition of coals, sands, silts, and shales of the Åre Formation (Figure 7). This formation is about 490 m thick. It is an important gas-prone source rock with possible oil generation potential and is further discussed later.

A marine transgression in the Pliensbachian Stage deposited the shallow marine sands of the Åre 2/Tilje equivalent. This unit is on average about 140 m thick and constitutes a reservoir objective in the Haltenbanken area. It is discussed in more detail later. Continued marine transgression culminated with the deposition of the Ror marine shale during the Toarcian Stage.

Middle Jurassic

Sandy deposition caused by clastic influx from regional updoming, both marine and fluvial, returned to the area in the Middle Jurassic. These sands are referred to as the Fangst Group and are a major reservoir objective equivalent in part to the Pelion/Sortehat in eastern Greenland (Figure 7). There is wide variation in the thickness of this unit as a result of the influence of synsedimentary faulting and erosion associated with the Cimmerian fault movements.

Upper Jurassic

The initial phases of the Cimmerian diastrophism caused partial or total erosion of the Fangst Group on the major highs and resulted in the formation of a widespread unconformity. This event was succeeded by deposition of the Melke shales (Callovian and younger), indicating fully marine conditions in Haltenbanken that persisted for the remainder of the Jurassic. The Late Jurassic transgression culminated with the deposition of the Spekk Formation (Draupne Formation equivalent "Hot Shale") in deeper water with anoxic bottom conditions. This unit is the main oil-prone source rock of the area.

Further Cimmerian block faulting in the Late Jurassic and Early Cretaceous brought about local erosion of the Spekk Formation (e.g., on Heidrun, the Nordland ridge extension, and the Trøndelag platform). In general, however, the unit is remarkably persistent, suggesting that much of the late Cimmerian structuring occurred within a fully marine environment.

Cretaceous

The rapid subsidence that occurred in the Cretaceous west of Haltenbanken is reflected in onlap of marine calcareous shales against the Cimmerian structures. The structurally higher features, such as the Nordland ridge extension, remained exposed until the Paleocene.

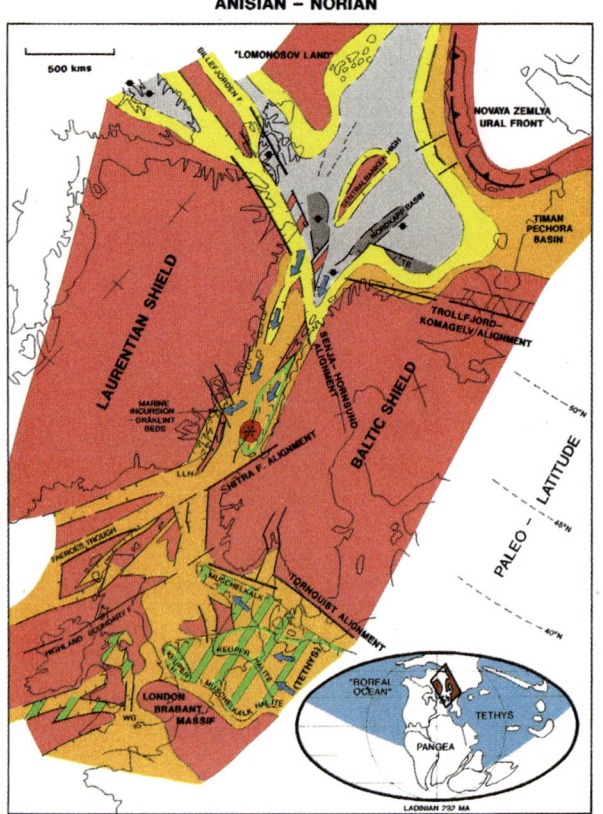
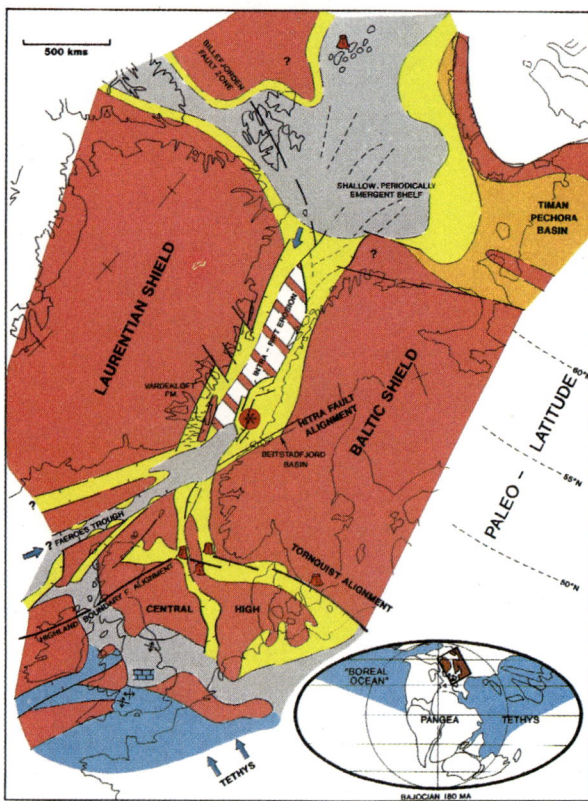

FIGURE 5. Paleotectonic maps of the early Mesozoic in northwest Europe (after Doré, 1990).

Tertiary-Holocene

Seafloor spreading commenced in the early Tertiary, and Haltenbanken subsequently developed as a passive continental margin (Doré and Gage, 1986).

The succeeding Paleocene sediments are also largely of marine shale. Tuffaceous sediments at the Paleocene-Eocene boundary form a regional seismic marker, the Balder Formation equivalent. The tuffs represent widespread vulcanism associated with the opening of the North Atlantic. A Paleocene intrusive center (Vest Brona) is present a short distance south of Haltenbanken (Bugge et al., 1980).

Post-Paleocene sedimentation was primarily of marine claystone, accumulating on a passively subsiding ocean margin. Glacially derived sands and boulder clays are present in the Quaternary. The Tertiary-Holocene section is more than 1980 m thick in the Haltenbanken designated area.

Status of the Exploration of Haltenbanken in 1983

Database

By 1983, eight wells had been completed in the Haltenbanken area and had confirmed the existence of the Jurassic rift section with the associated excellent reservoir source and seal. CNI had interests in, or had traded for, all but two of these wells. As stated above, both regional and detailed seismic surveys totaling 15,520 km had been purchased or traded for by CNI. All the open blocks for the eighth round were covered by a detailed 2 by 2 km seismic grid, except for the southern parts of blocks 6507/7 and 6507/8, which were covered by a 1 by 1 km grid. The data quality of the detailed surveys was fair to good.

By 1983, Haltenbanken was an area of proved potential for gas condensate. A moderate-size gas accumulation was discovered by well 6507/11-1 (Saga) and further delineated by well 6407/2-2 (Figure 8). Statoil's discovery of light, gassy oil with well 6407/1-2 in 1982 was a landmark in Norwegian oil exploration, producing the first liquid hydrocarbons from Haltenbanken.

Generally good reservoir development was interpreted in the open blocks at Middle Jurassic Fangst and Lower Jurassic Tilje/Åre levels, although variable erosion of the units had occurred on the edge of the Trøndelag platform and on the Nordland ridge. A wide range of porosities (15 to 34%) was expected in these units as a result of varying burial depths.

Definitive structuring was mapped in the designated eighth round blocks at base Cretaceous, Middle Jurassic, and Lower Jurassic levels (Figure 8). The structures reflected the over-all north-northeast–south-southwest tectonic grain imparted to the area by the Cimmerian

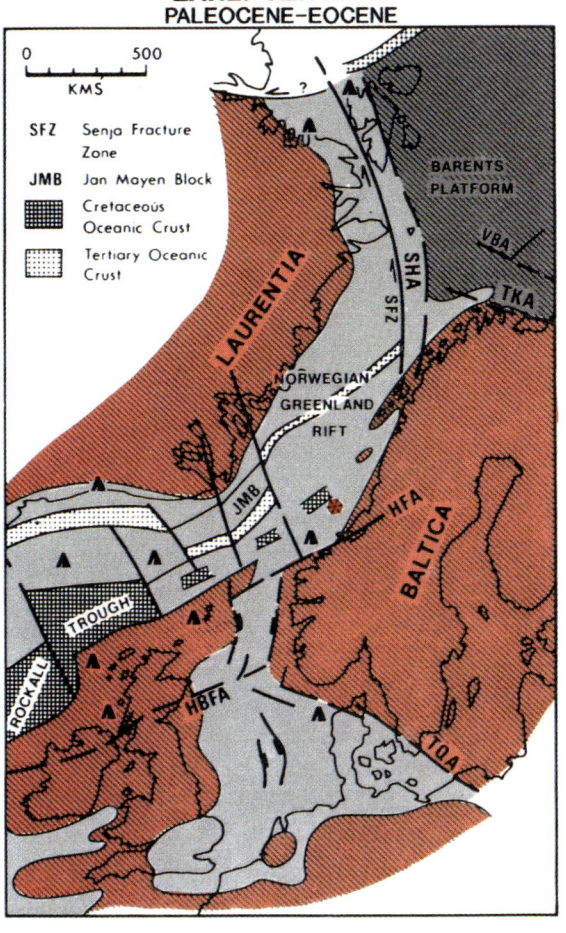
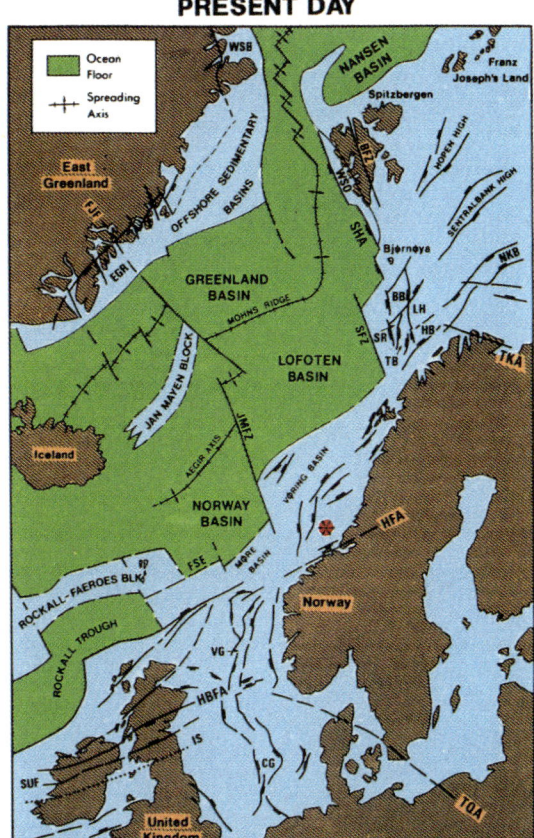

FIGURE 6. Paleotectonic maps of the Cenozoic-Holocene in northwest Europe (after Doré and Gage, 1986).

diastrophism. They included four-way dip closures, horsts, and tilted fault blocks. Top seal to all Fangst/Tilje/Åre prospects was provided by Spekk, Melke, or Lower Cretaceous shales. The sealing capacities of these units had not yet been tested. Lateral seal in all instances was provided by Upper Jurassic-Lower Cretaceous shales.

The prime source rocks were interpreted as being the Upper Jurassic Spekk Formation (Draupne Formation "Hot Shale" and Heather Formation equivalents), and the Jurassic-Triassic lower Åre Formation (Coaly Unit) (Figure 9). The Spekk Formation was viewed as an important source rock for oil, whereas the Coaly Unit was interpreted as a major gas source with possible oil potential. Both were considered immature on the Trøndelag platform and the Nordland ridge extension, but mature in the west of the Haltenbanken.

Therefore, CNI interpreted good potential for major hydrocarbon accumulations in the eighth round acreage. Substantial volumes of both oil and gas were likely to have been generated in the vicinity of the blocks in the last few million years.

Reservoir Quality

The primary reservoir rocks were interpreted as being the Fangst, Tilje, and Åre (Figure 9).

Porosity vs. depth plots were made for three separate levels—Tilje/Åre, and lower and upper Fangst (Figure 10)—and were combined to give an over-all porosity gradient, and this gradient was generally used in reserve estimates. High porosities were probable on strongly upfaulted horsts such as prospect B. Porosities in the more basinal areas, such as prospect C and the northwestern portion of prospect A, were almost certainly much lower (about 16%). However, well 6407/1-2 had demonstrated that oil could be produced from sands in this porosity range.

Source Rocks

The Jurassic Spekk Formation shales and Åre Formation coal beds were present over most of the area. However,

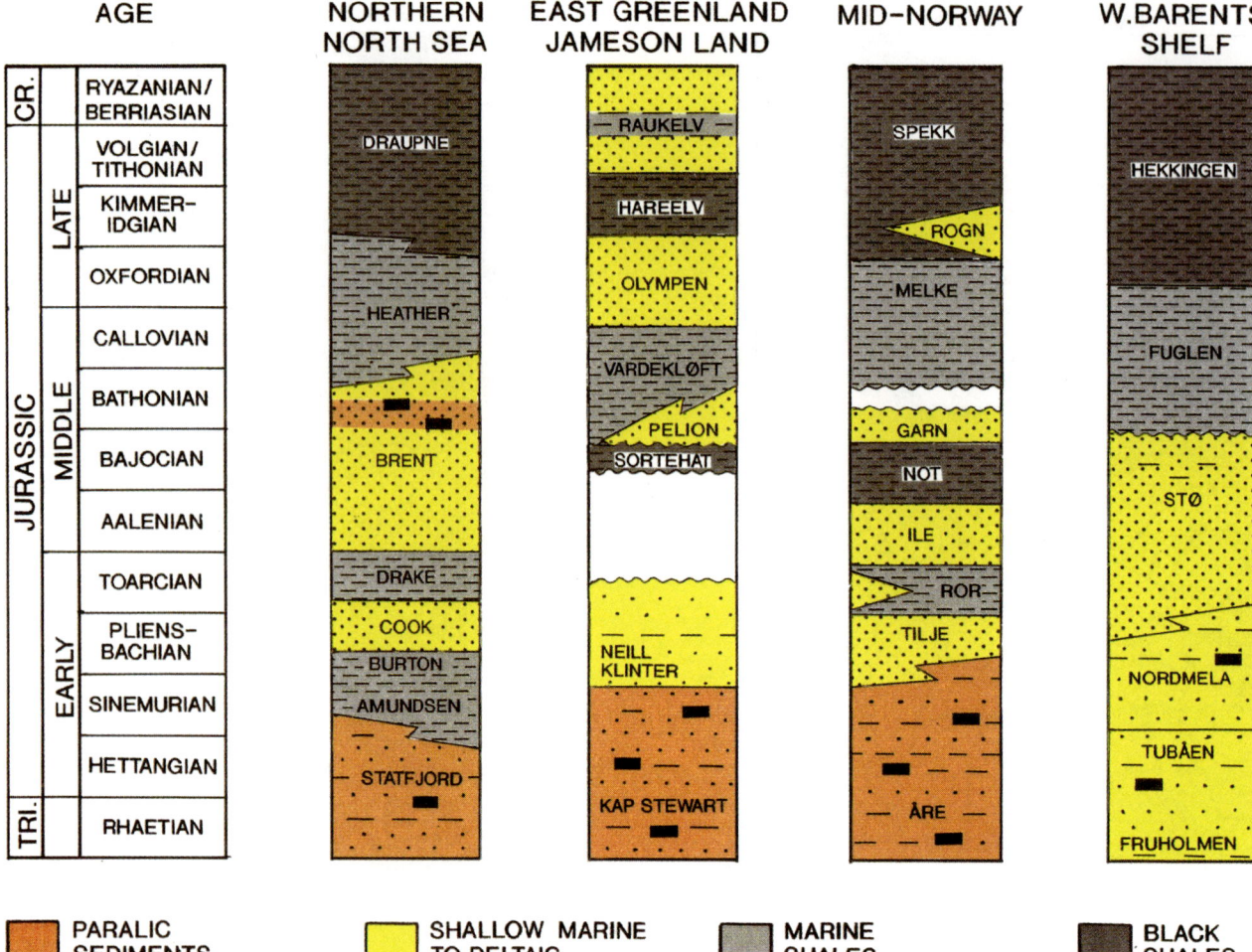

FIGURE 7. Regional stratigraphy of the proto-Atlantic rift (after Doré, 1990).

the Spekk Formation was thin or absent over some of the major highs as a result of Cimmerian erosion.

The Spekk Formation had the highest gamma ray readings in the Haltenbanken area and was a rich source rock with a total organic carbon content of up to 13% (average, 4%).

Most Haltenbanken geochemical reports indicated that the dominant kerogen type was amorphous type II, with some type III. The Spekk Formation was therefore expected to be oil prone, but to have significant additional gas-generating potential.

Maturity

Conoco had developed in-house techniques (Leadholm et al., 1985) for modeling vitrinite reflectance (R_o) from velocity data to assist in identification of oil kitchens in undrilled basins. Conoco also used maturation techniques such as burial history, heat flow analysis, and TTI as originally developed by Lopatin (1971) and Waples (1980).

The computed R_o values on the maturity maps shown in Figure 11 illustrate the maturity distribution in Haltenbanken. The maps indicate that the Spekk Formation is immature on the Trøndelag platform and on the southerly extension of the Nordland ridge. The western part of Haltenbanken is within the oil window (R_o, 0.6 to 1.2%). The Spekk Formation enters the initial wet gas generation stage (R_o, 1.2 to 1.5%) immediately west of the designated acreage. However, burial/maturation curves indicate that even in this area the dominant hydrocarbon generated in the last few million years was probably oil (Figure 12).

The westernmost blocks of Haltenbanken lie within the present Spekk Formation oil window, and the Jurassic reservoir rocks were interpreted as having received oil from the Spekk Formation in the last 5 million years. The Åre Coaly Unit, which was expected to be a prolific source, was generating gas in the same area. The migration paths were uncertain. The oil could be proximal to the reservoirs in the present-day "oil window area" (deeper than 0.8 R_o) in prospect A. Alternatively, the oil could have migrated 5 to 15 km upward and along faults and sand reservoir conduits, to the higher structural traps such as prospect B.

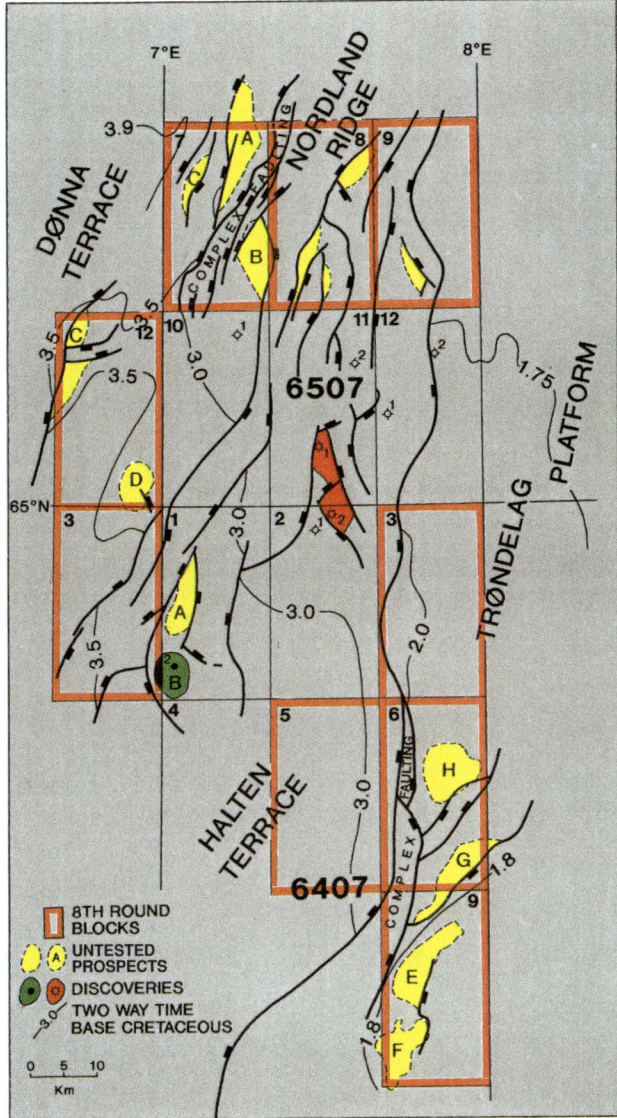

FIGURE 8. Time map (1983 interpretation) on base Cretaceous unconformity, showing prospect inventory (Haltenbanken area).

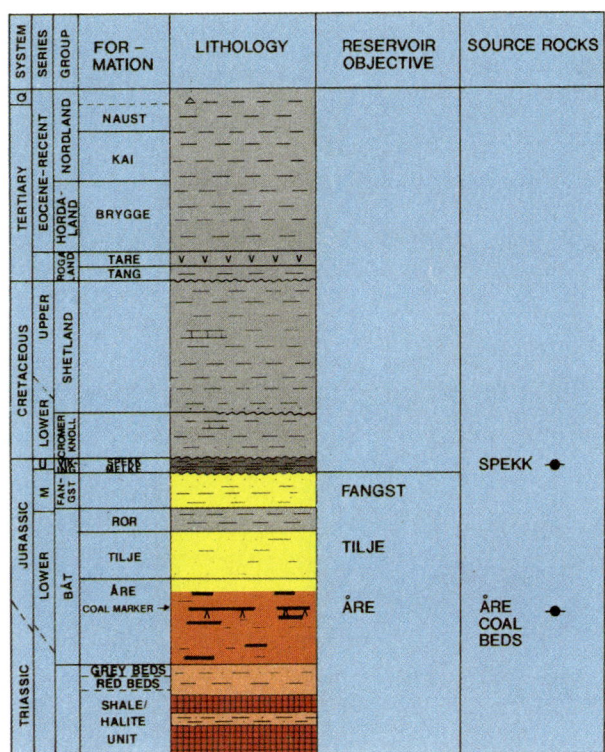

FIGURE 9. Stratigraphic column for Heidrun field.

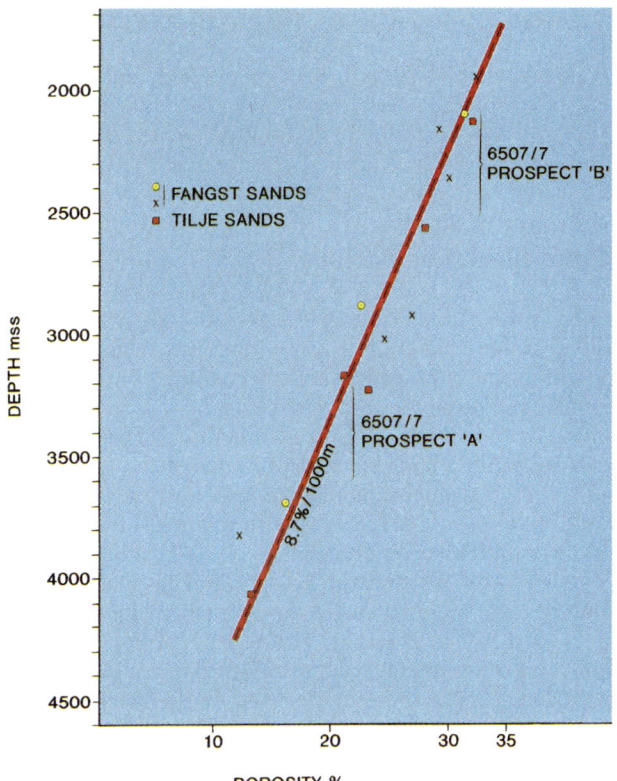

FIGURE 10. Porosity vs. depth composite plot for Haltenbanken Fangst and Tilje sands.

Structural Closures

All prospective hydrocarbon traps discussed in this paper were of the structural type. Closures were present in the area at all mapped levels. They included tilted fault blocks, horsts, and four-way dip closures.

Tilted fault blocks were the largest and most numerous structures. Major prospects of this type included prospects A, C, and D in the west of Haltenbanken (Figure 8). Prospect B was the most prominent trap of the horst type, although closure of this structure depended on a strong southerly plunge.

Prospects E and F on the Trøndelag platform were of low relief as a result of the gentler structuring that affected this province. Prospect G was a large, well-defined anticline at base Cretaceous level, formed by westward slump faulting above a major basement fault.

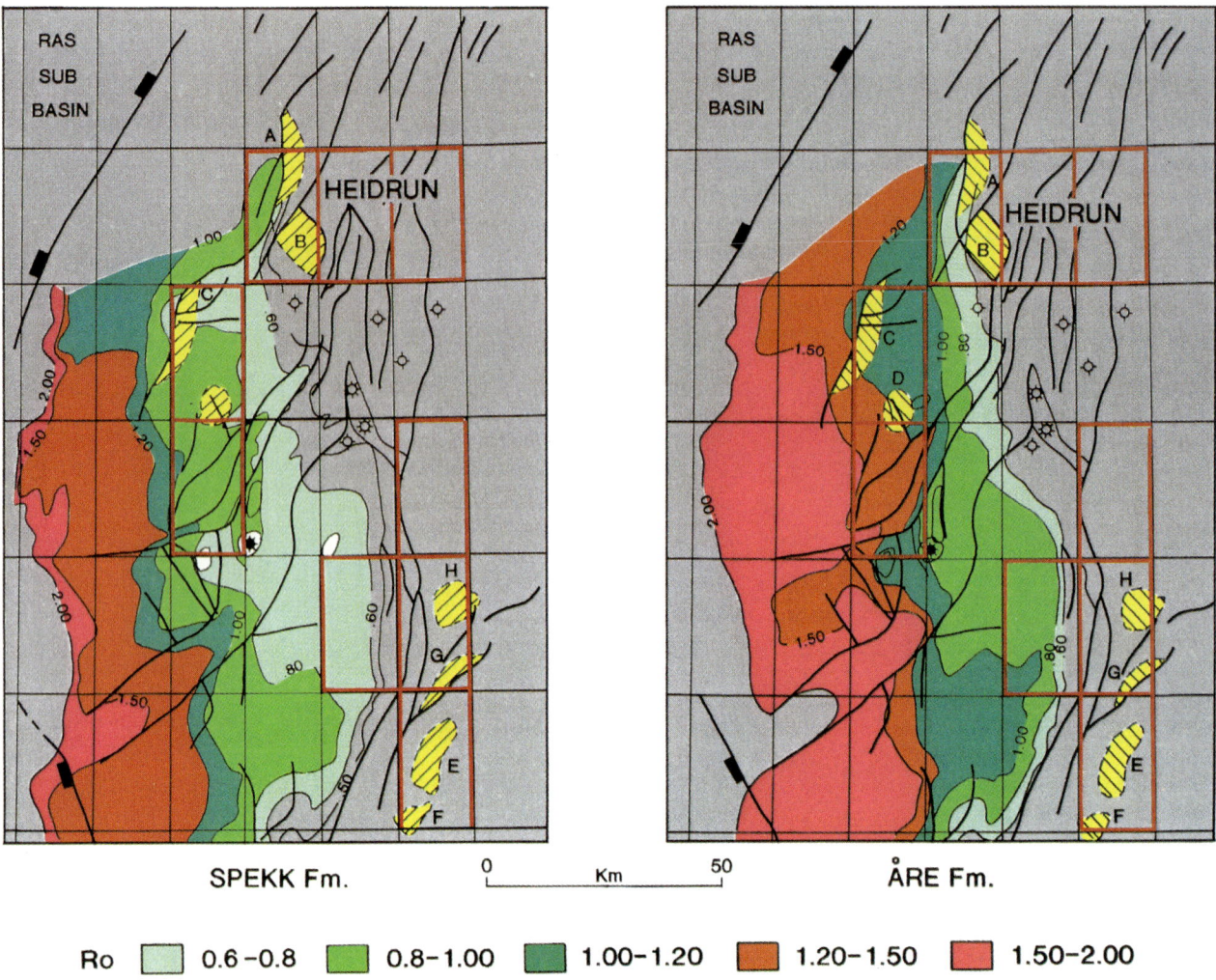

FIGURE 11. Maturity maps (R_o) of Spekk and Åre formations in Haltenbanken.

Summary of the Eighth-Round Block Evaluations

Conoco now had nearly all the ingredients for a major oil play in Haltenbanken, but it was not clear which blocks contained the most prospective structures. The seismic mapping indicated ten to 12 giant Jurassic structures with potential oil reserves of 0.5 to 1.0 billion bbl. These structures were tilted fault blocks and horsts of the trap door and simple anticlinal structural styles. Thus, the next step was to select those structures with mature source rocks and favorable reservoir quality.

It was obvious that the prospects in the north were adjacent to, or mapped inside, the oil kitchens for both the Spekk and Åre formations. The prospects in the south were, however, immature and might require long path migration. Conoco therefore focused on the blocks to the north and analyzed the most northwesterly block, 6507/7, in detail. This block had two giant structures—structures A and B.

But what about the reservoir? Porosity vs. depth curves (Figure 10) calibrated to well offset data for 6507/7 showed that prospect B could have high porosity (30 to 35%), whereas prospect A had medium porosity (15 to 20%) and the potential risk of lower porosity.

Thus, the exploration analyses showed that there were two world-class, billion barrel potential, giant structures in the same block (6507/7): prospect A had a risk of low porosity but was interpreted as having mature oil-prone source rocks, whereas prospect B had high porosity but immature source rocks. Long-path oil migration into prospect B was a strong possibility, because there were both open migration paths from the southwest oil kitchens feeding this prospect, and major faults that could act as conduits for vertical migration.

CNI upgraded block 6507/7 and was successful in its eighth-round application to the Norwegian government. Block 6507/7, containing the highly prospective A and B structures, was awarded to Conoco, Arco, Tenneco, and Statoil under PL 095 in 1984; CNI was designated as the operator.

Structure A was drilled by CNI in 1984 with well 6507/7-1. This well was plugged and abandoned after exhibiting only minor oil and gas shows. Well 6507/7-2

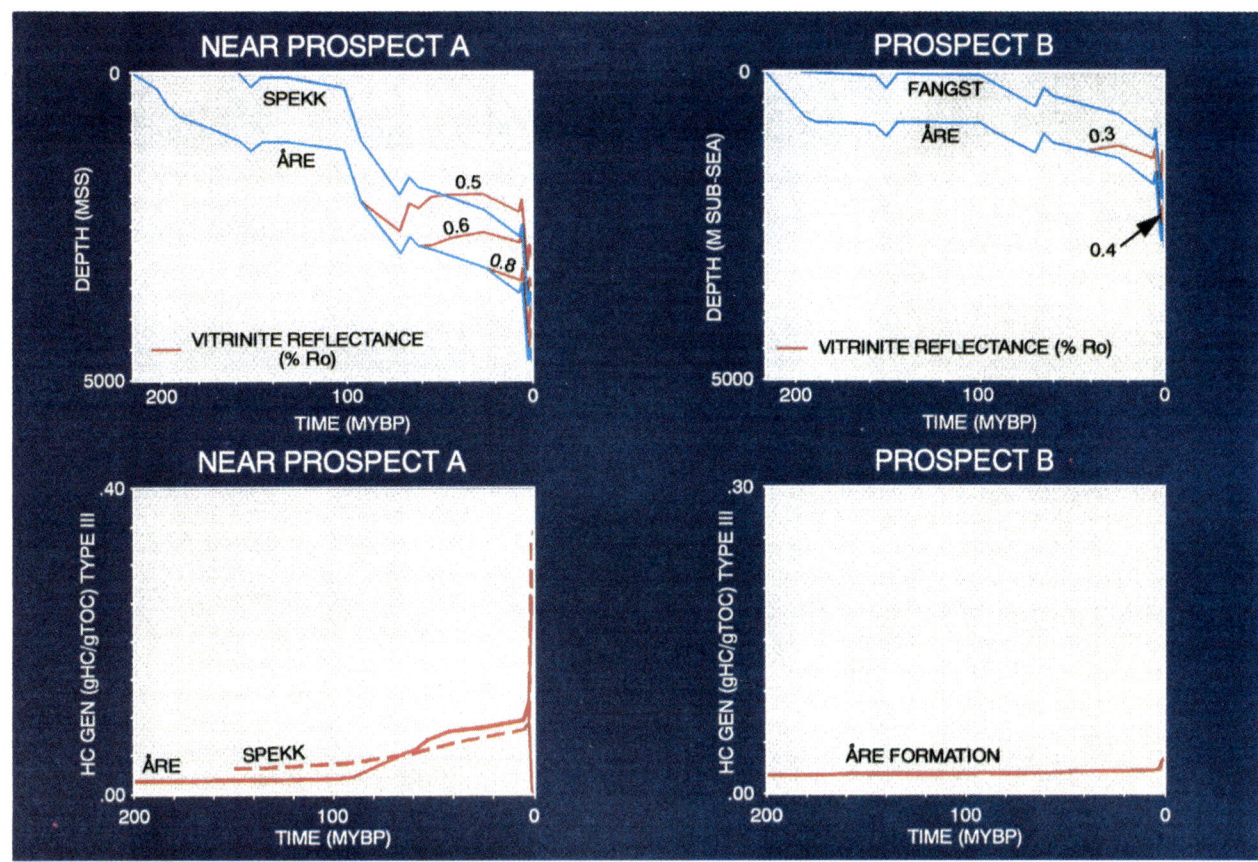

FIGURE 12. Maturation and yield plots for block 6507/7 prospects.

was spudded on structure B on February 25, 1985, and made the giant oil and gas discovery that was later named Heidrun, and reported by Koenig (1986).

PHASE II—APPRAISAL AND DEVELOPMENT

Overview

Continuous appraisal drilling of Heidrun was undertaken, and by mid-1987 an additional five wells had been drilled and a statement of commerciality issued. Statoil, as operator of block 6507/8, drilled well 6507/8-1 in 1986 and proved the extension of Heidrun into that block. The partners unitized Heidrun in 1989 with 75% of the field in block 6507/7 and 25% in block 6507/8 (Figure 13). Reservoir data for the Heidrun field is illustrated in Table 1.

In order to locate the appraisal wells, a 1015-km 2D survey with 450-m spacing, the CN 8502 survey, was acquired over the field in 1985. However, the data quality was not acceptable for field development. Therefore, during the summer of 1986, a 3D survey with 25-m spacing was acquired over the field, and state-of-the-art processing of the survey was completed in May 1987. The quality of the 3D seismic data is far superior to that of the 2D data over the area, as illustrated by line Z-Z' (Figures 14 and 15). Faults with 25 m of throw (such as that in

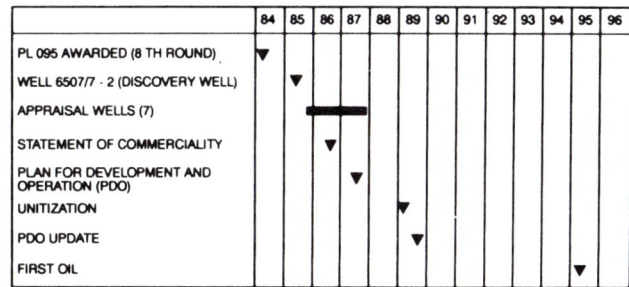

FIGURE 13. Appraisal and development (phase II) schedule for Heidrun field.

6507/7-8) that were indiscernible on the 2D seismic are often clearly imaged on the 3D seismic (Figure 15). The 3D survey provides the geophysical input for the geologic model that forms the basis of the current reservoir model.

The reservoir boundary that can be mapped with greatest confidence is the top of the Fangst Group, because the Melke/Fangst interface is represented by a large acoustic impedance contrast and a strong reflection that can be correlated with confidence. Near the crest of the structure, in the area where the Upper Jurassic shale has been eroded, the impedance contrast is weaker, although the pick is still quite good because it corresponds to the base Cretaceous unconformity.

TABLE 1. Reservoir data for Heidrun field.

Parameter or characteristic	Fangst reservoir	Tilje/Åre 2 reservoir	Åre 1 reservoir	Total
Oil reserves, million bbl:				
STOOIP	1141	606	1038	2785
Recoverable	575	104	71	750
Gas reserves (recoverable), tcf:*				
Free gas	—	—	—	1.32
Associated gas	—	—	—	0.45
Porosity (average), %	29	26	28	—
Permeability, md:				
Average	1450	360	1788	—
Range	670–19,500	90–10,000	30–11,000	—
Water saturation (average), %	15	42	24	—
Net to gross (average)	0.292	0.82	0.54	—
Oil column, m	195	160	100	—
Oil-water contact, m (subsea)	2468	2429–2450	2414	—
Productive closure, km^2	37.0	23.1	8.0	—
Gravity of oil, ° API	29	22–28.6	22–28.6	—
Gas cap	Yes	Yes	Yes	—
Gas/oil ratio, std. ft^3/STB	628	332–641	332–641	—
Pour point, °F	<−50	<−50	<−50	—
Water resistivity (R_w) at 60°F, ohm m	0.185	—	0.160	—
Reservoir pressure	Normal	Normal	Normal	—
Drive mechanism	Water; Gas cap expansion	Water; Gas cap expansion	Depletion(?)	—

*The distribution of gas reserves in the Heidrun reservoir is still under study. These figures refer to the total gas reserves from all Heidrun reservoirs.

Other reservoir boundaries (base Fangst, top Tilje, and top Åre 1) are mappable but are less reliable picks because of the relatively small acoustic impedance contrasts. Nonetheless, it has been important to map these horizons in order to delineate the fault pattern, the extent of reservoir truncation, and the gross variations in the thickness of unit reservoirs.

The intra-Åre Coal Marker has been a very important mapping horizon. The Coal Marker is a strong seismic reflector that is present over the entire area and has been very helpful in understanding the structural configuration of the field. Together with the base Cretaceous unconformity, the Coal Marker has been instrumental in determining the amount of truncation over the crestal part of the field.

The interpretation of the Heidrun 3D survey was carried out on the Geco Charisma II workstation. The final interpretation was transferred from the workstation to the Vax-based Zycor mapping package, on which all final reservoir maps have been generated.

Detailed Reservoir Structure

The Heidrun reservoirs are disrupted by a coeval, conjugate fault system that formed during the Late Jurassic to Early Cretaceous. The conjugate north- and northeast-trending fault sets on the top Fangst structure map (Figure 14) most likely developed in response to 3D strain caused by basement-controlled rifting (Schmidt, in press). Fault density increases in the northern portion of the structure, where two conjugate faults bounding the field converge.

The northeasterly faults are related, in part, to reactivation of Paleozoic faults with the same strike. A major extensional event occurred during the Late Jurassic to

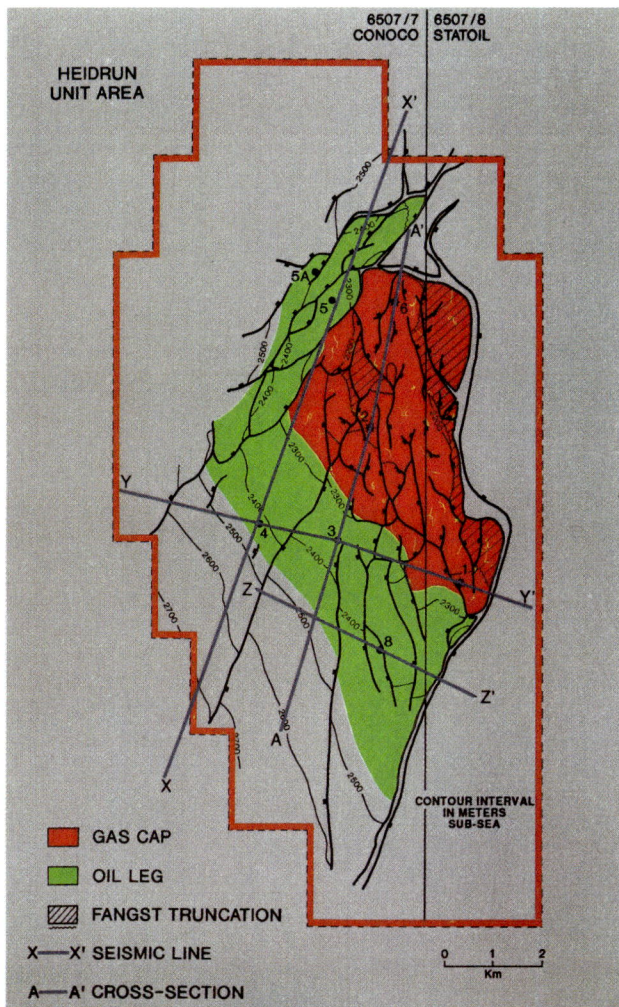

FIGURE 14. Depth map on top of Fangst Group, and seismic line locations.

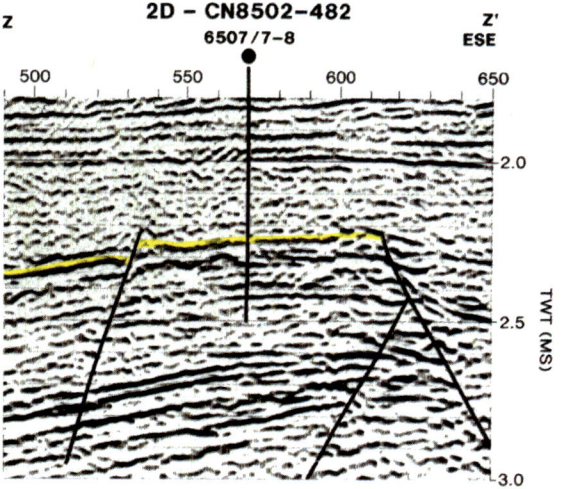

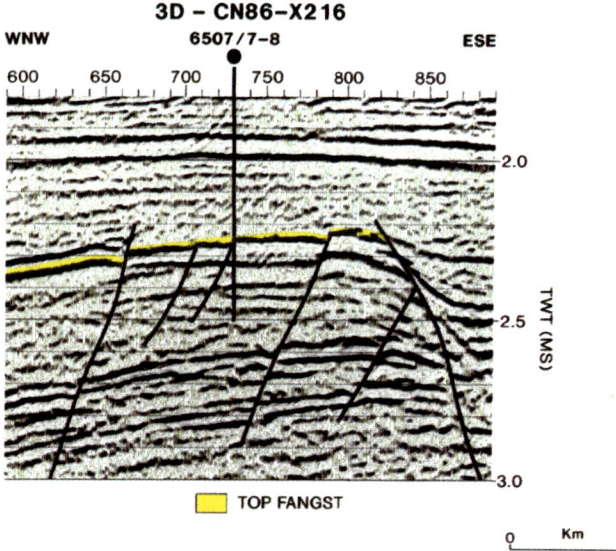

FIGURE 15. Comparison of 2D and 3D seismic data for Heidrun field.

Early Cretaceous (i.e., late Cimmerian). Some of the northeast-trending Paleozoic faults were reactivated during this event, and a north-trending fault set, which is often antithetic to the northeast-trending faults, also developed. Apparently, extension associated with this event could not be accommodated entirely by reactivation of the older northeast-trending faults.

The late Cimmerian event is the only deformational event that affected the Heidrun reservoir units. Faults within the field that are associated with this event characteristically have throws less than 50 m. The throws of faults seldom exceed the thickness of the Fangst reservoir and never exceed the thickness of the entire reservoir section (Fangst, Tilje, and Åre reservoirs) except on the large bounding faults of the field. Figures 16 and 17 illustrate the excellent data quality of the 3D seismic, and Figure 14 shows the locations of these lines.

Sedimentology and Stratigraphy

Zonation

The three Heidrun unit reservoirs are the Fangst Group, the Tilje Formation, and the Åre Formation. These are further subdivided into 12 zones (Figure 18). In order to establish a fieldwide zonation of these unit reservoirs, various techniques ranging from purely sedimentological to biostratigraphic and lithostratigraphic based on core and log responses have been applied (Pedersen et al., 1988). All of these techniques have been brought together to produce a geologic model that shows deposition through discrete moments in time (Figure 19). The fact that the zonation is honored by data from these different disciplines has permitted better modeling of reservoir

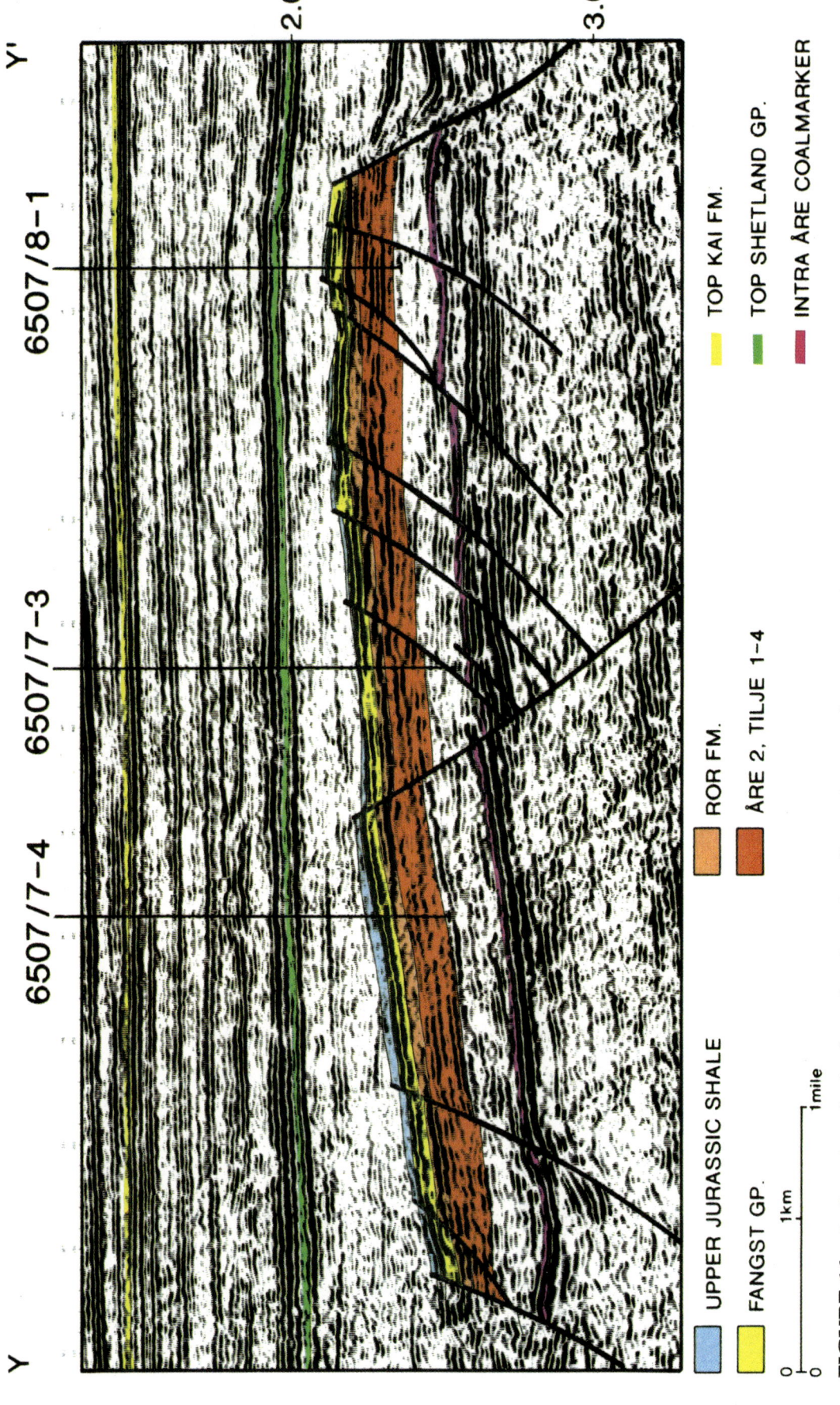

FIGURE 16. Strike-oriented seismic section through wells 6507/7-4, 6507/7-3, and 6507/8-1.

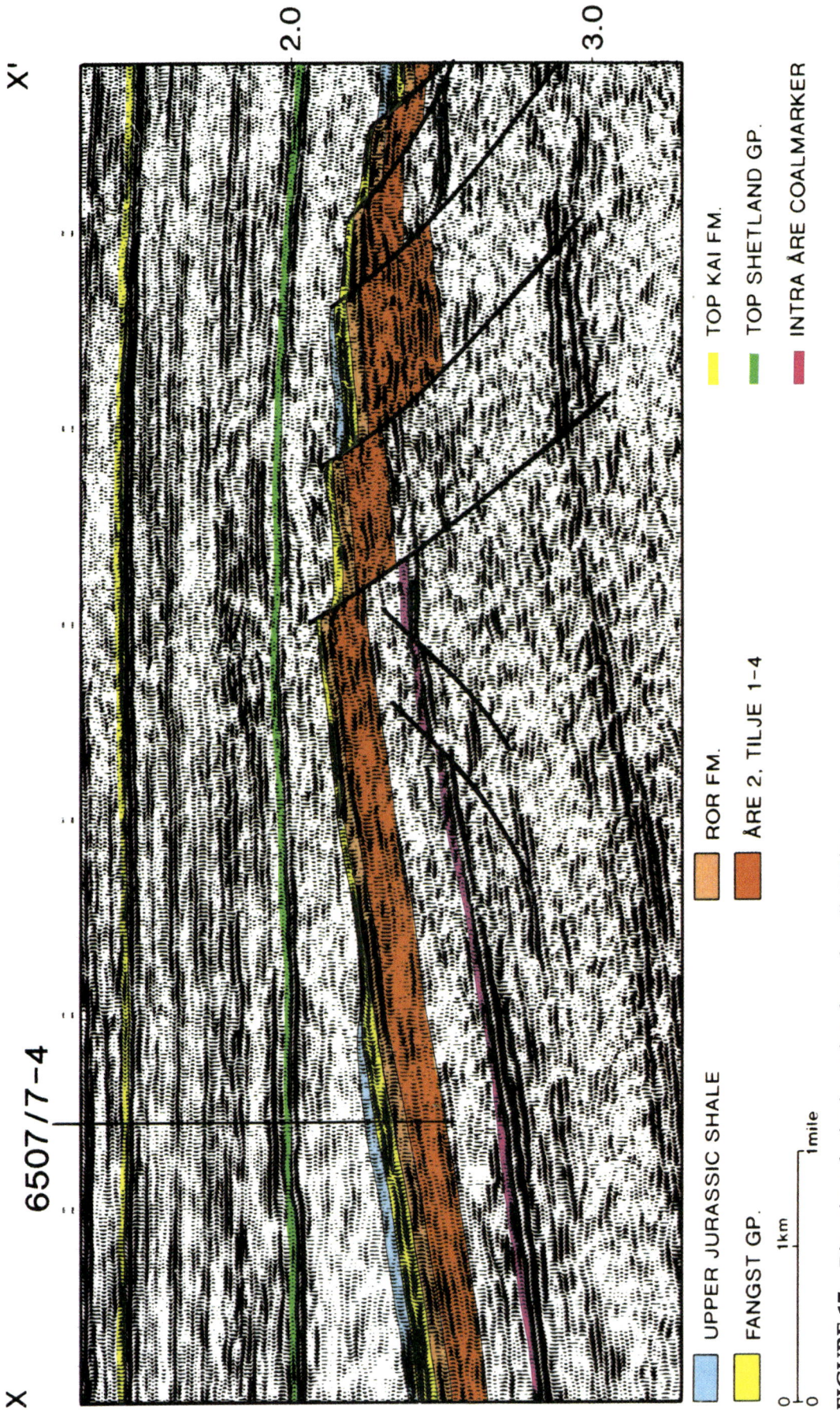

FIGURE 17. Dip-oriented seismic section through well 6507/7-4.

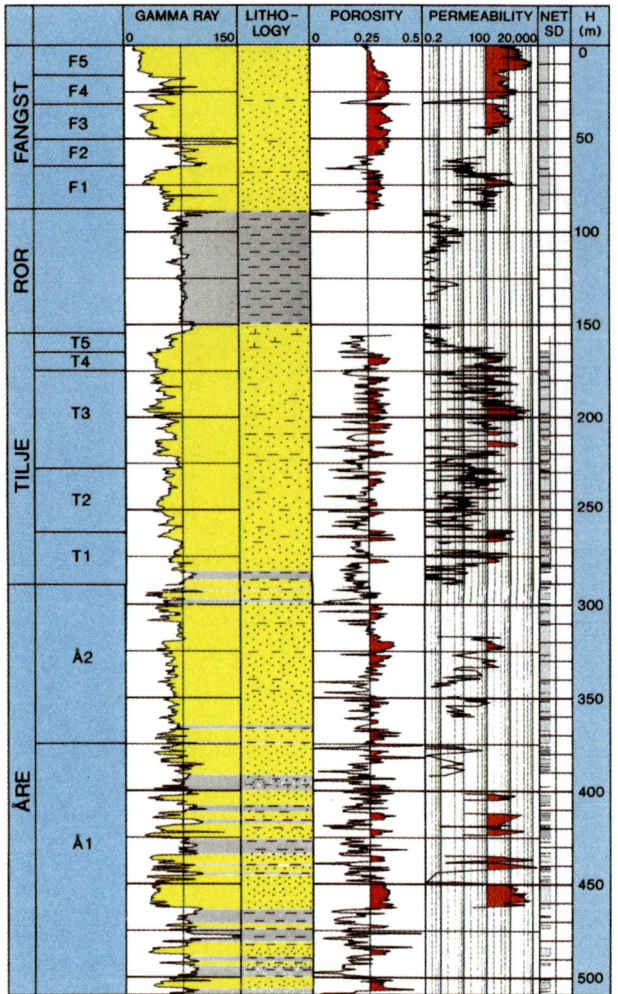

FIGURE 18. Heidrun field type well log section composite of wells 6507/7-4 and 6507/7-6.

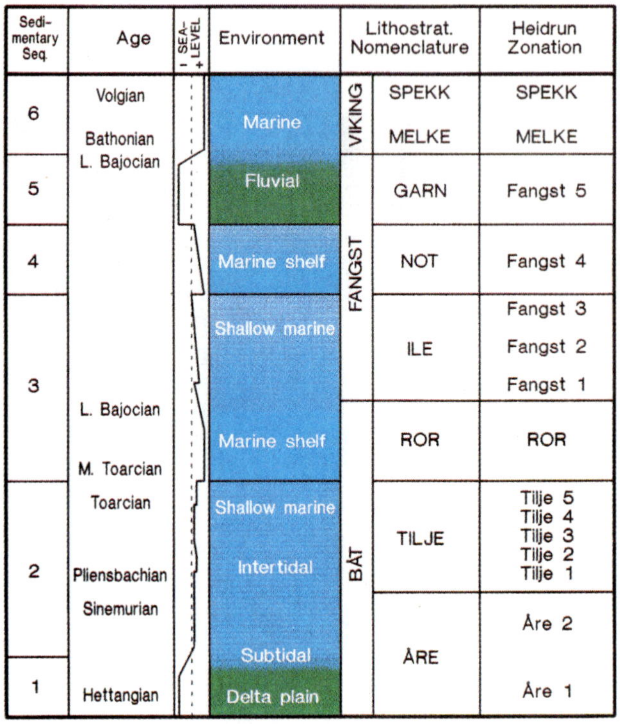

FIGURE 19. Early/Middle Jurassic paleoenvironments in Heidrun field (after Pedersen et al., 1988).

continuity. The current reservoir development plan and volumetric studies are based on the 12 reservoir zones.

The lithostratigraphic correlation within the Heidrun field is based on well logs and cores from the eight wells drilled within the field. A total of 1715 m of core has been logged and described in detail, including petrographic analysis (Pedersen et al., 1988; Harris, 1989).

Åre 1 Zone

The Åre 1 zone has been cored in only one of the Heidrun wells (6507/7-6) but has been penetrated and logged in two of the other wells (6507/7-2 and 6507/7-3).

Core interpretation indicates that the reservoirs are fluvial channel sands deposited on a delta plain or alluvial plain. The Åre 1 zone is composed of fine-grained vertical accretion deposits (75%) and sandy lateral accretion channel deposits (25%). These channel sands exhibit only minor diagenetic alterations and are excellent reservoirs.

The Åre 1 zone was deposited as a mud-dominated deltaic plain crossed by meandering channels, some of which were of considerable depth and lateral dimensions. Flow toward the south is indicated by dipmeter logs within cross-stratified channel intervals. The main reservoirs within this zone are point-bar sandstones within meander belts that should trend generally from north to south. Stream depths may have exceeded 15 m, implying large streams and meander belts likely to have been several kilometers wide. Muddy floodplain deposits with thin coal beds separate the channel sandstones and represent effective lateral and vertical seals. The coal beds and organic shales that are thermally mature may provide sources of hydrocarbons. Figure 20 (right) illustrates a typical Åre 1 channel sand.

The transition zone between Åre 1 and Åre 2 has been faulted out in the only well with available core. This transition zone is caused by a regionally controlled subsidence, and can be followed over the entire Haltenbanken area. The boundary is diachronous and becomes stratigraphically younger eastward.

Åre 1 Petrology

The channel sandstones in the Åre Formation contain quartz, potassium feldspar, minor amounts of mica, and virtually no clay. Garnet is a common accessory phase.

Åre 1 Diagenesis

Åre channel sandstones have generally undergone relatively minor diagenetic changes (Figure 21, right). Common diagenetic features include dissolution of potassium feldspar, alteration of mica to kaolinite, and precipitation of kaolinite books in open pore space. In general, alter-

 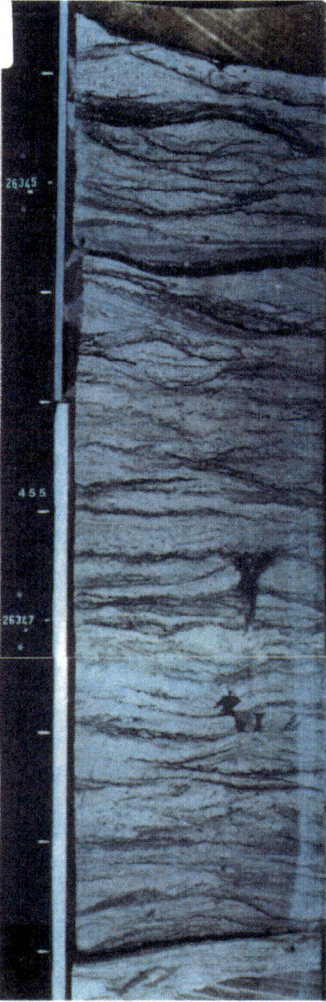 

FANGST GP.
(6507/7-4, 2456.8m C.D.)

TILJE FM.
(6507/7-4, 2634.6m C.D.)

ÅRE FM.
(6507/7-6, 2433.2m C.D.)

FIGURE 20. Core photographs of the Heidrun field reservoir. *Fangst Group (F5)* (left). Medium- to coarse-grained, poorly to moderately sorted, subrounded, cross-bedded fluvial sandstone, consisting predominantly of quartz with a significant percentage of lithic fragments. Environment of deposition is an alluvial plain, high-energy channel system. *Tilje Formation (T3)* (center). Fine- to very fine-grained sandstone, interlaminated with siltstone and claystone. Sand is ripple cross-laminated with moderate to intense bioturbation. Environment of deposition is tidal flat/shoal. *Åre Formation (Å1)* (right). Medium- to fine-grained, tabular cross-bedded sandstone overlain by siltstone and claystone interlaminated with very fine-grained sandstone, displaying laminae parallel to ripple cross-laminae. Lower sandstone is a fluvial channel that was subsequently abandoned and overlain by low-energy overbank deposits.

ation has had only a minor effect on sandstone porosity and permeability. Locally, the channel sandstones have been cemented by sparry dolomite. This cement is localized in discrete zones typically 1 m thick.

Sandstones deposited in an overbank setting are considerably richer in mica and clay. In part, the clay is kaolinite, derived from alteration of mica. These sandstones also contain considerably more siderite than the channel sandstones. The siderite replaces detrital mud clasts and fecal pellets; locally, siderite constitutes more than 50% of the rock volume.

Åre 2 and Tilje 1 to 5 Zones

This unit is the most heterogeneous of the Heidrun reservoirs. The total average thickness of the six zones is about 200 m. Environments of deposition are varied and range from intertidal mudflats to shallow marine shelf deposits, with the most marine influence in the middle and uppermost parts of the sequence. Figure 20 (center) shows a typical Tilje tidal sandstone.

Regionally, because deposition ranged from shallow marine along the axis of the basin to broad tidal and

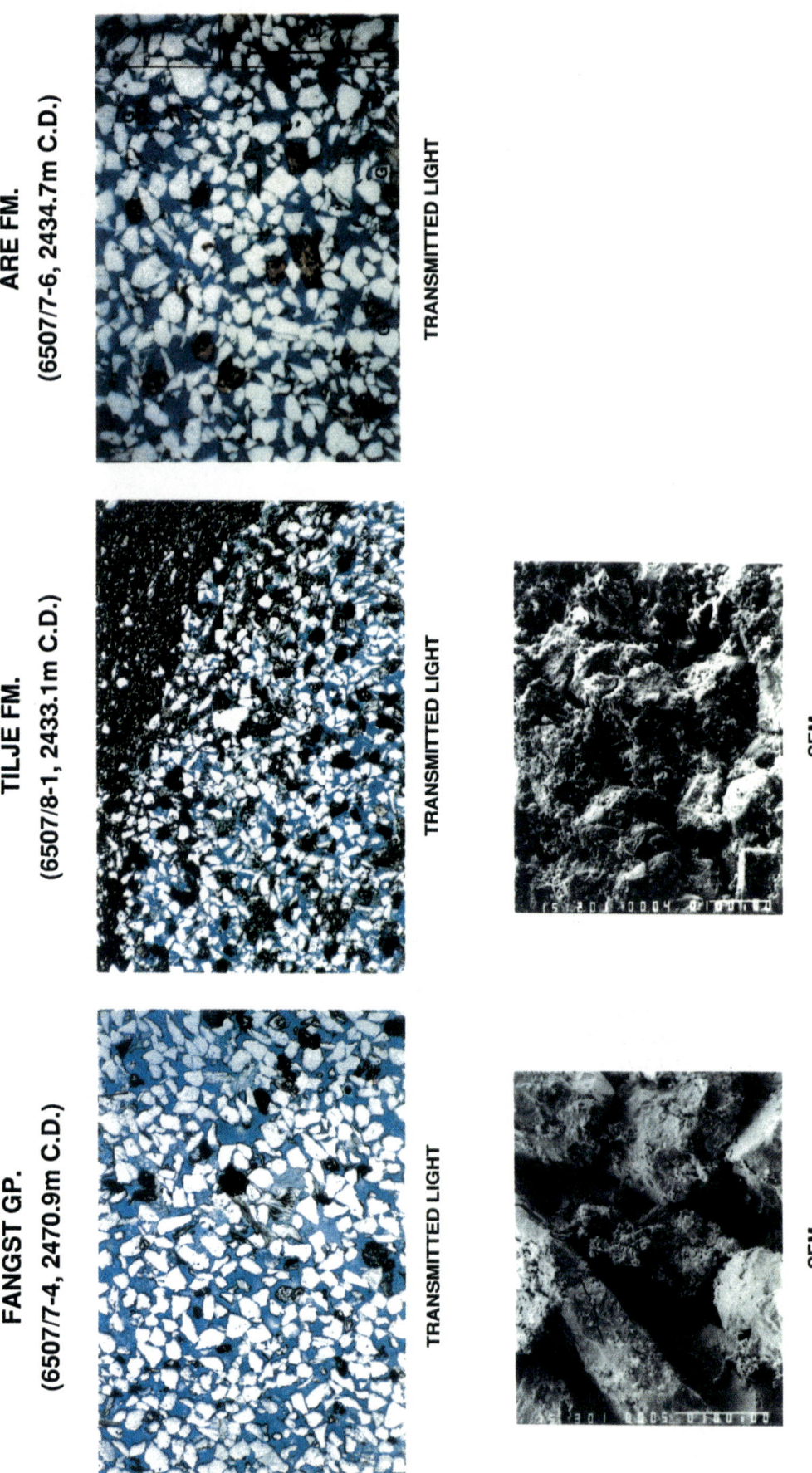

FIGURE 21. Thin section and S.E.M. micrographs of Heidrun field reservoirs. *Fangst Group* (left). The sample has a relatively open pore system with small amounts of clay appearing as pore-lining cement—e.g., the grain in the center in the center (top). The pores are large, and the pore system seems relatively open in all directions with no sedimentary structures. Porosity is unaffected by significant diagenetic alteration. *Tilje Formation* (center). The most common cementing phases are syntaxial quartz overgrowths, kaolinite, and illite. The sample consists of alternating thin layers of very fine-grained, well-sorted sandstone and matrix-rich siltstone. This sample contains pore-filling kaolinite cement—irregular masses (pseudomorphs?), booklets, and vermicules—and it seems as if dissolution of potassium feldspar is less common. *Åre Formation* (right). Thin section photomicrograph of clean, quartz-rich channel sandstone. Clear, white grains are quartz. Yellow or brown grains are potassium feldspar. Gray, high-relief grains (labeled "G") are garnet. Note the open packing of grains, the lack of detrital or authigenic clay, and the lack of cement.

deltaic at the margins, correlation is difficult. The end of the Pliensbachian was marked by an abrupt Toarcian transgression that deposited Ror shales conformably over the Tilje. This stratigraphic boundary, which may be nearly isochronous throughout the basin, is easily recognizable in most wells.

Åre 2 Zone

The lower part of the Åre 2 unit is sparsely represented in cores but contains flaser-bedded shale and rippled sand that is desiccation cracked and contains some root zones, implying exposure and growth of vegetation. These features suggest that deposition was mainly in the upper or landward part of the tidal zone. In the middle and upper parts of the Åre 2 interval, the deposits are either subtidal or in the lower, less frequently exposed parts of the intertidal zone. There are no root zones, cracks are less common, and burrow forms are more diverse. Some of the sandy units are wave or current rippled and contain thin, flat-bedded zones with ooids, implying deposition on wave-washed shoals. Well data suggest a more marine environment to the south or southwest within the Heidrun field area.

The upper boundary (between Åre 2 and Tilje 1) defines the start of the more open marine environment at the base of the Tilje 1 zone.

Åre 2 Petrology

The Åre 2 sandstone is an arkose with both potassium feldspar and plagioclase. Detrital muscovite flakes are abundant, especially in the finer-grained laminae and beds. Carbonaceous fragments and carbonized wood fragments are abundant in some beds. Mudstone rip-ups and siderite peloids (fecal pellets?) are also common in some beds. Authigenic clays are uncommon and are restricted to kaolinite that has replaced some detrital feldspar (?) grains.

Tilje 1 Zone

The lower part of the Tilje 1 zone contains sandy shale with diverse marine burrowing fauna. Some coarser sandstone occurs in well 6507/7-5 as isolated rippled zones, suggesting that the source of sand lay to the north. Ripped-up siltstone pebbles and coarse quartz granules attest to the high energies of the shallow marine environment in the lower Tilje 1.

The tidal flat sequence in the upper Tilje 1 indicates that the rate of sediment supply caught up with the rate of subsidence. The upper boundary of Tilje 1 probably was caused by a relative rise in sea level that produced the marine shelf setting of the Tilje 2. This boundary is regionally well defined and easily correlated across the study area.

Tilje 2 Zone

Tilje 2 is the zone with the most diverse marine burrow forms and the most intensive bioturbation within the Tilje Formation. These characteristics suggest that deposition in this zone occurred under more marine conditions and was somewhat slower than the rest of the formation. The unit is shaliest at the base, suggesting a transgression and local deepening that rather abruptly brought marine conditions over the previous subaerially exposed tidal flat. The interval becomes sandier upward, not regularly, but rather as a series of upward-coarsening cycles 5 to 10 m thick.

Basal Tilje 3 beds resting on the uppermost burrowed sandstone are tidal-flat deposits, indicating an upward change from marine to marginal marine environments. This boundary is caused by an increase in the rate of sediment supply relative to the rate of subsidence.

Tilje 3 Zone

The Tilje 3 zone has five subunits of flaser-bedded tidal-flat deposits in a section only about 50 m thick. Each of these units contains sand-filled cracks that indicate emergence and drying episodes. Between these muddy tidal-flat zones are sandy beds that represent both tidal channels and shallow subtidal shoals with marine burrows. The entire interval is interpreted as a fairly energetic tidally dominated shoreline deposit, where sandy subtidal shoals swept by currents and waves intermingle with more muddy intertidal flats crossed by tidal channels.

Sand is present within lens-shaped ripples, representing traction transport. Silty shale drapes these ripples as thin parallel laminae, representing deposition from suspension. The repeated alternation of traction and suspension deposits indicates dramatic fluctuations in flow intensity, a common attribute of tidal environments. Shale zones are commonly cracked, and the cracks are filled with sand, indicating exposure and desiccation.

The upper boundary is marked by a return to marine conditions and is probably caused by a relative rise in sea level. This boundary is regionally well defined and easily correlated across the field area.

Tilje 4 Zone

The Tilje 4 zone ranges in thickness from 6 to 8 m and is composed of an intensely burrowed sandstone that represents shallow shelf deposition.

The boundary toward the overlying zone is defined by the base of the siderite cementation of Tilje 5. This boundary is well defined in the Heidrun field. However, because it is defined by diagenetic processes, it has little bearing in a regional correlation.

Tilje 5 Zone

The Tilje 5 zone varies from burrowed shaly sandstone at the base to coarse-grained sandstone with pebbles of siderite and cemented sandstone or grainstone, carbonate ooids, and large-scale sets of tabular cross-strata at the top. This capping unit probably represents marine shoals that had relatively large migrating sand waves. However, these sandstones are very mixed by burrowing and modified by siderite cementation. Deposition was slow enough to allow formation of ooids (by wave agitation) and very thorough burrowing and leaching.

The top of Tilje 5 is marked by a change to marine, sandy shale settings. This boundary was caused by a eustatic rise in sea level that may represent the Toarcian highstand at about 182 m.y.a. observed throughout southern Europe and North Africa.

Tilje Petrology

The sandstones of the Tilje Formation are arkosic to subarkosic arenites. Quartz content varies from 26 to 79% of the bulk rock. When these sandstones are normalized to a composition of quartz and feldspar, the two dominant framework constituents, quartz makes up 56 to 89%. Quartz occurs as monocrystalline and polycrystalline grains, with lesser amounts of microcrystalline quartz and quartz/feldspar rock fragments. The dominant feldspar variety is potassium feldspar. Muscovite and carbonaceous fragments are common accessories and generally are concentrated in the finer fractions.

Siderite occurs as sand-sized grains or peloids in much of the Tilje Formation. They are commonly concentrated along laminae with carbonaceous fragments and micas, indicating hydrodynamic equivalency with the lighter grains and suggesting that the peloids were not siderite (density, about 11.5 gm/m^3) at the time of deposition. The peloids are presumed to have been composed largely of siliciclastic mud, perhaps originating as fecal pellets, and to have been sideritized diagenetically.

Tilje Diagenesis

The Tilje Formation is a very good reservoir and is uncemented except for kaolinite and some siderite. Siderite cement is abundant in the upper few meters of the Tilje and greatly reduces porosity. The Tilje porosity is dominantly primary and intergranular. Many of the detrital feldspar grains are partially dissolved; in some places, dissolution is nearly complete and only a thin rim (an overgrowth of different composition?) remains. The majority of the feldspars appear to be unaffected by diagenesis. Oomoldic porosity occurs in some beds where carbonate and/or clay cortices are partially dissolved. Porosity formed by the dissolution of framework grains comprises an insignificant percentage of the over-all porosity.

Pressure solution and quartz overgrowth/cementation are negligible to nonexistent in these sandstones, although Bjørlykke et al. (1985) observed these features in other cores in the Haltenbanken area that are buried to only 2000 m. No stylolites were observed in the Tilje Formation in Heidrun field.

Results of x-ray diffraction analyses of clay separates indicate that authigenic kaolinite is the dominant clay, with lesser to trace amounts of illite and chlorite. The chlorite and illite identified by x-ray diffraction are presumed to be mostly detrital in origin and will not affect production. As mentioned earlier, much of the detrital, siliciclastic mud is sideritized. The detrital mud occurs in the form of peloids, intraclasts, laminae, thin beds, and matrix. The diagenetic conditions for sideritization are unknown, but if the detrital muds contained a high percentage of iron-rich chlorite or biotite, a local source of iron may have facilitated it (Figure 21, center).

Ror Formation and Fangst Zones 1 to 3

The Ror Formation shales and the Fangst zones 1 to 3 sandstones together form a major sequence and represent a major Toarcian to lower Bajocian transgressive/regressive cycle.

Ror Formation

The Ror Formation consists predominantly of shale that becomes coarser upward, and is 45 to 60 m thick in the Heidrun area. The Ror Formation represents deposition in a low-energy shallow marine setting. Cores and logs show this formation to be shaly and finest-grained near the base, with increasing sand content upward, particularly in the upper 20 m. A thin conglomerate occurs at or near the top of the Ror and is composed of rounded, cemented sandstone pebbles.

A eustatic rise in sea level, and possible tectonic collapse in the Toarcian, resulted in widespread deposition of marine shelf muds over mid-Norway, especially on the Halten terrace. The effects of this Toarcian event are seen in other areas of northwest Europe and the Mediterranean. Within these shales, thin sandstone units are developed that were derived from local highs such as Heidrun. The upper part of the Ror Formation becomes coarser-grained, and the Ror-lower Fangst boundary is difficult to pick because both are part of the same over-all shoaling sequence. The boundary between Ror (Toarcian) shale and lower Fangst (Ålenian-lower Bajocian?) sandstones is simply a facies transition related to the influx of more sand. This regression is probably depositional and not a result of significant eustatic changes or tectonic activity.

Fangst Reservoir Zonation

The Fangst reservoir is illustrated in Figures 18 and 19. In the Heidrun field, the lower Fangst has been divided into F1 and F2 units for reservoir zonation; the upper part of this sandy sequence (F2) represents a prograding strand plain capped by beach deposits, so that the northern part of Haltenbanken was emergent in this time interval. The upper boundary of the lower Fangst is defined as the base of the zone F4 shales (Bajocian)—the Not Formation, which in Heidrun is a thin (<1 m thick) shale. The upper Fangst zones 4 and 5 are segmented by an erosional boundary.

Fangst 1 Zone

The Fangst 1 zone is a shallow marine sandy deposit, shaliest at the base and increasing in sand content upward. Three distinct lithofacies are included: (1) fine-grained, partly glauconitic sandstone with wave ripples and scattered shaly partings, slightly to thoroughly burrowed by diverse marine fauna; (2) medium-grained, slightly glauconitic sandstone with tabular- or trough-shaped cross-beds, sharp bases, and scattered shale clasts in units 15 to 38 cm (6 to 15 in.) thick; and (3) rarer thin, well-laminated, very fine-grained sandstone with hummocky cross-strata.

The upper boundary is defined by the shallowing of the water and is marked by the deposition of the hummocky cross-stratified Fangst 2 zone.

Fangst 2 Zone

The Fangst 2 zone is a lower shoreface to offshore deposit and is composed of hummocky, cross-stratified, fine- to very fine grained sandstone. The sandstone has a high thorium content, which gives a high gamma ray log response that could be misinterpreted as a shaly response. Similar high thorium concentrations and gamma ray log responses are noted in the Tilje 1 zone within hummocky beds. The thorium concentration appears to be related to very fine grained heavy minerals, particularly thorite, and therefore represents a hydraulic concentration process.

Fangst 3 Zone

The Fangst 3 zone is composed of (1) mostly trough cross-stratified fine-, medium-, or coarse-grained sandstone, with some layers rich in carbonaceous debris, interpreted as upper shoreface surf zone deposits, and (2) finely laminated, low-angle, finely cross-stratified, well sorted, fine- to medium-grained sandstone that is interpreted as beach deposits. The depositional interpretation would be compatible with progradation of a sandy delta in a wave-dominated setting.

The top of Fangst 3 is marked by the middle Fangst (Not Formation) transgression, which is followed by a regressive cycle. This boundary is regionally well recognized.

Fangst 4 Zone

The Fangst 4 zone is interpreted as a shallow marine deposit. In its lower part, it is shaly and has an abundant and fairly diverse marine burrow assemblage. Upward, the unit becomes more sandy, is largely burrowed by diverse marine fauna, and locally contains preserved cross-strata. Originally, the sediment was probably transported as ripples or larger bed forms, but rarely was the deposition rapid enough to prevent nearly complete burrowing. The Fangst 4 interval becomes notably more shaly in the southeastern part of the Heidrun area, suggesting a more offshore depositional environment.

The top of Fangst 4 is an erosional surface that has been caused by a local eustatic relative drop in sea level. The erosion is a result of the fluvial Fangst 5 sands prograding southward across the Heidrun area. Toward the south, the boundary becomes less erosional on a regional scale as the Fangst 5 sand becomes fluviomarine.

Fangst 5 Zone

The Fangst 5 zone in the Heidrun area is interpreted as a fluvial deposit on the basis of: (1) the dominance of trough cross-strata, (2) lack of any marine indicators such as burrows or glauconite, (3) the sharp erosional base, and (4) the tendency toward textural upward fining. The sandstone is fine- to medium-grained in the upper part of the unit and increases to medium- to coarse-grained near the base. Cross-strata are trough-shaped, and sets are 15 to 75 cm thick. There are no textural or stratification breaks that would separate the unit into any thinner distinctive facies (Figure 20, left).

The basal surface is sharp and apparently erosional. Cross-stratified sandstone rests abruptly on burrowed sandstone with a fully marine aspect. There are no common depositional models that would include a contact between these facies in a uniformly aggrading sequence. Additionally, a similar contact appears to be common at this stratigraphic level over a wide area of the Haltenbanken. The superposition of these discordant facies implies a relative drop in sea level, and erosion.

The top Fangst 5 boundary was formed as a result of a eustatic rise in sea level that caused marine shales to be deposited on top of Fangst 5. This boundary correlates well regionally.

Fangst Petrology

The Fangst sands are composed of quartz with minor to moderate amounts of potassium feldspar and mica, mostly muscovite. They are classified as subarkoses and lithic subarkoses on a standard quartz feldspar lithic (QFL) plot. Minor amounts of garnet are commonly present, as are trace amounts of hornblende and pink tourmaline. Thorite is present in trace quantities throughout the Fangst and significantly affects the log response, especially the gamma ray log. The basal part of Fangst 1 also contains up to 20% clay oolites (possibly chamosite) (Harris, 1989).

The Fangst zones vary in composition, most notably in their detrital mud and mica contents.

Fangst Sandstone Diagenesis

The Fangst sandstone is uncemented, contains little clay, and has excellent porosity and permeability (Figure 21, left). This is a function of the shallow depth of burial combined with favorable primary depositional factors such as coarse grain size and minor detrital clays and micas. Diagenesis has not played a significant role in the development of porosity and permeability. Diagenetic features in the Fangst sandstones are very minor and are described in Harris (1989).

Secondary porosity resulting from feldspar dissolution only contributes a small amount to the total porosity, typically 2 to 4% of the rock volume. It is likely that a lack of plagioclase in the sandstone is a diagenetic effect rather than a primary depositional one.

Thin zones of sandstone, 1 m or less in thickness, are tightly cemented by poikilotopic calcite and are known as "doggers."

The lower part of the Fangst 1 zone contains locally significant amounts of siderite and pyrite cement. In addition to filling pore space, the siderite also replaces detrital mud. Pyrite locally amounts to 20% of the rock and is a pore-filling cement.

Biostratigraphy and Palynofacies

An extensive biostratigraphic study of the eight wells in Heidrun field has been undertaken and is described in Pedersen et al. (1988) (Figure 19).

Geologic Model Summary

The Heidrun geologic model has been constructed from the 3D seismic-derived structure maps, the sedimentological subdivision of the stratigraphy, the average petrophysical parameters, the segmentation resulting from the complex faulting, and the hydrocarbon contact distributions.

The structural complexity of the Heidrun field has necessitated a joint geologic and geophysical approach to construction of the base Fangst, top Tilje, and top Åre 1 depth maps. The top Fangst depth map and the base Cretaceous depth map, both interpreted from the 3D geophysical interpretation, have formed the data from which all map construction has taken place. Three total isochores of the Fangst Group, the Ror Formation, and the Tilje 1 to 5 and Åre 2 zones were used to stack down from the above data in order to construct base Fangst, top Tilje, and top Åre 1 maps. Uneroded zone isochores, generated from predicted sediment distributions, were used to stack up from the base of the respective units (Fangst and Tilje/Åre 2) to construct the individual zone structural maps by clipping ("eroding") the respective units against the top map.

Geologic Model for Sand Distribution

Isochore maps for all zones have been drawn by taking into consideration regional sedimentology studies, sediment source direction, and timing of structural development of the field from the regional studies (Figure 22).

Both regional studies and intrafield studies (i.e., sedimentology, dipmeter, and biostratigraphy) indicate that the predominant direction of sediment supply for the Fangst reservoir zones of Heidrun is north to east. On a regional scale, there is evidence to suggest that there may have been a local high to the northwest of Heidrun that acted as a source area for parts of the Tilje and Åre reservoirs. This is supported by the facies distribution maps and isopach maps prepared for the Fangst, Tilje, and Åre reservoir units in the Haltenbanken area.

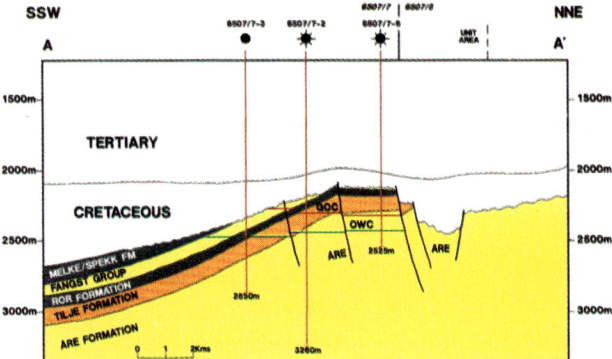

FIGURE 22. Hydrocarbon contact distribution model for Heidrun field.

Petrophysical Interpretation

Database

The petrophysical database includes well logs and core analyses for all the Heidrun wells. The special core analyses included permeabilities to oil and water, relative permeability, cation exchange capacity, formation resistivity factor, and resistivity indexes.

Formation water samples have been recovered from the Fangst Group in well 6507/7-4 and from Åre 1 in well 6507/7-2.

Porosity and net-to-gross maps were constructed using average values from the wells for each reservoir zone. Because sedimentary processes are interpreted as being the primary controls on reservoir quality, the depositional models established from detailed sedimentological descriptions were honored in the mapping of these parameters away from well control.

Average values of hydrocarbon saturation were obtained for each reservoir zone in each well. Oil saturation maps were constructed to honor the well data and to reflect the transition zone above the oil-water contact (OWC). Transition zone heights were determined with the aid of capillary pressure curves. Gas saturation maps in the gas cap were contoured to honor well data.

Fluid Contact Distribution

A fluid contact distribution model for Heidrun field (Figure 22) has been employed to illustrate the effect of fault segmentation on the fluid contacts in certain reservoir units.

The fault complexity varies across the field and is minimal in the southern and southwestern parts of the field, which contain a large portion of the Fangst oil. Greater complexity is observed over the crestal areas, in the gas cap areas, and in the northwest region of the field.

The segmentation of the Heidrun field is important for the fluid contact distribution in the Tilje and Åre zones. Faults may act as pressure boundaries in these formations, causing various contact levels in different fault blocks. In the Fangst Group, however, this does not appear to be the case. The Fangst is interpreted as having the same OWC and GOC across the field.

Source Rock Summary

Selected samples of the Lower Jurassic to Cretaceous intervals of several Heidrun wells were examined by Robertson Research.

In general terms, the sediments in this area are immature to a depth of about 2200 m, early mature from 2200 to about 3100 m, middle mature from 3100 to about 4100 m, and late mature below this depth. The relatively low maturity of the Jurassic interval on the Heidrun field indicates that these sediments are at their maximum depth of burial, and that this location has probably been a structural high since the Jurassic.

The Spekk ("Hot Shale") Formation in well 6507/7-1 is organically rich and sapropelic and is a probable source of the oils analyzed. The change in the "Hot Shale" Formation from sapropelic shale in well 6507/7-1 to vitrinitic shale/siltstone on the Heidrun field may be a function of the paleomorphology of the area.

The oils analyzed have been generated by the same middle mature source rock (or by very similar middle mature source rocks), which contains a mixture of algal and humic kerogen and shows some biodegradation. Oil was emplaced in the Heidrun field when it was 1000 m shallower than at present. An additional phase of oil emplacement occurred when the reservoirs were below the "biodegradation window."

Oil-Source Rock Correlation

The Spekk Formation "Hot Shale" in the location of well 6507/7-1 and in the Smørbukk areas is a likely source rock in terms of quality and maturity. It is also noted that some of the biomarkers of the oils may indicate a contribution from a coaly source rock, such as the Åre 1.

Development Plan

The revised plan for development and operation (PDO) was submitted to the Norwegian government in December 1989. First oil is scheduled for 1995. Statoil will assume the operatorship of the field within nine months after installation of the main platform.

The reservoir development plan includes both water and gas injection to optimize oil recovery. The plateau production rate is 200,000 BOPD. The current drilling program calls for 58 wells, including six subsea water injectors. The maximum drilling angle is 73°, which allows all the producers to be platform wells.

CNI and the unit owners propose to develop Heidrun field with an innovative concrete tension leg platform (TLP), incorporating concrete module support beams (Figure 23). This concept combines Norwegian expertise in concrete construction with Conoco's experience in building and operating TLPs elsewhere in the world.

The TLP will have a displacement of about 175,000 tons, an operating draft of 64 m, and a freeboard of 22 m. The use of concrete module support beams, which will be 140 m long, will shorten the over-all construction time and allow barge installation of large modules, thus providing considerable cost savings. The modules will be designed for maximum self-sufficiency to allow as much commissioning work as possible to be completed onshore.

Oil storage will be in either a floating storage and offloading vessel or a concrete subsea tank. Dedicated shuttle tankers will transport the oil to shore.

A pipeline from Heidrun to mid-Norway will be used to transport the gas for onshore manufacturing or power generation. Initially it will transport about 75 million standard ft^3 of associated gas per day, and after a few years it will transport gas from the gas cap. The potential participants in a regional pipeline, and its size and landing site, are all under active consideration at present. Conoco and Statoil have concluded that construction and operation of a methanol plant for the Heidrun gas would be both feasible and economical.

FIGURE 23. Schematic illustration of concrete tension leg platform and floating storage and offtake vessel.

SUMMARY

1. Heidrun is a giant oil field with 750 million bbl of oil contained in Jurassic reservoirs with porosities greater than 30% and permeabilities higher than 10 darcys in the cleanest Fangst sands.

2. The discovery of Heidrun in 1985 resulted from extensive regional exploration studies, which identified a Jurassic rift in the Haltenbanken. Heidrun is the largest oil field in Haltenbanken, and the most northerly oil field in Europe.

3. The application of systematic exploration mapping and use of modern exploration technology such as organic maturation contributed to the upgrading of prospective Jurassic traps.

4. The development phase required better seismic definition, and a 3D seismic survey acquisition improved dramatically the confidence in the reservoir description.

5. The challenge of developing Heidrun in 350 m of water is being met by the combination of Norwegian expertise in concrete construction and Conoco experience with tension leg platform technology.

ACKNOWLEDGMENTS

The author wishes to thank CNI, Statoil, Norsk Hydro, Neste OY, and DNO for permission to publish this paper. The author gratefully acknowledges the assistance of his many colleagues at Conoco Norway and Conoco Inc. ERS, on whose interpretations this paper was based. Also acknowledged are Dr. J. Harms for his sedimentological input and useful suggestions, Dr. A. G. Doré for critically reviewing the manuscript, R. H. Koenig and Walt Pusey for their helpful comments, Bruce Reid for his help in preparing and editing the manuscript, Marianne Kapstad for drafting the illustrations, and John Dutton for his technical writing and his patience.

REFERENCES CITED

Bjørlykke et al., 1985, Diagenesis and reservoir properties of Jurassic sandstones from the Haltenbanken area offshore mid Norway, in Habitat of hydrocarbons, Norwegian oil and gas finds: Norwegian Petroleum Society, p. 275-287.

Bugge, T., T. Prestvik, and K. Rokoengen, 1980, Lower Tertiary volcanic rocks off Kristiansund, mid Norway: Marine Geology, v. 35, p. 277-286.

Bukovics, C., and P. A. Ziegler, 1985, Tectonic development of the mid-Norway continental margin: Marine Petroleum Geology, v. 2, n. 1, p. 2-22.

Bukovics, C., E. G. Cartier, N. D. Shaw, and P. A. Ziegler, 1984, Structure and development of the mid-Norway continental margin, in Petroleum geology of the north European margin: Norwegian Petroleum Society, p. 407-423.

Campbell, C. J., and E. Ormassen, 1987, Geology of the Norwegian oil and gas fields, in A. M. Spencer et al., The discovery of oil and gas in Norway: an historical synopsis, p. 1-37.

Dalland, A., D. Worsley, and K. Ofstad, 1988, A lithostratigraphic scheme for the Mesozoic and Cenozoic succession offshore mid and northern Norway: Norwegian Petroleum Directorate, Bulletin n. 4, p. 65.

Doré, A. G., 1990, The structural foundation and evolution of Mesozoic seaways between Europe and the Arctic Sea, in Channell, Jansen, and Winterer (eds.), The volume of Tethyan paleogeography.

Doré, A. G., and M. S. Gage, 1986, Crestal alignments and sedimentary domains in the evolution of the North Sea, northeast Atlantic margin and the Barents shelf, in Petroleum geology of NW Europe, v. 2, p. 1131-1149.

Harris, N. B., 1989, The reservoir geology of the Fangst Group (Middle Jurassic), Heidrun field, offshore mid-Norway: AAPG Bulletin, v. 73, n. 11, p. 1415-1435.

Jørgensen, F., and T. Navrestad, 1981, The geology of the Norwegian shelf between 62°N and the Lofoten Islands, in Petroleum geology of the continental shelf of NW Europe, Institute of Petroleum, p. 407-413.

Koenig, R. H., 1986, Oil discovery in 6507/7; an initial look at the Heidrun field, in Habitat of hydrocarbons on the Norwegian continental shelf: Norwegian Petroleum Society, p. 307-313.

Leadholm, R. H., T. T. Y. Ho, and S. K. Sahai, 1985, Heat flow geothermal gradients and maturation modelling on the Norwegian continental shelf using computer methods, in Petroleum geochemistry of the Norwegian shelf: Norwegian Petroleum Society, Graham and Trotman, London, p. 131-143.

Lopatin, N. V., 1971, Temperature and geologic time as factors in coalification: Izvestiya Akademiia Nauk USSR, Seriya Geologicheskaya, n. 3, p. 95-106 (in Russian).

Oftedahl, C., 1975, Middle Jurassic graben tectonics in mid Norway, Jurassic northern North Sea symposium, Article 21: Norwegian Petroleum Society.

Payton, C. E., 1977, Seismic stratigraphy: its application to hydrocarbon exploration, AAPG Memoir 26.

Pedersen, T., J. C. Harms, N. B. Harris, R. W. Mitchell, and K. M. Tooby, 1988, The role of correlation in generating the Heidrun field geologic model, in Correlation in hydrocarbon exploration: Norwegian Petroleum Society, p. 327-338.

Price, I., and R. P. Rattey, 1984, Cretaceous tectonics off mid Norway: implications for the Rockall and Faeroes-Shetland troughs: Journal of The Geological Society, London, v. 141, p. 985-992.

Schmidt, W. S., in press, Structure of the mid Norway Heidrun field and its regional implications, in Structural and tectonic modelling and its application to petroleum geology: Norwegian Petroleum Society.

Waples, D. W., 1980, Time and temperature in petroleum formations: application of Lopatin's method to petroleum exploration: AAPG Bulletin, v. 64, p. 916-926.

Ziegler, P. A., 1988, Evolution of the Arctic North Atlantic and the Western Tethys, AAPG Memoir 43.

Chapter 25

The Snorre Field
A Major Field in the Northern North Sea

K. Jorde
G. W. Diesen

Saga Petroleum a.s.
Sandvika, Norway

ABSTRACT

The Snorre field is located in the northern North Sea and belongs to the prolific hydrocarbon province on the western margin of the Viking graben, which contains the Brent, Statfjord, and Gullfaks fields. The Snorre field was discovered in 1979. It has two main reservoirs—the Triassic Lunde Formation and the Triassic–Jurassic Statfjord Formation. Each of these reservoirs consists of a network of fluvial sand bodies in a mudstone matrix.

The reservoir properties range from fair to very good. Typical test permeabilities are 2000 md in the upper part of the Statfjord Formation and 125 to 380 md in the underlying units.

The Snorre field contains 490 million standard m^3 (3.1 billion bbl) of undersaturated oil in place, of which approximately 75% occurs in the Lunde Formation. The oil-water contact varies from 2561 m (8400 ft) on the crest to 2599 m (8530 ft) in the western region.

This field is a structural/stratigraphic trap formed by westward tilting and erosion of a major fault block. Both reservoirs are truncated by the Kimmerian unconformity and overlain by Jurassic and Cretaceous shales. Organic-rich Upper Jurassic shales constitute the major source rock.

The Snorre field will undergo a two-phase development. In phase 1, a tension leg platform (TLP) will be located in the southwestern part of the field. Production will start in 1992 with six wells predrilled from a template beneath the TLP. Additional TLP wells and another subsea template will complete phase 1. For phase 2, there are two different options—either relocating the TLP to the northern part of the field or leaving it in its original position and adding two more subsea templates.

The field development plan (FDP) calls for water injection and includes 73% of the oil in place. On this basis, it is estimated that 119 to 122 million standard m^3 (750 to 770 million bbl) of oil will be recovered, depending on which phase 2 alternative (TLP relocation or subsea extension) is chosen. Studies aimed at increasing field production, by developing portions of the oil that are not included in the current plan, are presently being performed.

INTRODUCTION

The Snorre field is located in the Norwegian part of the northern North Sea, approximately 240 km northwest of Bergen (Figure 1), and covers an area of approximately 100 km² (39 mi²). The field is situated within the southern part of block 34/4 and the northern part of block 34/7, representing production licenses PL 057 and PL 089, respectively.

On April 6, 1979, PL 057 was awarded to the following group of partners: Saga Petroleum a.s. (operator), 15%; Amoco Norway a.s., 10%; Amerada Hess (Norway) Ltd., 5%; Den Norske Stats Oljeselskap a.s. (Statoil), 50%; Deminex (Norge) A/S, 15%; and Texas Eastern Norway A/S, 5%.

On March 9, 1984, PL 089 was awarded to: Saga (operator), 10%; Statoil, 50%; Det Norske Oljeselskap (DNO), 1%; Deminex, 4%; Elf Aquitaine Norge A/S, 8%; Esso Norge a.s., 15%; and Norsk Hydro Produksjon A/S, 12%.

The field was unitized in 1987. A few transfers and acquisitions of field and license shares have taken place, and the present partners and their shares are as follows: Saga (operator), 11.3%; Statoil, 41.4%; Esso, 10.3%; Deminex, 10.0%; Idemitsu, 9.6%; Norsk Hydro, 8.3%; Elf Aquitaine, 5.5%; Amerada Hess, 1.5%; Enterprise, 1.5%; and DNO, 0.7%.

EXPLORATION HISTORY

The potential of the tilted fault block structures of the Tampen spur was proved in 1971 with the Shell/Esso discovery of the Brent field in U.K. waters (Bowen, 1975). The 1974 Mobil discovery of the Statfjord field (Kirk, 1980) verified the potential of this area and emphasized the possibilities of the Snorre prospect, which is located along the northern continuation of the Brent-Statfjord structural trend (Figure 2).

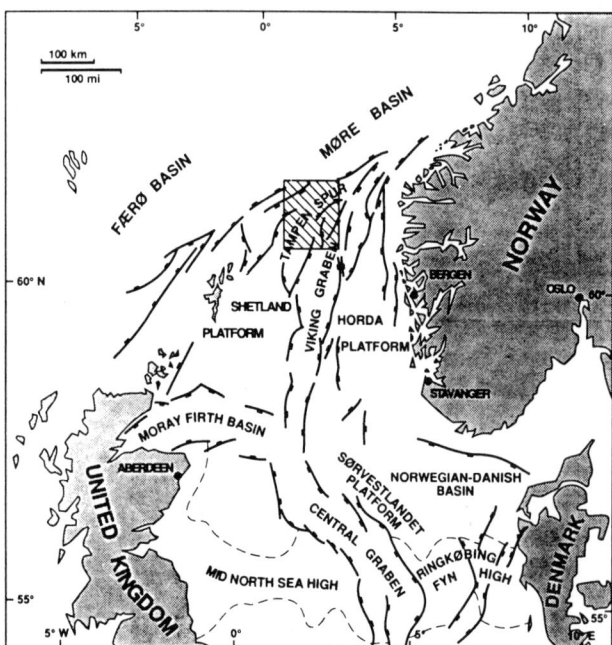

FIGURE 1. Tectonic map of the North Sea, showing the location (cross-hatched area) of the Snorre field (partly after Ziegler, 1988).

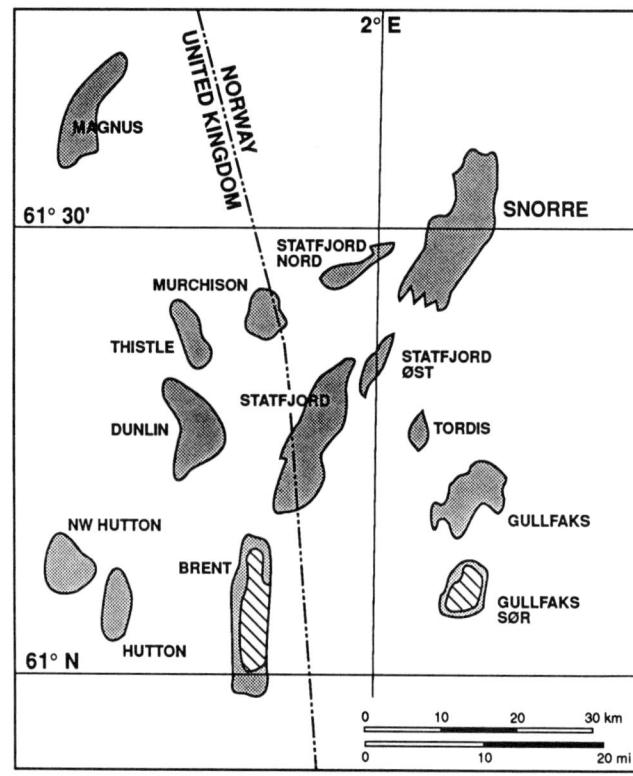

FIGURE 2. Oil fields of the Tampen spur. Gas caps are indicated by cross-hatching.

Several 2D seismic surveys with fairly wide spacing were shot in the area—e.g., a Saga survey in 1980, the Tampen Spur Group Shoot in 1981, and a joint Saga-Norsk Hydro-Statoil survey in 1983 (Hollander, 1987). These data were supplemented by rather extensive 3D surveys: Geco and Esso covered the southern part of the field in 1983; and Saga, as PL 089 operator, shot a survey over the northern part of the field in 1984. In all, 9000 km (5600 mi) of 3D seismic has been shot over the field.

DISCOVERY HISTORY

The first Snorre well, 34/4-1, was drilled in 1979 on the northeastern flank of the structure (Figure 3). It penetrated the stratigraphically deepest parts of the field—the middle and lower members of the Lunde Formation—as well as the underlying Lomvi and Teist formations (Figure 4). A 78-m (225-ft) oil column was encountered in the middle member of the Lunde Formation, with an oil-water contact of 2561 m (8402 ft).

Well 34/4-4, drilled in 1982, encountered oil in the upper member of the Lunde Formation. Well 34/7-1, drilled in 1984, proved that the oil in the Lunde Formation extended southward into PL 089. Well 34/7-3, drilled in 1984 in the southwestern part of the field, proved that the overlying Statfjord Formation was also an oil-bearing structure. Well 34/7-3 also proved a deeper oil-water contact of 2595 m (8514 ft), resulting in a total oil column of 300 m (980 ft).

A total of 11 exploration and appraisal wells have been drilled within the field boundaries, providing an extensive database comprising some 1635 m (5360 ft) of core and 29 production tests. Nine more wells were drilled within blocks 34/4 and 34/7 through July 1990, resulting in several discoveries. An extensive program of exploration for undrilled prospects is continuing.

STRATIGRAPHY

The oldest geologic unit identified in the Snorre wells is the Scythian-Anisian Teist Formation of the Hegre Group (Figure 4). It consists of reddish brown, silty mudstone and lesser amounts of siltstone and fine-grained sandstone, which appear to have been deposited on a distal alluvial plain, with local lacustrine and fluvial environments. The thickness of the Teist formation in the Snorre field is not known, but it exceeds the 576 m (1890 ft) penetrated in well 34/4-4.

The overlying unit, the Anisian-Ladinian Lomvi Formation, consists of fine-grained sandstones and subordinate siltstone and mudstone. The Lomvi is interpreted as fluvial deposits, in agreement with Vollset and Doré (1984), possibly with minor eolian components. The thickness of the unit increases southward from 77 m (253 ft) in well 34/4-6 to 93 m (305 ft) in well 34/7-6.

Although the Teist and Lomvi formations apparently were deposited during a period of sea level lowstand, the sea level appears to have risen sufficiently to result in a marginal marine environment for the basal parts of the overlying Ladinian-Rhaetian Lunde Formation.

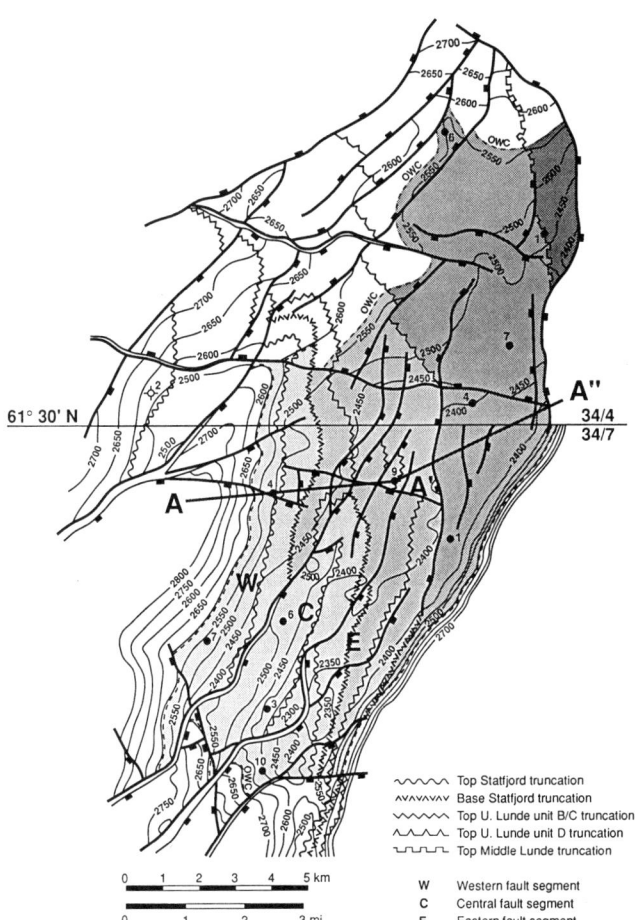

FIGURE 3. Top reservoir map for Snorre field (contour values are in meters subsea).

Chronostratigraphy			Lithostratigraphy		
System	Series	Stage	Group	Formation	Member
Jurassic	Lower	Pliensbachian	Dunlin	Amundsen	Calcareous
		Sinemurian		Statfjord	Upper
		Hettangian			Lower
Triassic	Upper	Rhaetian	Hegre Group	Lunde	Upper
		Norian			Middle
		Carnian			Lower
	Middle	Ladinian		Lomvi	
		Anisian			
	Lower	Scythian		Teist	Upper
					Lower

FIGURE 4. Lower Jurassic-Triassic stratigraphy of the northern North Sea.

The Lunde Formation, which has been subdivided into three informal members, also consists of alternating sandstones and mudstones of fluvial origin. The river systems appear to have varied from braided streams with wide, shifting channels and nonchannelized sheet flood deposits to low-sinuosity to meandering rivers. The middle and lower members also contain minor lacustrine deposits. The Lunde Formation is interpreted as having been laid down on a wide alluvial plain under a semihumid to semiarid tropical climate (Nystuen et al., 1989), resulting in red beds and paleosols with calcrete nodules.

The thicknesses of the lower and middle members increase northeastward, from 155 and 114 m (509 and 374 ft), respectively, in well 34/7-6, to 222 and 145 m (728 and 476 ft) in well 34/4-4. The thickness of the upper member of the Lunde Formation is rather constant over the field, varying from 824 to 854 m (2703 to 2802 ft).

The Rhaetian-Sinemurian Statfjord Formation also consists of alternating fluvial sandstones and mudstones, which are interpreted as having been deposited by braided rivers on an alluvial plain. The marine incursion that has been identified on the Statfjord field (Kirk, 1980) does not appear to have extended into the Snorre field. The thickness of the formation decreases northward from 99 m (325 ft) in well 34/7-3 to 72 m (236 ft) in well 34/4-2.

There are strong similarities between the upper part of the Lunde Formation and the lower part of the Statfjord Formation, and a gradational change from one to the other. In the upper part of the Statfjord Formation, however, more grayish-colored beds, coaly mudstones, and a higher K-feldspar/plagioclase ratio than in the red beds below indicate a more humid, subtropical to tropical climate (Nystuen et al., 1989).

STRUCTURE

The Snorre field is located on the Tampen spur, which forms a northward extension of the Shetland platform (Figure 1). The field is situated close to the eastern margin of the spur, toward the Viking graben. The Snorre area was part of a large sedimentary basin from the Permian to the Middle Jurassic. The basin was subjected to the Triassic-Jurassic rifting episode that strongly influenced the northern North Sea. For the Snorre area, this resulted in a major Late Jurassic uplift, with the Tampen spur forming a regional high during the Early Cretaceous. With its marginal position, Snorre was strongly affected by the associated erosion. Thus, more than 800 m (2600 ft) must have been removed from the crest of the field (Karlsson, 1986).

The Snorre field constitutes the easternmost and highest part of a rotated fault block with a northwestward dip of 8 to 10° (Figure 5). The field is cut by a series of faults striking north-northeast that subdivide it into fault segments (Figure 3). These faults have typical throws of 100 to 200 m (330 to 660 ft) in the southern part of the field, decreasing northward. Several east-west-striking faults also transect the field, further subdividing it.

RESERVOIR CHARACTERISTICS

The fluvial processes that deposited the Snorre field rocks resulted in a series of sandstone bodies heterogeneously distributed in a mudstone matrix.

For volumetric calculations and reservoir characterization, the upper Lunde Formation has been subdivided into five reservoir zones—units A, B/C, D, E, and F—on the basis of net/gross ratio variation (Figure 6). Correspondingly, the Statfjord Formation has been subdivided into three units—lower, middle, and upper. The reservoir properties of the different units vary widely, from the essentially nonreservoir upper Lunde unit A to the very good upper Statfjord Formation, which has a net/gross ratio of 0.57, a porosity of 25%, and permeabilities in the darcy range (Figure 6).

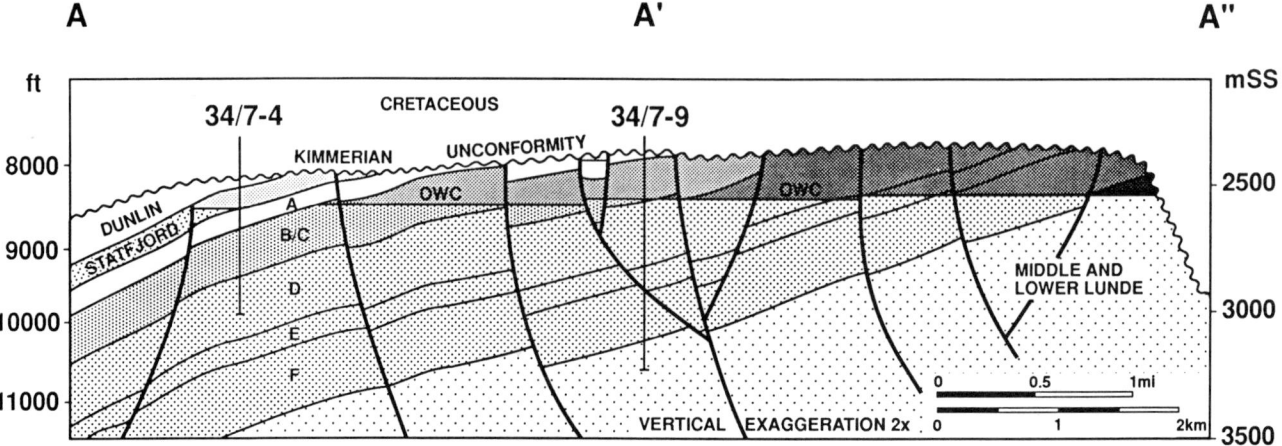

FIGURE 5. Cross section of the Snorre field (for location, see Figure 3).

Formation	Sedimentological environment	Reservoir unit		N/G	Porosity	Effective perm. (mD)
Statfjord Fm.	Braided streams	Upper Statfjord		0.57	0.25	1300-2000
		Middle Statfjord		0.47	0.23	250
		Lower Statfjord		0.19	0.21	250
Upper Lunde Fm.	Low-sinuosity rivers	Unit A		0.05	0.21	-
		Unit B/C		0.25	0.24	125
	Braided streams	Unit D	Lunde DEF	0.48	0.24	380
		Unit E		0.42	0.23	
		Unit F		0.64	0.24	
Middle Lunde Fm.	Low-sinuosity rivers	M. Lunde	M/L Lunde	0.25	0.18	60
Lower Lunde Fm.		L. Lunde		0.30	0.20	

FIGURE 6. Reservoir characteristics of the Snorre field.

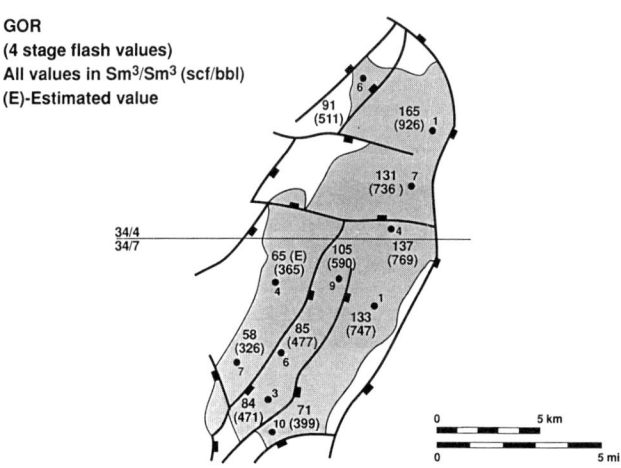

FIGURE 7. Gas-oil ratio distribution for the Snorre field.

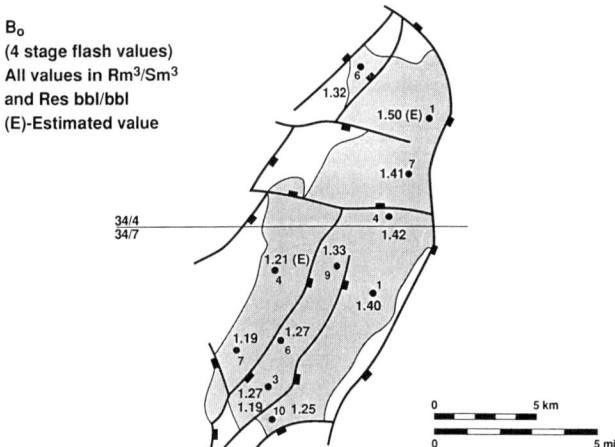

FIGURE 8. Formation volume factor distribution for the Snorre field.

FLUID PROPERTIES AND OIL-WATER CONTACTS

The Snorre reservoirs are overpressured. Both the Lunde and Statfjord reservoirs have a pressure of 383 bar (5554 psi) at 2475 m (8120 ft).

Analysis of well test samples shows that the fluid properties vary significantly and systematically over the field. Thus, the gas-oil ratio (GOR) decreases from an estimated value of 165 standard m³/standard m³ (926 ft³/bbl) in well 34/4-1 to 58 standard m³/standard m³ (326 ft³/bbl) in well 34/7-7 (Figure 7). A corresponding decrease is observed for the bubble point pressure, which varies from 220 to 90 bar (3190 to 1305 psi) in the same two wells, and in the formation volume factor, which decreases from 1.50 to 1.19 (Figure 8). Lateral variation is observed also in oil density, which varies from 654 to 764 kg/m³ (85 to 54° API) for the same two wells. The temperature shows a slight decrease, from 93°C (199°F) in the eastern part of the field to 90°C (194°F) in the western part. Vertical differences in fluid properties also are observed, but in general these differences are significantly smaller than the variations between wells.

The most likely explanation for the variations in the fluid properties is considered to be that hydrocarbons are currently migrating into the field. In this model, most of the oil (with a low gas content) is interpreted as having migrated into the field from the Møre Basin to the north and/or spilled out of the northern part of the Statfjord field (Figure 9). In addition to this, significant amounts of gas or gas-rich oil are believed to be migrating into the field from a local basin to the east (Figure 9).

The oil-water contact is another parameter that exhibits lateral variation, deepening from northeast to southwest. The variation is greatest for the Lunde Formation, which has contacts ranging from 2561 to 2595 m (8402 to 8514 ft) subsea, based on RFT measurements.

The contacts in the Statfjord Formation are estimated as 2599 m (8527 ft) subsea for the western fault segment and 2595 m (8514 ft) subsea for the rest of the field. Because the Statfjord Formation has been oil filled in all Snorre wells, these values are based on the Statfjord Formation oil gradient and the Lunde Formation water gradient.

Wells recently drilled adjacent to the Snorre field indicate that the water pressures in the Statfjord Formation within the field may be somewhat lower than those in the Lunde Formation, suggesting Statfjord Formation oil-water contacts potentially deeper than the current estimates.

RESERVOIR MODELING

Prior to development planning, a significant effort was undertaken to create a method of transforming the complex geometry of the fluvial reservoirs into a more simplified model that could be used as input for the reservoir simulators. This was done by creating synthetic representations of the fluvial sand body network. The representations were constructed by means of a computer program called "SISABOSA," developed through a joint effort by

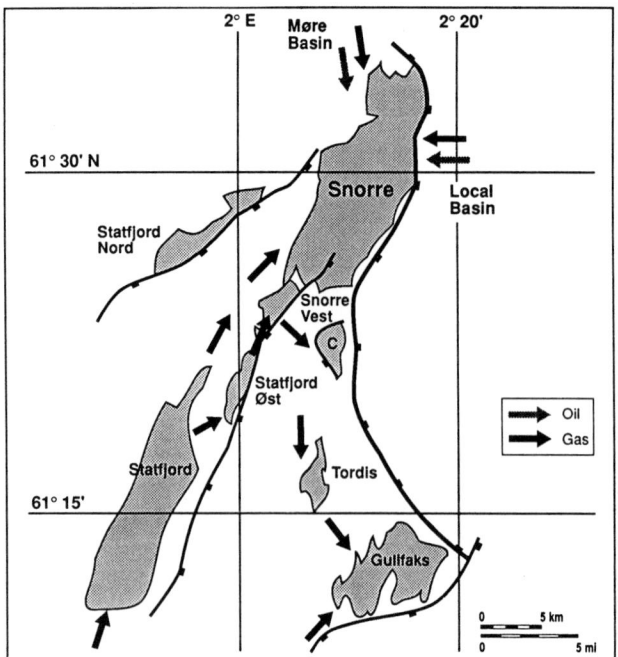

FIGURE 9. Possible routes of oil migration into the Snorre field area.

Saga and the Norwegian Computing Center (Augedal et al., 1986).

The SISABOSA program creates a 3D model of fluvial reservoirs by distributing parallelepipeds representing bodies of permeable sand in a volume representing impermeable mudstone (Figure 10). The distribution is performed as a stochastic process conditioned on predefined geometric constraints, and the process is continued until a prescribed sand-content variation is achieved.

The parallelepipeds constitute simplified approximations of channel sand bodies. Thicknesses and widths were assigned to the parallelepipeds on the basis of values determined from channel sands of analogous modern rivers (Stanley et al., 1990).

Average long-axis directions were determined from Snorre field dipmeter studies of cored intervals and from oriented cores; a northerly direction was indicated for Statfjord Formation sand bodies, and a northeasterly to easterly direction for the Lunde Formation. The dispersion in long-axis orientation was based on dipmeter data from cored intervals and from analyses of oriented cores.

Ten synthetic representations were generated for each model, using the most likely set of input parameters. In addition, five representations for highest-case parameters and five for lowest-case parameters were run to evaluate the variability in the models. From the ten runs with the most likely parameters, a normal distribution of sand body interconnectedness measurements was generated (Stanley et al., 1990). The median interconnectedness value was used to select the representation that was used as input for reservoir simulation.

For the reservoir simulation, a full-field model was run in order to investigate the effects of sealing or nonsealing major faults. However, most of the modeling was performed by means of element models, each of which typically covered 1.5 by 6 km (1 by 4 mi) (Figure 11). Grid cell size varied from 75 by 75 m (250 by 250 ft) for Lunde unit B/C to 100 by 250 m (330 by 820 ft) for the Statfjord Formation models, with a cell thickness of 3 m (10 ft) for all models.

In order to generate quantitative information on the restrictions to fluid flow imposed by closely spaced but

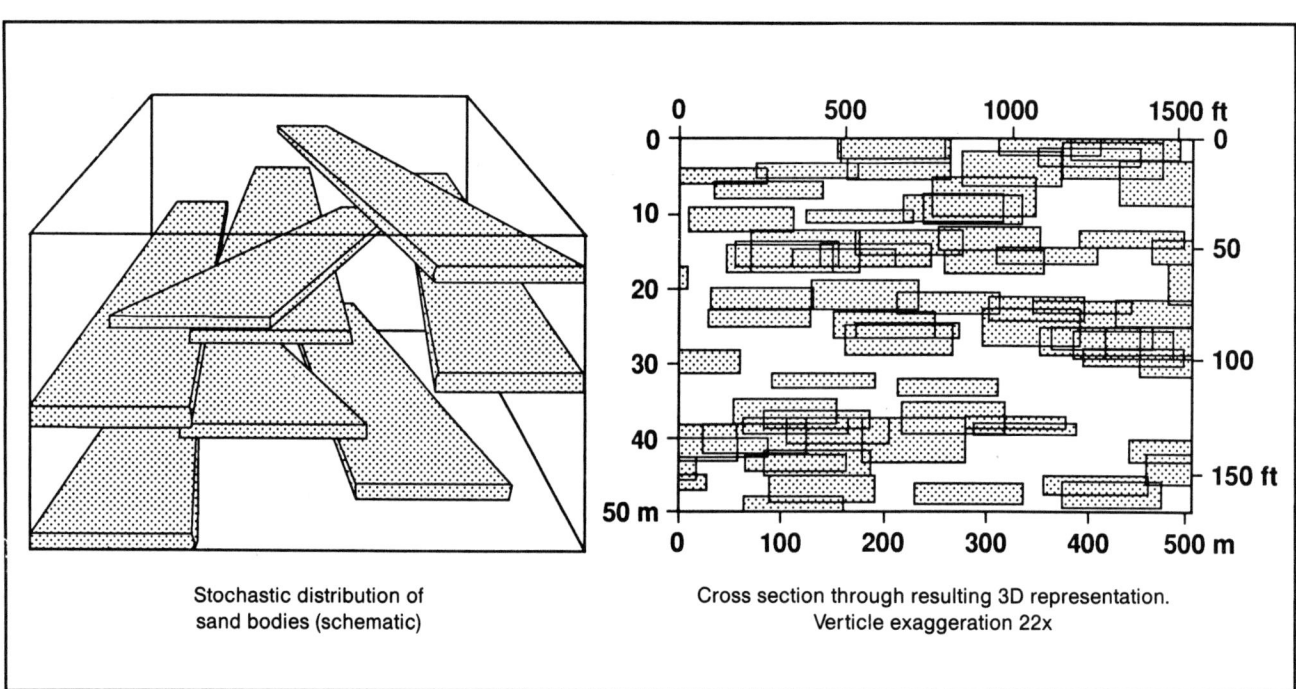

FIGURE 10. Computer modeling of fluvial channel sands.

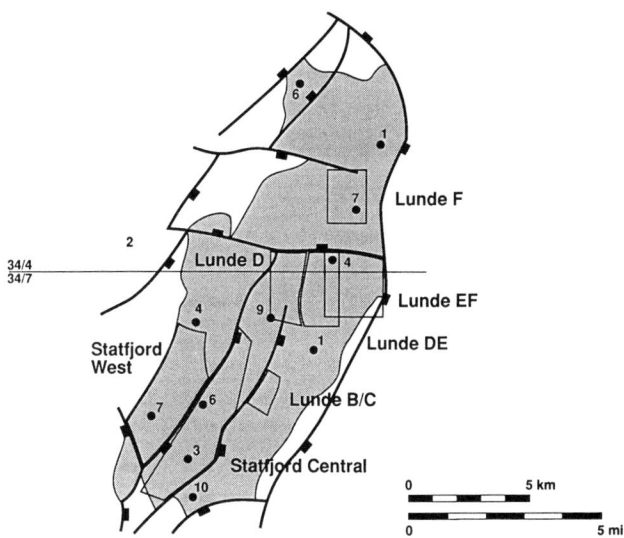

FIGURE 11. Snorre field element model areas.

age of the sand contents of the nine planes (Nybråten et al., 1990).

Although the resulting simulation models necessarily constituted simplified representations of the heterogeneous fluvial reservoir, they allowed a significant part of the available information on reservoir heterogeneity to be quantified and incorporated. Also, the stochastic aspect of the "SISABOSA" program means that, although the exact locations of individual sand bodies—and hence the production behavior of individual wells—cannot be predicted, average sand body distribution can be modeled. This method can provide estimates of average well behavior and identify potential problems in field development. In addition, early estimates of field producibility and recoverable reserves can be made for field planning purposes.

FIELD DEVELOPMENT

Several development scenarios were evaluated for the Snorre field. A major item was the relatively great water depth, which ranges from 295 m (970 ft) in the southern part of the field to 380 m (1250 ft) in the north. Also, the large areal extent of the field and the large number of wells necessitated by the heterogeneous, fluvial reservoirs constituted important constraints on the development scenarios.

nonconnected channel sand bodies, transmissibility values in the x, y, and z directions were calculated for all grid cells. These values were determined by introducing nine parallel planes between the centers of two adjacent grid cells (Figure 12) and computing the harmonic aver-

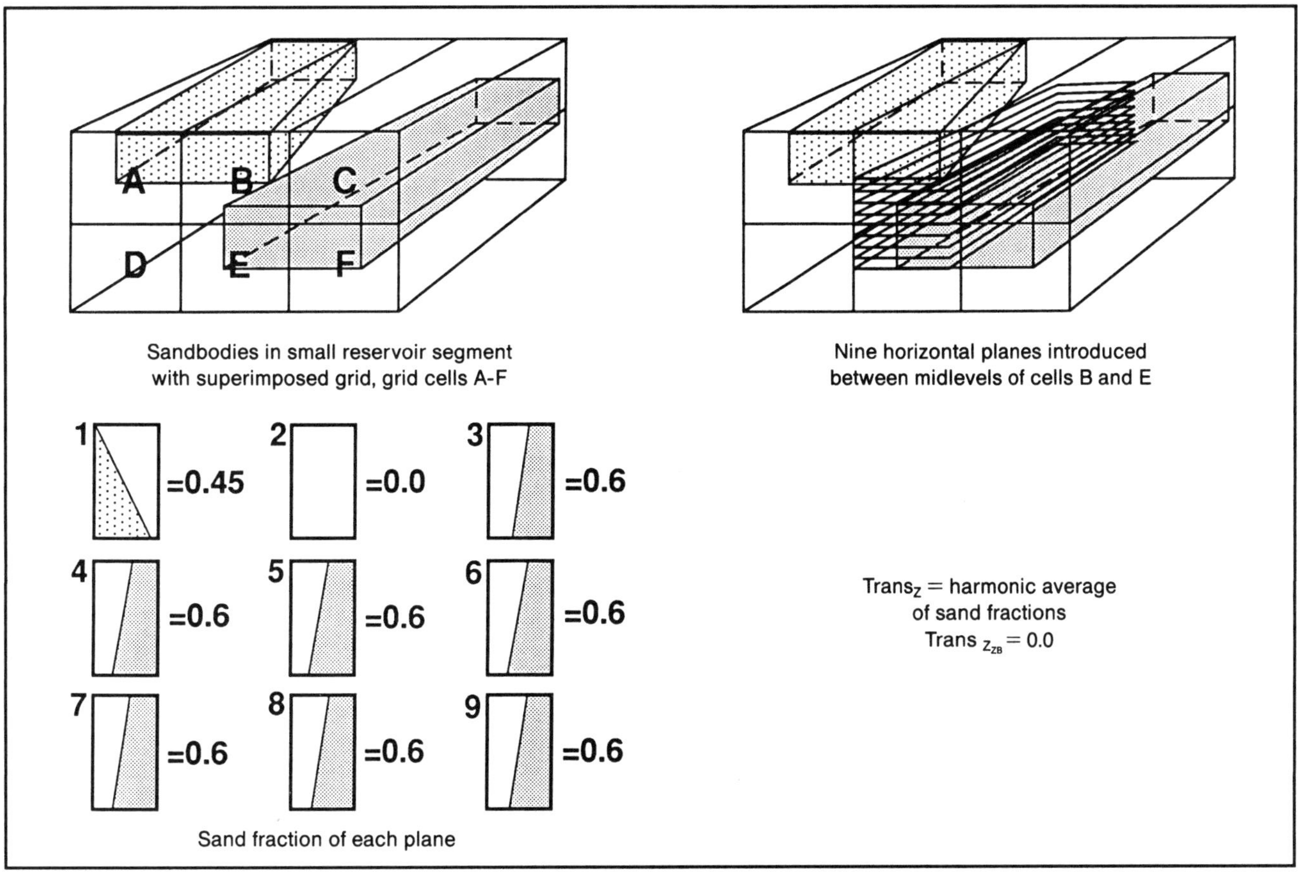

FIGURE 12. Calculation of vertical transmissibility in reservoir simulation grid.

The optimal solution was considered to be a two-phase development plan. In phase 1, a tension leg platform (TLP) will be located in the southwestern part of the field (Figure 13). It will be situated above a large subsea template and will be secured to the seabed by 16 thin-wall steel tube tethers (Figure 14). The tethers will be fastened to skirtlike concrete foundations that will penetrate some 18 m (59 ft) into the seabed.

In a predrilling program that started in September 1990, three production and three injection wells will be drilled from a semisubmersible rig through the TLP well template, allowing a fast production buildup when production begins in 1992.

A second subsea production system (template A) will be installed in the autumn of 1992 to the northeast of the TLP (Figure 13). Drilling from template A will start immediately, with first production in early 1993.

Two options have been documented for phase 2. One option involves a relocation of the TLP to the northern part of the field (Figure 15). The relocation would take place in 2004 and would be supplemented by the drilling of more template A wells. In the alternative option, the TLP would remain in its initial position, and two more subsea production systems would be used to produce the reserves in the northern part of the field (Figure 15). The two subsea templates, B and C, would be installed in 1998 and 1999, respectively. According to the current plans, production from the field would continue until about 2011 (for the extension option) or about 2014 (for the TLP relocation option).

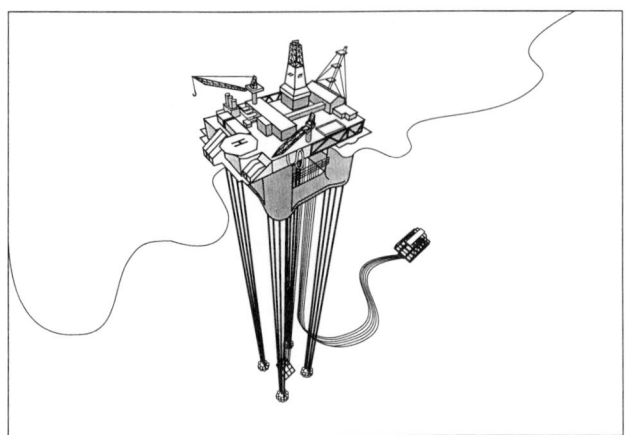

FIGURE 14. Snorre tension leg platform and subsea production system.

The TLP, as designed for the field development plan (FDP), will be a four-column platform with a ring pontoon (Figure 14). It will have an integrated deck measuring 92 by 136 m (302 by 446 ft). Total estimated topside weight is 36,250 MT, and estimated displacement is approximately 106,000 MT.

The TLP will have a two-stage processing unit with one separation train. The design oil-processing capacity is 27,300 standard m^3 (170,000 bbl) per calendar day, and the total liquid capacity is 40,900 standard m^3 (257,000 bbl) per calendar day, assuming a process regularity of 0.91. The resulting gas and partially stabilized crude will be transferred through separate pipelines to the Statfjord field. Final stabilization of the crude, as well as subsequent storage and tanker offtake, will be performed on the Statfjord A platform. The Snorre gas will also be further processed on the Statfjord A platform, and will be exported through the Statpipe system.

RESERVES

The total field STOOIP is estimated at 490 million standard m^3 (3080 million bbl) (Table 1). The FDP includes only some 73% of the total STOOIP—i.e., the oil occurring in the Statfjord Formation and in units D, E, and F of the upper Lunde Formation. During the first development phase, an estimated 68 million standard m^3 (428 million bbl) will be recovered from the Statfjord Formation and the southern part of the Lunde reservoir. Estimated recovery during the second phase is 51 million standard m^3 (321 million bbl) for the subsea extension option, or 54 million standard m^3 (340 million bbl) for the TLP relocation option.

The oil in Lunde unit B/C, with a total STOOIP of 105 million standard m^3 (660 million bbl), constitutes significant additional reserves. Owing to the low net/gross ratio of this unit (Figure 6), and hence its uncertain sand body interconnectedness, the unit B/C oil is not included in the current development plan. Plans have been made, however, for obtaining more information on the production potential of this unit through an extensive data acquisi-

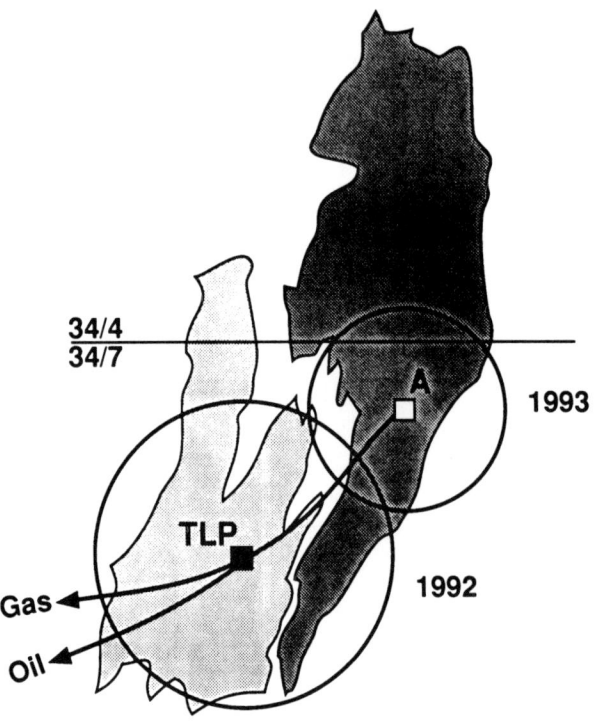

FIGURE 13. Snorre phase 1 development of Statfjord and Lunde D, E, and F reservoirs.

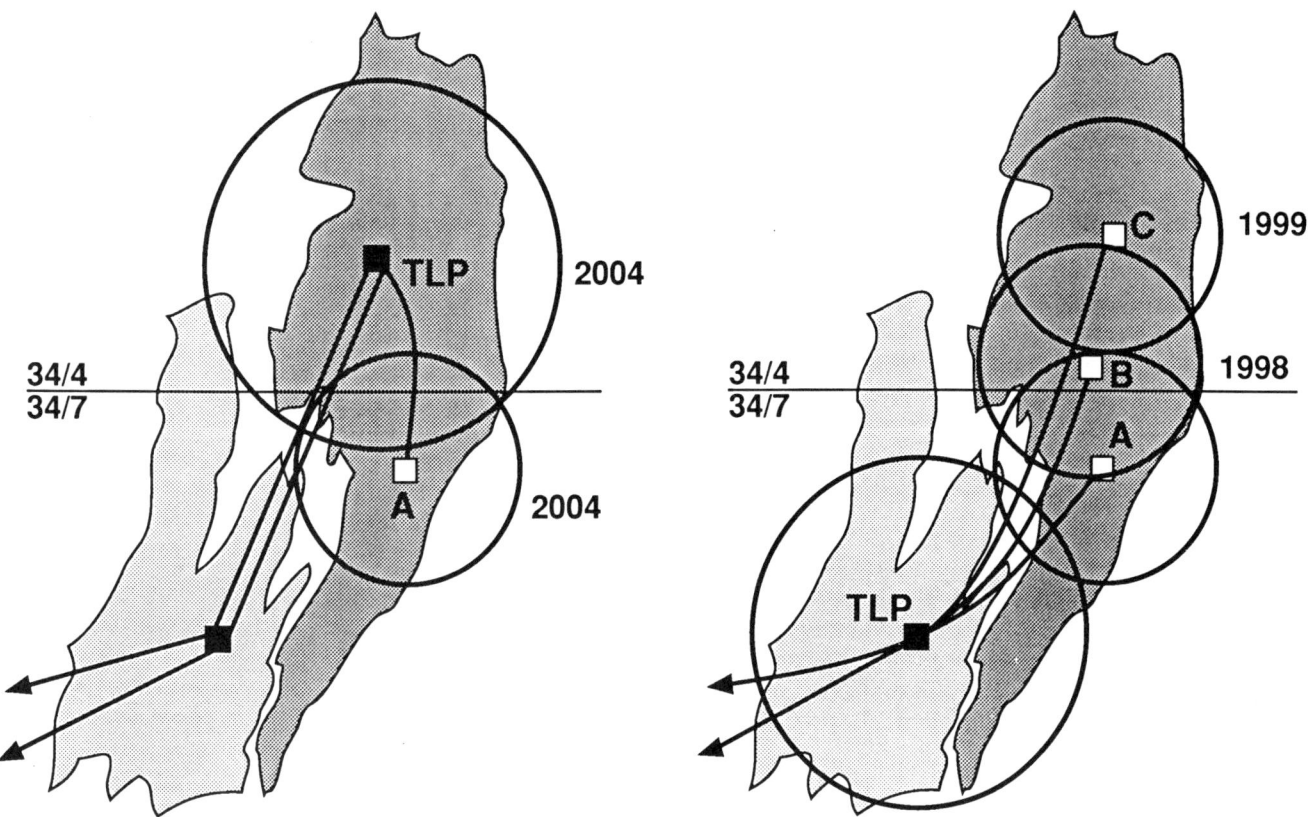

FIGURE 15. Snorre phase 2 development options: TLP relocation (left) and subsea extension (right).

TABLE 1. Snorre field oil reserves.

Reservoir unit	STOOIP* Million standard m³	Million bbl	Reserves** Million standard m³	Million bbl
Upper Statfjord	67	421		
Middle Statfjord	33	207	44	277
Lower Statfjord	23	145		
Upper Lunde:				
Unit A	(10)	(63)		
Unit B/C	(105)	(660)		
Unit D	94	591		
Unit E	58	365	75	472
Unit F	91	572		
Middle Lunde	(5)	(31)		
Lower Lunde	(4)	(25)		
Total	490	3080		
Included in FDP	366	2301	119	749

*Numbers in parentheses are not included in the field development plan.
**Assuming subsea extension option.

tion program that will include extensive coring and water injection supported test production.

Significant reserves are also present in the smaller structures adjacent to the Snorre field (Figure 9). The optimal development scenarios for these reserves are presently being evaluated.

CONCLUSIONS

With its heterogeneous, fluvial reservoir, its large areal extent, and its great water depth, the Snorre field provides several challenges. In order that these challenges can be met in reservoir planning and management, emphasis has been placed on extensive data collection, close cooperation between geologists and reservoir engineers, and detailed modeling. The technological challenges have been met by aiming at a high degree of flexibility, with a phased development program based on relocatable TLP and subsea production systems.

ACKNOWLEDGMENTS

The authors would like to thank the PL 057 and PL 089 partners for permission to publish this paper. Thanks are also extended to all our Saga colleagues on whose work this field presentation is based, and particularly to Grethe B. Røed, who prepared the illustrations.

REFERENCES CITED

Augedal, H. O., K. O. Stanley, and H. Omre, 1986, SISABOSA, a program for stochastic modelling and evaluation of reservoir geology, *in* J. Nitteberg (ed.), Conference on reservoir description and simulation with emphasis on EOR, Oslo, September 1986: Institute for Energy Technology.

Bowen, J. M., 1975, The Brent oil field, *in* A. W. Woodland (ed.), Petroleum and the continental shelf of NW Europe, v. 1: Barking, Geology Applied Science Publishers, p. 353–361.

Hollander, N. B., 1987, Snorre, *in* A. M. Spencer et al. (eds.), Geology of the Norwegian oil and gas fields: Graham and Trotman, p. 307–318.

Karlsson, W., 1986, The Snorre, Statfjord and Gullfaks oilfields and the habitat of hydrocarbons on the Tampen spur, offshore Norway, *in* A. M. Spencer et al. (eds.), Habitat of hydrocarbons on the Norwegian continental shelf: Norwegian Petroleum Society, Graham and Trotman, p. 181–197.

Kirk, R. H., 1980, Statfjord field—a North Sea giant, *in* M. T. Halbouty (ed.), Giant oil and gas fields of the decade 1968-1978: AAPG Memoir 30, p. 95–116.

Nybråten, G., E. Skolem, and K. Østby, 1990, Reservoir simulation of the Snorre field, *in* A. T. Buller et al. (eds.), North Sea oil and gas reservoirs-II: The Norwegian Institute of Technology, Graham and Trotman, p. 103–114.

Nystuen, J. P., R. Knarud, and K. Jorde, 1989, Correlation of Triassic to Lower Jurassic sequences, Snorre field and adjacent areas, northern North Sea, *in* J. D. Collinson (ed.), Correlation in hydrocarbon exploration: Norwegian Petroleum Society, Graham and Trotman, p. 273–289.

Stanley, K. O., K. Jorde, N. Ræstad, and C. P. Stockbridge, 1990, Stochastic modelling of reservoir sand bodies for input to reservoir simulation, Snorre field, northern North Sea, Norway, *in* A. T. Buller et al. (eds.), North Sea oil and gas reservoirs-II: The Norwegian Institute of Technology, Graham and Trotman, p. 91–101.

Vollset, J., and A. G. Doré, 1984, A revised Triassic and Jurassic lithostratigraphic nomenclature for the Norwegian North Sea: Norwegian Petroleum Directorate, Bulletin n. 3, 53 p.

Ziegler, P. A., 1988, Evolution of the Arctic-North Atlantic and the western Tethys: AAPG Memoir 43, 198 p. + 30 plates.

Chapter 26

Oseberg Field

Jens Hagen
Benedicte Kvalheim

Norsk Hydro a.s
Bergen, Norway

ABSTRACT

The Oseberg field is located in the Norwegian North Sea in blocks 30/6 and 30/9, approximately 140 km northwest of Bergen, Norway. The field came onstream in December 1988.

The Oseberg field comprises three major eastward-tilted fault blocks called Alpha, Gamma, and Alpha North. The hydrocarbons are contained in the sandstones of the deltaic Middle Jurassic Brent Group. The Brent Group exhibits excellent reservoir properties, including porosities from 20 to 25% and permeabilities ranging up to several darcys. The gross thickness of Brent ranges mainly between 50 and 190 m (Table 1). The maximum extension of the field is 26 km, and the maximum width is about 6 km. The structural dip is normally 6 to 10° east-northeast.

Gas caps are present in all structures. The Alpha structure has a vertical gas column of 380 m and an oil column of 210 m, for a total hydrocarbon column of about 600 m.

Reservoir engineering studies have shown that a greater ultimate volume of oil can be recovered by gas flooding than by waterflooding. Consequently, pressure maintenance by gas injection has been chosen as the production mechanism for the Alpha and Gamma structures. An adequate supply of gas for injection can be obtained only by importation of large quantities from the huge Troll field by means of a 48-km-long pipeline. The Alpha North structure will be produced mainly by water flooding.

On the Oseberg field itself, the development comprises a field center in the southern part of the field (onstream), a production platform in the northern part (in construction), and three subsea wells.

The current estimates of field reserves are 231.6 million standard m^3 of oil and 92 billion standard m^3 of gas.

TABLE 1. Oseberg field reservoir characteristics.

Trap/rock parameters

Trap type	Truncated fault block
Depth to crest	2120 m (6955 ft) subsea
Gas-oil contact	2497 m (8192 ft) subsea
Oil-water contact	2695–2719 m (8842–8921 ft) subsea
Gas column, approximate	380 m (1247 ft)
Oil column	203–222 m (666–728 ft)
Productive closure, approximate	115 km^2
Lithology	Sandstone
Gross thickness:	
Brent Group	46–187 m (151–614 ft)
Oseberg Formation	17–65 m (56–213 ft)
Net/gross ratio:	
Brent Group, approximate (range)	0.70 (0.50–0.83)
Oseberg Formation, approximate	0.98
Porosity (Oseberg Formation), average (range)	23.7% (20–27%)
Water saturation (Oseberg Formation), average (range)	15% (9–26%)
Permeability, approximate (range)	2 darcys (1–3.5 darcys)

Hydrocarbons (Alpha and Gamma)

Stock tank oil density	0.857 g/cm^3
Bubble point (at gas-oil contact)	280.7 bar
Solution gas/oil ratio	130–160 std. m^3/std. m^3

Volumes

Recoverable oil (1990 base estimate)	231.6 million std. m^3
Imported gas (TOGI project)	25.5 billion std. m^3
Recoverable gas (including reproduction of imported gas)	92 billion std. m^3
Drive mechanism	Pressure maintenance by gas injection

INTRODUCTION

The Oseberg field (named after the Oseberg long boat, a ninth century Viking ship discovered in 1904) is located in the Norwegian North Sea approximately 140 km northwest of Bergen, Norway (Figure 1). It was discovered in 1979 and was put onstream on December 1, 1988. The reserves are estimated to be 231.6 million standard m^3 (1457 million STB) of oil and 92 billion standard m^3 (3.25 trillion standard ft^3) of gas. A comparison of the reserve figures for the various Norwegian oil fields (Figure 2) shows that Oseberg is currently Norway's third-largest oil discovery.

EXPLORATION AND DISCOVERY HISTORY

Prior to any drilling on the Oseberg field, the huge hydrocarbon potential in the northern North Sea had already been established by a series of major discoveries in both the U.K. and Norwegian sectors. The Norwegian discoveries included giant fields such as the Statfjord field to the

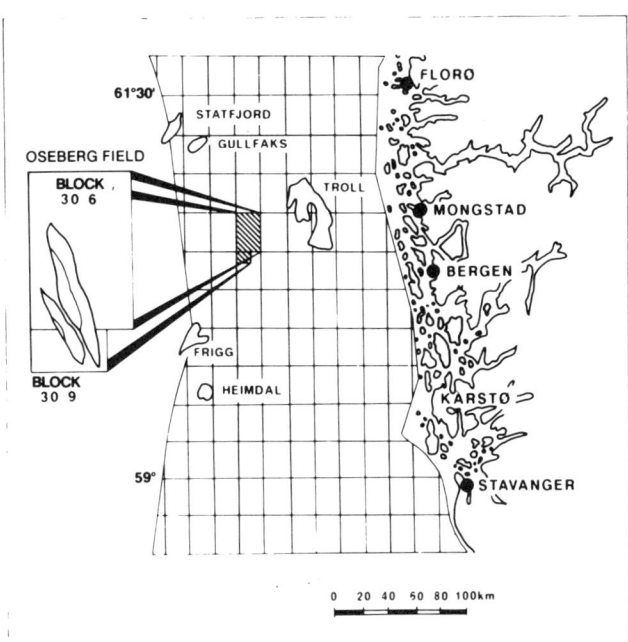

FIGURE 1. Location map for the Oseberg field.

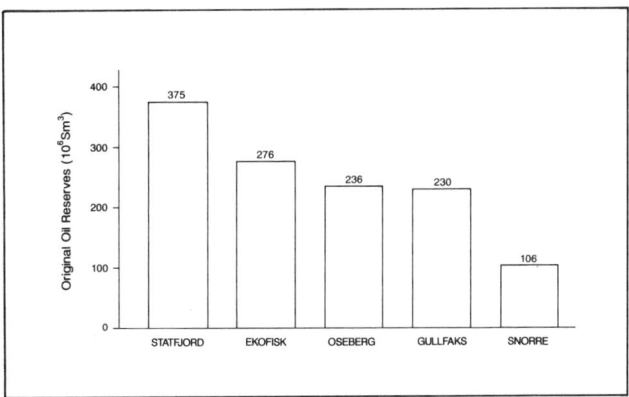

FIGURE 2. Comparison of oil reserves for Norwegian fields (from Norwegian Petroleum Directorate, Annual Report, 1989).

northwest of block 30/6 and the Frigg field to the southwest (Figure 1).

Several large eastward-dipping structures in block 30/6 were identified by seismic. The expectation of a major hydrocarbon discovery was great even before this block was awarded. Consequently, it often was referred to as "the silver block."

In 1979, block 30/6 was awarded to the following group of companies: Norsk Hydro a.s, 12.50%; Statoil, 50.00%; Saga Petroleum a.s, 7.50%; Elf Aquitaine Norge a.s, 13.33%; Total Marine Norsk a.s, 6.67%; and Mobil Exploration Norway Inc., 10.00%.

The results of the first well did not quite match the high expectations. This well was drilled in 1979 in a structurally high position on the largest fault block in the area, the Alpha structure (Figure 3). This well was intended to evaluate the potential of the Paleocene and Middle and Lower Jurassic strata, which were known to be hydrocarbon bearing in other fields. Gas-bearing sandstones were found in the Middle Jurassic Brent Group, but the gross thickness was only 68.5 m. In addition, the reservoir properties were moderate. Furthermore, the Paleocene had no significant sandstone development, and the Lower Jurassic Statfjord Formation was water bearing.

The first oil discovery was made in 1981 by the fourth well. This well, located in a downdip position on the Alpha structure, showed a significant thickening of the reservoir rocks. Gross thickness was 88 m. Reservoir quality was far better than that intersected by the first three wells. The free water level (FWL) was finally established in 1982 by the fifth well on the structure.

It was now evident that a significant oil and gas field had been discovered. The vertical hydrocarbon-bearing section measured approximately 600 m, from the crestal position at about 2120 m MSL to the FWL at about 2719 m MSL. The extension of the Alpha structure alone was about 22 km, and the width was as great as 5 km (Figure 3).

In 1982, the operatorship of the Oseberg field was transferred from Statoil to Norsk Hydro. An aggressive

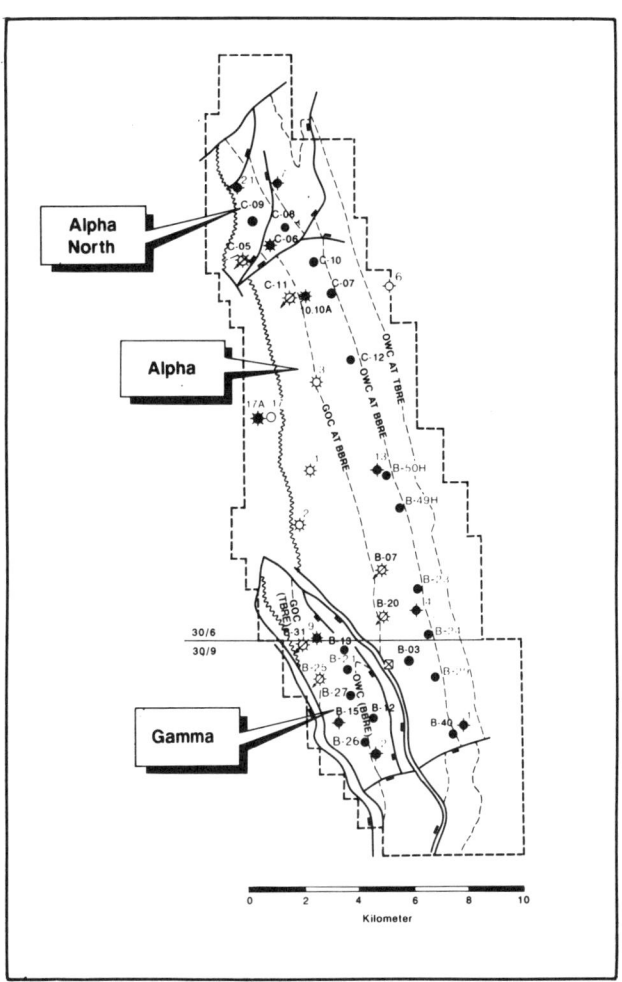

FIGURE 3. Well location map of the Oseberg field.

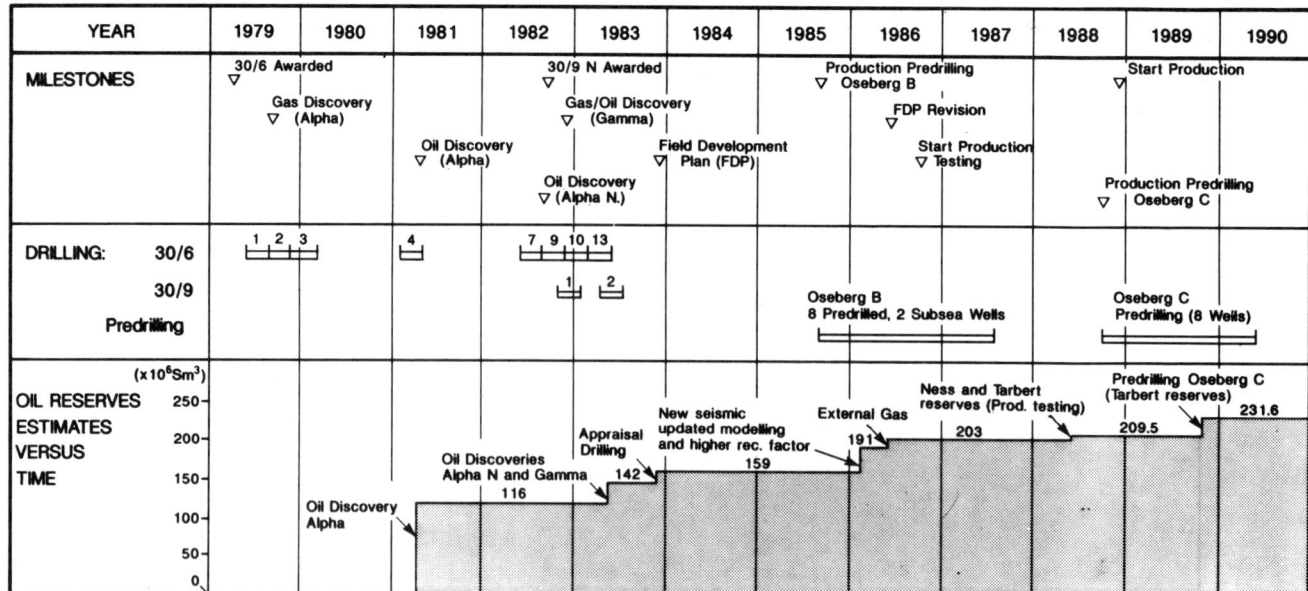

FIGURE 4. Summary of milestones, drilling activity, and reserve estimates for the Oseberg field.

exploration and appraisal drilling program was adopted to establish the hydrocarbon potential of the area and to prepare a sound basis for field development.

The two downfaulted blocks adjacent to the Alpha structure—the Alpha North and Gamma structures (Figure 3)—were both drilled in 1982, and both proved to be hydrocarbon bearing. These two structures had reservoir thicknesses significantly greater than that of the Alpha structure, and their reservoir properties were excellent.

It was now apparent that the Oseberg discovery extended into the neighboring block to the south, block 30/9. It was obvious that future development would require a unified approach across the two blocks. Thus, the northwest corner of block 30/9 was awarded in 1982 to the following companies: Norsk Hydro a.s (operator), 16.0%; Statoil, 73.5%; and Saga Petroleum a.s, 10.5%.

The final appraisal drilling was focused on verification of the southward extension of the field into block 30/9 and determination of the oil-water contact in the Gamma structure. A field development plan was issued in late 1983.

Figure 4 summarizes the exploration/appraisal drilling program and the subsequent reserve estimates. This diagram illustrates the progressive increase in the Oseberg reserve estimates with time, as the field went from a gas discovery with no known oil, to an assumed modest oil and gas discovery, and finally to what proved to be a giant oil and gas field.

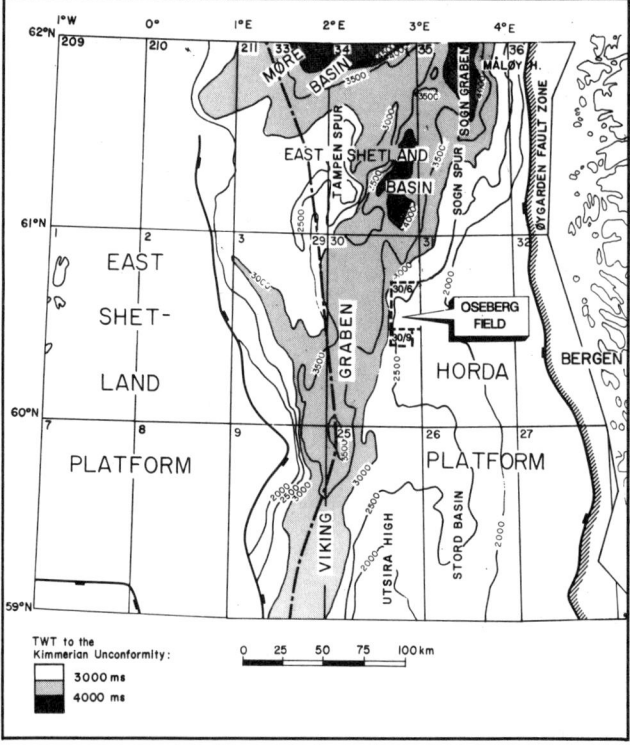

FIGURE 5. Location of the Oseberg field in relation to surrounding major structural features.

STRUCTURE

The Oseberg field is located on the margin of the Viking graben (Figure 5). The field consists of three major eastward-dipping fault blocks called Alpha, Alpha North, and Gamma (Figures 3 and 6). The reservoir rocks have a structural dip of about 6 to 10° east-northeast. In the crestal position, the Brent Group is eroded and capped by Cretaceous shales and limestones.

The main structure, the Alpha structure, is about 22 km long, including the Alpha South compartment. A few northeast-southwest faults are present in the southern

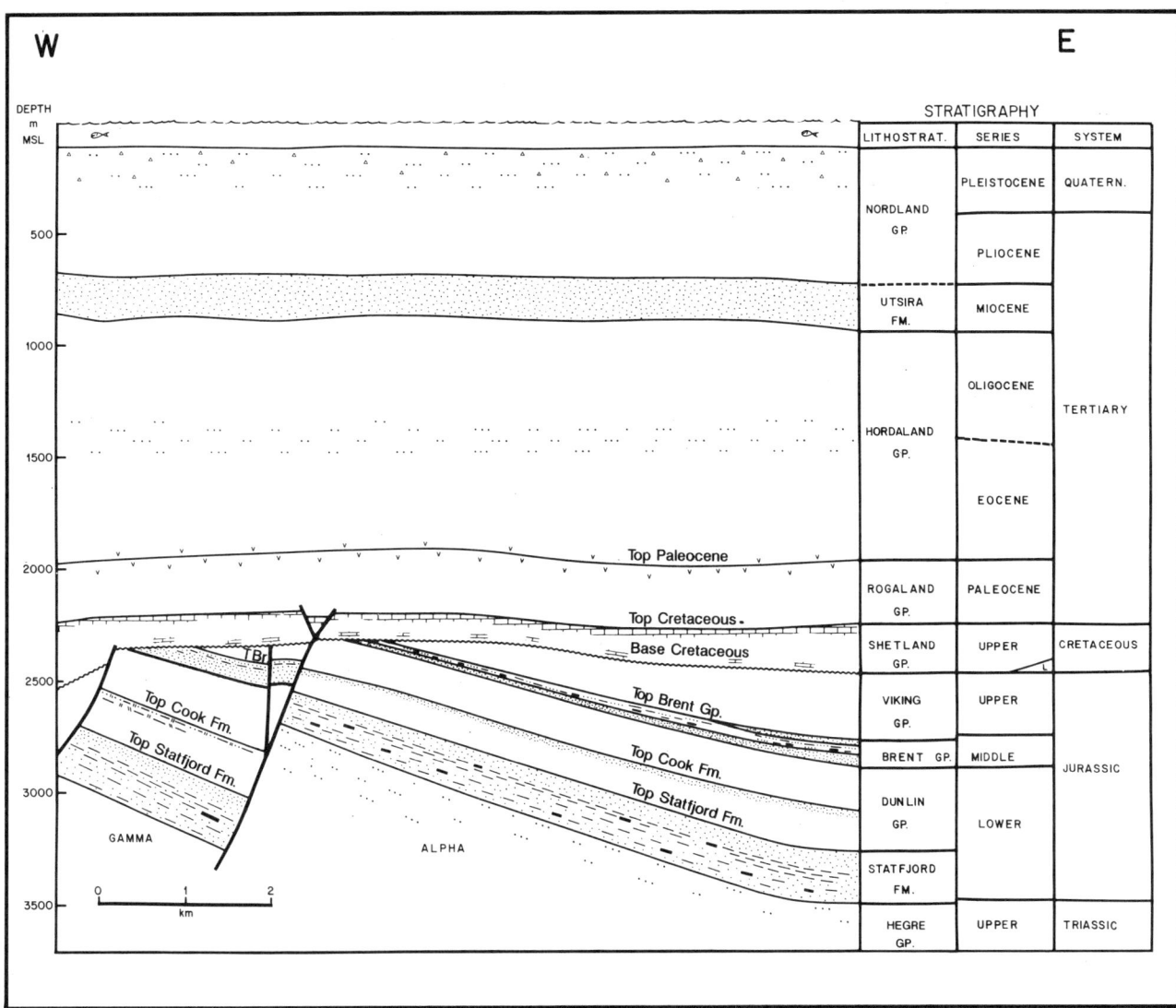

FIGURE 6. Schematic structural cross section through the Oseberg field, with generalized lithostratigraphy.

part of the structure. Apart from these, only minor discontinuous faults paralleling the main north-northwest–south-southeast fault direction have been mapped on the basis of 3D seismic.

The two downfaulted structures have somewhat more internal faulting, but faulting is not considered to be a problem for field development.

The major north-northwest–south-southeast faults generally exhibit long-ranging activity and also show signs of differential subsidence during Brent deposition. The northeast–southwest-trending faults exhibit more punctuated and generally later activity (Nipen, 1987).

The crestal position of the Alpha structure is at about 2120 m MSL. The gas-oil contact is at 2497 m MSL and the free water level is at 2719 m MSL, resulting in a total vertical hydrocarbon column of about 600 m. The field itself is approximately 26 km long by 6 km wide.

STRATIGRAPHY

A generalized stratigraphic column is shown in Figure 6. The Early to Middle Jurassic sandstones of the Brent Group constitute the main reservoir rocks. In addition, a minor discovery of oil and gas has been made in the underlying Early Jurassic Cook Formation. However, this find has not yet been included in the production plans for the Oseberg field.

The Brent Group has been subdivided into five formations—the Tarbert, Ness, Etive, Rannoch, and Oseberg. It consists of rocks of a mainly deltaic origin. Depositional subenvironments of the respective formations are summarized in Figure 7. In a regional context, deposition of Brent sediments consisted of three major phases (Graue et al., 1987):

1. Aalenian lateral infill of sandstones from the east. This

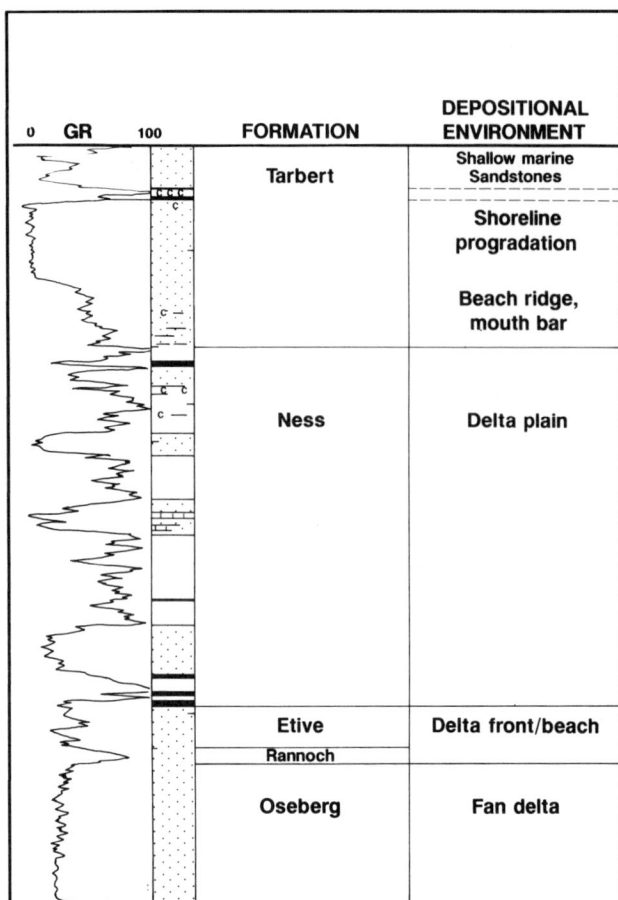

FIGURE 7. Stratigraphy and depositional environment of the Brent Group.

resulted in the deposition of the fairly localized but thick fan delta sandstones across the Oseberg area. These deposits are referred to as the Oseberg Formation.

2. Late Aalenian to early Bajocian progradation of the Brent delta from south to north. This led to the deposition of the delta front/beach sandstones of the Rannoch and Etive formations and the delta plain deposits of the Ness Formation.

3. Early Bajocian to early/middle Bathonian retreat of the Brent delta, resulting in deposition of delta plain deposits of the Ness Formation and finally deposition of beach and shallow marine sandstones of the Tarbert Formation. At least two major pulses of transgression have occurred: one took place before the onset of the Tarbert deposition, and the other was the final transgression of the Brent delta in this area.

The regional stratigraphic aspects of the Brent Group are illustrated in a simplified form in Figure 8.

SOURCE

The main source of the Oseberg oil is the highly organic shales of the Upper Jurassic Draupne Formation (Kimmeridge Clay Formation). To the west of the field, in the Viking graben, large quantities of mature source rock exist. Generation of hydrocarbons in the graben area may have started as early as the late Maastrichtian Stage (Dahl et al., 1987), although the main migration of oil into the field occurred somewhat later.

RESERVOIR GEOLOGY

This section will discuss in more detail the Oseberg, Ness, and Tarbert formations.

Oseberg Formation

The Oseberg Formation is the most important geologic unit for development of the Oseberg field, because it contains most of the recoverable oil. The Oseberg Formation is a fan delta deposit (Graue et al., 1987). Two partly overlapping lobes have been mapped (Figure 9). For both lobes, the inferred transport direction is from the Horda platform area situated to the east. The thickness is partly fault controlled, with crestal thinning (Figure 10) and syndepositional thickening across the major faults. On the downfaulted structures, sand isopachs of up to 65 m are observed. The main deposition of the Oseberg Formation took place in a fairly localized area around the field. There is evidence of a rapid depositional pinch-out toward the south.

When logs are inspected, the Oseberg Formation may appear as a massive and homogeneous sand, because basically no shales are present. When cores are described, however, it is evident that grain size variations occur, especially in the foresets, which tend to have higher degrees of small-scale heterogeneity than the rest of the sequence. This is reflected even in plots of permeability vs. depth for some of the wells (Figure 11).

The Oseberg sands are very permeable. Permeabilities are normally above 1 darcy and usually range from 1 to 4 darcys, although permeabilities higher than 10 darcys have been recorded. The average porosity is about 24%.

The permeability trends have been studied in detail for reservoir modeling purposes, because the Oseberg oil will be produced by gas flooding. Despite both somewhat higher permeabilities in many wells in the upper part of the Oseberg Formation and the greater heterogeneity observed in the foreset/bottomset sequence, the over-all high permeability will tend to gravity-stabilize the gas flooding process.

The reservoir quality of the Oseberg Formation, as it is for the rest of the Brent Group, is controlled mainly by primary depositional processes. The main variations in quality are related to differences in grain size and detrital clay content, although diagenetic processes have somewhat modified and enhanced the initial differences.

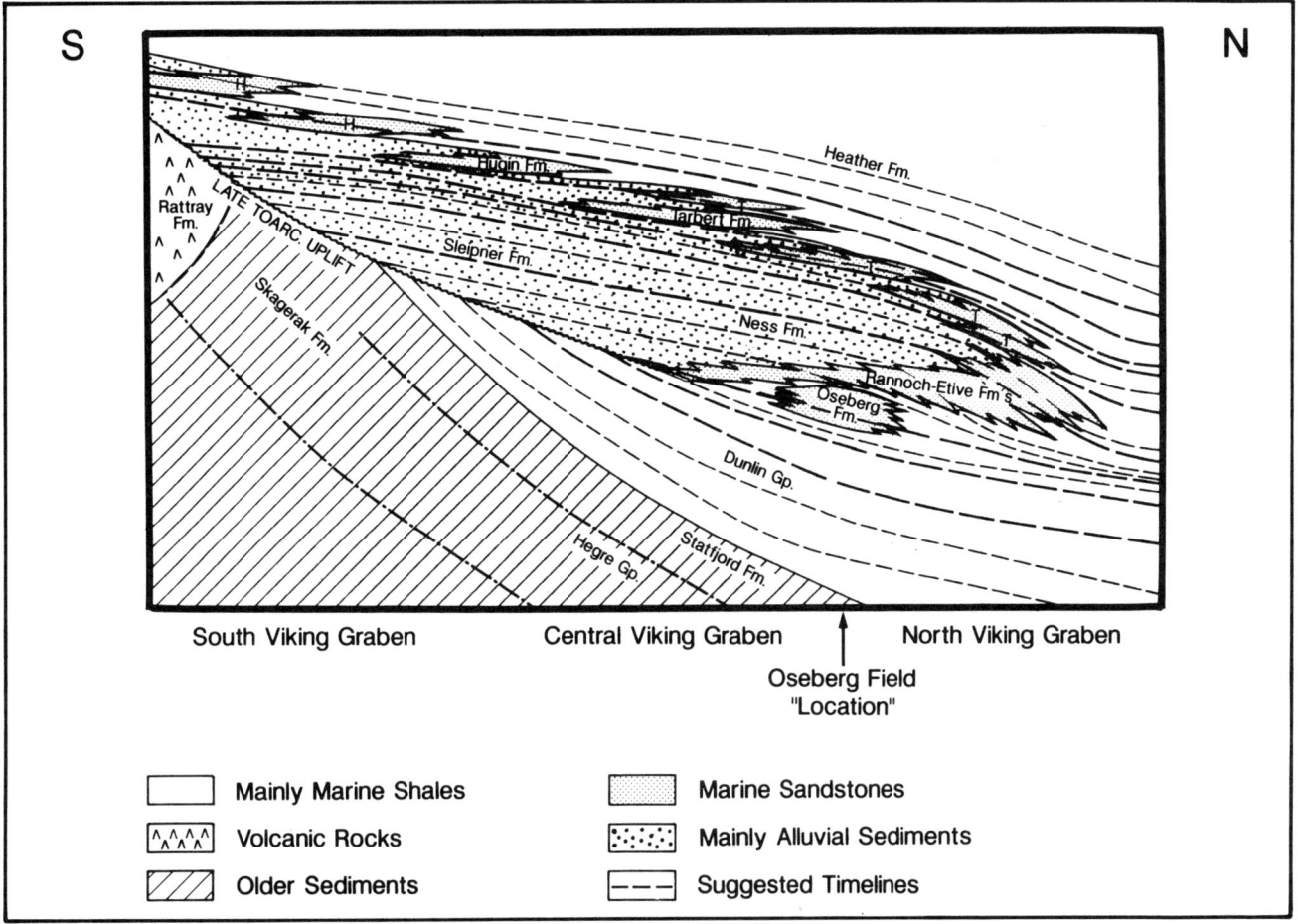

FIGURE 8. Regional Brent Group stratigraphy. Note the fairly localized distribution of the Oseberg Formation. (After Graue et al., 1987.)

This highly permeable formation has good reservoir communication, as demonstrated by its production history. The production from the southern and central parts of the Alpha structure has created pressure responses, which have been recorded in the northern part of the field in the predrilled production wells, indicating full reservoir communication along the Alpha main structure. Furthermore, in the individual wells, static pressure profiles recorded by RFT logs typically have exhibited uniform pressure depletion throughout the Oseberg Formation. Such data have proved that pressure communication exists between the overlapping fan lobes, and that there are no sealing faults or sedimentary features along the Alpha main structure (19 km long).

Ness Formation

The Ness Formation contains roughly 25% of the Oseberg field's total oil in place, but it accounts for only about 10% of the reserves. The low recovery presently estimated is a consequence of the complex reservoir geometry of this formation.

The Ness Formation consists of fluvial and fluvial-related sandstones interbedded with coals and shales. This reservoir complexity is illustrated by Figure 12, which is a schematic north-south cross section of the Alpha structure. The complexity of this formation is also illustrated by observations of different oil-water contacts within the Ness Formation on two of the structures that constitute the Oseberg field.

In studies of the Ness Formation, several different techniques have been used to improve reservoir mapping, including traditional core and log studies, stochastic sand modeling, seismic inversion techniques, and even long-term production testing. These studies have provided valuable information on the depositional environments, reservoir properties, volumetric estimates (statistical methods), and estimates of production performance. However, the major challenge at this stage of field development is deterministic mapping of the main sand systems. Such mapping will allow optimization of produc-

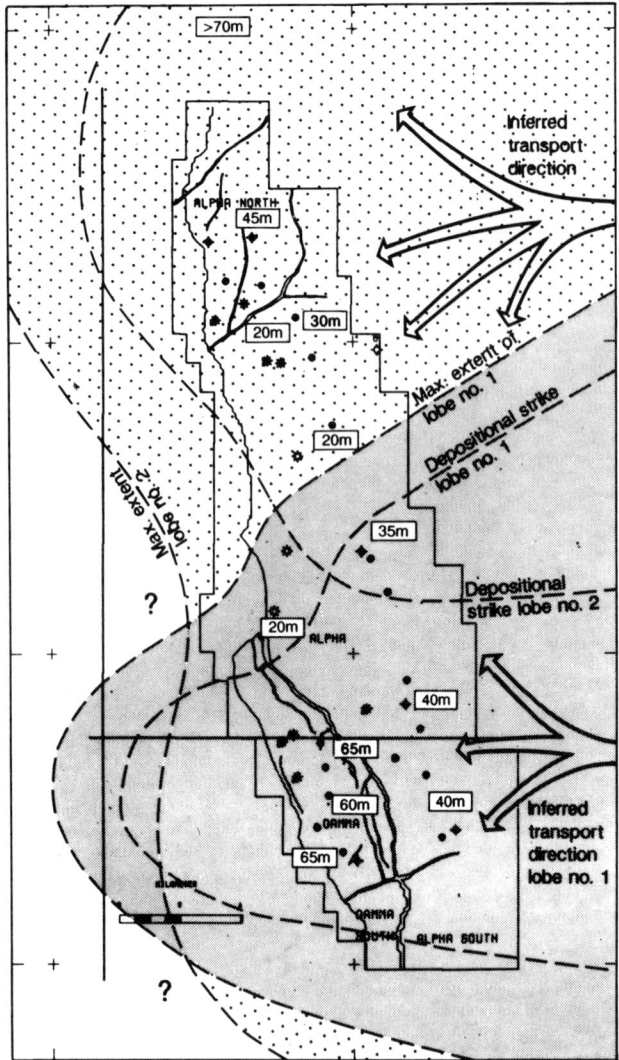

FIGURE 9. Oseberg Formation, showing suggested distribution of fan delta lobes.

tion from the Ness Formation, possibly involving dedicated production and injection wells in the major sand systems.

Tarbert Formation

The Tarbert Formation, which is present in both the southern and northern parts of the field, is a sheet sand reservoir (Figure 13). In the northern part of the field, it is an important reservoir with excellent properties, exhibiting sand thicknesses up to 42 m, porosities as high as 25%, and permeabilities up to 4 darcys.

The Tarbert deposition represents the over-all retreat of the Brent delta. Two main phases of sand deposition occurred. Following the initial transgression that marked the end of the Ness Formation was a constructional phase with progradation of Tarbert sheet sands backed by deposition of continental shales. A major transgression with sand deposition occurred during the late Bajocian, followed by a rotational uplift in the early to middle Bathonian that resulted in downflank deposition and upflank erosion. Final transgression of the entire area had taken place by the middle to late Bathonian (Figure 14).

The present distribution of the Tarbert sands, which are absent in the crestal and central parts of the field, is a result of the relationship between deposition and erosion. Given this reservoir distribution, the importance of the Tarbert reservoir in the north was first fully realized during the predrilling of production wells. It then became evident that locally this reservoir was more important than the Oseberg Formation, which had been the main target for the exploration and appraisal drilling.

FIELD DEVELOPMENT AND PRODUCTION

Fixed Installations

The Oseberg development consists of a field center with two platforms in the southern part of the field, and a single platform in the northern part (Figure 15). In addition, two subsea production wells are located in the central part of the field. They are both linked to the field center.

The field center comprises the Oseberg A and Oseberg B platforms. The Oseberg A platform is a concrete platform with processing equipment, injection modules, and accommodation facilities. Prior to production startup on December 1, 1988, the planned processing capacity was 240,000 bbl (38,000 standard m^3) per calendar day. In 1990, the processing capacity was upgraded to 320,000 bbl (51,000 standard m^3) per calendar day. Peak production has been approximately 357,000 bbl (57,000 standard m^3) for a single calendar day.

The Oseberg B platform is a steel platform for drilling and production. It is linked to the Oseberg A platform. The drilling platform has 48 well slots.

The produced oil is piped to a terminal at Sture, on the west coast of Norway. This is the first oil pipeline ever routed to the Norwegian mainland, although there has been offshore Norwegian oil production since 1971.

The Oseberg C platform is an integrated drilling, production, and accommodation steel platform that is currently under construction in the northern part of the field. This platform will have 26 well slots and a processing capacity of 110,000 bbl (17,500 standard m^3) of oil per calendar day. Production from the C platform will begin in the autumn of 1991. The distance between the C platform and the field center to the south is approximately 14 km. The produced oil will be piped to the field center and then to onshore facilities.

Predrilling

An important aspect of the field development as been the predrilling of production and injection wells through a

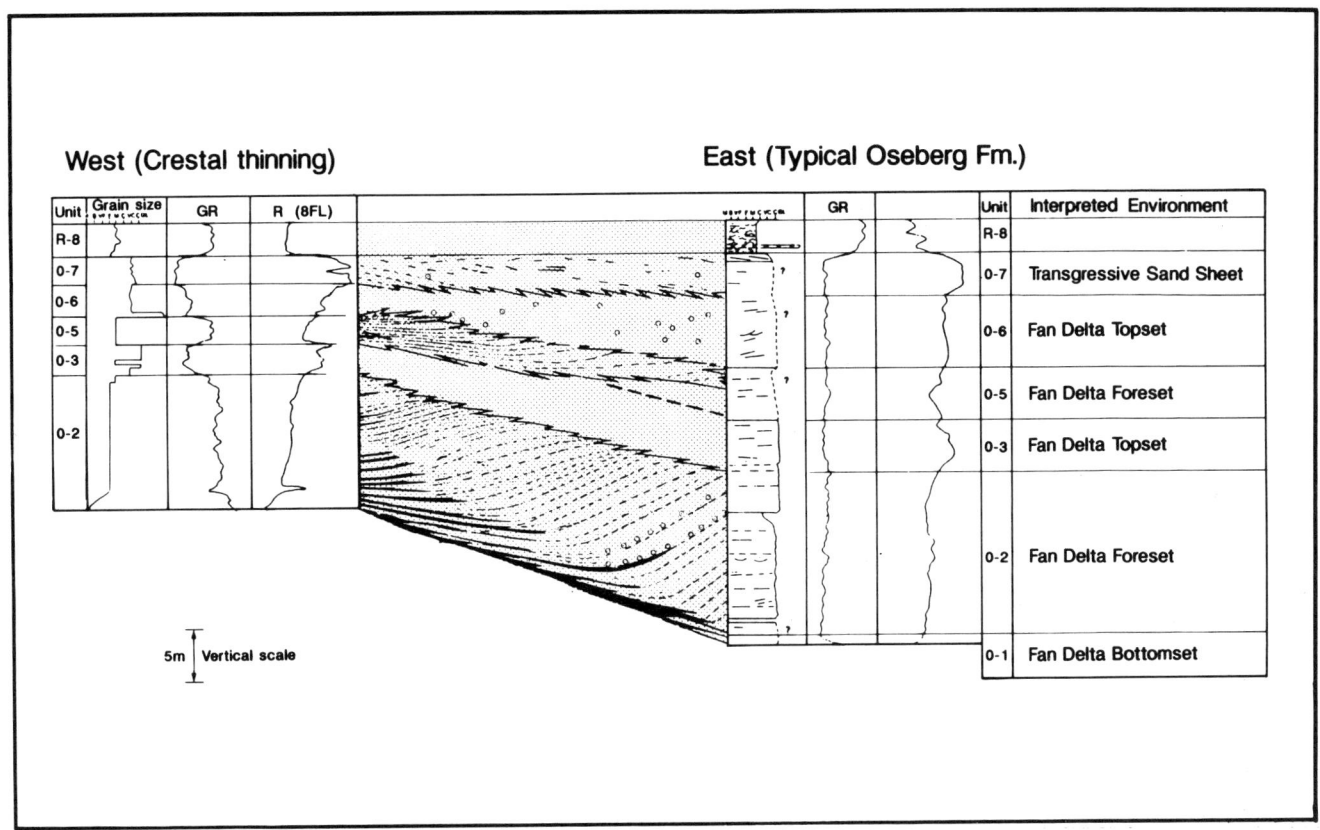

FIGURE 10. Depositional environment of the Oseberg Formation.

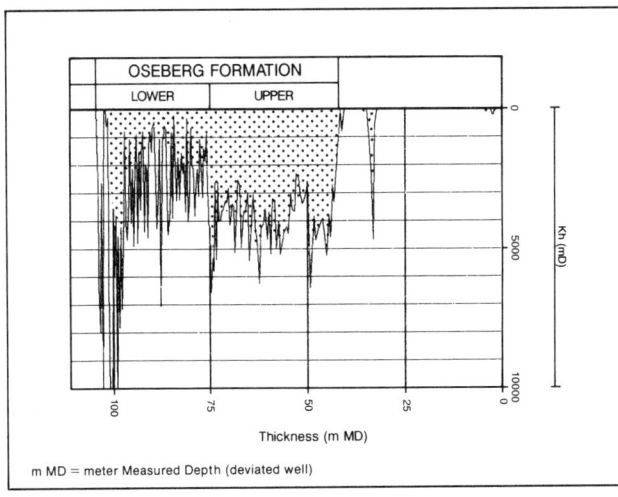

FIGURE 11. Permeability profile for the Oseberg Formation. The lower Oseberg section represents the foresets and bottomsets.

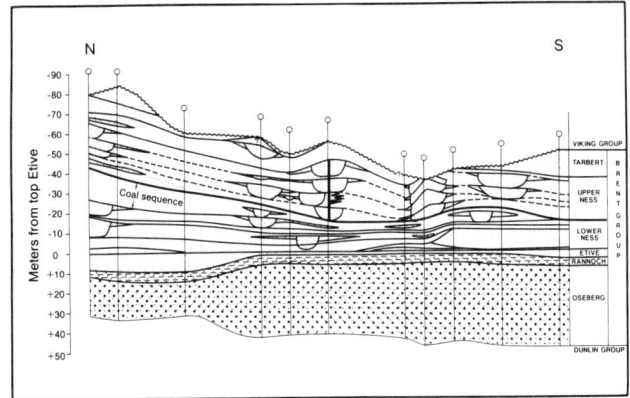

FIGURE 12. Schematic north-south profile of the Alpha structure, illustrating some of the reservoir complexity of the Ness Formation.

subsea template prior to the installation of the platforms. Ten wells have been predrilled in the south, including two subsea wells, and eight wells have been predrilled in the north.

Predrilling allows rapid production buildup (Figure 16). In addition, it provides early access to an increased geologic database, which can be used to improve production planning and optimize field development. This is best illustrated by the predrilling in the north, which proved the presence of significant Tarbert reserves, basically at the same time as the production capacity of the

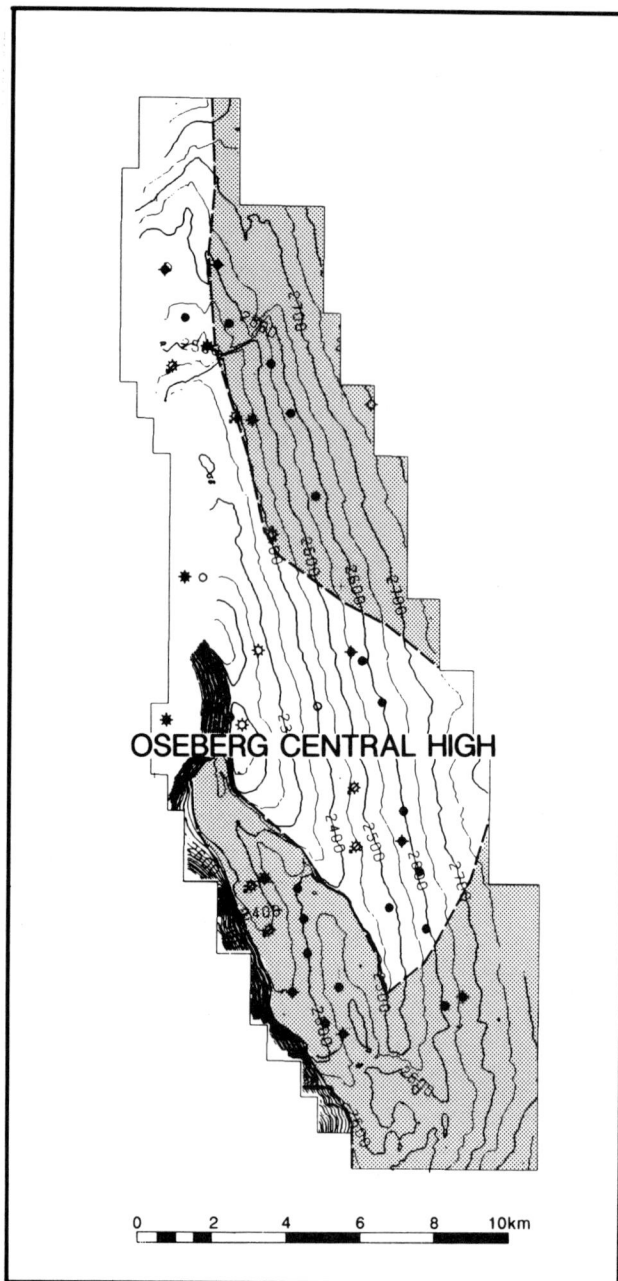

FIGURE 13. Distribution of Tarbert Formation sands (gray shading).

field center was upgraded. When both the increased reserve base for the northern platform and the upgraded production capacity in the south were considered, it was decided to build a pipeline that could transport unprocessed oil to the field center. This will allow both optimum utilization of production platforms and balanced production over the field.

Gas Injection

Production from the Alpha and Gamma structures will employ pressure maintenance, with updip gas injection and downdip oil production. Reservoir studies have concluded that oil recovery from the Oseberg, Rannoch, and Etive formations in the Alpha structure will be 12 million standard m³ (75 MMSTB) greater with pressure maintenance by gas injection than it would be with pressure maintenance by water injection. In addition, fewer dedicated injector wells will be needed for gas injection than would be required for water injection (saved 20 wells).

The gas for injection will come from several sources. All the associated gas not used for fuel will be reinjected. However, in order to maintain the reservoir pressure at a sufficient level, import of external gas is needed. This requirement has led to the TOGI (Troll-Oseberg Gas Injection) project, whereby gas will be produced by a subsea installation at the Troll field and exported to the Oseberg field through a 48 km long pipeline (Figure 15). Five TOGI wells have been drilled, and gas production commenced on January 22, 1991.

By means of the TOGI project, Norsk Hydro is linking two giant offshore fields and initiating the first gas production from the Troll field.

Some injection gas will also be imported from Gamma North, which is a small satellite field to the west of the main Oseberg field. Gamma North will be produced by a subsea well linked to the Oseberg C platform.

Production

Oil production from the field center began on December 1, 1988. Figure 16 summarizes both the historic production and the planned future production profile. Because the oil is produced by gas flooding, the profile consists of an early oil production phase followed by a gas production phase.

Figure 16 also illustrates how effective predrilling has been in ensuring rapid production buildup. Shortly after production start-up, the average oil rate reached the initial production capacity of about 38,000 standard m³ (240,000 STB) per day.

The present production rate of about 51,000 standard m³ (320,000 STB) per day is sustained by only 11 producers and four injectors (as of January 1991). The average initial well capacity has been as high as about 6000 standard m³ (37,000 STB) per day. These high well capacities illustrate the excellent reservoir properties of the Oseberg field.

ACKNOWLEDGMENTS

The authors thank their many colleagues whose work provided the information and ideas presented in this paper. Special thanks to go J. R. Johnsen, H. R. Moe, and S. Pollen for their assistance and criticism. The paper on the Oseberg field by Nipen (1987) has also been used as a general source of information.

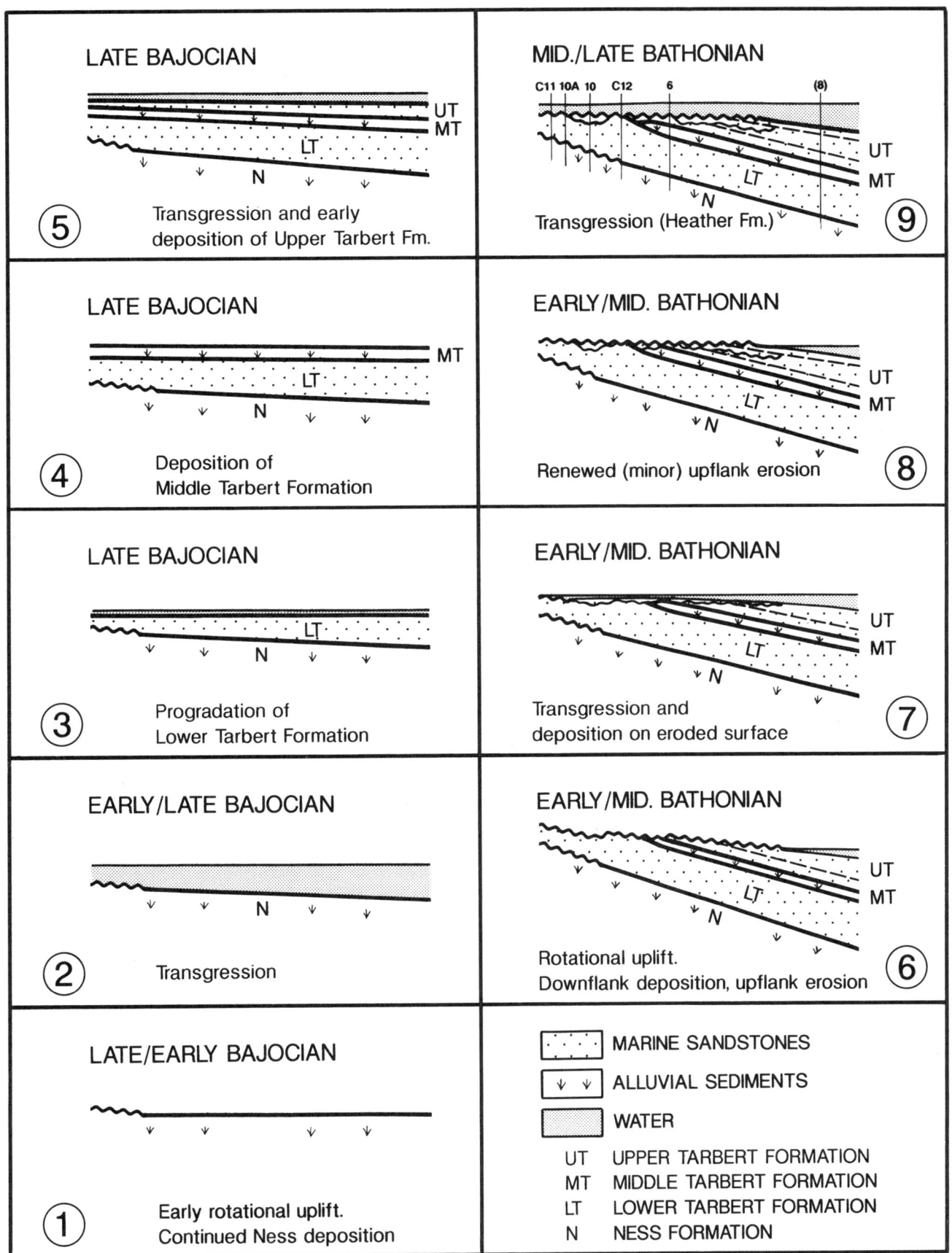

FIGURE 14. Suggested tectonosedimentary development of the Tarbert Formation across the Alpha structure.

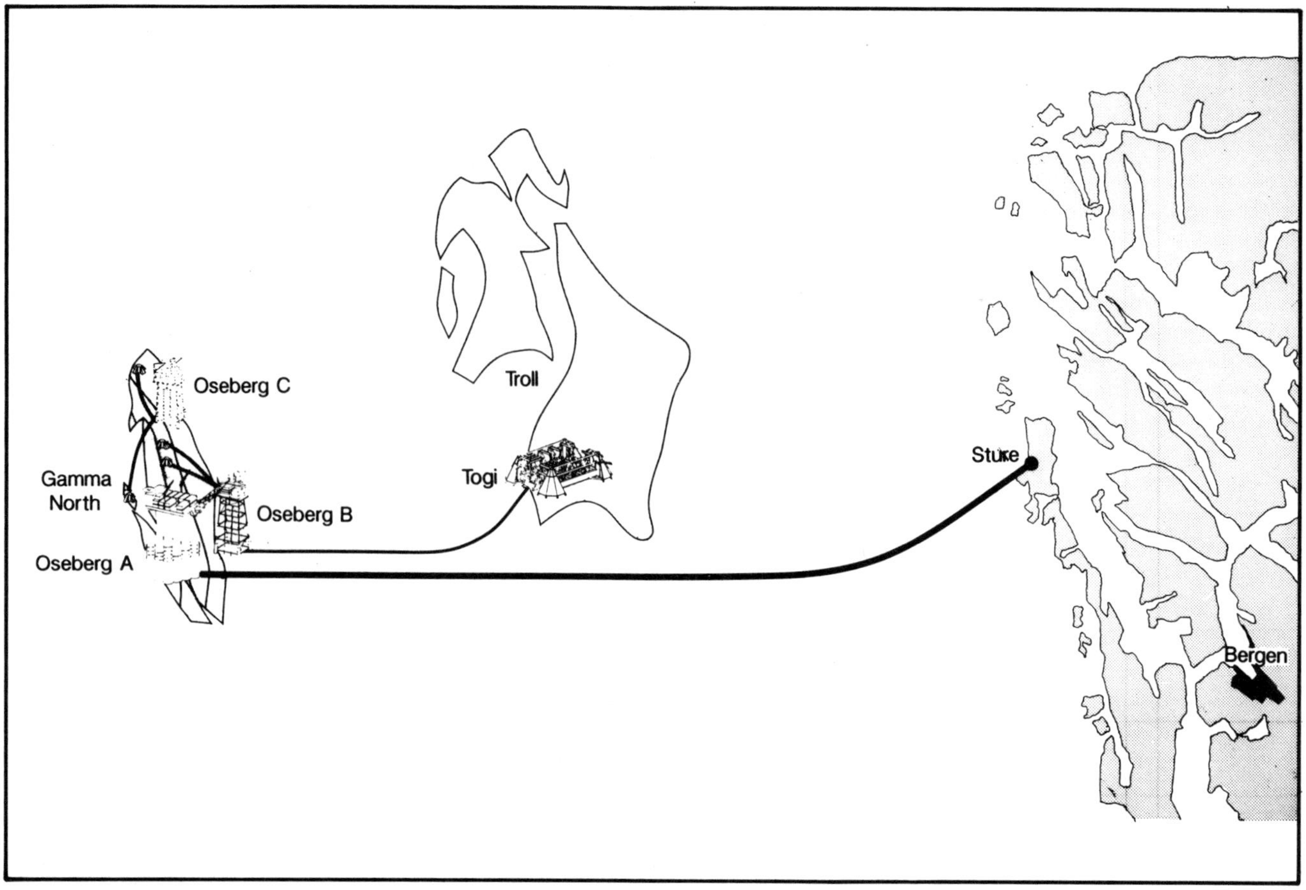

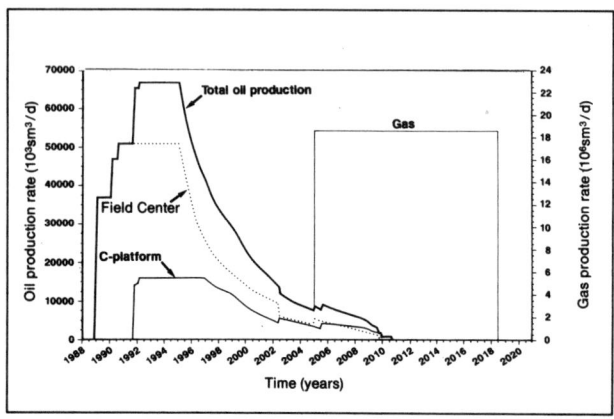

FIGURE 15. Oseberg field development, including the TOGI installations (installations not drawn to scale).

FIGURE 16. Production profile for the Oseberg field (planned production after 1991).

REFERENCES CITED

Dahl, B., E. Nysaether, G. E. Speers, and A. Yukler, 1987, Oseberg area-integrated basin modelling, in J. Brooks and K. W. Glennie (eds.), Petroleum geology of north west Europe: London, Graham and Trotman.

Graue, E., W. Helland-Hansen, J. Johnsen, L. Lømo, A. Nøttvedt, K. Rønning, A. Ryseth, and R. Steel, 1987, Advance and retreat of Brent delta system, Norwegian North Sea, in J. Brock and K. W. Glennie (eds.), Petroleum geology of north west Europe: London, Graham and Trotman.

Nipen, O., 1987, Oseberg, in A. M. Spencer et al. (eds.), Geology of the Norwegian oil and gas fields: London, Graham and Trotman.

Chapter 27

The Gullfaks Field

Ole Petterson
Arvid Storli
Eva Ljosland
Ole Nygaard

Statoil
Bergen, Norway

Ian Massie

Conoco

Henrik Carlsen

Statoil
Stavanger, Norway

ABSTRACT

The Gullfaks giant oil field in the Norwegian sector of the North Sea was discovered in 1978. The Gullfaks field contains oil reserves on the order of 230 million standard m^3.

Gullfaks represents the shallowest structural element of the Tampen spur. It was formed during the Upper Jurassic to Lower Cretaceous as a sloping high, with a westward structural dip gradually decreasing toward the east. The major north–south-striking faults, with eastward-sloping fault planes, divide the field into several rotated fault blocks. The central and eastern parts of the structure were eroded by the Early Cretaceous transgression.

The reservoir sands are the Middle Jurassic delta-deposited Brent Group, the Lower Jurassic shallow marine sandstones of the Cook Formation, and the Lower Jurassic fluvial channel and delta plain deposits of the Statfjord Formation.

The Brent reserves in the western part of the field are currently being developed from the Gullfaks A and B platforms, and the eastern part is being developed from a third platform, Gullfaks C. Water injection is the major drive mechanism, maintaining the reservoir pressure above the bubble point.

One of the most important factors in the reservoir development of the Gullfaks field has been the effect of fault transmissibilities on lateral and vertical pressure distributions.

INTRODUCTION

The Gullfaks field is located in the central part of the East Shetland Basin on the Viking graben (Figure 1). The field lies within the Norwegian offshore license PL 050 in block 34/10, which was awarded in 1978 to a group comprising Statoil (85%), Norsk Hydro (9%), and Saga Petroleum (6%), with Statoil designated as the operator. The block contains four separate hydrocarbon-bearing structures (Figure 2). The Gullfaks field, in the northeastern corner of the block, is the most important structure. It was declared commercial in 1980.

The areal extent of the Gullfaks reservoir is 50 km^2, which made it evident at an early stage that three platforms would be needed for full exploitation of the reserves. Gullfaks A and C are fully independent processing platforms, whereas the Gullfaks B platform has only first-stage oil separation. Partially stabilized oil is trans-

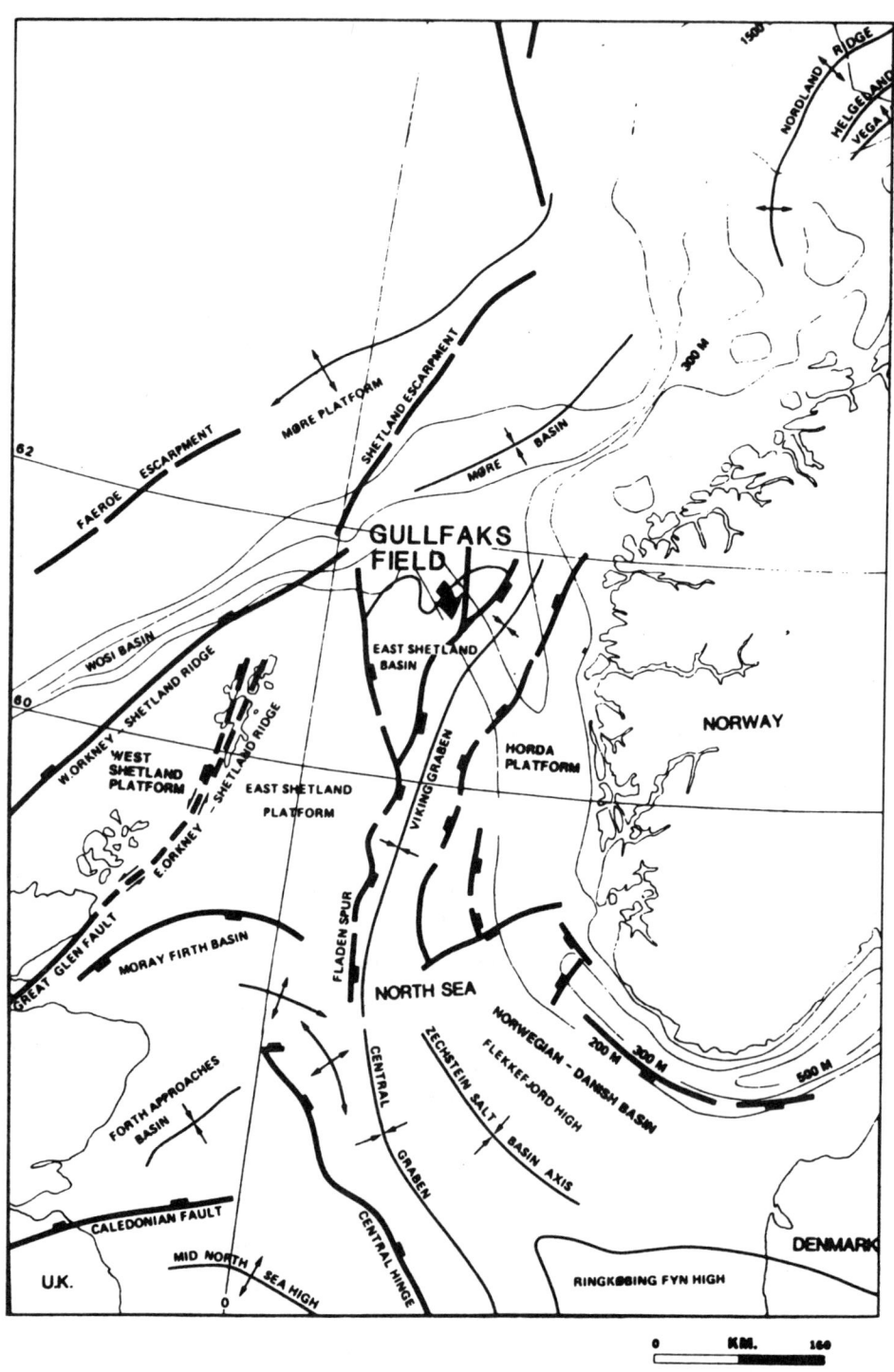

FIGURE 1. Location map for the Gullfaks field.

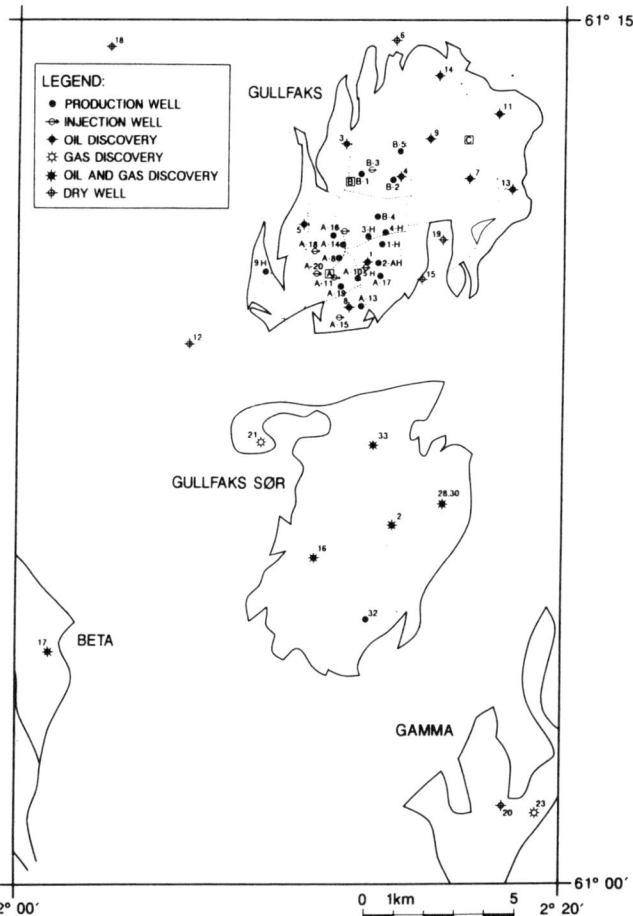

FIGURE 2. Fields in block 34/10, Norwegian continental shelf.

ferred from Gullfaks B to Gullfaks A or C for complete processing. The platform locations and pipeline infrastructure for the field are shown in Figure 3.

Crude oil is stored in the concrete bases of Gullfaks A and C and loaded into tankers by means of two loading buoys. Produced gas is exported through the Statpipe system to Emden, Germany. Gas injection requirements are met by employing combined oil production/gas injection wells so as to handle modulation of gas deliveries.

The A and B platforms have 42 drilling slots each, and Gullfaks C has 52 slots. In addition, a total of 21 subsea wells can be connected to the platforms. The current (1988) development plan includes 115 development wells for the entire field.

EXPLORATION HISTORY

The first well on the Gullfaks structure, well 34/10-1, was spudded on June 20, 1978, just after the license had been awarded (Figure 4). This well penetrated the Jurassic section and encountered 162 m of oil-filled Brent sandstone with oil down to the base of the Brent Group. The reservoir properties were found to be excellent, with porosities as high as 34% and permeabilities up to 4 darcys. A 60 m thick regressive sand in the Cook Formation of the Dunlin Group, as well as the Statfjord Formation, were dry. The next four exploration wells (34/10-3, -4, -5, and -6) were drilled in the western part of the field. All except one (34/10-6) proved oil-filled Brent sandstone. An oil-water contact (OWC) was established at 1947 m MSL. At this point, two issues had to be resolved:

1. The extent of the Brent sandstone in the eastern direction.

2. Whether or not, if the Brent sand was eroded or not present, oil could be preserved in the Cook Formation or in the deeper Statfjord Formation.

It was therefore essential that more and better seismic data be acquired prior to further exploration drilling in the easternmost part of the structure. Hence, a 3D seismic survey was shot over the structure in 1979. Improvements in basic data quality, together with the large increase in data quantity, meant that previously unmapped areas in the eastern part of the structure could now be mapped. Six wells (34/10-7, -8, -9, -11, -13, and -14) were then drilled in the eastern area and were all successful, and two appraisal wells (34/10-15 and -19) were drilled in the southeastern area in an effort to delineate the field in that direction. The Cook sandstone, which was the main objective, was found to be dry in well 34/10-15 and eroded in well 34/10-19.

At the end of 1983, the exploration phase was brought to an end after a total of 14 wells had been drilled on the structure. Ten of these wells were successful, three were dry, and one was abandoned because of problems with shallow gas. Three wells are essential to an understanding of the structural positions of the main reservoirs in the Gullfaks field (Figure 5):

1. Well 34/10-3 in the western area, which discovered the OWC in the Brent reservoir at 1947 m MSL.

2. Well 34/10-9 in the central area, which found oil-bearing Cook Formation sediments below the oil-bearing Brent sandstone. An OWC in the Cook Formation was measured at 2090 m MSL.

3. Well 34/10-11, which was drilled into a horst block in the eastern area and discovered oil-bearing Statfjord Formation sands structurally 300 m higher than the adjacent western fault block.

GEOPHYSICS

The Gullfaks field has been described as the most complex structure so far developed in Norwegian waters. There are several reasons for this description, but the main one is the complex fault pattern that intersects and divides the field into many small fault blocks. A total of 32 fault blocks have been mapped (Figure 6). This large number of fault blocks has required that a rather large number of exploration and appraisal wells be drilled, and high-quality seismic data be acquired, in order to achieve a reasonable

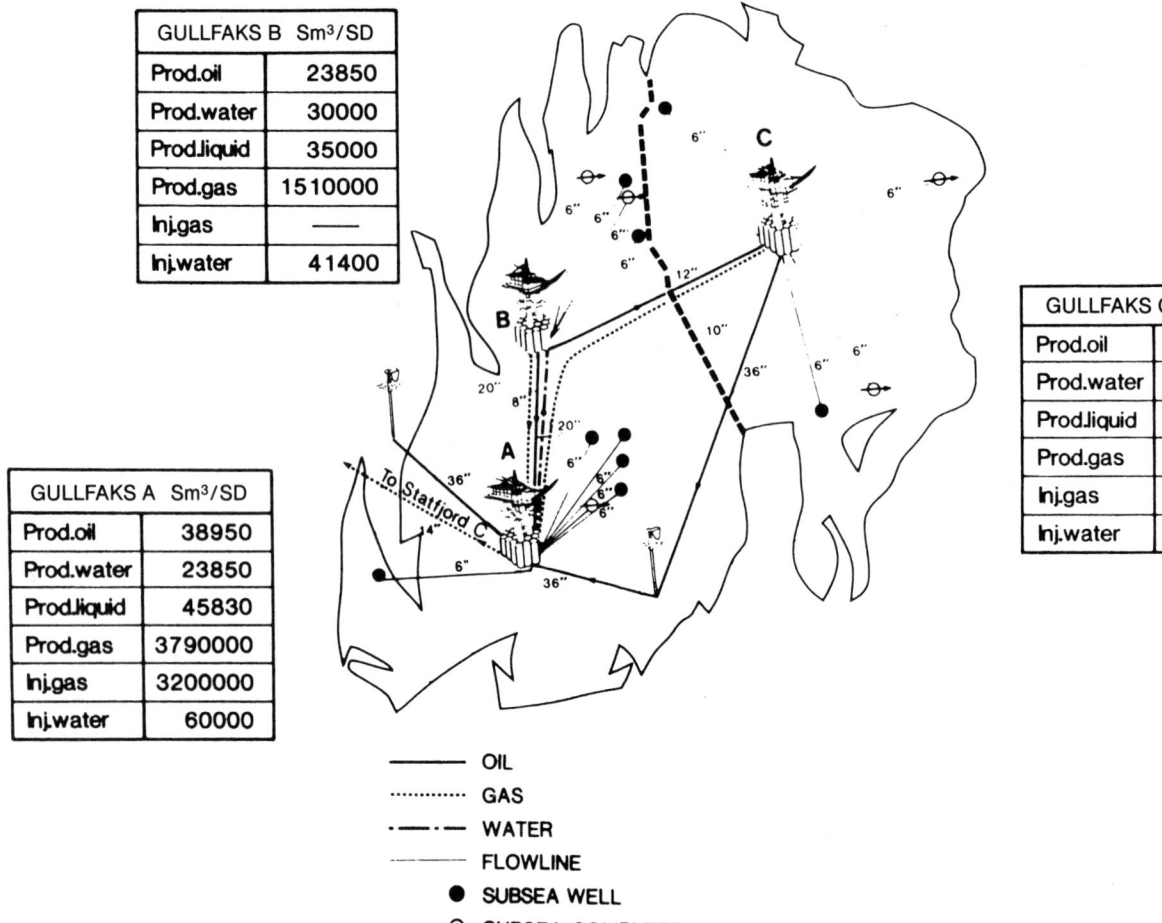

FIGURE 3. Platform locations and pipeline infrastructure for Gullfaks field.

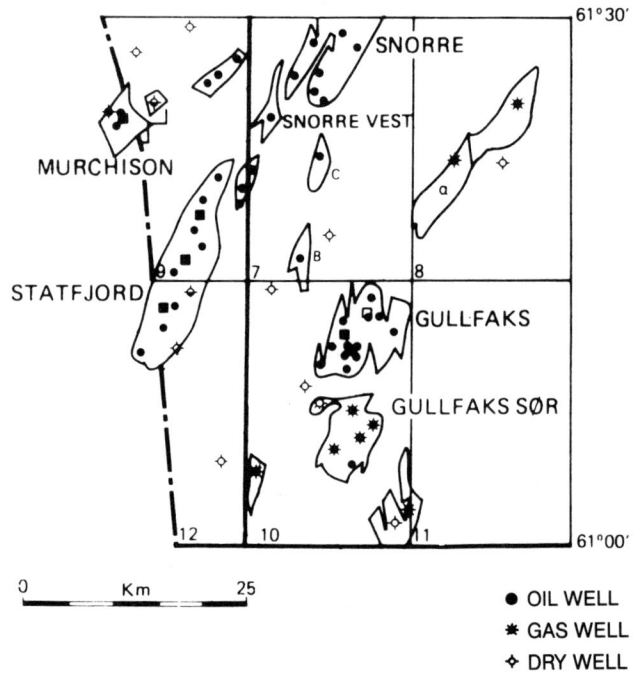

FIGURE 4. Exploration and delineation wells in the Gullfaks area.

degree of confidence in the interpretation and structural mapping. Several sets of seismic data have been acquired over the Gullfaks field. Since the initial interpretation, which was based on a 1 by 1 km conventional 2D seismic survey taken in 1974, two 3D seismic surveys have been collected (1979 and 1985).

In order to improve the quality of the data and the reliability of the structural maps prior to development drilling, the acquisition of a new 3D data set was carried out in 1985. This survey consisted of 8700 km of seismic data divided into 550 east-west lines with a spacing of 25 m. Processed data from this survey (1985) have proved to be of higher quality than the data collected in 1979 (Figure 7). In particular, the line spacing of 25 m has resulted in improvements in the cross lines. This has provided better control of fault patterns and much better definition of reflectors. Despite the observed improvements in the 1985 survey, it must be emphasized that the seismic data are still of relatively low quality in comparison with those acquired for many other fields in the region. The most important reasons for this are:

♦ Variation in water depth from 130 m in the southwest to 230 m in the northeast, with a steep sea bottom slope crossing the field.

♦ The presence of shallow gas to a depth of approximately 450 m MSL.

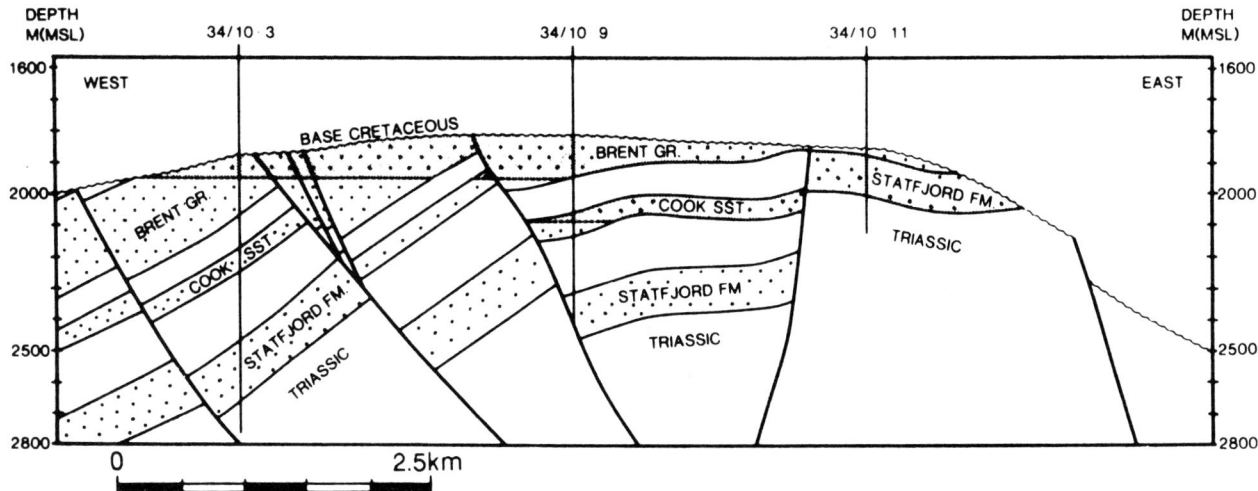

FIGURE 5. Structure of the Gullfaks field seen in cross section.

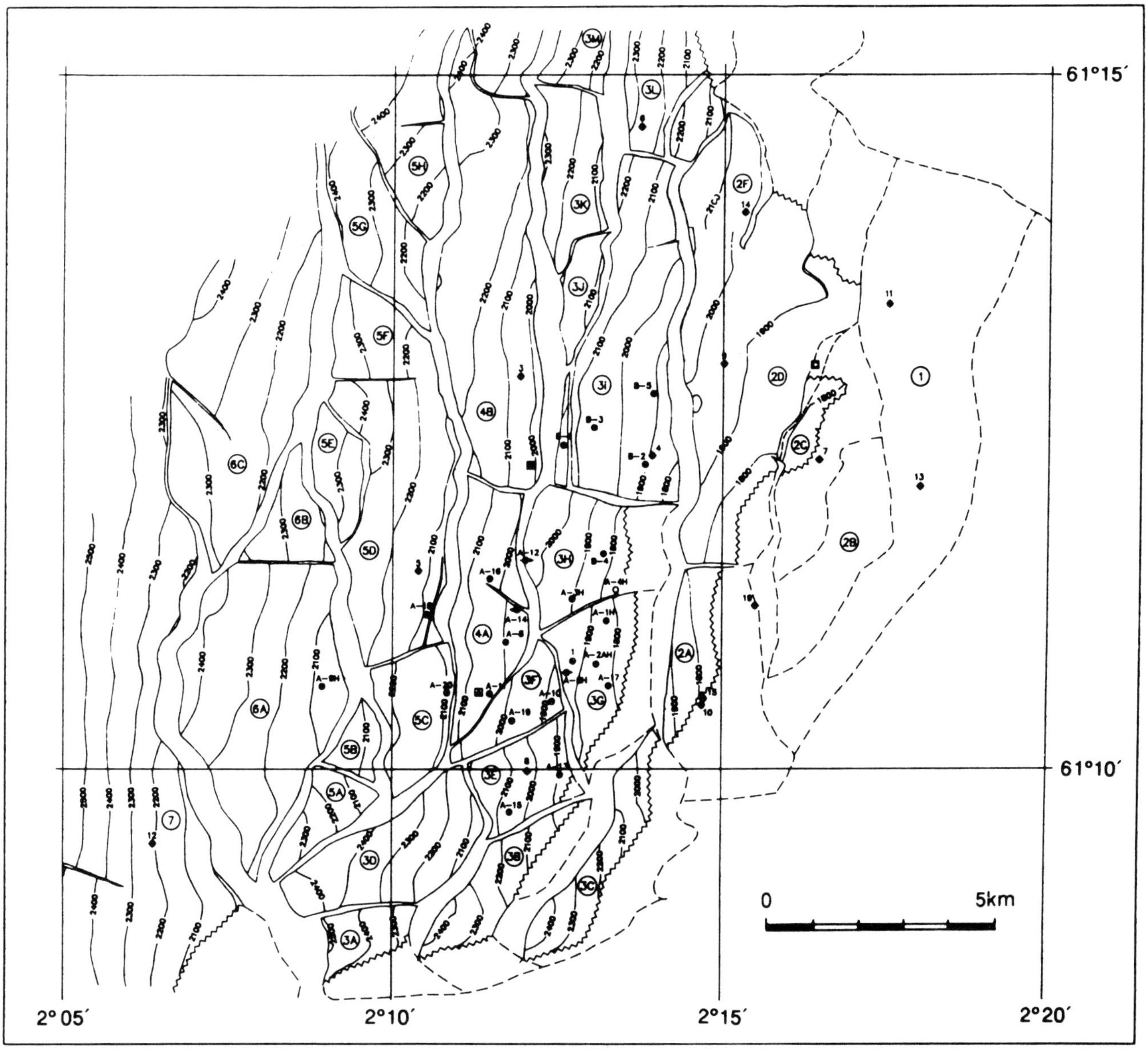

FIGURE 6. Structure of the Gullfaks field showing the numerous fault blocks.

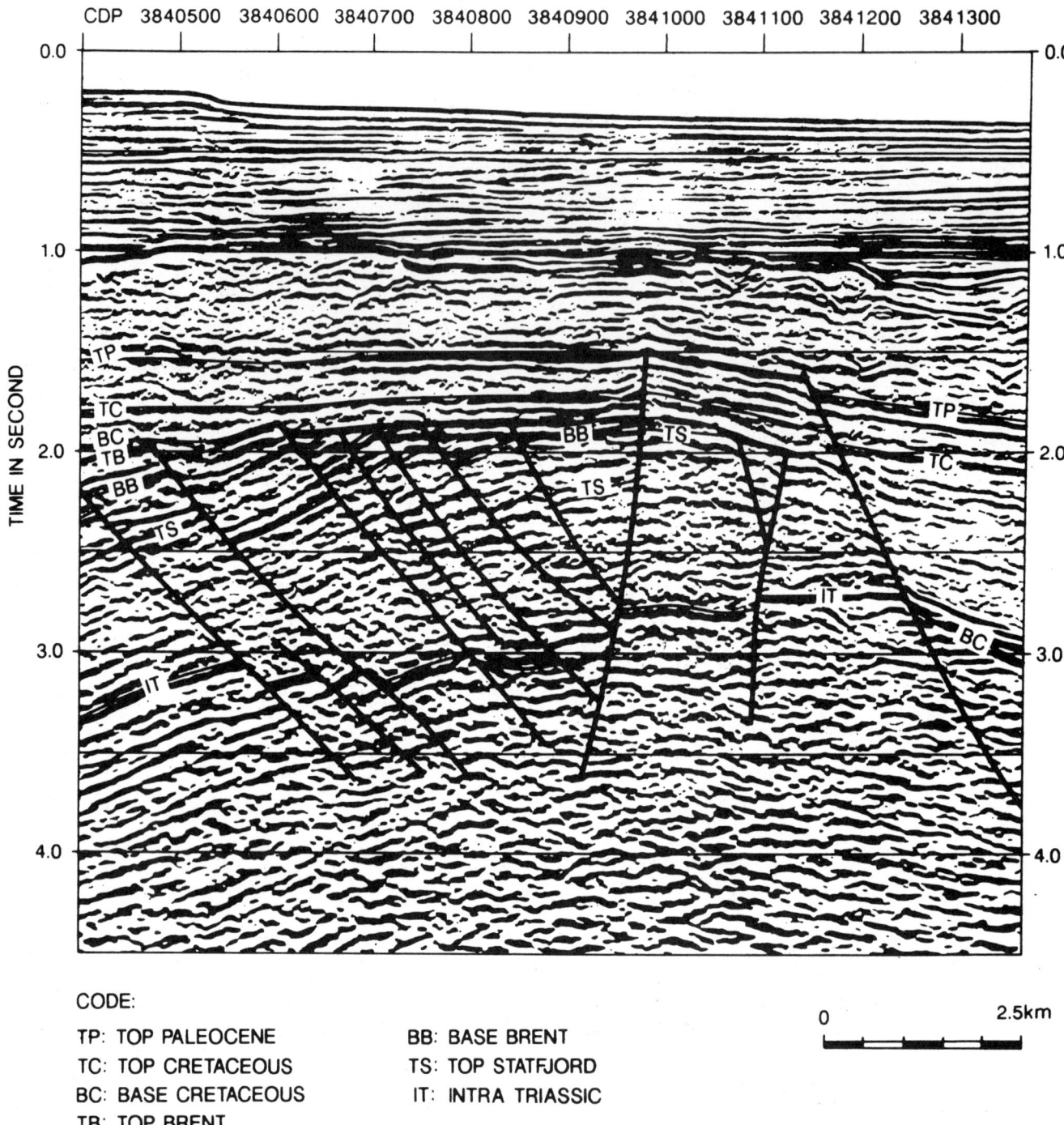

FIGURE 7. Seismic profile of Gullfaks field.

- The presence of gas throughout a large interval in the post-Jurassic sections.

- Very strong reflectors at the tops of the Paleocene and the Cretaceous, which reduce the amount of original seismic energy penetrating the pre-Cretaceous section. These two reflectors also generate strong multiples, which interfere with the pre-Cretaceous reflections.

- Deterioration of the seismic data as a result of the large number of fault blocks.

In spite of the above-mentioned problems, it can be concluded that the new 3D seismic data have provided important insights into the tectonic evolution of the Gullfaks field. In addition, borehole information from 13 exploration wells and 23 development wells has resulted in better control with respect to identification of the various reflectors and faults. For improved accuracy of mapping, most of the development wells have been deliberately deepened in order to penetrate potential faults and minimize uncertainties. However, areas in the central and eastern parts of the structure remain very difficult to interpret because of discontinuous reflectors, noise, and strong multiples within the seismic sections. This is particularly evident in the detection, orientation, and extension of faults.

STRUCTURAL DEVELOPMENT

The structural genesis of the Gullfaks field took place throughout the Jurassic, culminating during the Late Jurassic to Early Cretaceous in connection with regional tectonic movements.

The extensional tectonics that seem to have controlled the development of the Gullfaks structure resulted in sagging with normal faulting and block rotation in the western part of the structure. The central area underwent "keystone" faulting that resulted in a graben feature, whereas the eastern part remained as an elevated horst structure. The structural development took place within the framework of a half-graben with sliding and rotation of the fault blocks on a deep detachment zone (Figure 8). The intense faulting throughout the period formed a very marked north–south-oriented fault system. The widths of the fault blocks vary from 0.5 to 1 km, and the major fault throws vary from 50 to 150 m.

Most faults terminate at the Kimmerian unconformity, which seals off the Gullfaks reservoirs at approximately 1700 m MSL in the crestal area. To the north, east, and south, the structure is limited by major boundary faults that separate it from the Tampen spur rim (block 34/8), the Viking graben, and the Gullfaks Sør structure, respectively.

Most authors seem to favor the assumption that extensional stress prevailed throughout the Late Jurassic to Early Cretaceous. However, a new seismic/structural study (Fossen, 1989) argues that dextral strike-slip movement along the northeast-southwest fault system at the northeastern margin of the Tampen spur (Figure 9) accompanied the extensional phases. This movement resulted in compressional structures, involving the for-

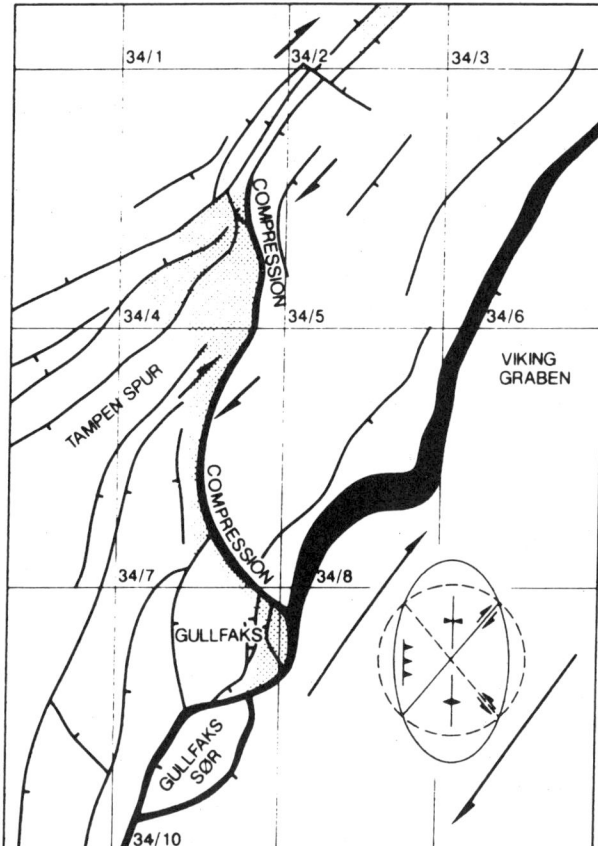

FIGURE 9. Structural model of Gullfaks field.

mation of thrusts, reverse faults, and compressional folds at the Gullfaks field. The transpressional tectonics took place in the Late Jurassic to Early Cretaceous. In a regional context, the strike-slip movement may be related to dextral strike-slip movements along the Møre-Trøndelag fault zone farther to the northeast.

The Gullfaks field can be divided into three structural regimes—the western, central, and eastern areas (Figure 8).

Western Area

The western area comprises a series of rotated fault blocks. The normal faults, which strike in a north-south direction, exhibit fault planes that dip at rather unusually low angles of approximately 30 to 40° to the east, whereas the formations dip at 10 to 15° to the west. Usually, there is good agreement between dips detected from seismic lines and from dipmeter logs, and normal faults develop with fault plane dips of about 60° (Figure 10A). In cases where block rotation yields fault plane dips of 30 to 40°, bedding dips should be expected to be on the order of 20 to 30° (Figure 10B), and not as on the Gullfaks field, where the observed dip is 10 to 15° unless other deformations exist.

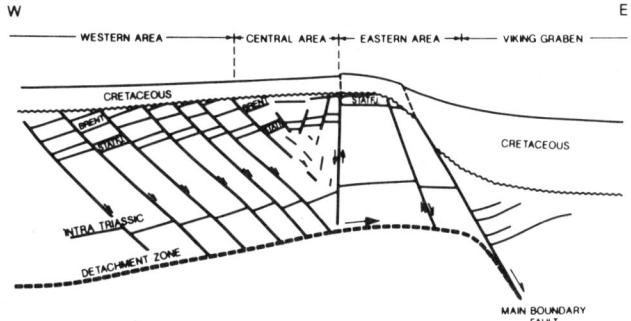

FIGURE 8. Structural regions of Gullfaks field.

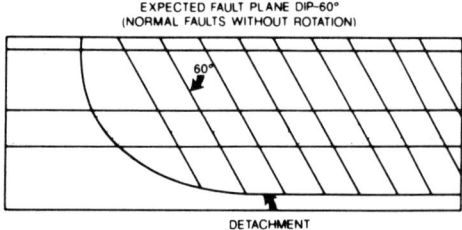
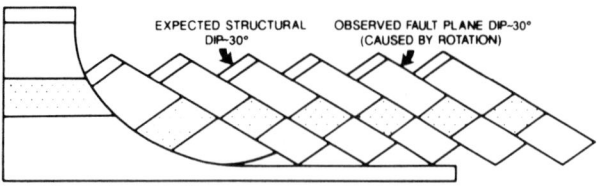
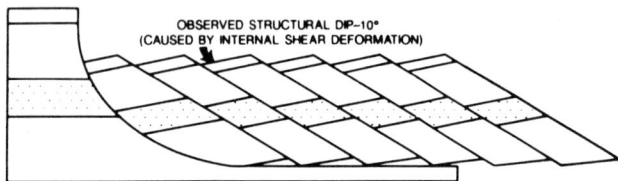

FIGURE 10. Structural model of the evolution of the western area of Gullfaks field.

This unusual feature of low fault dips in combination with low bedding dips has been explained by the presence of two types of deformation:

1. The "domino-type" deformation of several fault blocks, with rotational deformation of the blocks situated above a detachment (Figure 10B).

2. Internal shearing within the separate blocks, maintaining the low bedding dips observed (Figure 10C).

An analog to this type of deformation has been produced in a sandbox experiment by McClay and Ellis (1987).

Space problems close to the detachment are expected to be solved by ductile deformation. Ductile deformation in this context means a change in shape without observable discontinuities on the scale of the seismic resolution.

Central Area

The central area could be described structurally as a graben feature. The graben is interpreted as the result of a "keystone" collapse (Sæland and Simpson, 1982).

Seismic sections normally show a poorly defined reflector pattern that is believed to be the expression of a more complexly deformed structure.

Reflectors show dip variations from approximately 10° to more subhorizontal in the eastward direction.

Eastern Area

The eastern area forms a horst structure, where the Statfjord Formation is lifted approximately 300 m relative to the central area. The elevated position of the horst has resulted in complete erosion of the Brent and Cook reservoirs. At the northern and southern flanks of the horst, the erosion can be followed down into the Hegre Group of the Upper Triassic (Figure 11, cross section B-B'), resulting in erosion of more than 700 m of Jurassic strata.

The horst structure is bounded by an almost vertical fault from the central area (Figure 5) and by a major boundary fault with a throw on the order of 1000 to 1500 m in the Viking graben in the east.

STRATIGRAPHY AND ENVIRONMENTAL MODEL

The recoverable oil in the Gullfaks field is found mainly in three reservoir sandstones:

1. The Statfjord Formation. (For typical log presentation, see Figure 12.)

2. The Cook Formation. (For typical log presentation, see Figure 13.)

3. The Brent Group. (For typical log presentation, see Figure 14.)

Minor reserves are also present in the Upper Triassic Hegre Group. The lithostratigraphic nomenclature used for the Gullfaks field (Figure 15) follows the subdivisions of Vollset and Doré (1984).

Hegre Group

The oldest reservoir sandstones are the medium-grained Triassic fluvial sandstones of the Lunde Formation. Exploration well 34/10-13 has proved and tested reserves in this formation. However, the reservoir quality is considered to be moderate, and the reservoir potential to be low.

Statfjord Formation

In the Gullfaks area, the Statfjord Formation is subdivided into the Raude, Eiriksson, and Nansen members (Figure 12). The Statfjord Formation represents an alluvial environment that changed from a well-drained, semiarid setting with episodic sheet flood deposition to a more humid, alluvial plain with poorly drained, locally swampy overbank areas and channelized flows (Collinson, 1985, confidential contract report for Statoil).

Raude Member

The Raude Member is divided into the three reservoir units Raude 1, 2, and 3. The lowest unit, Raude 1, is a

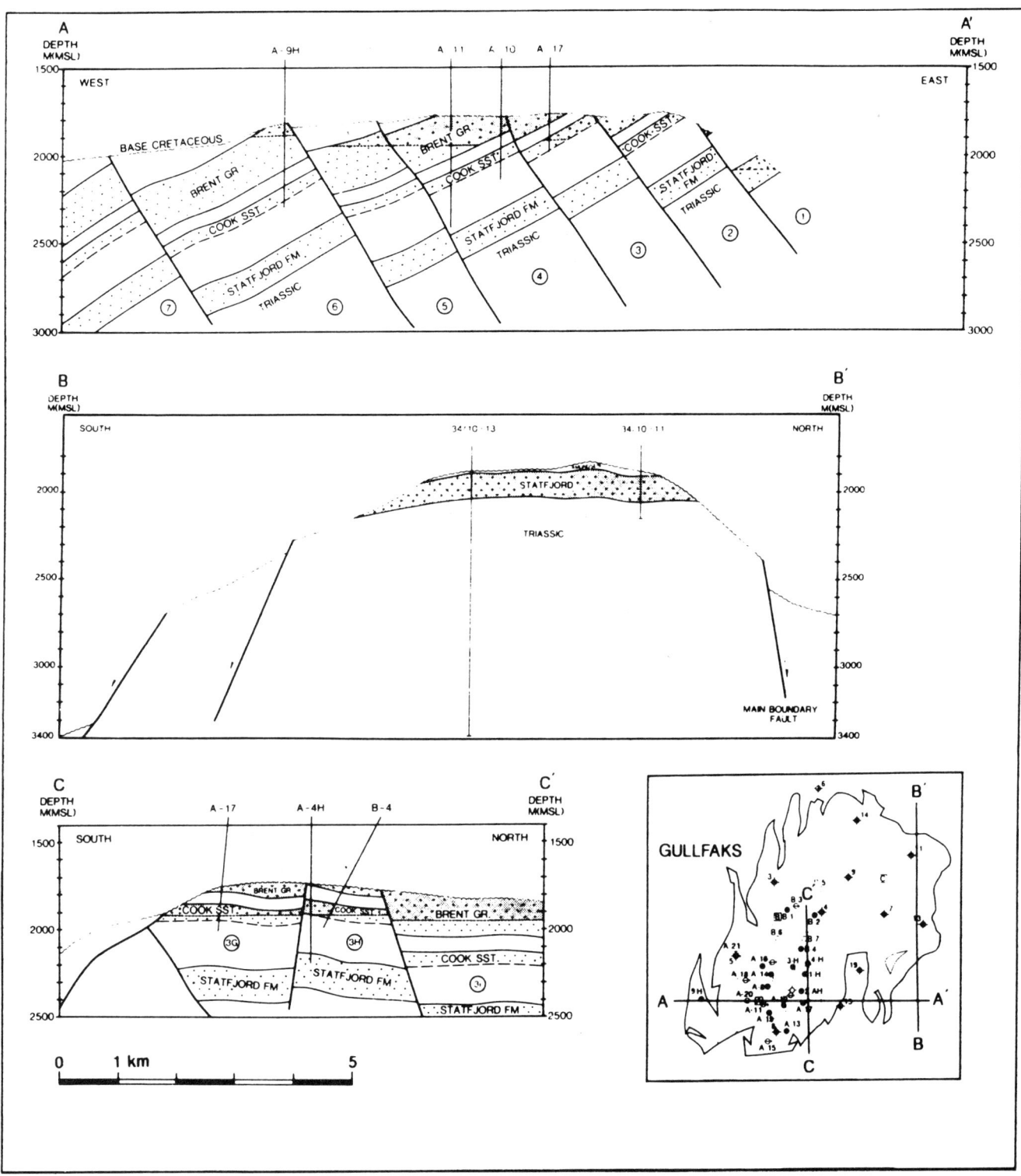

FIGURE 11. Structure of the eastern and central areas of the Gullfaks field seen in cross section.

coarsening-upward shale/silt unit with poor reservoir quality and a net/gross ratio of only 0.1. Raude 2 and 3 comprise several sheet flood sands and alluvial sands rapidly deposited during short periods of flooding in a generally dry climate. The sands and shales have been influenced by pedoturbation, and soil profiles as well as caliche are common. Raude 2 and 3 are of medium reservoir quality owing to a high kaolinite content in the pores. Thickness varies from 100 m in the west to 80 m in the east.

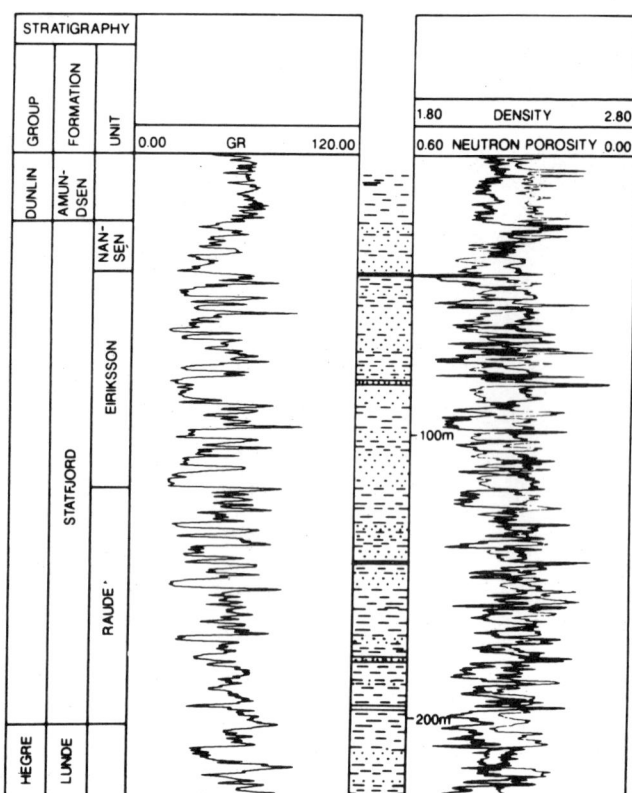

FIGURE 12. Typical log for the Statfjord Formation.

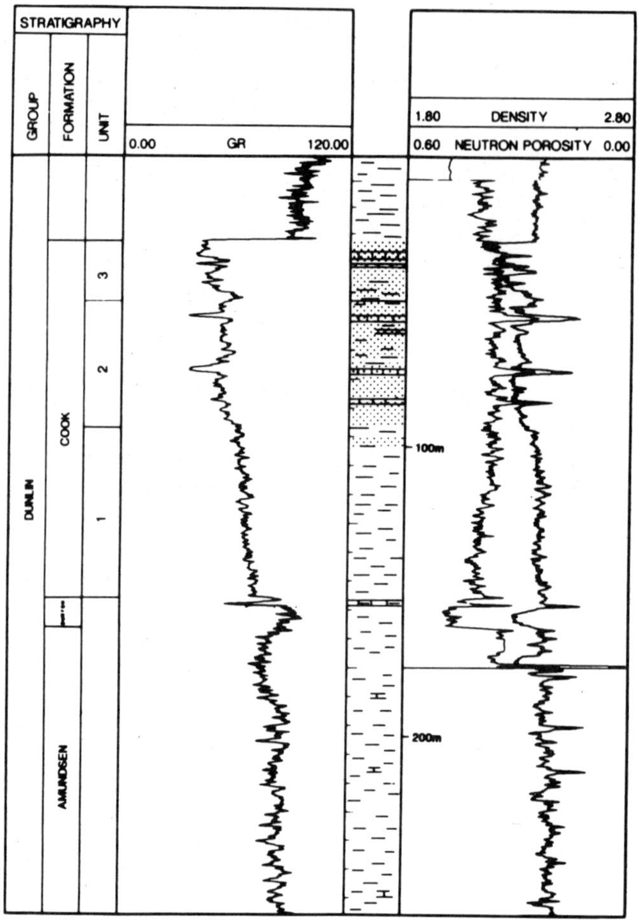

FIGURE 13. Typical log for the Cook Formation.

Eiriksson Member

The Eiriksson 1, 2, and 3 reservoir units were deposited in an environment with meandering channels and swampy overbank areas in a climate that was gradually becoming more humid. Each unit is defined by one or more stacked channels at the base, and a shaly top. The coarse-grained channel sandstones exhibit good to excellent reservoir properties, with porosities up to 28% and a net/gross ratio of 0.60 in the eastern oil-filled area. Thickness varies between 60 and 75 m. The vertical and lateral continuities are good as a result of the stacked fluvial channels.

Nansen Member

The advance of a transgressive low-energy shoreline terminated the Statfjord Formation with the deposition of the Nansen Member. In its lower part, the Nansen Member comprises a well-sorted, laterally extensive, medium-grained sandstone. The upper part is more heterogeneous, with a mixture of different shoreline and embayment sediments of more restricted areal distribution. Reservoir qualities of the Nansen Member are good to moderate, with a net/gross ratio of 0.6 in the eastern area. This 15 to 25 m thick sand is likely to be laterally extensive, but may be wholly or partly broken by shale developments.

Dunlin Group

The Dunlin Group is subdivided into the Amundsen, Burton, Cook, and Drake formations (Figure 13). The marine transgression that terminated the deposition of the Statfjord Formation progressed southward, and the 170 to 180 m thick claystones of the Amundsen and Burton formations were deposited.

Cook Formation

An intra-Dunlin regression, together with increased tectonic activity, caused silt and sand to be transported farther into the East Shetland basin, depositing the Cook Formation. On the Gullfaks field, the Cook Formation is subdivided into the Cook-1, Cook-2, and Cook-3 units (Figure 13). The lowermost unit, Cook-1, consists of open marine silty claystone that has no potential as a reservoir rock. The bioturbated muddy sandstones of Cook-2 were deposited on an inner marine shelf and generally have a thickness of 50 m over the field. The uppermost unit, Cook-3, consists of interbedded sand and shale. Its thickness is 40 m in the southern area, but decreases to 15 m in

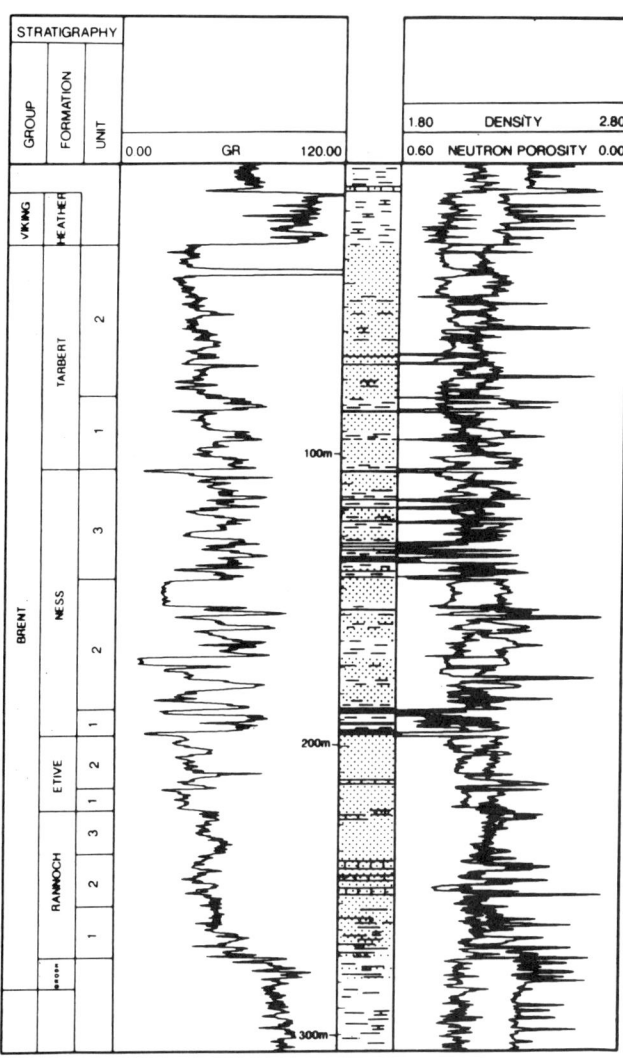

FIGURE 14. Typical log for the Brent Group.

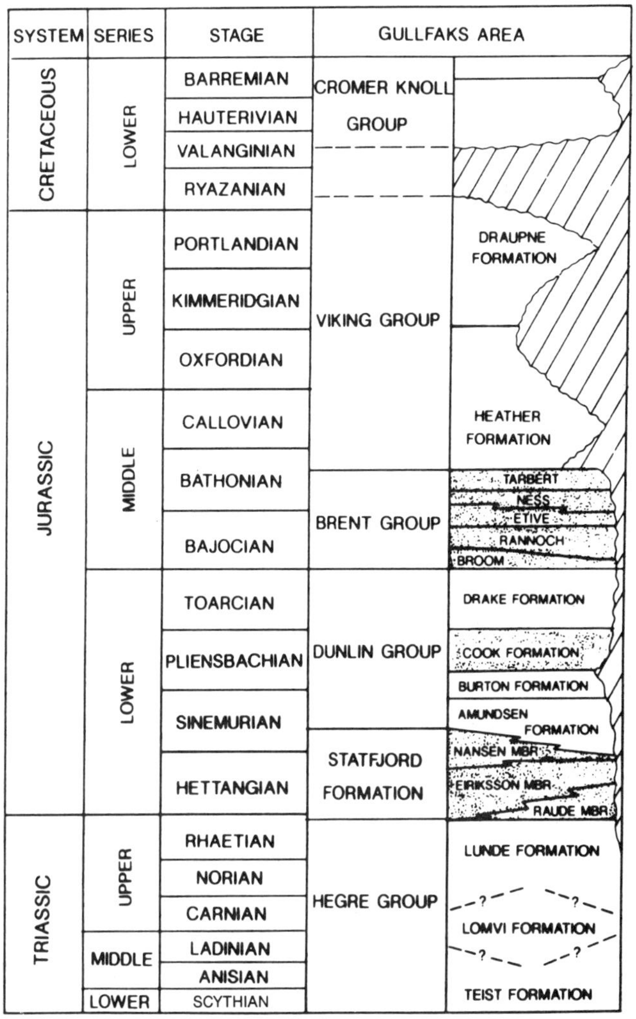

FIGURE 15. Lithostratigraphic nomenclature for Gullfaks field.

the northeast region of the field. The Cook-3 unit shows cleaner sands upward, forming an excellent reservoir at the top. It was deposited in a restricted marine setting, probably under anoxic bottom conditions. This is indicated by the absence of trace fossils, calcareous microfossils, and a high hydrogen index in the shales.

Tightly calcite-cemented sandstones, which generally are 0.5 to 2 m thick, occur rather frequently throughout the Cook Formation. They are relatively nonporous (2 to 8% porosity) and have permeabilities of less than 1 md. Both modular and layered zones are observed. Log examinations suggest that the calcite-cemented zones, particularly in the Cook-2 unit, occasionally form continuous, essentially stratigraphic layers that can be correlated between wells over distances of more than 4 km. These layers may constitute laterally continuous permeability barriers that will be important in the development of the Cook Formation.

The general transgression in this area continued after the deposition of the Cook Formation, and the Drake Formation was deposited as a marine shale with varying amounts of silt. The thickness in the Gullfaks area varies from 70 m in the south to 120 m in the north.

Brent Group

The Brent Group is subdivided into the Broom, Rannoch, Etive, Ness, and Tarbert formations (Figure 14). All the formations in the Brent Group have been classified as deposits associated with a river delta system.

The vertical succession of the Brent Group is divided into two separate delta systems. A wave/fluvial-dominated delta interpretation is preferred for the lower Brent delta, or "Bajocian" delta lobe, which comprises the Broom, Rannoch, Etive, and Ness formations. The upper Brent delta, or "Bathonian" delta lobe, is interpreted as a fluvial/tide-dominated delta, and comprises the Tarbert Formation.

An idealized core description of the Brent Group is shown in Figure 16, indicating that, in the Gullfaks field, all the "classical" Brent Group formations are present.

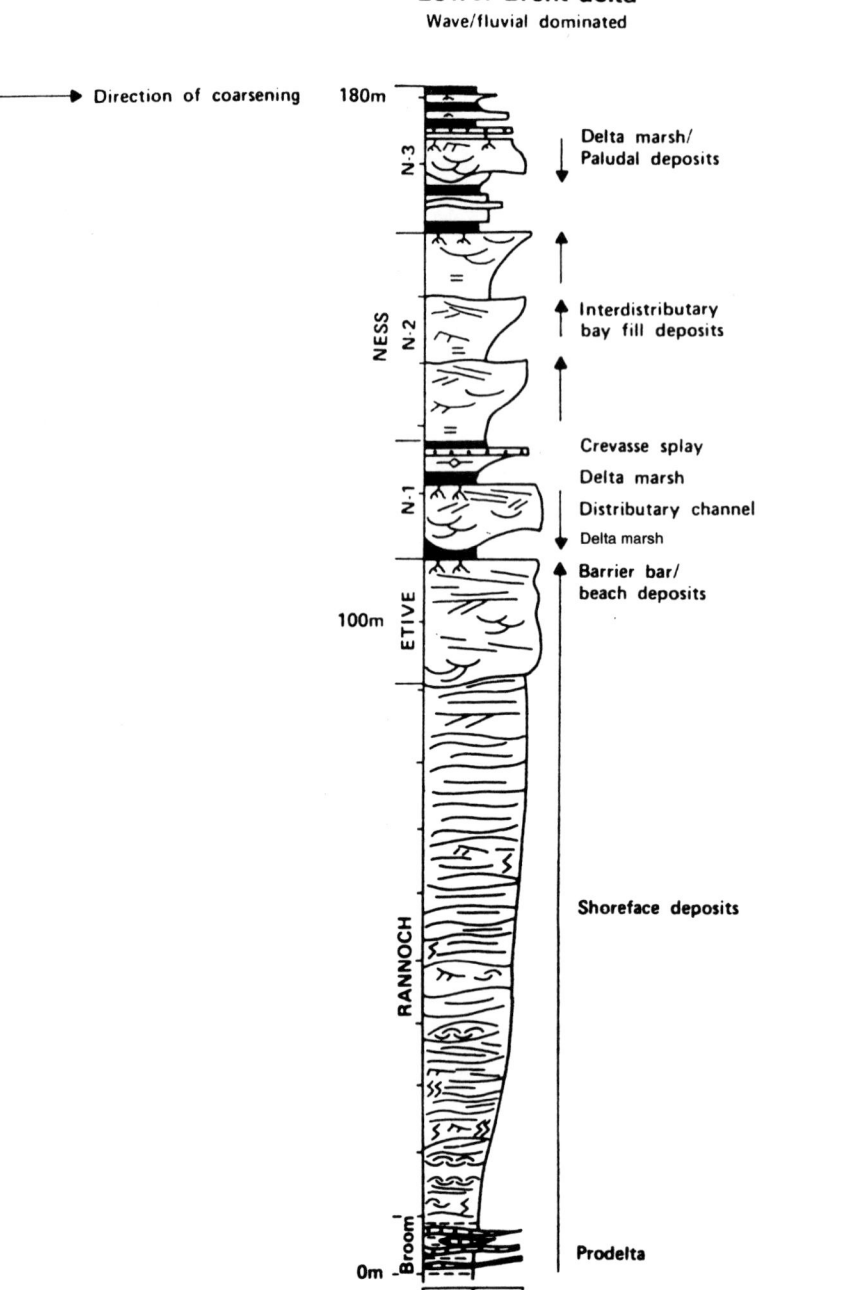
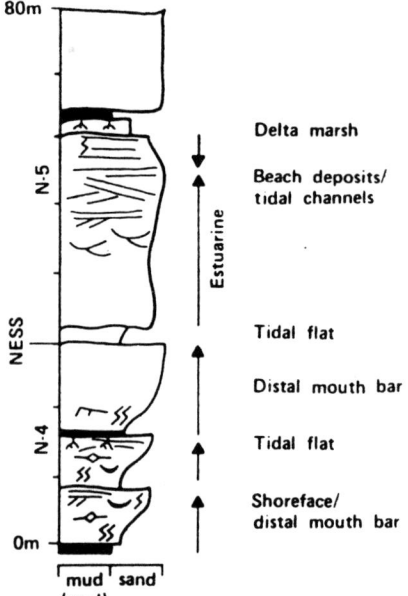

FIGURE 16. Core description of the Brent Group.

Broom Formation

The Broom Formation is a mudstone/shale formation 5 to 15 m thick, with thin layers of coarse sandstone and gravel beds, and is interpreted as prodelta deposits in an open marine environment. The thin sandstone beds usually are carbonate cemented and in restricted communication with the overlying sandstones of the Rannoch Formation.

Rannoch Formation

The Rannoch Formation, generally regarded as delta front shoreface micaceous sandstone, is subdivided into the three reservoir units Rannoch-1, Rannoch-2, and Rannoch-3. The Rannoch Formation comprises an over-all upward-coarsening sequence in a sheet-formed sandstone body that thins to the south, with a total thickness of 50 to 90 m. Abundant hummocky cross-stratification shows that wave processes were important during the deposition of this formation.

The lowest unit, Rannoch-1, is characterized by a decreasing mud content in a well-defined upward-coarsening trend. Bioturbation, small-scale ripples, micro-hummocky cross-bedding, and lenticular bedding are characteristic. Minor carbonate-cemented beds occur, and these beds may be laterally extensive. Units Rannoch-2 and Rannoch-3 have good to excellent reservoir properties in the fine- to medium-grained sandstones. Hummocky cross-stratification, minor burrows, and small-scale ripples are observed in cores. Carbonate-cemented beds are rather common but are not correlatable between wells, and therefore are probably laterally restricted. In the topmost few meters of the Rannoch Formation, the mica content may be locally high, particularly in the southern part of the field. This considerably reduces the vertical permeability in this part of the reservoir.

Etive Formation

Barrier bar complexes form the depositional environment of the Etive Formation. On the basis of low-angle large-scale cross-stratification, grain size, heavy mineral concentrations, and parallel laminations, a high-energy beach environment is suggested. The Etive Formation consists of medium- to coarse-grained sandstones with a thickness of 15 to 40 m. The lower part commonly reflects upward-fining sequences that may represent channel fill deposits. These sequences are interpreted as tidal inlets in the barrier bar systems. The reservoir quality is excellent owing to the clean, laterally persistent sandstones with minor mica and clay contents and little calcite cementation. High porosities (30 to 36%) and high permeabilities (1.0 to 4.5 darcys) reflect the superb reservoir qualities.

Ness Formation

The first occurrence of a coal bed above the clean sands of the Etive Formation is considered to be the base of the Ness Formation. On the Gullfaks field, Ness is subdivided into three units, Ness-1, Ness-2, and Ness-3, and has a total thickness of 85 to 110 m. A further subdivision into 11 subunits has also been carried out in a detailed correlation of this formation. A delta plain environment is suggested for the three units, predominantly in a lower delta plain facies.

The lowest unit, Ness-1, consists of interbedded coals, mudstones, and sandstones, occasionally with thick sandstone channel deposits. These channel deposits are interpreted as low-sinuosity distributary channels that are expected to be in local communication with the underlying Etive Formation. Unit Ness-2 is characterized by several upward-coarsening sandstone sequences having good reservoir quality. The depositional style suggests crevasse splay, crevasse channel, overbank flooding, and minor mouth bar sequences, representing interdistributary bay fill deposits.

The Ness-3 unit marks the end of the lower Brent delta progradation in the Gullfaks area. Typical of this unit are the common occurrences of soil profiles with siderite. The Ness-3 unit is dominated by siltstone/claystone and coal deposits, but fluvial channel sands and lacustrine deposits have been recorded also. Where uneroded, Ness-3 contains laterally persistent permeability barriers across the Gullfaks field. Only very locally can the fluviatile channel sands be in communication with the underlying Ness-2 sands.

Tarbert Formation

The boundary between the Ness and Tarbert formations represents a marine transgressive event that separates these two different phases of delta building. The Tarbert Formation is subdivided into two units and for reservoir purposes has been further subdivided into seven subunits. Owing to the Kimmerian erosion, core and log data are limited and are restricted to the western part of the field.

Sedimentary structures associated with tidal flat and shoreface deposits are found in the lower part of the Tarbert Formation. Lithology varies from shales, siltstones, and coal beds to medium- to coarse-grained sands in which calcite cementation is quite common. This lower part represents a permeability barrier that is likely to cover most of the field. Estuarine and bar complex deposits are suggested as the depositional environments of the upper part of the Tarbert Formation. The lithology of the upper unit is also variable, but it is generally more homogeneous than that of the lower unit. Heavy minerals and carbonate cementation occur. Thick, clean sands with excellent reservoir properties are found locally.

Where it is uneroded by the Kimmerian unconformity, the Tarbert Formation is underlain by the shales of the Heather Formation.

RESERVOIR GEOLOGY

The complexity of the Gullfaks field has been confirmed by drilling of development wells. These wells have contributed important knowledge about the reservoir geology, such as:

- The Gullfaks field is extremely faulted. Many of the faults are smaller than the resolution of the seismic data.

- The existence of several east–west-oriented faults has been established.

- The majority of the faults that have been mapped with sand-to-sand contact are nonsealing, and thus provide limited restriction to fluid flow between fault blocks.

- The existence of sealing faults in which sand-to-sand contact is mapped has been verified.

- Through a combination of remapping of fault locations and the finding that the Brent reservoir is deeper than previous interpretation had indicated, it has now been proved that the Brent reserves in the western area of the field are located mostly in the

Ness and Tarbert formations. Formerly, the Brent reserves were mapped in the Etive and Rannoch formations.

♦ Oil-bearing Cook Formation sandstone has been established in two fault blocks.

♦ In the western fault blocks, the Ness Formation is uneroded or only slightly eroded. Well control has shown that the Ness Formation can be correlated over wide areas. For use in reservoir models, a detailed correlation containing 12 sand beds has been carried out.

HYDROCARBONS

Source

The Upper Jurassic Draupne Formation (Figure 15) is the most important source rock for generation of hydrocarbons in the East Shetland Basin. Source rock analysis from wells in the northern North Sea indicates a total organic content on the order of 5 to 10%.

The Draupne Formation is often eroded on the structural heights, as it is at the Gullfaks structure, but reaches a thickness on the order of 200 to 400 m in the Viking graben. Supplementary source rocks of less importance are assumed to be the shales of the Heather Formation and the marine shales of the Drake Formation. The shales and coals within the Ness Formation of the Brent Group are also potential source rocks, but their volumes are restricted.

Three important "kitchens" for maturation of hydrocarbons are situated around the Tampen spur area (Figure 17): the "Oseberg kitchen" to the south, the "Troll kitchen" to the east, and the "Møre kitchen" to the north of Snorre field.

Migration

Hydrocarbon generation and migration most likely started during the Paleocene to Eocene. At that time, the Gullfaks structure was sealed off by the Upper Cretaceous shales and marls, which from other fields in the area are known to have formed efficient seals (Skarpnes et al., 1982).

Direct migration of hydrocarbons from the thick source rocks in the Viking graben is assumed to have made only a partial contribution to the oil filling of the structure. The reason for this is the major boundary fault that is believed to have restricted potential migration from the "Troll kitchen." However, the deepest oil is found within the Cook, Statfjord, and Hegre formations in the eastern area of the Gullfaks field, in wells 34/10-7, -9, -11, and -13. Together with the fact that two different types of oil are observed, this indicates that migration from the northeast (block 34/8), east, and south (Gullfaks Sør) might all be regarded as potentially important alternatives to the more general view of a "filling by spill

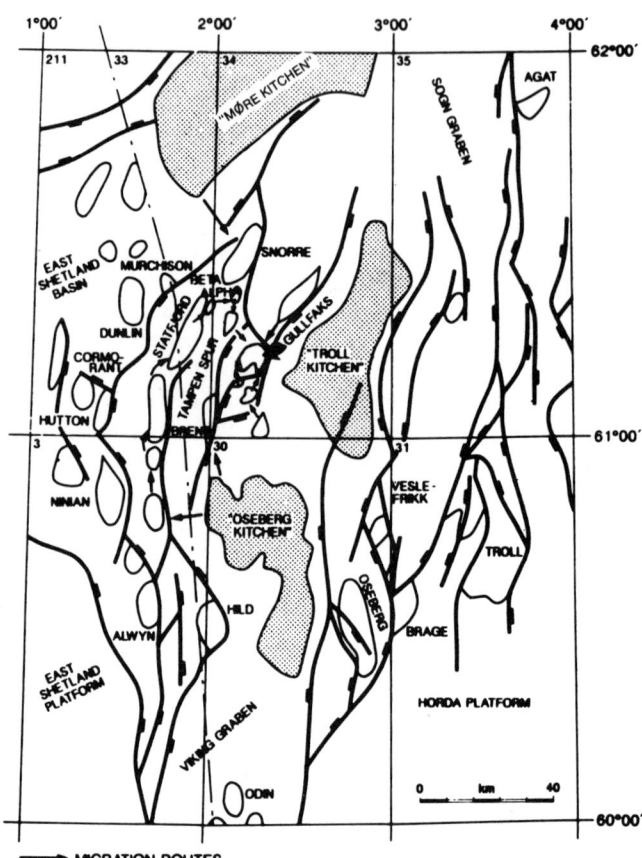

FIGURE 17. Migration map of the Gullfaks area.

only" migration from the filled neighboring structures to the northwest.

ORIGINAL RESERVOIR DEVELOPMENT PLAN

The commerciality report dated October 1980 proposed that the western area of the field be developed first, from Gullfaks A and B, and that the eastern part be developed later, from Gullfaks C (Figure 18). The proposed program for Gullfaks reservoir development described below is taken from the development plans for phase I, dated January 1986, and phase II, dated December 1984. An updated fieldwide development plan was completed in November 1988.

Early studies showed pressure support by waterflooding to be the most suitable recovery mechanism for the types of hydrocarbons and reservoir conditions present in all the Gullfaks reservoirs. Of the miscible flooding options investigated, only the injection of rich gas into the Cook Formation showed potential, but it was considered less favorable than waterflooding.

In the field development plan, the following formations were considered as separate reservoirs to be developed individually until field results proved communication with another reservoir:

♦ Upper Brent (Ness and Tarbert formations)

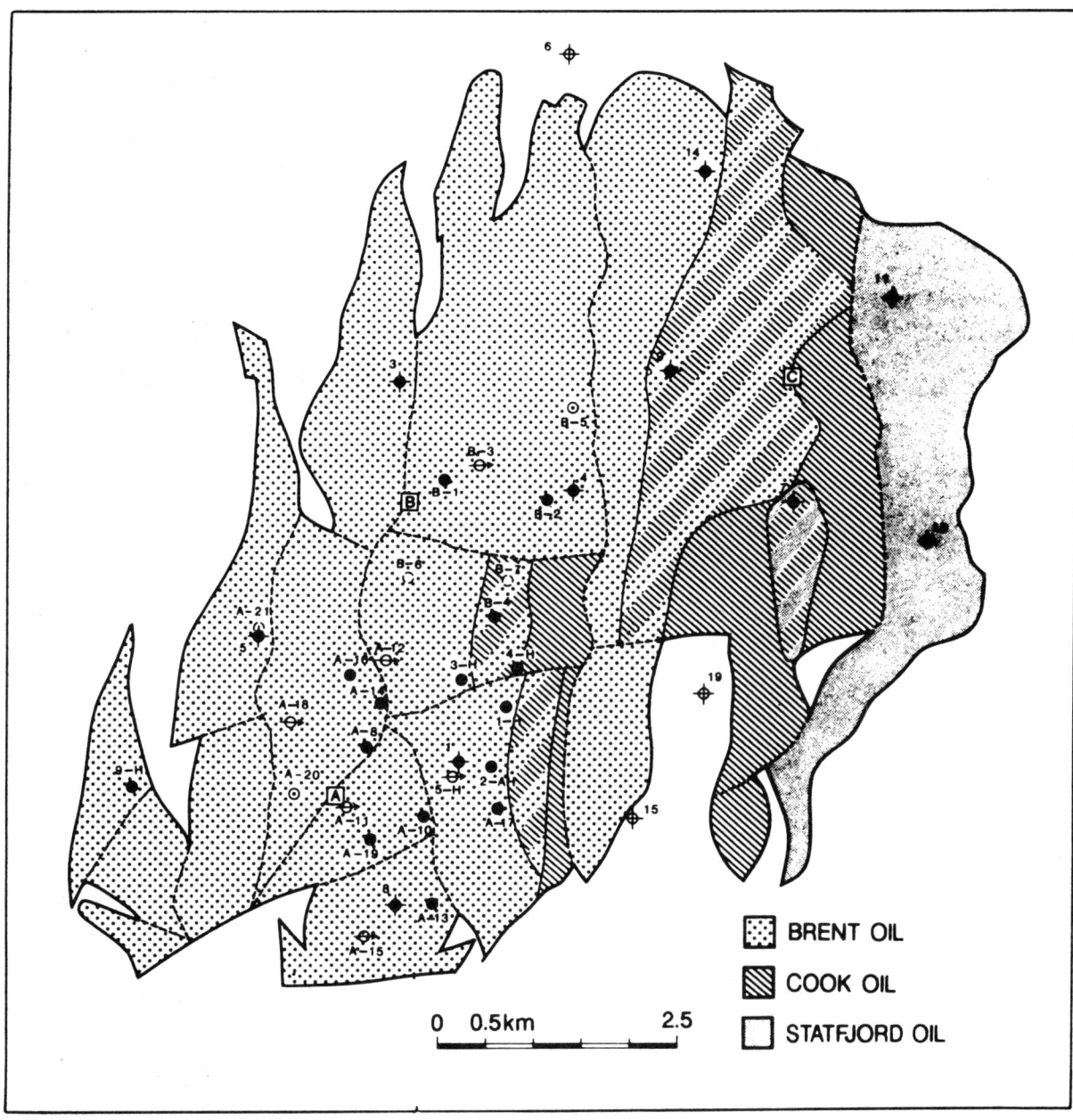

FIGURE 18. Locations of Brent, Cook, and Statfjord oils in the Gullfaks field.

- Lower Brent (Rannoch and Etive formations)
- Cook Formation
- Statfjord Formation.

Other important features of the early development plan were as follows:

1. Reservoir pressure would be maintained above the bubble point in each of the reservoirs. Each reservoir would be developed by the number of dedicated wells necessary to ensure maximum recovery within the economic constraints of the Gullfaks project. For wells that would penetrate two reservoirs, recompletion into the second reservoir would be carried out when considered favorable. Upper Brent sands would be developed after the lower Brent wells had been drilled in each region. By use of this development scenario, areal continuity and maximum information pertaining to the upper Brent sands would be obtained prior to the development of these sands.

2. Until communication had been observed in the field, each fault block would be developed as an isolated unit with the necessary numbers of production and water wells.

3. Lower Brent would be developed initially by production and injection wells perforated in the Rannoch sands only, in order to reduce water override in the more permeable Etive sands. Dedicated production and injection wells were to be completed in the Tarbert sands or the Ness sands in the upper Brent. Where possible, injection wells would be located below the OWC, but it was estimated that 60% of the Ness and Tarbert injection wells would have to be perforated above the OWC (Figure 19).

4. Selective perforation would be the preferred completion method. In sands with insufficient strength, gravel packing would be considered.

In the planning phase, it was realized that the development of Gullfaks would not be straightforward. The extent of the faulting already mapped for the field, together with the knowledge that fields in the U.K. sector were proving to be more complex than expected, reinforced the belief that the reservoir complexity should not be underestimated. In the field development studies carried out between 1980 and 1985, it became apparent that certain reservoir properties influencing production performance could be determined only through production experience. Such properties included:

- Fault transmissibility
- Vertical transmissibility within each reservoir
- Sand strength
- Effect and flow pattern of injected water.

The development plan was deliberately based on a pessimistic interpretation of these properties. Accordingly, production forecasts were conservative, taking into account the possibility of unpleasant surprises in the development phase.

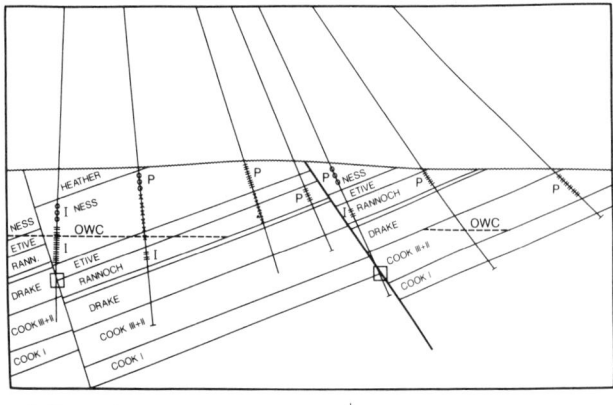

FIGURE 19. Locations of production (P) and injection (I) wells in the Gullfaks field.

RESERVOIR MANAGEMENT

Correct formulation of a reservoir management program was essential to the achievement of the objectives of the field development plan and the acquisition of strategic data as soon as the field came on production. Predevelopment planning studies highlighted activities that, if implemented at an early stage, would provide maximum information on communication routes within the field.

EARLY DRILLING SCHEDULE

The decision was made to accelerate production through early start-up of the subsea wells in the Gullfaks A area. In addition to improving the economic outlook by reducing the time to plateau specified in the field development plan, this decision also provided an earlier opportunity to reduce the degree of uncertainty in the field data.

Priority was given to early drilling of high-rate production wells in locations with minimum uncertainty. These wells were drilled systematically into previously undeveloped fault blocks so that the effects of production from adjacent fault blocks would be reflected in the measured pressures. Water injection wells were drilled when the bottom hole flow pressures in the production wells were at or just above the bubble point. The exact timing of pressure support was dependent on the depletion trends and production drawdowns, which varied from one fault block to another.

Radioactive tracers were injected into selected lower Brent water injection wells early in their injection lives. These wells were situated such that injected water could move into three different fault blocks and upper Brent sands. Analysis of produced water in neighboring wells in these adjacent fault blocks was used to identify water movement across faults.

VERTICAL COMMUNICATION

RFT pressures have been used extensively to establish vertical communication within each reservoir and fault block. Through a carefully planned drilling and coring program, good areal and vertical core coverage have been achieved. Integration of RFT pressures and core inspection and log evaluation results have been used to map the areal and vertical continuities of sands and barriers.

WATER INJECTION PROGRAM

The effects of the planned pressure maintenance program have been evaluated by monitoring reservoir pressure in all fault blocks simultaneously. In addition, introduction of radioactive tracers into early water injection wells has been important in gaining an understanding of the communication routes between the fault blocks early in the life of the field. The tracer content of produced water is employed as an input parameter for reservoir simulation models used to predict the water flow pattern. In wells where the sands to be investigated have not been perforated, the gamma spectroscopy tool (GST) is used to detect water movement behind casing in the well.

SAND STRENGTH

The knowledge of sand strength of each sand unit has been increased prior to production from each sand through the combined use of a numerical sand production prediction simulator, the multimineral log evaluation model, and core inspection. Through selective perforation, a balance between elimination of weak sands and achievement of high well productivity has been maintained. The success of this approach is monitored by use of sonic sand detectors on each platform well flow line together with downhole detectors run as parts of logging tools.

PRODUCTION

The actual production history of the Gullfaks field and the production forecast prior to start-up are shown in Figure 20. It can be observed that production was initially, and has remained, in excess of that expected. There have been setbacks caused by sand production, but oil production still has remained higher than originally forecast.

The Brent reservoir consists of reservoir rocks with superb quality (porosities in excess of 30% and permeabilities in the darcy range). These reservoir qualities, in addition to the fact that the reservoirs are relatively shallow, are factors that contribute to the poor consolidation of the sandstone reservoirs.

In the lower Brent reservoir (Rannoch Formation), sand production, generally associated with water breakthrough, results in a decrease in production rates.

Sand production occurs in wells producing from the upper Brent Formation but cannot be related to water production. To overcome the setbacks caused by sand production, several wells have been gravel packed (pr. 1 July 1990—8 wells). In addition to gravel packing of new wells in which sand production problems are anticipated, existing wells have been worked over and gravel packed.

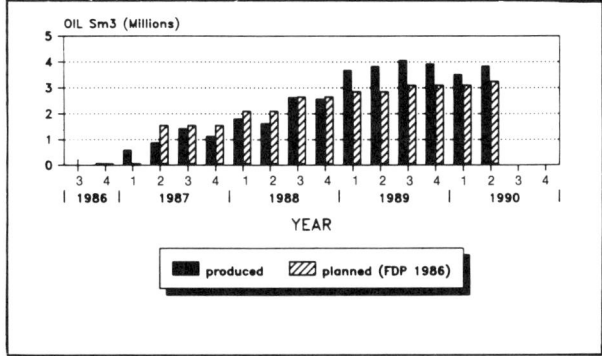

FIGURE 20. Production history vs. field development plan for Gullfaks field.

Drilling of the first wells from the Gullfaks C platform has indicated that the area around this platform is geologically more complex than had been anticipated. This is reflected by the fact that the first well drilled from the platform did not come in as expected, therefore resulting in a new interpretation of the area.

One of the most important factors in the reservoir development of the Gullfaks field has been the effect of fault transmissibility. A comprehensive data acquisition program, including permanent downhole pressure gauges, has allowed improved monitoring of the strategic injection and production plan. For characterization of fluid movement, radioactive tracers have been injected into selected water injection wells.

Long-term field development plans employing full-field simulation models have been supplemented successfully by smaller detailed fault block models. Well locations and drilling schedules are continuously updated, incorporating information from new wells and the latest production data.

As a supplement to water injection, Statoil is currently evaluating alternative methods of increasing oil recovery. A pilot test for surfactant injection will be carried out, and testing of alternating water/gas injection is also planned.

ACKNOWLEDGMENTS

The authors wish to express their thanks to Den Norske Stats Oljeselskap a.s, Norsk Hydro a.s, and Saga Petroleum a.s for their permission to publish this paper. The views and opinions expressed here are those of the authors, and are not necessarily shared by the Gullfaks field operator and partners. We also wish to express our thanks to everyone who has contributed in any way to the preparation of this paper, including colleague geologists and reservoir engineers, drafting personnel, and other technical staff.

REFERENCES CITED

Deegan, C. E., and B. J. Scull (compilers), 1977, A proposed standard lithostratigraphic nomenclature for the central and northern North Sea: Report of the Institute of Geological Science, n. 77/25; Norwegian Petroleum Directorate Bulletin n. 1.

Fossen, H., 1989, Indication of transpressional tectonics in the Gullfaks oilfield, northern North Sea: Marine and Petroleum Geology, n. 1, p. 22–30.

Karlsson, W., 1986, The Snorre, Statfjord and Gullfaks oilfields and the habitat of hydrocarbons on the Tampen spur, offshore Norway, in A. M. Spencer et al. (eds.), Habitat of hydrocarbons on the Norwegian continental shelf: Graham and Trotman, p. 181–197.

McClay, K. R., and P. G. Ellis, 1987, Geometries of extensional fault systems developed in model experiments: Geology, v. 15, p. 341–344.

Sæland, G. T., and G. S. Simpson, 1982, Interpretation of 3-D data in delineating a subconformity trap in block

34/10, Norwegian North Sea: AAPG Memoir 32, p. 217–235.

Skarpnes, O., E. Briseid, and D. I. Milton, 1982, The 34/10 delta prospect of the Norwegian North Sea: exploration study of an unconformity trap: AAPG Memoir 32, p. 207–216.

Thomas, B. M., P. Møller-Pedersen, M. F. Whitaker, and N. D. Shawn, 1984, Organic facies and hydrocarbon distribution in the Norwegian North Sea, *in* B. M. Thomas et al. (eds.), Petroleum geochemistry in exploration of the Norwegian shelf: Graham and Trotman, p. 3–26.

Vollset, J., and A. G. Doré, 1984, A revised Triassic Jurassic lithostratigraphy nomenclature of the Norwegian North Sea: Norwegian Petroleum Directorate, Bulletin n. 3, p. 1–53.

Chapter 28

Troll Field
Norway's Giant Offshore Gas Field

L. Bolle

A/S Norske Shell
Tanager, Norway

ABSTRACT

Troll field, a giant gas field underlain by an oil rim of variable thickness, is located approximately 80 km offshore Norway on the northwestern edge of the Horda platform and adjacent to the eastern margin of the Viking graben in waters between 300 and 355 m deep. The field extends over four Norwegian blocks (31/2, 31/3, 31/5, and 31/6) and covers an area of 710 km^2.

Hydrocarbons are trapped in three eastward-tilted fault blocks capped by clays of Upper Jurassic to Paleocene age. The reservoir comprises sediments of the Fensfjord, middle Heather, Sognefjord, and upper Heather formations of the Jurassic Viking Group. Deposition took place as a cyclic shallow marine sandstone sequence with alternations of transgressive sands and silts and progradational shoreface facies induced by regional sea level fluctuations.

The field is divided into three communicating hydrocarbon provinces that coincide with the major fault blocks. In Troll East, a gas cap of up to 250 m overlies a thin oil rim (0 to 4 m thick). In Troll West Gas, a maximum gas cap of 210 m covers an oil rim some 12 m thick. In Troll West Oil, a modest gas column of 43 m caps up to 28 m of oil. The reservoir rocks exhibit excellent properties, with porosities up to 35%, permeabilities in the darcy range, and no shale interbeddings.

Since the discovery of Troll in 1979, 25 exploration and appraisal wells have delineated initial in place hydrocarbon volumes of 1670 billion standard m^3 of free gas and 615 million m^3 of oil, as reported for planning purposes by the joint operators.

Phase I of the field development plan will constitute exploitation of the main gas accumulation in Troll East from an offshore concrete gravity structure together with onshore processing facilities. First gas is planned for 1996. Successive development of Troll West Oil could require horizontal wells feeding a floating or fixed production platform. The independent Troll-Oseberg Gas Injection (TOGI) project will export first gas in 1991 from a six-slot subsea template located in the south of Troll East for injection in the Oseberg field.

INTRODUCTION

The Troll field is a vast gas accumulation underlain by an oil rim of variable thickness. It is located offshore Norway in water depths of up to 355 m and about 80 km northwest of the city of Bergen (Figure 1). The field stretches over four Norwegian exploration blocks (31/2, 31/3, 31/5, and 31/6) held under two production licenses (PL 054 and PL 085), and encompasses an area of 710 km^2.

With its 1670 billion m^3 (59 tcf) of gas and 615 million m^3 (3.9 billion bbl) of oil initially in place, Troll is the second-largest gas field ever discovered offshore Europe, and it is to be classified as a "super giant gas field" (Fitzgerald, 1980). Norwegian mythology has inspired the field's name to reflect its "monstrous" dimensions, which will secure gas deliveries to the European continent well into the next century.

Phase I field development is currently under way, and first contractual gas deliveries are planned for October 1996. The Troll gas supply wells for the Troll-Oseberg Gas Injection (TOGI) project have been drilled and are expected to come onstream in early 1991. The phase II feasibility evaluation is ongoing: a horizontal well has been drilled in the thicker oil rim to assess its potential. Prolonged production testing is still ongoing, and preliminary results have been encouraging. A second horizontal well for evaluation of the feasibility of development of the thinner oil rim is planned for the end of 1990.

HISTORY

In 1972, a disconformable, nearly horizontal event (a "flatspot") was defined on seismic sections acquired over the unallocated offshore blocks 31/2, 3, 5, and 6, suggesting a possible hydrocarbon-water contact in a large areal closure, subsequently called the "Flathead structure" or "flatspot" (Figure 2). Furthermore, the subhorizontal reflector was clearly truncated by a dipping stratigraphic reflector and coincided well with the closure and spillpoint of the mapped prospect. Polarity reversals at the top reservoir reflector were observed along the edges of the assumed accumulation. Stacking velocities abruptly increased below the flatspot, implying an interval velocity increase of some 900 m/sec. This indicated the existence of gas or very light oil overlying water (Birtles, 1986). The assumed reservoir was interpreted as a Middle Jurassic Brent equivalent, although the considerable distance to the nearest well and the abundance of unconformities in pre-Cretaceous strata made predictions difficult.

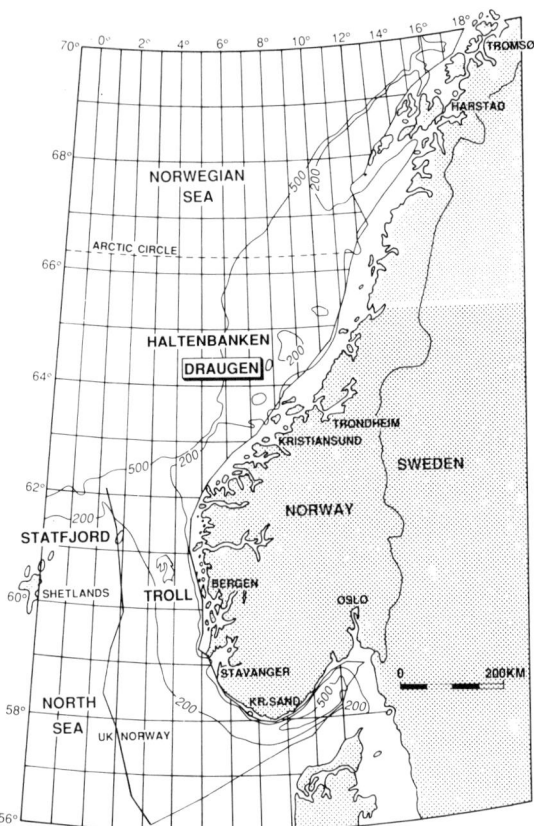

FIGURE 1. Location map for Troll field.

In April 1979, production license PL 054 for block 31/2 was granted to the following companies: Den Norske Stats Oljeselskap A/S, 50%; A/S Norske Shell (operator), 35%; Norsk Hydro Produksjon A/S, 5%; Norske Conoco A/S, 5%; and Superior Oil Norge A/S, 5%.

The discovery well 31/2-1 was drilled in the same year and confirmed that the flatspot coincided with a hydrocarbon-water contact. The well encountered 130 m of high-quality gas-bearing sands covering a 12 m thick oil column in Upper Jurassic Viking strata. The flatspot played an important role in the subsequent delineation of the field by acting as an invaluable calibration tool for depth conversions. An additional four exploration wells and nine appraisal wells were drilled on the structure in block 31/2 by the end of 1984. The "Statement of Commerciality" for Troll West was issued on November 15, 1983.

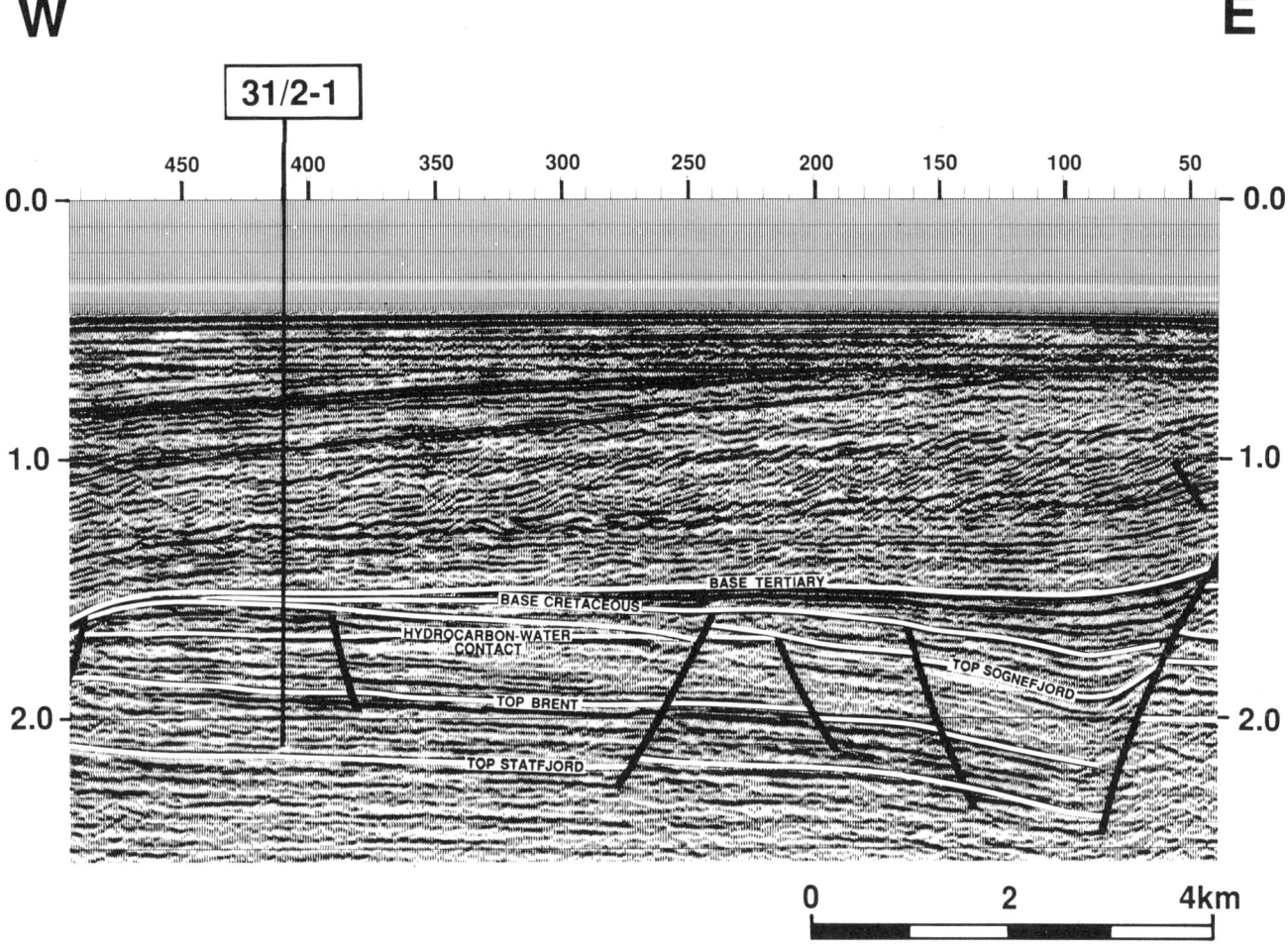

FIGURE 2. Seismic line illustrating the horizontal event (flatspot).

The adjacent blocks 31/3, 31/5, and 31/6 were awarded in July 1983 under production license PL 085 with a shared operatorship among all the licensees: Den Norske Stats Oljeselskap A/S, 85%; Norsk Hydro Produksjon A/S, 9%; and Saga Petroleum A/S, 6%.

The existence of an eastern gas structure (Troll East) was expected because of the areal extent of the flatspot. This was confirmed in late 1983 by well 31/6-1. A total of nine exploration and appraisal wells had been drilled on the structure in PL 085 by the end of 1985. Two of these wells appraised the southern flank of Troll West in block 31/5. Appraisal and delineation of the field were then complete.

In February 1984, the Norwegian authorities requested of the PL 054 and PL 085 licensees that development of the Troll accumulations be based on field unitization. In September 1986, these negotiations resulted in a full unitization agreement for the field, with an initial equity split of 32%/68% for PL 054 and PL 085, respectively. Together with some share reassignments, this resulted in Troll unit interests as follows: Den Norske Stats Oljeselskap A/S, 74.57600%; A/S Norske Shell, 8.28800%; Norsk Hydro Produksjon A/S, 7.68800%; Saga Petroleum A/S, 4.08000%; Elf Aquitaine Norge A/S, 2.35344%; Norske Conoco A/S, 2.01456%; and Total Marine Norsk A/S, 1.00000%. The Norwegian state's direct economic involvement is administered by Den Norske Stats Oljeselskap (Statoil) and amounts to 62.696%.

REGIONAL GEOLOGIC SETTING

Throughout the Permian–Triassic period, most of the North Sea was an area of continental deposition within a novel and rapidly subsiding, but later failed, intracratonic rift system of which the Central and Viking grabens and the Morray fault system are the remnants. Fluvial deposits accumulated near the rift margins, and finer alluvial and lacustrine sediments were laid down toward the basin centers.

Paralic and lacustrine conditions continued into the Lower Jurassic with deposition of the Statfjord Formation. A sea level rise established the marine environment in which the regionally extensive Dunlin shales developed. The Middle Jurassic updoming allowed the construction of deltaic systems of which one deposited the Brent sands sequence. Collapse of the updomed areas in the Callovian Stage resulted in drowning of the basin and gave rise to thick accumulations of deep marine Viking shales in the basin center. On the flanks and around intrabasinal highs, shallow marine conditions persisted or developed, and allowed deposition of a thick, sand-rich Viking sequence in the Troll region.

Contemporaneous with the opening of the Atlantic Ocean during the Kimmerian orogeny, North Sea rifting continued during the Upper Jurassic and into the Lower Cretaceous Aptian Stage. A series of megafault blocks was induced along the graben perimeters. Tilting and emergence (or semiemergence) caused periods of erosion and nondeposition on local highs. The Troll field is located near the northwestern edge of the Horda platform on the Sogn Spur high, close to the eastern margin of the Viking graben (Figure 3).

The Upper Jurassic Draupne Formation—a Kimmeridge clay equivalent—and the overlying Cretaceous marls and calcareous mudstones vary in thickness and distribution. This is caused by a series of unconformities that resulted from synsedimentary continuation of block faulting and tilting. Rifting activity ceased from the Aptian Stage onward, and was followed by a widespread subsidence.

Cretaceous or Jurassic sediments are overlain by a thick sequence of westward-dipping marine Tertiary and Quaternary strata that are unconformably covered by Pleistocene formations.

RESERVOIR STRATIGRAPHY AND DEPOSITIONAL MODEL

The oil and gas in the Troll structure are contained in medium- to coarse-grained, hardly consolidated sands and fine micaceous sandstones and siltstones of the Middle to Upper Jurassic Viking Group. In the central and northern North Sea, the Viking Group is characteristically developed as shales and claystones with some thin, areally restricted intercalations of sandstones. In the Troll area, the Viking Group contains a shaleless, stacked shallow marine sand sequence of the Krossfjord, Fensfjord, Sognefjord, and Heather formations (Figures 4, 5, and 6).

Deposition took place on a coast-attached shelf as a cyclic sequence with alternations of transgressive sands and silts and progradational shoreface facies proximal to coastal distributaries. The sequence architecture is controlled by minor regional sea level fluctuations that are framed in the major over-all late Callovian–early Volgian regional transgression. Tidal influence is indicated by the occurrence of "double drape" stratification. Coarse and poorly sorted storm deposits, together with strata that in parts are strongly bioturbated, underlie the shallow marine sands.

The *Fensfjord Formation* comprises a stacked series of small, coarsening-upward units, with a fine micaceous sand at the base grading upward to medium- to coarse-

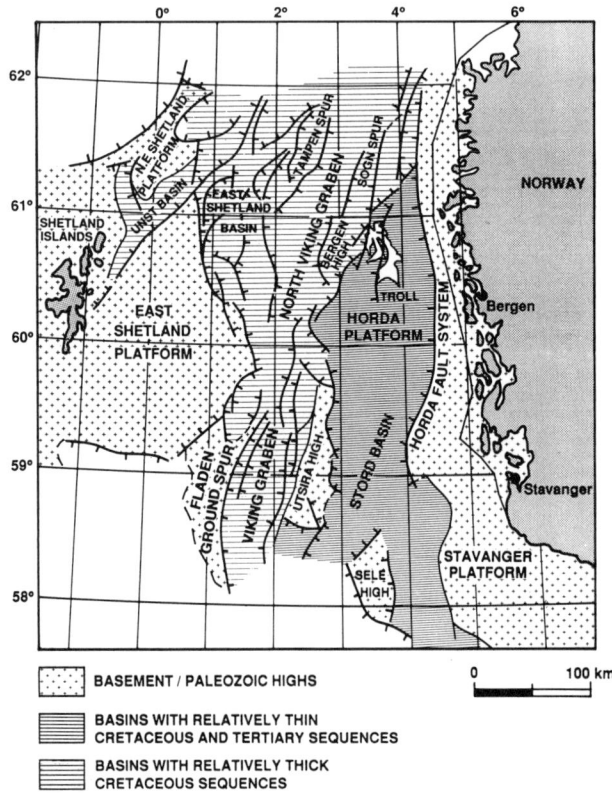

FIGURE 3. Map of regional tectonic elements.

CHRONOSTRATIGRAPHY		LITHOSTRATIGRAPHY	
PERIOD	STAGE	GROUP	FORMATION
EARLY TERTIARY	EOCENE	ROGALAND	BALDER
	PALEOCENE		SELE
			LISTA
			MAUREEN
LATE CRETACEOUS		SHETLAND	
EARLY CRETACEOUS	RYAZANIAN	CROMER KNOLL	
LATE JURASSIC	VOLGIAN KIMMERIDGIAN OXFORDIAN	VIKING	DRAUPNE U. HEATHER SOGNE FJORD MID HEATHER
MIDDLE JURASSIC	CALLOVIAN		FENSFJORD
	BATHONIAN		KROSS FJORD LOWER HEATHER
	BAJOCIAN AALENIAN	BRENT	TARBERT
			NESS
			ETIVE
EARLY JURASSIC	TOARCIAN	DUNLIN	DRAKE
	PLIENSBACHIAN		COOK
			U. AMUNDSEN
	SINEMURIAN		JOHANSEN
			L. AMUNDSEN
	HETTANGIAN		STATFJORD
TRIASSIC	RHAETIAN	HEGRE	

FIGURE 4. Stratigraphic table for the Troll area.

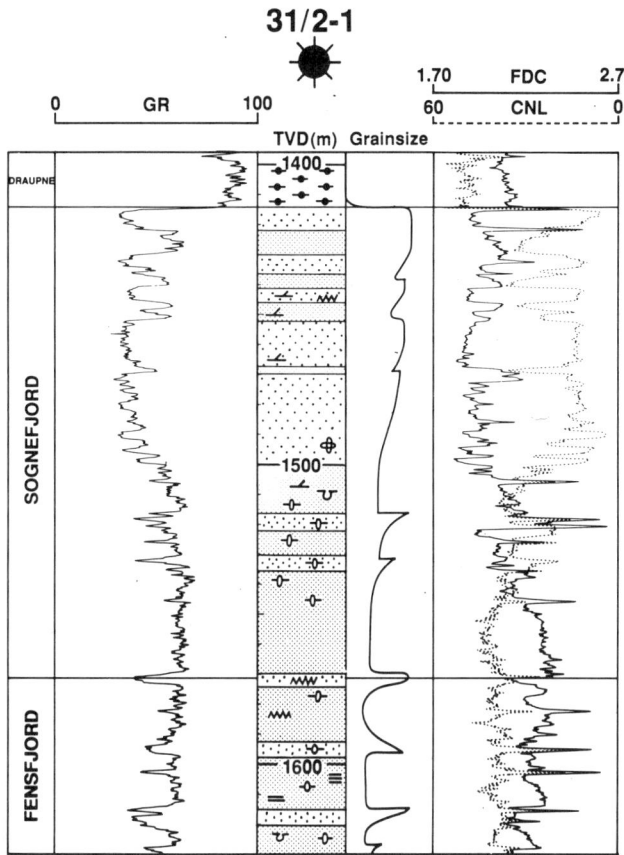

FIGURE 5. Type log for Troll West.

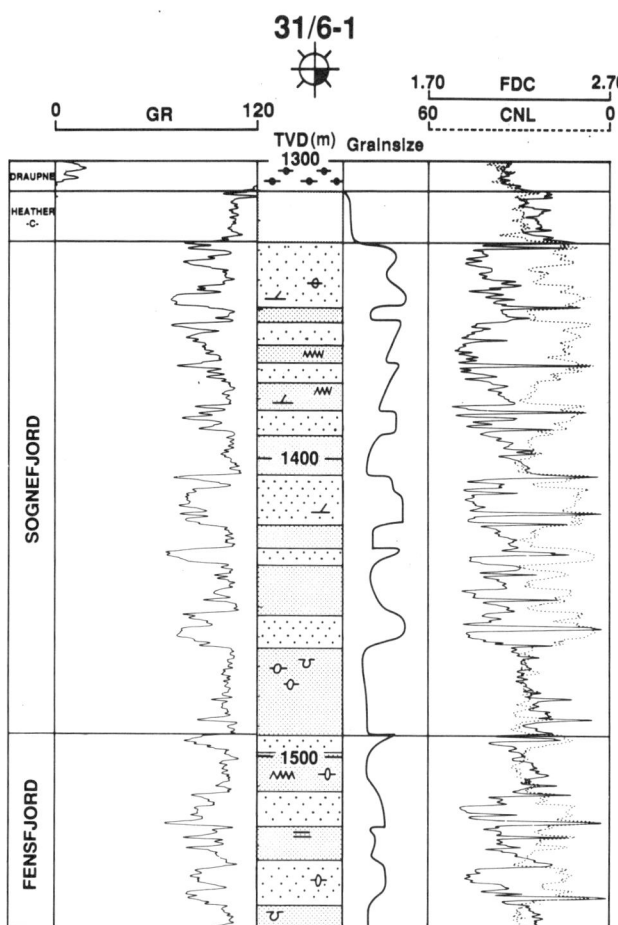

FIGURE 6. Type log for Troll East.

grained sands at the top. The sequence is interpreted as a progradational shoreface facies. It has a maximum gross thickness of 300 m and porosities from 25 to 30%. Only a small portion of the hydrocarbons is accommodated in the Fensfjord formation, but this formation contains the bulk of the underlying aquifer.

The *Heather-B and Sognefjord formations* comprise six depositional cycles, each marked by a rapid rise in sea level. Every complete cycle starts with a distal, low-energy, fine micaceous sand at the base and grades upward to a coarse, clean sand, representing the shoreface progradation, at the top. Subsequent transgressive phases often cover the sequence with coarse, reworked storm deposits from a more proximal setting. In the eastern block, cleaning is less effective owing to the lower energy level, and some finer sands containing mica represent the lateral equivalents of the clean sands in the western area. The reservoir rocks exhibit excellent properties in the clean sands, with porosities up to 35% and permeabilities in the darcy range. The lower-energy micaceous sands display porosities of about 26 to 32% but have dramatically lower permeabilities of less than 100 md. This package has a maximum thickness of about 220 m and contains the bulk of the hydrocarbons as well as an important segment of the flank aquifer.

Unit C of the *Heather Formation* is a particularly poor reservoir in comparison with the Sognefjord Formation. It is an open marine, low-energy siltstone facies that is rather well cemented. Hence, porosities are generally below 20%, and permeabilities are on the order of 10 md. This unit forms a rapidly westward pinching-out wedge with a maximum thickness of some 44 m.

Biostratigraphy based on palynology and palynofacies has been essential for the correlation of the sand cycles because of their inherent lateral variations and obvious similarities. Well-to-well continuity of individual sand units cannot readily be established. The Troll reservoir has been deposited over a period of 15 million years in a subsiding basin. The relative abundance of organic material has allowed the definition of six major palynozones within the Sognefjord Formation and one in the middle Heather Formation. The palynozones were found to coincide well with the fining-upward sedimentary cycles and aided the construction of seven lithostratigraphic reservoir zones. Recent research has revealed the link between the "regional" Haq eustatic curve and the Troll reservoir stratification (Figure 7). All allostratigraphic boundaries can be related to maximum flooding surfaces and sequence boundaries as defined from dynocysts and ammonites in the North Sea basin. Several distinct carbonate-cemented horizons occur within the reservoir, and many of them coincide with the above-mentioned sequence boundaries and maximum flooding surfaces. Diagenetic and mineralogical evidence suggests

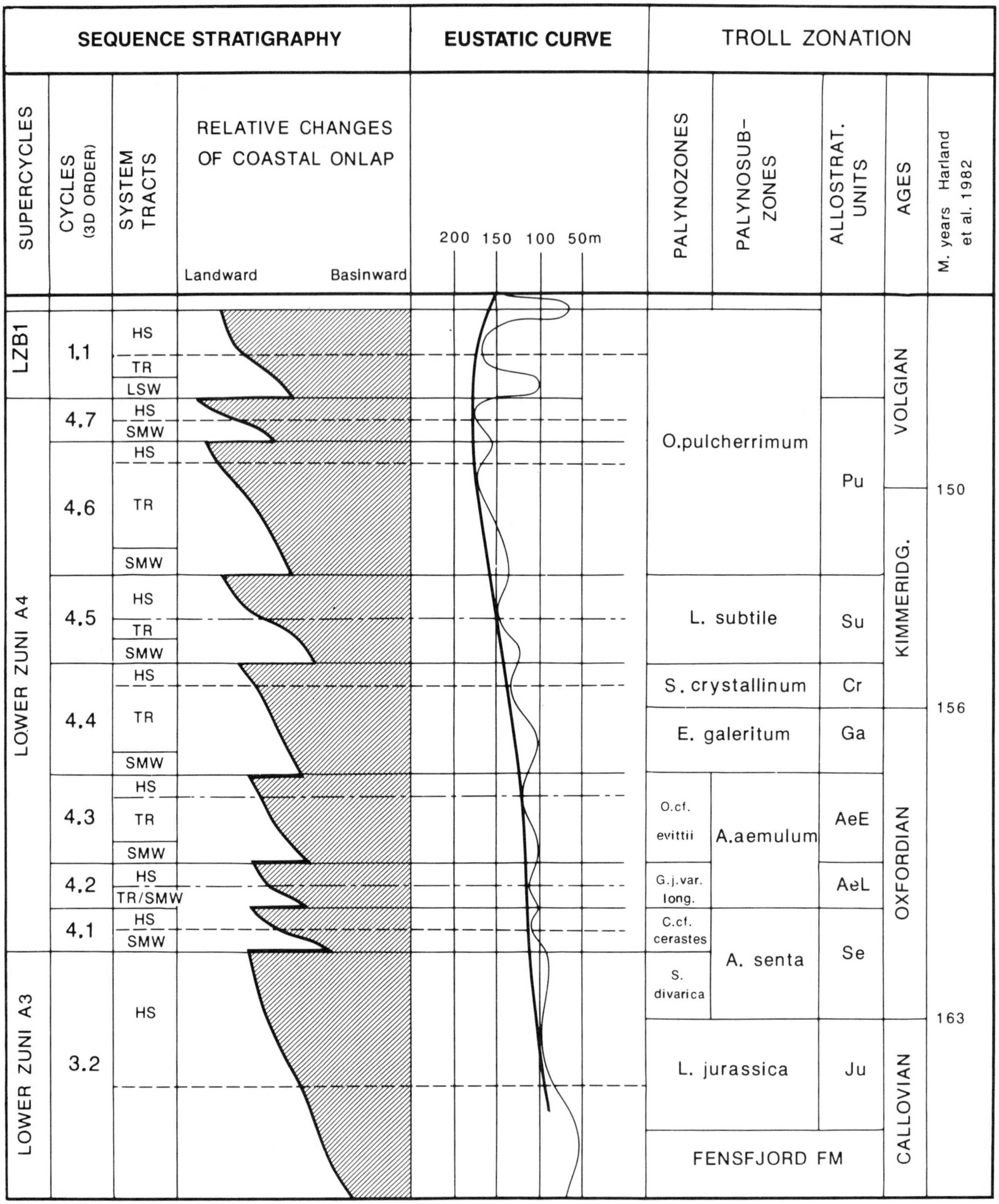

FIGURE 7. Correlation between eustatic curve and Troll deposits (from Haq et al., 1987).

a "hardground-like" near-surface generation for these cements, which hence may have an extensive distribution. Their undisturbed lateral continuity is uncertain and may range from a few centimeters to several kilometers. Thicknesses vary from a few centimeters to several meters. It is conceivable that these cements have a local influence on vertical communication, which is of particular importance for a possible oil development.

The Troll reservoir fabric consists mainly of quartz, feldspars, varying amounts of mica and lithic fragments, and minor amounts of accessory minerals. Textural and mineralogical evidence indicates that the sands are first-cycle sediments that were transported over relatively short distances. The source area is located to the northeast of the central fault block, as indicated by the "proximity to source" analyses of the palynofacies (Figure 8). It embodies high-grade metamorphics and granite intrusions.

STRUCTURE AND TRAP DEVELOPMENT

Hydrocarbons are trapped in three low-relief, eastward-tilted fault blocks that decrease in over-all crestal depth from east to west (Figures 9 and 10). The top of the eastern megafault block is just below 1300 m subsea, whereas the westernmost block has a crest some 200 m deeper. Coinciding with these major fault blocks and based on their respective hydrocarbon habitats, three communicating hydrocarbon provinces have been defined:

1. The Troll West Oil province (TWOP)
2. The Troll West Gas province (TWGP)
3. The Troll East (gas) province (TEP).

To the west, major normal boundary faults with throws of several hundred meters delineate the fault blocks and juxtapose the reservoir sands against clay formations. Dip closures are present to the north, east, and south, and also to the west in Troll West Oil, where the hydrocarbon contacts are found in an erosional, cliff-like feature. Owing to erosion and nondeposition, the cap rock is of varying age.

Troll East and the southern part of the Troll West Gas province are covered by an eastward-thickening local Kimmeridge shale equivalent—the Draupne shales. The Troll West Oil province is overlain by Cretaceous marls and shales. Claystones of Paleocene age cap the northern areas of Troll West. Pressure communication among the three hydrocarbon provinces through the aquifer is abundant, but hydrocarbon communication is limited to a few narrow channels.

The hydrocarbon trap developed in two stages. An early trap was formed in the west as a result of block faulting and tilting caused by ongoing rifting activity during the late Kimmerian Stage. Later on, faulting and block rotation extended eastward and gave rise to the formation of the central and eastern fault blocks. Intra-field faulting is most pronounced in the early developed Troll West Oil province fault block, where some faults have influenced the hydrocarbon distribution. Generally, the intrareservoir faults show small displacements—tens

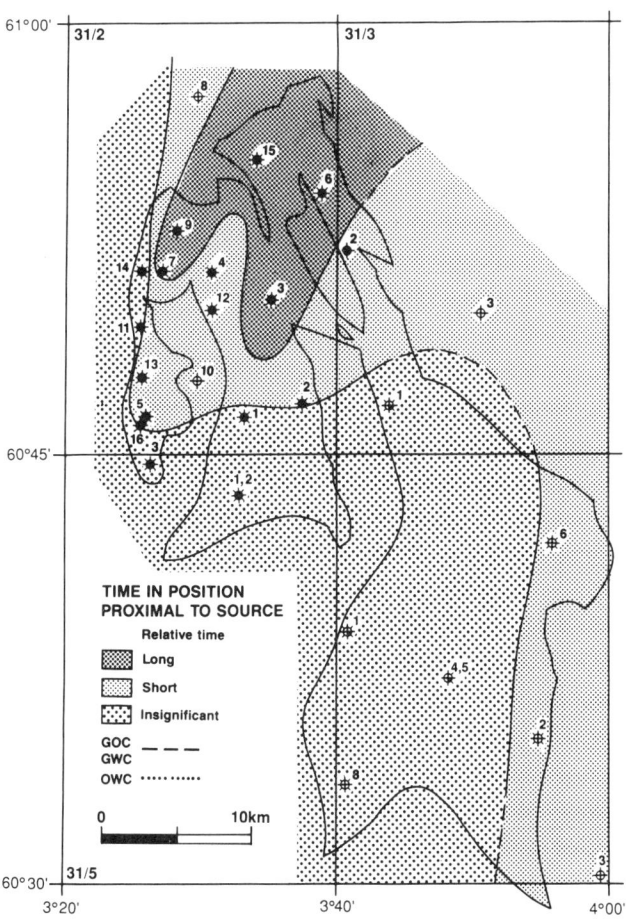

FIGURE 8. Map showing proximities to source for Troll field wells.

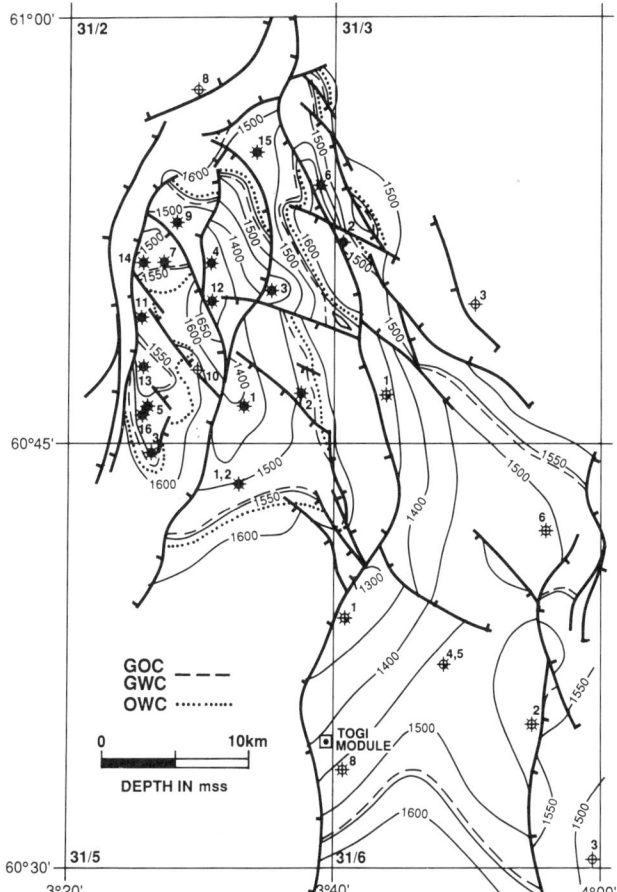

FIGURE 9. Map on top of the Troll reservoir.

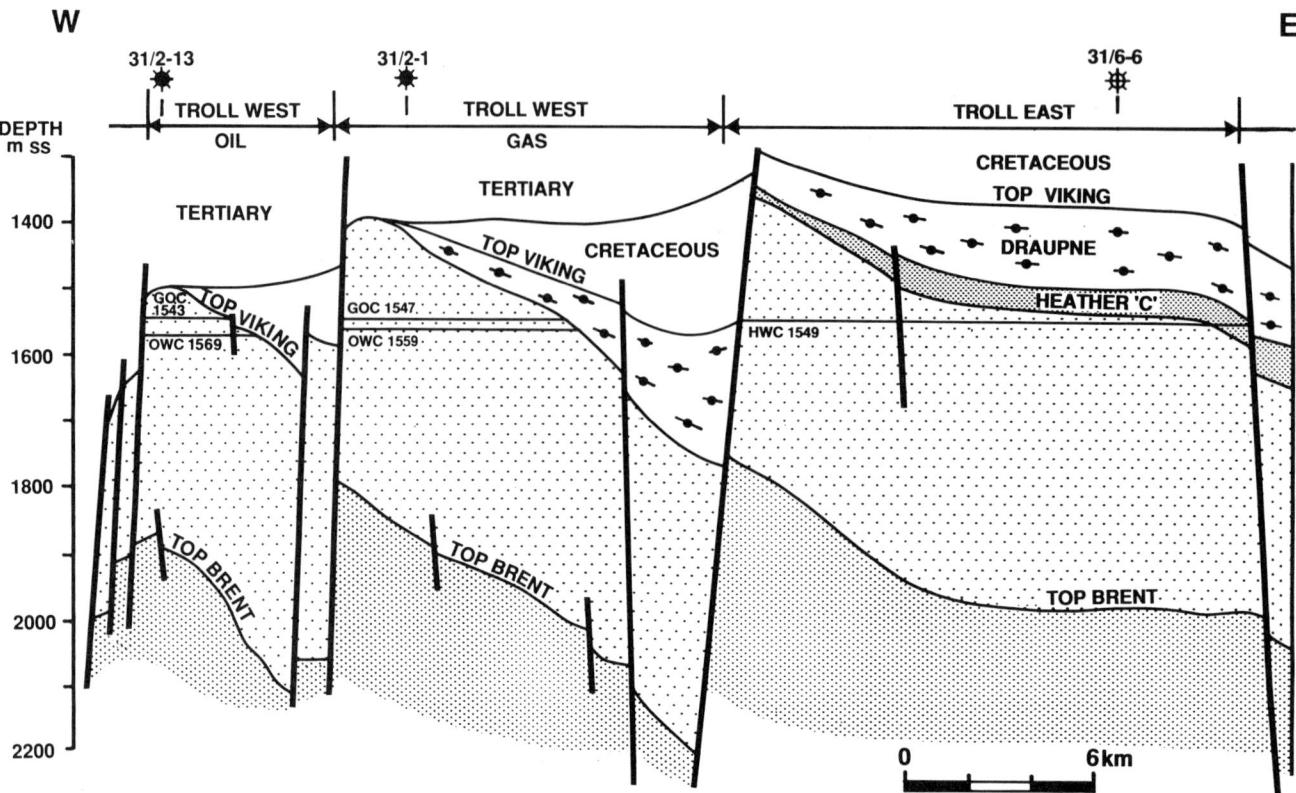

FIGURE 10. Structural cross section of Troll field.

of meters at the most—and are probably nonsealing. With no clay development in the Troll reservoir, there is no potential for clay smearing. Other diagenetic and tectonic effects—such as cataclasis, grain reorientation, and mineralizations—are possible, but have not yet been shown to effect communication in the Troll field.

Gas column thicknesses are structurally controlled. The thickest accumulations occur along the western flanks of the tilted fault blocks. Oil distribution is probably controlled by a combination of structure and stratigraphy. Juxtaposition of high- and low-permeability sands in the oil leg across faults may result in oil-water contact variations (Figure 11). The pore entry pressures of the low-permeability sands are considered sufficient to have restricted the free movement of oil, which would explain the differences among the three provinces. However, one cannot rule out the possibility that the free movement of oil has been restricted to a certain extent by a combination of sealing faults and uplift.

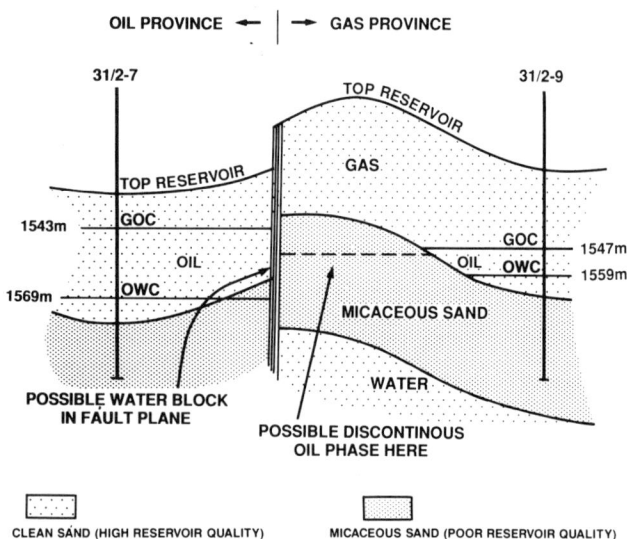

FIGURE 11. Possible explanation of permeability barriers for oil.

HYDROCARBON EMPLACEMENT AND DISTRIBUTION

The organic-rich shales of the Upper Jurassic Draupne Formation constitute the dominant source rock for the hydrocarbons in the Troll reservoir. Hydrocarbons of various maturities were generated, which indicates a kitchen of large areal extent. It is likely that this kitchen is located to the west (or northwest) of the field in the deeper parts of the Viking and Sogn graben systems. Migration pathways from these areas to the Troll field exist; they are uncomplicated but have been influenced by major faults. Migration into the present-day structure is possible only in the western and northwestern parts of Troll West, and is conceivably ongoing.

Source rock maturation modeling indicates that significant generation and migration of oil, together with minor gas expulsion, began during the Paleocene-Eocene. Gradual maturation of the source rock resulted in a period, during the Oligocene-Miocene, that consisted mainly of gas generation.

Some additional gas may have originated from other Jurassic source rocks, such as the Brent Group and Statfjord Formation coals. The Draupne Formation cap rock over the reservoir is immature and has not contributed to hydrocarbon generation.

The phased trap development and contemporaneous hydrocarbon generation allowed accumulation in the early western (and only) fault block prior to development of the central and eastern boundary faults. A nonhorizontal fossil oil-water contact in the Troll West Oil province and oil saturations in the gas cap are today's only remnants of this early stage. The position of the fossil oil-water contact suggests that the structure was not filled to the spillpoint at that time. The central and eastern fault blocks have since developed, and nowadays the field is filled to the spillpoint with both oil and gas. The respective fault blocks host hydrocarbons each of which has a different distribution (Figure 12).

In the Troll East province, a gas cap of up to 250 m overlies a thin oil rim from 0 to 4 m thick. The Troll West Gas province contains a maximum gas cap of 210 m, which covers an oil rim with an average thickness of 12 m and a maximum thickness of 16 m. In the Troll West Oil province, a modest gas column of some 43 m caps up to 28 m of oil in the northern compartments. The southern compartment contains an oil column 22 m thick.

The volumes and distributions of hydrocarbons initially in place (as reported for planning purposes by the joint operators) are presented in Table 1.

The Troll gas is a sweet, lean gas with a rather constant chemical composition over the field (Table 2). The gas contains about 93% methane and has a gravity of 0.61 relative to air. Condensates with gravities of 40 to 45° API are produced at a ratio of 29 m^3 per million m^3 of gas. The oil exhibits different properties throughout the field. Viscosities of some 1.4 cp are found in the southern oil province. The other oil rims usually contain more biodegraded oils, with viscosities up to 2.0 cp. Oil gravities vary from 24 to 29° API, indicating a process of biodegradation. Neither oil nor gas properties exhibit straightforward correlation patterns. The following multiphase emplacement history has been proposed to account for this. Early in the emplacement sequence, reservoir temperatures were still low, and biodegradation was able to act on all migrating hydrocarbons. Conditions gradually became harsher, and finally biodegradation nearly stopped and undegraded hydrocarbons were stored in the reservoir, mixing with the stock already present.

Over 90% of the hydrocarbons lie in the Sognefjord Formation. Excellent reservoir quality will allow about 75 to 80% of the gas to be recovered. Poorer recovery factors are expected for the rather tight Heather-C Formation silts. Oil recovery factors are heavily dependent on the thickness of the oil column. Two drive mechanisms are proposed:

1. Gas expansion provides direct energy for gas production and will translate into a gas cap drive for any possible oil production.

2. A large aquifer is in direct communication with all hydrocarbon accumulations and offers potential for water-driven production of both oil and gas.

DEVELOPMENT

Phase I

The Troll field will be developed in phases. Phase I comprises the development of the Troll East (gas) province. First gas deliveries are planned for October 1996. Norske Shell has been appointed operator for the planning and construction of the facilities. Den Norske Stats Oljeselskap A/S (Statoil) will take over operatorship for the drilling and production phases.

In 1986, a "plan for development and operation" was submitted to and approved by the Norwegian Ministry of Petroleum and Energy. This plan proposed construction of a fully integrated drilling, production, and living-quarters unit supported by a gravity-based concrete substructure. Gas was to be exported through the Zeepipe and Statpipe/Norpipe transport systems to the European continent, and it was assumed that condensate would be evacuated to Sture via the Oseberg field.

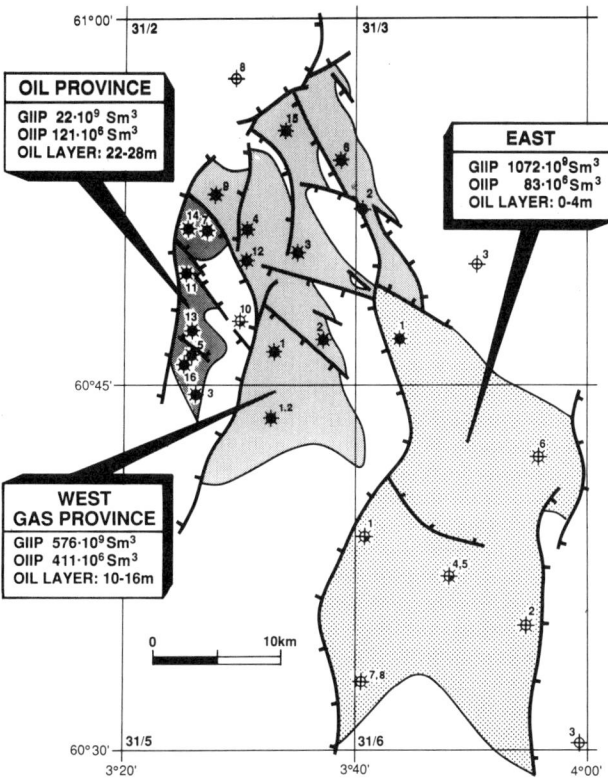

FIGURE 12. Hydrocarbon distributions for Troll field.

TABLE 1. Volumes and distributions of hydrocarbons initially in place in the Troll field.

Compartment	Oil-water contact (m subsea)	Gas-oil contact (m subsea)	Volumes Gas (billion m³)	Oil (million m³)
Troll East (gas) province	1549	1547	1072	83
Troll West Gas province	1559	1547	576	411
Troll West Oil province:				
North	1569	1543	13	81
South	1569	1547	9	40
Total	—	—	1670	615

TABLE 2. Reservoir data for Troll field.

Trap

Type	Structural, tilted fault blocks
Crestal depth (Troll East), approximate	1300 m subsea
Gas-oil contacts, average:	
Troll East	1547 m subsea
Troll West Gas	1547 m subsea
Troll West Oil South	1547 m subsea
Troll West Oil North	1543 m subsea
Oil-water contacts, average:	
Troll East	1549 m subsea
Troll West Gas	1559 m subsea
Troll West Oil South	1569 m subsea
Troll West Oil North	1569 m subsea
Gas column (Troll East), maximum	250 m
Oil columns, maximum:	
Troll East (Well 31/6-5)	4 m
Troll West Gas (well 31/2-9)	16.5 m
Troll West Oil (well 31/2-7)	28 m
Field area	710 km²
Reservoir pressure, at 1547 m subsea	158 bar

Hydrocarbon-bearing reservoir

Formations	Fensfjord, middle Heather, Sognefjord, upper Heather
Age	Callovian–Portlandian
Porosity range	17–36%
Permeability range	0–10+ darcys

Hydrocarbon properties

Gas gravity, relative to air	0.61
Gas quality	Sweet, dry, lean
Oil gravity	0.882–0.910 g/cm³ (29–24° API)
Oil quality	Slightly waxy
Gas/oil ratio	52–72 std. m³/std. m³
Condensate/gas ratio	29 std. m³/million std. m³
Condensate gravity	0.802–0.825 g/cm³ (45–40° API)

In early 1989, state-of-the-art technology for construction of multiphase pipelines justified reconsideration of the original project design. An alternative plan involving onshore location of the entire processing plant was developed. In this plan, two 67 km long, 36-in. carbon steel multiphase pipelines would be used to transport nearly untreated reservoir fluids onshore. Corrosion protection by free water knockout and glycol injection offshore, combined with an adequate passive corrosion margin, would offer an acceptable safety factor. Owing to the steep and irregular shore approach of the sea bottom, the pipelines would run through a tunnel starting some 130 m below sea level and traversing a distance of more than 3 km.

In May 1990, a revised development concept based on the onshore processing option was submitted to the Ministry of Petroleum and Energy. The offshore platform processing facilities and accommodations have been slimmed down. However, the complete structure will still be some 430 m high and will be one of the tallest constructions ever towed out to open sea (Figure 13).

The Troll phase I platform will stand in 303 m of water near the crest of the Troll East structure in block 31/6. The nearest exploration well is well 31/6-1, some 2.5 km due west. The position of the large basement-controlled boundary fault prohibits location of the platform closer to the structural crest. Studies have indicated a remote but existing potential for reactivation of the western boundary fault. The reservoir compaction and related sea bottom subsidence resulting from reservoir depletion are predicted to be on the order of 6 m. According to a finite element computer model, initiation of fault reactivation would require both a 13-m differential compaction across the boundary fault and a compressional stress regime of which the existence is not currently anticipated.

The northwestern edge of the Horda platform is known for its relatively frequent but low-amplitude basement-induced seismicity (Lindholm, 1987). It cannot be ruled out that future tectonic events may reactivate boundary faults, and it is therefore considered prudent to position the platform away from possible areas of deformation or fault subcropping. At the same time, drilling through large faults will be avoided.

Development wells will be targeted on two circles with radii of 500 m and 375 m, respectively. Gravel packing will be necessary because of the unconsolidated nature of the reservoir sands.

Troll-Oseberg Gas Injection (TOGI) Project

The Oseberg oil field, operated by Norsk Hydro, is located some 48 km west of the Troll field (see Chapter 26). Oil production will be supported by injection of gas produced from the Troll field but processed at Oseberg before compression. First gas deliveries are planned for early 1991.

In late 1989, a six-slot subsea well template—the TOGI Module—was installed on the southwestern flank of the Troll East (gas) province about 9 km southwest of the phase I platform location in 325 m of water. The module is designed to supply an annual offtake of 2.6 billion m^3 over a period of approximately 11 years. Gas will be exported through a 20-in. multiphase pipeline. All functions of the TOGI Module are operated by full remote control from the Oseberg platforms, and the entire construction is designed for diverless maintenance.

Future Oil Developments

In 1989, the Troll West Oil province was covered with a 3D seismic grid to enhance understanding of this structurally most complex part of Troll field. The technical feasibility of horizontal well technology applications was demonstrated at the end of 1989 by the drilling of well 31/2-16S in the southern compartment of the oil province close to exploration well 31/2-5 in a water depth of 322 m. The well was drilled with a 500-m horizontal production section targeted 4 m above the oil-water contact in the 22-m thick oil rim. Reservoir sand quality is excellent, with permeabilities ranging from 3 to 10 darcys. A 12-month production test using the Petrojarl production vessel is currently being undertaken to assess the viability of an oil development. Preliminary results have been very promising.

In light of the positive results from the horizontal well testing in the Troll West Oil province, the thin (12-m) oil rim in the Troll West Gas province represents another prospect. However, the sands in the oil leg of the gas province exhibit reservoir quality poorer than that of the sands in the oil province. This may be partly compensated for by increasing the length of the horizontal well section. Present plans call for test production from a horizontal well (31/5-4S) drilled near appraisal well 32/5-2. Drilling of an 800-m long horizontal section, with a vertical tolerance of some 2 m, has been completed.

Future Gas Developments

The Troll West Gas province will provide a second gas drainage point. Production timing will depend on contract gas sales and possible oil production scenarios.

In order fully to develop the Troll East (gas) province, an additional drainage point is foreseen in the northern

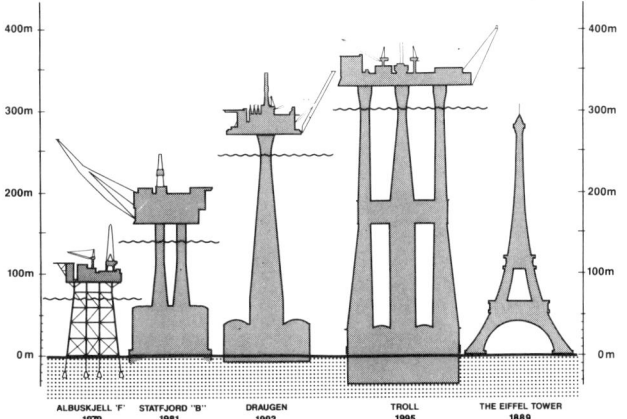

FIGURE 13. Evolution of platform size.

part of the province in block 31/3 close to well 31/3-1. Timing will be fully dependent on offtake rates and contract negotiations.

CONCLUSIONS

The Troll field is a major gas (and oil) field located in one of the most hostile environments currently being exploited. The development of this field in a safe, cost-effective, and environmentally friendly manner poses a substantial challenge to its operators. Front-end technology, such as multiphase pipelines, deep water platforms, and horizontal drilling, will play an important role in the required development scenarios. Nearly 25 years will have elapsed between the seismic discovery of the field in the early 1970s and first gas production in 1996, and yet the project has now matured and is well on its way to becoming Europe's largest single offshore gas supplier for the next half-century.

ACKNOWLEDGMENTS

This paper is a compilation of data selected from abundant reports and articles on the Troll field published over the past 15 years. Therefore, it does not necessarily represent the opinions of the operator or of any of the license partners. The author thanks all Troll licensees (Den Norske Stats Oljeselskap A/S, A/S Norske Shell, Norsk Hydro Produksjon A/S, Saga Petroleum A/S, Elf Aquitaine Norge A/S, Norske Conoco A/S, and Total Marine Norsk A/S) for permission to publish this paper.

REFERENCES

Anonymous, 1990, Multi phase flow concerns guide TOGI system design: Oil and Gas Journal, July 23, 1990, p. 38-42.

Birtles, R., 1986, The Seismic flatspot and the discovery and delineation of the Troll field, in A. M. Spencer et al. (eds.), Habitat of hydrocarbons on the Norwegian continental shelf: London, Graham and Trotman.

Fitzgerald, T. A., 1980, Giant field discoveries 1968-1978: an overview. AAPG Memoir 30.

Gibbons, K., et al., 1990, Sequence architecture, facies development and carbonate cemented horizons in the Troll field reservoir, offshore Norway, in Special publication volume, from the meeting Advances in Reservoir Geology, London, January 29-30, 1990.

Gray, D. I., 1987, Troll, in A. M. Spencer et al. (eds.), Geology of the Norwegian oil and gas fields: London, Graham and Trotman.

Haq, B. U., J. Hardenbol, and P. R. Vail, 1987, Chronology of fluctuating sealevels since the Triassic: Science, v. 235, p. 1156-1166.

Harland et al., 1982, Stratigraphy of the North Sea: Cambridge University Press.

Lien, S., K. Seines, S. Havig, and T. Kydland, 1990, Experience from the ongoing long-term test with a horizontal well in the Troll oil zone, 1990 SPE Annual Technical Conference and Exhibition, New Orleans, September 23-26, 1990.

Lindholm, C. D., 1987, Seismicity in the northern North Sea: Ph.D. thesis, University of Bergen.

Osborne, P., and S. Evans, 1987, The Troll field: reservoir geology and field development planning, in J. Kleppe et al. (eds.), North Sea oil and gas reservoirs: London, Graham and Trotman.

Chapter 29

The Suizhong 36-1 Oil Field, Bohai Gulf, Offshore China
Reservoir Delineation by Geophysical Methods

John B. Gustavson

*Gustavson Associates, Inc.
Boulder, Colorado, U.S.A.*

Xin Shi Gang

*Bohai Oil Corporation
Tanggu, Tianjin, China*

ABSTRACT

The Suizhong 36-1 oil field was discovered by Bohai Oil Corporation, a subsidiary of CNOOC, in 1987 in the Bohai Gulf, Liaodong Bay. The structure consists of Oligocene fluviodeltaic and lacustrine sandstones that contain heavy oil in a combination structural-stratigraphic trap. Seismic data were reprocessed and interpreted. Synthetic seismograms were used for correlation and wavelet processing of the seismic data. Seismic-petrophysical analyses were carried out to relate the measurable seismic parameters to the subsurface rock and fluid parameters.

Interpretations were carried out on a GeoQuest workstation and were iterated several times to obtain a consistent match with the wells. Seismic attribute generation and analysis were used primarily to reveal patterns associated with the depositional environments. The approximate inversion or pseudoinversion method was applied to the seismic data to obtain the subsurface acoustic impedance distribution that was the basis for the reservoir parameter estimates.

For estimation of hydrocarbon reserves, first the values of acoustic impedance were converted into estimates of porosity using the relationships derived from seismic-petrophysical analysis, and then the porosity distribution was transformed into maps of net reservoir thickness, permeability, and oil saturation. The results suggest that accurate predictions of oil in place are possible with this methodology.

INTRODUCTION

An integrated reservoir study of the Suizhong (SZ) 36-1 offshore oil field area was carried out, including comprehensive analyses of the geologic, petrophysical, geophysical, reservoir engineering, and production engineering aspects of the field. The Suizhong 36-1 field was discovered by Bohai Oil Corporation (BOC) in 1987 (Li, 1989). It is located in the offshore waters of the rifted, Mesozoic-Tertiary Liaodong basin of eastern China (Figure 1). The main portion of the field is situated on the Liaoxi horst block, and parts of the field extend into the adjacent Liaozhong and Liaoxi depressions.

The principal reservoirs are in unconsolidated sands of the Oligocene lower Dongying Formation; these shallow-water deltaic sediments were deposited in a lacustrine basin. Twelve wells have been drilled in the field area, and sizable reserves of low-gravity (19° API) oil are indicated. This paper discusses first the regional and field geology and then the use of seismic inversion methods to determine the oil in place.

REGIONAL GEOLOGY

The SZ 36-1 field is located in the Liaodong basin, an extensional or rifted basin in the northern part of the Bohai Gulf basin of eastern China. The Liaodong basin covers an area of approximately 27,000 km². It is essentially a faulted depression or crustal sag that trends northeastward, roughly parallel to the nearby continental margin. It is bordered on the west by the Liaoxi coastal plain and on the east by the Tan-Lu fault, a major right-lateral strike-slip fault.

The Liaodong basin formed where incipient Mesozoic intracontinental rifting thinned the crust, which in this region was composed of Precambrian metamorphic rocks and Paleozoic carbonates. A series of grabens and half-grabens developed along basement fracture zones (Figure 2). These depressions, which are bounded by normal faults, became subbasins that underwent successive episodes of extension and subsidence that continued into the Cenozoic. The fault trends are aligned parallel to shear directions caused by tectonic interaction between the Sino-Korean craton and the Pacific plate.

The stratigraphic section within the SZ 36-1 study area ranges in age from Precambrian to Holocene. Tensional stresses increased during the early part of the Tertiary Period. This resulted in successive episodes of fracturing and basin expansion. The tensional stresses may have been related to the uplift that occurred to the west of the study area, where the first phase of Himalayan mountain-building was underway. Two half-graben sags pertinent to this study—the Liaoxi and Liaozhong depressions—were formed at this time. These features served as local depocenters where thick sequences of fluviodeltaic and lacustrine sediments accumulated (Figure 3).

As lacustrine waters transgressed into the general area from northeast to southwest, Eocene-Oligocene Shahejie strata filled the topographic depressions and began to onlap the emergent faulted ridges. Deposition of the overlying Oligocene Dongying Formation buried any remaining topography; what previously had been a segmented basin and range province now gradually became a larger and unified sedimentary basin.

Lower Dongying sedimentation represents a peak stage of subsidence. Deltaic and bar sandstones were deposited locally in shallow lacustrine environments; dark organic muds with intercalated turbidites were deposited in deeper parts of the lacustrine basins.

By analogy to the greater Bohai Gulf basin, traps of various types in reservoir rocks of different ages are expected to exist in the Liaodong Bay basin area. In the SZ 36-1 field, which is a combination structural-stratigraphic trap, the Oligocene lower Dongying reservoir section contains numerous transgressive-regressive sedimentary cycles of fluviodeltaic-lacustrine origin. Although reservoir types vary somewhat, the dominant ones are porous distributary channel, distributary mouth bar, and delta-front sandstones; these facies interfinger in complex patterns. Individual sand thicknesses range from a few meters to more than 60 m; most individual sands do not carry from well to well. Average porosity values in these beds exceed 30%.

At least 12 wells have been drilled in the field area, all of which have penetrated the lower Dongying Formation or older sediments. The structurally highest wells have oil pay in four to six zones in the lower Dongying, whereas the flank wells have pay in as few as two zones. Across

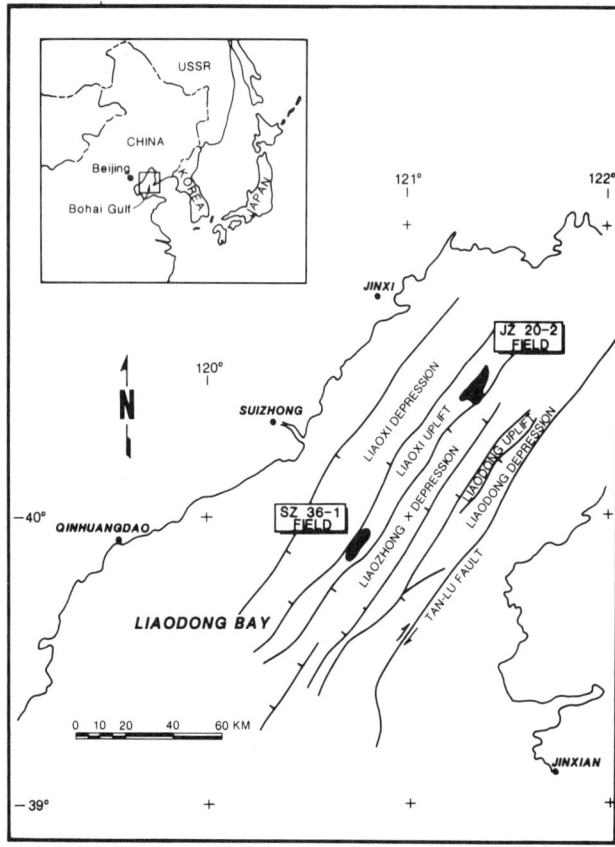

FIGURE 1. Tectonic map of Liaodong Bay and the SZ 36-1 field area.

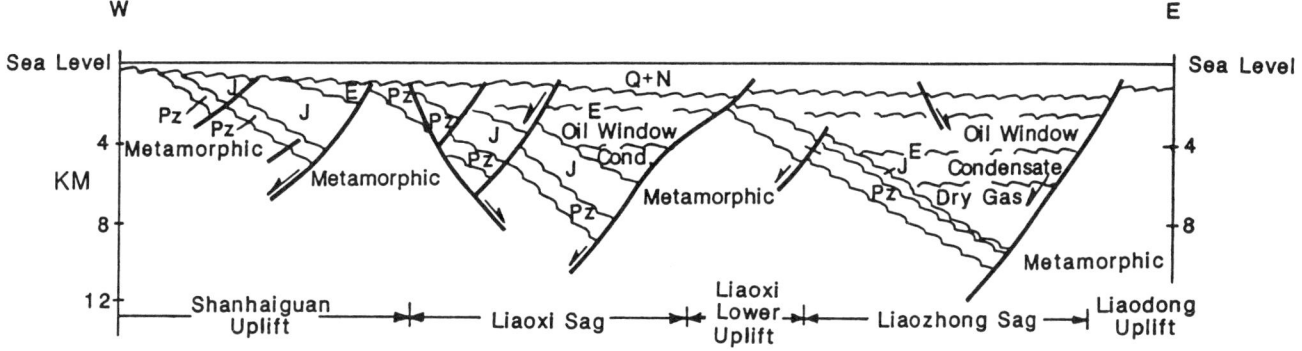

FIGURE 2. East-west schematic structure section, Liaodong Bay (from Bohai Oil Corp.).

AGE		FORMATION		LITHOLOGY	THICKNESS
NEOGENE	PLIOCENE	MINGHUAZHEN		SANDSTONE, SILTSTONE, SHALE, CONGLOMERATE, MUDSTONE	600–1500m
NEOGENE	MIOCENE	GUANTAO		SANDSTONE, SILTSTONE, SHALE, BASAL CONGLOMERATE	400–900m
PALEOGENE	OLIGOCENE	DONG – YING	UPPER	MUDSTONE, THIN SANDSTONE INTERBEDS	30–200m
PALEOGENE	OLIGOCENE	DONG – YING	LOWER	ALTERNATING SANDSTONE, SILTSTONE, MUDSTONE IN NUMEROUS COARSENING-UPWARD SEQUENCES	300–700m
PALEOGENE	EOCENE-OLIGOCENE	SHAHEJIE		CARBONATES, MUDSTONE, SHALE, SANDSTONE. OCCASIONAL VOLCANIC INTERBEDS	100–900m

FIGURE 3. Generalized Tertiary stratigraphic column of the SZ 36-1 field area.

the major bounding fault, on the downthrown side, pay has been found only in limited parts of the upper Dongying Formation.

FIELD GEOLOGY

The SZ 36-1 field covers an area of about 200 km^2. Only limited well control exists within this area (Figure 4). The geologic study concentrated on the stratigraphy and sedimentation of the lower Dongying Formation. The first major tasks were to establish a valid network of correlations within the lower Dongying and to develop a workable zonation system. The zonation was utilized not only in environmental mapping but also in geophysical mapping, petrophysical analysis, and reservoir studies. Correlations were easier to establish within the shaly upper part (zones 1A and 1B) of the lower Dongying than in the mixed lithologies of the underlying reservoir sequence. The first step in correlation within the reservoir section itself was to isolate the sand sequences by correlating the major shale intervals. After these shale intervals had been connected from well to well, the sandier intervals above and below the shale intervals were zoned and correlated by identifying "depositional cycles."

Zones within the reservoir sequence could be mapped seismically. Most of the reservoir sands are in zones 2A, 2B, and 2C; a lesser number are in the underlying, structurally deeper zones. The amount of sand varies considerably from zone to zone and from well to well.

Several of the reservoir zones (2A, 2B, 2C, 3, and 5 in the main part of the field) pinch out to the south between wells 2D and 6. These pinch-outs are interpreted as being limits of deposition rather than erosional edges; they can be seen on seismic and are of great significance for reserve estimates and field development planning.

A major high-angle normal fault, with hundreds of meters of throw at the reservoir level, separates the crest of the Liaoxi uplift from the downdropped Liaoxi depression to the west. On the downthrown side of the fault, the seismic indicates that some of the lower Dongying zones thicken into the fault. This demonstrates that this and other related faults were sporadically active during deposition of the lower Dongying Formation. This same fault, which has a throw of about 350 m at the zone 2A level, was also active in the post-Dongying period; it displaces the unconformity at the base of the Guantao Formation and can be seen to cut almost to the present-day sea floor. The thinnest sections of zone 1A are located along the high part of the upthrown fault block.

The interrelated objectives of the next major phase of the study were the development of a suitable depositional model for the lower Dongying and the preparation of a series of environmental maps. These maps would aid in the understanding of reservoir sand distribution. Wireline log patterns were used to identify environments of deposition. In the SZ 36-1 field study area, the wells are too few in number and too widely spaced to permit the construction of detailed paleoenvironmental and reservoir distribution maps from log patterns. Nevertheless, log patterns have been a useful tool in making more general environmental interpretations.

Log patterns believed to be diagnostic of five types of environments within the over-all deltaic framework have

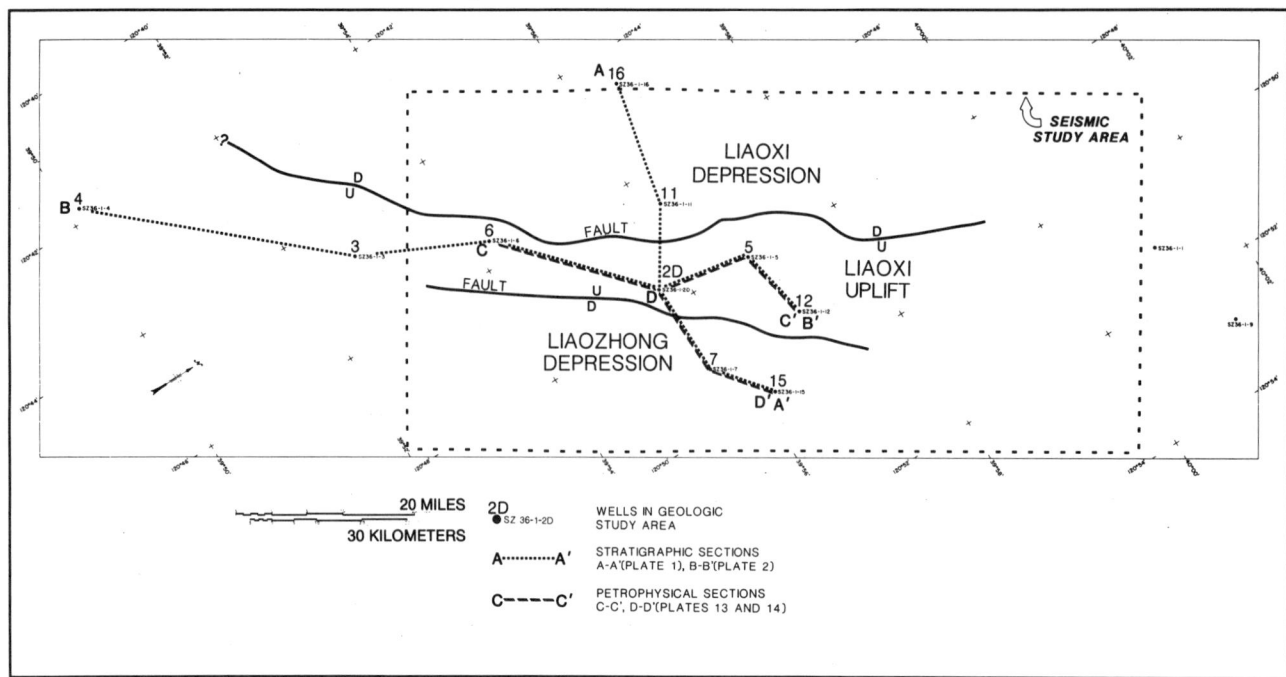

FIGURE 4. Index map of the SZ 36-1 field area.

been identified in the study area. The environments represented by these log patterns are:

1. Prodelta
2. Delta front
3. Barrier bar
4. Distributary mouth bar (stream mouth bar)
5. Distributary channel.

Typical examples of these log patterns, taken from wells in the study area, are shown in Figure 5. It should be emphasized that the pattern of a single log segment is, in and of itself, not diagnostic. It must be considered in relation to the underlying and overlying sediments in the well and in relation to adjacent wells.

Other sedimentary environments undoubtedly exist in the study area. Unfortunately, the sparse well control, and to some extent the extremely sandy nature of much of the lower Dongying Formation itself (if more shale were present, the log patterns might be somewhat more diagnostic), make recognition of the other environments very speculative. Some of the sandstones in the area are probably of point-bar origin, but this cannot be demonstrated by log patterns.

Dipmeter data from six wells in the field were processed and interpreted for direction of transport and paleoslope. Seismic lines were examined in an effort to locate any channellike anomalies that might be present. In an attempt to map the general distribution of channels, all of the seismic lines were scanned for anomalous-looking features. The two principal display formats examined were the variable intensity display and the Hilbert transform phase display. The locations of any anomalous channellike features (or barlike features in a few cases) were noted, and the stratigraphic zonation was determined. By analysis of both strike and dip lines, it was possible to map the approximate locations and trends of the channels.

After the information described above had been analyzed, it was concluded that the lower Dongying reservoir sequence consisted of a series of stacked, river-dominated lacustrine delta lobes (Figure 6). There is little evidence that wave or current action played a significant role in the formation of the deltas. Prodelta, delta front, barrier bar, distributary mouth bar, and distributary channel environments have been identified. The over-all direction of transport of sediments in the field area was from north-northeast to south-southwest. In the southern part of the study area, the direction of transport may, at times, have been from west to east (a different system of lobes?). The progradational phase of delta development consisted of zones 2A through 8; the abandonment (transgressive) phase—during which either the basin deepened or lobe switching occurred—included zones 1A and 1B.

Ten environmental maps of zones 1A through 6 were prepared. An example (zone 3) is shown in Figure 7. These maps show the general areal distribution of environments (delta front, distributary channel, etc.) during the deposition of a particular zone. Because the environments change vertically as well as laterally in a given zone, these maps are to some degree a compromise; particular emphasis was placed on the more reservoir-prone environments. The environmental maps show a series of similarly shaped prograding deltaic lobes. Progradation and

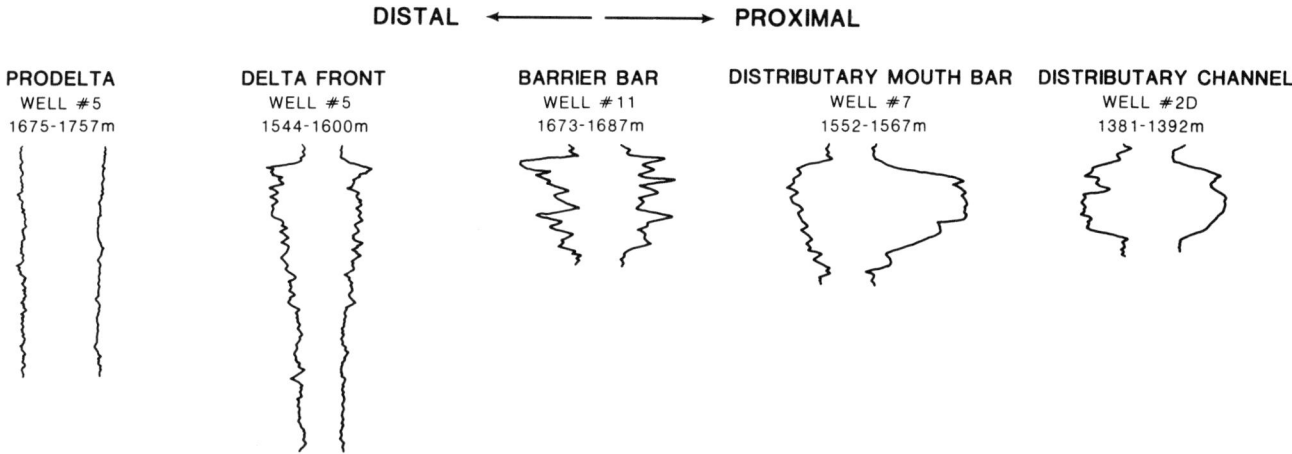

FIGURE 5. Wireline log patterns representative of depositional environments in the lower Dongying deltaic complex, SZ 36-1 field area. See Figure 6 for model of depositional environments.

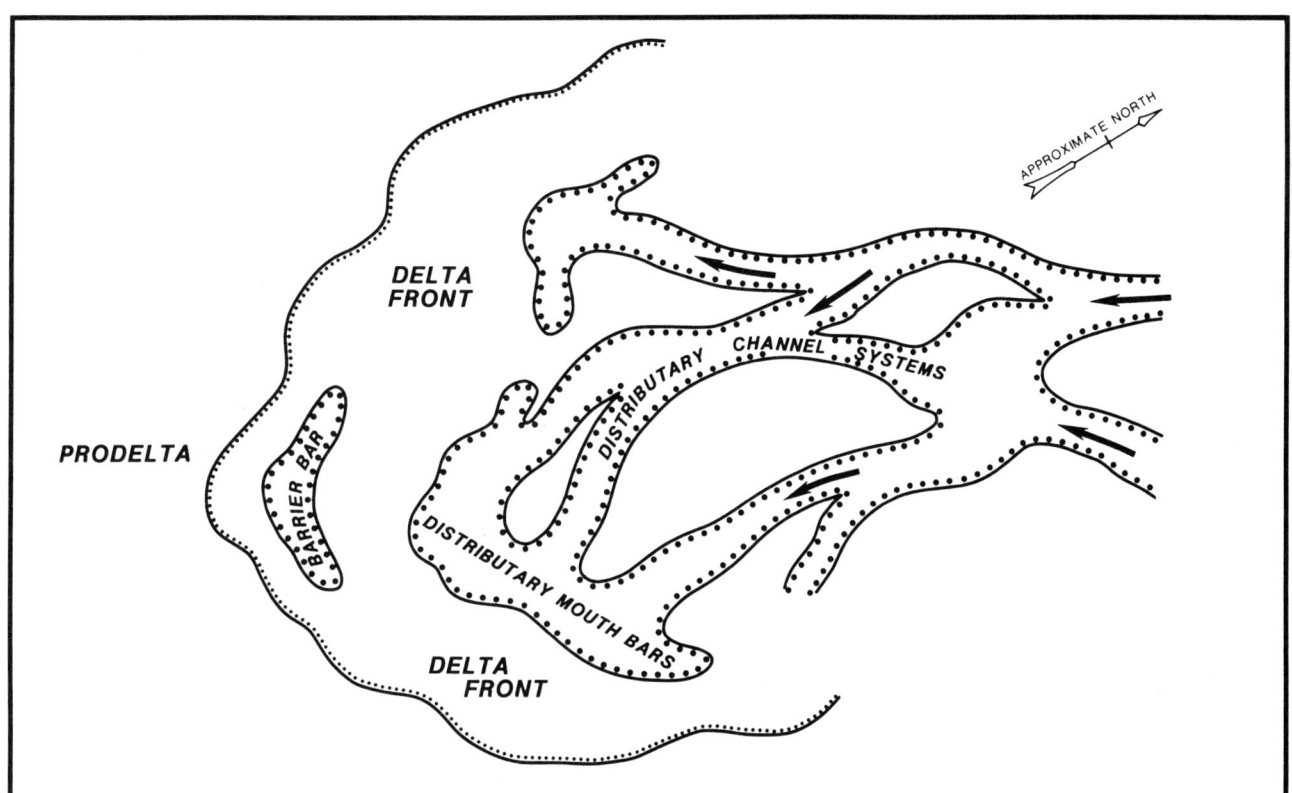

FIGURE 6. Deltaic depositional model of lower Dongying Formation.

the transport of sediment were from north-northeast to south-southwest in the main field area, and perhaps from west to east in the southern part of the study area.

Thin section, clay mineral, trace element, and heavy mineral data from several wells in the SZ 36-1 field area were tabulated and plotted graphically. The spotty distribution of the data precluded any detailed analysis. The data did suggest, however, that the source area for the lower Dongying was probably a metamorphic (or igneous-metamorphic) terrane not far distant from the site of deposition.

PETROPHYSICAL STUDY

The objectives of the petrophysical study were to establish a database for well log and core analysis; to analyze the log, core, and other data using up-to-date computer

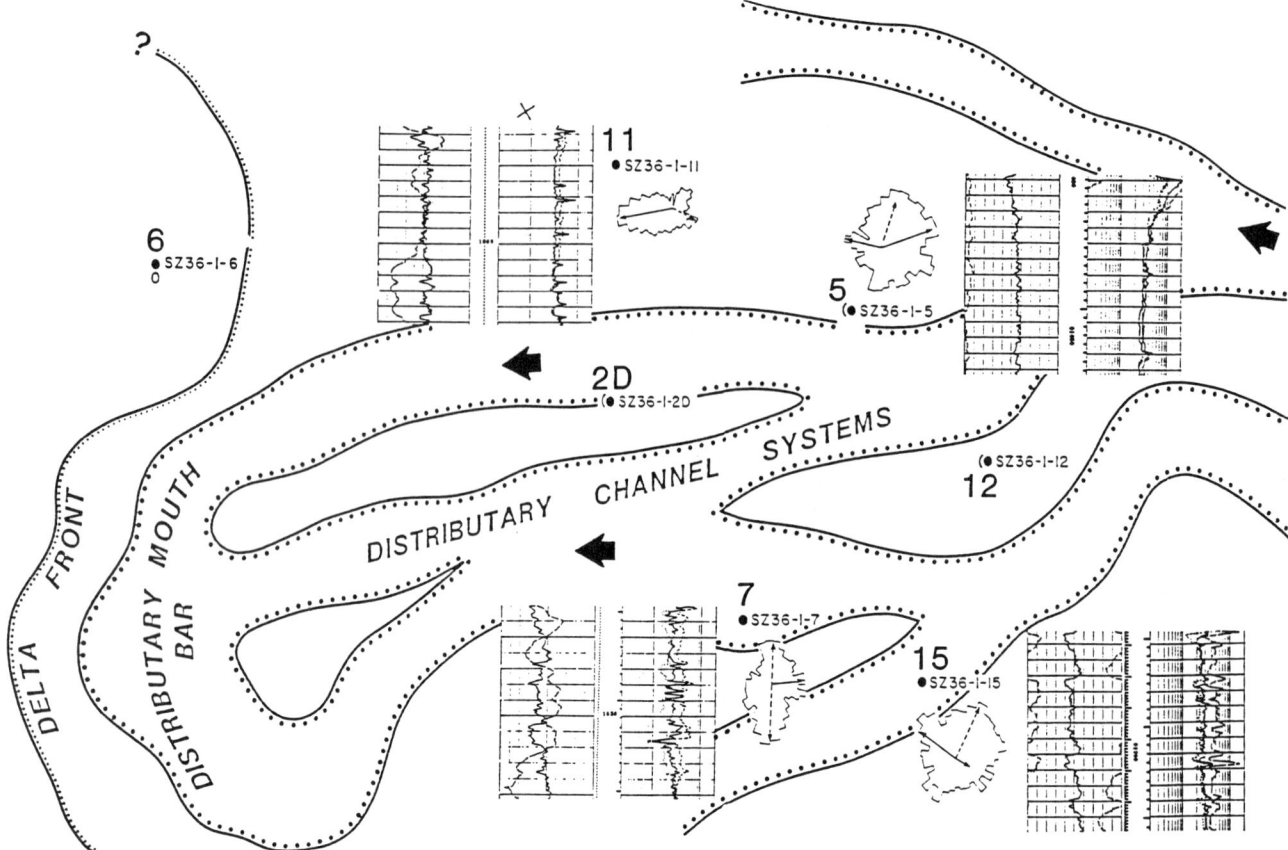

FIGURE 7. Generalized environmental map of zone 3, SZ 36-1 field.

technology; and to develop petrophysical parameters and models that would aid in the geologic, geophysical, and engineering phases of the over-all field study. Most of the core data were from the SZ 36-1-5 well; this well was to serve as a model throughout much of the study. Quality control procedures included depth shifting, borehole and environmental corrections, and normalization of the well log data.

Because the core data were too limited to provide information on the entire field area, it was necessary to calibrate the more extensive wireline log data with the core data. Density and acoustic log data were used to develop a modified log porosity equation, which agrees reasonably well with known core porosities. The match of log and core porosities was sufficiently good that clay corrections were found to be unnecessary. Grain size appears to be the major factor controlling the petrophysical properties of the porous and often unconsolidated sands that make up the reservoirs in the lower Dongying Formation.

When core permeabilities were compared with core porosities, an excellent correlation was observed; log permeabilities could therefore be derived directly from log porosities.

Lithologic studies had been carried out by BOC on the core samples in the field area. Because log porosities could be closely correlated with core lithologies, empirical lithology logs could be developed from the porosity logs.

Analyses of waters recovered during drill-stem and production tests were used to determine the average salinity and resistivity of formation waters in the field. These values were utilized, along with the porosity data, to calculate water saturations in the reservoirs.

BOC's classification of drill-stem test results, along with a production index, were compared with the porosity and water saturation data in order to develop criteria for the recognition and evaluation of oil zones in the field. Suitable cutoff values for water saturation and porosity were found to be 60 and 27%, respectively. The high porosity cutoff is considered typical for unconsolidated sands. A cutoff value of 10 ohm m for deep lateral resistivity measurements was deemed appropriate.

When oil-water contacts in the field wells were studied, the following conclusions were drawn: (1) not more than one oil-water contact is present in any individual well; and (2) at least two, and possibly three or more, oil-water contact systems are present in the field area. Wells 2D, 5, 6, and 12, all located along the "crest" of the structure, appear to have a common oil-water contact at about −1507 m. The downflank wells, 7 and 15, have a somewhat deeper contact at about −1580 m.

The reservoirs for the six wells mentioned above are in the lower Dongying Formation. The database for wells 11 and 16 is not sufficient to pick oil-water contacts accurately. Wells 11 and 16 are in a different fault block across a major fault from wells 2D, 5, 6, 7, 12, and 15, and their reservoir sands are in the upper Dongying. It would not be surprising to find, with more well control, that the oil-water contact(s) in this fault block is (are) different from those in the other parts of the field structure.

The results of the petrophysical study, when combined with the geologic and geophysical study of the reservoir sands, were used in calculating oil in place.

DEVELOPMENT GEOPHYSICS

A total of 350 km of seismic data (Figure 8) was reprocessed, and detailed interpretations and analyses were carried out on all but 50 km of these data. In addition, special processing was carried out on approximately 20 to 35 km of data in the central part of the field. All conventional processing and most special processing were done by outside contractors, whereas interpretations and all seismic-engineering analyses were carried out at BOC on a GeoQuest workstation with ADSEIS- and Zycor-augmented software. Seismic-petrophysical analyses were carried out on an 80386-class microcomputer using PCI software.

A serious attempt was made to reprocess vertical seismic profile (VSP) data so as to improve correlation of surface seismic data with subsurface reservoir parameters. Unfortunately, incomplete field data and/or documentation made such an effort infeasible.

Synthetic seismogram construction was standard, and the results were used for correlation and wavelet processing of the seismic data. Seismic-petrophysical analyses were carried out to relate the measurable seismic parameters to the subsurface rock and fluid parameters.

Surface seismic processing comprised conventional reprocessing of the seismic data; special processing of the data, primarily for hydrocarbon indicators; and three-dimensional velocity analysis. The reprocessing improved the quality of the seismic data over that of the original processed data. Amplitude vs. offset (AVO) processing revealed no obvious hydrocarbon indicators in the main reservoir zones. Velocity analysis produced a three-dimensional velocity field that was the key to the interpretation and engineering work that followed.

Seismic Interpretation

Layer and fault interpretations were carried out on a GeoQuest workstation and were iterated several times to obtain a consistent match with the wells. Seismic attribute generation and analysis were used to reveal patterns associated with the depositional environments. The phase component is shown in Figure 9. The phase component emphasizes the boundaries between reflection packets or subcycles. For the Tertiary, a probable interpretation is that closely spaced boundaries represent low-energy

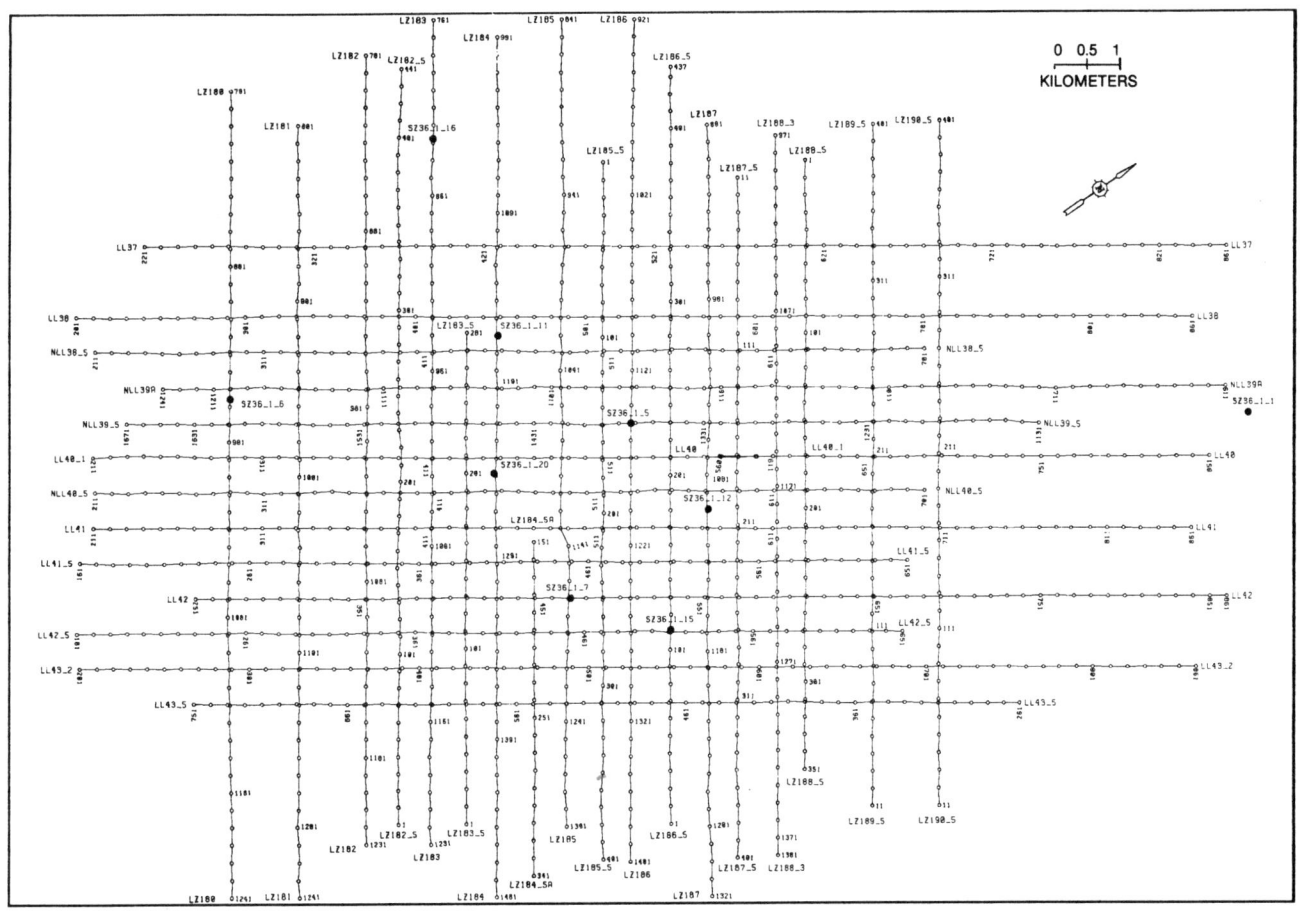

FIGURE 8. Seismic grid and well location map for SZ 36-1 field.

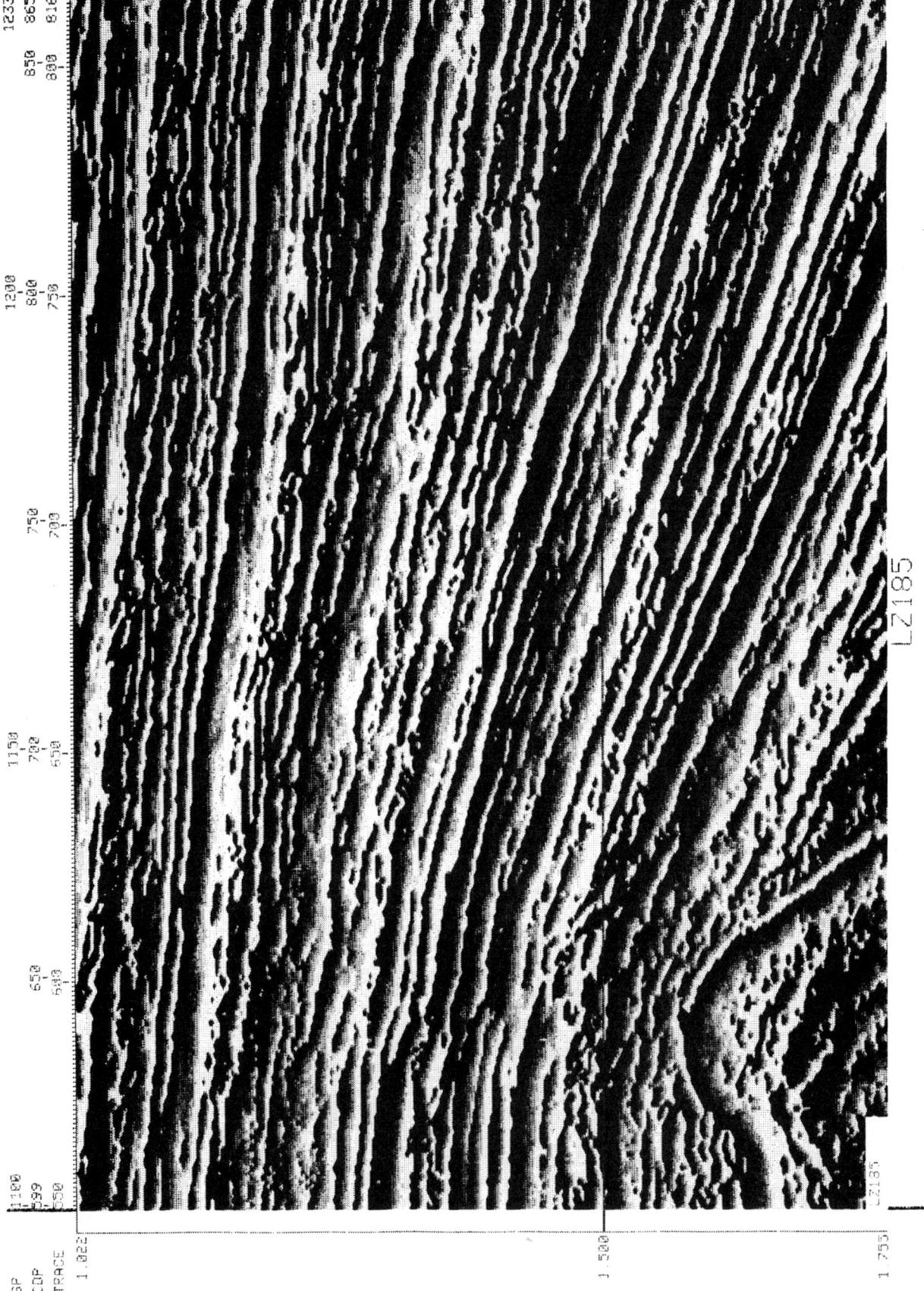

FIGURE 9. Phase component showing low-energy (closely spaced boundaries) and high-energy (widely spaced boundaries) depositional environment.

depositional environments, whereas widely spaced boundaries represent higher-energy environments.

A good deal of local information about the rocks and fluids could be deduced from analysis of the details of the reflected events themselves. From the areal pattern of changes in the seismic events, a significant amount of information could be derived about the environments of sedimentary deposition, and consequently about the lithologies of the rocks. Figure 10 shows examples of interval velocity patterns that clearly reveal a delta front with various coarse-grained to shaly lithologies.

The seismic method has evolved to the point where its integrated use with petrophysical and engineering information can lead to quite reliable estimations of reserves in places where the total amount of well control is very limited. Examples of this type of application have begun to appear in the literature (e.g., De Buyl and Guidish, 1986), but not all potential techniques have been exploited.

The starting point for the extraction of volumetric information using seismic methods is the generation of an absolute acoustic impedance curve. Acoustic impedance, as defined in exploration seismology, is the product of the bulk density of a rock and the compressional velocity of elastic waves through that rock. It is one of the few basic properties of the earth that can be extracted in detail from seismic data, because the strength of a reflected seismic event is directly proportional to the change in acoustic impedance from one type of rock to another.

Two kinds of seismic trace inversion processes were used to obtain quantitative estimates of the subsurface acoustic impedance distribution. One, the approximate inversion or pseudoinversion method developed by Lindseth (1979) and by Oldenburg et al. (1983), was applied to the seismic data and used as the basis for all the reservoir parameter estimates. The other, the generalized linear inversion (GLI) approach, was applied to 20 km of seismic data as part of a calibration effort.

The first step was to convert the acoustic impedance to a depth function in order to relate it to the well information. The velocity function that was used was the same one used to convert the seismic time horizons to depths so that the acoustic impedance values remained accurately registered to the rocks with which they were associated.

The next step was to introduce information derived from petrophysical examination of the logs, cores, and rock samples from the wells. This information was presented in the form of functional relationships or tables (see Table 1) relating, primarily, acoustic impedance and porosity, and, secondarily, acoustic impedance and saturation, permeability, lithology, etc. To the same extent that good functional relationships could be established between the acoustic impedance and the specific rock properties, the seismic could be used to extend these properties away from the wells.

Once the above relationships had been generated, it was possible to translate the acoustic impedance information into estimates of porosity, permeability, saturation,

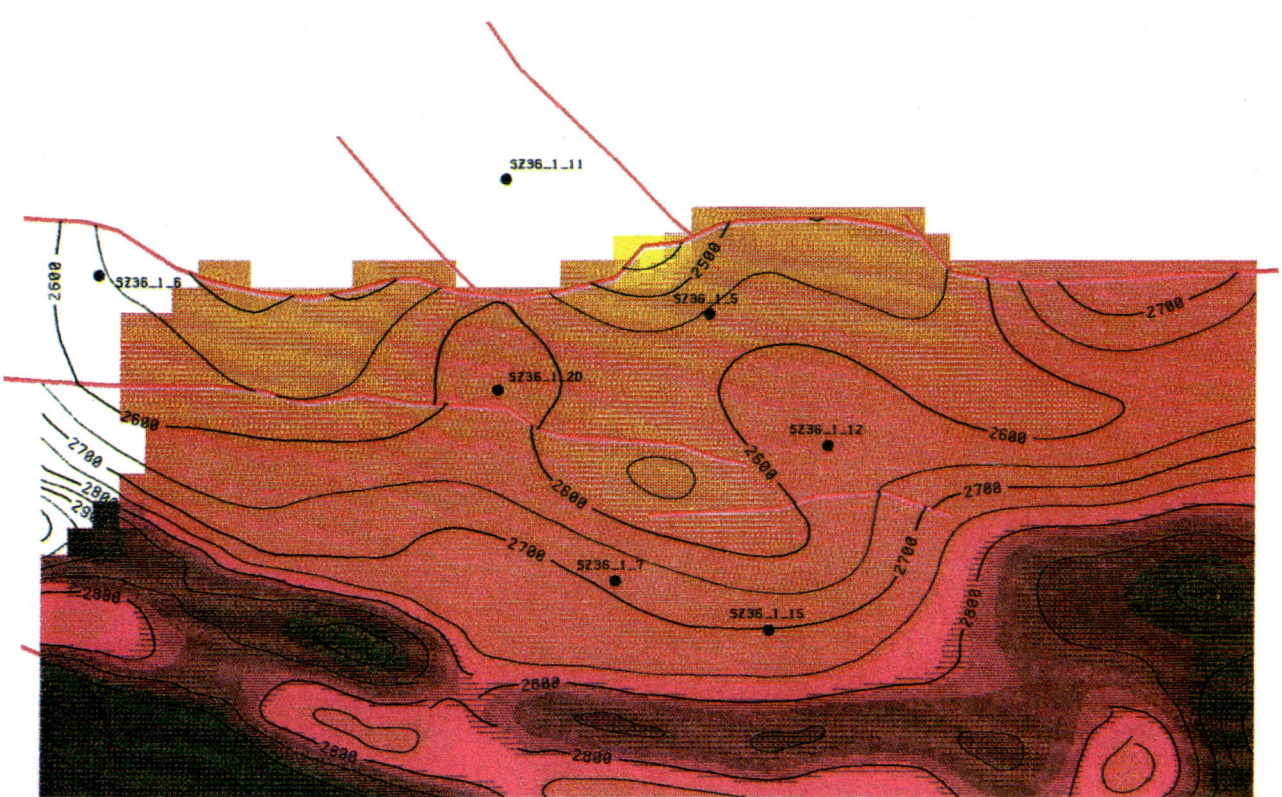

FIGURE 10. Interval velocity map, showing prograding delta front.

TABLE 1. Summary of regression results.

Reservoir parameter	Seismic or seismically derived parameter	Linear regression relationship	Total number of samples		Correlation coefficient*
			Before editing	After editing	
Porosity (ϕ)	Acoustic impedance (ρv)	$\phi = 0.789 - 0.258\,E^{-04}\,\rho v$	4435	3695	0.938
Porosity (ϕ)	Seismic velocity (v)	$\phi = 0.993 - 0.803\,E^{-04}\,v$	4435	3724	0.935
Porosity (ϕ)	Density (ρ)	$\phi = 1.464 - 0.532\,\rho$	4435	3600	0.882
Permeability (K)	Porosity (ϕ)	$\log K = -5.054 + 26.289\,\phi$	487	321	0.751
Water saturation (S_w)	Porosity (ϕ)	$S_w = 2.466 - 6.148\,\phi$	4435	3362	0.879

*Based on edited data.

net thickness of porosity within a reservoir interval, and so forth. The quality of these translations depended on several factors, including the amount of clay in the reservoir sands, the degree of sorting, the mineralogical content of the grains, etc. The detailed approach to dealing with these factors was quite variable and depended on the circumstances in each specific area. However, the principles were universal. From geology, the depositional environment and the diagenetic history of the sediments were known. Through the patterns thus observed, it was possible to gain considerable insights into the probable distribution of clays, grain sorting, and grain size distribution of the reservoir rocks. This was an interpretative process and was greatly facilitated by use of the workstation.

Direct detection of hydrocarbons was tried in an effort to map oil-water contacts in the field. Relative amplitude profile (RAP) sections were analyzed for amplitude anomalies. Amplitude vs. offset (AVO) profiles were also analyzed for anomalies. All efforts failed to detect hydrocarbons or oil-water contact levels conclusively. It is hypothesized that two factors combined to make hydrocarbon detection impossible: (1) the low gravity (and other physical characteristics) of the oil (about 19° API), and (2) the absence of gas in the hydrocarbons.

DISTRIBUTION OF OIL IN PLACE

The first step was to calculate a scale factor, which is a measure of the net fraction of a reservoir interval that is occupied by oil, for each x,y grid point in the field. This required an area-wide distribution of the ratio N/G, which is the net thickness of producible reservoir rock (i.e., rock having porosities above some threshold value) divided by the gross thickness of the total reservoir interval. Also needed were estimates of the average porosity (above threshold) and the average oil saturation distribution over the area. All of these estimates were derived directly from the acoustic impedance curves and seismic attribute patterns. Once they had been derived, their product was taken to obtain the scale factor.

In order to generate an estimate of the reservoir's distribution of oil in place, it was necessary to have, in addition to the scale factor, the distribution of the bulk rock volume of the total reservoir unit above the oil-water contact. Except at the wells, where direct measurements usually could be taken, this estimate came from structural mapping of the seismic data. Basically, seismic mapping of the top and bottom of the reservoir unit (or of sections as close to the top and bottom as could be differentiated seismically) was carried out. Because such mapping is usually done in terms of time, a conversion to depth had to be made.

From the petrophysics, the oil-water contact depth distribution was introduced to obtain the reference level from which to calculate, at each grid point, the net reservoir thickness.

The last step in the reserve estimation process was to multiply, for each point in the grid, the value of the scale factor by the net reservoir thickness to obtain the net column of oil associated with that grid point (Figure 11). A summation of all the oil columns over the field then yielded an estimate of the oil in place for the reservoir unit. Any number of maps and cross sections could also be printed out for development engineering purposes.

CONCLUSIONS

The process of deriving the distribution of the oil in place in a producing reservoir is fairly straightforward in principle. The procedure described above makes use of those properties most directly obtainable from seismic data—namely, acoustic impedance, average velocity, and areal attribute patterns. When these parameters are combined with the detailed information obtained from petrophysical analyses of logs and cores and the results of DST and other engineering tests, it becomes a relatively simple task to calculate reserves and other quantities of importance. These methods are of particular value in the economic evaluation of producing fields that either have very few wells in relation to their lateral extents or are suspected of having extensions that have not been defined by drilling.

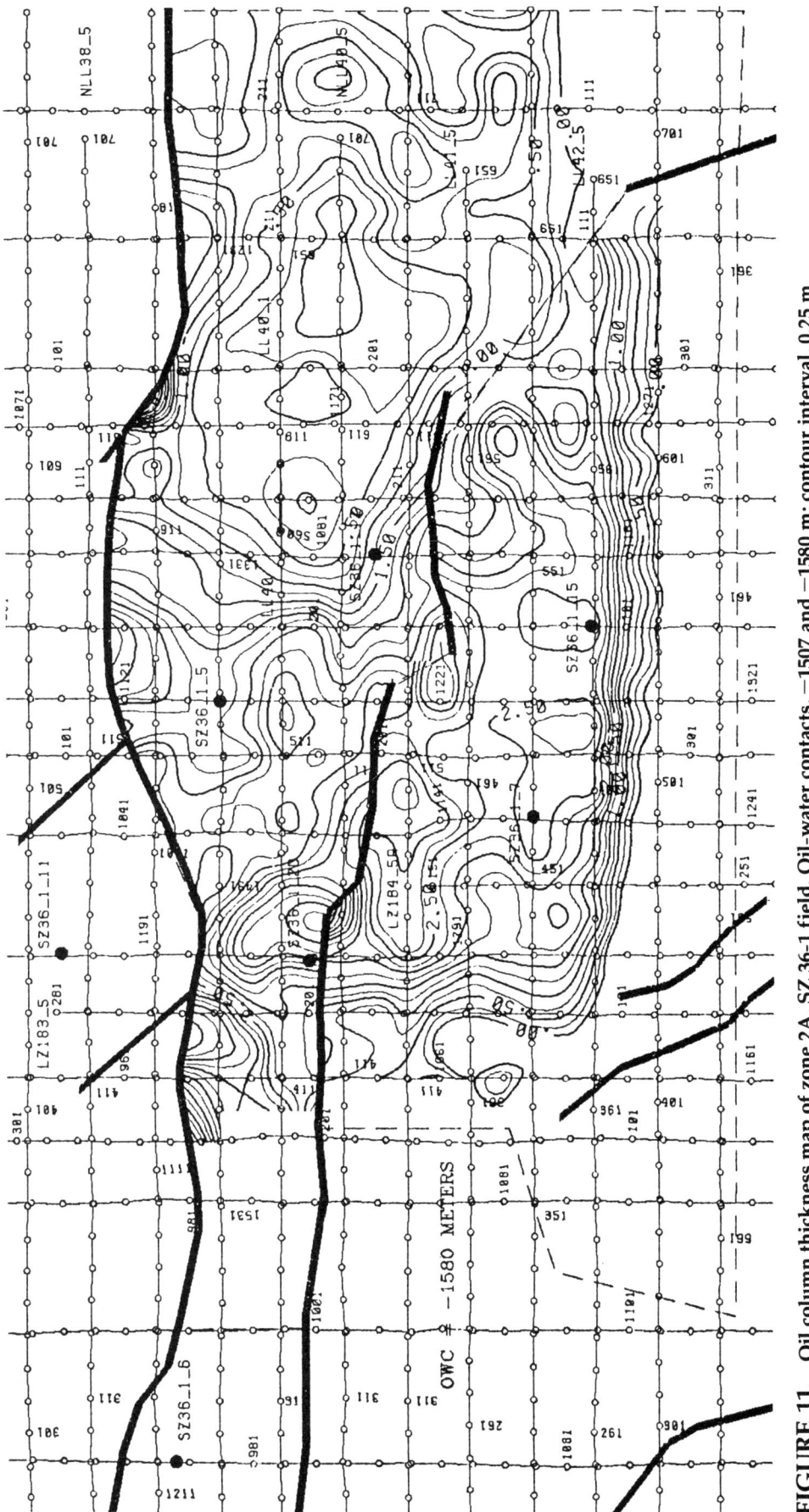

FIGURE 11. Oil column thickness map of zone 2A, SZ 36-1 field. Oil-water contacts, −1507 and −1580 m; contour interval, 0.25 m.

The accuracy of such predictions often can be estimated from a knowledge of the limitations of the methods employed in the calculation of the data being used. Particularly for seismic data, the quality of the inferences is very dependent on the care and completeness with which the data are prepared prior to the derivation of the acoustic impedance distributions. The quality of the final results is also heavily dependent on the integration of information derived from all sources and on the use of one kind of data for calibration of another.

When the Suizhong 36-1 oil field was discovered, the low gravity of the oil, the lack of reservoir energy, and the unconsolidated sands combined to make commercial field development questionable. However, as described in this case history, the delineation of the oil reserves by geophysical methods provided data on both the quantities and the detailed locations of the oil that were sufficiently accurate to allow a waterflood development program to be engineered.

ACKNOWLEDGMENTS

This feasibility study was conducted under a grant from the U.S. Trade and Development Programs. The authors wish to thank Bohai Oil Corporation for permission to publish this paper.

REFERENCES CITED

De Buyl, M., and T. Guidish, 1986, Reservoir description from seismic lithologic parameter estimation: Society of Petroleum Engineers, Paper 15505.

Li Bingquan, 1989, The discovery and development of SZ 36-1 oil field in Liaodong Bay: Proceedings, 21st Annual Offshore Technology Conference, p. 429-436.

Lindseth, R. O., 1979, Synthetic sonic logs—a process for stratigraphic interpretation: Geophysics, v. 44, n. 1, p. 3-26.

Oldenburg, D. W., T. Scheuer, and S. Levy, 1983, Recovery of acoustic impedance from reflection seismograms: Geophysics, v. 48, n. 10, p. 1318-1337.

Chapter 30

Iagifu/Hedinia Field
First Oil from the Papuan Fold and Thrust Belt

R. H. Matzke

Chevron Overseas Petroleum Inc.

J. G. Smith

Chevron International (Yemen) Limited

W. K. Foo

Arctic Onshore District CUSA Production Inc.

ABSTRACT

The Iagifu/Hedinia accumulation will support the first large-scale petroleum development in the Papuan fold and thrust belt. The discovery well, Iagifu-2X, was drilled in 1986 and flowed 45° API oil and gas from a thrust-cored anticlinal structure with closure enhanced by sealing across the sole fault. The primary reservoir is the Early Cretaceous (Berriasian) Toro sandstone. Well locations are selected on the basis of surface geology and existing well results only, because it has proved impossible to acquire seismic data of reasonable quality in a cost-effective manner. Delineation drilling continued until early 1990, with wells ultimately outlining a broad subsurface structure that extends beneath and is detached from the surface anticlines at Iagifu and Hedinia. A development plan was submitted to the Papua New Guinea government in May 1990, citing estimated reserves of 146.6 million bbl of oil in the Toro sandstone and proposing first production in 1992. The region has excellent petroleum potential. Oil and gas accumulations have now been discovered on nearby structures along a 70-km trend with total estimated potential in excess of 300 million bbl of oil and 4 tcf of gas.

INTRODUCTION

The Iagifu/Hedinia field is located in the petroleum prospecting license PPL 100 and is operated by Chevron on behalf of BP Australia, Ampolex, BHP Petroleum, Oil Search, and Merlin Petroleum. The field lies within the Papuan fold and thrust belt (Figure 1). It is about 350 mi (560 km) northwest of Port Moresby, the capital of Papua New Guinea (PNG). Nearest road access is at Poroma, 60 km to the northeast (Figure 2) across a karstic mountainous region covered by dense rain forest. Exploration operations have been entirely helicopter supported, and the construction of production facilities will be supported by air lift. Construction began in July 1990, and first oil production is projected for mid-1992.

Iagifu-2X, the discovery well, was named for a village at the foot of the prominent surface anticline. The prospect was defined solely on the basis of surface geology and topographic expression and interpretation of synthetic aperture radar images. In March 1986, production tests flowed an aggregate 6500 BOPD from the Early Cretaceous Toro sandstone. Delineation drilling began immediately and continued until February 1990. By that time, 22 exploratory wells, including sidetracks, had been drilled to establish the field limits and reserves prior to filing for a petroleum development license in May 1990. The large number of wells was required because (1) it had not been possible to acquire definitive seismic data across the karstic Miocene limestones, and (2) there was a major detachment in Cretaceous shales beneath the limestones that limited the degree to which surface geologic measurements could be projected to reservoir depth.

HISTORY OF EXPLORATION IN THE PAPUAN FOLD BELT

Overview

Early oil exploration in PNG was concentrated along the coasts and major rivers (Rickwood, 1990). Initial drilling was conducted by a precursor of BP in 1915 along the Papuan coast, northwest of Port Moresby. In contrast, the Western and Southern Highlands of PNG were so remote and rugged that they were thought to be uninhabited until Australian gold prospectors entered the region in

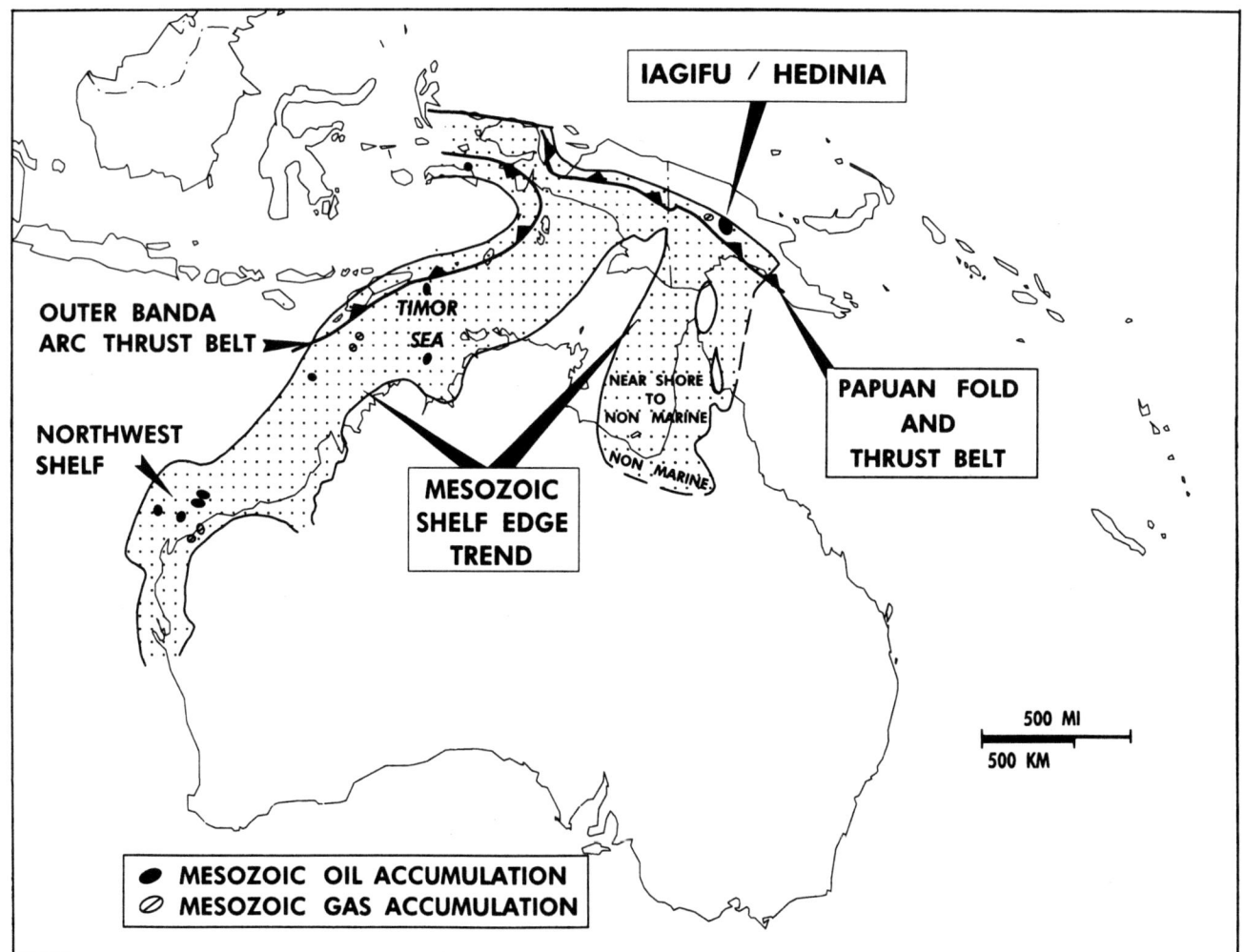

FIGURE 1. Map showing location of Iagifu/Hedinia field within Papuan fold and thrust belt, relative to Australia.

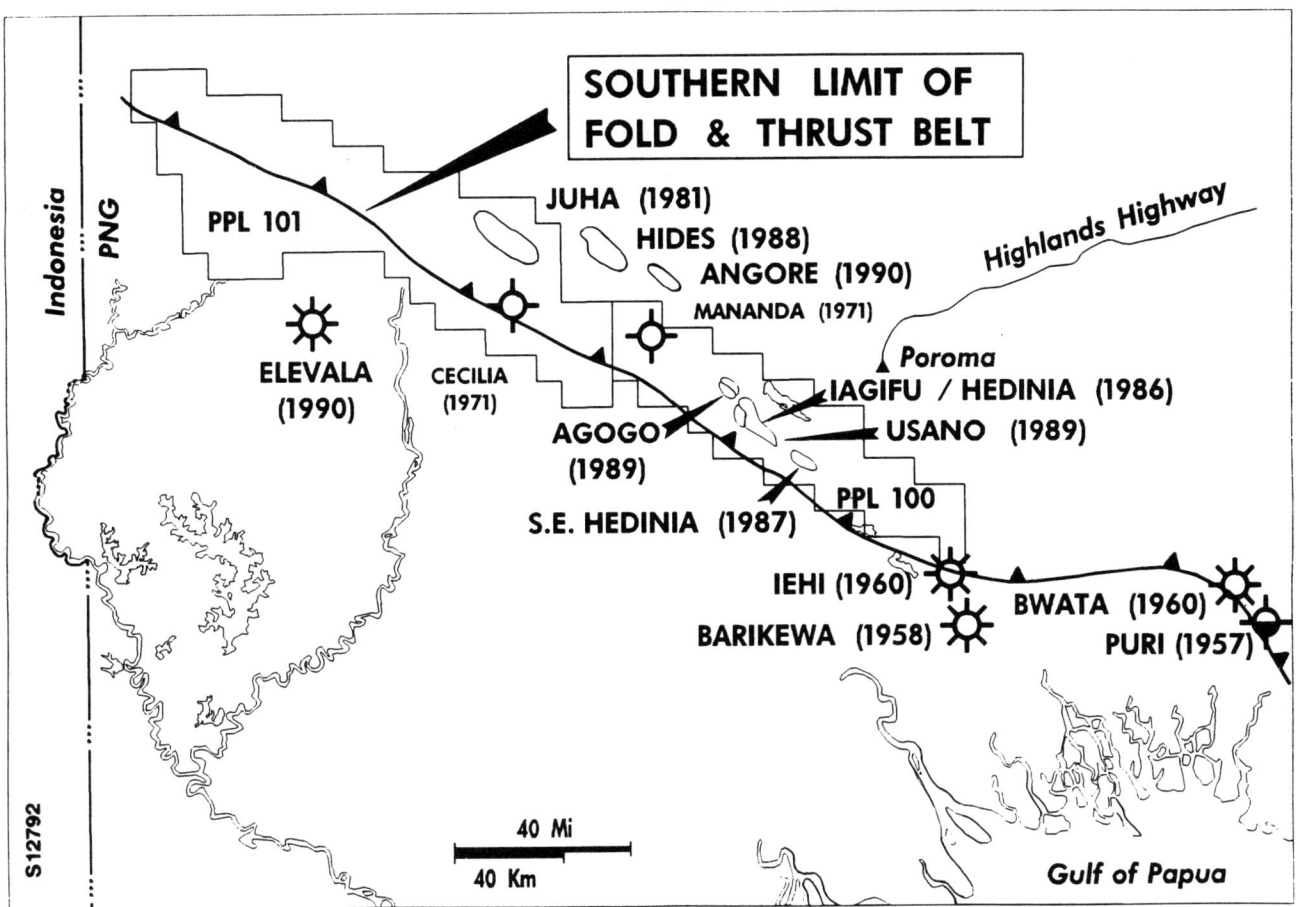

FIGURE 2. Generalized map of Chevron–joint venture acreage (PPL 100 and PPL 101), showing locations of discoveries to date and the position of the deformation front.

1930 and discovered a thriving population in excess of one million people.

Early geologic expeditions to the Western Highlands noted the abundance of oil and gas seeps, the presence of thick Jurassic–Lower Cretaceous sandstones, and the numerous well-defined folds, including Iagifu and Hedinia, that are outlined by resistant Miocene limestones of the karst plateau (Figure 3). Until heavy-lift helicopters became widely available in the early 1970s, drilling was not possible in the highlands of the fold and thrust belt.

Through the late 1950s, drilling on river-accessible foreland structures south of the fold and thrust belt resulted in a series of gas discoveries at Iehi, Barikewa, and Bwata in Berriasian and older Jurassic sandstones (Figure 2). Puri-1, operated by Australasian Petroleum Company (APC) in 1956–57, was the first well in the world to be supported entirely by helicopters. A small quantity of high-gravity oil was produced from Cenozoic basinal limestones at Puri-1 along the southeastern continuation of the fold and thrust belt. These noncommercial discoveries, as well as the promising surface geology, were compelling reasons to maintain exploration interest.

The first two wells in the Western Highlands fold belt were drilled during 1971 by APC and Texaco/Chevron. Gulf (now merged into Chevron) became operator of the APC licenses through a farm-in in 1978. In 1981, the Gulf-operated joint venture discovered the Juha gas/condensate field approximately 100 km northwest of Iagifu/Hedinia. Although there is no market for the gas reserves, condensate flows of up to 1700 BOPD at Juha proved that liquid hydrocarbons could be produced from Mesozoic reservoir sands at potentially commercial rates.

Iagifu/Hedinia Chronology

In March 1986, the Iagifu-2X well marked the first discovery of black oil in the Mesozoic section, flowing oil and gas from three zones at individual rates of up to 3000 BOPD. A total of 144 dry holes or noncommercial producers had been drilled in PNG prior to this initial success (IMPS, 1990).

When the likelihood that development of the Iagifu discovery would prove to be a commercial project was recognized, the licenses formerly issued as PPL 17 and PPL 18 were renegotiated as PPL 100 and PPL 101. PPL 100 carried a requirement that the joint venture submit a petroleum development license (PDL) for Iagifu by November 28, 1989, or show cause for declaring the field noncommercial. Between the discovery in March 1986 and

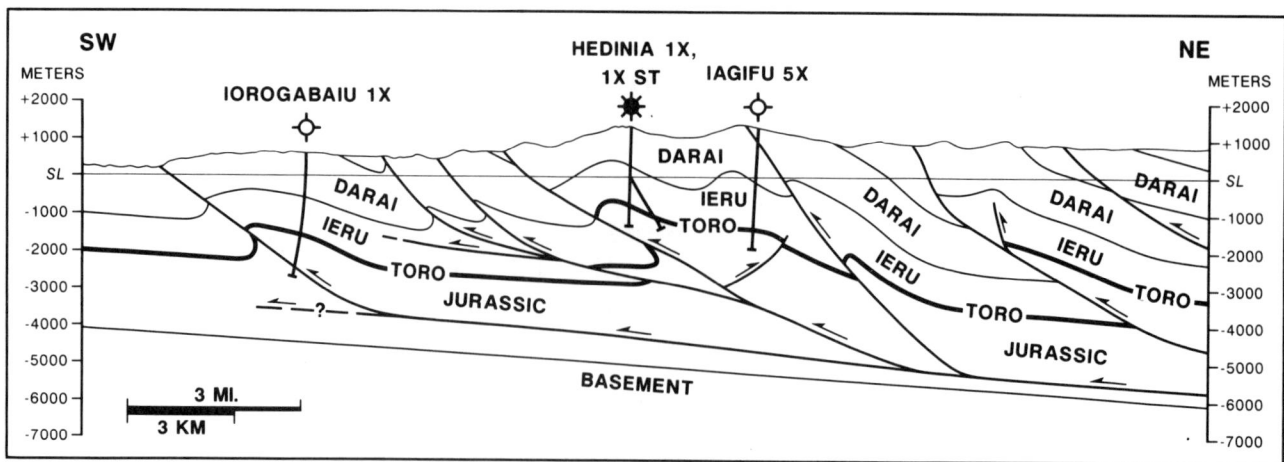

FIGURE 3. Regional cross section across PPL 100 through Iagifu/Hedinia field.

January 1988, three delineation wells were drilled at locations based on nearly 100 km of geologic traverses across the Iagifu structure. The results were a new pool discovery at Iagifu-3X and two dry holes at Iagifu-4X and -5X (Figures 4 and 5). In March 1988, Hedinia-1X was drilled on the adjacent Hedinia surface anticline 2.5 km south of Iagifu-2X. It encountered an extension of the Iagifu gas cap and was plugged back for an exploratory sidetrack, Hedinia-1XST, which discovered the continuation of the Iagifu oil leg in June 1988. Aggressive delineation drilling resulted in 13 more wells and exploratory sidetracks across the field between July 1988 and the November 28, 1989, deadline for PDL submission (Figure 4). An extension of the deadline was granted, and an additional three wells were completed prior to submission of the PDL application on May 1, 1990.

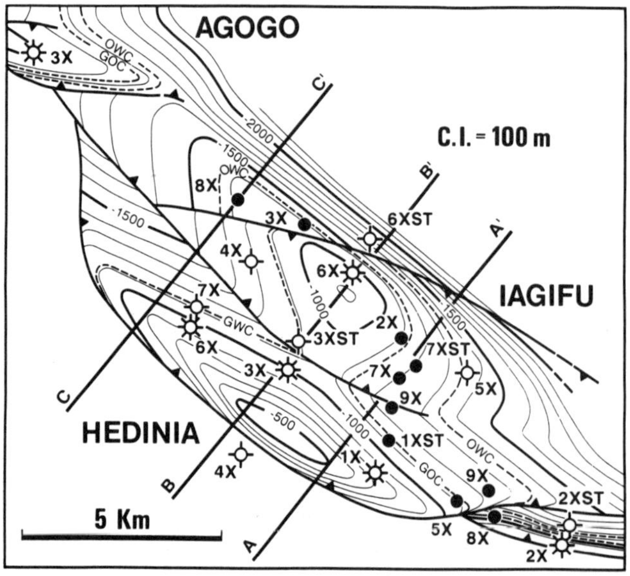

Oil discoveries in 1989 at the nearby Agogo and Usano structures added additional reserves, which will be developed as part of the initial Iagifu/Hedinia project.

GEOLOGY

Tectonic Setting

The Papuan fold and thrust belt (Figures 1 and 2) extends across the island of New Guinea from the Sulawati basin in Irian Jaya to the Gulf of Papua. Within the fold belt, structures are outlined by the topographic surface, which generally follows the top of the 1 km thick Miocene limestone section (Figure 3). Structural style varies considerably along strike, from broad, high-relief, basement-involved structures near the PNG–Irian Jaya border, to an assemblage of lower-relief, en echelon anticlines in the Iagifu/Hedinia area, to an imbricate stack of thrust sheets farther to the southeast, near the Gulf of Papua. Strata at least as young as Pliocene–Pleistocene are involved in the deformation, and analysis of recent seismic activity suggests that the area is currently undergoing shortening (Abers and McCaffrey, 1988; Davies, 1990). The present-day plate tectonic configuration of PNG has evolved in response to oblique Tertiary convergence between the Australian and Pacific plates, as recently summarized by Smith (1990).

Depositional History

The stratigraphy of the Papuan fold and thrust belt, as illustrated in Figure 5, is divided into two gross packages: a Mesozoic rift-drift passive margin succession of southern provenance, and a Cenozoic foreland basin succession of primarily northern provenance.

Deposition in the Papuan basin commenced with rifting and subsequent spreading along the northern

FIGURE 4. Structure map on top of Toro Formation, Iagifu/Hedinia field. Section lines indicate locations of Figures 7, 8, and 9. Seismic line shown in Figure 10 parallels section A-A'.

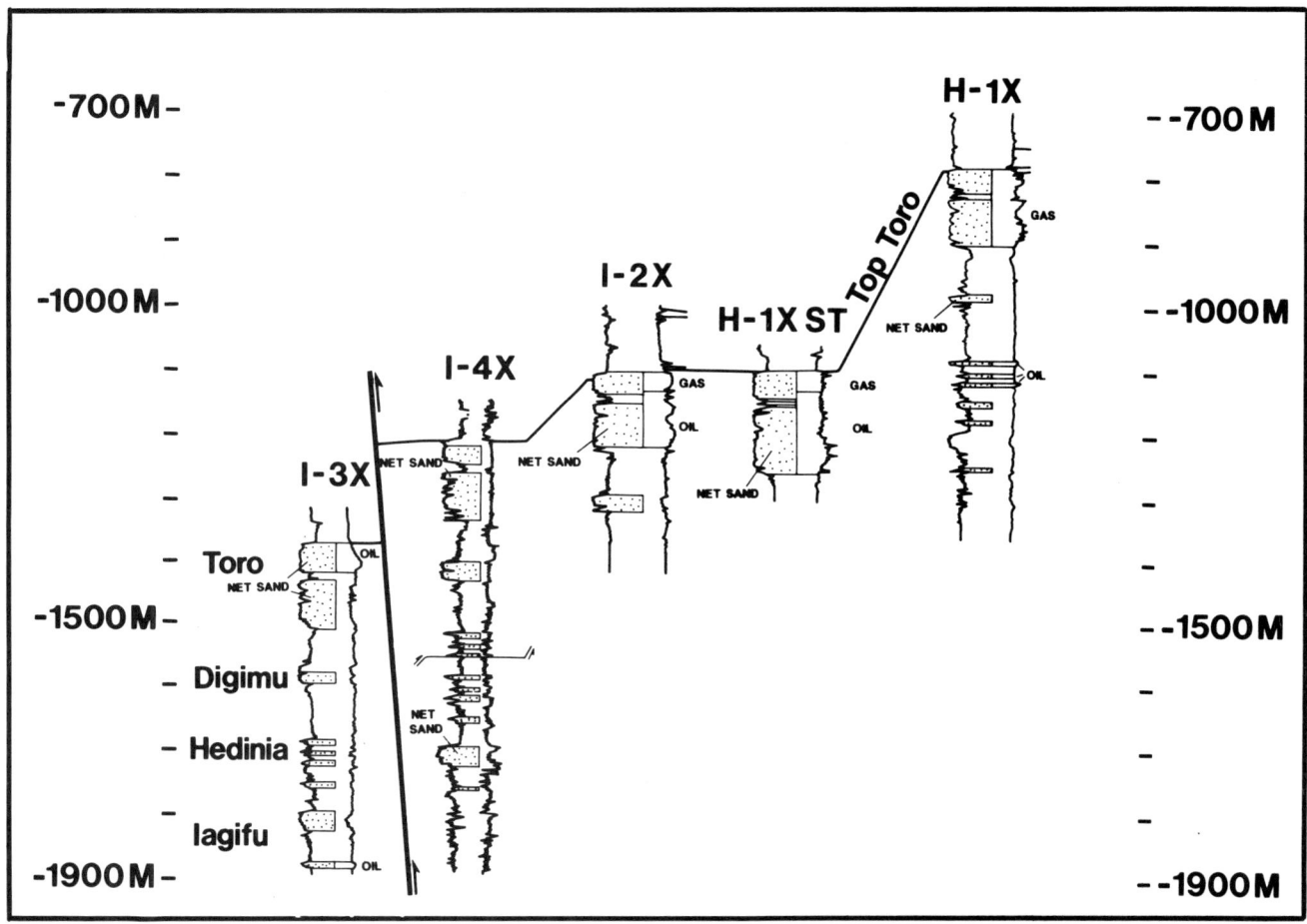

FIGURE 5. Schematic structural cross section showing early delineation history of Iagifu/Hedinia field.

margin of the Australian plate during the Triassic and Middle Jurassic, respectively (Home et al., 1990). The Jurassic-Upper Cretaceous section records a long interval of transgressions onto the Australian craton, punctuated by brief regressive episodes. Reconstructions of the early history of the basin have been presented by Home et al. (1990) and by Struckmeyer et al. (1990). The discussion below is limited to the stratigraphy penetrated within the license area, as illustrated in Figures 5 and 6.

Mesozoic Stratigraphy

The oldest strata penetrated in the Iagifu/Hedinia area comprise approximately 200 m of shallow (?) marine sandstones, siltstones, and shales of Callovian age in the upper Koi Iange Formation. The Koi Iange Formation is overlain with apparent conformity by the 600 m thick Oxfordian-Tithonian Imburu Formation (Figure 6). The basal Imburu section is predominantly open marine shale and siltstone, marking a significant transgression at the end of Koi Iange deposition. Thin intervals of fair- to good-quality source rocks are present in the lower Imburu. The uppermost 300 m of Imburu Formation comprises three coarsening-upward sand packages (Figure 5) termed the Iagifu, Hedinia, and Digimu members, in ascending order (Denison and Anthony, 1990). These sandstones are secondary reservoir objectives at Iagifu/Hedinia.

The Berriasian Toro Formation is the primary reservoir objective in the Papuan fold and thrust belt and contains the bulk of the oil reserves at Iagifu/Hedinia. It is approximately 100 m thick and can be subdivided into three coarsening-upward cycles in the field area. The lower contact of the Toro sands with Imburu shales is transitional and is placed at the base of the oldest coarsening-upward massive sand cycle.

The upper contact with the Ieru Formation is sharp but conformable. It records a major transgression at the base of the condensed, late Berriasian-Barremian sequence of shales and siltstones that comprises the Alene Member of the Ieru Formation. The 75 m thick Alene Member is the top seal for Iagifu/Hedinia and other accumulations in the Toro sandstone. Geochemical analysis suggests that the Alene and younger units within the Ieru Formation do not contribute significantly to generation of hydrocarbons in the PPL 100 area.

The upper Ieru section is dominated by shales, siltstones, and argillaceous sandstones that display an overall coarsening-upward trend, reflecting progradation of coarser shelf facies northward. In the Iagifu/Hedinia area, Ieru strata of upper Cenomanian-Turonian age are truncated beneath the pre-Cenozoic unconformity.

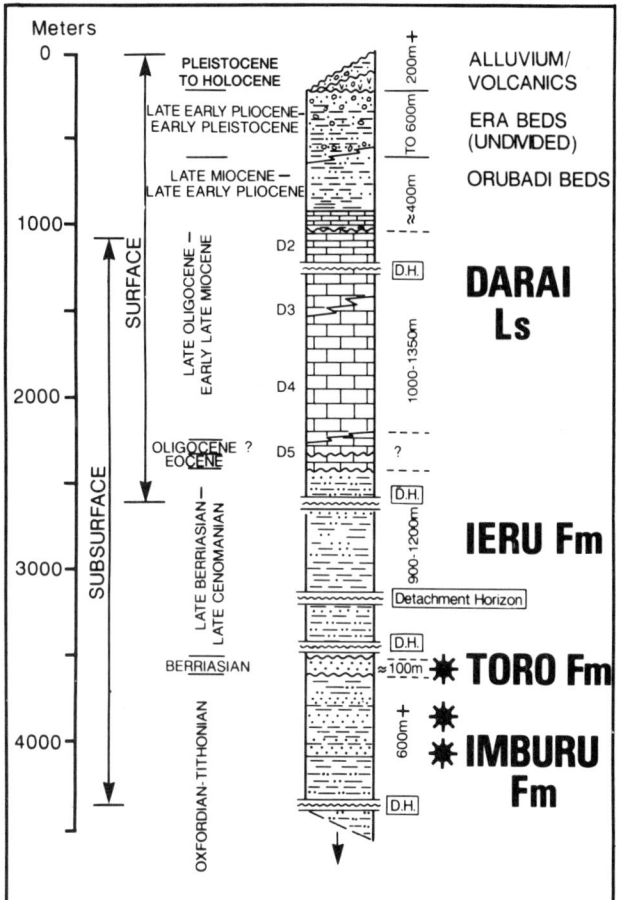

FIGURE 6. Stratigraphic section encountered by wells in PPL 100.

Cenozoic Stratigraphy

The Darai limestone of Eocene (?) to late Miocene age is a platformal unit approximately 1000 to 1350 m thick that marks the base of the foreland basin succession. Carbonate deposition was terminated diachronously by southward encroachment of siliciclastic deposition related to uplift of the fold and thrust belt. Pigram et al. (1989) have discussed the conceptual geometry of the Darai foreland basin succession.

The late Miocene-early Pliocene Orubadi beds are approximately 400 m thick, with interbedded marine shales and limestones grading upward through terrestrial sands and shales into a transitional contact with the Era beds. The Era beds comprise a synorogenic early Pliocene-Pleistocene unit up to 600 m thick. They consist of interbedded nonmarine sandstones, siltstones, and claystones, with thin lignites being locally present. The sandstones are often conglomeratic, with clasts derived from uplifted Tertiary beds. Toward the top of the Era beds, the unit also contains tuffaceous beds and agglomerates.

Pleistocene and Quaternary basaltic volcanic aprons from Mt. Bosavi and Mt. Sisa on the northwest and Mt. Murray on the southeast overlie the Darai, Orubadi, and Era beds with pronounced angularity. Thus, volcanic activity partly postdates the emplacement of major structures in this area of the thrust belt. However, coeval volcanic rocks are folded in conjunction with the frontal thrust belt structures, which indicates that active uplift has continued to the present time, with the deformation front migrating toward the foreland.

Surface Geology

Throughout PPL 100 and the Iagifu/Hedinia area, the dominant unit is the thick Darai limestone, which crops out in anticlines and in the hanging walls of thrust sheets. The Darai limestone crops out extensively throughout the fold and thrust belt, and is divided informally for mapping purposes into three unequally thick and poorly defined lithologic units: D2, D3, and D4 (Lamerson, 1990). The synorogenic Orubadi and Era beds overlie the Darai limestone, forming narrow strike valleys beneath succeeding thrust sheets in the hanging walls of which the Darai limestone is exposed. The Era beds are exposed only locally in PPL 100 and are generally confined to structural depressions in the footwall sections of major thrusts. Outcrops of the clastic succession are isolated, and exposures are generally poor in the rain forest and only somewhat better in the stream valleys. To the northeast of PPL 100, volcanics associated with Mt. Bosavi and Mt. Sisa unconformably overlie strata extending from the Darai limestone on the Mananda anticline to the Era beds in the footwall of the Mananda-Hedinia thrust.

Structure

The Iagifu/Hedinia complex comprises genetically related structures and hydrocarbon pools developed in the hanging wall of the Hedinia thrust, as shown in cross section in Figures 7 to 9. The Iagifu and Hedinia anticlines are distinct features at the surface, separated by a tight, deep syncline. At the Toro level, the structures merge into a broad, northwest-dipping panel, with only a minor reversal between Iagifu and Hedinia. The Iagifu surface anticline appears to be almost completely detached within the Ieru shales.

Surface control for the Iagifu and Hedinia structures comprises contact and bedding attitude data from several hundred kilometers of traverses along lines cut through the jungle. Subsurface control is limited to formation tops and dipmeter interpretations from the wells drilled to date, because attempts to obtain high-quality seismic data have been unsuccessful. Figure 10 portrays an 18-km, 192-fold CDP line acquired over Iagifu/Hedinia in 1988 at costs in excess of US $100,000 per kilometer. Data of reasonably good quality have been collected at reservoir depth in areas where the karstic Miocene limestone is covered by younger clastics, but the data from segments where limestone crops out are uninterpretable. The failure to define structure at depth by the seismic method is compounded by the nearly complete detachment of the Iagifu structure within the Ieru section. The technique of projecting the surface fold geometry downward to reservoir depth is hindered by this structural discordance.

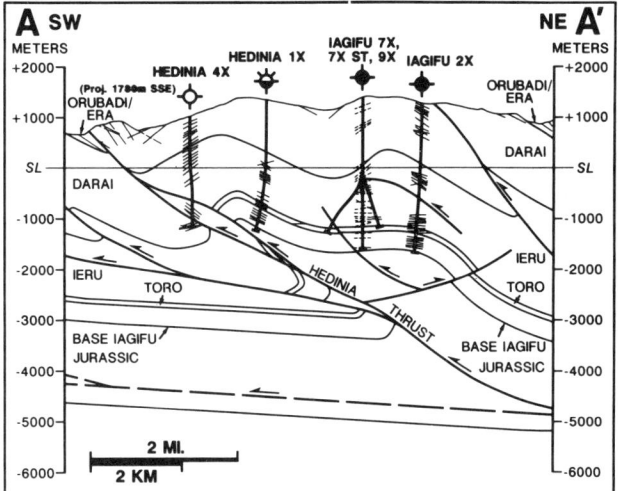

FIGURE 7. Dip-oriented cross section A-A' through the southeastern part of Iagifu/Hedinia field. Note detachment of Iagifu surface anticline within the Ieru shales, as demonstrated by Iagifu-7X dipmeter. Note continuity of oil and gas accumulation between Iagifu-2X and Hedinia-1X. For location, see Figure 4.

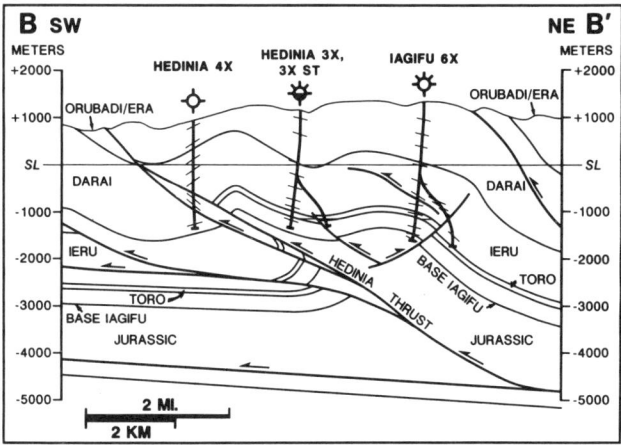

FIGURE 8. Dip-oriented section B-B' through the central part of Iagifu/Hedinia field. For location, see Figure 4.

The Hedinia thrust (or thrusts) has (have) been penetrated in four wells. As depicted in cross sections, the southwest limb of the Hedinia anticline is short as a result of truncation against the fault. It appears unlikely that the Toro unit is present to a depth sufficient to provide dip closure down to the base of the hydrocarbon column.

The Toro and subjacent reservoir units are juxtaposed across the Hedinia thrust with footwall strata ranging from Darai limestones to Ieru shales. Trapping at the level of the gas column is assumed to have resulted from a combination of dip reversal and juxtaposition of the reservoirs against impermeable strata in the footwall of the Hedinia thrust. The oil accumulation below the gas

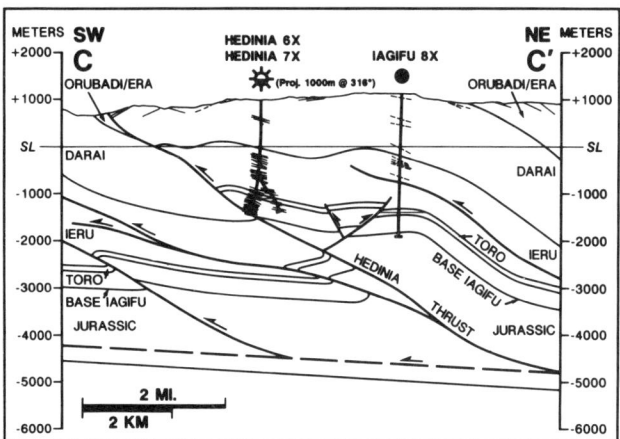

FIGURE 9. Dip-oriented section C-C' through the northwestern part of Iagifu/Hedinia field. Note accumulation at Iagifu-3X trapped on downthrown side of minor fault. For location, see Figure 4.

cap is asymmetric (Figure 4) as a result of the tilted oil-water contact described below.

Minor faulting appears to have been a significant factor in the isolation of individual pools within the complex. Given the level of subsurface control, it is possible to infer the presence of many minor structures. However, for simplicity in this discussion, only three minor faults are indicated on the map shown in Figure 4 and the sections shown in Figures 7, 8, and 9.

The backthrust depicted on the north flank of the structure locally juxtaposes the Digimu Member in its hanging wall with the Toro unit in its footwall, based on interpretation of pressure data and palynology. This implies a minimum stratigraphic separation of 60 m, which is sufficient to isolate the Toro unit and trap oil in the Iagifu-3X/8X accumulation, as depicted in Figure 9.

The southward-directed minor thrust in the saddle between Iagifu and Hedinia exhibits approximately 60 m of stratigraphic separation where it is penetrated at Iagifu-4X, which is insufficient to isolate completely the Toro unit between the Iagifu and Hedinia closures.

The final significant fault separates the main Iagifu/Hedinia accumulation from the Hedinia-8X/Usano pool. This fault is based on 400 m of repeated Ieru section in Hedinia-8X together with offset of the Toro contours between Hedinia-5X and Hedinia-8X, and on pressure data that indicate separation of both oil and gas columns between these two wells.

A few minor extensional faults have developed in the Ieru and Toro sections—possibly as a function of gravity sliding at the time of deposition, or of extension on the outer arc of concentric folds. These faults do not appear to have been significant factors in the trapping of the accumulations.

Geochemistry

Biomarker analyses demonstrate that the principal source rocks for Iagifu/Hedinia crudes are Oxfordian-Kimme-

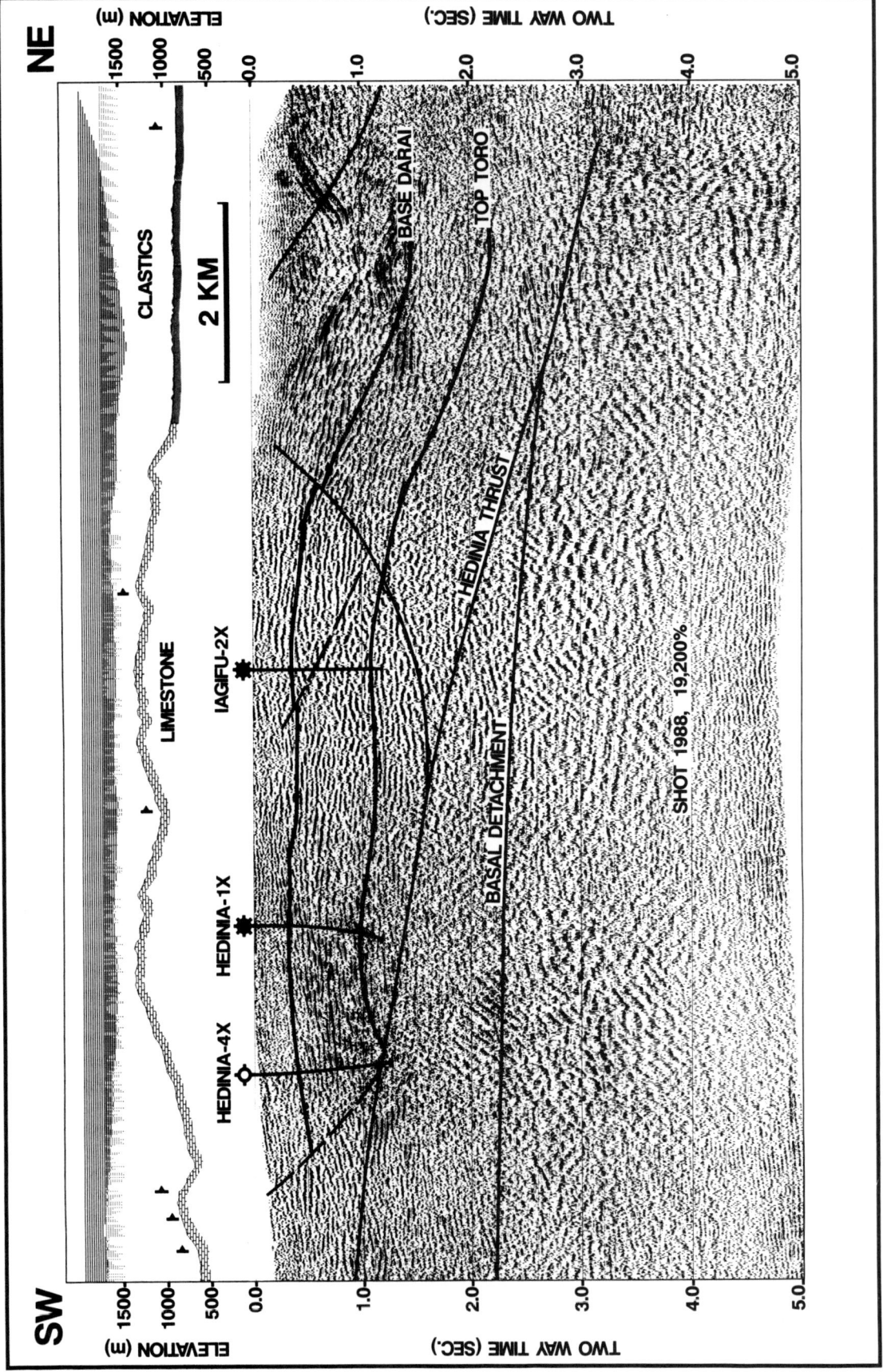

FIGURE 10. Seismic line PN-88-IAG-1, which parallels the line of section for Figure 7. Reasonable-quality data were recorded over clastic outcrops, but poor data or no data are evident where limestone crops out. Display is a stacked section, 192-fold CDP.

ridgian marine shales of the Imburu Formation. Where it is penetrated in "on-structure" locations within PPL 100, the Imburu section is immature to marginally mature. Generation occurred in synclinal areas downdip to the northeast. Maximum maturation levels were achieved as a result of burial beneath the Miocene–Pliocene foreland basin succession followed by tectonic loading associated with development of the fold and thrust belt.

RESERVOIR ROCK AND FLUID PROPERTIES

Toro Reservoir

The Lower Cretaceous (Berriasian) Toro sandstone consists of multiple cycles of coarsening-upward (cleaning-upward) quartz arenites varying in both thickness and completeness (Figure 11). This unit is interpreted as a series of stacked regressive barrier bar sand bodies. No facies higher than foreshore has yet been cored. At Iagifu/Hedinia, the Toro reservoir is divided into three members termed, in descending order, the A, B, and C sands, which are each separated by thin, transgressive breaks of bioturbated sand and claystone. Sands are friable and noncalcareous to slightly calcareous, and are composed of poorly to moderately sorted, fine- to medium-size, subrounded to rounded grains. The sediments are mineralogically very mature, with glauconite grains being the only volumetrically significant framework material other than quartz. Occasionally, minor amounts of kaolinite and pyrite are also present. The Toro net sand thickness averages approximately 90 m at Iagifu/Hedinia.

Reservoir quality in the thick, clean, upper shoreface sandstones is generally good, as summarized in Table 1 and Figure 11. Porosity is typically 12 to 15%, and permeability usually exceeds several hundred millidarcys (Table 1). Estimated connate water saturations are in the range of 15 to 25%.

Occlusion of porosity results almost entirely from syntaxial quartz overgrowth. Retention of permeability suggests that at least part of this cementation was early, thereby providing a rigid framework that resisted subsequent burial compaction. The lack of any other major diagenesis—especially authigenic clay formation—limits the potential for formation damage, such as sensitivity to freshwater completion fluids. Toro wells are typically capable of oil flows in excess of 7500 BOPD.

Digimu Reservoir

The Digimu sand is Late Jurassic (late Tithonian) in age and is separated from the overlying Toro unit by approximately 60 m of shale (Figure 11). It comprises two sand lobes separated by a thin shale. The sequence is interpreted as a regressive shallow marine bar system (Denison and Anthony, 1990). The sands are friable, fine- to medium-grained quartz arenites with minor amounts of glauconite, and are similar to the Toro sands in reservoir quality. Total net sand thickness increases northwestward across the area from a depositional limit near the southeastern boundary of the field to approximately 27 m at Hedinia-6X. Digimu wells are programmed for initial production rates of approximately 2500 BOPD.

Hedinia Sands

The Hedinia sands are Late Jurassic (early and middle Tithonian) in age and are separated from the overlying Digimu sand by 60 to 75 m of siltstone and claystone (Figure 11). The Hedinia unit comprises four coarsening-upward sand lobes, each 3 to 12 m thick, interbedded with claystones and siltstones. The sequence is interpreted as a series of stacked, regressive shoreface units, with lower shoreface silts and muds grading into shelf sands and better-sorted foreshore or bar sands (Denison and Anthony, 1990). Sands are quartzose, friable, very fine to fine grained, and moderately to well sorted. The total interval thickness is between 30 and 90 m. The only successful test of the Hedinia sands within the field area flowed 90 BOPD with several thousand pounds per square inch of drawdown at the sand face.

The Hedinia and underlying Iagifu sands have porosities that often exceed 20%, but routine core analysis shows that these sandstones are tight. Inspection by scanning electron microscopy demonstrates that pervasive early chlorite clay grain coatings protected quartz framework grains from later overgrowth. However, the clay coatings, in combination with finer grain size, resulted in much smaller and more rugose pore throats than those in the quartz-cemented Toro sands. In spite of the higher porosities of these sandstones, their permeabilities are significantly lower, probably averaging only a few millidarcys. Microporosity within the intricate chlorite structure is essentially inaccessible to migrating liquid hydrocarbons under normal reservoir buoyancy (capillary pressure) conditions. Irreducible water saturations in these rocks typically run more than 50 to 60%.

Iagifu Sands

The Iagifu member is Late Jurassic (Kimmeridgian) in age and is separated from the overlying Hedinia sands by 30 to 45 m of siltstone. The Iagifu interval consists of four generally coarsening-upward sand lobes separated by interfingering shales. The sequence is interpreted as representing a repetitive stack of regressive strandlines, or poorly developed barrier bars, deposited on a surface of low slope (Denison and Anthony, 1990). The uppermost sand, the Iagifu A sand, is usually the best developed and ranges up to 30 m in thickness. Sand quality diminishes downsection through the B, C, and D sands. The sands are very fine to coarse grained, subangular to subrounded, and poorly to moderately sorted, and are composed of 60 to 80% quartz with trace amounts of glauconite, mica, and carbonaceous material. Tests of the Iagifu C and D sands at Iagifu-3X flowed up to 3100 BOPD, although with severe drawdown.

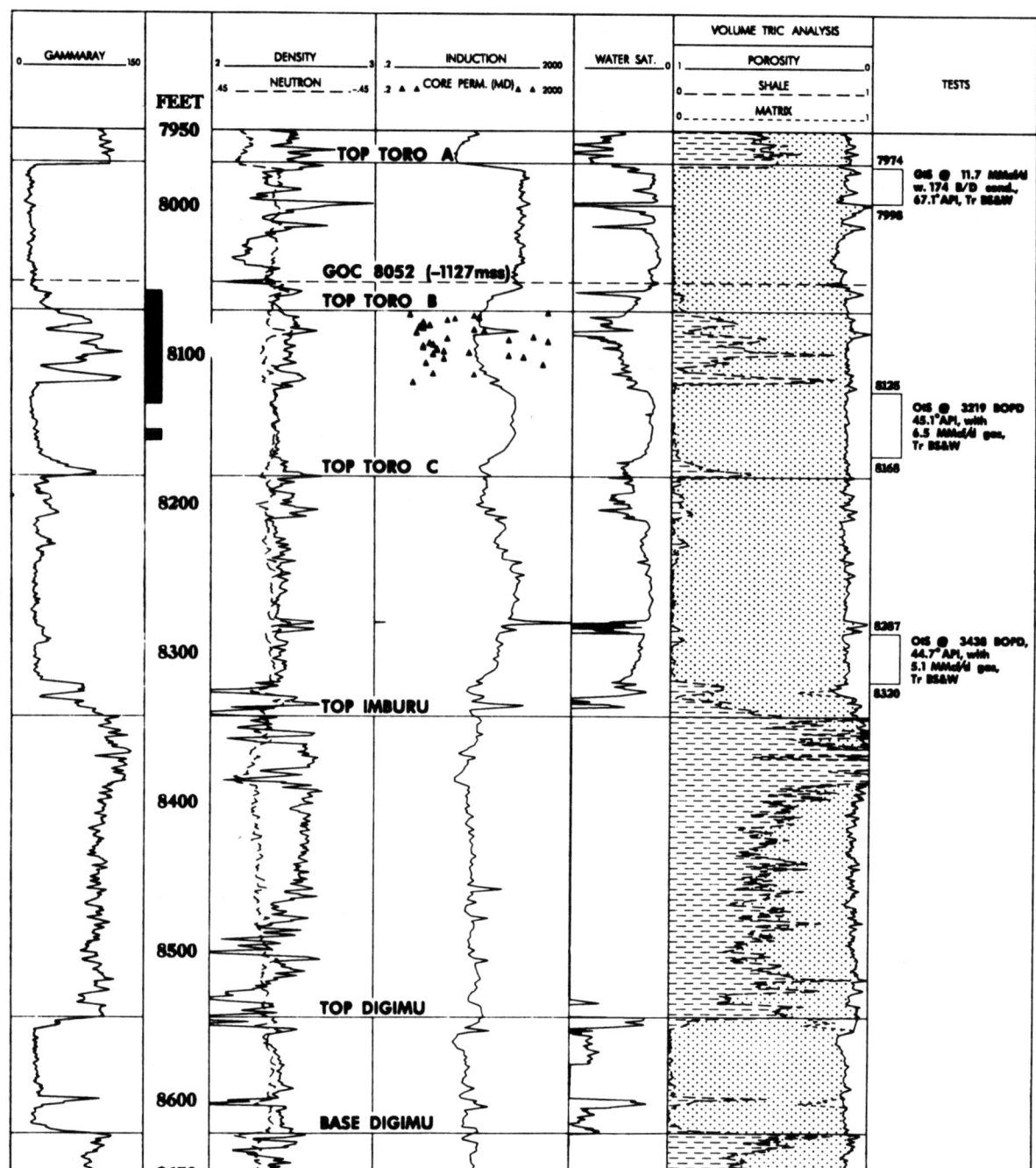

FIGURE 11. Type log for Iagifu-2X discovery well.

Reservoir Fluid Properties

Crude oils within the Iagifu/Hedinia field have a gravity of 45° API and a viscosity of 0.30 cp. Sulfur content averages 0.04 wt. %, and CO_2 content averages 1.5 mole %. Producing GOR is approximately 1000 standard ft^3/STB, and a formation volume factor of 1.55 was determined by PVT analysis. Solution gas gravity is 0.891. Condensate yield from the gas cap is approximately 10 bbl/mmcf. A summary of fluid parameters is presented in Table 2.

The formation water is brackish, containing total dissolved solids (TDS) of 11,300 mg/L. The predominant ionic constituents are Na^+ and Cl^-. Resistivity of formation water is 0.48 ohm m at 75°F (24°C).

Reservoir Temperature and Pressure

Temperatures in Iagifu/Hedinia wells are typically 140°F (60°C) for the Toro pools and 145°F (63°C) for the Digimu

TABLE 1. Reservoir rock properties for Iagifu/Hedinia field.

| | Reservoir | |
Property	Toro	Digimu
Gross thickness, average, m	100	8–27
Net sand thickness, average, m	90	6–25
Porosity (ϕ), average, %	13.4	15.5
Permeability (K), average, md	300	400
Water saturation (S_w), average, %	21.8	30.5
Temperature, °C	60	63
Expected flow rate, BOPD	8000	2000

TABLE 2. Reservoir fluid properties for Iagifu/Hedinia field.

| | Reservoir | | |
Property	Toro I-2X pool	Digimu pool	Torl I-3X pool
Oil gravity, °API	45	45	45
Reservoir gas-oil ratio, std. ft³/bbl	1000	1075	1000
Reservoir pressure, psia	2115	2151	2510
Bubble point pressure, psia	2050	2056	2437
Gas cap gravity (air = 1)	0.72	—	—
Methane content, %	0.83	—	—
Condensate-gas ratio, bbl/mmcf	10	—	—
H_2S content, %	Nil	—	—
CO_2 content, %	0.8–2.2	—	—
Water salinity (TDS), mg/L	11,300	—	—

pools. The geothermal gradient averages 1.1°F/100 ft (20°C/km). Reservoir pressures are summarized in Table 2; they range from 2115 to 2510 psi. The average pressure gradients are 0.29 psi/ft within the oil columns and 0.06 psi/ft within the gas columns.

Tilted Oil-Water Contact

On the southwest flank of the main block Toro pool, there is a pronounced tilt (up to 3°) of the oil-water interface. The northwest lobe of the oil accumulation was flushed during development of the tilted, asymmetric accumulation, leaving a residual oil column up to 90 m thick. There is also a mild tilt (0.33°) of the gas–residual oil interface between Hedinia-7X and Hedinia-3XST. The oil-water interface appears to be horizontal across both the Iagifu 3X/8X pool and the Hedinia Digimu pool.

DEVELOPMENT PLAN

The Iagifu/Hedinia field will be developed in three phases. Phase I will comprise installation of all facilities required for first oil export from the Iagifu/Hedinia and Agogo fields. Start-up at Iagifu/Hedinia is planned for mid-1992. Gas injection facilities will be installed in phase II and will be online one year after first oil. Reservoir data gathered in phases I and II will be used to determine subsequent pressure maintenance schemes. Phase III will encompass the construction of water injection facilities at Iagifu/Hedinia if required. Well requirements for the total project include as many as 20 producers, six gas injectors, and, if necessary, up to five water injectors. The Iagifu/Hedinia central production facility (CPF) will have a design capacity of 96,000 BOPD and 96 MMCFGD.

Figure 12 is a schematic illustration of the crude export system. A 261-km pipeline will be constructed from the Iagifu/Hedinia field to a marine terminal in the Gulf of Papua. The 508-mm (20-in.) diameter pipeline will operate with a capacity of 150,000 bbl per day under gravity flow conditions. Capacity can be increased to 270,000 bbl per day by use of two pump stations. The first 171 km of the pipeline will be buried onshore. The remaining marine pipeline segments will comprise 50 km of pipeline laid in the main river channel of the Kikori and Nakari rivers, and 40 km of offshore pipeline laid in the Gulf of Papua, ending at the marine terminal.

The offshore marine terminal will consist of a platform in 20 m of water and a single-point mooring buoy

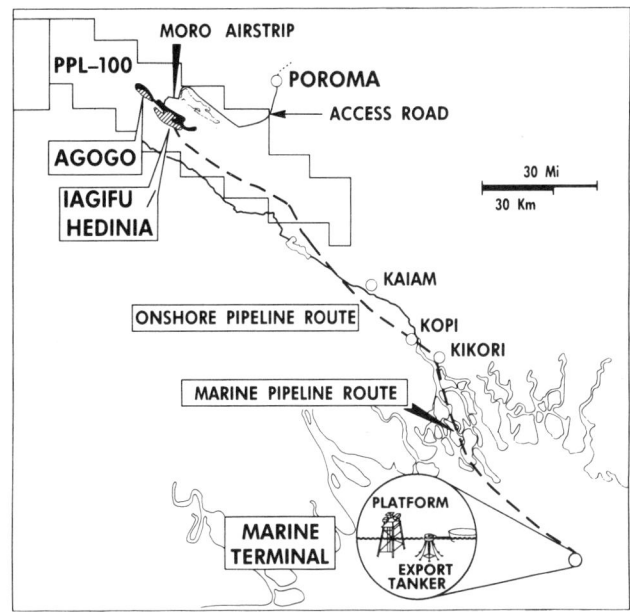

FIGURE 12. Schematic outline of Iagifu/Hedinia field facilities and export system. Construction commenced in May 1990, and first oil is projected for mid-1992.

located in a minimum water depth of 25 m to allow 50,000- to 150,000-DWT export tankers to approach, moor, and load. Export tankers will continuously occupy the mooring buoy and receive production from the pipeline. Three 100,000-bbl storage tanks located at the CPF will be used during change-out of export tankers and during loading delays. The marine terminal facilities will also include a workboat to assist export tankers, perform routine maintenance, and provide emergency standby and oil spill response capabilities.

Material and equipment for the field facilities and onshore infrastructure will be brought into PNG through the port of Lae, on the eastern coast. Material will be trucked to a staging area/airfield to the north of Lae and then flown to the field airstrip at Moro by Hercules aircraft. From the Moro airfield, material will be moved by road to the construction site.

Material for the pipeline will be brought into PNG by ocean barges and transshipped onto coastal lighters for transport to a staging area at the town of Kopi and to pipe yards along the route. The pipe will be moved along the right-of-way by truck.

ACKNOWLEDGMENTS

The authors would like to thank the members of the PPL 100 joint venture—Chevron Niugini Pty. Ltd., BP Petroleum Development Ltd., Ampol Exploration Pty. Ltd., BHP Petroleum (PNG) Inc., Oil Search Ltd., and Mitsubishi Oil Co. Ltd.—for permission to publish this paper. The interpretations presented here are those of the authors and do not necessarily correspond with the views of all parties within the joint venture. We would also like to thank the numerous Chevron employees who contributed their ideas and their time during preparation of the manuscript, especially Paul Lamerson, John Hebberger, Bob Ladd, and Roger Severson.

REFERENCES CITED

Abers, G., and R. McCaffrey, 1988, Active deformation in the New Guinea fold-and-thrust belt: seismological evidence for strike-slip faulting and basement-involved thrusting: Journal of Geophysical Research, v. 93, p. 13332-13354.

Davies, H. L., 1990, Structure and evolution of the Border Region of New Guinea, in Carman, G. J. and Z. Carman (eds.), Petroleum exploration in Papua New Guinea: Proceedings of the First PNG Petroleum Convention, Port Moresby, February 12-14, 1990, p. 245-269.

Denison, C. N., and J. S. Anthony, 1990, New Jurassic subsurface lithostratigraphic units, PPL-100, Papua New Guinea, in Carman, G. J. and Z. Carman (eds.), Petroleum exploration in Papua New Guinea: Proceedings of the First PNG Petroleum Convention, Port Moresby, February 12-14, 1990, p. 153-158.

Home, P. C., D. G. Dalton, and J. Brannan, 1990, Geological evolution of the western Papuan basin, in Carman, G. J. and Z. Carman (eds.), Petroleum exploration in Papua New Guinea: Proceedings of the First PNG Petroleum Convention, Port Moresby, February 12-14, 1990, p. 107-117.

IMPS, 1990, PNG wells directory: IMPS Research Pty. Ltd., Port Moresby, PNG, 2nd ed., January 1990.

Lamerson, P. R., 1990, Evolution of structural interpretations in Iagifu/Hedinia field, Papua New Guinea, in Carman, G. J. and Z. Carman (eds.), Petroleum exploration in Papua New Guinea: Proceedings of the First PNG Petroleum Convention, Port Moresby, February 12-14, 1990, p. 283-300.

Pigram, C. J., P. J. Davies, D. A. Feary, and P. A. Symonds, 1989, Tectonic controls on carbonate platform evolution in southern Papua New Guinea: passive margin to foreland basin: Geology, v. 17, p. 199-202.

Rickwood, F. K., 1990, Towards development—the long history of petroleum exploration in Papua New Guinea, in Carman, G. J. and Z. Carman (eds.), Petroleum exploration in Papua New Guinea: Proceedings of the First PNG Petroleum Convention, Port Moresby, February 12-14, 1990, p. 1-13.

Smith, R. I., 1990, Tertiary plate tectonic secting and evolution of Papua New Guinea, in Carman, G. J. and Z. Carman (eds.), Petroleum exploration in Papua New Guinea: Proceedings of the First PNG Petroleum Convention, Port Moresby, February 12-14, 1990, p. 229-244.

Struckmeyer, H. I. M., M. Yueng, and M. T. Bradshaw, 1990, Mesozoic paleogeography of the northern margin of the Australian plate and its implications for hydrocarbon exploration, in Carman, G. J. and Z. Carman (eds.), Petroleum exploration in Papua New Guinea: Proceedings of the First PNG Petroleum Convention, Port Moresby, February 12-14, 1990, p. 137-152.

Chapter 31

Fortescue Field, Gippsland Basin, Offshore Australia
Flank Potential Realized

J. H. Hendrich
I. D. Palmer

Esso Australia Ltd.
Sydney, Australia

D. A. Schwebel

Exxon Company International
Florham Park, New Jersey, U.S.A.

ABSTRACT

The Fortescue field is the last of the "giant" oil fields discovered to date in the offshore Gippsland basin, southeastern Australia. This field is a stratigraphic oil accumulation on the western flank of the giant Halibut-Cobia oil field. The Fortescue field discovery well, West Halibut-1, was drilled in 1978—some 11 years after the Halibut discovery—as a follow-up to the dry Fortescue-1 wildcat. The occurrence of a stratigraphic trap in a sand-prone section is a unique feature of the Fortescue field. Fortescue reservoirs are Eocene sandstones that are interpreted as having been deposited in coastal plain, upper shoreface, and lower shoreface environments. Fortescue reservoirs are stratigraphically younger than, and hydraulically separated from, those in the underlying Halibut-Cobia field. Pressure data have conclusively demonstrated that there are at least three separate hydraulic systems within Fortescue field. Reserves are estimated at 280 MMSTB on the basis of an original oil in place estimate of 415 MMSTB.

Fortescue field was developed by two 21-conductor platforms, Fortescue A and Cobia A, which also developed the underlying Cobia reserves. These platforms were commissioned in 1983. At the conclusion of development drilling in early 1986, 28 productive wells had been drilled into the field.

Peak production of 100,000 BOPD from the combined development facilities was achieved in 1984 and sustained until 1986. Production received a boost in 1989 from two infill wells that were drilled following interpretation of the available reservoir data.

Following these successes, a 3D seismic survey was acquired over Fortescue, Halibut-Cobia, and Mackerel fields in 1990. A revised detailed reservoir description based on these data is expected to allow identification of additional infill and workover opportunities and to help mitigate production declines in these mature fields.

INTRODUCTION

The Gippsland basin is located mainly offshore southeastern Australia (Figure 1) and is Australia's most prolific oil and gas province. Initial reserves are estimated at 3.6 billion bbl of crude and condensate, 0.6 billion bbl of LPG, and 8.3 tcf of dry gas. The sequence stratigraphy and habitat of hydrocarbons in the Gippsland basin have been described by Rahmanian et al. (1990).

Fortescue field is a stratigraphically trapped oil accumulation on the western flank of the giant Halibut-Cobia oil field in the offshore Gippsland basin, Australia. Fortescue field was discovered in 1978 and lies 65 km offshore in water depths ranging from 65 to 73 m. Development drilling proceeded between 1983 and 1986 from two platforms, Cobia A and Fortescue A.

Cumulative production to date is about 200 MMSTB. This represents 70% of the recoverable reserves, which are assessed at 280 MMSTB. The field is currently producing at a rate of 45,000 BOPD, down from a maximum rate of 100,000 BOPD.

In 1989, two successful infill wells were drilled following a geologic reinterpretation and the integration of all available data, including production history and production logging data. Seismic modeling performed on an experimental high-resolution seismic line was instrumental in the choice of one of the two locations. These wells are currently responsible for 50% of the field's total production. A 3D seismic survey was recently acquired over the Fortescue-Halibut-Cobia area and will play a key role in defining the remaining potential for these fields.

EXPLORATION HISTORY

The discovery of Fortescue field in 1978 was the last "giant" oil discovery made in the Gippsland basin. Figure 1 illustrates the commercial oil and gas discoveries made through 1990. The majority of the large discoveries were made during the highly successful exploration period in the late 1960s. The discoveries made since 1978 have been significantly smaller, reflecting the maturity of the conventional play types.

The exploration history of the Fortescue-Halibut-Cobia area is detailed in Thornton et al. (1980). Figure 2

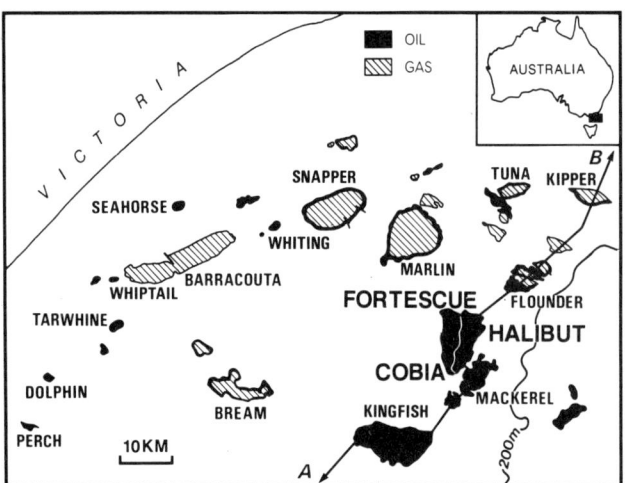

FIGURE 1. Gippsland basin discoveries to date. Cross section AB is shown in Figure 6.

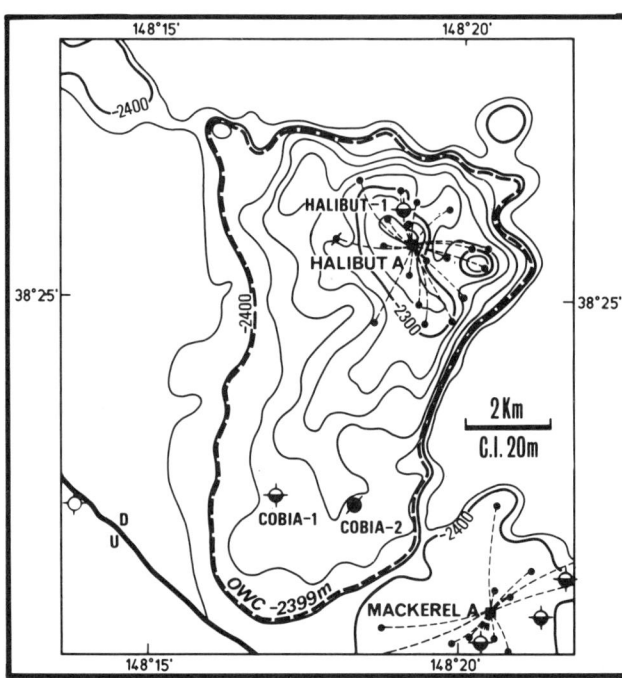

FIGURE 2. Structure map on top of Latrobe Group, showing early exploration and development wells in Halibut-Cobia area. For well legend, see Figure 17.

illustrates the early exploration history of this area. The Halibut-1 wildcat intersected a 120-m gross hydrocarbon column near the crest of the Halibut structure in 1967. On the strength of this discovery, the 24-conductor Halibut A platform was commissioned in 1969, from which 20 development wells were drilled between 1969 and 1970. To date, cumulative production from the 19 productive Halibut wells (including one plugged and abandoned, or P&A) is 729 MMSTB, which represents 85% of the recoverable reserves. Some Halibut wells have produced in excess of 50 MMSTB.

The Cobia-1 wildcat was drilled on the southern flank of the Halibut field in 1972 and confirmed the southerly extension of the field. Cobia-2, drilled in 1977, confirmed the Cobia area for development with an additional platform. Cobia-2 was itself completed as a subsea completion tied to the Mackerel platform by two 90-mm, 4.4-km flow lines. Cobia-2 pioneered the use of subsea completions in the Bass Strait and produced 1.7 MMSTB of oil between June 1979 and April 1983.

Figure 3 illustrates the exploration wells associated with the Fortescue discovery. Fortescue-1 was drilled in 1978 and intersected water-bearing, reservoir-quality sands that were stratigraphically younger than those of the underlying Halibut reservoirs. This well provided strong evidence that a stratigraphic trap might exist on the western flank of the Halibut-Cobia field. Later that year, West Halibut-1 was drilled updip of Fortescue-1 and intersected a 54-m oil column, thereby discovering the Fortescue field some 11 years after the Halibut discovery.

The Fortescue-2, -3, and -4 delineation wells, drilled in 1979, confirmed that Fortescue was a "giant" oil discovery. These wells established separate original oil-water contacts for the Fortescue field that were deeper than that of the underlying Halibut-Cobia field. RFT pressures also showed a pressure discontinuity across the shale base seal between the two fields.

STRATIGRAPHY

The generalized stratigraphy of the Gippsland basin is summarized in Figure 4. The economic basement is the Early Cretaceous Strzelecki Group. The Cretaceous-Eocene clastic sediments of the Latrobe Group were deposited in a rapidly subsiding east-west graben formed at the time of breakup between the Australian and Antarctic continents. Unconformably overlying the Latrobe Group are the marine shales of the Lakes Entrance Formation, which provide an excellent seal for the top of the Latrobe structures, and the calcareous mudstones of the Seaspray Group commonly known as the Gippsland limestone.

The top of the Latrobe Group is an unconformity surface cut by deep channels as a result of structuring and subsequent erosion during the Eocene. Figure 5 illustrates how the shale-filled Marlin channel contributes to the trapping mechanism for many of the large fields, including Fortescue and Halibut-Cobia, by providing a lateral seal. Figure 6 is a cross section through the major fields in the basin. Oil and gas accumulations occur primarily at the top of the Latrobe Group, which has been the main exploration target. Large structures in this play have been effectively exhausted, and current exploration focus has turned to traps within the Latrobe Group, of which the Flounder and Kipper fields are examples.

A detailed stratigraphic description of the Fortescue reservoirs is provided in Thornton et al. (1980), and the sequence stratigraphic framework is described in Rahmanian et al. (1990). Figure 7 illustrates the depositional sequences that encompass the Fortescue and Halibut-Cobia reservoirs. Reservoirs prefixed "M" are Halibut-Cobia reservoirs, whereas "FM" prefixes denote Fortescue reservoirs. The reservoirs shown in Figure 7 were depos-

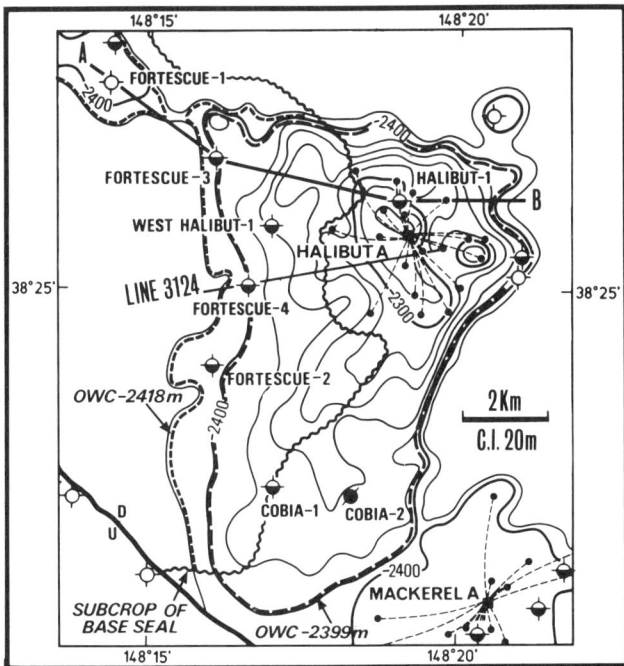

FIGURE 3. Structure map on top of Latrobe Group, showing discovery history of Fortescue field. Cross section AB is shown in Figures 7 and 8. For well legend, see Figure 17.

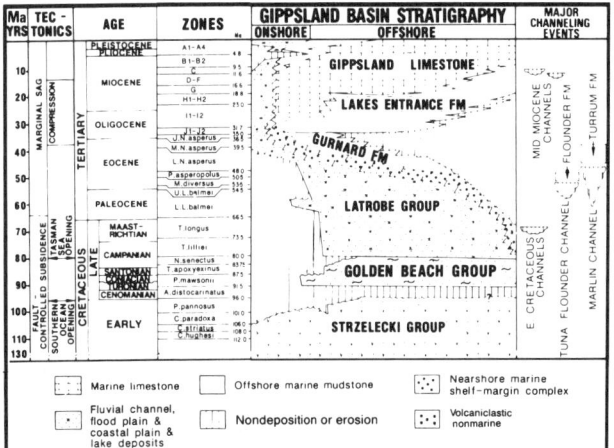

FIGURE 4. Stratigraphy and tectonics of the Gippsland basin.

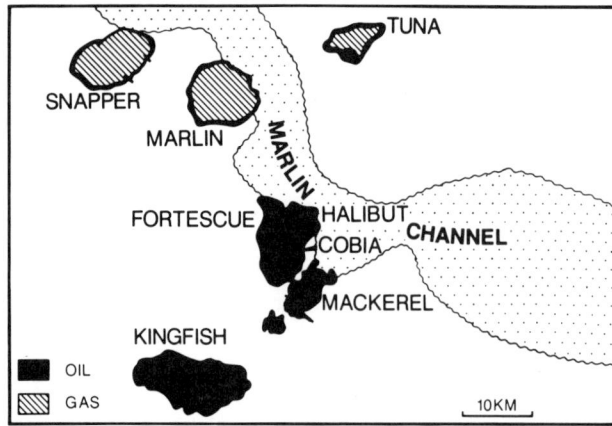

FIGURE 5. Eocene channeling in the Gippsland basin. Shale-filled Marlin channel contributes to trapping mechanism for many large fields at the top of the Latrobe Group.

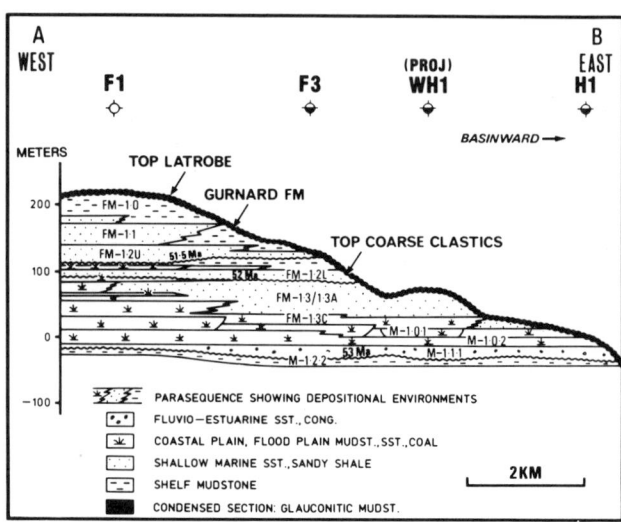

FIGURE 7. Stratigraphic cross section AB. Fortescue reservoirs are prefixed "FM"; Halibut-Cobia reservoirs are prefixed "M." For location of section, see Figure 3.

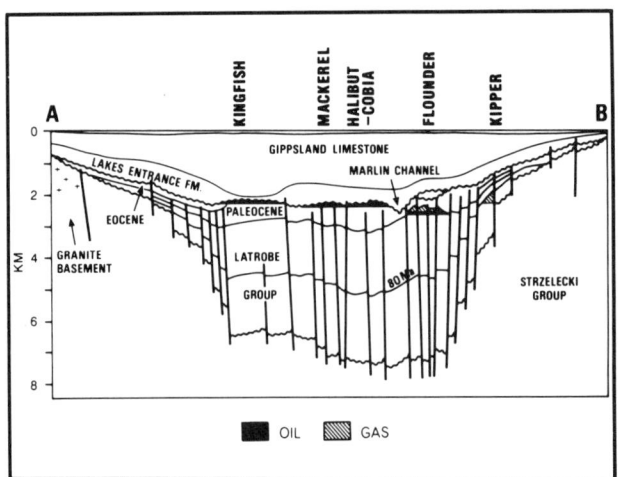

FIGURE 6. Regional cross section AB, showing major discoveries. For location, see Figure 1.

ited during the period represented by the lower to upper *Malvacipollis diversus* palynological zone. Fortescue reservoirs display an over-all retrogradational stacking pattern attributed to large, basin-wide accommodation that resulted from the combined effect of tectonic subsidence and eustatic sea level rise (Rahmanian et. al., 1990).

Interpreted reservoir facies from base to top include fluvial, coastal plain, nearshore marine, and offshore strata. Marine influence gradually increases up the section. The marine shales of the FM-1.2 reservoir and nonmarine shales of the FM-1.3A and FM-1.3C reservoirs act as semiregional and local internal seals that affect the production behavior and reservoir pressure distribution within the field.

STRUCTURE

The Fortescue and Halibut-Cobia fields are large stratigraphic traps at the top of the Latrobe Group. Entrapment is provided by truncation of westward-dipping strata by the erosional unconformity at the top of the Latrobe Group. The morphology of the structure at the top of the Latrobe Group is illustrated in Figure 3. The Halibut A platform is located near the crest of the structure, whereas Cobia A is situated on the low-relief southern flank of the structure. The steep northern and eastern flanks were formed by channeling during the Eocene. Figure 8 illustrates the internal geometry of the Fortescue and Halibut-Cobia reservoirs. Hydrocarbons occur in several pools, isolated by relatively thin, laterally continuous seals.

Figure 9 illustrates the seismic expression of Fortescue reservoirs as a wedge bounded by the top of the Latrobe Group unconformity and a westward-dipping (3 to 4°) internal base seal. The westward dip of the top of the Latrobe unconformity is shallower than the dip of the underlying strata, resulting in the preservation of progressively younger strata to the west. Two normal growth faults trending west-northwest intersect the field. However, growth had ceased by the onset of Fortescue reservoir deposition, and fault throw is interpreted as being insignificant in the lower Fortescue reservoir units (Thornton et al., 1980).

SEISMIC COVERAGE

The existing 2D seismic data form a grid, measuring approximately 0.5 × 1.0 km, of essentially pre-1985 data and do not have sufficient resolution for accurate mapping of any internal Fortescue reservoir horizons. A high-

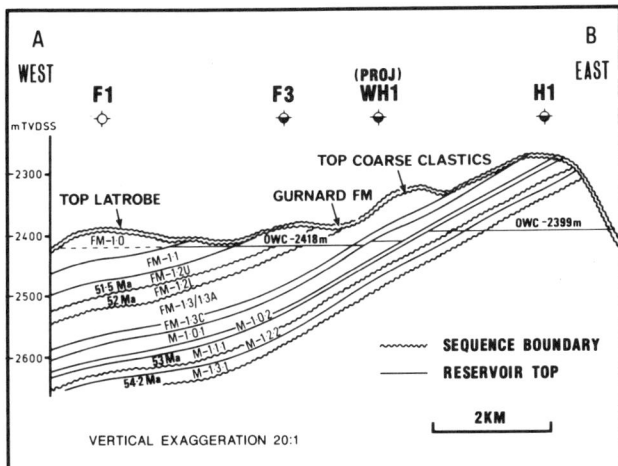

FIGURE 8. Structural cross section AB. Fortescue reservoirs are prefixed "FM"; Halibut-Cobia reservoirs are prefixed "M." For location of section, see Figure 3.

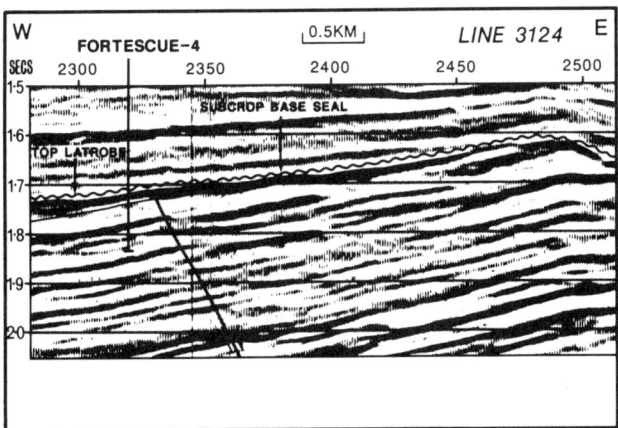

FIGURE 9. Seismic line 3124, illustrating seismic expression of Fortescue field on western flank of Halibut-Cobia field. Location of line is shown in Figure 3.

resolution 3D seismic survey was acquired over the Fortescue field, the Halibut-Cobia field, and most of the Mackerel field in 1990. This survey covered an area of 190 km^2 and consisted of 4350 km shot at a line spacing of 50 m. The 25-m shotpoint interval, together with the 50-m line spacing, yielded a bin size of 12.5 by 50 m.

The 60-fold data were recorded to 4 sec two-way travel time, or TWT, with the cable being towed at a depth of 6 m. This resulted in frequencies of up to 80 Hz at the target horizon (top of Latrobe Group) with a peak frequency of 35 Hz, compared with peak frequencies of 20 to 25 Hz in the best of the 2D data.

The data set is being interpreted on a Charisma workstation, which also enables the interpreter to perform attribute mapping. Early interpretive work indicates that the data have sufficient resolution to permit accurate mapping of internal reservoir horizons. This interpretation will form the basis for a more accurate reservoir description and further field development.

RESERVOIR CHARACTERISTICS

Porosity-Permeability

Table 1 summarizes the average porosities and net/gross ratios for the major Fortescue reservoir units. The FM-1.3C and FM-1.3A units are the oldest Fortescue reservoirs and have excellent average porosities but relatively low net/gross sand ratios. This reflects the marginal marine depositional environment of these units, highlighting the influence of facies changes from nearshore marine sands to coastal plain coals and shales within the field limits (Figure 7). These units contain 35% of the original oil in place.

The FM-1.3 and FM-1.2L units are generally thick and exhibit excellent reservoir properties. Figure 10 illustrates the interpreted maximum landward positions of paleoshorelines at various stages during deposition of Fortescue and Halibut-Cobia reservoirs. The paleoshoreline that existed during deposition of the FM-1.3 and FM-1.2L units is interpreted as having moved relatively landward with continuing transgression, compared with that at the time of FM-1.3C and FM-1.3A deposition. The FM-1.3 and FM-1.2L units contain nearly 50% of the original oil in place.

Because most of the exploration wells were drilled approximately along depositional strike down the field, the effects of the transgression on reservoir properties was not well defined until significant development drilling had taken place. As with many developments, the true geologic complexities were underestimated by the exploration drilling. For example, all the FM-1.3/1.3A and FM-1.3C sands were originally expected to comprise one large, relatively homogeneous unit throughout the reservoir.

TABLE 1. Average porosity and net/gross ratio for each major Fortescue reservoir unit.

Reservoir unit	Average porosity* (%)	Net/gross ratio
FM-1.0**	20	20
FM-1.1	12	30
FM-1.2U	21	27
FM-1.2L:		
Shaly	17	66
Clean	22	99
FM-1.3	21	71
FM-1.3A	20	34
FM-1.3C	19	43

*Porosity cutoff, 12%.
**Localized discrete sandstone body around West Fortescue-1.

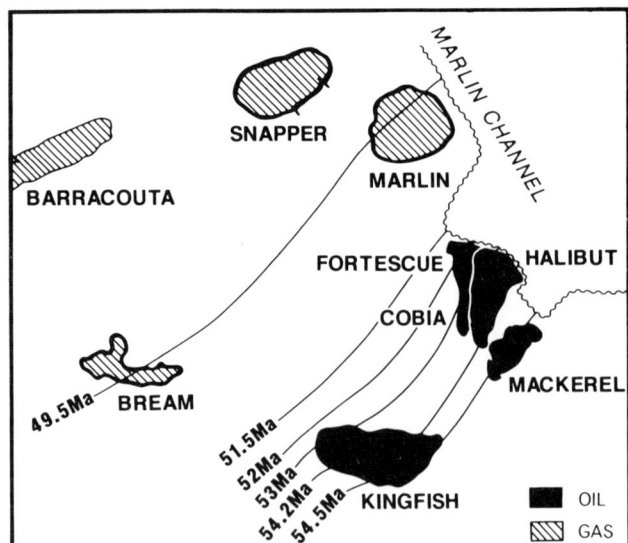

FIGURE 10. Interpreted maximum landward positions of paleoshorelines associated with deposits of the 54.5- to 49.5-Ma sequences.

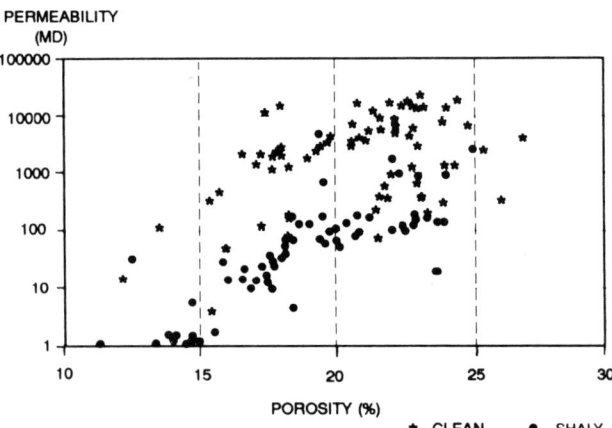

FIGURE 11. Core porosity vs. core permeability relationships for Fortescue reservoirs. Note clean and shaly trends.

A further landward shift in the paleoshoreline as a result of continued transgression resulted in a high proportion of marine shale during deposition of the FM-1.2U and FM-1.1 units within the field limit. A marine shale separates the FM-1.2 reservoir into upper (FM-1.2U) and lower (FM-1.2L) units. Pressure discontinuities of up to 150 psi (see Figure 15) have been recorded across this shale, indicating that it is acting as an effective seal during production. The FM-1.1 reservoir is of marginal quality within the field limit and has not been produced to date. The FM-1.1 reservoir properties improve markedly in a landward direction (Figure 7); however, the reservoir-quality FM-1.1 sands are mapped entirely within the aquifer.

The facies changes—particularly within the FM-1.3C/FM-1.3A and FM-1.2L/FM-1.2U units—have profound effects on the reservoir pressure dynamics and production performance.

Permeability

Figure 11 is a plot of core porosity vs. core permeability that illustrates two strong relationships that have been used in modeling the depletion of Fortescue field. Clean sands exhibit multidarcy permeabilities characteristic of Gippsland basin reservoirs at the top of the Latrobe Group. In contrast, shaly sands with similar porosities exhibit significantly lower permeabilities—in some cases two orders of magnitude lower. Shaly sands tend to be lower shoreface sands.

Although Fortescue reservoirs are of excellent quality and appear quite uniform on logs, significant permeability contrasts cause these reservoirs to behave in a highly stratified manner, resulting in uneven sweep of some reservoir units. Permeabilities decline markedly when core porosities fall below 15%. Reservoir rock is usually of excellent quality, which is reflected in the paucity of data points in the lower-porosity range.

Oil Properties

Table 2 summarizes the major properties of Fortescue oil. Fortescue oil is a light, undersaturated, low-viscosity crude with low bubble and pour points. The low gas-oil ratio (GOR), relative to the API gravity, is believed to be a function of methane having been stripped from the oil during migration. Hydrocarbon source, maturation, and migration are discussed in Rahmanian et al. (1990). A detailed geochemical analysis/comparison of the Fortescue, Halibut, and Cobia oils is documented in Thornton et al. (1980).

These oil properties, combined with excellent reservoir properties and strong aquifer support, ensure high sweep efficiencies and consequently high recovery factors estimated at 67% (Henzell et al., 1985). The strong aquifer support is attributed to the generally sand-prone Latrobe Group, which outcrops onshore and is continually recharged by meteoric waters (Kuttan et al., 1986).

TABLE 2. Major properties of Fortescue field oil.

Gravity, °API	41
Gas-oil ratio, std. ft³/STB	42
Viscosity, cp	0.67
Formation volume factor (B_o), rb*/STB	1.151
Bubble point, psig	364
Pour point, °C	5

**rb = reservoir barrels.

FIELD DEVELOPMENT

Fortescue field was developed from the 21-conductor Fortescue A and Cobia A platforms, which were commissioned in 1983. Recoverable reserves are estimated at 280 MMSTB on the basis of an original oil in place (OOIP) of 415 MMSTB. The productive field area of Fortescue field is approximately 31 km^2, and the maximum column height is 100 m.

Figure 12 illustrates the total net oil sand thicknesses of the Fortescue reservoirs. Reservoir quality and thickness are greatest in the north, where the top of the Latrobe unconformity has not cut as deeply into the Latrobe Group sediments.

Development drilling commenced from the Cobia A platform in March 1983 and from the Fortescue A platform in June 1983. At the conclusion of development drilling from Cobia A in May 1985 and from Fortescue A in April 1986, nine wells from Cobia A (including one redrill) and 23 wells from Fortescue A (including one redrill and one sidetrack) had been targeted for Fortescue reservoirs. Of these, 28 wells (20 Fortescue A and eight Cobia A wells) are productive from Fortescue field.

Three dry holes (including the last two wells in the original development program) were drilled from the Fortescue A platform, all from the No. 8 conductor. The last two development wells (A-8A and the A-8B sidetrack) targeted the updip pinch-out edge of the Fortescue reservoirs. Unexpected non-net rock thickness resulted in the truncation of the target reservoir, and these wells were plugged and abandoned as dry holes. At the conclusion of the original development drilling phase, development opportunities appeared to have been exhausted.

Two Cobia A wells (CF-8 and CF-17) originally interpreted as having intersected Fortescue reservoir units were recorrelated as having intersected a discrete accumulation within the underlying Halibut-Cobia M-1.0.1 reservoir (Bretherton, 1987). This accumulation (Figure 12) is interpreted as an erosional remnant formed by the top of the Latrobe unconformity in the low-relief southern area of the field.

PRODUCTION

Production from Fortescue field commenced in May 1983 from the Cobia CF-2 well, which came online at 12,000 BOPD. Production performance is illustrated as functions of time and cumulative production in Figures 13 and 14, respectively. Production reached a government-imposed ceiling of 100,000 BOPD (representing 12.5% estimated ultimate recovery, or EUR/yr) in early 1984. This rate was sustained until late 1987. Production decline was partially offset by the additional production capacity provided by two infill wells drilled in 1989.

Reservoir Pressures

Figure 15 shows datumned reservoir pressures of various Fortescue and Halibut-Cobia reservoirs vs. time. The majority of the data are RFT pressures that were obtained in six exploration and delineation wells and 20 development wells. These data are a valuable reservoir management tool and have been used extensively to aid geologic correlation.

Initial Fortescue reservoir pressures were slightly higher than initial Halibut pressures. The slight pressure decline (100 psi) in Halibut-Cobia reservoirs during the 21 years of production since 1969 reflects excellent reservoir continuity and highlights the excellent aquifer support in

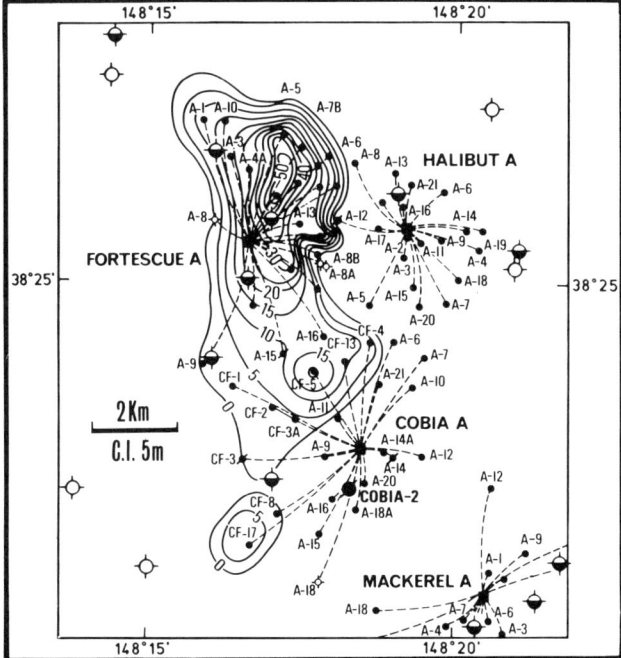

FIGURE 12. Total net oil sand thicknesses of Fortescue reservoirs, showing development wells. Note isolated erosional remnant in south. For well legend, see Figure 17.

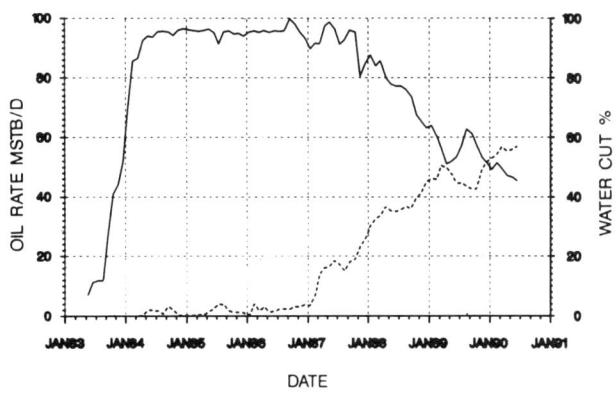

FIGURE 13. Fortescue field production performance vs. time. Note boost to performance provided by infill wells in 1989.

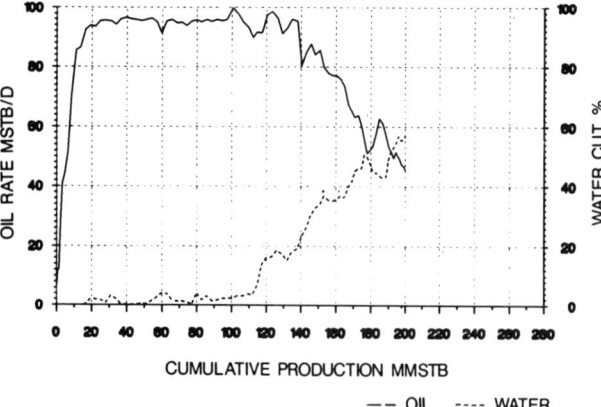

FIGURE 14. Fortescue field production performance vs. cumulative production.

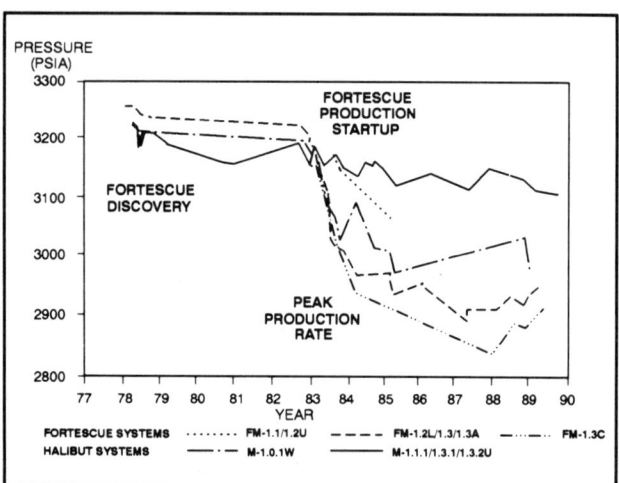

FIGURE 15. Reservoir pressure datumned to −2250 m subsea as a function of time. Note rapid pressure decline in Fortescue and M-1.0.1W reservoirs after Fortescue production start-up.

these fields. In contrast, Fortescue reservoir pressures declined rapidly following the start-up of production from the Fortescue field in 1983. Fortescue reservoir pressures stabilized and reached a steady state with the aquifer after production rates plateaued in 1984. After nearly seven years of production from Fortescue, several dynamic pressure systems are evident. Note in particular the pressure discontinuities among the FM-1.1/FM-1.2U, FM-1.2L/FM-1.3/FM-1.3A, and FM-1.3C units caused by semiregional and local seals.

Cumulative production from Fortescue field to date is 200 MMSTB, which represents approximately 70% of the recoverable reserves. Gas lift and water handling were commissioned in February 1987 on both the Cobia A and Fortescue A platforms. The current oil production rate is 45,000 BOPD from the combined facilities.

INFILL DRILLING

In 1989, the "gorilla" class jack-up drilling rig called the "Maersk Giant" was brought to Bass Strait waters to undertake development drilling on several new developments. This rig was also used to drill two infill wells on Fortescue field. Production performance in the northeast of the field suggested greater reservoir stratification than had initially been mapped after the end of development drilling. Analysis of pulsed neutron capture (TDT) logs suggested that thin (1-m) coals could be acting as local seals preventing efficient vertical sweep of the reservoirs. Production anomalies between areally close wells gave credence to this hypothesis. Indeed, history matching of a reservoir simulation model by means of the previously assumed less stratified geologic model proved to be impossible. Figure 16 is a schematic cross section that illustrates the incremental reserves postulated as being trapped beneath the unconformity surface updip of existing wells, provided that the sealing coals are of sufficient areal extent in an updip direction.

All wells in the area were recorrelated by use of a sequence stratigraphic framework and incorporation of all available production data. The refined geologic interpretation indicated that incremental reserves could exist in the FM-1.3A reservoir updip of the existing A-20 and A-14 wells (Figure 16). Figure 17 illustrates the distribution of these incremental reserves within the FM-1.3A reservoir and the locations of the A-8C and A-15A infill wells.

Seismic Modeling

The A-8C and A-15A infill wells were targeted on good-quality seismic lines (Figure 17). Line G88A-9902, passing through the A-8C location, is an experimental high-

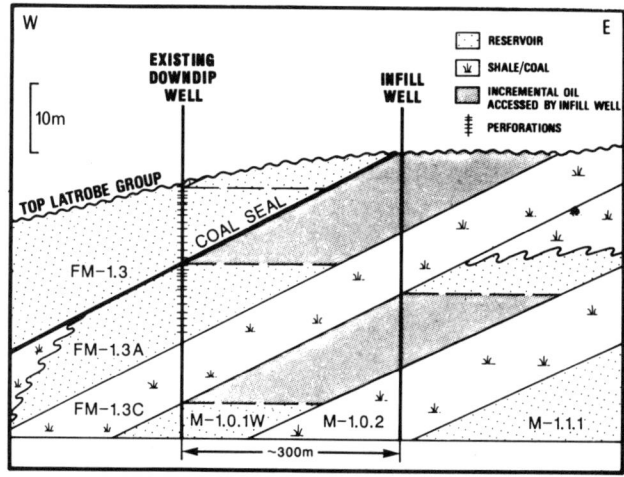

FIGURE 16. Schematic west-east cross section through proposed infill drilling location, illustrating postulated incremental reserves in the Fortescue FM-1.3A reservoir. Note incremental reserves in Halibut M-1.0.1W reservoir provided by updip stratigraphic pinch-out of sand.

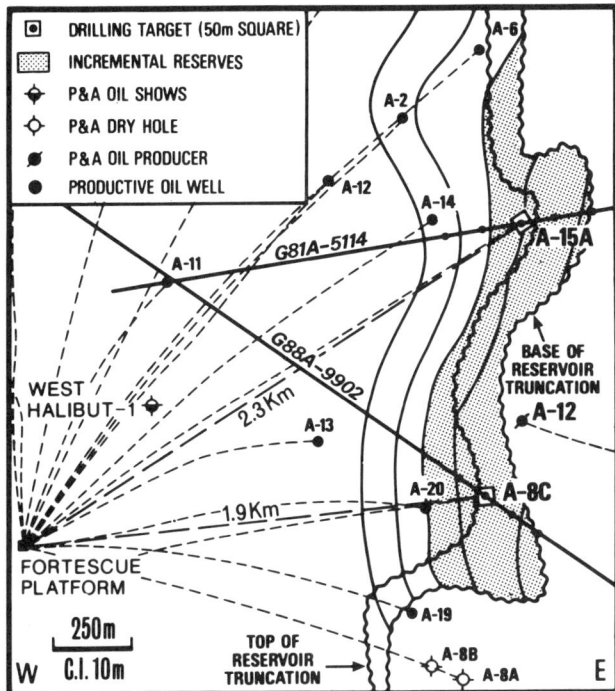

FIGURE 17. Schematic structure map on top of FM-1.3A reservoir on northeast flank of Fortescue field. Note infill drilling locations A-8C and A-15A near updip truncation edge of reservoir.

resolution line, whereas line G81A-5114 represents the best quality of the 2D vintage data.

Forward seismic modeling was performed on these lines to determine whether the updip truncation of the sealing coals by the top of the Latrobe unconformity could be identified with confidence. This modeling work indicated that a 1-m coal could in fact yield a recognizable change in the character of the seismic reflection at the top of the Latrobe Group close to the truncation edge, and played a key role in the choices of both well locations.

The A-8C location was targeted in a geologically uncertain area. Two dry wells (A-8A and A-8B) to the south and the abandoned Halibut A-12 well to the north defined the maximum easterly limit of any Fortescue reservoir in this area. Fortescue A-19 to the southwest had encountered abnormally thick (20 m, compared with an average of 2 to 3 m) Gurnard Formation sediments, which had truncated the FM-1.3A reservoir in this area. The A-19 well is producing from the underlying FM-1.3C reservoir, which was not expected to be developed in the A-8C infill location. The data from the experimental high-resolution seismic line (G88A-9902) were of sufficient quality for the A-8C infill well to be targeted with confidence despite the geologic uncertainties.

Conductor Availability

Only one conductor, the A-8 conductor, through which three dry holes had previously been drilled from the Fortescue A platform, was immediately available for drilling. A second conductor was provided when the A-15 well, which had watered out, was plugged and abandoned. The A-15 surface casing was reused to drill the A-15A infill well after cutting and pulling of the production casing. Conductor reclamation for infill drilling is described in detail by Glenton (1988).

Results

Both infill wells achieved their objectives and were completed as oil producers, giving production from Fortescue field a welcome boost. Well A-8C provided the additional surprise of unexpected sand development in the FM-1.3C reservoir. Initially, the A-8C well was completed in this FM-1.3C sand, but now that its water cut is rising rapidly, consideration is being given to commingling its production with that from the FM-1.3A sand (the original target), which is under a slightly higher pressure. Figure 18 illustrates the early production performance of the infill wells relative to the average performance of the other Fortescue wells, highlighting the significant role that these wells have played in restoring production rates in a declining field. The production performance of the infill wells is typical of the initial production performances of most Fortescue wells. The infill wells account for 50% of the current production capacity from Fortescue field.

RESERVOIR MANAGEMENT

Effective reservoir management requires close reservoir surveillance. Oil-water contact movement is monitored through the use of pulsed neutron capture logs. Relatively low formation water salinities (35,000 ppm) together with shaly sands make definitive log interpretation difficult. However, multiple passes and time lapse analysis are yielding satisfactory results.

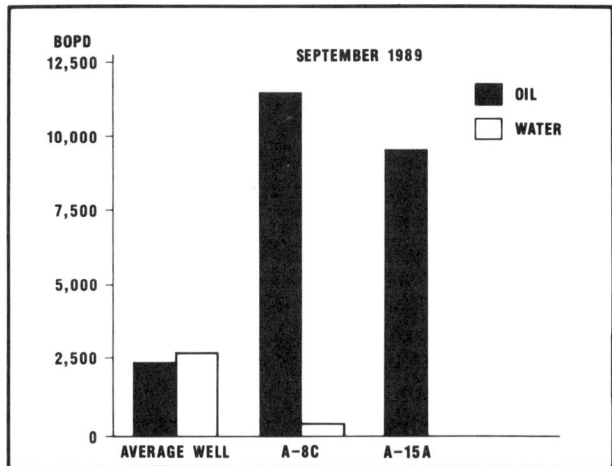

FIGURE 18. Early production performance of Fortescue infill wells A-8C and A-15A compared with average performance of remaining Fortescue producing wells.

Workover activity has evolved full circle and is continuing. Early activity was concentrated on additional perforation and reperforation opportunities as reservoir heterogeneity greater than that originally anticipated became increasingly evident. This heterogeneity has resulted in inefficient sweep of some reservoir units. Current workover activity is being concentrated on identification of water shutoff opportunities.

A 3D computer-based model of the porosity distribution within the Fortescue reservoirs has been built using an EXXON proprietary program called RESMAP. A detailed description of 3D modeling using the RESMAP program is given by Mudge and Thomson (1990). The 3D model describes the geologist's interpretation of the distribution of specific reservoir attributes (e.g., porosity) within the field. The resultant 3D model is interfaced with a MARS reservoir simulator to model depletion of the field.

CONCLUSIONS

The Fortescue field is a "giant" stratigraphic oil accumulation containing original reserves of 280 MMSTB. It was not discovered until 11 years after the discovery of the adjacent and even larger Halibut-Cobia field, largely because of the lack of recognition of the stratigraphic trapping component. The occurrence of a stratigraphic accumulation in a sand-prone section is a unique feature of the Fortescue field.

The current understanding of the geology and producing characteristics of the field is continually being refined as new data are acquired to maximize oil recovery and optimize the depletion strategy. Recent data indicate that Fortescue field is stratified to a degree far greater than that originally anticipated. Consequently, additional infill wells may be required for production of the remaining reserves.

Modeling work on the 1988 experimental seismic line and the success of subsequent infill drilling based on this work emphasize the value of high-quality seismic data. This recent infill drilling has helped to offset the decline in field production and accounts for 50% of the current production. Additional infill drilling and workover opportunities are expected to be realized through a more detailed reservoir description based on high-resolution 3D seismic coupled with data from an ongoing reservoir surveillance program. In particular, this new data is expected to define other infill drilling opportunities analogous to the stratigraphically isolated attic pools successfully targeted by wells A-8C and A-15A.

ACKNOWLEDGMENTS

The authors thank Esso Australia Ltd. and its joint venture partner, BHP Petroleum Ltd., for permission to publish this paper.

REFERENCES CITED

Bretherton, T. A., 1987, Fortescue field—post development geological study: Esso Australia in-house report, unpublished.

Glenton, P. N., 1988, The Snapper development, Gippsland basin: APEA Journal, v. 28, pt. 1, p. 29–40.

Henzell, S. T., H. R. Irrgang, E. J. Jenssen, R. A. M. Mitchell, G. O. Morrell, I. D. Palmer, N. W. Seage, G. J. Hicks, M. J. Hordern, and C. W. Kable, 1985, Fortescue reservoir development and reservoir studies: Australian Petroleum Exploration Association Journal, v. 25, n. 1, p. 95–106.

Kuttan, K., J. B. Kulla, and R. G. Neumann, 1986, Freshwater influx in the Gippsland basin: impact on formation evaluation, hydrocarbon volumes, and hydrocarbon migration: Australian Petroleum Exploration Association Journal, v. 26, n. 1, p. 242–249.

Mudge, W. J., and A. B. Thomson, 1990, Three dimensional geological modelling of the Kingfish and West Kingfish oilfields: the method and applications: APEA Journal, v. 30, pt. 1, p. 342–354.

Rahmanian, V. D., P. S. Moore, W. J. Mudge, and D. E. Spring, 1990, Sequence stratigraphy and the habitat of hydrocarbons, Gippsland basin, Australia, in J. Brooks (ed.), Classic petroleum provinces: Geological Society of London, special publication.

Thornton, R. C. N., B. J. Burns, A. K. Khurana, and A. J. Rigg, 1980, The Fortescue field—new oil in the Gippsland basin: Australian Petroleum Exploration Association Journal, v. 20, n. 1, p. 130–142.

Chapter 32

Petroleum Geology and Prospects of the Tarim (Talimu) Basin, China

Hu Boliang

Lanzhou Institute of Geology
The Chinese Academy of Sciences
Lanzhou, China

ABSTRACT

China's Tarim Basin is considered to be the last large onshore basin where giant oil and gas deposits remain to be discovered and developed. In 1977, the discovery well, Ke No. 1, located in what is now the Kekeya oil and gas field on the southwest margin of the basin, produced oil and gas at high flow rates. The initial rates were 1600 m^3 (10,000 bbl) of oil and 270,000 m^3 (9.5 mmcf) of gas per day. In the northern part of the basin, on the Shaya uplift, another discovery well, the Shacan No. 2, produced oil and gas from Ordovician dolostone at initial rates of 1000 m^3 (6290 bbl) of oil and 2 million m^3 (70 mmcf) of gas per day in 1984. Recently there have been still other discoveries of substantial volumes of oil and gas in various formations. The ages of the producing zones vary over a wide range—Cretaceous, Jurassic, Triassic, Carboniferous, and Ordovician. Especially encouraging was the Central No. 1 discovery well in the Central No. 1 structure, which yielded 576 m^3 (3620 bbl) of oil and 340,000 m^3 (12 mmcf) of gas per day in 1989. All these discoveries indicate that Tarim Basin is promising.

The Tarim Basin, with an area of 560,000 km^2, was a huge platform during the Paleozoic. Later, in the Mesozoic-Cenozoic, the platform subsided and became a basin. There were two major cycles of marine transgression and regression that resulted in thick sediments containing two marine facies favorable for oil generation—one in the Sinian-Devonian, and the second, after a hiatus, in the Carboniferous-Permian. In addition, the Triassic-Jurassic section contains good continental facies source rocks.

In two oil fields of the Tarim Basin—the Yiqikelike field in the Kuche depression and the Kelatuo field in the Southwest depression—the crude oils were derived primarily from Jurassic and Triassic sediments and have geochemical characteristics of the Jurassic continental coal measure strata. However, some production of Miocene oil and gas of the Kekeya field might be derived from Carboniferous–Permian source rocks. Most of the oil and gas in the four fields/blocks on the Shaya uplift was generated from Paleozoic marine source rocks, but part of it is from Triassic–Jurassic continental source rocks, resulting in an oil and gas accumulation from cosources. Separate oil and gas accumulations from Paleozoic marine facies and Mesozoic continental facies also have been found there. The oil and gas of the Central No. 1 structure may have been derived from Paleozoic source rocks.

In the Tarim Basin, the areas considered to have optimal conditions for giant oil and gas fields are the Shaya uplift, the Central uplift, and the southwest margin of the Southwest depression.

INTRODUCTION

The enormous resources of oil and gas in the Tarim (Talimu) Basin (Figure 1) make it the last large onshore basin in the world that remains to be developed. The basin lies in the southern part of Xinjiang, China. Its area is about 560,000 km². It is bounded by three mountain ranges: the Kunlun and Arking mountains to the south and the Tianshan Mountains to the north. The Tarim Basin is the largest intermontane basin in the world.

The occurrence of oil seepage in Kang Village, Kuche depression, Tarim Basin, was recorded in 2000 B.C. In 1942, a team led by Huang Jiqing made a geologic investigation of the petroleum potential there and put forward the idea that the petroleum might have been derived from continental facies. A petroleum reconnaissance survey began in 1950 in the basin. By 1958, a small oil field—the Yiqikelike oil field—was discovered in the Kuche depression. On May 17, 1977, the Kekeya oil and gas field was discovered in the south margin of the Southwest depression alongside the Xinjiang-Tibet Road. The initial rates of production of light oil and gas from the discovery well, Ke No. 1, were 1600 m³ (10,000 bbl) of oil and 270,000 m³ (9.5 mmcf) of gas per day and caused a boom in exploration for giant oil and gas fields in the Tarim Basin.

On October 22, 1984, in the northern part of the basin on the Shaya uplift, the Yakela oil and gas field was discovered. The discovery well, Shacan No. 2, produced from Ordovician dolomite. The initial production rates were about 1000 m³ (6290 bbl) of oil and 2 million m³ (70 mmcf) of gas per day. Great interest in exploration for giant oil fields in the Tarim Basin was once more aroused. Since 1988, more than ten wells on the Shaya uplift have yielded oil and gas at high flow rates. Table 1 presents information about some of these wells. The Central No. 1 well, in the hinterland of the basin, produced 576 m³ (3620 bbl) of oil and 360,000 m³ (12 mmcf) of gas on October 31, 1989. These new discoveries demonstrate the possibility of finding major and giant oil and gas fields in the Tarim Basin.

GEOLOGIC SETTING AND TECTONICS

The Tarim (Talimu) Basin, also called the Tarim platform, is located in the westward extension of the China platform (Figure 1). It is a Mesozoic-Cenozoic basin based on a Paleozoic platform. The basement is comprised of Proterozoic metamorphic rock overlain by Sinian-Paleozoic marine strata and, above these, Mesozoic-Cenozoic continental strata.

The southern margin of the Tarim Basin and the Kunlun Mountain geosyncline are separated by the Kunlun Mountain deep frontal fault and the Arking Mountain deep fault; the northern part of the basin borders the Tianshan Mountain geosyncline on the South Tianshan Mountain deep fault and the North Kuluketage fault. On the basis of seismic and geologic information, the basin can be divided into six tectonic units indicated by Roman numerals in Figure 2: (I) the Kuche depression, (II) the Shaya uplift, (III) the Awarti-Mankal depression, (IV) the Central uplift, (V) the Southwest depression, and (VI) the Southeast fault block. Around the periphery of the basin (and also on the platform) are four fault block uplifts: (VII) the Keping fault block in the northwest, (VIII) the Kuluketage fault block in the northeast, (IX) the Tikelik fault block in the southwest, and (X) the Arking fault block in the southeast.

Kuche Depression

The Kuche depression tectonic unit lies in the north margin of the basin (Figures 2 and 3). It measures 470 km from east to west and 20 to 75 km from north to south, with an area of about 23,000 km². The depression borders the

FIGURE 1. Map of China showing location of Tarim Basin.

TABLE 1. Some Tarim Basin discovery and exploratory wells with high flow rates.

Field/ block	Well	Age	Lithology	Depth (m)	Date	Flow rate Oil, m³/day (bbl/day)	Gas, m³/day (mmcf/day)
Kekeya	Ke No. 1	Lower Miocene	Sandstone	3500–3783	1977	1600 (10,000)	270,000 (9.5)
Yakela	Sha No. 2	Ordovician	Dolostone	5363–5391	1984	1000 (6300)	2,000,000 (70)
Yakela	Sha No. 4	Sinian	Dolostone	5424–5429	1986	13 (82)	12,000 (0.42)
		Jurassic	Sandstone	5375–5380	1988	154 (969)	26,500 (0.936)
Yakela	Sha No. 5	Lower Cretaceous	Sandstone	5324–5328	1988	500 (3100)	320,000 (11)
Yakela	Sha No. 7	Cambrian	Dolostone	5415–5450	1986	69 (430)	13,900 (0.491)
		Jurassic	Sandstone	5367–5371	1988	100 (600)	5000 (0.2)
Akekule	Sha No. 14	Ordovician	Limestone	5295–5380	1988	190 (1200)	1,000,000 (35)
	Sha No. 18	Paleozoic	Limestone	4950–5157	1989	1400 (8800)	420,000 (15)
Tazhong	Central No. 1	NA	NA	NA	1989	576 (3620)	340,000 (12)

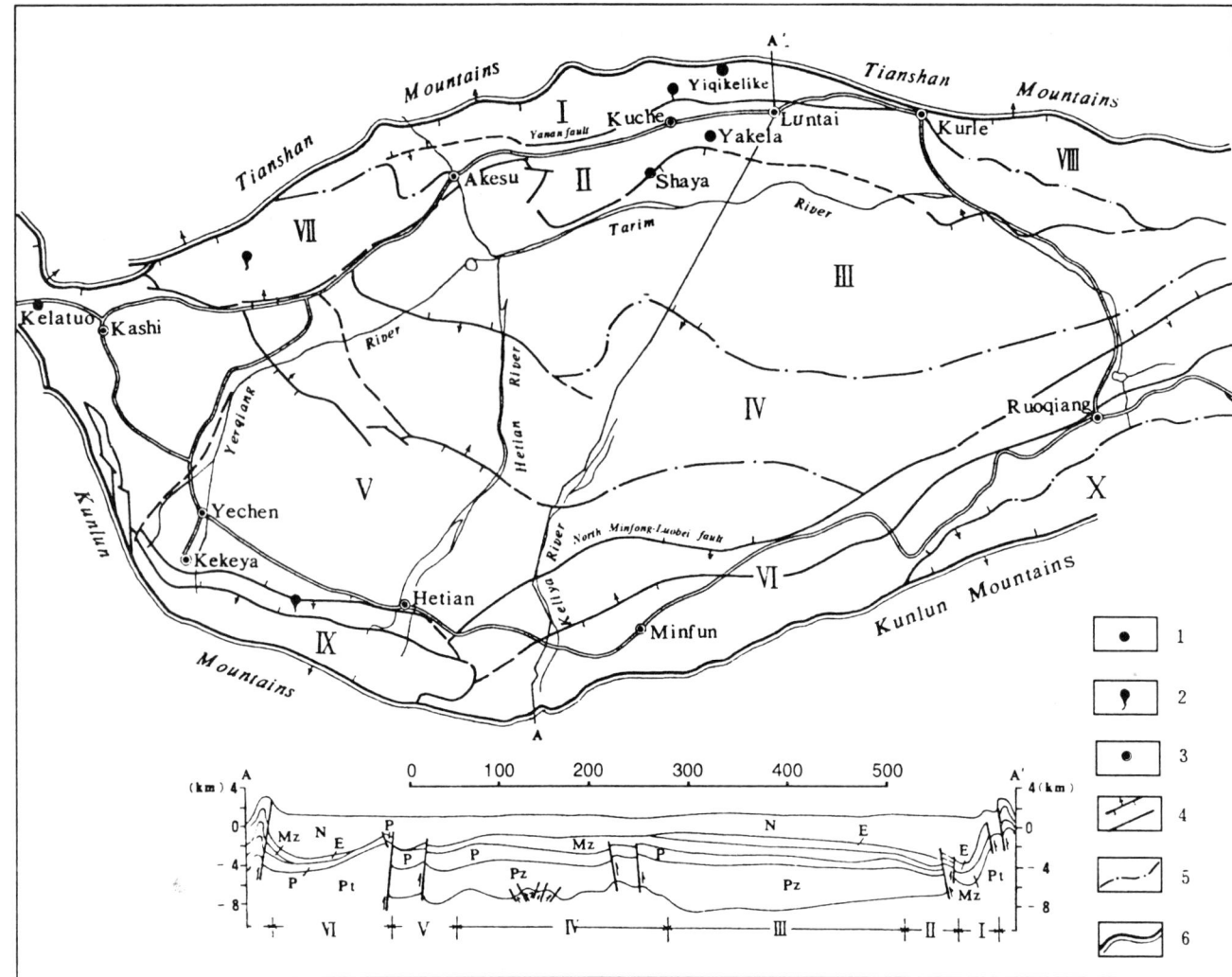

FIGURE 2. Tectonic and oil field location map of the Tarim Basin, China (modified after Xie Xiaoan, 1986): (I) Kuche depression, (II) Shaya uplift, (III) Awarti-Mankal depression, (IV) Central uplift, (V) Southwest depression, (VI) Southeast fault block, (VII) Keping fault uplift, (VIII) Kuluketage fault uplift, (IX) Tikelike fault uplift, and (X) Arking fault uplift. Symbols: (1) oil and/or gas field, (2) oil show, (3) county and road, (4) fault, (5) boundary of tectonic unit, and (6) boundary of the Tarim Basin. Chronostratigraphic symbols used here are shown in Figure 4. Approximate length of cross section, 740 km.

Tianshan Mountain geosyncline on a deep fault to the north, and the Shaya uplift on the Changmuziduk section and the Yanan fault to the south. It is a Mesozoic-Cenozoic fault depression with a sedimentary section more than 9500 m thick. Triassic sediments are mainly distributed over the middle part of the depression; Jurassic deposits extend east to west and gradually thin southward, then abruptly terminate. Total thickness of the Triassic-Jurassic strata is about 3000 m. Dark mudstone source beds about 1300 m thick are dispersed in the middle part of the Triassic section, the upper part of the Lower Jurassic section, and the Middle Jurassic section. On the north are four rows of east-west anticlinal trends that obviously are structurally related to the Tianshan Mountains. The north part of the depression is heavily folded. The terrain slopes gently and openly to the south, where more than 80 seepages are known and 60 structures have been mapped. Sixty-three wells have been drilled on 12 of these structures, resulting in the discoveries of the Yiqikelike field and the Tugalming oil-bearing structure. Oil and gas shows were encountered in the wells on six of the structures.

Shaya Uplift

The east-west-oriented Shaya uplift tectonic unit is situated between the Kuche depression and the Awarti-Mankal depression, covering an area of about 30,000 km² (Figures 2 and 3). The uplift, in fact, is a paleoanticline with a length of 500 km and a width ranging from 50 to 70 km. The core of the anticline, consisting of phyllite of Sinian age, is in Xinhe-Luntai County. The south limb of

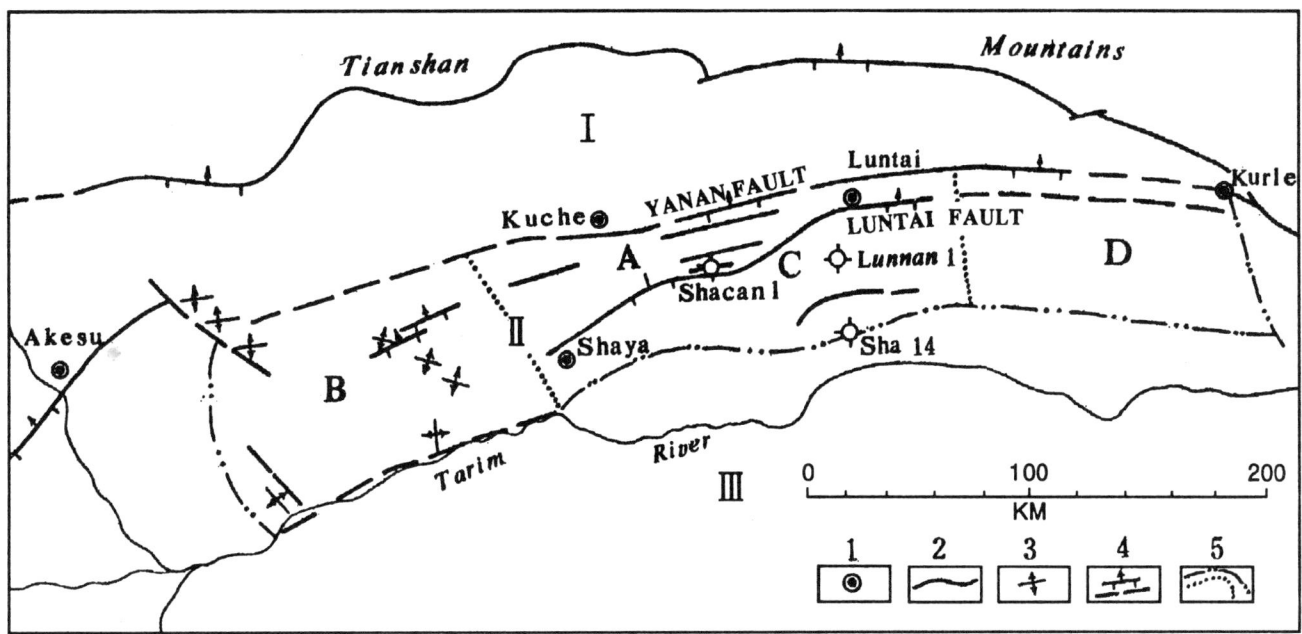

FIGURE 3. Tectonic regions of the Shaya uplift in the Tarim Basin: (I) Kuche depression, (II) Shaya uplift, and (III) Awarti-Mankal depression. Tectonic features: (A) Luntai fault uplift, (B) West Shaya fault-fold, (C) Akekum slope, and (D) Kurle nose uplift. Symbols: (1) county, (2) stream, (3) anticline, (4) fault, and (5) boundary.

the anticline ranges in age from Sinian to Permian; the north limb is complicated by the long-active Yanan fault.

The Shaya uplift began to appear with the onset of the Tarim orogeny (Figure 4). As a result of the Tarim orogeny, the Sinian section thins northward. The uplift rose slowly during the Cambrian-Ordovician, but at the end of the Ordovician it began to rise more rapidly. A result of this uplift was that Silurian-Devonian sediments ceased to be deposited on the south limb. During the Hercynian orogeny, uplift continued and then ceased. Finally the area subsided and was buried beneath Mesozoic and Cenozoic sediments. The uplift, plunging to the west, now lies beneath northward-dipping monoclinal Cenozoic strata.

The Shaya uplift contains three fault suites with different orientations. Most of these faults are oriented east-west or west-southwest-east-northeast; a few have southeast-northwest orientations. Most are compressional faults; a few are tension-shear faults.

The fault planes generally dip to the north. They are steep in their upper parts but become less steep at depth. Faults with large throws or other displacements were active over long geologic times. The tension-shear faults are steep and generally small. The three fault suites cut the uplift into a series of uneven blocks that have different forms and sizes. The faults are paralleled or crossed by folds.

The Shaya uplift is a structural complex with an optimum paleogeographic position and requisite structures for the generation, migration, and accumulation of oil and gas. Kang Yuezhu (1989) further divided the Shaya uplift into four secondary structural zones according to tectonics and sediment characteristics (A, B, C, and D in Figure 3).

Luntai Fault Uplift Zone

The Luntai fault uplift zone is bounded on the west by Xihe county, on the east by Yangxia County, on the north by the Yanan fault, and on the south by the Luntai fault. Paleozoic sediments were deposited only in the southwest of this area, whereas a large area of pre-Sinian phyllite underlies the great Mesozoic deposit. The east-northeastward faults have cut pre-Mesozoic strata into several fault blocks that are concealed beneath undisturbed mid-Cenozoic sediments. The Yakela oil and gas field is associated with such a structural unit.

West Shaya Fault-Fold Zone

Mesozoic and Paleozoic strata are fully developed in the West Shaya fault-fold zone. The Permian System, including volcanic strata, is widely distributed. Some exploratory wells encountered many formations with oil and gas shows in the Miocene and Paleocene, and oil was obtained at a high flow rate in the lower Paleozoic section.

Akekum Slope

The Akekum slope is situated along the Luntai fault (its northern boundary) and extends southward to the Tarim River. To the west and east are Shaya County and the Kurle nose uplift zone, respectively. During Paleozoic Erathem deposition, the area was tilted from north to south, and consequently the sediments thicken to the south. Large volumes of oil and gas have been produced from the wells in the Akekum and Akekule structures.

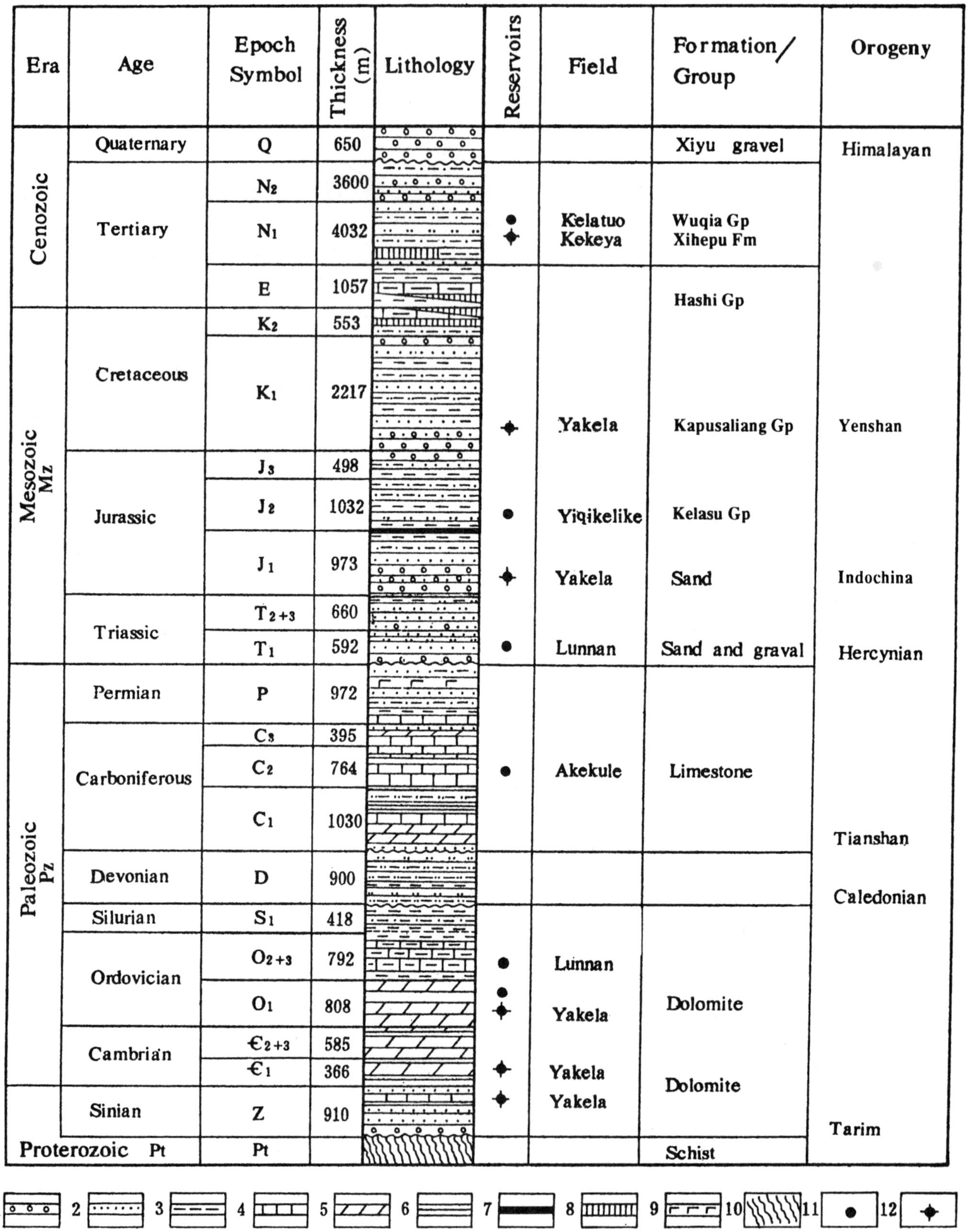

Era	Age	Epoch Symbol	Thickness (m)	Lithology	Reservoirs	Field	Formation/ Group	Orogeny
Cenozoic	Quaternary	Q	650				Xiyu gravel	Himalayan
Cenozoic	Tertiary	N_2	3600				Wuqia Gp	Himalayan
Cenozoic	Tertiary	N_1	4032		●◆	Kelatuo Kekeya	Wuqia Gp Xihepu Fm	Himalayan
Cenozoic	Tertiary	E	1057				Hashi Gp	Himalayan
Mesozoic Mz	Cretaceous	K_2	553					Yenshan
Mesozoic Mz	Cretaceous	K_1	2217		◆	Yakela	Kapusaliang Gp	Yenshan
Mesozoic Mz	Jurassic	J_3	498					Yenshan
Mesozoic Mz	Jurassic	J_2	1032		●	Yiqikelike	Kelasu Gp	Yenshan
Mesozoic Mz	Jurassic	J_1	973		◆	Yakela	Sand	Indochina
Mesozoic Mz	Triassic	T_{2+3}	660					Indochina
Mesozoic Mz	Triassic	T_1	592		●	Lunnan	Sand and graval	Hercynian
Paleozoic Pz	Permian	P	972					Hercynian
Paleozoic Pz	Carboniferous	C_3	395					Hercynian
Paleozoic Pz	Carboniferous	C_2	764		●	Akekule	Limestone	Hercynian
Paleozoic Pz	Carboniferous	C_1	1030					Tianshan
Paleozoic Pz	Devonian	D	900					Tianshan
Paleozoic Pz	Silurian	S_1	418					Caledonian
Paleozoic Pz	Ordovician	O_{2+3}	792		●	Lunnan	Dolomite	Caledonian
Paleozoic Pz	Ordovician	O_1	808		●◆	Yakela	Dolomite	Caledonian
Paleozoic Pz	Cambrian	ε_{2+3}	585					Caledonian
Paleozoic Pz	Cambrian	ε_1	366		◆	Yakela	Dolomite	Caledonian
Paleozoic Pz	Sinian	Z	910		◆	Yakela		Tarim
Proterozoic Pt		Pt					Schist	Tarim

FIGURE 4. Composite stratigraphic section showing principal reservoirs in the Tarim Basin. Lithologic symbols: (1) gravel, (2) sandstone, (3) mudstone, (4) limestone, (5) dolomite, (6) shale, (7) coal, (8) anhydrite, (9) basalt, (10) schist, (11) oil reservoir, and (12) oil and gas reservoir.

Kurle Nose Uplift Zone

The Kurle nose uplift zone lies in the eastern part of the Central uplift—that is, in the area of Yangxia County and Kurle. Cenozoic and thinner Mesozoic strata overlie lower Paleozoic deposits, pre-Sinian deposits, and granite. The Kurle nose uplift zone is a westward-dipping uplifted nose related to the Kuluketage uplift (VIII, Figure 2).

Awarti-Mankal Depression

The Awarti-Mankal depression tectonic unit (the Awarti and Mankal depressions considered together) is the largest depression in the Tarim Basin, with an area of about 150,000 km² (Figures 2 and 3). Quaternary desert pavement overlies the entire area, and the basement rock is buried to a depth of more than 10 km; between these, the formations are warped in a double structure. The Paleozoic section, about 3000 m thick, and the Mesozoic section, about 2500 m thick, are depressions. The Triassic section alone is nearly 1000 m thick. The Cenozoic section is a northward-dipping monocline.

Central Uplift

The Central uplift tectonic unit traverses the mid-basin (Figures 2 and 5). It is bounded on the north by the Awarti-Mankal depression and on the south by the Southwest depression and the Southeast fault block. It has a total area of more than 100,000 km². The Bachu fault uplift, which is in the western part of the Central uplift, has been an uplift since the Paleozoic and was not buried until the late Neogene. Twelve wells have been drilled, and a small quantity of oil has already been obtained from the Lower Carboniferous section in the Qui No. 1 well, on the Bachu fault uplift. The area from the eastern part of the Central uplift to Luobupo Lake is slightly uplifted. The surface is desert. Some anticlinoria have been discovered, one of which—the Center No. 1 structure—has an area of 8200 km². Oil and gas at high flow rates have been obtained from the Center No. 1 well. The Central uplift has the most nearly optimal conditions for hydrocarbon accumulations of the magnitude of a giant oil and gas field.

Southwest Depression

The Southwest depression tectonic unit consists of the north slope, the Maigeti slope, and the Kashi and Yechen depressions, which comprise the deep, elongate Southwest depression itself. The depression has a total area of about 150,000 km² (Figures 2 and 5). The subsidence center involves Kashi, Yechen, and Hetian in front of the Kunlun Mountains. Upper Paleozoic deposits reach a thickness of about 12,000 m. In the Mesozoic section, most Triassic sediments are missing and the Jurassic section is distributed in a very narrow fault trough in front of the mountains. The Upper Cretaceous-Eogene section is a shallow water marine platform facies deposit where the

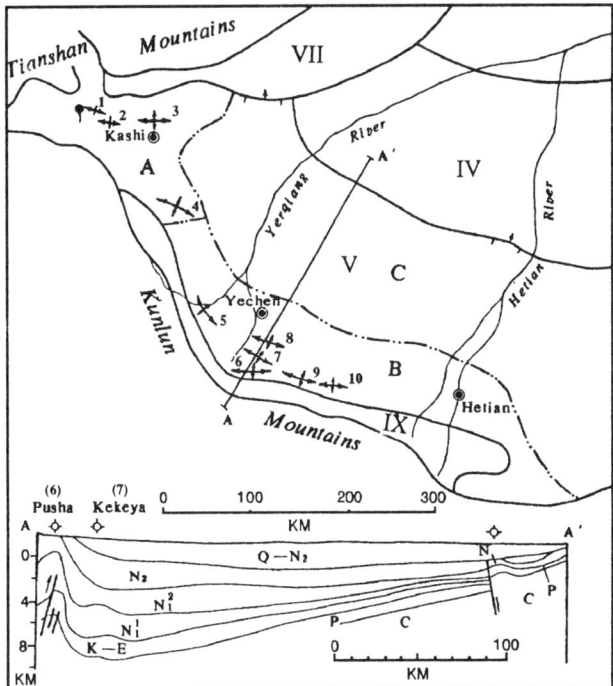

FIGURE 5. Tectonic regions and relationships of the Southwest depression in Tarim Basin: (IV) Central uplift, (V) Southwest depression, (VII) Keping uplift, and (IX) Tikelike fault uplift. Tectonic features: (A) Kashi depression, (B) Yechen depression, and (C) Maigeti slope. Major anticlines: (1) Kelatuo, (2) Mingyaolu, (3) Kashi, (4) Yingjiesa, (5) Qipan, (6) Pusha, (7) Kekeya, (8) Guman, (9) Yueliqung, and (10) Sonzhu. Formations and symbols are shown in Figure 4.

Tethys sea transgressed along the northern foot of the Kunlun Mountains eastward into the area around Hetian. The upper Tertiary section is a series of continental facies deposits. The Neogene System consists of a molasse accumulation, more than 6000 m thick, in front of the mountains.

Spread out in front of the West Kunlun Mountains are three tiers of Cenozoic structures. The south tier is sharply folded; the steep folds gradually become smooth anticlines northward. The three sets of Cenozoic structures comprising the tiers include 35 anticlines. Seventy-four wells have tested 18 of the structures in the Southwest depression, but the only discovery has been the Kekeya oil and gas field. Nonetheless, the possibility remains that some giant oil and gas fields will be found in the Southwest depression in the future.

Southeast Fault Block

The Southeast fault block tectonic unit lies in the southeastern part of the basin (Figure 2). Extending to the west to Chele, it is bounded to the east by Luobupo Lake, to the north by the North Minfong-Luobei fault, and to the south by the foot of the Kunlun and Arking mountains. It stretches east-northeastward over a distance of 1000 km, has a width ranging from 70 to 120 km, and encompasses an area of nearly 100,000 km². The Southeast fault block, which involves the Minfong-Lobuzhuang fault uplift and the Minfong-Nurqiang fault depression, is a very long uplift zone. In its higher parts, the Triassic strata rest directly on the Proterozoic section. There are Jurassic coal-bearing strata along the front margin in the fault depression, and above these strata are Cenozoic continental facies deposits.

GEOLOGIC HISTORY AND STRATIGRAPHY

The Tarim Basin is part of the China platform. A relatively complete sequence of Paleozoic and Mesozoic-Cenozoic rocks overlies the basement of sub-Sinian metamorphic rock. The thickness of the sedimentary section reaches 30,000 m. The geologic history of the Tarim Basin (Figure 4) involves three stages: the pre-Sinian geosyncline development stage, the Paleozoic platform development stage, and the basin development stage.

Pre-Sinian Geosyncline Development Stage

During the Archeozoic-Proterozoic, about 20,000 m of clastic, carbonate, and volcanic rocks was deposited. In the last phase of the Tarim movement, some fold metamorphism occurred. An enormous acidic igneous mass was erupted and ultimately became the basement of the Tarim platform.

Paleozoic Platform Development Stage

In the Paleozoic, two great marine transgressions and two transgressive and regressive sedimentary cycles occurred, and the Sinian-Devonian and Carboniferous-Permian strata were deposited.

Sinian

In Kuluketage, Mankal, and Keping, the Sinian System is more than 1000 m thick. The lower series is made up mainly of coarse clastic rock and morainal material in which are embedded neutral and acidic volcanic rocks; this series covers the sub-Sinian metamorphic rock in angular discordance. The middle and upper series consist primarily of carbonate and clastic rocks that were deposited in a shallow marine environment.

Cambrian-Ordovician

Except on the Southeast fault block, neritic carbonates with embedded clastic rock were deposited over the entire area (Figure 6). As a result of the Caledonian orogeny, parts of the area lack Upper Ordovician deposits. Folding began in Cambrian-Ordovician rocks, which thin from northeast to southwest in the basin. The thickness of the Cambrian-Ordovician deposits ranges from 1000 to 4000 m.

Silurian-Devonian

The area covered by Silurian-Devonian sediments is more limited than that covered by Cambrian-Ordovician deposits. In the southeastern region and part of the northern region of the basin, the Cambrian-Ordovician section is not present. The Silurian deposits consist of green sandstone interlayered with mudstone and shale facies of a regressive shallow marine environment. The Devonian System is made up of red clastic rocks. The total thickness of the Silurian-Devonian sediments ranges from 900 to 3000 m.

Carboniferous-Permian

At the end of the Devonian Period, the early Hercynian orogeny (Tianshan orogeny) caused folding and faulting, and the older strata were overlain by Carboniferous System strata in angular unconformity. The combined Carboniferous-Permian System is widely distributed except in the Southeast fault block and the Shaya uplift to the north (Figure 7). The Southwest depression was the subsidence center during the Carboniferous, and deposits there consist mainly of shallow marine carbonates interbedded with mudstone and shale. In the western part of the basin, the Lower Permian strata are composed of marine and continental carbonate and clastic rocks; the eastern region contains continental clastic rocks; and volcanic rocks are found in the area of Bachu and in the northern part of the basin. The Upper Permian strata are composed of red continental clastic rocks that are covered by Mesozoic Erathem strata in angular unconformity. The Carboniferous-Permian sediments are thin in the northeastern part of the basin but thicken to 4500 m in the Southwest depression.

Basin Development Stage

The later Hercynian orogeny that occurred at the end of the Paleozoic Era caused substantial growth of the Tianshan and Kunlun mountains. Since then, the Tarim Basin has become a Mesozoic-Cenozoic continental sedimentary basin. The sub-Mesozoic rocks were much folded and faulted, and the Triassic strata were overlaid in angular unconformity.

Triassic

The Triassic deposits of the Tarim Basin comprise a sedimentary series of fluvial, lacustrine, and piedmont clastic facies. The thickness of this series of strata is greatest in the Kuche depression in front of the Tianshan Mountains, ranging from 1500 to 2000 m, whereas in the Awarti-Mankal depression the Triassic section is only 500 to 1000 m thick (Figure 8). Triassic rocks had never been found elsewhere in the basin until Triassic sediments were recently reported to exist in front of the Kunlun Mountains.

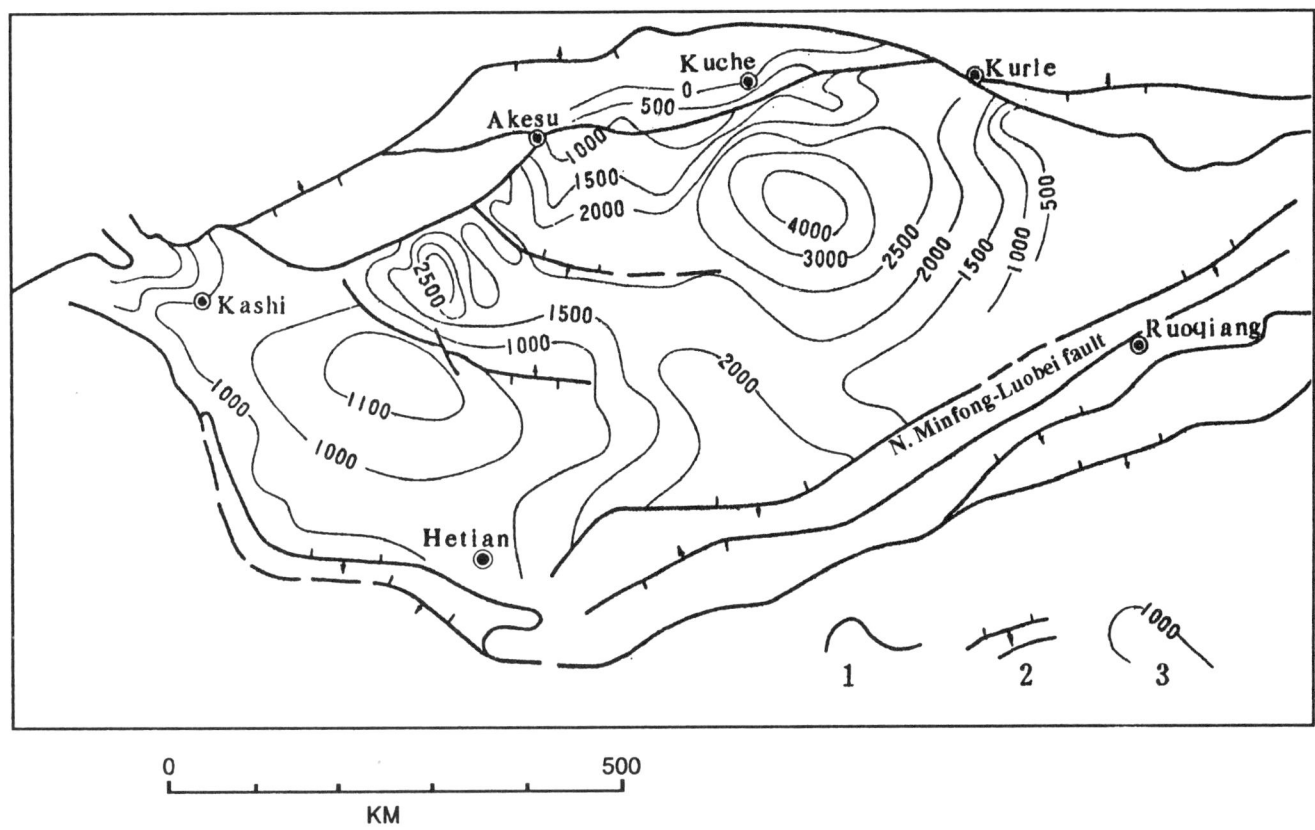

FIGURE 6. Isopach map of the Cambrian–Ordovician deposits in the Tarim Basin (after Kang Yuezhu, 1989). Symbols: (1) boundary, (2) fault, and (3) isopach line. Contour intervals, 500 and 1000 m.

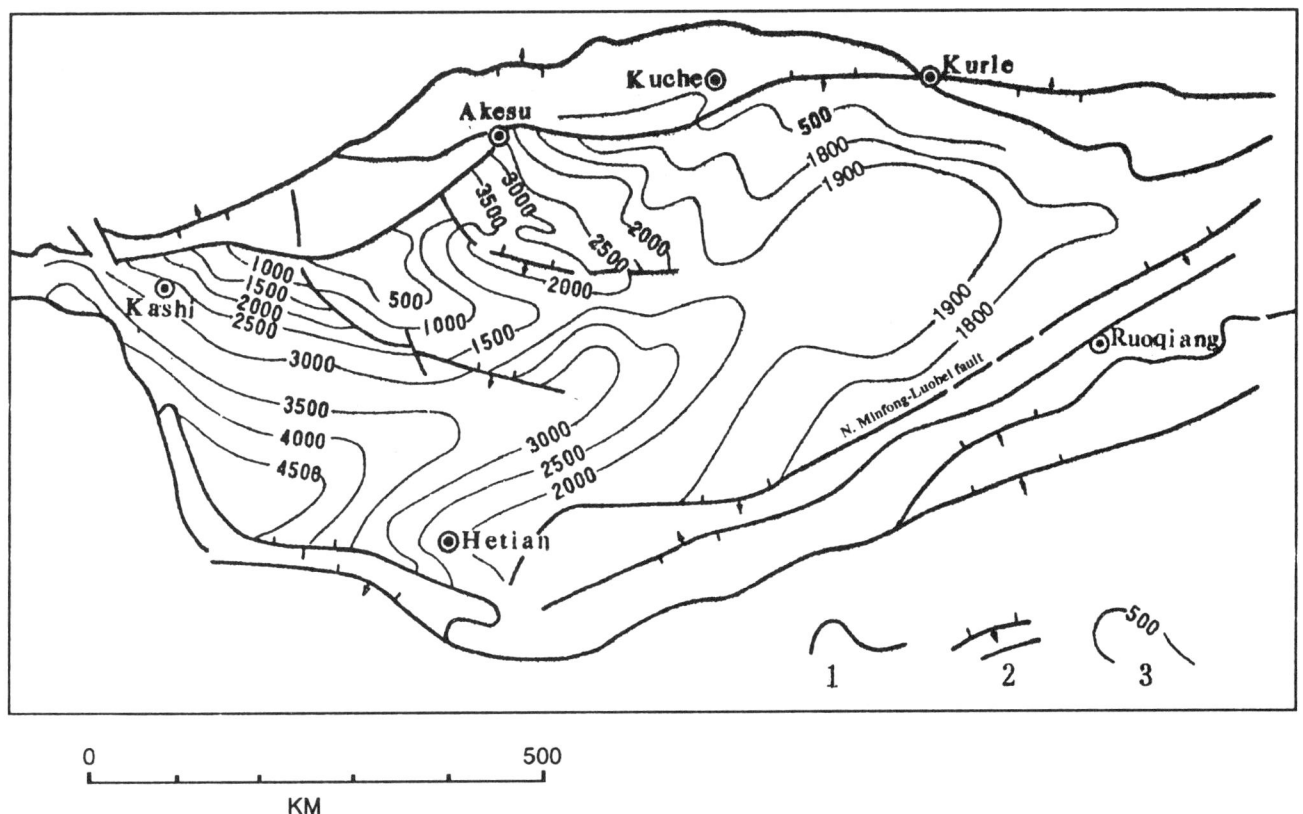

FIGURE 7. Isopach map of the Carboniferous–Permian sediments in the Tarim Basin (after Kang Yuezhu, 1989). Symbols: (1) boundary, (2) fault, and (3) isopach line. Contour intervals in meters.

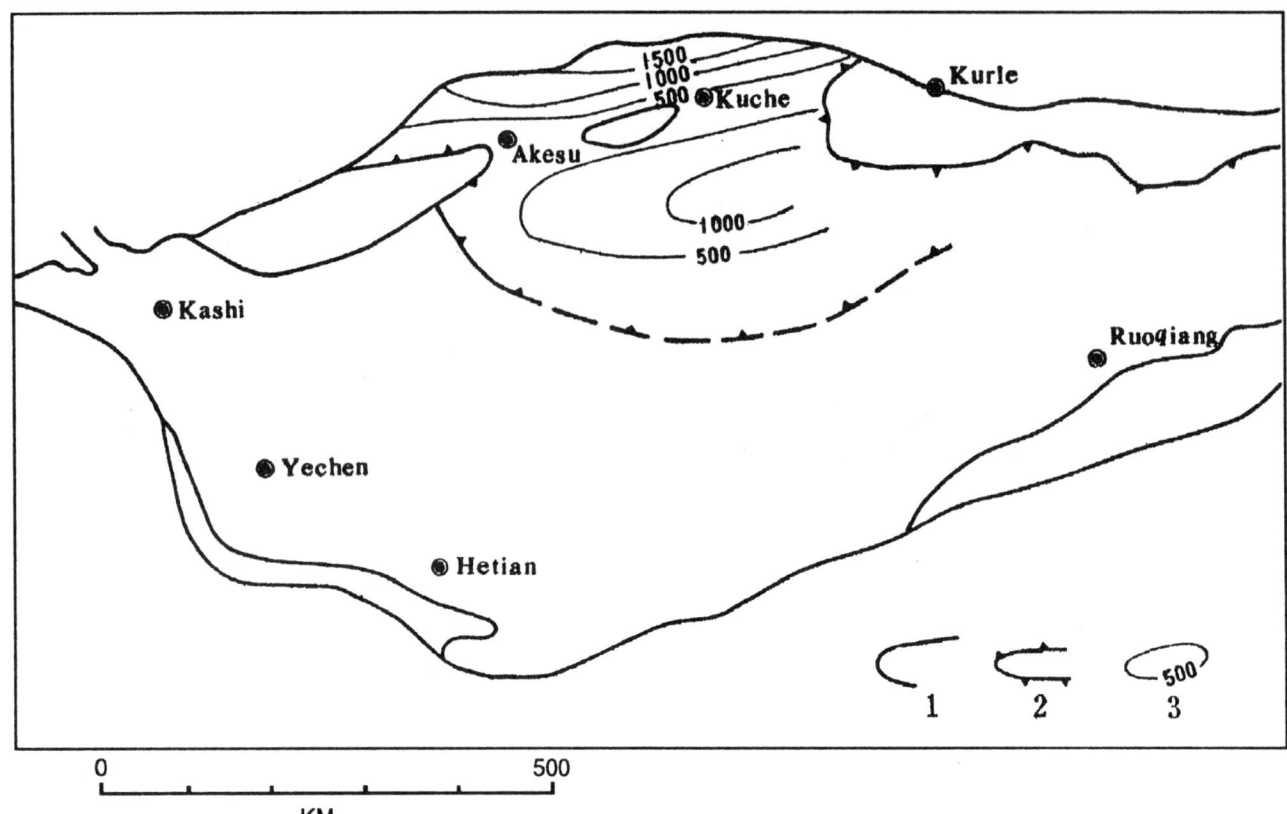

FIGURE 8. Isopach map of the Triassic deposits in the Tarim Basin (modified after Cheng Rongling, 1988). Symbols: (1) boundary, (2) fault, and (3) isopach line. Contour interval, 500 m.

Jurassic

The Jurassic deposits consist of shallow lake, marsh, and river facies that are distributed over the Kuche and Awarti-Mankal depressions in thicknesses from 1000 to 1500 m (Figure 9). In contrast to that of the Triassic, the sedimentary center of the Jurassic section in the Awarti-Mankal depression moved distinctly eastward. The Jurassic sediments were also distributed in the narrow depression in front of the Kunlun and Arking mountains; here, the sedimentary center is to the west of Kashi and consists of a suite of coal-bearing sediments interlayered with thin marine facies, ranging in thickness from 1400 to 3800 m and thinning from northwest to southeast. In the Southeast fault block, controlled by the northeast-southwest-oriented fault to the north, the Jurassic strata appear in several isolated depressions in front of the mountains, where rock character and facies are markedly changed. Thicknesses in these depressions range from a few hundred to 1000 m.

Cretaceous

In the Early Cretaceous, the area of sedimentary deposition enlarged a little. Red sandstone, mudstone, and mudstone-sandstone sediments were deposited in the Kuche, Awarti-Mankal, and Southwest depressions in thicknesses ranging from 183 to 2217 m. In the Late Cretaceous–early Tertiary, owing to the eastward transgression of the Tethys sea, a suite of shallow marine, nearshore, and lagoonal facies, including carbonates, anhydrites, and mudstone, were deposited in thicknesses ranging from 39 to 1580 m. The sea transgressed into the eastern part of the Tarim Basin through Tangusbas gorge, resulting in a marine facies about 300 to 500 m thick. Figure 10 is an isopach map of the Upper Cretaceous sediments, and Figure 11 is an isopach map of the Paleocene–Eocene deposits.

Tertiary Miocene sediments are brackish lagoonal deposits that precipitated after the recession of the seawater. The Himalayan orogeny caused substantial growth of the mountains surrounding the basin, and the deposition center moved continuously to the inner part of the basin; these conditions resulted in the deposition of tremendously thick, brown-red upper Tertiary clastic rocks. The thickest sediments (8812 m) were deposited in the Southwest depression.

Quaternary

Quaternary deposits are primarily distributed around the perimeter of the basin in front of the mountains, but also include lower Pleistocene conglomerates in the western region. The maximum thickness of the Quaternary section is about 3000 m.

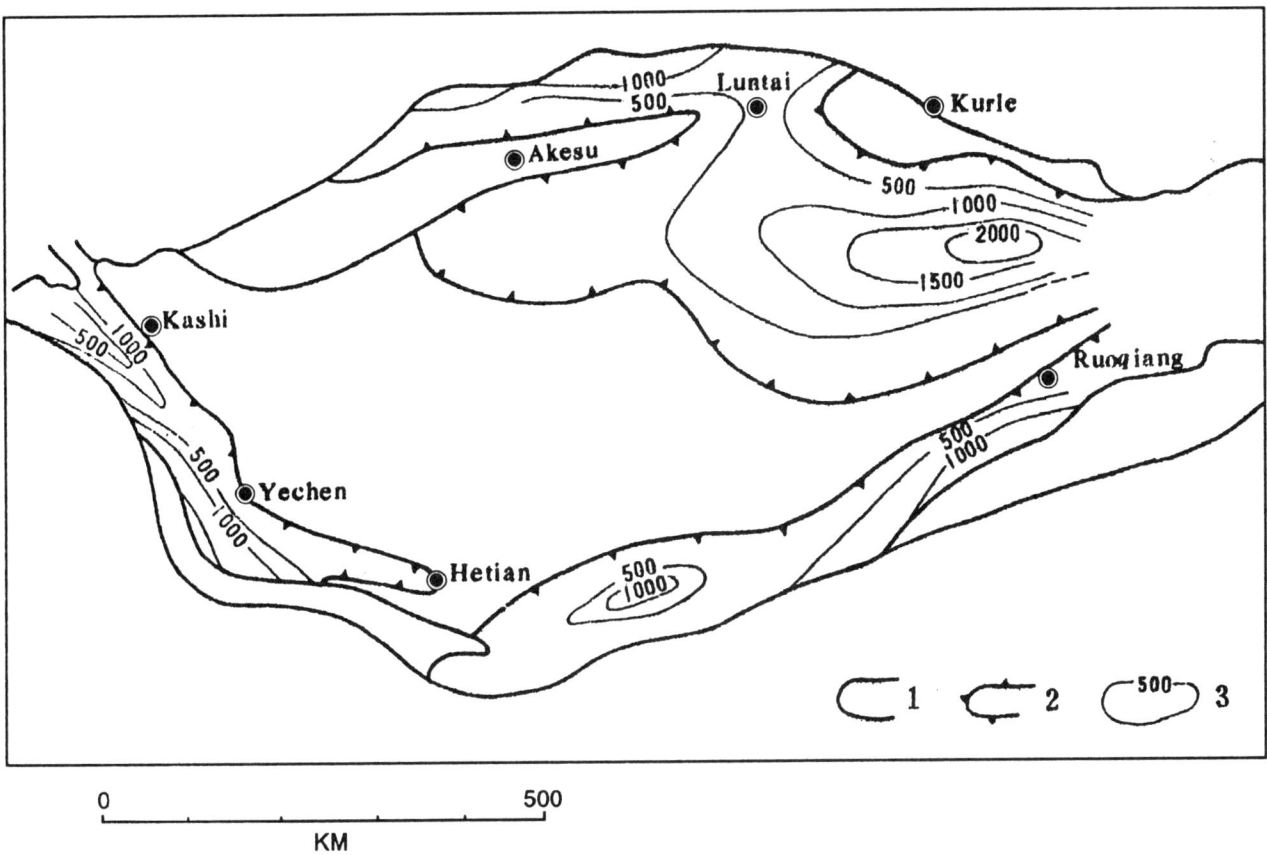

FIGURE 9. Isopach map of the Jurassic deposits in the Tarim Basin (modified after Cheng Rongling, 1988). Symbols: (1) boundary, (2) fault, and (3) isopach line. Contour interval, 500 m.

GIANT OIL AND GAS FIELDS

So far, many wells have been drilled in the Tarim Basin, and oil and gas have been produced at high flow rates from some of them. In this basin, however, only one giant oil and gas field is in the stage when exploration and appraisal are finished and development is under way—the Kekeya oil and gas field. Besides the Kekeya field, the Yakela, Lunnan, Akekule, and some other oil and gas fields discovered on the Shaya uplift on the north side of the Tarim Basin are being explored. Considering the results to date, there is no doubt that they will be numbered among the giant oil and gas fields. The Central No. 1 structure on the Central uplift, with an area of 8200 km², is certain to be the most optimal site for another giant oil and gas field.

Kekeya Oil and Gas Field

The Kekeya oil and gas field, which is located in an anticlinal zone in the Southwest depression along the south margin of the Tarim Basin (Figure 5), currently contains the largest quantity of remaining gas reserves in China. The anticlinal zone is composed of three rows or tiers of anticlines in front of the Kunlun Mountains, extending nearly east-west. The axes of the anticlines are en echelon. The first row of structures consists of anticlines with very steep flanks; the second and third rows become progressively gentler and deeper toward the basin. The Kekeya field lies on a second-row anticline, paralleling the Pusha anticline in the first row and the Guman anticline in the third row (Figure 12). The Kekeya anticline was formed during the first pulse of the Himalayan orogeny at the end of the Oligocene, was more strongly folded during the second pulse of the Himalayan orogeny in the Miocene, and, during the third pulse in the Pliocene, was still more strongly folded and uplifted so that it had an amplitude of 450 m.

The Kekeya anticline has a steep north flank (dipping 12°30′) and a slightly less steep south flank (11°30′); the long axis is three times the length of the short one (Figure 13). The upper part of the structure is the Atushi Formation of the Pliocene Series; the lower part is the Miocene Wuqia Group, also known as the Xihepu Formation (N_1X), which contains the oil and gas reservoirs of the Kekeya field. The main part of the anticline is not cut by the fault. Structural closure at the eighth member of the Xihepu Formation is 300 m, and the area of closure is 33 km². According to the seismic information, the number of faults increases at the depth of the anticline, and fault throws become longer. For example, there is only one fault within the Xihepu Formation on the north flank of the structure, whereas there are five faults deep in the

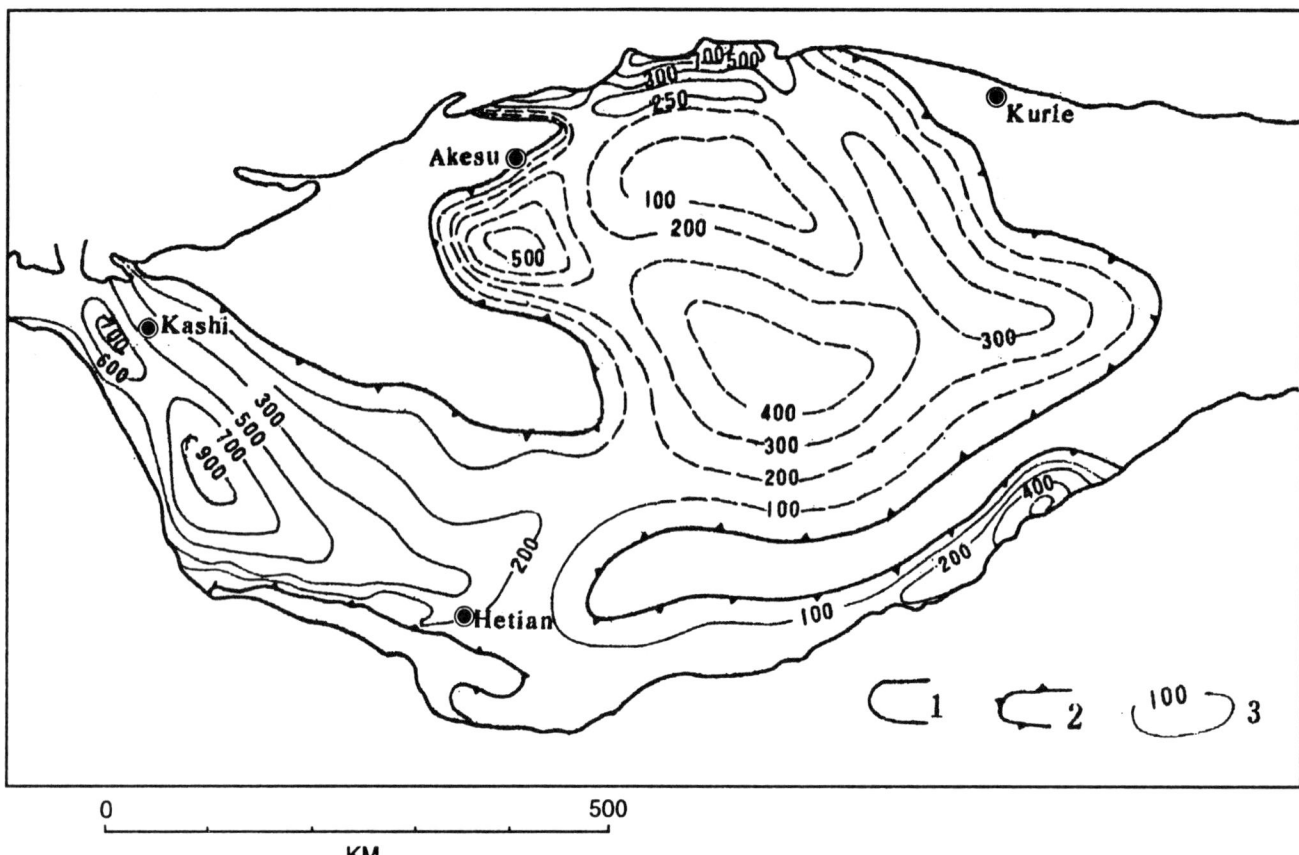

FIGURE 10. Isopach map of the Upper Cretaceous sediments in the Tarim Basin (modified after Yong Tianshou, 1984). Contour intervals are 50, 100, and 200 m.

Permian section. These faults provided upward migration routes for oil and gas from what are now secondary oil and gas accumulations in the Kekeya structure.

Most of the Xihepu Formation reservoirs of the Kekeya field consist of fine sandstone, whereas the others consist of siltstone. The effective porosity is 13.3 to 15.9%, and the permeability is 11.2 to 90.6 md. Oil saturation is 57.5 to 69.3%. Kekeya field is a complex anticlinal-type oil, condensate, and gas trap that consists of the Xihepu Formation and five other oil and gas pools of different properties in which the accumulation of the oil, gas, and water is controlled by local structure, the character of the reservoirs, and the reservoir seals. The Xihepu reservoirs are as follows:

1. Gas condensate pool in the third member (X_3) of the Xihepu Formation. The X_3 reservoir is a layered gas condensate pool with bottom water. It is not very thick.

2. Gas condensate pool in the fourth member (X_4^1) of the Xihepu Formation. The X_4^1 pool is composed of five sand-layer suites (seven single layers). The largest of these has a net thickness of 36.3 m, the average thickness of the layers is 20.5 m, and the greatest gas column height is 177 m. The gas-bearing area is 13.54 km².

3. Gas condensate pool with an oil ring in the fourth and fifth members of the Xihepu Formation. The X_4^2-X_5 interval includes nine sand-layer suites (ten single layers) and is the major oil and gas pool in the Kekeya structure. The condensate and gas fill the structure down to 1425 m below sea level, and beneath this there is a light crude oil ring (with bottom water beneath it) in the interval from 1425 to 1490 m below sea level. The thickest pay section is about 93.3 m thick. Structural closure is 247.5 m, and the oil- and gas-bearing area is 24.8 km². The oil column has a height of 65 m and a width of only 350 to 600 m.

4. Gas condensate pool in the seventh member (X_7^2) of the Xihepu Formation. The X_7^2 pool consists of four sand-layer suites (seven single layers). The crude oil distribution is ringlike, with a condensate and gas cap and bottom water. The thickest pay section has a maximum thickness of 45 m, and the oil ring is about 600 m wide.

5. Gas condensate pool in the eighth member (X_8) of the Xihepu Formation. The X_8 pool, which consists of four

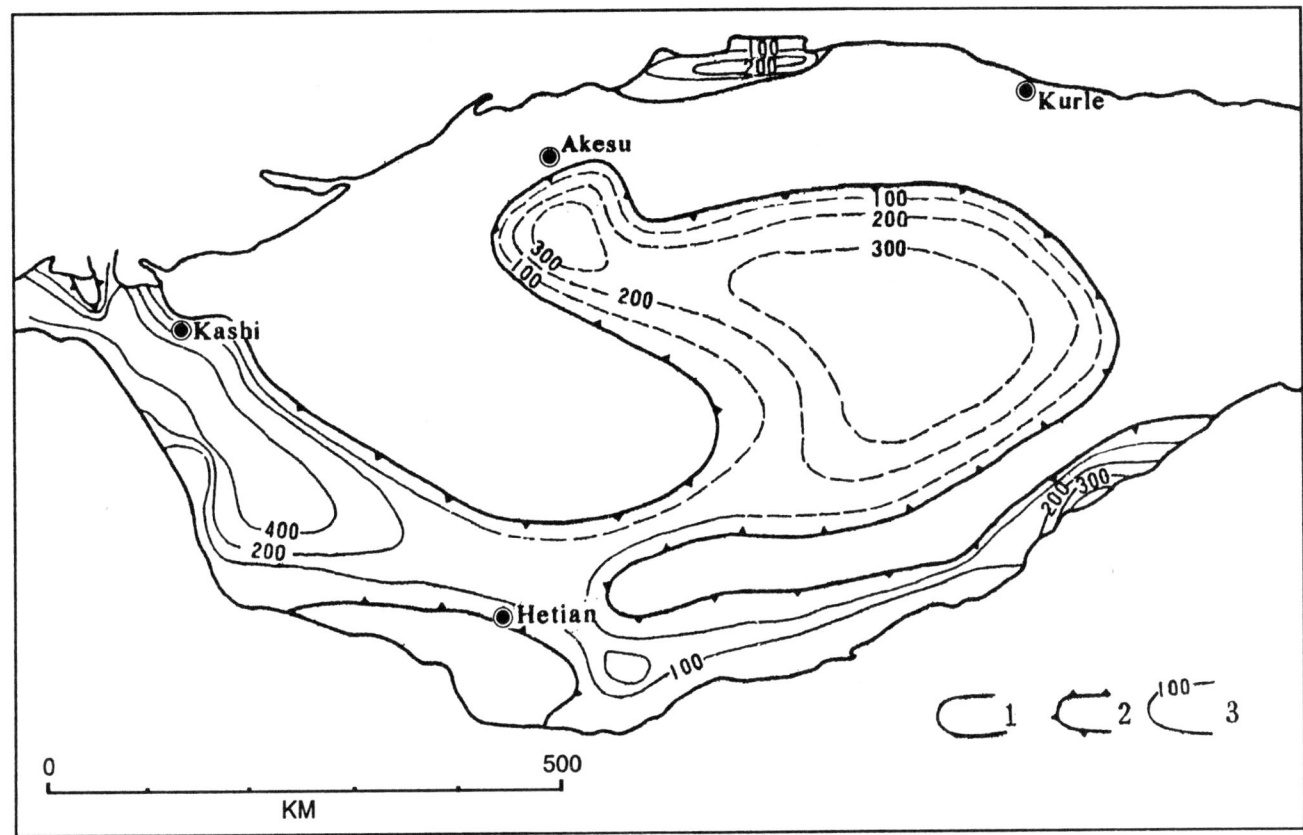

FIGURE 11. Isopach map of the Paleocene-Eocene deposits in the Tarim Basin (modified after Yong Tianshou, 1984). Contour interval, 100 m.

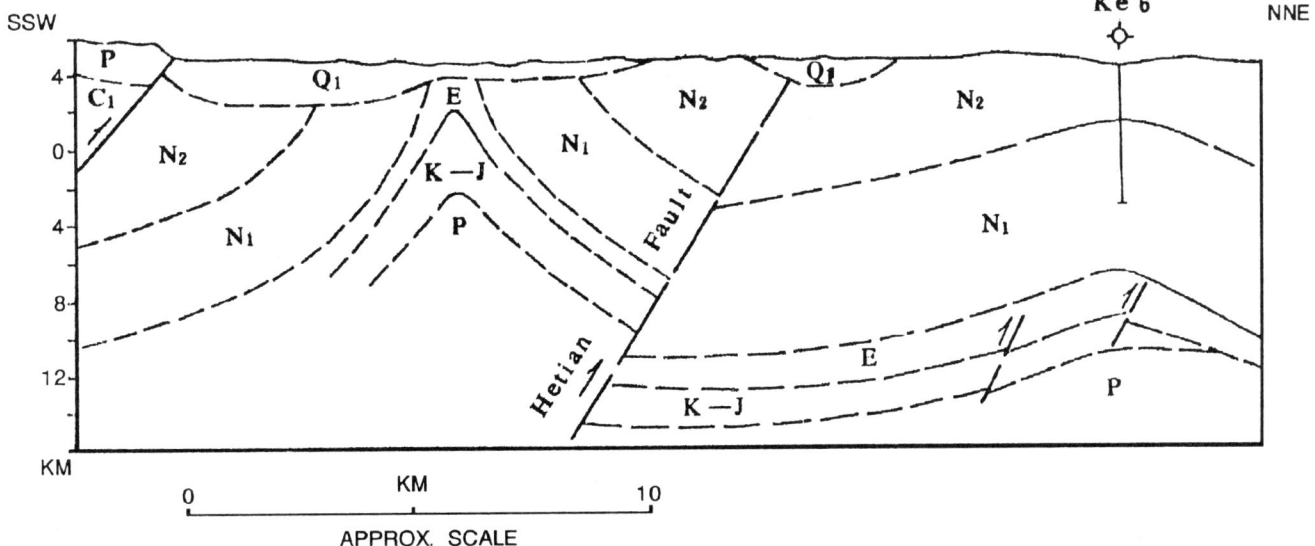

FIGURE 12. Section across the Pusha and Kekeya anticlines, Yechen County, along the Xinjiang-Tibet Road. The location of this cross section is approximately coincidental with the southern part of line A-A' in Figure 5. Stratigraphic symbols are shown in Figure 4. (After Song Lixung, 1989.)

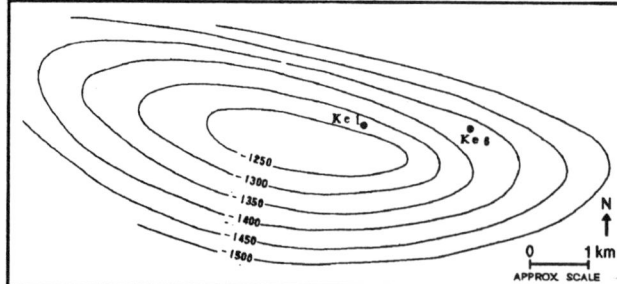

FIGURE 13. Kekeya field structure contour map on the top of Xihepu second member in the southwest part of the basin. Subsea contour intervals, 50 m. (After Qiu Zongtian, 1984.)

sand-layer suites (seven single layers), is a gas condensate pool with bottom water. The X_8 reservoir is highly overpressured: average reservoir pressure is over 494 atm, and the maximum pressure encountered is 649.6 atm.

In the Southwest depression there are quite a few structures similar to the Kekeya anticline. Wells have been drilled on some of these structures, and although oil and gas shows have been discovered, so far no commercial oil and gas flows have been found. Perhaps the failure has been one of technology. It is expected that with the progress of exploration technology, some giant oil and gas fields like the Kekeya field will be found in the future.

Yakela Oil and Gas Field

The Yakela oil and gas field lies in the midsection of the Luntai fault uplift in the Shaya uplift (Figures 2 and 3). This area is cut by the four east-northeast faults and is divided into three fault blocks; among these, the Yakela structure is the highest, with an area of 100 to 150 km². Its south boundary is the large overthrust Luntai fault, and its north limit is the Ya No. 1 normal fault (Figure 14). Beneath the unconformity at the base of the Triassic section, the lower part of the structure is a buried paleomountain. The strata dip to the southwest at 25 to 30°, decreasing in age from northeast to southwest (Figure 15). The overlying Mesozoic–Cenozoic sediments, which are more than 5000 m thick, completely conceal the Paleozoic mountain. The Yenshan orogeny at the end of the Cretaceous made the Triassic-Cretaceous structure a gentle anticline and a good oil and gas trap. In the Tertiary, a tension fault developed on the north flank of the structure (Ya No. 1 fault).

Yakela Field Stratigraphy

In the Yakela field, taking the Shacan No. 2 well as an example (Figure 14), the strata encountered are as follows (after Kang Yuezhu, 1985):

Quaternary (Q). Gray and yellow-gray silty clay and clayey silt, 113.5 m thick.

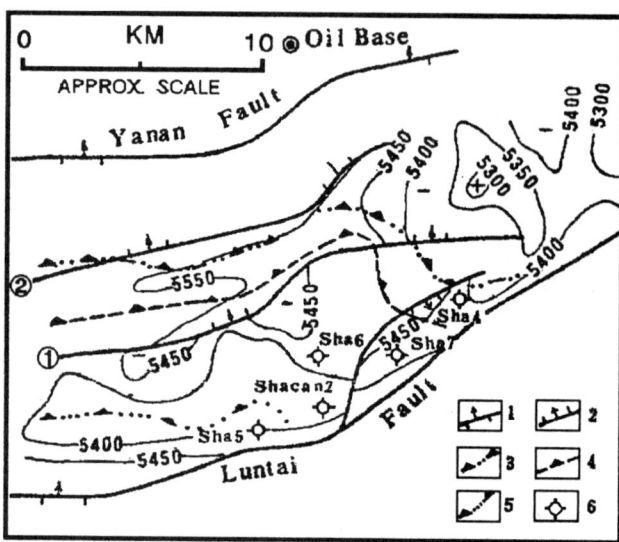

FIGURE 14. Yakela field structure contour map on the erosional surface of the Paleozoic section on the northern edge of the Tarim Basin. The location of the field is shown in Figure 2. Subsea contours are in meters, at various intervals as indicated. (After He Haiquan et al., 1989.) Symbols: (1) thrust fault, (2) normal fault, (3) unconformity boundary between Sinian and pre-Sinian, (4) unconformity boundary between Cambrian-Ordovician and Sinian, (5) unconformity boundary between Silurian-Devonian and Cambrian-Ordovician, and (6) exploratory well. Circled numerals 1 and 2 denote Ya No. 1 and No. 2 faults, respectively.

Neogene Pliocene Kuche Formation (N_1K). Upper part: gray silt and fine sand interbedded with light-yellow mudstone. Lower part: light-brown, fine-grained quartz sandstone and gray-white silt to fine sandstone, interbedded with light-brownish (small amount of light-greenish) gray mudstone. Total thickness, 2647 m.

Miocene Kangcun Formation (N_1K). Gray-brown mudstone, siltstone, and light-gray siltstone, 817 m thick.

Miocene Jidik Formation (N_1J). Top: blue-gray mudstone and silty mudstone (about 54 m thick). Upper part: light-brown and gray-brown silty mudstone. Lower part: brown-gray mudstone embedded in red-brown gypsum mudstone. Thickness reaches 547 m.

Miocene Suwiyi Formation (N_1S). Upper part: yellow-brown, medium- to fine-grained sandstone and dark-gray "fine-grained" gravel interbedded with light-brown mudstone, in part with gypsum. Lower part: light-brown mudstone, gray siltstone, and yellow, medium- to coarse-grained sand. Total thickness, 273 m.

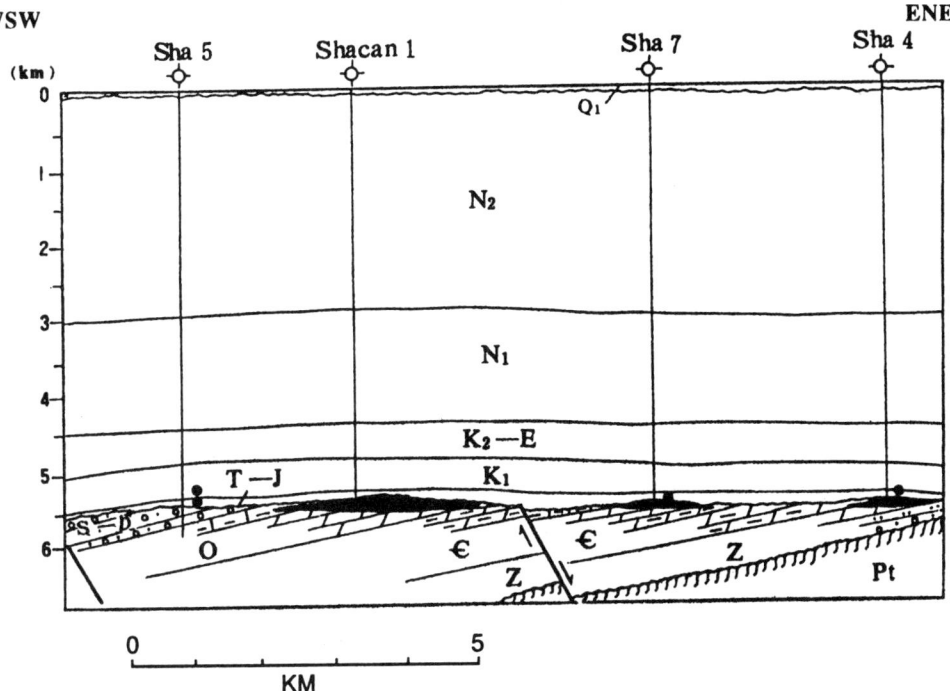

FIGURE 15. Section across the Yakela field. Cross section parallels the Luntai fault (Figure 14), about 1 km north. The location of Shacan No. 1 is shown in Figure 3. (Modified after He Haiquan et al., 1989.)

Eogene Kumgelie Formation (Ek). Fine-grained sandstone interbedded with red-brown silty mudstone and siltstone, 360 m thick.

Lower Cretaceous (K_1). Brown, fine-grained silty sandstone interbedded with red-brown silty mudstone and dark-brown mudstone, 480 m thick.

Lower Jurassic–Upper Triassic ($J_1 + T_{2+3}$). Upper part: gray-purple siltstone, fine-grained sandstone, and gray, medium- to coarse-grained sandstone and gravel interbedded with dark-brown mudstone, 109.5 m thick.

Ordovician (O). Chert strata interbedded with gray, medium-grained dolomite, siliceous dolomite, and dark-gray limestone, 36 m thick.

When the Shacan No. 2 well was drilled to a depth of 5391 m, a strong blowout occurred. The estimated initial production rates were more than 1000 m³ (6300 bbl) of oil and 2 million m³ (70 mmcf) of gas per day. After that, below the Paleozoic section, algal dolomite of the upper Sinian and microcrystalline and fine crystalline dolomite of the mid-Cambrian were encountered in drilling. The Sha No. 4 and Sha No. 7 wells produced substantial amounts of oil and gas; then Lower Jurassic sandstone yielded oil and gas at high flow rates. In the Sha No. 5 well, Silurian–Devonian (or, as some believe, Carboniferous) clastic sediments and Ordovician dolomite were encountered during drilling in the pre-Mesozoic strata. A small amount of heavy oil was obtained at the top of the Cretaceous, but oil and gas at high flow rates were tested in the Lower Cretaceous (Table 1). Besides this, good oil and gas shows were found in the Kidik and Kumglemu formations of the lower Tertiary.

Yakela Reservoirs

Yakela field reservoirs can be classified into three types:

1. *Porous clastic reservoirs*. Mesozoic reservoirs are primarily porous clastics, and most of them are concentrated in the Upper Triassic, Lower Jurassic, Lower Cretaceous, and Tertiary (lower Miocene) strata. The average thickness of single reservoir units is 5 to 20 m; sometimes the thickness can exceed 100 m. The porosity is 6 to 21%, and permeability ranges from 11 to 1700 md.

2. *Fractured, karstic carbonate rock reservoirs* have been found mainly in the lower Paleozoic Cambrian–Ordovician carbonate rocks in which primary and secondary porosity, fractures, and caves are well developed. The average width of the fractures and pores is about 1 mm; the maximum width can be more than 3 mm.

3. *Weathered zone secondary porosity reservoirs*. At the end of the Hercynian orogeny, the Shaya uplift continued to rise for a long time. The exposed Paleozoic rocks were weathered, and a thick weathered zone was devel-

oped. In carbonate rocks, a great quantity of secondary pores and many fractures and openings gave rise to characteristics such as high permeability in the weathered zone. The conditions were optimal for oil and gas accumulations.

The major rock units overlying the lower Paleozoic oil and gas pool of the Yakela oil field are those of the lower part of the unconformable Mesozoic structure (Figure 15). These overlying impermeable beds sealing the lower Paleozoic oil and gas pool beneath them are primarily mudstones and gypsum mudstones. The reservoirs of the lower Paleozoic strata have formation pressures slightly lower than normal and form an underpressured hydrodynamic system, whereas pressures in oil and gas reservoirs in the Mesozoic section are somewhat higher than normal and form a high-pressure hydrodynamic trap. These conditions favor the preservation of the lower Paleozoic oil and gas pool.

Lunnan Oil Field

The Lunnan oil field, about 40 km to the south of Luntai County, is on the Akekum slope south of the Shaya uplift (Figure 3). On September 5, 1987, in the lower Jurassic clastic rocks, the discovery well, Lunnan No. 1, at a depth of about 4770 to 4800 m, recovered highly commercial oil and gas flows in drill stem tests. Later, the lower Ordovician weathered zone, at a depth of 5030 to 5052 m, yielded a certain amount of heavy oil with a relative density of 0.96 and a viscosity of 7800 MPa s at 30°C.

The Lunnan No. 1 well is 46 km from the Shacan No. 2 well. The well location had been set at the western high point of the Lunnan No. 1 geologic anomaly. This discovery well was followed by several exploratory wells that produced oil and gas at high flow rates from different formations of the Ordovician carbonate rocks and from clastic rocks of the Triassic, Jurassic, and Lower Cretaceous sections.

According to the latest report (Yong Tianshou et al., 1990), comparison of the lithology and mineral composition of the Mesozoic rocks encountered during drilling of the Lunnan No. 1 well with those of the Kuche depression rocks, along with the identification of plant fossils, show these Mesozoic rocks to be Cretaceous and Jurassic, but not Triassic.

The Cretaceous rocks (3916 to 5039 m deep) are those of the Kapusaliang Group (K_1). They are massive, friable, brick-red sandstones, with gravel at the bottom, overlain by the lower Tertiary Suwiyi Formation.

The Jurassic rocks (4214 to 5039 m deep) comprise the following five formations:

1. *Qigu Formation.* Cherry-colored mudstone and sandy mudstone interbedded with azure and blue-gray-green silty sandstone and sundry rocks, with a thickness of 119 m.

2. *Qiakemak Formation.* Gray-green, blue-gray-green, and purple-gray sandstone interlayered with brown mudstone and sandy mudstone in layers of unequal thicknesses. The mudstone includes abundant calcium nodules. This formation reaches a thickness of 75 m.

3. *Kezilenul Formation.* The upper part is a suite of gray-green and dark gray-green sandy mudstone and gray-green and light gray-green sandstone interbedded in unequal thicknesses. The lower part is a suite of gray, dark-gray, and gray-green sandy mudstone, carbonaceous shale, siltstone, and grayish-white to white, poorly sorted sandstone interbedded with beds bearing spore-pollen fossils. The thickness exceeds 246.5 m.

4. *Yangxia Formation.* The upper part is a suite of dark-gray, flat-bedded, lacustrine mudstone facies. The middle and lower parts are gray and dark-gray silty sandstone, sandy mudstone, and grayish-white to white, irregularly interbedded, poorly sorted sandstone containing the index fossil Coniopteris hymenophylloides of the Lower and Middle Jurassic. The thickness of this formation is 170.5 m.

5. *Ahe Formation.* A dark-gray mudstone interlayered with grayish-white to white gravelly sandstone. This formation contains a great quantity of fossils and reaches a thickness of 214 m. It is the major oil-productive bed of the Lunnan No. 1 well.

In the Lunnan field, the Paleozoic section is a "Y-type" structure composed of small-scale compressional "pop-up" blocks. In fact, it is a small rebound thrust block. The lower part of Figure 16 is a section across the Lunnan Akekum structure. Under the Sha No. 9 well is a rebound thrust block of Paleozoic rock overlain by a gentle capping anticline composed of Mesozoic strata. Much oil and gas generated from Paleozoic source rocks accumulated in the block, and even more of them migrated upward into the capping anticline.

Other Giant Oil and Gas Fields

Besides the Kekeya, Yakela, and Lunnan oil and gas fields, the Akekule, Yingmaili, Central No. 1, and other structures have produced oil and gas at high flow rates. Perhaps these structures will be found to be giant oil and gas fields waiting to be appraised and developed.

The Akekule structure is about 10 km south of the Lunnan structure. As with the Lunnan structure, the shallower part of the Akekule structure is a capping anticline composed of Mesozoic strata, and beneath it is a rebound thrust block in Paleozoic rock strata. The discovery well Sha No. 14 and the exploratory well Sha No. 18 encountered oil and gas at high flow rates, especially in the latter well, whose estimated initial potential exceeded 1400 m³ (8800 bbl) of condensate and 420,000 m³ (14.8 mmcf) of gas per day.

The Yingmaili structure is located in the Shaxi faulted fold zone of the Shaya uplift. Substantial amounts of oil have been produced from Ordovician limestone in this structure.

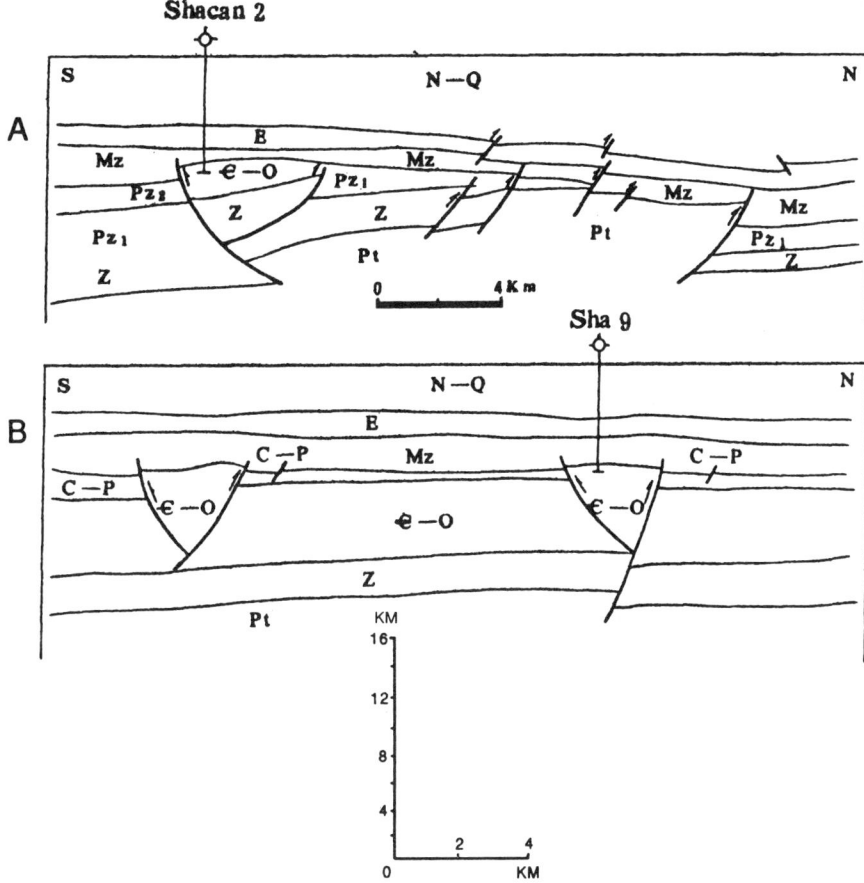

FIGURE 16. Sections across the compressional "pop-up" blocks of the Yakela structure (A) and of the Akekum and Akekule structures (B). The sections are normal to the Luntai fault (Figure 14). (After Tang Liangjie, 1989.)

The Central No. 1 structure lies in the Central uplift and covers an area of 8200 km². It is the most prospective area. In the Central No. 1 well, high oil and gas flow rates have already been obtained; the initial production potential is 576 m³ (3620 bbl) of oil and 340,000 m³ (12 mmcf) of gas per day.

ACKNOWLEDGMENTS

The author thanks Professor Liu Shuying of Lanzhou Commercial College, Lanzhou, China, for translating this paper into English.

Editor's note: In the translation into English, certain phrases required restating to ensure clarity and the continuity of expression. The editor has necessarily exercised the privilege of such restatement.

REFERENCES CITED

Cheng Rongling, 1988, The prospects for oil and gas and the sedimentary sequences from Triassic to Jurassic in the Tarim Basin: Experimental Petroleum Geology, v. 9, n. 3, p. 253–257.

Cheng Ziyung, 1988, Local structures related to oil and gas in the northern Tarim Basin: Xinjiang Petroleum Geology, v. 9, n. 1, p. 1–8.

He Haiquan, Zhongxian Zhang, and Tang Yidan, 1989, On the formation conditions of the Yakela oil and gas field in the Tarim Basin: Experimental Petroleum Geology, v. 11, n. 1, p. 1–5.

Hu Boliang, 1979, Distribution of carbon isotopes of crude oil in the Tarim Basin, *in* Collected works on petroleum geology: Geological Press, 1982, p. 130–135.

Hu Boliang, 1981, On the organic geochemical characteristics and oil source rock correlations in the Tarim Basin: Oil & Gas Geology, v. 2, n. 4, p. 359–368.

Hu Boliang and Shen Jianzhong, 1990, Geochemical characteristics of hydrocarbons in the Yakela structure of the Shaya uplift in northern Tarim Basin and their generation, migration, and accumulation: Experimental Petroleum Geology, v. 12, n. 3, p. 232–247.

Hunt, J. M., 1979, Petroleum geochemistry and geology: San Francisco, W. H. Freeman, 617 p.

Kang Yuezhu, 1985, Discovery of high rate of oil and gas flow in the Shaya uplift, Tarim Basin: Xinjiang Geology, v. 3, n. 1, p. 34–41.

Kang Yuezhu, 1986, Characteristics of the tectonic movements of the Tarim platform: Xinjiang Geology, v. 6, n. 1, p. 48–59.

Kang Yuezhu, 1989, Enormous discoveries resulting from hydrocarbon exploration in the Tarim Basin: Oil & Gas Geology, v. 10, n. 3, p. 276-282.

Qiu Zongtian, 1984, Main characteristics of the Kekeya condensate gas field in the Tarim Basin: Xinjiang Petroleum Geology, v. 5, n. 3, p. 68-78.

Ren Jishun, 1989, Prof. Huang Jiqing and China's oil exploration: Oil & Gas Geology, v. 10, n. 3, p. 233-246.

Song Lixung, 1989, The formation, evolution, and prospecting of oil in the Southwest depression, Tarim Basin: Xinjiang Petroleum Geology, v. 10, n. 1, p. 20-34.

Tang Liangjie, 1989, Fault types and hydrocarbon potential of the northeastern Tarim Basin: Oil & Gas Geology, v. 10, n. 1, p. 45-51.

Tissot, B. P., and D. H. Welte, 1984, Petroleum formation and occurrence: Heidelberg, Springer, p. 93-129.

Xie Xiaoan, 1986, On geological conditions for the formation of giant oil and gas fields in the northern Tarim Basin: Xinjiang Petroleum Geology, v. 7, n. 2, p. 19-28.

Yan Yuegui, Yang Shizhuo, Hu Boliang, Wen Changqing, and Jin Huijuan, 1983, Some problems concerning petroleum geology of the Tarim Basin: Scientia Sinica (Series B), v. 26, n. 11, p. 1201-1215.

Yong Tianshou, 1984, The marine stratum from Late Cretaceous to the Eogene period in the west Tarim Basin: Xinjiang Petroleum Geology, special issue, 75 p.

Yong Tianshou, Lixiong Song, Yuede Yu, and Yu Xinqi, 1990, The Mesozoic stratum of the Lunnan 1 well in Tarim Basin: Xinjiang Petroleum Geology, v. 11, n. 2, p. 132-135.

The Hydrocarbon Potential of the Norwegian Continental Shelf

Finn Roar Aamodt
Norwegian Petroleum Directorate
Stavanger, Norway

ABSTRACT

Exploration for hydrocarbons on the Norwegian continental shelf began in 1965. The first commercial discovery (Ekofisk), made in 1969, led to first production of hydrocarbons in 1971. Subsequently, many oil and gas fields were discovered, 19 of which were in production by June 1989. Numerous fields will be developed in the next decade. It is anticipated that by mid-1990 total Norwegian oil and gas production will be about 100 million and 30 million TOE, respectively. Oil production is still growing, and gas production is expected to grow significantly from the turn of the century. Proved resources for the entire Norwegian continental shelf as of June 1989 were estimated at 5.26 billion TOE—2.16 billion TOE of oil and 3.1 billion TOE of gas. Most of these resources (60%) lie in giant fields. All the discovered giants are either in production or in the development phase. Most of them were discovered from 1978 to 1985.

Exploration costs are increasing because most of the recent discoveries are smaller than previous ones. However, exploration costs in Norway are still lower than in most other areas. The R/P (resource/production) ratios for Norway are approximately 20 for oil and 120 for gas. The remaining unproved potential of the Norwegian continental shelf is expected to be at least 4 to 5 billion TOE, of which 70% or more is anticipated to be gas. New resources will have to be found in the North Sea, offshore mid-Norway, and in the Barents Sea.

The future challenges for Norway will be to discover new oil fields, to make economically viable the smaller existing fields, to enhance oil recovery from existing fields, and to find markets for its large gas resources. Norway will remain a major oil-producing nation well into the next century, and a major gas-producing nation for a much longer period.

Editor's note: An oral presentation on the hydrocarbon potential of the Norwegian continental shelf was made at the conference in Stavanger; however, a full paper was not written. The illustrations that accompanied the oral presentation are sufficiently self-explanatory that it is appropriate to include them in this volume, along with the abstract from the conference program.

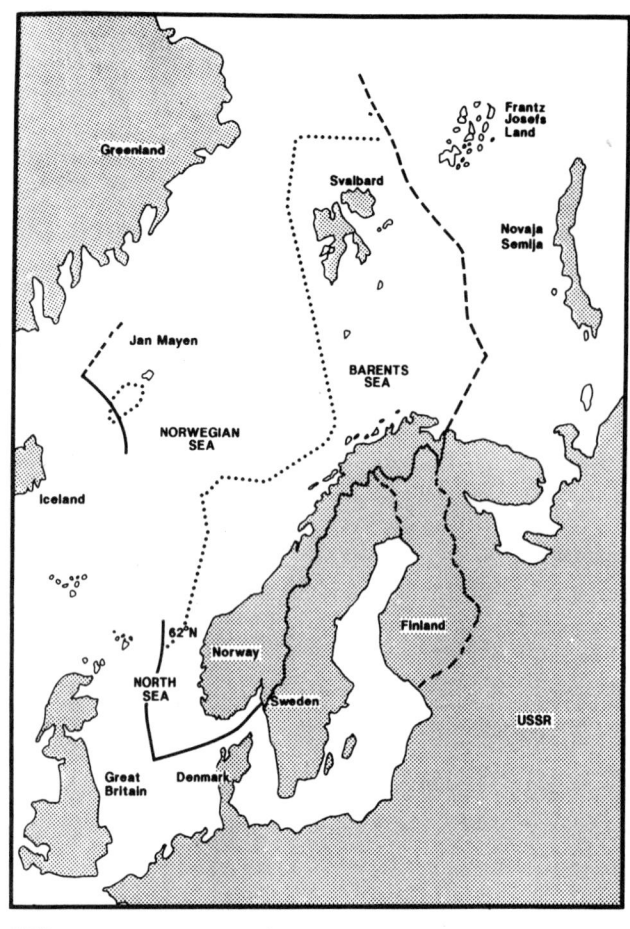

FIGURE 1. Norwegian offshore areas covered by sedimentary rocks.

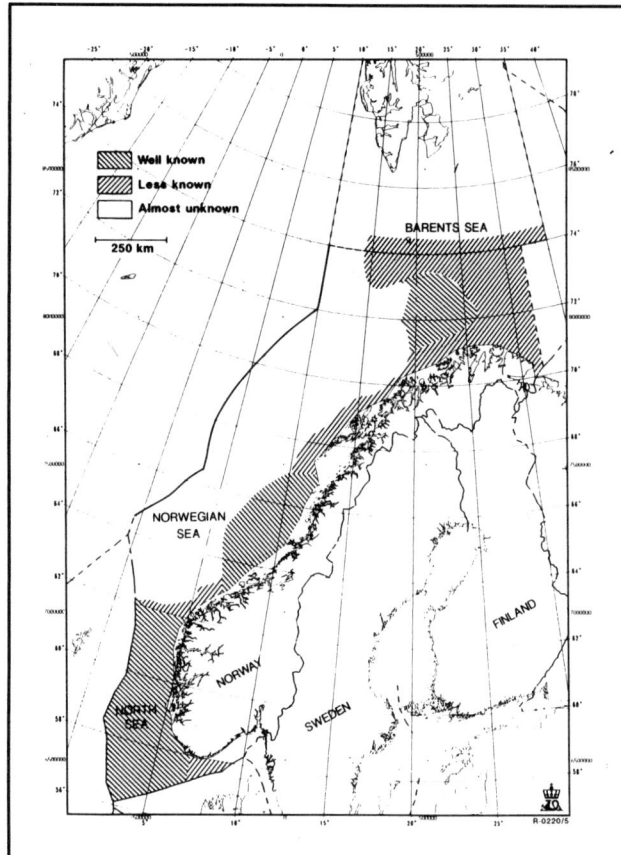

FIGURE 3. Exploration maturity map.

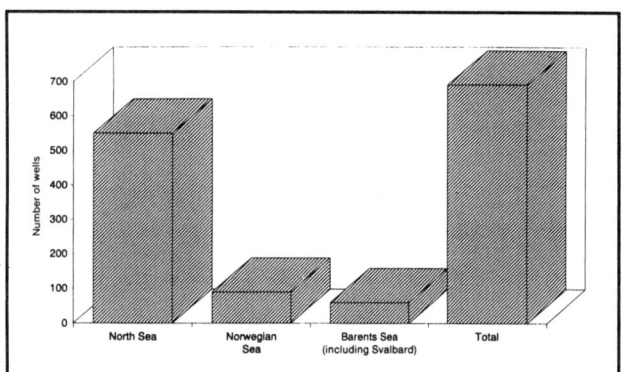

FIGURE 2. Norwegian exploration wells.

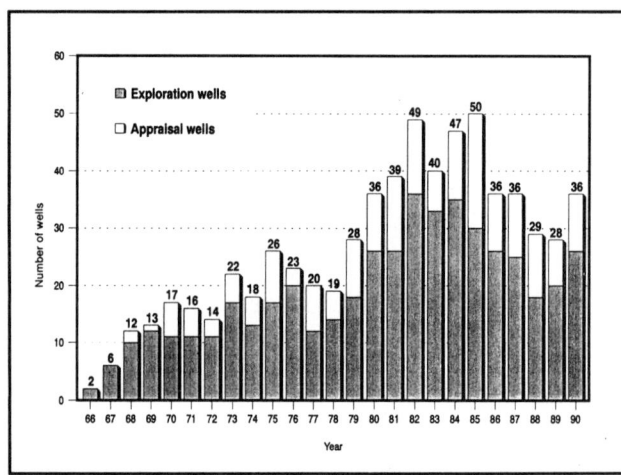

FIGURE 4. Exploration drilling on the Norwegian continental shelf, 1966–90.

FIGURE 5. Geographic distribution of proved resources (billion TOE).

	Oil/condensate	Gas	Total
North Sea	1.94	2.58	4.52
Mid-Norway	0.21	0.26	0.47
Barents Sea	0.01	0.26	0.27
Total	2.16	3.1	5.26

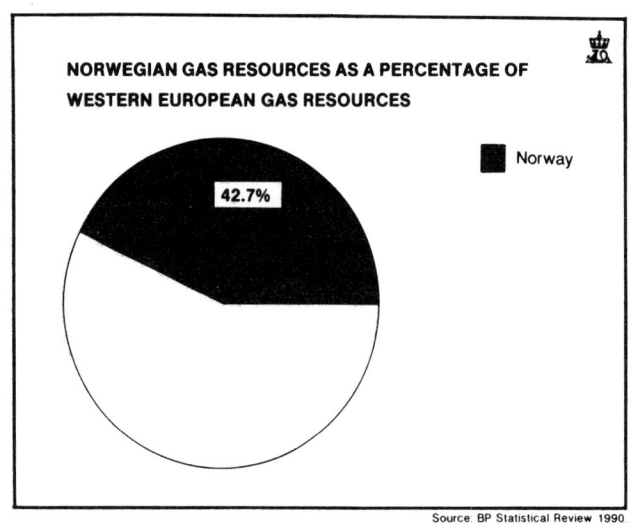

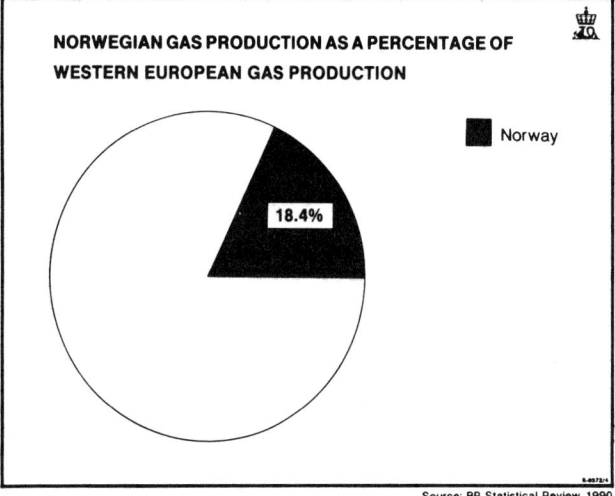

FIGURE 7. Norwegian gas resources and production as percentages of Western European gas resources and production.

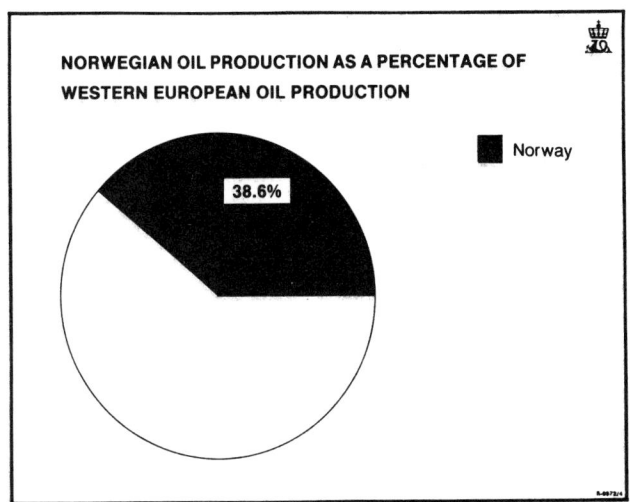

FIGURE 6. Norwegian oil resources and production as percentages of Western European oil resources and production.

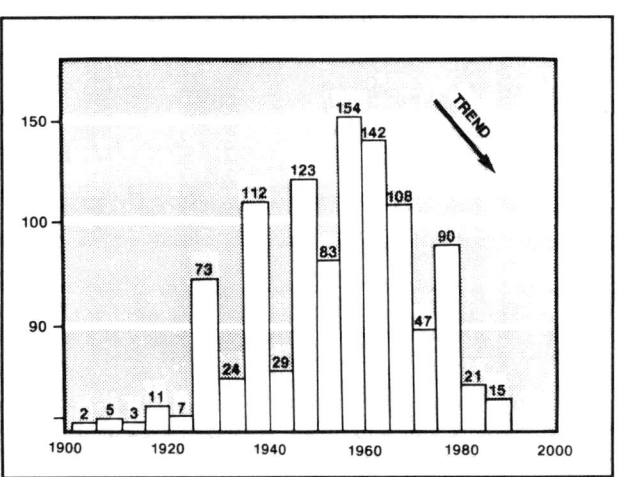

FIGURE 8. World giant oil discoveries.

FIGURE 9. Proved Norwegian oil and gas resources.

Field size	No. of fields	Resources (billion std. m³)	
		Oil	Gas
<10	29	0.09	90
10–20	15	0.13	130
20–40	19	0.34	300
40–60	16	0.57	730
>100	8	1.34	1850
Total	87	2.47	3100

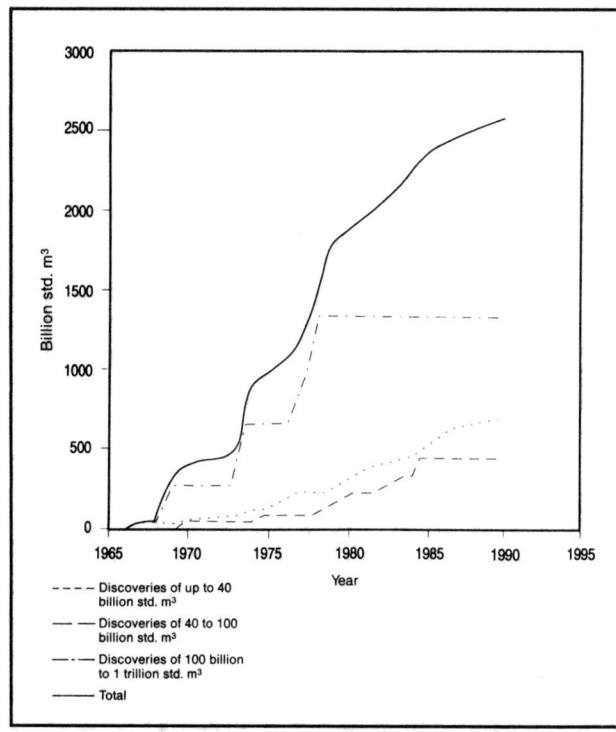

FIGURE 10. Growth in recoverable Norwegian oil resources since 1965.

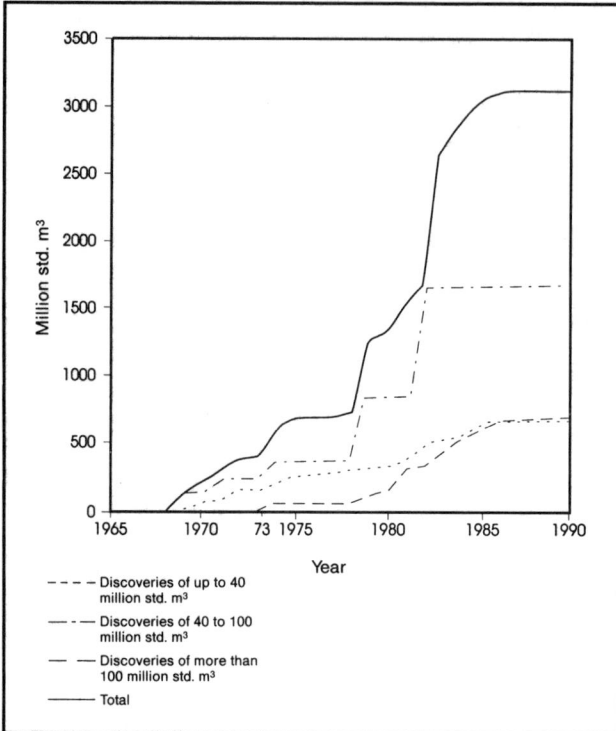

FIGURE 11. Growth in recoverable Norwegian gas resources since 1965.

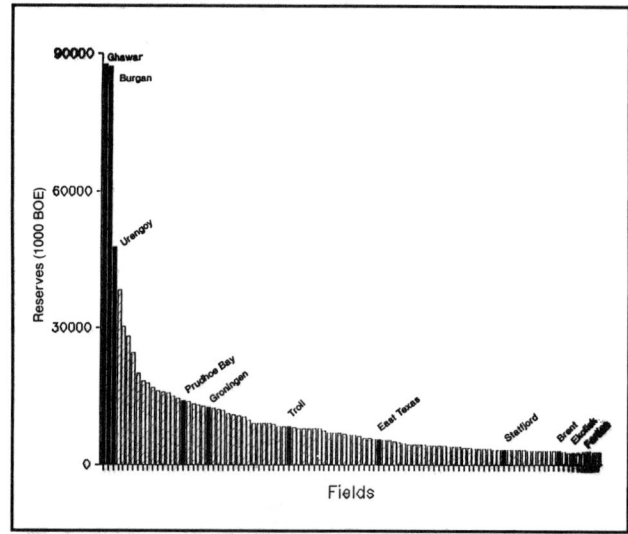

FIGURE 12. The world's 100 largest oil and gas fields.

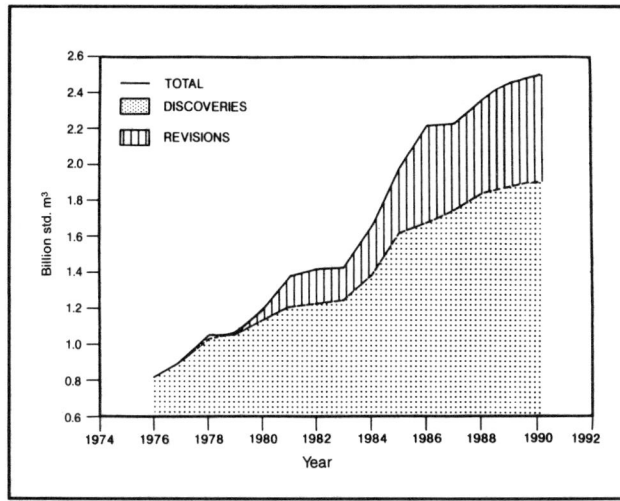

FIGURE 13. Revisions in estimates of Norwegian oil resources.

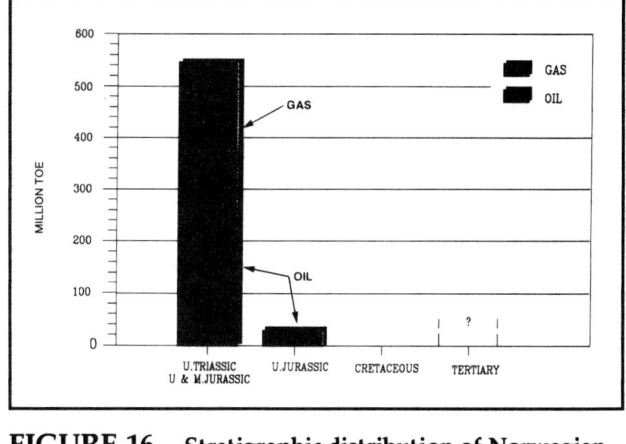

FIGURE 16. Stratigraphic distribution of Norwegian oil and gas resources discovered in the last five years, 1986–90.

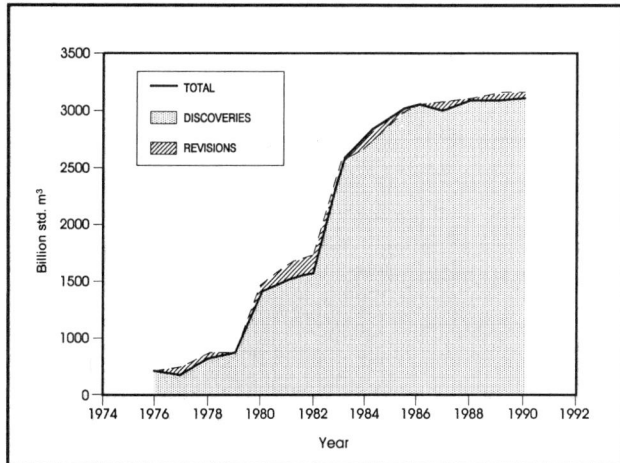

FIGURE 14. Revisions in estimates of Norwegian gas resources.

FIGURE 17. Future oil and gas potential of the Norwegian continental shelf.

Improved oil recovery potential

Estimated by NPD to be on the order of 400 to 800 million MT of oil.

Undiscovered potential

Estimated by NPD to be on the order of 3.5 to 4.0 billion TOE.

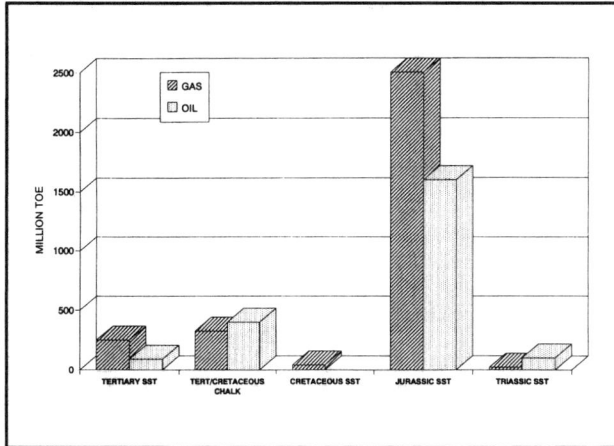

FIGURE 15. Stratigraphic distribution of Norwegian oil and gas resources.

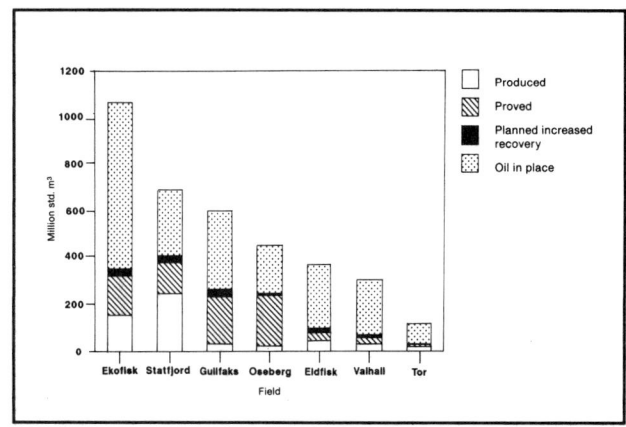

FIGURE 18. IOR potential of existing fields on the Norwegian continental shelf.

FIGURE 19. Oil in place in thin reservoirs on the Norwegian continental shelf.

No. of fields	Oil in place (10⁶ MT)	Thickness (m)
10	1000	5–25

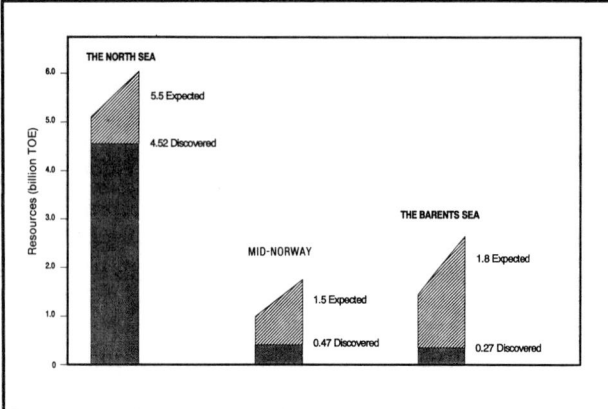

FIGURE 20. Geographic distribution of oil and gas resources on the Norwegian continental shelf.

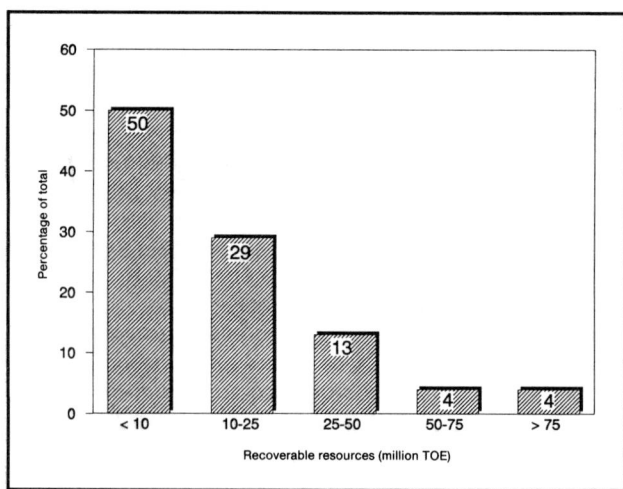

FIGURE 21. Undrilled prospects on the Norwegian continental shelf.

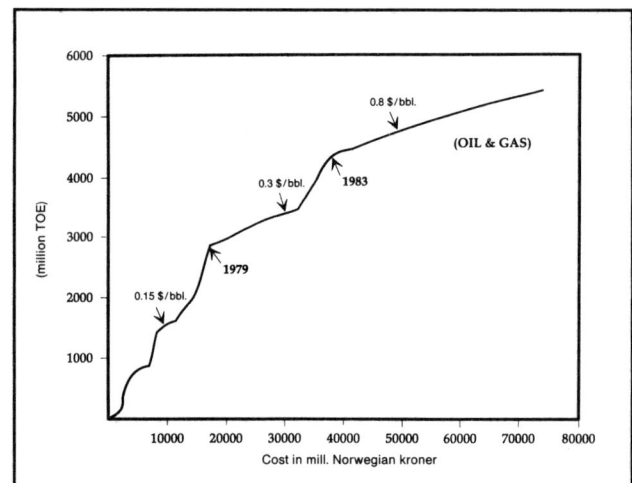

FIGURE 22. Unit exploration costs for Norwegian wildcat and appraisal wells, 1967–91.

The Gorgon Gas Field

L. J. Clegg
M. J. Sayers
A. M. Tait

West Australian Petroleum Pty. Limited
Perth, Australia

EXTENDED ABSTRACT

The Gorgon gas field is situated 130 km offshore on the Northwest Shelf of Western Australia. The field was discovered by West Australian Petroleum Pty. Limited (WAPET) in 1980, and two successful appraisal wells were drilled in 1982 and 1983. The Gorgon structure is an elongate horst block trending north-northeast at the southwestern end of the Rankin Trend, which also contains Woodside Offshore Petroleum's Goodwyn and North Rankin gas fields.

The Rankin Trend forms the western margin of the Barrow and Dampier subbasins. It was formed in the Early Jurassic during active rifting, in response to downfaulting along the eastern margins of these half-grabens. The extensional faulting of the Rankin Trend follows two dominant trends. Northeasterly faults parallel the rift axis, and north-northeasterly faults follow a reactivated Paleozoic/Precambrian trend. At the close of the Jurassic, uplifting took place to the south of the Barrow Subbasin in association with the breakup of Australia and India. This provided large volumes of sediment, which moved north and spread out across the Exmouth Plateau and the Barrow Subbasin, forming the deltaic Barrow Group. In the Early Cretaceous, continental breakup took place as India separated from Australia. This cut off the sediment supply to the Barrow Group. The marine Muderong shale was then deposited as a regional seal across the Rankin Trend and the Barrow and Dampier subbasins and adjoining shelves. Cretaceous and Tertiary sediments overlying the Muderong Shale were deposited as prograding slope wedges on the northwest Australian continental margin.

The Triassic Mungaroo Formation, which hosts the gas at Gorgon, is a prerift sequence of meander belt sandstones and overbank siltstones and shales.

The Triassic sandstones dip north within the Gorgon horst, which is unconformably overlain and sealed by Cretaceous shale of the Barrow Group. The discovery well, Gorgon 1, encountered 98 m of net gas pay in a 489-m gross Triassic interval above a gas-water contact at 4111 m subsea. North Gorgon, 23 km north of Gorgon 1, encountered 127 m of net gas pay in an 884-m gross interval above a lowest known gas at 4135 m subsea. Central Gorgon 1, situated between the previous two wells, encountered 35 m of net gas pay in a 443-m gross interval above a gas-water contact at 4049 m subsea.

Most sands in the Gorgon field have tested at over 850,000 m^3 (30 million ft^3) of gas per day. The sandstones consist mainly of quartz, with kaolinite occurring as diagenetic replacements of feldspar grains. Porosity of the reservoir sandstones decreases with depth from 20 to 12% owing to the growth of authigenic quartz cement.

The Gorgon gas field is at least 45 km long by 5 km wide. It is estimated to contain 312 billion m^3 (11 tcf) of gas in place.

Editor's note: An oral presentation on the gas potential of the Gorgon gas field was made at the conference in Stavanger; however, a full paper was not available for publication. The authors have kindly provided an extended abstract for inclusion in this volume.

Index

Abkatun field, 73, 83, 85, 91-92
Agat field, 291
Agave field, 73, 81, 82, 84, 86
Akekum slope, 497
Albacora field. *See also* Campos Basin.
  cross section, 130
  development and production projects, 132-134
  discovery history, 126-128
  geology, 128-129
  isopach maps, 131
  location maps, 124, 139
  oil pool limits, 125
  reservoir characteristics, 131
  reservoir facies, 133
  signatures, local, 127
  stratigraphy, 128-129
  structural contour maps, 129
  structure, 128
Alba field, 297-305
  depositional processes, 302
  development, 305
  exploration history, 298
  facies, 302
  geologic setting, 298
  location map, 298
  petrographic description, 304
  reservoir characteristics, 304
  reservoir structure, 303-304
  sandstone relationships, 300-302
  stratigraphy, 298-302
  structural features surrounding, 298
  structure, 302-304
Amberjack high, Monterey Formation isopach map, 13
Aptian-Albian Ben Nevis sandstones, 50
Avalon sandstones. *See* Ben Nevis/Avalon sandstones.
Awarti-Mankal depression, 499

Bacab field, 73, 83, 85, 92, 93
Barbara field, 265-276
  development scheme, 276
  geologic model, 273
  geology, 267
  geophysics, 272
  history, 267
  location map, 266
  petrophysical and thermodynamic parameters, 272, 276
  quaternary depositional environment, 267
  reservoir and fluid parameters, 276
  stratigraphy, 272
  tectonic sketch, north Adriatic area, 271
  time maps, "A" level, 269, 270
  trap, 267, 272
  type log, 268
Barremian unconformity, 48
Batab field, 73, 83, 85, 92-93
Bellota field, 73, 82, 84, 86, 87
Ben Nevis/Avalon sandstones
  appraisal, 41
  Aptian-Albian Ben Nevis sandstones, 50
  Barremian unconformity, 48
  core log, 49
  depositional model, 47-51
  development, structural and stratigraphic, 51
  post-Barremian unconformity—Avalon sandstones, 48-50
  pre-Barremian unconformity—Avalon sandstones, 47-48
  reservoir characteristics, 50-51
  reservoir geometry and quality, 50
Brent field, 286
Brent Group, 422, 423, 439-441
Broom Formation, 440

Caan field, 73, 83, 85, 93-94
Cabinda. *See* Takula field and Greater Takula area.
Campos Basin, 123-153. *See also* Albacora field, Carapeba field, Linguado field, Marimbá field, Marlim field, Vermelho field.
  cross section, 126
  discovery history, 126-128
  evolution, 125
  fields, comparative histogram, 140
  fields, production characteristics, 139
  fields, volumes, 139
  geology, 124-126, 128-129, 138-142
  petroleum potential, 125-126
  seismic regional section, 142
  stratigraphic column, 141
  stratigraphy, 124
  structural features, 124-125
  tectonic framework, 141
Caño Limon field, 175-195. *See also* Llanos basin.
  characteristics, 190
  cross sections, 194
  location map, 178

sedimentary sequence, 195
stratigraphic column, 183
stratigraphy, 188–190
structure, 188, 189, 190
Carapeba field. *See also* Campos Basin.
   development and production, 150
   geologic sections, 145, 147
   location map, 139
   productive sandstones, 146
   reservoirs, 139, 143, 145–147
   structural map, 146
   volumes and production characteristics, 139
Cardenas field, 73, 82, 84, 86–87
Ceuta-Tomoporo field, 163–173. *See also* Maracaibo basin.
   exploration and discovery history, 167–169
   generation and migration of hydrocarbons, 167
   geologic and thermal modeling, 169
   Miocene sedimentation on the Ceuta high, 168
   production parameters for reservoirs, 164
   reserves, 172, 173
   reservoirs, 169–172
   seismic section, 165
   stratigraphy, 166–167
   structural setting, 164–166
Chiapas-Tabasco area
   API gravity variation, 99
   field characteristics, 82
   field descriptions, 81, 86–91
   field production data, 84
   location, 73
   seismic line in central part, 81
   stratigraphy, regional, 75–79
   tectonic aspects, 80
Chuc field, 73, 83, 85, 94, 95
Cook Formation, 438–439

Diatomite, unaltered, SEM photomicrograph, 17
Digimu reservoir, 479
Dongying Formation, 463
Douala basin. *See* Sanaga Sud field.
Draugen field, 371–382
   development planning, 379–380
   exploration and appraisal, 372
   fluid properties and volumetrics, 379
   Garn Formation, 378
   geologic setting, 372–374
   location map, 372
   overview, schematic, 381
   reservoir characteristics, 375–378
   reservoir geology, 374–375
   Rogn Formation, 376–378
   stratigraphy, regional, 372, 374
   structure, 372
Dunlin Group, 438

Eastern Venezuelan basin. *See also* El Furrial field.
   cross section, regional, 158
   depositional systems, 157
   exploration history, 156–157
   location map, 156
   sedimentary cycles, 158
   stratigraphic chart, generalized, 158
   stratigraphy, 157–158
   structural framework, 158–159
Ek field, 73, 83, 85, 94, 95, 96
El Furrial field, 155–161. *See also* Eastern Venezuelan basin.
   development program, 161
   exploration history, 156–157
   location map, 156
   petrophysical parameters, 160
   production data, 159–160
   production profile, 161
   reservoir characteristics, 159, 160
   seismic line, structural interpretation, 157
   stratigraphic cross section correlating wells, 160
   stratigraphy, 157–158
   structural framework, 158–159
   tectonic framework, regional, 159
Espada Formation, 9–10
Etive Formation, 441

Fladen Ground spur, 298, 299
Fortescue field, 483–492
   development, 489
   drilling, infill, 490–491
   exploration history, 484–485
   oil properties, 488
   oil sand thicknesses, net, 489
   permeability, 488
   porosity-permeability, 487–488
   production, 489–490
   reservoir characteristics, 487–488
   reservoir management, 491–492
   reservoir pressures, 489–490
   seismic coverage, 486–487
   seismic modeling, 490–491
   stratigraphy, 485–486
   structure, 486

Garn Formation, 378
Gippsland basin, 484–486
Giraldas field, 73, 82, 84, 87, 88
Glassy chert, quartz phase, SEM photomicrograph, 18
Gorgon field, 517–518
Gullfaks field, 287, 429–446
   Brent Group, 439–441
   Broom Formation, 440
   Cook Formation, 438–439
   drilling schedule, early, 444
   Dunlin Group, 438
   Etive Formation, 441
   exploration history, 431
   geophysics, 431–435
   Hegre Group, 436
   hydrocarbons, source and migration, 442
   lithostratigraphic nomenclature, 439
   location map, 430
   Ness Formation, 441
   pipeline infrastructure, 432
   platform locations, 432
   production, 445
   Rannoch Formation, 440–441
   reservoir development plan, original, 442–444
   reservoir geology, 441–442

reservoir management, 444
sand strength, 445
seismic profile, 434
Statfjord Formation, 436-438
stratigraphy and environmental model, 436-441
structure and structural development, 433, 435-436
Tarbert Formation, 441
vertical communication, 444
water injection program, 444-445
wells, exploration and delineation, 432

Halibut-Cobia field, 484-490
Haltenbanken province. See Draugen field, Heidrun field.
Hammerfest Basin. See also Snøhvit field.
   history, 350-352
   stratigraphy and deposition, 352-354
Hedinia sands, 479
Hegre Group, 436
Heidrun field, 383-406
   appraisal and development, 393-405
   block evaluations, eighth round, 392-393
   development plan, 405
   exploration, 384-393
   geologic model, 404
   location map, 385
   maturity, 390
   maturity maps, 392
   paleoenvironments, Early/Middle Jurassic, 398
   paleogeographic evolution, 386-388
   paleotectonic maps, 388, 389
   petrophysical interpretation, 404
   porosity vs. depth plot, 391
   reservoir data, 394
   reservoir quality, 389
   reservoir structure, 394-395
   sedimentology and stratigraphy, 395-403
   seismic sections, 396, 397
   source rocks, 389-390, 404-405
   stratigraphic column, 391
   stratigraphy, regional, 390
   structural closures, 391
   structural features map, 386
   tectonic development, 386
   time map showing prospect inventory, 391
   type well log section composite, 398
Hibernia field, 35-54
   appraisal, 40-41
   depositional model—Ben Nevis/Avalon sandstones, 47-51
   depositional model—Hibernia sandstones, 43-46
   development plan, 51-52
   exploration history, 39-40
   geology and stratigraphy, regional, 36-39
   location map, 37
   Mesozoic rift basins and principal positive areas, 38
   production system, principal components, 53
   reservoir data, 52
   structure, 41-43
   trap, migration, and source of hydrocarbons, 51
   wells tested prior to and drilled since discovery, 39
Hibernia sandstones
   appraisal, 40-41
   bay and crevasse splay sequence, 45
   core log, 44
   delta front deposits, 44
   delta plain sequence, 44-45
   depositional model, 43-46
   depth structure map, 42
   fault block configuration, 43
   reservoir characteristics, 46
   reservoir geometry and quality, 45-46
   stratigraphic cross section, north-south, 47
Huldra field, 290-291

Iagifu/Hedinia field, 471-482
   chronology, 473-474
   cross section, regional, 474
   cross section, structural, 475
   cross sections, dip-oriented, 477
   depositional history, 474-475
   development plan, 481-482
   Digimu reservoir, 479
   discovery well, type log, 480
   exploration history, 472-474
   facilities and export system, schematic outline, 481
   geochemistry, 477-479
   geology, 474-479
   Hedinia sands, 479
   Iagifu sands, 479
   location maps, 472, 473
   reservoir rock and fluid properties, 479-481
   reservoir temperature and pressure, 480-481
   stratigraphy, 475-476
   structure, 476-477
   structure map, 474
   surface geology, 476
   tectonic setting, 474
   Toro reservoir, 479
Iagifu sands, 479
Iris field, 73, 82, 84, 87, 88

Jalama Formation, 10
Jeanne d'Arc Basin
   lithostratigraphy, generalized, 40
   schematic illustration, 41
Jujo field, 73, 82, 84, 87-89

Kekeya field, 503-506
Kribi block. See Sanaga Sud field.
Kuche depression, 494-496
Ku field, 73, 83, 85, 94-95, 96
Kurle nose uplift zone, 499

Lagunillas Formation, 170-171
Laiodong Bay, 460, 461
La Rosa Formation, 171
Latrobe Group, 484-486
Linguado, Carapeba, Vermelho, and Marimbá fields, 137-153. See also Linguado field, Carapeba field, Vermelho field, Marimbá field, Campos Basin.
Linguado field. See also Campos Basin.
   development and production, 145
   geologic sections, 142, 144
   location map, 139

productive rock, principal, 142-143, 144
reservoirs, 139, 142-145
structural map, 144
volumes and production characteristics, 139
Llanos basin. *See also* Caño Limon field.
drilling history, 179
evolution, 178-180
exploration history, 176-178
exploration program, 184-188
geographic setting, 176
geologic section, 181
geologic setting, 178
index map, 178
land blocks, original, 185
oil source, 182-184
seismic sections, 186, 187, 188
stratigraphy, 180-181
structure, 181-182, 184
Tertiary isopach map, 182
Louisiana and Texas, biostratigraphic chart, 28
Luna field, 73, 82, 84, 89
Lunnan field, 508
Luntai fault uplift zone, 497

Maloob field, 73, 83, 85, 95-97
Maracaibo basin. *See also* Ceuta-Tomoporo field.
exploration and discovery history, 167-169
hydrocarbons, generation and migration, 167
oil fields, 164
source rock types, 169
stratigraphy, 166-167
structural map, 164
Marimbá field. *See also* Campos Basin.
amplitude seismic map, 152
development and production, 151-152
geologic sections, 151, 152
location map, 139
productive sandstone, 151
reservoirs, 139, 143, 150-152
structural map, 152
volumes and production characteristics, 139
Marlim and Albacora fields, 123-135. *See also* Marlim field, Albacora field, Campos Basin.
Marlim field. *See also* Campos Basin.
blanket sandstones, 132
cross section, 130
development and production projects, 134-135
discovery history, 126-128
geology, 128-129
isopach map, 132
location maps, 124, 139
oil pool limits, 125
reservoir characteristics, 131
signatures, local, 127
stratigraphy, 128-129
structural contour map, 130
structure, 128
Mexico's giant fields, 73-99. *See also* Chiapas-Tabasco area, Sonda de Campeche area.
characteristics, 82, 83
field descriptions, 80-97
geologic provinces, southeast Mexico, 74

history of discoveries, 73-74
location, 73
production data, 84, 85
stratigraphy, regional, 75-79
tectonic events in southeast Mexico, 80
tectonic frameworks and schematic sections, 79
Miller field, 307-322
biostratigraphic framework, 314
delineation history, 310
depositional model, 314-316
development, 320
exploration history, 309-310
facies descriptions, 314
geologic setting, regional, 310-314
hydrocarbons in place, 320
location map, 308
reserves, 320
reservoir geology, 317-320
seismic interpretation, 317
source rock, 320
stratigraphic column, 311
stratigraphy, 310-316
structure, 316-317
Monterey Formation
isopach map, 13
reservoirs, percent sulfur vs. API gravity for oils, 20
Rock-Eval characteristics, 22
Rock-Eval hydrogen and oxygen index plot, 22
structure maps, 7, 10, 15
Muspac field, 73, 82, 84, 89, 90

Nauchlan sands, 299, 301, 303, 304
Ness Formation, 423-424, 441
Northern Viking graben
depositional environment, compositional effects, 283-284
geochemistry of crude oils, 285-293
geochemistry of Jurassic source rocks, 279-285
geochemistry of oils, 277-296
geologic framework, 278-279
hydrocarbons, expulsion and migration, 293
isotope profile data for oils, 288-291
oils in various fields, 286-292
petroleum generation and expulsion, 284-285
reservoir rocks and sample locations, 279
source characteristics of oils, 285-286
stratigraphic nomenclature, 280
structural features and oil fields, 279
thermal maturity of oils, 285
North Sea
lower Jurassic-Triassic stratigraphy, 409
tectonic map, 408
Norwegian continental shelf, hydrocarbon potential, 511-516

October field, 231-249
African rift basins, 233
central Gulf of Suez accommodation zone area, 237
Egyptian oil production history, 233
exploration and discovery history, 238-240
Gulf of Suez dip domains, 236
location map, 233
Middle East tectonics, 234

Nubia Formation characteristics, 242–245
  oil characteristics, 246
  production and pressure history, 246–247
  reservoir distribution, 239, 240
  source rocks, generation, and migration, 245–246
  stratigraphic column, 238
  stratigraphy and producing horizons, 236–238
  structure, 240–242
  structure map, pre-Miocene, 239
  tectonic and structural setting, regional, 232–236
Oseberg field, 287, 290, 417–428
  development and production, 424–426
  drilling activity, 420
  exploration and discovery history, 419–420
  gas injection, 426
  installations, fixed, 424
  location maps, 419, 420
  milestones, 420
  Ness Formation, 423–424
  oil reserves for Norwegian fields, 419
  Oseberg Formation, 422–423
  predrilling, 424–426
  production, 426
  reserve estimates, 420
  reservoir characteristics, 418
  reservoir geology, 422–424
  source rocks, 422
  stratigraphy, 421–422
  structure, 420–421
  Tarbert Formation, 424
  well location map, 419
Oseberg Formation, 422–423

Papuan fold and thrust belt, 472–474
Paredon field, 73, 82, 84, 90
Paris basin. *See also* Villeperdue field.
  block diagram, regional, 252
  history and geology, 252–254
Pinda reservoir, 205–206
Point Arguello field, 3–25
  development, 20–21
  discovery well test results, 9
  exploration and discovery history, 4–7
  hydrocarbons, 19–20
  location maps, 4
  Monterey Formation isopach map, 13
  Monterey oil sample, gas chromatogram, 21
  Monterey reservoir, fracture permeability model, 19
  Neogene depocenter, geohistory diagram, 23
  regional setting and geologic history, 7–8
  reservoir, 16–18
  seismic sections near wells, 16
  stratigraphy, 8–14
  structural cross section along axis, northwest-southeast, 12
  structure, 14–16
  trap, 15–16
Point Sal Formation, 10–11
Pol field, 73, 83, 85, 97
Porcelanite, SEM photomicrograph, 17
Post-Barremian unconformity, 48–50
Pre-Barremian unconformity, 47–48

Pre-Caspian basin, cross section, generalized, 104
  geologic structure, 102–103
  stratigraphic column, generalized, 103
  tectonic features, 103

Rannoch Formation, 440–441
Rogn Formation, 376–378

Sable Island area
  significant discoveries, 57
  wells with geopressure, 61
Sanaga Sud field, 217–230
  Bouguer gravity, 219
  completion results for wells, 227–229
  depositional environments, 220–221
  development and production, 229
  drilling information, 228
  drill-stem test results, 229
  exploration history, 219
  geochemistry, 226–227
  geology, regional, and basin development, 218–219
  geophysics, 227
  hydrogen index and paleobathymetry vs. depth, 226
  isomaturity profile, 227
  lithofacies descriptions, 221–222
  lithofacies distribution, 221
  location map, 218
  paleontology, 223–225
  petrographic analysis of reservoir quality, 222–223
  petrophysical reservoir properties, 228
  reservoir potential and reserves, 229
  stratigraphic units, major, 221
  stratigraphy, 219–220
  structural cross section, 225
  structure map, detailed, 226
  structure and trap configuration, 225
  vitrinite reflectance, 227
Santa Barbara Channel offshore, stratigraphic section, 6
Santa Maria basin offshore
  location maps, 5, 9
  stratigraphic section, 6
  tectonic elements, major, 11
Scotian Shelf
  location map, 56
  stratigraphic section, generalized, 60
  tectonic elements, 58
  tectonic setting, regional, 58, 59
Sen field, 73, 82, 84, 90, 91
Shaya uplift, 496–499
Smørbukk field, 323–348
  development plan, 347
  diagenesis and reservoir quality, 337–342
  drill-stem test results, 336
  exploration and discovery history, 324
  hydrocarbons, 342–343
  hydrocarbon volumes, in-place, estimated, 346
  mercury injection capillary pressure data, 343
  mid-Norwegian shelf, tectonic elements, 331
  perspective view, 324
  porosity and permeability vs. core depth, 342
  reserves, 343–347
  reservoir fluid compositions, 344

reservoirs, 333–337
reservoir simulation, 343–344
reservoir zones between wells, correlation, 328–329
residual gas and water saturations, 343
stratigraphic column, 327
stratigraphy, 324–330
structural contour map, 326
structure, 333
tectonic setting, 330–333
time and lithostratigraphic section, 326

Snøhvit field, 349–370. *See also* Hammerfest Basin.
aquifer, 358
borehole stress and fault sealing capacity, 363–365
brittleness of seal, 362
burial history, 366
core descriptions, 357
development, 367
discovery history, 350–352
erosion and uplift, 365–367
exploration history, 350
geochemical analysis of seal, 363
geochemistry, 359–361
geologic section, north-south, 353
hydrocarbons, 359–360
leakage from reservoir, 361–362
lithology of seal, 362
location map, 351
mass fragmentograms of oil, 360
maturation, 361
migration/filling of reservoir, 361
mineralogy, 354
permeability of seal, 363
pore pressure of seal, 363
reservoir characteristics, 354–358
reservoir layering, 356
reservoir properties, 358
rock strength of seal, 362–363
seal, 362–365
sedimentology, 354–358
source rocks, 360–361
stratigraphy and deposition, 352–354
structural depth map, 352
structural elements, southwestern Barents Sea, 354
structural geology, 354

Snorre field, 287, 407–416
cross section, 410
development, 413–414
discovery history, 409
exploration history, 408–409
fluid properties and oil-water contacts, 411
fluvial channel sands, computer modeling, 412
formation volume factor distribution, 411
gas-oil ratio distribution, 411
location map, 408
oil migration routes, 412
reserves, 414–416
reservoir characteristics, 410
reservoir modeling, 411–413
stratigraphy, 409–410
structure, 410
tension leg platform and subsea production system, 414
top reservoir map, 409

vertical transmissibility calculation, 413

Sonda de Campeche area
API gravity variation, 99
field characteristics, 83
field descriptions, 91–97
geologic column, typical, 76
location, 73
production data for fields, 85
seismic line over Taratunich structure, 81
stratigraphy, regional, 75–79
structural geology, regional, 79–80
tectonic aspects, 81

South Viking graben, 311, 312, 313
Statfjord field, 286, 287
Statfjord Formation, 436–438

Suizhong 36-1 field, 459–470
deltaic depositional model, lower Dongying Formation, 463
development geophysics, 465–468
environmental map, zone 3, 464
geology, field, 461–463
geology, regional, 460–461
index map, 462
interval velocity map, 467
oil column thickness map, 469
oil in place, distribution, 468
petrophysical study, 463–465
regression results, 468
seismic interpretation, 465–468
stratigraphic column, 461
tectonic map, 460
well location map, 465
wireline log patterns, 463

Takula field and Greater Takula area, 197–215
Aptian salt deposition, 202
development program, 208–209
exploration history, 198–200
geology, regional, 200–203
Pinda reservoir, 205–206
post-salt period, 202–203
pre-salt period, 202
production and export facilities, 209–210
reservoir fluid properties, 207–208
source rocks and hydrocarbon migration, 208
stratigraphy and reservoir geology, 205
structure, 203–205
Toca reservoir, 205
Vermelha reservoir, 206–207

Talimu Basin. *See* Tarim (Talimu) Basin.
Tarbert Formation, 424, 441
Tarim (Talimu) Basin, 493–510
Akekum slope, 497
Awarti-Mankal depression, 499
Central uplift, 499
development stages, 500–502
geologic history and stratigraphy, 500–502
geologic setting and tectonics, 494–500
giant oil and gas fields, 503–509
isopach maps, 501, 502, 503, 504, 505
Kekeya field, 503–506
Kuche depression, 494–496

Kurle nose uplift zone, 499
location map, 495
Lunnan field, 508
Luntai fault uplift zone, 497
reservoirs, principal, stratigraphic section, 498
Shaya uplift, 496–499
Southeast fault block, 499–500
Southwest depression, 499
tectonic and oil field location map, 496
wells, discovery and exploratory, 495
West Shaya fault-fold zone, 497
Yakela field, 506–508
Tecominoacan field, 73, 82, 84, 90–91
Tengiz field, 101–122
discovery history, 103–106
formation fluids, 118–119
geochemistry, 119–120
geologic structure of pre-Caspian basin, 102–103
location map, 102
Lower Permian P_1 base salt, configuration, 104
oil and gas properties, 119
porosity distribution, 118
reserves, 118
reservoir rock features, 114–118
seismic sections, 105, 107
seismogeologic model, 110–114
stratigraphic column, 108
stratigraphic model, 117
stratigraphy, 106–110
tectonics, 110
time-stratigraphic cross section, 117
well No. 102, gas chromatogram of oil, 119
Texas and Louisiana, biostratigraphic chart, 28
Toca reservoir, 205
TOGI Project, 426, 457
Toro reservoir, 479
Tranquillion Volcanics, 10
Troll field, 291, 447–458
cross section, structural, 454
development, 455–458
future oil and gas developments, 457–458
geologic setting, regional, 449–450
history, 448–449
hydrocarbon emplacement and distribution, 454–455
location map, 448
reservoir data, 456
reservoir map, 453
reservoir stratigraphy and depositional model, 450–453
stratigraphic table, 450
structure and trap development, 453–454
tectonic elements, map of regional, 450
Troll-Oseberg Gas Injection Project, 457
type logs, 451
wells, map of proximities to source, 453
Troll-Oseberg Gas Injection Project, 426, 457

Uech field, 73, 83, 85, 97, 98

Vaqueros sandstone reservoir, structure map, 8
Venture field, 55–71
composite section, 61–62
diagenetic history, 66–67
gas analysis, average, 71
geochemical data, 67–69
geopressure, 59–61
location maps, 56, 57
porosity vs. depth, 65–66
porosity vs. depth plot, composite, 68
pressure and temperature vs. depth, 65
regional setting, 56–57
reservoir characteristics, 69–70
reservoir parameters for sands, 66
sand 3A structure, 63
seismic control, 62–63
seismic sections, north-south, 64
stratigraphic cross section, 67
stratigraphic section, composite, 62
stratigraphic section, general, 57–59
structure, 65
structure on top of 3A sand, 63
Venture sand sequence, 63–65
well B-43, No. 2 and No. 6 sands, photomicrographs, 69
well B-43, porosity vs. depth plot, 69
well B-43, resistivity and interval transit time for shale, 68
well B-43, temperature and pressure vs. depth, 68
well D-23, No. 2 sand, photomicrograph, 69
well H-22, modified Van Krevelen diagram, 70
well H-22, vitrinite reflectance data, 70
Vermelha reservoir, 206–207
Vermelho field. *See also* Campos Basin.
development and production, 150
geologic sections, 148, 150
location map, 139
productive sandstone, principal, 148
reservoirs, 139, 143, 147–150
seismic section, 147
structural map, 150
volumes and production characteristics, 139
Veslefrikk field, 287
Villeperdue field, 251–263. *See also* Paris basin.
contour maps, 253, 254
discovery, and subsequent evolution of entrapment configuration, 254–256
exploration history and geology, 252–254
lithology, sedimentology, and porosity distribution of lower Callovian carbonate reservoir, 256–257
migration history and trapping mechanism, 259
oolitic facies, 256, 257
production data, 259–260
profile, NNW-SSE, 259

West Chalkley prospect, 27–33
area map, 29
discovery well log, 32
drilling, 31
economics, 30
exploitation, 31–33
lease acquisition, 30
log evaluation and testing, 31
Miogyp marker seismic structure maps, 30, 31
Miogyp stratigraphic cross section, 29
Miogyp trend map of oil and gas fields, 28
prospect sales, 30–31

seismic control, regional, 29
seismic database, 30
seismic lines, 30
seismic program map, proprietary, 32
stratigraphic control, 29
study area map, 28

top Miogyp sand structure maps, 32
trend analysis, 28–29
West Shaya fault-fold zone, 497
Witch Ground graben, 298, 300

Yakela field, 506–508